W9-BWM-869

ENCYCLOPEDIA OF HUMAN BIOLOGY

VOLUME 6 Pi–Se

ENCYCLOPEDIA OF HUMAN BIOLOGY

VOLUME 6 Pi–Se

Editor–in–Chief
Renato Dulbecco
The Salk Institute
La Jolla, California

ACADEMIC PRESS, INC. *Harcourt Brace Jovanovich, Publishers*
San Diego New York Boston London Sydney Tokyo Toronto

This book is printed on acid-free paper. ∞

Academic Press, Inc.
San Diego, California 92101

United Kingdom Edition published by
Academic Press Limited
24–28 Oval Road, London NW1 7DX

Library of Congress Cataloging-in-Publication Data

Encyclopedia of human biology / [edited by] Renato Dulbecco.
p. cm.
Includes index.
ISBN 0-12-226751-6 (v. 1). -- ISBN 0-12-226752-4 (v. 2). -- ISBN 0-12-226753-2 (v. 3). -- ISBN 0-12-226754-0 (v. 4). -- ISBN 0-12-226755-9 (v. 5). -- ISBN 0-12-226756-7 (v. 6). -- ISBN 0-12-226757-5 (v. 7). -- ISBN 0-12-226758-3 (v. 8)
1. Human biology--Encyclopedias. I. Dulbecco, Renato, 1914-
[DNLM: 1. Biology--encyclopedias. 2. Physiology--Encyclopedias.
QH 302.5 E56]
QP11.E53 1991
612'.003--dc20
DNLM/DLC
for Library of Congress 91-45538
CIP

PRINTED IN THE UNITED STATES OF AMERICA
91 92 93 94 9 8 7 6 5 4 3 2 1

CONTENTS OF VOLUME 6

HOW TO USE THE ENCYCLOPEDIA

We have organized this encyclopedia in a manner that we believe will be the most useful to you and would like to acquaint you with some of its features.

The volumes are organized alphabetically as you would expect to find them in, for example, magazine articles. Thus, "Food Toxicology" is listed as such and would *not* be found under "Toxicology, Food." If the first words in a title are *not* the primary subject matter contained in an article, the main subject of the title is listed first: (e.g., "Sex Differences, Biocultural," "Sex Differences, Psychological," "Aging, Psychiatric Aspects," "Bone, Embryonic Development.") This is also true if the primary word of a title is too general (e.g., "Coenzymes, Biochemistry.") Here, the word "coenzymes" is listed first as "biochemistry" is a very broad topic. Titles are alphabetized letter-by-letter so that "Gangliosides" is followed by "Gangliosides and Neuronal Differentiation" and then "GangliosideTransport."

Each article contains a brief introductory Glossary wherein terms that may be unfamiliar to you are defined *in the context of their use in the article*. Thus, a term may appear in another article defined in a slightly different manner or with a subtle pedagogic nuance that is specific to that particular article. For clarity, we have allowed these differences in definition to remain so that the terms are defined relative to the context of each article.

Articles about closely related subjects are identified in the Index of Related Articles at the end of the last volume (Volume 8.) The article titles that are cross-referenced within each article may be found in this index, along with other articles on related topics.

The Subject Index contains specific, detailed information about any subject discussed in the *Encyclopedia*. Entries appear with the source volume number in boldface followed by a colon and the page number in that volume where the information occurs (e.g., "Diuretics, **3:** 93"). Each article is also indexed by its title (or a shortened version thereof) and the page ranges of the article appear in boldface (e.g., "Abortion, **1: 1–10**" means that the primary coverage of the topic of abortion occurs on pages 1–10 of Volume 1).

If a topic is covered primarily under one heading but is occasionally referred to in a slightly different manner or by a related term, it is indexed under the term that is most commonly used and a cross-reference is given to the minor usage. For example, "B Lymphocytes" would contain all page entries where relevant information occurs, followed by "*see also* B Cells." In addition, "B Cells, *see* B Lymphocytes" would lead the reader to the primary usages of the more general term. Similarly, "*see under*" would mean that the subject is covered under a subheading of the more common term.

An additional feature of the Subject Index is the identification of Glossary terms. These appear in the index where the word "defined" (or the words "definition of") follows an entry. As we noted earlier, there may be more than one definition for a particular term and, as when using a dictionary, you will be able to choose among several different usages to find the particular meaning that is specifically of interest to you.

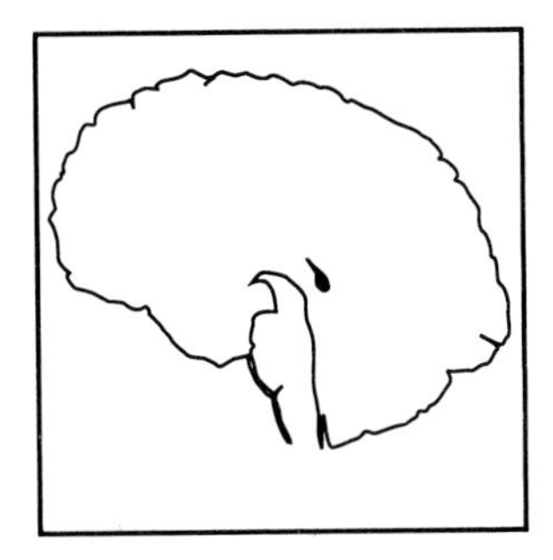

Pineal Body

RUSSEL J. REITER, *The University of Texas Health Science Center at San Antonio*

Glossary

Hydroxyindole-*O*-methyltransferase Melatonin-forming enzyme in the pineal gland; its activity is high and does not exhibit a change over the light–dark cycle

Melatonin One end product of serotonin metabolism in the pineal gland and the major secretory product, or hormone, of the gland; its production increases markedly at night

***N*-acetyltransferase** Enzyme in the pineal gland that metabolizes serotonin; its activity increases at night and may determine the quantity of melatonin formed

Norepinephrine Catecholaminergic neurotransmitter in the sympathetic neurons which end in the pineal gland; its release at night stimulates the production of melatonin in the pinealocytes

Pinealocyte Major cellular element of the pineal gland; these are the cells that produce melatonin

Serotonin Biogenic amine that is distributed widely in the brain and is in especially high concentrations in the pineal gland; it is a precursor of melatonin

Suprachiasmatic nuclei Paired groups of neurons in the hypothalamus of the brain that, at night, send neural signals to the pineal gland, initiating and maintaining melatonin synthesis

Tryptophan Amino acid that is taken up by the pinealocytes, where it is, in part, converted to serotonin and eventually to melatonin

THE PINEAL BODY, or pineal gland, is a small organ near the center of the brain that functions as an organ of internal secretion. Although its chief hormonal product seems to be the indoleamine *N*-acetyl-5-methoxytryptamine (melatonin), it may release other endocrine products as well. Melatonin production in and secretion from the pineal gland is much greater at night than during the day and, as a result, melatonin levels in the blood are usually five- to tenfold higher at night. After its entrance into the blood, melatonin passes into other bodily fluids as well, e.g., into saliva and cerebrospinal fluid (CSF) in which its levels also rise at night. The 24-hr rhythm of melatonin production and secretion is directly related to the light–dark environment that an individual witnesses. Light, perceived by the eyes, prevents the pineal gland from producing melatonin; thus, exposure to light at unusual times (i.e., during the night), rapidly suppresses melatonin production and causes a drop in blood levels of the compound. In general, the longer the night, the longer the duration of elevated melatonin. As day length (and therefore night length) varies seasonally, the pineal gland, because of the secretion of melatonin, provides information concerning time of year to all other organs in the body. The pineal gland is part of the biological clock that assists the organism in adapting to the external environment, both daily and seasonally.

I. Anatomical Location and Morphology

The pineal gland, a part of the epithalamus, is an unpaired organ that is an embryological outgrowth of the dorsal aspect of the developing brain. It eventually comes to lie near the center of the central nervous system (Fig. 1). The pineal gland is roughly shaped like a pine cone, which gave rise to its name.

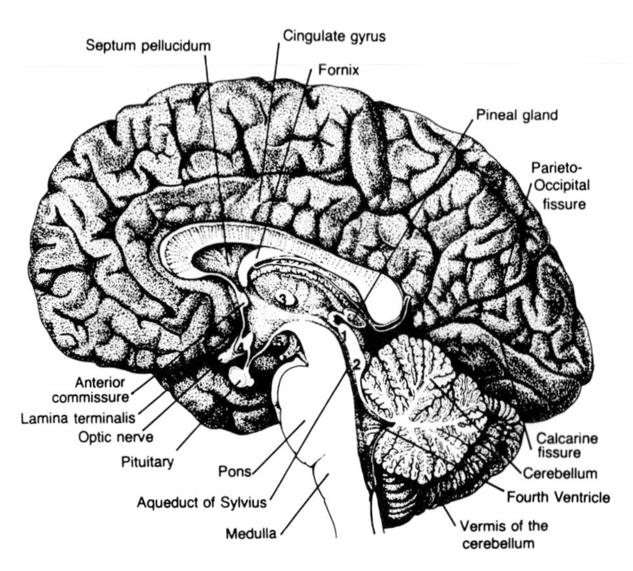

FIGURE 1 Midsagittal section of the human brain showing the location of the pineal gland and its relation to other neural structures. 1, Superior colliculus; 2, Inferior colliculus; 3, Massa intermedia; 4, Optic chiasm; 5, Mammillary body.

Although its weight varies widely among individuals, the pineal gland in adults is usually between 100 and 150 mg. The pineal gland is richly perfused with blood vessels derived from the posterior cerebral arteries. The venous drainage of the gland is directly into the large venous sinuses, which surround the organ. Microscopically, the pineal gland consists of modified neurons, referred to as pinealocytes, and other supportive elements, especially glial cells. The number of pinealocytes may decrease in advanced age, when calcium deposits, which are visualized radiologically, also form in the gland.

II. Innervation

The innervation of the pineal gland is essential for its function. The primary innervation originates in the eyes. The eyes send retinohypothalamic projections through the optic nerves to the hypothalamus, where the fibers synapse on neurons in the suprachiasmatic nuclei (SCN) (Fig. 2). The SCN output, through a neural pathway, which probably includes several additional synapses, reaches neurons in the intermediolateral cell column of the upper thoracic cord. These neurons belong to the sympathetic nervous system. Their axons leave the central nervous system in the ventral roots of the spinal cord and pass up to the superior cervical ganglia in the neck, where they make synaptic contacts with postganglionic sympathetic neurons. The axons of these postganglionic neurons accompany blood vessels into the skull, and their fibers eventually terminate in the vicinity of the pinealocytes in the pineal gland. These nerve terminals are rich in the catecholaminergic neurotransmitter norepinephrine (NE). NE is released from the nerve endings primarily at

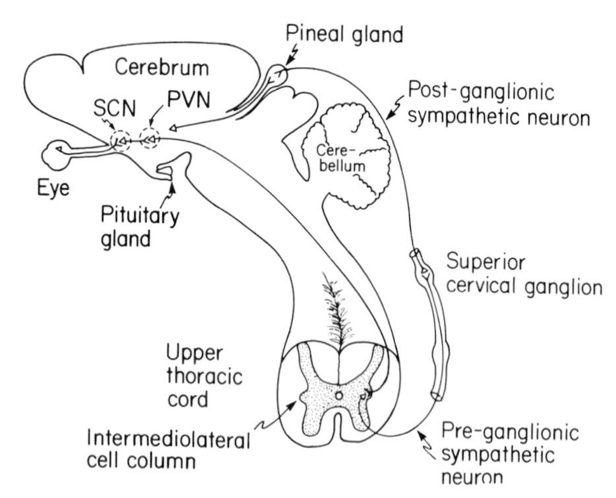

FIGURE 2 Neuroanatomical connections between the eyes and the pineal gland. The neurons between the suprachiasmatic nuclei (SCN) and the pineal gland must be intact for the gland to function normally. Neurons in or near the paraventricular nuclei (PVN) may be a relay point between the SCN and the intermediolateral cell column of the upper thoracic cord. [From R. J. Reiter, 1989, The pineal gland, *in* "De Groot's Endocrinology," Vol. 1, 2nd ed. (L. J. De Groot, ed.), pp. 240–253, Saunders, Philadelphia.]

night and interacts with specific receptors in the cell membranes of the pinealocytes.

The release of NE at night is initiated by a neural signal originating in the SCN of the hypothalamus. During the day, the activity of the SCN is suppressed by light perceived by the retinas and, therefore, NE is not released onto the pinealocytes. If the neural connections between the SCN and the pineal gland are interrupted, either surgically or by disease processes, the pineal cannot receive neural messages from the SCN and, as a result, even during the night, NE is not released from the postganglionic sympathetic fibers that end in the pineal gland; as a result, no nighttime rise in melatonin occurs.

Besides the sympathetic innervation of the pineal gland, fibers originating from neurons in various areas of the brain may also enter the pineal through the blunt stalk by which it is anatomically connected to the central nervous system. Which neurotransmitters are released by these fibers and their functional importance for the pineal gland remain unknown.

III. Melatonin Synthesis

The release of NE from postganglionic sympathetic neurons onto the pinealocytes at night activates the cells to produce its chief hormonal product, melatonin. After its discharge from the sympathetic axons, NE binds to $\beta 1$- and $\alpha 1$-adrenergic receptors in the pinealocyte membranes. The activation of the β-receptors is by far the most important stimulus for promoting melatonin production. β-receptors are linked to an enzyme adenylate cyclase via a G-protein. NE binding to the β-receptors activates adenylate cyclase, which then converts adenosine triphosphate to cyclic adenosine monophosphate (cAMP), an important intrapinealocyte second messenger. Cyclic AMP promotes the synthesis of new RNA and protein, which eventually leads to an increase in the activity of another enzyme, serotonin-*N*-acetyltransferase, which is important in controlling melatonin production. [*See* ADENOSINE TRIPHOSPHATE (ATP).]

Melatonin is the metabolic product of the amino acid tryptophan. Tryptophan is taken up by the pinealocytes from the blood; it is quickly acted upon by the enzyme tryptophan hydroxylase, which converts it to 5-hydroxytryptophan. 5-Hydroxytryptophan is metabolized to 5-hydroxytryptamine or serotonin in the presence of the enzyme L-aromatic acid decarboxylase. Serotonin, which is produced in many regions of the brain, is in especially high concentrations in the pineal gland. In the pineal, the bulk of the serotonin is located in the pinealocytes with a much lower amount being located in the sympathetic neurons that are found in the gland. Serotonin is metabolized via a number of pathways including deamination by monoamine oxidase, *O*-methylation by hydroxyindole-*O*-methyltransferase (HIOMT), and *N*-acetylation by *N*-acetyltransferase. In the pineal gland, NAT is nocturnally activated by the β-receptor–cAMP mechanisms and rapidly acetylates serotonin. NAT is located in the cytoplasm and it functions by transferring the acetyl group from the cofactor acetyl coenzyme A to acceptor amines, mainly serotonin. The activation of NAT is believed to be, in part, responsible for the drop in pineal serotonin levels during darkness (Fig. 3).

The product of serotonin *N*-acetylation is *N*-acetylserotonin, which accumulates in the pineal gland at night (Fig. 3). In the pinealocyte, *N*-acetylserotonin in turn is rapidly converted to melatonin by the enzyme HIOMT, which transfers a methyl group from the cofactor *S*-adenosylmethionine to 5-hydroxyindoles. The activity of HIOMT does not vary over the light–dark cycle but, because ample enzyme is present, most of the melatonin produced occurs at night, when its precursor, *N*-acetylserotonin, is being formed (Fig. 3). The increased production of melatonin at night has led to the general conclusion that darkness activates the pineal gland whereas light inhibits it; however, this conclusion is based solely on the metabolism of serotonin to melatonin and may not reflect what is happening to other metabolic pathways within the pineal.

IV. Melatonin Levels in Blood

Melatonin is a highly lipid-soluble molecule, which, once produced, rapidly escapes from the pinealocytes. The mechanisms of release, however, and its controls remain to be defined. Release occurs directly into the numerous capillaries, which perfuse the gland; as a consequence, blood titers of melatonin rise at night, with maximal values five- to tenfold greater than daytime levels (Fig. 3). In most

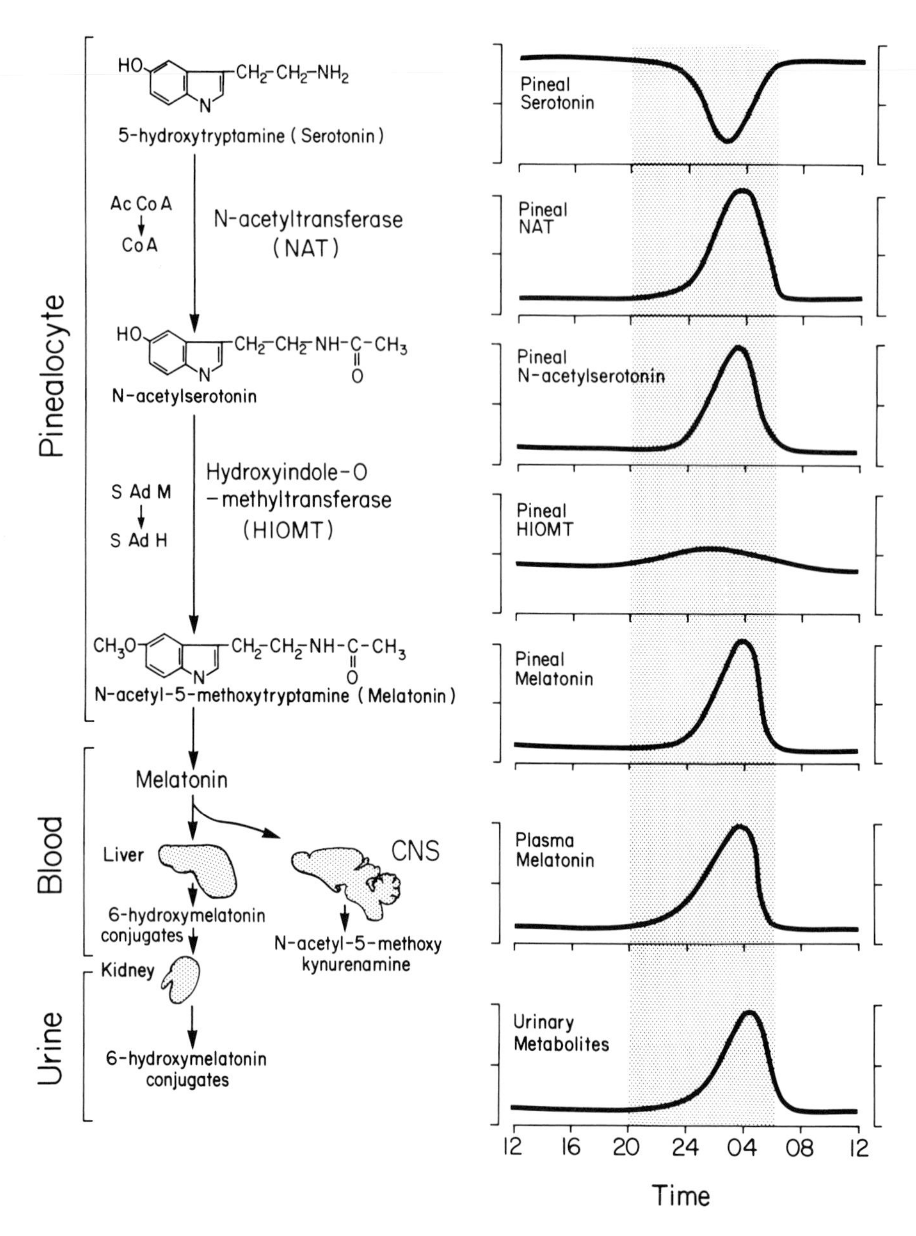

FIGURE 3 Left, metabolic conversion of serotonin to melatonin in the pinealocyte as well as its sites of degradation and excretion. Right, the daily fluctuations in the various constituents over a 24-hr period; the shaded area is the daily period of darkness. (Modified from R. J. Reiter, 1986, Funktionelle Morphologie und Pathophysiologie der Zirbeldrüse, *in* "Endokrinologie der Kindheit und Adoleszenz" (D. Gupta, ed.), pp. 53–57, Georg Thieme, Stuttgart.)

humans, blood melatonin values usually begin to rise at lights out in the evening, reach peak values between 2:00 and 3:00 A.M., and drop during the latter half of the dark phase to reach daytime levels early in the morning. Males and females have similar 24-hr rhythms of blood melatonin.

Newborn infants lack a blood melatonin rhythm; the retina–brain–pineal system matures rapidly during the first year after birth, and in 1-yr-old infants the day–night variation in blood melatonin is very apparent. Whereas daytime levels of melatonin remain uniformly low throughout life, high nocturnal values vary with age. In 1- to approximately 8-yr-old individuals noctural levels of blood melatonin are the highest, and then they decrease rapidly during puberty. In advanced age the day–night difference in blood melatonin may be very slight.

Because melatonin is highly soluble in lipid, which is a major constituent of the outer membrane of cells, it is able to enter into all cells and fluids in the body. Thus, when blood melatonin levels increase so do titers of this compound in, for example, the saliva and CSF. Usually, salivary melatonin levels are about one-third what they are in the blood whereas CSF titers may be nearly equivalent to those in the blood. [*See* SALIVARY GLANDS AND SALIVA.]

V. Melatonin Metabolism and Excretion

Melatonin is metabolized primarily in the liver and, to a lesser extent, in the brain (Figs. 3 and 4). In the liver, melatonin is converted primarily to 6-hydroxymelatonin, which is then coupled to sulfuric acid to form 6-hydroxymelatonin sulfate; a much smaller quantity of 6-hydroxymelatonin is coupled to glucuronide acid and appears in the urine as 6-hydroxymelatonin glucuronide. In the brain, melatonin is metabolized to a hallucinogenic agent *N*-acetyl-5-methoxykenurenamine. Both the hepatic and neural metabolites are discharged into the blood and are eventually excreted in the urine along with small quantities of unchanged melatonin (Fig. 4). Because blood melatonin values are higher at night, its by-products are also in highest concentrations in the first urine sample collected in the morning (Fig. 3). If the pineal gland is either surgically removed or otherwise destroyed due to a disease process, melatonin and its metabolites virtually disappear from bodily fluids.

VI. Role of Light in Melatonin Production

Light perceived by the eyes clearly is important in synchronizing the 24-hr melatonin production and secretion rhythm. Hence, in blind subjects (without visible light perception) the melatonin cycle persists but the duration of its rhythm is no longer 24 hr in duration, rather the period of the cycle is around 25 hr and, as a result, each day the melatonin rhythm moves 1 hr out of phase with that of the previous day. Because of this, blind humans may experience their highest melatonin levels at any time throughout the day or night. Sighted humans always have their melatonin peaks at night because the rhythm is synchronized by the prevailing light–dark environment. In general, the longer the night (such as during the winter months) the longer the duration of elevated melatonin.

Although the amplitude of the nocturnal melatonin peak varies among individuals, within a given subject it usually is highly stable and difficult to perturb. A factor known to alter the 24-hr pattern of melatonin production is light. Thus, light exposure at unusual times (e.g., during darkness) rapidly suppresses pineal melatonin synthesis with an associated drop in circulating levels. If an individual is awakened at night and is subjected to a bright light, blood melatonin titers are depressed to low daytime values within 30–60 min. Dim light exposure (e.g., moonlight, starlight, very low-intensity room light) will not have this suppressive effect; however, normal or slightly brighter room light inhibits pineal melatonin production and blood levels fall. The minimal brightness of light to which the human pineal will respond is apparently about 200 lux. The pineal also seems to respond in a graded manner such that the brighter the light, the greater the suppression of blood melatonin levels. Light exposure has been used to treat several clinical conditions (e.g., seasonal affective disorder or winter depression); this treatment is referred to as phototherapy.

Besides the brightness of the light, the color or wavelength of light is also important in determining the response of the pineal gland. Blue wavelengths of light (around 500–520 nm) seem to be maximally suppressive, suggesting that the blue-sensitive retinal photopigment rhodopsin may be the mediator of the inhibitory effects of light on pineal melatonin synthesis. Other wavelengths (e.g., near ultraviolet, green, yellow, red) have no or little effect.

Humans are exposed to "unusual" light–dark cycles during travel across time zones (e.g., a flight from the United States to Europe), which may cause a subject to be exposed to only a brief night or an unusually long day. These alterations are not followed by an immediate synchronization of the melatonin rhythm to the light–dark cycle at the new destination; rather, the melatonin rhythm can take up to 10 days to re-acclimate, with older individuals requiring a longer time. During this period of reacclimation, it is usual to experience a phenomenon commonly known as jet lag; this feeling of fatigue, insomnia, etc., has been speculated to be, in part,

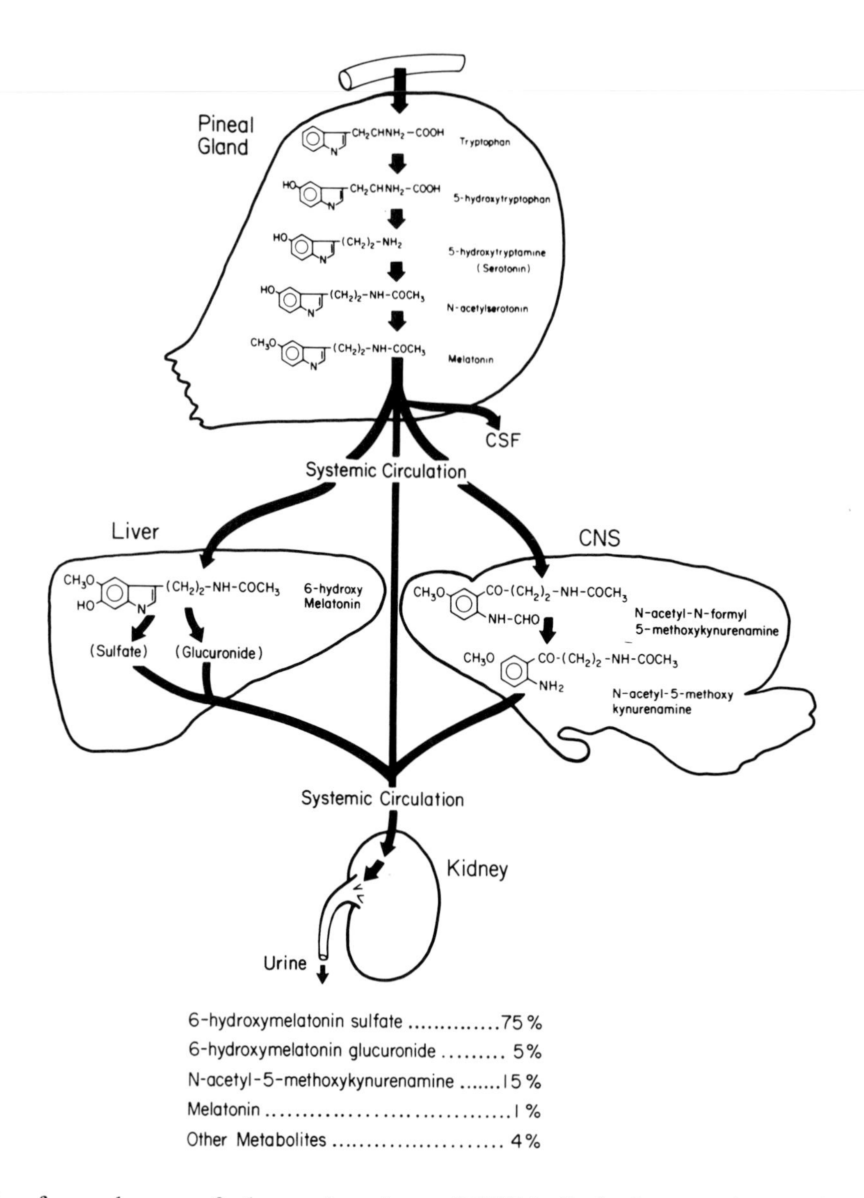

FIGURE 4 Production, secretion, metabolic fates, and urinary excretion of melatonin and its metabolites. Besides its release into the blood, melatonin may be released directly or secondarily into the cerebrospinal fluid (CSF). Melatonin is metabolized in both the liver and in the central nervous system (CNS). 6-hydroxymelatonin sulfate, 75%; 6-hydroxymelatonin glucuronide, 5%; N-acetyl-5-methoxykynurenamine, 15%; Melatonin, 1%; Other Metabolites, 4%. (From R. J. Reiter, 1986, Impact of photoperiodic information on pineal metabolism and physiology, *in* "Biometeorology 10," part 2, (R. W. Gloyne, Y. Inaba, M. Iriki, M. Kikuchi, and M. M. Yashino, eds.), pp. 178–197, Swet and Zietlinger, Amsterdam.)

related to lack of synchrony of the melatonin rhythm with the sleep–wake cycle. Melatonin is being tested as a treatment for jet lag.

VII. Concluding Remarks

The pineal gland, by producing melatonin during darkness, signals all body organs as to time of year, because, regardless of where an individual is on Earth, day lengths change seasonally. The higher the latitude, either north or south, the greater the seasonal variations in day length; within the Arctic and Antarctic circles (roughly 66.5° north or south latitude, respectively), there are actually periods of constant light and darkness. These seasonal variations in photoperiod exposure are transduced into a chemical messenger (melatonin) in the pineal gland, which provides the internal organs with a link to the external environment. In this sense, the pineal gland provides both daily and seasonally useful information. [*See* CIRCADIAN RHYTHMS AND PERIODIC PROCESSES.]

Bibliography

Arendt, J. (1986). Assays of melatonin and its metabolites: Results in normal and unusual environments. *J. Neural Transm.*, Supplement **21**, 11–34.

Cardinali, D. P., Vacas, M. I., Rosenstein, R. E., Etchegoyen, G. S., Keller Sarmiento, M. I., Solveyra, C. G., and Pereyra, E. N. (1987). Multifactorial control of pineal melatonin synthesis: An analysis through binding studies. *In* "Advances in Pineal Research," Vol. 2 (R. J. Reiter and F. Fraschini, eds.), pp. 51–66. John Libbey, London.

Gupta, D., and Attanasio, A. (1988). Pathophysiology of pineal function in health and disease in children. *In* "Pineal Research Reviews," Vol. 6 (R. J. Reiter, ed.), pp. 262–300. Alan R. Liss, New York.

Lewy, A. J., Sack, R. L., and Singer, C. M. (1985). Immediate and delayed effects of bright light on human melatonin production: Shifting "dawn" and "dusk" shifts the dim light melatonin onset. *Ann. N. Y. Acad. Sci.* **453,** 253–259.

Reiter, R. J. (1985). Action spectra, dose–response relationships and temporal aspects of light's effects on the pineal gland. *Ann. N.Y. Acad. Sci.* **453,** 215–230.

Reiter, R. J. (1986). Normal patterns of melatonin levels in the pineal gland and body fluids of humans and experimental animals. *J. Neural Transm.*, Supplement **21,** 35–54.

Reiter, R. J. (1988). Neuroendocrinology of melatonin. *In* "Melatonin—Clinical Perspectives" (A. Miles, D. R. S. Philbrick, and C. Thompson, eds.), pp. 1–42. Oxford University Press, Oxford.

Reiter, R. J. (1989). The pineal gland. *In* "De Groot's Endocrinology," Vol. 1, 2nd ed. (L. J. De Groot, ed.), pp. 240–253. Saunders, Philadelphia.

Waldhauser, F., Steger, H., and Vorkapic, P. (1987). Melatonin secretion in man and the influence of exogenous melatonin on some physiological and behavioral variables. *In* "Advances in Pineal Research," Vol. 2 (R. J. Reiter and R. Fraschini, eds.), pp. 207–222. John Libbey, London.

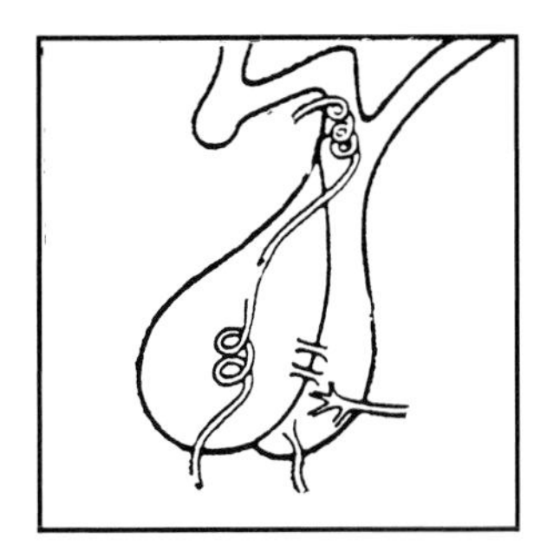

Pituitary

WILLIS K. SAMSON, *University of Missouri School of Medicine*

Glossary

ACTH Adrenocorticotropic hormone. Major posttranslational product of pro-opiomelanocortin gene transcription

AVP Vasopressin or antidiuretic hormone. A neuropeptide produced in unique neurons in the hypothalamic paraventicular and supraoptic nuclei and released from nerve terminals in the posterior pituitary (neural lobe)

FSH Follicle-stimulating hormone, produced in gonadotrophs

GH Growth hormone, produced in somatotrophs

LH Luteinizing hormone, produced in gonadotrophs

OT Oxytocin, a neuropeptide produced in unique hypothalamic neurons in the paraventricular and supraoptic nuclei, released from nerve terminals in the posterior pituitary (neural lobe)

Portal vessels Venous plexus, which connects the hypothalamus and anterior lobe of the pituitary gland. Delivers humoral agents of neural origin, which control hormone production and release

PRL Prolactin, produced in lactotrophs

TSH Thyroid-stimulating hormone, produced in thyrotrophs

THE PITUITARY is an aggregation of neural and endocrine tissue located on the ventral surface of the brain on the midline, lying beneath and in part continuous with the hypothalamus (Fig. 1). The posterior (neural) lobe is a ventral extension of the axons of hypothalamic neurons. From these terminals are released the neuropeptide hormones OT and AVP. The neural lobe therefore is important for the hormonal control of fluid and electrolyte homeostasis and lactation. The intermediate lobe of the pituitary is only distinguishable during embryological development in the human, its cells merge with both the neural and anterior lobe cells after birth. The largest portion of the pituitary gland is the anterior lobe. It originates from ectodermal derivatives on the roof of the developing oral cavity. Cells migrate dorsally toward the expanding neural tube and establish a position in front of and partially surrounding the neural lobe and its connection with the hypothalamus, the infundibular stem. A peculiar vascular connection, the hypophyseal portal vessels, develops connecting the anterior lobe with the hypothalamus through which trophic factors gain access to the hormone-producing cells of the gland. Hormones released from these cells control a wide variety of physiologic functions including growth, metabolism, and reproduction.

I. Development and Structure

A dorsal evagination of the ectodermal lining of the stomodeum just anterior to the endoderm of the pharynx initiates during week 4 of gestation in humans. Cells separate from the roof of the presumptive buccal cavity forming a spherical mass called *Rathke's pouch.* These cells will develop into the hormone-secreting elements of the anterior pituitary gland or adenohypophysis. By week 6 they have approximated the ventral outgrowth of neural tissue, which will constitute the neurohypophysis or neural lobe. At the time of apposition, Rathke's pouch undergoes considerable structural rearrangement with the collapse of its hollow center. The anterior wall proliferates to become the glandular tissue characterizing the bulk of the anterior lobe, the pars distalis. The thin posterior wall contacts

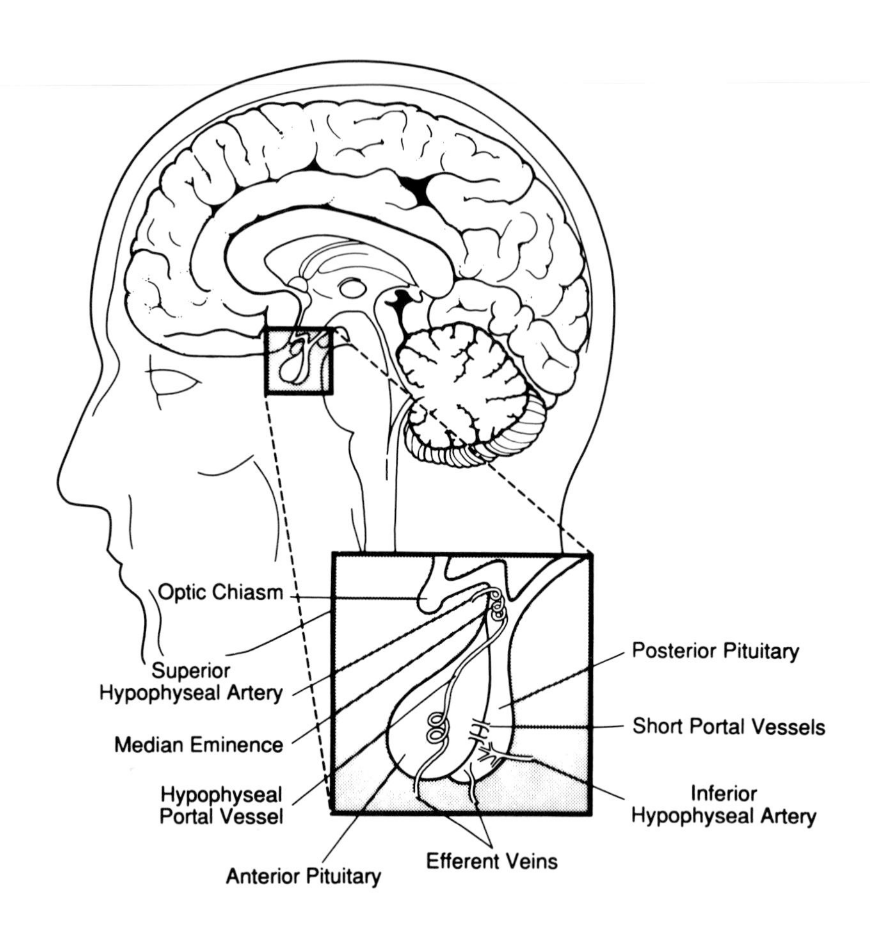

FIGURE 1 Diagrammatic illustration of location of the pituitary gland beneath the hypothalamus. Positions of the adenohypophysis (anterior pituitary) and the neurohypophysis (posterior pituitary) are demonstrated, as well as the arterial supply of these tissues. The hypophysial portal vasculature is schematized for simplicity.

the developing neurohypophysis and fuses with it. In the embryo this tissue remains distinguishable as the pars intermedia. Lateral extensions of the developing pouch surround the infundibular stem as the pars tuberalis. The neurohypophysis or neural lobe has been formed by the ventral extension of axons from nerve cells located in the developing hypothalamic paraventricular and supraoptic nuclei. As these axons reach the base of the hypothalamus in the midline, they form a ridge of tissue in the median eminence and, progressing ventrally as the infundibular stem around which the pars tuberalis adheres, end in secretory terminals in the infundibular process or neural lobe.

During the sixth week of gestation, mesenchymal elements have aggregated into the portal vessel system, which will provide the vascular link between the hypothalamus and the developing adenohypophysis. The superior hypophysial artery, most commonly a branch of the internal carotid artery, supplies tissue at the base of the hypothalamus surrounding the separation of the infundibular stem, and the stem itself. This region of the hypothalamus is the median eminence, which appears as a vascularized ridge of tissue. Capillaries eminating from the superior hypophysial arteries in the median eminence form a dense network and are characterized by fenestrated endothelial elements through which neural factors released locally diffuse. These capillary loops give rise to long portal veins that descend along the infundibular stem and ramify into the sinusoids of the adenohypophysis. Via this unique portal vessel system neurally derived agents gain access to the hormone-producing cells of the anterior lobe. The infundibular process (neurohypophysis) receives arterial supply via the inferior hypophysial artery. In addition to the venous drainage of this tissue, a short portal system of vessels connects the sinusoids of the neurohypophysis with those of the anterior lobe, providing another potential pathway for the delivery of trophic (stimulatory and inhibitory) factors to the endocrine cells of the adenohypophysis.

The adenohypophysis, or anterior lobe, is comprised then of the large pars distalis, the pars tuberalis, and the indistinct pars intermedia. The neurohypophysis is mainly the infundibular process, or neural lobe, and the infundibular stem. Some authors consider the median eminence part of the neurohypophysis, and although the internal layer of the median eminence does contain hypothalamo-neurohypophyseal fibers of the OT and AVP neural systems projecting toward the infundibular stem, the majority of the tissue in the median eminence is comprised of nerve terminals terminating in the vicinity of the hypothalamo-hypophyseal portal vasculature; therefore this tissue is more related to anterior pituitary lobe function. For the sake of this review, the median eminence will be considered hypothalamic tissue. In the adult human, the pituitary gland (both adenohypophysis and neurohypophysis) measures 1.0–1.5 cm in length and width and 0.5 cm in depth. It weighs approximately 0.5 g, except during pregnancy when the gland may weigh as much as 1.0 g. The gland sits protected in a bony cavity of the sphenoid bone, the sella turcica. It is separated from the overlying nervous tissue of the hypothalamus by a layer of dura mater, the diaphragma sella, through which the infundibular stem passes. The major venous drainage of the pituitary gland is via the cavernous and transverse sinuses and the internal jugular veins.

II. Anterior Lobe Function

A. Morphology

Adenohypophyseal cells are encapsulated in a dense collagenous matrix and are arranged in sinusoidal fashion in close apposition to the thin-walled vascular elements of the gland. Until recently the glandular cells of the adenohypophysis were classified on the basis of their affinity for routine histologic dyes. Two classes of cells were identified: chromphils and chromophobes. Modern techniques such as immunocytochemistry using antibodies directed against specific proteins produced by the cells have enabled more accurate identification of cell types. Six distinct, glandular cell types have been identified, and their characterization is described in Table I.

B. Control of Hormone Release

Anterior pituitary function is primarily controlled by the action of trophic substances, releasing and inhibiting factors, delivered to the gland by the long and short portal vessels. These factors originate for the most part in nerves located in the hypothalamus and rostral diencephalon. The majority of the trophic factors are small peptides, although other factors (e.g., more classical neurotransmitters) do exert potent effects on adenohypophyseal hormone synthesis and secretion. The primary controller of pituitary function then is the hypothalamus. This tissue, which lies below the thalamus on the base of the diencephalon, is a heterogeneous population of neurons receiving abundant afferent input from other brain areas. Because of the absence of the blood-brain barrier at some locations, the hypothalamus also receives afferent stimuli that are humoral in nature. The hypothalamic cell groups, called *nuclei*, process incoming signals, integrating this information before sending out efferent neural messages controlling a wide variety of essential functions including cardiovascular, renal, and gastrointestinal funciton, as well as autonomic nervous

TABLE I Adenohypophyseal Morphology: Cell Types and Hormone Content[a]

Cell Type	Staining Affinity	Hormone Content	Secretory Granule Size	% of Total
Somatotroph	Acidophilic	Growth hormone	150–600 nm	50%
Lactotroph	Acidophilic	Prolactin	150–350 nm or 300–800 nm	10–25%
Corticotroph	Basophilic	Pro-opiomelanocortins (ACTH, lipotropins, endorphins)	250–400 nm	15–20%
Thyrotroph	Basophilic	Thyroid-stimulating hormone	100–200 nm	<10%
Gonadotroph	Basophilic	Luteinizing hormone and follicle-stimulating hormone	150–250 nm or 350–450 nm	10–15%
Mammosomatotroph	Acidophilic	Prolactin and growth hormone	400–1200 nm	

[a] Other cell types have been identified on the basis of secretory granule content and affinity or avidity for acidic and basic dyes; however, the identity of their secretory product(s) has not yet been elucidated.

TABLE II Hormones of the Adenohypophysis

Hormone	Origin	Structural Identity	Major Trophic Factors
Growth hormone	somatotrophs and mammosomatotrophs	191 amino acids (21,000 MW)	somatostatin (−) growth hormone–releasing hormone (GHRH, +)
Prolactin	lactotrophs and mammosomatotrophs	198 amino acids (22,000 MW)	dopamine (−) thyrotropin-releasing hormone (TRH, +) vasoactive intestinal peptide (VIP, +) oxytocin (+)
Pro-opiomelanocortins	corticotrophs		corticotropin-releasing hormone (CRH, +)
ACTH		39 amino acids (4,500 MW)	vasopressin (+) norepinephrine (+)
Alpha-MSH		13 amino acids (1,800 MW)	
Beta-lipotropin peptides			
gamma-lipotropin		58 amino acids (6,000 MW)	
beta-endorphin		31 amino acids (3,500 MW)	
Thyrotropin (TSH)[a]	thyrotrophs	alpha subunit (92 amino acids) beta subunit (112 amino acids)	thyroid-stimulating hormone (TRH, +)
Luteinizing hormone (LH)[b]	gonadotrophs	alpha subunit (92 amino acids) beta subunit (121 amino acids)	luteinizing hormone–releasing hormone (GnRH)
Follicle-stimulating hormone (FSH)[c]	gonadotrophs	alpha subunit (92 amino acids) beta subunit (118 amino acids)	GnRH (+) activin (+) inhibin (−)

[a] TSH is a glycoprotein with about 14% by weight carbohydrate (28,000 total MW).
[b] LH is a glycoprotein with about 16% by weight carbohydrate (30,000 MW).
[c] FSH is a glycoprotein with >20% by weight carbohydrate (32,000 MW).

system activity and brain-derived behaviors. In coordination with these neural functions, the hypothalamic integrators also control such varied functions as metabolism, growth, and reproduction via their delivery of trophic factors to the median eminence where on release they diffuse into the portal vessels and gain access to the pituitary gland. Thus the pituitary sees a constant variety of humoral cues, neuroendocrine in nature, which control the release of its hormones into the peripheral circulation. [*See* HYPOTHALAMUS.]

A secondary level of control of adenohypophyseal function is exerted directly at the level of the anterior pituitary gland. Hormones released from the anterior lobe exert a variety of actions in distant tissues, some of which result in the elaboration and secretion of target tissue-specific hormones into the blood (e.g., estrogens from the ovary in response to gonadotropin stimulation) (Table II). These peripherally derived humoral agents not only reach the integrative centers of the hypothalamus where they can exert positive (stimulatory) or negative (inhibitory) effects on the hypothalamic control of adenohypophyseal function, but also exert selective effects on the anterior lobe cells themselves. These long-loop feedback actions of target tissue-specific factors play an important role in fine-tuning the ability of the anterior pituitary to respond to hypothalamic signaling. In this manner the synthetic and secretory activity of the gland can be accelerated, limited, or synchronized in an appropriate physiologic fashion.

Two other levels of control involve feedback effects of the anterior lobe hormones themselves. There is evidence that these hormones may gain access to the hypothalamus and exert direct effects on the hypothalamic neurons, elaborating the releasing and inhibiting factors controlling their secretion. This is termed *short-loop feedback* and is exemplified by the ability of prolactin to gain access to the brain and affect the activity of dopaminergic neurons. Hormones have also been implicated in a form of autofeedback in which they control the activity of other pituitary cell types (paracrine control) or of their own cells of origin (autocrine control). This phenomenon has been termed *ultrashort-loop feedback*, and although these feedback loops have yet to be conclusively demonstrated in humans, abundant data derived from animal studies suggest the presence of such control mechanisms in humans as well.

C. Cell Types: Hormone Production, Actions, and Control of Release

1. Somatotrophs

Growth hormone (GH)–producing cells are located mainly in the lateral wings of the pars distalis and are medium in size with a spherical or oval shape. They possess a centrally located, spherical nucleus, well-developed rough endoplasmic reticulum, and extensive Golgi complexes. These profiles, together with the presence of abundant secretory granules averaging about 300 nm in diameter, reveal the endocrine nature of these cells. The abundance of these somatotrophs accounts for the fact that the human pituitary contains about 10% by weight GH. Surprisingly, the number or activity of the somatotrophs is not influenced by age, suggesting then that the increased effectiveness of GH during development is due not to more hormone being secreted but instead to a more receptive tissue response to GH. GH has a plasma half-life of approximately 20 min and circulates in adults at normal levels of 2–4 ng/ml (5–8 ng/ml in adolescents). Secretion occurs in a pulsatile fashion with more exaggerated pulsatility present during puberty. Levels of GH remain fairly constant throughout the day with the exception of minor surges 3–4 hr after meals and a major (often 10-fold) surge during the initial period of deep sleep (at the onset of stages III and IV, before REM sleep).

The actions of GH are best described by conditions present during hyper- or hyposecretion. Hyposecretion is characterized by dwarfism, a condition in which a symmetrical reduction in growth occurs. If hypersecretion occurs early in life, the resulting gigantism is reflected in a symmetrical overgrowth of all tissues including bones, muscles, connective tissues, and visceral organs. Hypersecretion after fusion of epiphyseal plates cannot stimulate bone growth, and as a result only the soft tissues of the body (connective tissue, skin, visceral tissues) and the cancellous bone (hands, jaw) respond, a condition known as *acromegaly*. GH receptors are present in the liver and on fibroblasts, adipocytes, and lymphocytes. GH administration results in decreased plasma amino acid composition, reflecting a stimulation of amino acid transport into cells, particularly liver and muscle, and increased protein synthesis (anabolism). The effects on protein synthesis are thought to occur at a pretranslational level. GH infusion also results in positive nitrogen balance as well as the retention of calcium, phosphorus, potassium, and magnesium. The hormone antagonizes the action of insulin, resulting in a decreased use of carbohydrates and the promotion of hepatic and muscular glycogenesis. With the diversion of amino acids for protein synthesis and carbohydrates for glycogen formation, lypolisis and fatty acid oxidation are stimulated, resulting in ketogenesis. Mitosis increases in the hematopoetic system.

The most striking effect of GH is growth of cartilage and bone. To exert these effects, the hormone must first act in the liver to stimulate the production and release of factors called *somatomedins*. These peptides (approximately 7,500 MW) share significant structural similarity with proinsulin and therefore have been termed *insulin-like growth factors* (IGFs). IGFs circulate in plasma associated with a binding protein or proteins, and several subclasses of IGF receptors have been characterized. Many of, if not all, the biological activities of GH can be ascribed to the action of the somatomedins. They stimulate collagen synthesis, sulfate incorporation into cartilage, and mitosis. Long bones (e.g., the tibia and femur) grow by increases in ossification of the diaphyseal and epiphyseal plates. The IGFs stimulate thickening of the plates, resulting in increased ossification and long bone growth. This cannot occur even in the presence of excess GH after epiphyseal plate closure has occurred in adolescence. [*See* GROWTH, ANATOMICAL; PEPTIDES.]

Metabolic, hormonal, and neural factors control somatotroph activity. Insulin-induced GH release is caused not by a direct effect of insulin *per se*, but instead by the hypoglycemic action of insulin. Delayed postprandial GH secretion, particularly after a high-protein meal, is caused by the direct stimulatory effects of certain amino acids (e.g., arginine and leucine). Falling levels of plasma-free fatty acids also result in GH secretion. Metabolic inhibition of GH secretion occurs in hyperglycemia, in obese individuals, and when free fatty acid levels are high. Hormonal effects on GH secretion include stimulatory effects of estrogens, glucagon, and vasopressin and inhibitory effect of GH itself by an autocrine action. Plasma levels of GH are also low in hypothyroidism. By far the most significant regulation of somatotroph function is exerted by the hypothalamus and other central nervous system structures.

Two peptides of hypothalamic origin are the primary regulators of secretion. Somatostatin, a peptide that exists in both a 14 and an N-terminally extended 28-amino-acid form, inhibits basal and stimulated GH secretion *in vitro*. *In vivo* administration of the peptide in humans results in little effect on the already low basal levels of hormone; however, the sleep-related surge can be suppressed, and hypersecretion of GH in acromegalics can be reversed. Somatostatin-producing neurons are found adjacent to the walls of the third cerebroventricle in the periventricular region of the hypothalamus. Fibers containing the peptide project to the median eminence, where somatostatin is released into the vicinity of the capillary loops of the portal vasculature. The mechanism of action of somatostatin at the level of the somatotroph is thought to be via an inhibition of stimulated adenylate cyclase activity and to an action downstream from the formation of cAMP, perhaps caused by an action on calcium influx. GH-releasing hormone (GHRH) is a 44-amino-acid peptide produced in cells of the infundibular region of the hypothalamus. Short axonal processes deliver the peptide to the adjacent median eminence. It is active *in vitro* at doses similar to those present in portal blood of rats and is an effective stimulator of GH secretion when infused in humans. GHRH stimulates secretion of GH via adenylate cyclase–dependent and –independent [probably via increases in calcium influx and changes in phosphatidylinositol (PI) turnover] mechanisms. It synergizes with other agents known to stimulate these two signal transduction pathways (e.g., prostaglandins). The releases of both somatostatin and GHRH are controlled by numerous neural and hormonal factors including the somatomedins, which have been hypothesized to feedback in a negative fashion at both the pituitary and hypothalamic levels to dampen GH secretion. Interactive effects of somatostatin and GHRH on their own release have been described. Alpha-adrenergic agents act centrally to stimulate GH secretion, perhaps explaining stress-induced elevations in plasma GH levels. Beta-adrenergic agents exert inhibitory control. Finally, in lower species there is evidence for short-loop negative feedback of GH on the hypothalamic cells responsible to GHRH release or on those neurons *per se*.

2. Lactotrophs

Prolactin (PRL)-producing cells are located mainly in the lateral posterior aspect of the pars distalis, adjacent to the posterior lobe. The adenohypophysis contains approximately 0.1 mg PRL, although the hormone undergoes rapid postmortem degradation, so this might be a low estimate. Two types of lactotrophs exist. The most abundant cells are small and angular, often arranged in contact with each other. Although they possess well-developed endoplasmic reticulum and Golgi complexes, they have sparse, small granules (150–350 nm). the Golgi complexes are stacked near the cell surface, giving the appearance that these cells are highly active and that the newly synthesized PRL is rapidly secreted. The second class of lactotrophs represent those cells in which stored PRL accumulates. They appear in isolation throughout the pars distalis, are large and often irregular in shape, and possess abundant, large secretory granules (300–800 nm). [*See* GOLGI APPARATUS.]

The number of lactotrophs varies greatly during life. The fetal pituitary is producing PRL already by the tenth week of gestation, and under the influence of maternal estrogens, both fetal and maternal pituitaries demonstrate a hyperplastic population of lactotrophs. Numbers of cells fall then in childhood as they also do postpartum in the absence of lactation. In the nonpregnant state, adult female pituitaries have approximately the same number of lactotrophs as found in males. Plasma PRL levels in adult, nonpregnant, nonlactating females are about 10 ng/ml and are slightly lower in males and children. Levels in plasma progressively rise during pregnancy to reach third trimester levels of as high as 300 ng/ml. The plasma half-life of PRL is 20–30 min, and its high metabolic clearance rate suggests a secretion rate of about 0.2 mg/day, indicating the high synthetic activity of the lactotroph. Other than in pregnancy, the most remarkable secretion of PRL occurs in response to suckling, with plasma levels rapidly increasing to as high as 10 times basal with mechanical stimulation of the nipples. This mechanoreceptor-mediated stimulation of PRL secretion also occurs in the nonlactating female but not in males, indicating a sexual dimorphism with regard to the control of PRL secretion. Under basal conditions, PRL secretion is pulsatile in nature, with only minor fluctuations except during sleep when one or more major surges occur, resulting in a doubling or tripling of plasma levels. Unlike nocturnal GH surges, these PRL surges are not synchronized with the sleep cycle and normally occur toward the end of the period of sleep. Virtually any form of stress, whether caused by physical intervention or a manifestation of a perceived threat (psychogenic stress), results in PRL secretion. Additionally, stimulation of mechanoreceptors in the vagina and on the uterine cervix during intercourse results in PRL secretion. The pathways involved in these reflex releases of PRL are thought to involve brain opiate, serotonergic, dopaminergic, and peptidergic systems. Although present in rodents, the existence of a midcycle PRL surge in human females is still controversial.

The mammary gland is the primary site of action of PRL. The breasts rapidly develop during both puberty and pregnancy. The growth and differentiation is called *mammogenesis*, and although PRL is thought to play an important role, its actions appear to be secondary to those played by estrogen and progesterone. Ductile development is stimulated by increasing estrogens, and lobulo-alveolar growth in the estrogen-primed gland is induced by progesterone. Permissive effects of cortisol, GH, and insulin are also involved. PRL's effects are thought to be mediated via specific membrane receptors, which result in an activation of phospholipase A activity and eventually increased transcription and translation. The proliferative effect then is not only one involving mitosis, but also the *de novo* synthesis of enzymes such as protein kinases and milk proteins. During pregnancy, the high levels of estrogens and progesterone not only stimulate mammogenesis, but also actually inhibit milk production. After parturition, levels of these steroids fall, and if suckling ensues, the lactogenic effects of PRL are uncovered. These effects appear to require the permissive actions of adrenal steroids, as well as GH, insulin, thyroxine, and parathyroid hormone. [*See* STEROIDS.]

Prolactin acts in the adrenal gland and ovary to increase the availability of steroid precursors and in the presence of luteinizing hormone (LH) stimulates the secretion of progesterone. Effects on the growth and development of accessory sex glands have been described. Receptors are present in the prostate and testis, suggesting a role for PRL in growth and development of these tissues. In situations of hypersecretion, an antigonadotrophic effect is expressed, probably at the level of the hypothalamus, which in part explains the relative infertility associated with lactation. Finally, although controversial, a renotropic action of PRL is suggested by clear antinatriuretic effects demonstrated in lower species and the presence of PRL receptors in the kidney.

Neural and hormonal factors control the release of PRL. As mentioned previously, estrogen stimulates hormone secretion, probably by both a direct effect on the lactotroph and effects exerted on hypothalamic control mechanisms. The major, recognized, neural factor regulating factor PRL release is dopamine, which exerts potent inhibitory effects on the lactotroph. Dopaminergic neurons originating in the infundibular region project either to the median eminence or neural lobe, delivering the catecholamine to the long and short portal vessels. Much of the magnitude of PRL secretion observed under a variety of physiologic circumstances can be explained on the basis of reduction of dopaminergic neuronal activity and the ensuing removal of inhibitory tone. Inactivation of adenylate cyclase appears to be one of the cellular mechanisms for dopamine's PRL-inhibiting effect. However, dopamine withdrawal alone cannot completely describe the timing or extent of PRL secretion, leading to the hypothesis that PRL-releasing factors must also be present, which either act in the absence of dopamine's inhibitory tone or which can stimulate PRL release even in the presence of dopamine. Several peptides of hypothalamic origin have been demonstrated to act as PRL-releasing factors in humans. Thyrotropin-releasing hormone (TRH), vasoactive intestinal peptide (VIP), and oxytocin (OT) all are capable of stimulating PRL secretion; however, the exact role played by each of these peptides in the physiologic regulation of PRL release is still unclear. In the case of TRH, an involvement of phospholipase C and PI hydrolysis resulting in inositol triphosphate and diacylglycerol formation has been reported. The resulting increase in intracellular calcium and in levels of protein kinase C are thought to be the mechanisms by which the peptide stimulates PRL release. VIP's PRL stimulatory effect is thought to be caused by a stimulation of adenylate cyclase activity. Recently, evidence has been presented for the presence of novel peptides with potent PRL-releasing and -inhibiting activities, suggesting that the identities of the primary factors responsible for the physiologic control of lactotroph function remain unidentified.

3. Corticotrophs

Located primarily in the center of the pars distalis and at the border adjacent to the neural lobe, the corticotrophs are medium to large cells with a prominent perinuclear body, thought to be lysosomal in origin. The rough endoplasmic reticulum is well-developed, Golgi complexes are abundant, and typical secretory granules are 250–400 nm in diameter. Corticotrophs appear early in fetal life, and the number of these cells does not seem to be affected by age or sex. Hyperplasia is observed in certain disease states. Corticotrophs that were originally part of the fetal intermediate lobe migrate during development into both the posterior aspect of the pars distalis and into the neural lobe. These cells differ not only in size from those of the adenohypophysis, being smaller and more cuboidal, but also in the manner in which they modify the major posttranslational secretory products.

The human pituitary contains about 0.2 mg of ACTH, which is one of the primary secretory products of the corticotroph. ACTH is posttranslationally derived from a large glycoprotein, proopiomelanocortin (POMC), which is the primary product of the POMC gene transcription and translation. Posttranslational processing of the POMC gene product results in the formation of a variety of bioactive peptides, which may undergo further posttranslational processing including glycosylation and phosphorylation, although the significance of these final two steps in the human remains obscure. The POMC gene encodes a large protein, which in addition to ACTH contains the sequences of beta-lipotropin (LPH) and, at the N-terminus, a signal peptide followed by a large N-terminal fragment, for which no clear bioactivity has been demonstrated. Along the peptide backbone of the POMC gene product exist numerous arginine and lysine doublets, which provide cleavage sites for the generation of multiple peptide fragments. In the human adenohypophysis, the two major, bioactive fragments are the 39-amino-acid peptide ACTH and the 91-amino-acid LPH at the carboxy terminus. ACTH can undergo further processing to yield alpha-melanocyte–stimulating hormone (MSH), a 13-amino-acid peptide, and the 22-amino-acid corticotropin-like intermediate peptide (CLIP). This process probably does not occur in the pars distalis, but instead in the cells comprising the vestiges of the intermediate lobe. Within the beta-LPH peptide reside the sequences for gamma-LPH and beta-endorphin. These 58- and 31-amino-acid peptides, respectively, are liberated from the beta-LPH molecule within the corticotroph. Cells remaining from the fetal intermediate lobe, now present at the boundary between the anterior and posterior lobes and dispersed within the neural lobe, have the capacity to process gamma-LPH further to form the 18-amino-acid peptide, beta-MSH, and to generate the 12-amino-acid peptide, gamma-MSH from the N-terminal fragment of POMC.

The plasma half-life of ACTH is approximately 20 min, and only 10–20% of total content is secreted each day. Under normal, nonstressed conditions levels of ACTH in plasma are low, >10 pg/ml. ACTH is secreted in a pulsatile fashion, with pulses of only minor amplitude occurring at intervals of about 20 min throughout the day. A major increase in pulse amplitude is responsible for elevated, surge-like levels present at night. Peak levels of ACTH are present in plasma between 2 and 8 AM. In addition to this circadian fluctuation in hormone secretion, ACTH release is stimulated by external factors including stress (pain or anxiety), pyrogens, and hypoglycemia.

The primary action of ACTH is exerted in the adrenal gland, where the peptide stimulates the conversion of cholesterol to pregnenolone. After binding to its specific adrenal receptor, ACTH increases PI turnover, resulting in an activation of adenylate cyclase. The ensuing increase in phosphorylation of key enzymes eventuates both a stimulation of cholesterol esterase activity, releasing free cholesterol, and an activation of the rate-limiting 20–22 desmolase reaction, thus initiating the enzymatic pathway for the biosynthesis of adrenal steroids in general and, in particular, cortisol. Because ACTH assays have proven laborious, plasma cortisol determinations are often used to access ACTH secretion. Alpha-MSH stimulates melanocytes, and the hyperpigmentation of ACTH excess might be caused by this action. Beta-endorphin exerts profound effects on neuronal activity, yet it is unclear whether peptide of pituitary origin accesses the brain. Potentially significant effects of beta-endorphin on cardiac and renal function are now being examined in animal and human studies. [*See* CHOLESTEROL.]

The 41-amino-acid peptide, corticotropin-releasing hormone (CRH), is the primary hypothalamic factor regulating corticotroph activity. Specific CRH receptors have been localized on the corticotroph, and activation of these receptors results in increased adenylate cyclase activity, which can be correlated with ACTH secretion. A downstream effect mediated via calcium influx has also been described. Most of the neural pathways involved in the central control of ACTH release converge on the CRH-producing neurons of the hypothalamus, located predominantly in the periventricular regions. Cholinergic, serotonergic, adrenergic, GABAergic, and histaminergic influences on CRH release in the median eminence have been described. Glucocorticoids are thought to exert long-loop negative feedback on ACTH secretion by actions within the hypothalamus and directly at the level of the corticotroph. Other neuropeptides have been described to have ACTH-releasing activity as has norepinephrine. Vasopressin exerts CRH actions via unique receptors, not characteristic of either the V-1 (vascular) or V-2 (renal) binding subclasses. OT and VIP have been implicated in the control of corticotroph activity; however, their roles in the control of ACTH secretion in the human are unclear.

4. Gonadotrophs

Gonadotrophs are medium-sized cells found throughout the pars distalis, often adjacent to capillaries. Their frequent close apposition to lactotrophs has suggested paracrine interactions between these two cell types. Although most gonadotrophs, which make up 10–15% of the cells in the adenohypophysis, produce both LH and follicle-stimulating hormone (FSH), ultrastructural evidence for subpopulations of gonadotrophs producing only LH or FSH exists. Physiologic and pharmacologic studies also support the concept of distinct subpopulations. The rough endoplasmic reticulum and Golgi complexes are well-developed, and the spherical nucleus is often eccentric in location. Secretory granules containing FSH have been reported to be larger (350–450 nm) than those containing LH (150–250 nm). In the absence of circulating gondal steroids, large gonadotrophs called *castration cells* are present. These cells are characterized by dilated and enlarged cytoplasmic organelles, indicating enhanced synthetic activity.

The gonadotropins are glycoproteins comprised of alpha and beta subunits. They share a common alpha subunit, which is also similar to the alpha subunit of thyrotropin. Biologic specificity is contributed by their unique beta subunits, which do share some structural homology. Owing to the wide differences in purity of the available standard reference preparations, levels of LH and FSH in plasma and pituitary gland have been expressed in terms of an immunologic preparation and quantitated as International Units of activity (IU). Levels present in the pituitaries of adult men and menstruating women are approximately 700 IU for LH and 200 IU for FSH. These levels would be approximately 100–300 ng of hormone. Plasma levels are in the low ng/ml range and, with the exception of during the periovulatory period, are stable in adult men and women. Plasma gonadotropins are low in prepubertal children. Gonadotropins are secreted in a pulsatile fashion with peaks of LH occurring every 1–2 hr. This type of secretion is called *circhorial* because of its approximately hourly nature. Peaks of FSH are less evident because of its greater half-life in plasma than that of LH. Diurnal, or daily, fluctuations in LH and FSH secretion are minor in the adult; however, such changes do appear during sleep in pubertal females. FSH appears to be preferentially secreted before puberty. A third mode of gonadotropin secretion is related to the ovarian cycle of adult, premenopausal women. These surges are the gonadotropin response to positive gonadal steroid feedback. In postmenopausal women, the low circulating levels of estrogen and to some degree inhibin result in a rise in gonadotropin secretion, often to concentrations 10–15 times greater than those present before ovarian failure.

In the testes, FSH stimulates spermatogenesis via an action on the Sertoli cells. It also induces LH receptors on the Leydig cells, thereby promoting the action of LH to stimulate testosterone production. In the ovary, FSH stimulates follicular growth past the early antrum stage. It also acts on the granulosa cells to cause the conversion of androgens to estrogen, via an action on aromatase activity, and synergizes with estrogen to increase LH receptors on the granulosa cells. Follicular growth is also stimulated by LH. Additionally, LH stimulates androgen production by the thecal cells, thereby synergizing with FSH to stimulate estrogen production. Follicular rupture during ovulation and corpus luteum formation is stimulated by unknown mechanisms by LH, which thereafter exerts a luteotropic effect stimulating estrogen and progesterone formation.

Neural and hormonal factors control gonadotroph activity. The decapeptide hormone, gonadotropin–releasing hormone (LHRH), is produced in neurons located in the infundibular region and the rostral hypothalamus. These neurons project to extrahypothalamic sites where they are thought to be involved in the generation of stereotypic sexual behaviors and to the median eminence for release of GnRH into the vicinity of the capillaries of the portal vessels. GnRH binds to specific receptors on the gonadotrophs and stimulates cAMP formation; however, it is not clear that this is the second-messenger system involved in the stimulatory effect of the decapeptide. Instead the role of extracellular calcium has been supported by cell culture studies, which revealed a calmodulin-sensitive pathway. The mechanism of action of the calcium effect of GnRH seems to involve phospholipid turnover via activation of phospholipase A-2 and arachidonic acid formation with subsequent protein phosphorylation. Other byproducts of PI breakdown might also be involved in GnRH activation of the gonadotroph because diacylglycerol formed by PI hydrolysis activates protein kinase C, which synergizes with calmodulin.

Although the existence of a separate hypothalamic, FSH-releasing hormone has been speculated on for many years, it has failed to be identified. Instead a glycoprotein of gonadal origin, originally named *inhibin*, acts at both the hypothalamic and pituitary levels to inhibit FSH secretion. Two forms of inhibin have been isolated from porcine follicular fluid, each comprised of alpha and beta subunits linked by disulfide bridges. Inhibin A differs from inhibin B only in the structure of the beta subunit. Recent studies in humans point to the stimulatory effect of FSH on inhibin release, suggesting a physiologically significant long-loop negative feedback from gonads to the hypothalamo-pituitary unit. Recently it has been reported that a protein formed by disulfide linkage of the respective beta subunits of inhibin A and inhibin B has potent FSH-releasing activity. This protein has been named *activin*.

In addition to the novel glycoprotein-mediated gonadal influences on the gonadotroph, gonadal steroids exert profound negative and positive effects on the release of the gonadotropins. Mainly exerted by estradiol, the negative feedback effects of gonadal steroids are a result of dampening of the hypothalamic pulse generator, which controls episodic GnRH secretion, and a result of a diminution of pituitary responsiveness to GnRH. Thus, in the follicular phase of the human menstrual cycle when estradiol and progesterone levels are low, the frequency of LH pulses is higher than during the luteal phase, when progesterone and estradiol secretion is maintained. At midcycle in the human female, rising estrogen levels from the developing follicle begin exerting a positive feedback effect by increasing GnRH pulsatility and sensitizing the gonadotroph to the action of the decapeptide. The midcycle surges of LH and FSH are then terminated by the high secretion rates of estrogen and progesterone from the corpus luteum.

5. Thyrotrophs

Thyrotrophs make up only a small proportion of adenohypophyseal cells (5–10%), are located in the center of the pars distalis, and are characterized by their large size and the presence of small secretory granules. These cells produce the glycoprotein hormone thyroid-stimulating hormone (TSH), which like the gonadotropins consists of alpha and beta subunits. These subunits are products of separate genes and are assembled only after addition of the carbohydrate side chains. The alpha subunit is structurally homologous to that of LH and FSH. The adenohypophysis contains approximately 0.1 mg TSH. The plasma half-life of the glycoprotein is about 50 min. Normal plasma levels are 1–4 ng/ml, which are often still referred to in terms of International Units (normal mean, 1.8 μU/ml). These low levels often complicate accurate determination of secretory activity of the thyrotroph, necessitating measurement of plasma thyroid hormones as an indicator of thyrotroph function.

Adenylate cyclase activity of the thyroid is stimulated as a consequence of TSH binding to a specific receptor. The increased cAMP levels result in protein phosphorylation and numerous changes in thyroid cell activity. Morphologic changes occur, the cell becoming cylindrical in shape losing its cuboidal profile and the amount of colloid in the follicles declines. TSH stimulates iodide transport, thyroglobulin synthesis, iodotyrosine and iodothyronine formation, thyroglobulin proteolysis, and thyroxine (T4) and triiodothyronine (T3) release.

Neural and hormonal factors control the release of TSH. A small (3-amino-acid) peptide known as TRH is produced in neurons of the preoptic and parvocellular paraventricular hypothalamic areas and delivered to the median eminence. TRH binding results in increased cytosolic calcium, probably caused by mobilization from mitochondrial stores. This free calcium binds to calmodulin and activates protein kinases, leading to eventual protein phosphorylation and granule extrusion. Continued TRH exposure, however, down-regulates the number of TRH receptors at the cell surface, resulting in decreased thyrotroph responsiveness. Estrogens up-regulate the TRH receptor, although T4 causes a decrease in the number but not affinity of TRH receptors. T4 can be converted within the thyrotroph to T3 by the action of 5′-monodeiodinase. In addition to down-regulating the number of TRH receptors, thyroid hormones decrease TSH synthesis. Finally, although there long has been speculation that thyroid hormones can exert long-loop negative feedback effects on hypothalamic mechanisms controlling TSH release, only recently have studies in monkeys demonstrated this possibility and also has the transcription of the thyroid hormone receptor gene been demonstrated in the hypothalamus.

III. Posterior Lobe Function

A. Morphology

The posterior lobe, often called the *neurohypophysis*, weighs approximately 0.10–0.15 g in the adult. As discussed above, the lobe is made up primarily of axonal processes that originate in the supraoptic and paraventricular hypothalamic nuclei. These fibers are unmyelinated and possess terminal swellings characteristic of neurosecretory endings. The lobe also contains corticotrophs that were originally part of the fetal pars intermedia and glial elements called *pituicytes*. The hypothalamo-neurohypophyseal fibers deliver OT and vasopressin (AVP) to the neural lobe in association with specific proteins, the neurophysins, which were once thought to function as carrier molecules but are now known to be posttranslational products of the OT and AVP prohormones. Their biologic functions remain undetermined. Neurons producing OT and AVP also project to brain sites other than the neural lobe, including regions associated with stereotypic behaviors that have been associated with the peptides, and to the median eminence. Therefore, these peptides are communicated via either the long portal vessels from the median eminence or the short portal vessels from the neural lobe to the adenohypophysis. Releasing factor activities for OT at the level of the corticotroph and lactotroph and for AVP at the levels of the corticotroph and thyrotroph have been described, although their role as trophic agents in human physiology has yet to be convincingly established.

Both hormones are peptides consisting of nine amino acids, containing an internal disulfide linkage between cystine residues at positions 1 and 6. The presence of this disulfide bridge is necessary for the peptides to maintain full bioactivity. They differ in structure from each other only at the 3 and 8 positions. As mentioned above, each hormone is synthesized as part of a larger prohormone, the neurophysin molecule (approximately 10,000 MW) being cleaved off during posttranslational processing. Estrogen treatment stimulates the production of OT and its associated neurophysin. Thus this neurophysin has been termed the *estrogen-stimulated neurophysin*. AVP-associated neurophysin production is stimulated by nicotine, so this glycoprotein is known as the *nicotine-stimulated neurophysin*. Individual neurons of the magnocellular supraoptic and paraventricular nuclei produce either OT or AVP, never both. The prohormones are transferred to the Golgi apparatus after synthesis on the ribosomes and then packaged into neurosecretory granules, in which posttranslational cleavage of the hormone from its associated neurophysin takes place. Delivery down the axon to the neurohypophysis occurs with flow rate of 8 mm/hr (faster than normal axoplasmic flow). These hypothalamo-neurohypophyseal cells are neurons capable of generating and propagating action potentials that on arrival at the terminals result in depolarization and exocytosis of the neurosecretory granules. This release phenomenon is called *stimulus-secretion coupling*. Both hormones circulate in blood unbound to carrier proteins and are rapidly removed from the circulation by the kidney, liver, and brain. Although the older literature suggests a plasma half-life for OT and AVP of only 5 min, recent studies using physiological dose levels of the hormone suggest half-lives of less than 1 min. Plasma levels in adults under resting conditions are approximately 1–2 pg/ml for AVP and 10–100 pg/ml for OT. OT levels are slightly higher in adult, nonpregnant, nonlactating females than in males and are known to fluctuate during the menstrual cycle in correlation with estrogen levels.

B. Posterior Pituitary Hormones

1. Vasopressin

Two classes of AVP receptors have been established. V2 receptors are present on the peritubular (serosal) surface of cells in the distal convoluted tubules and medullary collecting ducts in the kidney. Adenylate cyclase activity is stimulated in these cells by AVP, and the cAMP formed activates a protein kinase on the luminal membrane, which results in protein phosphorylation and an enhancement of permeability of the cell to water. The change in permeability stimulated by AVP permits back diffusion of solute-free water down an osmotic gradient from hypotonic urine to the hypertonic renal medullary interstitium. The end result is an increase in urine osmolality (relative to glomerular filtrate or plasma) and a decrease in urine flow. A second class of AVP receptor exists on vascular smooth muscle. Activation of these V1 receptors results in vasoconstriction, shunting of blood away from the periphery, and elevation of central venous pressure. This is demonstrated by the observation that total peripheral resistance increases in a linear fashion with increases in plasma AVP levels. Normally, the increase in total peripheral resistance caused by AVP is buffered by baroreceptor mechanisms, which reflexively cause a decrease in cardiac output and sympathetic tone, so that other than regional perfusion effects, the increase in arterial pressures is not great.

Release of AVP can be stimulated by osmotic and nonosmotic factors. These are best typified by water deprivation and hemorrhage-induced release of AVP, respectively. Water restriction results in increased plasma osmolality as the water lost in urine and via evaporation and respiration is not replaced. This increase in plasma osmolality is sensed by specialized cells called *osmoreceptors*, located in two vascular regions of the central nervous system where the blood-brain barrier is missing: the vascular organ of the lamina terminalis (OVLT) and the subfornical organ (SFO). Increases in plasma osmolality result in a loss of cellular water from these osmoreceptors and, via neuronal relays probably involving acetylcholine-containing neurons, a stimulation of AVP release. The sensitivity of the osmoreceptor to changes in plasma osmolality can be enhanced by circulating angiotensin II. This mode of stimulation of AVP release is exquisitely sensitive, AVP being released in response to changes in plasma osmolality as small as 1%. Water loading, however, results in decreased plasma osmolality, a net gain of water in the osmoreceptors and an inhibition of AVP release.

Any stimulus resulting in a decrease in central blood volume (hemorrhage of greater than 8%, orthostatic hypotension induced by quiet standing, positive pressure breathing) is a potent stimulus for AVP secretion. However, maneuvers that increase total blood volume (isotonic saline or whole blood infusion, cold water immersion) suppress AVP release via an alpha-adrenergic mechanism. Low (left atrial) and high (carotid and aortic) pressure baroreceptors detect small alteration in blood volume (pressure) and, via afferents in the ninth and tenth cranial nerves, communicate volume–pressure changes to the central nervous system. Medullary structures receiving this input send efferents to the hypothalamus, which normally inhibit AVP release. Decreases in blood pressure result in a lowering (unloading) of neuronal activity in the afferents from the baroreceptors, less inhibition of AVP release, and increased circulating levels of the hormone. Renin released from the juxta-glomerular apparatus of the kidney in response to hypovolemia initiates the formation of angiotensin II and synergizes with it to sensitize the osmoreceptors leading to additional stimuli for AVP release. This phenomenon explains the increased sensitivity of the osmoreceptors to changes in plasma osmolality in volume-depleted states. This resetting of the osmoreceptor is thought to play an important role in the maintenance of normal fluid and electrolyte balance. A variety of other factors can modulate or alter AVP release. Secretion is induced by increased $PaCO_2$ or reduced PaO_2, pain, stress, heat, beta-adrenergic agents, gonadal steroids, opiates, barbiturates, nicotine, and prostaglandins. Hormone release is inhibited by cold, alpha-adrenergic agents, ethanol, and the cardiac hormones.

Secondary actions of AVP have been described. AVP appears to act within the central nervous system to facilitate the consolidation and retrieval of memory. Studies in humans have indicated a beneficial effect of AVP on short-term memory. As mentioned above, under certain circumstances, AVP can act as a CRH within the adenohypophysis. This effect is thought to represent in physiological terms a potentiative effect to that of CRH during stress. The most common example of altered AVP secretion in humans is the syndrome of deficient secretion called *diabetes insipidus*. Because of some deficiency in the hypothalamo-neurophypophyseal system, AVP secretion is low or nonexistent. This syndrome is characterized by an inability to concentrate urine, frequent urination, and excessive thirst. Patients are treated with a AVP analogue that has no V1 activity, thus the renal effects only predominate.

2. Oxytocin

The primary action of OT is on the myoepithelial cells surrounding the alveoli and ducts of the mammary gland, stimulating contraction of these cells and milk ejection. OT also stimulates rhythmic contraction of myometrial cells in the uterus, aiding in delivery of the fetus. Its uterine effects are not mandatory for the initiation of labor; however, it proceeds more slowly in the absence of OT. A commercial preparation of OT, called *pitocin*, is often employed postpartum to sustain contractions and decrease bleeding. Secondary actions of OT include a central effect to initiate maternal, instinctive behaviors and a pituitary action to stimulate PRL and ACTH secretion.

Activation of touch receptors in the nipples or vaginal stimulation during intercourse or delivery result in increases in afferent input to the OT-producing cells of the supraoptic and paraventricular nuclei. These spinothalamic inputs are thought to use acetylcholine or dopamine as final neurotransmitters. Estrogens secreted by the developing follicle in the periovulatory period also stimulate OT release. Mild stress (restraint, novel environment, apprehension, and fear) and hemorrhage also stimulate OT secretion. Inhibition of OT release is seen during extreme pain or high-temperature exposure. Disorders in OT secretion are often undetected. Deficient OT secretion results in difficulty nursing because of inadequate milk ejection.

Bibliography

Daughaday, W. (1985). The anterior pituitary. *In* "Williams Textbook of Endocrinology" (J. W. Wilson and D. W. Foster, eds.). W. B. Saunders, Philadelphia.

Ganten, D., and Pfaff, D., ed. (1985). "Neurobiology of Vasopressin." Springer-Verlag, Heidelberg.

Ganten, D., and Pfaff, D., ed. (1986). "Neurobiology of Oxytocin." Springer-Verlag, Heidelberg.

Imura, H., ed. (1985). "The Pituitary Gland." Raven Press, New York.

Muller, E. E., and Nistico, G., ed. (1989). "Brain Messengers and the Pituitary." Academic Press, San Diego.

Reichlin, S. (1985). Neuroendocrinology. *In* "Williams Textbook of Endocrinology" (J. D. Wilson and D. W. Foster, eds.). W. B. Saunders, Philadelphia.

Samson, W. K., and Mogg, R. J. (1990). Oxytocin as part of stress response. *In* "Behavioral Aspects of Neuroendocrinology" (D. Ganten and D. Pfaff, eds.). Springer-Verlag, Heidelberg.

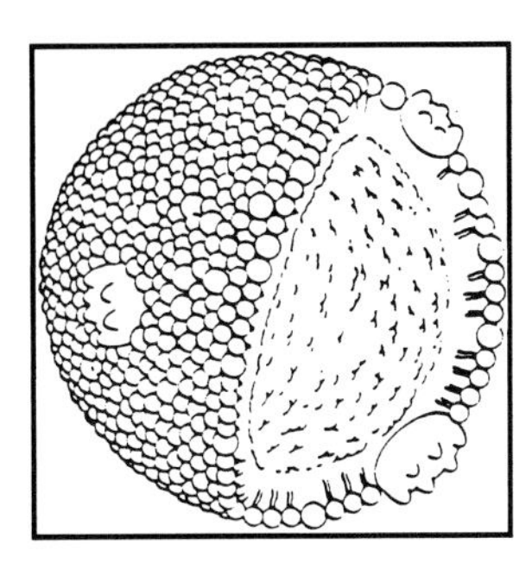

Plasma Lipoproteins

THOMAS L. INNERARITY, *Gladstone Foundation Laboratories for Cardiovascular Disease, Cardiovascular Research Institute, University of California, San Francisco*

Glossary

Apoprotein (or apolipoprotein) Protein component of a lipoprotein that transports lipids and acts as a ligand for lipoprotein receptors and as a cofactor for enzymes; the apoproteins are apoA, -B, -C, -D, and -E

Apoprotein B-100 (or apoB-100) Huge apoprotein (M_r = 550,000) that is the sole protein component of low-density lipoproteins and mediates the binding of low-density lipoproteins to their receptor; it is also present in very low-density and intermediate-density lipoproteins

Apoprotein E (or apoE) Apoprotein component of very low-density and intermediate-density lipoproteins and chylomicron remnants that mediates their binding to the low-density lipoprotein and chylomicron remnant receptors; it is also a component of a subspecies of high-density lipoproteins

Cholesteryl ester transfer protein Enzyme that catalyzes the transfer of cholesteryl esters from high-density to very low-density lipoproteins in exchange for triglycerides

Chylomicron remnant receptor Hepatic receptor that binds and removes chylomicron remnants from the circulation (also known as the apoprotein E receptor)

Chylomicron remnants Chylomicrons in which the triglycerides have been hydrolyzed by lipoprotein lipase; these lipoprotein particles are readily cleared from the circulation by hepatic lipoprotein receptors

Chylomicrons Huge triglyceride-rich lipoproteins (d < 0.95 g/ml) synthesized by the small intestine after a fatty meal; the ability of these large chylomicron particles to scatter light gives plasma from nonfasted subjects its milky appearance

High-density lipoproteins (d = 1.063–1.21 g/ml) Smallest and most protein-rich of the lipoproteins; plasma high-density lipoprotein concentration is inversely correlated with the risk of atherosclerosis; they promote the efflux of cholesterol from cells

Intermediate-density lipoproteins (d = 1.006–1.019 g/ml) Intermediate in density (between very low-density and low-density lipoproteins) and formed by the action of lipoprotein lipase on very low-density lipoproteins

Lecithin–cholesterol acyltransferase Enzyme that catalyzes the transfer of a fatty acid from the β position of phosphatidylcholine to the 3-β-hydroxy position of cholesterol, forming a cholesteryl ester

Lipoprotein lipase Enzyme responsible for the hydrolysis of triglycerides on chylomicrons and very low-density lipoproteins

Low-density lipoprotein receptor Membrane-bound receptor responsible for clearing low-density lipoproteins and other lipoproteins from the plasma, thereby supplying cells and tissues with cholesterol; their ligands are apoproteins B-100 and E

Low-density lipoproteins (d = 1.019–1.063 g/ml) End product of the lipolysis of very low-density and intermediate-density lipoproteins by lipoprotein lipase; they are cholesterol-rich lipoproteins that deliver cholesterol to cells via the low-density lipoprotein receptor; plasma low-density lipoprotein concentration is positively correlated with coronary heart disease

Space of Disse Small space between endothelial cells of the liver and hepatocytes that is exposed to the plasma

Very low-density lipoproteins (d < 1.006 g/ml) Synthesized by the liver and transport triglycerides to peripheral cells and tissue

LIPIDS ARE FATTY SUBSTANCES that are normally insoluble in an aqueous environment but are solubilized and transported in the plasma as water-soluble macromolecular complexes (pseudomicellar particles) known as lipoproteins. The major lipid components of lipoproteins are triglycerides, cholesterol, and phospholipids. The protein components are referred to as apolipoproteins, or apoproteins. Lipoproteins are spherical particles that comprise a nonpolar core (composed of triglycerides and cholesteryl esters) surrounded by a shell of polar lipids (composed of phospholipids and unesterified cholesterol) and apoproteins (see Fig. 1 and Tables I and II). Triglycerides are used to produce energy and to synthesize other lipids. Cholesterol is an essential component of cell membranes and is a precursor of steroid hormones and bile acids. Phospholipids are the principal component of cell membranes and serve as intra- and intercellular messengers. Lipoproteins transport lipids from their origin (liver or small intestine) to peripheral cells, tissues, and organs. Lipoproteins have a pathogenic role in atherosclerosis and heart disease: Numerous epidemiological studies have shown that elevated concentrations of low-density lipoprotein (LDL) cholesterol and decreased concentrations of high-density lipoprotein (HDL) cholesterol in the plasma are associated with an increased risk of coronary heart disease.

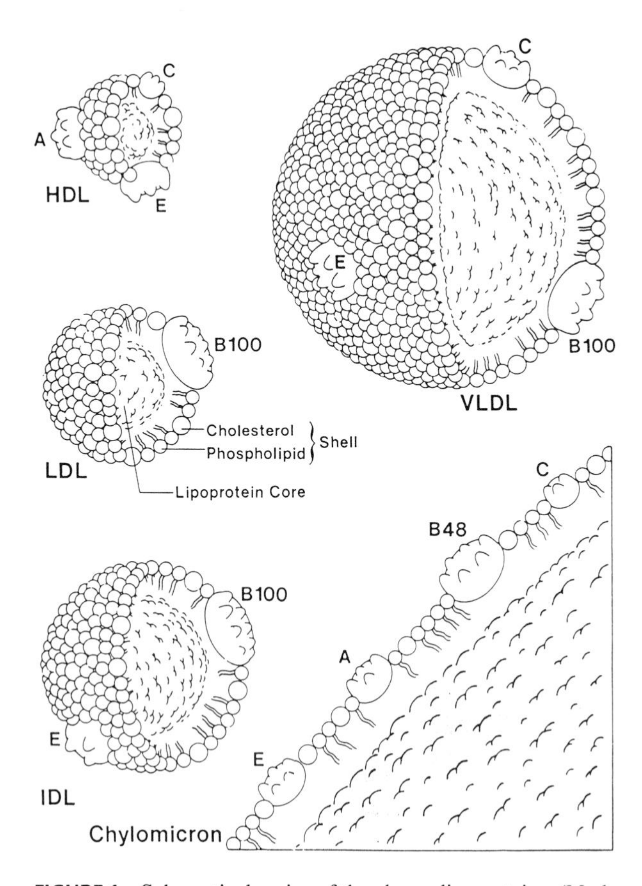

FIGURE 1 Schematic drawing of the plasma lipoproteins. (Modified from *Scientific American Medicine,* Section 9, Subsection II. © 1989 Scientific American, Inc. All rights reserved.)

I. Lipoproteins

A. General Concepts

Plasma lipoproteins are the primary carriers of lipids in the circulation. Lipoproteins can be divided into six major classes and two specialized classes (Table I). Four of the major classes of lipoproteins—very low-density lipoproteins (VLDL), intermediate-density lipoproteins (IDL), LDL, and HDL—are derived from the liver and are present in plasma from both fasted and nonfasted subjects. The other two major classes—chylomicrons and chylomicron remnants—are derived from the small intestine and are found in the plasma only after a fatty meal (postprandially). Like LDL, the specialized classes of lipoproteins, β-VLDL and lipoprotein(a), are significant because they are positively correlated with coronary heart disease and atherosclerosis.

Lipoproteins are classified according to the density of the salt solution in which they are separated by ultracentrifugation. The lipoproteins separate as they do because each class of lipoproteins is composed of a different percentage of lipids and proteins. The distinct composition and the disparity in density between lipids and proteins explain the differential flotation characteristics of lipoproteins. For example, as compared with water (1.0 g/ml) and normal saline (1.006 g/ml), tripalmitin has a density of 0.875 g/ml, cholesterol oleate 0.98 g/ml, dipalmitoyl lecithin 1.03 g/ml, and cholesterol 1.067 g/ml, whereas the density of the apoprotein moiety of these lipoproteins ranges from 1.33 to 1.37 g/ml.

TABLE I Characteristics of Plasma Lipoproteins

Lipoprotein class and density of flotation (g/ml)	Origin	Major functions	Major apoproteins	Principal lipids
Endogenous lipoproteins				
VLDL (d < 1.006)	Liver	Transport triglycerides to various tissues. Fatty acids hydrolyzed by LPL are used by the tissues for energy or stored as triglycerides.	B-100, E, C-I, C-II, C-III	Triglycerides
IDL (d = 1.006–1.019)	Derived from the lipolysis of VLDL by LPL	Some are precursors of LDL; some are taken up by the liver and provide cholesterol for the liver.	B-100, E	Cholesterol, triglycerides
LDL (d = 1.019–1.063)	Derived from IDL following lipolysis by LPL	Provide cholesterol to various tissues by receptor-mediated endocytosis. Increased levels of LDL cholesterol correlate directly with coronary heart disease.	B-100	Cholesterol
HDL (d = 1.063–1.21)	The apoproteins are derived from the hydrolysis of chylomicrons and VLDL and from *de novo* synthesis by the liver and intestine	Facilitate removal of cholesterol from extrahepatic tissues for transport to the liver. HDL cholesterol correlates inversely with coronary heart disease.	A-I, A-II, C-I, C-II, C-III	Phospholipid, cholesterol
Exogenous lipoproteins				
Chylomicrons (d < 0.95)	Small intestine	Transport triglycerides and cholesterol from intestine to the plasma.	B-48, A-I, A-II, A-IV, E, C-I, C-II, C-III	Triglycerides
Chylomicron remnants (d < 0.95)	Derived from chylomicrons following lipolysis by LPL	Transport cholesterol and triglycerides to the liver. The uptake by the liver is mediated by lipoprotein receptors.	B-48, E	Cholesterol, triglycerides
Specialized lipoproteins				
β-VLDL (d < 1.006)	Two subclasses: I. Intestine and II. Liver	Potentially atherogenic lipoproteins: subclass I represents cholesterol-rich chylomicron remnants; subclass II cholesterol-rich VLDL remnants	B-100 or B-48, E	Cholesteryl ester
Lp(a) (d = 1.05–1.12)	Liver	Concentration and phenotypes of Lp(a) correlate with accelerated atherogenesis.	B-100, (a)	Cholesterol

TABLE II Characterization of the Major Apoproteins

Apoprotein	Plasma concentration (mg/dl)	Molecular weight (number of amino acids of the mature protein)	Major sites of synthesis	Chromosomal location of the gene	Functions	Clinical disorders due to genetic variants or mutations
A-I	100–130	28,100 (243)	Small intestine, liver	11q	Structural protein of HDL; LCAT activator; tissue cholesterol efflux	Apo A-I–C-III deficiency Apo A-I deficiency A-I multiple mutants Tangier disease
A-II	30–50	17,400 (77)	Small intestine, liver	1q	Structural protein of HDL	
A-IV	15	43,000 (376)	Small intestine	11q	Associated with triglyceride transport in chylomicrons	
B-100	80–120	550,000 (4536)	Liver	2p	Necessary for VLDL biosynthesis and secretion; ligand for the LDL receptor	Abetalipoproteinemia Familial hypobetalipoproteinemia Familial defective apoB-100
B-48	<5	250,000 (2152)	Small intestine	2p	Necessary for chylomicron biosynthesis and secretion	Abetalipoproteinemia Chylomicron retention disease
C-I	5–7	6,600 (57)	Liver	19q	Modulates LCAT activation	
C-II	3–7	8,000 (79)	Liver	19q	Activates LPL	Familial hyperchylomicronemia
C-III	9–13	8,750 (79)	Liver	11q	Modulates receptor uptake of chylomicron remnants	
D	6–7	33,000 (169)	Adrenal, kidney, brain, liver, small intestine, etc.	3q	Unknown	
E	3–6	34,200 (299)	Liver, macrophages in various organs, astrocytes in the brain	19q	Ligand for lipoprotein receptors	Type III hyperlipoproteinemia Isoforms correlated with plasma cholesterol levels
(a)	0–100	350,000–750,000 (4,529)	Liver	6q	Binds to plasminogen receptors on endothelial cells	Levels and phenotypes of Lp(a) are correlated with coronary heart disease

Thus, lipoproteins with a greater proportion of lipids, especially neutral lipids (triglycerides and cholesteryl esters), are less dense and float at a lower density upon ultracentrifugation. In practice, lipoproteins from the plasma of fasted individuals are usually isolated by sequential ultracentrifugation using increasingly dense (more concentrated) salt solutions. For example, VLDL (d < 1.006 g/ml) float to the top of the tube when plasma is ultracentrifuged. After the VLDL are removed, the remainder is raised to d = 1.019 g/ml with salt and the ultracentrifugation is repeated to isolate IDL (d = 1.006–1.019 g/ml), which will now float to the top of the tube. The process is repeated to isolate LDL (d = 1.019–1.063 g/ml) and HDL (d = 1.063–1.21 g/ml). The density of lipoprotein particles is also inversely related to their size. The larger lipoprotein particles (i.e., chylomicrons, chylomicron remnants, and VLDL) contain a larger nonpolar core (triglycerides and cholesteryl esters) than do LDL and HDL (see Fig. 1). The classification of lipoproteins is further defined on the basis of electrophoretic mobility and apoprotein content. [*See* LIPIDS; PROTEINS.]

Although the lipoprotein classes are treated as distinct structural and functional entities, they should also be viewed as micellar structures in a state of dynamic equilibrium with each other and with cellular membrane components exposed to the circulation. For example, unesterified cholesterol rapidly equilibrates between various lipoprotein particles and cell membranes. Many of the apoproteins readily exchange between certain lipoproteins. Other components, such as the insoluble neutral lipids, exchange between lipoproteins with the help of transfer proteins, such as cholesteryl ester transfer protein, a protein that catalyzes the exchange of cholesteryl esters and triglycerides between certain lipoprotein classes.

The apoproteins act as detergents in that they solubilize the lipids for transport in an aqueous environment. These proteins have amphipathic helical regions that are well suited for this function. These regions contain both polar and nonpolar amino acid residues that line up on opposite sides of an α-helix. Thus, the helix has a hydrophobic (nonpolar) face and a hydrophilic (polar) face. Acidic and basic residues are located at the center or edge of the polar face, often as charged pairs. The nonpolar hydrophobic residues on the opposite face are thought to interact with the phospholipid surface of the lipoprotein shell. The combination of polar and nonpolar amino acid residues in this arrangement gives apoproteins their detergent properties and their ability to transport hydrophobic lipids in an aqueous environment.

B. Chylomicrons (d < 0.95 g/ml)

The largest of the lipoproteins (>1,000 Å in diameter) are synthesized by the mucosa of the small intestine to transport dietary lipids from the site of absorption into the circulation. These huge lipoproteins are composed mainly of lipid (~99%), of which 90% is triglycerides (Figs. 1 and 2 and Table I). The small amount of protein is made up of apoB-48, apoA-I, apoA-II, apoA-IV, and the C apoproteins.

The distinctive apoprotein that is a "marker" for intestinally derived lipoproteins is apoB-48. This form of apoB (approximately the amino-terminal 48% of apoB-100) is synthesized only in the intestine (in humans) and, therefore, is unique to chylomicrons and chylomicron remnants. Moreover, apoB-48 is synthesized by a novel biological process that has only been observed for this protein. Although both the hepatic apoB (apoB-100) and the intestinal apoB (apoB-48) are derived from the same gene, apoB-48 is not produced by alternate splicing of the RNA transcript or proteolytic processing of the polypeptide chain. Instead, this apoprotein is the result of a unique modification of the apoB mRNA that occurs only in the intestine: A nucleotide, cytosine, at a single site (apoB nucleotide 6666) is changed to a uracil. This substitution changes apoB codon 2153 from a CAA (a codon for glutamine) to the translational termination codon UAA. As a result of this termination of protein translation, a shortened apoB (apoB-48) is produced in the intestine. The two B apoproteins also differ in function. Apoprotein B-100 is involved in receptor-mediated clearance of LDL because it has a receptor-binding domain in its carboxy-terminal segment. Because apoB-48 does not contain this portion of apoB, it does not bind to lipoprotein receptors. Finally, both forms of apoB are required for the synthesis of different lipoproteins: chylomicrons (apoB-48) or VLDL (apoB-100).

Chylomicrons are synthesized by epithelial cells throughout the length of the small intestine; the bulk are produced by the jejunum. Digested dietary fats (fatty acids and cholesterol) in the form of bile acid micelles induce the synthesis of chylomicrons in the brush-border membranes of endothelial cells.

The rate of secretion of chylomicrons is a direct function of the rate of fat absorption. In experimental animals, large quantities of dietary fat increase the amount of triglycerides secreted by a factor of 20, whereas the amount of apoB is increased by only about twofold. These data are explained by the incorporation of many more molecules of triglyceride than apoB into chylomicron particles. The fatty acids that result from the hydrolysis of dietary triglycerides and monoglycerides serve as substrates for triglycerides that are being resynthesized in the smooth endoplasmic reticulum of the absorptive cell. Other lipid components (cholesterol, phospholipids) are either directly absorbed into the cell or are synthesized *de novo*. Nascent chylomicrons appear to be assembled and to accumulate in the Golgi apparatus, where glycosylation of the apoproteins occurs. The chylomicrons are released into the mesenteric lymph, proceed to the thoracic duct lymph, and finally enter the general circulation. Newly secreted chylomicrons possess apoB-48, apoA-I, apoA-II, and apoA-IV (apoproteins synthesized by the intestine) and later acquire apoE and apoC-I, apoC-II, and apoC-III when exposed to other lipoproteins (mainly HDL) in the lymph and plasma (Fig. 2). Apparently, apoB-48 is the only apoprotein that has to be synthesized for chylomicron secretion to occur.

C. Chylomicron Remnants ($d < 0.95$ g/ml)

Chylomicrons in the plasma are a substrate for lipoprotein lipase (LPL), an enzyme that hydrolyzes much of the core triglycerides, releasing free fatty acids (Fig. 2). The fatty acids are taken up by various tissues and oxidized as an energy source or are re-esterified to triglycerides. The triglycerides are stored in adipose tissue and are used as an energy source as needed. The triglyceride-depleted chylomicrons are referred to as chylomicron remnants. [*See* FATTY ACID UPTAKE BY CELLS.]

The initial step in the hydrolysis of chylomicrons is the direct binding of the lipoprotein particle to LPL on the endothelium of the capillary beds of adipose tissue, cardiac and skeletal muscles, and several other tissues (Fig. 2). Apoprotein C-II is a necessary cofactor in this process, greatly increasing the catalytic rate of triglyceride hydrolysis. Genetic mutations of either LPL or apoC-II result in the accumulation of chylomicrons and triglycerides in the plasma.

The depletion of core triglycerides results in the transfer of excess surface components such as phospholipids, A apoproteins, and C apoproteins from the chylomicron remnants to HDL (Fig. 2). At the same time, the remnants are becoming enriched in apoE acquired from HDL. With the loss of apoC-II and other changes in their composition, chylomicron remnants are a poorer substrate for LPL and the rate of hydrolysis of the triglycerides slows. Because only the triglycerides are hydrolyzed, the shrunken core of the remnants is relatively enriched in cholesterol and cholesteryl esters.

Chylomicron remnants are cleared from the circulation by the liver. Lipoprotein receptors on the sinusoidal surface of hepatocytes exposed to the blood recognize apoE on the surface of the remnant particles. This receptor-mediated process is extremely efficient: In humans, remnant particles have circulation half-life of less than 10 min.

Two lipoprotein receptors mediate the catabolism of the remnants: the LDL receptor and the chylomicron remnant receptor (Fig. 2). Although the chylomicron remnant receptor has not been biochemically identified or characterized, physiological and genetic data indicate that such a receptor exists and plays the principal role in removing these remnants from the circulation. The LDL receptor most likely removes a portion of the remnants. Because the remnant receptor has not been characterized, the intracellular pathway of this catabolic process remains vague. Current evidence indicates that remnants are degraded in the lysosomes of hepatocytes because the lipolysis and proteolysis of the remnant components are virtually complete. Lipids from the chylomicron remnants can then be reused for VLDL synthesis. For example, the triglycerides are hydrolyzed, and the resulting free fatty acids are reesterified into triglycerides and resecreted as the lipid core of VLDL. The cholesterol from the remnants is also used for VLDL formation, but it can have several additional fates: It can be directly secreted from the hepatocyte into the bile, or it can be oxidized to bile acids before being secreted into the biliary canaliculi.

D. Very Low-Density Lipoproteins ($d < 1.006$ g/ml)

The primary lipoprotein class synthesized by the liver is the VLDL. These large lipoproteins (300–700 Å in diameter) are responsible for transporting

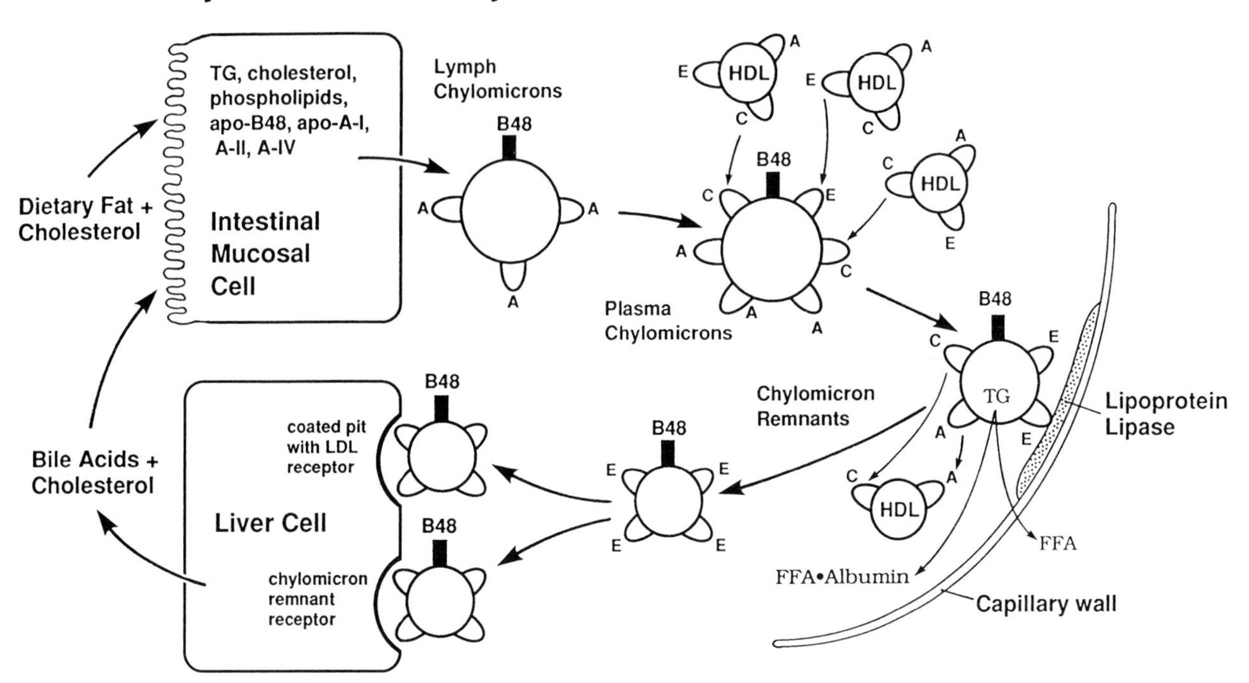

FIGURE 2 Synthesis and catabolism of intestinally derived lipoproteins. FFA, free fatty acids; TG, triglycerides. (Modified, with permission, from R. J. Havel, 1982, Approach to the patient with hyperlipidemia, *Med. Clin. North Am.* **66,** 319.)

triglycerides from the liver into the circulation. They are 88–90% lipid (~55% triglycerides, ~20% cholesterol, and ~15% phospholipid) and 10–12% protein. The apoproteins include apoB-100, apoE, the C apoproteins, and trace amounts of apoA-I. Their essential and distinctive apoprotein is apoB-100, which is necessary for the assembly and secretion of nascent VLDL. Each VLDL particle has only one copy of apoB-100.

Very-low density lipoproteins are synthesized in the liver. In cultured hepatocytes, the translation of apoB-100 requires about 10–15 min, its transport to the Golgi apparatus 5–10 min, and its secretion from the cell another 10 min. The apoB-100 molecules remain attached to the endoplasmic reticulum (possibly within the membrane) until the formation of nascent VLDL particles. Phospholipids and cholesteryl esters newly synthesized by the rough endoplasmic reticulum then combine with the apoB-100. Alternatively, cholesterol is derived from the receptor-mediated uptake and degradation of lipoproteins by the liver. Triglycerides used in VLDL assembly are synthesized in a specialized portion of the smooth endoplasmic reticulum.

The apoB-100, phospholipids, cholesterol, cholesteryl esters, and triglycerides are thought to form a complex at the smooth-surfaced termini of the rough endoplasmic reticulum. In fact, nascent VLDL particles were first visualized by electron microscopy in the transitional elements between the rough and smooth endoplasmic reticulum before they entered the Golgi apparatus. Nascent VLDL appear to be transported to the sinusoidal cell surface in large membrane-bound secretory vesicles, where they are released into the space of Disse, from which they enter the plasma. Nascent VLDL contain apoB-100, apoE, and only small amounts of the C apoproteins (Fig. 3). In the plasma, VLDL acquire additional E and C apoproteins, primarily from HDL. The nascent VLDL become mature plasma VLDL as they acquire cholesterol and lose phospholipids in exchanges with other lipoproteins.

E. Intermediate-Density Lipoproteins (d = 1.006–1.02 g/ml)

The catabolism of VLDL is very similar to that of chylomicrons. The triglycerides in VLDL are hy-

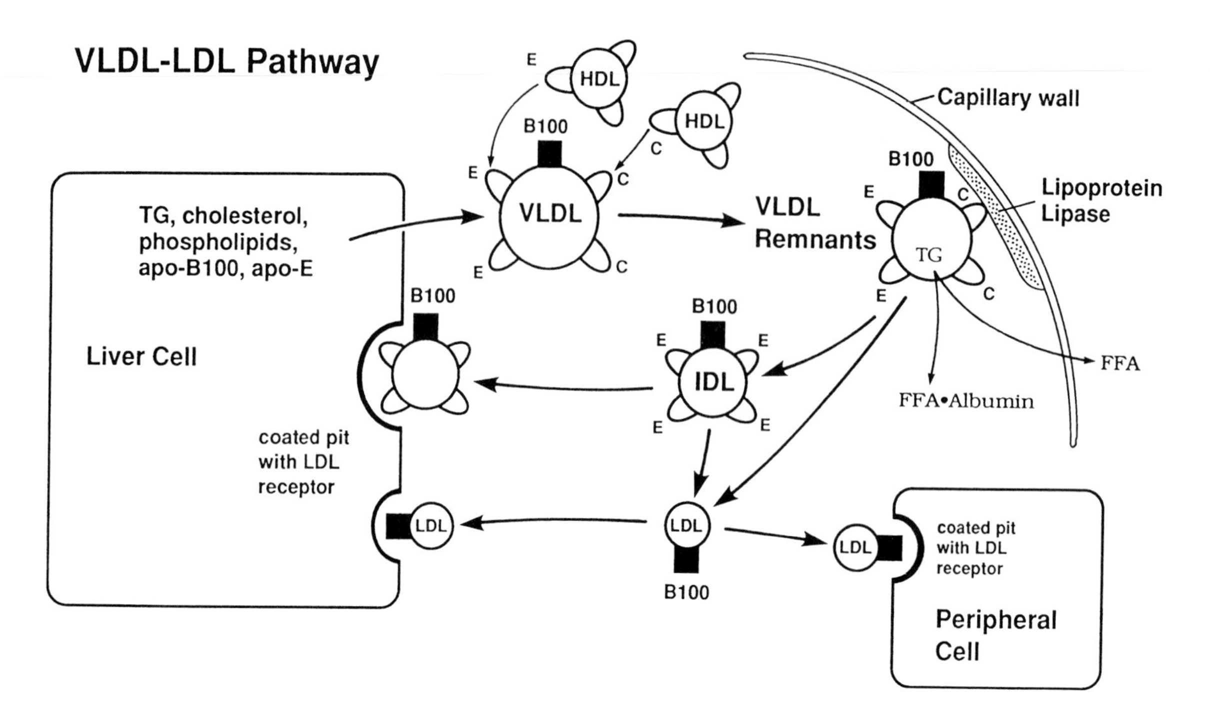

drolyzed by LPL and to a lesser extent by hepatic triglyceride lipase, resulting in the formation of VLDL remnants or IDL, or both (Fig. 3). The rate of hydrolysis of VLDL triglycerides is slower than that of chylomicron triglycerides: The normal residence time in the plasma for VLDL triglycerides is 15–60 min, whereas for chylomicron triglycerides it is 5–10 min. Depletion of the triglyceride core of VLDL, concurrent with the loss of the C apoproteins, decreases the size of the lipoprotein particles. The IDL are smaller and more cholesterol-rich than the VLDL and contain only apoE and apoB-100. Some of the VLDL remnants and IDL are cleared from the plasma via hepatic LDL receptors, and the fraction that escapes clearance by the liver is converted to LDL.

Even though newly synthesized VLDL contain apoB-100, these particles do not bind to the LDL receptor. Presumably, the receptor-binding domains of neither apoB-100 nor apoE are exposed or properly arranged in newly formed VLDL. Partial lipolysis by LPL and the concurrent loss of C apoproteins probably result in the activation or exposure of the receptor-binding domain of apoE, thereby enabling these particles to be bound and taken up through the interaction of apoE with LDL receptors. The VLDL remnants and IDL bind to the LDL receptors with a very high affinity because the multiple copies of apoE on their surface enables them to bind to several receptors simultaneously.

FIGURE 3 Synthesis and catabolism of hepatically derived lipoproteins. FFA, free fatty acids; TG, triglycerides. (Modified, with permission, from R. J. Havel, 1982, Approach to the patient with hyperlipidemia, *Med. Clin. North Am.* **66,** 319.)

Consequently, they are rapidly cleared from the circulation (where their half-life is ~15–20 minutes). After additional lipolysis, all of the apoE is lost and the receptor-binding domain of apoB-100 becomes functional: The resultant LDL particle binds to the LDL receptor via apoB-100 (Fig. 3). Compared with VLDL remnants and IDL, LDL have a lower affinity for LDL receptors because each LDL particle contains only a single copy of apoB-100, which interacts with only one receptor. Consequently, LDL remain in the circulation longer (a half-life of 2–3 days) than do IDL. In humans, about half of the VLDL remnants and IDL are taken up by the hepatic LDL receptor pathway; the remainder are converted to LDL. In general, a high expression of hepatic LDL receptors promotes the removal of IDL and VLDL remnants via this pathway and decreases the fraction converted to LDL.

F. Low-Density Lipoproteins (d = 1.019–1.063 g/ml)

Low-density lipoproteins are the major cholesterol-transporting lipoproteins in humans, carrying about two-thirds of the total plasma cholesterol. The lipid composition of LDL is ~35% cholesteryl ester, ~12% cholesterol, ~8% triglycerides, and 20% phospholipid. Lipids constitute ~75% of the LDL molecule, and apoB-100 constitutes most if not all of the remaining 25%. The LDL molecule is a spherical particle, 200 Å in diameter, and apoB-100 protein crisscrosses its surface. Because LDL possess β electrophoretic mobility, they were once referred to as β-lipoproteins.

The apoB-100 on the surface of these particles is recognized by hepatic and extrahepatic LDL receptors. This interaction is responsible for ~75% of the clearance of LDL from the plasma, principally through the liver (Fig. 3). Elevated plasma cholesterol concentrations, especially LDL concentrations, are strongly correlated with atherosclerotic coronary heart disease. Low-density lipoproteins are believed to be the major atherogenic lipoprotein class in humans; epidemiological studies have shown that even a slight reduction of serum LDL cholesterol can significantly decrease the incidence of coronary heart disease and the rate of progression of coronary lesions.

G. High-Density Lipoproteins (d = 1.063–1.21 g/ml)

High-density lipoproteins, with a particle diameter of 80–130 Å, are the smallest lipoproteins. Protein constitutes about one-half the weight of HDL, more than in any other lipoprotein class. The remainder of HDL particles are made up of various lipids: ~25% phospholipid (mainly phosphatidylcholine), ~16% cholesteryl ester, ~5% cholesterol, and ~4% triglycerides. High-density lipoproteins possess α electrophoretic mobility and were once referred to as α-lipoproteins. The major apoproteins of HDL (C and A apoproteins) are synthesized in the intestine and the liver. Both of these organs secrete nascent HDL particles, but their assembly and the form in which they enter the blood are poorly understood.

Nascent HDL produced by the liver appear as discoidal particles of about 200 Å in diameter and with a width of 45 Å. They have an outer bilayer composed of phospholipid and apoproteins, primarily apoE and apoA-I. The flattened disks are converted to spherical particles when the enzyme lecithin–cholesterol acyltransferase (LCAT) esterifies the HDL cholesterol. The intestine also appears to produce nascent discoidal HDL and small spherical HDL. In addition, a fraction of HDL, or at least major surface constituents of HDL, arise from the release of surface components from chylomicrons and VLDL during the hydrolysis of core triglycerides by LPL. These surface components include most of the A apoproteins, a large fraction of the C apoproteins, phospholipids, and some free cholesterol.

High-density lipoproteins are commonly subdivided into HDL_2 (larger, more lipid-rich HDL that float at d = 1.063–1.125 g/ml) and HDL_3 (smaller, denser particles that float at d = 1.125–1.21 g/ml). These subspecies can be remodeled by the acquisition of surface components from triglyceride-rich lipoproteins. For example, the smaller HDL_3 can acquire phospholipid and cholesterol from the lipolysis of chylomicrons and VLDL by the action of LPL, which converts them to the larger, more lipid-rich HDL_2. The concentration of HDL cholesterol, and in particular HDL_2, correlates inversely with coronary heart disease.

II. Specialized Lipoproteins

A. Lipoprotein(a)

Lipoprotein(a) [Lp(a)] closely resembles LDL in lipid composition and by the presence of apoB-100 but is clearly distinguished from LDL by the presence of a glycoprotein known as apolipoprotein(a) [apo(a)]. A single apo(a) molecule is bound to the apoB-100 molecule of LDL, most likely by a disulfide linkage. Lipoprotein(a) migrates with $\alpha 2$ electrophoretic mobility and floats at a density (1.05–1.08 g/ml) between that of LDL and HDL upon ultracentrifugation.

The highly glycosylated apo(a) exhibits a genetically determined size heterogeneity; seven isoforms migrate in the range of ~200–700 kDa on sodium dodecyl sulfate–polyacrylamide gel electrophoresis. It is encoded at a single gene locus by multiple alleles, present in different individuals, which produce proteins of various sizes. Apolipoprotein(a) is strikingly similar to plasminogen, the enzyme that proteolytically dissolves fibrin clots. The structure of apo(a) consists of an inactive protease domain and two plasminogenlike kringle domains. (Kringles are cysteine-rich sequences containing three internal disulfide bridges producing a pretzellike structure resembling the Danish kringle cake.) The size heterogeneity is due to the number of copies of kringle 4 present in the isoform. Plasma Lp(a) levels vary among individuals, from undetectable levels ≤100 mg/dl, and are related to the type of apo(a) isoform. In general, subjects with alleles that produce the smaller isoforms possess the highest concentrations of Lp(a) in their plasma.

High plasma levels of Lp(a) are strongly associated with atherosclerosis and are an independent risk factor for myocardial infarction. The risk of coronary heart disease shows a twofold increase when the level of Lp(a) is above 30 mg/dl, as it is in about 20% of the population. When LDL and Lp(a) are both elevated, the relative risk for early coronary heart disease increases up to sixfold. The mechanism by which Lp(a) accelerates atherosclerosis is unknown, but recent studies have shown that Lp(a) can bind to plasminogen receptors on macrophages and endothelial cells and accumulates in atherosclerotic lesions. Thus, Lp(a) may interfere with fibrinolysis by inhibiting plasminogen binding (and thereby formation of plasmin, the enzyme that breaks down fibrin) and/or by contributing to the formation of the foam cells characteristic of atherosclerotic lesions.

B. β-Very Low-Density Lipoproteins

β-very low-density lipoproteins (β-VLDL) are cholesteryl ester-enriched lipoproteins that float at $d < 1.006$ g/ml upon ultracentrifugation but migrate with β electrophoretic mobility. They are prominent in the plasma of animals fed a diet high in cholesterol and lipids and are present in the plasma of humans with type III hyperlipoproteinemia. β-Very low-density lipoproteins are a mixture of cholesteryl ester-enriched VLDL remnants of hepatic origin and chylomicron remnants of intestinal origin. Their major apoproteins are apoE and apoB-48 (chylomicron remnants) or apoB-100 (VLDL remnants). As in chylomicron and VLDL remnants, triglycerides constitute the predominant lipid. However, β-VLDL contain a much higher content of cholesteryl ester than do normal chylomicron and VLDL remnants.

In cholesterol-fed animals, β-VLDL accumulate because of an increased synthesis and decreased catabolism of intestinal chylomicrons and hepatic VLDL. In addition, their hepatic LDL receptors are downregulated in response to increased levels of plasma lipoproteins, preventing the remnant lipoproteins from being effectively cleared from the circulation. In subjects with type III hyperlipoproteinemia, β-VLDL accumulate because the remnant lipoproteins possess a genetic variant of apoE that is defective in its ability to bind lipoprotein receptors.

When β-VLDL are incubated *in vitro* with mouse peritoneal macrophages or human monocyte–macrophages, they are avidly taken up by these cells and cause marked intracellular cholesteryl ester accumulation. The morphological appearance of these atherosclerosis-associated cells gives them their name—foam cells. β-VLDL are the only naturally occurring lipoproteins that can cause foam-cell formation in macrophages. [*See* MACROPHAGES.]

III. Lipoprotein Receptors

Lipoprotein receptors play a critical role in lipoprotein metabolism. The best-characterized lipoprotein receptor is the LDL receptor. In 1985, Drs. M. S. Brown and J. L. Goldstein received the Nobel Prize in Physiology or Medicine for their discovery and characterization of this receptor. Their work demonstrated the significance of the LDL receptor pathway in cholesterol homeostasis and atherosclerosis and defined how this receptor influences both intracellular cholesterol concentrations and plasma cholesterol levels. The intracellular LDL receptor pathway was first defined in normal human fibroblasts and has served as a model for the study of receptor-mediated endocytosis for a number of receptor systems. [*See* CHOLESTEROL.]

The LDL receptor is a transmembrane protein residing in the plasma membrane of most cells. Low-density lipoproteins or apoE-containing lipoproteins bind to the LDL receptor via its amino-terminal ligand-binding region. Lipoproteins bound to the receptor are internalized via clathrin-coated pits on the plasma membrane. The coated pits "pinch off" to form coated vesicles, then fuse together, lose their clathrin protein coat, and become endosomes. Proton pumps acidify the endosomes, causing the receptor to release the lipoprotein(s). The receptor is then recycled to the cell surface while the lipoproteins are delivered to lysosomes, where enzymes degrade the apoproteins into their constituent amino acids and hydrolyze the cholesteryl esters to unesterified cholesterol. The cholesterol released from lysosomes is used for the synthesis of cell membranes, and in certain specialized cells it is a precursor for sterol end products. For example, in the adrenal gland and ovary, cholsterol is converted into steroid hormones; in the liver, it is converted into bile acids. [*See* BILE ACIDS.]

An oversupply of lipoprotein-derived cholesterol to cells has three major metabolic effects. First, it reduces the production of cellular cholesterol by turning off the synthesis of hydroxymethylglutaryl–coenzyme A reductase, the major rate-limiting enzyme involved in cholesterol biosynthesis. Second, the excess cholesterol promotes its own storage by activating the enzyme acyl coenzyme A–cholesterol acyltransferase. This enzyme re-esterifies cholesterol into cholesteryl esters, which are stored in the cytoplasm as lipid droplets. Third, excess cholesterol downregulates the synthesis and expression of the LDL receptor. The cell thereby adjusts the number of its cell-surface receptors so that the amount of intracellular cholesterol is sufficient to perform various functions, but not so high as to overload the cell.

In addition to controlling intracellular cholesterol, receptors regulate plasma LDL levels and determine, in part, how much VLDL is converted to LDL. Low-density lipoprotein receptors, which are expressed in most tissues, bind and remove about three-fourths of the LDL cleared from the plasma. The liver, which has a high concentration of LDL receptors, is responsible for about 70% of the receptor-mediated removal of LDL. The rate of removal of LDL from the circulation has been shown to depend on the number of LDL receptors expressed on the hepatocytes. About 25% of the LDL is cleared from the plasma by a poorly understood nonspecific mechanism.

The importance of receptor-mediated LDL clearance is demonstrated by two genetic defects: familial hypercholesterolemia, in which the LDL receptor is defective, and familial defective apoB-100, in which the ligand (apoB-100) is defective. In familial hypercholesterolemia, individuals who inherit one copy of the mutant LDL receptor gene possess LDL levels about twice that of normal. These familial hypercholesterolemic heterozygotes are much more susceptible to heart attacks than normal individuals are. Individuals who inherit two defective LDL receptor alleles (homozygotes) have circulating LDL levels six to eight times higher than normal and usually have heart attacks before the age of 20. Because normal hepatic LDL receptors efficiently remove LDL from the blood, the lack of these receptors results in the massive accumulation of LDL in the blood. In familial hypercholesterolemic homozygotes, LDL particles survive in the bloodstream about two and one-half times as long as in normal individuals because they must be removed by a metabolic pathway that is not as efficient as the LDL receptor pathway. In addition to the slower removal of LDL, about twice as much LDL is produced in familial hypercholesterolemic homozygotes. Thus, the build-up of LDL in the plasma is due to both a slower utilization and increased production of LDL particles.

In the second genetic disorder resulting in elevated plasma LDL levels, familial defective apoB-100, a single amino acid substitution in the receptor-binding domain of apoB-100, abolishes the ability of LDL to bind to the LDL receptor. Individuals heterozygous for this disorder, who possess a normal apoB-100 allele and one mutant allele, have one population of normal LDL and one that binds poorly, if at all, to LDL receptors. The abnormal LDL are not cleared efficiently and accumulate in the plasma to levels 50–70% above normal. In contrast to familial hypercholesterolemic heterozygotes, however, there is no overproduction of LDL; consequently, the LDL concentrations are not as elevated. These two genetic disorders of the LDL receptor system clearly illustrate the physiological significance of the LDL receptor pathway in regulating plasma LDL levels.

Several additional lipoprotein receptors are involved in lipoprotein metabolism and pathology: the chylomicron remnant, or apoE, receptor; the HDL receptor; and the acetyl LDL, or scavenger, receptor.

The chylomicron remnant receptor is a hepatic lipoprotein receptor that is responsible for the rapid clearance of chylomicron remnants from the circulation. The binding of the remnants is mediated by apoE (hence its other name, the apoE receptor), but otherwise this receptor is poorly characterized. Nevertheless, physiological and genetic evidence indicate that it plays a central role in chylomicron metabolism.

The HDL receptor has not yet been isolated or characterized but has been extensively studied in cultured cells. The binding of HDL to this receptor promotes the efflux of cholesterol from the cells to HDL. If these data obtained from tissue culture studies can be extrapolated to the whole organism, one would expect the HDL receptor to facilitate the transfer of cholesterol from peripheral cells and tissues back to the liver for excretion from the body, a process known as reverse cholesterol transport.

The acetyl LDL, or scavenger, receptor recognizes, binds, and internalizes LDL in which the epsilon amino groups of lysine have been chemically modified by acetylation, acetoacetylation, or reaction with malondialdehyde. The scavenger receptor is found on the surface of monocyte–macrophages and certain endothelial cells. This scavenger receptor does not appear to play a major role in normal lipoprotein metabolism but may be involved in atherosclerosis. Fatty atheroma plaques contain macrophages loaded with cholesteryl esters (foam cells), and foam-cell formation is strongly associated with elevated concentrations of LDL in the plasma. However, because macrophages normally possess limited numbers of LDL receptors, which are subject to normal downregulation by high concentrations of LDL, these normal LDL receptors do not appear to be significantly involved in the accumulation of massive amounts of cholesterol in the cytoplasm of foam cells. Rather, it is thought that LDL are modified by oxidation or some other process and are then recognized by the acetyl LDL receptors; these, unlike the LDL receptors, are not regulated by the intracellular content of cholesterol. Because the synthesis and expression of the acetyl LDL receptor is not downregulated after the uptake of modified LDL, macrophages exposed to these lipoproteins become increasingly filled with cholesteryl esters. Although the exact mechanism of chemical modification is still unknown, several lines of evidence indicate that LDL can be oxidized *in vivo,* and the subsequent uptake of oxidized LDL by the acetyl LDL receptor is a likely mechanism for atherogenesis and foam-cell formation. [*See* ATHEROSCLEROSIS.]

IV. Enzymes and Transfer Factors Involved in Lipoprotein Metabolism

Investigators have established the identity of three enzymes and one lipid transfer factor involved in plasma lipoprotein processing and metabolism: LPL, hepatic lipase, LCAT, and cholesteryl ester transfer protein (CETP).

LPL is the enzyme responsible for most of the hydrolysis of the triglycerides on chylomicrons and VLDL. It catalyzes the hydrolysis of the ester bonds of triglycerides, thereby releasing free fatty acids. This enzyme is synthesized in the parenchymal cells of a number of tissues and is transported to the endothelial surface of blood capillaries, where it is bound to cell-surface glycosaminoglycans (e.g., heparan sulfate). Under normal conditions, very little lipoprotein lipase circulates free in the plasma, but it can be released from endothelial cells by intravenous injections of heparin. The initial step in the hydrolysis of chylomicron and VLDL triglyceride is the binding of the lipoprotein particle to the lipase molecule on the surface of the endothelial cells. The initiation of LPL activity requires the presence of apoC-II on the surface of the lipoproteins. Genetic mutations in either LPL or apoC-II can result in triglyceride accumulation in the plasma.

Hepatic lipase resembles LPL in that it is synthesized by hepatic parenchymal cells and is transported to the hepatic sinusoidal endothelial cells (presumably while bound to heparan sulfate). Its function is not as precisely defined as LPL, but it appears to function in the later stages of the hydrolysis of IDL triglycerides. Unlike LPL, hepatic lipase is not activated by apoC-II.

Almost all of the cholesteryl ester transported by lipoproteins is formed by the action of lecithin–cholesterol acyltransferase (LCAT) in the blood. The substrates of this enzyme are phosphatidylcholine and cholesterol, and the products are lysolecithin and cholesteryl ester. LCAT is activated by apoA-I, which is the major protein component of HDL. Because the major physiological substrate of LCAT is HDL, the cholesteryl ester products produced in HDL can be rapidly transferred to acceptor lipoproteins by the action of CETP.

Cholesteryl ester transfer protein (CETP) facilitates the transfer of cholesteryl esters from HDL to VLDL, in exchange for triglycerides. The CETP is synthesized primarily in the liver and is secreted into the plasma. Each molecule of CETP can bind one molecule of triglyceride or cholesteryl ester.

Bibliography

Breslow, J. L. (1988). Apolipoprotein genetic variation and human disease. *Physiol. Rev.* **68,** 85.

Brown, M. S., and Goldstein, J. L. (1986). A receptor-mediated pathway for cholesterol homeostasis. *Science* **232,** 34.

Gotto, A. M., Jr., Pownall, H. J., and Havel, R. J. (1986). Introduction to the plasma lipoproteins. *Methods Enzymol.* **128,** 3.

Havel, R. J., and Kane, J. P. (1989). Introduction: Structure and metabolism of plasma lipoproteins. *In* "The

Metabolic Basis of Inherited Disease," 6th ed. (C. R. Scriver, A. L. Beaudet, W. S. Sly, and D. Valle, eds.). pp. 1129–1138. McGraw-Hill, New York.

Mahley, R. W. (1990). Biochemistry and physiology of lipid and lipoprotein metabolism. *In* "Principles and Practice of Endocrinology and Metabolism" (K. L. Becker, and C. R. Kahn, eds.). pp. 1219–1229. J. B. Lippincott, Philadelphia.

Mahley, R. W., Innerarity, T. L., Rall, S. C., Jr., and Weisgraber, K. H. (1984). Plasma lipoproteins: Apolipoprotein structure and function. *J. Lipid Res.* **25**, 1277.

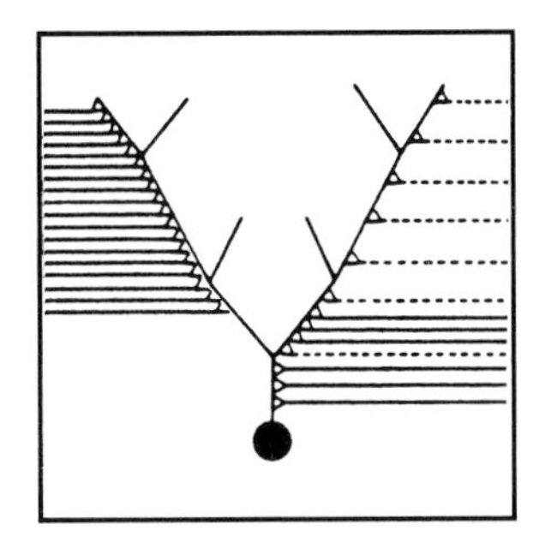

Plasticity, Nervous System

FREDRICK J. SEIL, *VA Medical Center, Portland and Oregon Health Sciences University*

Glossary

Axon Nerve cell process (fiber) that conducts electrical impulses to other cells. Most neurons have a single axon, but the axons may branch many times

Dendrite Neuron process that originates from the soma and that, along with the soma, receives the projections of multiple axon endings (terminals). Dendrites are generally multiple or complexly branched and function as integrators of incoming impulses

Denervate To deprive a target of its incoming axonal projections, such as by severing or otherwise injuring the projecting axons

Innervate To project axons to a target

Neuron Nerve cell, including the cell body (soma) and all its processes

Neurotransmitter Chemical substance, often an amino acid, that is released into the synaptic cleft from the presynaptic terminal in response to an electrical impulse. After crossing the synaptic cleft, the neurotransmitter interacts with a receptor on the postsynaptic membrane to produce a change in either an excitatory or inhibitory direction

Nucleus Anatomically defined, generally small collection of neurons whose projecting axons tend to be functionally similar

Synapse Site where two neurons communicate. A synapse is composed of a presynaptic element, usually an axon terminal, a postsynaptic element consisting of a specialized membrane on the surface of a nerve cell body, dendrite, or occasionally another axon, and a synaptic cleft, or space between the pre- and postsynaptic elements

BY CURRENT USAGE, *neural plasticity* is a term that refers to any change in the structure or function of the nervous system. The change may occur spontaneously [e.g., the normal turnover of nerve terminals connecting to muscle fibers (neuromuscular junctions)] or may be induced by some alteration of internal or external conditions (e.g., an injury or disease). Changes may occur at the level of a single cell or may involve the complex neural organization of an organism. Plasticity is a characteristic of the nervous system at any stage of life, including development, maturity, and aging. Plastic changes that occur during development and result in a departure from the normal developmental pattern are usually more dramatic than later occurring changes. The developing nervous system, with its interconnections more "diffusely" organized, has a greater capacity for modification of its connection patterns. Structural changes involve growth of new nerve fibers or processes or formation of new synapses, resulting in a reorganization of the neural circuitry. Functional changes can vary from expression of a different neurotransmitter by a cell to a local reflex change to a change in behavior. In essence, neural plasticity is the modifiability of the nervous system at a cellular, multicellular, or organismal level to permit an adaptive response to an ongoing or changing set of internal or external circumstances.

I. Structural Plasticity

A. Axon Collateral Sprouting

A common structural plastic change in response to an injury to the nervous system is collateral sprouting of axons, or the growth of collateral branches from intact, uninjured axons. Sprouted axon collaterals innervate target neurons, muscle fibers, or sensory receptor organs denervated because of damage to or loss of their primary axons. For example, lizards losing tails as a defense or escape reaction may regenerate new tails, but they do not regenerate the sensory ganglia that innervate the tails. Rather, the regenerated tails receive sensory innervation from collateral axonal branches that sprout from the most caudal of the surviving intact ganglia. The surviving ganglia thus extend the territory to which they project by growing additional axonal branches that substitute for the lost axons, and thus allow the development of sensation in the regenerated tails.

Another example of axon collateral sprouting is found in the septal nucleus, one of the deep nuclei of the forebrain. The neurons in this nucleus receive innervation from other parts of the brain via two major incoming nerve fiber bundles [i.e., the fimbria and the medial forebrain bundle (MFB)]. The fimbrial axons synapse only on dendrites, whereas a substantial portion of the MFB axons synapse on neuron cell bodies (somata). If the MFB is experimentally cut, the fimbrial axons undergo collateral sprouting and synapse on the cell somata as well as on the dendrites of septal nucleus neurons, thus maintaining innervation of the cell bodies (Fig. 1).

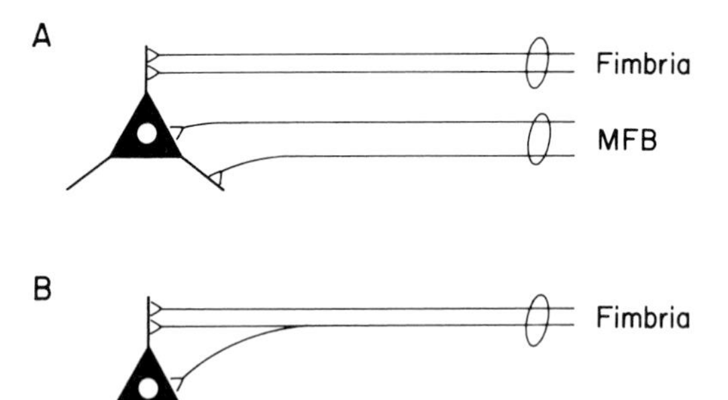

FIGURE 1 Collateral sprouting. (A) Neurons in the septal nucleus receive input from other brain areas by way of two fiber bundles, the fimbria and the medial forebrain bundle (MFB). Fimbrial axons terminate exclusively on dendrites, and MFB axons project to both dendrites and somata of septal nucleus neurons. (B) If the MFB is cut, fimbria axons undergo collateral sprouting and project to somata as well as dendrites of neurons in the septal nucleus. [From Seil, J. F., ed. (1989). "Neural Regeneration and Transplantation." Alan R. Liss, New York, with permission.]

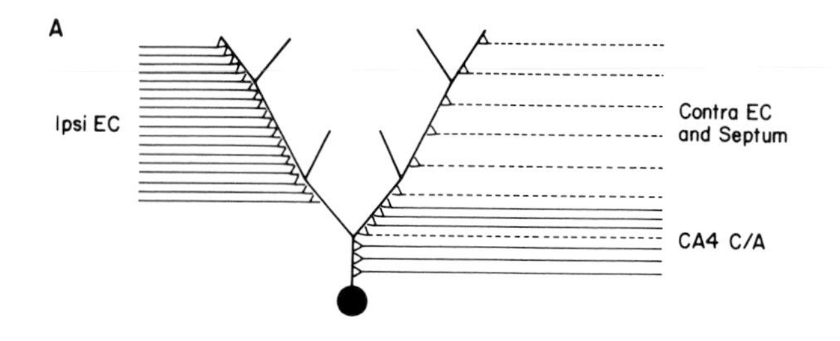

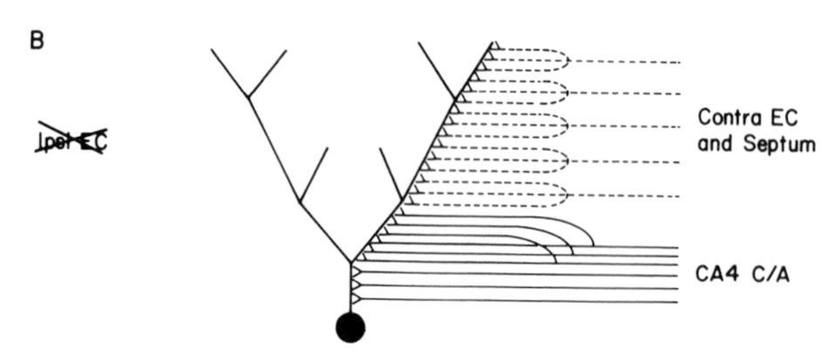

FIGURE 2 Synaptic plasticity. (A) Dendrites of granule cell neurons of the hippocampus form multiple synapses (represented as —<) with axons from the ipsilateral entorhinal cortex (Ipsi EC) of the brain and from the CA4 area of the hippocampus bilaterally (CA4 C/A), and fewer synapses with axons from the contralateral entorhinal cortex (Contra EC) and the septal nuclei. Entorhinal cortex fibers project to the outer 75% of the granule cell dendritic tree, and CA4 fibers project to the inner 25%. (B) If the ipsilateral entorhinal cortex is destroyed (X), CA4 fibers, septal fibers, and fibers from the contralateral entorhinal cortex all undergo collateral sprouting and form new synapses on granule cell dendrites to replace lost synapses. CA4 fibers extend their projection to the inner 35% of the dendritic tree in this scheme of synaptic reorganization. [From Seil, F. J., ed. (1989). "Neural Regeneration and Transplantation." Alan R. Liss, New York, with permission.]

B. Synaptic Plasticity

Synaptic plasticity usually refers to a modification of synapses (e.g., a change in the number or size of the terminals converging on a target) or a change in the source of the projecting axons, with or without a change in the neurotransmitter of the projecting terminals, the end result of which is an alteration of the original neural circuitry. The structural alteration may be correlated with a functional change (see Section II,B). In parts of the nervous system where synapses turn over normally, the result of the turnover is a maintenance of the existing circuitry. When the synapse turnover is in response to a lesion, then the usual consequence is a change in the synaptic organization of the involved part of the nervous system.

One of the best studied examples of a lesion-induced change in synaptic organization is in the hippocampus, a part of the brain that has been associated with memory. The dendrites of the granule cell neurons of the hippocampus are densely innervated by fibers from a neighboring structure, the ipsilateral (same side) entorhinal cortex (Fig. 2A). The dendrites also receive a dense projection from another area of the hippocampus (CA4) bilaterally, and a sparse projection from the contralateral (opposite side) entorhinal cortex and the septal nuclei. The entorhinal cortex axons project to the outer three-quarters of the granule cell dendrites, whereas the CA4 fibers project to the inner quarter. [*See* HIPPOCAMPAL FORMATION.]

If the ipsilateral entorhinal cortex is destroyed (Fig. 2B), more than 60% of the synapses on the granule cell dendrites are lost. Intact CA4 axons, septal axons, and axons from the contralateral entorhinal cortex all undergo collateral sprouting and form new synapses on the granule cell dendrites that replace the lost synapses. The CA4 fibers extend their territory to innervate the inner 35% of the dendritic tree, whereas septal nucleus and contralateral entorhinal cortex fibers retreat from this area and only reinnervate the denervated territory. The new synapses formed by contralateral entorhinal cortex axons are like those of the ipsilateral entorhinal cortex that they replace (homotypical), whereas the new synapses formed by CA4 and septal nucleus axons are different from the original synapses (heterotypical). Both homotypical and heterotypical synapses are apparently functional, as determined by electrophysiologic studies. The result is a reorganization of the original circuitry in such a way as to compensate for the deficit produced by the lesion.

II. Functional Plasticity

A. Neurotransmitter Plasticity

Developing sympathetic neurons are plastic with regard to neurotransmitter expression. Such neurons are normally adrenergic (use norepinephrine as a neurotransmitter) but can be induced to become cholinergic (using acetylcholine as a neurotransmitter) by manipulation of their environment. This can be done *in vivo* by transplanting sympathetic neurons so that they innervate target tissue normally receiving cholinergic innervation, or *in vitro* by culturing the sympathetic neurons in the presence of nonneuronal cells or in medium conditioned by nonneuronal cells. The expression of one or the other neurotransmitter, norepinephrine or acetylcholine, profoundly changes the functional properties of the sympathetic neurons. This is a form of neuroplasticity that is the property of a single cell, as opposed to other kinds of changes (e.g., organizational changes) that involve at least a group of neurons.

B. Lesion-Induced Functional Changes

The red nucleus in the brainstem represents an area where inputs from two different parts of the brain converge. Axons from one of the deep nuclei of the contralateral cerebellum (hindbrain) project to the somata of the red nucleus neurons, whereas fibers from the ipsilateral sensorimotor cortex project to the outer portions of the dendritic trees. Red nucleus neurons are large cells that are subject to intracellular electrophysiologic recording. Stimulation of the contralateral deep cerebellar nucleus produces a fast-rising excitatory postsynaptic potential (EPSP) characteristic of a synapse on the cell body and near the recording electrode. Cortical stimulation produces a slow-rising EPSP, consistent with synapses located well out on the dendrites, at a distance from a recording electrode placed in the cell soma. If the deep cerebellar nucleus is lesioned, cortical stimulation 10 or more days later elicits an EPSP with an intermediate rise time, indicative of cortical axons having sprouted and formed new synapses on the inner parts of dendrites or on cell somata. The morphological findings support the electrophysiologic results and provide an elegant example of how structural and functional plastic changes may correlate. The new synapses that are formed in response to lesioning are heterotypical but, as also indicated by the electrophysiologic results, are functional.

C. Latent Synapses

Under some conditions of lesioning or of temporarily blocking the function of somatosensory pathways, rapid changes in the receptive fields mediated by such pathways may occur. The changes may be too rapid or too extensive to be explained by axonal sprouting and formation of new synapses. These kinds of data have raised the possibility of the existence of latent or "silent" synapses, which may be anatomically present but functionally ineffective. A temporary or permanent lesion could induce a change in effectiveness of such structurally static synapses, resulting in a functional change that is not based on a morphological reorganization of the underlying neural circuitry. Thus it is possible that structural changes in synaptic connectivity are not always necessary for the occurrence of functional plasticity.

D. Behavioral Plasticity

Behavioral plasticity is a broad term that encompasses such phenomena as a change of behavior in response to selective experience, conditioning, habituation, sensitization, imprinting, associative learning and memory formation, adaptation to unusual motor or sensory conditions (e.g., the wearing of distorting or inverting lenses), and others. Also generally classified under behavioral plasticity are the morphological changes that are induced in various areas of the brain by experience. Rats raised in an enriched laboratory environment as opposed to standard colony or impoverished conditions have reported increases in brain weight, glial density, cortical thickness, dendritic branching, numbers of dendritic spines and synapses, and increases in areas of individual synapses. Such changes occur at whatever age the enriched environmental condition is imposed. Similar changes occur after application of formal learning procedures.

The substrates for some kinds of behavioral plasticity are known. Clearly, a change such as a structural synaptic reorganization in response to injury along with a circuit change as determined electrophysiologically may have an associated change in behavior. Habituation or a decrement in response to a repeated inconsequential stimulus can, in the gill-withdrawal response of the sea slug, *Aplysia*, be associated with a depression of the EPSPs of motor neurons caused by a decrease in the quanta of neurotransmitter released after successive impulses by sensory neuron terminals at their junctions with motor neurons. In the same model, sensitization or intensification of the gill-withdrawal reflex by presentation of a strong stimulus elsewhere on the body surface, is associated with an enhancement of neurotransmitter release by sensory neurons. Imprinting, or the behavior of following the first large moving object seen, as occurs in birds (or other animals born in a relatively advanced state of development), is associated with an increase in postsynaptic receptive areas. Memory formation is believed by some to involve a mechanism such as long-term potentiation (LTP). LTP is an increase in the synaptic response of neurons induced by stimulation of fibers projecting to the neurons with moderate to high-frequency bursts (10 Hz or more). LTP develops after seconds and endures for hours or days. Evidence is available for functional changes at both pre- and postsynaptic sites, as well as for morphological changes at synapses after induction of LTP.

III. Plasticity After Transplantation

Plasticity associated with transplantation of the nervous system is included under a separate heading not because there is anything basically different about plastic changes associated with transplantation, but because (1) transplantation represents a totally contrived or artifactual phenomenon that does not occur in nature and (2) the changes that occur are generally in the opposite direction to those that occur in response to injury (i.e., the purpose of transplantation is to restore the normal structural/functional state, as opposed to the reorganization of normal patterns that occurs in response to injury or disease).

An initial question that can be raised is whether a nervous system that has undergone reorganizational changes as a consequence of injury, with the formation of new synapses and the establishment of alternate neurotransmitters, is capable of undergoing a second round of reorganization if the missing elements are restored. If so, can the original circuitry be faithfully reconstructed? The answers to these questions are in the affirmative, as animal studies with lesioned adult hippocampus and visual systems, as well as other systems, indicate that damaged and reorganized neural circuitry can be reconstructed by transplantation with fetal tissue. The axons of transplanted neurons find and synapse with appropriate targets in host tissue, and the grafts also receive host fibers. Electrophysiologic and behavioral studies attest to the appropriate functionality of at least some of the connections between graft and host nervous system.

Some insight into how circuit reconstruction might occur is gained from studies with neural tissue cultures. If cerebellar explants derived from neonatal mice are treated with a drug that eliminates one group of cortical neurons, the cerebellar granule cells, one of the surviving neuronal groups, the Purkinje cells, undergoes a remarkable sprouting of axon collaterals that form synapses on the dendritic sites normally occupied by the axon terminals of granule cells. The Purkinje cell neurotransmitter is not only different from the granule cell neurotransmitter, but is inhibitory rather than excitatory, which represents a radical change from the normal condition. These abnormal synapses are functional and inhibitory, as indicated by electrophysiologic studies.

If cerebellar granule cells from another source are now introduced into this reorganized system, the sprouted Purkinje cell axon collaterals degenerate or are withdrawn, and the heterotypical synapses that they had formed are replaced with normal synapses formed by the terminals of granule cell axons; normal function is restored. It appears that primary presynaptic elements have priority for the synaptic sites for which they were originally programed and that substitute terminals give way to primary presynaptic elements when the latter appear, even if the heterotypical synapses have become functional. This kind of hierarchical order in the nervous system facilitates the reconstruction of normal circuitry after transplantation.

Reconstruction of circuitry is not always necessary, however, for the restoration of function after neural transplantation. In some cases, grafting a population of neurons that produce a missing neurotransmitter near a target group of neurons, with resultant diffusion of the target neurons by neurotransmitter, may be sufficient for functional recovery.

Replacement of damaged parts of the nervous system by transplantation is possible because the nervous system remains plastic throughout life and is amenable to repeated rounds of reorganization or readaptation as new conditions arise. This plastic capability of the nervous system is the basis for hope that further developments in neural regeneration and transplantation are possible and can be extended from the experimental laboratory to humans.

Bibliography

Cotman, C. W., ed. (1985). "Synaptic Plasticity." The Guilford Press, New York.

Jenkins, W. M., and Merzenich, M. M. (1987). Reorganization of neocortical representations after brain injury: A neurophysiological model of the bases of recovery from stroke. *In* "Neural Regeneration," Progress in Brain Research, vol. 71. (F. J. Seil, E. Herbert, and B. M. Carlson, eds.). Elsevier, Amsterdam.

Rosenzweig, M. R., and Leiman, A. L. (1989). "Physiological Psychology," 2nd ed. Random House, New York.

Seil, F. J., ed. (1989). "Neural Regeneration and Transplantation," Frontiers of Clinical Neuroscience, vol. 6. Alan R. Liss, New York.

Steward, O. (1989). "Principles of Cellular, Molecular, and Developmental Neuroscience." Springer-Verlag, New York.

Zucker, R. S. (1989). Short-term synaptic plasticity. *In* "Annual Review of Neuroscience," vol. 12. (W. M. Cowan, E. M. Shooter, C. F. Stevens, and R. F. Thompson, eds.). Annual Reviews, Palo Alto, California.

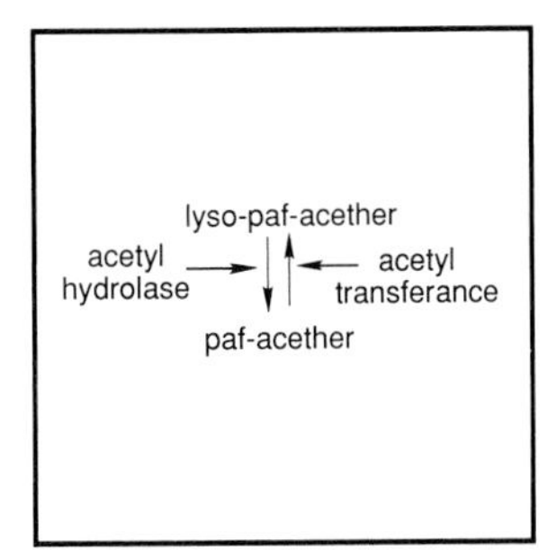

Platelet-Activating Factor Paf-acether

JACQUES BENVENISTE, *INSERM U 200*

Glossary

Atopic Greek word meaning "unusual," by and large equivalent to "allergic." The predisposition to develop allergic reactions in response to allergens that are harmless in the general population.

Chemotactic activity Ability of a chemical (natural or artificial) substance to attract cells. Complement components, leukotriene B_4, and paf-acether, are known examples of such substances.

Cytokines Soluble factors released by lymphocytes (lymphokines) or macrophages/monocytes (monokines). Such factors are implicated in cell-mediated immunity and, more generally, in the immune response.

Hypersensitivity reaction Exposure to an antigen generally induces protection. In some cases, immunopathological phenomena can appear instead: this is hypersensitivity. Gell and Coombs have described four types: type I is responsible for atopic diseases; type II generates cytotoxic reactions with or without implication of complement factors; type III corresponds to immune complex diseases; type IV is delayed hypersensitivity (i.e., cell-mediated immunity).

IgE Immunoglobulin produced upon exposure with antigens (allergens). They are heavier than IgG, and rich in carbohydrates. They are present in plasma in trace amounts but are capable, following their interactions with membrane high-affinity receptors, of sensitizing specialized cell types, mast cells in tissues, and polymorphonuclear basophils in blood. Upon a new encounter with the allergen, the bridging of adjacent IgE triggers cell degranulation and release of mediator.

Inflammation Major component of numerous, if not all, disease states characterized by pain, heat, redness, swelling. Dilation of vessels, emigration of cells and fluids toward the tissue explain the symptoms. A constantly operative normal means of defense, it often ends with local and sometimes generalized injury. Allergy is an immunologically triggered form of inflammation, with usually abrupt onset

Kidney mesangial cells One of the cell types in glomeruli of the kidney cortex. They serve both as a matrix for the glomeruli and as scavenger cells analogous to circulating or tissue macrophages.

Lymphocytes Small leukocytes present in blood (about 3,000/mm^3), tissues, and specialized organs (e.g., thymus, spleen, and lymph nodes). There are two types of lymphocytes: the T (for thymus) and the B (from the avian bursa of fabricius)

Phospholipids Lipids constituting the cell membranes (plasma membrane limiting the cell, inner membranes limiting organelles inside the cells). Phospholipids are made of a polar (hydrophlic) head and two "legs," apolar (hydrophobic) fatty acids, bound to a glycerol backbone. The link between the COOH of the acid and the CHOH of the glycerol is an ester bond. When the fatty chain is an alcohol, the link is an ether bond

Platelets Small cells, devoid of nucleus, present in blood (about 300,000/mm^3). Instrumental in limiting bleeding from injured vessels, in coagulation, and in thrombosis. Contain numerous active molecules that make them the major reservoir of physiologically important but also, in case of massive release, injurious factors that are implicated in vessel damage and more generally in inflammation

Polymorphonuclear leukocytes White cells present mainly in blood (about 5,000/mm^3) but also in inflamed or infected tissues. Their function is nonspecific scavenging of debris and bacteria (phagocytosis). About 95% of these cells are neutrophils, stained by neutral dyes. They contain numerous granules that carry a vast array of substances, the general function of which—on degranulation of the cell—is to fight microbes and foreign or autologous deposits. In the process, they may create local injury. About 1% and 1–5% of these cells are basophils and eosinophils stained with basic and acidic dyes, respectively. The latter cells are instrumental in allergic processes.

Secretagogue Natural or artificial substance that triggers secretion from cells or organs

PAF-ACETHER (paf, first described as—and often still named—platelet-activating factor) is one of the most potent mediators of allergy and inflammation. It is the clearest example of such a mediator belonging to the phospholipid class. Its role was initially thought to be restricted to immediate hypersensitivity reactions because it was first described as originating from IgE-sensitized basophils. Basophil degranulation and release of paf were then suspected to intervene in acute kidney diseases, exemplifying the relation between allergic reactions and inflammatory processes. Paf is now recognized as having a wider role, as a mediator of inflammation, because it is released from a variety of inflammatory cell types, including lung macrophages, a possible factor in the pathogeny of human asthma. Paf acts *in vitro* on various cell types including polymorphonuclear leukocytes and on numerous tissues and organs. It exhibits a strong chemotactic activity on proinflammatory cells, especially eosinophils. It causes the release of secondary mediators from various cell types and organs. *In vivo,* paf induces numerous pathological effects such as bronchoconstriction associated with pulmonary, cardiac, and vascular changes in guinea pigs, baboons, and humans, and nonspecific bronchial hyperreactivity in animals as well as in humans. Paf also increases skin vascular permeability, creates gastrointestinal damage, and exhibits an antihypertensive property. Paf represents a good example of a mediator released from and acting on inflammatory cells and organs, thus enhancing and/or perpetuating inflammation, notably in human lung diseases. Development of anti-paf drugs is a major goal. [*See* INFLAMMATION.]

I. Structure and Characterization

The structure of paf was elucidated as 1-*O*-alkyl-2-acetyl-sn-glycero-3-phosphocholine (Fig. 1). Hence the name *paf-acether,* illustrating both the main function of the molecule and its critical structural aspects (e.g., the ether link at position 1 and the acetate at position 2). It has been chemically synthesized. There are two main variations according to the length of the first chain: paf is most often a mixture of about 80% 16-carbon and 20% 18-carbon chains, notably in human cells and plasma. Phospholipids structurally close to the above depicted two molecules and endowed with some paf-like activity must be categorized as paf analogues.

Paf is routinely measured using bioassays (e.g., monitoring washed rabbit platelets based on either aggregation or serotonin release). The unspecific nature of these methods makes it necessary to distinguish paf from other platelet-activating agents such as arachidonic acid, thrombin, adenosine diphosphate (ADP), and thromboxane. Also, some paf analogues can trigger cell activation, although they are generally much weaker than paf itself. Paf

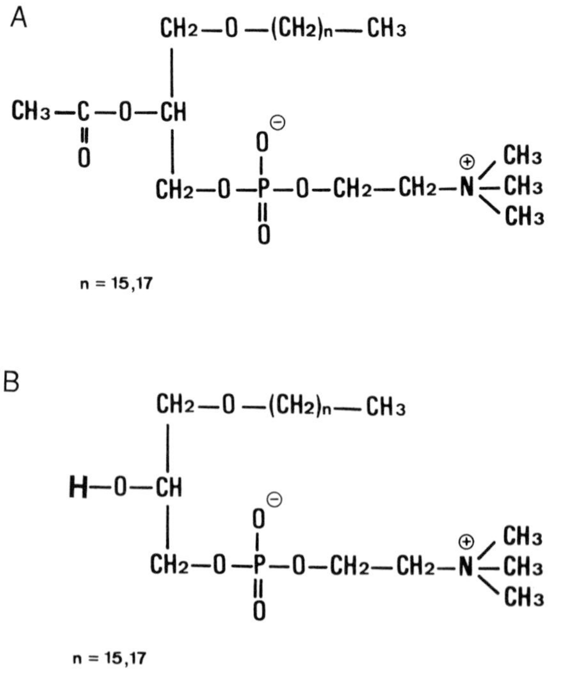

A

$CH_2—O—(CH_2)_n—CH_3$

$CH_3—C(=O)—O—CH$

$CH_2—O—P(=O)(O^{\ominus})—O—CH_2—CH_2—N^{\oplus}(CH_3)_3$

n = 15,17

B

$CH_2—O—(CH_2)_n—CH_3$

$H—O—CH$

$CH_2—O—P(=O)(O^{\ominus})—O—CH_2—CH_2—N^{\oplus}(CH_3)_3$

n = 15,17

FIGURE 1 Structure of paf-acether, (A); lyso paf-acether, (B).

is identified by the following criteria: (1) paf aggregates platelets (or releases serotonin) in the presence of arachidonic acid cascade blockers (aspirin or indomethacin) and of ADP scavengers; (2) specific paf-induced aggregation must be inhibited by paf antagonist drugs; (3) it is inactivated by phospholipases A2, C, and D and resistant to lipase from *Rhizopus arrhizus;* (4) it exhibits physicochemical characteristics (e.g., gas and liquid chromatography with mass spectrometry or electron capture as detectors) identical to that of synthetic paf. Using these criteria, the bioassays are fast, specific, sensitive, and accurate. Sensitive and specific radioimmunoassays are now commercially available.

II. Cell and Organ Sources

Paf is synthesized and released by IgE-sensitized rabbit basophils, but its origin from human basophils is still controversial (Table I). Calcium-dependent paf synthesis and release from human polymorphonuclear neutrophils and eosinophils is well-documented. Rabbit and human neutrophils synthesize paf when stimulated with various stimuli, among which are the ionophore A 23187 (which allows free entry of Ca^{2+} into cells), complexes of antigen and antibodies, the complement component C5a, phagocytozable particles such as serum-treated (yeast) zymosan (STZ), formyl-methionyl-leucyl-phenylalanine (FMLP), phorbol myristate acetate (PMA), tumor necrosis factor (TNF), or paf itself. About 50% of the synthesized paf and derivatives is released from neutrophils. However, when the cells are stimulated in high-concentration (e.g., above 5×10^6/ml) with or without subsequent decrease of the medium pH, the release and the overall level of paf synthesis is strongly reduced. This is probably due to paf adsorption to cell membranes and to conditions unfavorable to cell metabolism. This phenomenon may regulate the mediator output in inflammatory lesions.

In neutrophils, paf release can be dissociated from degranulation and superoxide anion produc-

TABLE I Paf-Acether Synthesis and Effects in Cells

Cells	Synthesis	Agonists	Effects
Basophils	+	Antigens (IgE)	*In vitro:* increased adhesion to endothelial cells *In vivo:* basopenia
Mast cells	+	Ionophore, IgE	–
Neutrophils	+	Phagocytosis, C_{5a}, PMA, FMLP, ionophore, TNFα, paf	*In vitro:* aggregation, degranulation, chemotaxis, adhesion, calcium mobilization, superoxide anion production, phosphoinositide metabolism, complement receptor increase *In vivo:* neutropenia
Eosinophils	+	IgE, ionophore	Degranulation, *in vitro* and *in vivo* chemotaxis, transformation in hypodense, superoxide anion production
Monocytes/macrophages	+	Phagocytosis, ionophore, immune complexes, allergen (IgE), LPS	Aggregation, superoxide anion production, IL-1 and TNFα production, calcium mobilization, oncogene modulation
Endothelial cells	+	Ionophore, bradykinin, IL-1, thrombin, histamine, LPS, leukotrienes C and D	Calcium mobilization, increased adhesion of cells, plasminogen activator synthesis
Platelets	±	Ionophore, thrombin	Aggregation, vasoactive amine release, calcium mobilization, phospholipase C/phosphoinositides stimulation, thromboxane A2 formation
Lymphocytes	?		
T cells	Only leukemia cell lines	—	Down-regulates CD3/TcR pathway Up-regulates CD2 alternate pathway
B cells	and natural killer cells	—	Modulates B-cell proliferative response
Skin fibroblasts	+	Ionophore	—
Keratinocytes	+	Ionophore	—
Kidney mesangial cells	+	Ionophore	Contraction, calcium mobilization

tion, which is a characteristic response of these cells. Eosinophils from patients with eosinophilia synthesize and release large amounts of paf on stimulation with the ionophore A 23187, STZ, and the specific antigen bound to cell IgE (but not IgG) antibodies in a dose-, time-, and calcium- and magnesium-dependent manner. Besides neutrophils and eosinophils, many cells are capable of synthesizing paf on stimulation with a specific secretagogue and/or via an IgE-dependent mechanism [platelets, monocytes/macrophages, natural killer cells, alveolar and peritoneal macrophages, vascular endothelial cells, mast cells, skin fibroblasts and keratinocytes, renal (mesangial) cells]. It is noteworthy that human alveolar macrophages, recovered by bronchoalveolar lavage from asthmatic patients, release paf *in vitro* when exposed to the specific antigen to which the patient is allergic. [*See* MACROPHAGES; NEUTROPHILS.]

The semisynthesis (i.e., acetylation of the paf precursor) and release of paf has recently been evidenced in the bacteria *Escherichia coli,* a result of possible importance in view of the pathogenetic role of enterobacteria. Moreover, paf thus appears to be an ancient and functionally conserved molecule and the competence to form and release this phospholipid an early phylogenic development. Paf has been obtained, after stimulation with suitable stimulatory substances, from numerous organs, notably the lung, heart, kidney, brain, liver, spleen, thymus, skin, ovary, embryo, and retina. Paf was present in low amounts in the alveolar space in chronic human inflammatory lung diseases. However, it was not found in bronchoalveolar lavage fluid from atopic patients, whereas lyso paf, the precursor metabolite of paf (see Section III), levels increased in fluid from occupational inflammatory lung diseases. Large quantities of paf were detected in bronchial mucus from children with chronic inflammatory lung diseases (e.g., cystic fibrosis, chronic obstructive pulmonary disease, and bronchopulmonary dysplasia). A potent role for paf in the development and maintenance of cutaneous inflammation and allergy has been suggested because of its identification in lesioned tissue. For instance, paf has been found in psoriatic scales and, after allergen stimulation, in skin blister fluids from allergic volunteers. The antigenic release was inhibited by pretreatment with substances that block the H-1 receptor for histamine. Paf was associated with primary acquired cold urticaria. Recently, ionophore-stimulated fibroblasts and keratinocytes from normal human skin were shown to synthesize and release paf.

It is noteworthy that free paf is absent from normal plasma and serum. However, paf—and its non-acetylated precursor/metabolite lyso paf—is bound in large amounts to lipoproteins in human plasma and in various cell types. The latter form of paf may represent a storage pool. The mechanism for its putative release from the bound form is unknown, yet lipoprotein-paf may play an important role in pathology.

III. Pathways of Biosynthesis

In many cell types, paf is newly synthesized in two successive steps (phospholipid remodeling pathway): (1) deacylation of membrane choline-containing ether lipids by a calcium-dependent phospholipase A2 (PLA2) enzyme, yielding lyso paf, and (2) acetylation of lyso paf into paf by a calcium-dependent enzyme, the acetyltransferase. Because in phospholipids arachidonic acid is often bound to the second carbon of the glycerol, the above depicted first step can yield both the paf precursor lyso paf and arachidonic acid, the precursor of prostaglandins and leukotrienes (Fig. 2). Therefore, the two pathways responsible for the synthesis of the major mediators of allergy and inflammation can evolve from one molecular species, the alkyl ether (long chain, i.e., mostly arachidonic acid) glycerophosphocholine, on stimulation of one enzyme (PLA2). The acetyltransferase activity increases severalfold in neutrophils stimulated by various agonists. Numerous experimental works suggest the involvement of enzyme(s) of the protein kinase C (PKC) family in acetyltransferase activation and paf synthesis. Essentially identical results have been obtained in different cell types including mast cells, macrophages, and platelets. Another (*de novo*) pathway for paf biosynthesis involves a choline phosphotransferase enzyme capable of transfering a phosphocholine group to the alkyl-acetyl-glycerol. Whether it operates in stimulated inflammatory cells is doubtful. Finally, it is conceivable, although not yet demonstrated, that large amounts of preformed paf be freed from the plasma and cell paf–lipoprotein complex.

Inactivation of paf depends on a calcium-independent acetylhydrolase enzyme, present in cells and plasma, that yields lyso-paf. The latter will be

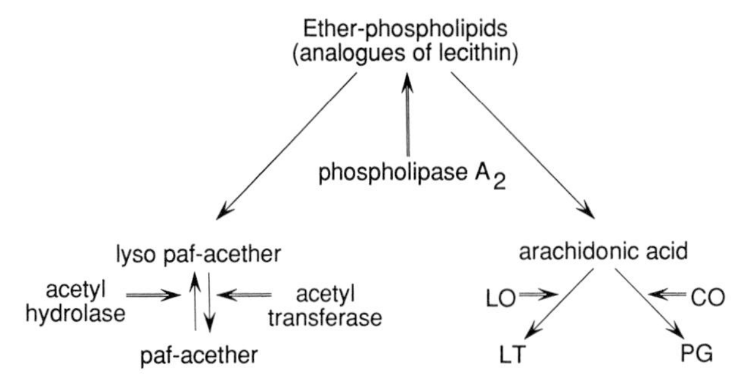

FIGURE 2 Diagrammatic representation of synthesis of paf-acether and arachidonic acid derivatives from membrane ether-phospholipids. LO, lipoxygenase; CO, cyclo-oxygenase; LT, leukotrienes; PG, prostaglandins; ⇒, enzyme activity.

either reacetylated into the active paf or acylated into the inactive ether phospholipid that is a component of the cell membranes. Acetylhydrolase is released from stimulated platelets and possibly other cells. This latter process may contribute to the plasma enzyme pool associated with the low-density lipoproteins. Plasma albumin also binds and inactivates paf.

IV. Effects on Cells

A. Platelets

Paf was first described in view of its capability of activating rabbit platelets (i.e., triggering calcium-dependent aggregation and release). Most mammalian platelets (with the notable exception of the rat and mouse) are sensitive to paf, which is in numerous species more potent than any other platelet agonist. Paf activates human platelets in the nanomolar range. Platelet stimulation in plasma requires about 10–100 times more paf than for cells suspended in artificial medium, because of the presence in plasma of acetylhydrolase and of large amounts of albumin which binds paf. Aggregation of platelets by paf is preceded by release of their intracellular content (e.g., the vasoactive amines histamine and serotonin, proteases, platelet-factor 4, and other coagulation factors) and by production of thromboxane A2. Platelet activation by paf is independent from other pathways such as ADP or arachidonic acid metabolites. Thrombopenia (an index of the activation of circulating platelets) readily follows the intravenous administration of paf, owing to platelet aggregation.

B. Polymorphonuclear Neutrophils, Eosinophils, and Basophils

Paf stimulates most of the neutrophil functions because it induces *in vivo* and *in vitro* aggregation, adherence, degranulation, chemotaxis, release of superoxide anions and leukotriene B4, and an increase in complement (C3) receptor expression. Paf induces similar biological effects on the eosinophil, but its most potent action on this cell is chemoattraction, which could be one of its major functions in inflammatory diseases.

The aggregating effect of paf on neutrophils was initially suggested by experiments showing that its intravenous injection into rabbits (and later into baboons) induced not only thrombocytopenia, reflecting its platelet-aggregating activity, but also neutropenia (and basopenia), because of sequestration of neutrophil aggregates in capillaries, mostly in the lung. In the guinea pig, eosinophils were detectable in lung parenchyma after paf challenge. *In vitro,* paf aggregated rat peritoneal and human blood neutrophils. Paf synthesized by human endothelial cells stimulated by leukotriene C4, D4, interleukin-1 (IL-1), or thrombin was shown to mediate adhesion of neutrophils to human endothelial monolayer cells. This neutrophil adherence was also the consequence of paf release by neutrophils. When the latter cells were exposed to low doses of paf, an anti-CDw18 (leukocyte surface glycoprotein complex) inhibited neutrophil adherence and an increase of complement receptors was noted. Paf seems to interact with antigens of the integrin family that are involved in the adhesion and subsequent emigration of neutrophils from the vascular lumen into the tissues, an essential step of the inflammatory process. Closely similar data have been obtained from the other polymorphonuclear cells, the eosinophils, and the basophils. These facts provide several interesting paf-mediated amplifying loops between proinflammatory cells and the vascular endothelium, which could be instrumental in inflammatory damage of the vessels and surrounding tissues, but also in physiological passage of cells from inside the vessels out.

Paf in the nanomolar to the micromolar range induces a calcium-dependent release of granule-associated enzymes, lysosyme, β-glucuronidase, peroxidase, and acid phosphatase from neutrophils and eosinophils.

In vitro, paf induces human neutrophil chemotactic but not chemokinetic response. Lyso paf is 1,000 times less potent than paf. Surprisingly, paf and lyso paf deactivated each other, whereas their chemotactic activity was unaffected by other chemotactic factors. The cross-desensitization between paf and lyso paf suggests that, at least in the neutrophil, they share a common receptor structure, which differs from the membrane receptor site for other factors. A stronger chemotactic activity of paf is reported for eosinophils compared with neutrophils. The eosinophil chemotactic activity of paf is 100-fold higher than the classical chemotactic factors (e.g., leukotriene B4 or eosinophil chemotactic factor of anaphylaxis). The paf chemotactic activity was also detectable *in vivo* in humans. The potent action of paf on eosinophils was described in human skin windows or chambers implanted on the skin and in guinea pig skin during subcutaneous inflammation. It is inhibited by some antihistamine drugs. Intratracheal or intravenous activity results in alveolar eosinophilia in baboon lungs, which persists 2 weeks after a unique paf challenge.

Paf appears to be less potent than other agents in inducing superoxide anion production by neutrophils. However, paf exhibits a remarkable enhancing effect on zymosan-induced superoxide production, suggesting that the sites of action of paf and of the particles are independent. Paf, at higher concentration than for degranulation, induces superoxide production in eosinophils. Higher sensitivity of eosinophils from atopic patients compared with normal subjects was reported. This finding may reflect a presensitization of the cells to paf, which could be relevant to the pathogenesis of human asthma.

C. Other Cells

Monocytes and macrophages are sensitive to paf. It induces cell aggregation and (weakly) superoxide production. Specific immunological effects will be described below. In vascular endothelial cells, paf initiates calcium mobilization and the release of plasminogen activator. Paf powerfully contracts and mobilizes calcium in kidney mesangial cells, thus possibly modulating renal blood flow. It induces differentiation at low dose and toxicity at high concentration on a neural cell line.

D. Paf and the Immune Response

Among immunocompetent cells, monocytes/macrophages are known as a major source of paf, whereas B and T lymphocytes are unable to produce it. However, some human B and T leukemic cell lines, as well as lymphoblastoid cells obtained by infecting B cells with Epstein-Barr virus, release paf-like material after nonspecific stimulation.

Monocytes/macrophages play a critical role in the onset and the development of both inflammatory and immune reactions by virtue of their ability to synthesize and release soluble products in response to membrane-perturbing or phagocytic stimuli. Among these secretory products are peptide mediators such as IL-1, IL-6, or TNFα and bioactive lipids including eicosanoid derivatives and paf. The monocyte/macrophage could not only be a paf-producing cell but also a target for paf. Indeed, paf has been shown to modify monocyte/macrophage functions such as phagocytosis, calcium influx, release of superoxide anions, formation of arachidonate metabolites, and TNFα. Moreover, paf can modulate c-*fos* and c-*myc* gene expression, indicating for the first time a paf-initiated differentiation signal at a genetic level. It also modulates adhesion molecules such as members of the integrin family, a group of extracellular matrix, and cell-adhesion receptors, which integrate the extracellular environment with the cytoskeleton. Regarding the functional relations between paf and other proteinic monokines, it has been recently shown that paf regulates IL-1 and TNFα production, which in turn can act on monocytes to produce more paf. This amplification loop during the course of the immune response (e.g., after bacterial infection) may play a role in tissue repair. Paf interactions with TNFα will be further described below.

In addition to their phagocytic, cytotoxic, or bactericidal activities, macrophages/monocytes present foreign antigens to T lymphocytes. Indeed, one of the earliest steps in T-cell activation is the engagement of the T-cell receptor (TcR) by antigen complexed with protein of the major histocompatibility complex (HLA) with the participation of soluble lymphokines and additional cell surface molecules such as the CD2 and CD3 molecules present on T cells. Recently, it was shown that paf modulates CD2 and CD3 cell surface expression on human T cell. This modulation does not reflect a general alteration of T-cell membrane components because no modification of HLA class I antigens is observed. Paf also differently modulates CD3- and CD2-dependent pathways of human T-cell activation. Indeed, when T cells are triggered via the CD3 pathway, paf decreases their proliferation by inhibiting IL-2 receptor expression. By contrast, on stimulation of the CD2-dependent pathway, paf potentiates T-cell proliferation, IL-2, IL-4 production, and IL-2 receptor expression. One possible interpretation of these findings, although hypothetical, is that paf affects CD2-triggered T-cell proliferation, in part by enhancing IL-2 and IL-4 production. The potentiation of IL-4 production may have profound implications in the regulation of IgE production and therefore in the pathogenesis of atopic diseases. This dichotomy of paf effects on T-cell activation indicates for the first time that a naturally occurring cytokine differently regulates T-cell activation triggered via either the CD2 or CD3 pathways. Such a control mechanism might be essential in determining the overall quality and the magnitude of the immune response. In this regard, it is interesting to note that a recent report demonstrates that human thymus contains large amounts of paf as well as paf precursors. Because the thymus microenvironment is essential in the proper development of bone marrow progenitors committed to the T-cell lineage into thymocytes capable of emigrating to the periphery as functional T lymphocytes, it is conceivable that paf may be a novel candidate involved in the regulation of intrathymic T-cell differentiation. Although specific paf binding sites were recently described in a Burkitt B-cell lymphoma line, the precise site of action of paf on T lymphocytes is unknown. Yet these data, when considered together, indicate a more fundamental role for paf than the now classical ones in allergy and inflammation and point it out as an important candidate in the regulation of immune function. [*See* IMMUNE SYSTEM; LYMPHOCYTES; T-CELL RECEPTORS; THYMUS.]

V. Mechanisms of Cell Activation

A. Receptors for Paf

The existence of receptors for paf was suggested by the specificity of paf action as compared with structurally related phospholipids and by cross-desensitization studies. Indeed, target cells are activated when paf binds to specific sites on the plasma membrane. This is the case for platelets from humans, rabbit, guinea pig, dog, cat, and horse but not from mouse and rat, and also human neutrophils and eosinophils, macrophages, human endothelial cells, and animal tissue homogenates from lung and hypothalamus. Studies with paf analogues have clarified the structural requirements for specific paf binding as being the stereospecificity, the etheroxide at the position 1 of the glycerol, a short fatty acid at position 2 and the choline polar head group. However lyso paf, devoid of 2-acetyl, interferes with paf binding in neutrophils. Because paf-receptor antagonists do not operate equally on all cell types and organs, different types of paf receptors have been suspected to exist but are not identified. The presence of high- and low-affinity binding sites specific to paf was reported in human neutrophils. The dissociation constant (K_d) calculated for the high-affinity binding sites varied from 0.1 to 0.4 nM. Between 1×10^3 and 5×10^6 sites per cell were determined, but this variation could be caused by different experimental conditions used (e.g., whole cells or membrane fractions). Paf receptors on human and guinea pig eosinophil were analyzed by eosinophil peroxidase release, calcium mobilization, superoxide production under paf stimulation, and the use of the paf antagonist WEB-2086 (see Section VIII). The importance of the specific paf binding was supported by the fact that different antagonists inhibited the paf-mediated platelet aggregation and the labeled paf binding with nearly identical dose–response curves. The paf receptors have been extensively studied in intact cells and plasma membrane preparations but have not been isolated from cell membranes. Moreover, specific binding sites for paf have not yet been evidenced in all target cells and tissues.

B. Transmembrane Signaling

Paf is instrumental in transmembrane signaling. In human neutrophils and many other cell types, the binding of paf to its specific site induces rises in cytosolic calcium and subsequent activation of protein kinase C (PKC), which may lead to superoxide anion production and degranulation. It was suggested that PKC activation is involved in the decrease of high-affinity paf binding sites, possibly a negative feedback regulation mechanism of paf-induced activation. In fact, paf acts on target cells via the ubiquitous signaling system that triggers functional stimulation in most cells: activation mediated by receptor-linked guanine nucleotide proteins of the phosphoinositide-specific phospholipase C, which generates endogenous calcium mobilizing agents and PKC activators.

VI. Effect on Tissues and Organs

A. The Lung

1. *In Vitro* Effects

The constrictor effect of paf on isolated bronchial smooth muscle is still controversial, whereas this effect is well-established on guinea pig lung parenchymal strips. An amount of paf injected in isolated ventilated lung perfused with saline solution in the absence of blood induces a vasoconstriction and a bronchoconstriction. In rabbit lung, paf has a greater effect in pulmonary vasculature than on airway pressure, whereas the reverse is true in guinea pig. These data indicate that paf can act on lung tissue in the absence of blood platelets and/or leukocytes.

2. *In Vivo* Effects

a. Immediate Reactions Intravenous administration of paf induces in guinea pigs an immediate bronchoconstriction associated with thrombocytopenia. In nonhuman primates or in rabbits, a rapid and significant bronchoconstriction is associated with a concomitant increase of pulmonary inspiratory pressure. The bronchoconstrictor effect of paf is maximal at 20 sec after challenge and is spontaneously reversible, recovering the baseline value 40–45 min later. This bronchial effect is associated with a rapid systemic thrombocytopenia and neutropenia, which occur less than 1 min after paf challenge. The blood platelets are transitorily accumulated into pulmonary vasculature. Modifications of blood gas are observed 10 min after paf challenge: a fall of arterial O_2 and an increase in CO_2. All modifications described in experimental animal models also appear in humans.

b. Late Reactions In the baboon, an eosinophil accumulation appears in the alveolar space 1 hr after paf challenge, with no changes in the number of other leukocytes. This specific chemotactic effect is inhibited by antiasthma drugs. Eosinophils are still present 14 days later. This lung eosinophil accumulation was also observed in guinea pig lung parenchyma. Eosinophil lung accumulation has not been described in humans, yet paf induces a skin eosinophil infiltration in normal and atopic subjects. A nonspecific bronchial hyperreactivity to metacholine appears 1 hr after paf inhalation or intratracheal deposition. In humans or in the baboon, the bronchial hyperreactivity is still persistent 14 days later.

3. Effect of Chronic Administration

Chronic administration of low doses of paf to guinea pigs is unable to induce the major effects of paf described above. However, hyperreactivity to histamine is observed. Lung tissue is congestive with some changes in the epithelial membrane, and eosinophil number is increased compared with normal tissue.

4. Paf and Asthma

Paf is the only mediator that can mimic the main characteristics of human asthma: increase in lung airway resistance; influx of inflammatory cells, predominantly eosinophils, in the tissue and in the alveolar space; long-lasting bronchial hyperresponsiveness; and mucus secretion (Table II).

TABLE II Paf-Acether and Asthma

Paf-acether
is released from allergen-stimulated alveolar macrophages
induces bronchoconstriction in various mammalian species including humans
promotes alveolar eosinophil infiltration
creates a long-lasting (up to 4 weeks) nonspecific bronchial hyperreactivity in humans and experimental animals
triggers mucus secretion

B. The Heart

The involvement of paf on heart physiology was first demonstrated on isolated heart from normal guinea pigs. A concentration-dependent reduction by paf of coronary flow and contractile force has been observed. The deacylated compound lyso paf or paf-related molecules do not affect the tissue, indicating the specificity of paf receptors and/or target cell(s). Paf-receptor antagonists attenuate the acute myocardial ischemia. The action of paf on the electrical and mechanical activities of isolated auricles of guinea pig heart suggests that this mediator changes the calcium and potassium conductance on the cell membrane of cardiac cells.

C. The Kidney

Paf infusion induces a dose-dependent decrease in glomerular filtration rate, electrolyte excretion, and urinary and renal blood flow. *In vivo* renal blood pressure is not affected by paf administration. In various experimental diseases (e.g., unilateral ureteral obstruction or obstructed hydronephrotic kidney), paf dose-dependently stimulates the release of prostaglandins and thromboxane A2. The latter release is inhibited by paf-receptor antagonists.

D. The Skin

Paf is a potent *in vivo* agonist capable of inducing inflammatory events in the skin. When injected into human or animal skin, it increases vascular permeability and neutrophil infiltration. Paf elicits in normal individuals a wheal-and-flare response with a perivascular neutrophil infiltrate. In allergic patients, a typical hypersensitivity reaction occurs, with an intense eosinophil infiltration similar to that induced by the specific allergen. Plasma protein exudation follows intradermal injection of low doses of paf, whereas higher doses induce platelet and then red blood cell accumulation. The dissociation between protein exudation and blood cell emigration is in agreement with a direct effect of paf on vessel wall permeability, independently from platelet activation.

E. Paf and Shock

Shock is characterized by different paf-like symptoms including systemic and cardiac hypotension, pulmonary hypertension, and an increase of vascular permeability. *In vivo* systemic administration of paf, of bacterial endotoxins, or of more purified products such as lipopolysaccharides from bacterial membranes can induce shock in animals. The analogies between endotoxin shock and the effects of systemic paf in rats have raised the hypothesis that paf might be an important mediator of endotoxin shock. Thus many pharmacological studies using different types of paf antagonists, either from natural origin or obtained by synthesis, have been done on animals submitted to endotoxin injection, and their results seem to corroborate the hypothesis. Moreover, in another experimental model, rats exhibit necrotizing bowel lesions associated with shock after an injection of paf, or a combination of paf and endotoxin under conditions in which either one is inactive. Recent investigations have shown that TNFα, a peptide cytokine endowed with a pivotal role in modulating acute and chronic disease states, is also able to induce shock and necrosis of the gastrointestinal tract with morphological changes indistinguishable from those induced by paf. In fact, the lesions are associated with paf production in bowel tissue and can be prevented by pretreatment with paf antagonists. The latter data suggest a sequential secretion of TNFα and paf in the pathogenesis of shock associated with digestive tract injury. Finally, paf has been credited with potent ulcerogenic and spasmogenic effects on rat stomach.

F. Reproduction

Embryo-derived paf may be involved in the nidation of the fertilized egg, in the fetal lung maturation, and in the parturation, the latter being related to its myometrum-contracting effect.

VII. Paf Antagonists

The first paf-receptor antagonists described were analogues of paf [e.g., CV-3988 (Takeda Chemicals, Japan) or SRI63-441 (Sandoz Chemicals, USA)]. Then naturally occurring substances were chemically defined [e.g., kadsurenone (Merck Sharp and Dohme, USA) from Haifenteng and ginkgolide B from Ginkgo biloba (compounds of the BN series, Institut Henri Beaufour, France)]. The synthetic triazolobenzodiazepines [e.g., alprazolam, triazolam, WEB-2086, and 2170 (Boehringer Ingelheim, GFR)] also express potent and specific paf-receptor antagonistic properties. Other compounds are ONO-6240, L-652-731 (Merck), Rhône-Poulenc-48740, and the SRI series from Sandoz (United States of America). Some of the calcium antagonists (e.g., diltiazem or verapamil), membrane-active drugs (e.g., xylocain), drugs that increase the intracellular level of cyclic AMP (e.g., prostacyclin), antiallergic drugs (e.g., ketotifen and nedocromil sodium), and the paf-adsorbing serum albumin are nonspecific paf antagonists. Diets rich in fish liver and onion oils have also receptor-independent inhibitory effects, which are mediated by a decrease of paf and thromboxane A2 synthesis. Thus these nonspecific paf inhibitors might be helpful as additional treatments of diseases in which paf is involved.

VIII. Conclusions

The vast array of *in vitro* and *in vivo* effects of paf makes it difficult to precisely delineate its physiological and/or pathological role. There is little doubt that many of these characteristics, obtained in experimental situations far from the human pathophysiology, will not survive time. It remains that given the potent vasopermeant and cell-attracting properties of this mediator, it is most probably involved in many situations in which an acute inflammation is created. Its role in the onset and maintenance of chronic inflammation is purely hypothetical at present. However, paf likely intervenes in asthma and other inflammatory lung diseases, local and generalized shock, burns, gastrointestinal diseases, and possibly rheumatic diseases. In fact, it might play a pivotal role as part of the intracellular cascade responsible for functional stimulation of cells. It may also act as a major regulator of cell differentiation—as shown in lymphocytes and neuronal cells—if not, when present in large amounts, as an inducer of cell injury. The use of the specific anti-paf drugs now under development should help in clarifying the pathophysiological role of this mediator and hopefully yield important benefits for the understanding and management of numerous pathological states.

Acknowledgments

The contribution of Annie and Bernard Arnoux, Ruth Korth, Ewa Ninio, and Yolène Thomas and the help of Françoise Lamarre for manuscript handling are gratefully acknowledged.

Bibliography

Benveniste, J. (1988). Paf-acether, an ether phospholipid with biological activity. *In* "Abberration in Membrane Structure and Function" (A. Leaf, L. C. Bolis, and M. L. Karnovski, eds.), pp. 73–85. Alan R. Liss, New York.

Denizot, Y., Dassa, E., Kim, H. Y., et al. (1989). Synthesis of paf-acether from exogenous precursors by the prokaryote *Escherichia coli*. *FEBS Lett.* **243,** 13–16.

Doebber, T. W., and Wu, M. S. (1987). Platelet-activating factor (PAF) stimulates the PAF-synthesizing enzyme acetyl-CoA: 1-alkyl-sn-glycero-3-phosphocholine O^2-acetyltransferase and PAF synthesis in neutrophils. *Proc. Natl. Acad. Sci. U.S.A.* **84,** 7557–7561.

Dulioust, A., Duprez, V., Pitton, C., et al. (1990). Immunoregulatory functions of paf-acether. III. Down-regulation of high-affinity interleukin 2 receptor expression. *J. Immunol.* **144,** 3123–3129.

Handley, D. A., Saunders, R. N., Houlihan, W. J., and Tomesch, J. C. (1990). "Platelet-Activating Factor in Endotoxin and Immune Diseases." Marcel Dekker, New York.

Kornecki, E., and Ehrlich, Y. H. (1988). Neuroregulatory and neuropathological actions of the ether-phospholipid platelet-activating factor. *Science* **240,** 1792–1794.

Leyravaud, S., Bossant, M. J., Joly, F., Bessou, G., Benveniste, J., and Ninio, E. (1989). PMA-induced paf-acether biosynthesis and acetyltransferase activation in human neutrophils. Comparison with a physiologic stimulus: Opsonized zymosan. *J. Immunol.* **143,** 245–249.

McIntyre, T. M., Zimmerman, G. A., and Prescott, S. M. (1986). Leukotrienes C_4 and D_4 stimulate human endothelial cells to synthesize platelet-activating factor and bind neutrophils. *Proc. Natl. Acad. Sci. U.S.A.* **83,** 2204–2208.

O'Flaherty, J. T., and Wykle, R. L. (1983). Biology and biochemistry of platelet-activating factor. *Clin. Rev. Allergy* **1,** 353–367.

Shalit, M., Von Allmen, C., Atkins, P. C., and Zweiman, B. (1988). Platelet-activating factor increases expression of complement receptors on human neutrophils. *J. Leukocyte Biol.* **44,** 212–217.

Sun, X. M., and Hsueh, W. (1988). Bowel necrosis induced by tumor necrosis factor in rats is mediated by platelet-activating factor. *J. Clin. Invest.* **81,** 1328–1331.

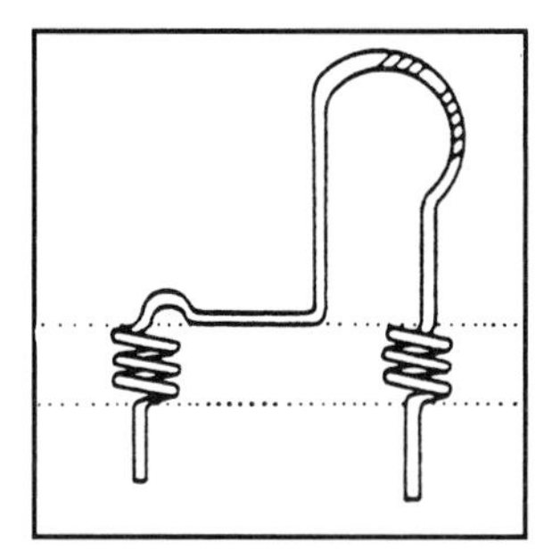

Platelet Receptors

G. A. JAMIESON, *American Red Cross*

Glossary

Adhesion Attachment of a monolayer of platelets to adhesive substrates such as collagen, laminin, or von Willebrand factor

Aggregation Commonly, the fibrinogen-dependent platelet–platelet interaction induced by soluble mediators such as thrombin or adenosine diphosphate; alternatively, the piling up of layers of platelets on the adherent or anchored monolayers

Anchorage Subsequent state involving the activation and spreading of adherent platelets resulting in a more tenacious interaction with the substrate

PLATELETS ARE NOT TRUE CELLS but are fragments of their precursor, the megakaryocyte, and they lack a nucleus. They can become activated because, on their surface, they have a variety of receptors that interact with activating molecules (agonists), either soluble substances formed in the blood or adhesive proteins exposed following damage to the vascular endothelium. Platelet activation results in the loss of discoid shape, secretion of intracellular components such as serotonin, the formation of long processes (pseudopodia) on the platelet surface, and the rearrangement of surface components, which then form secondary binding sites for adhesive proteins, such as fibrinogen, leading to the formation of hemostatically effective platelet thrombi.

Each of the receptors for which reasonably detailed information is available will be discussed in the alphabetical order of the agonists with which they interact.

I. Adenosine

Adenosine receptors are coupled to adenylate cyclase, the enzyme that in cells produces cyclic adenosine monophosphate, a secondary messenger in signal transduction. These receptors are of two main types: the A1 receptors, which inhibit the cyclase, and the A2 receptors, which stimulate it. The adenosine receptors in platelets are exclusively of the A2 subtype. This conclusion is based on the relative ability of a series of adenosine analogues to stimulate the cyclase. The most potent of these bind to isolated platelet membranes with high affinity [dissociation constant (Kd), 12 n*M*; maximum number of receptor sites (B_{max}), 1.1 pmol/mg protein at 37°C].

II. Adenosine Diphosphate

Adenosine diphosphate (ADP) has an important role as a platelet-aggregating agent. Exogenous ADP from red cells or other cellular sources can cause both platelet activation and the inhibition of stimulated adenylate cyclase. This activation, whether induced by ADP itself or by other agonists, then induces the secretion from platelets of ADP packaged in the dense granules that constitute an intracellular pool separate from the pool of metabolically active nucleotides. This secreted ADP can then promote further aggregation and recruitment of platelets to the thrombus.

Despite the importance of ADP in platelet function, the mechanisms by which it activates platelets are imperfectly understood. Essentially, any modification in the structure of ADP, whether in the purine, ribose, or phosphate moieties, results in the loss of platelet-aggregating activity. The only exception is ADP analogues substituted in the C2 position of the adenine ring, which are also able to induce full platelet activation. Certain other structural analogues are partial or complete antagonists and block the interaction of ADP with its receptors. Adenosine triphosphate (ATP) is the most powerful antagonist and is effective at the same concentration as ADP itself. [*See* ADENOSINE TRIPHOSPHATE (ATP).]

The ADP receptor on platelets does not closely resemble other adenine nucleotide receptors. In other tissues, including endothelial cells, both ADP and ATP are agonists, and there are no known reversible antagonists. Nucleotide receptors in other cells also generally show much lower stereoselectivity than does the ADP receptor of platelets, which has an absolute requirement for the natural form of the nucleotide.

A major difficulty in obtaining kinetic binding data has been the rapid metabolism of ADP and its analogues by platelets and the complicating effects of ADP secretion from the activated platelets. These problems have been circumvented by the use of platelets fixed with formaldehyde. The affinity of ADP (Kd 0.35 μM; 160,000 sites/platelet) and of some 20 analogues for the receptor were, generally, in the range expected from the potency of these substances as agonists and antagonists, except in the case of C2-substituted analogues, which showed a weaker-than-expected binding.

This difference may relate to the question of whether platelet activation is mediated by a single type of receptor coupled to two pathways, one inducing activation and the other affecting adenylate cyclase, or by two different receptors, each separately affecting one of the pathways. The first, or single-receptor hypothesis, is supported by the constant ratio in the activities of a wide range of structurally diverse ADP analogues in their effects as agonists or antagonists of platelet activation and inhibition of stimulated adenylate cyclase. The two-receptor hypothesis is supported by the fact that the adenosine analogue 5′-fluorosulfonylbenzoyl adenosine (FSBA), which has been used to label nucleotide-binding sites in a number of purified enzyme systems, inhibits ADP-induced platelet aggregation but does not inhibit adenylate cyclase. ^{3}H-FSBA labels a single membrane component (M_r 100,000) that has been termed "aggregin," but the mechanism by which this protein mediates ADP-induced platelet activation has not been determined.

Labeling of the α-chain of glycoprotein (GP)IIb (M_r 120,000) by ADP analogues suggests that it may be part of a platelet ADP receptor. This observation is consistent with the decreased activation induced by ADP with platelets from patients with Glanzmann's thrombasthenia, which lack GPIIb/IIIa.

III. Collagen

When the integrity of the endothelial cell monolayer lining the blood vessels is disrupted, platelets rapidly adhere to components exposed in the blood vessel wall, become activated, secrete granule contents, and form a hemostatic plug (thrombus) formed of platelets and other blood constituents. Fibrillar collagen has been identified as the most effective macromolecular constituent of the vessel wall in forming a thrombus. Direct binding of platelets to collagen without the interpolation of other adhesive proteins appears to occur in regions of the vasculature where blood is flowing relatively slowly. [*See* COLLAGEN, STRUCTURE AND FUNCTION.]

Numerous candidates have been proposed as the platelet collagen receptor, but few have been adequately characterized. Therefore, seven criteria have been suggested for establishing the role of a putative collagen receptor for platelets. These seven criteria are as follows. (1) The purified receptor should bind to native but not denatured collagen or other adhesive proteins. (2) The isolated receptor should compete with membrane-bound receptor in platelet activation by collagen but not by other agonists. (3) The isolated receptor should neutralize inhibition of adhesion by anti-whole platelet antibodies. (4) Anti-receptor antibodies should inhibit adhesion of unactivated platelets to collagen but not to other adhesive proteins, and this inhibition should be seen at the earliest stages of platelet adhesion. (5) The antibodies should also inhibit activation (including shape change) by collagen but not by other agonists. (6) Induced expression in nonadhesive cells should confer the property of collagen adhesion. (7) Receptor-associated defects should be

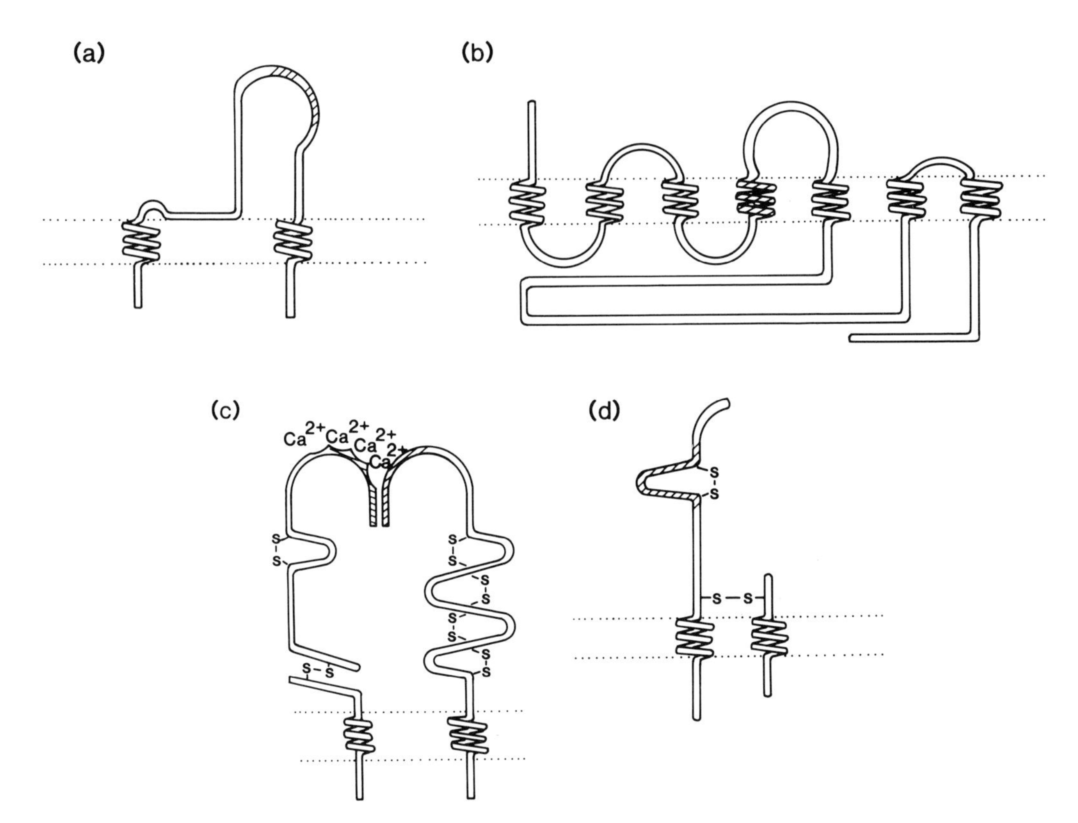

FIGURE 1 Probable structures of known platelet receptors. Amino- and carboxyl-termini are indicated together with disulfide linkages and Ca^{2+}-binding domains thought to be important to receptor function. Bound carbohydrate is not shown. Ligand-binding domains are indicated by the cross-hatching on the peptide backbone. The dotted lines indicate the boundaries of the lipid bilayer, and the extracellular space is oriented at the top of the figure in each case. (a) The primary collagen receptor GPIV; (b) the platelet α_2-adrenergic receptor; (c) the integrin GPIIb/IIIa, which in activated platelets serves as a binding site for the adhesive proteins fibrinogen, vWF, and fibronectin; (d) GPIb, the high-affinity receptor for α-thrombin and the primary (recognition) receptor for vWF.

identifiable in the platelets of patients with specific defects in adhesion or collagen-induced activation.

Platelet GPIV is the only platelet component that at present meets these seven criteria for a collagen receptor. GPIV also occurs in a variety of other cells, including endothelial cells, where it mediates the attachment of red cells infected with *Plasmodium falciparum* malaria, and it may function as the platelet receptor for thrombospondin.

GPIV (M_r 88,000) has been isolated from platelets and shown to bind avidly to fibrillar type I collagen (Kd 0.35 nM). The cDNA for the molecule has been cloned, and the deduced primary amino acid sequence indicates two transmembranous domains with a third hydrophobic domain, which may remain in close proximity to the lipid bilayer and account for the high degree of proteolytic resistance of GPIV in platelets (Fig. 1). Two domains demonstrating partial homology with collagen-binding domains of other proteins have also been identified.

Another membrane glycoprotein suggested as a platelet collagen receptor is the VLA-2 integrin (GPIa/IIa). Monoclonal antibodies against VLA-2 inhibit platelet adhesion and lipid vesicles (liposomes) containing purified GPIa/IIa on their surface adhere to collagen.

Available data are consistent with the interpretation that recognition of, and adhesion to, collagen by platelets is mediated by GPIV. This leads to membrane rearrangement and the expression of adhesive binding sites in the GPIa/IIa complex, which mediates the subsequent stages of spreading and anchorage as secondary events.

IV. Epinephrine

The α_2-adrenergic receptor is probably the best characterized of all the platelet receptors, although it has not yet been unequivocally established whether epinephrine is a platelet agonist in its own right or it merely potentiates the activation induced by other agonists.

The platelet α_2-adenergic receptor has been isolated, and the corresponding cDNA has been cloned. The deduced primary amino acid structure shows that it is a member of the family of G-protein-coupled receptors, which includes the β-adrenergic and muscarinic receptors and rhodopsin, the receptor for light in the cones of the eye. The peptide backbone contains seven regions of 20–25 predominantly hydrophobic amino acids in an α-helical conformation, which loop back and forth through the plasma membrane and are numbered 1–7 from the amino-terminal domains (Fig. 1). There are a total of six loops of hydrophilic amino acids connecting the membrane-spanning regions, three projecting into the cytoplasm and three into the extracellular space.

The fourth membrane-spanning domain is involved in ligand-binding to human platelet α_2-adrenergic receptors.

V. Fibrinogen

The binding of fibrinogen to platelets following activation is a prerequisite for their aggregation. Extensive studies in numerous laboratories have shown that the binding site for fibrinogen becomes expressed in the GPIIb/IIIa complex after platelet activation. This conclusion is based on the ability to bind fibrinogen covalently to GPIIb/IIIa using cross-linking agents, the inhibition of fibrinogen-binding by monoclonal antibodies to GPIIb/IIIa, and the lack of aggregation seen with platelets from patients with Glanzmann's thrombasthenia, which lack GPIIb/IIIa.

Structural features of GPIIb and GPIIIa have been deduced from biochemical and molecular studies and have established that the two molecules are members of the integrin superfamily of adhesive binding proteins. GPIIb is a 135 kDa integral membrane protein composed of 100 kDa α-chain disulfide-linked to a 23 kDa β-chain and contains four sequences of amino acids homologous to the Ca^{2+}-binding domains of calmodulin. GPIIIa is a single chain of 95 kDa containing 21 intrachain disulfide bonds in four cysteine-rich repeats (Fig. 1). GPIIb is synthesized as a larger precursor. It undergoes posttranslational processing prior to forming a complex with GPIIIa, which is inserted into the plasma membrane. Both GPIIb and GPIIIa contain a single transmembrane-spanning region and a short cytoplasmic domain.

The molecular basis of fibrinogen receptor exposure following platelet activation remains unknown but extracellular divalent cations play an important role in the stability and function of GPIIb/IIIa, and the complex may serve as a channel for the influx of extracellular Ca^{2+}. The majority of high-affinity binding sites for Ca^{2+} on the platelet surface have been localized to GPIIb/IIIa, possibly within the calmodulin-like domains of GPIIb. Ca^{2+} is required to maintain GPIIb and GPIIIa as a complex in solution. Although it is possible that Ca^{2+} functions as a bridge between GPIIb and GPIIIa, it seems more likely that it functions to maintain one or both of these glycoproteins in a conformation required for their stable association. Full fibrinogen-binding function of the complex has been observed only when it is present in its native membrane environment within the intact platelet.

VI. Laminin

Platelets adhere to laminin, a consituent of the basal lamina that lies directly under the endothelial cells lining blood vessels. In contradistinction to their attachment to other adhesive proteins such as collagen and von Willebrand factor (vWF), platelets adherent to laminin under static conditions do not spread or undergo other morphological changes characteristic of activation. Platelets will also adhere directly to the laminin-derived peptide Tyr-Ile-Gly-Ser-Arg. Anti-laminin antibodies inhibit the adhesion of platelets to segments of vascular subendothelium in flowing blood, indicating the physiological relevance of this interaction and, under these conditions, platelets do become activated, possibly due to the added effects of fluid flow.

A 67 kDa laminin-binding protein, identical to that found in human breast carcinoma tissue and a variety of other normal and transformed cells, has been affinity-isolated from platelets. Monoclonal antibodies to this receptor inhibit the binding of platelets to laminin. Monoclonal antibodies to the VLA-6 integrin, corresponding to the GPIc/IIa complex, also inhibit platelet adhesion to laminin. The relationship between adhesion mediated by the

67 kDa receptor and that mediated by GPIc/IIa is not clear, but this may be another example, like that seen with collagen and vWF, of different receptors being involved in the adhesion and anchorage phases of platelet attachment.

VII. Platelet-Activating Factor

The structure of platelet-activating factor (PAF) is 1-0-alkyl-2-acetyl-SN-glycero-3-phosphocholine with hexadecyl and octadecyl moieties comprising the major alkyl constituents, and it is the paradigm of a class of biologically active phospholipids of the general structure alkylacylglycerophosphocholine. It causes aggregation of washed human platelets at low concentrations (1–10 n*M*) but may have a wide range of biological functions, as it is also a potent activator of a variety of cells, including neutrophils, monocytes, endothelial cells, smooth muscle cells, mesangial cells, and neurons. PAF on human platelets binds with high affinity (Kd 0.5–5 nM) to a specific receptor coupled to a GTP-binding protein. PAF-binding proteins of M_r 160,000–180,000 have been identified and may, in fact, be on the inner surface of the membrane but may be accessible to PAF due to its structural similarity to membrane phospholipids. [*See* PLATELET-ACTIVATING FACTOR PAF-ACETHER.]

VIII. Thrombin

Thrombin is the most potent platelet agonist and can cause full aggregation at very low concentrations (0.3 n*M*). Several lines of evidence indicate that α-thrombin activates platelets by two distinct pathways corresponding to high affinity (Kd 0.3 nM, 53 sites/platelet) and moderate affinity (Kd 11 nM, 1,700 sites/platelet) receptors. The two pathways also differ in their requirements for receptor occupancy and in the role of guanine nucleotide regulatory proteins. Furthermore, prior treatment of platelets with the proteolytic enzyme chymotrypsin separates two main effects of platelet activation: It blocks the thrombin-induced inhibition of adenylate cyclase and the activation of phospholipase A2, whereas the activation of phospholipase C and protein kinase are unaffected.

GPIb has been identified as the high-affinity thrombin receptor, as demonstrated most directly by the effect of monoclonal antiGPIb antibodies, which decrease binding of, and activation by, low concentrations of thrombin (<0.4 n*M*). Similar effects are also observed when GPIb is selectively cleaved from platelets by a proteolytic enzyme (*Serratia marcescens* metalloprotease). Other evidence supporting GPIb as the thrombin high-affinity receptor is that it may be chemically cross-linked to thrombin in intact platelets and that there are concomitant decreases in thrombin-binding, platelet responsiveness, and altered GPIb content in platelets from patients with Bernard–Soulier syndrome, patients with myeloproliferative disorders, or platelets treated with chymotrypsin or elastase.

GPIb (M_r 170,000) consists of two disulfide-linked subunits GPIbα (M_r 143,000) and GPIbβ (M_r 22,000) but, in the membrane, it probably exists as a multimolecular complex of about 900,000 Da. The thrombin-binding site of GPIb has been located in the M_r 45,000 carbohydrate-poor, amino-terminal domain of the GPIb α-chain and may overlap with, but is not identical to, the vWF-binding domain on GPIb for vWF (another clotting factor). cDNAs for both α- and β-chains of GPIb have been cloned, and each possesses a single membrane-spanning domain (Fig. 1).

The nature of the moderate-affinity thrombin receptor is at present unknown.

IX. von Willebrand Factor

In the regions of the microvasculature where the blood moves with high velocity, the adhesion of platelets to subendothelium appears to be mediated by vWF bound to collagen or, possibly, to other subendothelial components. This initial interaction is dependent on the presence of GPIb on the platelet surface, and the vWF-binding domain of GPIb has been localized as being in the M_r 45,000 distal fragment, which also contains the binding site for thrombin. Agglutination of formalin-fixed platelets by vWF in the presence of the antibiotic ristocetin has been a useful laboratory test for this GPIb-dependent interaction. The binding domain in vWF involved in this GPIb-dependent interaction is a Val-Thr-Leu-Asn-Pro-Ser-Asp-Pro-Glu-Cys-Gln sequence.

Platelets that have undergone this initial attachment mediated by the GPIb–vWF interaction then undergo activation and anchorage to GPIIb/IIIa at the sequence Arg_{1744}-Gly-Asp_{1746} present in the carboxy-terminal third of the vWF molecule. These steps provide the clearest evidence in platelets of different receptors mediating adhesion and anchorage to the same adhesive substrate.

Bibliography

Jamieson, G. A. (ed.) (1988). "Platelet Membrane Receptors: Molecular Biology, Immunology, Biochemistry, and Pathology." Alan R. Liss, Inc., New York.

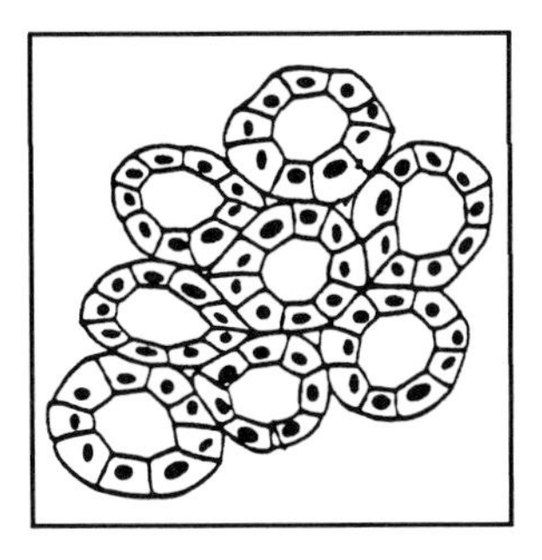

Polarized Epithelial Cells

ENRIQUE RODRIGUEZ-BOULAN, *Cornell University Medical College*

W. JAMES NELSON, *Stanford University*

Glossary

Cytoskeleton Structural and motile elements of the cell; classified as cytoplasmic and submembranous. The former is formed by filaments that are classified according to their thickness. These filaments provide the cell's "scaffold" (intermediate filaments, 10 nm), "muscle" (thin, 5 nm, actin filaments), and "tracks" for the transport of organelles (microtubules, 20 nm)

Endoplasmic reticulum Main site of protein and lipid synthesis of secretory material in cells specialized in secretion

Epithelial phenotype All the apparent features that distinguish the epithelial cell

Junctional complex In epithelial cells, the typical junctional complex is formed by three distinct junctions that may be distinguished at the electron microscope level: the tight junction, the zonula adherens, and the desmosome

Morphogenesis Generation of form or shape, the process by which the embryo or its parts acquire their definitive shape

Secretion Externalization of material produced or synthesized by the cell (e.g., proteins, glycoproteins, lipoproteins)

Transcytosis Process by which material is transferred from one side to the other of the cells

THIS ARTICLE describes the structure and function of polarized epithelial cells and their distribution in the embryo and adult. The highly asymmetric structure of the polarized epithelial cell is shown to be the origin of a variety of unidirectional functions that constitute the basis of its active barrier role. The polarized organization of the cytoskeleton fulfills a role in embryonic tissue morphogenesis and in the establishment of the polarized epithelial phenotype. The mechanisms involved in the maintenance of epithelial cell polarity in mature epithelia are discussed.

I. Distribution and General Organization

Polarized epithelial cells form tightly apposed superficial monolayers that line all body cavities and tubular structures. They play a crucial role in tissue and organ morphogenesis during embryo development. In addition, they form selective permeability barriers between biological compartments in both the embryo and the adult. These two critical functions explain the widespread distribution of the epithelial cell phenotype (Fig. 1).

Characteristically, epithelial tissues display closely apposed cells with minimal amounts of intercellular material. They lie on protein meshworks designated *basement membranes*, which are synthesized in part by epithelial cells and in part by fibroblasts in the connective tissue. The polarized epithelial cell is characterized (Fig. 2) by (1) the distribution of plasma membrane proteins and lipids to three distinct surface domains, termed *apical, lateral*, and *basal*; (2) specialized tight junctions

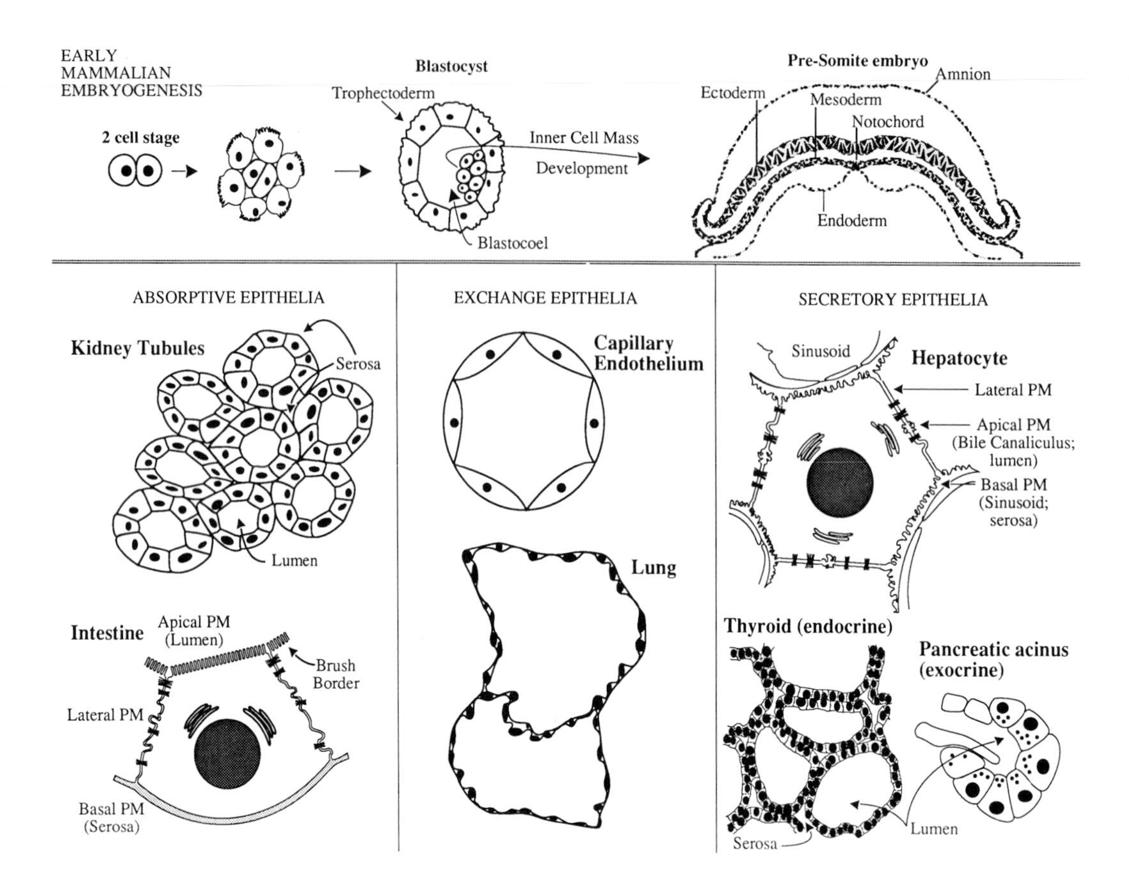

FIGURE 1 Tissue distribution of polarized epithelial cell phenotype. **Top:** Formation of a polarized epithelial cell phenotype is the first overt sign of cellular differentiation in early mammalian embryogenesis. Cleavage of the fertilized egg generates polarized outer blastomeres and unpolarized inner blastomeres. Outer blastomeres form the trophectoderm, an ion-transporting epithelium. Apolar blastomeres form the inner cell mass that develops into three major germ layers (i.e., ectoderm, endoderm, and mesoderm) from which all adult organs and tissues are formed. **Bottom:** Polarized epithelial cells are the major cell type in a variety of organs and tissues in the adult organism where they form selective permeability barriers between luminal and serosal compartments of different ionic milieu. Three classes of epithelia can be defined (i.e., absorptive, exchange, and secretory) according to their major functional activity.

that separate apical and lateral surface domains and form barriers to the intercellular diffusion of macromolecules and ions; (3) cohesive cell-cell attachments via cell-adhesion molecules (CAMs) and the junctional complex; (4) strong attachments to the basement membrane via specific receptors or substrate attachment molecules (SAMs); and (5) the polarized distribution of cytoplasmic organelles and the cytoplasmic and cortical cytoskeleton.

II. Biological Roles of Epithelial Cells: Morphogenesis and Active Barrier

The structural characteristics of the polarized epithelial cell are responsible for two major biological roles: morphogenesis during embryo development, and active barrier both in the embryo and in the adult organism.

A. Cytoskeleton and the Morphogenetic Role of Epithelial Monolayers

Components of the cytoskeleton contribute to the organization of fully polarized epithelial cells and the roles of these cells in tissue and organ morphogenesis (Fig. 3). Some components interact specifically with junctional elements that interlink cells of the epithelium and attach cells to the substratum. Other cytoskeletal components are involved in organization of apical and basolateral membrane proteins. Together, these cytoskeletal organizations contribute to the surface and cytoplasmic polarity of epithelial cells and transmit surface polarity cues elicited by interactions with other cells and with the substratum. This "conversation" between surface

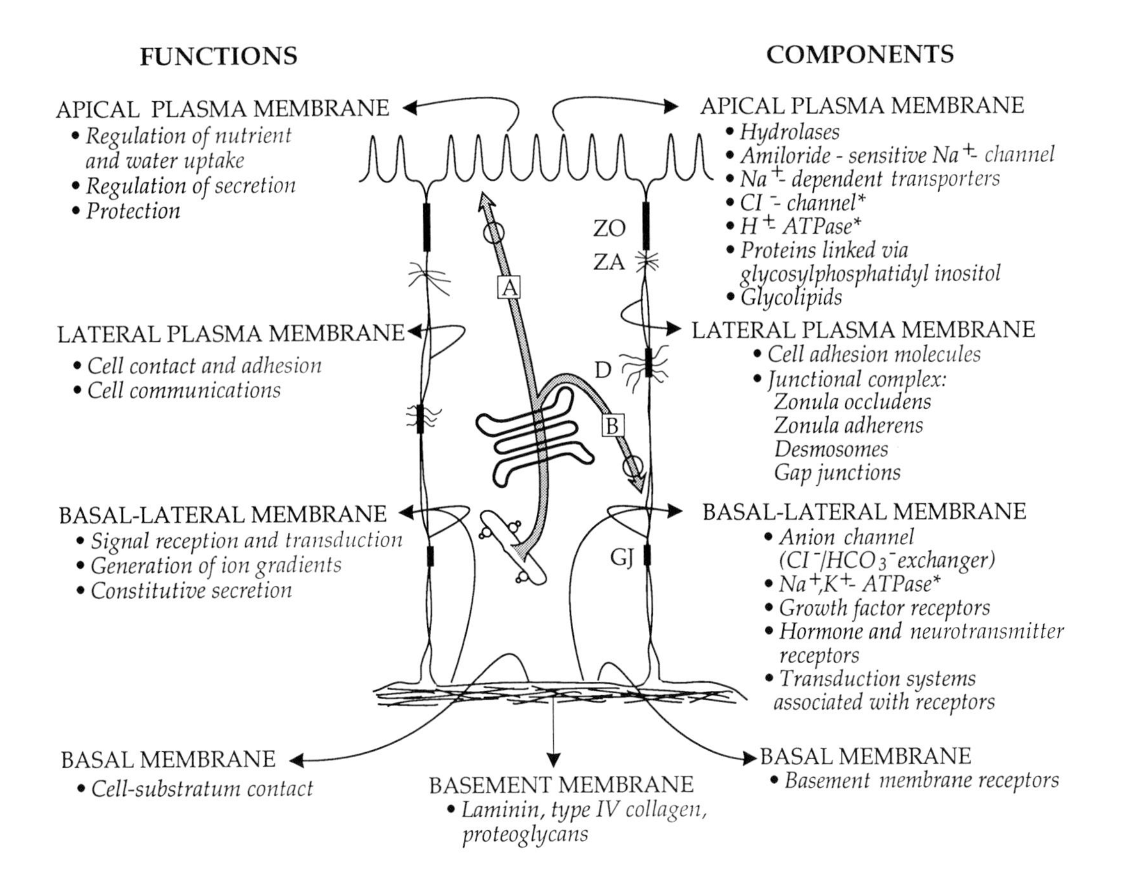

FIGURE 2 Functional surface domains in the polarized epithelial cell. Different ion transport and receptor systems are located in apical and basolateral domains of plasma membrane. Apical membrane is enriched in glycolipids, GPI-anchored proteins, and hydrolases; in exocrine glands, it is the site of regulated secretion (pathway A). Lateral membrane contains the junctional complex, which includes the tight junction or zonula occludens (ZO), the zonula adherens (ZA), and numerous desmosomes (D); it also displays gap junctions (GJ) that allow communication, via small messenger molecules and ions, between neighboring cells. Basal membrane is enriched in receptors for substrate molecules, whereas both lateral and basal membranes have receptors for blood-borne hormones and growth factors. [Reproduced with permission from Rodriguez-Boulan, E., and Nelson, W. J. (1989). *Science* **245,** 718–725.]

molecules and the cytoskeleton results in a dynamic structural and functional cellular reorganization during development of a polarized epithelium.

The important roles of epithelial monolayers in organ morphogenesis are based on the cohesive structure of the epithelium in which cells are linked together through the junctional complex and the contractile cytoskeleton. The latter comprises filaments of actin that circumscribe the apical portion of each cell and are attached to a junction, termed the *zonula adherens*, that links together apposing cells. Contraction of these filaments draws the apices of connected cells together like a "purse string", resulting in an infolding (invagination) of the epithelium at that point. Other, noncontractile cytoskeletal filaments (e.g., intermediate filaments) play a more static role in tissue organization by forming an intracellular scaffold of filaments from one side of the cell to the other at sites of specific intercellular junctions, termed *desmosomes*. This scaffold maintains the structural continuum of the epithelium against shearing stress. The microtubule component of the cytoskeleton is organized into an apical web of filaments that extend in longitudinal bundles to the base of the cell. These microtubules play roles in the spatial organization of organelles in the cytoplasm (e.g., endoplasmic reticulum and Golgi complex) and as tracks for the movement of vesicles to and from different domains of the plasma membrane. Finally, components of the intermediate filament and actin filament cytoskeletons are linked to SAMs at the base of the cell.

The characteristic structural and functional organization of apical and basolateral membrane domains is regulated, in part, by distinct cytoskeletons. The apical domain has numerous membrane extensions, each termed a *microvillus* (~1 μm in length and 0.1 μm in diameter), collectively termed the *brush border*, that project into the extracellular milieu and create an increased surface area for absorption and secretion. Each microvillus contains a central core of actin filaments and associated proteins (i.e., villin, fimbrin). At the base of the microvillus, the actin core is embedded in a matrix (i.e., the terminal web) of contractile proteins (e.g., tropomyosin, actin), which is linked laterally to the zonula adherens region of the apical junctional complex by other proteins (e.g., α-actinin, vinculin). The basolateral membrane domain contains a different cytoskeletal framework that comprises several proteins described in human red cells. The principal protein, termed *fodrin* (i.e., the nonerythroid form of spectrin), is a long, flexible, rod-shaped molecule (200 nm long by 10 nm wide). Individual fodrin molecules may be linked together by actin and other associated proteins to form a protein meshwork on the cytoplasmic face of the plasma membrane. The meshwork is linked to specific membrane proteins (e.g., Na^+, K^+-ATPase, a specific basolateral membrane protein) through a protein termed *ankyrin*, which binds to fodrin. A high molecular weight complex (i.e., the membrane cytoskeleton) of Na^+, K^+-ATPase, ankyrin, and fodrin has been identified in certain kidney cells. Another component of the complex appears to be the epithelial CAM termed *uvomorulin* or *E-cadherin*. Lateral membrane contacts between cells, which are regulated by this CAM, is believed to trigger the polymerization of the membrane cytoskeleton into a protein network that becomes decreasingly extractable by membrane solubilizing agents (e.g., detergents). Interactions between membrane proteins and this network cause progressive immobilization of constituent

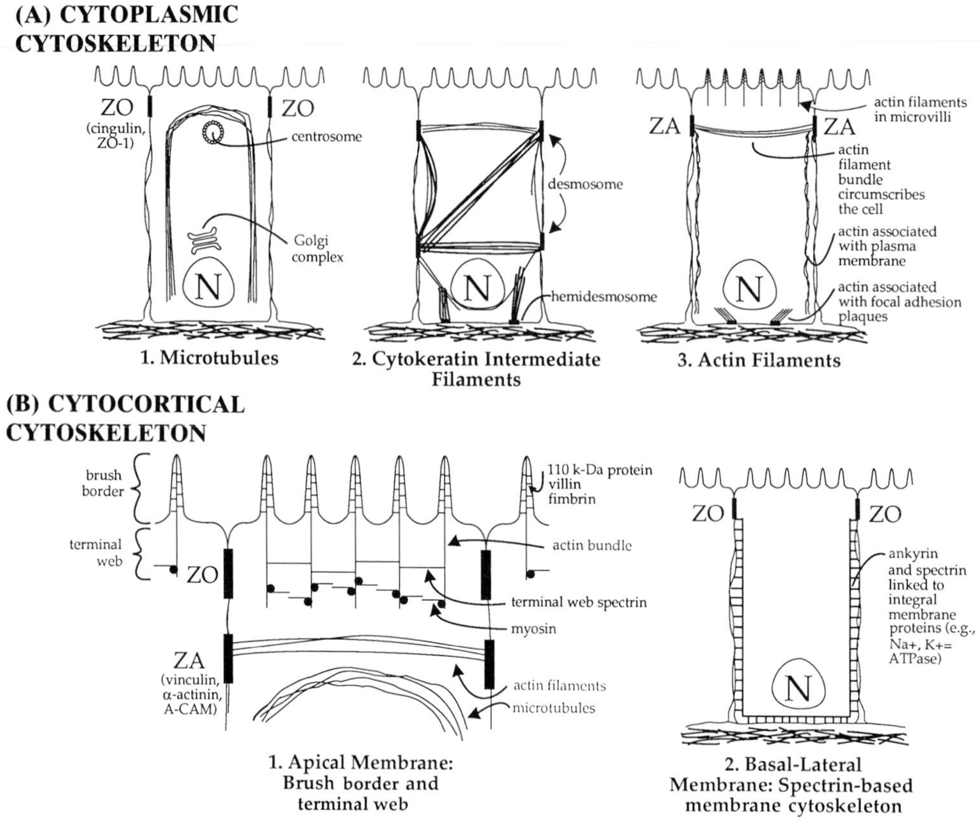

FIGURE 3 Polarized organization of the cytoplasmic and cytocortical cytoskeleton. Connections between cellular junctions and the cytoplasmic cytoskeleton are schematized **(A)**. Different apical and basolateral domain-specific submembrane cytoskeletons are found in epithelial cells **(B)**. The subapical cytoskeleton is actin based: Bundles of actin filaments form the cores of microvilli that end basally in the terminal web; this filamentous network ends laterally in the zonula adherens (ZA). Basal cytoskeleton is composed of fodrin, ankyrin, and actin; on cell–cell interaction, "subunits" of this cytoskeleton polymerize into a large network. ZO, zonula occludens.

proteins, resulting in their accumulation at the forming basolateral membrane domain.

B. Roles of Epithelial Cells as Passive and Active Barriers

Epithelial monolayers behave both as passive and active barriers between biological compartments (e.g., blood and the lumen of the gut or kidney tubules) (Fig. 1). The role as a passive barrier is dependent on the presence of tight junctions, a characteristic epithelial structure. Tight junctions are formed by parallel strands and grooves between closely apposed cells, as seen by freeze-fracture electron microscopy; in thin-section electron microscopy, they appear as regions of fusion between the membranes of neighboring cells, where the normal intercellular gap is occluded. Two peripheral membrane proteins, ZO-1 and cingulin, have been identified at the cytoplasmic aspect of the tight junction, but integral membrane proteins have not yet been found. The tight junction blocks the passage of macromolecules between cells (i.e., paracellular route) and confers a selective permeability to cations over anions across the epithelium (presumably due to the possession of a net negative charge in the sealing elements of the junction). Tight junction permeability may be regulated by contraction of the apical cytoskeleton (see Section II,A).

The role of polarized epithelial cells as active barriers is given by the metabolic activities of the individual epithelia. Polarized epithelial cells modify their environment by secretion, ion and nutrient transport, and transport of macromolecules from one biological compartment to the other across the epithelium, termed *transcytosis*. Secretion may occur at either the apical or the basolateral membrane (i.e., polarized secretion) and may be either continuous (such as the production of basement membrane components, i.e., laminin, proteoglycans, and type IV collagen) or regulated by hormones or neurotransmitters. Ion transport is essential for the maintenance of the salt composition of the internal medium and is carried out by a variety of ion channels and transporters. Transcytosis of macromolecules (usually immunoglobulins) plays a role in the immunological defense against infectious agents (e.g., transcytosis of maternal antibodies across the placenta to the fetus). Energy for secretion, ion transport, and transcytosis is usually needed and is provided by hydrolysis of ATP.

III. Generation of the Polarized Epithelial Phenotype

The first overt sign of multicellular organization in the mammalian embryo occurs after about four cell divisions of the fertilized egg, when an outer layer of polarized epithelial cells, termed the *trophectoderm*, is formed (Fig. 1). The trophectoderm surrounds the developing embryo and eventually forms the placenta. Further cell divisions within the embryo give rise to distinct germ layers of epithelial cells from which all tissues and organs are formed. Morphogenesis of a variety of organs (e.g., intestinal tract, urinary and genital tubes) requires folding of these germ layers into hollow linings or tubes. Later in development, epithelial cells continue to appear at critical stages. Many body organs (e.g., lung, liver, kidney, glands) develop from branched, hollow linings of epithelial cells that arise from small epithelial buds and surrounding mesenchymal cells. The latter are induced to differentiate into epithelial cells, aggregate, and form the epithelial linings of organs. Contacts with the basement membranes is important for the conversion of mesenchymal cells into epithelial cells, although the specific genes that are activated in this process are still uncharacterized. In addition, there are some instances in development when loss of the epithelial phenotype results in the formation of a different tissue. An example of this is the formation of mesoderm from primitive streak ectoderm at the two–germ layer stage. [*See* EMBRYO, BODY PATTERN FORMATION.]

Basement membrane components are specific for different epithelial tissues and somehow maintain the characteristic differentiated phenotype of that tissue (e.g., determine that tracheal epithelium remains tracheal, intestinal epithelium remains intestinal). A variety of receptors for basement membrane components may be involved in this effect. The cellular mechanisms of control are still unknown.

In vitro model systems using kidney and thyroid epithelial cells in culture have been employed to study the stages in the development of cell polarity. These studies indicate that the formation of the apical and the basolateral pole can be uncoupled both in time and space. The generation of the apical pole is a rapid process initiated by either cell-cell or cell-substratum contact. The generation of the basolateral pole, instead, is more gradual and requires extensive cell-cell contact and the development of intercellular junctions. The generation of these membrane domains requires removal of misplaced proteins and is coordinated with the assembly of domain specific submembranous cytoskeletal networks (see Section II,A).

IV. Maintenance of Epithelial Cell Polarity

The different protein compositions of apical and basilateral surfaces of epithelial cells can be demonstrated either by purification and analysis of these two membranes or by application of molecular "labels" or "tags" to each surface to track the distributions of newly synthesized proteins. These studies show that once the polarized epithelial phenotype is established, maintenance of the two distinct sets of proteins in the apical and basolateral membranes is a dynamic process that depends on the coordinate insertion of new proteins in the appropriate surface domain.

Initial studies on the biogenesis of plasma membrane proteins made use of the observation that enveloped RNA viruses bud in a polarized fashion from the surfaces of epithelial cells. Influenza virus buds from the apical surface, whereas vesicular stomatitis virus (from the same family as rabies) buds from the basolateral surface. Studies carried out with the envelope glycoproteins of these viruses have provided insight into the process of sorting of surface molecules by epithelia (Fig. 4), because these viral glycoproteins are synthesized and transported to the cell surface in a manner identical to that of cellular plasma membrane proteins. Synthesis of both apical and basolateral membrane proteins is carried out by ribosomes bound to the endoplasmic reticulum. Insertion of the nascent polypeptide chain into the endoplasmic reticulum membrane is mediated by a signal sequence similar to that employed by cellular secretory proteins. Irrespective of their final destination, both sets of proteins are processed similarly by endoplasmic reticulum and Golgi enzymes and follow an identical route until the distal (or *trans*) cisterna of the Golgi apparatus. This is a special sorting compartment designated the *trans* Golgi network (TGN) or Golgi-endoplasmic reticulum-lysosome (GERL). [*See* Golgi Apparatus.]

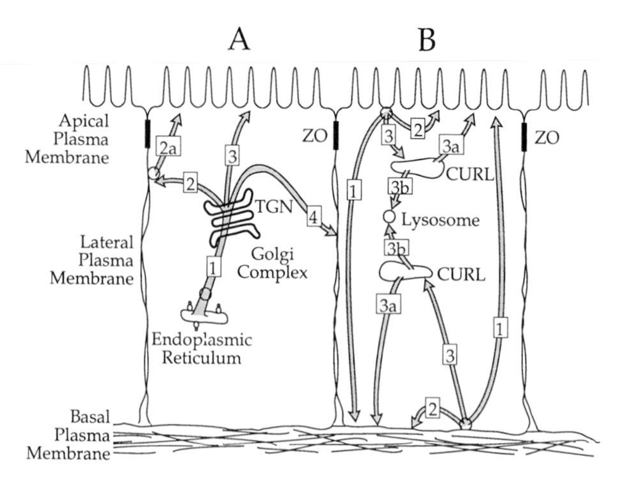

FIGURE 4 Protein transporting pathways in epithelial cells. *A*: Biosynthetic pathways. After synthesis in the endoplasmic reticulum, apical and basolateral membrane proteins are transported together to the Golgi complex (pathway 1). In intestinal and liver cells, there is evidence for initial delivery of both groups of proteins to the basolateral membrane (pathway 2); basal proteins are retained, whereas apical proteins are endocytosed and delivered to the apical membrane (pathway 2a). Studies with viral glycoproteins in cultured cell lines of kidney epithelia have demonstrated intracellular sorting in a distal compartment of the Golgi apparatus, the *trans* Golgi network (TGN). At this point, apical and basolateral membrane proteins are sorted into different vesicles that are targeted to the corresponding plasma membrane domain. *B*: Recycling pathways. Ligands endocytosed via apical membrane receptors are delivered to apical endosomes (pathway 1), whereas ligands endocytosed via basal membrane receptors are delivered to a different endosome population, the basal endosomes (pathway 2). From either set of endosomes, receptors may recycle back to the same surface (pathways 1c and 2a), transcytose to the other surface (pathways 1a and 2c), or be transferred to lysosomes (pathways 1b and 2b). [Reproduced with permission from Rodriguez-Boulan, E. and Nelson, W. J. (1989). *Science* **245,** 718–725.] ZO, Zonula occludens; CURL, compartment of uncoupling of receptor and ligand.

In the TGN, proteins destined for the apical and basolateral membranes are sorted from one another and incorporated into different vesicles that are then targeted to and that fuse with the corresponding plasma membrane domain. Transport of vesicles from the TGN to the plasma membrane may be mediated by tracks of microtubules. The mechanism(s) involved in the sorting and targeting of different membrane proteins is not well understood. It is thought that receptors in the TGN may recognize different signals on proteins and sort them into different vesicles. However, another possibility is that only one pathway is signal-mediated and the other is default. The sorting of proteins into vesicles destined for the apical membrane is accompanied by the enrichment in the vesicle membrane of glycolipids of the ceramide type (i.e., glucosyl and galactosyl-ceramide). Concomitant incorporation of apical proteins and glycolipids into vesicles may constitute a sorting mechanism through noncovalent interaction between apical proteins and these glycolipids. Significantly, proteins that are linked to the membrane by a glycolipid, glycosyl phosphatidylinositol (GPI), are also targeted to the apical membrane. Transfer of GPI to proteins normally secreted through the basolateral membrane results in their redirectioning to the apical membrane, suggesting a sorting role for GPI. The tendency of these glycolipids to form patches and thus segregate themselves from other membrane lipids may be important in this sorting process.

Studies on the sorting and targeting of cellular plasma membrane proteins have not only confirmed that these proteins use the same pathways as viral envelope proteins, but have also uncovered additional pathways (Fig. 4A). In intestinal cells, some apical proteins reach their destination through an indirect, "transcytotic" pathway, similar to that employed by the IgA receptor (discussed below). Part of or all the newly synthesized apical proteins are initially delivered to the basolateral surface, together with basolateral membrane proteins, where they are selectively endocytosed and then transferred to the apical surface. Apical membrane proteins of the hepatocyte (liver cell) studied so far also follow a similar transcytotic route. The signals that direct proteins (e.g., IgA receptor) along this complex pathway appear to be located in different domains of the protein; different signals may become dominant at different stages of the pathway as a result of posttranslational modifications (e.g., phosphorylation), thus determining the complex pathway of this molecule.

After proteins reach their respective plasma membrane domain, they may remain there until they are degraded, or they may be recycled through intracellular compartments (Fig. 4B). The latter happens in the case of hormone or growth factor receptors located in the basolateral membrane (e.g., transferrin receptor). These proteins are cycled through distinct populations of vesicles termed *endosomes* that are associated with either the apical or the basolateral membrane domain. Proteins cycling through these compartments are usually returned to the membrane domain from which they originated.

V. Restriction to the Diffusion of Plasma Membrane Proteins in Polarized Epithelial Cells

Once proteins reach their appropriate plasma membrane domain, their ability to diffuse in the plane of the lipid bilayer must be restricted, otherwise randomization of proteins over the whole cell surface would rapidly take place. Proteins in each plasma membrane domain appear to be partially immobilized by the cytoskeleton. However, substantial fractions of proteins are free to diffuse. In addition, lipid compositions appear to be different in each domain. Experiments with fluorescent lipid probes have shown that when they are incorporated to one surface domain, their diffusion is prevented by the tight junction. However, when the lipid probe has the ability to translocate from the outer to the inner leaflet of the bilayer ("flip-flop"), it quickly equilibrates with the other surface domain. This experiment indicates that tight junctions restrict diffusion of lipids in the outer, but not in the inner, leaflet of the bilayer. Consequently, there are greater differences in the lipid composition between domains in the outer leaflet of the bilayer than in the inner leaflet.

Experiments have been carried out to study whether tight junctions prevent diffusion of proteins. (This is certainly expected because proteins are larger than lipids.) Strategies that disrupt tight junctions (e.g., removal of Ca^{2+} from the medium) result in progressive randomization of surface protein compositions. However, the interpretation of these results is made difficult by the fact that the same experimental strategies also result in loss of all lateral membrane contact and disruption of the organization of the cytoskeleton, which may be equally responsible for the randomization of surface components.

VI. Diseases with an Alteration of the Polarized Epithelial Phenotype

Polarized epithelial cells play a fundamental role in the regulation of the ionic environment of an organism. Each of the stages involved in the generation and maintenance of the polarized epithelial phenotype described above may be altered by genetic mutation. These alterations may have dire consequences for normal tissue function and organismal survival.

The most frequent type of human tumors is the carcinoma (85% of all cases), which is derived from epithelial tissues. Most carcinomas are characterized by various degrees of loss of cell polarity. Some genetic diseases appear to result from alterations in the establishment and maintenance of surface polarity. For example, polycystic kidney disease is characterized by the development of numerous kidney cysts and the apical localization of the Na,K-ATPase (usually a basolateral membrane enzyme). Microvillar inclusion disease (i.e., a familial enteropathy of the small intestine) is characterized by the lack of cell surface expression of microvillar enzymes, which are stored abnormally in intracellular vesicles rich in microvilli. Efforts to understand how the polarized epithelial cell phenotype is generated and maintained may lead to important insights into the etiology and treatment of these and other diseases. [*See* NEOPLASMS, ETIOLOGY.]

Bibliography

Fawcett, D. W. (1986). "A Textbook in Histology," 11th ed. W. B. Saunders, Philadelphia.

Lisanti, M. P., and Rodriguez-Boulan, E. (1990). Glycolipid membrane anchoring provides clues to the mechanism of protein sorting in polarized epithelial cells. *Trends Biochem. Sci.* **15,** 113–118.

Matlin, K. S., and Valentich, J. D., eds. (1989). Functional epithelial cells in culture. *In* "Modern Cell Biology" (B. H. Satir, series ed.). Alan R. Liss, New York.

Nelson, W. J. (1989). Topogenesis of plasma membrane domains in polarized epithelial cells. *Curr. Opinion Cell Biol.* **4,** 660–668.

Rodriguez-Boulan, E., and Nelson, W. J. (1989). Morphogenesis of the polarized epithelial cell phenotype. *Science* **245,** 718–725.

Special topic: Polarity of epithelial cells. (1989). *In* "Annual Review of Physiology," vol. 51, (D. C. Eaton, section ed.), pp. 727–810. Annual Review, Palo Alto, California.

van Meer, G., and Simons, K. (1989). Lipid polarity and sorting in epithelial cells. *J. Cell Biochem.* **36,** 51–58.

Wandinger-Ness, A., and Simons, K. (1989). The polarized transport of surface proteins and lipids in epithelial cells. *In* "Intracellular Trafficking of Proteins" (J. Hanover and C. Steel, ed.). Cambridge University Press, Cambridge, England.

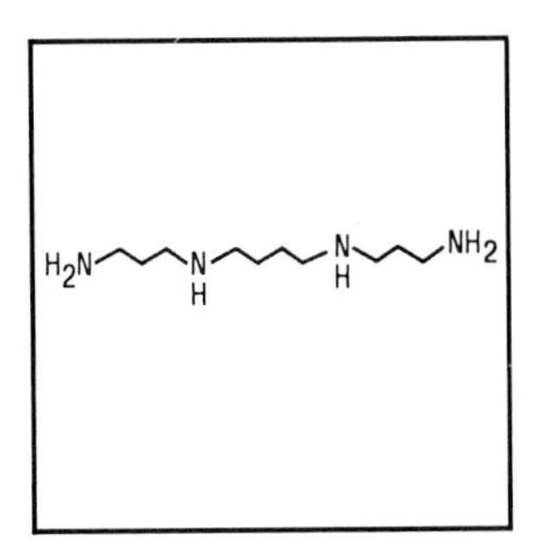

Polyamine Metabolism

PETER P. McCANN, *Marion Merrell Dow Inc.*

ANTHONY E. PEGG, *Milton S. Hershey Medical Center, Pennsylvania State University*

Glossary

α-difluoromethylornithine Enzyme-activated, irreversible inhibitor of ornithine decarboxylase, which blocks the formation of putrescine

Ornithine decarboxylase First enzyme in the general polyamine biosynthetic pathway, converting ornithine to the diamine putrescine

Putrescine 1,4-Diaminobutane, the first amine in the biosynthetic pathway for polyamines

Spermidine Product of the biosynthetic reaction which adds an aminopropyl group to putrescine

Spermine Product of the addition of an aminopropyl group to spermidine

***S*-adenosylmethionine decarboxylase** Enzyme that produces decarboxylated *S*-adenosylmethionine as an aminopropyl donor for both spermidine and spermine formation

THE BIOCHEMISTRY AND CELLULAR PHYSIOLOGY of polyamine metabolism (the synthesis and interconversion of putrescine, spermidine, and spermine from ornithine) has been a steadily growing field of research, and although the function of polyamines is still not well understood at the molecular level, a great deal of information attests to the importance of polyamines in cellular function. More recently, there has been a growing awareness that polyamine metabolism provides a useful target for the design of inhibitors, which have value as pharmacological agents. The availability of such inhibitors has led to a rapid increase in knowledge of the importance of polyamines in a wide variety of living organisms, and the clinical utility of these inhibitors in certain situations is now clearly evident. In view of the ubiquitous distribution of polyamines and their key role in a variety of cellular processes, further uses for these inhibitors as chemotherapeutic agents is expected.

I. General Outline of Polyamine Metabolism

A. Biosynthesis and Retroconversion

The polyamine biosynthetic pathway (Fig. 1) has been well studied in mammalian and other eukaryotic cells. The precursor diamine putrescine is the result of the decarboxylation of ornithine by the pyridoxal phosphate-dependent enzyme ornithine decarboxylase (ODC). The polyamines spermidine and spermine are then sequentially formed by aminopropyl groups added to putrescine. Two quite distinct enzymes, spermidine synthase and spermine synthase, which both require an aminopropyl donor in the form of *S*-adenosylmethionine, catalyze the formation of these two polyamines. *S*-adenosylmethionine decarboxylase (AdoMetDC), a pyruvoyl enzyme, is responsible for the formation of *S*-adenosylmethionine (AdoMet).

Although no other pathway is responsible for the *de novo* formation of polyamines in mammalian cells, a specific membrane transport system exists for putrescine, spermidine, and spermine; it is highly regulated and controls the uptake of exogenous sources of these amines. There is also a retroconversion of spermidine and spermine to putrescine controlled by both spermidine–spermine-N^1-acetyltransferase (SAT) (polyamine acetylase) and polyamine oxidase. SAT converts spermidine to its N^1-acetyl derivative, which in turn is cleaved by polyamine oxidase to putrescine and 3-acetamidopropanal (Fig. 1). There is an analogous retrocon-

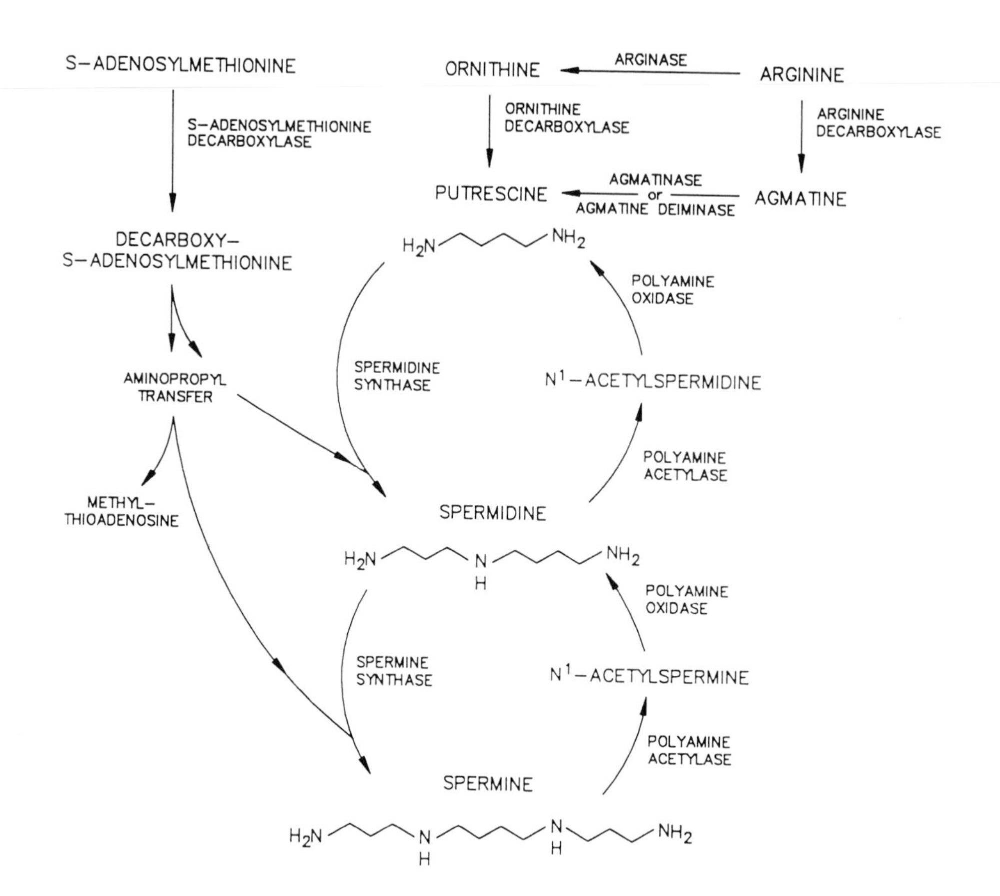

FIGURE 1 General overview of polyamine metabolism. It should be noted that not all of these reactions occur in all species. For example, the pathway to putrescine via arginine decarboxylase is not present in mammalian cells.

version of spermine into spermidine. Putrescine, the N^1-acetylspermidine derivative, and in some cases spermidine are excreted from various mammalian cells. Polyamines can also arise from exogenous sources such as diet. Serum oxidases may, in fact, degrade such extracellular amines. [*See* CELL MEMBRANE TRANSPORT.]

B. Incorporation of Polyamines into Other Cellular Molecules

Posttranslational modification of specific proteins is one clearly defined and specific function of polyamines. For example, transglutaminase-catalyzed incorporation of spermidine and spermine into proteins occurs in higher eukaryotes, although the specific physiological function of such incorporation is not well understood.

The formation of hypusine in the initiation factor eIF-4D for protein synthesis is another example of polyamines serving as precursors of protein modification. Hypusine, N^{ε}-(4-amino-2-hydroxylbutyl)lysine, originates from the conversion of putrescine to spermidine and to deoxyhypusine, although only the 1,4-diaminobutane fragment of spermidine is incorporated into hypusine itself.

Although not present in bacteria, hypusine is found in many mammalian cells and in yeast as part of eIF-4D protein, suggesting its importance in some general, well-conserved biological process.

II. Polyamine Metabolism in Mammalian Cells

A. Ornithine Decarboxylase

ODC is a pyridoxal phosphate-dependent enzyme that is found only at very low levels in resting cells. Cellular levels of ODC can be significantly increased, sometimes by a 100-fold or more, by stimulation with such things as growth factors, hormones, drugs, and other regulatory molecules; however, ODC still represents only a minor portion of the total cellular protein (e.g., from 0.01% of the cytosolic protein in androgen-stimulated mouse kidneys to 0.00012% in thioacetamide-stimulated rat liver). A macromolecular inhibitor of ODC, termed antizyme, has been isolated in a number of cells and may be another regulatory factor controlling ODC activity as well.

The use of the competitive inhibitor α-methylornithine conclusively demonstrated that formation of putrescine by ODC was essential for DNA replication and, consequently, mammalian cell growth. Later synthesis of the more potent inhibitor α-difluoromethylornithine (DFMO; eflornithine, Ornidyl®) allowed extensive work to explore the relationship between polyamine depletion caused by ODC inhibition, its consequent effects on cell replication, and, in some cases, cell differentiation.

DFMO (an enzyme-activated, mechanism-based, irreversible inhibitor of ODC) and some other substrate and product-related inhibitors, which have somewhat improved biochemical properties, have been widely used as biochemical and pharmacological tools. In fact, DFMO has been extensively explored as therapy for several types of cancer and protozoan infections. A number of studies have shown that treatment of cells in culture with DFMO significantly decreases intracellular putrescine and spermidine with little or no effect on spermine, whereas an enormous increase occurs in the overall levels of decarboxylated AdoMet. Other more potent inhibitors such as (2R, 5R)-δ-methyl-α-acetylenic putrescine had greater effects on spermine levels but did not achieve complete depletion as was the case with spermidine.

B. *S*-adenosylmethionine Decarboxylase

Mammalian AdoMetDC is the enzyme that provides decarboxylated AdoMet as an aminopropyl donor in the formation of spermidine from putrescine. Once it has been decarboxylated, AdoMet is committed to polyamine production, as no other reactions utilizing decarboxylated AdoMet at any physiologically significant rate are known. Therefore, the production of decarboxylated AdoMet is kept low and constitutes the rate-limiting factor in spermidine formation. Mammalian AdoMetDC is activated by putrescine and repressed by spermidine, linking the supply of decarboxylated AdoMet to the need for spermidine and the availability of the other substrate (putrescine) for spermidine synthesis. AdoMetDC has an enzyme-bound pyruvate as cofactor and, analogous to ODC, is only present in cells at very low levels (e.g., equal to 0.015% of the soluble protein in ventral prostate and to 0.0007% in liver). Similarly, AdoMetDC is also regulated by a number of external growth factors and stimuli.

Mammalian AdoMetDC is a dimer of two pairs of subunits with molecular weights of 30,621 and 7,681. These subunits are formed by the cleavage of a proenzyme chain (M_r about 38,000) in a reaction that forms the pyruvate prosthetic group at the amino-terminal end of the larger subunit. The processing–cleavage step may be an autocatalytic reaction of the proenzyme chain and is accelerated in the presence of putrescine. Because this step is essential for the production of the active sites of the enzyme, it provides an attractive target for the future design of therapeutically useful inhibitors of AdoMetDC.

Methylglyoxal bis(guanylhydrazone) (MGBG) is a potent inhibitor of AdoMetDC and was used in early experiments to prevent formation of decarboxylated AdoMet and, thus, synthesis of spermidine. Although such experiments suggested the importance of polyamines for cell growth, they were not conclusive because MGBG is nonspecific and has a variety of effects on mitochondrial and other physiological functions unrelated to effects of inhibition of polyamine biosynthesis. MGBG can be considered as a structural analogue of spermidine and is taken up by cells via a specific polyamine transport mechanism. Thus, the reversal of the effects of MGBG by spermidine may be as much due to interference of its cellular uptake or binding as to its replacement of intracellular polyamines.

Even though other inhibitors of AdoMetDC have been utilized, only recently have newer specific and more potent compounds become available such as 5′-deoxy-5′[*N*-methyl-*N*-(aminooxyethyl)]-aminoadenosine, 5′-deoxy-5′[*N*-methyl-*N*-(3-hydrazinopropyl)]aminoadenosine, and *S*-(5′-deoxy-5′-adenosyl)-methyl-thioethylhydroxylamine. These irreversible inhibitors of AdoMetDC apparently bind to the active site of the enzyme and form a covalent bond with the enzymatic pyruvate prosthetic group. An enzyme-activated inhibitor of AdoMetDC has also been described—5′-{[(*Z*)-4-amino-2-butenyl]methylamino}-5′-deoxyadenosine.

These inhibitors produce profound but expected changes in intracellular polyamine levels and consequently inhibit cell growth. For example, they reduce spermidine and spermine as well as decarboxylated AdoMet and 5′-methyl-thioadenosine and cause an enormous increase of putrescine in L1210 cells. Following depletion of spermidine and spermine, cell growth is arrested in spite of the large concentration of available putrescine; addition of either one depletes polyamine fully restored normal cell growth. Utilization of DFMO and an AdoMetDC inhibitor together completely block the increase of all three amines, demonstrating the potential utility of inhibition of AdoMetDC as a target for drug design.

C. Aminopropyltransferase

Spermidine synthase is responsible for the transfer of the aminopropyl group from decarboxylated AdoMet to putrescine. A second enzyme, spermine synthase, carries out a similar reaction and adds another aminopropyl group to spermidine. Although the mechanisms of action of these two aminopropyl transferases are analogous, both spermidine synthase and spermine synthase are specific and discrete enzymes, each having its own unique substrate. The cellular amounts of both these aminopropyl transferases are normally much higher than either ODC or AdoMetDC and are apparently regulated by the levels of their substrates (e.g., decarboxylated AdoMet). However, apparently the disposition of available decarboxylated AdoMet toward spermidine or spermine is probably determined by the relative amounts of the two synthases, and marked changes in spermidine synthase activity have been observed in response to hormones, tissue regeneration, and cell growth factors.

The multisubstrate analogue, *S*-adenosyl-1,8-diamino-3-thioctane (AdoDato), designed as an inhibitor of spermidine synthase, strongly inhibits this enzyme in mammalian cells. Another potent inhibitor of the same enzyme, cyclohexylamine, is competitive with respect to putrescine with a K_i of about 0.2 μm.

Although AdoDato is an effective *in situ* inhibitor of spermidine synthase in mammalian cells, its overall utility is limited because of the large increase in decarboxylated AdoMet resulting from the compensating rise in AdoMetDC after the inhibition of spermidine synthase. This decarboxylated AdoMet, in turn, can be used for the synthesis of spermine via spermine synthase, and thus the total polyamine pools are not significantly altered. Cyclohexylamine, however, does decrease overall polyamine levels in murine tumors.

Design and synthesis of specific spermine synthesis inhibitors has been of recent interest because of the relative ineffectiveness of ODC inhibitors in decreasing spermine levels and because the physiological significance of spermine synthase, present in mammalian cells but not found in many microorganisms, is not clearly understood at present.

The first inhibitor shown to appreciably block spermine synthase *in vitro* was *S*-methyl-5′-methylthioadenosine. Another compound, *S*-adenosyl-1,12-diamino-3-thio-9-azadodecane (AdoDatad), structurally related to AdoDato by addition of an aminopropyl moiety, was designed and found to be a potent, specific, multisubstrate analogue inhibitor of spermine synthase. This enzyme step is also blocked effectively by n-butyl-1,3-diaminopropane. However, neither AdoDatad nor n-butyl-1,3-diaminopropane had any effect on cell proliferation, although spermine was significantly depleted, suggesting either no requirement for spermine or that the increased amounts of spermidine found could compensate for the lack of spermine.

D. Polyamine Transport

Besides having the ability to synthesize polyamines, cells also possess a specific membrane transport system for the uptake of exogenous polyamines. Although its biochemical mechanism is not well understood, this uptake system is regulated by the intracellular polyamine content. Thus, when polyamines are depleted by use of specific inhibitors, the transport system responds to increased uptake and vice versa.

Of particular import is the fact that the transport system reduced the therapeutic effectiveness of polyamine biosynthetic inhibitors such as DFMO because extracellular polyamines (e.g., from diet, intestinal microorganisms, cell turnover) can be utilized via the transport system of polyamine depleted cells. For example, DFMO is significantly more active against a polyamine transport mutant L1210 cell line *in vivo* than against the parent cell line. DFMO is also more effective against tumors in rodents when intestinal polyamine oxidase is inhibited, thus preventing formation of polyamines from their N^1-acetyl derivatives.

Many polyamine analogues are actually substrates for the transport system and thus act as competitive inhibitors in relation to the natural polyamines. In fact, a number of carrier systems likely have somewhat overlap specificities. One system is Na^+-dependent and can be regulated by compounds, which change Na^+ flux. Characterization of such systems should ultimately allow the design and synthesis of specific inhibitors to regulate entry and efflux of the polyamines.

MGBG toxicity is quite obviously related to its transport and ultimate accumulation within the cell. Such high levels of MGBG cause inactivation of mitochondria and finally inhibition of macromolecular biosynthesis. Although MGBG accumulation is via the polyamine transport system, it does not seem to mimic the natural polyamines and other polyamine analogues in downregulating the polyamine transport system. As the system continues to transport MGBG into cells that already contain very high levels of the drug, MGBG's lack of repression of such transport may be a factor in its significant cytotoxicity.

Various *N*-alkyl polyamine analogues have been synthesized and are taken up by the polyamine transport system. For example, N^1,N^8-bis(ethyl)-spermidine reduced ODC levels and intracellular putrescine, spermidine, and also spermine in L1210 cells, whereas N^1,N^{12}-bis(ethyl)spermine was even more active in that it also reduced AdoMetDC levels and caused almost complete depletion of all the polyamines. Such effects are produced in part because these analogues are recognized by the polyamine regulatory systems for ODC and AdoMetDC biosynthesis. In addition, efflux of intracellular polyamines increases by some yet undefined mechanism, which normally regulates intracellular polyamine levels.

The bis(ethyl)polyamines are also potent inducers of SAT, and enzyme increases of several 100-fold are seen in cells within hours after exposure to the analogues. Such induced increases in SAT may also affect the rapid excretion of intracellular polyamines because such acetylation facilitates conversion of nonexcreted spermine to spermidine and putrescine, which are both excreted from the cell.

III. Polyamine Metabolism in Disease-Causing Microorganisms and Viruses

A. Protozoa

The polyamine metabolism of disease-causing protozoa has been of particular interest because of the pronounced sensitivity of several species to the effects of several polyamine inhibitors, primarily DFMO.

The original observations of the effects of DFMO on protozoa were made in African trypanosomes when the inhibitor was found to cure acute, lethal infections of *Trypanosoma brucei brucei* in mice. These initial results led to the rapid clinical use of DFMO in Africa against what would have been fatal cases of late-stage, arsenic-resistant West African sleeping sickness caused by *Trypanosoma brucei gambiense*.

Numerous experiments have been conducted concerning the role of polyamines in trypanosomal growth and differentiation. The African trypanosome appears to be unusually sensitive to DFMO, due in part to both its rapid doubling time and lack of spermine. ODC in the trypanosome, however, appears to be kinetically similar to the mammalian enzyme. Again, analogous to mammalian cells, DFMO enters the trypanosome by passive diffusion rather than facilitated transport. An intact host-immune system is necessary for cures of trypanosome infections, and other evidence indicates that polyamines are required for the shift of antigenic determinants on the parasite membrane. Furthermore, DFMO and consequent polyamine depletion induces morphological and biochemical shifts of entire trypanosome populations from parasitic bloodstream forms to "short, stumpy" forms, which do not replicate. This differentiation event is independent of an earlier actual block of bloodstream form replication.

Trypanosomes also contain a novel spermidine-containing cofactor, which is necessary for glutathione reductase activity. This cofactor, trypanothione, uniquely present in trypanosomatids, is greatly reduced in *T. b. brucei* after DFMO treatment and is probably yet another target for polyamine depletion in these protozoa.

Recent experiments on molecular aspects of ODC from *T. b. brucei* have demonstrated that the trypanosomal ODC has a much longer half-life than the mammalian enzyme. Although the trypanosomal enzyme is very homologous to the mammalian enzyme, a likely explanation for this is that ODC from *T. b. brucei* lacks a specific 36 amino acid carboxy-terminal peptide containing a "PEST" (proline–glutamate–aspartate–serine–threonine-containing) region, which promotes rapid protein degradation.

The other polyamine biosynthetic enzymes have

been less studied in trypanosomes and, although the AdoMetDC of *T. b. brucei* is inhibited by such drugs as Berenil®, pentamidine, and MGBG, the effects of such inhibition contribute little or nothing to the antiprotozoal effects of these agents. As is the case with the mammalian enzyme, spermidine synthase from *T. b. brucei* is quite sensitive to both cyclohexylamine and AdoDato. Neither one has any *in vivo* effect on trypanosome infections as might be predicted, because neither one actually alters trypanosome polyamine levels.

DFMO will also inhibit the growth of promastigotes of the trypanosomatid *Leishmania donovani* but has no effect on its intracellular-infective cousin *Trypanosoma cruzi,* the cause of American trypanosomiasis (Chagas' disease). Unexpectedly, α-difluoromethylarginine, an irreversible inhibitor of arginine decarboxylase (an enzyme not present in mammalian cells) significantly inhibited *T. cruzi* infection of macrophages as well as actual replication of the parasite. These findings indicate the apparent presence of arginine decarboxylase in *T. cruzi* and a possible unique target for this now incurable disease.

A number of the sporozoea class of protozoa are quite sensitive to the effects of DFMO, including *Eimeria* spp. and *Plasmodia* spp., the latter causing various forms of malaria, the most common parasitic disease worldwide. DFMO inhibits the *in vitro* replication of *Plasmodium falciparum* and will cure exoerythrocytic infections of *Plasmodium berghei* in mice and also block the sporogonous cycle of *P. berghei* in mosquitoes. While not curative of erythrocytic infection of *P. berghei,* DFMO will significantly lower parasitemia and, in combination with a bis(benzyl)polyamine analogue, will actually cure *P. berghei*-infected mice. [*See* MALARIA.]

Pneumocystis carinii, the cause of the lethal opportunistic pneumonia in acquired immunodeficiency syndrome patients, has been thought to be related to the protozoal class *Sporozoea,* subclass *Coccidia,* although recently it has been shown to have ribosomal RNA more homologous to certain fungi. In any case, DFMO has been successfully used in numerous clinical instances as therapy for *P. carinii* pneumonia. Experiments in cell culture and immunosuppressed rats have shown that *P. carinii* is inhibited by DFMO, and its growth-inhibitory effects can be reversed by putrescine. [*See* PROTOZOAL INFECTIONS, HUMAN.]

B. Bacteria

In bacteria, as well as a number of prokaryotic organisms, an alternate pathway exists for the synthesis of putrescine via the decarboxylation of arginine to form agmatine, which then yields putrescine (Fig. 1). As bacteria have both ODC and arginine decarboxylase, inhibition of either enzyme alone induces the other to increase and thus maintain a specific intracellular concentration of putrescine.

Although bacteria, in general, only contain putrescine and spermidine and not spermine, it has been difficult to deplete intracellular polyamines and thus have specific effects on cell growth and replication.

Remarkably, several bacteria such as *Escherichia coli* and *Klebsiella pneumoniae* have ODC enzymes that are wholly unreactive to DFMO. Other analogues, however, such as α-difluoromethylputrescine and α-monofluoromethylornithine do completely and irreversibly inhibit the enzymes from these organisms. α-Difluoromethylarginine as well as even other more potent arginine and agmatine analogues were found to irreversibly inhibit the arginine decarboxylases found in both bacteria and plants.

AdoMetDC from *E. coli* is inhibited by MGBG in a cell-free system, but this drug is not effective on intact cells. Some analogues of AdoMet as well as pentamidine and Berenil® are also *in vitro* inhibitors of cell-free bacterial AdoMetDC. With regard to spermidine synthase, both cyclohexylamine and AdoDato are potent inhibitors of the *E. coli* and *Pseudomonas aeruginosa* cell-free enzymes. However, cyclohexylamine is taken up by both organisms and will inhibit growth specifically due to depletion of spermidine, unlike AdoDato, which is not taken up.

The most significant growth inhibition of bacteria (e.g., *E. coli, P. aeruginosa,* and *Serratia marcescens*) was demonstrated with combinations of α-monofluoromethylornithine, α-difluoromethylarginine, and cyclohexylamine, wherein all three polyamine biosynthetic enzymes (ODC, arginine decarboxylase, and spermidine synthase) were simultaneously inhibited. Although appreciable growth inhibitory effects were noted, they were not as profound as in a number of protozoa and fungi. Therefore, the above-mentioned inhibitors of polyamine biosynthesis will unlikely be useful as therapy for any bacterially caused infections.

C. Fungi

Filamentous fungi and most yeast synthesize putrescine solely via ODC, unlike bacteria. Although only limited experimental work has been done with zoophilic and disease-causing fungi and yeast, the ODC in *Saccharomyces cervisiae* and *Saccharomyces uvarum* is sensitive to DFMO. Furthermore, polyamines are depleted by DFMO in several human pathogenic *Candida* spp., whereas *Candida tropicalis* growth was also significantly inhibited by DFMO as well. α-Monofluoromethyldehydroornithine methylester was shown to be a potent inhibitor of two species of dermatophyte fungi, *Microsporum* and *Trichophyton*. In fact, it was 25 times more effective than DFMO itself, which also significantly arrested the growth of these organisms.

D. Viruses

Viruses as intracellular parasites depend completely on their metabolically active host cells for replication. However, in some cases, viral enzymes necessary for replication are encoded in the viral DNA or RNA. The pox viruses such as vaccinia apparently encode both ODC and AdoMetDC and have been shown to respond to polyamine inhibitors such as α-methylornithine, DFMO, and MGBG. DNA viruses in the herpes family, such as herpes simplex virus and human cytomegalovirus, have been shown to be sensitive to the effects of DFMO and MGBG as well. A correlation was made as well between the ability of herpes simplex virus to induce the synthesis of polyamines and the antiviral effects of the inhibitors. It was also found that depletion of polyamines was necessary in the host cells prior to infection with cytomegalovirus to have an effect on viral replication by DFMO. Overall, although polyamine biosynthesis is required for viral replication, more experimental work is needed to effectively determine any possible utility of inhibitors for viral diseases. [*See* VIROLOGY, MEDICAL.]

Bibliography

McCann, P. P., Pegg, A. E., and Sjoerdsma, A. (eds.) (1987). "Inhibition of Polyamine Metabolism: Biological Significance and Basis for New Therapies." San Diego, Academic Press.

Pegg, A. E. (1986). Recent advances in the biochemistry of polyamines in eukaryotes. *Biochem. J.* **234,** 249–262.

Pegg, A. E. (1988). Polyamine metabolism and its importance in neoplastic growth and as a target of chemotherapy. *Cancer Res.* **48,** 759–774.

Pegg, A. E., and McCann, P. P. (1982). Polyamine metabolism and function. *Am. J. Physiol.* **243,** C212–C221.

Pegg, A. E., and McCann, P. P. (1988). Polyamine metabolism and function in mammalian cells and protozoans. *ISI Atlas of Science: Biochemistry* **1,** 11–18.

Porter, C. W., and Bergeron, R. J. (1988). Enzyme regulation as an approach to interference with polyamine biosynthesis—An alternative to enzyme inhibition. *Advan. Enzyme Regul.* **27,** 57–82.

Porter, C. W., and Sufrin, J. R. (1986). Interference with polyamine biosynthesis and/or function by analogs of polyamines or methionine as a potential anticancer chemotherapeutic strategy. *Anticancer Res.* **6,** 525–542.

Sjoerdsma, A., and Schechter, P. J. (1984). Chemotherapeutic implications of polyamine biosynthesis inhibition. *Clin. Pharmacol. Ther.* **35,** 287–300.

Tabor, C. W., and Tabor, H. (1984). Polyamines. *Annu. Rev. Biochem.* **53,** 749–790.

Tyms, A. S., Williamson, J. D., and Bacchi, C. J. (1988). Polyamine inhibitors in antimicrobial chemotherapy. *J. Antimicrob. Chemoth.* **22,** 403–427.

Polymorphism of the Aging Immune System

BERNHARD CINADER, *University of Toronto*

Glossary

B cells Bone marrow-derived cells that differentiate, acquire receptors, and can finally interact with antigens and synthesize antibodies
T cells Thymus-derived cells, also coming from the bone marrow, that differentiate in the thymus and then leave
Tc cells Cytotoxic T cells that can kill target cells
NK cells Nonspecific killer cells
LAK cells Lymphokine-activated killer cells
Economic correction Designates a relation between youthful activity and middle-age decrease in that activity

AGING OF ANY PARTICULAR FUNCTION or synthetic activity occurs at different rates in different individuals (i.e., the rate of progression is polymorphic, or variable). Functionally different tissues of the same individual age at different rates, i.e., aging is not synchronized, it is compartmentalized. In many compartments, a direct relationship exists between the density of receptors, concentration or activity in youth, and rate of decrease in later life—an economic correction of youthful synthetic exuberance. Genetic programs, wear and tear, repair capacity, and feedback controls are involved in the progression of aging. In the immune system, involution of the thymus is an important factor in aging of the T-cell compartment; different T-cell types change at different rates in different individuals.

I. Progression of Aging

In the influenza epidemics of 1957–1958 and 1962–1963, deaths occurred more frequently among the elderly than among the middle-aged. These deaths were due to many causes, including cardiovascular diseases and secondary bacterial pneumonias, but some evidence indicates that an age-related decline in immune response was a factor in the relatively high influenza infections among the elderly. In fact, this is a general feature of aging: The concentration of natural antibodies to flagella (bacterial surface appendages) is, on the average, twice as high in individuals in their thirties to forties than in those in their sixties. As the amount of antibody to foreign antigens decreases, the incidence of antibodies to self-antigens (i.e., of the individual's own body), such as DNA, increases. These changes in immune responsiveness occur at different ages in different individuals and in some individuals cannot be detected, even at a very advanced age. Thus, individual differences (i.e., polymorphism) are considerable in the rate and severity of age-related changes. In the same individual, age-related changes of different cell populations are initiated at different ages and progress at different rates. These differences in immune response depend on multiallelic genes and environmental influences.

II. Aging of the Immune System

The progression of age-related changes depends on the ability to repair biochemical damage and on the ability to generate precursors from stem cells and generate end cells from precursors. Throughout life, the immune system is dependent on processes of differentiation from precursors to functionally specialized lymphocytes of two types: committed thymus-derived (T) and bone marrow-derived (B) cells. Because precursor cells replicate and their descendants differentiate to become mature cells, the immune system is profoundly affected by wear and tear of precursor cells and less by wear and tear of mature cells than are such tissues as muscle and brain. Therefore, to what extent do precursor cells lose their ability for self-renewal? There appears to be a correlation between the number of *in vitro* cell division of fibroblasts and age of donor. Argument for a biological link between this type of *in vitro* and *in vivo* aging can be found in the reported correlation between average lifespan of different species and the number of *in vitro* divisions their fibroblast can undergo. Nevertheless, the effect of age on the *in vivo* capacity for self-renewal remains a controversial subject. It may depend on the developmental distance from the primitive stem cell. The further the precursor cell differentiates toward a fully committed cell (end cell), the greater the probability that self-renewal declines. The polymorphism of these changes remains to be explored and may contribute to the resolution of contradictory results. [*See* LYMPHOCYTES.]

In early postnatal life, T cells acquire their functional specialization in the thymus. Involution of the thymus is a programmed feature of postnatal development and reaches completion in middle age (in humans, at about 50 yr of age). In men it occurs at an earlier age than in women; hormonal controls can affect the rate of thymus regression. The pituitary gland is implicated in some of the changes. Furthermore, the age-related decrease in the secretion of thymus hormones, such as thymulin, may be due to a decrease in synthesis of thyroxine, the thyroid hormone. Whether or not initiation of thymus regression is triggered by endocrine controls is not definitively established. Events in the thymus itself are multicentric, which results in different types of precursor cells, i.e., precursors of suppressor cells for humoral and cell-mediated immunity age at different rates and even in different directions (one class decreases while another increases with age in some inbred mouse strains). [*See* THYMUS.]

A decreasing supply of thymic peptides and cytokines may be pacemakers of the impact of involution on the immune response and contribute to the peripheral consequences of involution. The involution of the thymus initiates peripheral age-related changes in immune responsiveness during the second half of life. However, precursor cells, which leave the thymus, settle in spleen and lymph nodes and are sources for renewal of T-cell function in later life. Finally, the involution of the thymus affects peripheral function if or when the peripheral pools have exhausted their ability for self-renewal.

The general view on aging-progression, deduced from observations on humans, can be extended and confirmed in appropriate animal model systems, particularly with animals of different homogeneous genetic background (i.e., with strains of inbred mice). The validity of extension from animal to man depends on the validity of the assumptions that the factors involved in aging are common to all mammals. In fact, there is indirect evidence for this view. A positive correlation exists between lifespan and concentration in the liver of carotenoids, α-tocopherol, ceruloplasmin, and ascorbate; an inverse correlation is evident between lifespan and concentration of cytochrome, gluthathione transferase, and catalase. It is, therefore, reasonable to conclude that aging of different mammals depends on similar processes and that it is legitimate to extrapolate from short-lived to long-lived mammals. In general, aging is apparently dramatic in some, but not all, compartments of the immune system. We shall briefly consider the polymorphism of some of the following all cell populations:

1. T-suppressor capacity for antibody response
2. T-suppressor capacity for cell-mediated immunity
3. Cytotoxic T cells (CTL)
4. B cells and T-cell helper capacity
5. Nonspecific killer (NK) and lymphocyte-activated killer (LAK) cells

III. Polymorphism of Aging

In some inbred strains of mice, a marked decrease is noted in suppressor capacity for the humoral response with a simultaneous increase of suppressor capacity for cell-mediated immunity. In other strains of mice, these processes occur very slowly, and in yet others suppressor capacity for cell-mediated immunity either does not change with age or increases to middle age and then decreases.

A great diversity exists in age-related changes of CTL cells and of their inhibition by suppressor cells. The extent of LAK cell activity differs in young animals of different strains and, as animals age, the level decreases to a greater extent the greater the youthful activity (i.e., economic correction).

B cells undergo age-related changes in some individuals (e.g., inbred mouse strains), which apparently renders the ability to respond to antigens less dependent on T-cell help as the animal ages. Relatively few age-related changes are found in NK cells of most strains of mice; in some there even appears to be an increase with age in middle age followed by a decline in old age. In short, polymorphism occurs in the rate at which age-related changes occur and in the cell type that is predominantly affected by these changes. This can be easily established with inbred mice, because young and old animals of the same genetic background can be compared. This type of comparison cannot be made in the study of human aging; the heterogeneity observed in the rate at which the human immune system ages can be assumed to be a reflection of the differences in the type of aging that occurs in different individuals (i.e., with different genetic backgrounds).

IV. Compartmentalized Aging and Polymorphism

Because individual differences of age-related changes will be illustrated in terms of suppressor capacity in various contexts, we will give a brief summary of how suppressor capacity of two classes (one affecting humoral immunity, the other cell-mediated immunity) can be assayed.

Suppressor capacity in the thymus, affecting humoral response (antibody production), can be measured by reconstitution experiments, in which lethally irradiated mice are given either lymphocytes of young donors or lymphocytes from the same young donors but mixed with thymus cells from donors of different ages. The reconstituted mice are then immunized, and antibody is measured. In some strains of mice, thymus cells from young donors reduce the humoral response of the lymphocytes, showing a suppressor effect, whereas thymus cells from old donors increase it, showing reduced suppression.

Changes in suppressor capacity for cell-mediated immunity can be measured by the ability of thymus cells to prevent lymphocytes to be sensitized by targets and to kill them. Results obtained by this method have been confirmed by a variety of other tests for cell-mediated immunity.

Differences within the same strain in age-related changes of these two classes of suppressor capacity are remarkable. The independence of the aging changes occurring in the two classes has been demonstrated by nutritional, hormonal, and pharmacological interventions, which can affect aging of one class without affecting aging of the other.

Reference to the rate of age-related changes modified by various external factors has already been made. These factors can be used to determine whether or not aging in different cellular subsets is under independent genetic controls (i.e, whether or not aging is compartmentalized). This distinction between independence and interdependence can be analyzed by interventions, which change the rate of one process without changing that of another, and by selecting different types of interventions, which act on fundamentally different metabolic or developmental processes. To this end, hormones, purine analogues, and different diets have been employed.

The rate of increase in precursors of suppressor cell for CTL can be accelerated by administration of β-estradiol and progesterone; this treatment does not bring about changes in humoral response. The rate of decrease in precursors for suppression of humoral response can be delayed with purine analogues, which are readily available as antiviral drugs. This type of drug may affect progression differentially because adenosine deaminase and purine nucleosidase deficiency have different effects on humoral and cell-mediated immunity. So far, two such antiviral compounds have been tested and found to modify age-related progression of the humoral response. A third strategy for modification of age-related progression is based on feeding diets, which differ in fatty acid composition. This is a strategy of choice, because lipid composition of membranes changes during differentiation and the ratio of polyunsaturated to saturated fatty acid (P/S) affects membrane fluidity. In fact, diets induce changes in suppressor capacity for humoral as well as for cell-mediated immunity: diets low in P/S prevent or delay resistance against down regulation (i.e., against tolerance induction) and delay loss of thymus suppressor activity for antibody formation; diets high in P/S prevent or delay the age-related increase of suppressor activity for cell-mediated immunity.

It follows that: (1) suppressor capacity for humoral and cell-mediated immunity can age at different rates and even in different directions (i.e., one decreasing, the other increasing) and (2) the rate of aging can be changed in one of these types of suppressor capacity without changing it in the other. Therefore, a reasonable to conclusion is that age-related progression in these two compartments is under the control of different genes, at least in individuals with certain types of genetic backgrounds.

Age-related changes in suppressor capacity are only one of several instances of aging, occurring at different rates in different compartments. Other instances of compartmentalized aging have been found in differential progression of aging of antibody responsiveness in the mucosal and peripheral immune system, in different isotypes, and in age-related progression of antigen-presenting cells. In short, aging within the immune system is compartmentalized; in the same individual, different compartments, as well as same compartments of different individuals, age at different rates. These conclusions are based on studies of aging in rodents but are clearly applicable to human aging. It is well known that a man may lose his ability to divide hair follicle (i.e., become bald), while retaining the ability to divide precursor cells of sperm (i.e., remain fertile). By extension, compartmentalization of the immune system, analyzed in rodents, can be assumed also to apply to man, though this remains to be rigorously demonstrated.

V. Economic Correction: An Intracompartmental Control Process

Individual differences in the quantity of a given gene product at different ages is yet another aspect of the individuality of aging to consider. The range of this polymorphism is great early in life, decreases in middle age, and then increases again in senility. Between early age and middle age, the synthesized quantity of various types of molecules and cell types decreases; the magnitude of this decrease is greater the greater the quantity of a given gene product produced early in life. This relationship between youthful quantity and magnitude of age-related decrease is referred to as economic correction. As a consequence of this process, the quantity of many macromolecules, produced by different individuals, varies over a much larger range in youth than in middle age. The level, attained in middle age, remains constant over a considerable portion of the second half of life and shows an individually variable decrease toward the end of life.

It is apparent from the foregoing survey of polymorphism and compartmentalization of aging processes, that this type of data represents the fine structure of the heterogeneity of aging. Insight into the genetics of polymorphism is not only of biological but also of clinical interest. The initiation and progression of diseases of old age depend on a variety of controlling processes (including T-cell regulation), which prevent immune responses that might damage self; the effectiveness of this controlling process changes dramatically in some individuals, although not in others. Preventive medicine for degenerative diseases of old age can only be developed if the genetic individuality of the polymorphism is known. The ability to identify populations at risk would allow us to develop strategies that delay or prevent changes that predispose to a particular type of degenerative disease. The growing proportion of the elderly in the world's population makes this task an important target for improving the quality of life and preventing excessive pressure on our health services.

Bibliography

Bergener, M., Ermini, M., and Stahelin, H. B. (ed.) (1988). "Crossroads in Aging," *1988 Sandoz Lecture in Gerontology*. Academic Press, London.

Chandra, R. K. (ed.) (1985). "Nutrition, Immunity and Illness in the Elderly." Pergamon Press, New York, Oxford, Toronto, Sydney, Frankfurt.

Cinader, B. (ed.) (1989). Symposium 2.4.2: Genetics of Aging, XVIth International Congress of Genetics. Genome, National Research Council, Ottawa, Ontario.

Cinader, B., and Kay, M. M. B. (1986). Differentiation and regulatory cell interactions in aging. *Gerontology* **32,** 340–348.

Courtois, Y., Faucheux, B., Forette, B., Knook, D. L., and Treton, J. A. (eds.) (1986). "Modern Trends in Aging Research." John Libby Eurotext, London, Paris.

de Weck, A. (ed.) (1984). Lymphoid cell functions in aging. *In* Topics in Aging Research in Europe." Eurage, Rijswijk.

Goidl, E. A. (ed.) (1986). "Aging and the Immune Response." Marcel Dekker, New York.

Maddox, G. L., and Busse, E. W. (eds.) (1987). "Aging: The Universal Human Experience." Springer Publishing Co., Inc., New York, New York.

Yamamura, Yuichi and Tomio Tada (eds.) (1983). "Progress in Immunology V." Academic Press Inc., San Diego, California.

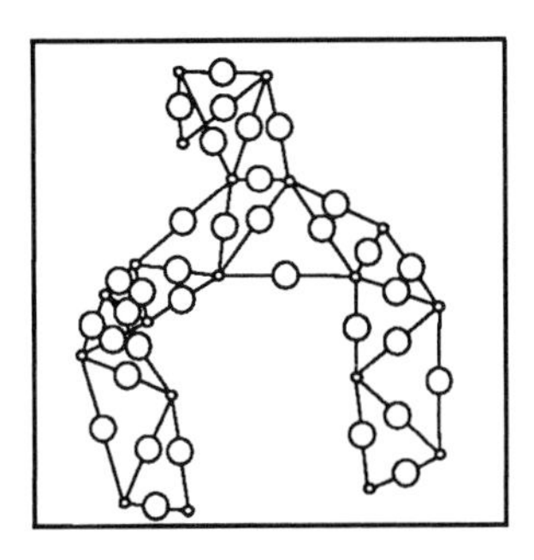

Polymorphism, Genes

ANDRÉ LANGANEY, *Université de Genève*

DANIEL COHEN, *Centre d'Etude du Polymorphisme Humain*

Glossary

Admixture Effect on genetic pools of the interbreeding of two different populations

Base pair Unit of information of the DNA molecule: base pairs are the "letters" of the genetic code

Chromosome Linear structure carrying the genes inside the cell nucleus

DNA Deoxyribonucleic acid: chemical constituent of the genes

Founder effect Sampling effect, in which a limited number of emigrants will take away only a part of the genes of their population of origin and will have different frequencies of the same genes

Genes Units of genetic information transmitted from parents to children according to Mendel's laws

Genetic drift Random change of gene frequencies from generation to generation in a single population. Genetic drift is fast in small populations and slow in large populations

Genome Whole genetic information of an individual. The human genome is made of 46 chromosomes carrying 3 billion base pairs of DNA

Mutation Random change of DNA structure. Some mutations change genes product, some others (silent mutations) do not

Recombination Random association of parental genes during meiosis (cell divisions producing sex cells) or fecundation

THE MOST STRIKING discovery of population biology is that in any sex-reproducing population, all individuals are different one from another except, in some respect, identical twins. This variation within populations and between individuals was already studied by Darwin more than one century ago. But ignoring the biology of sex and genetics, Darwin did not succeed when he tried to explain it. Nevertheless, he understood that natural variation of populations and individuals was the basis of natural selection, permitting some lineages to transmit their "types" while some others disappeared.

Today, variation between individuals and between populations of a same biological species is named *polymorphism*. Polymorphism is the fact that for any characteristic, individuals of the same species will show different types (*poly* = several, *morph* = type). *Polymorphism* is less general a word than *variation*, which is used for changes of a single individual or differences between species.

The human species is one of the most polymorphic among living species. Most characteristics vary from one individual to another within populations, and some of them have different ranges of variation from one population to another. Polymorphism can be split into two parts, according to the concerned characteristics and to their variation pattern.

I. Quantitative Polymorphism

The first type of polymorphism is the variation of the outside look of an individual: face and body shapes, skin and hair colors, height or body measurements, and so on. This type of polymorphism, easily recognized by the human sight or easy mea-

Encyclopedia of Human Biology, Volume 6.

surements, can be said to be "quantitative." This means that the variation between individuals is continuous and that the measures made in a population will be real numbers of a continuous interval of variation. For instance, somebody is not short, medium, or tall, but more or less tall. People are not black, white, or yellow skinned, but they are more or less brown, the extremes being possible.

There is no doubt that variation between individuals or between populations for quantitative polymorphisms is, at least partly, under genetic control. For example, family heredity of skin color within population, or in cases of admixture, proves that some genes are responsible for being more or less dark- or light-skinned. But despite intensive investigations for more than a century of genetics, no one has yet identified the genes that make somebody more or less something. Many sophisticated statistical approaches did not prevent "quantitative genetics" from remaining a "black box" of modern biology.

So the most interesting result about quantitative polymorphism is its pattern of variation within and between human populations, whatever its genetic background will be. First, for all quantitative characteristics, there is an enormous variation within populations, between individuals. Measures of a quantitative polymorphism (e.g., height or skin color), usually vary according to a normal (Laplace Gauss) probability distribution, which means that measures around the mean of the population are more common than those that lay far aside, but these still exist. For example, height measures vary currently up to 10 inches away from a population mean for one sex. This makes that mean differences between sexes and between populations commonly less than extreme differences between individuals within a population. Also, measure distributions of a quantitative polymorphism in a set of human populations usually overlap between populations. This is especially true in the case of skin color, which can be shown to vary continuously worldwide from the lightest white-skinned people to the darkest blacks, through all "yellow-skinned" or brown intermediary grades.

In some cases variation of quantitative characteristics has been significantly related to environmental factors, even if the ultimate reasons of this geographical pattern remains puzzling. For example, it is well known that mean skin colors of human populations vary according to their latitude of origin. All dark-skinned populations originated from the tropics, whereas all populations from temperate or arctic areas are light-skinned; gradients of skin colors are observed from the equator to the pole in Africa and Europe, in Asia, in North and even South Amerindians. Rigorously speaking, mean skin color is more correlated with sunlight than with latitude.

Height and body shape distributions are more complicated, being commonly correlated with the type of environment rather than with latitude. People are short and stocky in the arctic, in high altitude (Andes or Himalayas) but also in equatorial forests. They are middle-sized in temperate forest and savannah, and their mean height increases up to the cold temperate zone, whereas they are commonly tall and thin in hot deserts of Africa.

II. Genetic Polymorphism

It is well known since the beginning of the century that blood transfusion is conditioned by the ABO blood groups and that these blood groups are determined by mendelian genes. Each individual of any population belongs to one of the four main ABO blood groups (A, B, AB, and O) and not to anyone else. So the ABO blood groups are a qualitative polymorphism in which an individual belongs or not to each category. Such a polymorphism is also called *discontinuous* or, in a mathematical sense, *discreet*. Its measure for an individual is "present" or "lacking" for each type and not a continuous real number as for quantitative polymorphism. At the population level, the measure will be the frequencies of the various possible types [either phenotypes (e.g., the four ABO blood groups) or genotypes-gene combination of each individual] or, when they are computable, the frequencies in the population of the genes that settle the possible genotypes (genes A, B, and O for the ABO blood groups).

Mostly in the past 25 years, and much faster during the past 10 years, the accelerating acquisition of knowledge in immunology, biochemistry, and molecular biology has permitted us to discover hundreds of new qualitative genetic polymorphisms concerning other red cell blood groups, seric blood groups, cell histocompatibility groups, and protein and DNA molecular variations. All these genetic systems split human populations in various numbers of phenotypes, themselves determined by systems of genes [alleles = genes located at the same spot (locus) of a peculiar human chromosome], which current numbers vary from one to more than 40 (for the HLA-B gene, for example). [*See* GENES.]

The current estimations of the number of functional genes in humans vary from 40,000 to more than 100,000, that is to say, 100 million DNA base pairs among 3 billion in the human genome. Despite much difficulty to check it, it is usually thought that at least a fifth of the human functional genes are polymorphic, with two or more gene variants at the protein level. This would mean that in every human population, the number of theoretically possible genetically different individuals is much greater than 3 exponent 8,000 (3 = number of genotypes for two alleles, 8,000 = minimal number of polymorphic functional genes among the 23 human chromosome pairs). Such a number is of a much greater order of magnitude than the estimated number of the chemical atoms of the whole universe! And it is a low underestimation because many genes have many more than two alleles and because we did not take into account the by far more numerous "silent mutations," which change the DNA without changing the gene product or which change noncoding sections of the DNA (more than 90% of our gene pool).

This means that human polymorphism is almost infinite in every human population. The existing individuals at one generation are only a tiny sample of the enormous potential diversity. Because of the sex mechanisms, two human people never have the same genetic background if they are not identical twins produced by a split of a human egg or embryo. Each parent transmits for each genetic system one gene out of two that he or she received from his or her own parents. So each sperm or ovocyte is produced by a "recombination lottery" among billions of possible combinations of the alleles of thousands of genetic systems. So during the meiosis (cell division that produces germ cells) a parent never produces the same ovocyte or the same sperm, and then each fecundation combining an original ovocyte with an original sperm produces a completely new human being. Each of us has a unique genetic constitution that appeared at conception and will disappear at death, without unusual and sophisticated genetic engineering processes. Any human being possesses a kind of unique chemical identity card in its DNA, written in the genetic code as a 3-billion-base-pair word.

The main consequence of this genetic diversity is that from generation to generation, human populations are made of different sets of individuals. So they cannot prevent changes from generation to generation, which means that sexual mechanisms (meiosis and fecundation) oblige human populations to evolve from generation to generation. Human evolution is not a scientist dream or hallucination; it is a trivial fact established by our precise knowledge of genetics. [*See* EVOLUTION, HUMAN.]

III. The Origins of Human Polymorphism

Human genetic diversity is amplified by genetic recombination, but it takes its roots in mutations that create new genes (new alleles) by chance errors in gene duplications. Because modern humans are likely to belong to a recent species (about 1,000 centuries old), most of present human genes were produced by old mutations that occurred in our human or nonhuman ancestors, and few of our genes are modern human specific alleles (those are commonly rare in populations). It has been demonstrated that human beings share some of their genes (those basic in cell biology) with species like fruit flies or even wheat or fungi. Many of our genes are common with other vertebrates. Most of them are shared by mammals and even more by other primates, our nearest animal cousins. Sophisticated molecular biology techniques permit us to estimate the age of some of our alleles, and the results vary from more than 1 billion years to few generations for some rare mutations. Nevertheless, we share a large part of our blood group or enzyme alleles with the nonhuman primates, and most of our genes differ only by neutral mutations from those of the other mammals.

IV. Variation Between Populations

Meanwhile, genetic polymorphism within human populations has proved to be unexpectedly important; genetic differences between human populations or races were found to be much less than expected according to physical diversity.

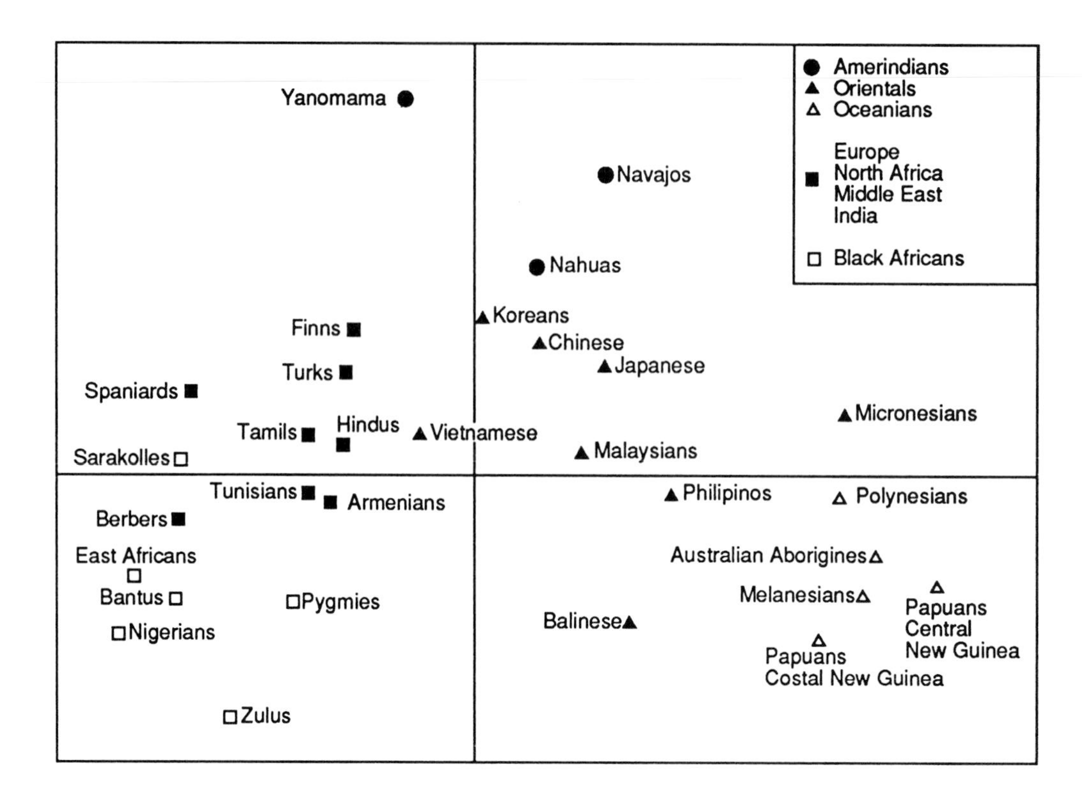

FIGURE 1 Genes and geography: this map is a genetic map of gene frequencies of the human major histocompatibility system (HLA). Without introducing any geographic data, this analysis gives back a map of the world: geography is the major factor of human genetic differentiation. [(Data from A. Sanchez Mazas, Laboratoire de Génétique et Biométrie, University of Geneva)].

For most of the studied part of the human genome, the same alleles of the genes are present in almost all human populations. The differences between populations are variations of frequencies of the same alleles and usually not the fact that an allele would be found here and not there. There are some exceptions for some rare alleles and/or for some polymorphic systems with numerous alleles. But if it is easy to tell that an allele we find is present in a population, it is almost impossible to tell that it is absent in another population. Cost reasons usually make it impossible to study large samples of individuals, and small samples "miss" rare genes. Anytime that a population was studied by a large sample, alleles that were reputed to be characteristic of "other races" were found. For example, large samples of European blood donors commonly show, at low frequencies, alleles that had been thought once to be only black African or Oriental, even when any recent admixture is excluded.

When we study the variation of genetic polymorphism frequencies throughout the world, the first evidence is that this variation is absolutely continuous from the west of Europe to East Asia and further Indian America via the Bering Strait and toward Oceania through Southeast Asia (Fig. 1 and 2). The same can be said from South Africa up to Scandinavia. We cannot find large gaps in gene frequencies between races or populations, except in small isolated populations (because of founder effects and genetic drift) or when a long-range migration without admixture has faced populations formerly isolated for a long while.

So genetic differentiation of human beings is mostly geographic (Fig. 3), the "genetic distance" between populations following the geographic distance between their original locations and sometimes the structure of communication paths (shape of continents, isolation by mountains, oceans, glaciers, and so on). Detailed continental studies have also pointed out that there is commonly a narrow correlation between genetic diversity and languages diversity whereas there are few if any correlations between genetic diversity and quantitative polymorphism. For example, New Guinea Papuans physically look like some African Bantous, whereas they are quite different from a genetic point of view.

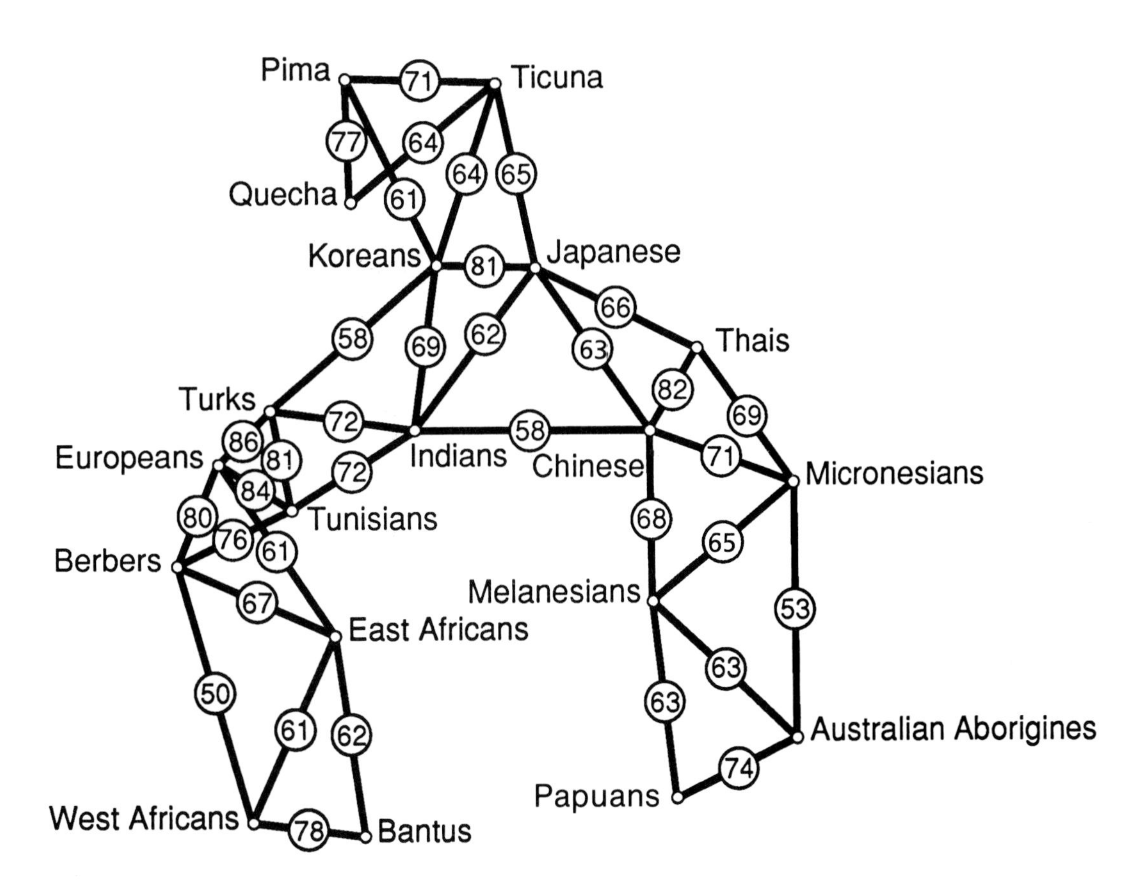

FIGURE 2 Human genetic network: populations are linked according to percentage of gene frequency they share for three very polymorphic genetic systems (Rhesus blood groups, Gm seric groups, and HLA). (Data from A. Sanchez-Mazas and A. Langaney L. G. B., University of Geneva.)

Papuan genes are definitely Oriental genes, closer to those of the Chinese than to those of the Africans, whatever their skin color is. [*See* POPULATION GENETICS.]

V. Polymorphism and Prehistory of Modern Humans

The first fossils of anatomically modern humans have been shown recently to be about 100,000 years old, and these people were dwelling between the Middle East (Qafzeh) and East Africa (Omo Kibish). The present genetic structures of human populations show limited differentiation between populations. Such results support a recent common origin of all modern human populations, between 100,000 and 150,000 BP, and important migrations between populations during prehistoric and historic past. This theory is supported by independent data from immunology and molecular biology and is compatible with fossil and archaeological data. So most human population geneticists think that the ancestors of modern humans originated about a thousand centuries ago in a single population, which could have dwelled somewhere between East Africa and western Asia. Then, this population split and spread itself throughout the Old World until 70,000 BP, reaching Oceania and America later on. Peripheral populations of South and West Africa, north and western Europe, eastern Asia, Oceania, and Amerindians possess different subsets of the gene pool commonly observed from East Africa to India. So we can think that when small groups of hunter gatherers reached these areas, they lost a part of the original gene pool, which they did not carry with them (the founder effect) and/or genetic drift, the normal process of variation and random accumulation of new differences between populations. This settled the present differences between human gene frequencies. A thousand centuries is not enough to obtain noticeable frequencies of many new mutations occurring during that colonization of the

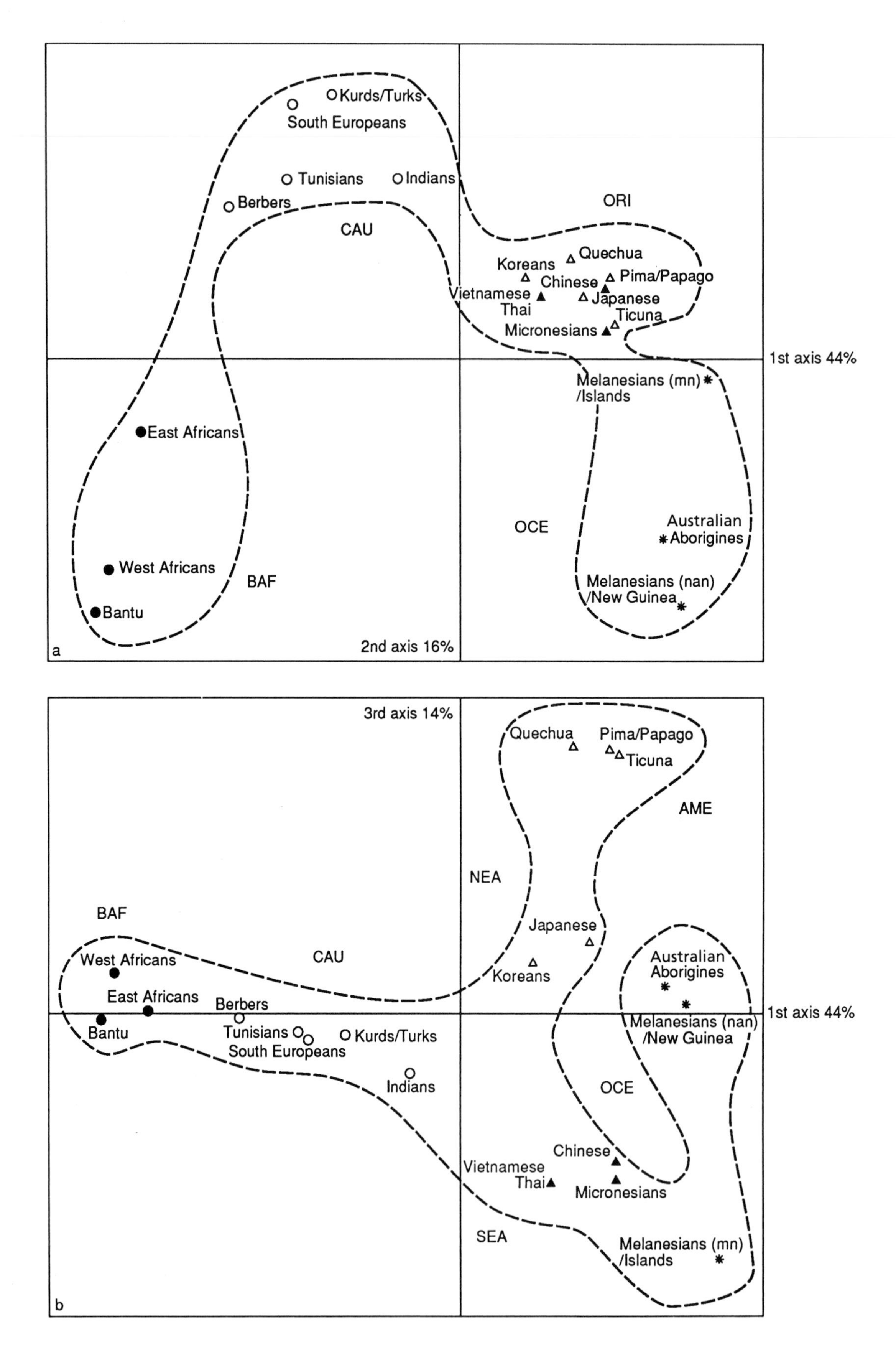

FIGURE 3 Multivariate analysis of the human genetic network BAF, Black Africans; CAU, Caucasoids; ORI, Orientals NEA, North East Asia; SEA, South East Asia; AME, Amerindians; OCE, Oceanians (same polymorphisms as Fig. 2). [From Sanchez-Mazas, A., and Langaney, A. (1988). *Hum. Genet.* **78,** 161–166, with permission.]

world. So it is believed that founder effects, genetic drift, and migration were the main factors of modern human genetic evolution.

The high level of discordance between genetic polymorphism and quantitative polymorphism must also be stressed. Although genetic polymorphism evolves according to human pedigrees and migrations, quantitative polymorphism seems to evolve fast, adapting to local environment conditions. Body shapes and measurements can vary in few generations as they did in Europe since the middle age. Skin color seems to be able to change drastically in less than 50,000 years, because Orientals and Melanesians share the same ancestors at this time scale. So the outside characteristics of the human body (shape, size, color), which commonly define the racial groups, have little to do with the remote history of human genealogy; we must consider that race is an empirical concept, not a genetical or historical one.

Bibliography

Cann, R. L., Stoneking, M., and Wilson, A. C. (1987). Mitochondrial DNA and human evolution. *Nature* **325,** 31–36.

Cavalli-Sforza, L. L., Kidd, J. R., Kidd, K. K., et al. (1986). DNA markers and genetic variation in the human species. *Cold Spring Harbor Symp. Mol. Biol.* **LI,** 411–417.

Excoffier, L., and Langaney, A. (1989). Origin and differentiation of human mitochondrial DNA. *Am. J. Hum. Genet.* **44,** 73–85.

Excoffier, L., Pellegrini, B., Sanchez-Mazas, A., Simon, C., and Langaney, A. (1987). Genetics and history of sub-Saharan Africa. *Am. J. Phys. Anthrop.* 151–194.

Langaney, A. (1988). "Les Hommes: Passé, Présent, Conditionnel." Armand Colin, Paris.

Langaney, A., Sanchez-Mazas, A., Pellegrini, B., and Excoffier, L. (1988). "From the Language of Genes to the Genesis of Languages." ISI Meeting, May 23–26, Torino. Stanford University Press (in press).

Piazza, A. (1988). "Modern Man History: Methodology, Results and Hypotheses." ISI Meeting, May 23–26, Torino. Stanford University Press (in press).

Sanchez-Mazas, A., and Langaney, A. (1988). Common genetic pools between human populations. *Hum. Genet.* **78,** 161–166.

Wainscoat, J. S., Hill, A. V. S., Boyce, A. L., et al. (1986). Evolutionary relationships of human populations from an analysis of nuclear DNA polymorphisms. *Nature* **319,** 491–493.

Polypeptide Hormones

MICHEL CHRÉTIEN, ROY A. SIKSTROM, *Clinical Research Institute of Montreal*

Glossary

Binding assay Test of the interaction between a hormone and a receptor
Endocytosis Process by which material such as a bacteria, a hormone, or a toxin, after making contact with the cell surface, is taken into the cell
Endosome Intracellular vesicles into which endocytosed hormone–receptor complexes are transported
Exocytosis Process by which a secreted protein is transported to the cell membrane and released
Feedback control Regulation of the flux through a biological pathway such that a given point in the pathway is able to influence earlier points
Hormone affinity Tightness of binding between a hormone and a receptor or other binding element
Hormone specificity Uniqueness of fit between a hormone and a receptor or other binding molecule
Receptor Cell protein able to strongly bind a hormone or other extracellular signaling molecule
Second messenger Intracellular signaling molecules which result from hormone–receptor binding at the cell surface
Target cell Cell upon which a hormone acts

HORMONES BELONG TO A CLASS of substances known to have a powerful action in minute amounts. They are important in biological processes as intercellular messengers. The term hormone was coined from a Greek root to describe an agent able to excite or activate. The concept, first expressed by Bayliss and Starling in 1902, was based on an active substance present in a grandular secretion that could be mediated by way of the blood to produce a physiologic effect. This has led to the current view of a hormone as an agent produced by any tissue of the body that upon secretion may act locally or be transported by the blood to various organs and tissues, where its reactions may affect the organism as a whole. In this chapter, we will consider only hormones that are protein or peptide in their structure, conventionally designated peptide hormones. A rapidly growing group of active peptide agents called growth factors will not be discussed here, nor will the steroid prostanoid or catecholamine hormones.

I. The Nature of Peptide Hormones and Their Biological Effect

Historically, peptide hormones were thought to belong to a single class of hormones secreted by endocrine or ductless glands, but they are now recognized to be produced and released by a large variety of cells—for instance, atrial natriuretic factor (ANF) is produced by the heart. Peptide hormones are produced at a variable rate in response to environmental or other stimuli. A key element in the cell receiving the chemical message is the receptor specific for each hormone. Insufficiency or an excess of a particular hormone, or a defect in the receptor or its means of signal transmission, may cause clinical disorders, which in many cases have helped us to understand the normal endocrine system functions. [*See* ATRIAL NATRIURETIC FACTOR.]

Peptide hormones stimulate and control many metabolic processes of widely varying natures, but they have several properties in common.

(1) They resemble other crucial biological agents such as enzymes or vitamins in that they are effective in tiny amounts and are not consumed in the

course of their action, although they may be degraded in lysosomes, after acting, as a form of cellular regulation.

(2) In most cases, they are produced in a cell or tissue other than that in which they ultimately act.

(3) They are secreted into the circulation, which transports them throughout the body. Thus, their concentration in the blood at any given time gives at least an indication of the activity of the gland in question and the degree of exposure to and interaction with the target cells. Because hormone effects are highly specific, only small amounts are necessary, and the blood levels are normally extremely low, usually 10^{-10}–10^{-12} *M*. Peptide hormones have half-lives ($t_{1/2}$) of 5–60 min.

(4) Structurally, peptide hormones are rather small proteins having molecular weight (MW) of 30,000 or less; they may be as small as enkephalin and thyrotropin-releasing hormone, which contain five and three amino acids, respectively. They are water-soluble and, hence, do not usually require carrier proteins in circulation. The variety in nature and activity of peptide hormones is shown in Table I.

A number of peptide hormones are synthesized with equal ease by cells in the gastrointestinal tract or the central nervous system. Indeed, perhaps any known cell type has some degree of endocrine character; thus, in Table II we have arranged some peptide hormones that do not seem to fit so well into existing categories. The role and the sources of peptide hormones are likely to continue to extend in this way, because our advanced techniques in protein chemistry and molecular biology find us isolating and characterizing peptides in ever greater numbers, often before we know their cellular role or whether or not they have biological activity. [*See* PEPTIDE HORMONES OF THE GUT.]

TABLE I Active Peptide Hormones

Hormone[a]	Properties	Physiological effects
	Pituitary	
ACTH (adrenocorticotropin)	39 AA; MW 4,500; plasma level 50 pg/ml; $t_{1/2}$ = 25 min; acts via cAMP	Regulates adrenal function Increases steroidogenesis and synthesis of RNA and new adrenal proteins Promotes lipolysis in fat cells Induces insulin and growth hormone secretion
β-endorphin (β-END)	31 AA	Produces analgesia and euphoria
α-MSH (α-melanocyte-stimulating hormone)	13 AA	Promotes an increase in pigmentation
β-MSH (β-melanocyte-stimulating hormone)	18 AA	Promotes an increase in pigmentation
TSH (thyroid-stimulating hormone)	201 AA; MW 32,000; α, β-subunits, glycosylated	Regulates thyroid function Increases thyroid size and vascularity Increases iodide uptake, thyroglobulin synthesis Increases T_3 and T_4 release
FSH (follicle-stimulating hormone)	204 AA; MW 32,000; $t_{1/2}$ ~60 min; α, β-subunits, glycosylated	Stimulates follicular development in the ovary and gametogenesis in the testes
LH (luteinizing hormone)	204 AA; MW 32,000; $t_{1/2}$ ~30 min; α, β-subunits, glycosylated	Promotes luteinization of the ovary and stimulates Leydig cell function of the testes
GH (somatotropin)	191 AA; MW 22,650; single peptide chain with 2 S—S bonds	Regulates intermediary metabolism Increases protein synthesis and amino acid transport Accelerates fat mobilization and stimulates fatty acid oxidation Inhibits carbohydrate uptake by muscle Decreases sensitivity to insulin
Prolactin	198 AA; MW 23,510; single peptide chain with 3 S—S bonds	Promotes lactation
Vasopressin (AVP)	9 AA; single peptide chain with 1 S—S bond	Raises blood pressure Increases water resorption in kidney tubules

TABLE I *(continued)*

Hormone[a]	Properties	Physiological effects
Oxytocin	9 AA; single peptide chain with 1 S—S bond	Promotes contraction of the uterus Stimulates mammary glands
	Insulin	
Insulin	51 AA; (chain A, 21 AA, chain B, 30 AA); MW 6,000; 2 S—S interchain bonds	Stimulates metabolic activity for synthesis and/or storage of carbohydrates, fats, proteins, and nucleic acids Promotes glucose uptake, glycogenesis, and lipogenesis Promotes synthesis of protein, ATP, DNA, and RNA
Glucagon	29 AA; MW 3,485	Regulates carbohydrate metabolism mainly in liver by promoting glycogenolysis
Pancreatic polypeptide	36 AA	Stimulates basal acid secretion Increases gut motility
	Gastrointestinal Tract	
Gastrin	17 AA and 34 AA	Promotes secretion of gastric acid and pepsin Increases growth of gastric mucosa Increases secretion of pancreatic enzyme as well as insulin, somatostatin, and pancreatic polypeptide Increases water and electrolyte secretion by stomach, pancreas, liver, and Brunner's glands Increases gastric motility
CCK (cholecystokinin)	Heterogeneous size; 33 AA, 12 AA, and 8 AA; $t_{1/2}$ 3 min; also found in the central nervous system	Regulates processes in the pancreas, the biliary system and gut Stimulates the release of pancreatic digestive enzymes
Secretin	27 AA; MW 3,055; $t_{1/2}$ 4 min; linear peptide	Acts on pancreatic acinar cells to release water and bicarbonate Acts to augment the release by cholecystokinin of pancreatic enzymes
VIP (vasoactive intestinal peptide)	28 AA; MW 3,326; $t_{1/2}$ 1 min; linear peptide; also found in the central nervous system	Promotes relaxation of the smooth muscles of the circulation system and of the gut Increases secretion of water and electrolytes from pancreas and gut
Bombesin	14 AA; MW 1,620; $t_{1/2}$ 5 min; also found in nerves	Induces hypothermia Stimulates hormone and pancreatic enzyme secretion and smooth muscle contraction
Motilin	22 AA; MW 2,698	Increases motility of the gastrointestinal tract Decreases acid secretion
Neurotensin	13 AA; MW 1,692; also found in hypothalamus	Promotes vasodilation and vascular permeability Decreases plasma volume and promotes hypotension
Substance P	11 AA; MW 1,528; also found in many nerve cells	Often acts through the nervous system as a sensory neurotransmitter, analgesic, and excitatory neurotransmitter Promotes contractions of gastrointestinal smooth muscle A potent vasoactive agent
GIP (gastric inhibitory polypeptide)	42 AA isolated from small intestine	Inhibits gastric acid secretion Promotes insulin release

TABLE I *(continued)*

TABLE I *(continued)*

Hormone[a]	Properties	Physiological effects
Galanin	29 AA isolated from intestine; also found in the central nervous system	Inhibits insulin release Promotes release of growth hormone and prolactin
	Thyroid and Parathyroid	
CT (calcitonin)	32 AA; a single chain peptide with 1 S—S bond	Lowers serum calcium
PTH (parathyroid hormone)	84 AA; single peptide chain	Stimulates the rate of resorption of calcium from the glomerular filtrate Promotes calcium resorption from bone Enhances the absorption of calcium from the gastrointestinal tract
	Hypothalamus	
TRH (TSH-releasing hormone)	3 AA	Stimulates TSH release from anterior pituitary Stimulates PRL release
LHRH (LH-releasing hormone)	10 AA	Releases LH and FSH
GRH (GH-releasing hormone)	40 AA and 44 AA	Stimulates growth hormone secretion
CRH (corticotropin-releasing hormone)	41 AA; found throughout brain	Stimulates the release and synthesis of POMC products from the pituitary
Somatostatin	14 AA and 28 AA; also found in pancreas and gut	Inhibits secretion of growth hormone and TSH and many extrapituitary hormones
	Central Nervous System	
Enkephalins	5 AA	Inhibits the release of neurotransmitter Induces analgesia
BNP (brain natriuretic peptide)	26 AA; also found in heart	Natriuretic-diuretic Hypotensive
Dynorphin	17 AA; also found in pituitary and gut	Induces analgesia
Neuromedin B	10 AA isolated from spinal cord; also found in gastrointestinal tract and pituitary	Promotes muscle contraction
NPY (neuropeptide Y)	36 AA isolated from brain; carboxy-terminal tyrosine amide; also found in placenta	Vasoregulator Inhibits secretion of LH and GH
CGRP (calcitonin gene-related peptide)	37 AA; 1 S—S bond; carboxy-terminal amide; also found in the peripheral nervous system	Mediates smooth muscle contraction Alters blood pressure and heart rate

Abbreviations used: AA, amino acid(s); cAMP, cyclic adenosine monophosphate; MW, molecular weight in daltons; pg/ml, picograms per milliliter; S—S, a disulfide bond; $t_{1/2}$, half-life.

[a] Hormones are listed in a more commonly used full or abbreviated name with the alternative form in parentheses. Hormones have been classed with the tissue from which they were first isolated or where they are found in the greatest amount, although many hormones have been found in more than one tissue.

A. Regulation of Hormone Levels

The effect of a peptide hormone mainly depends on its concentration in the circulation. The producing cells have a basal or steady secretion and respond to need by increasing the secretion from the basal to a stimulated rate.

1. Basal Peptide Hormone Secretion

Basal secretory levels are not well known because of the difficulty of measuring secretion rates directly. Due to the development of the radioimmunoassay and the radioreceptor assay, the blood levels of most peptide hormones can be measured fairly accurately; these levels reflect the overall balance among secretion, plasma degradation, and target cell association. In most instances, they correctly reflect the state of the actual secretion. [*See* RADIOIMMUNOASSAYS.]

2. Stimulated Hormone Secretion

Different forms of stimulation can lead to enhanced secretion, as shown, for instance, by the

TABLE II Diverse Peptide Hormones

Name	Characteristics	Source	Physiologic effects
Inhibin	MW 31,000–32,000; α-subunit 18,000 Da and β-subunit 14,000 Da joined by 1 S—S bond; α-subunit is glycosylated	Originally isolated from follicular fluid	Potent suppressor of FSH release in males and females, whereas LH and other pituitary hormones are unaffected
	Time of action 4–18 hr; 70% similarity between β-subunit and TGFβ	Also found in gonads and placenta	Acts on gonads to influence steroidogenesis
ANF (atrial natriuretic factor)	28 AA; $t_{1/2}$ ≈5 min; precursor is processed at release from the cell	Isolated from heart atria; also found in brain and lung	Promotes natriuresis and diuresis and has vasorelaxant properties Lowers blood pressure
Ang II (angiotensin II)	8 AA; $t_{1/2}$ ~1 min; plasma concentration 10–30 pg/ml; precursor is synthesized in liver and processed in circulation to Ang II	Isolated from blood	Potent vasoconstrictor Stimulates the formation of aldosterone Acts on kidney to induce sodium retention
Endothelin	21 AA synthesized in endothelial cells; 2 S—S bonds	Isolated from cell culture medium	Potent and long-lasting vasoconstrictor
PL (placental lactogen)	191 AA; MW 21,700; single peptide chain; 2 S—S bonds	Isolated from placenta	Promotes growth Stimulates lactation
HCG (human chorionic gonadotropin)	236 AA; MW 46,000; α, β-subunits, glycosylated	Found in placenta	Maintains the corpus luteum once pregnancy has occurred

Abbreviations used: AA, amino acid(s); MW, molecular weight in daltons; pg/ml, picogram per milliliter; S—S, disulfide bond; $t_{1/2}$, half-life.

secretion of insulin by the β cell of the pancreas. The principal factor controlling insulin secretion is the concentration of glucose in the blood. A rise in glucose level causes insulin to be released. Other agents, including amino acids such as arginine, glucagon, and other hormones, can also release insulin but are less effective and are dependent on the presence of a glucose effect. Certain α-adrenergic agents and somatostatin can decrease insulin release. Thus, stimulated secretion responds to a complex hierarchy of agents and effects. [*See* INSULIN AND GLUCAGON.]

B. Neurosecretion and the Hypothalamus

Peptide hormone secretion under neural regulation may be visualized by two general mechanisms. A nerve terminal may make direct contact with a secreting cell, or a neuron may itself secrete a peptide, which may act directly to alter target cell function or pass via the circulation to act upon a second secretory cell. Such neurons act as both nerve cells and endocrine cells and are called neurosecretory cells. An example are cells in the hypothalamus that, when stimulated by other nerve cells in the higher regions of the brain, secrete a specific peptide into the portal blood vessels of the pituitary stalk. Such agents have been called releasing factors, or hormones. As shown in Table I, several releasing factors have been characterized. In some respects, this portrays a hierarchical system of hormone action with the hypothalamus as the main regulator tissue.

C. Hormone Levels in the Blood

The level of a particular hormone in the blood and its effect on target cells is the product of a complex set of dynamic interacting components.

(1) The level of hormone available in the blood is dependent on its rate of synthesis and the secretion of the hormone from the tissue of origin.

(2) The actual transport in the circulation is believed to take place in an unaided fashion, but certain hormones such as vasopressin and oxytocin may have carrier proteins, which act as specific transport systems.

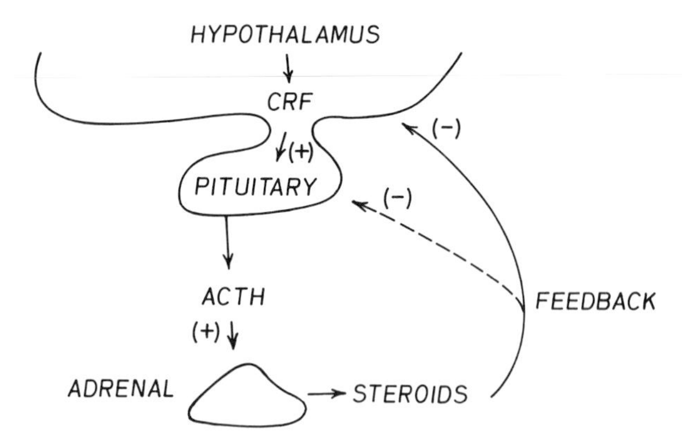

FIGURE 1 Feedback regulation. The regulation by the hypothalamus of the pituitary, and in turn the adrenal is shown schematically. CRF secreted by the hypothalamic tissues stimulates the pituitary to release ACTH, which acts on the adrenal gland. The adrenal cells produce steroid hormones, which feed back negatively on the hypothalamus and, to a lesser extent, on the anterior pituitary, tending to decrease the secretion of CRF and ACTH, respectively.

(3) Also in the blood in some cases (angiotensin I → II), an enzymatic activation produces a hormone from an inactive precursor.

(4) Finally, the hormone levels in circulation may be decreased by degradation to restore a proper balance. Many peptide hormones are degraded via lysosomes in their target cells after biological action has been effected. In addition, peptide hormones are rapidly cleared from the circulation by being taken up by the liver or kidney. Variation of any of the above factors will alter the blood concentration of the hormone and therefore its activity. There is also a dynamic balance of the various hormones in the organism. For example, release by the hypothalamus of corticotropin-releasing factor acts upon special cells (corticotrophs) in the pituitary causing the release of adrenocorticotropin (ACTH), which in turn acts upon the adrenal to induce the release of steroid hormones (Fig. 1). These hormones then feed back on the hypothalamus and the pituitary with a negative stimulus, which tends to balance the whole process. [*See* HYPOTHALAMUS.]

D. Target Cell Effects

The range of responses of cells to a particular hormone is intrinsic to the target cell and largely depends on the level and nature of the differentiated state of the cell. A given peptide hormone may act on several different types of target cells and exhibit different responses in each of them. For instance, when insulin acts on fat cells, it is known to induce glucose transport and lipid biosynthesis, whereas in liver it stimulates amino acid transport and glycogen synthesis and, if administered into the cerebrospinal fluid of an experimental animal, it acts upon the brain cells to bring about a decrease in food intake and a weight loss.

Essentially all cells in the body are a target for one or more hormones. A single hormone may induce several responses in target cells, and, conversely, several different hormones can provoke the same function in a particular target cell. As a general rule, most hormones have a wide range of effects, which are recognized as our ability to perceive and measure subtle changes within cells increases.

The interaction of the hormone with a target cell is brought about by a receptor inserted in the plasma membrane, which has high affinity and specificity for the hormone. Each target cell has on its surface a limited number of receptors for a particular hormone, and not all of them (10% or more) need be bound to the hormone to provide a maximum biological effect. In fact, many target cells tend to regulate the number of receptors on their surface up or down, dependent on whether the level of the corresponding hormone is low or high.

1. Receptors and Hormone Binding

Through the blood, all the cells in the body are exposed to a low and roughly equal concentration of each of the hormones at any given time. The receptor provides the cells' specifically unique ability to recognize and strongly bind the bioactive hormone from among the many diverse substances in the blood. Even though the hormone may be present among other proteins at very low levels and among peptides at much higher concentrations, it can be distinguished by a specific receptor and attracted to it effectively at a chance of one in one million to one in one billion. At the heart of this interaction is the strong binding of a region of the receptor with a complementary segment of the hormone. Thus, the tightness of fit or the affinity observed in the hormone–receptor complex is a crucial characteristic intrinsic to this form of intercellular communication.

A schematic representation is given in Fig. 2. The characteristics of the hormone–receptor interaction are determined by binding assays, in which the ability of a hormone or analogue to bind to a particular receptor is tested by exposing it to membranes or

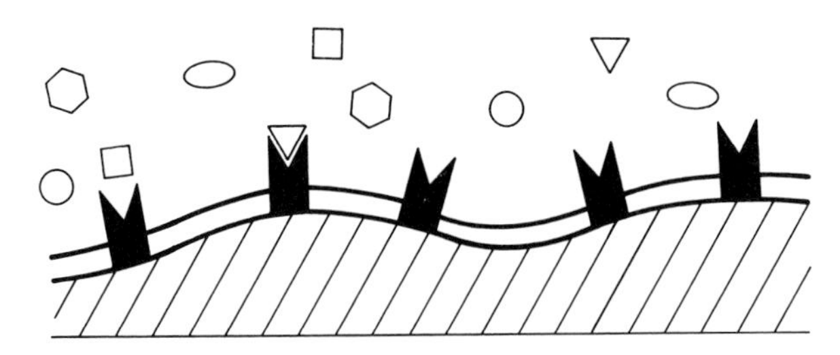

FIGURE 2 Receptor specificity. Shown schematically are the target cell receptors being exposed to the blood containing their corresponding hormone (triangles) and to other hormones represented by the other geometric figures.

cells known to contain that receptor. The same test can be used to search for the presence of a certain receptor. Usually the hormone is labeled with a radioisotope ^{125}iodine (^{125}I) or tritium, or other quantifiable agents. After a binding incubation, the bound hormone–receptor complex is separated from the free hormone, and both the free and the bound hormone are measured. A control is usually carried out with an identical amount of labeled hormone plus a vast excess of unlabeled hormone. The excess of unlabeled hormone decreases to zero the binding of the labeled hormone to the receptor. This is called a nonspecific control. Such a binding experiment is illustrated in Fig. 3, which shows radioautographs of ^{125}I-ACTH binding in adrenal tissue.

The receptor as the key element in hormone action can be seen by the fact that the receptor for insulin is very closely conserved in its specificity and affinity throughout all vertebrate species, whereas the insulin molecules of different species do differ from one another in their affinity for the receptor. Moreover, a close correlation is observed between the ability of a given insulin hormone to bind to the insulin receptor and the extent of biological activity elicited by that same insulin ligand. In general, the affinity of a peptide hormone for its receptor closely parallels the biological activity.

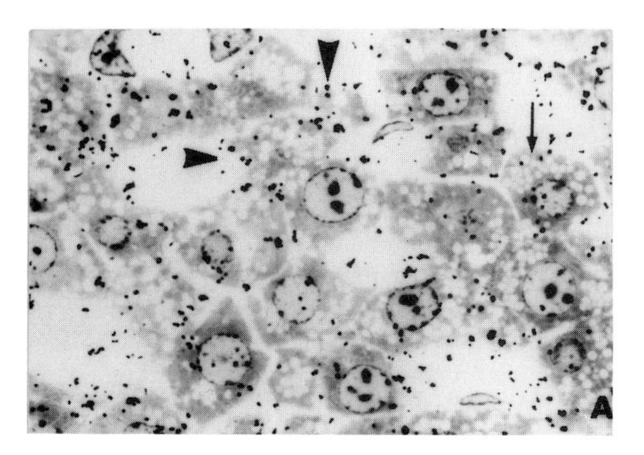

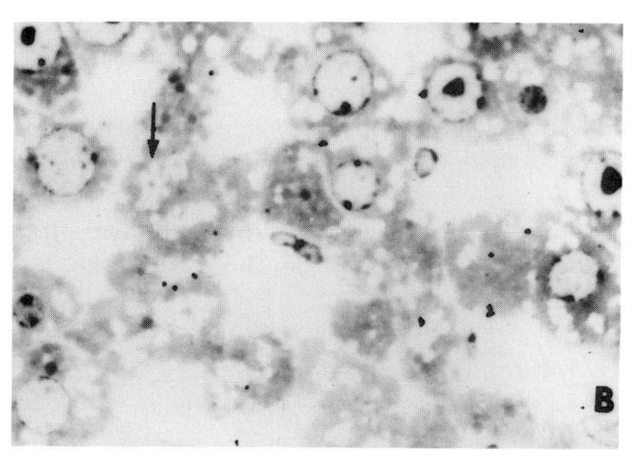

FIGURE 3 Radioautograph of ACTH binding in adrenal tissue. A. ACTH labeled with ^{125}iodine (^{125}I) -ACTH, which has been bound to adrenal cells. B. The nonspecific control radioautograph obtained from a binding experiment in which an amount of ^{125}I-ACTH identical to A, plus a 200-fold excess of unlabeled ACTH interacted with an equivalent adrenal tissue section. The excess of cold ACTH greatly dilutes the binding of ^{125}I-ACTH, thus producing many fewer grains. After the binding incubation, the tissue is preserved with a fixative and placed in direct contact with a photographic emulsion. Upon development, silver grains are produced from the disintegrations of the ^{125}I-ACTH, thus revealing the presence of binding sites. Arrowheads indicate grains overlying ^{125}I-ACTH bound to adrenal cells. Small arrows show lipid droplets characteristic of these cells. Magnification ×1,100.

2. Hormone Binding and Biological Activity

It is necessary to understand the relationship between the hormone–receptor binding and the ability to produce the biological effects. All active hormones or hormone analogues having activity are designated as agonists. Other analogues, called antagonists, bind to the receptor, sometimes with an affinity even greater than that of the native hormone, but do not produce any biological activity. Studies involving hormone analogues have shown that the binding region of the hormone is usually different from the region of the hormone that provides the biological response. Thus, antagonist molecules can readily attach to the hormone-binding site on the receptor molecule but are unable to trigger the cellular activity; however, their presence on the receptor-binding site prevents the binding of the bioactive hormone, if it be present. Some hormones have affinity for the receptor of other hormones, though usually less effectively than the intrinsic hormone, and thus act as partial agonists. For example, growth hormone, prolactin, and placental lactogen have partial binding affinity on the lactogenic receptor. The same is true of many synthetic analogues of shorter peptide hormones such as luteinizing hormone-releasing hormone, somatostatin, etc.

When a hormone binds to its receptor, the hormone–receptor complex serves to activate the cellular response. The receptor appears to possess the full complement of information necessary to activate the cell response; the hormone serves mainly to induce the receptor to express this information. This has been shown in the case of insulin using antibodies directed against the insulin receptor, which are able to block insulin binding; conversely bound insulin is able to block the binding of the antibodies. These antibodies can mimic the range of biological responses of insulin in the absence of hormone. Also, when several structurally related hormones bind to one another's receptors, the effect elicited is that expected from the receptor and not from the hormone.

3. Internalization of Peptide Hormone–Receptor Complexes

The binding of the hormone to its receptor initiates a series of intracellular events that lead to the biological effect of the hormone. Having triggered the initiation of the endocrinal message, the hormone generally does not dissociate from the receptor, but the complex is engulfed in membrane-coated pits and internalized as such into vesicles called endosomes, which are free in the cytoplasm. Although cellular events subsequent to internalization of hormone–receptor complexes are not well understood, clearly in many systems the hormone is separated from the receptor in the endosome and delivered to a lysosome to be degraded, while in some cases the receptor is recycled to the cell surface and in other cases it is also degraded in lysosomes. Thus, the problem of the meaning of receptor internalization in the effect of the hormone in the cell arises.

4. Receptor Structure

The known receptors for peptide hormones are large molecules with hydrophobic domains to permit their partial immersion in the plasma membrane. Although some receptors consist of a single polypeptide chain, others have several subunits and globally may have molecular weights of several hundred thousand to more than 1 million daltons. The study of receptors has progressed more slowly than that of the peptide hormones because receptors are present in low concentrations in cells, and, being rather hydrophobic, they tend to be insoluble in aqueous solutions. Hence, detergents or other agents are often required for their isolation, and then alter the native characteristics of the receptor.

The insulin receptor is one of the best-studied receptors and is considered here in some detail. Analysis of isolated receptor and cloning and sequencing of the cDNA complementary to the messenger RNA expressing the receptor led to the following structural description (Fig. 4). This receptor is a heterodimer with the two α- and two β-subunits. The α-subunits (M_r ~125–135,000 Da) are linked to each other and to the β-subunits (M_r 95,000 Da) by disulfide bonds. The β-subunits have a short hydrophobic sequence, which spans the cell membrane, and a substantial carboxy-terminal region extending into the cell cytoplasm. DNA studies suggest that one α- and one β-subunit are synthesized as a single polypeptide chain separated by a short sequence of four basic amino acid residues

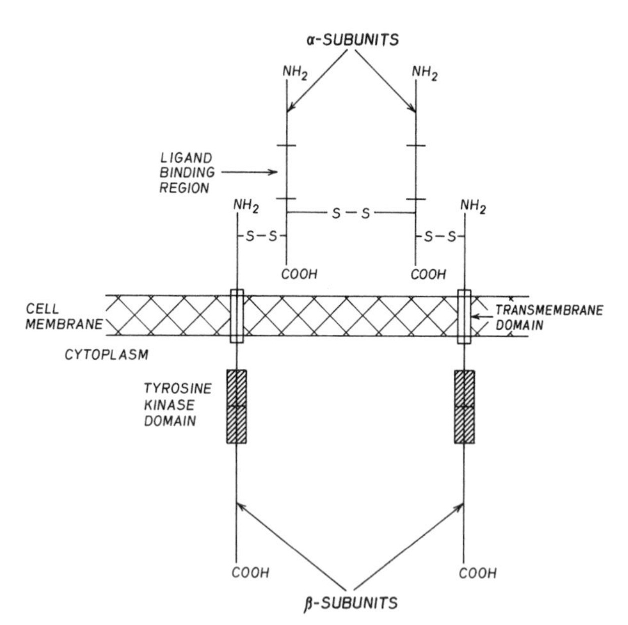

FIGURE 4 Structure of the insulin receptor. The receptor is composed of two α polypeptide chains joined by a disulfide bond, joined in turn via disulfide bonds to the two β-chains. Each of the β-subunits is anchored to the cell membrane and has an open box representing the transmembrane region. The carboxy-terminal region of each β-chain is in the cell cytoplasm, and the tyrosine kinase domains are indicated with cross-hatched boxes. The ligand-binding region is shown on each α-subunit in the extracellular space.

(Arg-Lys-Arg-Arg), which is cleaved during receptor assembly. Insulin binding to its receptor leads rapidly to an autophosphorylation of a tyrosine residue on the β-subunit. This result shows that the receptor possesses enzymatic activity as protein kinase. How ligand binding to the α-subunit outside the cell can generate a kinase action in the cytoplasmic extension of the receptor and how this fits into the mechanism of hormone action is unclear. Although hormone action will not be dealt with in any detail here, it will be of great interest to see whether the kinase activity is linked merely to hormone binding or to hormone–receptor aggregation and internalization.

II. Biosynthesis of Peptide Hormones

One of the interesting features in the biosynthesis of peptide hormones is that the active form of the hormone found in circulation is not the form in which it is initially synthesized. Usually, precursors are first formed and are subsequently processed to one or several bioactive mature hormones. Because the discussion of transcription and translation is covered in detail elsewhere, we will turn directly to the processing of a newly synthesized hormone precursor. During synthesis, the signal sequence, which mediates the penetration of the polypeptide through the membrane, is removed, leaving a resultant precursor, or prohormone, form, which usually requires further processing to yield an active hormone. The precursor penetrates into the cisternae of the endoplasmic reticulum (ER). Probably, the final processing takes place in the ER, the Golgi complex, and one or more secretory vesicle-storage granules until its ultimate secretion from the cell into the circulation.

A. Precursors

The knowledge of the DNA sequence corresponding to a given protein hormone obtained with the technology of molecular genetics makes it possible to deduce the structure of the precursor protein, sometimes revealing other potential peptide hormones derived from the same precursor, even if they have not yet been characterized. Pro-opiomelanocortin (POMC), a polyprotein originally isolated from the pituitary, is an example of a precursor that contains several active hormones including ACTH, α- and β-melanocyte-stimulating hormones (α-MSH and β-MSH), and β-endorphin (β-END) within its sequence (Fig. 5). Although most studies on biological activities have been focused on ACTH, the MSH's, and β-END, the possibility that other segments of POMC may come to show activity cannot be excluded.

B. Processing

Processing is generally understood to include any transformation of the initial polypeptide, which may take place during the total course of biosynthesis before the final mature and active hormone is produced. Clearly a key modification involved in the conversion of a prohormone to its active component is the cleavage of the peptide chain at the appropriate site(s), which generally occurs at a pair of basic amino acids (Table III). Processing at such sites, however, is not obligatory. For example, ANF retains an arginine-arginine sequence after processing to the active hormone (Table III), and, in a few prohormones, cleavage sites are marked by either a single basic amino acid or even the absence of a basic residue. Cleavage at the paired basic amino acid residues is believed to be carried out by a uniquely specific endopeptidase, a maturation enzyme. However, at present, only yeast indicates something about the nature of such a maturation enzyme or enzyme(s). Because active peptides do not terminate with arginines or lysines, a second carboxypeptidaselike enzyme would be required to remove them. A carboxypeptidase E has been described that can remove terminal basic residues.

The proteolytic cleavages are usually not the only events necessary to obtain a bioactive peptide from a precursor. Specific amino acid modifications are also involved, such as glycosylation, amidation, acetylation, phosphorylation, sulfation, covalent attachment of lipid, disulfide-bond formation, and even asymmetric center inversion, which ultimately lead to the peptide form most commonly found secreted.

C. Diversity of Peptide Hormones

A wide range of sizes and characteristics is seen among known peptide hormones, suggesting that these properties are adaptations that enhance function and survival at the cellular and organism level. By being able to create a large number and variety of peptide hormone products, each with the potential to act on target tissues, the cell has enhanced its

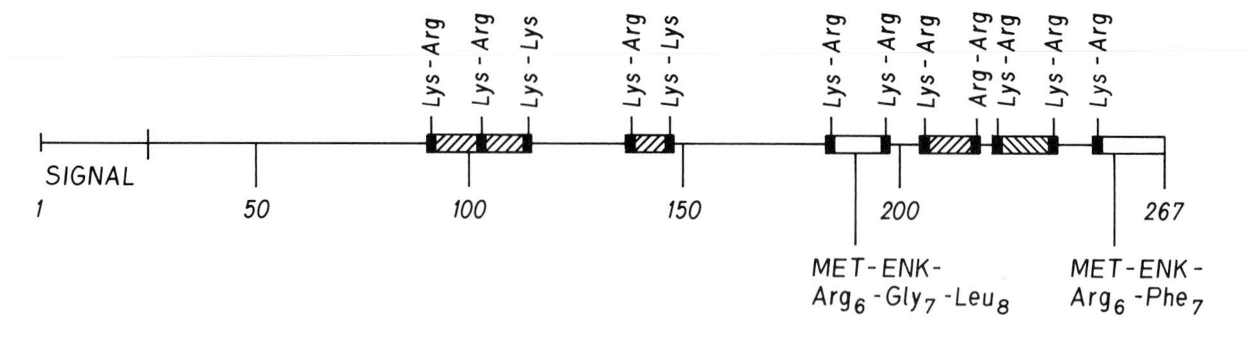

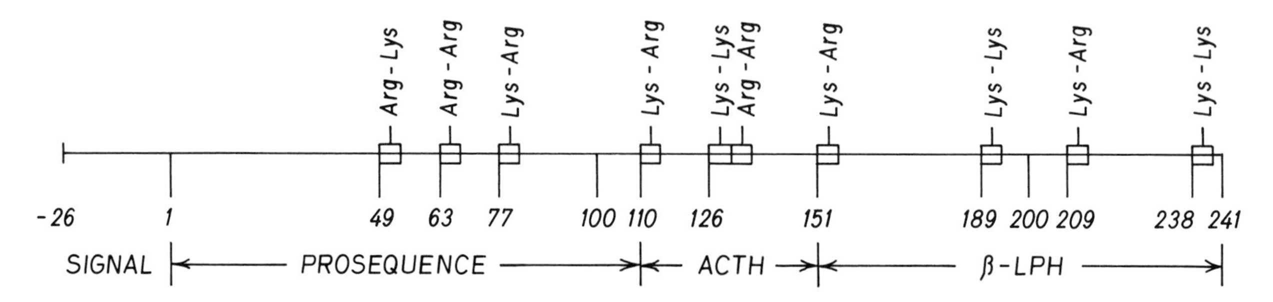

chances to control and adapt. This molecular diversity is known to be expressed at the level of precursor processing and in the synthesis of mRNA.

FIGURE 5 Structures of precursors. The upper diagram represents the protein precursor preproenkephalin, which contains four copies of methionine enkephalin (MET ENK), one copy of leucine enkephalin (LEU ENK), a MET ENK heptapeptide, and a MET ENK octapeptide. The lower diagram illustrates features of the precursor prepro-opiomelanocortin (pre-POMC), which gives rise to ACTH, β-END, and α- and β-MSH. The region of the amino-terminal prosequence, ACTH, and the intermediate precursor β-LPH are shown. The signal sequence is shown for both precursors, and the potential cleavage sites are marked by pairs of basic amino acids.

1. Diversity at the Level of Transcription

In thyroid tissue, the calcitonin gene gives rise to the protein precursor preprocalcitonin, whereas in the hypothalamus, as the result of an alternative form of splicing of the RNA transcript, the same gene produces a distinct peptide called calcitonin gene-related peptide. Thus a posttranscriptional event expresses two distinct species of mRNA (coding for functionally distinct bioactive peptides) from a single gene. [*See* DNA AND GENE TRANSCRIPTION.]

In a similar fashion, two species of substance P precursors were derived via alternative RNA splicing from a single gene. In this case, tissue-specific splicing generates two distinct mRNAs encoding a precursor for substance P alone or substance P together with substance K. All current data suggest that substances P and K have different cellular roles, different receptors, and different tissue distributions. These examples illustrate the importance of posttranscriptional mechanisms for providing hormonal cellular diversity.

2. Peptide Hormone Diversity from Protein Processing

The key modification in the conversion of a prohormone to its active component is the cleavage of the peptide chain at its appropriate site(s). This is another method used by the cell for producing more than one hormone from a single transcript. The described case of POMC (Fig. 5) is one of the best examples of how the cell has amplified its choices leading to adaptation and viability. A second good model is the precursor of enkephalin, which has 12 potential sites of cleavage (Fig. 5). This adds tremendously to the expression of cellular diversity.

TABLE III Precursors

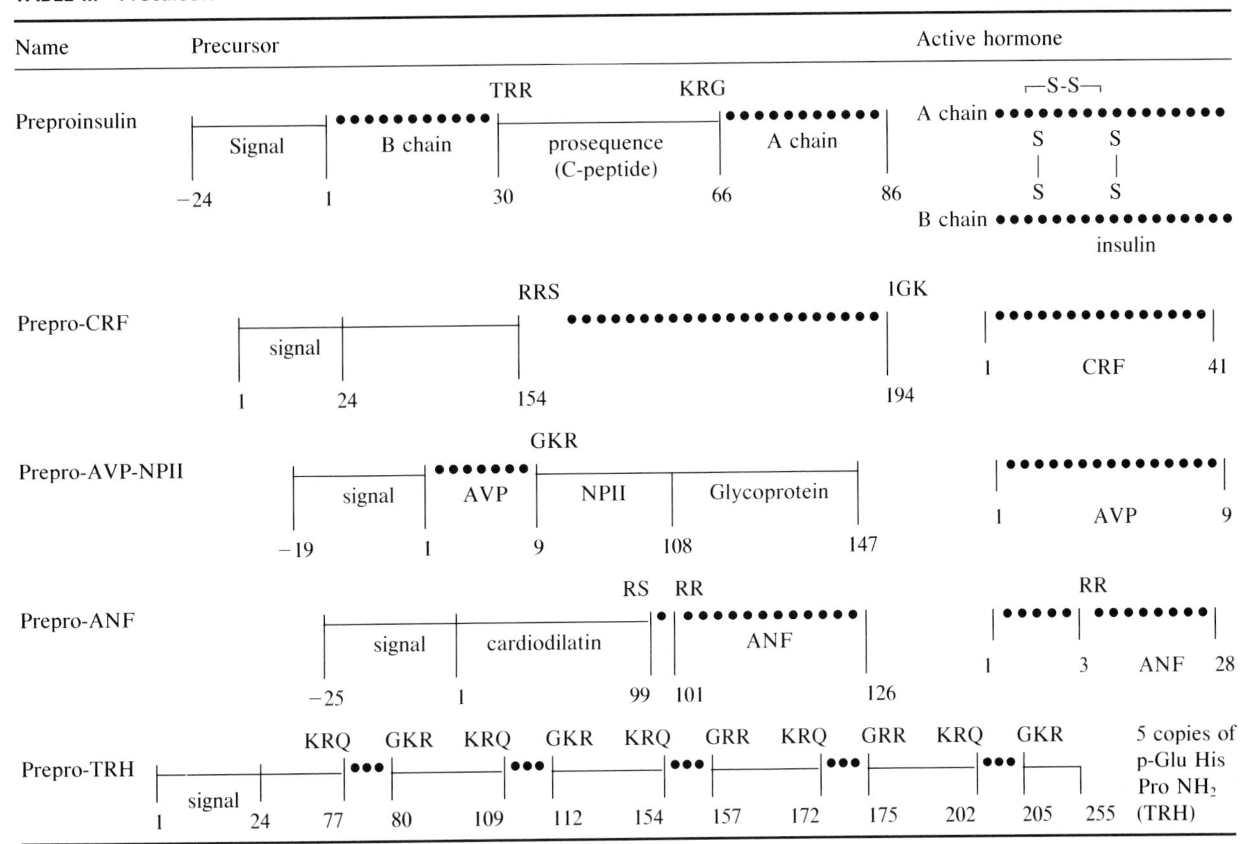

Abbreviations: ANF, atrial natriuretic factor; AVP, arginine vasopressin; CRF, corticotropin-releasing factor; G, glycine; His, histidine; I, isoleucine; K, lysine; NPII, neurophysin II; p-glu, pyroglutamate; $ProNH_2$, proline amide; Q, glutamine; R, arginine; S, serine; TRH, thyroid-stimulating hormone-releasing hormone.

3. Tissue-Specific Processing

The precursor POMC is synthesized in the pituitary as well as in other sites, such as the hypothalamus. Differences in the nature and/or extent of enzymatic processing in the different tissues result in the production and secretion of different sets of end-product peptides. The pattern of POMC processing in brain appears to be distinct from that in either the anterior or the intermediate pituitary lobe. Further modification of α-MSH- and β-END-related peptides via acetylation also appears to be specific to pituitary intermediate lobe cells as well as to the hypothalamus. In the light of studies showing that α-*N*-acetylation inactivates the opioid properties of β-END_{1-31} and potentiates the behavioral effects of $ACTH_{1-13}$-NH_2 (MSH), the selective production of the acetylated peptides could represent another mechanism for regulation of biological actions of neuronal POMC derivatives.

Similarly, the enkephalins, another family of opioid peptides, are generated biosynthetically via a precursor found both in adrenal medulla and brain tissue. In these two tissues, the processing occurs very differently. In brain, cleavage appears to begin from the N-terminus to produce free synenkephalin and enkephalin, whereas in the adrenal processing is less thorough, which produces larger peptides. Additionally, prodynorphin, a precursor that has repeating copies of leukenkephalin and other peptides, is also processed with marked variation in several different regions of the brain. Therefore, even if a precursor derived from a single gene is synthesized in many cellular locations, alternative pathways of processing may lead to different products or differing proportions of products in each of the different tissues, thus creating additional functional diversity.

4. Tissue-Specific Biological Activity

As already illustrated above with insulin, a peptide hormone may express a different bioactivity in different target cells. For instance, many hormones active in peripheral tissues have also been identified in regions of the brain. Immunohistochemical mapping of brain tissue has shown a wide distribution of the precursors POMC, proenkephalin, and prodynorphin. Conceivably, active hormones from these precursors and others might exhibit a different kind of activity at target cells in different locations within the brain and the whole organism. In this manner, through the multiplicity of peptide hormone action, an increased level of cellular diversity is realized.

III. Summary

Multicellular organisms require communication to organize and coordinate the activities of development, cell division and growth, metabolism and maintenance, behavior, and emotional expression. The biochemical communication, which brings about such diverse cellular activities, is effected in part by many hormones, one class of which is the peptide hormones. To completely describe the peptide hormone as a messenger, we must consider the characteristics of the secreting cells and the cells that are subject to the hormone effects—the target cells. Together they act as sending and receiving components in a biological communication system.

Although varying in size, peptide hormones are usually of low molecular weight and are generally water-soluble, so that once secreted from the producing cell they are transported in the circulation to their site of action. A critical entity in the communication system is the receptor situated in the plasma membrane of the target cell. The receptor has an extremely high specificity and affinity for its intrinsic hormone, which allows recognition and binding to the target even though the hormone may be only present in the blood at concentrations of 10^{-8}–10^{-12} M or lower, and blood contains many other proteins. The receptor is the inherently more complicated component of the hormone–receptor couple, and it tends to be more conserved in evolution. Furthermore, the major burden of conveying the endocrine message falls on the receptor. In the course of transmitting the biological action, the hormone–receptor complex usually seems to be internalized within the cell, where degradation or recycling may follow.

In the course of evolving into very complex entities, cells have adapted and developed many communication choices. The flexibility of choices available for a cell to cope with and respond to external stimuli or to control its environment are reflected, in part, in the diverse ways of synthesizing and modifying peptide hormones for any need and in response to temporal changes.

Bibliography

Alberts, B., Bray, D., Lewis, J., Raff, M., Roberts, K., and Watson, J. D. (1983). "Molecular Biology of the Cell." Garland Publishing Inc., New York.

Chrétien, M., Sikstrom, R. A., Lazure, C., Mbikay, M., Benjannet, S., Marcinkiewicz, M., and Seidah, N. G. (1989). Expression of the diversity of neural and hormonal peptides via the cleavage of precursor molecules. *In* "Peptide Hormones as Prohormones" (J. Martinez, ed.). Ellis Horwood Ltd., Chichester, England.

Chrétien, M., Sikstrom, R. A., Lazure, C., and Seidah, N. G. (1988). Neuronal and endocrine peptides. *In* "Neurotransmitters and Cortical Function" (R. W. Dykes and P. Gloor, eds.). Plenum Publishing Corp., New York.

Imura, H., Shizume, K., and Yoshida, S. (eds.) (1988). "Progress in Endocrinology." Excerpta Medica, Amsterdam.

Krieger, D. T., Brownstein, M. J., and Martin, J. B. (eds). (1983). "Brain Peptides." John Wiley & Sons, New York.

Rosen, O. M. (1987). After insulin binds. *Science* **237,** 1452–1458.

Williams, R. H. (1981). "Textbook of Endocrinology." W. B. Saunders Co., Philadelphia.

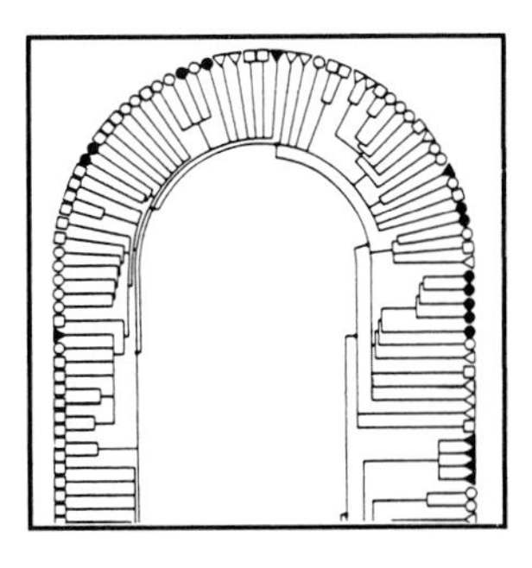

Population Differentiation and Racial Classification

MICHAEL D. LEVIN, *University of Toronto*

Glossary

Cline Graded variation in gene frequency along a line of geographic transition

Geographical race Broad, geographically delimited collections of races (populations)

Local race Breeding population, or population isolate

mtDNA Mitochondrial DNA inherited through females. Differences in individual mtDNA allow hypotheses on evolution

Population All the organisms of a species; within a species, a community of individuals where mates are usually found; a local or breeding group; the inhabitants of a place

Race Group of *populations* of a *species* distinct from other groups of the same species in at least a few characteristics; a subspecies

POPULATION DIFFERENTIATION is the result of genetic change. Populations of a species may be denoted as races, or subspecies if the differences in the gene frequencies in each are regarded as significant. The use of the term *race* to refer to human populations is, however, controversial scientifically and politically. The political use of the term, which is usually part of a racist ideology, has affected the scientific discussion about its use. It is possible to separate political and scientific questions to some extent, but not absolutely, as political arguments tend to build on scientific information and scientific questions are often stimulated by political statements. The distinction between scientific uses of the term and social and political uses of the term *race* is therefore important.

Two major questions depend on an understanding of the processes of population differentiation: the origin of modern humankind, *Homo sapiens sapiens*, and the maintenance of human genetic diversity. Both are connected to the use of the term *race*. Many scientists believe that answers to these questions do not depend on a division of humankind into races. The influence of language, distance, culture, and other nonbiological differences on evolutionary processes and the uses of genetic variation in the study of population histories are among the questions of current scientific interest.

The history of racial classification, a different problem, is marked by literary influences and scientific and political controversy and often appears to have been based on and to be motivated by notions of group superiority. Attempts to establish associations between various morphological and other biological differences and to correlate these differences with psychological and social characteristics are controversial not only politically but scientifically. Many scientists do not consider these associations legitimate. These attempts to extend the question of human variation beyond differences in genetic variation have generally been regarded as unscientific. The association of morphological characteristics with moral and behavioral characteristics, moreover, has been criticized because it supports racism, the ranking of subpopulations of the human species, and has been used to justify discrimination against individuals.

I. Terminology and the History of Classification

A. Terminology of Population Differentiation

The differentiation of population as *races* implies that the differences among the human beings are sufficient to subdivide the species into subspecies. The purpose of racial studies is the definition of relevant and significant differences that would allow this kind of classification. Racial studies have gradually been exhausted as criteria on which group differences rest are found to be inadequate for logical reasons (e.g., arbitrariness, failure to distinguish one group from another, and lack of correspondence to linguistic or cultural uniformities) and for statistical reasons (e.g., greater variance within the populations, defined as races, than between them, the inability to discover a single characteristic that can be associated exclusively with one racial population).

As different criteria were found inadequate, qualifications of the simple term *race* were developed. The terms *geographical race* and *local race* were suggested to associate distinct populations with specific places. These places of origin defined aboriginal populations retrospectively to a relatively fixed world population as of 1492. Clines were suggested as an alternate for race and for population as an object of study. Clines, although they did organize valuable data, have not replaced the concept of population, but have made understandings of its use more sophisticated. Use of clines has demonstrated that differences in single genes between populations were gradual along a spatial dimension. Isolation was suggested as a criteria of racial separation. With research revealing the clinal variation of genetic frequencies, the argument for isolation, or separation, of human populations was severely weakened. The failure to demonstrate isolation and the clear interdependence of environment and behavior in genetic changes are significant arguments against the use of the term *race* for humankind populations. More recent research, dependent on more sophisticated statistical multivariate techniques described below, indicates a range of boundaries of greater or lesser degrees of difference between populations. The scientist who have done this research recognize that clustering of populations into broad categories is possible but have not used *race* to describe these categories.

B. History of Classifications

The discoveries of the New World and the expansion of contact with Africa and Asia had made Europeans aware of the great diversity of the human population. The growth of science led to attempts to understand and incorporate in existing ideas this diversity. The apparent isolation of and morphological differences among human populations seemed to be most significant. What had seemed, from a European perspective, to be a process of increasing differentiation of relatively isolated human populations came to an end in the 15th century with the voyages of discovery, which dramatically increased communication, migration, and change. The contrast makes the rate of change before 1492 seem slow and the populations seem isolated. The combination of the humanness and differences of New World and African and Asian peoples was a problem that demanded understanding. It is only recently that the Eurocentric dimension of this view has been recognized.

The first classification of humankind into races was probably that of François Bernier, a French physician, who in 1684 suggested four: Europeans, Far Easterners, "blacks," and Lapps. Despite criticisms, most notably of Leibniz in 1737, who wrote that "there is no reason why men who inhabit the earth should not be of the same race, which has been altered by different climates . . ." (a view with truly modern echos), race classification was elaborated over the centuries.

In the 18th century, three classifications of humankind were suggested that have continued to influence the form of racial thinking. In 1737, as part of his classification of all living organisms, Linnaeus "divided human kind into four varieties: *Homo Europaeus*, *Homo Asiaticus*, *Homo Afer*, and *Homo Americanus*." Linnaeus based his classification on the idea of the fixity of the species, that species were immutable. Varieties, however, were members of species that differed because of environmental conditioning. The immutability, or adaptability, of organisms is an underlying question in all race classifications as it raises questions of which factors are the basis of human differentiation. George Buffon in his *Natural History* (1749–1804) suggested five races: Lapland, Tartar, Southern Asiatic, European, and American. He believed the white race was the norm and others were exotic variations, but of one species. In 1775 in *The Natural Variety of Mankind*, Johann Friedrich Blumenbach, a professor of medicine at the University of Göttingen, who despite believing that classifications are arbitrary, subject to exceptions and representations of continua, suggested five categories: Caucasian, Mongolian, Ethiopian, American, and Malay. This fivefold classification has had a long life in the literature on race in the more general form of white, yellow, black, red, and brown. Blumenbach coined the term *Caucasian* to describe the "white race." Buffon and Blumenbach are regarded as founders of physical anthropology, and Blumenbach was known as the "father of craniology," the study and use of skulls for scientific purposes. These fathers of the scientific study of human variation, especially Blumenbach, were highly critical of attempts to rank races. Among those who argued for the separate origin and for superiority and inferiority of races were Voltaire, Thomas Jefferson, Lord Kames, and an English physician, Dr. Charles White, who wrote *An Account of the Regular Gradation in Man* (1799). The questions of the unity or plurality of human origins (monogenesis or polygenesis) touched on questions of religion, politics, particularly the legality of slavery and later, immigration policy, as well as science.

The 19th and 20th century classifications were marked by debate on the formal or adaptive character of races, whether there were fixed characteristics that separated the races, as a result of separate origin or of adaptation, or whether human differentiation was a result of continuing adaptive processes affecting genetic and morphological characteristics differentially so that major taxonomic distinctions of *Homo sapiens* into subspecies or races were not useful for scientific thinking and work. The main expression of this debate was the question of types, their validity, and their use. The question of "pure" racial types arose as part of the debate on typology, and variations of the question of purity, miscegenation or mongrelization, race preservation, and dilution became scientific and popular issues. In 1899, Wm. Z. Ripley used cranial data in *The Races of Europe* in which he distinguished the Nordic, Alpine, and Mediterranean races, each characterized by a definite head shape, other morphological characteristics, and particular mental and temperamental traits. Modern classifications include that of Deniker (1900) of six categories, which, for example, defined Europeans as having fair, wavy or straight hair and light eyes, and that of E. A. Hooton (1946) who suggested three primary races (White, Negroid, and Mongoloid) and four other general kinds of races (primary subrace, composite race, composite subrace, and residual mixed types), each comprising various numbers of races. The typologies of Deniker and Hooton indicate some of the problems of racial typologies. Typologies as classificatory systems tend to be rigid. Accommodation of variant individuals in the schema often is difficult, resulting in increasing numerous and complex sets of categories. The basic problems of establishing a typology become more difficult (e.g., which criteria are fundamental, which secondary, the number of criteria that should be used) and are all questions that cannot be determined in the absence of a theory. Cranial measurement and brain size became important criteria of differentiation. Whatever criteria were given prominence, none could be consistently applied and would differentiate all humans into categories. Neither could pure types be found, nor could a single or even a few characteristics be specified that would allow discrete categorization of all humans.

S. M. Garn has offered the most recent classification based on three categories: geographical races, which are defined as "a collection of populations whose similarities are due to long-continued confinement within set geographical limits"; local races, which "correspond more nearly to the breeding populations themselves . . . isolated by distance, by geographical barriers or by social prohibitions, . . . are totally or largely endogamous"; and micro races, which are related to "very real differences in the genetic makeup of cities and continual changes in the frequencies of genes," which create regional differences. Garn names nine geographical races: American, Polynesian, Micronesian, Melanesian, Australian, Asiatic, Indian, European, and African. He names 32 local races but does not name micro races, although he suggests they may be defined in terms of cities. This extension of the concept *race* to all magnitudes of populations, based on the concept of breeding, can be seen in retrospect as an attempt to preserve the concept by making it a synonym for population.

II. Evolutionary Aspects of Population Differentiation

The failures of the attempts to establish a typology of human differentiation based on morphology were compounded as genetic and molecular information was brought to bear on the problems of human differentiation. It came to be recognized that what had been described as *racial characteristics* were adaptations to environments, or the result of other evolutionary phenomena, drift, migration, and admixture. Black skin of Australians and Africans, for example, is a similar bodily feature that evolved independently, these populations being, in terms of their histories as shown by genetic research, as far apart as possible. Garn, in the same work in which he proposed his nine geographical races, acknowledged that "the distinctive characteristics of every race may now be understood in terms of the special environments in which they have lived." The confirmation of the variation of genes with environment weakened significantly any concept of absolute distinctiveness of populations within the human population, any notion of separateness between subpopulations in the human species. Moreover, to continue to define characteristics that were recognized to be adaptations as *racial* begs the question.

The study of variations of single characteristics, by use of clines, maps of the spatial variation in morphology, gene frequencies, or other biological traits, has demonstrated that these characteristics vary more gradually than do linguistic, or cultural, or "racial" definitions of groups would lead us to believe. Although he did not use the term *cline*, G. E. Morant, in his *Races of Europe*, demonstrated that morphological differences were gradual and less than linguistic and cultural differences on which the Nazis based their irredentist and racial supremacist claims. The study of variation by plotting gene frequencies as clines has opened the study of human variation to evolutionary theory. An interesting example of polymorphism is the presence of tawny hair among Australian aborigines as children, which is a maximum of 90% in the western Australian desert, declining from this center. Fig. 1 illustrates the clinal distribution of tawny hair among Australian aborigine children.

Other genes are known to be indicative of local environments, notably those that determine the diseases thalassemia and sickle cell anemia, and to vary in ways well-represented in clinal distributions. Both thalassemia and sickle cell disease, although strongly associated with certain regions (the Mediterranean for the former and Africa and India for the latter), vary independently of "racial" categories. Such variations are examples of microevolution and suggest emphatically that one characteristic cannot be used to define a subspecies or races. Africans are more likely to have sickle cell disease,

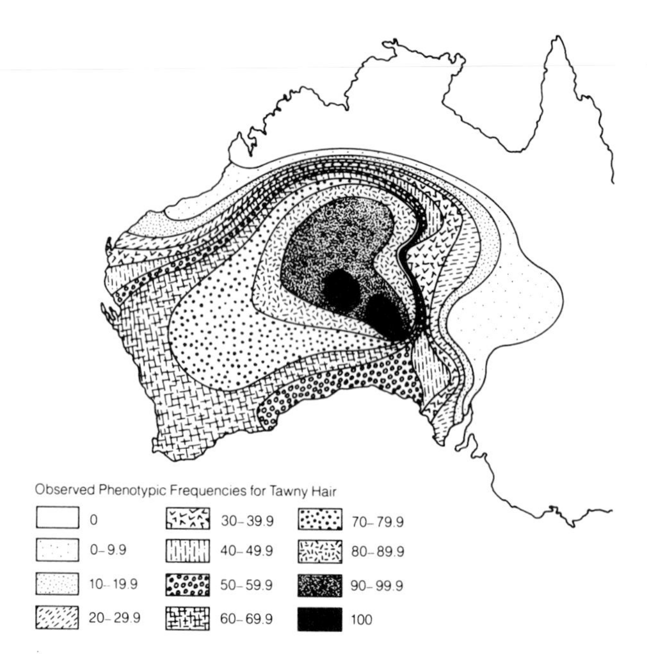

FIGURE 1 Gene distribution, tawny hair in Australia. Clines illustrate the decreasing gradient in distribution of the concentration of tawny hair.

but the disease is a result of a combination of cultural, environmental, and genetic factors. It has evolutionary advantage, resistance to malaria, when an individual is heterozygous but is fatal for homozygous individuals. African environments led to the evolution of certain morphological features; some African environments led to the evolution of the sickle cell gene. The association is real, but the processes are separate.

Multivariate studies have been used to test directly differentiation due to race. Leowontin began with the classification of major geographical races and local races. Using categories suggested as races by previous authors, he examined the degree to which these categories explained human diversity. His conclusion was that the populations, described in the literature as "races," explain little of human diversity. "The mean proportion of the total species diversity that is contained within populations is 85.4%, . . . Less than 15% of all human diversity is accounted for by differences between human groups! Moreover, the difference between populations within a race accounts for an additional 8.3%, so that only 6.3% is accounted for by racial classification." The continued accumulation of evidence on genetic variation continues to indicate the continuity of variation in the human species, rather than discrete, separate groupings.

Multivariate studies have also demonstrated other adaptive effects and have been used to test racial classifications. Studies have shown that gene frequency distribution varies with the advance of agriculture in Europe and with climate. Language change and differences have been shown to be associated with genetic variations and population differentiation. These studies strongly support the possibility that evolutionary change and its interdependence with cultural or political factors (e.g., language) determined genetic boundaries and forms of migration.

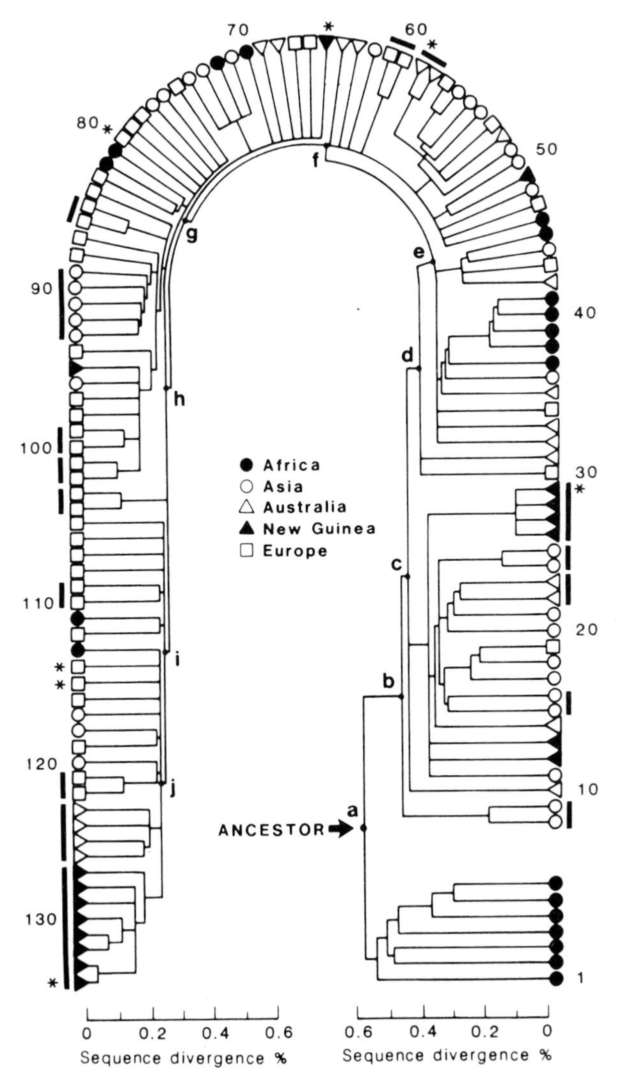

FIGURE 2 Genealogical tree for 134 types of human mtDNA (133 restriction maps plus reference sequence). The multiple origins of all populations represented except the African implies that each area was repeatedly colonized. On the basis of this data, *Homo Sapiens* thus has a common African origin, and present-day populations were created by repeated migrations from Africa, replacing any earlier populations.

III. Population History

Recent research has been establishing increasingly sophisticated measures of population differentiation and has been suggesting hypotheses to explain differentiation. New techniques offer answers to particular aspects of the questions of the origin, or evolution, of modern humans and of the factors that created and maintain the differentiation of the present human population. The study of the population history of humankind is an area of active and extremely interesting and often controversial research.

A. Mitochondrial DNA

Competing claims are made for the evidence from studies of mitochondrial DNA (mtDNA). The original research drawn from five geographical populations demonstrates both the unity and likely African origins of the human species and the diversity (i.e., multiple origins) of all the populations except the African. Fig. 2 shows the divergence of mtDNA of 147 people from five geographical populations. The distribution of members of these geographical populations across the dendogram indicates the biological diversity within geographical populations and the difficulty in attributing bases of unity to them other than place. This study indicates that the early migrations of *Homo sapiens* from Africa was not a continuous radiation, but multiple movements, not unique in direction. A reinterpretation of this data challenges the idea of unique origins and suggests the evolutionary origins of modern humans in three or more regions. This argument is known as "the regional transition hypothesis."

B. Geography, Language, and Culture

Spatial and cultural variables have been used to suggest answers for the question of origin and of population differentiation. Linguistic, archaeological, and genetic evidence has allowed a classification of human language phyla and genetic clusters that suggests six or seven main groups. This study suggests genetic groupings and language families have similar patterns of divergence and form similar genealogical trees. This classification is compatible with a common origin of humankind.

The study of prehistoric migrations from Asia to Melanesia and the Americas, the spread of agriculture in Europe, and the migrations and separations of culturally distinct groups have been aided by genetic comparisons. For Europe and New Guinea, language boundaries have been shown to coincide with boundaries showing sharp genetic change. The emphasis in these studies is on the degree of morphological and genetic differentiation and the explanation for the maintenance of these differences, differences that depend on isolation allowing differentiation and reducing gene flow, which would have reduced variability. Both language differences and geography are factors in creating isolation and, thus, differentiation. The relative importance of each and the exact relation to genetic variation is still to be established. It should be emphasized that in these studies linguistic and genetic divergence is not an exact match and boundaries do not always coincide, nor is change from group to group ordered in a particular way.

Variables affecting genetic change and differentiation in population include geography, gradual "demic" population movement and more rapid mass migration, language, and culture. Evidence of population separation includes DNA and non-DNA polymorphisms, dentition, and other paleoanthropological data. These studies can outline speculative histories of human evolution and of populations (e.g., estimated times of migration, population affinities), but the fundamental problems of analysis are as yet unresolved. Among these are questions of how to standardize measures of differences in frequencies of genetic distribution, linguistic differences and classification, and other characteristics (especially dentition; the rate of change of language, of DNA, rates of genetic drift and mutation; the effect of and rate of change of environmental change on gene frequencies). Empirical problems also exist: If, as is likely, small populations separated, genetic changes would have likely been more rapid than in large populations. Thus, the two populations would not only as expected diverge genetically, but would have done so at different rates. Broader issues are also in question: the shape and basis for deciding the form of the human genealogical diagram. Methodological problems caused by the wide range of possible error also limit accuracy of hypotheses of origin.

Studies attempting to trace population histories and origins will, nevertheless, continue to fascinate us as scientists and as human beings. Although the answers to these questions of origin are likely to remain speculative, the processes of population differentiation (i.e., differences in gene frequencies), it has become clear, are dependent not only on neutral processes, mutation, selection, gene flow, and genetic drift, but also on culturally determined marriage preferences, mass migrations, and gradual population expansions. The demonstration of cultural and spatial factors affecting genetic change and in creating and preserving population differences is extremely important in establishing the complexity and historical character of population differentiation.

IV. Meaning and Difference: Terminology and Politics

A. Changing Terminology in Anthropology

The use of the term *race* has carried with it the notion of fixed characteristics, much like the notion of fixity of species, which was problematic in the development of evolutionary theory. When morphological and anatomical characteristics were the variables of human differentiation and the data discontinuous, the possibility of discrete human subspecies could be entertained. Before archaeological and molecular biological data indicated the unity of the human species, multiple origins of *Homo sapiens* could be argued. The main characteristics of human differentiation, the continuity of change, its gradualness, and the relative lack of difference between groups, which continue to be demonstrated with the advances of research, indicate that race as a category of evolution or a major division in the human species is unsupported. The repeated attempts to demonstrate the separation of races by reduction of morphological characteristics to genetic differences has largely failed. After this work is considered, only the differences in bodily features and aboriginal geographical status (i.e., as of the 15th century) remain as the basis of racial classification.

The language of discussion of human differentiation and diversity in textbooks in physical anthropology has changed significantly. Chapter headings have changed from "Human Races" or "Races of Man," to "Human Diversity" or "Human Variability." Terms such as "racial characteristic" and "racial admixture" have either disappeared in favor of "bodily features" or been abandoned in recognition that they implied an identification between a race and a specific characteristic. The remaining defenses of the concept of race are common sense (i.e., based on the evidence of our eyes) or simply that such distinctions and concepts are useful and interesting. The alternative to discussion of population differentiation sometimes takes the form of describing *Homo sapiens* as a polytypic species. The scientists studying the evolutionary impact of language and geography use the terms *group*, *phyla*, *genetic entity*, *population cluster*, and the most general, *population*.

B. Use of Biology for Political Purpose

The use of the term *race* has often implied unbridgeable differences between individuals. The attempts to demonstrate differences between racial groups have been used as the basis of beliefs about the superiority and inferiority of racial groups (i.e., racism). Racism has led to the implementation of policies and practices of discrimination and genocide in the 20th century. Racism depends on the belief that morphology or bodily features allow conclusions about behavior, intelligence, and morality, that in some way appearance and behavior derive from a common, in this case, racial, source. Despite scientific evidence to the contrary, people cling to these prejudices. What is described as biological is often cultural, as is the belief of superiority. [*See* STEREOTYPES AND PREJUDICE.]

Modern research makes it increasingly clear that human populations have had differing histories but that at the level that racial differences have been noted, the differences do not establish separate population groups. The arbitrary and general concept population is useful for scientific purposes as its limits and definition can be specified, whereas race, not only because of its political implications, is not. Moreover, group membership and individual identification is not biological, but cultural.

Bibliography

Ammerman, A. J., and Cavalli-Sforza, L. L. (1984). "The Neolithic Transition and the Genetics of Populations in Europe." Princeton University Press, Princeton, New Jersey.

Bateman, R., Goddard, I., O'Grady, R., Funk, V. A., Mooi, R., Kress, W. J., and Cannell, P. (1990). Speaking of forked tongues: The feasibility of reconciling human phylogeny and the history of language. *Curr. Anthropol.* **31,** 1–24.

Birdsell, J. B. (1985). "Human Evolution," 3rd ed. Houghton Mifflin, Boston.

Cann, R. L., Stoneking, M., Wilson, A. C. (1987). Mitochondrial DNA and human evolution. *Nature* **325,** 31–36.

Cavalli-Sfroza, L. L., Piazza, A., and Menozzi, P. (1988). Reconstruction of human evolution: Bringing together genetic and archaeological, and linguistic data. *Proc. Natl. Acad. Sci. U.S.A.* **85,** 6002–6006.

Garn, S. M. (1971). "Human Races," 3rd ed. Charles C Thomas, Springfield, Illinois.

Gossett, T. F. (1963). "Race: The History of an Idea in America." Southern Methodist University Press, Dallas, Texas.

Gould, S. J. (1979). "The Mismeasure of Man." Norton Publishers, New York.

Greenberg, J. H., Turner, C. G., II, and Zagura, S. L. (1986). The settlement of the Americas: A comparison of the linguistic, dental and genetic evidence. *Curr. Anthropol.* **27,** 477–496.

Kirk, R., and Szathmary, E., eds. (1985). "Out of Asia: Peopling the Americas and the Pacific." The Journal of Pacific History, Canberra.

Lewontin, R. C. (1972). The apportionment of human diversity. *Evolutionary Biol.* **6,** 381–398.

Menozzi, P., Piazza, A., and Cavalli-Sfroza, L. L. (1978). Synthetic maps of human gene frequencies in Europeans. *Science* **201,** 786–792.

Nelson, H., and Jurmain, R. (1985). "Introduction to Physical Anthropology," 3rd ed. West Publishing, St. Paul, Minnesota.

Piazza, A., Mennozzi, P., and Cavalli-Sfroza, L. L. (1981). Synthetic gene frequency maps of man and selective effects of climate. *Proc. Natl. Acad. Sci. U.S.A.* **78,** 2638–2642.

Sokal, R. R., Oden, N. L., Legendre, P., et al. (1990). Genetics and language in European populations. *Am. Naturalist* **135,** 157–175.

Spuhler, J. N. (1988). Evolution of mitochondrial DNA in monkeys, apes and humans. *Yearbook Physical Anthropology* **31,** 15–48.

Washburn, S. L. (1963). The study of race. *Am. Anthropologist* **65**(3, Pt. 1), 521–531.

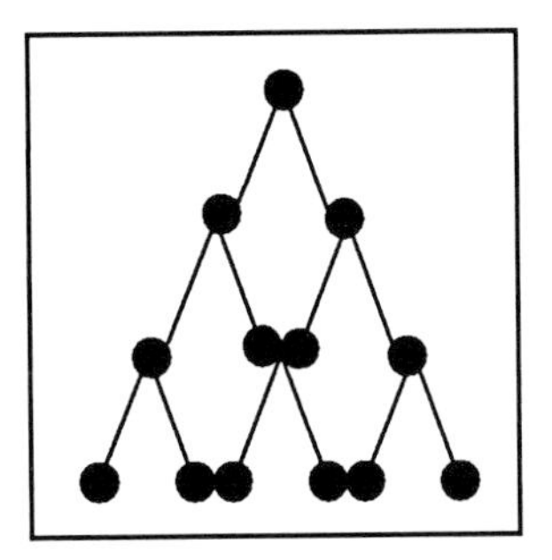

Population Genetics

RICHARD LEWONTIN, *Harvard University*

Glossary

Allele frequency Relative proportion of all genomes in a population that carry a particular allele at a locus

Assortative mating Mating in which the choice of partner is based on some genetic, social, or geographical criterion. It may be positive, in which like mates with like, or negative, in which there is an avoidance of similar mates

Genetic drift Random fluctuation in allele frequency at a locus arising from sampling error in breeding of a finite population

Natural selection Differential survival and reproduction of genotypes as a result of functional and anatomical differences in the organisms that carry them

Polymorphism Presence in a population of more than one allele at a locus, so that the frequency of the most common allele is not more than 99%

Reproductive fitness Expected number of offspring produced by individuals of a given genotype including the probability that the individuals survive to reproductive age

INDIVIDUAL HUMAN beings differ from each other genetically, as do geographical populations and ethnic and linguistic groups. Population genetics is concerned first with the description of that individual and group variation, second with the population forces that are acting to maintain or alter that variation at the present time, and third with the reconstruction of the forces in the past that have resulted in the current genetic composition of the human species. Because the human species is better known biologically and historically than any other species, human population genetics also provides models for the study of genetic variation in living organisms in general.

I. Individual Variation

Surveys of the genotypes of individuals within human populations have shown an immense variation at the level of individual nucleotide positions, genes, gene complexes, whole chromosomes, and genomes of cell organelles (e.g., mitochondria). No two individuals, even identical twins, are genetically identical because mutational errors in DNA replication are sufficiently frequent to guarantee that base pair differences will exist between the genomes of daughter cells of a single cell division. So, for example, it two genomes are surveyed for nucleotide differences by restriction endonucleases, which cut DNA at specific well-defined sequences, they will differ from each other at about one in every 200 base pairs. The functional significance of these differences is unknown because the restriction sites are not localized to particular genes nor to particular base positions within codons. Much more revealing surveys of genetic variation come from studies of particular genes and gene complexes such as those specifying enzymes, blood group antigens, histocompatability systems, and specific developmental features. [*See* GENES; GENOME, HUMAN.]

Surveys of more than 100 enzyme-coding loci, using gel electrophoresis to detect amino acid substitutions caused by mutations, have shown that approximately one-quarter are polymorphic within a major racial group and one-third are polymorphic within the species as a whole. A locus is classified as polymorphic if the most common allele at the locus has a frequency of 0.99 or less. Another measure of genetic diversity of a population is the proportion of heterozygotes (i.e., in which the two genes of a pair are not identical) averaged over loci. For a typical human population, this proportion is roughly 0.065 for enzyme-coding loci. Both in the proportion of loci polymorphic and average heterozygosity, humans are typical of vertebrates in general.

In some cases one allele at a polymorphic locus is frequent, but in others there are two or more allelic forms with roughly equal frequencies. Table I shows the frequencies of the homozygotes (in which the two genes of the same pair are identical) and of the heterozygotes for the 15 most polymorphic enzymes in the English population. Polymorphisms of a similar degree are found for red cell antigens. Of 33 known red cell antigen genes, about one-third are polymorphic with an average heterozygosity of 0.16. Table II shows the allelic frequencies at the polymorphic blood group loci in the English population.

By far the most extraordinary polymorphism in human populations is for the HLA gene complex of tightly linked loci specifying histocompatability proteins. In Caucasians, for example, there are 14 alternate alleles at the A locus, 17 alleles at the B locus, 6 at the C, and 8 at the locus. The alleles at all the loci are generally in intermediate to low frequency, and none of the loci has a very common form (usually referred to as "wild type").

The consequence of the large amount of genic polymorphism is that the probability that two individuals chosen from a European population are genetically identical, considering only the 35 most polymorphic loci, is about 8×10^{-14}. Nor do these highly polymorphic loci exhaust the known variation. About two per 1,000 individuals carry a rare variant, recognizable electrophoretic allele for some enzyme-coding locus. "Inborn errors of me-

TABLE I Frequencies of Various Enzyme Variants Found in the English Population

	Homozygotes			Heterozygotes		
Enzyme	1	2	3	1/2	2/3	1/3
Red cell acid phosphatase	0.13	0.36	0	0.43	0.05	0.03
Phosphoglucomutase 1	0.59	0.06	—	0.35	—	—
Phosphoglucomutase 3	0.55	0.07	—	0.38	—	—
Placental alkaline phosphatase	0.41	0.07	0.01	0.35	0.05	0.12
Peptidase A	0.58	0.06	—	0.36	—	—
Adenylate kinase	0.90	0.01	—	0.09	—	—
Adenosine deaminase	0.88	0.01	—	0.09	—	—
Alcohol dehydrogenase 2	0.94	—	—	0.06	—	—
Alcohol dehydrogenase 3	0.36	0.16	—	0.48	—	—
Glutamate-pyruvate transaminase	0.25	0.25	—	0.50	—	—
Esterase D	0.82	0.01	—	0.17		
Malic enzyme	0.48	0.09	—	0.43	—	—
Phosphoglycolate phosphatase	0.68	0.03		0.29		
Glyoxylase 1	0.30	0.21		0.49		
Diaphorase 3	0.58	0.05		0.36	—	—

TABLE II Blood Type Frequencies in the English White Population

System	Type	Frequency
ABO	A	0.447
	B	0.082
	AB	0.034
	O	0.437
MNS	MS	0.201
	Ms	0.093
	MNS	0.260
	MNs	0.236
	NS	0.060
	Ns	0.149
Rh	r	0.147
	R_1	0.535
	R_2	0.150
	R_1R_2	0.129
	R_0	0.022
	R′	0.011
	r″	0.006
P	P_1	0.266
	P_1P_2	0.499
	P_2	0.234
Secretor	Se^+	0.773
	Se^-	0.227
Duffy	Fy^a	0.177
	Fy^aFy^b	0.462
	Fy^b	0.301
Kidd	Jk^a	0.583
	Jk^aJk^b	0.361
	Jk^b	0.056
Dombrock	Do^a	0.664
	Do	0.336
Auberger	Au^a	0.857
	Au	0.143
Xg	Xg^a	0.894
	Xg	0.106
Sd	Sd^a	0.912
	Sd	0.088
Lewis	Le^a	0.224
	Le	0.776

tabolism," in which some enzymatic function is deficient or absent, are usually the consequence of homozygosity for a low-frequency mutant allele at a locus and thus are individually rare. Heterozygotes for these alleles are much more common, however. In a sample from the United States examined for 14 inherited disorders, it was estimated that 13% of newborns were heterozygous for a mutant allele. It is clear that all human individuals are heterozygous for scores of mutations that would be deleterious in homozygous condition. [*See* GENETICS, HUMAN.]

II. Variation Between Groups

Human geographical populations are obviously differentiated for some genetically influenced traits (e.g., skin color hair, form, facial characters, and height). These are not typical of all genes, however. For two-thirds of the human genome estimated to be monomorphic, all human individuals, irrespective of geographical population, are genetically identical with the exception of those individuals carrying rare mutant alleles. For the polymorphic one-third of the genome, the degree of differentiation between populations varies. Table III shows the three most and three least geographically differentiated protein-coding loci known. Even for the highly differentiated Duffy, Rhesus, and P loci, there is no allele that is 100% in one racial group but absent in another, although there are large frequency differences. Currently, 17 loci have been studied in a sufficiently wide geographical sample to make an estimate of the relative variation between and within groups. Table IV shows for each locus the proportion of the heterozygosity that occurs within local ethnically and linguistically homogenous populations (i.e., Kikuyu or French or Japanese), between such populations within major geographical race divisions (Black Africans, Caucasians, Asians, Oceanians, etc.), and between these races. Although there is some variation from gene to gene, on the average 85% of human genetic diversity occurs between individuals within a local population and the remaining 15% is split roughly evenly between population differentiation within major races and major racial differentiation.

TABLE III Examples of Extreme Differentiation and Close Similarity in Blood Group Allele Frequencies in Three Racial Groups

Gene	Alleles	Caucasoid	Negroid	Mongoloid
Duffy	Fy	0.0300	0.9393	0.0985
	Fy^a	0.4208	0.0607	0.9015
	Fy^b	0.5492	—	—
Rhesus	R_0	0.0186	0.7395	0.0409
	R_1	0.4036	0.0256	0.7591
	R_2	0.1670	0.0427	0.1951
	r	0.3820	0.1184	0.0049
	r′	0.0049	0.0707	0
	others	0.0239	0.0021	0
P	p_1	0.5161	0.8911	0.1677
	P_2	0.4839	0.1089	0.8323
Auberger	Au^a	0.6213	0.6419	
	Au	0.3787	0.3581	
Xg	Xg^1	0.67	0.55	0.54
	Xg	0.33	0.45	0.46
Secretor	Se	0.5233	0.5727	
	se	0.4767	0.4273	

III. Forces Controlling Variation

Given the observed genetic variation within and between populations, how is the amount and nature of this variation to be explained? Genetic variation within a species is created, maintained, and altered by a combination of six forces: mutation and recombination, which create variation, and assortative

TABLE IV Proportion of Genetic Diversity Accounted for Within and Between Populations and Races

Gene	Within Races: Within Populations	Within Races: Between Populations	Between Races
Hp	0.893	0.051	0.056
Ag	0.834	—	—
Lp	0.939	—	—
Xm	0.997	—	—
Ap	0.927	0.062	0.011
6PGD	0.875	0.058	0.067
PGM	0.942	0.033	0.025
Ak	0.848	0.021	0.131
Kidd	0.741	0.211	0.048
Duffy	0.636	0.105	0.259
Lewis	0.966	0.032	0.002
Kell	0.901	0.073	0.026
Lutheran	0.694	0.214	0.092
P	0.949	0.029	0.022
MNS	0.911	0.041	0.048
Rh	0.674	0.073	0.253
ABO	0.907	0.063	0.030
Mean	0.854	0.083	0.063

mating, natural selection, random genetic drift, and migration, which modulate that variation.

A. Mutation

Ultimately all genetic variation comes from the random errors that occur in DNA replication either spontaneously or as a result of chemical and physical mutagens. There are no really good estimates of spontaneous mutation rates per gene in humans, but if they are typical of mammals, a rough estimate of 10^{-5} per gene (approximately 10^{-9} per nucleotide) per generation can be used. Because the rate of increase of an allele in a population from repeated mutations alone is equal to the mutation rate times the frequency of the allele *from* which it is mutating, mutation is a weak force in changing allele frequencies. The human species as a whole is probably not more than 250,000 years old, and present major geographical races probably diverged about 35,000 years ago. Given a human generation time of 25 years, only 10,000 generations have elapsed since the species was formed and 1,500 generations since the divergence of the races. Then, by mutation alone, an allele could only have increased in frequency by a maximum of 0.10 since the founding of the species or 0.015 since the divergence of the races. Clearly other forces must operate on the mutant genes as they arise. [*See* MUTATION RATES.]

B. Recombination and Sexual Reproduction

In the absence of sexual reproduction, genetic diversity among individuals would depend on the serial occurrence of different mutations in a given ancestral line. Sexual reproduction and the consequent recombination of genetic material from the maternal and paternal lines allow historically independent mutations to be combined in the same individuals, giving rise to vastly greater total variation and allowing variants that have arisen in different lines in the recent past to appear in a single genome. One consequence of nuclear genome recombination is that separate family lines do not preserve a trace of their ancestral history because they are constantly mixing. Because there is no recombination in mitochondria, however, which are entirely maternally inherited, it is possible to make reconstructions of their ancestral history and so trace the common ancestry of individuals and groups.

C. Natural Selection

Different genotypes may have different physiologies and morphogenesis as a consequence of different gene products. In turn, these physiological and anatomical differences may cause differences in the fertility and probability of survival of the different genotypes. In such a case, allele frequencies will change within the population. It is by no means certain, however, that genetic variation will have such effects on reproductive fitness. Many nucleotide changes occur in intergenic regions and in silent positions in codons and may have no effect at all on the organism. Changes in introns may be selectively neutral but, if they occur at splice junctions, may prevent normal protein synthesis. This is the case for the *thalassemias,* a collection of mutations that interfere with normal globin synthesis because they interfere with exon splicing. Even if nucleotide changes cause amino acid substitutions in proteins, these substitutions do not necessarily affect the biochemical activity or structural role of the protein. Moreover, even if physiological or anatomical changes occur, some may be irrelevant to differential reproduction or survivorship during reproductive ages. Genetic changes that influence the probability of survivorship during middle or old age have little or no effect on offspring production. Finally, genetic changes that do affect reproduction and survival may do so in a way that is sensitive to environment so that fluctuating environments may result in no clear trend in allelic frequency change. In sum, it cannot be assumed that a genetic variation will be subject to natural selection in the absence of clear experimental evidence on reproductive fitness. For example, attempts to find different reproductive rates among different blood group genotypes have failed, except for the fetal death associated with Rh incompatability. Various statistical associations have been claimed between blood groups or HLA genotypes and various diseases (e.g., the association between ABO blood types and peptic ulcer). But none of these associations, if real, have been shown to result in different reproductive rates. Do people who eventually get peptic ulcers leave fewer (or more) children? It has been claimed, for example, that darker-skinned populations occur where higher ultraviolet irradiation would cause skin cancer in light-skinned people, whereas very light skin in higher latitudes is an adaptation to making Vitamin D with low ultraviolet irradiation. But it has never been shown that Finns would get a significant amount of skin cancer in Africa or that Africans would suffer from rickets in Finland or that, even if they did, a difference in reproductive schedules would follow. Natural selection has virtually never been demonstrated in the human species except for clearly deleterious genes, which are generally rare. Natural selective explanations for major polymorphisms are lacking. The only exception is the sickle cell polymorphism, in which the protection that heterozygotes have from malaria is established and does explain the maintainance of the sickling allele in the population (discussed below).

There are essentially three forms of natural selection that may be relevant to human genetic variation. In the first form, the loss of reproductive fitness caused by an allele is either recessive or partly manifested in the heterozygote. If there are repeated mutations from the normal allele to the deleterious form, a, an equilibrium between the loss of the allele because of selection and the recruitment of the allele from new mutations will occur such that the frequency, q, of the deleterious allele will be

$$q = \sqrt{\mu/(1 - w_{aa})}$$

where μ is the rate of mutation to the deleterious allele and w_{aa} is the reproductive fitness of the deleterious homozygote. If there should be any deleterious effect in heterozygotes, homozygotes are so rare as not to make any difference to the result and

$$q = \mu/(1 - w_{Aa})$$

where w_{Aa} is the fitness of heterozygotes. In either case the equilibrium frequency of the allele will be low. For example, if homozygotes die before reproduction, $w_{aa} = 0$ and

$$q = \sqrt{\mu}$$

which for realistic mutation rates will be smaller than 1%. Because homozygotes only occur at a frequency q^2, the frequency of affected persons in the population will only be one in 10,000 for this lethal gene. Presumably the inborn errors of metabolism owe their characteristically low frequencies to this mutation–selection balance.

A second form of selection, in which heterozygotes between two alleles have a higher reproductive fitness than either homozygote, leads to a stable intermediate frequency of the alleles independent of the mutation rate. The frequency of the allele a will be

$$q = \frac{w_{AA} - w_{Aa}}{w_{AA} + w_{aa} - 2w_{Aa}}$$

where w is the reproductive fitness of the genotypes *aa, Aa,* and *AA*. Sickle cell anemia conforms to this prediction. Homozygotes for the sickling gene, *SS*, usually do not survive to adulthood ($w_{SS} = 0$), because of their severe anemia. In West Africa where there is a high frequency of falciparum malaria, homozygotes for the normal hemoglobin A die of malaria at a moderate rate, whereas heterozygotes seem to be totally protected against malaria ($w_{SA} = 1$). The observed incidence of the sickle cell allele is about 10%, which would be predicted from the formula for equilibrium if the malarial death rate was about 10% ($W_{AA} = 0.9$). A confirmation of this theory is that the frequency of the allele S has decreased among the descendants of West African slaves in North America at a rate more than twice that expected from the rate predicted simply from the interbreeding with Caucasians.

A third form of selection occurs when the heterozygote has a *lower* reproductive fitness than either homozygote, as in the case of the maternal–fetal incompatibility for Rh alleles. In such cases, an unstable equilibrium is expected and one or the other of the alleles should become fixed in the population, depending on the initial frequency of the alleles. This has not happened in human populations, all of which are polymorphic for Rh alleles in what appears to be a long-term polymorphism predating the present geographical divergence of human groups. The explanation of this contradiction remains unknown.

The failure to find satisfactory selective explanations for the widespread polymorphisms in the human species means that reproductive differentials between genotypes are small, about 1% or less. To detect such selection, age-specific mortality and fertility schedules would have to be compiled for extremely large samples of individuals of different genotypes, on the order of hundreds of thousands.

D. Genetic Drift

In a population of finite size, the genetic composition of the population will not be reproduced exactly from generation to generation but will be subject to sampling error because each generation is a sample of the gamete pool of the previous generation. The smaller the population size, N, the greater this sampling error. As a consequence, even in the absence of any force of selection or mutation, allele frequencies in a population will fluctuate from generation to generation, the fluctuations being inversely proportional in size to the population number. This process of random change of allele frequencies is *genetic drift*. The consequences of genetic drift are

1. New mutations as they arise are almost always lost within a few generations. The probability of loss is $(2N\text{-}1)/2N$. Even a slightly favorable new mutation, with selective advantage s, has a probability of only $2s$ of being retained. Thus, new favorable mutations must occur over and over again before they are incorporated.
2. Polymorphisms that are not being maintained by selection within populations will be lost as one of the alleles randomly drifts to fixation at a frequency of 100%.
3. Different populations will diverge from each other in genetic composition as one allele increases in frequency in some populations and decreases in others. So, the variation that is seen in allele frequencies of polymorphic loci in human populations may simply be the result of random divergence since the separation of the populations.

Eventually some alleles will be totally lost from some populations, and this process is accelerated if new populations are founded by only a few individuals or pass through a severe bottleneck in population size. This is the probable explanation, for example, for the absence or very low frequency of the B blood type in American Indians who presumably came over the Bering Strait in small numbers during the last glacial period and then were isolated from the ancestral populations in Asia when the ice melted. Eventually, if populations remain isolated from each other for a long enough period, they may become totally differentiated at a locus, with no alleles in common. This has not happened for any genes in the human species, presumably because human populations have not been totally isolated from each other for long periods. Again, it must be remembered that the major geographical races are only about 1,500 generations old.

E. Migration

Populations exchange genes as a consequence of migrations, invasions, and conquests. Such gene exchanges reduce the genetic differences between populations that have built up as a result of isolation or different selection in different environments. If two populations with the frequencies q_1 and q_2 of an allele exchange a proportion m of their members in one generation, the difference in their allele frequencies in the next generation is simply

$$(q_1 - q_2)' = (1 - m)(q_1 - q_2)$$

This linear relation can be used to estimate how much gene flow has occurred between two populations over a few generations, because the total migration will be roughly the sum of the one-generation rates. For example, the Duffy blood group allele Fy^a has a frequency of about 0.40 in Europeans and up to 70% in American Indians, but is absent in West Africans. Among blacks in New York the frequency is 0.081. Then, assuming that only Europeans had a significant genetic input into black populations, an estimate of the proportion of European ancestry in American blacks from New York is

$$m = 1 - \frac{(0.40 - 0.081)}{(0.40 - 0)} = 0.189$$

The estimates for other northern cities are similarly high (0.26 for Detroit and 0.22 for Oakland), but southern samples show much lower values (0.037 for Charleston, SC, and 0.106 for rural Georgia). These figures are the total migration of European genes into the black population since slaves were first introduced in large numbers about 10 generations ago. The estimates would be more difficult to make if blacks have a significant amount of American Indian ancestry, which they do in some regions.

G. Human Population Structure

Human populations are not isolated islands of randomly mating individuals that occasionally receive migrants from other populations, nor is the human species one large interbreeding unit. There is a hierarchical mating structure. At the local level, even within small villages and tribes, there is assortative mating by clan, religion, caste, and appearance. This is sometimes a positive assortment, as in the tendency to mate within religions, castes, and social classes, or there may be negative assortment, when there are taboos against mating within totem groups or families. Assortative mating by skin color is strong in the United States, with "cross-racial" marriages making up a fraction of 1% of all unions.

At the next level, there is selective mating by geographical locality. Despite a strong bias toward mating between individuals born within a small geographical radius, there are also traditions of marrying out to other villages and locales. There are also national and linguistic barriers that correspond to large geographical areas and finally major geographical separations by continent or widely spaced island groups. But none of these is absolute, and there is a continuous flow of migrants fluctuating in numbers depending on unique historical events. Because of asymmetrical social definitions of group membership, gene passage usually appears to be one-way. Any person with known black ancestry is described as "black" in the United States, so it is impossible to estimate the flow of genes from Africans into the "white" population.

In summary, human populations have a structure that allows genetic drift to cause local differentiation between castes, religions, regions, ethnic groups, etc., but that structure is not one of rigidly isolated islands of population of a fixed size. Rather, there is partial isolation by geographical and social distance, differing in degree from region to region and historical epoch to historical epoch.

F. Synthesis of Forces

Mutation, migration, selection, and genetic drift are all operating simultaneously, and the genetic diversity within and between populations is a consequence of opposing tendencies of the different forces. Mutation introduces variation into populations and in the short run causes differentiation between populations because the same mutations will not have occurred in all local groups. In the long run, however, recurrent mutation will give the same spectrum of changes for all populations. Migration increases the diversity within populations by mixing of local differentiated groups but decreases the differentiation between populations. Selection has different effects depending on the nature of the fitness differences. Selection in favor of heterozygotes maintains variation within populations. Directional selection favoring one genotype causes populations to become homozygous and makes all local populations genetically similar unless different alleles are favored under different environmental conditions. There is no direct evidence in human populations for directional selection of different alleles in different populations, although that may be the case for skin color or body fat. Selection against heterozygotes, as in the case of maternal–fetal incompatability, causes populations to become homozygous and to differentiate from each other. Finally, random genetic drift causes a loss of genetic variation within populations and a divergence between populations.

The importance of random drift as opposed to the deterministic directional forces of selection, migration, and mutation depends on the product of population size, N, and the parameters of the deterministic forces of selection (w), migration (m), and mutation (μ). Generally, if the product Nw or Nm or $N\mu$ is greater than 1, the deterministic force predominates and little random differentiation between populations will occur by drift. If the product is smaller than 1, however, the deterministic force is insufficient to prevent random differentiation. So, if local populations of Brazilian Indians are approximately a few hundred in size, selection intensities of less than 1% will be insufficient to determine the frequencies of polymorphic genes, and the populations will drift in their allele frequencies in a random way. If, however, such local populations exchange even one migrant individual per generation with each other ($m = 1/N$), $Nm = 1$ and they will not diverge randomly from each other, although they may diverge as a whole from some other more distant population.

The pattern of genetic variation within and between human populations is then the result of some combination of these diversfying and homogenizing

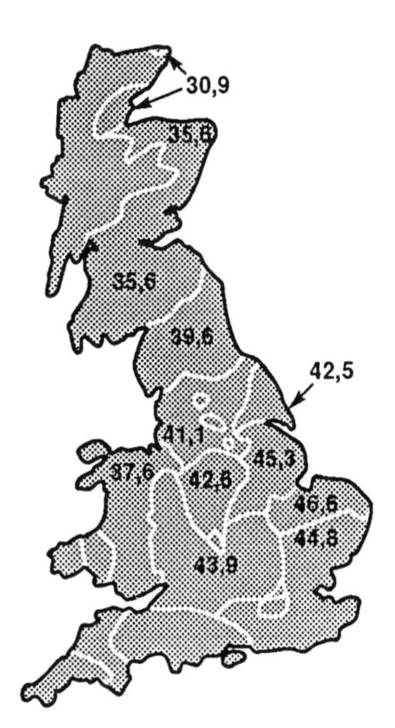

FIGURE 1 Cline of frequency of the blood type A in Britain. [Reproduced, with permission, from R. C. Lewontin, Human Diversity. *Sci. Am.*, p. 19.]

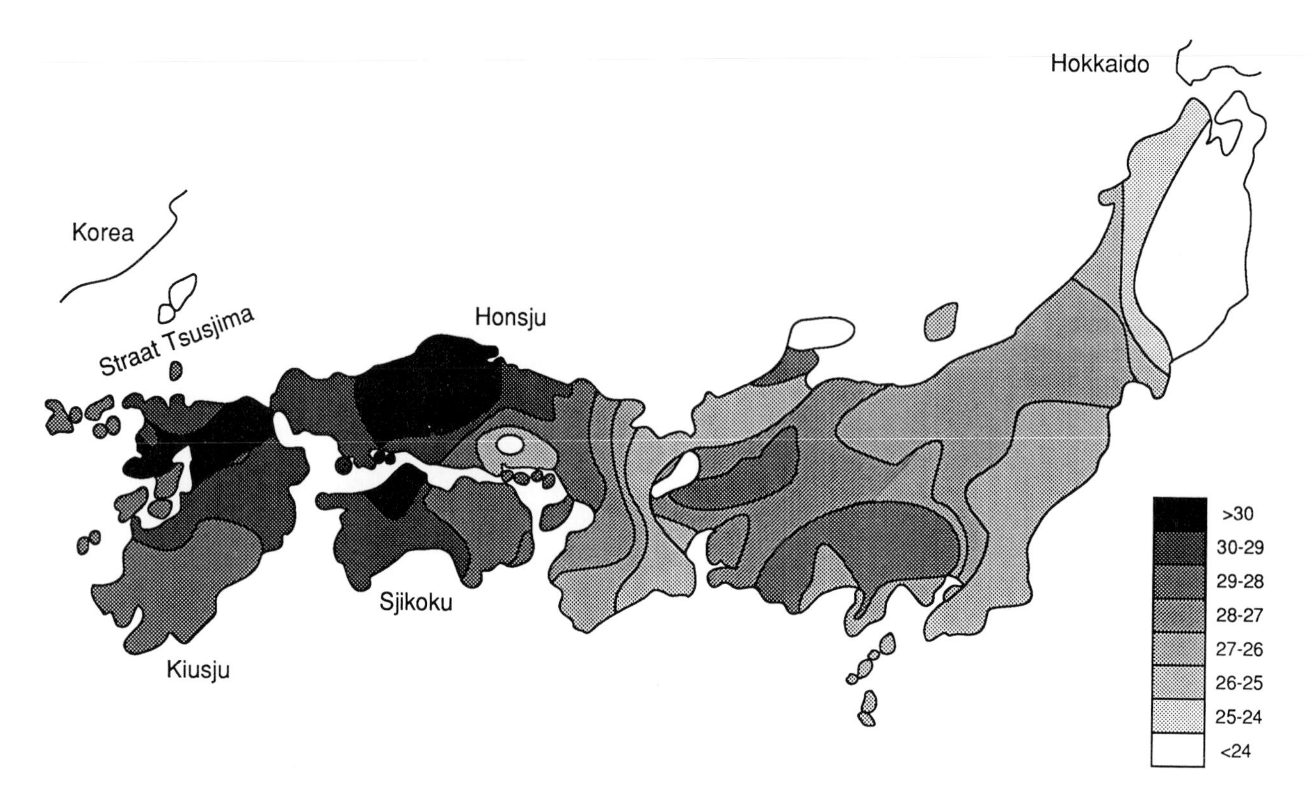

FIGURE 2 Cline of frequency of blood type A in the Japanese archipelago. [Reproduced, with permission, from R. C. Lewontin, Human Diversity. *Sci. Am.*, p. 19.]

forces. Unfortunately, it has not been possible to measure with sufficient accuracy the rates of selection, mutation, migration, and population size to make a clear assignment of the forces for most human variation. This problem is immensely complicated by the rapid historical changes that have occurred in population sizes, migration patterns, environmental conditions, and social structures in relatively few generations of human history.

IV. Reconstructing History

Although we do not know what forces were responsible for establishing the widespread polymorphisms in human populations or the original differentiation of geographical groups, it is clear that the forces of selection and random drift have not been strong compared with the role of migration during the Past 10,000 years. Geographical patterns of allele frequency distribution on national and continental scales show the clear traces of migrations and invasions whose histories are otherwise known from historical records. Figures 1 and 2 show examples of these genetic traces of migration. Figure 1 shows the percent of blood type A in Britain. The decreasing cline northward and westward from the original invasion sites of the Danes on the east coast (so-called Danelaw) reflects the high frequency of A among Scandinavians and the low frequency among the Celts and Picts of Scotland. Although not shown on the map, there is a similar low frequency of A in Ireland and in Liverpool, which is the chief port of entry into Britain from Ireland. Reciprocally, there is an island of high A frequency in far western Ireland as a consequence of a long-time garrisoning of British troops there.

Figure 2 shows a cline in A-type frequencies from the northeast to the southwest of the Japanese archipelago with a concentration of high A in Honshu and Kyushu bordering the straits of Tsushima. This corresponds to the known repeated invasions of Japan by mainland peoples of China through the closest point in Korea. The low frequency of A in the far north presumably is that of the aboriginal peoples (perhaps the modern-day Ainu) who mixed with the mainland invaders. What is extraordinary about these two cases is that the historically determined clines have persisted for more than 2,500 years despite the high mobility and urbanization of Britain and Japan. In general, in Europe, differ-

FIGURE 3 Clines of the first three principal components of allele frequencies in Eurasia, based on HLA and blood group polymorphisms. Each of the principle components if represented by a different color. The more intesne the color, the value of the principle component. [Reproduced, with permission, from P. Menozzi, A. Piazza, and L. L. Cavalli-Sforza. (1978). Synthetic maps of human gene frequencies in Europeans. *Science* **201,** 786.]

ences in allele frequency between populations is well-correlated with geographical distance but not with linguistic family differences.

The parallel between allele frequency clines and known history for small regions makes it likely that large continental clines could be used to trace large-scale migrations. Information from a large number of different polymorphic systems can be combined into multivariate statistical indices, the *principal components*, and when these are mapped for Eurasia, as in Fig. 3, there is clear evidence of a cline from central Asia fanning out into Europe paralleling the postulated spread of agricultural people between 9,000 and 5,000 years ago. The possibility of reconstructing human migration patterns from gene frequency patterns is a consequence of the relative weakness of selective forces and of the maintenance of local breeding populations over short periods of time so that history is not obscured by mixing.

Bibliography

Cavalli-Sforza, L. L., and Bodmer, W. F. (1971). "The Genetics of Human Populations." W. H. Freeman, San Francisco.

Harris, H. (1980). "Human Biochemical Genetics," 3rd ed. rev. Elsevier/North Holland, Amsterdam.

Lewontin, R. C. (1972). The apportionment of human diversity. *Evolutionary Bio.* **6,** 381.

Menozzi, P., Piazza, A., and Cavalli-Sforza, L. L. (1978). Synthetic maps of human gene frequencies in Europeans. *Science* **201,** 786.

Race, R. R., and Sanger, R. (1975). "Blood Groups in Man," 6th ed. Blackwell, Oxford.

Spiess, E. B. (1989). "Genes in Populations." John Wiley, New York.

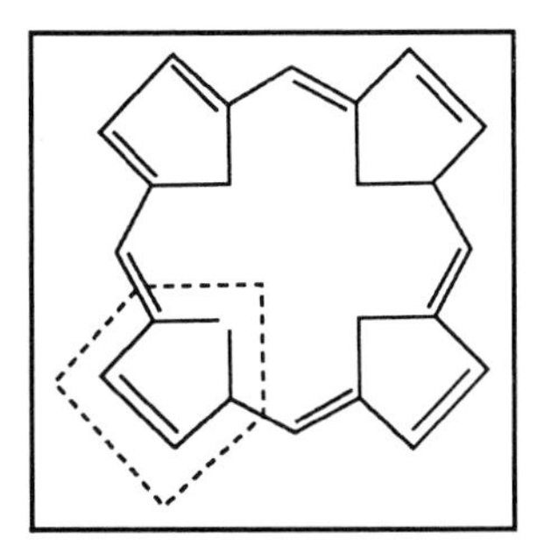

Porphyrins and Bile Pigments, Metabolism

DAVID SHEMIN, *Professor Emeritus of Biochemistry and Molecular Biology, Woods Hole, Massachusetts*

Glossary

Carboxyl group —COOH
Isomer Compound having the same percentage of composition and molecular weight as another compound but differing in chemical structure
Isotope Any of two or more atoms that have the same number of protons but different atomic weights, (e.g., ^{14}N has 7 protons and 7 neutrons whereas ^{15}N has 7 protons and 8 neutrons)

PORPHYRINS ARE heterocyclic compounds composed of four pyrrole rings linked to each other by four methene bridge carbon atoms (Fig. 1). These cyclic compounds are extremely stable, having been found in petroleum shale oil and in fossilized excrements millions of years old. The basic structures of porphyrins are found in the heme molecules of hemoglobin, myoglobin, cytochromes, catalases, and the like, in the chlorophyll structure of plants and photosynthetic organisms, and in the corrin ring of vitamin B_{12} (Fig. 2). The porphyrin ring of heme is linked by a coordinate-covalent bond to iron, whereas chlorophyll contains magnesium and the corrin ring is linked to cobalt. The heme proteins are concerned with oxygen transport (hemoglobin), oxygen storage (myoglobin), oxidation (cytochromes and peroxidases), and nitrogen fixation (leghemoglobin). Whereas heme is concerned primarily with the release of available energy, the role of chlorophyll is photosynthesis, by which solar energy builds up reducing and oxidizing potentials, so as to store chemical energy.

As discussed in Section I, the biosynthesis of porphyrins involves the formation of a "parent" porphyrin structure having an acetic acid (A) and a propionic acid (P) side chain in each of the B positions of each of the pyrrole rings. With an A and a P side chain in each of the pyrrole rings, one can construct four structural isomers. The parent structures of all biological functioning porphyrins or porphyrin-derived compounds have the following distribution of side chains, starting from ring A (Fig. 1): AP, AP, AP, PA; they are designated isomer III. In particular pathological or metabolic altered states, isomer I (AP, AP, AP, AP) is formed as well. The unique structures of the side chains eventually found in the functioning porphyrin (e.g., protoporphyrin of heme) arise by subsequent alteration of the A and P side chains of the parent isomer III; for example, the methyl side chains of protoporphyrin are derived by decarboxylation of the four acetic acid (A) side chains, and the two vinyl side chains (rings A and B) are derived by decarboxylation and oxidation of the propionic acid (P) side chains (Figs. 1 and 3).

I. Biosynthesis and Porphyrins

The first committed step in the synthesis of the porphyrin structure, and thereby in the synthesis of heme, chlorophyll, and the corrin ring, is the formation of an aminoketone, δ-aminolevulinic acid (ALA). In humans as well as other members of the animal kingdom and also in photosynthetic bacteria, the aminoketone is enzymatically synthesized from the simple amino acid glycine and from succinic acid with the participation of two coenzymes—coenzyme A and pyridoxal phosphate—which are derivatives of two essential human

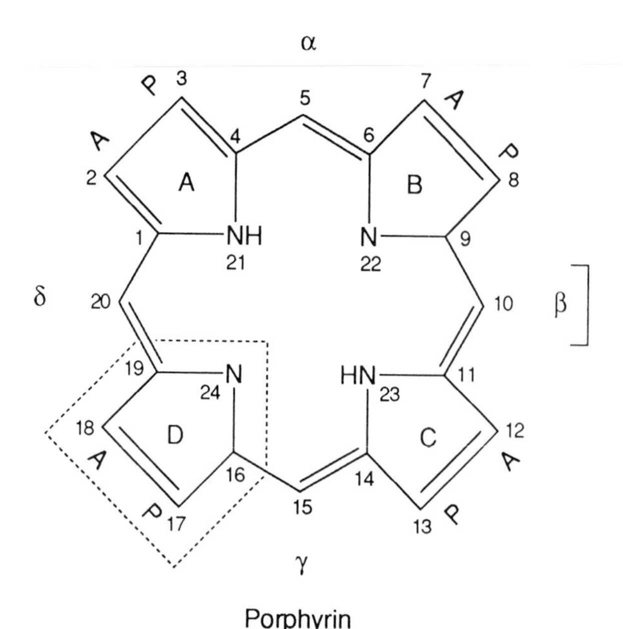

FIGURE 1 Uroporphyrin III. α, β, γ, δ, Methene bridges; Side chain abbreviations: A, $—CH_2COOH$; P, $—CH_2—CH_2—COOH$.

dietary compounds: panthothenic acid and vitamin B_6, respectively (Fig. 3). It was recently found that vertebrates have two separate genes that encode for the enzyme which catalyzes this reaction (i.e., δ-aminolevulinate synthase). The erythroid gene is expressed exclusively in the red blood cell, whereas the hepatic form of the enzyme is probably expressed ubiquitously. It is of further interest that plants, algae, and other microorganisms (e.g., *Cyanobacteria, Chromatium,* and *Methanobacterium*) synthesize the aminoketone from glutamic acid, an amino acid.

Subsequently, two molecules of ALA are enzymatically condensed to form a monopyrrole, porphobilinogen (PBG). The enzyme catalyzing this reaction (i.e., PBG synthase) is a zinc-containing enzyme markedly inhibited by the presence of lead. Four molecules of PBG are then condensed enzymatically with the participation of two enzymes (PBG deaminase and uroporphyrinogen III cosynthase) to form uroporphyrinogen III (reduced uroporphyrin III). (In the absence of cosynthase, the nonbiological functioning isomer I is formed.) The A side chains of uroporphyrinogen III are enzymatically decarboxylated to methyl groups to form coproporphyrinogen III and subsequently the P side chains of rings A and B are decarboxylated and dehydrogenated to vinyl groups to form protoporphyrinogen. The latter is oxidized to protoporphyrin, which is enzymatically converted to heme by the addition of iron. It is worth noting that the first step (in Fig. 3) and last three enzymatic steps occur in the mitochondria, whereas the intermediate enzymes are found in the cytosol. The rate-limiting step in the synthesis of porphyrins appears to be the synthesis of ALA, partially regulated by the concentration of heme. Chlorophyll and the corrin ring of vitamin B_{12} are synthesized from intermediates in the outlined scheme (Fig. 3).

II. Porphyrias

There are approximately seven well-defined human inborn errors in the metabolic process concerned with the synthesis of porphyrins. The genetic inheritance is usually autosomal dominant, except for congenital erythropoietic porphyria and ALA dehydratase deficiency porphyria, which are autosomal recessive disorders. Two of these disorders are briefly described here.

A. Congenital Erythropoietic Porphyria

Congenital erythropoietic porphyria is a rare disease in which the homozygote is characterized by an increase of porphyrins in the blood, urine, and feces. Since the abnormality lies in the deficiency of uroporphyrinogen III cosynthase, the excretion is that of uroporphyrin I and coproporphyrin I. The accumulation of the porphyrins in the tissues, bones, and teeth also causes the individual to be photosensitive, exposure to sunlight causing blisters which readily become infected. The bones and the teeth of individuals with this disorder are reddish. A similar disorder has been found in cattle and in the fox squirrel.

B. Acute Intermittent Porphyria

Acute intermittent porphyria is an autosomal dominant disorder in which the metabolic defect is in the deficiency of the enzyme porphobilinogen deaminase. Approximately 50% of the normal enzymatic activity is found in the subjects' tissues. Most of the subjects who inherit this genetic deficiency do not exhibit any clinical symptoms or biochemical abnormalities. However, when exposed to some drugs (e.g., barbiturates or sulfonamides) or some stresses, they accumulate PBG and ALA and ex-

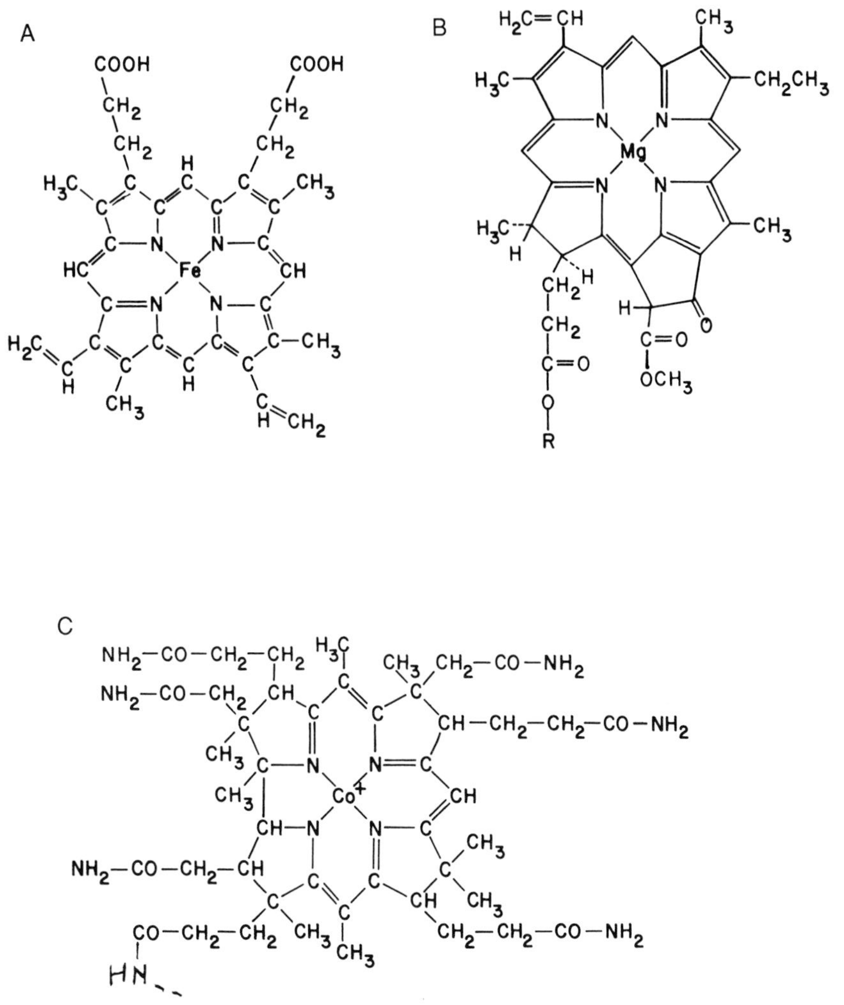

FIGURE 2 Structures of heme, chlorophyll, and the corrin ring of vitamin B_{12}.

press clinical symptoms. Along with marked increases in PBG and ALA in the urine, the subjects can have acute abdominal pains, neurological symptoms, paralysis, muscle weakness, and some abnormal behavioral patterns. It is of interest that, before being recognized as a metabolic disorder, many subjects, in the acute phase, have received appendectomies and even been assigned to mental institutions. The incidence of the disease is apparently more common in Scandinavia, South Africa, Great Britain and Lapland. The occurrence in some instances has been traced to the original individual carrying the defective gene. The injection of hemin (heme–Fe^{3+}), which inhibits ALA synthesis, appears to lessen the symptoms occurring in the acute phase.

III. Life Span of the Red Blood Cell

Knowledge of the pathway by which porphyrins are synthesized enables one to specifically label the heme of hemoglobin by the ingestion, for example, of glycine tagged with the stable isotope of nitrogen (^{15}N) and to determine the life span of the red blood cell under normal physiological conditions. By plotting the ^{15}N concentration of the hemin samples isolated from blood samples against time, one can determine the average life span of the red blood cell, since the hemoglobin in the cell is not in a dynamic state of synthesis and degradation. The average life span is found to be about 120 days (Fig. 4). This survival time of the red blood cell in patients with congenital porphyria or polycythemia vera is simi-

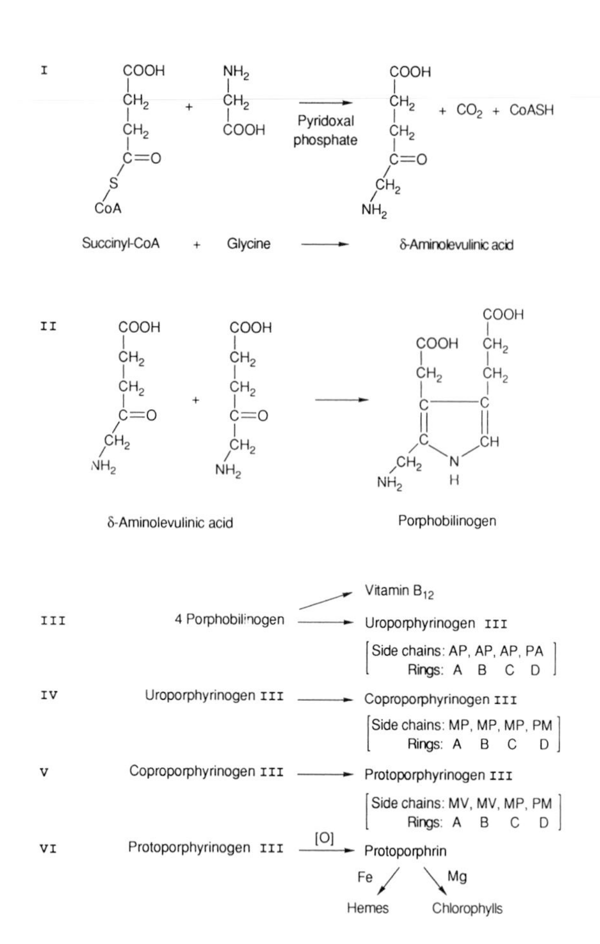

FIGURE 3 Outline of the biosynthesis of a porphyrin. CoA, Coenzyme A; Side chain abbreviations: M, —CH_3; A, —CH_2—COOH; V, —CH=CH_2; P, —CH_2—CH_2—COOH.

lar to that of normal subjects, whereas in sickle cell anemia or untreated pernicious anemia the red blood cell seems to be destroyed indiscriminately.

IV. Catabolism of Porphyrins

The heme moieties of hemoglobin and other heme proteins are eventually metabolized to bile pigments (Fig. 5) and eliminated in both the stool and the urine. Approximately 80% of the bile pigment is derived from hemoglobin.

The heme molecule is metabolized by the reticuloendothelial system by microsomal heme oxygenases to yield the bile pigment biliverdin, which is then reduced by a reductase to the orange pigment, bilirubin (Fig. 5). In this process the α-methene bridge carbon atom is removed as carbon monoxide. The release of carbon monoxide, a poisonous gas, poses a problem, for it combines with hemoglobin and myoglobin and blocks oxygen transport. The binding of carbon monoxide by heme is about 25,000 times stronger than the binding of oxygen. However, the binding affinities of hemoglobin and myoglobin for carbon monoxide are only about 200 times greater than that for oxygen. [*See* RETICULOENDOTHELIAL SYSTEM.]

This reduction of the affinity for carbon monoxide in hemoglobin and myoglobin is brought about

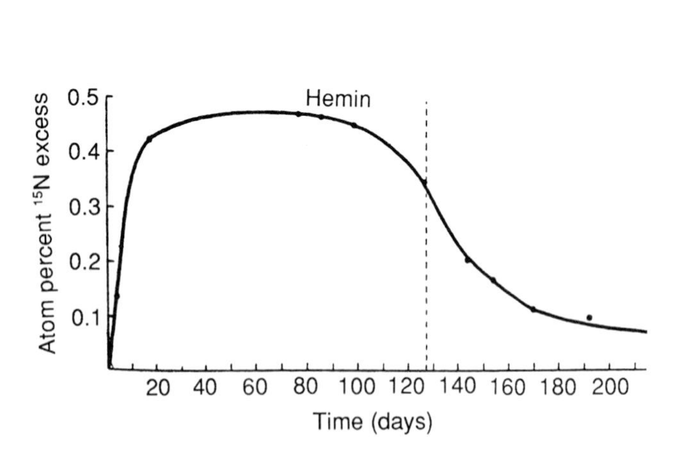

FIGURE 4 Life span of the human red blood cell. Plot of the ^{15}N concentration of the heme of hemoglobin versus time after the ingestion of ^{15}N-labeled glycine.

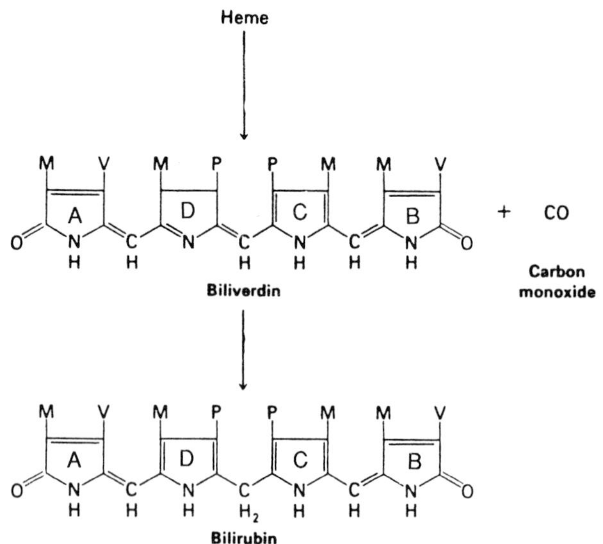

FIGURE 5 Structures of bile pigments in the conversion of heme to biliverdin and bilirubin. CO, Carbon monoxide; M, —CH_3; V, —CH=CH_2; P, —CH_2—CH_2—COOH.

by bending the binding angle of the carbon monoxide interaction with heme by a distal histidine of the protein. In complexes of carbon monoxide with isolated heme, iron, carbon, and oxygen atoms are in a linear array (Fe—C≡O), whereas the carbon monoxide axis is at an angle to the Fe—C bond (—Fe—C⫻O) in hemoglobin. The linear binding of carbon monoxide is prevented mainly by the steric hindrance of the distal histidine. This decreased affinity of hemoglobin and myoglobin for carbon monoxide is a biological advantage, since the endogenously produced carbon monoxide only interferes with about 1% of the sites, instead of having a more poisonous effect.

The rather insoluble bilirubin is released into the circulation, where it binds to the serum albumin and subsequently is cleared by the liver. (The liberated iron is recycled for heme synthesis.) In the liver the bilirubin is esterified with glucuronic acid and thereby converted to a more soluble derivative. The esterified bilirubin and any unesterified pigment are secreted into the bile and subsequently into the intestines, where it is further acted on by bacteria to form urobilinogens. The urobilinogens are, in part, resorbed and reexcreted into the bile and, in part, excreted in the stool and the urine. Oxidized derivatives of urobilinogens are partially responsible for the color of these excretions. The absence of urobilinogens in the stool and in urine is indicative of obstruction of the bile ducts. It is of interest that it is advantageous to have bilirubin as the circulating bile pigment rather than biliverdin, which is the end product of heme catabolism in birds and reptiles, for bilirubin is a strong antioxidant. Bilirubin scavenges peroxyradicals and protects fatty acid, transported by albumin, from oxidation. In this process the bilirubin is oxidized to biliverdin, which is then rapidly reduced to bilirubin.

V. Disorders in Bilirubin Metabolism

Bilirubin is potentially a toxic compound, having diverse effects, such as inhibition of protein synthesis, carbohydrate metabolism in the brain, ATPase activity, and many other enzymatic reactions; therefore, hyperbilirubinemia and the resulting jaundice are indeed a serious matter. Hyperbilirubinemia and jaundice can occur as a result of abnormalities in the process of formation, excretion, and metabolism of bile pigments. Obviously, liver disorders such as hepatitis, cirrhosis, and cancer play a role in causing an increased amount of unconjugated serum bilirubin, as does an abnormal degree of hemolysis of red blood cells. The lack of the functioning liver enzyme which catalyzes the esterification of bilirubin by glucuronic acid also results in an increased concentration of water-insoluble unconjugated bilirubin in the circulation. A rather significant percentage of newborns have hyperbilirubinemia, due to a combination of increased bilirubin production and insufficient glucuronyl transferase activity for bilirubin.

Bibliography

Baldwin, J. M. (1980). The structure of human carbonmonoxy hemoglobin at 2.7 Å resolution. *J. Mol. Biol.* **136,** 103–128.

Baldwin, J., and Chothia, C. (1979). Hemoglobin: The structural changes related to ligand binding and its allosteric mechanism. *J. Mol. Biol.* **129,** 175–220.

Friedman, J. M. (1985). Structure, dynamics, and reactivity in hemoglobin. *Science* **228,** 1273–1280.

Kappas, A., Sassa, S., and Anderson, K. E. (1983). The porphyrians. *In* "The Metabolic Basis of Inherited Disease" (J. B. Stanbury, J. B. Wyngaarden, D. S. Fredrickson, J. L. Goldstein, and M. S. Brown, eds.), 5th ed., pp. 1301–1384. McGraw-Hill, New York.

Shaanan, B. (1983). Structure of human oxyhaemoglobin at 2.1 Å resolution. *J. Mol. Biol.* **171,** 31–59.

Shemin, D. (1982). From glycine to heme. *In* "From Cyclotrons to Cytochromes" (N. O. Kaplan and A. Robinson, eds.), pp. 117–129. Academic Press, New York.

Stocker, R., Yamamoto, Y., McDonagh, A. F., Glazer, A. N., and Ames, B. N. (1987). Bilirubin is an antioxidant of possible physiological importance. *Science* **235,** 1043–1046.

Stocker, R., Glazer, A. N., and Ames, B. N. (1987). Antioxidant activity of albumin-bound bilirubin. *Proc. Natl. Acad. Sci. U.S.A.* **84,** 5918–5922.

Wolkoff, A. W., Chowdhury, J. R., and Arias, I. M. (1983). Hereditary jaundice and disorders in bilirubin metabolism. *In* "The Metabolic Basis of Inherited Disease (J. B. Stanbury, J. B. Wyngaarden, D. S. Fredrickson, J. L. Goldstein, and M. S. Brown, eds.), 5th ed., pp. 1385–1420. McGraw-Hill, New York.

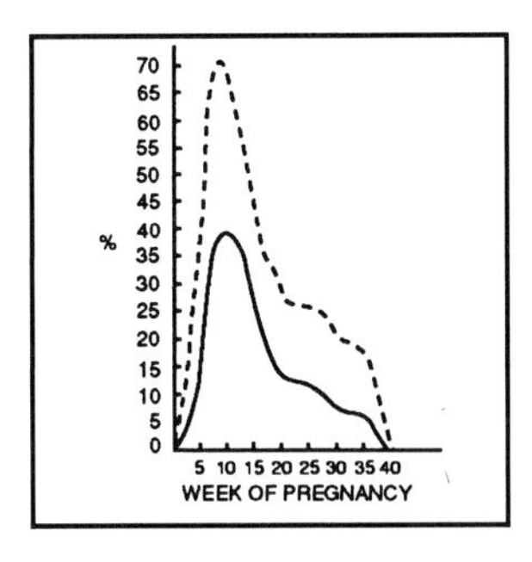

Pregnancy, Nausea and Vomiting

FORREST D. TIERSON, *University of Colorado at Colorado Springs*

Glossary

Emesis gravidarum Nausea and vomiting of pregnancy (NVP) primarily in the first trimester (morning sickness)

Hyperemesis gravidarum Severe NVP extending beyond the 16th week, interfering with maternal nutrition and fluid balance

NAUSEA AND VOMITING of pregnancy (NVP) is a ubiquitous complex of symptoms. Although the relation between NVP and pregnancy is well known (and is often mentioned in the popular literature), relatively little is known about the cause of NVP or about its epidemiology. NVP is often used as an indicator of pregnancy well-being. Many studies have demonstrated a positive association between the presence of early NVP (morning sickness) and favorable pregnancy outcome. In addition, epidemiological studies have reported associations between the presence of early NVP and smoking, between the presence of early NVP and increased infant birthweight, and between the absence of NVP early in pregnancy and increased risk of spontaneous abortion (usually early in pregnancy as well).

I. Historical Background

Nausea and vomiting during early pregnancy is so common that it is considered to be a normal and expected part of pregnancy. Indeed, nausea of pregnancy often provides the first indication that a woman is actually pregnant. NVP has been of interest during most of recorded history, with the first known description of vomiting in pregnancy coming from a papyrus source dated about 2000 BC. The earliest explanations for the cause of NVP relied mostly on explanations having to do with excessive food intake. Supposedly, this excessive intake of food resulted in nerve compression, which, in turn, activated NVP. By the end of the 19th century, psychological causes were advanced to explain NVP. It was thought that vomiting of pregnancy was a manifestation of neurosis and that, therefore, vomiting was readily amenable to suggestive treatment to affect a cure. Women with NVP were said to exhibit significant psychopathology of emotion—especially anxiety and tension, depression, and resentment. Positive correlations were found between the severity and duration of symptoms of NVP and the degree of emotional disturbance experienced by the pregnant woman. In addition, it was claimed that NVP was used by pregnant women to avoid work and to avoid or refuse intercourse. NVP was also thought to be the manifestation of a wish that the pregnancy be aborted.

Although a wealth of evidence now exists that implies organic causes for NVP, these psychologically based explanations have remained in vogue until recently. Actually, an overreliance in the past on psychogenic explanations still detrimentally influences our understanding and perception of NVP today.

Despite popular interest in NVP, little scientific inquiry has been expended in developing an understanding of this complex of symptoms. During the past 40 years, no more than about 45 papers have been published on NVP, and the majority of these papers have been concerned almost solely with hyperemesis. Although several studies have reported some data on prevalence of symptoms of NVP, little information has been presented on time of onset or duration of symptoms. In addition, partly because of the nature of when pregnancies are ascertained, most of the descriptive studies have been retrospective in nature, reporting observations in women ascertained late in pregnancy or even well after pregnancy termination.

II. Prevalence of NVP in Modern Populations

Based on retrospectively obtained data, many investigators have reported that nausea of pregnancy affects between 50 and 70% of pregnant women. In a relatively recent prospective study of 414 pregnant women of high socioeconomic status in Albany, New York, 89.4% reported having nausea and/or vomiting of pregnancy. The total number of women reporting vomiting of pregnancy represented 56.5% of the study population, whereas 32.9% had nausea alone. Only 10.6% of the women had no symptoms of NVP.

Among women who ever had nausea, the mean and median week of onset was the sixth week of gestation. Women who had nausea with vomiting did not differ in the mean week of onset of symptoms from women having only nausea. Figure 1 displays the pattern of nausea and vomiting over the course of pregnancy by showing the percentage of women in the entire sample who were experiencing nausea with or without vomiting (dotted line) and vomiting (solid line) at any specific period during gestation.

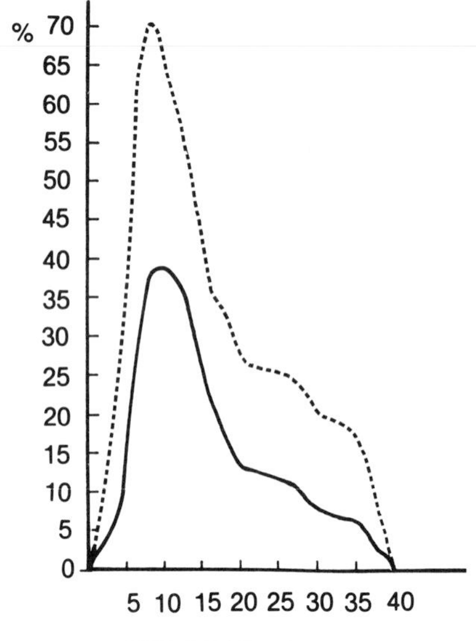

FIGURE 1 Pattern of nausea (*dotted line*) and vomiting (*solid line*) of pregnancy. Displayed is percentage of women in entire sample of 414 women experiencing symptoms of nausea and vomiting of pregnancy by time of pregnancy when symptoms were experienced. [From Tierson, F. D., Olsen, C. L., and Hook, E. B. (1986). Nausea and vomiting of pregnancy and association with pregnancy outcome. *Am. J. Obstet. Gynec.* **155**(5), 1019, with permission.]

As seen in Fig. 1, only about 20% of all women had developed symptoms of nausea with or without vomiting by the fourth week of gestation. Thereafter, the number of women developing symptoms of nausea increased sharply, so that by the eighth week nearly 80% of the women in the sample had developed nausea, and 70% were still actually experiencing NVP. By week 16, 98% of all women who ever developed nausea (about 86% of the total sample) had begun having symptoms. A total of 52% of all women in the study (91.5% of the women who ever developed vomiting) had developed vomiting by the 12th week—at which time about 35% were still experiencing vomiting of pregnancy.

Of those women who had nausea, 30% had stopped having symptoms by the 12th week. Fifty percent of the women had stopped having symptoms by the 15th week. Even so, by the 20th week of gestation, 25% of the women ever having nausea were still experiencing symptoms. Of those women expressing vomiting of pregnancy, 50% had stopped having symptoms by the 15th week. Nausea persisted longer among women whose nausea was accompanied by vomiting than among women with nausea alone.

The incidence of NVP determined in the Albany study is somewhat higher than reported by others, although nausea and vomiting are rarely treated as separate symptoms. Much of the difference between this study and previous ones is most likely due to the greater ascertainment of women with only minor symptoms of nausea in this prospective study (in which pregnancies were ascertained by the 13th week of gestation). Again, several retrospective studies have reported that about 70% of the women in the study population experienced NVP.

III. NVP and Pregnancy Outcome

Hyperemesis gravidarum, which is usually accompanied by maternal weight loss, has been associated with low birthweight (small for gestational age) infants because of the fact that hyperemesis interferes with maternal nutrition and fluid balance. Particularly severe cases can be life-threatening—both to mother and fetus. The effects of "normal" NVP are not as dramatic. Initially, women experiencing vomiting of pregnancy gain weight more slowly than women with no NVP or women with only nausea. However, after their symptoms of vomiting decrease in severity, women who previously experienced vomiting of pregnancy gain weight faster than women who had only nausea and faster than women who had no symptoms of NVP. As a consequence, by about the 25th week of pregnancy, no significant difference exists in maternal weight gain between women with symptoms of NVP and women with no symptoms.

In the Albany study, women without any symptoms of NVP accounted for a significantly larger proportion of the fetal deaths (because of spontaneous abortion or miscarriage). This finding is in line with results obtained from other studies. The generally accepted reason for this association is that a low level of steroidal hormones (progesterone and estrogens) is not enough to induce NVP—but at the same time the low level is not enough to maintain pregnancy and, as a consequence, the pregnancy is spontaneously aborted. [*See* ABORTION, SPONTANEOUS; STEROIDS.]

IV. Origins of NVP

Little information is available that would allow comparison of hormone levels with NVP in early pregnancy because most studies of nausea and vomiting of pregnancy have been concerned with hyperemesis gravidarum. The relation between levels of steroidal hormones in the third trimester (obtained from studies of hyperemesis) and hormone levels early in pregnancy (especially as they influence NVP) is not known.

Although the etiology of NVP is still not well-defined, a hormonal influence is suspected. Because some women experience nausea similar to NVP with the use of oral contraceptives or when they are on estrogen medication, increased levels of sexual steroids (progesterone and estrogens) during pregnancy are thought to be related to the development of NVP. In addition, NVP and nausea during the use of oral contraceptives are both usually confined to the first few months—either after conception or after first starting to take oral contraceptives. In both cases, there is a rapid increase in hormone levels at this time. As mentioned earlier, low levels of steroidal hormones during early pregnancy would not trigger the development of NVP, but the same low levels would also be insufficient for proper pregnancy development and the pregnancy would be aborted.

It has been suggested that women who experience early NVP may have lower functional liver capacities. As a result of the lower capacities, these women would be overly sensitive to estrogens or their metabolites and would be more likely to develop NVP. The functional load on the liver would be similar during early pregnancy and when oral contraceptives were first administered—both conditions when the production of estrogens rapidly increases.

A correlation between increased levels of human chorionic gonadotropin (HCG) and NVP has been reported. Such a correlation would certainly be expected, because levels of HCG rapidly increase in early pregnancy. However, several studies fail to demonstrate any causative connection. Also, immunological factors may possibly be involved in the origin of NVP, but again, the actual nature of this relation is not clear.

Bibliography

Jarnfelt-Samsioe, A. (1987). Nausea and vomiting in pregnancy: A review. *Obstet. Gynecol. Surv.* **42**(7), 422–427.

Jarnfelt-Samsioe, A., Samsioe, G., and Velinder, G.-M. (1983). Nausea and vomiting in pregnancy—A contribution to its epidemiology. *Gynecol. Obstet. Invest.* **16,** 221–229.

Kasper, A. S. (1980). Nausea of pregnancy: An historical medical prejudice. *Women Health* **5**(1), 35–44.

Tierson, F. D., Olsen, C. L., and Hook, E. B. (1986 and 1989). Nausea and vomiting of pregnancy and association with pregnancy outcome. *Am. J. Obstet. Gynecol.* **155,** 1017–1022 and **160,** 518–519.

Primate Behavioral Ecology

PAUL A. GARBER, *University of Illinois*

Glossary

Behavioral ecology Study of relations between ecological factors, social organization, and mating systems in animal species

Behavioral strategies Alternative patterns of behavior that influence reproductive success, access to resources, and/or social position

Female-bonded group Social groups composed mainly of female matrilines and unrelated males. In these species, females generally remain in the group in which they are born, males emigrate from their natal group, and kinship appears to play a significant role in female social interactions

Food patch The spatial and temporal clumping of food items; generally defined as the crown of a tree or an area in which food resources are distributed such that an animal can continue to feed without having to switch to a new food type. The size of the food patch sets some upper limit on the number of animals that can feed simultaneously

Mating system Patterns of sexual interactions, mate preferences, and sexual competition among reproductively active individuals residing in a social group. The mating system differs from the social system in that not all adults within a group may be reproductively active

Phylogeny Evolutionary history of a species or taxonomic group

Primate community Species that live in the same environment, overlap in their use of resources, and interact such that the activities of each species is likely to have some effect on the activities of the other species

Social group Aggregation of individuals who are spatially cohesive, interact regularly, form social bonds, and travel together. Individuals within a social group tend to have more frequent interactions with each other than individuals from different social groups

THE ORDER PRIMATES represents a diversified group of some 200 species of prosimians, monkeys, apes, and humans that are generally characterized by a single infant at birth, a relatively long period of infant dependence, and complex behavioral and social interactions. With few exceptions, primates live in social groups composed of individuals of all ages and both sexes. The structure and form of these groups vary considerably, however, and recent studies in the behavior and ecology of nonhuman primates have identified a number of primary factors that influence the size, composition, and cohesiveness of social groups. These include (1) the threat of *predation,* (2) *food distribution and availability,* (3) *kinship,* (4) *infant care,* and (5) *intrasexual aggression* and *mating patterns.*

The study of primate *behavioral ecology* has centered on the comparative approach. Comparisons of the *social organization* and feeding ecology of closely related species living in the same ecological community indicate that although phylogeny and body size constrain the ways in which a species responds to changes in the environment, interspecific differences in social organization are often explained in terms of subtle differences in diet, food quality, and the distribution of preferred food resources. It has generally been assumed that even among members of the same social group, individual competition for access to important feeding sites

Encyclopedia of Human Biology, Volume 6.

tends to limit the size of the social unit. When resources are scarce or distributed in small scattered feeding sites, within-group *feeding competition* can result in a decrease in group size, a decrease in group cohesion, the formation of foraging subgroups, or a reduction in the health and nutritional status of low-ranking group members. In contrast, larger groups may be favored under conditions of high food density or in areas where increased vigilance and opportunities for predator detection play a significant role in reducing *mortality*.

Factors that are important in understanding the behavioral ecology of nonhuman primates are also important in understanding the behavior of human primates. Relations among behavioral, ecological, and social patterns found in our closest living relatives provide explanatory models for identifying ancestral and derived character traits and adaptations that have shaped human biology and evolution. In this review the behavioral ecology of a select group of primate species is examined and compared, in an attempt to focus on the ways in which group size and composition influence individual *behavioral strategies* and *reproductive success*. A list of the scientific names of primate species discussed in this article is presented in Table I.

TABLE I List of the Common and Scientific Name of Primate Species Discussed in this Article

Common name	Scientific name
Black spider monkey	*Ateles paniscus*
Bonnet macaque	*Macaca radiata*
Brown capuchin monkey	*Cebus apella*
Chacma baboon	*Papio ursinus*
Chimpanzee	*Pan trogloydes*
Dusky titi monkey	*Callicebus moloch*
Gelada baboon	*Theropithecus gelada*
Hanuman langur	*Presbytis entellus*
Humans	*Homo sapiens*
Japanese macaque	*Macaca fuscata*
Lar gibbon	*Hylobates lar*
Long-tailed macaque	*Macaca fascicularis*
Mountain gorilla	*Gorilla gorilla*
Moustached tamarin	*Saguinus mystax*
Night monkey	*Aotus trivirgatus*
Olive baboon	*Papio cynocephalus anubis*
Orangutan	*Pongo pygmaeus*
Pigtailed macaque	*Macaca nemistrina*
Red colobus	*Colobus badius*
Red howler monkey	*Alouatta seniculus*
Rhesus macaque	*Macaca mulatta*
Ring-tailed lemur	*Lemur catta*
Saddle-back tamarin	*Saguinus fuscicollis*
Siamang	*Hylobates syndactylus*
Squirrel monkey	*Saimiri sciureus*
Talapoin monkey	*Miopithecus talapoin*
Vervet monkey	*Cercopithecus aethiops*
Woolly spider monkey	*Brachyteles arachnoides*
Yellow baboon	*Papio cynocephalus cynocephalus*
Yellow-handed titi monkey	*Callicebus torquatus*

I. Predation

In addressing questions concerning the advantages of social group living, it has been suggested that the threat of predation is a major factor promoting social cohesion, group stability, and increased group size in diurnal primates. For small-bodied arboreal monkeys, raptors such as hawks and eagles are likely to pose the greatest predatory threat. Defense against *aerial predators* commonly involves concealment, vigilance, a cessation of movement, or flight. In the case of larger-bodied and more terrestrial primates, mammalian carnivores present the greatest danger. *Antipredator tactics* against these threats include vigilance, retreat to an arboreal haven, frequent alarm calls, and a variety of branch-shaking and arboreal threat displays. If group size is a direct response to predator activity, associated patterns of behavior (both protective and avoidance) and grouping should exist in areas of high predator density that are not commonly observed in areas of lower predator pressure. For example, although the night monkey exhibits a noctural activity pattern throughout most of its range, this primate has been observed to travel and forage frequently during the day in areas where large predatory owls are common. Similarly, in areas free of predators, some populations of long-tailed macaques live in smaller social groups than populations in areas where the threat of predation is greater. However, in the absence of systematic information on predation or predatory attempts on these species, conclusions regarding the relation between group size and group protection must be viewed with caution.

Many species of primates exhibit characteristic *alarm calls* and behavioral responses that are specific to particular classes of predators. The existence of these calls suggests that predators have exerted a strong influence on at least some aspects of primate behavior. In a series of taped playback experiments, for example, it was determined that vervet monkeys display acoustically distinct alarm calls for six different types of predators. In this species there is evidence that females give alarm calls more frequently in the presence of kin than in the presence of nonkin. However, because high-ranking animals call more frequently than low-ranking animals, it is difficult to separate the effects of kinship, social status, and spatial position on calling behavior.

Data on a population of ground-living yellow baboons in Amboseli, Kenya, indicate that at least 25% of all the known or presumed deaths occurring during a 14-month period were the result of predatory attacks. Although there have been reports that adult male baboons may defend the group from predators, such behavior rarely occurs and its effect on individual survivorship is unknown. In contrast, red colobus males have been observed to defensively attack and chase chimpanzees attempting to prey on group members. In many cases these males were effective in reducing the hunting success of the chimpanzees. Although other factors such as injury, disease, and starvation may contribute more to mortality than predation, the threat of predation appears to have a significant effect on behavior and the development of antipredator tactics.

If group size is a primary response to predator pressure, primate species living in the same ecological community are expected to have groups of similar overall size. This, however, is often not the case. For example, in a community of 10 species of neotropical primates in Peru, group sizes appear to reflect more closely the breeding system and feeding behavior of each species than the degree of predator pressure. Species exhibiting a *monogamous mating pattern* had the smallest group size, followed by *polyandrous*, and *polygynous* species. Although body size is an important factor in vulnerability to predators, small-bodied species were characterized by both the smallest and largest group sizes. In this and many other primate communities, the availability of resources during food-limited periods of the year may have a more direct effect on group size than does predation.

II. Food Distribution and Availability

The abundance, availability, and distribution of food resources exert a strong influence on the structure and size of primate foraging groups. A number of hypotheses have been proposed to explain the adaptive basis of group size and *foraging behavior*. In species for which the costs of foraging increase rapidly with additional group size, smaller groups are likely to exploit their environment more efficiently than larger groups. When resources are distributed in small, scattered patches, individuals living in larger groups may be forced to travel greater distances, visit more feeding sites, and compete more aggressively over limited food rewards than members of the same species living in smaller groups. However, when exploiting highly productive, clumped, or rapidly renewing feeding sites, individuals foraging together in a larger social unit may obtain feeding advantages by cooperative hunting (e.g., chimpanzees and humans) or collectively disturbing concealed prey (squirrel monkey), traveling together to minimize return times to exploited food patches (e.g., mixed species troops of tamarins), or by successfully defending major feeding sites from smaller neighboring groups (e.g., female-bonded groups of Old World monkeys). Although many researchers have assumed that individuals in smaller groups have lower net *foraging costs* than individuals in larger groups, this is likely to depend on the number and distribution of both large and small food patches as well as the costs of *subgroup formation*. In species characterized by flexible subgrouping patterns, individuals may avoid competition by temporarily separating from the main group and exploiting smaller resource patches. This foraging pattern requires high levels of social communication, an ability to quickly reestablish social bonds, and an exchange of information about resource availability in the environment. Among nonhuman primates, such extreme flexibility in foraging behavior is reported in chimpanzees, black spider monkeys, woolly spider monkeys, and some populations of long-tailed macaques.

If subgroup formation is a general response to resource scarcity and patchiness, intragroup feeding competition can have an important influence on group stability and individual *reproductive success*. Among chacma baboons, within-group feeding competition for limited resources can result in loss of body weight, inadequate nutrition, and death for adult females and infants. This reflects the ability of larger and more aggressive adult and subadult males to displace females and young from feeding sites. Similarly, in the brown capuchin monkey the aggressive behavior of the dominant male is the most important factor affecting the feeding success of other group members. Subordinates avoid competing directly with higher ranking individuals by foraging independently and at lower-quality feeding sites. When forced to feed apart from the main group, low-ranking group members suffer a decrease in energy intake as well as a higher risk of predation. In this brown capuchin population, the high cost of subgroup formation and the ability of dominant animals to control access to high-quality resources limited the size of the social group to between 5 and 14 individuals.

Given their large body size, substantial daily food requirements, and absence of major predators (except humans), the great apes (gorilla, chimpanzee, and orangutan) are expected to forage in small groups. And although small foraging parties (one–five animals) are common in orangutans and chimpanzees, the heaviest of all living primates, the mountain gorilla, lives and forages in stable, cohesive social groups of 10–20 animals. This apparent anomaly is explained by the fact that these African apes exploit superabundant patches of terrestrial herbaceous vegetation. The plant species eaten by gorillas are among the most common in their habitat. These large patches can accommodate larger feeding groups, and aggressive feeding competition between group members is infrequent.

III. Kinship

There is evidence among several species of primates that patterns of cooperative behavior such as coalition formation, hunting, food sharing, resource defense, and assistance in caring for young occur mainly between family members and close kin. However, because precise information on the genetic relatedness of group members is generally unknown or restricted to individuals related through the female line, attempts to examine patterns of kin recognition and kin-associated behavior in primates have primarily focused on matrilineal relations. In Japanese macaques, for example, female relatives are found to groom, rest, feed, and travel together more frequently than expected based on the number of potential social partners present in the group. Similar mother–mature-offspring associations and sister associations have been documented among female rhesus macaques, bonnet macaques, pigtailed macaques, olive baboons, gelada baboons, Hanuman langurs, vervet monkeys, squirrel monkeys, ring-tailed lemurs, and chimpanzees. In many cases, females of one matriline socially dominate females of a second matriline. During agonistic encounters or at limited feeding sites, support provided by matrilineal kin may play a key role in determining priority access to resources.

Among male primates, there is only limited evidence of patrilineal-based kin affiliations. This may reflect the general mammalian pattern of intense intrasexual competition among reproductively active males or parental uncertainty through polygynous matings, as well as the fact that in many primate species dispersal patterns result in social groups that are largely composed of a single or unrelated set of adult males. In species in which related males are reported to reside in the same group (e.g., the gorilla, red howler monkey, and chimpanzee), male coalitions appear to play a critical role in preventing aggressive and infanticidal attacks by nonresident males, as well as maintaining the integrity of the breeding unit.

IV. Infant Care

There is considerable evidence that matrilineal relatives directly contribute to the care and protection of infants in female-bonded primate societies. However, with few exceptions, information on male–infant interactions in primate species has failed to identify a direct link between probable paternity and caregiving behavior. In the case of olive baboons, for example, protective and affiliative interactions between adult males and infants are often better explained in terms of the maintenance of a social bond between the male and the infant's mother than by kin-related behavior. In this species, recent immigrant males spend more time grooming and interacting with infants than do resident males who are probable fathers.

Among Old World primates in which paternity is more certain, extreme variation exists in the nature and frequency of male–infant interactions. In species exhibiting a haremlike mating pattern (harem polygyny), fathers tend to interact minimally with their offspring. Although it is difficult to quantify the importance of vigilance behavior in protecting young, at present there is no evidence that the single breeding male in a harem group invests more in infant care than do reproductively active adult males in multimale polygynous breeding groups.

In New World species characterized by a nuclear family social group, a single infant at birth, and a pair-bonded mating pattern (e.g., dusky titi monkey, yellow-handed titi monkey, night monkey), fathers substantially contribute to infant care. In these primates, the adult male is the primary infant caretaker and devotes considerable time and effort in carrying and sharing food with the young. Among monogamous lesser apes, however, patterns of male caregiving are more variable. Whereas siamang males may spend up to 78% of the day transporting and grooming a single young, lar gibbon males provide little direct infant care. Thus, despite the fact that paternal certainty can be an important factor influencing infant care, that alone is not a sufficient condition from which to predict male caregiving behavior. It is likely that additional factors such as the relation between nonmaternal care and *infant survivorship,* the cost of reproduction to the female, and the role of male–infant alliances in establishing a *social–sexual bond* with males and females have contributed importantly to the evolution of male caregiving behavior.

Recent information on the social and mating patterns of tamarin and marmoset monkeys provides additional support for a critical relation between the high cost of reproduction to the female and patterns of male caregiving behavior. Tamarins and marmosets are unusual among higher primates in their production of twin offspring, a polyandrous mating system in some species, the fact that females can produce two litters per year, and the degree to which adult males cooperatively participate in infant care. Studies on both captive and free-ranging groups indicate that although virtually all group members assist in caring for the young, adult males are the principal caretakers. In the case of moustached and saddle-back tamarins, groups commonly contain two or three adult males. There is evidence that the number of adult male helpers residing in groups is positively correlated with infant survivorship. Although it remains uncertain whether adult helpers are related to the breeders and act to increase the kinship component of their reproductive success or are unrelated and act to increase the individual component of their reproductive success, the size and structure of tamarin social groups is strongly influenced by the high costs of reproduction to the breeding female and the need for nonmaternal infant care. In the absence of at least two adult nonmaternal helpers, it is unlikely that a female tamarin or marmoset could successfully rear twin infants.

It has been proposed that a monogamous mating pattern and adult male provisioning of females and young were an important and early component of the behavioral ecology of human ancestors. However, based on a broader understanding of primate mating and social patterns, it is now apparent that the bases of male–infant interactions are extremely complex and may reflect factors other than *paternity*. Moreover, among modern humans there exists extreme variation in the degree to which adult males provide direct infant care. In some societies fathers may spend up to 14% of their day interacting with young (e.g., !Kung San Bushman). However, in most cases, male–infant interactions are infrequent and limited to small amounts of play behavior. Although fathers do provide food for their offspring, it remains uncertain whether this behavior developed as a mechanism to maintain a social bond between an adult male and an adult female or whether it evolved as a mechanism to increase the survivorship of a male's offspring.

V. Intrasexual Aggression and Mating Competition

Among mammals, females invest more heavily in the production of offspring than do males. This begins with the initial expenditure of energy required to produce an egg and continues through the period of gestation and lactation. Given that the cost of reproduction to the female is significantly greater than it is for a male, female reproductive success is constrained by the number of offspring that survive (not produced), whereas male reproductive success is limited by availability and access to fertile females. Males often compete aggressively for access to reproductive partners, resulting in extreme variance in male reproductive opportunities and reproductive success. In contrast, although females can exercise some level of mate choice and select male partners based on particular physical or behavioral qualities, at present there is little direct evidence that a female's mating preferences have a significant impact on her lifetime reproductive success.

In primates characterized by a polygynous mating system, intrasexual competition among males has a direct effect on patterns of social tolerance, aggression, and migration. In the case of unimale breeding groups, intense competition for *reproductive sovereignty* between the harem leader and individuals or bands of nonresident males often exists. Depending on the size of the harem, the age of the harem leader, and the number of extragroup males in the population, attempted takeovers of the group can occur rarely or every few months. These takeovers are extremely aggressive and result in severe injuries and sometimes death.

Social interactions among males residing in multimale polygynous groups are highly complex and include elements of *social cooperation*, alliances, and physical aggression. In many baboon and macaque species, males compete with each other for positions in a dominance hierarchy. These hierarchies are unstable, however; as resident males leave the group, new males enter, and changing alliances alter dominance relations. In these groups a single male rarely is able to monopolize access to all reproductive females. Rather, a number of males may mate with a given female, although some males may mate more frequently and at a more optimal time for conception. In these primates, there exists a variety of tactics or alternative behavioral strategies by which males are able to gain access to reproductive partners (e.g., immigration into a new social group, coalition and alliance formation, development of a social–sexual bond with particular females, formation of consort relationships).

Females can also compete for reproductive partners. This can occur through direct solicitation of copulations, transfer into another social unit, or directed aggression against other females. In the case of tamarins and marmosets discussed in the previous section, it has been proposed that females mate preferentially with males that provide direct care for their offspring. Alternatively, dominant female rhesus macaques, yellow baboons, and red howler monkeys have been observed to attack subordinate females who were in the process of copulating. In addition, female primates engage in a variety of behavioral and reproductive tactics that serve to confuse paternity, reduce the opportunities for *infanticide,* and limit the amount of aggression directed against their offspring.

There is increasing evidence that in some primate species, dominant females can impair or *suppress the reproductive cycle* of subordinate group members. Although the physiological mechanisms through which this occurs are not well understood, both stress and *pheromone* cues can disrupt normal endocrine function. In captive female talapoin monkeys, for example, subordinates that receive high levels of aggression exhibit a decrease in sexual activity and show steroid hormone levels indicative of neuroendocrine disfunction. Similarly, social stress can result in low levels of conception and increased rates of abortion in vervet monkeys. Finally, among free-ranging tamarins and marmosets, regardless of the number of adult females in a group, only a single female produces offspring. Data from captive tamarin and marmoset groups indicate that subordinate females are frequently anovulatory. Ovulation inhibition may be related to scent marking and the resultant pheromone signals given by the dominant female. Thus, factors such as social dominance, stress, hormone production, and nutritional condition all appear to influence individual reproductive success and the behavioral interactions of group-living primates.

The form and structure of a primate group represents a compromise between the threat of predation and predator protection, foraging costs and access to high-quality feeding sites, reproductive opportunities and mating exclusivity, and the importance of group members in providing care for young. These factors interact in complex ways and have produced the variety of behavioral, mating, and social patterns found in living primates. Although there exist certain qualitative relations among diet, feeding ecology, group size, and group cohesion, detailed information on the distribution, patch size, seasonality, renewal rate, and nutritional quality of foods exploited by primates is needed before more quantitative and predictive models of behavioral ecology can be tested. In particular, studies must focus on differences in the costs and benefits of particular foraging patterns for males and females, the manner in which female foraging patterns change during pregnancy and lactation, and whether subordinates, migrants, and extragroup individuals suffer higher foraging costs and greater risks of predation. In those species for which the advantages of social group living are not directly related to feeding and predation, access to mating partners and assistance in caring for young may play a more formidable role in the structure and form of the social group.

Bibliography

Fleagle, J. G. (1988). "Primate Adaptation and Evolution." Academic Press, New York.

Garber, P. A. (1987). Foraging strategies among living primates. *Annu. Rev. Anthropol.* **16,** 339–364.

Jolly, A. (1985). "The Evolution of Primate Behavior," 2nd Ed. MacMillan Press, New York.

Richard, A. F. (1985). "Primates in Nature." W. H. Freeman, New York.

Smuts, B. B., Cheney, D. L., Seyfarth, R. M., Wrangham R. W., and Struhsaker, T. T. (1986). "Primate Societies." University of Chicago Press, Chicago.

Terborgh, J., and Janzen, C. H. (1986). The socioecology of primate groups. *Annu. Rev. Ecol. Systematics* **17,** 111–136.

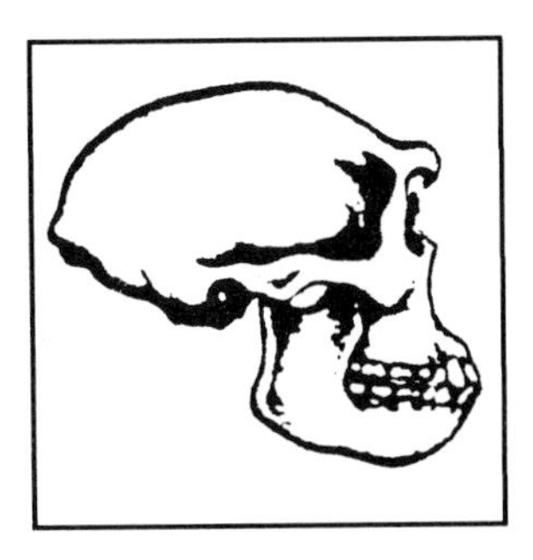

Primates

ERIC DELSON, *Lehman College and Graduate School, City University of New York, and American Museum of Natural History*

IAN TATTERSALL, *American Museum of Natural History*

Glossary

Cathemeral Activity pattern of an animal wherein activity is distributed over the entire 24-hr cycle, rather than limited to the daytime (diurnal), night-time (nocturnal), or the times around dawn and dusk (crepuscular)
Morphology Visible attributes; anatomy
Phylogeny Evolutionary history; a statement of evolutionary relationships
Systematics Study of the diversity of life and of the relationships among taxa
Taxonomy Theory and practice of classifying organisms; derived from the noun "taxon" (pl. taxa), which denotes a named unit at any rank of the Linnean hierarchy

PRIMATES IS THE ORDER of mammals containing the lemurs, lorises, tarsier, monkeys of the Old and New Worlds, apes, and humans, totaling approximately 200 living species. Also usually regarded as primates, besides the fossil relatives of extant species, are the members of an early Tertiary group of "archaic" forms known as Paromomyiformes.

I. Definition and Classification

Attempts to define the order Primates in terms of morphological characteristics uniquely shared by all of its members have routinely been frustrated by the fact that phylogeny, rather than morphology, gives the group its unity. However, if we disregard the archaic forms of the Paleocene and early Eocene, we can follow W. E. Le Gros Clark in identifying several "progressive trends" that have some value in demarcating developments in primate evolution from those in other mammalian lineages. These include a decreased significance of olfaction compared with stereoscopic vision; enhancement of grasping and manipulative capabilities, associated with the replacement of claws by nails on all or nearly all digits; and a tendency toward relative enlargement of the "higher" centers of the brain, notably the association areas of the cortex.

A diagram of primate evolutionary relationships at the family level is shown in Fig. 1. Many of its details, perhaps most importantly the affinities of the tarsier, are actively debated, and we do not regard this statement as definitive. However, it does serve as the basis for the outline classification presented in Table I. Geological ranges of the families identified in Figure 1 are shown in Fig. 2.

II. "Archaic" Primates

The primate order originated prior to about 65 Myr (million years) ago, and early primates flourished in the northern continents during the Paleocene epoch, around 64–54 Myr ago. The most successful of these archaic forms were *Plesiadapis* (Figs. 3 and 4) and its relatives, which were characterized by small braincases, long faces, clawed feet, and an enlarged, specialized anterior dentition. "Archaic" primates lacked the postorbital bar, a strut that delineates the orbit laterally in all living primates, and are included in the order principally on the bases of the structure of their bony ear and postcranial bones and their molar tooth morphology. Generally, in early primates, the molars show some lowering of cusp relief, which has been taken to suggest that the origin of the primate order lay in a dietary shift

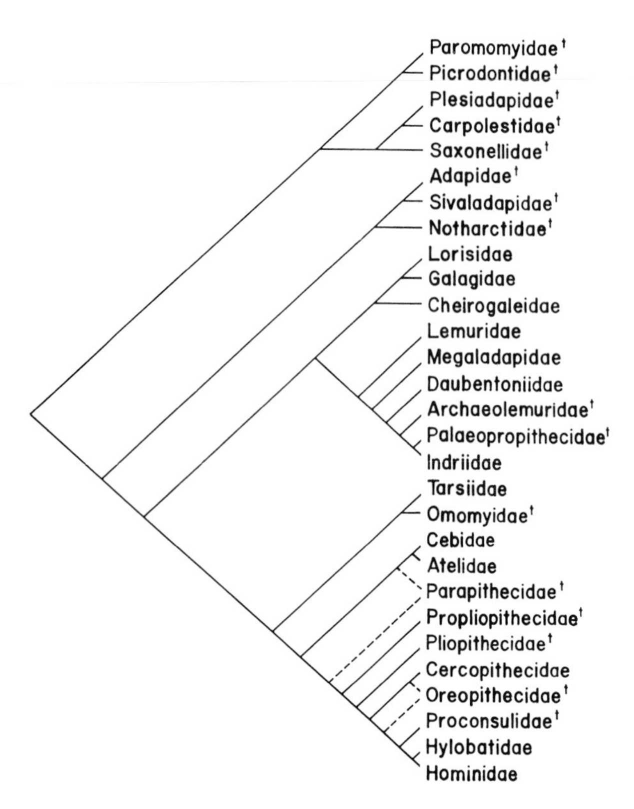

FIGURE 1 A provisional phylogeny of the primates. † denotes extinct taxon. Dotted lines indicate alternative placement of disputed taxa.

away from insects and toward plant foods. Postcranial remains are rare but indicate that *Plesiadapis* itself, though heavy-boned, was arboreal.

III. "Lower" Primates of Modern Aspect

A. The Living "Lower" Primates

The extant strepsirhine (or "lower") primates (suborder Strepsirhini; Fig. 3) are most abundantly represented on the island of Madagascar, where higher primates were absent before the recent arrival of humans. All living strepsirhines are united in possessing a "grooming claw" on the second pedal digit and a "tooth comb," in which the slender, elongated front lower teeth lie horizontally and are closely pressed together. They also retain a primitive form of the nasal apparatus, with a "tethered" upper lip (fixed to the upper jaw in the midline, rather than fully mobile) and a moist rhinarium (or "wet nose," as in dogs). In comparison with the "higher" primates, the sense of smell, as expressed by these anatomical retentions and the presence of olfactory marking behaviors using feces, urine, or the exudations of specialized glands, is relatively important. On the other hand, the diurnal visual sense is less developed, as reflected in weak or absent color vision. Notably, however, in contrast to the higher primates, many strepsirhines are nocturnal.

Surviving strepsirhines in Africa and Asia are all classified as lorisoids, belonging to the families Lorisidae (the quadrumanous lorises and pottos) and Galagidae (the leaping bushbabies); so also are the dwarf lemurs of Madagascar (family Cheirogaleidae). All lorisoids are nocturnal primates, which, although social, are relatively nongregarious; as noted, scent-marking provides an important channel for intraspecific communication among these forms.

The Malagasy families Lemuridae (the lemurs, ruffed lemurs and bamboo lemurs), Indriidae (the sifakas, babakotos, and woolly lemurs), Daubentoniidae (the aye-aye), and Megaladapidae (sportive lemurs) are more diverse. Of these families, the latter two are exclusively nocturnal, and the last probably lost a diurnal genus only recently; the other two (the principally quadrupedal Lemuridae and the vertical-clinging and leaping Indriidae) are chiefly composed of diurnal or cathemeral species and show a great deal of variation in social organization. Groups range from bonded pairs (permanent associations of an adult male with an adult female) with immature offspring through small multimale, multifemale units to large heterosexual groups of 30 individuals or more. These groups, as in most higher primates, appear to be based around female nuclei, and male exchange is common, especially during the brief annual breeding season. Diets appear to vary widely by season and locality, and despite their generally lower ratios of brain to body size, these diurnal strepsirhines seem to be every bit as opportunistic in the exploitation of their habitats as are the higher primates.

B. The Fossil Record of Lower Primates

By the end of the Paleocene, new kinds of primates were beginning to appear alongside the archaic forms, and in the following Eocene epoch, between about 54 and 36 Myr ago, these more modern-looking forms rapidly replaced the archaic types. The fossil record, as currently known, provides no clear ancestry within the Paleocene primate radiation for that of the Eocene; it is possible that the African record, as yet unknown, holds the key to this relationship. Eocene primates are generally classified

TABLE I Partial Classification of the Order Primates

Rank	Taxon	"Common" name
Semiorder	Paromomyiformes†	"Archaic" forms
Suborder	Plesiadapiformes†	"Archaic" forms
Superfamily	Paromomyoidea†	"Archaic" forms
Superfamily	Plesiadapoidea†	"Archaic" forms
Genus	*Plesiadapis*†	
Semiorder	Euprimates	"Primates of modern aspect"
Suborder	Strepsirhini	"Lower" primates, strepsirhines
Infraorder	Adapiformes†	Extinct "lemurlike" forms
Genus	*Notharctus*†	
Genus	*Smilodectes*†	
Genus	*Leptadapis*†	
Infraorder	Lemuriformes	Lemurs, lorises
Superfamily	Lemuroidea	"True" lemurs
Family	Lemuridae	
Superfamily	Indrioidea	Sifakas, aye-aye, etc
Family	Indriidae	Sifakas
Family	Daubentoniidae	Aye-aye
Family	Megaladapidae	Sportive lemurs
Superfamily	Lorisoidea	Lorises, galagos, mouse lemurs
Family	Lorisidae	Lorises
Family	Galagidae	Galagos, bushbabies
Family	Cheirogaleidae	Mouse and dwarf lemurs
Suborder	Haplorhini	"Higher" primates and tarsiers
Hyporder	Tarsiiformes	Tarsier, fossil relatives
Family	Tarsiidae	Tarsiers
Family	Omomyidae†	Extinct "tarsierlike" forms
Genus	*Necrolemur*†	
Hyporder	Anthropoidea	"Higher" primates
Infraorder	Platyrrhini	New World monkeys, platyrrhines
Family	Cebidae	Marmosets, capuchins
Subfamily	Cebinae	Capuchins, squirrel monkeys
Subfamily	Callitrichinae	Marmosets, tamarins
Family	Atelidae	Howler, spider, saki
Subfamily	Atelinae	Howler, spider monkeys
Subfamily	Pitheciinae	Sakis, titis, owl monkeys
Genus	*Cebupithecia*†	
Infraorder	Catarrhini	Old World "higher" primates, catarrhines
Superfamily	Parapithecoidea†	Extinct Egyptian "monkeys"
Genus	*Apidium*†	
Superfamily	Pliopithecoidea†	Early modern catarrhines
Genus	*Propliopithecus*†	
Genus	*Pliopithecus*†	
Superfamily	Cercopithecoidea	Old World monkeys
Family	Cercopithecidae	Old World monkeys
Subfamily	Cercopithecinae	Cheek-pouched Old World monkeys
Subfamily	Colobinae	Leaf-eating Old World monkeys
Genus	*Mesopithecus*†	
Superfamily	Hominoidea	Apes, humans
Family	Proconsulidae†	Extinct apes
Genus	*Proconsul*†	
Family	Hylobatidae	Gibbons
Family	Hominidae	Great apes, humans
Subfamily	Dryopithecinae†	Extinct early hominids
Subfamily	Homininae	African apes, humans
Genus	*Australopithecus*†	
Genus	*Homo*	
Subfamily	Ponginae	Orangutans, extinct relatives
Genus	*Sivapithecus*†	

† Denotes extinct taxon; genera listed are those illustrated or mentioned in the text.

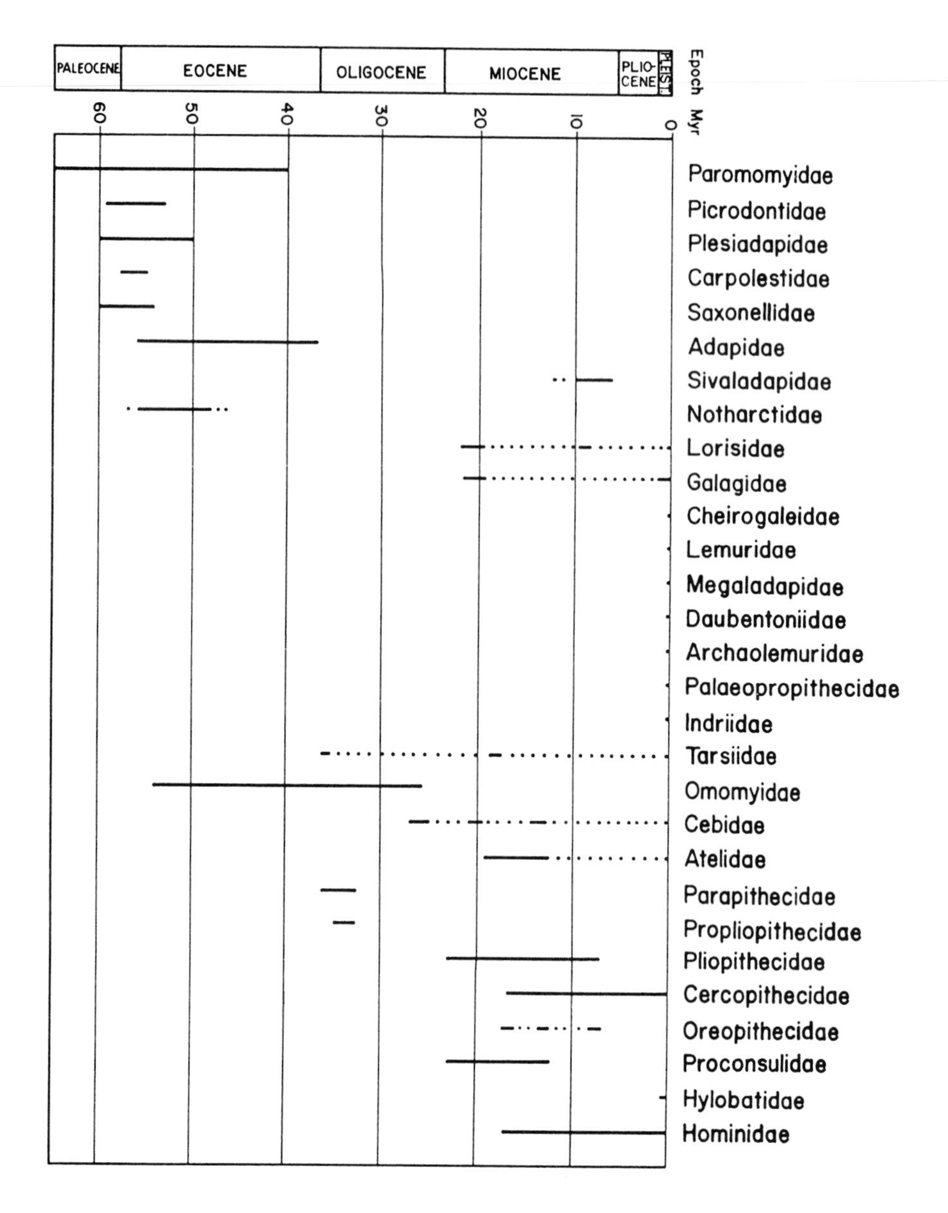

FIGURE 2 Temporal distribution of the primate families shown in Figure 1. Dotted lines represent presumed range continuity between known occurrences.

into the lemurlike Adapiformes and the tarsierlike Omomyidae, although this simple division probably obscures a situation of considerably greater complexity, and some of today's lemurs may bear more than general affinities to elements of the Eocene primate fauna. Whatever the exact geometry of the relationships between the lower primates of today and those of the Eocene may ultimately prove to be, however, as functioning organisms, the latter were already clearly similar to today's lower primates. Undoubtedly arboreal in habit, Eocene primates had grasping hands and feet, with an opposed hallux and sensitive digital pads backed by flat nails; braincases were relatively large whereas the face was reduced, presumably in correlation with a de-emphasis of the sense of smell. Conversely, the eyes faced forward to enhance overlap of the visual fields, thus promoting stereoscopic vision, and were ringed by bone (see Figs. 4 and 5).

As climates cooled subsequent to the Eocene, primates began to disappear from the northern continents, and later fossil evidence, principally from the Miocene of Africa and Asia, reveals strepsirhine primates closely related to those that now live in those areas. Late Holocene deposits in Madagascar

FIGURE 3 Representative living lower primates. Clockwise from top left: forkmarked lemur (Cheirogaleidae), bamboo lemur (Lemuridae), sportive lemur (Megaladapidae), avahy (Indriidae), potto (Lorisidae), bushbaby (Galagidae), tarsier (Tarsiidae). Center: aye-aye (Daubentoniidae). Not to scale. (Illustration by Don McGranaghan.)

indicate the survival until recently, on that island, of a much more varied lemur fauna than exists there today; extinct genera include several large-bodied and specialized forms.

IV. "Higher" Primates and Tarsiers

The living primates have been subdivided alternatively into the Prosimii (=Plesiadapiformes, Strepsirhini, and Tarsiiformes here) versus Anthropoidea or into the Strepsirhini versus Haplorhini, with the main difference involving the position of the tarsier. Most current research, both genetic and morphological, supports the phylogenetic linkage of tarsiers to the other "dry-nosed" or haplorhine primates (monkeys, apes, and humans), as opposed to the "wet-nosed" strepsirhines characterized, as noted above, by a tethered upper lip and greater sensory emphasis on olfaction. The resulting division is usually recognized at the taxonomic level of the suborder, as in our classification. However, as universally accepted today, the platyrrhines (New World monkeys) and the catarrhines (Old World monkeys, apes, humans, and their extinct relatives) shared a common ancestry distinct from all other primates, thus forming a natural group—the "higher" primates, or anthropoids. This group is formally recognized as the Anthropoidea and was previously ranked as a suborder, when tarsiers were considered prosimians. To recognize both of these important systematic evolutionary concepts, we include the Anthropoidea as a subunit within the Haplorhini, giving it the named but seldom-used rank of hyporder.

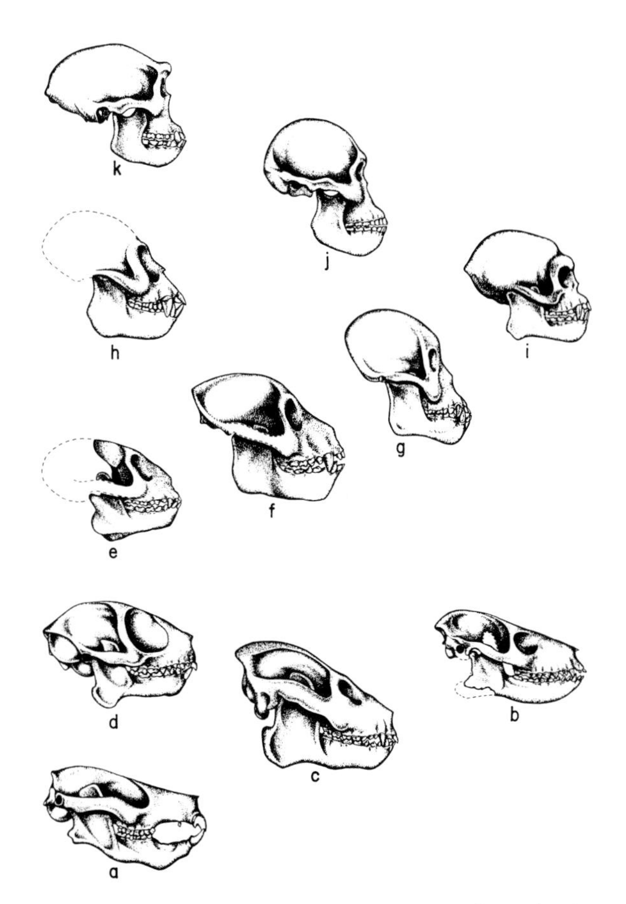

FIGURE 4 Skulls in lateral view of various extinct primates. a. *Plesiadapis tricuspidens*; b. *Notharctus tenebrosus*; c. *Leptadapis magnus*; d. *Necrolemur antiquus*; e. *Apidium phiomense*; f. *Propliopithecus* (=*Aegyptopithecus*) *zeuxis*; g. *Proconsul africanus*; h. *Cebupithecia sarmientoi*; i. *Pliopithecus vindobonensis*; j. *Australopithecus africanus*; k. *Homo erectus*. Not to scale. (Illustration by Don McGranaghan.)

FIGURE 5 Skeletons of various extinct primates. a. *Plesiadapis tricuspidens*; b. *Smilodectes gracilis*; c. *Propliopithecus* (=*Aegyptopithecus*) *zeuxis*; d. *Proconsul africanus*; e. *Pliopithecus vindobonensis*; f. *Mesopithecus pentelicus*; g. *Australopithecus afarensis*. Not to scale. (Illustration by Don McGranaghan.)

A. The Living Haplorhines

Modern tarsiers (Tarsiidae) occur as four species of the genus *Tarsius*. They are among the smallest primates, restricted to islands of southeast Asia and the Phillippines, where they live in pair-bonded groups. Their huge eyes are an adaptation to their principally nocturnal activity pattern, and their elongated leg and foot bones allow a rapid-leaping locomotion (Fig. 3).

The South American platyrrhines are considered by many to be the most diverse primate superfamily, in terms of diet, locomotion, and morphology. The family Cebidae includes the subfamilies Callitrichinae (marmosets and tamarins) and Cebinae, both characterized by a short face and lightly built masticatory system designed for a diet of insects, fruits, and gums. The marmosets are the smallest living anthropoids; they are generally pair-bonded, and the male takes an active role in the care of the young, often twins. The cebines include both large- (capuchins) and small-bodied (squirrel monkey) species, which live in large groups and have relatively large brains for platyrrhines. The family Atelidae includes mostly larger platyrrhines with a heavily built masticatory apparatus (jaws, teeth, projecting face, chewing muscles) adapted to a frugivorous (fruit-eating) or folivorous (leaf-eating) diet. The uniformly large atelines (including the spider and howler monkeys) are characterized by prehensile (grasping) tails and suspensory locomotion and feeding postures, although some species are acrobatic, whereas others move cautiously; most atelines live in moderately large social units. By contrast, the pitheciines may form pair-bonded social groups (the titi and the nocturnal–cathemeral owl monkey) or small heterosexual units (the somewhat larger-bodied sakis and uakaris). The latter forms have adapted to a diet of hard-shelled or hard-seeded fruits, for which they have developed a specialized dentition (Fig. 6).

FIGURE 6 Representative living higher primates. Clockwise from upper left: tamarin (Callitrichinae), spider monkey (Atelinae), orangutan (Ponginae), colobus (Colobinae), chimpanzee, human (Homininae), saki (Pitheciinae), macaque (Cercopithecinae). Not to scale. (Illustration by Don McGranaghan.)

The Old World monkeys (superfamily Cercopithecoidea) are the most numerous living primate group, with some 75 species in about 15 genera, and among the most widely distributed. They share a distinctive dental pattern and the presence of ischial callosities, tough rump patches on which they sit in trees (also found in gibbons and possibly an ancient catarrhine feature). One family, Cercopithecidae, is divided into two subfamilies, Cercopithecinae and Colobinae. The former is characterized by cheek pouches for temporary food storage, a simple stomach, and generally a somewhat to greatly elongated snout; the diet is eclectic, including fruits, some leaves and shoots, nuts, insects, and even rare vertebrate prey. A partly terrestrial quadrupedal pattern of locomotion probably is typical for the cercopithecines, with several groups independently either strongly terrestrial or more fully arboreal. The colobines, on the other hand, are more acrobatic arboreal runners and leapers, with only one living species spending much time on the ground. Their diet is concentrated on leaves and shoots, as mirrored in their enlarged and sacculated stomach. Social organization varies widely among the cercopithecids, including bonded pairs, unimale "harems" with several females, and groups with roughly equal numbers of both sexes and between 15 and 200 members. African cercopithecines include mangabeys, baboons, mandrills, guenons, and patas monkeys, while the macaques occur in Northwest Africa and range from Pakistan to Japan and Indonesia; they live in environments from desert fringe through savannah, rain forest, deciduous forest to high mountains. Colobines are less tolerant, with the African colobus varieties and the Asian langurs and doucs found mainly in deciduous or evergreen forest, although "golden monkeys" range to high elevations in China and proboscis monkeys live in mangrove swamps.

To humans, the most intriguing primates are the apes, grouped (along with people) in the superfamily Hominoidea. The lesser apes, or gibbons (*Hylobates*), form one family found only in Southeast Asia. These pair-bonded fruit-eaters are most famous for their brachiating pattern of locomotion, involving ricochetal arm-swinging and below-branch suspension, as contrasted with the above-branch running of most monkeys. Gibbon males and females both defend their home ranges or territories by means of "song" calls of several types, which indicate their position to neighbors and cement the pair bond.

In the past, great apes (Asian orangutan, African chimpanzee and gorilla) were often classified together in the family Pongidae, as opposed to humans in Hominidae, but a variety of mainly genetic studies have demonstrated that humans and great apes are extremely close to each other in evolutionary terms and that African apes and humans form a phyletic unit or evolutionary subgroup as contrasted with orangutans. Thus, by analogy to other primates (and mammals in general), only a single family Hominidae is recognized to reflect the low diversity of this group, and it is in turn divided into the subfamilies Ponginae for the orangutan and its extinct relatives and Homininae for humans, chimpanzees, and gorillas. The orangutan is, almost like lorisoids, relatively nongregarious: the large adult males patrol broad ranges overlapping those of females and their offspring, but the two sexes seldom travel together. Gorillas live in social groups with one or two adult males and several females, whereas chimpanzees usually form small groups of males, females, and young or consort pairs, which fluidly "fuse" into larger bisexual groups and divide again every few days. Diets of most hominoids consist mainly of fruits, although gorillas also eat other plant parts, and chimpanzees have been seen to actively hunt various animals including monkeys. Chimpanzees also use twigs to obtain termites from their mounds and leaves to get water from tree cavities; thus, they are considered to make and use tools on a regular basis, and they teach this behavior to their offspring. Great apes do not truly brachiate (like gibbons do) as adults: orangutans use hands and feet interchangeably when climbing and walk quadrupedally on the ground; chimpanzees and gorillas "knuckle-walk" on the ground but also climb actively and swing manually in trees (except for adult gorillas).

B. The Fossil Record

The living tarsiers represent only a small fraction of the past diversity of the Tarsiiformes. The earliest tarsierlike primates, known as the Omomyidae, first appeared in the early Eocene, along with the lemurlike adapiforms (see above). This group was diverse throughout the Eocene in North America, with several species often found together. Additional species, including several in a distinct subfamily, occurred in Europe, and rare fossils possibly belonging to the Omomyidae have been recognized in Asia and northern Africa. Recently, fragmentary remains of possible close relatives of *Tarsius* itself have been recovered from the Oligocene Fayum deposits of Egypt and from the Miocene of Thailand. Most likely, an animal broadly similar to the Eocene omomyids was ancestral to the anthropoids, but no known form is a good candidate for such an ancestor.

The earliest New World monkeys appeared in South America late in the Oligocene, about 27 Myr ago, with no trace of earlier primates south of Texas. During the early Cenozoic, South America was an island continent, isolated from Africa by the expanding South Atlantic Ocean and from North and Central America by a broad Caribbean–Pacific connection, and the source area of the platyrrhines is, thus, a continuing problem in primate evolution. Both North America (with Eocene omomyids) and Africa (with somewhat earlier anthropoids; see below) have been championed, with the latter view more widely but not universally accepted. In one of the most fascinating aspects of New World monkey evolution, fossils that can be readily identified as belonging to each of the two living families occur by 20 Myr ago, and by 15 Myr ago, fossils that belong to each of the four living subfamilies have been identified; some of those fossils can even be identified as close evolutionary relatives or ancestors of certain living genera. This very early diversification of living lineages is uncommon among primates (see Figs. 1, 4, and 5).

By contrast, the history of catarrhine primates is best seen as a series of radiations, each replacing its antecedents after a brief overlap interval. The 30–35-Myr-old Fayum beds in northern Egypt provide an important "window" onto early catarrhine evolution, revealing members of two main groups or superfamilies. The Parapithecoidea were morphologically monkeylike, sharing general features of locomotor ability, relative brain size, and probably lifeways with living monkeys of both the Old and New Worlds. Although once proposed as ancestors of the cercopithecoids, they are today viewed as an early side-branch radiation of either the catarrhines or the anthropoids as a whole. They were mainly arboreal fruit-eaters; at least one species was unique among primates in having lost all lower incisor (front) teeth (Figs. 4 and 5).

Contemporary with the parapithecoids were several species of *Propliopithecus* (one is often called *Aegyptopithecus*). In tooth and skull morphology, these animals are more similar to the living catarrhines and may be similar to the common ancestor of all later Old World anthropoids. More fragmentary remains of possible early catarrhines have been reported from the later Eocene of Burma and recently from Algeria, Oman, and the earliest levels in the Fayum. Additional genera of "primitive" catarrhines, not closely related to either monkeys or apes, are known from the Miocene of Africa, Europe, and Asia. These forms, along with the older *Propliopithecus*, offer insight into the morphology and diversity of early catarrhines and may be simplistically grouped as the pliopithecoids, although several lineages are likely conflated in this group.

The oldest definite Old World monkeys (cercopithecoids), recognized by their distinctive tooth morphology, are known from the earlier Miocene (15–20 Myr) of eastern and northeastern Africa. The two modern subfamilies probably differentiated in Africa between 15 and 12 Myr ago, but they are not documented until about 10 Myr ago. Today, the leaf-eating colobines are almost entirely arboreal and generally of rather small size, but large, terrestrially active forms lived in Europe and Africa between 5 and 2 Myr ago and possibly extended into Mongolia. The same interval (the Pliocene) saw a great radiation of cercopithecine monkeys in Africa, the spread of macaques across Eurasia, and the development of a number of apparently independent terrestrial lineages.

The ape–human superfamily Hominoidea is also first represented in the early Miocene of Africa, where *Proconsul* and related forms occur between 23 and 14 Myr ago. This group is now usually thought to be similar to the common ancestor of the two living hominoid families: the gibbons (Hylobatidae) and the humans and great apes (Hominidae). No gibbon ancestors have been discerned in the fossil record, although previous researchers suggested certain pliopithecoids for that role; a major problem is that modern gibbons combine very conservative craniodental morphology with (secondarily?) small body size and unique locomotor adaptations, so that recognizing a "proto-gibbon" that lacks these unique features is operationally difficult.

The earliest members of Hominidae (the dryopithecines) apparently occurred in Africa and Arabia between about 17 and 14 Myr ago, but they cannot be linked with either modern subfamily. Later dryopithecines lived in Eurasia and Africa from 15 to 9 Myr ago; they are somewhat more like modern hominids in general, but still do not show any characters linking them unequivocally to either living subfamily. The orangutan lineage is the first to be documented in the fossil record, with members of the genus *Sivapithecus* known from Turkey, Pakistan, and India (and several other genera from China and Eastern Europe) between 12 and 7 Myr ago. *Sivapithecus* shares with orangutans several specialized features of the face and teeth (such as frontal sinuses derived from the maxillary sinus, narrow interorbital region, ovoid orbits, outwardly rotated upper canines, and constricted incisive canal [the passageway linking the floor of the nose to the roof of the mouth]), which can be readily recognized in even fragmentary fossils.

The hominine lineage is far less well documented in the 10–4 Myr interval, with only a few teeth or jaws in eastern Africa representing potential early relatives of the African apes and humans. No later fossils are linked to the chimpanzee or gorilla, but human evolution is rather well known after 4 Myr ago. *Australopithecus* includes five species characterized by brains slightly expanded relative to body size as compared with those of apes, small canine teeth incorporated into the incisor complex, and a variety of features related to their bipedal locomotor adaptation. By about 2 Myr ago, the earliest members of the genus *Homo* can be diagnosed by an increase in brain size, an apparently more modern postcranium and locomotor system, and the presence of stone tools. Humans occupied Eurasia by at least 1 Myr ago, but anatomically modern fossils are not known until around 100,000 years ago, in Africa and then the Near East. By about 30,000 years ago, modern humans had spread over the Old World including Australia, replacing more archaic varieties such as the European Neanderthals, and reached the New World by at least 12,000 years ago. [*See* COMPARATIVE ANATOMY.]

Bibliography

Clark, W. E. Le Gros (1971). "The Antecedents of Man," 3rd ed. Edinburgh University Press, Edinburgh.

Fleagle, J. G. (1988). "Primate Adaptation and Evolution." Academic Press, New York.

Napier, J. R., and Napier, P. H. (1985). "The Natural History of the Primates." MIT Press, Cambridge.

Richard, A. (1985). "Primates in Nature." W. H. Freeman, New York.

Szalay, F. S., and Delson, E. (1979). "Evolutionary History of the Primates." Academic Press, New York.

Tattersall, I. (1982). "The Primates of Madagascar." Columbia University Press, New York.

Tattersall, I., Delson, E., and Van Couvering, J. A. (eds.) (1988). "Encyclopedia of Human Evolution and Prehistory." Garland, New York.

Wood, B. A., Martin, L., and Andrews, P. (eds.) (1986). "Major Topics in Primate and Human Evolution." Cambridge University Press, Cambridge.

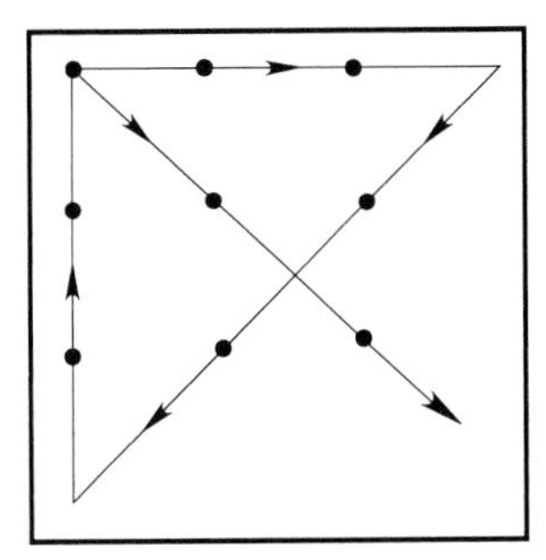

Problem Solving

K. J. GILHOOLY, *Aberdeen University*

Glossary

Problem reduction representation Representation of a problem as a set of goals, subgoals, sub-subgoals, and so on, which need to be accomplished before solution

State space representation Representation of a problem as a set of possible states into which the problem material can be manipulated

I. General Characterization

A. Defining Problem Solving

What is a problem? The definition offered by the Gestalt psychologist Karl Duncker many years ago is still serviceable. He wrote that "a problem arises when a living organism has a goal but does not know how this goal is to be reached."

This is a useful initial formulation that signals a number of points. First, that a "task" set by an experimenter is not necessarily a problem for a given individual. Whether a task is a problem depends on the person's knowledge and on his or her ability to locate relevant knowledge, should he or she have it. Second, a problem may vanish or be dissolved if the person changes his or her goals. A third point is that a problem does not effectively exist until the person detects some discrepancy between his or her goals and the current situation.

B. Dimensions of Problems

Most psychological studies of problem solving have dealt with well-defined problems. If we accept the useful proposal that problems in general can be viewed as having three components (viz., a starting state, a goal state, and a set of processes that may be used to reach the goal from the starting state), a problem is well-defined if all three components are completely specified. Problems in mathematics, in logic, and in board games tend to be well-defined. Although well-defined, such problems can be difficult, and the psychologist is faced with the task of explaining how we humans, with our various limitations, manage to solve geometry, chess, and similar scale problems in reasonable times. Of course, it will be still more difficult to explain how we tackle those ill-defined problems that are more typical of real life than the well-defined variety.

Ill-defined problems leave one or more components of the problem statement vague and unspecified. Problems can vary in degree of definedness, ranging, for example, from "make a silk purse out of a sow's ear" (which leaves possible methods unspecified) to "make something useful from a sow's ear" (which has a vague goal and unspecified methods) to "make something useful" (which leaves vague the goal, the starting state, and the methods available).

It seems a reasonable strategy for psychologists to start by studying people's ways of handling apparently well-defined problems and then to move on to consider ill-defined tasks. Research suggests that people tackle ill-defined tasks by seeking a well-defined version of the problem, which they then work within until the problem is solved or a new definition is tried. Thus studies of how well-defined problems are solved will be relevant to part of the process of solving ill-defined problems.

Another useful distinction is that between adversary and nonadversary problems. In an adversary problem, the person is competing with a thinking opponent, whereas in nonadversary problems, the struggle is with inert problem materials (real or symbolic) that are not reacting to what the person does with a view to defeating him or her (despite what the problem solver may feel).

A third distinction that has become increasingly important in problem-solving research is that between knowledge-rich and knowledge-lean problems. To a large extent, this distinction refers to the solver's view of the problem. A problem is knowledge-rich for solvers who bring considerable relevant knowledge to the problem. For example, if someone has just been told the basic rules of a game that they have never encountered before, a problem in that game would be knowledge-lean for that person, but the same problem would be knowledge-rich for an expert player of the game. Many artificial puzzles used in studies of problem solving are knowledge-lean for most people. Until relatively recently, most studies of problem solving focused on knowledge-lean puzzles, particularly in the case of nonadversary problem solving. More recently there has been considerable interest in the study of knowledge-rich nonadversary tasks (e.g., computer programming and physics problem solving). Such studies frequently involve contrasts between the behavior of people for whom the problems concerned are knowledge-rich (experts in the area) with people for whom the problems are knowledge-lean (novices in the area).

C. Outline of Main Stages in Problem Solving

A few steps are generally discerned in problem solving. These are the following:

1. Detecting that a problem exists; realizing that there is a discrepancy between the current situation and a goal and that no way is known to reach solution without search. In the course of working toward the major goal, numerous specific subgoals may arise and provide further targets for problem solving.

2. Formulating the problem more completely. How are the starting conditions to be defined and represented? How are the goals to be defined and represented? What relevant actions are available? A problem formulation is an internal representation (or mental model) of the problem components, and solution attempts are made within this representation.

Two broad ways of representing problems are (a) in state space terms and (b) in problem reduction terms. In the state space approach, the solver typically works forward from the starting state and explores alternative paths of development as these branch out from the starting state. Within the state space approach, the solver may also work backward by inverse actions from goal states toward the starting state. A resulting successful sequence discovered by mental exploration could then be run forward to actually reach a goal state from the starting state. Indeed, both forward and backward exploration may be tried in various mixtures within the same problem attempt. In the problem reduction approach, the overall problem is split up into subproblems. For example, if the goal state involves a conjunction of conditions (achieve X *and* Y *and* Z), problem reduction would seek to achieve each condition separately (either by further problem reduction or by state space search). A key method within problem reduction is means–ends analysis. In this method, subgoals are set up for achievement such that those subgoals would lead to achievement of higher level goals. These subgoals correspond to subproblems, which can themselves be reduced. Problem and subproblem reduction continue until subgoals are reached that can be immediately solved because a method is already known for them.

3. Given a representation and a choice of approach (state space or problem reduction), solution attempts can begin. Although in the case of well-defined problems, computers could in theory continue until solution or until all possible move sequences have been exhausted, humans usually give up—or at least set problems aside temporarily—long before all possible state action or problem reduction sequences have been explored. That is to say, stop rules are invoked. Different types of stop rules can be readily imagined. For example, one type of rule would be to stop exploring within a particular representation if solution has not been reached by a certain time and return to the problem formulation stage to produce a revised formulation for further exploration. A second type of rule would be to accept a "good enough" problem state rather than continue exploration for an optimal state (sometimes called *satisficing*). A third type of rule would be to only stop if an optimal solution is reached (i.e., a "maximizing" rule). A further type of stop rule would be an "abandon" rule (i.e., a judgment that further reformulation and/or exploration would not be worthwhile and the problem should be left unsolved). Intuitively, this would seem to involve rather complex assessments of the chances of succeeding or failing and of the cognitive and other costs of continuing work or abandoning the problem. Again, this type of stop rule appears to have been little researched.

II. Common Blocks to Problem Solving

A. Set

"Sets" are fixed approaches to solving certain classes of problems, and these fixed approaches can be powerful blocks to solution. A robust demonstration of such effects can be obtained using water jar problems. In these tasks, the subjects are asked to say how they could get exactly a specified amount of water using jars of fixed capacities and an unlimited source of water [e.g., give three jars (A, B, C) of capacities 18, 43, and 10 units, respectively, how could you obtain exactly 5 units of water?]. The solution may be expressed as B − A − 2C. After a series of problems with that same general solution, subjects usually have great difficulty with the following problem: Given three jars (A, B, and C) of capacities 28, 76, and 3 units, respectively, how could you obtain exactly 25 units of water? In fact, the solution to this problem is very simple (i.e., A − C), but when it comes after a series of problems involving the long solution (B − A − 2C), many people fail to solve or are greatly slowed down compared with control subjects.

Similar results have also been found with other types of problems such as anagrams. If subjects have a run of anagrams with the same scrambling pattern, they will be slowed down when an anagram with a different scrambling pattern appears in the sequence. Again, if there is a run of "animal" solutions to a sequence anagram, solving will be slowed down when a "nonanimal" solution is required.

As well as training effects, the layout of a problem can induce strong sets. For example, if subjects are given a pattern of nine dots as below

and asked to draw four straight lines to connect up all the dots without raising the writing tool from the paper, few solve it within 20 min. Most subjects are misled by the square layout into keeping their efforts within the square (although no such restriction was imposed in posing the problem). The problem cannot be solved while this "set" is dominant. Solution

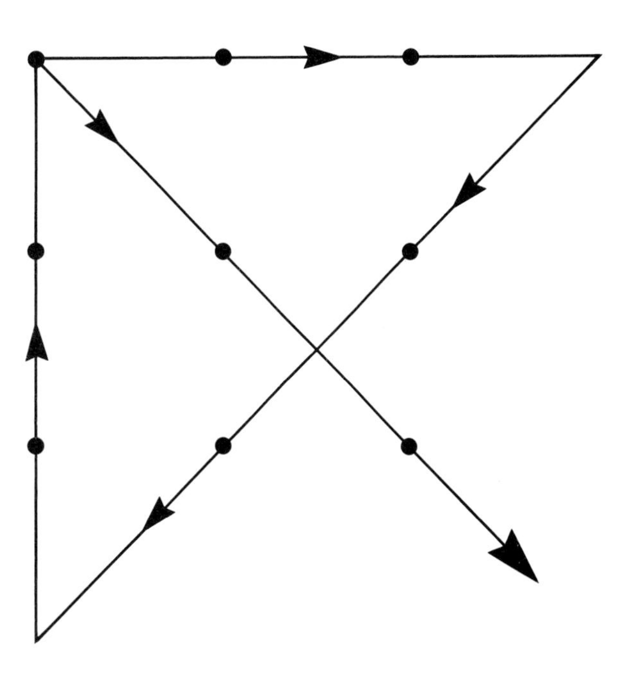

requires the lines to go outside the square shape.

B. Functional Fixity

A related block to effective problem solving, known as *functional fixity*, has also been identified. Functional fixity tends to arise when an object has to be used in a new way to solve some problem. The classic study of functional fixity was carried out by Duncker in 1945 using the "box" (or "candle" problem). In this task, subjects were presented with tacks, matches, three small boxes, and three candles. The goal was to put the candles side by side on a door, in such a way that they could burn in a stable fashion. For one group of subjects the boxes were empty, but for the other group (experimental group) the boxes were used as containers and held the matches, tacks, and candles. The solution is to use the boxes as platforms and fix them to the door using the tacks. It was found that the solving rate was much higher in the control group than in the experimental group. Duncker explained this result in terms of a failure to perceive the possible platform function of the boxes when they were presented as containers. Functional fixity has been independently demonstrated and further investigated in a large number of later studies. The phenomenon of functional fixity is a robust one and, along with "set," is doubtless a major source of difficulty in real-life problem solving.

III. Expertise in Problem Solving

It is a striking fact that an expert in a given area can come up with much better and often faster solutions than can a beginner. How is this done? Does the expert, for instance, mentally run through a much larger number of possible actions more rapidly than the beginner?

Studies of chess skill have been informative about the nature of expertise; chess is a good area for this type of research because differences in skill level are easily measured. Early studies by the Dutch psychologist De Groot reported that experts and less-skilled players differed relatively little in the amount of mental searching they did abut possible moves and counter moves. De Groot was able to keep track of the mental search patterns of the players by having them think aloud as they chose moves. Both his expert and less-expert groups of players looked ahead a maximum of about six steps and considered totals of about 35–50 moves and countermoves before deciding on their own best move. However, the more-skilled players always chose much better moves than did the less-skilled. A clue as to how the chess masters differed from the less-skilled players came from studies in which the subjects were shown chess positions for short times (5 sec). After such short exposures, master level players could recall the board with greater than 90% accuracy, whereas less-skilled players performed with around 40% accuracy. When the chess positions were made up by placing pieces at random on the board, however, both the master level and less-skilled players scored about 40% correct recall of the boards after short exposures. So it was inferred that the master level players had built up in their permanent or long-term memories a large number of familiar chess patterns that they could then recognize when a new but realistic position appeared. If pieces were arranged at random, few if any familiar patterns would appear, and performance would fall back to amateur levels. Similarly, in playing the game and choosing moves, familiar patterns will be recognized and will guide the search more effectively. More recent research, using larger numbers of subjects and a wider range of chess skill, has also shown that more skilled players do in fact tend to search somewhat more extensively and more rapidly than less-skilled players and are better at evaluating the potential of intermediate positions. The amount of searching is, however, but a fraction of that carried out by the best computer chess programs.

Evidence from biographies of leading chess players indicates that about 10 years of concentrated study is required to reach expert levels. An early start also seems beneficial. Studies of other areas of mental skill likewise indicate about a 10-year period of intensive practice before high-level performance is reached. Thus experts' apparently effortless solving of new problems in their fields is purchased with many years of prior effort. Prior experience has built up the extensive background knowledge that experts use to describe new problems effectively, to fill in implicit details, and to anticipate likely directions of solution. Results similar to those in chess have been reported now in a wide range of problem areas, including physics, political science, mathematics, and programming. However, when experts encounter problems that exceed their expertise, they too must revert to the more general methods of search outlined in Section I of this article.

Bibliography

Gilhooly, K. J. (1988). "Thinking: Directed, Undirected and Creative," 2nd ed. Academic Press, London.
Gilhooly, K. J. (ed.) (1989). "Human and Machine Problem Solving." Plenum Press, New York.
Holding, D. H. (1985). "The Psychology of Chess Skill." Lawrence Erlbaum, Hillsdale, New Jersey.
Sternberg, R. J., and Smith, E. E. (1988). "The Psychology of Human Thought." Cambridge University Press, Cambridge, Massachussetts.

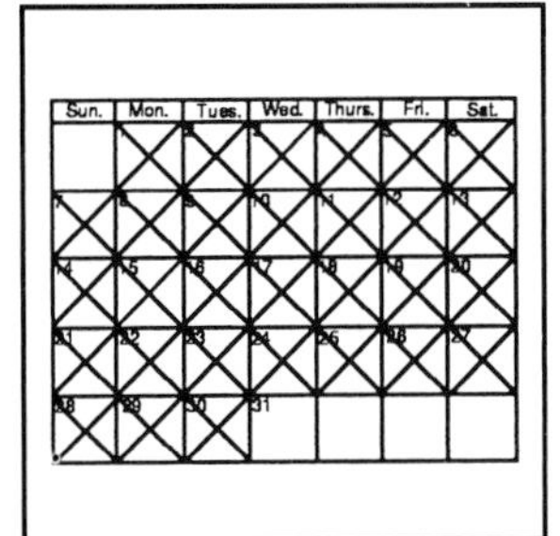

Procrastination

NOACH MILGRAM, *Tel-Aviv University*

Glossary

Frustration tolerance Ability to sustain goal-directed behavior on a particular task in an effective manner despite aversive situational features and associated unpleasant emotional responses

Locus of avoidant behavior Variables that affect task-avoidant behavior may be found either in the person (internal) in the form of traits or generalized proactive avoidant tendencies, or in the task (external) in the form of aversive stimuli that evoke reactive avoidant tendencies; internal variables predict differences among people, and external variables differences among tasks

Procrastination Life pattern that is characterized by inefficient behavior in decision-making, setting of priorities, scheduling, and adherence to schedule in one or more life areas and is a source of personal distress

Task motivation Procrastinator may approach any given task with one of the following motivations: (1) to do one's best, (2) to avoid the possibility of success, (3) to avoid the possibility of failure, (4) to guarantee partial or complete failure

Time frame Span of time whose starting point is when one first becomes aware that a given issue should be resolved or a given task performed and whose end point is the deadline for accomplishing it or shortly thereafter; may be objective (e.g., from the date when an examination is announced until the day of the examination itself) or subjective (e.g., from the time one notes that the laundry should be done until one runs out of clean clothes)

PROCRASTINATION is a maladaptive behavior with regard to a ubiquitous, essential cognitive and executive process—the setting of task priorities, determining when a given task is done and how well it is done, the selection from innumerable choices of a particular course of action, and following through with one's decision. This paper attempts to examine several conceptually important distinctions within procrastination and to formulate an integrative approach to understanding the diverse phenomena subsumed under the term. It surveys the major etiological and intervention perspectives and summarizes recent research on personal–social characteristics associated with procrastination.

I. Procrastination—A Puzzling Phenomenon

Procrastination, loosely defined as putting off for tomorrow what one should do today, is an extremely common, puzzling phenomenon. Between 10 and 40% of high school and college students display academic procrastination; i.e., they fail to complete school assignments on time, they prepare for examinations at the last minute, and, as a consequence of these actions, they do less well in school. A more disruptive form of procrastination is chronic indecision about major adult privileges–responsibilities (e.g., vocational choice, job search, marriage, parenting) and/or failure to follow through on one's decisions. A third, subtle, self-defeating form of procrastination is chronic indecision and inefficiency in scheduling and performing the various routines of daily living (e.g., paying bills, preparing meals, doing the laundry, meeting social obligations).

Because procrastination is counterproductive and self-defeating, why do so many people procrastinate? One answer, typically given with regard to questions about other human errors of commission or omission, is that procrastination, like any bad habit, gratifies some underlying needs of the procrastinator. It is, however, a most peculiar bad habit. On the one hand, it can be highly detrimental to one's well-being (like drinking, smoking, or eating to excess). On the other hand, it does not have an equally damaging social reputation: It is not regarded as embarrassing or shameful.

II. Inadequacies in the Layperson's Behavioral Definition

The layperson's definition of procrastination refers to points in a time frame. *When* does someone make a decision and/or do a particular task—promptly, as soon as one becomes aware of the issue or the task; somewhat later; toward the end of the time frame; at the last minute; or after the deadline, if at all? To define procrastination exclusively by time frame, however, is problematic for methodological and conceptual reasons.

A. The Methodological Problem

It is difficult to formulate a uniform operational definition for different points (e.g., prompt, later, still later, last minute, after deadline) in the time frame of a given task for a given person, much less for a number of tasks and people. Some tasks have short time frames (e.g., when to get up in the morning to go to school or work has a time frame ranging from minutes to an hour or two). When to do the laundry has a time frame of several hours to several days, but usually not several weeks. Scheduling a dental appointment has a still longer time frame, weeks or even months.

Length of time frame for the same task may vary, moreover, from person to person. The time frame for a dental appointment is different for people with or without dental problems. A person with no dental problems may skip a 6-month checkup with impunity; a person with chronic dental difficulties may skip a 3-month checkup at his or her own risk. Similarly, the time frames for deciding which job to accept or which apartment to rent or whether or not to marry someone have highly idiosyncratic temporal lengths.

Tasks with definitive starts and deadlines may be found at school and at work (e.g., submitting college applications, passing in a written report, preparing for and taking an exam, paying bills). The time frames of these tasks are uniform for large numbers of people because of the practice of formally dating when an assignment is given and when it must be completed. On the other hand, except for doing the task immediately (promptly, a counterindication of procrastination) or doing it past the deadline (late, a clear indication of procrastination), all other points in the time frame are open to differing subjective assessments as to degree of procrastination. We may choose to accept each individual's assessment of his or her promptness or procrastination on each task or we may rely on group norms computed by averaging the subjective ratings of many people on each task. In either case, problems of objective measurement and statistical analysis abound.

B. The Conceptual Problem

There is also a conceptual obstacle to using time frame ratings as indices of procrastination. When a woman calmly decides to do some tasks fairly late in a given time frame—because they are less important than other tasks or because the other tasks have more stringent time pressures—and does so in an efficient manner, she is not procrastinating, but merely establishing priorities for what gets done and when, and following through. Similarly, when a person has no intention of doing a particular task but engages in apologetic postponements in preference to outright noncompliance, she is simply using a well-known social strategy to avoid unpleasant arguments. As we shall see, procrastination is defined by a number of interrelated criteria, some objective and others subjective.

III. Necessary Criteria for Comprehensive Definition

To better understand procrastination defined here as a disruptive, bad habit, it is necessary to define the phenomenon in a comprehensive and concise manner. Four interrelated criteria are considered necessary for a comprehensive diagnosis of the phenomenon.

A. Inefficient Behavioral Process

We live in a world in which we are beset by countless issues and tasks, major and minor, unusual and routine. These issues and tasks demand that we (1) make decisions whether to do or not to do them, (2) establish priorities about when to do them, (3) select problem-solving strategies about how to do them, and (4) carry out our decisions. We may engage in this decision-making, priority-assigning, and problem-solving sequence of behaviors in an efficient manner. Or we may procrastinate.

Procrastination is an inefficient behavioral sequence of starts and stops. The procrastinator may spend hours or days in a closed loop on any of the above phases (e.g., postponing decisions, failing to follow through, and repeating the scheduling–aborting sequence again and again). This inefficient behavioral process may be applied to decision-making and problem-solving on major issues and tasks in life as well as to inefficient processing of the many minor routines of daily living.

Delays in making decisions and in acting upon them are also costly. By delaying, we may be unable to take advantage of a particular option. For example, someone else bought the house we wanted; by the time we decided to have children, we found that we were unable to do so safely, if at all. A lengthy, drawn-out decision-making process reduces the amount of time and energy available for investment in other pursuits. Finally, inefficiency in handling the routines of daily living contributes to a sense of futility, the feeling that we are drowning in a sea of details, that we never have the time and opportunity to do the really important things. This sense of futility contributes to what has been termed a high hassle index (feeling hassled by life routines) and is regarded by some as a major stressor in modern life with highly adverse effects on mental and physical health.

B. Consequent Substandard Behavioral Product

As a consequence of inefficient scheduling and failure to adhere to one's schedule, many people complete tasks at the last minute, but at a price in quality: They do them less well. In other words, a consequence of an inefficient behavioral process is an inadequate behavioral product. Studying for an exam at the last minute usually means poorer preparation and receiving a lower grade than if one studied for the examination earlier with opportunity for additional review as needed. This criterion, substandard product, is met, however, only if the poor performance is due to the inefficient behavioral process. If the student does poorly because of lack of ability or interest in the course of study, then we are dealing with a different phenomenon.

C. Perceived Legitimacy of Task Demand

Behavioral inefficiency and associated substandard behavioral product are not enough. Procrastination is restricted to tasks that we wish to do or feel compelled to do. We acknowledge the legitimacy of the claim of the task in question on our time and effort. Two men may be engaged in identical delaying behavior. The one acknowledges that he should do the task in question, do it well, and go about it in a reasonably efficient manner, but he fails to do so. The other has no desire or intention to do the particular task at all or to do it well. The former is procrastinating, the latter is simply marching to the tune of a different drummer. In other words, procrastination is defined in part by subjective self-report. Or to coin a phrase, "Procrastination is in the mind of the performer and not in the eyes of the beholder."

D. Consequent Emotional Upset

The fourth criterion that closes the circle described by the other criteria is emotional upset. People who typically procrastinate become upset in varying degrees about their behavior because they acknowledge the legitimacy of the task demands and are painfully aware that they have not met their commitments in a proper manner. This kind of inefficient, unproductive, undisciplined behavior is open to criticism, both from others and from oneself, with consequent adverse effects on one's self concept.

In summary, the current consensus in the field is that a comprehensive conceptualization of procrastination must necessarily deal with the following issues: the person's acknowledgement that the task should be done, that it should meet some personally defined standard, and that it should be completed with reasonable attention to deadlines; individual versus group standards for orderly behavior, promptness, adequate preparation, and product quality; the observed efficiency of the behavioral process and the quality of the product; and personally experienced distress associated with perceived behavioral inefficiencies and inadequacies.

IV. Three Theories of Etiology and Intervention

Three theoretical schools have dealt with procrastination: the psychodynamic or psychoanalytic, the behaviorist, and the cognitive behavior modification perspective.

A. The Psychodynamic School

This school, based on the psychoanalytic principles of Sigmund Freud, emphasizes largely unconscious, irrational, and unrealistic psychic events—wishes and fears, and/or traumatic childhood experiences—as responsible for the self-defeating behaviors defined by procrastination. By means of depth (uncovering) interviews, clients become aware of these unconscious factors, learn to evaluate them more realistically, integrate them into an increasingly positive self-concept, and eventually dispense with the maladaptive procrastinatory behaviors that were motivated by these psychic events. For example, a person may be unconsciously motivated by fear of success. To succeed is equivalent to committing an unacceptably aggressive act against parents or siblings and to risk loss of their love or support. Hence, one strives to avoid success whether in school or at work. As a consequence, one engages in the kind of avoidant, inefficient behaviors that we have been describing as procrastination and thereby guarantees that one will not succeed. [*See* PSYCHOANALYTIC THEORY.]

B. The Behaviorist School

The behaviorist school is based on the work of Ivan Pavlov, John Watson, B. F. Skinner, and Joseph Wolpe. According to this approach, a procrastinatory pattern develops when a neutral stimulus acquires aversive properties because it occurs in temporal and spatial contiguity to a genuinely aversive stimulus (classical or Pavlovian conditioning). Thereafter, one continues to behave as if the original aversive episode were about to recur when exposed to the conditioned stimulus, and attempts to avoid these anxiety-producing situations (operant or Skinnerian conditioning). Avoidant responses are less anxiety-provoking than confrontational responses and are by definition reinforcing: Behaviors that reduce anxiety are more likely to recur than behaviors that increase anxiety. For example, a man may have been conditioned to avoid the possibility of a devastating failure experience by employing procrastinatory avoidant behaviors. Because failure is typically associated with a failure to achieve a desired result because of lack of native ability, the procrastinator is insulated from this experience and, instead, will attribute his substandard achievement to lack of time to do better.

Treatment consists of analyzing situational cues and contingencies that maintain maladaptive habits, introducing new cues and incentives to reinforce adaptive habits, and utilizing techniques of relaxation training and systematic gradual exposure to anxiety-arousing situations to achieve desensitization (Wolpe) and permit disciplined coping with aversive, stressful situations.

C. Cognitive Behavior Modification

This school is based on the work of Albert Ellis, Donald Meichenbaum, and Cyril Franks. It emphasizes the identification of those cognitive activities—images, ideas, expectancies, and self-initiated verbalizations—that contribute to the acquisition and maintenance of bad habits such as procrastination. For example, a young woman may have been raised in a home in which the major way she was able to assert her independence in the face of parental domination was by indirect subtle, passive resistance. This defensive style is later generalized to routine demands made by society in general, her employer and husband in particular. Procrastination lends itself admirably to a passive–aggressive orientation. By procrastinating, the woman is not refusing to do a particular task, but she is asserting that she has not had the opportunity to do so as yet. She may state again and again her desire to do what is required but, in point of fact, never completes the task in question. Her sense of adequacy as a person is bolstered paradoxically by her substandard decision-making and problem-solving behavior and work quality, because she interprets the situation in self-assertive terms: No one is going to push me around.

Another example of a counterproductive cognition relevant to procrastination is the overgeneralized truism: "Anything worth doing is worth doing (very) well or not at all." In point of fact, the opposite may be true. Many things are not worth doing as a matter of personal preference, but we may acknowledge that they must be done because of social convention or other external demand. In these circumstances, we will do them, not necessarily to the best of our ability, but rather to meet the minimal requirement set by those imposing the task upon us. Moreover, given our priorities, we may freely assign to a particular task a low investment priority and give a minimum of time and skill in it, so as to be free to devote more effort to other tasks.

Treatment consists of identifying these unrealistic, exaggerated self-defeating generalizations that one transmits to oneself and replacing them with more appropriate messages to be administered to oneself in future decision and task adherence situations.

The three schools differ markedly in terminology, underlying assumptions, and formal therapeutic techniques, but some of the differences are more apparent than real, especially in the case of the latter two. A given therapist or treatment center may use concepts and techniques from all three, depending on the particular kind of procrastination presented by the client. When procrastination is confined to a circumscribed area, as in academic procrastination, symptomatic group intervention is recommended, i.e., workshops and study groups providing instruction and encouragement to use behavioral techniques to replace bad study habits with good ones. Conversely, when procrastination is generalized to many life areas and disrupts work or home commitments or both, intervention is more intrusive and designed to uncover the covert verbalizations or the unconscious motivations that contribute to procrastination.

V. Survey of Research

A. Clinical Research on Procrastination

Most of our knowledge about procrastination and procrastinators has been obtained from clinical interviews of clients in psychotherapy and counseling. Certain personal–social characteristics were found to be conducive to the phenomenon—fear of failure; conversely, fear of success or motivated underachievement; passive–aggressive orientation toward performing socially required tasks; and low frustration tolerance and poor self-regulation skills. The latter terms refer to the difficulty some people experience in coping with disagreeable decisions or tasks. Often, people must choose between two unattractive alternatives, a difficult situation that has been termed an avoidance–avoidance conflict. Less difficult is deciding between two attractive alternatives (an approach–approach conflict). Most difficult of all is deciding between two or more alternatives each characterized by attractive as well as unattractive features. Consequently, people characterized by low frustration tolerance and poor ability to control or regulate their upset feelings, obsessive thoughts, and impulsive, maladaptive actions are considered to be at high risk for procrastination.

B. Survey of Recent Empirical Research

Given the widespread nature of the procrastination, it is instructive that few empirical research studies deal with the general phenomenon or with its personal–social correlates in particular. A brief summary of major findings follow.

Two behavioral manifestations of procrastination have been discussed: procrastination as defined by *when* a task is completed (points on a time frame) and as defined by *how* it is done (inefficient behavioral process). The question may be raised whether these behavioral phenomena are wholly independent or highly correlated. If the former is found, this means that a great many people are comfortable in electing to do things relatively late or toward the end of their respective time frames and, in fact, adhere to their schedules without undue difficulty. Psychometrically reliable estimates of time frame and of process procrastination were obtained on a series of daily life routines. The high positive correlation between the two (nearly 0.70) indicated that people who complete tasks early in the time frame are highly efficient in their scheduling and adherence behavior; conversely, the small number of people who acknowledge completing tasks near the end or at the last possible minute almost invariably report inefficient scheduling and inefficient adherence behaviors with regard to these tasks.

From these studies, we also learn that persistent procrastination in routine life tasks is associated with a perception of life routines as unpleasant impositions from the outside for men and women, and with low scores on life satisfaction, learned resourcefulness (the ability to employ behavioral strategies to regulate one's interfering cognitive and emotional responses), and Type A time pressures to work rapidly for men only.

Procrastinators are more sensitive to the attractiveness of a given project than nonprocrastinators, giving it more time if they like it and less if they do not. Nonprocrastinators, by contrast, are more disciplined or more confirming, depending on one's point of view, and doggedly persist on projects regardless of their personal appeal. Procrastinators are less likely to complete stressful projects, whereas nonprocrastinators tend to regard stressful projects as challenges and positive experiences they are likely to complete. Procrastinators are less rational than others because, while they elect to do things at the last minute, they lack the skills requisite to accomplish this goal: They tend to underestimate the amount of time needed to complete projects and are lacking in organizational tendencies and organizational skills.

Recent research calls attention to different kinds of procrastinators as well as to the possibility of different kinds of nonprocrastinators. With regard to the former, two types of procrastinators have been identified by factor analysis of procrastination and other trait scales: Those characterized primarily by high neurotic disorganization and rebelliousness who fail to invest adequate time in their projects and regard them as difficult and stressful; and those characterized by high neurotic disorganization and low organizational skill and energy level, whose projects are low in difficulty and stress level. The latter group may be regarded as classic underachievers.

On the positive side, we find procrastinators high in breadth of interests and in self-engaged independent thinking who remind us of procrastinators cited earlier—people who are not upset about their failure to make or meet deadlines because they are marching to the tune of a different drummer, whether out of rebelliousness against or indifference to social pressures. Some of these people are able to find or to create unusual work and home settings that permit them procrastinatory license and afford them numerous satisfactions.

With regard to nonprocrastinators, it is worthwhile to differentiate between those who are apprehensive about the possibility and consequences of being late and those who are not. The former group is likely to be characterized by high social sensitivity and fear of social rebuff, and the latter chiefly by concerns about behavioral efficiency.

VI. Integration

To integrate many of the concepts relevant to procrastination, a 3-×-4 model of procrastination has been formulated, with source of task avoidant response on the vertical axis and one's task motivation on the horizontal. The former refers to the primary source of the avoidant behaviors—traits or generalized response tendencies within the individual (internal), aversive stimulus features of tasks or classes of tasks (external), or both (internal–external). Trait measures permit us to predict the behavior of individuals across tasks, so-called state or task measures permit us to predict the relative aversiveness of tasks across groups of people, and a combination of the two permit us to predict the behavior of people in different tasks.

The horizontal axis refers to one's motivation with regard to the task at hand. One's motivation or goal ranges from the positive pole of wishing to complete the task in a creditable manner (success striving) to the negative pole of wishing to fail (failure seeking). Intermediate points on this axis would be the desire not to achieve success (success avoidance) and the desire to avoid the appearance of failure (failure avoidance). Success avoidance and failure avoidance are not equivalent, because the former is closer to the positive end of the continuum and the latter to the negative end.

These levels exhaust all theoretical possibilities. A person without procrastinatory tendencies (internal) is by definition not likely to procrastinate. A task (external) without aversive stimulus properties does not elicit procrastinatory responses. Finally the absence of any goal with regard to a task (suc-

TABLE I Procrastination Model

	Task motivation			
Locus of avoidant behavior	Success striving	Success avoiding	Failure avoiding	Failure striving
Internal	1	2	3	4
Internal–external	5	6	7	8
External	9	10	11	12

cess or failure) means that whatever the quality of one's task-directed behavior, it does not meet the third criterion of procrastination, perceived legitimacy.

The 3-×-4 model yields 12 cells, each representing a different avoidance source × task motivation situation. We would predict the least procrastinatory behavior in cells 1, 5, and 9, and the most in cells 4, 8, and 12. Cells 5–8 would yield higher procrastination scores than 1–4 or 9–12 because high-avoidant people procrastinate more on a highly aversive task than low-avoidant people, or high-avoidant people on a less aversive task. Cell 8 would procrastinate more than any of the others. The model permits investigation of these hypotheses and offers treatment or intervention implications. Procrastination determined primarily by the aversive nature of the particular task may be better treated by a behaviorist orientation, whereas procrastination determined primary by generalized traits within the person by a psychodynamic or cognitive one. The differentiation between the nuances of success avoidance, failure avoidance, and failure striving would have distinct implications for the appropriate treatment.

Bibliography

Burka, J. B., and Yuen, L. M. (1983). "Procrastination: Why You Do It, What To Do About It." Addison-Wesley, Reading, Massachusetts.

Ellis, A., and Knaus, W. J. (1975). "Overcoming Procrastination." Institute for Rational Living, New York.

Lay, C. H. (1986). At last, my research article on procrastination. J. Res. Pers. **20,** 474–495.

Lay, C. H. (1987). A modal profile analysis of procrastinators: A search for types. Pers. Individ. Diff. **8,** 705–714.

Milgram, N. A. (1988). Procrastination in daily living. *Psychol. Rep.* **63,** 752–754.

Milgram, N. A., Sroloff, B., and Rosenbaum, R. (1988). The procrastination of everyday life. *J. Res. Pers.* **22,** 197–212.

Rorer, L. G. (1983). "Deep" RET: A reformulation of some psychodynamic explanations of procrastination. *Cog. Ther. Res.* **7,** 1–10.

Rosenbaum, M. (1982). Learned resourcefulness as a behavioral repertoire for the self-regulation of internal evens: Issues and speculations. *In* "Perspectives in Behavior Therapy in the Eighties" (M. Rosenbaum, C. M. Franks, and Y. Jaffe, eds.). Springer, New York.

Sabini, J., and Silver M. (1982). "Moralities of Everyday Life." Oxford University Press, New York.

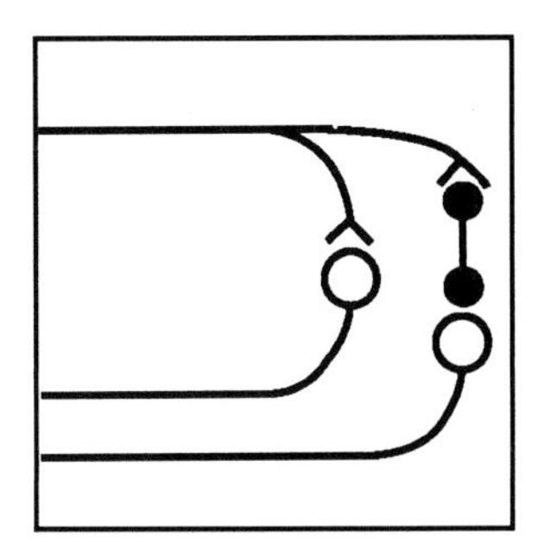

Proprioceptive Reflexes

S. C. GANDEVIA, DAVID BURKE, *The Prince Henry and Prince of Wales Hospitals and School of Medicine, University of New South Wales*

Glossary

Antagonistic muscles Muscles with actions that are anatomically opposite (e.g., flexion and extension of a joint)
Corticopontine Fibers that arise in (or close to) the motor cortex and are destined for the brainstem
Corticospinal Fibers that arise in (or close to) the motor cortex and reach the spinal cord
Disynaptic Involving two synapses
Homonymous Homonymous motoneurons (or receptors) are those innervating (or in) the same muscle
Interneuron A relay cell within the spinal cord
Latency Time taken between a stimulus and a response
Monosynaptic Involving only one synapse
Motoneuron Cell within the brainstem or spinal cord that sends an axon to a skeletal muscle and innervates a number of muscles fibers
Oligosynaptic Involving a few synapses (perhaps three to five)
Postsynaptic inhibition Mechanism for reducing the excitability of a neuron via a neurotransmitter, which hyperpolarizes the neuron
Presynaptic inhibition Mechanism for reducing the efficacy of a synapse by reducing the amount of neurotransmitter released at the synapse
Proprioceptive input Input to the nervous system from receptors in muscles, joints, and skin
Proprioceptive reflex Relatively stereotyped automatic motor response to a proprioceptive input
Reticulospinal pathways Descending pathways from the brainstem to the spinal cord
Synapse Specialized connection between one neuron and a second neuron
Synergistic muscles Muscles that act together to produce a particular action

ALL HUMAN MOVEMENTS, except perhaps extremely rapid ones, can be modified during their performance. These modifications may be deliberately initiated by the subject once they are aware that the initial movement is no longer appropriate or has run into an unanticipated obstacle. In addition, however, movement is continuously modulated by activity in reflex pathways, and the subject may be quite unaware that these modifications are occurring. Reflex activity forms an integral part of normal voluntary movement: Only rarely does one occur without the other. Apart from voluntary motor acts and their reflex modulation, some acts can be programmed voluntarily to occur only in response to a particular stimulus (triggered reactions). Some skilled motor acts (e.g., signing one's name, driving a car) can become so learned and automatic that the intrusion of consciousness rarely occurs. By convention, these are not considered reflex acts. The following discussion considers some of the circuits responsible for the reflex modulation of movement.

Reflexes depend on information detected by specific receptors and transmitted to the central nervous system to modify either the motor program (including the motor command reaching neurons in the motor cortex) or the effects of the descending corticospinal drive onto the motoneuron pools in the brainstem and spinal cord. The feedback that acts to modify movement may involve the special senses (e.g., vision, balance mechanisms in the inner ear); or may come from the periphery as a consequence of the movement itself (e.g., from receptors in the skin, joint and muscles of the moving limb). Only the latter (i.e., proprioceptive reflexes) will be considered in this article.

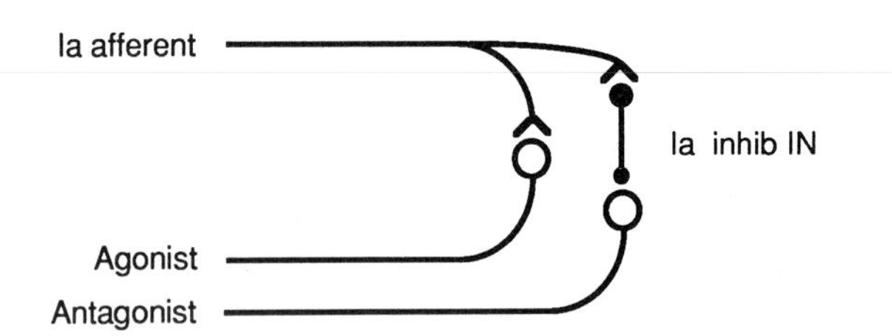

FIGURE 1 The monosynaptic excitatory connection between muscle spindle afferents (Ia afferents) and the agonist (homonymous) motoneuron pool, and the disynaptic inhibitory connection to the antagonist muscle via the Ia inhibitory interneurons (Ia inhib IN). Filled symbols used for inhibitory interneurons.

I. Reflex Circuits

A. Spinal Circuitry

The spinal cord, particularly at the cervical and lumbar enlargements (which contain the innervation for the limbs), contains complex circuitry, the function of which is only now being fully dissected in experimental animals such as the cat. In human subjects, a number of simple circuits have been defined physiologically and their role in the control of movement studied. Only those circuits that have been defined for human subjects will be discussed in detail: monosynaptic Ia excitation, oligosynaptic Ia excitation, reciprocal inhibition, group Ib inhibition, recurrent inhibition, and presynaptic inhibition of the Ia afferent terminals. Ia and Ib afferents innervate primary muscle spindle endings and Golgi tendon organs, respectively. [*See* PROPRIOCEPTORS AND PROPRIOCEPTION.]

In addition, although the precise circuitry has not been determined for human subjects, cutaneous and joint afferents exert reflex effects on motoneurons, and this circuit is subject to supraspinal control. [*See* SPINAL CORD.]

1. Reflexes Involving Muscle Afferents

The simplest spinal circuit is the direct (monosynaptic) connection between group Ia afferents from muscle spindles and the motoneurons of their own (homonymous) muscle (Fig. 1). There are also monosynaptic connections to the motoneurons innervating synergistic muscles and a disynaptic inhibitory pathway mediating reciprocal inhibition to motoneurons innervating antagonistic muscles. The monosynaptic influence on synergists is usually weaker than the homonymous connection and can even affect muscles that act at nearby joints. Experimental evidence indicates that each primary muscle spindle afferent makes synaptic connections with virtually all motoneurons in the homonymous muscle. However, the absolute strength of the synaptic connection between a single spindle afferent and motoneuron is relatively weak compared with the synaptic drive required to discharge a motoneuron. Any movement or disturbance that synchronizes the muscle spindle input will produce a proportionately greater effect on the motoneuron pool. The reciprocal inhibition is mediated via an interneuron termed the Ia inhibitory interneuron. Each such interneuron projects to a small fraction of the antagonist motoneuron pool.

Apparently, as in the cat, another excitatory connection exists between group Ia afferents and their homonymous motoneurons, exerted through an oligosynaptic pathway. This has been documented for some upper and lower limb muscles in humans. The relevant interneurons (and the Ia inhibitory interneuron) are the site of convergent input from cutaneous afferents and from descending pathways activated during a voluntary contraction (presumably including the corticospinal pathway). Group II muscle afferents innervating secondary muscle spindle endings probably also exert an excitatory effect on homonymous motoneurons in humans although the synaptic details are not yet clear.

The monosynaptic Ia pathway contains no interneurons, but the level of activity in this circuit can still be modulated through the mechanism of presynaptic inhibition (Fig. 2). This is mediated by in-

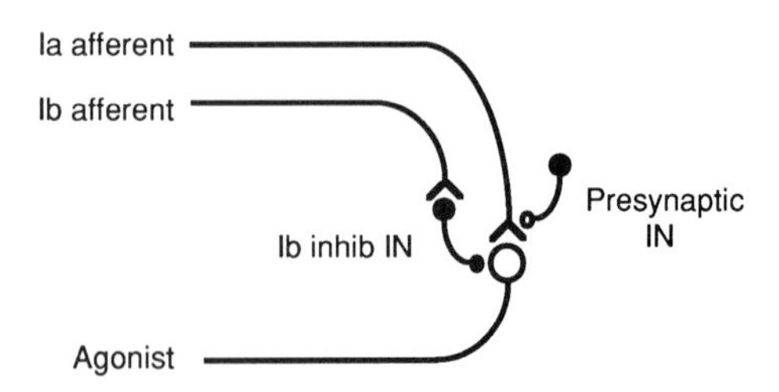

FIGURE 2 The inhibitory connection between Golgi tendon organ afferents (Ib afferent) and the agonist motoneuron pool via the Ib inhibitory interneuron (Ib inhib IN). Also shown is a presynaptic interneuron (Presynaptic IN), which can reduce the effectiveness of synaptic transmission.

terneurons that synapse on the terminals of the Ia afferent fibers before they reach the motoneuron, thus forming an axo–axonic synapse that is presynaptic to the Ia–motoneuron synapse. Activity of this interneuron diminishes the amount of excitatory transmitter released by an impulse traveling along the Ia afferent fiber and so decreases the excitatory effect of that impulse on the motoneuron. The presynaptic inhibitory interneuron receives powerful inputs from cutaneous and muscle afferents from the periphery and is also subjected to supraspinal control, its activity being depressed during contraction of the homonymous muscle. Although the full extent and functional importance of presynaptic inhibition are not established, it provides the central nervous system with a potent means of controlling the input to spinal and, possibly, supraspinal sites. Thus, when standing unsupported, presynaptic inhibition may be increased for some muscles (soleus) and decreased for others (quadriceps).

Because fusimotor neurons (both gamma and beta) are activated in a deliberate voluntary contraction of a muscle, the activity of group Ia afferents is increased, and this activity enters spinal reflex circuits that are already "primed" by the actions on interneurons described above. Hence, voluntary contraction of a muscle is associated with enhancement of the reflex influence of group Ia afferents from the contracting muscle on its homonymous alpha motoneurons and on the alpha motoneurons of synergists. There is also parallel inhibition of antagonists via the disynaptic reciprocal inhibitory pathway. By contrast with their powerful reflex effects on alpha motoneurons, group Ia afferents have relatively weak reflex effects on gamma motoneurons, which may be just as well because otherwise the Ia input could drive gamma motoneurons to produce more Ia activity, and this might create an unstable oscillation. [*See* FUSIMOTOR SYSTEM.]

Group Ib afferents from Golgi tendon organs convey a message related to the force of contraction and have an inhibitory effect on homonymous and synergistic motoneurons through a disynaptic pathway (Fig. 2). The group Ib effect via the Ib inhibitory interneuron has a widespread influence within the spinal cord, possibly more diffuse than the group Ia effect. The Ib inhibitory interneuron receives other peripheral inputs, for example, from cutaneous and joint afferents (see below). In the cat, it also receives some excitatory input from homonymous group Ia afferents, with the result that group Ia activity can theoretically both excite and inhibit homonymous motoneurons. This apparently paradoxical arrangement is one of several means by which the central nervous system may modulate the excitatory effect of muscle spindle inputs to motoneurons. It is not known whether or not this connection exists in humans.

2. Reflexes Involving Cutaneous and Joint Afferents

The afferent fibers from mechanoreceptors in the skin have complex actions on both alpha and gamma motoneurons, the effect being dependent on the skin region stimulated. The spinal reflex effects are mediated by oligosynaptic pathways, containing at least two interneurons, both of which receive inputs from other sources, both peripheral (e.g., group Ib and group II muscle afferents) and supraspinal (e.g., corticospinal and reticulospinal pathways).

Relatively little is known about the specific reflex effects of the different classes of cutaneous mechanoreceptors. Clearly, however, afferents from specialized cutaneous receptors can exert complex oligosynaptic reflex effects within a single motoneuron pool, as illustrated by the following examples. First, cutaneous afferents may exert differential effects onto the different types of motoneurons within the motoneuron pool. Thus, low-threshold, early-recruited motoneurons of the first dorsal interosseous muscle (an intrinsic muscle of the hand) may be inhibited by cutaneous afferents from the index finger (the digit the muscle moves), while higher-threshold, later-recruited motoneurons are excited. This reflex arrangement favors the activation of motoneurons that are capable of rapid, forceful contractions. Cutaneous inputs from functionally related digits (e.g., the thumb) probably have a similar reflex effect. Secondly, there is a highly synchronized afferent input from cutaneous afferents within the fingers and hand when a grasped object slips fractionally against the skin surface. This input leads to a rapid reflex increase in force, which is probably mediated through an oligosynaptic spinal pathway. [*See* SKIN AND TOUCH.]

Similarly, specialized mechanoreceptors in joints have oligosynaptic pathways to motoneurons, in part mediated by convergence on the Ib inhibitory interneuron. The overall pattern of activity from joint afferents can cause a marked change in the activity of muscles operating across a joint. As a joint moves into full flexion, the extensor muscles are facilitated at a spinal level; conversely, as the joint moves into full extension, the flexor muscles are facilitated. Such a reciprocal arrangement, documented for the knee and elbow joints in the cat, serves to protect a joint from the potentially damaging effects of excessive flexion or extension. The detailed neuronal basis for this arrangement has not been defined although the Ib inhibitory interneuron is probably involved. There is evidence for reflex inhibition of muscles acting about the human knee when the joint capsule is distended by an effusion. [*See* PROPRIOCEPTORS AND PROPRIOCEPTION.]

In summary, cutaneous and joint afferents almost certainly have excitatory and inhibitory connections to the same motoneuron pool, the reflex operating being the one "chosen" by descending commands to be appropriate for the particular movement. In spinal animals and in humans with spinal cord injury, this supraspinal control is lost, and the dominant reflex effect appears to be inhibition of motoneurons innervating extensor muscles and facilitation of those innervating flexor muscles, with the result that even innocuous stimuli can provoke "flexor spasms" in such patients.

3. Recurrent Inhibition

The output of the motoneuron can be modulated by an intraspinal feedback circuit using the Renshaw cell to produce recurrent inhibition (Fig. 3). A collateral branch coming off the axon of the alpha motoneuron excites nearby Renshaw cells, which directly inhibit the homonymous motoneuron pool. Renshaw cells inhibit both alpha and gamma motoneurons and, in addition, the Ia inhibitory interneuron directed to antagonistic motoneurons. Hence, activity in the Renshaw circuit tends to curtail a contraction and decrease the inhibition of the antagonist. In addition, the Renshaw cell circuit may help to focus the motor output on a restricted group of muscles. Like all interneurons, the Renshaw cell is subject to peripheral and descending control. During strong voluntary contractions, its excitability is turned down, so that there is less recurrent inhibition of the contracting muscle and the reciprocal inhibitory pathway to its antagonist is open. During weak voluntary contractions, Renshaw cells are excited to provide graded inhibition of the agonist motoneuron pools and indirect facilitation of the antagonist.

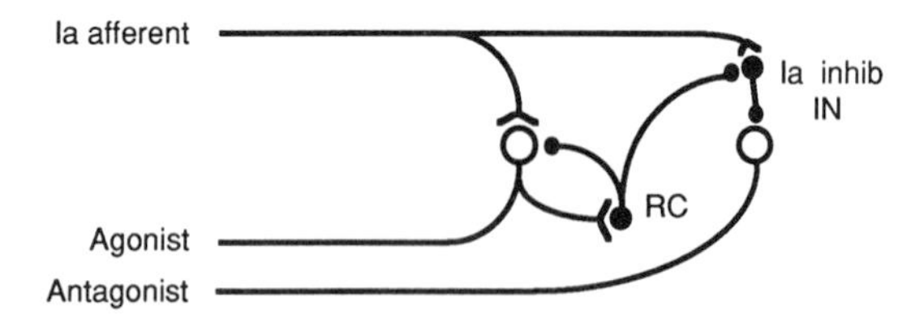

FIGURE 3 The Renshaw cell (RC) is excited by a collateral of the motor axon to provide recurrent inhibition of the motoneuron pool. The Renshaw cell also inhibits the Ia inhibitory interneuron (Ia inhib IN). These actions curtail activation of the agonist and favor activation of the antagonist. Filled symbols used for inhibitory interneurons.

B. Transcortical (Long-Loop) Circuits

Rapidly conducting afferents from mechanoreceptors in skin, joint, and muscle (particularly the muscle spindle and the Golgi tendon organ) have oligosynaptic projections to the primary motor and somatosensory areas of the cerebral cortex. This afferent information reaches the cortex by many pathways: The major ones traverse the posterior columns and, for muscle afferents from the lower limbs, the dorsal spinocerebellar tract to the dorsal column nuclei (and nucleus Z), where, after a relay, the input enters the medial lemniscus, only to relay again in the thalamus on the way to the cortex. At each level—spinal, thalamic, and cortical, there may be intramodality convergence (within the muscle afferent or cutaneous afferent types) and intermodality convergence (between the cutaneous and muscle afferents). In addition, the ascending volley in the sensory pathways can be modified at these three levels by descending motor pathways or local influences.

Within the primary motor cortex, the ascending sensory information has excitatory synapses on the upper motor neuron, which projects via corticospinal pathways to spinal motoneuron pools. The corticospinal fibers pass through the medullary pyramids, forming the pyramidal tract. These descending pathways make direct (monosynaptic) connections with their target motoneurons. Individual corticospinal axons may synapse with many motoneurons within a pool and even with many functionally related motoneuron pools. In addition, evidence in cats and monkeys shows indirect connections via interneurons at a local segmental level or via propriospinal neurons, which are located in the high-cervical spinal cord. The relative importance of these indirect connections in humans is not known, although recent evidence supports the presence of analogous propriospinal interneurons, which are activated in voluntary contractions and on which afferents from skin and muscle also converge. The presence of interneurons in the pathways from motor cortex to motoneuron provides an additional means by which sensory feedback from the periphery or the activity of other descending pathways can influence the resulting muscle contraction.

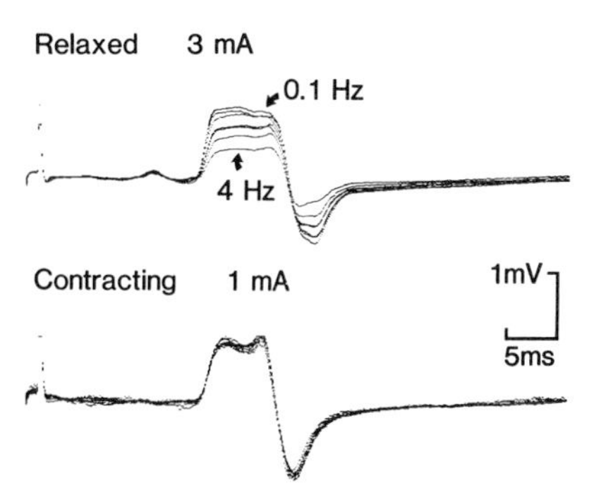

FIGURE 4 H reflex of a flexor muscle of the wrist produced by electrical stimulation at a range of frequencies (0.1, 0.2, 0.5, 1, 2, 3, and 4 Hz). When the muscle is relaxed (upper records) the reflex response attenuates as stimulus rate is increased; however, when the muscle contracts, the reflex can be obtained at a lower stimulus intensity (1 mA rather than 3 mA), and its amplitude is not affected by the stimulus rate. ms, msec. [Reproduced, with permission, from D. Burke, R. W. Adams, and N. F. Skuse, 1989, *Brain* **112,** 417–433.]

II. Proprioceptive Reflexes

A number of reflexes from mechanoreceptors in muscle and skin can be readily demonstrated in humans, the most easily elicited being the tendon jerk. However, the majority of these reflexes depend on subjecting the nervous system to an intense afferent volley generated by an artificial stimulus and, though clinically useful, such reflexes reveal little about how reflex activity is integrated into the control of movement. Nevertheless, reflexes such as the tendon jerk are useful because they enable physicians to check the integrity of spinal circuits and to quantify the changes in reflex intensity in different neurological diseases.

In normal human subjects, any reflex that is elicited repeatedly will habituate (i.e., the same stimulus will produce a smaller response). The habituation can be minimized by using irregularly repeated stimuli and keeping the repetition rate low, approximately 1/5–10 sec. The degree of habituation is greater for cutaneous reflexes such as the glabellar tap (the reflex blinking that occurs when the skin of the forehead is tapped), possibly because cutaneous reflexes contain more interneurons. However, if the test muscle is contracted voluntarily, the degree of habituation is much less, whether the reflex is primarily of cutaneous or muscle origin, and stimulus rates of >1/sec are then feasible (Fig. 4).

A. Tendon Jerk

If the tendons of a number of muscles are tapped briskly, a brief twitch contraction will occur in the relevant muscle. The time interval (latency) between the tendon tap and the onset of the reflex contraction depends on the distance of the muscle from the spinal cord, but it is short (e.g., 30–40 msec for the calf muscles (Fig. 5) and 20–25 msec for the thigh muscles. An abrupt tap on the tendon excites not only mechanoreceptors in the appropriate muscle but also those in nearby muscles and skin and produces a number of impulses from the more sensitive receptors. It is generally accepted that the major afferent input produced by a tendon tap is group Ia activity from the percussed muscle, although group Ib and cutaneous afferents are also excited. The major excitatory pathway from the tendon jerk is the group Ia monosynaptic pathway. However, the excitation produced in the motoneurons lasts so long that other spinal pathways and other afferent fibers can also affect the reflex discharge. The duration of this excitation is much longer for the tendon jerk than for the H reflex (See Section II,C) in human motoneurons (Fig. 5).

The tendon jerk can be potentiated in normal subjects by a number of different maneuvers; this phenomenon is known as reflex reinforcement. Perhaps the simplest method of reinforcement is to ask the subject to contract the relevant muscle weakly because, as discussed above, this potentiates transmission in group Ia excitatory pathways and suppresses relevant inhibitory interneurons. However, if the contraction is too strong, the muscle stiffens and it can be difficult to see the enhanced reflex. In practice, most physicians use a technique called the Jendrassik maneuver, in which the subject contracts a remote muscle rather than the one being tested. The potentiation produced by this maneuver is less than is produced by contraction of the test muscle and occurs through activation of descending pathways to alter reflex transmission and the excitability of motoneurons.

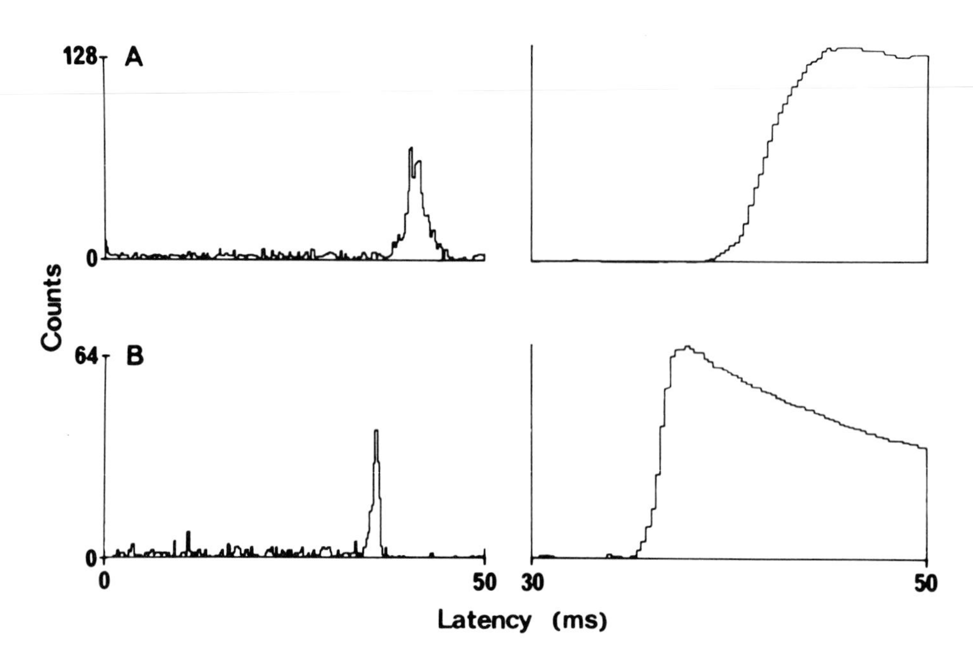

FIGURE 5 A histogram of the discharge of a voluntarily activated motor unit in soleus in response to percussion on the Achilles tendon (A) and to electrical stimulation of the tibial nerve to produce the H reflex (B). The occurrence of a discharge in the motor unit produces a single count in the histogram. The tendon tap and H-reflex stimuli are delivered at time 0. On the left, the increase in probability of discharge probably reflects the differential of the rising phase of the responsible excitatory event within the motoneuron (excitatory postsynaptic potential [EPSP]). On the right, the cumulative increase in probabilities of discharge are therefore presumed to represent the rising phases of the EPSPs. In these plots latency has not been corrected for the trigger delay (2.5 msec). Note that with the tendon tap (A), the reflex response occurs approximately 5 msec later than with H-reflex stimuli (B). Also note that the duration of the excitation is much longer with the tendon tap. Numbers of triggers used in the histograms were about 2,500. ms, msec. [Reproduced, with permission, from D. Burke, S. C. Gandevia, and B. McKeon, 1984, *J. Neurophysiol.* **51,** 185–194.]

B. Unloading Reflex

If a subject contracts a muscle steadily against a resistance that is suddenly removed, the contracting muscle abruptly shortens. Recordings of the electromyogram of the contracting muscle reveal that the muscle contraction is transiently interrupted at short latency after the removal of the load. The unloading reflex is produced by the abrupt removal of ongoing reflex support to the contracting motoneurons, particularly from group Ia afferents from spindle endings in the contracting muscle. The latency of the silent period in the contracting muscle is similar to the latency of the tendon jerk of the same muscle; the unloading reflex can be considered the inverse reflex to the tendon jerk.

The physiological significance of the unloading reflex is that it reveals the presence of supportive excitation to the contraction from peripheral reflex sources. Clearly, the ongoing activity in reflex pathways must have had a significant influence on motoneuron firing for its removal to be capable of silencing motoneurons, even if only transiently.

C. H Reflex

This reflex is named after Paul Hoffmann, who first described that a reflex muscle contraction can be produced by electrical stimulation of the afferent fibers coming from the muscle (Fig. 4). Group I muscle afferents are activated by low-intensity stimulation of the relevant nerve. The latency of the reflex contraction is approximately 5 msec less than the latency of the tendon jerk for the same muscle, the difference being attributed to the fact that the electrical stimulus does not require activation of the muscle spindle receptor and conduction along the distal part of the afferent fibers (Fig. 5). The H reflex has been considered the electrically induced analogue of the mechanically induced tendon jerk, and differences between the two reflexes in different experimental situations have been attributed to receptor (muscle spindle) mechanisms. Thus, comparisons of these two reflexes have been used as measures of the degree of sensitization of spindle endings by fusimotor drive. [*See* FUSIMOTOR

SYSTEM]. However, the afferent volleys for the two reflexes differ in a number of respects, all of which could alter the resulting reflex. Perhaps the major difference is that the electrically evoked volley will contain significant Ib activity, whereas the mechanically induced volley will not.

Group Ia afferents are responsible for the H reflex largely via a monosynaptic pathway. However, the group Ib contamination of the afferent volley produces disynaptic inhibition in the homonymous motoneurons, and this curtails the duration of the Ia excitation. Changes in the H reflex can be produced by affecting the Ib inhibitory interneuron. The H reflex provides a probe with which the function of the human spinal cord can be investigated. Virtually all the studies of reflex circuitry within the human spinal cord have depended on demonstrating changes in the H reflex. As a final example, recent studies of the H reflex of soleus during walking suggest that the reflex excitability may not parallel that of the motoneuron pool. The reflex responsiveness is high during the stance phase of walking, and even higher during standing, both circumstances in which active maintenance of body position against gravity is required.

It is often (erroneously) stated that the H reflex is more restricted in distribution than the tendon jerk. However, if the subject contracts the test muscle, both the H reflex and tendon jerk can be elicited from many limb muscles (Fig. 4).

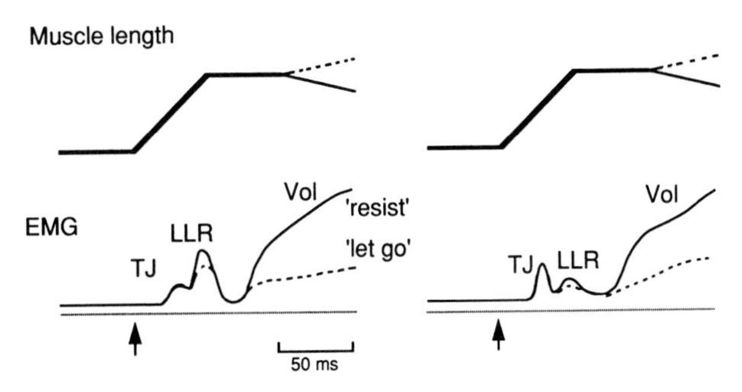

FIGURE 6 Diagrammatic representation of the responses of human muscles to stretch. Upper panels show muscle lengthening (with stretch as an upward deflection), and lower panels show the electromyographic responses (EMG) when the stretches are given during a steady contraction. On the left, the pattern obtained for some distal muscles are depicted, and on the right, for some proximal muscles. Following the muscle stretch, there are three phases of the response: a short-latency spinal response, equivalent to the tendon jerk (TJ), the long-latency response (LLR), and the voluntary response (Vol). The long-latency response is bigger when the subject is instructed to resist the applied stretch than when instructed to let go immediately after the stretch. In the former case, the muscle length begins to return toward the initial value. ms, msec.

D. Stretch Reflexes

These reflexes refer to the responses observed following stretch of a contracting muscle (Fig. 6). The stretch is usually applied rapidly and produces a burst of discharges from dynamically sensitive receptors (particularly muscle spindle endings) within the muscle. However, the disturbance is not highly localized so that intramuscular receptors from adjacent muscles and nearby cutaneous and joint receptors are also activated. For most human muscles, the response to abrupt stretch of a contracting muscle shows three components. The first occurs at a latency consistent with a monosynaptic reflex and is probably equivalent to the tendon jerk (see above). The second occurs at a longer latency and is thus consistent either with a more slowly conducting afferent volley reaching the spinal cord or with a longer neuronal pathway within the central nervous system. The latencies for the tendon jerk and long latency response are about 25 and 50 msec, respectively, for the long flexor muscle of the thumb. The third, and often blending temporally with the second component, is the deliberate voluntary reaction to the stretch.

The second component of the stretch reflex has aroused considerable debate because of the suggestion that it may represent the operation of a long-loop, possibly transcortical reflex (see above). Evidence supporting this conclusion includes the relationship between the latency of the second component and the distance between the stretched muscle and the brain, and observations that the component is reduced or absent in patients with lesions interrupting the pathways to and from the motor cortex. However, not all findings have supported this explanation, and it is clear now that other factors must also be considered, including the oligosynaptic activation produced by more slowly conducting group II afferents from the secondary endings of muscle spindles, and the continuation of excitation due to multiple discharges of primary muscle spindle endings and even from cutaneous receptors. Different mechanisms acting together probably are responsible for the second component of the stretch reflex, and the relative importance of any one mechanism may differ for different muscle groups (Fig. 6).

Not only does rapid muscle stretch (or shortening) produce excitation (or inhibition) of the relevant muscles, but extremely slow perturbations may have a similar effect. Much as described above, the reflex responses produced by slow stretch can be guided according to the subject's intent to resist or to relax during the perturbation. It is as if reflex circuits can be preset prior to a voluntary contraction or prior to an externally induced adjustment to movement. In other words, the overall gains of the stretch reflex response, and thus the effective stiffness of the muscle, can be controlled. It must be emphasized that the resting (background) discharge of muscle and cutaneous mechanoreceptors provides a "tonic" input to the central nervous system. Small changes in the discharge rates of muscle spindle afferents from a muscle group will affect motoneuron excitability. Thus, while phasic or transient stimuli (e.g., sudden stretch, H-reflex inputs) are often used to assess reflex pathways, the slowly changing tonic input accompanying normal movement will continuously modulate motoneuronal output. Furthermore, many proprioceptors, including cutaneous receptors, respond disproportionately to small mechanical disturbances, such as occur during natural movement, rather than to large mechanical disturbances, such as a tendon tap.

E. Reflex Effects of Muscle Vibration

Vibration activates many mechanoreceptors, in muscle and skin, but receptors in the muscle spindle are particularly sensitive and, when vibration is applied directly to a muscle or tendon, the reflex effects are generally attributed to group Ia afferents from the muscle spindle. If a muscle or its tendon is vibrated, two phenomena may occur: The tendon jerk and the H reflex will be inhibited, and a slowly increasing contraction develops in the vibrated muscle. The former phenomenon (vibratory inhibition) is largely due to presynaptic inhibition of the monosynaptic group Ia pathway due to the vibration-induced activity in muscle (and cutaneous) afferents. The slow-developing contraction (tonic vibration reflex) is perhaps due to group Ia afferent activity transmitted to the homonymous and synergistic motoneurons along nonmonosynaptic pathways. At least in some muscles, however, cutaneous afferents contribute to this nonmonosynaptic excitation. The tonic vibration reflex is subject to activity in descending motor pathways: On the one hand, it can be completely overridden voluntarily and, on the other, it can be potentiated by a weak voluntary contraction of the test muscle.

Muscle vibration also results in inhibition of the motoneurons of antagonistic muscles, presumably a reflection of Ia reciprocal inhibition. In addition, vibration of the Achilles tendon can disturb postural equilibrium in standing human subjects. When the Achilles tendons are vibrated, the subject sways backwards as if to compensate for apparent stretch of the calf muscles, and this may be sufficient to provoke a fall.

III. Conclusion

This chapter has focused initially on some of the spinal and suprapinal circuitry that is available to mediate reflex contributions to muscle performance, and then on some of the proprioceptive reflexes that can be evoked in human subjects. The latter emphasis reflects the need for neurologists and neurophysiologists to have means of monitoring particular neuroanatomical pathways. Clinically, the proprioceptive reflexes are often evoked by stimuli, which are more intense and synchronized than those occurring naturally. Thus, perhaps not surprisingly, much current research is shifting toward examining the coordinated responses of many muscles within a limb, or in muscles acting over many body segments, to smaller and more natural stimuli. Examples include the responses to upper limb movement, which can be recorded throughout the lower limb and trunk, and the response of neck, trunk, and leg muscles to small disturbances of stance. Analyses of these more complex situations are providing a picture of a less stereotyped and more modifiable reflex organization. Such an organization would fit with the increasing number of classes of interneurons that have been characterized within the spinal cord of experimental animals.

Bibliography

Baldissera, F., Hultborn, H., and Illert, M. (1981). Integration in spinal neuronal systems. *In* "Handbook of Physiology," Sect. I, Vol. II, Part 1 (V. B. Brooks, ed.), pp. 509–597. American Physiological Society, Bethesda, Maryland.

Burke, D., Adams, R. W., and Skuse, N. F. (1989). The effects of voluntary contraction on the H reflex of human limb muscles. *Brain* **112,** 417–433.

Burke, D., Gandevia, S. C., and McKeon, B. (1983). Afferent volleys responsible for spinal proprioceptive reflexes in man. *J. Physiol. (London)* **339,** 535–525.

Colebatch, J. G., and McCloskey, D. I. (1987). Maintenance of constant arm position or force: reflex and volitional components in man. *J. Physiol. (London)* **386,** 247–261.

Fournier, E., Pierrot-Deseilligny, E. (1989). Changes in transmission in some reflex pathways during movement in humans. *News Physiolog. Sci.* **4,** 29–32.

Hultborn, H., Meunier, S., and Pierrot-Deseilligny, E. (1987). Changes in presynaptic inhibition of Ia fibres at the onset of voluntary contraction in man. *J. Physiol. (London)* **389,** 757–772.

Jankowska, E., and Lundberg, A. (1981). Interneurones in the spinal cord. *Trends Neurosci.* **4,** 230–233.

Marsden, C. D., Rothwell, J. C., and Day, B. L. (1983). Long-latency automatic responses to muscle stretch in man: Origin and function. *In* "Motor Control Mechanisms in Health and Disease" (J. E. Desmedt, ed.), pp. 509–539. Raven Press, New York.

Neilson, P. D., and Lance, J. W. (1978). Reflex transmission characteristics during voluntary activity in normal man and in patients with movement disorders. *In* "Cerebral Motor Control in Man: Long Loop Mechanisms," Vol. 4 (J. E. Desmedt, ed.), pp. 263–299. Karger, Basel.

Rack, P. M. H. (1981). Limitations of somatosensory feedback in control of posture and movement. *In* "Handbook of Physiology," Sect. I, Vol. I, Part 1 (V. B. Brooks, ed.), pp. 229–256. American Physiological Society, Bethesda, Maryland.

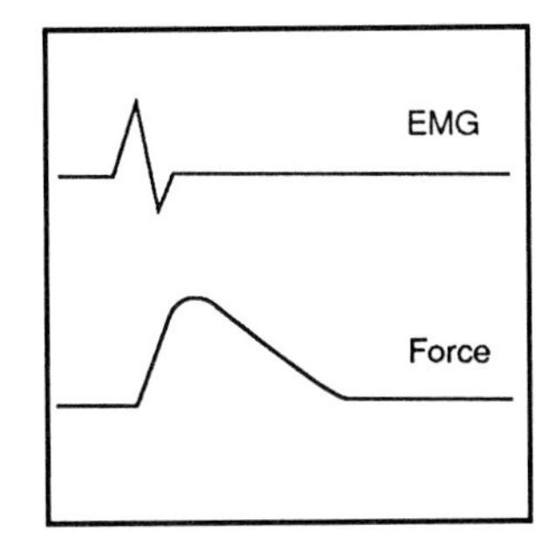

Proprioceptors and Proprioception

S. C. GANDEVIA, *The Prince Henry Hospital and School of Medicine, University of New South Wales*

Glossary

Golgi tendon organs Sensory receptors within musculotendinous junctions capable of signaling the force of muscle contraction and its rate

Joint receptors Sensory receptors within joint capsules and ligaments, which respond to local stress or pressure changes within the joint

Muscle spindles Complex sensory and motor structures located within muscles; afferent fibers from muscle spindle endings can signal the length of the muscle and its velocity of movement

Nerve fibers Afferent and efferent nerve fibers convey information to and from the central nervous system, respectively

Perceived motor commands Signals of motor command generated within the central nervous system, which can directly influence proprioceptive sensation

Proprioception Global term including sensations of the position and movement of the limbs and trunk, together with sensations of heaviness and force; loosely used synonyms include kinesthesia and joint position sense

Somatosensory cortex Area(s) of the cerebral cortex that receive ordered inputs from muscle, joint, cutaneous, and other afferent fibers

THE PHYSIOLOGICAL PROPERTIES of the major classes of proprioceptor located in muscles, tendons, joints, and skin are the focus of this chapter. The proprioceptive sensations evoked by these afferents are examined together with the roles for perceived motor commands. It is emphasized that proprioception encompasses a group of sensations including limb position and movement, sensations of force and heaviness, and sensation of the timing of muscle contraction.

I. Background

This chapter is concerned mainly with the physiological properties of proprioceptors (i.e., the peripheral sensory receptors, which provide signals about proprioception) and, secondly, on their contributions to specific aspects of proprioception. Emphasis has been placed deliberately on the properties of proprioceptors in human subjects, partly because there are differences in proprioceptors among species, but also because, based on neurophysiological studies in humans, the principles governing proprioception have been established in the last two decades. In addition, it is necessary to consider briefly the role of motor signals generated internally within the central nervous system in proprioception; these are loosely termed signals of motor command, corollary discharges, or perceived motor commands. Unlike some other senses, such as smell, vision, and hearing, voluntary movement and muscle contraction are usually a prerequisite for generation of proprioceptive signals. For this reason alone, not surprisingly signals of central motor command are important potential contributors to proprioception.

Since early in this century, the terms proprioception and kinesthesia have coexisted with little attempt to define them explicitly. The role of the relevant peripheral receptors and the central neural mechanisms have, at times, been hotly debated for different aspects of proprioception. However, in the last two decades, a number of psychophysical and electrophysiological studies have led to greater agreement about the principal neural mechanisms. The terms proprioception and kinesthesia are commonly used synonymously; they encompass a group of sensations that *includes* (1) sensations of limb (or joint) movement and position, (2) sensations of force and heaviness, and (3) sensations of the timing of muscular contraction.

Specialized sensory receptors within muscles and tendons, and in joint capsules and ligaments, can provide useful information about the "movement status" of any body part. In addition to their role in cutaneous sensation, even receptors within the skin can assist proprioceptive sensation, partly because some are inevitably excited as a result of joint movement. On this basis, no major class of specialized peripheral receptor in skin, joint, or muscle can be denied a role in proprioception. To determine their potential role requires a knowledge of their response properties, i.e., the specific peripheral events, which they can transmit to the central nervous system. It is then necessary to confirm that role by appropriate psychophysical experiment.

II. Properties of Proprioceptors

A. Muscle Spindle Endings

The muscle spindle is an intricate apparatus consisting of both sensory and motor components and is found in almost every skeletal muscle (Fig. 1). Each human muscle spindle usually contains 3–15 specialized muscle fibers, termed intrafusal fibers. The spindles lie parallel to the main (extrafusal) muscle fibers but do not produce significant contribution to total output of muscle force. The intrafusal fibers may be more numerous in human muscles than in the muscles of cats, the species in which much of the physiological work has been undertaken. There are at least three types of intrafusal fiber within each spindle: two types of nuclear bag fiber (dynamic [bag 1] and static [bag 2]) and the nuclear chain fiber. [*See* SKELETAL MUSCLE.]

There are two types of specialized sensory endings associated with muscle spindles: the primary spindle ending, which has spiral sensory terminals on the central portion of all the intrafusal fibers from a spindle (particularly the dynamic bag fiber in human spindles), and the secondary spindle ending with terminals on all but the dynamic bag intrafusal fiber. The density of spindles is highest for small muscles (of the trunk and distal extremities) when measured as the number of spindles per unit of muscle weight, but it is more uniform if measured per number of motoneurons innervating the muscle. The detailed anatomy of muscle spindles may vary within and between muscles.

Afferent fibers innervate sensory receptors and conduct signals to the spinal cord. They are often classified according to their diameter: the largest and most rapidly conducting afferents innervate primary spindle endings (group Ia afferents) and Golgi tendon organs (Ib afferents; see Section IIB). Secondary endings are innervated by smaller diameter myelinated fibers (group II afferents). While the basic anatomy of muscle spindles is similar in different species, some differences do exist. For instance, in humans the majority of group I muscle afferents in the upper and lower limbs have a conduction velocity similar to that of the specialized cutaneous and joint afferents (60–70 m/sec). In contrast, in the cat and monkey the fastest cutaneous afferents conduct more slowly than group I muscle afferents. Human spindles have greater attachments to the adjacent extrafusal muscle fibers than those in the cat, and this may render their afferents more sensitive to forces transmitted "in series" from adjacent muscle fibers. [*See* HUMAN MUSCLE, ANATOMY.]

The primary and secondary spindle endings signal the length, and changes in length, of the muscle in which they reside. The discharge frequency of the primary spindle ending encodes mostly the velocity of movement, although the length and, to a lesser extent, the acceleration are also encoded (Fig. 2A; see also Fig. 3). The discharge of the secondary ending encodes preferentially the static muscle length or joint position. Properties of the different proprioceptors are summarized in Table I. The traditional way to quantify the responses of spindle endings is to stretch the muscle, increasing its length over about 1 sec; the primary ending shows an abrupt increase in firing frequency and then a slower increase throughout the stretch. The firing frequency declines abruptly once the new length is reached. In contrast, the discharge frequency of the secondary ending follows the profile of the length change more closely. Thus, the primary spindle ending shows a greater dynamic response to the change in length than does the secondary. Both have a static, or steady, discharge, which increases as the muscle length increases, but the secondary ending has a greater sensitivity (i.e., greater discharge frequency per unit length change).

Over the full physiological range of joint excur-

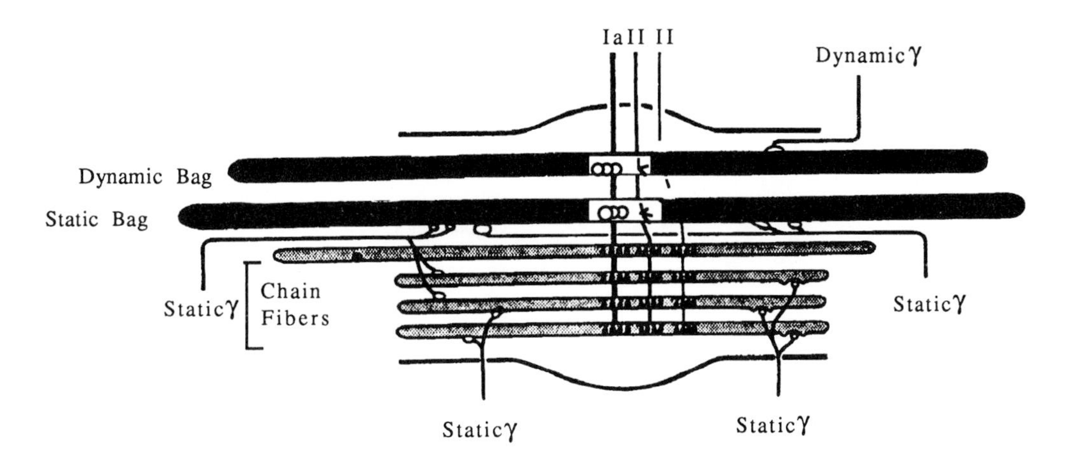

FIGURE 1 Simplified diagram of the structure and innervation of a typical cat muscle spindle. Two nuclear bag fibers and four nuclear chain fibers are shown. There is one dynamic nuclear bag intrafusal fiber (dynamic bag), and one static nuclear bag fiber (static bag). The diagram shows one primary spindle ending (Ia), which has sensory spiral terminals of a group Ia axon round every intrafusal fiber. It shows two secondary spindle endings (II); one secondary ending (left) has terminals on both chain fibers and nuclear bag fibers, and the other secondary ending (right) has sensory spirals on chain fibers only. Note the selective innervation of dynamic bag fiber by dynamic γ axon, and both selective and nonselective innervation of the static bag fiber and the chain fibers by static γ axons. [Simplified from I. A. Boyd, 1985, Muscle spindles and stretch reflexes, *in* "Scientific Basis of Clinical Neurology" (M. Swash and C. Kennard, eds.), Churchill Livingston, London.)

sion, the discharge frequency of primary spindle endings is not linearly related to the length change. The primary ending has a small "linear range," being exquisitely sensitive to very small increments in length (up to 100–200 μm in the cat), but is much less sensitive to larger increments. In contrast, the secondary ending responds in a more linear way over a much wider range of lengths. This means that the primary spindle ending responds more readily than other intramuscular receptors when vibration at high frequency and low amplitude is applied to the tendon. The discharge may be locked to the phase of vibration with one discharge per vibratory cycle. This susceptibility to vibratory stimuli has been exploited in psychophysical experiments (see Section IV) as a means to stimulate (preferentially) one class of proprioceptor. Due to mechanical properties of the intrafusal (dynamic bag) fiber, the primary ending's sensitivity to small length changes can be "reset" once a new operating length is achieved, so that its sensitivity is preserved over the physiological range of joint positions. The sensitivity of spindle endings is such that they respond to local events within the muscle, including the twitch contraction of nearby extrafusal muscles fibers and the mechanical disturbance set up by the arterial pulse. Responsiveness also varies depending on the

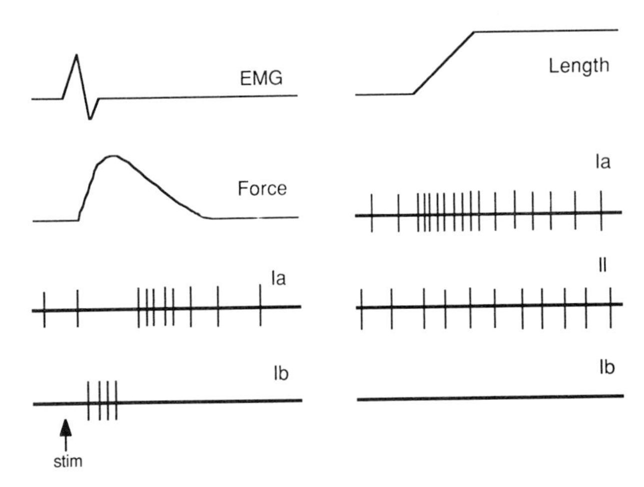

FIGURE 2 Left, Diagrammatic representation of the procedures used for physiological identification of muscle spindle afferent (Ia) and Golgi tendon organ afferent (Ib). Following electrical stimulation to produce a large muscle twitch, the Ia afferent fires during the falling phase of the twitch force, whereas the Ib afferent fires during the rising phase. The electromyographic response (EMG) and force profiles are shown. The discharge of the single afferent fibers are shown by the action potentials (represented as vertical lines). (Ib) Right, Diagrammatic representation of the response of muscle spindle endings (group Ia [primary] and group II [secondary]) and tendon organ afferent to a passive increase in length. There is a dynamic response in the Ia afferent to the onset of stretch. The discharge of the group II spindle afferent has a higher discharge frequency at the longer muscle length. The Ib afferent shows no response to the passive length change although it discharges to dynamic active changes in muscle force.

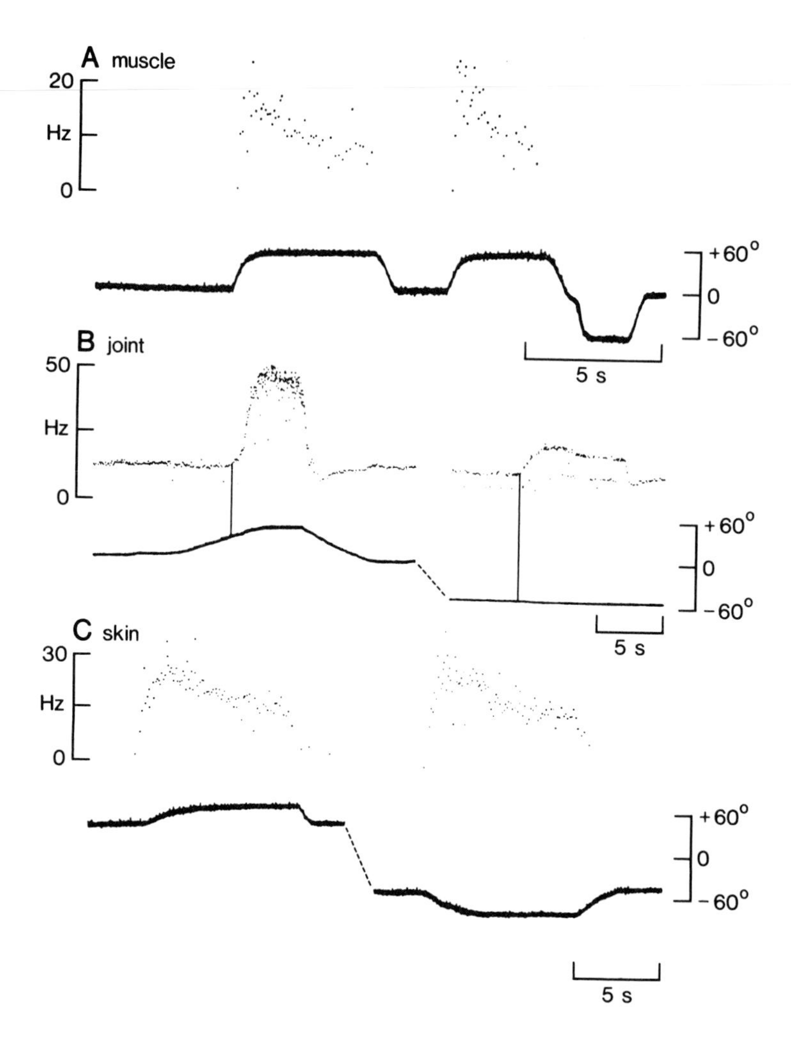

FIGURE 3 Typical responses of human proprioceptive afferents to passive movements. In each of the three panels, the upper record shows the discharge frequency of the afferent (plotted in Hertz or impulses per second), and the lower record shows the goniometer signal of joint angle. The time calibration is shown for each panel. A. Response of a muscle spindle afferent from the adductor of the ring finger to passive stretch and shortening applied by passive flexion and extension of the proximal interphalangeal joint. In all panels, the discharge frequency is plotted against time. This goniometer record shows stretch as an upward deflection. The afferent shows a slowly adapting response to the lengthening movements away from the rest position. B. Response of a typical joint afferent from the proximal interphalangeal joint of the index finger. The afferent had a background discharge across the full angular range (120°). This discharge increased only with hyperextension (left) *and* hyperflexion (right) of the joint (see goniometer record below). This is termed a bidirectional response. C. Response of a slowly adapting cutaneous afferent innervating skin near the proximal interphalangeal joint. The afferent discharged toward both ends of the range of passive movement. [Panel B is modified from D. Burke, S. C. Gandevia and G. Macefield, 1988, Responses to passive movement of receptors in joint, skin and muscle of the human hand, *J. Physiol.* (*London*) **402,** 347–361.]

spindle's exact location in the muscle, presumably because the stretch "seen" by each spindle varies slightly according to its specific location.

Output from the central nervous system to intrafusal muscle fibers on which the spindle's sensory endings lie can alter both the resting discharge of the ending and its responsiveness to changes in length. This output consists mainly of small-diameter γ motoneurons, which synapse toward the poles of the dynamic nuclear bag fibers (γ dynamic) or the static nuclear bag and the nuclear chain fibers (γ static). In addition, some spindles receive a second motor output from β motoneurons, which have a shared output to both the spindle and to extrafusal fibers. This form of innervation is the sole intrafusal innervation in amphibia. It is also present in primates and has been visualized in human muscles. It may be especially common in muscles of small size such as the intrinsic muscles of the hand.

Activation of different elements of the fusimotor

TABLE I Summary of Properties of Human Joint, Muscle Spindles, and Cutaneous Receptors in the Hand to Passive Movements[a]

	Response across middle of range	Unidirectional response	Multi-axial response	Proprioceptive signal
Muscle spindle endings	100%	100% unidirectional (i.e., increases with muscle lengthening)	73%	signals muscle lengthening and rate of lengthening
Joint receptors (80% of population)	0%	60% unidirectional	70%	detection of limit to movement
Joint receptors (20% of population)	100%	34% unidirectional	66%	minor signal of joint angle and movement
Slow-adapting cutaneous receptors	0%	67% unidirectional	24%	minor signal of joint angle and movement
Rapid-adapting cutaneous receptors	0%	18% unidirectional	27%	event detector

[a] The term response denotes a change in discharge frequency of the afferent.

system (through γ and β motoneurons) can produce changes in the background discharge rate and in the response of the ending to imposed length changes. Dynamic fusimotor inputs (via the dynamic nuclear bag fiber) may produce only a small change in background discharge of primary spindle endings but a marked increase in the response to length changes (increase in the dynamic response). Static fusimotor inputs via the static nuclear bag fiber markedly increase the background discharge rate of primary endings (and less so for the secondary ending), with little effect on the primary ending's response to length changes. Static fusimotor input via the nuclear chain fiber increases the background discharge and the static response to muscle stretch, and it can cause the primary (but not the secondary) endings to fire in one-to-one relation with the fusimotor-induced contraction of the chain fiber (a phenomenon termed driving). When both dynamic and static inputs to a primary spindle ending are activated, the effect of the latter may be dominant. [*See* FUSIMOTOR SYSTEM.]

B. Golgi Tendon Organs

Golgi tendon organs have been studied in less detail than muscle spindle endings, although their role in the control of movement may be equally important. Golgi tendon organs lie at the musculotendinous junctions throughout the muscle, in-series with the extrafusal muscle fibers. They are innervated by large diameter (group Ib) afferent fibers. They encode the contractile force developed by the extrafusal muscle (and its rate of change) with little adaptation over time and are relatively insensitive to passive lengthening of the muscle (Fig. 2). A number of muscle fibers from different motor units insert into the capsule of each tendon organ. The high sensitivity of the tendon organ to active muscle force is indicated by the ability of individual muscle fibers innervated by a single motoneuron (i.e., a motor unit) to increase the discharge of the tendon organ. Indeed, this response to contraction forms the critical basis for distinction between the spindle and tendon organs during experimental studies (Fig. 2). Thus, the tendon organ discharges during the rising phase of contractile force, whereas the muscle spindle falls silent (i.e., is "unloaded") during this phase and discharges preferentially during the falling phase of the contraction as the muscle lengthens slightly. Motor units of different size may produce a similar discharge from a single Golgi tendon organ suggesting that these receptors are designed to sample the force from different regions of the muscle. They retain an ability to signal fluctuations in force when many motor units are recruited or even during shortening contractions. The ultimate contractile force "seen" by the tendon organ is a complex mixture of the direct in-series and in-parallel component of force.

Both individual tendon organs and muscle spindle endings are extremely sensitive transducers. They can respond with high discharge rates to increments in local active force of about 10 mg and length changes of about 10 μm, respectively. However, both tend to saturate and reach a maximal response with the much larger static force and length changes that occur during normal behavior. This has led to the suggestion that these proprioceptors are especially important in transducing the local events according to their microenvironment within the muscle. The overall "average" of the activity in these endings may provide an accurate picture of the "average" force or length of the parent muscle. However, the sensitivity of the intramuscular afferents is such that the central nervous system also receives information about the dynamic performance of subvolumes of muscle. The discharge of these receptors in response to relatively large force and length changes contributes to conscious proprioception, and it is also likely that the microsensitivity of the endings to small mechanical changes has a perceptual role.

C. Joint Receptors

Receptors within joints were once thought to provide the only signals required for sensation of limb position and movement; indeed, this sensation was often termed *joint* position sense. One basis for this view is the presence of specialized receptors within joints and ligaments. The most important joint receptors for proprioception are probably the Ruffini endings, located within the joint capsule. Golgi endings, histologically similar to those at musculotendinous junctions, are found within ligaments. Paciniform endings are also associated with joints. Controversy has surrounded the capacity of specialized joint receptors to signal the angle at a joint to the central nervous system. Classic studies in the 1950s suggested that individual (slowly adapting) joint afferents were "tuned" to signal a particular region of the joint angular range and that they responded to both the static angle of the joint and the velocity of joint movement. Such properties would clearly make joint receptors powerful candidates to provide proprioceptive information. However, whereas each joint probably contains some receptors with these properties, the large majority of human joint afferents respond preferentially as a joint is moved toward and beyond the limits of its usual range (Fig. 3). Some joint receptors discharge at both ends of the movement range, e.g., at the extremes of flexion *and* extension (i.e., bidirectional response), and thus would signal joint position ambiguously. Furthermore, their discharge may increase at the extremes of movement in more than one axis, e.g., flexion and extension, and abduction and adduction (i.e., multiaxial response). These properties would make the receptors unambiguous detectors of movements, which are going outside the usual range of joint movement and, therefore, may harm the joint. Coupled with this, muscles whose contractions would continue to move the joint out of its physiological range are probably reflexly inhibited by these "limit detectors" within the joints, thus preventing damage to the joint.

D. Cutaneous Receptors

There are at least four specialized types of cutaneous receptors that respond to mechanical events involving the glabrous or nonhairy skin surface. The afferent fibers associated with two of these specialized receptors respond with a rapidly adapting discharge to mechanical indentation of the skin; they innervate the Pacinian and Meissner's corpuscles. Afferents innervating the Merkel–cell neurite complex and Ruffini endings are slowly adapting in their response to mechanical stimuli. Details about these receptors and the specific contribution of these afferent fibers to specialized aspects of cutaneous sensibility are addressed elsewhere. Theoretically, these afferents could have an important role in sensation of limb position and movement. Firstly, the rapidly adapting afferents may signal in a nonspecific way that a disturbance has occurred at or near a joint (i.e., event detector); secondly, slowly adapting cutaneous afferents may signal that the skin overlying a particular joint has been stretched or distorted; and thirdly, the generalized facilitation of central nervous system pathways produced by the continually changing input from cutaneous afferents has an indirect role in proprioception. This is particularly so for the hand with its high density of cutaneous innervation in the tips of the fingers. Thus, anesthesia of the fingertip impairs proprioceptive judgements from proximal finger joints and even adjacent fingers. This is less evident for skin overlying large joints in the lower limb. [*See* SKIN AND TOUCH.]

Direct recordings of the discharge of cutaneous afferents innervating the human hand show that all receptor classes discharge during active and passive movements of the fingers. The background discharge of these afferents may be minimal when the hand has been in a rested position for some minutes. Single, slowly adapting cutaneous afferents innervating the skin of the fingers, commonly discharge, like typical joint receptors, at both ends of a normal angular range of movement (Fig. 3). Despite this potential ambiguity, signals from cutaneous afferents may help to resolve difficulties that may exist in the central nervous system when interpreting the discharge from muscle and joint proprioceptors. These may arise for signals of muscle length and velocity from spindle endings in muscles that cross several joints.

III. Proprioceptive Mechanisms

A. Sensations from Proprioceptors

Given the response properties of the proprioceptors, it is possible to state which of them have the potential to contribute to a particular sensation. For example, when a joint is passively moved across its normal angular range, muscles will be stretched (exciting muscle spindle endings), the joint capsule will be stressed (exciting some joint receptors), and the skin near the joint will be disturbed (exciting slowly and rapidly adapting cutaneous receptors). Thus, theoretically, muscle spindle, joint, and cutaneous receptors could provide the sensorium with information about what happened at the joint. However, this is dependent on central perceptual mechanisms having access to the neural activity generated by the different classes of proprioceptor. If the movement is made actively, then Golgi tendon organs will also be activated, and there will also be signals related to the central commands for movement.

For some decades, it was denied that the discharge of specialized intramuscular receptors, particularly muscle spindle endings, could be perceived. It was argued that signals in these endings would be hard to decode because of the interaction between the fusimotor input and the changes in muscle length, both of which affect the muscle spindle discharge. Furthermore, it proved more difficult to demonstrate a projection to the sensorimotor cortex from primary muscle spindle endings than from cutaneous and joint receptors. However, now both psychophysical and electrophysiological evidence indicates that muscle spindle endings in primates project to the sensorimotor cortex and play a major role in sensation of limb position and movement sense. This evidence is the following.

(1) Subjects experience illusory movements (and changes in limb position) when a muscle is vibrated in a way that excites the muscle spindle endings within it. The direction of these illusory movements is consistent with a perceived elongation or stretch of the vibrated muscle. These illusory movements can also be detected during a voluntary movement. Furthermore, electrical stimulation of muscle afferents can also produce similar illusory movements. These illusions are best explained by the perception of the discharge of muscle spindle afferents. They are quantitatively influenced by the number of spindle afferents activated and their discharge frequency. By contrast with sensations derived from some single cutaneous afferents innervating the finger tip, the discharge of many muscle receptors may be required to generate perceived proprioceptive signals. The illusions produced by muscle spindles are powerful, but labile, being less obvious if vision of the relevant limb is permitted. This presumably indicates that there is a dynamic (as well as a static) representation of the limb within the central nervous system, which can be influenced by many cues.

(2) The ability to detect changes in limb position and movement remains when the contribution from joint and cutaneous receptors is abolished by anesthesia. Although this procedure does not impair proprioceptive acuity for large joints in the leg, acuity is impaired for joints in the hand. Clearly, for the former joints, proprioceptive acuity can be entirely normal (across the usual angular range) when reliance is placed solely on specialized intramuscular receptors. Consistent with this view is the observation that surgical removal of a joint and associated tissues and its replacement with a prosthesis does not produce an overt deficit in detection of changes in joint position and movement in the hand as well as in the hip and knee.

(3) Some joints can be positioned so that the muscles that normally move them are unable to do so. Movement at these joints can no longer stretch these muscles, and the ability to detect joint rotation is then severely impaired.

(4) Muscle spindle endings (both primary and secondary) and tendon organs have anatomically documented projections to the sensorimotor cortex in experimental animals. Based on several techniques, the projections to the somatosensory cortex have been determined in human subjects for rapidly conducting muscle afferents from distal and proximal muscles (Fig. 4), and even from muscles of the trunk. It is likely that the signals from intramuscular receptors undergo significant central processing at or before reaching a cortical level. This may be re-

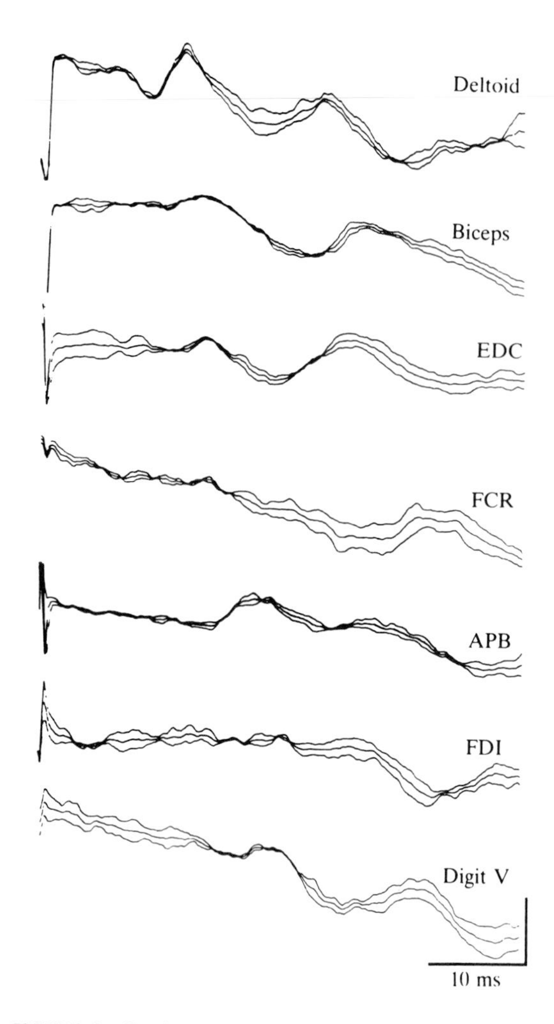

FIGURE 4 Cerebral potentials recorded from the parietal scalp overlying the somatosensory cortex in one subject. Potentials produced by intramuscular stimulation of the motor point of anterior deltoid (top traces), short head of biceps brachii, extensor digitorum communis (EDC), flexor carpi radialis (FCR), abductor pollicis brevis (APB), first dorsal interosseous (FDI), and by stimulation of the digital nerves of the little finger (bottom traces). Stimuli were delivered at three times the relevant motor or sensory threshold. The number of responses in each average and vertical calibration were as follows: deltoid n = 1,024, 1.4 μV; EDC 1,024, 1 μV; FCR 3,000, 1 μV; APB 2,048, 1 μV; FDI 2,048, 1 μV; little finger–hand area electrode 1,024, 2 μV. Potentials were somatotopically organized with those from distal muscles (e.g., APB, FDI) more lateral on the scalp than those from proximal muscles (e.g., deltoid). ms, msec. [From Gandevia and Burke, 1988, Projection to the cerebral cortex from proximal and distal muscles in the human upper limb, *Brain* **111**, 389–403.]

quired (a) to determine useful signals of angular position, velocity, and other derivatives of muscle length; (b) to determine which particular joint is moving, given that some muscles span several joints; and (c) to decode the component of muscle spindle discharge that results from peripheral events rather than fusimotor activity.

While available experimental evidence suggests that joint receptors have only a minor role to play in proprioceptive sensations when the joint is in its mid-range, they likely play a greater role when the joint approaches the end of its usual angular range. Thus, both the proprioceptive acuity and reflexes evoked by joint afferents are enhanced when a joint is distended following a traumatic effusion or affected by some forms of arthritis.

Golgi tendon organs respond to active muscle force, and signals from them project to the somatosensory cortex in the cat and presumably also in human subjects. However, while it is likely that their signals can directly influence force sensations, definitive evidence for their role in the perception of muscle force has not been easy to obtain. As argued below, under many circumstances, subjects appear to use signals of centrally generated motor command rather than peripheral signals of tension in the estimation of forces and tensions.

B. Signals of Perceived Motor Commands

Additional components of proprioceptive sensation depend, in a less direct way, on signals derived simply from peripheral afferents. The best known example of this is the sensation of muscular force (or heaviness). Whereas signals from skin, joint, and tendon organ afferents would register the force used to support or move an object, when subjects are formally required to signal the perceived force (or heaviness) they do so by reference to a centrally generated signal of the effort required in the task. Thus, as muscle fibers fatigue, a greater effort is required, and so the perceived force increases. This phenomenon has been studied in many situations in which the relationship between the central motor command or effort and the achieved force is altered (Fig. 5). This preferred reliance on a partly centrally generated signal for estimates of force, effort, and heaviness may have a teleological basis. For example, it means that the sensorium becomes aware of the increased central motor command required for a fatiguing task rather than the overt tension failure that necessitated it. This would ensure that associated changes in proprioceptive reflexes are generated in proportion to the increase in central command. Indeed, it is as if subjects can check on the status of a particular motoneuron pool and its muscle fibers by reference to a central signal. Accuracy

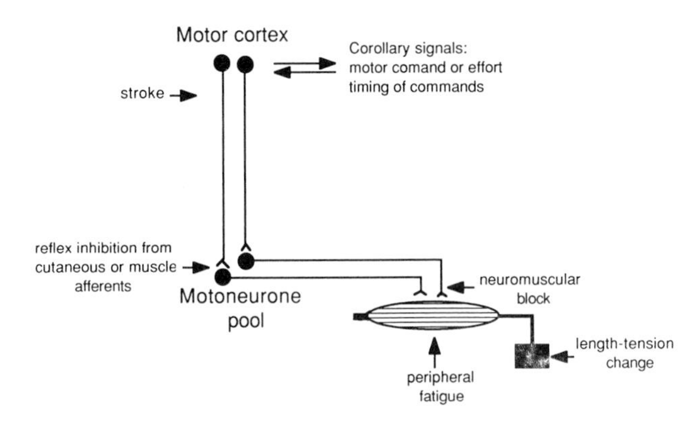

FIGURE 5 Simplified presentation of descending motor pathways, the motoneuron pool, and muscle. The modifications that produce an increase in perceived force and heaviness are marked. Signals related to the motor outflow are involved in the perception of motor command or effort and in perception of the time at which motor commands are dispatched. They do not give rise to the sensation that a limb has moved.

of judgement about force, as for movement sensations, may be relatively uniform for different joints within a limb. Certainly, the marked distal acuity for cutaneous sensation is not paralleled by marked acuity for some proprioceptive sensations.

Some limited evidence suggests that the neural mechanisms underlying the generation of the relevant signal of motor command for force perception probably involve the primary motor cortex. However, the interaction of such internal signals with those from peripheral receptors warrants further study. It is, nonetheless, now well established that signals of motor commands do not provide directly a sense that the relevant body part has moved: Attempts to move a paralyzed limb are associated with a feeling of great effort or weight, not with a sense that the limb actually moved.

Signals related to centrally generated motor commands probably have a number of other proprioceptive functions apart from simple movement production. They act principally to allow evaluation and interpretation of ongoing signals from proprioceptors. Some additional functions follow. First, they influence the transmission of somatosensory inputs to the sensorimotor cortex just prior to and during movement. Second, they influence the potency of many reflexes that affect motoneurones during movement. Third, they provide signals that can be used to decode the signal from muscle spindles (and presumably other proprioceptors) into that component due to an external perturbation to a movement and that expected based on the descending voluntary drive to the fusimotor system. Such decoding is especially relevant given that the fusimotor system can alter selectively the bias and the gain of muscle spindle signals. Fourth, some signals that rely on force *and* length (e.g., compliance and stiffness) may be biased by the level of perceived motor command. Finally, the perceived *time* at which a muscle contraction is commanded occurs before movement onset and must also represent perceptual access to a centrally generated motor signal. [*See* FUSIMOTOR SYSTEM.]

IV. Conclusions

Many of the proprioceptive mechanisms discussed above have related to simple movements or perceived responses to passive movements. Although neither of these situations occurs routinely in daily life, the relevant neural mechanisms are likely to be based on those revealed under simpler experimental conditions. Proprioception covers a range of sensations associated with the central commands needed to produce force and movement, and the monitoring of the peripheral consequences of the movement. The importance of peripheral proprioceptors in motor control is revealed by the deficits following loss of proprioceptive input in the ability to execute complex movements, to learn movement sequences, and to sustain a constant level of muscle contraction.

Acknowledgment

The author's work is supported by the National Health and Medical Research Council of Australia. Comments on the manuscript by D. Burke and G. Macefield are gratefully acknowledged.

Bibliography

Boyd, I. A., and Gladden, N. H. (eds.) (1985). "The Muscle Spindle." Stockton Press, New York.

Burgess, P. R., Wei, J. Y., Clark, F., and Simon, J. (1982). Signalling of kinesthetic information by peripheral sensory receptors. *Ann. Rev. Neurosci.* **5,** 171–187.

Burke, D., Gandevia, S. C., and Macefield, G. (1988). Responses to passive movement of receptors in joint, skin and muscle of the human hand. *J. Physiol.* (*London*) **402,** 347–361.

Gandevia, S. C., and Burke, D. (1988). Projection to the cerebral cortex from proximal and distal muscles in the human upper limb. *Brain* **111,** 389–403.

Goodwin, G. M., McCloskey, D. I., and Matthews, P. B. C. (1972). The contribution of muscle afferents to kinaesthesia shown by a vibration induced illusions of movement and by the affects of paralysing joint afferents. *Brain* **95,** 705–748.

Hulliger, M. (1984). The mammalian muscle spindle and its central control. *Rev. Physiol. Biochem. Pharmacol.* **101,** 1–110.

Matthews, P. B. C. (1972). "Mammalian Muscle Receptors and Their Central Actions." Arnold, London.

Matthews, P. B. C. (1988). Proprioceptors and their contribution to somatosensory mapping: complex messages require complex processing. *Can. J. Physiol. Pharmacol.* **66,** 430–438.

McCloskey, D. I. (1981). Corollary discharges: Motor commands and perception. *In* "Handbook of Physiology, Sect. I, The Nervous System," Vol. 2, Part 2, pp. 1415–1447. American Physiological Society, Bethesda, Maryland.

Proske, U., Schaible, H.-G., and Schmidt, R. F. (1988). Joint receptors and kinaesthesia. *Exp. Brain Res.* **72,** 219–224.

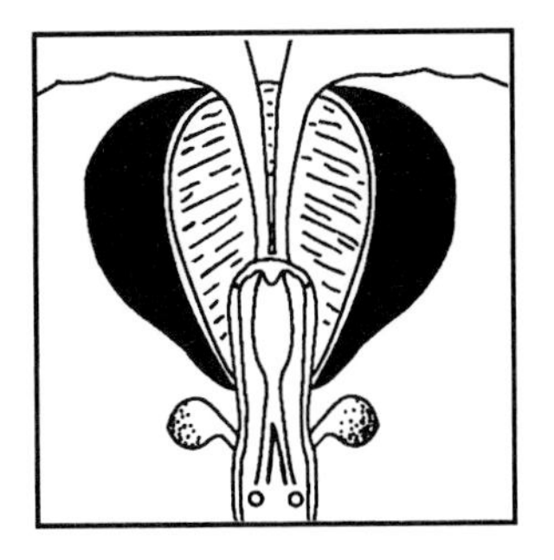

Prostate Cancer

MAARTEN C. BOSLAND, *New York University Medical Center*

Glossary

Adenocarcinoma Neoplasm derived from epithelial cells, which grows essentially in a glandular fashion and is malignant (invades into surrounding tissues and disseminates to form metastases)

Androgen receptor Intracellular protein that binds the active androgenic hormone 5α-dihydrotestosterone; the hormone–receptor complex interacts with DNA to regulate expression of genes that are involved in androgen-specific responses of the target cell

Androgens Steroid hormones that, either directly or after metabolic conversion in target cells to 5α-dihydrotestosterone, bind to the androgen receptor, thereby having trophic effects on target cells; they are produced by the testes and, in both sexes, the adrenal cortex

Antiandrogens Compounds that antagonize the action of androgens (e.g., by binding to the androgen receptor without having biological activity) thereby competitively inhibiting the action of 5α-dihydrotestosterone on the target (cancer) cell

Hormonal therapy Therapy of hormone-sensitive cancers that involves either (1) reduction of the production of the hormone in question or removal of its source, or (2) inhibition of its action on the target (cancer) cell by a compound that blocks the receptor for that hormone (e.g., antiandrogens) or that inhibits the formation of the active form of the hormone (e.g., 5α-reductase inhibitors)

Latent prostate cancer Form of prostate cancer that is very common in aging men and does not cause symptoms within the lifetime of most men

Luteinizing hormone-releasing hormone analogues Peptides that are analogues of luteinizing hormone-releasing hormone (LH-RH), which is produced in the hypothalamus; they bind to the LH-RH receptor in the pituitary and either have LH-RH activity and exhaust the production of LH by the pituitary, or have no biological activity, thereby competitively inhibiting the action of LH-RH

Male accessory sex glands Exocrine, androgen-dependent glands that are part of the male genital tract and produce secretum that is part of the ejaculate; besides the prostate, they include the seminal vesicles and bulbourethral glands

Tumor heterogeneity Presence of tumor cells within the same neoplasm that have different biological and phenotypical characteristics, e.g., differ in hormone-sensitivity

THE BIOLOGY AND PATHOLOGY of cancer of the prostate gland is the focus of this article[1]; its clinical aspects are briefly summarized. The causes of prostate cancer are discussed in more detail, including epidemiology and studies in animal and tissue culture systems.

1. This work was supported in part by Grant No. 43151 from the National Cancer Institute.

I. Introduction

Prostate cancer occurs frequently in the United States and in much of western Europe. It is the second-most frequent cause of death due to cancer in these countries, after lung cancer and before cancer of the large bowel and rectum. In 1990, prostate cancer will surpass lung cancer as the most frequently diagnosed cancer in U.S. males. In the United States, an estimated 5–6 out of every 100 men will develop prostate cancer in their lifetime, and 3–4 will die from it. Because prostate cancer is such a widespread disease, a large amount of information exists about diagnosis, treatment, and prognosis. Furthermore, the prostate is the major target organ for male sex hormones and, therefore, much research has been conducted on aspects of endocrinology and biochemistry of this gland. However, little is known about the causes (etiology) of prostate cancer, unlike other major human cancers, such as lung cancer, which is known to be caused predominantly by smoking.

II. Anatomy and Biology of the Prostate Gland

The prostate is one of the accessory sex glands in male mammals; the other major male accessory sex glands are the seminal vesicle and the bulbourethral gland. The human prostate is a single gland that completely surrounds the urethra at the base of the urinary bladder; the larger part of the gland is dorsally located between the urethra and the rectum. The gland is approximately 2.5–3 cm long and 3–4 cm in diameter in young adult males, and it weighs 18–25 g. It is composed of 30–50 compound tubuloalveolar glands that share 16–32 excretory ducts that open out into a part of the urethra that is surrounded by the prostate. The ducts of the paired seminal vesicles merge with the paired deferent ducts to form the ejaculatory ducts, which project dorsally through the prostate and open into the lumen of the prostatic part of the urethra via a structure that projects into this lumen termed the colliculus seminalis or verumontanum. The glands comprising the prostate are embedded in firm connective tissue (stroma) consisting mainly of fibroblasts and smooth muscle cells. A fibroelastic capsule surrounds the prostate.

In the mammalian fetus, the prostate develops as four paired groups of endodermal epithelial buds growing from the urogenital sinus, which derives from the primitive hindgut, into the surrounding endodermal mesenchyme. This mesenchyme plays a decisive role in determining the future development of the bud epithelium. In the male, the Müllerian duct regresses almost completely, and the Wolffian (mesonephric) duct gives rise to the deferent ducts and seminal vesicles. In the human fetus, one of the paired groups of prostatic buds does not further develop, but in rodents this group forms into the ventral lobe of the prostate. The other three groups of buds develop into separate prostate lobes in rodents—dorsal, lateral, and anterior lobes—the latter is synonymous with the coagulating gland. In humans and several other species, however, these three groups merge into a single glandular mass. Various subdivisions of the human prostate have been proposed based on the embryology of the gland and on morphological and pathological considerations. From a clinical point of view, it is most useful to think of the prostate as composed of two glandular zones: the glands that directly surround the urethra and the glands that are located more in the periphery.

The prostatic ducts are lined by simple or pseudostratified columnar epithelium, whereas the prostatic glands contain two different cell types: the glandular or secretory cell and a nonsecretory basal cell. The secretory cell contains the intricate apparatus for producing the prostatic secretum: rough endoplasmatic reticulum, mitochondria, Golgi structures, and secretory granules surrounded by a single membrane. The mechanism of secretion is of the merocrine type (secretion by expulsion of the contents of secretory granules: exocytosis) and, possibly, also the apocrine type (secretion by expulsion of membrane-bound vesicles containing secretum). The basal cell may function as a reserve cell that proliferates to make up for losses of secretory cells. The latter cell type, however, can also divide.

The major function of the prostate and seminal vesicle is to produce part of the seminal fluid during ejaculation (approximately 30% of this fluid is produced by the prostate and 70% by the seminal vesicles). The function of the bulbourethral gland is unclear at present. The seminal fluid plays a critical role in maintaining sperm viability and sperm motility after ejaculation. The secretum of the prostate contains many constituents; quantitatively most important are citric acid, zinc, and spermine (a polyamine), which all probably protect the sperm cells in some manner, and the enzyme acid phosphatase, whose function is unclear; the major ion in prostatic secretum is sodium. Some of the major constituents of the seminal vesicle secretum are fructose (provides energy for the sperm cells) and protective substances such as protease inhibitors and uric acid. The synthesis and secretion of the prostatic fluid is under nervous as well as hormonal control (predominantly androgens).

The considerable fibromuscular portion of the human prostate also acts as a sphincter of the urethra, in conjunction with the pelvic musculature that controls micturation. Furthermore, this fibromuscular tissue plays a major role in the process of the ejaculation of the seminal fluid and, thereby, of the semen. This function is neuronally controlled.

The prostate is not an endocrine gland in the real sense, because it does not synthesize hormones but produces the active androgen metabolite 5α-dihydrotestosterone (DHT) from the male sex hormone testosterone. Testosterone is in males predominantly produced in the testes by the Leydig cells and is, to a much lesser extent (5%), derived from other sources, e.g., the adrenal cortex. The conversion from testosterone to DHT is irreversibly effectuated by the enzyme 5α-reductase in the prostate epithelial cells. The formation of DHT is under tight control, as is its further metabolism in the prostate epithelium by oxidoreductase enzymes (reversibly) to metabolites, which are further enzymatically hydroxylated (irreversibly) to inactive steroids; this is necessary because DHT is the active hormone that binds to a DHT-specific androgen-receptor protein in the prostate epithelial cell. Whether this binding occurs in the cytoplasm, in the nucleus, or in both compartments is unclear. The hormone–receptor complex ultimately binds to acceptor sites within the nucleus, probably to specific regions in the DNA that regulate transcription of genes involved in prostate cell responses to androgens. These DNA regions may be comparable to the estrogen-responsive elements, which have been shown to play a role in the response to estrogens in estrogen-responsive tissues. [*See* DNA and Gene Transcription.]

Control of cell proliferation in the prostate is under primary control of androgenic steroids, but estrogens almost certainly also play a role in this respect. The pituitary peptide hormone prolactin probably facilitates the effect of androgens. The human and rodent prostate contains estrogen receptors and possibly prolactin receptors. Synthetic and secretory activities of the prostate epithelium are under the control of the same hormones but are also regulated by the autonomic nervous system. The secretion of the steroid hormones that regulate prostate function is under control of the hypothalamo–pituitary system. The secretion of testosterone by the testes, for example, is regulated by the pituitary peptide luteinizing hormone (LH), and the secretion of LH is regulated by the hypothalamic substance LH-releasing hormone (LH-RH). The secretion of LH-RH, in turn, is controlled by feedback influence of the blood concentration of testosterone and by the central nervous system. This delicate regulating system maintains the level of testosterone in the blood at a very constant level, thereby maintaining the prostate gland. An additional regulating system is the serum protein sex hormone-binding protein, which reversibly binds most testosterone in the blood, thereby determining the fraction of free testosterone that is available to enter tissues and cells. Cell proliferation and secretory function of the prostate epithelium are also regulated by the stromal cells that surround it, possibly via secretion of humoral factors (paracrine acting growth factors that diffuse to the epithelial cells). The stromal cells of the prostate also have been shown to be under hormonal control; they can form DHT from testosterone and have androgen receptors.

III. Pathology and Tumor Biology of Prostate Cancer

The great majority of prostate tumors in men is of the adenocarcinoma type, i.e., a malignant neoplasm derived from epithelial cells that grows in a glandular fashion. Prostate cancer should be distinguished from benign hyperplasia of the prostate (BPH), which is very common in aging men. BPH is a joint proliferation of the glandular epithelium and the fibromuscular stroma that does not display malignant properties; it is predominantly localized in the inner zone of the prostate surrounding the urethra and usually causes obstruction of the urinary flow. Prostate cancer, on the other hand, is a proliferation of prostatic epithelial cells, and it is predominantly located in the peripheral zone of the prostate. Prostate cancer is a malignant disease: the tumor cells invade into surrounding tissues, they penetrate the prostatic capsule and blood and lymph vessel walls, and, once in the blood or lymph fluid, disseminate and lodge in other tissues, forming metastases. These metastases most frequently occur in the pelvic, spinal, and other bones, in the pelvic lymph nodes, and in the lungs. The reason for the predilection of prostate cancer to form skeletal metastases is not understood. Prostate carcinomas produce a number of substances in abnormal amounts, eventually (usually in advanced stages of the disease) leading to elevated levels of these substances in the blood. Of these, the enzyme prostate-specific acid phosphatase and prostate-specific antigen are clinically most important, because they can be used as markers to detect prostate cancer and to monitor the response to treatment. [*See* METASTASIS.]

One form of prostate cancer does not cause any clinical symptoms and usually does not progress to clinical cancer within the lifetime of a man. This so-called latent prostate cancer occurs very frequently in aging men. At least 50% of all men are estimated to develop this type of prostate cancer. Some investigators estimate that it occurs in as much as 80% of men >80 yr of age. This cancer is small, often occurs multifocally, and does not penetrate the prostate capsule or metastasize. It is only detectable by histologic examination of prostate tissue that is removed surgically (for treatment of BPH) or that is collected after death by autopsy.

Most prostate cancers are androgen-dependent; i.e., they will regress when the production of androgens in the body is stopped by some means or when the action of androgens on the prostate (cancer) cell is blocked in some way. Eventually, however, almost every prostate carcinoma that initially regressed upon hormonal treatment will relapse to a hormone-insensitive state and grow in the absence of androgens. Why this happens is not completely clear—the cancer cells could adapt to the absence of androgen; the tumor could be the result of a confluence of multiple small tumors, some hormone-sensitive and others not; or small numbers of hormone-insensitive cells could develop in a tumor that was initially 100% androgen-sensitive and the hormonal treatment may select for these. The last possibility seems most likely to be the predominant mechanism of the relapse. The development of tumor heterogeneity (a mix of androgen-sensitive and -insensitive cells) is thought to be caused by a genetic instability of the cancer cells.

Two important characteristics of prostate cancer are the very slow growth rate of this tumor and the large variability in biological properties, such as degree of hormone-sensitivity, production of acid phosphatase, and degree of histological differentiation (or grade, distinguished as well, moderately, or poorly differentiated based on morphologic characteristics of the cancer cells and their nuclei, and on the growth pattern of the carcinoma). Both properties create a problem for the treatment of prostate cancer as slow-growing tumors are relatively resistant to chemotherapy, and the variability obviously has a negative effect on the reliability of diagnostic and therapeutic procedures.

IV. Clinical Aspects of Prostate Cancer

Prostate cancer is usually diagnosed when a patient encounters symptoms related to obstruction of urinary flow or pain, usually from metastatic processes. Palpation of the dorsal aspect of the prostate is possible by a digital rectal examination. This rectal exam can detect a large percentage of prostate carcinomas, but unfortunately only some 20% of them are in an early stage of their development, and only these are potentially curable by total prostatectomy (surgical removal of the prostate). The only other method by which small, early cancers can be detected is by transrectal (via the rectum) ultrasound techniques, but this method is not suitable for screening purposes. Thus, the majority of prostate cancers have metastasized and/or are at an advanced stage at the time of detection. One other way some prostate carcinomas are detected is by histological examination of tissue that is surgically removed from the prostate as treatment for BPH; this is usually done by a transurethral (via the urethra) resection procedure. Measurement of the blood concentration of markers such as prostate-specific antigen and acid phosphatase will often only detect advanced stage cancers. Confirmation or rejection of the suspicion of prostate cancer upon a rectal exam is usually done by histologic examination of a needle biopsy (either taken by regular biopsy needle or spring-loaded biopsy gun method or by aspiration of cells via a fine needle, usually carried out via the rectum) and recently by ultrasound examination.

Prostate cancers are clinically divided into groups according to their extent, the presence of metastases, and the elevation of serum acid phosphatase. This procedure is called staging and will determine the treatment that is proposed. Staging often involves the surgical removal of the regional pelvic lymph nodes that drain the prostate to examine them for the presence of metastases, as well as methods to detect bone metastases (e.g., X-ray methods or a bone scan examination). A summary of the method of staging for prostate cancer that is most widely used in the United States is given in Table I. Stages A1 and A2 are found incidentally in transurethral resection material; stages A and B are usually without symptoms.

Stage A1 is usually not treated but followed. Stages A2 and B are treated with prostatectomy or radiation. Radiation can be applied by implantation of needles containing a source of radiation, most often iodine-125, or by external radiation methods that can be accurately focused at the prostate. External radiation is also applied to stages C and D1. Hormonal therapy is used for all stages beyond stage A, and chemotherapy is usually reserved for patients that have stage-D2 disease, which is resistant to hormone therapy. Combinations of different modes of therapy are common. Because the disease is essentially incurable beyond stage A, most patients receive one type of treatment (e.g., hormonal treatment) until it is not effective any more; then, other therapy modes will be applied (e.g., chemotherapy).

Hormonal treatment consists of some form of lowering the blood concentration of testosterone to very low levels or of blocking the effect of testosterone on the androgen-sensitive cancer cell. Lowering the testosterone concentration can be achieved by castration (orchiectomy), by treatment with estrogens (particularly diethylstilbestrol), or by treatment with LH-RH analogues. The latter treatment either competitively inhibits the effects of natural LH-RH on the pituitary (analogues with no biological activity) or exhausts the LH production by the pituitary following an initial increase due to constant hyperstimulation of LH secretion (analogues with agonistic activity). These treatments will take away or inhibit testicular androgen production, but they will not suppress androgen production by the adrenal gland. Therefore, antiandrogenic compounds have been used that will block the binding of DHT to the androgen receptor in the prostate epithelial (cancer) cells. A combination of inhibition of testicular androgen production and application of antiandrogens to block away the effect of adrenal androgens, so-called total androgen blockade, is thought to be more effective than either treatment alone, but this has not (yet) been proven. The androgen dependence of most prostate cancers and the effectiveness of castration and estrogen treatment was first demonstrated in the early 1940s by Dr. Charles Huggins and coworkers, who received the Nobel Prize for his work.

Only surgery, and sometimes radiation therapy, can be curative. However, because of the old age of most patients, any treatment that will prolong life will be of great value, particularly when symptoms are suppressed and side effects are minimal or absent (palliative therapy). Radiation therapy improves survival somewhat in most cases, depending on stage and grade of the tumor; 10-yr survival rates in the range of 50% have been reported for stage-C and -D disease. Hormonal treatment will effectively suppress symptoms in 80–85% of patients for variable periods, ranging from a few months to many years, depending on the stage of the disease; the average response duration is approximately 1 yr. However, overall survival is not improved in most cases, and survival after failure to respond to hormonal therapy is only 6–12 mo. The results of chemotherapy trials of hormone-insensitive stage-D patients have been largely unsuccessful. Clearly,

TABLE I Staging of Prostate Cancer[a]

Stage	Local lesion	Elevation of prostatic acid phosphatase	Regional lymph node metastases	Distant metastases	% of patients[b]
A1	Not palpable, focal	No	No	No	5–10
A2	Not palpable, diffuse	No	No	No	
B	Palpable, localized	No	No	No	10–15
C	Local extension beyond capsule	No	No	No	35–40
D1	Any of above	Yes	No	No	
	Any of above	No	Yes	No	40–45
D2	Any of above	Yes/No	Yes/No	Yes	

[a] According to the American Urologic Society
[b] Percentage of patients that have that stage of prostate cancer at the time of first diagnosis.

better therapeutic approaches to control hormone-insensitive prostate cancers are necessary.

V. Descriptive Epidemiology of Prostate Cancer

Prostate cancer is not diagnosed before the age of 70–75 yr in most men, and it is thereby typically a disease of old age, more so than any other type of cancer. However, the occurrence of prostate cancer rises more steeply with increasing age than any other cancer and is the most frequent cancer in men over 80 yr of age in many countries.

The occurrence of prostate cancer among different countries varies considerably. In general, prostate cancer incidence and mortality rates are high in North America and western Europe; lower rates are found in eastern and southern Europe and in some Asiatic and most South American countries; and prostate cancer is infrequent in some Central and South American countries, in southeastern Asiatic countries, and in Japan. In fact, the ratio of occurrence in the highest risk population (U.S. Blacks) to the risk in the lowest risk populations (e.g., Chinese living in Shanghai) is larger for prostate cancer than for any other type of cancer. Some examples of this variability are given in Table II. An impression of the actual numbers of new cases (incidence) and deaths (mortality) from prostate cancer is given in Table III. These figures and most data that epidemiologists use to compare (prostate) cancer rates in different groups of people are age-adjusted, i.e., mathematically corrected for differences in the age distribution of the populations studied. This is necessary because as the proportion of older individuals increases, so does the number of cancer cases.

Large-scale migration to the United States occurred at the end of the last century and the first half of this century, predominantly from Asiatic, Central American, and European countries. Many of the countries from which these migrants originated are the same areas that have an intermediate or low risk for prostate cancer (Table II). These migrant populations in the United States appear to have a risk for prostate cancer that is higher than the risk prevailing in their homeland, and that approaches the risk in the general U.S. population. This change in prostate cancer risk has been best studied among Japanese living in Hawaii (see Table III) and the San Francisco Bay area. Clearly, the environment, and not some genetic factor, determines the risk for prostate cancer.

Nevertheless, prostate cancer is less frequent among these migrant populations in the United States, including Mexicans, Japanese (Table III), Chinese, and Filipinos, than among U.S. white men. The indigenous people in the United States (American Indians and Alaskan Inuit) also have a lower prostate cancer risk than white men. These observations might suggest racial (genetic) differences in susceptibility to the factors that cause prostate cancer; on the other hand, not all U.S. men share the same environment (e.g., their life-style, which is an environmental factor, may differ considerably), and this may be related to these differences in prostate cancer risk.

Furthermore, prostate cancer is very frequent among U.S. Blacks (see Table III), whereas this disease is not very frequent among the predominantly black populations of some Caribbean islands; it is probably also infrequent in Central and southern Africa. In fact, U.S. Blacks have the highest risk for prostate cancer in the world. The differences in prostate cancer occurrence among black men living in different parts of the world also indicates that environment is a more important determinant of prostate cancer risk than is genetic makeup.

TABLE II Examples of the Geographic Variation in Prostate Cancer Occurrence

High-risk countries	Intermediate-risk countries	Low-risk countries
Sweden	Chile	Mexico
United States	Spain	Ecuador
Norway	Poland	Singapore
Federal Republic of Germany	Paraguay	Shanghai
France	India	Japan
Hungary	Greece	Thailand

TABLE III Examples of Age-Adjusted Incidence and Mortality Rates for Prostate Cancer[a]

Annual number of new cases (per 100,000) of prostate cancer in	
United States (Whites, 1971)	59.0[b]
United States (Whites, 1977)	70.4[b]
United States (Blacks, 1971)	94.1[b]
United States (Blacks, 1977)	115.8[b]
United States (Whites in Hawaii, 1978–1982)	58.3[c]
United States (Japanese in Hawaii, 1978–1982)	31.2[c]
Japan (Osaka, 1978–1982)	5.1[c]
Shanghai (1978–1982)	1.8[c]
Annual number of prostate cancer deaths (per 100,000) in	
Sweden (1976)	21.0[c]
United States (1976)	14.5[c]
Paraguay (1976)	8.8[c]
Japan (1976)	2.3[c]

[a] Actual rates depend on which population is used as the standard for the age-adjustment.
[b] Age-standardized to the 1970 U.S. population.
[c] Age-standardized to the "Standard World Population."

The occurrence of prostate cancer, in both mortality and incidence, has increased, albeit slowly, over the past several decades in almost every country that has been studied. Because the method of age-adjustment has been used in these studies to correct for differences in life expectancy, this increase in prostate cancer is real and not related to the increasing length of life that has occurred in many countries. Increases in the quality of medical diagnostic procedures has not led to increased detection of prostate cancer as cause of death but probably has contributed to the increased number of diagnosed new cases. The increased occurrence of the disease over time suggests that the presence and influence of environmental determinants of prostate cancer have increased. Prostate cancer occurrence has increased much more rapidly in Japan during the last two decades than in, for example, the United States; this has been ascribed to the rapid westernization of the Japanese life-style. The increase in prostate cancer frequency in the United States has been much larger among Blacks than among Whites. In fact the Black–White ratio in prostate cancer death rates has changed from 0.6 in 1930 to 1.8 in 1970. Interestingly, this change coincided with a large-scale migration of Blacks from rural to urban areas, and this change in environment may have contributed to the change in prostate cancer occurrence.

Old men frequently have a latent form of cancer of the prostate, as pointed out earlier. The occurrence of this form of prostate cancer does not vary much in different countries. Even in populations in which the occurrence of prostate cancer as a disease differs widely (e.g., Japanese living in Japan and Hawaii, and U.S. white and black men), latent prostate cancer is about equally frequent. However, there is a type of latent cancer that is larger and grows more aggressively than average, and this type does show the same geographic variation as clinical prostate cancer. Based on these observations, it is believed that the environmental factors that appear to determine prostate cancer risk influence the progression from small, not very invasively growing prostate cancer to larger, more invasive cancer (which ultimately develops in a tumor that causes clinical symptoms).

Cancer patterns in some populations that differ from the general population in their life-style environment have been studied to determine the influence of life-style on cancer development. Best studied in this regard are groups of Mormons in Utah and Seventh-Day Adventists in California. Because of religious reasons, these groups usually do not smoke or drink alcoholic beverages, tea, and coffee, have rather strict sexual mores, and, the Adventists in particular, refrain from a rich diet, and some are vegetarians. Cancer of the breast (in females) and colon (in both sexes) occurs less frequently in both groups, and this has been ascribed to their moderate life-style. Prostate cancer, however, is equally frequent in these two religious groups and in the rest of the U.S. population. The differences in prostate cancer occurrence among countries and, with several exceptions, among ethnic groups in the United States are rather similar to the differences in occurrence found for breast and colon cancer. Also, the changes in cancer patterns in migrants in the United States are very comparable for breast, colon, and prostate cancer. Thus, there are probably some similarities in the causes of these three important human cancers, but there are likely also major differences in the etiology of prostate cancer on the one side and of breast and colon cancer on the other. Both colon and breast cancer are thought to be related to dietary risk factors, and breast cancer also to hormonal factors. Prostate cancer, however, probably has a more complex etiology. [*See* BREAST CANCER BIOLOGY.]

To estimate if factors related to social and economic status are associated with cancer risk, cancer patterns of groups that differ in income, occupation, or level of education are often compared. No consistent relation has been found between prostate cancer risk and such indicators of socioeconomic status. A consistent finding, however, is that in many European countries prostate cancer is more frequent in urbanized than in rural areas. In the United States, this pattern has only been found among Blacks, but not Whites. This excess risk for prostate cancer in urban areas may be related to pollution or an increased likelihood of occupational exposure to carcinogens, but it may also be related to urban–rural differences in, for example, life-style.

Men with blood-relatives that have (had) prostate cancer have a higher risk for developing prostate cancer than men who do not. In this regard, a few studies have shown an excess risk of as high as three- to fourfold (there are also some contradictory reports). These findings indicate that genetic factors can contribute to a person's risk for developing prostate cancer, but they do not necessarily explain the differences found between, for example, U.S. Blacks and Whites.

Prostate cancer and benign hyperplasia of the prostate (see earlier) are probably unrelated processes. The epidemiology of the two diseases is rather different, and the morphology and site of occurrence in the prostate gland are very dissimilar, as pointed out earlier.

In general, married men have a higher risk for prostate cancer than single men, and widowed and divorced men have a higher risk than married men. Among U.S. Blacks, however, single men are at the highest risk, followed by married and then widowed or divorced men. In fact, single U.S. black men are probably the population at highest risk for prostate cancer worldwide at the present time. These associations between risk and marital status might be related to sexual behavior, as discussed later.

In summary, the descriptive epidemiology has provided the following information on the probable causes of prostate cancer:

1. Environment rather than genetic factors seems to determine a person's risk of developing prostate cancer (geographic variation and migrant studies).
2. These environmental determinants probably influence the progression from very slow-growing, not very malignant latent cancers to more malignant cancers that will eventually metastasize (lack of geographic variation for latent cancers).
3. The presence and influence of these environmental factors has probably increased over the past several decades (increasing occurrence of prostate cancer).
4. Aspects of life-style (in particular, dietary habits [similarity with the epidemiology of breast and colon cancer]), exposure to environmental or occupational carcinogens (higher urban than rural occurrence of prostate cancer), and aspects of sexual behavior (marital status) may be be related to the development of prostate cancer.
5. Notwithstanding the probable importance of environmental determinants, some genetic factors may influence prostate cancer risk on an individual level (familial aggregation of risk).
6. For some unknown reason, black men in the United States have a higher risk for developing prostate cancer than any other population in the world.

VI. Animal and Tissue Culture Models of Prostate Cancer

Animal models are extremely useful for research on the causes and treatment of many types of cancer. Tissue culture (*in vitro*) methods are helpful in elucidating the mechanisms by which certain factors cause cancer (e.g., chemical carcinogens) and control the growth of cancer cells (e.g., hormones). *In vitro* models can also be used to screen the potential usefulness of anticancer drugs.

Animal models for cancer can be distinguished into two basic categories: (1) induction models and spontaneous models and (2) transplantation models. The spontaneous models and, particularly, induction models can give information about factors that cause (induce) cancer (e.g., chemical carcinogens) and factors that enhance or inhibit the formation of cancer (e.g., diet or hormone treatment). The transplantation models involve transplantation of small pieces of tumor tissue or of a suspension of tumor cells under the skin or at a blood vessel-rich location, such as under the kidney capsule. These models are primarily used to study aspects of tumor biology (e.g., hormone sensitivity and tumor heterogeneity) and to test the effectiveness of anticancer treatments, such as chemotherapeutic agents.

Induction and spontaneous animal models of prostate cancer are still in development, and no systems are available that can be readily used. The systems that have been described in the literature are summarized in Table IV. Human prostate cancer is an adenocarcinoma that metastasizes and is usually dependent for some time on the presence of androgens for growth. Clearly from Table IV, several potential models do not have these characteristics and hormone sensitivity has not been studied in many cases. Furthermore, the prostate tumors in some models are located in the ventral prostate, which does not exist in men, as pointed out earlier (Section II). Nevertheless, there are some promising systems, particularly those that combine treatment with a chemical carcinogen and enhancement of cell proliferation (by hormone treatment to make the tissue sensitive to the action of the carcinogen), and those that combine treatment with a carcinogen and long-term treatment with testosterone.

Transplantation models of cancer are distinguished in two categories: (1) those that use tumor tissue from a specific, inbred animal strain and transplant it into animals of the same strain (which do not reject the tumor tissue because of inbreeding), and (2) those that use human tumor tissue transplanted into so-called nude mice, which lack the thymus, and are therefore partly immunodeficient and will not reject the transplants. The first category has been used to study aspects of hormone (androgen) sensitivity of prostate cancer, and the development of insensitivity of prostate tumors to hormones that occurs in prostate cancer patients after prolonged treatment with estrogens or antiandrogens. Both the first and second categories have been used to study the therapeutic effects of chemotherapeutic agents, (anti)hormones, and combinations thereof. Various aspects of the metastasizing of prostate cancer have been studied, particularly in transplantation models of animal origin. A fair number of transplantation systems derived from human prostate carcinomas have been reported. These sys-

TABLE IV Spontaneous and Induction Models of Prostate Cancer in Animals

Model system (method)	Location	Hormone sensitivity	Other major characteristics
Lobund–Wistar rat; spontaneous	Dorsolateral prostate	Not well established	Develops only in very old rats; adenocarcinoma; metastasizes; low incidence
ACI/SegHapBR rat; spontaneous	Ventral prostate	Not well established	Develops only in very old rats; adenocarcinoma; does not metastasize; high incidence
Induced in rats by long-term treatment with testosterone, alone or in combination with estrone or estradiol-17β	Dorsolateral prostate	Unknown or variable	Adenocarcinoma; metastasizes; low incidence
Induced in rats by *N*-nitrosobis(2-oxopropyl)amine (BOP)[a]	Ventral prostate	Unknown	Squamous cell carcinoma; metastasizes
Induced in rats by combination of BOP[b] and long-term testosterone treatment	Dorsolateral prostate	Unknown	Adenocarcinoma; metastasizes; high incidence
Induced in rats by *N*-methyl-*N*-nitrosourea (MNU)[b]	Dorsolateral prostate	Probably androgen-dependent	Adenocarcinoma; metastasizes; low incidence
Induced in rats by combination of MNU[b]and long-term testosterone treatment	Dorsolateral prostate	Unknown	Adenocarcinoma; metastasizes; high incidence
Induced in rats by 3,2′-dimethyl-4-aminobiphenyl (DMAB)[a]	Ventral prostate	Probably androgen-dependent	Adenocarcinoma; does not invade in surrounding tissue; does not metastasize; low incidence
Induced in rats by combination of DMAB[a] and cycles of hormone treatment[c]	Ventral prostate	Unknown	Adenocarcinoma; no invasion; does not metastasize; high incidence

[a] This chemical carcinogen is injected repeatedly (10–20 times).
[b] This chemical carcinogen is injected once or on 3 consecutive days, when cell proliferation is enhanced by hormone treatment (sequential treatment with an antiandrogen and then testosterone).
[c] Just before DMAB injection, estrogens are administered and then withheld, stimulating cell proliferation in the prostate.

TABLE V Examples of Transplantable Prostate Carcinoma Systems of Animal Origin

		Major characteristics			
Model system	Most important sublines	Androgen-sensitive	Growth rate	Degree of differentiation	Metastasizes
Dunning tumors[a]	R-3327H	yes	slow	well	rarely
	R-3327-PAP	partly	slow	well	rarely
	R-3327HI	no	slow	well	rarely
	R-3327AT1&2	no	fast	poor	rarely
	R-3327MAT-LyLu	no	fast	poor	yes (lung and lymph nodes)
	R-3327MAT-Lu	no	fast	poor	yes (lung)
Noble tumors[b]	Androgen-sensitive	yes	moderate	moderate	yes (lung and liver)
	Androgen-insensitive	no	moderate	poor	yes (lung and liver)
Pollard tumors[c]	PAI	no	moderate	poor	yes (lung and lymph nodes)
	PAII	no	moderate	poor	yes (lungs, liver, bone marrow, etc.)

[a] Developed from spontaneous adenocarcinoma found in dorsolateral prostate of an old COP rat in 1961 by Dr. W. Dunning; it was originally androgen-dependent.
[b] Developed from hormone-induced dorsolateral prostate adenocarcinomas in NBL rats by Dr. R. L. Noble.
[c] Developed from spontaneous dorsolateral prostate carcinomas in aged Lobund–Wistar rats by Dr. M. Pollard.

tems display a wide variety of characteristics, such as hormone sensitivity and growth rate. Basically, three systems are derived from animal prostate carcinomas; their characteristics are summarized in Table V. This table demonstrates that these systems allow studies on differences in hormone sensitivity, growth rate, morphology (degree of histological differentiation), and metastatic capacity and pattern (via the bloodstream and lymphatic system). Of these, the Dunning tumors have been most widely used in prostate cancer research.

In vitro models can be distinguished in three ways: (1) systems of human and of animal origin; (2) systems that use tissue or cells directly derived from normal tissue (primary cultures, which will eventually die) and systems that are derived from tumors or that consist of cells that are immortal, both of which will grow indefinitely (established cell lines); and (3) systems that culture cells or small pieces of tissue (explants). Explant systems of normal prostate, mostly of rodents, were among the first to be used in studies on the effects of chemical carcinogens in tissue cultures in the early 1950s. Explant studies have provided useful information on effects of chemical carcinogens, retinoids, and hormones and on aspects of the metabolism of hormonal steroids. A variety of established cell lines of prostate epithelial origin exist; many have been derived from the earlier-mentioned transplantable rat tumor systems, and some have been developed from human prostatic carcinomas. Those derived from rat tumors have been the most widely used to study the effects of hormones and chemotherapeutic agents.

VII. Causes (Etiology) of Prostatic Cancer

Although prostate cancer is a very common cancer in men, little is known with certainty about the causes of this disease. Studies in animal models and in tissue culture and, in particular, epidemiological investigations have provided some insights into the possible causes of prostate cancer.

Epidemiological methods can be distinguished into four categories: (1) *Correlation studies* determine the correlation between, for example, the estimated consumption of a particular food in different countries (international) or in different areas within one country and the mortality or incidence of a type of cancer in those countries or areas. (2) *Case-control studies* compare a group of cancer patients (usually 50–200 individuals) with one or more appropriate control groups and determine if certain factors occur more or less frequently among cases than in controls; e.g., if occupational exposure to certain chemicals or the consumption of certain foods is more or less frequent. These studies can be based on information on death certificates (retrospectively) or on information gathered by interviews or questionnaires. In the latter cases, not only frequency of occurrence of factors can be determined, but sometimes also more exact quantitative information, such as amounts of certain foods consumed. (3) *Prospective studies* collect information from a large group (cohort) of individuals (usually many thousands) by interviews or questionnaires, and then follow the cohort until cancer occurs in a sufficient number of persons to allow a comparison between them and a group of persons from the cohort that did not develop this cancer. (4) A special type of epidemiology, sometimes called *metabolic epidemiology*, determines the correlation between the presence of cancer and parameters that are related to the causation of the cancer, such as the concentration of certain hormones, indicators of nutritional status, or markers of exposure to a specific chemical carcinogen. Factors such as consistency and degree of the associations found, as well as dose–effect relationships, coherence of associations among studies of different types, and biological plausibility all play a role in the evaluation of the results of epidemiological studies.

A. Environmental Factors

As pointed out earlier, environment is thought to play a major role in the causation of prostate cancer. Aspects of life-style (e.g., diet and nutrition and sexual behavior) and occupational and environmental exposure to carcinogenic factors all probably play a role in this respect. Smoking, which is a major factor in the causation of several human cancers, is not related to prostate cancer risk.

1. Diet and Nutrition

Most indications and evidence that dietary factors are involved in the etiology of prostate cancer are derived from epidemiological studies, but there are also some animal and *in vitro* investigations that have addressed this question. Rather than giving details of these various studies, a summary of their results is presented in Table VI. Notwithstanding the many difficulties in interpreting eipdemiological results, a few associations are consistent and coherent: (1) A strong positive association (more consumption is associated with higher risk) has been found for edible fats and oils and total fat; (2) a moderate, positive association has been established for eggs, animal (saturated) fats, total and, possibly also, animal protein; and (3) there is no association between risk and the consumption of alcoholic beverages. It is often not clear which of these factors are causally related with prostate cancer risk, because they are often interrelated, e.g., the consumption of animal fat and animal protein usually go hand in hand. [*See* FATS AND OILS (NUTRITION).]

The findings for vitamin A and β-carotene are very contradictory. *In vitro* studies have shown that retinol and other retinoids counteract the effects of chemical carcinogens on the morphology of rodent prostate explants. The results of some epidemiological studies, however, indicate that vitamin A and/or carotenes may increase risk for prostate cancer, whereas others suggest a protective effect. Most studies that showed increased risk found this only for patients 70 yr of age and older, but not for younger patients, and they estimated total dietary intake of vitamin A and carotene. Most studies that found a negative or no association determined only intake of one or a few vitamin A–β-carotene-rich foods, and they often did not distinguish between younger and older cases. There are as yet no reports of animal studies. The possible influence of retinoids and carotenes on prostatic carcinogenesis is presently unresolved. However, vitamin A has been shown to be essential for normal function of the prostate and normal differentiation of the prostate epithelium, both in studies on human vitamin A deficiency cases and in animal and organ culture studies. Furthermore, vitamin A has been shown to protect against the development of cancers other than prostate in a number of animal models, although in some cancer induction models vitamin A has enhanced carcinogenesis. [*See* VITAMIN A.]

Zinc is important in the functional activity of the prostate gland, and it is one of the major constituents of the prostatic fluid, as mentioned earlier. Whether or not this trace mineral plays a role in the development of prostate cancer is, however, not known.

One possible mechanism by which dietary factors can influence the development of prostate cancer is via effects on the endocrine system. Indeed, a number of studies in animals and humans have shown that changes in the composition of the diet can affect the concentrations of circulating testosterone, estradiol-17β, and prolactin in males. The data are, however, too fragmentary at present to draw firm conclusions, but they do indicate that both the production of hormones and their metabolic clearance from the blood can be influenced by diet. Interestingly, obesity, which is in part related to the diet, is associated with an increased risk for prostate cancer, and this condition has been shown to influence

TABLE VI Summary of Epidemiological, Animal Model, and Tissue Culture Studies on the Relation between Dietary Factors and Prostate Risk

Dietary factor	Associations[a] found in: Correlation studies	Case-control studies	Prospective studies	Animal studies	*In vitro* studies
Foods					
Meat, all/cattle	++	0	0		
Pork	+	0?			
Poultry	0	0?			
Edible fats/oils	+++	++			
Eggs	+?	++	+		
Fish	0	−	0		
Cereals	−−				
Sugars	++				
Vegetables, all	0	?			
Green-yellow (Japanese)		−	?		
Carrots		−−			
Pulses/nuts/seeds	−−	−			
Nutrients					
Fat, total	+++	++			
Animal/saturated	++	++			
Vegetable/unsaturated	0	?		+?	
Protein, total	++	+?			
Animal	++				
Carbohydrates					
Simple	++				
Complex	0				
Caloric intake (energy)	+				
Vitamin A–carotenes	0	???	?		−−−
Vitamin C	0	0?	0		
Selenium	−				
Zinc	+?				
Beverages					
Coffee	++				
Tea	0				
Alcohol, total	?	0	0		
Beer	+?	0			
Wine	0	?			
Hard liquor	0	0			

[a] Associations are indicated as: +++, strongly positive; ++, moderately positive; +, weakly positive; 0, no association; −, weakly negative; −−, moderately negative; −−−, strongly negative; ?, inconclusive results; ? after a + or − sign, limited evidence for the association; ???, very contradictory results.

the production and metabolism of sex steroid hormones in men. [*See* ENDOCRINE SYSTEM; OBESITY.]

2. Sexual Factors, Venereal Disease, and Prostatitis

The results of some case-control studies have suggested that prostate cancer risk is related to aspects of sexual behavior, such as sexual drive and ejaculatory activity. Although it is plausible to assume that such relations exist, it is far from clear what sexual factors are in fact related with increased risk. A consistent positive association has been found in case-control studies between prostate cancer risk and a history of venereal disease. It is not clear, however, which venereal diseases are more important in this respect and by what mechanism they would influence prostatic carcinogenesis. Prostatitis, inflammation of the prostate, is very common, and venereal diseases are frequently associated with prostatitis. There is some suggestive, but not conclusive, evidence that a history of prostatitis is associated with increased prostate cancer risk.

3. Occupational Factors, Environmental Pollution, and Radiation

Increased occurrences of prostate cancer have been found among workers in a number of occupations; these are summarized in Table VII. There are no indications that exposure to specific chemicals is related to increased prostate cancer risk in men, and animal studies have shown that a variety of chemical carcinogens can produce prostate cancer under favorable conditions. Cadmium has often been reported as a potential prostate carcinogen; however, at present absolutely no epidemiological evidence proves this. On the other hand, animal and *in vitro* studies provide some indication that cadmium can produce neoplastic changes in the prostate; in animals (rats) these changes have only been observed in the ventral prostate lobe.

Some evidence indicates that prostate cancer is more frequent in areas with a high level of air pollution. This is consistent with the earlier-mentioned higher prostate cancer rates in urban than in rural areas.

The rat ventral prostate, and probably also the human prostate (studied *in vitro*), can accumulate, secrete, metabolize, and bind a variety of chemical carcinogens. Some of these processes can produce active carcinogens in the prostate cells and, thereby, contribute to the formation of prostate cancer.

Recently, a slightly increased frequency of prostate cancer was found among men that had been exposed to heavy radiation from the Hiroshima and Nagasaki atomic bomb explosions. X-irradiation of the pelvis can cause prostate carcinomas in rats. There are a number of reports of increased prostate cancer risk in workers exposed to radiation on the job, particularly in the nuclear industry, but there are also studies that did not find this. Thus, exposure to γ and α radiation may be a risk factor for prostate cancer.

B. Endogenous Factors

As pointed out earlier, hormones, particularly androgenic steroids, regulate normal function of the prostate gland, and the majority of human prostate cancers is, at least for some time, dependent on the presence of androgens; therefore, it is plausible that hormones, and particularly androgens, are involved in the development of prostate cancer.

Differences in hormonal parameters between prostate cancer patients and controls have not yielded insights in this respect, because such differences are related to the presence of the disease, but not necessarily to its development. A relation between changes in hormonal status with aging and the development of prostate cancer has been suggested, particularly a decline in testosterone and an increase in estrogens in the blood.

A number of studies have compared hormonal parameters in groups of men that differed in risk for prostate cancer, e.g., South African Blacks, U.S. Blacks, and U.S. Whites. The only somewhat consistent finding was that the concentration of testosterone and LH were often slightly higher, and never lower, in the high-risk than in the low-risk populations. A number of animal studies (Section VI) now show that testosterone enhances the formation of prostate cancer, both by increasing the sensitivity of the prostate tissue to chemical carcinogens and

TABLE VII Summary of Associations between Occupations and Prostate Cancer Risk

Occupational group	Association
Farmers and farmworkers	Increased prostate cancer risk is a consistent finding in case-control and cohort studies from several countries; pesticide exposure is possibly related to this finding.
Armed forces personnel	Increased risk has consistently been observed in this very diverse group in England and the United States.
Rubber industry workers	Increased risk has been found in workers in the mixing and batch preparation division in some plants, but not in others; some animal studies found prostate cancer after exposure to chemicals used in these divisions.
Nuclear industry workers	Increased risk has been observed in several studies, but not in some others.
Iron–steel foundry workers	Increased prostate cancer risk was found in one study, but not in several others.

by a tumor promotorlike action when administered after treatment with carcinogens. Testosterone stimulates cell proliferation in the prostate, thereby increasing the probability that lesions created in the DNA by a carcinogen become permanent, inheritable, genetic DNA alterations in a cell (mutations). It probably also acts by enhancing the further proliferation of cells with such mutations, which allows them to form clones of neoplastic transformed cells that will grow out to become a cancer.

As mentioned earlier, men with blood relatives that had prostate cancer are at increased risk for the disease. This suggests that genetic predisposition may play a role in the development of some prostate cancers. In one study, the blood concentration of testosterone has been shown to be lower in such men than in appropriate controls. This finding raises the possibility that genetic predisposition to prostate cancer is mediated via other mechanisms than those that are responsible for the differences in prostate cancer risk between, for example, U.S. white and black men or between U.S. Blacks and African Blacks (testosterone, see previous paragraph).

VIII. Summary and Perspective

Prostate cancer is a frequently occurring cancer in men in western countries. Hallmark features of prostate cancer are its dependence on androgenic steroids to grow, its sensitivity to hormonal treatment, and its relapse after initial response to hormonal treatment. Descriptive epidemiology indicates that environmental factors determine the risk for developing prostate cancer. Life-style factors, particularly western dietary habits (e.g., high fat and protein foods, obesity), are probably related to the development of prostate cancer. Further environmental risk factors include a history of venereal disease and, as yet unspecified, exposures to chemicals and ionizing radiation in certain occupations. Hormones are most likely involved in the development of prostate cancer. Particularly high circulating levels of androgens are likely to be associated with increased risk for cancer of the prostate.

Little is known with certainty about the etiology of prostate cancer. Thus, more epidemiological research is needed to define risk factors, and soon we will have the animal models available to substantiate epidemiological findings. These studies must concentrate on the role of hormones and life-style factors, such as diet and sexual behavior, and need to focus on reasons for the extremely high prostate cancer rates among U.S. black men. Very little is known about the mechanisms of prostate cancer development, e.g., how androgens may stimulate prostate carcinogenesis. The application of *in vitro* models and molecular biological methods is required to resolve such questions. A major challenge for the future is to find better methods for the early detection of prostate cancer and therapies that will kill the androgen-insensitive cell populations in human prostate cancers.

Bibliography

Bosland, M. C. (1986). Diet and cancer of the prostate: Epidemiologic and experimental evidence. *In* "Diet, Nutrition, and Cancer: A Critical Evaluation" (B. S. Reddy and L. A. Cohen, eds.), p. 125. CRC Press, Boca Raton, Florida.

Bosland, M. C. (1988). The etiopathogenesis of prostate cancer with special reference to environmental factors. *Adv. Cancer Res.* **51,** 1.

Carter, H. B., and Coffey, D. S. (1990). The prostate: an increasing medical problem. *Prostate* **16,** 39.

Catalona, W. J. (1984). "Prostate Cancer." Grune and Stratton, Orlando.

Coffey, D. S. (1988). Androgen action and the sex accessory tissues. *In* "The Physiology of Reproduction" (E. Knobil, J. Neill *et al.,* eds.), p. 1081. Raven Press, New York.

Coffey, D. S., Bruchovsky, N., Gardner, W. A., Resnick, M. I., and Karr, J. P. (eds.) (1987). "Current Concepts and Approaches to the Study of Prostate Cancer." *Prog. Clin. Biol. Res.* Vol. 239. Liss, New York.

Page, H. S., and Asire, A. J. (1985). "Cancer Rates and Risks," 3rd ed. US DHHS Publ. No. (NIH) 85-691, National Institutes of Health, Bethesda, Maryland.

Perez, C. A., Fair, W. R., Ihde, D. C., and Labrie, F. (1985). Cancer of the prostate. *In* "Cancer. Principles & Practice of Oncology," 2nd ed. (V. T. DeVita, S. Hellman, and S. A. Rosenberg, eds.), p. 929. Lippincott, Philadelphia.

Scardino, P. T. (1989). Early detection of prostate cancer. *Urol. Clin. N. Am.* **16,** 635.

Spring-Mills, E., and Hafez, E. S. E. (eds.) (1980). "Male Accessory Sex Glands. Biology and Pathology." Elsevier/North Holland Biomedical Press, Amsterdam.

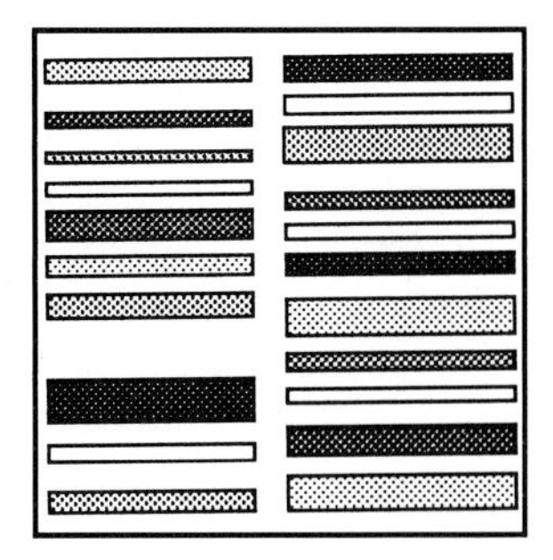

Protein Detection

CARL R. MERRIL, *National Institute of Mental Health*

Glossary

Autoradiography Method of using emitted rays from radioactively labeled material, such as labeled proteins, to expose photographic film to detect the labeled substance

Coomassie blue Organic stain that binds to proteins in a noncovalent manner, providing detection of proteins in the microgram range

Fluorography Detection method that uses fluors to convert the energy from radioactive decay into photons of light

Lowry procedure One of the most popular methods developed for the quantitative detection of proteins, it can detect as little as 0.1 μg protein

Operationally constitutive proteins Subset of proteins that have constant intragel density ratios in all of the gels used in an experiment; this subset can be used as a set of endogenous reference standards for density normalization for intragel protein comparisons

Protein silver stains Procedures developed for the detection of proteins by the reduction of ionic silver to metallic silver in the presence of proteins

Schlieren Patterns or shadows caused by variations in the refractive index in nonhomogeneous protein solutions; these patterns may be used to monitor protein separations

SOME PROTEINS, SUCH AS HEMOGLOBIN, may be detected by their color; however, most proteins do not absorb light in the visible range of the spectrum, and other methods have been required to detect their presence. Development of these protein-detection methods has progressed in parallel with the ability to resolve proteins from tissues and body fluids. Although most proteins can be visualized by their absorption of ultraviolet light, greater sensitivity has been achieved with organic and inorganic stains and color-producing reactions. Currently, the most commonly used stains are the organic stains, such as Coomassie Blue, and the silver stains, which depend on the reduction of ionic silver to metallic silver in the presence of proteins. In some cases, proteins may be monitored by labeling them with radioactive isotopes. However, to achieve a significant signal from a radioactive protein, it must be labeled to a high specific activity. This requirement limits the use of radioactive labeling to *in vitro* studies for the detection of the less-abundant proteins, as prohibitively large amounts of expensive radioactive precursors would be needed to label human proteins *in vivo*. In addition, the radiation hazards of such studies would be unacceptable.

I. Underlying Principles of Protein Detection

A. General Perspective

Certain proteins, such as those that make up human hair, are in sufficient abundance that their detection can often be made by a glance. Others, such as hemoglobin, are colored and can be detected by eye when they are present in milligram quantities. However, most proteins are not colored. Furthermore, proteins are generally present in complex mixtures in tissues and body fluids. The ability to detect and study the proteins contained in these complex mixtures depends on the ability to separate and visualize them. Currently, the most powerful separation methods are based on high-resolution, two-dimensional electrophoresis, a technique that separates the proteins by charge in the first dimension and by mass in the second dimension. Application of this technology has revealed that the concentrations of individual protein species, in most tissues and body fluids, varies by more than a millionfold. The capacity to detect the less-abundant, or trace, proteins, which may include proteins that function as vital enzymes and receptor molecules, has required the development of more sensitive detection methods, such as fluorescent dyes and the silver stains. [*See* PROTEINS.]

B. Historical Perspective

Proteins were first recognized in 1838 by the Dutch chemist Gerardus J. Mulder as "albuminoid substances" of biological origin, which often contain nitrogen, sulfur, and phosphorus. Many of the earliest protein detection methods, such as the Kjeldahl and Folin methods, made use of the presence of these elements or other reactive groups generally present in proteins. The Kjeldahl method is based on the determination of the organic nitrogen contained in the protein. This is accomplished by boiling the protein in concentrated sulfuric acid in the presence of a catalyst such as cupric sulfate. The resulting carbon dioxide and water are discarded, and the amount of protein is determined by measuring released ammonia. In the Folin method, copper and phenol are reacted with the phenolic side groups of the proteins, such as tyrosine, to produce a blue color. One of the micro modifications of the procedure, known as the Lowry method, is capable of detecting as little as 0.1 μg of protein. This method is still one of the most commonly used for the determination of protein concentration. While this method is about 10-fold more sensitive than detection of protein by the ultraviolet absorption of light, it still cannot detect <20,000 billion protein molecules (MW 30,000). A similar sensitivity can be achieved with the organic stain Coomassie Blue, which binds noncovalently to proteins. This stain is currently used in a general protein assay and for the visualization of proteins separated electrophoretically on polyacrylamide gel. More recently, silver stains have been demonstrated to have the capacity to detect as little as 0.01 ng protein, or as few as 200 million protein molecules. Although attaining even greater sensitivities would be useful, our ability to detect proteins has increased >100 millionfold since they were first recognized in the middle of the nineteenth century.

II. Detection with Light

The earliest studies of proteins relied on direct observations utilizing either the naturally colored proteins, such as myoglobin, hemoglobin, ferritin, and cytochrome *c*, or chemical methods, such as by precipitating relatively large amounts of proteins. Although most protein molecules cannot be observed directly with visible light, early investigators were able to study their electrophoretic properties by observing the migration of quartz microspheres with the proteins adsorbed to their surface in electric fields.

Tiselius demonstrated the use of ultraviolet light for the detection of noncolored proteins in the 1930s. Most of the proteins, with the exception of the protamines, contain aromatic amino acids, which provide them with a 280-nm absorption band. While it is possible to use this ultraviolet absorption to estimate the concentration of proteins, the absorption is not uniform for all proteins, as the content of aromatic amino acids varies considerably. The use of ultraviolet absorption for the detection of proteins during and after separation by electrophoresis requires a special light source, filters, and optical components that are transparent to ultraviolet light.

As the addition of protein to a solution will cause a change in the refractive index of the solution, determining protein concentration with a refractometer is possible. However, the solvent without the protein must be available as a blank. Tiselius demonstrated that the shadows, or schlieren, created by boundaries between regions with different refractive indices, due to the varying concentrations of proteins in electrophoretic systems, could be used to monitor the electrophoretic separation of proteins. The complex optical systems required for the use of schlieren systems has limited its application.

III. Detection with Organic Stains

A. Background

The use of solid support media, ranging from moist filter paper to polyacrylamide gels, stimulated the introduction and use of organic stains, particularly those that had already proven their use in histological staining applications for the detection of proteins. These stains eliminated many of the complications inherent in the ultraviolet and schlieren detection systems. The organic stains also provided increased sensitivities. These stains are often bound to the proteins with noncovalent bonds. This property requires that they be used to detect proteins after they are separated by procedures such as electrophoresis. Some of the first organic proteins employed were Bromophenol Blue and Amido Black. Some of these stains have proven to be useful because of their ability to differentially stain proteins; e.g., Oil Red 0 preferentially stains lipoproteins.

B. Coomassie Blue Stains

The most sensitive of the organic stains are the Coomassie Blue stains, with their capacity to detect as little as 0.1 μg protein. These organic stains were originally developed in the middle of the nineteenth century as acid wool dyes. They were named Coomassie dyes to commemorate the 1896 British victory in the battle for the occupation of the Ashanti capital, Kumasi, or Coomassie, in Africa. These triphenylmethane Coomassie stains have superior protein-staining abilities.

The first of triphenylmethane Coomassie Brilliant Blue stains introduced was Coomassie Brilliant Blue R250 (the letter R stands for a reddish hue, and the number 250 is a dye strength indicator). It can detect as little as 0.1 μg protein and gives a linear response up to 20 μg. However, the relationship between stain density and protein concentrations varies for each protein. Another Coomassie stain, Coomassie Brilliant Blue G250 (G indicates a greenish hue) is used for rapidly staining gel. Its capability as a rapid stain is based on its limited solubility in trichloroacetic acid, which permits its use as a colloidally dispersed dye. This dye does not penetrate gels and only stains the protein on or near the surface. Another Coomassie stain, Coomassie Violet R150, has gained some favor by virtue of its ability to rapidly stain proteins separated by isoelectric focusing (a procedure that separates proteins by their charge) on polyacrylamide gels, while not staining the carrier ampholytes (which are used to create the pH gradient in these gels).

While the mechanism of Coomassie Brilliant Blue staining is not fully understood, it is known that these dyes require an acidic medium for electrostatic attraction to be exerted between the dye molecules and the amino groups of the proteins. The binding is fully reversible, as has been demonstrated by the loss of the stain following dilution under appropriate conditions. The relatively high-staining intensity of Coomassie Blue stains, compared with other organic dyes, appears to be due to their ability to form secondary bonds between the dye molecules, permitting the accumulation of additional dye molecules. The basic amino acids have been proposed to supply the major binding sites for the Coomassie dyes. This proposal is based on the following observations: polypeptides, which are rich in lysine and arginine, are aggregated by Coomassie G stain, suggesting that the dye interacts with the basic groups in the polypeptides and a significant correlation exists between the intensity of Coomassie Blue staining and the number of lysine, histidine, and arginine residues in the protein.

C. Other Common Organic Protein Stains

Amido Black (Acid Black 1) and Fast Green (Food Green 3) are also commonly utilized for protein detection; however, Coomassie Blue staining exhibits three times the intensity of Fast Green and six times the intensity of Amido Black. The staining intensities of these dyes are approximately proportional to their relative molar adsorption coefficients.

D. Organic Stains that Provide Fluorescent Protein-Detection Methods

Anilinonaphthalene sulfonate was the first fluorescent stain used to visualize proteins in gels. It is used as a post-electrophoretic stain. It appears to form a fluorescent complex with the protein's hydrophobic sites. It has a sensitivity limit of about 20 μg protein. Higher sensitivity has been achieved with pre-electrophoretic fluorescent stains, such as dansyl chloride. This stain reacts with proteins to form fluorescent derivatives in 1–2 min at 100°C, with a sensitivity limit of 8–10 nm.

A number of fluorescent stains were designed to increase the detection limits of amino acid analyzers. One such compound, fluorescamine, is nonfluorescent prior to its reaction with a protein. This compound reacts with the primary amines of the basic amino acids within each protein and the N-terminal amino acids to yield a fluorescent derivative. It has proven capable of detecting as little as 6 ng protein. The protein derivative of a related compound, 2-methoxy-2,4-diphenyl-3(2H)-furanone (MDPF) has the same speed and simplicity of reaction as fluorescamine, but it is 2.5 times as fluorescent as a fluorescamine-labeled protein. As little as 1 ng protein has been detected with MDPF, and it has a linear response from 1 to 500 ng. As with most other protein stains, the relationship between the relative fluorescence and the protein concentration varies for each protein, depending on the number of reactive groups in the protein.

Most of the fluorescent stains form covalent bonds with the proteins. Reaction conditions are generally best provided prior to the separation of the protein. Pre-electrophoretic staining has certain advantages: the possibility of performing stoichiometric reactions with proteins without the diffusion limitations imposed by staining within a gel matrix, the feasibility of following the process of electrophoresis visually with prestained proteins, and the absence of background problems due to dye-trapping or reaction of the dye with the gel. However, the covalent bonds formed between the proteins and these stains may alter the charge of the protein and affect its resolution in methods that separate proteins by their charge, such as with isoelectric focusing. Such charge alterations are generally not of significant consequence for electrophoretic techniques that separate proteins on the basis of molecular weight. The addition of a few relatively small dye molecules will generally have an insignificant effect on the mass of most proteins. Furthermore, even in purification techniques that may be affected by these stains, as long as the stains react with the proteins in a stoichiometric manner, the shifts in protein patterns should be highly reproducible, permitting the construction of valid protein maps and protein identifications.

Although fluorescent stains may achieve a greater sensitivity than most other organic stains, they require ultraviolet light for visualization, and direct quantitation requires fairly sophisticated equipment. These problems, coupled with the altered mobility of the proteins during isoelectric focusing, have inhibited the utilization of the fluorescent stains.

IV. Detection with Metallic Stains

A. Background

In the twelfth century, Count Albert von Bollstädt reported that silver nitrate could blacken human skin. However, silver was not used as a stain in modern scientific applications until Krause adapted it to stain fresh tissues for histological examination in 1844. By 1873, Golgi, followed by Cajal, utilized silver stains to revolutionize the understanding of the anatomy of the central nervous system. Silver was introduced as a general stain for proteins separated by polyacrylamide gel electrophoresis by Merril and Switzer in 1979. This stain was adapted from a histological silver stain. The silver stains have permitted the detection of as little as 0.01 ng protein. The gain in sensitivity afforded by silver staining has stimulated the recent widespread use of this method for the detection of proteins separated by polyacrylamide gel electrophoresis. The first silver stains used for the detection of proteins were adapted from histological stains, and they were often tedious, requiring hours of manipulation and numerous solutions. However, during the 10 years since the introduction of silver staining as a general method for the detection of proteins, simplified staining protocols have been developed. These silver staining protocols can generally be divided into three main categories: the diamine, or ammoniacal silver stains; the nondiamine chemical stains, based on photographic chemistry; and stains based on the photodevelopment or photoreduction of silver ions to form a metallic silver image.

B. Chemistry of the Silver Stains

Diamine silver stains were first developed for the visualization of nerve fibers by Cajal. They rely on ammonium hydroxide to form silver diamine complexes to stabilize the silver ion concentrations. Image production is initiated by acidifying the stain solution, usually with citric acid in the presence of formaldehyde, to lower the concentration of free ammonium ions, thereby permitting the liberation of silver ions. The silver ions are reduced with the formaldehyde to form a metallic silver image of the protein pattern. An optimal concentration of citric acid is necessary to provide a controlled rate of silver ion reduction and thus prevent the nonselective deposition of silver. Diamine stains may become selectively sensitive for glycoproteins if the concentration of silver ions is too low. This specificity can be minimized by maintaining a sufficient sodium–ammonium ion ratio in the diamine solution. However, some investigators have enhanced this tendency of the diamine stains to develop a protein stain that selectively stains neurofilament polypeptides.

The nondiamine chemical-development silver stains were developed by adapting photographic photochemical protocols. These use silver nitrate to supply silver ions. Image formation is initiated by the selective reduction of the silver ions to metallic silver by formaldehyde under alkaline conditions. Sodium carbonate and/or hydroxide and other bases are used to maintain an alkaline pH during development. The formic acid, produced by the oxidation of formaldehyde, is buffered by these bases.

The photodevelopment stains utilize energy from photons of light to reduce ionic to metallic silver. Proteins enhance the photoreduction of ionic to metallic silver in gels impregnated with silver chloride. The use of photoreduction provides for a rapid, simple silver stain method for detecting proteins.

In some cases, combining the photodevelopment and chemical-development methods has been advantageous. One such stain permits the detection of proteins in the nanogram range in <15 min. This stain uses silver halide to provide a light-sensitive detection medium and to prevent the loss of silver ions from membranes or thin-layer gels; photoreduction to initiate the formation of silver nucleation centers; and chemical development to provide a high degree of sensitivity by depositing additional silver on the silver nucleation centers. The stain's rapidity of action and ability to stain samples spotted on membranes, such as cellulose nitrate, have permitted its use as a quantitative protein assay.

The basic mechanism underlying all of the above protein detection protocols with silver-staining involves reduction of ionic to metallic silver. Detection of proteins in gels or on membranes requires a difference in the oxidation-reduction potential between the sites occupied by proteins and the adjacent sites in the gels or on the membranes. If a protein site has a higher reducing potential than the surrounding gel or matrix, then the protein will be positively stained. Conversely, if the protein site has a lower reducing potential than the surrounding gel or matrix, the protein will appear to be negatively stained. These relative oxidation-reduction potentials can be altered by the chemistry of the staining procedure.

The reactive groups that are necessary for the silver stains are the sulfur and amino groups. Evidence also indicates that the silver stains require the cooperative effects of several intramolecular functional groups to form complexes with the silver prior to the reduction of ionic to metallic silver. The importance of the basic and the sulfur-containing amino acids has been corroborated by observations with purified peptides and proteins of known amino acid sequence. For example, leucine enkephalin, which has neither sulfur-containing nor basic amino acids, does not stain with silver, whereas neurotensin, which also has no sulfur-containing amino acids but does have three basic amino acid residues (one lysine and two arginines), does stain. The importance of the basic amino acids has been further substantiated by evaluations of the relationship between the amino acid mole percentages of proteins and their ability to stain with silver.

Some proteins have proven to be difficult to stain with silver, for as yet undetermined reasons. Calmodulin and troponin C require pretreatment with glutaraldehyde, whereas histones require pretreatment with formaldehyde and Coomassie Blue. However, even with this pretreatment the sensitivity for histones is decreased 10-fold as compared with detection of neutral proteins.

C. Color Effects with Silver Stains

Whereas most proteins stain with monochromatic brown or black colors, certain lipoproteins tend to stain blue and a number of the glycoproteins appear yellow, brown, or red. This color effect has been shown to be analogous to a photographic phenomenon first described by Thomas Seebeck in 1810 when he noted that if the spectrum of visible light obtained by passing sunlight through a prism was projected onto a silver chloride-impregnated paper, some of the colors of the spectrum appeared on the paper. The production of color by both the photographic experiment conducted by Seebeck and certain proteins separated on polyacrylamide gels is caused by the size of the metallic silver particles, produced by the photoreduction effects of the light; the refractive index of the photographic emulsion or electrophoretic gel; and the distribution of the silver particles. The smaller silver grains (<0.2 μm in diameter) transmit reddish or yellow-red light, whereas grains >0.3 μm give bluish colors, and larger grains produce black images.

The production of color by proteins separated on polyacrylamide gels may be enhanced by lowering the concentration of reducing agent in the image development solution, prolonging the development time, adding alkali, or elevating the temperature during staining. These colors may aid in identification of certain proteins.

D. Sensitivity of Protein Detection with Silver Stains

Currently, silver stains offer one of the most sensitive nonradioactive methods for detecting proteins, particularly for proteins separated by gel electrophoresis. They are generally 100-fold more sensitive than the Coomassie stains, and they can often detect as little as 0.01 ng protein. The chemical-development silver stains are, in general, more sensitive than photodevelopment silver stains. However, the loss in sensitivity by photodevelopment silver stains may be compensated for by their ability to produce a protein image within 10–15 min after gel electrophoresis. High sensitivities with silver stains requires care in the selection of reagents, including the water that is used to make up the solutions. The water should have a conductivity of <1 mho/cm. Contaminants may cause a loss of sensitivity and result in staining artifacts. One common contaminant, a keratinlike protein, produces artifactual bands with molecular weights ranging from 50 to 68 kDa.

V. Detection of Specific Proteins

A. Detection with Stains

Silver stains have been developed that demonstrate considerable specificity for nucleolar proteins and neurofilament polypeptides. Other silver-staining protocols produce colors with certain types of proteins: sialoglycoproteins may be stained yellow, whereas lipoproteins tend to stain blue. Lipoproteins are also preferentially stained with the stain Oil Red O. With other protocols, the glycoproteins appear yellow, brown, or red, whereas other proteins are stained brownish-black. Glycoproteins may also be detected by the red color they produce following oxidation with periodic acid and treatment with fuchsin sulfurous acid (Schiff's reagent).

B. Detection of Specific Enzymes

Enzymatically active proteins may be detected and quantified by monitoring their enzymatic activity. A number of enzymatic reactions can be manipulated to produce a color. In some cases, secondary reactions must be coupled to the initial enzymatic reaction. These colors can be used to detect the proteins in solution and in some cases directly in the electrophoretic support media. For example, the enzyme acid phosphatase can catalyze the modification of phenophthalein to produce a pink color, whereas amylase can be detected by placing a gel containing this enzyme in contact with a starch plate. Subsequent treatment of the starch plate with iodine will produce a purple color, except in regions that were in contact with amylase. These regions will appear as clear areas. A number of these assays use electron transfer dyes, such as methyl thiazolyl tetrazolium, to detect enzymes involved in electron-transfer reactions. Over 100 specific enzymes can now be detected by specific color reactions in a manner similar to those mentioned above. For this type of detection to be successful, electrophoretic parameters, buffers, and temperature must be optimized so that the proteins maintain their enzymatic activities. However, when pancreatic proteins were separated by the relatively harsh conditions of high-resolution two-dimensional, electrophoresis, which involves denaturation and treatment with detergent, 15 pancreatic proteins still displayed enzymatic activity.

C. Detection of Protein by Specific Antibody Binding

Directly detecting proteins separated in gels with antibodies is difficult, due mainly to the inability of the antibodies to diffuse into the gel matrix. However, antibody detection can be enhanced by transferring the proteins to a thinner matrix, such as nitrocellulose paper, diazo-modified cellulose, cyanogen bromide-activated paper, or nylon membranes. This transfer is usually facilitated by electroblotting or Western blotting. By carefully selecting the transfer buffer and electrical parameters, >90% of the protein from the gel on the membrane matrix can be captured. Many of the protein stains developed to visualize protein in gels can be applied to proteins bound to membranes. Detection of specific proteins with immunostains have been achieved with both polyclonal and monoclonal antibodies. It has been possible to detect subnanogram quantities of a specific protein by reacting it with antibodies that are complexed with peroxidase. The antigen–antibody is visualized by immunoperoxidase staining. It is also possible to use antibodies that have been complexed with enzymes, fluorescein, or rhodamine. Some investigators have been able to detect as little as 1 pg of a specific protein with these techniques.

VI. Detection of Radioactive Proteins

A. Background

Radioactively labeled proteins may be visualized without staining by autoradiographic methods. These autoradiographic methods were first introduced by Saint-Victor in 1867 and Becquerel and Curie in 1896 when they used photographic plates to search for elements and mineral crystals that emit rays that could penetrate the paper and expose and darken the photographic plates in a manner similar to those discovered by Roentgen. Becquerel discovered that uranium crystals emitted such rays, whereas Curie found that similar rays could be emitted by crystals containing other elements. She named this phenomenon radioactivity. While autoradiography has been used successfully to detect ^{14}C-labeled proteins, fluorographic techniques were introduced by Wilson in 1958 to study tritium-(^{3}H) labeled compounds.

Proteins need to be labeled to a high specific activity with radioactive isotopes to provide for the detection of the trace proteins. For example, 0.1 ng of a 40,000-MW protein labeled to a high specific activity of 1,000 Curies per millimole produces only 5 disintegrations per minute. Proteins labeled to a high specific activity may be detected with sensitivities equal to, and often better, than those obtained by most stains. However, the use of such radioactively labeled proteins is limited, as it is difficult to achieve these high specific activities in animal studies and unethical to utilize such high specific activities in research involving humans.

Proteins are generally labeled during their synthesis in cells with either ^{3}H, ^{14}C, ^{35}S, ^{32}P, or ^{125}I. These isotopes may be incorporated in a ubiquitous amino acid, such as methionine or leucine, to label most of the proteins in a cell culture or tissue slice, or they may be incorporated in a specific precursor, such as glucosamine, to label only the glycoproteins. Labeling the proteins *in vitro* is also possible with such reagents as iodinated *p*-hydroxyl-phenylpropionic acid or *N*-hydroxysuccinimide ester (Bolton-Hunter reagent). Although, these labeling methods often affect the electrophoretic migration of the protein, other procedures such as reductive methylation with boro-^{3}H-hydride do not appear to affect the migration.

B. Detection with Autoradiography

Autoradiographic detection depends on the formation of a latent image in the film following exposure of photographic film to ionizing radiation from beta, gamma, or X-rays. The image is made visible by normal photographic development. In general, a film's sensitivity to ionizing radiation increases linearly with temperature until 60°C. Dehydrating gels prior to autoradiography is generally useful to reduce the path length from the emitting isotope to the film. Certain isotopes such as ^{3}H emit such weak beta rays that they provide inefficient labels for autoradiographic detection, even when the gels are dehydrated. It is possible to enhance the sensitivity of the film by bathing the preexposed film in dilute solution of silver nitrate or by baking the film for 2 hr in a hydrogen atmosphere at 66°C.

C. Detection with Fluorography

It is possible to improve the detection of proteins labeled with ^{3}H, ^{14}C, and ^{35}S by converting some of the energy from the radioactive decay to light by the addition of a fluor or scintillation agent to the gel or membrane containing the labeled protein. Dispersal of the fluor throughout the gels is necessary when the gel contains proteins labeled with ^{3}H, as the path length of the weak beta particles emitted by this isotope is often too short to penetrate the gel to reach the photographic film. The sensitivity of this method will be enhanced if the film's spectral sensitivity matches the emission spectra of the fluor used. In contrast to autoradiography, low temperature exposure (−70° to −80°C) during fluorography increases the sensitivity. This low temperature enhancement may be due to reduced energy loss of the fluor molecules through intramolecular vibrations, thereby making more energy available for light production. The low temperature may also prolong the lifetime of unstable latent images, until they can be stabilized by reduction of four or more silver ions by multiple photon hits. By maintaining fluorographs at low temperatures, sensitivity can increase about 10-fold. A further threefold increase in sensitivity may be achieved if the film is prefogged to a uniform optical density of 0.15. This prefogging reduces the threshold for the creation of stable latent images. Fluorography may also enhance the detection of isotopes, which are high-energy emitters, such as ^{32}P or ^{125}I. These isotopes have emissions that generally pass through the film with energy to spare. By placing the film between the gel and an image intensification screen (containing phosphors or fluors), any radiation passing through the film will strike the screen, releasing photons of light, which will provide for a fluorographic image in addition to the autoradiographic image produced by a particle's original traverse of the film.

VII. Quantitative Detection of Proteins

A. Background

Many protein-detection techniques may be employed quantitatively, provided that their methodological limitations are respected. The major limitation for almost all of the methods, including stain and autoradiographic methods, is that most proteins exhibit protein-specific quantitative responses. These protein-specific variations are indicative of a dependence of these methods on the content of specific groups within each type of protein that permits detection. In the case of radioactively labeled proteins, detection is determined by the mole percent of the labeled amino acids; for the detection of protein with ultraviolet light, it depends on the number of aromatic amino acids; whereas for silver stain detection, it depends primarily on the mole percent of the basic amino acids. Because each type of protein has a specific amino acid arrangement, they will each respond to a detection method in a unique manner. Although these protein-specific detection variations may prove troublesome, they may also be utilized to differentiate proteins. In intergel quantitative studies, they limit quantitative comparisons to homologous proteins.

B. Quantitation with Specific Protein-Detection Methods

One of the most common methods used for the quantitative detection of proteins, the Lowry method produces specific staining curves when the density of the color reaction is plotted against the concentration of the protein. Because this method is based on the Folin reaction, in which copper ions and phenol are reacted with the phenolic side groups of the proteins, such as tyrosine, to produce the blue color, the slope of the staining curve will mainly depend on the number of phenolic side groups in the protein.

Proteins detected with Coomassie Blue also display protein-specific staining curves. The variation in Coomassie Blue staining has been shown to be related to the mole percent of the basic amino acids in the protein. These curves are generally linear over a protein range of 0.5–20 μg.

Proteins detected with silver staining display a similar correlation between the intensity of the staining reaction and the number of basic amino acids in each protein. The linear portion of the silver-staining curve generally begins at 0.02 ng and extends over a 40-fold range. It is possible to extend the useful quantitative range of the protein detection methods beyond the linear range by using curve-fitting techniques.

Protein concentrations of >2 ng usually cause a saturation in the silver image, which results in nonlinearity above that concentration. On analysis of gel images, saturated bands or spots can be recognized by either a plateau of the staining density in the center of the bands or spots or the presence of a center that is less intensely stained than the regions near the edges. This effect is similar to the "ring-dyeing" noted with some of the organic stains. In ring-dyeing, the stain concentration is less in the center of a band or spot than at the edge—an artifact due to an insufficient diffusion of dye molecules into the protein band or spot.

C. Quantitation Standards

Because most of the protein-detection methods depend on specific reactive groups within each protein, the standards chosen should reflect this bias. For example, because there is a correlation between the intensity of Coomassie Blue staining and the number of basic amino acids in each protein, it is critical that the standard chosen should have an equivalent number of basic amino acids. If a general protein standard is needed, it should contain about 13 mole percent basic amino acids, as the basic amino acid content of proteins ranges between 10 and 17 mole percent, with a modal content of 13 mole percent. The common choice of bovine serum albumin as a protein standard may be flawed in this respect, as it has a basic amino acid content of 16.5 mole percent. Similar consideration must also be made for silver-staining standards, as the silver stains also have a dependence on the presence of basic amino acids.

D. Quantitative Intergel Protein Comparisons

The occurrence of protein-specific staining curves with most staining protocols and the specific emissions of radiolabeled proteins labeled with specific radioactive amino acids requires the limitation of quantitative intergel comparisons to homologous protein bands or spots on each gel. For example, the actin spot on one gel can be compared with an actin spot on another gel, but not with a transferrin spot. Furthermore, gels to be compared must have been run under similar conditions (percentage acrylamide, stacking gel specifications, etc.) because migration distance affects band or spot compression, which in turn may influence the dyeing reactions. Quantitative intergel variations may influence the dyeing reactions; quantitative intergel comparisons also require the presence of reference proteins for the normalization of spot or band densities. Endogenous reference proteins may be found by searching each gel for a subset of proteins that have constant intragel density ratios. This subset of spots can be defined as a set of "operationally constitutive proteins." These proteins can be used to calculate specific normalization factors for each of the gels. This scheme corrects for variations in the protein-detection procedure, in image digitization, and in initial protein-loading.

Bibliography

Gershoni, J. M., and Palade, G. E. (1983). Protein blotting: principles and applications. *Anal. Biochem.* **131,** 1.

Merril, C. R., Harasewych, M. G., and Harrington, M. G. (1986). Protein staining and detection methods. *In* "Gel Electrophoresis of Proteins" (M. J. Dunn, ed.). Wright, Bristol.

Young, E. G. (1963). Occurrence, classification, preparation and analysis of proteins. *In* "Comprehensive Biochemistry" (M. Florkin and E. H. Stotz, eds.). Elsevier Publishing Company, New York.

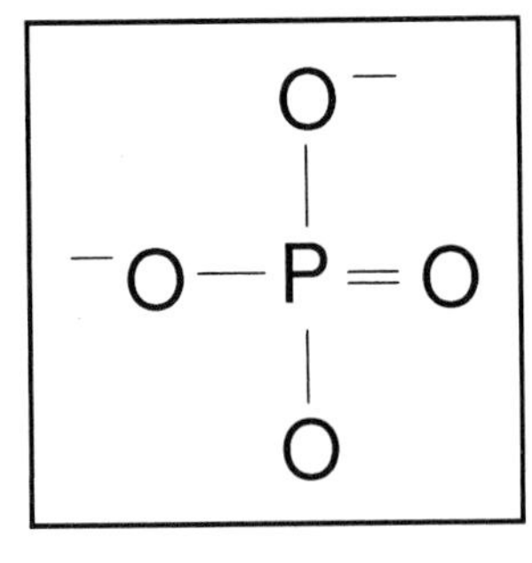

Protein Phosphorylation

CLAY W SCOTT, JITENDRA PATEL, *ICI Pharmaceuticals Group, ICI Americas Inc.*

Glossary

Calmodulin Intracellular calcium-binding protein, which mediates many functions of calcium in eukaryotes

Phosphoprotein phosphatase Class of enzymes that removes the phosphoryl moiety from phosphoproteins

Protein kinase Class of enzymes that catalyzes the transfer of a phosphoryl moiety from a donor molecule onto a specific amino acid residue of a substrate protein

Second messenger Intracellular molecule generated in response to an extracellular signal, which modulates the behavior of the cell

PROTEIN PHOSPHORYLATION is a principal mechanism by which a cell can modulate the biological activity of particular proteins. The phosphorylation of proteins is a reversible modification regulated by protein kinases and phosphoprotein phosphatases. Protein kinases catalyze the transfer of the γ-phosphoryl moiety from a donor adenosine triphosphate (ATP) molecule onto the hydroxyl group of a specific amino acid residue in the recipient protein (Fig. 1). This phosphorylation event can either increase or decrease the function of a protein, depending on the particular protein and the actual site being phosphorylated. The covalent attachment of a phosphoryl group (PO_4^{2-}) with its high charge density onto a protein induces a conformational change in the protein. The consequences of phosphorylation of an enzyme can be an increase or decrease in kinetic properties (K_m or V_{max}) or a response to allosteric effectors. The phosphorylation of a cell-surface receptor can alter its affinity for ligand. Many proteins are phosphorylated at multiple sites by distinct protein kinases with differing effects on biological activity (discussed below). Phosphoprotein phosphatases catalyze the removal of phosphate groups from these phosphoproteins, causing the protein to return to its former functional state. Both protein kinases and phosphatases are under regulatory control, and therefore the relative activities of these enzymes will generally determine the degree of phosphorylation of a given protein and, hence, its functional activity. In some circumstances, however, the availability or accessibility of the substrate protein will dictate its phosphorylation state.

The profound effect of phosphorylation on protein function is utilized by the cell to coordinately regulate a number of diverse biological processes including metabolism, secretion, gene expression, electrical excitability, and contractility. Protein phosphorylation plays an especially prominent role in the transduction and amplification of extracellular signals into intracellular responses. Hormones, neurotransmitters, and growth factors bind to cell-surface receptors, initiating a cascade of biochemical events that generates an intracellular second messenger (Fig. 2). These second messengers activate specific protein kinases, which continue the transduction pathway by phosphorylating various cellular proteins. These phosphorylated proteins ultimately cause a change in the behavior of the cell—i.e., the response to the extracellular signal. Some cell-surface receptors contain within their cytoplasmic domain a protein kinase activity. Agonist binding to these receptors stimulates the kinase activity and propagation of the external stimuli. In these

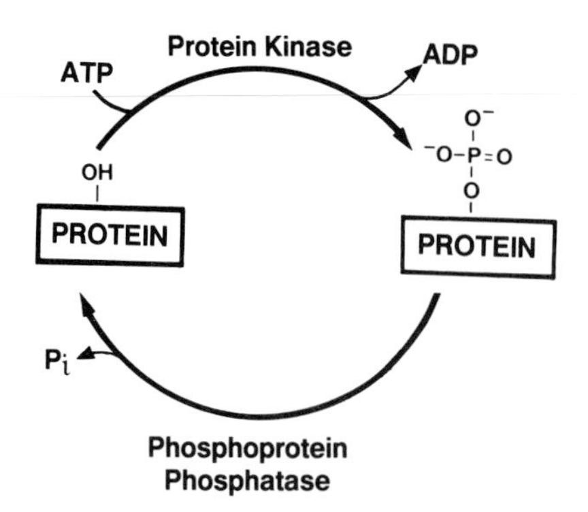

FIGURE 1 Protein phosphorylation is the enzymic transfer of a phosphoryl group onto a specific amino acid residue (usually serine, threonine, or tyrosine). Although ATP serves as the primary phosphate donor molecule, some protein kinases can utilize GTP. P_i, inorganic phosphate.

circumstances, the requirement for a second messenger may be abrogated.

I. Protein Kinases

Protein kinases can be categorized by the amino acid residue they phosphorylate (serine and threonine versus tyrosine) and are further divided on the basis of the effector system that controls their activity. The following section will describe particular protein kinases emphasizing their mechanism of regulation. This section is confined to those protein kinases that are activated by second messenger systems. Other protein kinases are not directly regulated by second messengers. These effector-independent protein kinases can be modulated by substrate availability, although they might be modulated by other, unidentified second messengers.

A. Cyclic Adenosine Monophosphate-Dependent Protein Kinase

Cyclic adenosine monophosphate (cAMP)-dependent protein kinase is the only known intracellular receptor for cAMP in mammalian cells and presumably mediates the effects of the many hormones and neurotransmitters that use cAMP as a second messenger. Cyclic AMP-dependent protein kinase exists as an inactive holoenzyme composed of a regulatory (R) dimer and two catalytic (C) monomer subunits. Both R (monomer) and C have relative molecular masses (M_r) in the 40,000–45,000 range. There are two cAMP-binding sites on each R subunit or four binding sites per dimer. Activation of the holoenzyme occurs as:

$$R_2C_{2(\text{inactive})} + 4\text{cAMP} \rightleftharpoons R_2(\text{cAMP})_4 + 2C_{(\text{active})}$$

Binding of four molecules of cAMP to the regulatory dimer promotes dissociation of the holoenzyme complex. The dissociated catalytic subunits are active and phosphorylate appropriate substrate proteins. Hydrolysis of cAMP by phosphodiesterases shifts the equilibrium toward the inactive holoenzyme, thereby terminating the actions of the catalytic subunit. The two cAMP-binding sites in each R monomer have extensive amino acid homology, suggesting they were formed via a gene duplication event. Two distinct forms of R, designated RI and RII, have been identified based on differences of physical properties, in apparent molecular mass, amino acid sequence, and susceptibility to autophosphorylation by C subunit. Recently, molecular cloning techniques have identified four different genes encoding R subunits (two for RI and two for RII) as well as two genes for C. Although the biological significance of these different proteins is not yet understood, the selective tissue distribution of the different gene products suggests they may play distinct roles in regulating various cellular events.

B. Cyclic Guanosine Monophosphate-Dependent Protein Kinase

Cyclic guanosine monophosphate (cGMP)-dependent protein kinase exists as a dimer with each monomer (M_r 80,000) containing two cyclic nucleotide-binding sites and a catalytic domain. These domains show extensive homologies to the corresponding regions in cAMP-dependent protein kinase, although for cGMP-dependent protein kinase the cyclic nucleotide-binding sites and catalytic domain are linked by peptide bond and do not separate upon binding cGMP. Activation of the holoenzyme occurs by binding four molecules of cGMP:

$$4\text{cGMP} + E_{2(\text{inactive})} \rightleftharpoons 4\text{cGMP} \cdot E_{2(\text{active})}$$

As is the case with cAMP-dependent protein kinase, dissociation and hydrolysis of cyclic nucleotide causes a shift in the equilibrium toward the inactive holoenzyme. The two cGMP-binding sites on each monomer share homology in terms of amino acid sequence yet are distinct in that cGMP-

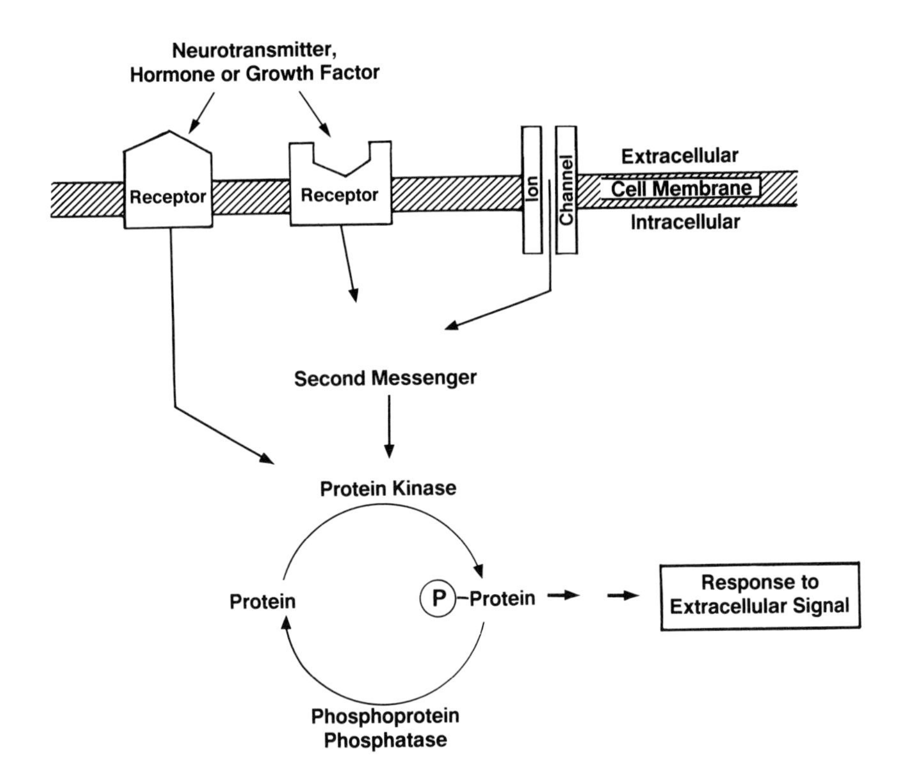

FIGURE 2 Protein phosphorylation is a common mechanism used to transduce extracellular stimuli into intracellular responses. Various extracellular molecules bind to their cell-surface receptors and regulate protein kinases either directly (when the receptor itself is a kinase) or indirectly (through changes in levels of second messengers). Activation of a specific protein kinase produces an increase in the phosphorylation of selective proteins, which leads to a change in the behavior of the cell (e.g., contraction, secretion, gene expression, metabolism). The second messengers can be generated by activating membrane-bound enzymes (in the case of cAMP, cGMP, diacylglycerol, inositol triphosphate, and Ca^{2+}) or by passage through receptor–voltage-activated ion channels (Ca^{2+}).

related molecules bind with different affinities to the two sites.

Although cGMP-dependent protein kinase has been extensively characterized in terms of physicochemical properties, the physiological substrates for this enzyme as well as its biological function remain relatively obscure. A role for cGMP-dependent protein kinase in cellular metabolism has been most clearly defined in the regulation of vascular smooth muscle contractility. In this tissue, substances that relax the muscle, both nitrovasodilators and atriopeptins, stimulate guanylate cyclase resulting in increased synthesis of cGMP and activation of cGMP-dependent protein kinase. The mechanism by which cGMP-dependent protein kinase relaxes the muscle, however, is not understood.

C. Calcium–Calmodulin-Dependent Protein Kinases

The family of protein kinases that are regulated by calcium plus calmodulin include calcium–calmodulin-dependent protein kinases I, II, and III, myosin light-chain kinase, and phosphorylase kinase. With the exception of calcium–calmodulin-dependent protein kinase II (CaM-kinase II), these enzymes have very limited substrate specificities. CaM-kinase II will be described as an example of this class of protein kinases, with the understanding that the other enzymes, although similar in terms of second messenger requirements, display distinct physicochemical properties.

CaM-kinase II is present in most mammalian tissues but is most abundant in brain, where it can account for up to 1% of total protein in some regions. The inactive holoenzyme is an oligomer with a molecular weight of 540,000–650,000 comprised of varying ratios of two or three distinct subunits each of M_r 50,000–60,000. Each subunit contains all of the functional domains that comprise the oligomer, namely an ATP-binding domain, a calmodulin-binding domain, and a catalytic domain. Other domains identified within each monomer include two autophosphorylation domains and an inhibitory domain. Although the regulation of CaM-kinase II activity is not completely understood, a number of points have been recently clarified. In the inactive state, the inhibitory domain blocks the ATP-binding domain. An increase in intracellular calcium, due to an influx through activated calcium channels or release from intracellular storage sites, results in the binding of calcium to calmodulin. The calcium–calmodulin complex binds to CaM-kinase II, inducing a conformational change in the enzyme, which alleviates the interaction of the inhibitory domain with the ATP-binding site. The calcium–calmodulin-enzyme complex is now in an active state and upon binding ATP can phosphorylate exogenous substrates as well as undergo autophosphorylation. The autophosphorylated enzyme loses its dependency on calcium–calmodulin; therefore, a decrease in intracellular calcium levels—which would cause dissociation of calcium–calmodulin from the enzyme—does not abolish CaM-kinase II activity. This represents a potential amplification step whereby small changes in calcium concentrations could produce sustained activation of CaM-kinase II and substantial changes in cellular function. It should be noted that the existence of autophosphorylated CaM-kinase II has been difficult to measure *in vivo* and is therefore still open to some speculation.

D. Protein Kinase C

Many receptors to extracellular signals are coupled to phospholipase C, which upon activation catalyzes the hydrolysis of phosphatidylinositol-4,5-bisphosphate (PIP2) within the plasma membrane, producing the two second messengers inositol 1,4,5-trisphosphate (IP3) and 1,2-diacylglycerol (DAG). IP3 initiates the mobilization of calcium from intracellular stores, thereby continuing the signal transduction pathway. The second messenger DAG stimulates the Ca^{2+}-activated phospholipid-dependent protein kinase, protein kinase C. DAG increases the affinity of protein kinase C for calcium such that activation of the kinase can occur with calcium concentrations found in resting cells. The localization of both DAG and phospholipid to the plasma membrane suggests that protein kinase C must interact with this membrane for activation. Indeed, many studies have shown that protein kinase C moves from the cytoplasm to the plasma membrane after cells are treated with agents that stimulate PIP2 hydrolysis. Activation of protein kinase C by diacylglycerol and phospholipid affords a second limb for conversion of extracellular signals into intracellular responses. Inactivation of protein kinase C occurs as a result of the conversion of DAG to phosphatidic acid. However, under some circumstances inactivation may occur by proteolytic degradation—initially to the active fragment termed protein kinase M and subsequently to inactive fragments.

Protein kinase C exists as a monomer with a molecular mass of about 85,000. It is present in all tissues, although most abundant in brain. It is now clear that a number of different isoforms of protein kinase C exist; currently seven isoforms have been identified using molecular cloning techniques. The biological significance for this diversity is not understood. Subtle differences in various biochemical properties as well as differences in tissue distribution suggest that the different isoforms have distinct functions in the modulation of or response to various physiological stimuli.

Phorbol esters (which are present in croton oil) are tumor-promoting agents that appear to elicit their biological effects through activation of protein kinase C. The phorbol ester 12-O-tetradecanoylphorbol-13-acetate (TPA) has structural similarity to diacylglycerol and can stimulate protein kinase C activity. The inappropriate activation of protein kinase C by phorbol esters is believed to be the basis for their tumor-promoting effect.

E. Growth Factor Receptor Tyrosine Kinases

A family of receptors for certain serum growth factors have been shown to contain intrinsic protein kinase activities. These receptor kinases specifically phosphorylate tyrosine residues and include the receptors for insulin, insulinlike growth factor 1 (IGF-1), epidermal growth factor (EGF), platelet-derived growth factor (PDGF), and colony-stimulating factor (CSF-1).

The insulin and IGF-1 receptors are structurally homologous oligomers composed of two α-subunits, which are located extracellularly and linked by disulfide bonds to each other, and to two β-subunits. The β-subunits transverse the plasma membrane and contain the intracellular tyrosine kinase domain. The α-subunits compose the ligand-binding domain. In contrast to these receptors, the EGF, PDGF, and CSF-1 receptors are monomeric proteins containing extracellular, transmembrane, and intracellular domains.

In intact cells, insulin stimulates the phosphorylation of its receptor on tyrosine as well as serine and threonine residues. The purified receptor, however, phosphorylates itself as well as exogenous substrates exclusively on tyrosine residues. It is now accepted that the insulin receptor is a tyrosine-specific protein kinase and, in turn, a substrate for at least two different kinases: the receptor kinase itself (autophosphorylation) and a serine–threonine protein kinase. All growth factor receptor kinases show similar selectivities for tyrosine residues and, therefore, serine–threonine phosphorylation of these receptors is the consequence of other protein kinases. [*See* INSULIN AND GLUCAGON.]

Stimulation of the receptor tyrosine kinase activities requires binding of the growth factor (the agonist). How these receptors transduce the ligand-binding signal across the plasma membrane to activate the tyrosine kinase is unclear. A currently favored hypothesis is a ligand-induced dimerization of the receptors. Ligand-induced aggregation appears to be an intrinsic function of the EGF receptor and may occur with all of the tyrosine receptor kinases. It is interesting to note that antibodies to the insulin receptor, which cause receptor cross-linking, also stimulate the receptor tyrosine kinase activity. In addition, both PDGF and CSF exist as disulfide-linked dimers and may be capable of interacting with the agonist-binding domain of two monomers, thereby promoting dimerization.

Both the insulin and EGF receptors undergo rapid intramolecular autophosphorylation reactions, which further enhance their tyrosine kinase activity. The autophosphorylated receptor kinase phosphorylates substrate proteins, thereby propagating the response to the agonist. In contrast to these receptors, the role of autophosphorylation in regulating the PDGF and CSF-1 receptor kinases has not yet been established.

A large body of evidence supports the hypothesis that the receptor kinase activities play crucial roles in mediating the actions of the various growth factors, although the substrates for these receptor kinases are just beginning to be identified. Unfortunately, relating the phosphorylation of one specific protein by a receptor tyrosine kinase with a particular biochemical response is not yet possible.

The tyrosine kinase activities of the insulin receptor and EGF receptor are regulated by serine–threonine phosphorylation. Protein kinase C phosphorylates both receptors *in vitro*, inhibiting ligand binding, and decreasing tyrosine kinase activity. Similar results are observed in cells treated with phorbol ester, suggesting that the protein kinase C effects are physiologically significant. Although cAMP-dependent protein kinase does not appreciably phosphorylate the insulin receptor *in vitro*, activating cAMP-dependent protein kinase in intact cells results in phosphorylation of the receptor along with decreased insulin binding and diminished insulin receptor tyrosin kinase activity. Whether these effects are due to phosphorylation of the receptor by the cAMP-dependent protein kinase itself, or they occur via another kinase activated by this enzyme is not clear. In either case, these observations provide direct evidence of crosstalk between different effector systems and demonstrate the diverse pathways that can be utilized to modulate a particular biochemical response.

F. Oncogene Protein–Tyrosine Kinases

Certain retroviruses induce the malignant transformation of cells through the expression of a viral oncogene. For several of these retroviruses, the oncogene products are tyrosine-specific protein kinases. These transforming proteins are structurally altered variants of normal cellular protein–tyrosine kinases. Examples of such viral oncogenes include v-src, v-fps, v-abl, and v-yes. Because these variants induce neoplastic transformation, their cellular homologues are probably important regulatory enzymes that control normal cell growth and metabolism. However, the mechanisms that regulate the cellular protein–tyrosine kinases remain undefined. In addition, the physiological substrates for the enzymes have not been delineated. A few proteins have been identified as substrates for the viral tyrosine kinases, and it is assumed that phosphorylation of these proteins by the cellular homologues could have physiological consequences. [*See* ONCOGENE AMPLIFICATION IN HUMAN CANCER.]

Vinculin is a cytoskeletal protein localized to adhesion plaques in cells. It is thought to act as a bridge to connect actin filament bundles with the plasma membrane. The product of the src oncogene, $pp60^{v\text{-}src}$, also localizes to the adhesion plaque and can phosphorylate vinculin *in vitro*. Cells transformed by the v-src oncogene display altered actin bundling and a distorted morphology, as well as an increase in tyrosine phosphorylation of vinculin. It has not been proven, however, that these morphological changes are due to the phosphorylation of vinculin by $pp60^{v\text{-}src}$.

P36 is the major phosphotyrosine-containing protein in cells transformed with the v-src oncogene.

P36 and p35 (a substrate for the EGF receptor kinase) belong to the calcium, phospholipid, and actin-binding protein family, calpactins. These two proteins are found in the cortical skeleton of endothelial and epithelial cells, a protein layer that underlies the plasma membrane. P35 and p36 may link the plasma membrane with the cytoskeleton (an intracellular system composed of various filaments) and/or intracellular vesicles and thus be involved in endocytosis (uptake of substances into cells) and exocytosis (excretion of substances). It has recently been discovered that p35 and p36 belong to a family of proteins collectively termed lipocortins, which inhibit the enzyme phospholipase A_2, and are induced by hormones of the adrenal cortex (corticosteroids). At present, how phosphorylation of these proteins modulates their activities is unclear.

G. Consensus Sequences for Protein Kinases

Protein kinases recognize substrates on the basis of their amino acid sequence. Studies with synthetic peptides as substrates have shown that changing amino acids within the vicinity of a phosphorylation site can dramatically affect the ability of the peptide to undergo phosphorylation. Cyclic AMP-dependent protein kinase, for example, recognizes the sequence -Arg-Arg-*X*-Ser(Thr)-X-, where *X* represents any amino acid. Peptides, which lack either of the two basic arginines, or in which the arginines are displaced from the phosphorylatable serine, exhibit significant increases in K_m and/or decreases in V_{max}. Although the primary sequence of a protein is an essential determinant for recognition by protein kinases, higher orders of structure resulting from the folding of the polypeptide chain can influence substrate–enzyme interaction. Small peptides, for example, which contain identical primary sequence to natural substrate proteins, typically display poorer phosphorylation kinetics. In addition, some proteins that are not substrates for protein kinases can be phosphorylated after denaturation. Higher orders of structure can, therefore, have either a positive or a negative effect on substrate specificity.

II. Phosphoprotein Phosphatases

In contrast to protein kinases, of which over 80 distinct enzymes have been identified, phosphoprotein phosphatases appear to be relatively few in number and promiscuous in activity. Indeed, the broad substrate specificity of these enzymes can create difficulty in defining distinct enzyme forms. Currently, the classification of phosphatases are based on preferences of substrate (α- or β-subunit of phosphorylase kinase) and susceptibility to inhibition by two thermostable proteins (inhibitor-1 and inhibitor-2). The identification of enzymes with differing properties within these class of phosphatases has led to further subcatagories (Table I).

A. Phosphatase-1

Phosphatase-1 has been purified in at least three distinct forms: an inactive MgATP-dependent complex, an active catalytic monomer of M_r 32,000–38,000, and an active glycogen-bound form. The MgATP-dependent complex is composed of the catalytic subunit bound to inhibitor-2. Activation of the complex is not completely understood but is accomplished *in vitro* by incubating with MgATP and a protein kinase named F_A/GSK-3. This kinase phosphorylates inhibitor-2, thereby removing the inhibitory action on the catalytic subunit. The catalytic subunit–phospho inhibitor-2 complex undergoes autodephosphorylation, returning to the inactive conformation, thus completing the phosphorylation–dephosphorylation cycle.

In skeletal muscle, the phosphatase-1 catalytic subunit is bound to a M_r 103,000 glycogen-binding (G) component in a 1 : 1 complex. This complex is attached to glycogen via the G component. The phosphatase-1 catalytic subunit is active in this configuration, although physically constrained to glycogen particles. Within this conformation, the catalytic subunit is afforded protection from the actions of inhibitor-1 and -2. Phosphorylation of G component by cAMP-dependent protein kinase promotes release of the active catalytic subunit into the cytoplasm. [*See* MUSCLE, PHYSIOLOGY AND BIOCHEMISTRY.]

TABLE I Classification of Phosphoprotein Phosphatases

Type	Inhibited by inhibitor-1 and -2	Specificity for phosphorylase kinase
1	yes	β-subunit
2A	no	α-subunit
2B	no	α-subunit
2C	no	α-subunit
3	no	?

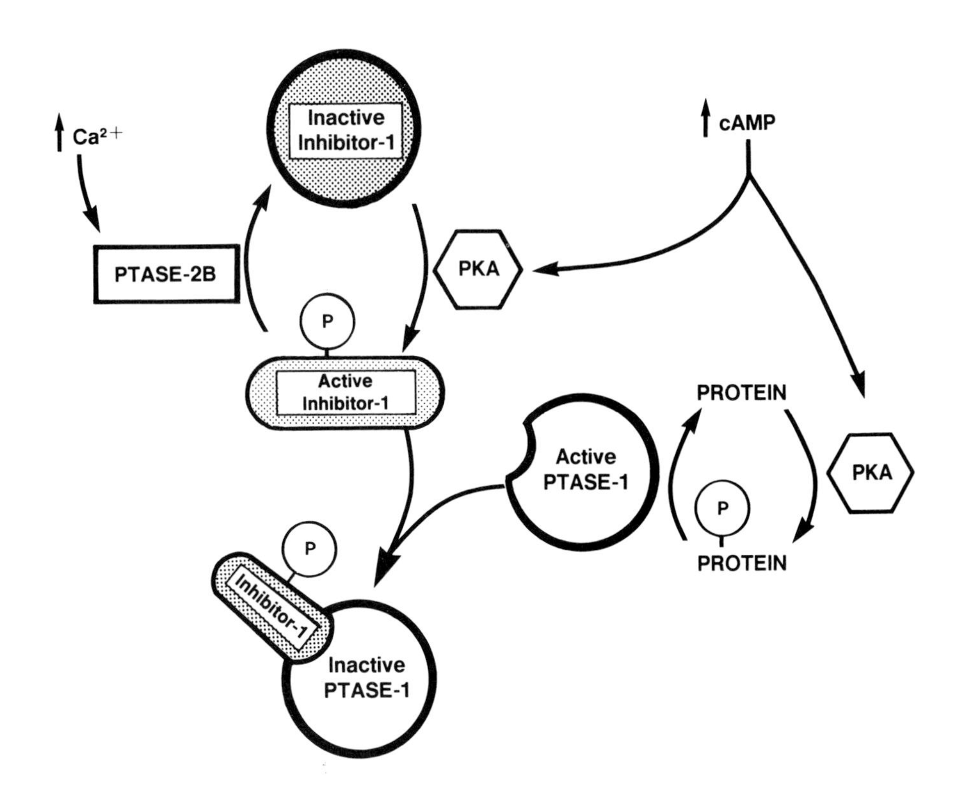

FIGURE 3 Protein phosphorylation as a site for cross-communication between different second messenger systems. Stimulation of cAMP synthesis and consequent activation of cAMP-dependent protein kinase (PKA) results in phosphorylation–activation of inhibitor-1. Phospho inhibitor-1 binds to phosphatase-1 (PTASE-1) causing inactivation of the phosphatase, resulting in a relative increase in phosphoproteins. An increase in intracellular calcium would attenuate the effects of cAMP by activating phosphatase-2B (PTASE-2B) and dephosphorylating inhibitor-1.

The free catalytic subunit can be inhibited by the activated form of inhibitor-1 (Fig. 3). Inhibitor-1 is a monomer with a M_r 18,600. Activation of inhibitor-1 occurs through its phosphorylation at threonine-35 by cAMP-dependent protein kinase. Phospho inhibitor-1 binds to the phosphatase-1 catalytic subunit, forming an inactive complex. Phosphorylation of inhibitor-1 by cAMP-dependent protein kinase results in a decrease in phosphatase activity and, therefore, an amplification of the cAMP-induced phosphorylation cascade.

Neural tissue contains an additional inhibitor of phosphatase-1 that has many of the same properties as inhibitor-1. DARPP-32 (for dopamine and cAMP-regulated phosphoprotein of M_r 32,000) is phosphorylated by cAMP-dependent protein kinase on a threonine residue within a region homologous to the phosphorylation site on inhibitor-1. Phospho DARPP-32, but not its dephospho form, can abolish phosphatase-1 activity in a manner analogous to inhibitor-1.

B. Phosphatase-2A

Phosphatase-2A consists of a central catalytic subunit of M_r 34,000–38,000 complexed with differing ratios of one or two additional distinct subunits (subunit A of M_r 60,000 and B of M_r ~54,000). The catalytic subunits of phosphatase-1 and -2 are distinct gene products, although they share a significant level of amino acid homology and antibody cross-reactivity.

The type 2A phosphatases have a broad substrate specificity but show a preference for the α- rather than the β-subunit of phosphorylase kinase. This class of phosphatases are stimulated by basic proteins and polyamines. The biological function of the A and B subunits are not clear, although reconstitution experiments have shown that they alter the enzymatic properties of the catalytic subunit.

C. Phosphatase-2B

Phosphatase-2B, or calcineurin, is a calcium–calmodulin-binding protein found primarily in brain. It exists as an inactive heterodimer of A (M_r 61,000) and B (M_r 16,000) subunits. The A subunit contains the catalytic domain as well as a calmodulin-binding domain. The B subunit has amino acid homology with calmodulin and can bind four molecules of calcium. Thus, the A subunit interacts with two related but distinct calcium-binding proteins: the B subunit, which does not dissociate from the A subunit, and calmodulin, which interacts with A subunit only transiently. Understanding the activation of phosphatase-2B has been difficult because the enzyme can exist in a number of different conformational states with differing activities. Evidence suggests that calcium binding to the B subunit allows for the A subunit to interact with calmodulin, which results in a 10-fold increase in its catalytic activity. In the absence of calmodulin, the calcium-bound enzyme is only minimally active. Divalent cations such as Mn^{2+}, Co^{2+}, and Ni^{2+} can bind to the A subunit in the presence or absence of bound calmodulin and further enhance activity.

Unlike other classes of phosphatases, phosphatase-2B has a rather narrow substrate specificity. Most of its substrates have functions that regulate protein kinases or phosphatases. For example, phosphatase-2B dephosphorylates inhibitor-1 and DARPP-32, thereby abolishing their inhibitory actions on phosphatase-1. An increase in intracellular calcium could, therefore, initiate a phosphatase cascade by activating phosphatase-2B and consequently phosphatase-1. This response represents a pathway to antagonize the actions of cAMP (Fig. 3) and is another example of cross-communication among different second-messenger systems.

D. Phosphatase-2C

Phosphatase-2C is a monomer with M_r 46,000. It is present in almost all mammalian tissues. It is completely dependent on Mg^{2+} or Mn^{2+} for activity and has a broad substrate specificity. Like the other type 2 protein phosphatases, phosphatase-2C prefers the α-subunit of phosphorylase kinase and is not regulated by inhibitor-1 or inhibitor-2.

E. Phosphatase-3 (Phosphotyrosyl–Protein Phosphatases)

Very little is known concerning the physical characteristics and substrate specificities of these phosphatases; the reasons for this are many: the discovery of reversible phosphorylation on tyrosine residues is quite new relative to serine–threonine phosphorylation, the cellular content of phosphotyrosine is 3,000-fold less than phosphoserine, and the substrates required to detect this activity are few in number and difficult to prepare in large quantities with high phosphorylation stoichiometry. Several laboratories have identified phosphotyrosyl–protein phosphatases that are comparable in size to the catalytic subunits of types 1, 2A, and 2C, although higher molecular weight enzymes have also been reported. This class of phosphatases is not affected by inhibitor-1 or -2.

III. Substrates for Protein Phosphorylation

Because protein phosphorylation has been implicated in regulating a vast number of cellular functions, not surprisingly the number of proteins undergoing reversible phosphorylation is quite large. In cells metabolically labeled with [^{32}P]phosphate, over 100 phosphoproteins can be detected. However, relatively few of these ^{32}P-labeled proteins have been positively identified, as well as their functions characterized with respect to changes in phosphorylation. The proteins described below represent some well-characterized examples and demonstrate the breadth of involvement of protein phosphorylation in cell function.

A. Cell-Surface Receptors

Cell-surface receptors are critical interactive elements between extracellular biomolecules and intracellular machinery. Any mechanism that modulates this interaction will have dramatic effects on the response of the cell to extracellular stimuli. Receptor phosphorylation represents one such mechanism. It is becoming increasingly apparent that a number of receptors for hormones, neurotransmitters, growth factors, and essential nutrients are subject to phosphorylation. Examples include the α_1-adrenergic and β-adrenergic receptors, the glucocorticoid receptor, the tyrosine-specific protein kinase receptors (see above), and the transferrin receptor. Phosphorylation appears to regulate both receptor function and subcellular distribution.

The β-adrenergic receptor binds catecholamines and activates adenylate cyclase via the stimulatory GTP-binding protein G_s. This results in increased synthesis of cAMP and activation of cAMP-dependent protein kinase. Sequential activation of the β-

receptor produces diminished response—a phenomenon termed homologous or agonist-specific desensitization. The biochemical basis for this effect may be due to the phosphorylation of the β-receptor by an enzyme termed the β-adrenergic receptor kinase (βARK). βARK phosphorylates the agonist-bound form of the receptor causing a functional uncoupling with G_s. Subsequent to phosphorylation, the receptor is sequestered within an intracellular compartment, unable to interact with extracellular agonists. Presumably, dephosphorylation and redistribution to the plasma membrane complete the resensitization of the β-receptor. Recent data have shown that βARK can phosphorylate other receptors *in vitro* and, therefore, may play a more general role in receptor modulation.

Cyclic AMP-dependent protein kinase can phosphorylate the β-receptor *in vitro* and is implicated in the heterologous desensitization of this receptor. In heterologous desensitization, the exposure of a cell to a specific agonist results in a diminution in subsequent responses to agonists acting through other receptor populations. As is the case with βARK, phosphorylating the β-receptor by cAMP-dependent protein kinase reduces the coupling between receptor and G_s. In contrast to βARK, however, cAMP-dependent protein kinase can phosphorylate the β-receptor independent of agonist occupation. This suggests a negative feedback loop, wherein stimulation of adenylate cyclase by any positively coupled receptor causes increased cAMP production and activation of cAMP-dependent protein kinase with subsequent phosphorylation of the β-receptor. In effect, this would dampen any potential stimulation of adenylate cyclase by the β-receptor pathway. Whether or not cAMP-dependent protein kinase can act in a similar fashion on other adenylate cyclase-coupled receptors is not yet apparent.

B. Structural Proteins

A wide variety of structural proteins are known to undergo reversible phosphorylation, from contractile proteins in muscle cells to cytoskeletal proteins in virtually all mammalian cell types. MAP2 and tau are neuronal proteins that bind to microtubules *in vivo* and stabilize microtubule structures. Both proteins promote the polymerization of tubulin into microtubules *in vitro*. MAP2 resides in the dendrites of neurons where it can bind neurofilaments and actin as well as microtubules. MAP2 appears to act as a coupling agent to form crossbridges between these various cytoskeletal proteins. Tau localizes primarily, if not exclusively, to the axonal compartment of neurons. In Alzheimer's disease, however, tau is found in the cell bodies as an integral component of neurofibrillary tangles, a pathological hallmark of this disease. Immunocytochemistry studies using antibodies that recognize phosphorylation sites on tau have shown that the neurofibrillary tangle-form of tau differs in its phosphorylation pattern compared with axonal tau. Whether these changes in tau phosphorylation promote the accumulation of tau into neurofibrillary tangles, or altered phosphorylation is a consequence of its abnormal neuronal location is not yet known.

Both MAP2 and tau when isolated contain varying amounts of endogenous phosphate. Both proteins can be phosphorylated *in vitro* by cAMP-dependent protein kinase and CaM-kinase II; in each case, phosphorylation inhibits their microtubule assembly-promoting activity. In addition, MAP2 can be phosphorylated by the EGF receptor tyrosine kinase and by protein kinase C, resulting in a loss of microtubule assembly-enhancement activity. These observations suggest that phosphorylation–dephosphorylation of cytoskeletal-associated proteins may be one mechanism by which changes in cell shape, structure, or function can be regulated. [*See* PHOSPHORYLATION OF MICROTUBULE PROTEIN.]

C. Ion Channels

Protein phosphorylation has been indirectly linked to modulation of both voltage-regulated and ligand-regulated ion channels. In some cases, most notably the nicotinic acetylcholine receptor and voltage-sensitive calcium channels, a direct link between channel activity and phosphorylation of the channel (or of channel-associated regulator proteins) have been made. All four major messenger-dependent protein kinases (cAMP-, cGMP-, calcium–calmodulin-, and calcium–phospholipid-dependent) have been shown to modulate the electrical excitability of different cells. Phosphorylation can regulate different characteristics of ion channels, from altering the time for opening or closing of a channel to regulating the number of functional channels capable of opening upon membrane depolarization.

The nicotinic acetylcholine receptor (nAchR) is a neurotransmitter-regulated ion channel located on the postsynaptic membrane of neuromuscular junctions. The biochemical properties of this receptor–channel have been established using receptor purified from *Torpedo* electric organ. The purified receptor consists of four subunits in an $\alpha_2\beta\gamma\delta$ complex. The receptor exists as a phosphoprotein *in vivo* and can be phosphorylated under physiological conditions by three protein kinases present on the postsynaptic membrane. Each of these kinases, cAMP-dependent protein kinase, protein kinase C, and a protein tyrosine kinase, phosphorylate multiple subunits of the receptor at a total of seven distinct sites. Phosphorylation of the purified receptor by cAMP-dependent protein kinase or the tyrosine kinase increases the rate of receptor desensitization sevenfold. This results in diminished ionic conductance through the postsynaptic membrane, i.e., a decrease in cholinergic synaptic transmission. Protein kinase C has not been shown to have a direct effect on purified receptors, although phorbol esters increase nAchR desensitization in cultured cells. Calcitonin gene-related peptide, a neuropeptide coreleased into the neuromuscular junction with acetylcholine, stimulates nAchR phosphorylation by promoting cAMP synthesis and activating cAMP-dependent protein kinase. The extracellular signal that regulates tyrosine-specific phosphorylation of the nAchR is not known. With the nAchR, protein phosphorylation acts as a mechanism by which three different extracellular signals can be integrated into a common response, namely the desensitization of the cell to a fourth extracellular signal.

D. Metabolic Enzymes

Glycogen synthase, the rate-limiting enzyme in glycogen synthesis, was one of the first phosphorylatable enzymes identified, yet the phosphorylation mechanisms that affect this enzyme are not completely understood. Glycogen synthase is a classical example of regulation through multisite phosphorylation and demonstrates the complexity involved in deciphering the effects of multiple phosphorylation events on enzyme activity. This complexity is due to the number of residues phosphorylated on glycogen synthase *in vivo* (minimally seven) and the fact that almost all serine–threonine protein kinases can phosphorylate glycogen synthase *in vitro*. Obviously, identifying the protein kinase(s) activated by an extracellular signal and responsible for *in vivo* phosphorylation is a difficult challenge. Figure 4 depicts the current understanding of the multisite regulation of glycogen synthase.

Phosphorylation of glycogen synthase at site 1, 2,

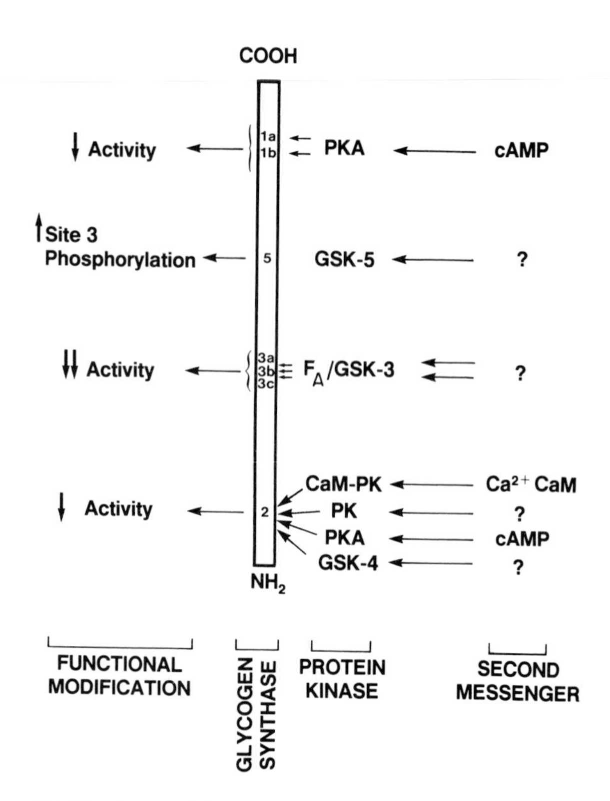

FIGURE 4 Multisite regulation of glycogen synthase activity. Phosphorylation of various sites is catalyzed by cAMP-dependent protein kinase (PKA), calcium–calmodulin-dependent protein kinase II (CaM-PK), phosphorylase kinase (PK), and several glycogen synthase kinases (F_A/GSK-3, GSK-4, GSK-5). Functional modification is indicated by vertical arrows. Question marks denote unidentified effectors. [Adapted from J. Patel, 1988, Protein phosphorylation: A convergence site for multiple effector pathways, *in* "Neuronal and Glial Proteins: Structure, Function and Clinical Application" (P. J. Marangos, I. Cambell, and B. M. Cohen, eds.), Academic Press, New York.]

or 3 inhibits enzyme activity, although site 3 produces a greater inhibitory effect. Phosphorylation of site 3 by F_A/GSK-3 requires prephosphorylation at site 5 by GSK-5 (also known as casein kinase II). The type 1 and 2A protein phosphatases are the predominant phosphatases that act on glycogen synthase. Sites 1, 2, and 3 are dephosphorylated at comparable rates by either phosphatase, whereas site 5 is dephosphorylated at a much slower rate.

Epinephrine stimulates the phosphorylation of glycogen synthase in skeletal muscle (resulting in diminished glycogen synthase activity), whereas insulin stimulates the dephosphorylation of glycogen synthase (enhancing enzyme activity). Epinephrine stimulates adenylate cyclase and thereby activates cAMP-dependent protein kinase, resulting in phosphorylation of sites 1a and 1b. Site 2 is also phosphorylated, either by cAMP-dependent protein kinase or phosphorylase kinase (which itself is activated by cAMP-dependent protein kinase). Also, enhanced phosphorylation of site 2 is possibly due to inhibition of protein phosphatase activity via phosphorylation of inhibitor-1 by cAMP-dependent protein kinase. Site 3 is also phosphorylated in response to epinephrine, presumably by F_A/GSK-3, although regulation of F_A/GSK-3 activity by epinephrine has not been directly demonstrated.

Insulin treatment causes a selective dephosphorylation of site 3, either by inhibiting F_A/GSK-3 activity or activating a protein phosphatase. Insulin also may alter the conformation of glycogen synthase such that it is a poorer substrate for F_A/GSK-3 (or better substrate for phosphatase). Thus far a direct connection between the insulin receptor tyrosine kinase and phosphorylation–dephosphorylation of glycogen synthase has not been established.

IV. Multisite Phosphorylation as an Integration Mechanism

An increasing number of proteins appear to be multiply phosphorylated with varying consequences on biological activity. Multiple site phosphorylation represents one approach by which different extracellular signals, activating distinct pathways and generating various second messengers, converge to regulate the function of a single protein. In this respect, multisite phosphorylation represents a convergence mechanism for coordinated regulation of cellular function. The consequences of multisite phosphorylation can be minimally separated into the following four categories.

(1) Phosphorylation of a common site by multiple effector-dependent protein kinases. Different extracellular signals can act in an additive fashion to regulate the function of a single focal protein, thus producing a concerted biochemical response. Examples include site 5 of glycogen synthase and serine-40 of tyrosine hydroxylase.

(2) Phosphorylation of distinct sites by multiple effector-dependent protein kinases producing similar functional responses. This phenomenon is observed with the nicotinic acetylcholine receptor as well as a number of cytoskeletal proteins.

(3) Phosphorylation of distinct sites by multiple effector-dependent protein kinases with opposing functional responses. The receptor tyrosine protein kinases are prototypic examples of this type of regulation.

(4) Phosphorylation of one site generates a recognition site for a second protein kinase. Although phosphorylation by the first kinase may or may not produce a change in protein function, it is a prerequisite for secondary phosphorylation, which does alter protein function. This mechanism is exemplified by phosphorylation of sites 5 and 3 on glycogen synthase (see above). Sequential phosphorylation may be a prerequisite for F_A/GSK-3 activity; in addition to its actions on glycogen synthase, sequential phosphorylation of phosphatase-1 by F_A/GSK-3 has been demonstrated using cAMP-dependent protein kinase as the initiating protein kinase. Sequential phosphorylation of glycogen synthase by cAMP-dependent protein kinase and casein kinase I has also been described.

Bibliography

Boyer, P. D., and Krebs, E. G. (eds.) (1986). "The Enzymes," Vol. 17. Academic Press, Inc., Orlando, Florida.

Boyer, P. D., and Krebs, E. G. (eds.) (1987). "The Enzymes," Vol. 18. Academic Press, Inc., Orlando, Florida.

Cohen, P. (1988). Protein phosphorylation and hormone action. *Proc. R. Soc. Lond.* **B234,** 115–144.

Edelman, A. M., Blumenthal, D. K., and Krebs, E. G. (1987). Protein serine/threonine kinases. *Ann. Rev. Biochem.* **56,** 567–613.

Yarden, Y., and Ullrich, A. (1988). Growth factor receptor tyrosine kinases. *Ann. Rev. Biochem.* **57,** 443–478.

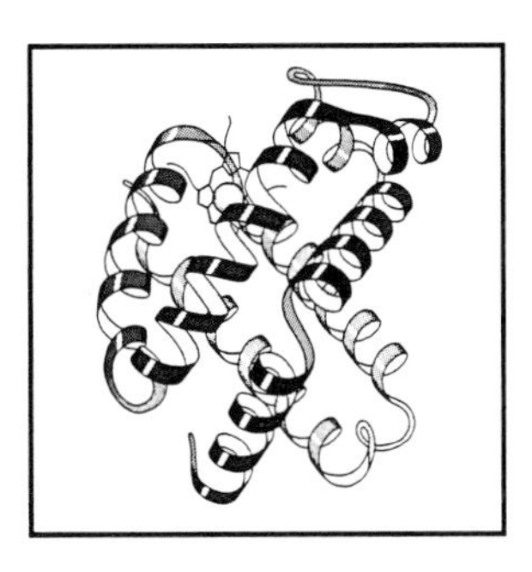

Proteins

THOMAS E. CREIGHTON, *Medical Research Council Laboratory of Molecular Biology*

Glossary

Polypeptide chain Structural unit of a protein; a linear chain of amino acids linked together by peptide bonds
Primary structure Sequence of amino acids in the polypeptide chain of a protein
Quaternary structure Assembly of individual polypeptide chains in large, multisubunit proteins
Secondary structure Regular local conformations of the polypeptide chain, especially α-helices, β-strands, and β-turns
Tertiary structure Three-dimensional structure of the polypeptide chain in a functional protein

PROTEINS ARE VITAL CONSTITUENTS of the human body and are essential for life as we know it. They have structural roles in giving our multicellular bodies their shape; are responsible for catalysis of all the metabolic reactions that take place to keep us alive; serve as hormones; protect us from infection; convert chemical energy into mechanical energy to permit us to move; detect light to permit us to see and movements of air to permit us to hear; and coordinate the nervous system to maintain us as perceptive individuals. An immense number of proteins in a human (about 100,000 at the best estimate) perform these varied functions, yet they all share the same basic architecture. The structures of proteins are the key to understanding how they work and why human beings and other living organisms are possible.

I. Functions

A. Structural

Many proteins are simply architectural, particularly the fibrous proteins. Keratins are the proteins that largely make up hair, nails, and skin. Collagen constitutes almost a quarter of the dry weight of the human body and plays a major role in holding it together. Some collagens form ropes and straps in tendons and ligaments, whereas others make woven sheets in skin, filtration membranes in kidneys, and calcium-reinforced skeletal frameworks in bone and teeth. [*See* COLLAGEN, STRUCTURE AND FUNCTION.]

Another fibrous structural protein is elastin, a rubberlike protein that occurs in most connective tissues of the body, especially in the walls of blood vessels and in ligaments. It is the major component of elastic fibers, which can stretch several times in length and then rapidly return to their original size and shape when the tension is released. [*See* ELASTIN.]

Proteoglycans are proteins with carbohydrates attached that form much of the matrix between cells in connective tissue. They are important in determining the viscoelastic properties of joints and other parts of the body subject to mechanical deformation. [*See* PROTEOGLYCANS.]

B. Informational

The human body can function in a coordinated manner only because its different parts communicate. Information is transmitted rapidly by the nervous system, and proteins are involved in all steps in transmitting and propagating nerve impulses. They do this by sensing the presence of nerve transmitters and by controlling the permeabilities of nerve membranes to different ions. The photoreceptor protein of the eye, rhodopsin, absorbs light, undergoes a change in its shape, and then triggers a nerve impulse, all so efficiently that a single photon is sufficient to activate a rod cell of the eye. Proteins are also involved in our other senses—touch, hearing, smell, and taste—but less is known about the molecular details of these processes.

The other type of communication involves release of a hormone into the blood by a gland in one part of the body and its detection by a different target organ. Many hormones are proteins, including insulin, glucagon, growth hormone, and corticotropin. The receptors that recognize the presence of all hormones, and trigger the appropriate response, are also proteins.

C. Immunochemical

The body's defenses against microorganisms and all other extraneous agents depend on proteins known as antibodies, or immunoglobulins, which specifically recognize and bind foreign molecules. A huge number of different antibody molecules, with different specificities, are possible, yet they all have very similar structures.

D. Transport

Many molecules are transported through the body by binding to specific proteins with an appropriate affinity. The O_2 required for metabolism is transported through the blood from the lungs by binding to the protein hemoglobin in the red blood cells and is stored in muscle cells bound to myoglobin. Serum albumin carries a variety of drugs and other small molecules. Iron is a necessary nutrient but is also toxic in its free form, so it is transported in the blood by transferrin and stored in tissues inside another protein, ferritin, which has a cavity capable of holding up to 4,500 ferric ions.

E. Catalysis

All the myriad of chemical reactions that occur within the human body are catalyzed by proteins, which in this case are known as enzymes. Even simple chemical reactions that occur spontaneously are catalyzed by enzymes, such as the hydration of CO_2 by carbonic anhydrase.

F. Pumping Across Membranes

Membranes are sheetlike assemblies of lipids that form impermeable barriers between cells and their environments and between various organelles within cells. Within membranes are many proteins, some of which act as highly selective pumps or gates for various ions and other small molecules. Protein pumps concentrate small molecules on one side of a membrane, against a concentration gradient. They function by binding the molecules tightly on one side of the membrane, and then using chemical energy to release them on the other side. In contrast, gates simply regulate the flow of specific molecules through membranes in both directions, depending on the concentration difference across the membrane. [*See* Ion Pumps; Membranes, Biological.]

G. Movement

The proteins of muscle, principally actin and myosin, convert chemical energy into mechanical energy. These two proteins accomplish this by binding to each other, changing shape so that one moves relative to the other, and then releasing and repeating the cycle.

II. Structures

The diverse functional properties of proteins can be understood in terms of their structures. At the simplest level, proteins are simply linear, unbranched polymers of amino acids. All proteins are made up of the same backbone and the same 20 amino acids. The sequence of amino acids linked together distinguishes one protein from another, and this level of structure is called the *primary structure*.

A. Primary Structure

1. Amino Acids

Of the 20 amino acids normally used to build proteins, 19 have the general structure

$$H_2N-\overset{\displaystyle R}{\overset{|}{C}}H-CO_2H$$

and differ only in the chemical structure of the side chain R (Fig. 1). Proline, the twentieth natural

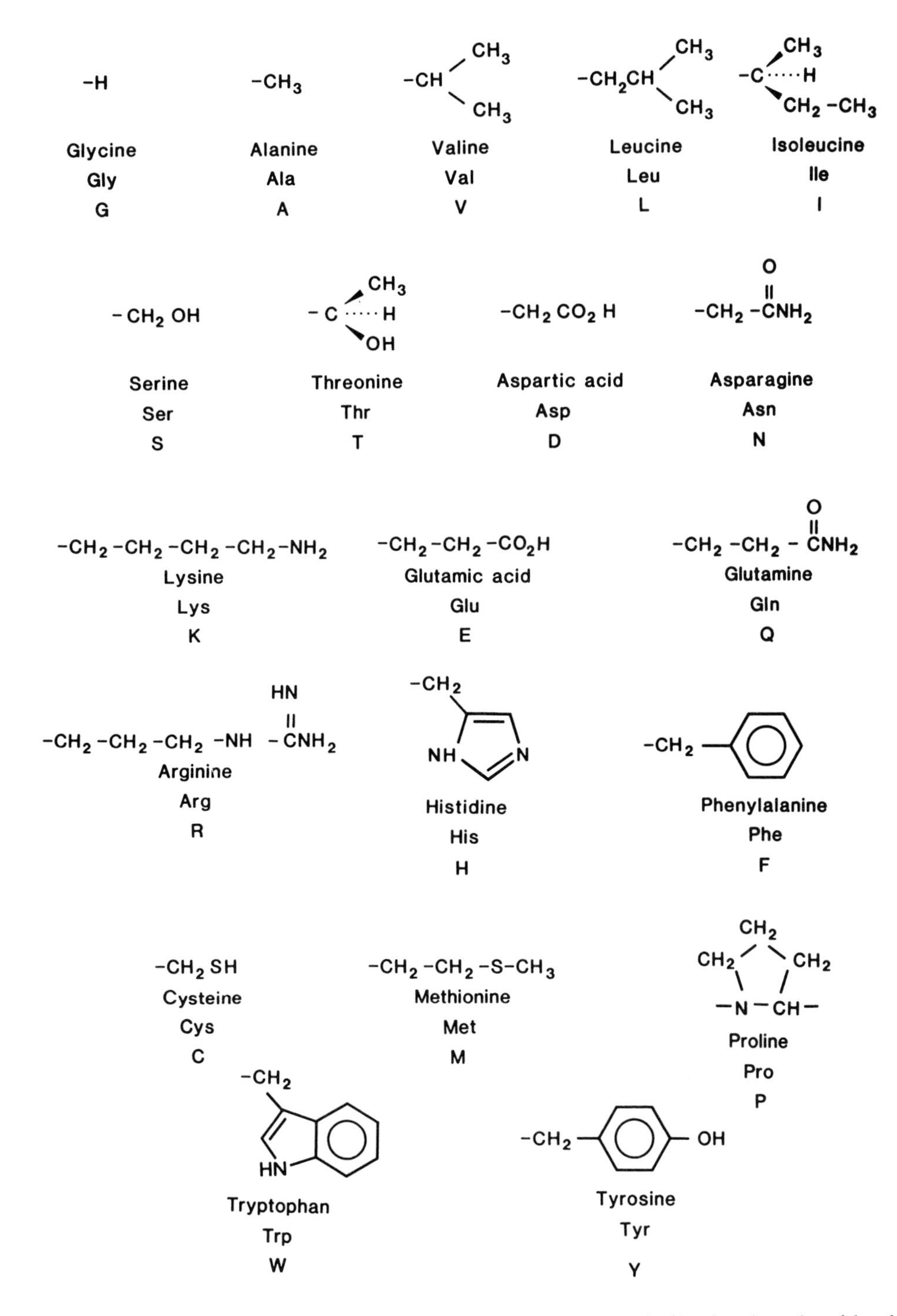

FIGURE 1 Chemical structures of the side chains of the 20 amino acids used to synthesize proteins. Below each side chain is the full name of the corresponding amino acid, plus the three- and one-letter abbreviations generally used for residues in proteins. The side chains of Ile and Thr have asymmetric centers, and only the isomer illustrated occurs naturally. The C^{α} and N atoms of the backbone are also included in the unique case of Pro.

amino acid, is similar but has the side chain bonded to the nitrogen atom:

H_2
C
H_2C CH_2
HN — CH—CO_2H

Except in the amino acid glycine, where the side chain is simply a hydrogen atom, the central carbon atom (C^α) is asymmetric and always of the L-isomer:

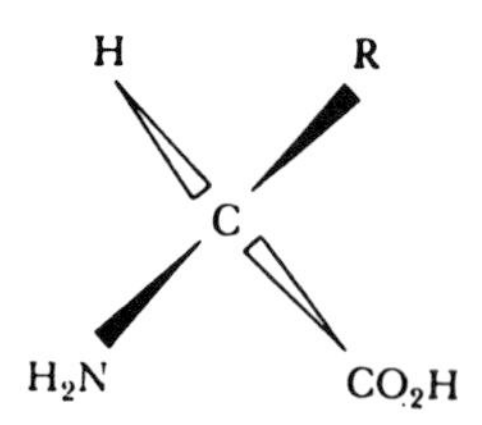

The 20 different amino acid side chains possess a variety of chemical properties that, when combined on a single molecule, give proteins properties far beyond those possible with smaller molecules or simpler polymers. Some side chains are simple nonpolar hydrocarbons, with no functional groups, but they have the important property of not interacting favorably with water, i.e., they are hydrophobic. The other side chains have polar functional groups, particularly amino, carboxyl, hydroxyl, thiol, and amide groups, which can participate in a variety of chemical reactions.

2. Polypeptide Backbone

Amino acids are assembled into proteins by linking them together via peptide bonds, expelling one water molecule per peptide bond in the process:

$$\text{—NH—}\overset{\displaystyle R_n}{\overset{|}{\text{C}}}\text{—CO}_2\text{H} + \text{H}_2\text{N—}\overset{\displaystyle R_{n+1}}{\overset{|}{\text{CH}}}\text{—CO}_2\text{H} \xrightarrow{\nearrow H_2O}$$

$$\text{—NH—}\overset{\displaystyle R_n}{\overset{|}{\text{C}}}\text{—}\overset{\displaystyle O}{\overset{\|}{\text{C}}}\text{—HN—}\overset{\displaystyle R_{n+1}}{\overset{|}{\text{CH}}}\text{—CO}_2\text{H}$$

Many amino acids (usually between 50 and 1,000, on average 300, but occasionally as many as several thousand are linked together in this way to form a linear polypeptide chain.

The polypeptide backbone is simply a repetition of the basic amino acid unit, comprising three atoms (the C^α and the amide N and carbonyl C′ of the peptide bond). The polypeptide chain has a free amino group at one end, the N-terminus, and a free carboxyl at the other, the C-terminus. The group of atoms originating from the same amino acid is designated as a residue, and individual residues are generally numbered sequentially, starting at the N-terminus.

The number and sequence of amino acids in the polypeptide chain distinguishes one protein from another. The amino acid sequences of approximately 10^4 proteins are known at the present time and contain some 2 million residues. Most protein sequences have no striking patterns and the distribution of residues is almost random. Only in certain structural proteins, with repetitive structures, are the amino acid sequences remarkable.

With any of 20 amino acids at each position, an immense number of different protein sequences are possible. For example, an average polypeptide chain of 300 residues could exist with any of 20^{300}, or 10^{390}, different sequences. One molecule of each such sequence would fill the entire universe 10^{287} times over. Obviously, not every protein sequence can exist today, nor could it have existed during the lifetime of the universe.

3. Biosynthesis

Natural proteins are synthesized within the cytoplasm of cells, using the genetic information and the individual amino acids. Of the 20 amino acids incorporated into proteins, only 11 are synthesized by humans; the remainder (histidine, isoleucine, leucine, lysine, methionine, phenylalanine, threonine, tryptophan, and valine) must be supplied by the diet, generally by the breakdown of proteins consumed.

Biosynthesis of proteins is a complex process in which the sequence of amino acids assembled into the polypeptide chain is determined by the nucleotide sequence of the corresponding gene (Fig. 2). The nucleotide sequence of the gene is first transcribed into a complementary sequence of nucleotides in the messenger RNA (mRNA). The individual amino acids are attached to specific adaptor molecules known as transfer RNA (tRNA). On a very large molecular apparatus of proteins and RNA known as the ribosome, the tRNA molecules recognize the sequence of nucleotides in the mRNA, three at a time (one codon), to insert the appropriate amino acid into the growing polypeptide chain. Synthesis starts at the amino end of the polypeptide chain, always with the residue Met, and progresses stepwise toward the C-terminus. The genetic code that is used to translate the nucleotide gene sequence into amino acid sequence of the protein is presented in Fig. 3. [*See* DNA SYNTHESIS.]

Protein synthesis is usually regulated at the initial stage of transcription of the gene into mRNA. Regulatory proteins bind at the 5′ end of the gene and determine the level of its transcription. In this way, proteins are only synthesized in the appropriate cell

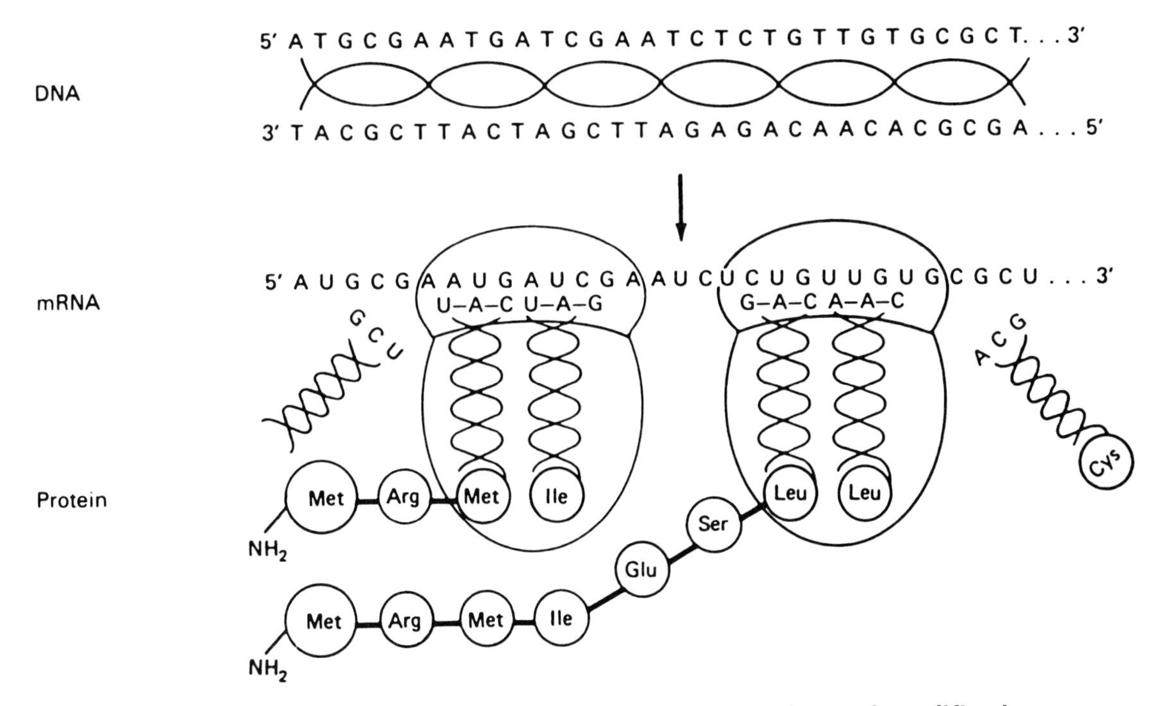

FIGURE 2 Biosynthesis of a polypeptide chain. At the top, the double-stranded DNA is shown, with the complementary nucleotide sequences of the two antiparallel strands containing the information for the protein primary structure. The nucleotide A is always paired with T, and C with G. One of the DNA strands is transcribed into a complementary mRNA strand, with the nucleotide U replacing T. The mRNA is translated by ribosomes, the two large acorn-shaped structures. Starting at the first AUG codon from the 5′ end, the triplet codons are recognized by hairpin-shaped tRNA molecules with the complementary anti-codon, to which the appropriate amino acid is attached. On the ribosome, the amino acids are linked together by peptide bonds. The ribosome on the left is adding the fourth residue, Ile, to the nascent chain, whereas that on the right is further along the mRNA and adding the eighth residue. Upon reaching a termination codon, the completed polypeptide chain is released from the ribosome.

type and under the appropriate circumstances. [*See* DNA AND GENE TRANSCRIPTION.]

Most genetic diseases of humans are caused by the absence or alteration of a particular protein, due to mutation of its gene. For example, the well-known mutation that produces sickle-cell anemia causes the sixth residue of the β chain of hemoglobin to be changed from the normal Glu to Val. This causes the protein to aggregate under certain conditions to produce the "sickling" of the red blood cells that results in the physiological symptoms. The hundreds of genetic diseases of humans known to result from absent or abnormal proteins include the hemophilias, the thalassemias, phenylketonuria, Duchenne muscular dystrophy, and color blindness. [*See* GENETIC DISEASES.]

4. Posttranslational Modifications

Only the 20 amino acids of Figure 1 are included in the genetic code and are incorporated into the initial polypeptide chain synthesized. In some proteins, the polypeptide chain undergoes alterations

First position \ Second position	U	C	A	G	Third position
U	Phe	Ser	Tyr	Cys	U
	Phe	Ser	Tyr	Cys	C
	Leu	Ser	Terminate	Terminate	A
	Leu	Ser	Terminate	Trp	G
C	Leu	Pro	His	Arg	U
	Leu	Pro	His	Arg	C
	Leu	Pro	Gln	Arg	A
	Leu	Pro	Gln	Arg	G
A	Ile	Thr	Asn	Ser	U
	Ile	Thr	Asn	Ser	C
	Ile	Thr	Lys	Arg	A
	Met	Thr	Lys	Arg	G
G	Val	Ala	Asp	Gly	U
	Val	Ala	Asp	Gly	C
	Val	Ala	Glu	Gly	A
	Val	Ala	Glu	Gly	G

FIGURE 3 The genetic code used to specify the amino acid residues incorporated during biosynthesis of proteins. Each codon of the mRNA consists of three of the nucleotides A, U, C, or G. The first nucleotide of each codon is given on the left, the second at the top, and the third on the right. Of the 64 possible codons, three normally code for no amino acid but terminate the polypeptide chain.

to the primary structure, usually in a very specific fashion, in which all the molecules of a particular protein are modified in the same way. Residues may be removed from either end of the polypeptide chain, the chain may be cleaved by proteolytic enzymes, and the terminal amino and carboxyl groups may be modified. A wide variety of specific alterations of amino acid side chains may also occur. The most common are addition of carbohydrate units to Asn, Thr, and Ser residues; phosphorylation of Ser, Thr, His, Lys, and Tyr residues, and attachment of various prosthetic groups that are involved in function, such as heme groups. Cys residues can also be cross-linked by oxidative formation of disulfide bonds between their thiol groups.

After biosynthesis on ribosomes in the cell cytoplasm, a protein must be directed to its appropriate place in the cell. Proteins destined to be secreted from the cell, or incorporated into membranes, have special sequences at their amino terminus (*signal peptides*) that serve this purpose. The signal peptides are usually cleaved off subsequently. The signals that send proteins to other compartments such as the nucleus, mitochondria, lysosomes, and peroxisomes are also special sequences, some of which are still being sought.

5. Evolutionary Origins of Proteins

The vast number of amino acid sequences possible with a polypeptide chain implies that whenever two protein sequences are substantially more similar than expected from chance, they must be related and must have arisen evolutionarily from a common ancestor, i.e., they are *homologous*. Two random sequences are expected to have about 6% of their residues identical, just by chance. Residues can be inserted and deleted by genetic mutations, so gaps may be introduced in either sequence when they are compared to maximize their similarities, but this decreases the significance of any similarities. Consequently, two proteins are generally taken to be homologous if more than 20% of their residues are identical when aligned with a minimum number of gaps.

The sequences of proteins provide a great deal of information about the evolutionary relatedness of the species from which they come. Generally, the more closely related the species, the more similar the amino acid sequences of homologous proteins. For example, the sequences of many human proteins, including cytochrome *c*, the hemoglobin α, β, and γ chains, and the fibrinopeptides, are identical to those of chimpanzees. Myoglobin and the hemoglobin δ chain from the two species each differ at a single residue.

Differences between proteins presumably have arisen during evolution by mutation of their genes after separation from their last common ancestor. The extent of the differences depends on both the time since the genes diverged and the rate at which mutations are permitted to occur. Some proteins change very slowly, some very rapidly. Most of the changes that have occurred appear to be neutral and to have no significant effect on the function of the protein. Most mutations that alter the function presumably are selected against during evolution, so the rate of change is usually the greater the fewer the functional constraints on the protein.

New proteins almost invariably arise by duplication of an existing gene and subsequent mutation. For example, the polypeptide chains of myoglobin and the α, β, γ, δ, and ε chains of hemoglobin are all homologous and undoubtedly arose by duplications of an ancestral globin gene. Whether all present proteins ultimately arose by duplication of one or a very few common ancestors is not clear, for over long periods of evolutionary time homology becomes undetectable due to the large number of mutations that have accumulated.

Some proteins have been elongated by duplication of all or part of the gene, so different segments of the protein sequence are homologous to each other. In other cases, parts of different genes have been duplicated and rearranged to generate new mosaic proteins.

In most cases, similarity of protein amino acid sequences implies similarity of function, although exceptions exist. The biological function of a new gene or protein is often inferred from any homology to genes or proteins of known function.

B. Folding of the Polypeptide Chain

The polypeptide chains of natural proteins differ from unnatural polypeptides and other polymers in that they fold back upon themselves to adopt compact, well-ordered, three-dimensional structures, in which the positions of most atoms are fixed in space by interactions with neighboring atoms. These folded conformations are essential for the biological functions of proteins.

1. Polypeptide Conformation

The three-dimensional conformations of proteins result from the ability of single covalent bonds to undergo rotation. For each residue (except Pro), rotations are possible around two single bonds of the backbone; the angles of rotation are given the designations ϕ and ψ (Fig. 4). In the case of Pro, its cyclic side chain restricts the value of ϕ to be about $-60°$. The other backbone bond, the peptide bond, has partial double-bond character and is constrained to be planar, but two different forms are possible. The *trans* form of the planar peptide bond (with the C^α atoms on opposite sides of the bond) is intrinsically favored, unless the following residue is Pro, when the *cis* isomer (with the C^α atoms on the same side of the bond) has a similar intrinsic stability. In addition to the backbone, the side chains of most of the amino acids have intrinsic flexibility comparable to that of small molecules.

Not all rotations about single bonds are possible in a fully unfolded polypeptide chain, the so-called random coil state, for clashes between atoms would occur in some cases. Local restrictions on flexibility of the backbone vary with the nature of the side chain; for example, 61% of the combinations of ϕ and ψ are possible with Gly residues, which have no side chain, but only 5% are feasible with the most

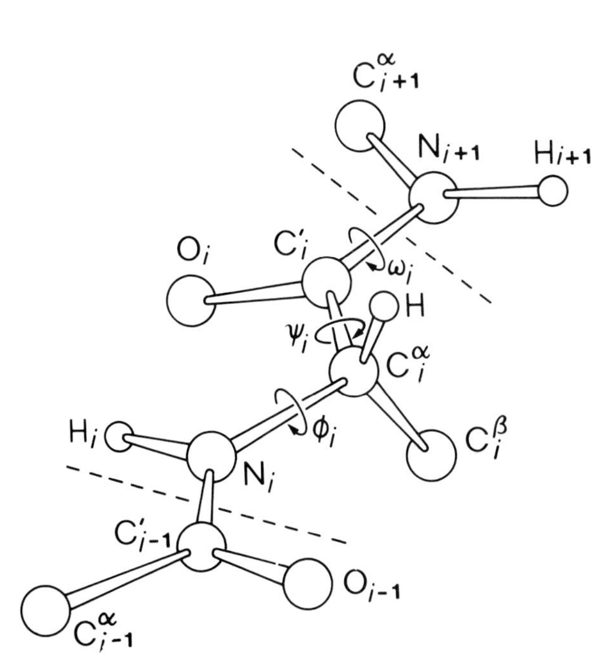

FIGURE 4 A short segment of polypeptide backbone, with the limits of one residue indicated by the dashed lines. Rotations about the bonds of the backbone produce different three-dimensional conformations. The angles of rotation ϕ and ψ have considerable flexibility, but ω must be close to either 0° or 180°, because the peptide bond has partial double-bonded character and must be planar. The polypeptide chain is illustrated in its fully extended conformation, where $\phi = \psi = \omega = 180°$.

restrictive side chains of Val and Ile residues. For most amino acids, 30% of the possible combinations of ϕ and ψ are possible, so the polypeptide backbone has substantial flexibility.

Each residue in a flexible polypeptide chain can adopt a number of different conformations, perhaps eight on average. In this case, an average-sized protein chain of 300 residues might be expected to adopt 8^{300} (10^{270}) different conformations. Not all of these would be feasible, because atoms distant in the primary structure would clash in some, but even if only one in 10 billion of these conformations were possible, the total number would be reduced to only 10^{260}.

The enormous number of conformations possible with an unfolded protein could not be encountered on a practical time scale. Rotations about single bonds to produce a different conformation occur about every 10^{-11} sec. Therefore, to sample 10^{260} conformations would require, on average, 10^{249} sec. This is vastly greater than the age of the Earth, about 10^{17} sec. Proteins obviously cannot fold to their tertiary structures by a random search, because they are observed to fold within a few seconds after completion of biosynthesis.

Completed proteins also can be unfolded and will usually refold on a similar time scale. Refolding occurs so quickly because unfolded proteins rapidly adopt a limited number of nonrandom conformations after being placed under refolding conditions. The slowest step is at a late stage of folding, when the protein gets very close to the final folded conformation.

Another way in which large proteins manage to fold so quickly is by folding initially into smaller structural units of some 50–200 residues, called domains, that are independently stable folding units. The domains fold sequentially during biosynthesis, as the polypeptide chain is being extended from the N-terminus.

2. Protein-Folded Conformations

The overall three-dimensional conformation of a protein is known as its *tertiary structure*. It can be determined in great detail by X-ray crystallography or, more roughly, by nuclear magnetic resonance analysis. The former gives the structure in the solid, crystalline state; the latter in solution. Fortunately, a single folded conformation for most proteins is not altered substantially by crystallization. Although nearly 200 different protein tertiary structures are known, they have a number of characteristics in common; one folded protein is illustrated in Color Plate 1.

The overall shapes of small folded proteins are roughly spherical, but larger proteins are folded into multiple domains. Individual domains rarely contain more than 200 amino acid residues.

The interior of a folded protein domain is remarkably compact, with very few cavities the size of a water molecule. In some proteins, a few water molecules are present within the protein interior, but they are involved in hydrogen bonding to the protein and are essentially part of the protein structure. The close-packing of atoms within a protein is such that about 75% of the interior volume is occupied by atoms, as in crystals of small molecules, but considerably greater than the 44–58% packing densities of liquids. Yet this close-packing in proteins is accomplished within the constraints of the covalent structure of the polypeptide chain, with relatively little strain due to unfavorable bond lengths, angles, or rotations.

The polypeptide backbone generally pursues a moderately straight course across the domain, turns on the surface, and then returns across the domain in a more-or-less direct path. The impression is of segments of somewhat stiff polypeptide chain interspersed with relatively tight turns or bends, which are almost always on the surface of the protein.

Marked differences are evident between the surfaces and interiors of proteins, depending on whether the protein occurs in aqueous solution or embedded within membranes. With water-soluble proteins, any amino acid can occur on the surface, but there are restrictions on which occur in the interior. Virtually all the side chains that are normally ionized (Asp, Glu, Lys, Arg residues) occur only on the surface, where they interact favorably with water, and not in the interior. Gly and Pro residues also tend to be on the surface, because they tend to be used in reverse turns of the polypeptide chain. The interior is composed almost entirely of nonpolar side chains (Ala, Val, Leu, Ile, Met, Phe, Tyr, Trp); virtually all the buried polar groups, principally the —NH—, and —CO— groups of the backbone, are paired in hydrogen bonds. With proteins that are embedded in the hydrophobic interiors of membranes, nonpolar side chains are on the protein exterior and interacting with the membrane. Polar side chains are less frequent than in water-soluble proteins and tend to be in the interior, where they are believed to have functional roles, such as guiding ions through channels in the membrane.

a. Secondary Structures The relatively straight segments of polypeptide backbone in a folded protein often adopt relatively regular conformations known as secondary structures. The α-helix is one, the β-strand the other (Fig. 5). Their primary com-

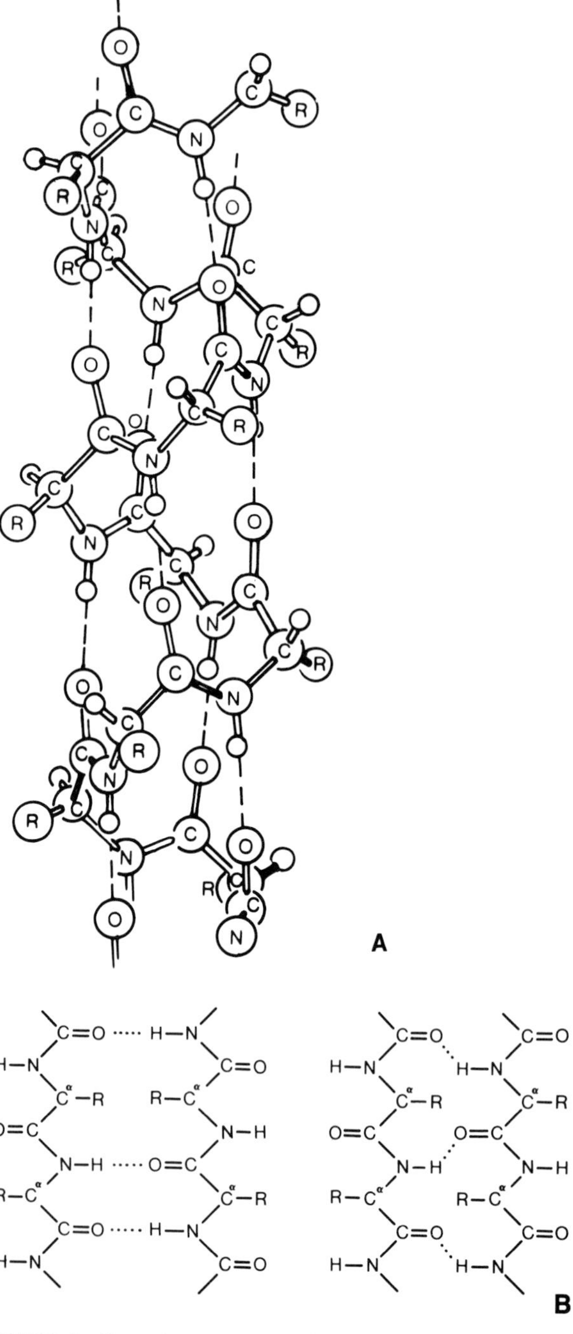

FIGURE 5 Secondary structures found commonly in proteins. A. An α-helix. B. β-sheets, antiparallel (left) and parallel (right). The side chains of the residues in β-strands alternately project above and below the plane of the figure.

mon characteristic is that all polar groups of the polypeptide backbone are paired in hydrogen bonds.

The α-helix is a right-hand helix with 3.6 residues per turn and a translation along the axis of 1.50 Å (0.15 nm) per residue. Most conspicuous is the hydrogen bond between the —CO— oxygen of the backbone of each residue and the backbone —NH— of the fourth residue along the chain. The polypeptide main chain forms the inner part of the rodlike helix, and the side chains radiate outward in a helical array. The side chains of residues three and four apart in the primary structure are near each other spatially.

In the β-strand conformation, the polypeptide chain is nearly fully extended, with the —CO— and —NH— groups of each residue pointing to one side, those of the adjacent residue to the other. The side chains alternately point up and down, so those of residues two apart in the sequence are near each other. Multiple β-strands aggregate side by side, forming hydrogen bonds between the —CO— of one strand and the —NH— of another to form a β-sheet. Adjacent strands may be either parallel or antiparallel, but the geometry of the chain is slightly different in the two cases, so sheets are often all of one type or the other. β-sheets in proteins are not flat but are almost always twisted, with each strand having a right-hand twist.

Approximately one-third of residues in proteins tend to be in helices and another third in β-sheets. Most of the interiors of proteins are composed of β-sheets, α-helices, or both packed together. Some proteins, such as myoglobin and hemoglobin (Fig. 6), are composed only of α-helices and turns, others consist entirely of β-sheets and turns, while many contain both β-sheets and α-helices. Simplified representations of proteins often depict α-helices as cylinders or coiled ribbons and β-strands as broad arrows pointing from the amino to the carboxyl end of the polypeptide chain (Fig. 6).

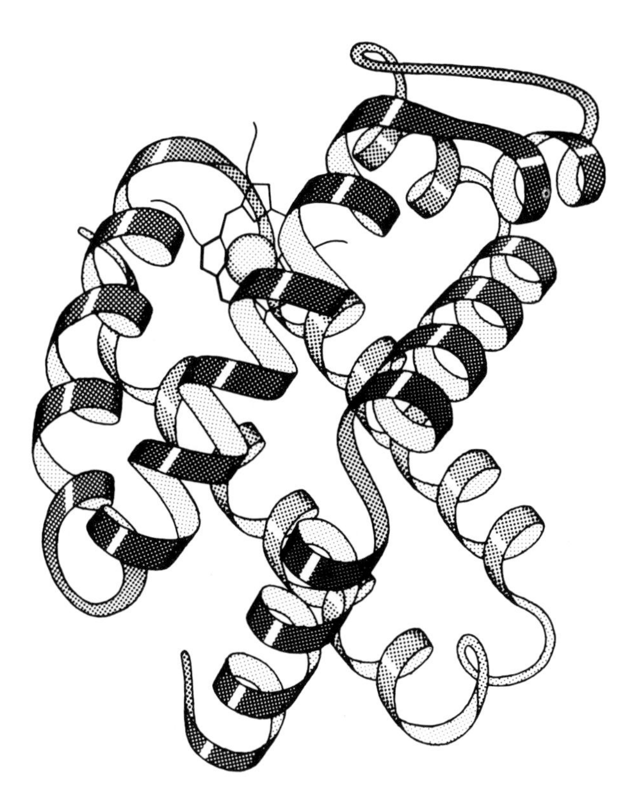

FIGURE 6 Schematic drawing of the topology of the polypeptide chain of the β-subunit of human hemoglobin. Only the polypeptide backbone is depicted (as a ribbon), plus the heme prosthetic group. The coiled ribbon depicts the α-helical conformation. (Drawing supplied by Jane Richardson.)

b. Fibrous Proteins Fibrous proteins, such as collagen and keratins, differ from the globular types of proteins described above in that they have repetitive, extended structures, without the reverse turns to give a compact structure. In collagen, three polypeptide chains of about 1,000 residues each are in helical conformations and wound about each other to give a molecule 14 Å in diameter and 3,000 Å in length (0.14 nm × 300 nm). This unusual architecture results from the repetitive sequence of collagen, in which every third residue in the sequence is Gly and many of the remainder are Pro. The absence of a side chain on the Gly residues permits the chains to coil about each other closely, whereas the restricted flexibility possible with Pro residues imparts rigidity to the structure.

Keratins consist of two or three polypeptide chains, each in an α-helical conformation, twisted about each other in a long, extended structure. This regular structure also results from repetitive features of the amino acid sequence.

The silk that is woven into cloth is also a fibrous protein of polypeptide chains containing 3,000–4,000 residues. It exists largely as antiparallel β-sheets interrupted by irregular segments of 100–200 residues and stacked on top of each other. This β-sheet structure also results from a regular pattern in the amino acid sequence, in which Gly residues tend to alternate with Ala or Ser. Alternate residues have their side chains on opposite sides of the β-sheet, so the silk β-sheets have Gly residues on one side and Ala and Ser on the other. The sheets stack with the same type of faces in contact.

C. Quaternary Structure

Many proteins normally function as part of larger structures made up of multiple copies of the same or different polypeptide chains. A wide variety of different structures are found, from single polypeptide chains to large complexes containing hundreds of polypeptides. Spherical viruses are made with an outer shell of many identical polypeptide chains interacting specifically to generate a regular icosahedron structure; a typical number is 180. This higher level of structure is designated as quaternary structure. When part of a larger aggregate, an individual polypeptide chain is referred to as a subunit or protomer and is often designated by a Greek letter.

Proteins interact with each other specifically by having surfaces on their folded conformations that are complementary, both in shape and in physical properties. The surfaces of interacting proteins generally fit together so that most of the surface atoms are in close, energetically favorable contact. Moreover, polar groups are usually paired in hydrogen bonds. Less frequently, ionized groups (which are electrically charged) are paired, one positive and one negative, to avoid burying any net charge. The interfaces between associated proteins are similar to the interiors of the individual folded subunits. The strengths of the interactions vary enormously, depending on both the degree of complementarity and the area of the interacting surfaces.

When two different protein molecules each have a complementary surface, a dimer with one molecule of each naturally results. Higher orders of structure require that more than one interacting surface be present on each protein subunit. In this case, aggregation will usually occur indefinitely, forming a long polymer, unless growth is inhibited by steric repulsions between subunits due to the geometry of the structure generated.

Aggregation of identical polypeptide chains requires that the interacting surface of each protein subunit be self-complementary. If the two complementary surfaces are not contiguous on the surface of the individual monomers, the association is said to be heterologous. An indefinite head-to-tail association will result, unless the geometry of association results in a closed structure. Many filamentous protein structures are generated by linear head-to-tail aggregation of globular subunits. Stable trimers, pentamers, hexamers, and heptamers of identical subunits are also made in this way, by fixing the geometry of interaction such that closed rings of the appropriate size result.

A more common type of association, isologous, is observed between identical molecules when a strictly limited aggregation is required. In this case, two molecules associate using the same surface on each. This requires that the single interacting surface on each subunit have a mirror plane of complementarity, so that the two identical surfaces are complementary. The result is a stable dimer in which the two monomers are equivalent. The dimer has a twofold symmetry axis through the junction of the two subunits, such that rotating the dimer by 180° about the axis reproduces the same molecule as the original (rotational symmetry). No further aggregation is possible, unless additional interacting sites are present. Tetramers are often formed by two different types of isologous association, so they are dimers of dimers.

D. Relationship Between Structural Levels

What secondary, tertiary, and quaternary structures are adopted by a protein depends entirely on its primary structure. Proteins with related amino acid sequences invariably have essentially the same secondary and tertiary structures, so a particular structure does not require a unique primary structure. In particular, the relative importance of the primary and tertiary structures in determining the secondary structure is not certain. Whereas the various amino acids tend to occur with different frequencies in α-helices, β-sheets, reverse turns, and irregular conformations, the stabilities of the various conformations depend on the presence of the tertiary structure. Do not assume that the primary structure determines the secondary structure, which then defines the tertiary structure.

The quaternary structure is determined by the residues within the interacting regions. Related proteins need not have the same quaternary structure, because changing only a few amino acid residues within the interacting region can alter dramatically the interaction with other proteins. For example, myoglobin functions as a monomer, whereas the homologous α, β, γ, δ, and ε chains of hemoglobin exist as $\alpha_2\beta_2$, $\alpha_2\gamma_2$, $\alpha_2\delta_2$, and $\alpha_2\varepsilon_2$ complexes.

E. Stability of Protein Folded Conformations

Whether or not the folded conformation of a protein is stable depends on both the protein's primary structure and its environment, i.e., the nature of the solvent, the pH, salt concentration (ionic strength), temperature, pressure, etc. Proteins can be unfolded, often reversibly, by high or low temperatures, extremes of pH (either very acidic or very basic), or addition of substances that promote unfolding (denaturants), such as urea or guanidinium chloride. Even under optimal conditions, the folded state of a protein is only marginally stable relative to the unfolded conformation. The many interactions holding together the folded protein are only barely able to compensate for the energetic cost of fixing the flexible polypeptide chain in a single conformation. Unfolding is usually cooperative in that once unfolding starts, it proceeds to completion; partially folded conformations are relatively unstable. This implies that the interactions stabilizing the folded state are also cooperative in that the stabilizing contribution of each interaction depends on the simultaneous presence of the others. Only by having a multiplicity of intrinsically weak interactions present simultaneously can folded conformations be stable.

A major contribution to stability of water-soluble globular proteins is the hydrophobic effect, which results from the shielding from the solvent of nonpolar groups within the interior. The nonpolar groups do not interact with other especially favorably, but they interact less well with solvent water in the unfolded state. Other contributions to the stability of the folded state are the hydrogen bonds between polar groups and the van der Waals interactions between nonpolar atoms that occur within the protein. Although similar interactions occur between the unfolded state and the solvent, those within the folded protein are intramolecular and, consequently, usually more favorable energetically. The cooperativity and complexity of these interactions, however, has thus far precluded a rigorous accounting for protein stability.

III. Interactions With Other Molecules

The functional properties of proteins become apparent only when they interact with other molecules.

A. Binding of Specific Ligands

1. Binding Sites

Every protein has a binding site for at least one other molecule, or ligand, with which it interacts during its function. Such binding sites are sufficiently specific so that the protein can distinguish between the appropriate ligand and all other molecules that it is likely to encounter. For most ligands, the binding site is a specific, relatively small area on the surface of the protein that is complementary, sterically and physically, to the ligand. Such interactions have been described here already in terms of quaternary structure (Section IIC). For very small ligands, such as oxygen atoms and electrons, that would be difficult to recognize specifically on such a basis, additional groups (prosthetic groups) are incorporated into the protein to add to the specificity. For examples, heme groups are used for binding electrons and oxygen molecules in cytochromes and hemoglobins, respectively. The protein first binds the appropriate prosthetic group, and together they can recognize specifically the correct small ligand. Nevertheless, many proteins are able to distinguish between many small ligands without prosthetic groups. For example, calcium-binding proteins can bind Ca^{2+} selectively in the presence of 1,000-fold higher concentrations of Mg^{2+}, a very similar ion.

Specific binding is accomplished by having within the binding site a constellation of several protein groups of intrinsic affinity for the ligand, arranged in the appropriate stereochemistry so that the ligand interacts with all the groups simultaneously. The binding sites on a protein usually preexist, formed by the folding of the protein. Binding of the ligand usually has only minor effects on the structure of the protein binding site.

Some proteins have only a single binding site at which they bind a single ligand, whereas others have binding sites for two or three ligands. Only in very large, complex proteins are there likely to be a greater number of binding sites. Each polypeptide chain generally has no more than one binding site for any ligand. In a multimeric protein, with *n* identical polypeptide chains, there should be *n* equivalent binding sites for each ligand. In some cases, however, the binding sites bridge the symmetry axes relating different polypeptide chains, so only a fraction of the *n* sites can be occupied simultaneously. [*See* DNA BINDING SITES.]

2. Binding Affinities

The binding affinity is usually expressed as a dissociation constant, which is the concentration of free ligand at which the protein binding site is occupied half the time at equilibrium; greater binding requires higher ligand concentrations. The affinities of ligands for different binding sites vary enormously. Some may be so great that binding is essentially irreversible. At the other extreme, many examples of binding are so weak as to be of questionable significance. If the value of the dissociation constant is substantially greater than the concentration at which the ligand is likely to be encountered naturally, that binding is generally considered not to be of functional significance.

B. Interactions Between Binding Sites: Allostery

Many proteins bind multiple ligands, and very often the binding of one ligand does not affect significantly the binding of another at a distant binding site. If the binding sites overlap, of course, binding of one ligand will inhibit binding of the other. If the binding sites are adjacent, binding of one ligand can increase or decrease the affinity for the other by affecting directly its binding site.

Much biological regulation occurs by the binding of one ligand altering the binding of another at a totally separate site; such proteins are said to be allosteric. The structural basis of the interaction between binding sites is best understood in the protein hemoglobin, where O_2 binding by each of the four subunits, $\alpha_2\beta_2$, is cooperative, in that binding at one site increases the affinity of the other sites. In addition, other ligands bind at other sites and also affect the O_2 affinity. These effects are crucial for hemoglobin to bind O_2 efficiently in the lung and to deliver it to other parts of the body.

The effects on O_2 affinity occur because the four subunits of the hemoglobin tetramer can exist in two somewhat different quaternary structures. One, designated T for "tense," is characterized by low affinity for O_2; the other, designated R for "relaxed," has high affinity. T is the predominant structure in the absence of O_2, but binding of O_2 shifts the equilibrium toward the R state, and thereby increases the O_2 affinity of the remaining sites. Other ligands affect O_2 affinity by binding preferentially to the R or T state.

By this simple mechanism, the binding of O_2 can be regulated very closely. Similar mechanisms probably occur with other allosteric proteins.

C. Immunoglobulins

Related proteins need not bind the same ligands, for changing only a few residues in a binding site can alter dramatically its specificity. Excellent examples are the immunoglobulins of the immune system, which can recognize virtually any molecules foreign to the body. Yet all immunoglobulins have remarkably similar three-dimensional structures. Their binding sites are composed of only 10–15 residues in six loops on their surface, out of a total of nearly 700 for an antibody molecule, so changing only these residues can alter drastically the antigen specificity.

IV. Chemical Catalysis By Enzymes

A. Enzyme Function

Enzymes are proteins that bind ligands but then cause them to undergo a specific chemical reaction. In this case, the ligands are known as substrates, and the region of the binding site is known as the active site. Only a single reaction generally occurs on any enzyme, even though its substrate in solution in the absence of the enzyme might undergo a number of different reactions at similar rates. An enzyme generally causes unimolecular reactions to occur at rates some 10^6–10^{14} times faster than would occur under the same conditions in the absence of the enzyme. The enzyme is not altered by the reaction and is only a catalyst. Therefore, the enzyme does not alter the normal equilibrium constant for the reaction and must also increase the rate of the reverse reaction to exactly the same extent under the same conditions. The products of the reaction can also be used as substrates in the reverse reaction, if the equilibrium permits.

A characteristic of enzyme-catalyzed reactions is that the rate of the reaction is proportional to the concentration of substrate only at low concentrations. With increasing concentration, the rate plateaus and approaches a maximum value (V_{max}) at high substrate concentrations. This kinetic behavior is a consequence of the enzyme binding the substrate molecules at a finite number of active sites. Maximal activity is obtained when the enzyme is saturated with substrate. The substrate concentration at which the rate is half maximal, the Michaelis constant, or K_m, is a function of both the affinity of the substrate for the enzyme and the rate of the reaction on the enzyme.

Enzymes increase the rates of chemical reactions in a variety of ways. The most fundamental is that the enzyme does not have optimal affinity for either the substrate or product but for the transition state for the reaction. The transition state for any chemical reaction is that species along the reaction pathway that is least stable, with the highest free energy. The rate of the reaction is determined by the free energy of the transition state, relative to that of the substrate; the lower the free energy of the transition state, the more rapid the reaction. By binding the transition state for a particular reaction most tightly, an enzyme lowers its free energy relative to the substrate and product and thereby increases the rate of the reaction on the enzyme. Enzymes catalyze specific reactions by binding tightly only specific substrates and transition states.

Enzymes are especially effective in catalyzing reactions between two or more reactants, because they can be bound simultaneously on the enzyme, at neighboring sites within the active site. The reaction between multiple substrates on the enzyme is effectively unimolecular, whereas the same reaction in the absence of the enzyme would require that the multiple reactants encounter each other simultaneously in solution, which is very improbable at low concentrations. The effective concentration between reactants bound to a protein can be very high. In some reactions, in which a chemical group is transferred from one substrate to another, the enzyme can be used as an intermediary, accepting temporarily the group from one substrate and subsequently transferring it to the second substrate.

Although no enzymatic reaction is understood entirely, the basic principles of enzyme catalysis appear to be relatively straightforward. The importance of enzymes for life cannot be overstated, because they cause the chemical reactions of metabolism to occur rapidly under physiological conditions at 37°C. [*See* ENZYMES, COENZYMES, AND THE CONTROL OF CELLULAR CHEMICAL REACTIONS.]

B. Regulation of Enzyme Activity

Enzymes do not function in isolation, and it is crucial that each enzyme be an integral part of the chemical reaction pathways that take place. Its substrate must be available at the appropriate time and concentration, and its reaction must be catalyzed only when there is a need for its product. Accordingly, the activities of enzymes are linked together in a complex network of chemical reactions, by which metabolism occurs, and they must be regulated.

One common method of integrating the activities of enzymes is to make them responsive to the availability of substances other than just their substrates. This is accomplished by having a binding site for the regulatory metabolites on the appropriate enzymes. Just as binding of a ligand at one site can affect binding at another by allostery (Section IIIB), so can catalysis at the active site be affected. Many enzymes are allosteric, and this is a common way of controlling the activity of an enzyme. For example, the end product of a biosynthetic pathway usually inhibits the first enzyme of that pathway, a phenomenon known as feedback inhibition. In this way, the end product is produced by the enzymes of the pathway only when its level falls and biosynthesis is required.

Another very common method of regulating the activity of an enzyme is to alter its functional properties by modifying it covalently, most often by phosphorylation of Ser, Thr, or Tyr residues. A variety of protein kinases that catalyze such reactions are known, with varying specificities for the proteins that they phosphorylate. It is important that such covalent modifications be reversible, so a variety of protein phosphatases may remove the phosphoryl groups from the target enzymes. The activities of the protein kinases and phosphatases are usually regulated by various physiological stimuli, such as hormones, growth factors, etc. Often complex cascades of phosphorylation of several kinases and phosphatases exist, resulting in a wide range of such control, and very small signals can be amplified to have substantial physiological effects.

Phosphorylation of proteins is a major means by which growth of the body is regulated. Many types of cancer are now known to be caused by malfunction of this regulatory system. [*See* PHOSPHORYLATION OF MICROTUBULE PROTEIN.]

C. Energy Transduction

Proteins catalyze many processes other than chemical reactions in solution; in particular, they can convert one form of energy to another. The ubiquitous adenosine triphosphate (ATP) is the usual currency of chemical energy and is involved in most energy interconversions, being either utilized or generated. For example, the chemical energy of ATP is converted to mechanical energy by moving one protein (myosin thick filaments) relative to another (actin thin filaments) in muscle. This movement of one protein relative to another is caused by a transient interaction between them, a conformational change in one (myosin) to move it relative to the other, followed by dissociation of the two and reassociation at a different point. In the process, the two proteins have moved relative to each other by about 100 Å (10 nm). Muscle contraction is simply the result of numerous such cycles, in each of which one filament slides past the other within the muscle. [*See* ADENOSINE TRIPHOSPHATE (ATP).]

The chemical energy for the movement comes from the hydrolysis of ATP, one molecule of which is consumed per myosin unit in each cycle. The details of the crucial conformational change in myosin are not known, but the complete cycle requires that the affinities of the myosin and actin for each other be closely controlled.

In general, transduction of the chemical energy of ATP requires only that two different forms of the enzyme (E and E*) be interconverted only by hydrolyzing ATP to adenosine diphosphate (ADP) and inorganic phosphate (P_i) or by carrying out the second reaction ($A \rightarrow B$) to which the first is to be coupled:

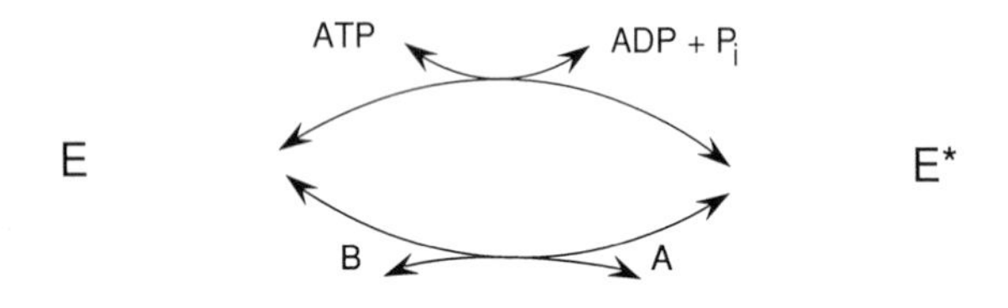

As long as E and E* can be interconverted only by these two reactions, the chemical energy of ATP hydrolysis will drive the reaction $A \rightarrow B$. That reaction can be anything capable of being carried out by a protein: mechanical movement by a conformational change, different affinities for ligands (to pump molecules across membranes against a concentration gradient), or emission or absorption of light.

Pumping ions and other metabolites across membranes, against concentration gradients, is accomplished by protein pumps embedded in the membrane, with a suitable channel for permitting passage of the appropriate small molecule through the membrane. In the best-understood ion pump, that for pumping Na^+ and K^+ ions, the chemical energy of ATP is used to phosphorylate the protein. Both phosphorylated and nonphosphorylated forms of the protein must exist in two conformational states, in which the channel is open to opposite sides of the membrane. By only permitting certain interconversions of the various forms of the protein pump, the chemical energy of ATP is used to transport Na^+ and K^+ ions against concentration gradients.

Light detection in the eye occurs upon its absorption by the pigment retinal, attached to the protein opsin; the complex is known as rhodopsin. Absorption of light causes the retinal to alter its shape, which affects the structure of rhodopsin. The rhodopsin then binds to another protein, transducin, and alters its affinity for its ligands. By a complex cascade of subsequent events, the absorption of a single photon of light by rhodopsin causes the closure of hundreds of Na^+ and K^+ channels in the membrane of the rod cell of the eye, which produces a electrochemical signal that is transmitted to other neurons of the retina.

V. Degradation

Proteins have only limited life spans, which vary markedly from one protein to another. Their ultimate fate is to be degraded to the constituent amino acids, and there is extensive turnover of proteins within most cells. Cells need mechanisms for surviving temporary conditions of starvation, by degrading "luxury" proteins in order to utilize their amino acids for biosynthesis of those more essential for survival. Maturation of reticulocytes into oxygen-carrying erythrocytes of the blood is accompanied by the selective degradation of mitochondria, ribosomes, and many proteins no longer required once the cell has synthesized its complement of hemoglobin. There is also a need for degrading abnormal proteins, either those that have succumbed to old age by chemical modifications or those that were synthesized in an incorrect form because of mistakes in the biosynthetic process. Abnormal proteins are readily recognized by most cells and are rapidly degraded. Degradation of some proteins is also necessary to permit regulation of their function. A major control over protein function in humans is at the stage of protein biosynthesis, and the level of any protein can be decreased only if there is a means by which it is degraded.

The proteins that turn over most rapidly are those that catalyze rate-determining metabolic reactions. For example, the most rapidly degraded protein in the liver is ornithine decarboxylase, with a half-life of only 11 min. It catalyzes the rate-determining step in polyamine biosynthesis and is not subject to any known regulation by allosteric control or by covalent modification. Instead, its rate of biosynthesis can vary more than a 1,000-fold, so decreases in its activity can only occur if there is a mechanism for degrading the protein.

In contrast, many proteins are long-lived, especially structural proteins. Proteins of the eye lens are not degraded, so proteins synthesized in the embryo are still present in the human eye lens 70 or more years later. During that time, chemical modifications of the proteins are inevitable and result in pigmentation, decreased solubility, and formation of cataracts. Other proteins exist for the life of the cell they are in, such as hemoglobin, which is stable throughout the 3-mo life-span of an erythrocyte. At the end of that time, the entire erythrocyte is degraded.

The factors that determine the rate of degradation of different proteins are still being sought, as is elucidation of the mechanism. Degradation of individual molecules of a protein is random, and newly synthesized molecules are just as likely to be degraded as old, unless chemically modified significantly. A single event is also sufficient to cause degradation; in some systems, covalent attachment of another protein, ubiquitin, is believed to be the event that marks a protein for degradation, but what regulates its attachment is not known.

Bibliography

Creighton, T. E. (1983). "Proteins: Structures and Molecular Properties." W. H. Freeman, New York.

Darnell, J., Lodish, H., and Baltimore, D. (1986). "Molecular Cell Biology." Scientific American Books, New York.

Dickerson, R. E., and Geis, I. (1983). "Hemoglobin: Structure, Function, Evolution, and Pathology." Benjamin-Cummings, New York.

Fersht, A. (1985). "Enzyme Structure and Mechanism." W. H. Freeman, New York.

Oxender, D. L., and Fox, C. F. (1987). "Protein Engineering." Alan R. Liss, New York.

Stryer, L. (1988). "Biochemistry." W. H. Freeman, New York.

Proteins (Nutrition)

CAROL N. MEREDITH, *University of California, Davis*

Glossary

Digestibility Percent of total protein intake that is absorbed by the gut

Protein Chains of amino acids found in all cells, containing an average 16% nitrogen. Dietary protein is often expressed as nitrogen intake

Protein quality Capacity of a dietary protein to promote growth in the young or nitrogen retention in the adult. Quality is determined by digestibility and the pattern of indispensable amino acids, which ideally is similar to the pattern of tissue proteins

Protein requirement Minimum protein intake that can sustain normal growth in the young or no net loss of body nitrogen in the adult. In healthy young adults it is 0.57 g/k when the diet provides adequate levels of all other nutrients

Protein sparing Effect that adequate or excess energy intake has on increasing the efficiency of nitrogen retention in the body, thus decreasing the protein requirement

Protein turnover Rate of new protein synthesis and of breakdown of body proteins, occurring at about three times the rate of protein intake in adults

Recommended dietary allowance Protein intake meeting the nutritional needs of practically all healthy persons, equal to 0.8 g of protein per kilogram body weight per day

Reference protein Protein of known good quality with a pattern of essential amino acids that can be used as a standard of comparison. Examples include whole egg protein for the adult and human milk protein for the infant

IN THE AVERAGE American diet, 12–14% of energy comes from protein, providing adults with more than 75 g of protein per day. Much of the dietary protein in developed countries is from animal products (e.g., meat, milk, eggs, and cheese), supplying the highest quality protein but also substantial and undesirable amounts of fat. In less-developed countries, most protein is provided by one or more staple foods of plant origin (e.g., rice in Asian countries and maize and cassava in sub-Saharan African countries). The rice-based diet of rural people in Thailand, for example, supplies 8% of energy as protein, which is only marginally enough for an adult. As countries become more prosperous and urbanized, protein intake increases, especially from animal foods. When income does not limit dietary choice, people tend to eat a diet with about 12% of total calories as protein, coinciding with the amount recommended by the U.S. National Research Council.

I. Chemical Composition of Dietary Proteins

Proteins in nature are made from different combinations of 20 amino acid building blocks. Each amino acid has at least one nitrogen-containing amine group attached to a hydrocarbon backbone of 2–11 carbons. The 20 amino acids function like letters in an alphabet, and the genetic code of each living organism determines how they will be put together to form a vast variety of proteins, analogous to the words and sentences in the books of a library. When we consume and digest proteins in food, the free amino acids enter the tissues and are knitted together through peptide bonds to make new proteins that are as different as hair, the transparent lens of the eye, or blood-clotting enzymes. Protein synthesis cannot occur at a normal rate unless each and every amino acid is present in the cell in the amounts demanded by the genetic code.

A. Indispensable or Essential Amino Acids

Eight amino acids cannot be made by animals and are considered essential or indispensable. In approximate order of abundance in animal muscle protein, they are lysine, leucine, valine, isoleucine, threonine, phenylalanine, methionine, and tryptophan. If a young animal is fed a diet in which one indispensable amino acid is absent, normal growth cannot occur. There are almost no natural food proteins that are totally lacking in one or more indispensable amino acids, with the exception of gelatin, which lacks tryptophan. However, many natural proteins supply only small amounts of one or more of the indispensable amino acids. When an essential amino acid is present in an amount too low to sustain normal growth, the protein is called *incomplete* and that amino acid is termed the *limiting amino acid*. The poor quality of many plant proteins is due to a pattern of indispensable amino acids that is markedly different from the pattern needed to make body proteins. It should be noted that incomplete proteins do not *lack* an essential amino acid, they simply provide a disproportionately small amount with respect to the body's needs.

B. Dispensable or Nonessential Amino Acids

Twelve amino acids can be made from other amino acids and carbon compounds and are considered dispensable or nonessential. In approximate order of abundance in animal muscle proteins, they are arginine, glutamate, glutamine, aspartate, asparagine, alanine, glycine, serine, tyrosine, histidine, proline, and cystine.

The ability to make all the dispensable amino acids at a rate allowing rapid growth is limited. The first intravenous feeding solutions made from mixtures of crystalline L-amino acids provided all the essential amino acids but only one or two dispensable amino acids, usually glycine. These formulas, leading to toxic levels of ammonia in infants and not efficiently used for growth, were improved by adding other dispensable amino acids such as arginine, alanine, glutamic acid, and proline. The most efficiently used artificial amino acid mixtures provide nearly all the dispensable amino acids, similar to natural food proteins.

II. Sources of Protein

A. Quantity in Foods and Diets

The amount of dietary protein is usually calculated from its nitrogen content multiplied by a conversion factor of 6.25, based on the fact that most proteins contain 16% nitrogen and nearly all dietary nitrogen is in the form of proteins or amino acids.

Protein is present in all unprocessed foods, and only highly purified foods (e.g., cooking oil or white sugar) lack protein. Table I shows the content and quality of protein in some foods. Prepared foods, containing more than 70% water, usually have between 1 and 12% of protein. Vegetables and fruit are almost protein-free, but plant seeds have more protein. Foods of animal origin tend to be richer in proteins. The most concentrated natural sources of dietary protein are lean meats (e.g., fish or skinned poultry).

B. Quality in Foods and Diets

Because the dispensable amino acids can be made in the body, quality is mainly determined by the amount and pattern of indispensable amino acids in the diet. Quality is tested in relation to growth or nitrogen retention. In the United States, the principal method for experimentally measuring the nutritive value of proteins is the Protein Efficiency Ratio (PER), which relates the weight gain of young male rats to the amount of protein eaten. In humans, quality is measured as the percentage of protein intake that is retained in the body. Nitrogen balance, or nitrogen retention, is determined by subtracting nitrogen excretion from nitrogen intake, after adapting to the test diet for 5–10 days. Most minor nitrogen losses (skin, hair, nails, secretions) are estimated and not usually measured directly. The balance method overestimates nitrogen retention and thus protein quality, as it is hard to measure all nitrogen losses accurately.

1. Amino Acid Patterns

By knowing the amount of each indispensable amino acid in a food protein, its quality can be pre-

TABLE I Content and Quality of Some Protein Foods[a]

Food	Protein concentration (g/100 g) Raw	Protein concentration (g/100 g) Ready to eat	Dig. (%)	Nut. Val. (%)	Lim. AA
Meat	25	25	100	70	Methionine
Egg	12	12	98	98	None
Cow's milk	3	3	98	83	Methionine
Human milk	1	1	100	100	None
Soy milk	3	3	92	60	Methionine
White beans	22	5	77	40	Methionine
Bread	9	9	97	37	Lysine
Rice	7	2	97	55	Lysine, threonine
Corn grits	9	1	86	51	Tryptophan, lysine
Potatoes	2	2	82	60	Methionine

[a] Dig., digestibility in rats = (N absorbed/N intake) × 100; Nut. Val., nutritive value in rats = (N retained/N intake) × 100; Lim. AA, Limiting amino acid, or the indispensable amino acid present in lowest amounts with respect to the need for that amino acid.

dicted by calculating its "chemical score." This expresses the concentration of each indispensable amino acid as a percentage of the concentration in an "ideal" or "reference" protein (e.g., egg protein). The lowest percentage obtained identifies the amino acid that is "limiting" and is a measure of the biological usefulness of the food protein. There is generally good agreement between the chemical score and biological measurements of protein quality. In the case of infants, the reference protein is human milk protein.

2. Complementarity

Most children and adults eat a variety of foods with different amounts and qualities of protein. Eating different proteins is beneficial, as proteins containing low amounts of one indispensable amino acid are eaten together with those proteins having abundant amounts of that amino acid, and the mixture absorbed from the gut has a more balanced composition.

In most of the world, cereals are the source of both protein and calories. Although the protein content of prepared cereals is not much lower than in milk-based formulas, the pattern of indispensable amino acids of cereals makes them a poor protein source during early childhood. Cereals are limiting in lysine (wheat) and often low in threonine (rice) and in tryptophan (maize). There are two ways of improving the quality of cereal proteins. One is by *supplementing* the cereal food with a small amount of an animal product rich in high-quality protein. For example, a dish prepared from rice, vegetables, and some egg provides good-quality protein. However, in poor rural communities, foods such as eggs, meat, or milk are expensive and scarce. Another method is by *complementing* the protein in a cereal food with other plant proteins that are richer in the cereal's limiting amino acid. In poor communities or for strict vegetarians, this is the only way to improve the protein quality of the diet. Compared with cereals, legumes provide more and better protein and are richer in lysine, thus increasing the lysine content of the mixture to the point that it does not limit protein utilization. As Table I shows, the quality of corn protein is limited by low lysine and tryptophan, whereas beans are limited by methionine and cystine. Experiments in young rats show that corn by itself has an efficiency of utilization of 56%, white beans by themselves have an efficiency of utilization of 76%, but a mixture of half beans and half corn, as in succotash, provides a protein that has an 82% efficiency of utilization (using the milk protein casein as a reference protein). Soybeans are a particularly good protein source and are used to make infant foods. Their high lysine content supplements the quality of cereals.

3. Amino Acid Digestibility

The amino acids in food proteins must enter the circulation before they can be used. Plant proteins are less easily broken down to amino acids because their structure is more resistant to digestive enzymes. Some plant foods (e.g., legumes) contain enzymes that inhibit protein digestion. Cooking improves digestibility by inactivating toxic enzymes and facilitating the attack of enzymes, by loosening protein structure.

4. Amino Acid Availability

Although moderate heating increases the availability of amino acids, the hot temperatures of toasting can destroy cystine and reduce the availability of lysine. In dry, nonacid foods, storage in hot climates or heat treatment favors a reaction between lysine and free sugars, forming yellowish or brown compounds, where lysine cannot be digested or absorbed. The loss of available lysine becomes nutritionally significant when the food is already limiting in that indispensable amino acid (i.e., in cereal products) or it is a major dietary source of lysine (i.e., powdered milk for infants and children).

C. Commercial Supplements of Proteins and Amino Acids

In recent years, a thriving market has been developed for protein and amino acid supplements, directed especially to strength-trained athletes. Weight lifters can spend hundreds of dollars on these products, although studies show that strength gain and muscle enlargement occur with training even if energy and protein intake are curtailed. The argument is that persons who are increasing muscle size and strength through resistance training need more protein to ''bulk up'' and that commercial protein supplements are more effective than dietary protein. Most supplements are expensive mixtures of milk protein and soy protein, not likely to be harmful in healthy persons consuming a varied diet and adequate amounts of fluids, but neither have they been proven effective. More highly touted and expensive are the pure amino acid supplements, sold as enhancers of protein synthesis and promoters of growth hormone release. Single amino acid supplements are potentially more hazardous and less useful for protein synthesis, as they can disrupt the normal entry of other amino acids into tissues.

III. Digestion and Distribution

Adults and children tend to eat during the day and fast during the night, whereas infants are usually constantly in the fed state. After a protein-containing meal, proteins are degraded by proteolytic enzymes secreted by the lining of the stomach and by the pancreas, producing free amino acids and small peptides made up of strands of two to four amino acids. The cells lining the upper small intestine take these up, and further break them down to free amino acids before they enter the blood. The concentration of free amino acids in the blood flowing from the gut to the liver increases several-fold. On reaching the liver, some of the amino acids are used for protein synthesis, some are broken down to small carbon compounds and urea, and the remainder are released to the general circulation. The liver contains enzymes for degrading all the amino acids except for leucine, isoleucine, and valine, which are mainly broken down in muscle. Insulin release following meals favors the uptake of amino acids from the blood, especially by muscle, which is the largest tissue in the body. All tissues constantly take up and release amino acids, but after meals a net uptake of amino acids and net increase in protein content is caused by greater synthesis than breakdown of tissue proteins.

In the fasted state, the flow of amino acids from the gut and liver to peripheral tissues diminishes, whereas the flow from the peripheral tissues to the visceral organs increases. The rate of protein synthesis decreases, and there is a net breakdown of proteins and net release of amino acids from muscle. Alanine and glutamine account for more than 50% of the amino acids released by muscle. These two dispensable amino acids are sources of carbon chains for glucose synthesis in the liver. During fasting, the glucose supplied by the liver feeds the tissues that have an obligatory and constant need for glucose as an energy source (e.g., the nervous system and red blood cells).

IV. Elimination of Excess Protein

The main function of amino acids is to provide units for the synthesis of protein, and other pathways are less important. Amino acids react with the enzymes of protein synthesis 20–100 times more avidly than with the enzymes of amino acid breakdown. This allows a highly efficient retention of amino acids present in low concentrations and provides a mechanism for breaking down excessive amounts of each amino acid. The activity of the enzymes that break down amino acids also adjusts to the amount of each amino acid in the diet. For example, an animal fed a low lysine diet (e.g., wheat flour) reduces the activity of the enzyme that oxidizes lysine, conserving it for protein synthesis. However, the efficiency of indispensable amino acid or protein conservation is not total, and a certain amount must be supplied daily in the diet to prevent a gradual loss of body proteins. Infants born with a deficiency of any of the amino acid breakdown enzymes accumulate toxic amounts of that amino acid and its breakdown products in the blood. The tissue that suffers the greatest damage is the central nervous system. The brain is susceptible because a large excess of a single amino acid can competitively inhibit the uptake of other amino acids, preventing the entry of the full range of amino acids needed for normal synthesis of proteins and neurotransmitters. In phenylketonuria, for example, the liver enzyme that converts phenylalanine to tyrosine is lacking, and unless a diet low in phenylalanine is provided, the child will become mentally retarded.

The enzymes that break down amino acids produce short-chain carbon compounds that can be used for energy metabolism or to make fats, glucose, and glycogen, while the potentially toxic nitrogen molecule is made into urea by the liver. The enzymes that make urea rapidly adjust their activity, increasing within hours of a protein-rich meal, or decreasing in response to a high carbohydrate, low-protein meal. Urea is excreted in the urine.

In newborn infants the capacity to make urea is limited. The diet should supply enough protein for growth, but not so much that amino acids and ammonia accumulate and produce toxic effects. In adults, a large protein intake with formation and excretion of urea is not harmful unless there is kidney damage, in which case a high-protein diet accelerates the loss of renal function.

V. Utilization of Dietary Proteins

A. Protein Synthesis and Breakdown

In infants, protein synthesis and breakdown occur at rates more than eight times greater than the rate of protein intake, and rapid growth results from a greater synthesis than breakdown. There are periods of enhanced protein synthesis in the development of various organs. The timing of these growth spurts determines which tissues will be most affected by dietary deficiencies. For example, the high rates of decay in the "baby teeth" of poor 6-year-old children may be a consequence of malnutrition during weaning that coincides with the formation of deciduous teeth.

Meals do not greatly affect the rate of whole body protein breakdown, but protein synthesis increases after meals to an extent that depends on the adequacy of dietary protein and energy.

B. Dietary Protein Requirements

There is a daily need for total protein and each of the essential amino acids. There are no stores of amino acids. If the diet provides too little protein, protein of poor quality, or insufficient calories so that amino acids are diverted toward energy production, there is growth faltering in young children and gradual loss of lean tissue in adults (negative nitrogen balance).

How much protein is enough? The adult's need for dietary protein is established as the smallest amount that allows all the nitrogen consumed as protein to equal all the nitrogen excreted, thus allowing nitrogen equilibrium. In healthy young American men fed egg or other high-quality protein, nitrogen equilibrium is achieved with a daily intake of 0.57 g of protein per kilogram of body weight. In developing countries, men consuming mixed animal and cereal proteins typical of the local diets achieve nitrogen equilibrium with about 0.75 g of protein per kilogram of body weight. The Recommended Dietary Allowance of protein in the United States is 0.8 g/kg, an amount that considers the lower digestibility and quality of mixed dietary proteins and the varied needs of a healthy population.

For the young child, satisfactory growth and good health are the criteria for dietary adequacy. The safe level of protein intake is based on the amount of cow's milk formula consumed by infants growing at a maximal rate. However, breast-fed infants consume much less protein, tend to grow more slowly, yet are healthier than formula-fed infants. In unsanitary and poor communities where infectious diseases are common, children who are no longer breast-fed may need more protein to allow rapid growth during periods of convalescence.

C. Effects of the Consumer

Age: The rate of growth, protein synthesis, and protein breakdown are greatest in newborns, coinciding with the highest need for dietary protein per unit of body weight. Protein needs decline with age, as

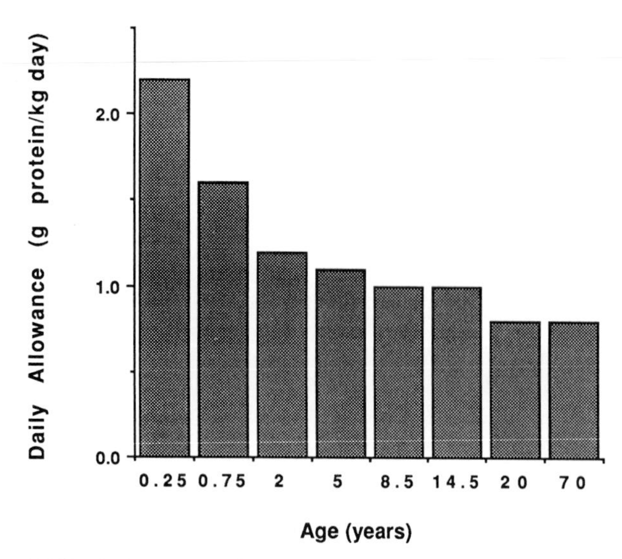

FIGURE 1 Recommended dietary allowance for protein. (Data from the U.S. National Research Council, 1989.)

reflected by the "safe levels of protein intake" in Fig. 1. Elderly men and women seem to need as much protein as young people, despite their lower lean body mass.

Sex: There is no difference in the dietary protein needs of men and women per unit of body weight, although men have greater body protein stores. During pregnancy, a woman is advised to increase protein intake by 6–30 g/day, to sustain the growth of her own tissues and of the fetus. The placenta actively abstracts amino acids from the mother's blood, providing an abundant supply for fetal protein synthesis. During lactation, women are advised to eat an additional 12–16 g of protein per day, by consuming a more protein-rich diet, although in most cultures women do not change the foods they eat during pregnancy or lactation. Animal studies show that dietary protein quantity and quality affect the rate of milk protein synthesis and total volume of milk secreted. Women consuming more than 100 g of protein per day produce more milk than those consuming low or marginally adequate amounts of protein.

Nutritional status: Dietary protein is more efficiently retained after a period of low total food intake, low protein intake, or low intake of an indispensable amino acid. During recovery from undernutrition, children gain weight at more than 20 times the normal rate for their age, as long as they get sufficient energy and protein. This is called *catch-up* growth. In adults, nutritional recovery is shown by positive nitrogen balance.

Physical activity: Activity can range from nearly complete immobility to rigorous physical training. Extremes of complete bed rest or exhaustive exercise each lead to a loss of protein stores, whereas moderate activity improves dietary protein use. Endurance athletes have a higher protein requirement (0.95 g/kg/day), but their large food intake generally supplies enough protein. Beginning a program of moderate exercise while consuming an adequate diet increases nitrogen retention in young or old adults and increases linear growth in young children. A resistance exercise program enlarges muscles by markedly increasing protein synthesis and slightly increasing protein breakdown, but strength-trained men do not appear to need more dietary protein to maintain nitrogen equilibrium.

Disease: Bed rest, infection, surgery, trauma, and even emotional distress lead to a loss of protein stores, shown by negative nitrogen balance and muscle wasting. There is increased protein breakdown, whereas protein synthesis can decline or increase, depending on whether protein intake is low or high. Enhanced breakdown of muscle proteins provides an increased supply of amino acids to the liver, white blood cells, and other tissues needing substrates to fight infection.

D. Effects of Diet

Energy intake: Changes in energy intake affect body weight, but effects on nitrogen retention are even more rapid. When energy intake is lower than energy expenditure, nitrogen retention progressively decreases. Obese adults on weight loss diets lose fat but also lose lean body mass (i.e., protein). Conversely, nitrogen retention increases when more calories are consumed, for reasons that are not clear. In infants, nitrogen retention is poor if either protein intake or energy intake are too low; above 70 kcal/kg/day, the use of protein for growth improves as a function of both energy and protein intake. Additional energy as carbohydrate appears to be more protein-sparing than fat in short experiments. [*See* ENERGY METABOLISM.]

Vitamins are cofactors in many enzymatic reactions involving amino acids. The need for thiamin depends on protein intake because thiamin is a cofactor in amino acid breakdown. Pyridoxine is needed for the synthesis of dispensable amino acids. Folate and vitamin B12 are involved in the metabolism of methionine and cystine, and B12 is also important for the oxidation of valine, methionine, and isoleucine. Minerals such as potassium and zinc are cofactors in the synthesis of nucleic acids and proteins and thus can limit protein utilization for growth.

VI. Protein Undernutrition

The most likely victims of protein undernutrition are poor children younger than 2 years of age, living in unsanitary urban slums or poor rural areas, and, most especially, the victims of war. In developed countries, protein undernutrition is less common or severe. It accompanies socioeconomic deprivation among families, often where there is alcoholism, drug addiction, or psychiatric disease. [*See* MALNUTRITION.]

A. Children

Protein and energy undernutrition worldwide is most serious in infants and young children, as they have a large daily requirement for protein and a rapid growth rate. In 1981, the World Health Organization estimated that malnutrition caused impaired growth in about 300 million children. It is caused by a combination of low supply and increased needs. Poverty and ignorance may lead a mother to feed her child with food that is too dilute, not sanitary, or providing low-quality protein. Food intake may be low because of an impaired appetite or the inappropriate withdrawal of food during infection or diarrhea. Protein needs may be increased because of frequent disease.

Two extremes of protein-energy malnutrition are described: kwashiorkor and marasmus. Kwashiorkor occurs when a young child is weaned from breast milk to a low-protein, high-starch, or high-sugar gruel in an unsanitary environment. The kwashiorkor child is irritable and lethargic, has thin limbs, a pot belly caused by edema, skin lesions reminiscent of pellagra, thin discolored hair, and a fatty liver. Death is common, mainly from severe diarrhea and bacterial infections. Mild to severe marasmus is more common. The child with marasmus is starved of all nutrients, including protein. In response to food deprivation, the child stops growing and uses up its reserves of fat and muscle as fuel to keep alive. The child with marasmus has a "skin and bones" body and the wizened face of an old man.

After treating life-threatening infections and imbalances of fluids and electrolytes, the refed undernourished child can gain weight rapidly, but it is more difficult to attain normal height.

B. Adults

In adults, protein-energy malnutrition is less severe or common. In the United States it can be found in persons with psychiatric problems (anorexia, senile dementia) or in hospitalized patients maintained for long periods on low-protein foods or amino acid–free intravenous feeding solutions. Underlying protein-energy malnutrition is often associated with prolonged convalescence after surgery or illness, hip fracture in the elderly, and the disorders caused by alcoholism and drug addiction.

Bibliography

FAO/WHO/UNU Expert Consultation. (1985). "Energy and Protein Requirements." WHO Technical Report Series 724, WHO, Geneva.

Munro, H. N., and Crim, M. C. (1988). The proteins and amino acids. *In* "Modern Nutrition in Health and Disease" (M. E. Shils and V. R. Young, eds.). Lea & Febiger, New York.

Torun, B., and Viteri, F. E. (1988). Protein-energy malnutrition. *In* "Modern Nutrition in Health and Disease" (M. E. Shils and V. R. Young, eds.). Lea & Febiger, New York.

Waterlow, J. C. (1984). Protein turnover with special reference to man. *Q. J. Exp. Physiol.* **69,** 409–438.

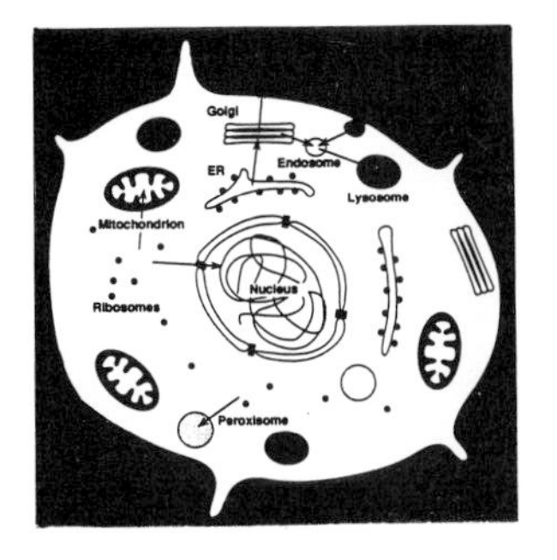

Protein Targeting, Basic Concepts

GUNNAR VON HEIJNE, *Department of Molecular Biology, Karolinska Institute Center for Biotechnology*

Glossary

Endocytosis Mechanism by which particles and molecules from the extracellular medium are internalized by the cell

Endosome Vesicular structure that serves as a relay station for protein transport to lysosomes

Golgi apparatus Vesicular structure that receives proteins from the endoplasmic reticulum and moves them toward lysosomes or the plasma membrane

Lysosome Vesicular structure that contains hydrolytic enzymes responsible for the digestion of macromolecules and other endocytosed materials

Mitochondrion Vesicular structure that contains the enzymes of the citric acid cycle and the respiratory chain where fatty acids and pyruvate are oxidized to carbon dioxide and energy-containing ATP

Peroxisome Vesicular structure that contains enzymes to carry out various oxidative reactions; also contains catalase, an enzyme that converts potentially toxic hydrogen peroxide to water and free oxygen

Plasma membrane Outer membrane of a eukaryotic cell

THE EUKARYOTIC CELL contains a number of distinct membrane-delimited compartments or organelles (e.g., the nucleus, the endoplasmic reticular network, the Golgi apparatus, lysosomes, mitochondria, and peroxisomes). These organelles import most or all of their constituent proteins from the cytosol. This cell can also secrete proteins to the surrounding medium, constitutively as well as in a regulated fashion. In addition to soluble proteins, integral membrane proteins can also be directed to their correct locations, be it the plasma membrane or one of the organellar membranes. These sorting processes depend on specific targeting sequences present in the proteins and on transport machineries in the target membranes.

I. The Sorting Problem

Except for a handful of proteins encoded in mitochondrial DNA and made inside this organelle, all proteins of the human cell are synthesized in the cytosol. Nevertheless, many of these proteins have to be transported to other locations inside or outside the cell to perform their function (Fig. 1). [*See* PROTEINS.]

Histones and other DNA-binding proteins (e.g., transcription factors, DNA and RNA polymerases, and splicing enzymes) must go into the nucleus. Most of the polypeptides involved in respiration must find their way to mitochondria, as must the polymerases and ribosomal proteins required for organelle-specific protein synthesis and DNA replication. [*See* HISTONES AND HISTONE GENES.]

The secretory pathway starts in the endoplasmic reticulum (ER), continues through the Golgi stacks (where a subsidiary route leading to the lysosomes branches off), and ends at the plasma membrane. Peroxisomal proteins are imported directly from the cytosol into the organelle. Proteins and other molecules can also be ingested by the eukaryotic cell through endocytosis, a pathway that leads from the plasma membrane via endosomes to the lysosomes.

Over the past 10–15 years our knowledge of the basic mechanisms responsible for these phenomena

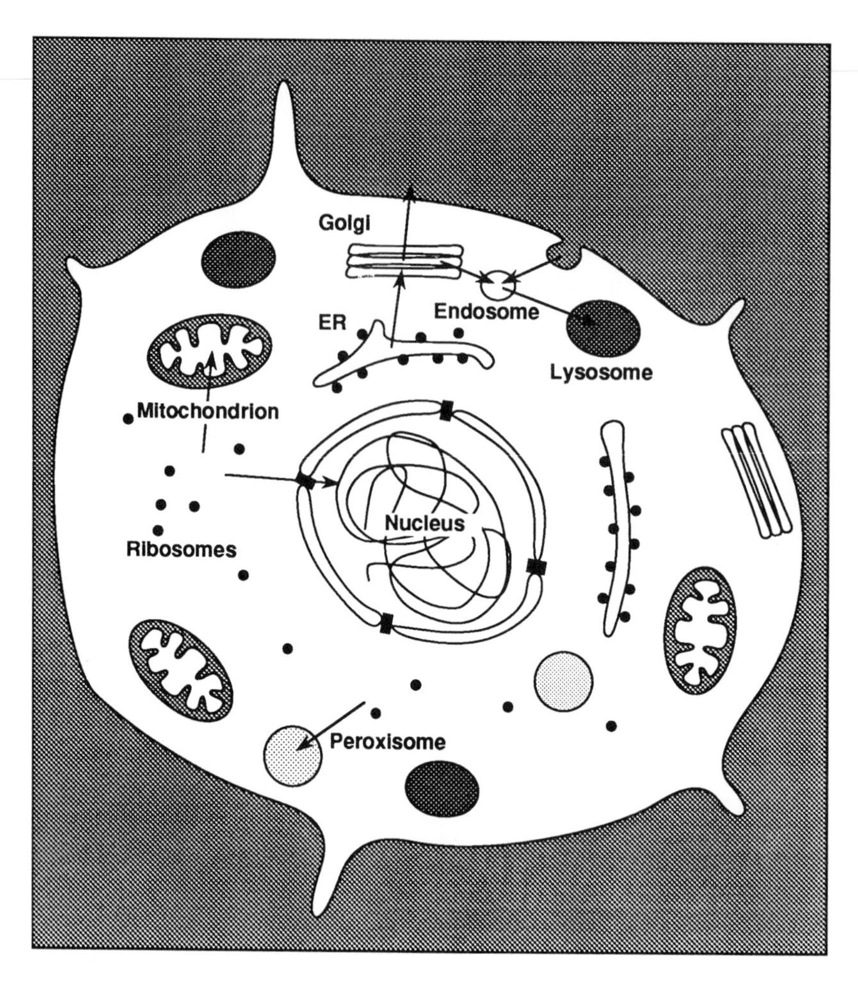

FIGURE 1 Protein-sorting pathways in a eukaryotic cell. ER, Endoplasmic reticulum.

has grown substantially, mainly as a result of rapid developments in recombinant DNA technology, *in vitro* reconstitution of organellar protein import, and genetic and biochemical analyses of the import apparatuses. It has been discovered that a sharply delimited segment(s) of a polypeptide chain can serve as a targeting signal, routing the molecule to the appropriate organelle, where membrane-bound protein complexes recognize the targeting signal and translocate the protein through one or more membranes to its final location. The basic designs of many of the classes of targeting signals are also known.

II. The Secretory Pathway

The secretory pathway is arguably the most important and the most complicated sorting pathway in the eukaryotic cell. During its passage from the ER to the plasma membrane, a protein travels through many organelles with different internal environments and is subject to the actions of a number of modifying enzymes. Among the most common posttranslational modifications are the removal of the targeting, or signal, peptide, the addition and trimming of oligosaccharides, and the addition of lipid moieties to the protein chain.

At the level of the Golgi apparatus, some proteins are shunted from the main pathway to the lysosomes. Of those that continue along the main branch, some are stored in secretory granules that only release their contents in response to an extracellular signal, whereas others are secreted directly into the surrounding medium in a constitutive fashion. [*See* GOLGI APPARATUS.]

The transit from one compartment to the next (i.e., from the ER to the *cis* Golgi, from one Golgi compartment to the next, and from the *trans* Golgi network to the plasma membrane) takes place in small "shuttle" vesicles that bud off from one mem-

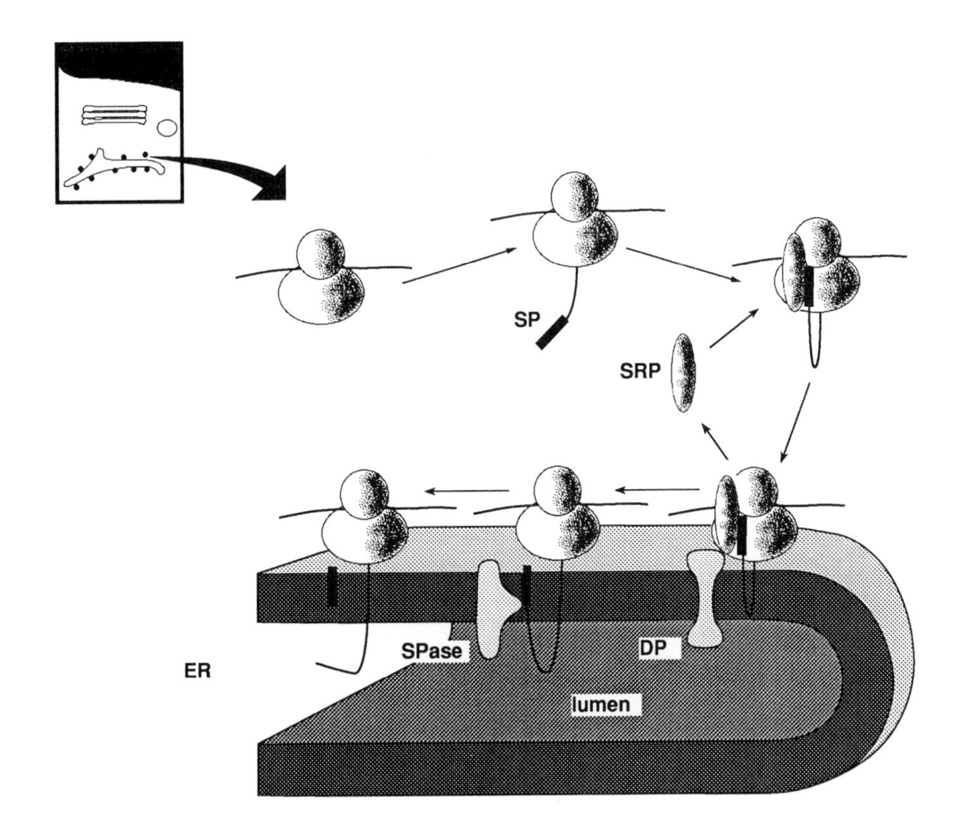

FIGURE 2 The signal recognition particle (SRP) cycle. As the ribosome starts translating the mRNA, an amino-terminal signal peptide (SP) on the nascent chain emerges. The SRP binds to the SP and the ribosome. This complex is recognized by the docking protein (DP) in the endoplasmic reticulum (ER). The SRP recycles, the growing nascent chain is translocated into the lumen of the ER, and the signal peptide is removed by the signal peptidase (SPase).

brane only to fuse with the next one in the series. Cytosolic proteins are apparently involved in both budding and fusion processes, thus directing the shuttle vesicles to their target membranes.

A. Events at the Endoplasmic Reticulum

1. Signal Peptides and the Signal Recognition Particle Cycle

The steps involved in the initial targeting to and translocation across the ER membrane are shown in Fig. 2. The amino-terminal 15–20 residues of a secretory protein contain the targeting signal. As soon as this signal peptide (SP) emerges from the ribosome, it is recognized by the so-called signal recognition particle (SRP), a complex consisting of six protein chains and one RNA molecule. The SRP binds to the ribosome, and the whole ribosome–SRP complex, then binds to the SRP receptor or docking protein, an integral component of the ER membrane. The interaction with the docking protein breaks the ribosome–SRP association. It is not known how the nascent protein chain is translocated across the membrane, but shortly after this process has begun, the signal peptide is removed by the action of signal peptidase, another integral membrane protein. Finally, the completed chain falls off the ribosome and folds into its native structure in the lumen of the ER. [*See* RIBOSOMES.]

Amino acid sequences are known for many signal peptides from different organisms. They lack strong sequence homology, but they all conform to the same basic pattern (Fig. 3): a positively charged amino-terminal region, a central apolar region, and a carboxy-terminal, more polar, region, including the cleavage site.

Gene fusion experiments have shown that the presence of a signal peptide is necessary, but not always sufficient, for translocation into the ER. In addition to having an SP, it seems that the nascent chain must be prevented from folding into a stable tertiary structure prior to translocation, or else the process will be aborted. Recent data indicate that so-called heat-shock proteins found in the cytosol

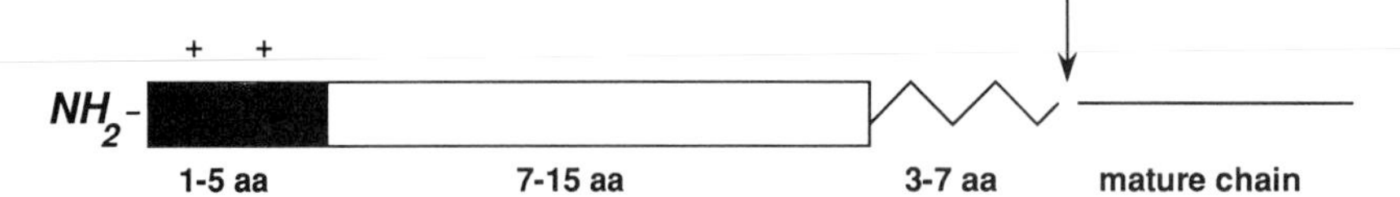

FIGURE 3 A secretory signal peptide. Typical lengths [i.e., the numbers of amino acids (aa)] of the three regions that make up the signal peptide are indicated.

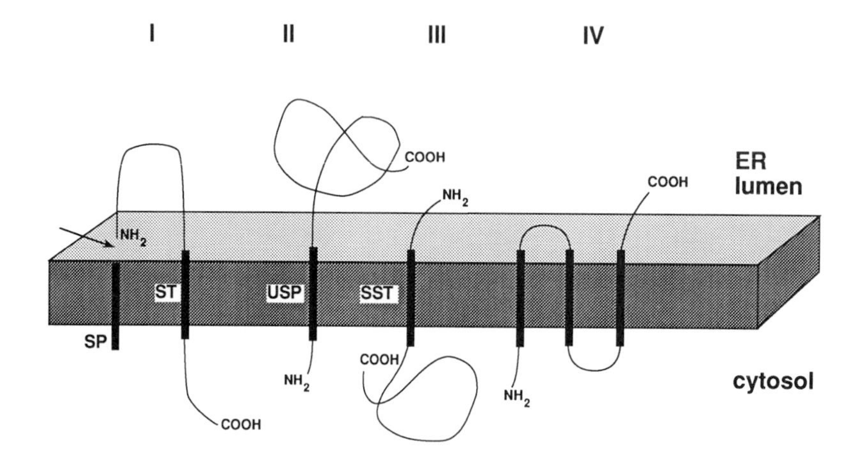

FIGURE 4 Class I membrane proteins are made with a cleavable signal peptide (SP) and an apolar stop–transfer sequence (ST). Class II proteins are anchored by an uncleaved SP (USP). Class III proteins are anchored by an uncleaved amino-terminal start–stop translocation sequence (SST). Class IV proteins have multiple membrane-spanning regions and might be oriented with their amino and carboxy termini on either side of the membrane.

help keep the nascent chain in an unfolded state. Other heat-shock proteins in the lumen of the ER are thought to catalyze proper folding of the translocated protein.

The early steps of glycosylation of the secretory proteins also take place in the ER. Core oligosaccharides are added to acceptor sites in the nascent chain during or immediately after translocation into the luminal compartment.

2. Membrane Protein Assembly

Proteins that span a membrane at least once are called integral membrane proteins. Most integral membrane proteins use the secretory machinery described above in the initial phases of the membrane insertion process, but additional elements in their sequences serve as stop–transfer signals that prevent the more distal parts of the chain from being translocated. This results in a transmembrane topology.

The four classes of integral membrane proteins are shown in Fig. 4. Class I proteins are made with a standard cleavable SP and have a stretch of 15–25 apolar amino acids that constitutes the stop–transfer signal (ST) and ends up spanning the ER membrane, with its carboxy-terminal end facing the cytosol. Class II proteins can be thought of as originating from secreted proteins, in which the cleavage of the SP has been blocked. The resulting uncleaved SP thus anchors a class II protein in the membrane, leaving the amino terminus facing the cytosol.

Class III proteins also lack a cleavable SP. They have a long apolar region at their amino terminus, but they do not have the typical cluster of positive charges found on the amino terminus of normal signal peptides. This allows them to insert into the membrane in the orientation opposite that of the class II proteins.

Class IV proteins, finally, have multiple apolar regions and span the membrane many times. Many transport proteins and ion pumps belong to this group.

B. Golgi Apparatus

The Golgi stacks serve a number of important functions in the secretory pathway: They contain enzymes involved in the trimming and modification of the oligosaccharide chains attached to secretory and membrane proteins in the ER, they sort apical from basolateral proteins (see Section II,E), and they shunt lysosomal proteins away from the main pathway.

Lysosomal targeting is signaled by a specific oligosaccharide modification: mannose 6-phosphate. A receptor with affinity for this moiety is located in the Golgi apparatus and shuttles appropriately modified proteins to endosomes and, ultimately, to lysosomes.

C. Constitutive versus Regulated Secretion

Many cells store some of their secretory proteins in dense secretory granules. When stimulated by the appropriate signal, these granules fuse with the plasma membrane and release their content into the medium. Other proteins are secreted directly from the *trans* Golgi network in an unregulated constitutive fashion. The signals for these two alternative pathways of secretion are not known.

D. Endocytosis

Endocytosis, the uptake of particles, nutrients, and other molecules from the surrounding medium, is intimately coupled with the secretory pathway. Solutes can be internalized nonspecifically through bulk phase uptake, but many peptides and other compounds are first bound to specific receptors on the cell surface. These receptors are transmembrane proteins with both extracellular and intracellular domains. After binding the ligand they cluster in invaginations of the plasma membrane, called coated pits. The coated pits pinch off from the membrane to form coated vesicles, which in turn can fuse with endosomal vesicles. In the low pH milieu of the endosome the ligand dissociates from the receptor, and the latter in many cases recycles to the plasma membrane.

Many enveloped viruses also enter the cell via this route. In the acidic environment of the endosome, proteins in the virus envelope undergo a conformational change that exposes fusinogenic regions, and the viral membrane fuses with the endosomal membrane, expelling the viral genome into the cytosol.

E. Sorting in Polarized Cells

Many cells are polarized in the sense of having two sides, with distinct cell surface properties. Thus, epithelial cells have an apical surface facing the external milieu and a basolateral surface facing the internal milieu of the organism. Many soluble proteins are known to be secreted preferentially from one side or the other, and many integral membrane proteins also display preferential sorting. It is not known what the signals for this kind of intracellular sorting are, but the two pathways are thought to diverge in the *trans* Golgi network or shortly after passage through the Golgi compartment. [*See* Polarized Epithelial Cell.]

III. Import into the Nucleus

The nucleus is surrounded by a double membrane studded with nuclear pores that allow the diffusion of small molecules of less than about 60,000 Da. Thus, small proteins can enter and exit the nucleus by passive diffusion, and their distribution between the nucleus and the cytosol depends on whether they bind to other molecules specifically localized to either compartment.

Larger proteins can only be imported into the nucleus by active transport through the nuclear pores. The details of this mechanism have not been elucidated, but it has been shown for a number of nuclear proteins that the signals for nuclear localization typically consist of short stretches of positively charged amino acids. Unlike most other targeting signals, the nuclear localization signals are not cleaved after import, which could be related to the fact that nuclear proteins have to reenter the newly formed nuclei after mitosis. Nuclear localization signals also seem to work cooperatively, such that multiple signals enhance the import efficiency when present in the same protein. [*See* Nuclear Pore, Structure and Function.]

IV. Import into Mitochondria

Mitochondria are surrounded by two membranes: outer and inner. Proteins are imported into the matrix compartment through contact sites, where the two membranes are in close apposition (Fig. 5). Many precursor proteins are made as full-length molecules in the cytosol and are imported posttranslationally; others may be imported cotranslationally.

A number of distinct steps on the import pathway have been identified. First, the precursor binds to a receptor on the mitochondrial surface. The amino-

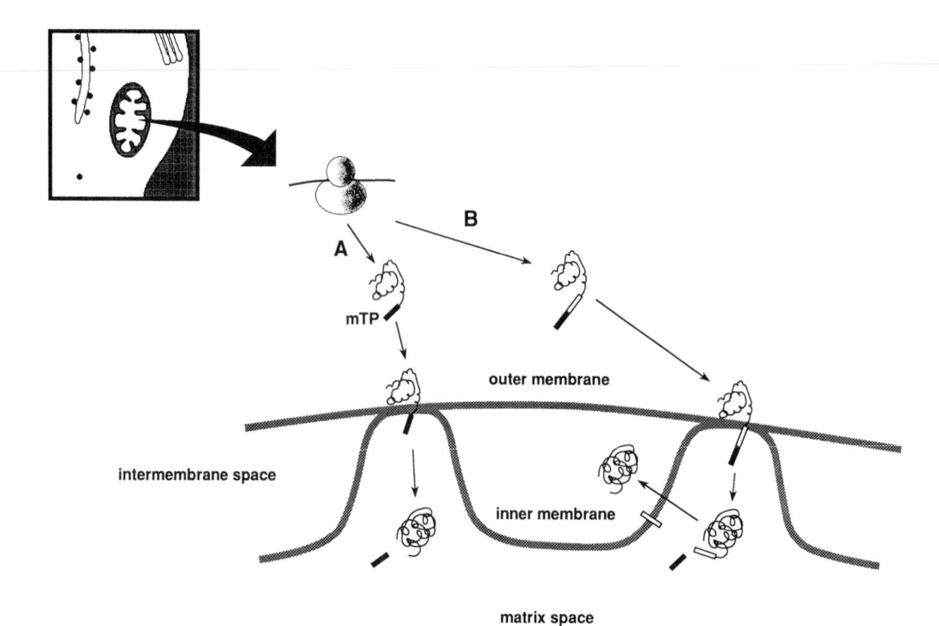

FIGURE 5 Mitochondrial protein import. (A) The unfolded precursor protein, with its mitochondrial targeting peptide (mTP), binds to the outer membrane of the mitochondrion and is imported through a contact site, where the outer and inner mitochondrial membranes come close together. (B) Proteins destined for the intermembrane space are made with a composite targeting peptide and are first imported into the matrix, only to be reexported through the inner membrane.

terminal part of the precursor, including the targeting peptide, is then translocated across a contact site in a process that requires the chain to be in a more or less unfolded conformation. This initial translocation step also requires an electrical transmembrane potential across the inner membrane. When the amino terminus has penetrated into the matrix space, the remaining parts of the molecule can enter the organelle, even in the absence of a membrane potential. Finally, the targeting peptide is removed by soluble proteases located in the matrix.

Cytosolic heat-shock proteins are thought to be important for maintaining the precursor in an unfolded state during the translocation process.

Mitochondrial targeting peptides are rich in positively charged residues, arginine in particular, and lack negatively charged acidic residues. They can fold into so-called amphiphilic α-helices (Fig. 6), with one charged and one apolar face. Such structures bind tightly to the surface of lipid bilayers, and it is conceivable that a direct protein–lipid interaction of this kind is an important step on the import pathway.

Proteins targeted to the intermembrane space are believed first to be imported into the matrix of the mitochondrion and then to be reexported across the inner membrane. In this case the targeting signal is a composite structure with an amino-terminal matrix-targeting part, followed by a stretch with many of the properties normally found in secretory SPs. The matrix-targeting signal is cleaved upon entry into the matrix, exposing the SP-like structure that guides the protein back through the inner membrane.

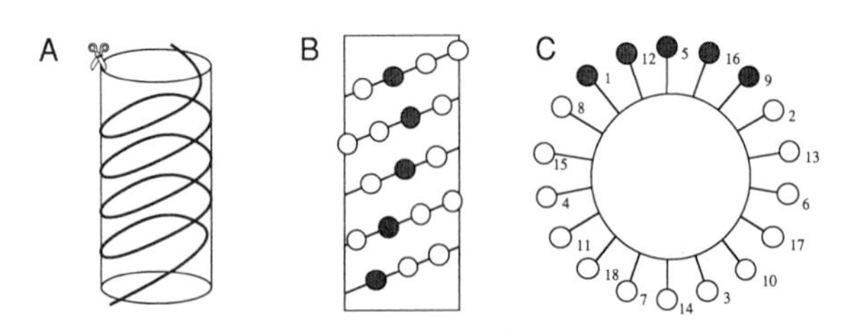

FIGURE 6 A typical mitochondrial matrix-targeting peptide, the amphiphilic helical structure of which is shown in three representations. (A) The backbone of an idealized α helix. (B) A helical net diagram in which the cylindrical helix has been cut open and folded flat. The positions of the side chains are indicated. (C) A helical wheel diagram looking down the axis of the helix. Numbering refers to the position in the linear sequence. Note that the positively charged residues (shaded areas) line up on one face of the helix in B and in C.

V. Import into Peroxisomes

Peroxisomes are small organelles that contain enzymes involved in various oxidative reactions. Protein import into peroxisomes is posttranslational and requires ATP. Gene fusion studies suggest that both internal and carboxy-terminal regions in the nascent protein might be active in the import process, and a highly conserved carboxy-terminal targeting signal with the sequence Ser–Lys–Leu–COOH is found in many, but not all, of the peroxisomal proteins.

VI. Evolution of Sorting Pathways

In bacteria, SPs direct proteins into or across the inner membrane. The SPs, as well as the mechanisms of targeting and translocation, are similar to what is found in the secretory pathway of eukaryotic cells. Indeed, bacterial proteins can often be secreted when made in eukaryotic cells and vice versa. Thus, it seems that protein secretion is a highly conserved process in evolution.

Sorting to the intermembrane space of mitochondria provides a nice example of this evolutionary conservation. Mitochondria, being descendants of bacteria, seem to have retained the capacity to translocate proteins from the matrix across their inner membrane into the intermembrane space in response to a SP-like structure. Since most of the genes originally present in the genome of the protomitochondrion have been transferred to the nuclear genome during evolution, the specific import pathway (i.e., cytosol to matrix) must have been added at an early stage. Intermembrane space proteins that presumably were first made inside the mitochondrion with a secretory SP to ensure correct targeting are now made in the cytosol with a matrix-targeting signal, followed by a secretion signal. Once inside the organelle the ancient secretory pathway is engaged, and the protein is routed to its final location.

The early eukaryotic cell must also have faced the general problem of mistargeting of its proteins. This could have been avoided in at least two ways: by increasing the specificity of targeting and by preferential proteolytic degradation of missorted proteins. So far, specific targeting, rather than specific degradation, seems to be the rule, although quantitative data are lacking.

The problem of mistargeting is intimately connected with the origin of the targeting signals. Targeting to both the secretory pathway and mitochondria involves sorting signals based on structures with one positively charged and one apolar part, and both kinds of peptides are known to interact strongly, though in different ways, with lipid bilayers. Thus, we might look for the origins of protein sorting in the simple physical chemistry of protein–lipid interactions.

Bibliography

Alberts, B., Bray, D., Lewis, J., Raff, M., Roberts, K., and Watson, J. D. (1983). "Molecular Biology of the Cell." Garland, New York.

Attardi, G., and Schatz, G. (1988). Biogenesis of mitochondria. *Annu. Rev. Cell Biol.* **4,** 289–333.

Borst, P. (1986). How proteins get into microbodies (peroxisomes, glyoxysomes, glycosomes). *Biochim. Biophys. Acta* **866,** 179–203.

Dingwall, C., and Laskey, R. A. (1986). Protein import into the cell nucleus. *Annu. Rev. Cell Biol.* **2,** 367–390.

Eilers, M., and Schatz, G. (1988). Protein unfolding and the energetics of protein translocation across biological membranes. *Cell (Cambridge, Mass.)* **52,** 481–483.

Goldstein, J. L., Brown, M. S., Anderson, R. G. W., Russell, D. W., and Schneider, W. (1985). Receptor-mediated endocytosis. *Annu. Rev. Cell Biol.* **1,** 1–39.

Griffiths, G., Hoflack, B., Simons, K., Mellman, I., and Kornfeld, S. (1988). The mannose 6-phosphate receptor and the biogenesis of lysosomes. *Cell (Cambridge, Mass.)* **52,** 329–341.

Hubbard, S. C., and Ivatt, R. J. (1981). Synthesis and processing of asparagine-linked oligosaccharides. *Annu. Rev. Biochem.* **50,** 555–583.

Rose, J. K., and Doms, R. W. (1988). Regulation of protein export from the endoplasmic reticulum. *Annu. Rev. Cell Biol.* **4,** 257–288.

Simons, K., and Fuller, S. D. (1985). Cell surface polarity in epithelia. *Annu. Rev. Cell Biol.* **1,** 243–288.

von Heijne, G. (1988). Transcending the impenetrable: How proteins come to terms with membranes. *Biochim. Biophys. Acta* **947,** 307–333.

Walter, P., and Lingappa, V. R. (1986). Mechanism of protein translocation across the endoplasmic reticulum membrane. *Annu. Rev. Cell Biol.* **2,** 499–516.

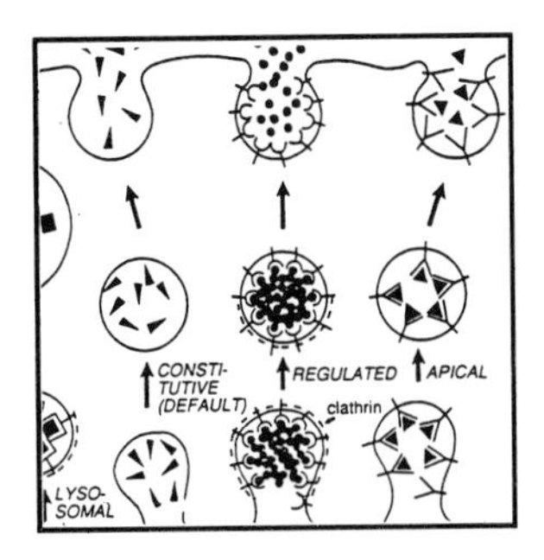

Protein Targeting, Molecular Mechanisms

ANTHONY P. PUGSLEY, *Institut Pasteur*

Glossary

Clathrin Heterooligomeric protein complex which forms a blanketlike cage around some vesicles budding from the *trans* Golgi network and around coated pits, which form endocytic vesicles at the plasma membrane

G proteins GTP-binding proteins which play central, but poorly understood, roles in signal transduction and in various aspects of protein transport

Heat-shock proteins Proteins whose synthesis is increased by a transitory shift to growth at high temperatures; many such proteins are present at relatively high basal levels and prevent proteins from prematurely folding or aggregating

Secretory pathway Pathway leading from the endoplasmic reticulum to the plasma membrane, or to lysosomes via the Golgi cisternae and the *trans* Golgi network

***Trans* Golgi network** Vesicular section of the Golgi apparatus located farthest from the endoplasmic reticulum and the site at which the different branches of the secretory pathway diverge

THE TARGETING of cytoplasmically synthesized noncytosolic proteins is initiated by a series of receptor interactions which direct these proteins to a target organelle [e.g., the endoplasmic reticulum (ER), mitochondria, peroxisomes, or the nucleus]. Translocation of the polypeptide across the proteolipid membranes of these organelles requires complex translocation machinery, which is different in each case, and in many cases the unfolding and refolding of the polypeptide prior to and after translocation. Proteins crossing the ER membrane might be retained in the ER, or transported via the Golgi apparatus and the *trans* Golgi network (TGN) to endosomes, lysosomes, secretory storage granules, the plasma membrane, and the external milieu. The transport of secretory proteins through the secretory pathway is mediated by vesicles that bud from donor organelles and fuse with recipient organelles or the plasma membrane.

I. General Principles

One of the most impressive features of cell architecture is the specific manner in which proteins are distributed. Proteins are rarely, if ever, found where they are not required; mitochondrial proteins, for example, are never found in the cytoplasm and are never secreted. One way to achieve specific localization would be for proteins to be randomly distributed throughout the cell, incorrectly localized proteins being degraded by "housekeeping" proteases. There is, however, no evidence for such a high level of protein turnover, and such a mechanism would be an extraordinary drain on energy resources. Thus, the distribution of proteins is more likely to depend on efficient and specific localization of proteins to their required sites (i.e., protein targeting). In human cells there are four protein targeting pathways: secretory, mitochondrial, peroxisomal, and nuclear. [*See* PROTEINS.]

In broad terms protein targeting can be divided into four stages: (1) Routing and (2) translocation are the stages at which targeted proteins are distinguished from proteins that, by default, are to remain in the cytoplasm. (3) Sorting and (4) transport occur after the protein has crossed the membrane of the target organelle, directing proteins to specific sites along the pathway and ensuring that they remain there. Each pathway has a terminus to which tar-

Encyclopedia of Human Biology, Volume 6.

geted proteins will be carried if they are not subject to sorting and retention. Individual steps in the pathways are energetically driven and depend on a complicated series of signal–receptor interactions, which ensure specificity and direction.

Many aspects of protein targeting have been extensively characterized in cells other than those of human origin. However, most targeting pathways seem to be more or less universal, at least among eukaryotic organisms, and it is therefore reasonable to extrapolate to human cells from studies with nonhuman cells and, in particular, cells from other mammals. Another important aspect of the universality of protein targeting pathways is that all proteins destined to a particular site almost invariably use a single common or "public," pathway. [*See* PROTEIN TARGETING, BASIC CONCEPTS.]

II. Routing and Translocation

The decision as to where a protein is to be routed is made each time a polypeptide is released from the ribosome. Only in rare cases do noncytosolic polypeptides remain in the cytosol for more than a short period, and, indeed, the routing of proteins into the secretory pathway almost invariably occurs during polypeptide elongation, shortly after the nascent polypeptide emerges from the ribosome [i.e., translation and routing (translocation) are tightly coupled]. This might also be the case for nuclear-encoded mitochondrial proteins. Such highly efficient protein routing is made possible by the fact that the sequence of individual targeted polypeptides including a segment (the routing signal) containing the information required for specific routing into one of the four target organelles (Table I). The only exceptions appear to be a small number of nuclear proteins which do not have routing signals. These proteins are routed to the nucleus by associating with other proteins (carriers) which have routing signals. These carriers could even be recycled, thus acting as shuttles between the nucleus and the cytoplasm. [*See* RIBOSOMES.]

A. Routing Signals

Comparisons of the large numbers of secretory, mitochondrial, and nuclear routing signals on nuclear-encoded proteins which have been characterized have led to some general conclusions regarding their structure and function. Two types of secretory routing signals (i.e., signal peptides and signal sequences) have been identified (Table I). Both comprise a core of 15–30 hydrophobic or neutral amino acids which is uninterrupted by polar or charged residues and which, according to structure predictions and as determined by physical measurements in apolar environments, forms an α-helix. Signal peptides may be distinguished from signal sequences by their closer proximity to the amino terminus of the polypeptide and by the frequent presence of positively charged arginines or lysines among the two to six residues that precede the core region (Table I). Most importantly, signal peptides are proteolytically processed (by signal peptidase) at a relatively well-conserved site which is itself separated from the core region by approximately three residues, with a strong tendency to form a β turn. Some authors, however, consider the terms "signal sequence" and "signal peptide" to be interchangeable.

The low level of primary sequence conservation observed among several hundred characterized secretory routing signals almost certainly means that they are recognized as structural motifs, rather than as specific sequences. Mitochondrial routing signals also exhibit a low degree of sequence conservation and, like secretory routing signals, are usually located at the amino terminus and are often proteolytically processed [by prepeptidase(s)]. However, mitochondrial routing signals have positively charged amino acids (usually arginines) evenly distributed throughout their length. If this segment of the polypeptide is folded as a helix, as suggested by structure predictions and by some physical measurements, then the positively charged residues would be aligned along one face of the helix. Sequences similar to mitochondrial or secretory routing signals are not found at the amino termini of cytoplasmic, nuclear, and peroxisomal proteins.

Nuclear and peroxisomal routing signals are not necessarily amino terminal; indeed, there is evidence that these signals are sometimes located at the extreme carboxy terminus or within the central part of the polypeptide. There are no known examples of processed nuclear or peroxisomal routing signals. Nuclear routing signals (karyophilic signals) are usually very short and contain several lysine residues (Table I); their sequences are not highly conserved, but they are obviously quite different from other routing signals. Some nuclear-targeted proteins may have more than one routing signal. This appears to increase the efficiency of routing or uptake into the nuclei. The sequence Ser–Lys–Leu (or closely related sequences) lo-

TABLE I Routing Signals in Noncytoplasmic Proteins Synthesized in the Cytoplasm of Human Cells[a]

Pathway	Signal	Length (residues)	Location in polypeptide	Structural/sequence motif	Processed	Receptors
Secretory	Signal peptide	15–30	Extreme amino terminus	Basic amino terminus, 15–20 residue α helix, hydrophobic core	+	SRP and ER membrane receptor
	Signal sequence	12–20	Usually amino terminal	Hydrophobic, α helix	–	SRP and ER membrane receptor[b]
Mitochondrial	Prepeptide	20–40	Extreme amino terminus	Basic amphipathic helix; serine, threonine, and arginine rich	+	Mitochondrial outer membrane[c]
	Outer membrane routing signal	20–40(?)	Extreme amino terminus	Similar to pre-peptide with lysine in place of arginine	–	Mitochondrial outer membrane[c]
	Others	—[d]	Amino terminal or internal	Not characterized	–	Mitochondrial outer membrane[c]
Nuclear	Karyophilic signal	5–17	Amino or car-carboxy terminal or internal	Lysine rich; often includes proline residue	–	Nuclear pore and possibly cytoplasmic[e]
Peroxisomal	Peroxisomal routing signal	≥20	Internal or carboxy terminal	Often carboxy-terminal Ser-Lys-Leu	–	Peroxisomal surface(?)[f]

[a] SRP, Signal recognition particle; ER, endoplasmic reticulum.
[b] Signal peptides and signal sequences are recognized by the same receptors.
[c] Evidence from experiments with mitochondria from fungi suggest that there are at least three classes of mitochondrial routing signal receptors corresponding to each of the three classes of mitochondrial routing signal.
[d] Not determined; few of these signals have been characterized.
[e] Competition experiments suggest that there may be two different receptors in the nuclear pore, and other experiments suggest the existence of a cytosolic receptor.
[f] Not characterized.

cated at the carboxy terminus of some peroxisomal enzymes functions as a routing signal for this class of protein.

Two lines of evidence show that routing signals are the major, if not only, determinants ensuring that polypeptides destined for an organelle do not remain in the cytoplasm. First, mutations that alter the structural motif of the signal cause such polypeptides to remain in the cytosol; second, normally, cytoplasmic proteins to which routing signals are coupled (either by genetic engineering or by chemical crosslinking) are routed to the organelle specified by the routing signal.

B. Routing Signal Receptors

Routing signals must be recognized by receptors, presumably located on the outer face of the target organelle. Indeed, receptor proteins for such signals have been detected on the surface of the ER and the mitochondria, and there is some evidence for receptors in nuclear pores and on the surfaces of peroxisomes.

The routing signals of the vast majority of secretory proteins are recognized by a cytosolic ribonuclear protein complex, the signal recognition particle (SRP), which is composed of six different proteins assembled on a 7 S RNA backbone. One protein in the complex recognizes the routing signal while another protein recognizes the heterodimeric SRP receptor located in the membrane of the ER. The SRP acts as a carrier which binds the routing signal as it emerges from the ribosome, thereby connecting it to the ER; it also reduces the rate of translation until it binds to its receptor in the ER. After binding the SRP is released from the routing signal (which can then bind to the second receptor on the ER membrane), and translation resumes. At the same time the ribosome binds to its own receptor, which, although unidentified, must also be located on the surface of the ER (Fig. 1).

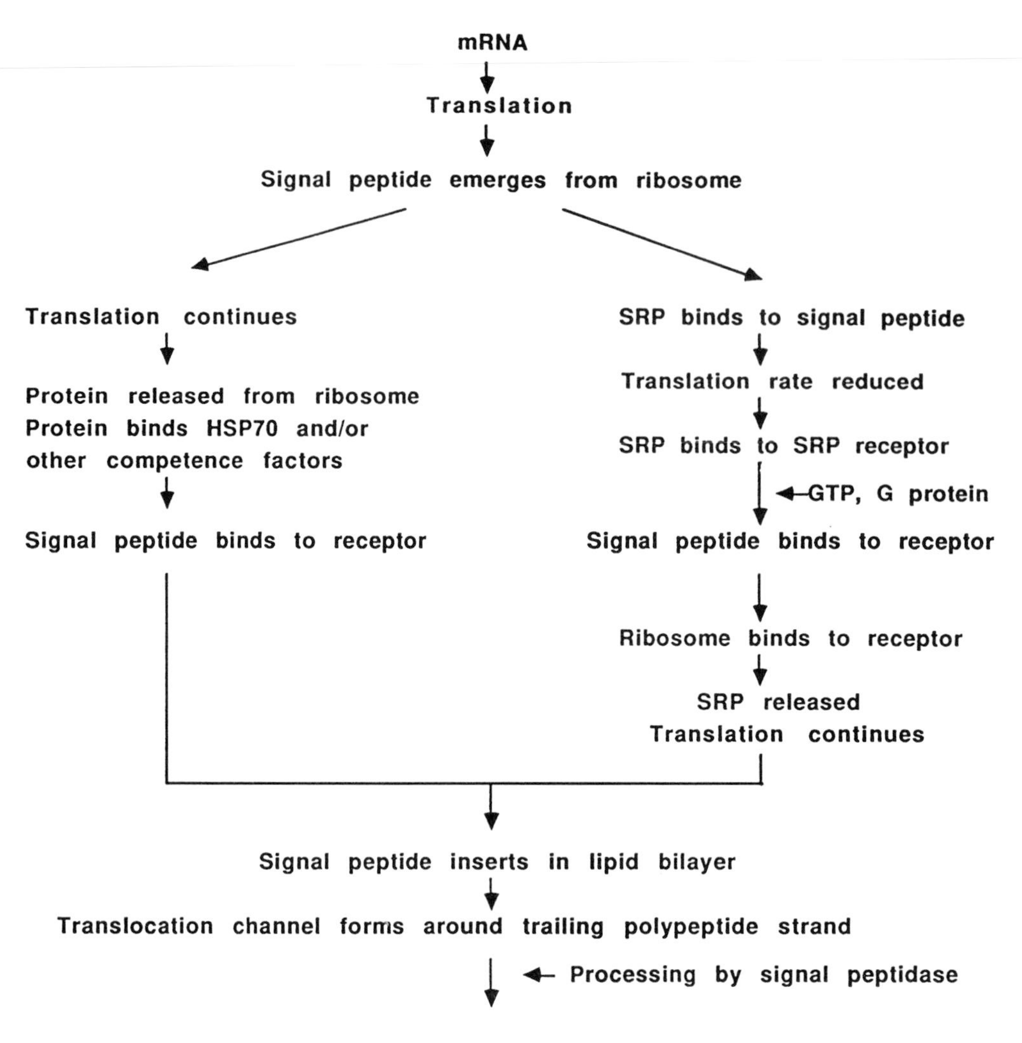

FIGURE 1 Early stages in the translocation of a secretory protein across the membrane of the endoplasmic reticulum. SRP, Signal recognition particle; HSP70, 70-kDa heat-shock protein.

This two-stage receptor system for routing secretory proteins to the ER is responsible for the tight translation–translocation coupling mentioned above. This feature of secretory protein routing could be vitally important for the next stage in protein targeting (i.e., protein translocation across the membrane), because it reduces the possibility that significant lengths of the polypeptide are made before they are translocated (see Section II,F). The presence of mRNA and associated ribosomes synthesizing secretory proteins on the surface of the ER imparts the characteristically "rough" appearance of this organelle.

It is conceivable that a double-receptor system could be involved in the routing of mitochondrial, nuclear, and peroxisomal proteins. Indeed, there is evidence for cytosolic receptors for imported mitochondrial and nuclear routing signals, in addition to receptors known or thought to be present on the target organelles. Mitochondrial protein routing could occur cotranslationally, since the routing signal is the first part of the polypeptide to emerge from the ribosome, especially if a two-stage receptor system operates in this case as well. Cotranslational routing of peroxisomal and nuclear proteins has not been reported, which is not surprising, in view of the fact that the routing signal is often in the central or carboxy-terminal part of the polypeptide.

C. Insertion of the Routing Signal

Translocation is the process by which proteins cross the membrane(s) of the target organelle. Except for nuclear proteins, which cross the nuclear envelope through preexisting nonspecific pores, translocation can be divided into two stages: the insertion of the routing signal into the membrane, followed by the linear translocation of the rest of the polypeptide. [*See* CELL MEMBRANE TRANSPORT.]

Protein translocation is initiated shortly after the routing signal binds to its cognate receptor, but it is not clear whether the routing signal must be immediately released from the receptor or, indeed, whether the insertion site is formed from lipids or proteins. The routing signal receptor could, in fact, form part of the insertion site. However, in view of the observed affinity of secretory and mitochondrial routing signals for lipids and, in particular, the uniformly hydrophobic character of the secretory routing signal core, it seems more probable that initial contact between the routing signal and the surface receptor is followed by insertion of the routing signal into the lipid bilayer of the membrane.

Irrespective of which of these two alternatives is correct, it is clear that the two types of routing signal (i.e., mitochondrial or secretory) insert in different ways. The major class of mitochondrial routing signals, the prepeptides, is proposed to insert "head (amino terminus)-first" into the mitochondrial envelope at sites where the two membranes are in close juxtaposition. The entire prepeptide can be exposed in the mitochondrial matrix, where it is processed by prepeptidase, under conditions that prevent the rest of the polypeptide from crossing the envelope (see Section III,C).

In contrast, studies with secretory proteins in which the extreme nonhydrophobic amino terminus of the signal peptide has been artificially extended by the addition of readily detectable foreign sequences show that the amino terminus remains exposed on the cytosolic side of the ER membrane, indicating that signal peptides penetrate into the target membrane in a "tail-first," or loop, configuration. The different insertion modes of the mitochondrial and secretory routing signals undoubtedly stem not only from their different structures (Section II,A), but also from differences in receptor interaction and the nature of the insertion sites (lipids in bilayer or nonbilayer conformation, or membrane proteins).

It is not known whether peroxisomal routing signals insert into the peroxisomal membrane; their role might, indeed, be restricted to routing signal receptor recognition. Further work is needed to determine how peroxisomal proteins are imported. Studies of certain patients with peroxisomal enzyme deficiencies (Zeeweger syndrome) suggest that soluble peroxisomal enzymes and membrane proteins could be targeted into the peroxisome by different routes.

Little is known about how nuclear envelope proteins are inserted into the nuclear membrane, although since the nuclear and ER membranes are contiguous and have similar, though not identical, protein compositions, nuclear membrane proteins could be routed first to the ER membrane and then sorted into the nuclear envelope.

D. Protein Translocation

Insertion of the routing signal into the target membrane is followed by the translocation of all or part of the remainder of the polypeptide across the same membrane. The site at which translocation occurs has not been precisely determined, although the two sites are likely to be adjacent, if not identical. There is evidence that mitochondrial proteins are imported at sites where the two membranes are fused, possibly by a translocation channel which spans both membranes. Since imported proteins are generally hydrophilic, it is assumed that the translocation channel is probably water filled, so that the polypeptide does not come into contact with the membrane lipids. This would avoid the excessively high energy expenditure required to transport charged amino acids through the apolar lipid environment. Similar considerations apply to the translocation of secretory proteins across the ER membrane and also to the import of proteins into peroxisomes. It is generally assumed that the walls of the translocation channels are formed from one or more transmembrane proteins, but lipid head groups, which are hydrophilic, could form the channel. Translocation channels are presumably closed when not actively involved in protein translocation, in order to prevent leakage from the lumen, or matrix, of the organelle.

Proteins are translocated linearly (i.e., completely unfolded) across the membrane of the ER. Proteins devoid of segments of high overall hydrophobicity (other than the routing signal) will be extruded into the lumen of the ER. A hydrophobic segment located downstream from the routing signal can act as a stop–transfer/membrane anchor signal, preventing further translocation and causing the dissociation of the translocation channel to allow the hydrophobic segment to become embedded in the membrane in a *trans* bilayer configuration. A third hydrophobic segment located further downstream from the stop–transfer signal could act as a secondary routing signal which will insert into and remain embedded in the membrane, initially as a loop structure, leading to the complex membrane topology of many of the integral membrane proteins of the ER, Golgi, lysosomal, and plasma membranes. It is important to note that the topology of the membrane proteins is established during insertion into and translocation across the membrane of the ER and is not substantially altered during transport to the final destination.

Mitochondrial proteins are also imported linearly, but, unlike secretory proteins, their translocation is not usually arrested by long hydrophobic segments. In mitochondria an additional complexity is represented by the presence of two membranes: inner and outer. Most nuclear-encoded proteins of the mitochondrial inner membrane and of the space between the two membranes probably transit via an intermediate that reaches the matrix, the space enclosed by the inner membrane. They could contain hydrophobic segments which eventually cause them to be reexported (i.e., sorted) to their final destination and/or to become anchored in the inner membrane (see Section III,C).

Mitochondrial outer membrane proteins seem to be the only exceptions to this rule (other than cytochrome *c*, which uses a private pathway to reach the intermembrane space). Although the routing signal of mitochondrial outer membrane proteins might channel them into the same insertion site as other imported proteins, they do not reach the matrix, but instead dissociate from the insertion site/translocation channel and fold into the outer membrane.

Nuclear proteins are transported into the nucleus as fully folded structures through nuclear pores. The diameter of the pore (4.5 nm) is sufficient to allow small molecules, including proteins with a low molecular radius, to diffuse into and out of the nucleus relatively freely. The role of the nuclear routing signal (the karyophilic signal) seems to be to trigger the further opening of the pore to permit the entry of larger molecules. [*See* NUCLEAR PORE, STRUCTURE AND FUNCTION.]

E. Proteolytic Processing of Routing Signals

The signal peptide and prepeptide classes of routing signals are specifically processed by proteases (e.g., signal peptidase and prepeptidase) located in the ER membrane and in the mitochondrial matrix, respectively.

Signal peptidase acts shortly after the looped tail of the signal peptide is exposed on the luminal side of the ER membrane. If the processed protein contains no stop–transfer signal (see Section II,D), it will be released into the lumen of the ER. Signal sequences, which are not processed, remain anchored in the ER membrane, with the carboxy-terminal end of the polypeptide exposed in the lumen of the ER. Secondary secretory routing signals are rarely processed by signal peptidase. Signal peptidase is composed of at least two different subunits and could be part of a larger protein complex involved in routing signal recognition, insertion, and translocation.

Two distinct prepeptidases have been identified in the mitochondrial matrix. Single-step processing by one enzyme or two-step processing at two different sites by two enzymes have been reported, depending on the precursor protein and the origin of the mitochondria. The better characterized of the two prepeptidases is a zinc-dependent heterodimer. One subunit seems to have catalytic activity, while the other could be involved in protein translocation and presentation of the processing site to the catalytic subunit. Processing is not required for complete import into the matrix, because some matrix proteins with prepeptidelike routing signals are not processed, and zinc depletion blocks prepeptidase action without preventing import. However, complete inactivation of the zinc-dependent prepeptidase by gene disruption causes unprocessed precursors of mitochondrial proteins to accumulate outside the mitochondrion, implying that prepeptidase (or, specifically the noncatalytic subunit) could be involved in protein translocation.

F. Additional Requirements for Routing and Translocation

Efficient routing and translocation of noncytoplasmic proteins require a number of factors in addition to the soluble and membrane-associated routing signal receptors and the translocation apparatus described above. *In vitro* assays show that these factors include soluble cytosolic proteins and nucleoside triphosphates.

A point of great interest is the conformation of secretory and mitochondrial proteins before they are translocated. The routing signal must remain exposed so that it can bind to its receptor(s), and folding of the peptide chain must be severely limited to allow translocation through the translocation channel (i.e., translocation competence). SRP is crucial in this respect, because it binds to the secretory routing signal as it emerges from the ribosome and slows translation to limit the length of the chain and consequently folding. Even when translation is almost completed, the presence of the ribosome at the carboxy terminus of the nascent secretory polypeptide seems to restrict secondary structure formation and to facilitate translocation.

A small group of secretory proteins less than 70 residues in length does not require SRP for *in vitro* translocation into ER vesicles. In these cases translocation requires at least two cytosolic proteins, one of which is HSP70 (a 70-kDa member of the so-called heat-shock family of proteins), whose level of production is increased when the cells are exposed to abnormally high temperatures. HSP70 presumably binds to these secretory proteins (possibly to regions of relatively high hydrophobicity) and must be released prior to translocation at the surface of the ER. ATP hydrolysis is required for HSP70 dissociation and perhaps also to drive translocation. It is not known whether HSP70 is required to maintain the translocation competence of secretory proteins which also require SRP (Fig. 1). [*See* ADENOSINE TRIPHOSPHATE (ATP).]

The translocation of proteins into mitochondria also appears to require HSP70, again explaining the ATP requirement for this process. ATP hydrolysis might also be required for the action of an "unfoldase" enzyme to remove residual secondary structures from precursor proteins immediately prior to import or for the release of other competence factors. However, the major driving force for mitochondrial protein import seems to be provided by a respiration-generated energy potential across the inner membrane.

Nucleoside triphosphates are also required for protein import into peroxisomes and nuclei. Peroxisomal enzymes presumably remain loosely folded prior to translocation, but a requirement for HSP70 or other cytosolic proteins has not been established. ATP may be required for routing-signal-triggered dilation of nuclear pores or for other, unresearched, aspects of nuclear protein import (e.g., import-coupled phosphorylation).

III. Sorting and Transport

Noncytoplasmic proteins reaching the nucleoplasm, the mitochondrial matrix, and, above all, the membrane or the lumen of the ER have not necessarily arrived at their final destinations. The following sections consider various aspects of the later stages of the targeting pathways.

A. Folding and Oligomerization

Most proteins, including noncytoplasmic proteins, interact to form homo- or heterooligomers. Folding and oligomerization (i.e., the formation of complexes of two or more polypeptides) of proteins which are linearly translocated across the ER, mitochondrial, and probably peroxisomal membranes occurs during or immediately after translocation to the luminal (*trans*) side of the membrane. Correct folding and oligomerization are not necessarily determined by the primary sequence alone; prosthetic groups (e.g., carbohydrates and fatty acids), which are added to the polypeptide chain during translocation or later might also affect protein folding. Furthermore, disulfide bond formation between cysteine residues in secretory proteins is catalyzed by an enzyme located in the ER, and mitochondria have a proline isomerase which is also important for protein folding.

Heat-shock proteins, other than those whose role in protein translocation has already been discussed, are important in protein folding, because they remain bound to incorrectly folded proteins (or proteins for which the correct partner is unavailable), which may eventually be degraded. As discussed below, this has important consequences for the subsequent movement of proteins out of the ER into the later stages of the secretory pathway or out of the mitochondrial matrix. [*See* HEAT SHOCK.]

B. Reexport

Proteins imported into the ER are subsequently unable to cross any other membrane which they encounter during transport through the cell, and thus remain encapsulated in organelles or transport vesicles or embedded in the membranes of these organelles or vesicles. Proteins imported into the mitochondrial matrix from the cytosol can be reexported to the inner membrane or intermembrane space. Interaction with HSP60 appears to be required in this sorting phenomenon, which could, in part, be determined by sequences located immediately downstream from the prepeptide, which are generally absent from the mature proteins. These "matrix export signals" could allow these imported proteins to be channeled into the same export pathway as that used by intermembrane space and inner membrane proteins encoded by the mitochondrial genome and synthesized within the organelle.

Most proteins which are specifically transported into the nucleus remain there, unless the nuclear envelope is mechanically punctured, implying that protein traffic is unidirectional. Some proteins are reexported from the nucleus, however, as ribonucleoprotein complexes which are too large to diffuse through nuclear pores, unless they are dilated in a way analogous to that which allows nuclear protein import. This implies the existence of a nuclear export signal in at least one of the proteins in the complex. This signal is presumably unmasked when the proteins and RNA associate in the nucleus.

C. Sorting and Transport in the ER–Golgi Complex

Further discussion in this section is entirely devoted to the later stages in the secretory pathway. The default destinations for secretory proteins are the plasma membrane or the external milieu, which are reached via the Golgi apparatus, the TGN, and the secretory vesicles. Sorting signals are required to shunt proteins into other branches of the secretory pathway which lead to the lysosome or to secretory granules in endocrine and exocrine cells, and to discriminate between proteins destined for the apical and basolateral membranes in polarized epithelial cells. Specific signals are also required for protein retention in the ER or the Golgi apparatus. [*See* GOLGI APPARATUS.]

Current evidence indicates that all proteins imported into the lumen of the ER, including "resident" ER proteins, are subsequently transported to the *cis* Golgi cisterna, or at least to the so-called transition element, which is believed to link the ER and the Golgi apparatus (Fig. 2). Endogenous ER proteins must then be recycled to the ER in salvage vesicles which bud from the *cis* Golgi or be in continual movement against the bulk flow of protein through the transition element. Luminal proteins of the ER have "salvage signals," which, in most cases studied so far, comprise the tetrapeptide Lys–Asp–Glu–Leu (K-D-E-L) at the extreme carboxy terminus of the polypeptide.

"KDEL" is apparently recognized by a receptor protein in the membrane of a compartment between the ER and Golgi. In addition, there must be a mechanism for segregating the salvage signal receptor from other Golgi membrane proteins, and a receptor located specifically in the ER membrane must recognize some component of the salvage vesicle to ensure that these vesicles are efficiently recycled. ER membrane proteins do not have the KDEL signal, but presumably rely on another mechanism for retrieval from the *cis* Golgi elements, if, indeed, they are not actually retained in the ER.

With the possible exception of these ER membrane proteins, the only proteins which do not join the bulk flow to the *cis* Golgi element might be incorrectly folded proteins that remain associated with heat-shock proteins and are subsequently degraded. It is not clear, however, whether these complexes are continually recycled to the ER from the *cis* Golgi element in the same way as free heat-shock protein. It is important to note that the time required by different proteins to fold correctly determines the length of time they spend in the ER. This, in turn, is the major factor determining the time required for secretory proteins to reach their final destination.

Secretory proteins other than those retained in or recycled to the ER are transported unidirectionally through the Golgi apparatus. Modification of secretory proteins beginning with core N-glycosylation of mannose residues and thioester linkage of fatty acyl groups in the ER is continued in the Golgi apparatus, with the trimming of mannose residues, terminal glycosylation, fatty acylation, phosphorylation, and sulfation. These different reactions occur in a specific order, and many of the enzymes which perform them are segregated into different cisternae of the Golgi apparatus and in the TGN. This implies the existence of mechanisms for the specific localization of these enzymes, but it is not known whether they are prevented from joining the bulk flow of proteins and membranes through the secretory pathway or are recycled to their specific site of residence in the same way that ER proteins can be recycled from the *cis* Golgi cisternae.

Secretory proteins are vectorially transported among Golgi cisternae in transport vesicles. Cytosolic and membrane proteins are required for vesicle release from the ER or donor Golgi cisternae and for their transport to, recognition by, and fusion with recipient cisternae (Fig. 2). This complicated transport system is still poorly understood. The intra-Golgi transport vesicles have a protein coat that is different from the clathrin coat present on some vesicles budding from the TGN (see below). Components of this coat, or other proteins present on the vesicle surface, must recognize a specific recep-

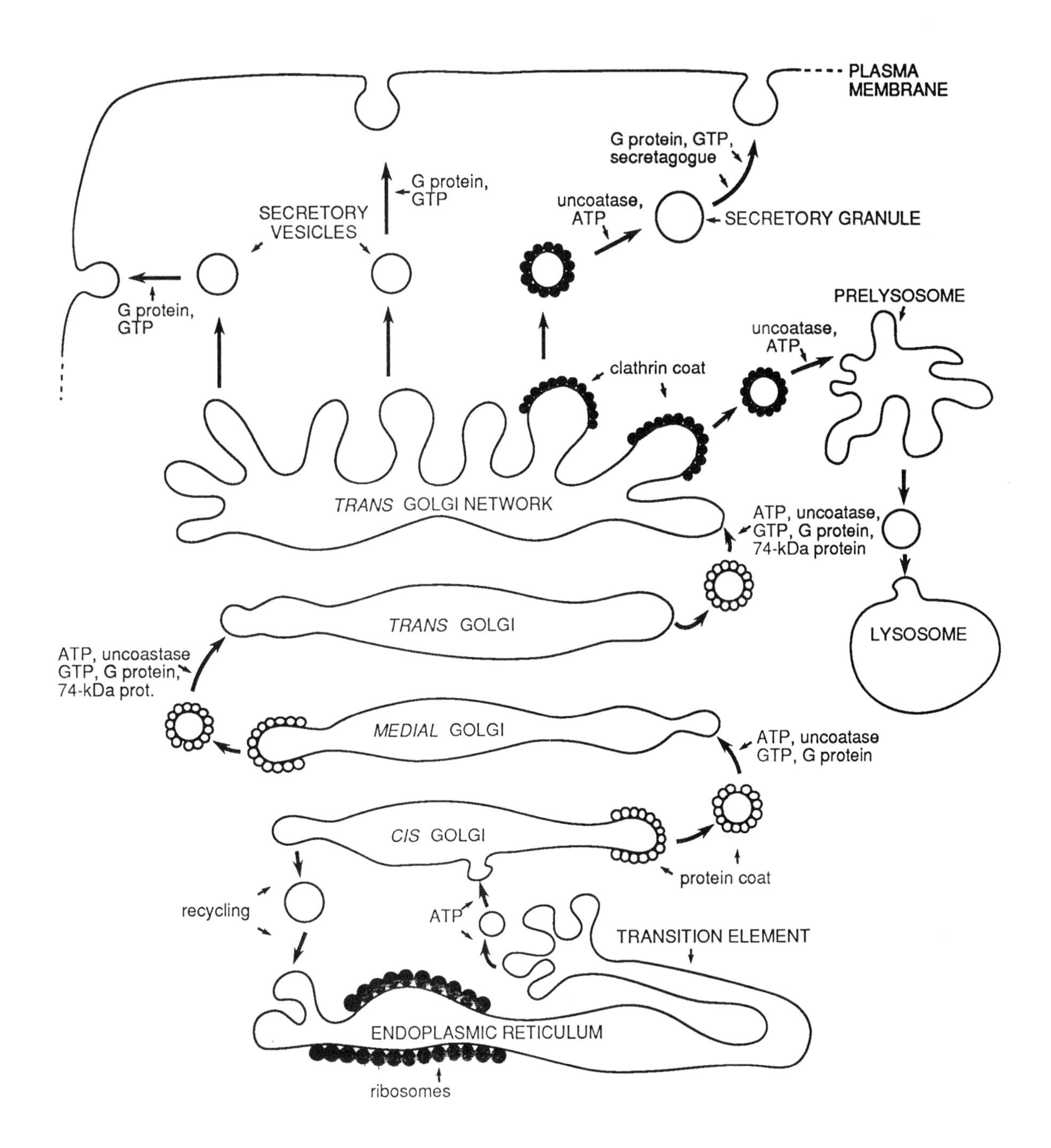

FIGURE 2 Model of post-endoplasmic reticulum sorting and transport of secretory proteins to the cell surface and lysosome, showing the requirements for each stage in the transport pathway. The recycling of proteins in vesicles from the plasma membrane to the prelysosome, from the prelysosome to the *trans* Golgi network (TGN), and from the TGN to the Golgi apparatus or between Golgi cisternae is omitted for clarity.

tor in the membrane of the recipient Golgi cisternae. At least one cytosolic protein is required for intra-Golgi transport, and G proteins and GTP might also play an important, but as yet unspecified, role in maintaining unidirectionality. Nucleoside triphosphate (ATP or GTP) hydrolysis might provide some of the energy required for vesicle fission or fusion and for the "uncoatase" which removes the protein coat prior to fusion (Fig. 2).

Many of the modifications incurred by secretory proteins in the Golgi apparatus and the TGN induce further conformational changes which are necessary for transport through the terminal stages of the secretory pathway. Incorrectly folded proteins can be retained for abnormally long periods in the Golgi

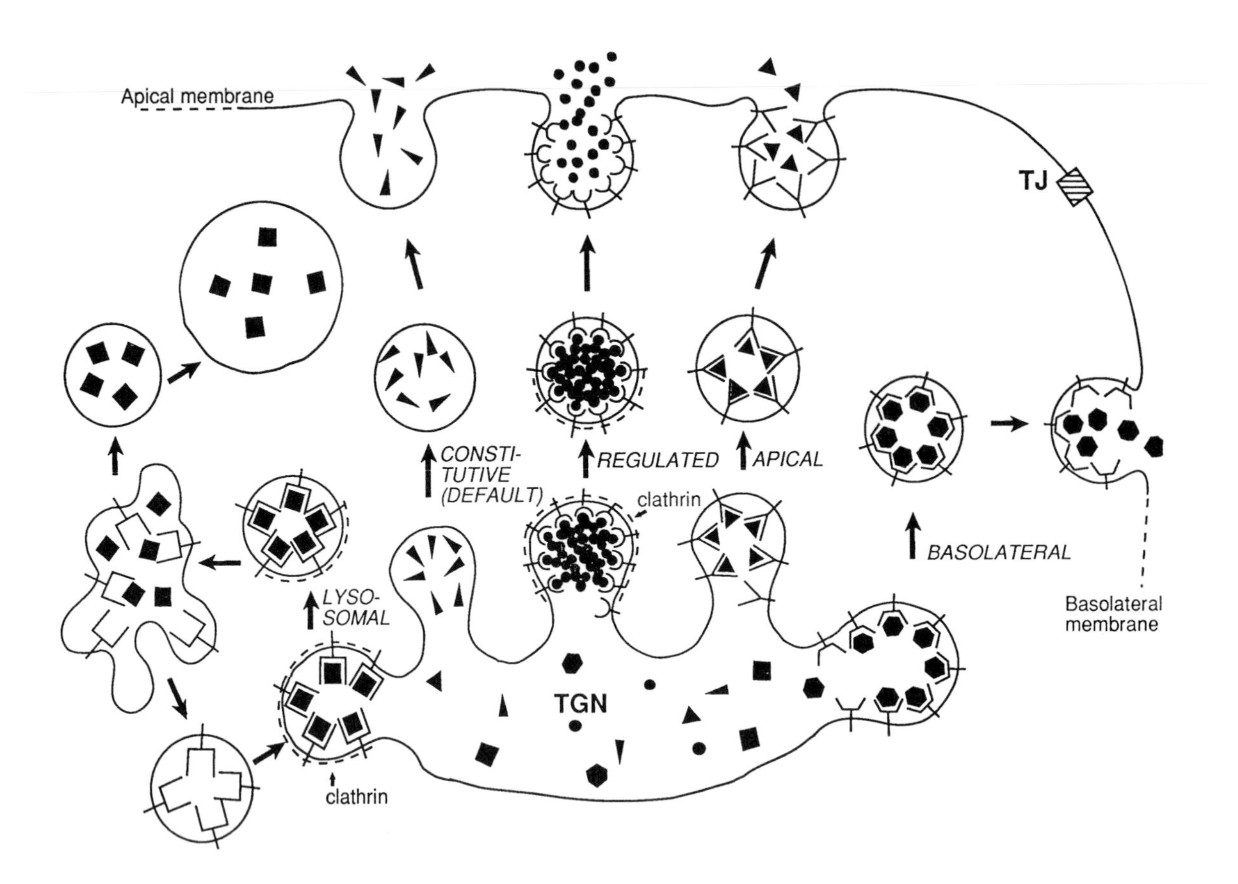

FIGURE 3 Receptor-mediated post-Golgi apparatus sorting of soluble secretory proteins into the lysosomal, regulated, apical, and basolateral terminal branches of the secretory pathway in a hypothetical polarized cell. Note that constitutive default sorting is not receptor-mediated and that sorting into the regulated secretory branch of the pathway may result from protein aggregation and granule formation, rather than from receptor interactions. TGN, *trans* Golgi network; L, lysosome; PL, prelysosome; TJ, tight junction separating outer leaflets of apical and basolateral membranes and linking adjacent cells in epithelial sheets.

cisternae or the TGN. Only in one case, that of the phosphorylated mannoses found exclusively on lysosomal enzymes, is there evidence that prosthetic groups constitute specific sorting signals. Mannose residues on proteins that are core-glycosylated in the ER can be phosphorylated in a sequence of two reactions in separate compartments of the *cis* Golgi network.

The enzyme catalyzing the first reaction, a phosphotransferase, must be able to recognize a signal in the lysosomal protein precursors which distinguishes them from all other secretory proteins. This signal is unlikely to be a continuous stretch of amino acids similar to routing signals or KDEL. Instead, it is likely to be a "patch signal" composed of amino acids from different segments of the polypeptide chain, which are brought together in the folded protein. The mannose 6-phosphate residues on lysosomal proteins are recognized by either of two receptor proteins, to which they remain bound until they have been sorted from other secretory proteins in the TGN.

The TGN is the site at which different classes of secretory proteins are segregated into the various teminal branches of the secretory pathway. Lysosomal enzymes are encapsulated in vesicles coated with the protein clathrin which bud from the TGN and migrate to the prelysosome/secondary endosome. Endocytosed proteins, imported from the outside of the cell and from the plasma membrane, are also transported to this compartment en route to the lysosome or before being recycled to the plasma membrane. The lower pH of the prelysosome relative to the TGN causes lysosomal enzymes to dissociate from the mannose 6-phosphate receptors, which are recycled to the TGN, while the enzymes are transported to the lysosome (Fig. 3).

Less is known about the sorting signals that cause other classes of secretory proteins to segregate into different vesicles budding from the TGN. Specific signals and receptors might not be required for proteins to enter the default branch of the secretory pathway. Secretory vesicles carrying mixtures of soluble and plasma membrane proteins continually bud from the TGN and migrate to the cell surface, with which they fuse to become an integral part of the plasma membrane and to release their contents into the medium. Sorting signals (probably patch signals) presumably distinguish between proteins that are destined to the apical or basolateral surfaces in polarized epithelial cells. Polarized transport might be facilitated, in part, by the movement of vesicles carrying these classes of proteins along microtubules to the apical or basolateral surfaces (Fig. 3).

The membranes of secretory granules budding from the TGN are coated with clathrin, but the coat dissociates as the granule matures, due to the action of an ATP-dependent uncoatase. The characteristically electron-dense appearance of secretory granules is due to the extremely high concentrations of protein they contain. The evidence supports a model in which protein condensing in the TGN forms a core which is enveloped by a clathrin-coated domain of the TGN membrane.

Many questions concerning the sorting and transport of secretory proteins remain to be answered. For example, while secretory proteins may have signals that allow them to be packaged into different classes of vesicles budding from the TGN, the vesicles themselves must carry the information for recognizing and fusing with their target membrane and for excluding proteins not destined to these sites. The complicated series of receptor–signal interactions which appears to be necessary to maintain the specificity and efficiency of sorting and transport could be supplemented by other factors (e.g., the preferential association of some classes of secretory proteins with lipids whose distribution through the cell is similarly organized.) This is a particularly attractive hypothesis in the case of polarized cells whose basolateral and apical membranes differ radically in both protein and lipid composition. Finally, the events that cause vesicles to bud from donor membranes and to fuse with recipient membranes, or the mechanisms that prevent secretory granules from fusing with the plasma membrane, unless specifically stimulated, remain poorly understood.

Bibliography

Burgess, P., and Kelly, R. B. (1987). Constitutive and regulated secretion of proteins. *Annu. Rev. Cell Biol.* **3,** 243–293.

Matlin, K. S. (1980). The sorting of proteins to the plasma membrane in epithelial cells. *J. Cell Biol.* **103,** 2565–2568.

Nicholson, D. W., and Neupert, W. (1988). Synthesis and assembly of mitochondrial proteins. *In* "Protein Traffic and Organelle Biogenesis" (R. C. Das and P. W. Robbins, eds.), pp. 677–746. Academic Press, San Diego, California.

Pugsley, A. P. (1989). "Protein Targeting." Academic Press, San Diego, California.

Rose, J. K., and Doms, R. W. (1988). Regulation of protein export from the endoplasmic reticulum. *Annu. Rev. Cell Biol.* **4,** 257–288.

Silver, P. A., and Hall, M. N. (1988). Transport of proteins into the nucleus. *In* "Protein Traffic and Organelle Biogenesis" (R. C. Das and P. W. Robbins, eds.), pp. 747–770. Academic Press, San Diego, California.

von Heijne, G. (1988). Transcending the impenetrable: How proteins come to terms with membranes. *Biochim. Biophys. Acta* **947,** 307–333.

Walter, P., and Lingappa, V. R. (1986). Mechanism of protein translocation across the endoplasmic reticulum membrane. *Annu. Rev. Cell Biol.* **2,** 499–516.

Proteoglycans

THOMAS N. WIGHT, *University of Washington*

Glossary

Aggregation Phenomenon that occurs when specific proteoglycans bind to a single molecule of hyaluronic acid via noncovalent bonds

Basement membrane Specialized form of extracellular matrix that contains type IV collagen, laminin, and heparan sulfate proteoglycan; it is present beneath several types of epithelial cells and surrounds other cells such as muscle and nerve

Biglycan Form of small interstitial proteoglycan (160–240 kDa) containing two chains either chondroitin sulfate or dermatan sulfate chains that is similar but not identical to decorin

cDNA Refers to the cloning of DNA: therefore, a cDNA molecule made by copying a specific mRNA molecule

Decorin Small interstitial proteoglycan (90–140 kDa) containing either chondroitin sulfate or dermatan sulfate chains that is intimately associated with collagen fibrils

Glycosaminoglycans Unbranched chains of repeating disaccharide units in which one of the monosaccharides is an amino sugar and the other is usually a hexuronic acid

Hyaluronic acid High-molecular-weight glycosaminoglycan consisting of a repeat disaccharide pattern of D-glucuronic acid and D-glucosamine

Proteoglycans Family of charged molecules containing a core protein to which are covalently attached one or more glycosaminoglycan chains

Syndecan Proteoglycan that is present on the surface of epithelial cells and interacts with collagen

PROTEOGLYCANS (PGs) are a group of complex and diverse macromolecules that are present in almost all tissues and are synthesized by a variety of cell types. They can be found in four distinct locations: (1) throughout the extracellular matrix (ECM), where they are referred to as interstitial PGs; (2) associated with specialized structures of the ECM such as basement membranes and basal laminae; (3) part of or associated with the plasma membrane of cells; and (4) in intracellular structures such as secretory storage granules and synaptic vesicles. These macromolecules consist of a core protein to which at least one glycosaminoglycan chain is covalently bound. Glycosaminoglycans are unbranched chains of repeating disaccharide units in which one of the monosaccharides is an amino sugar (hexosamine) and the other is usually a hexuronic acid. These chains vary in length and may contain 100 or more repeating disaccharide units except hyaluronic acid which can have tens of thousands of repeats. All glycosaminoglycans are highly negatively charged and with the exception of hyaluronic acid all are O-sulfated and in the case of heparan sulfate and heparin N-sulfated as well. Four basic types of glycosaminoglycans are recognized and defined according to the type of hexosamine (either *N*-acetylglucosamine or *N*-acetylgalactosamine) and/or the conformation of the uronic acid residues (either glucuronic or iduronic).

The four basic glycosaminoglycan types include hyaluronic acid, chondroitin/dermatan sulfate (CS/DS), heparin/heparan sulfate (Hep/HS), and keratan sulfate (KS). The chemical structure of the disaccharide patterns of the different glycosaminoglycans is diagrammed in Fig. 1. In addition to glycosaminoglycans, both N- and O-linked oligosaccharides may be covalently linked to the protein core.

I. Structure

The complexity and diversity of these macromolecules are derived largely from the number of different protein cores within specific proteoglycan families and from the variety in the complex carbohydrate structures produced by a large number of posttranslational modifications involved in the synthesis of the completed molecule. Usually,

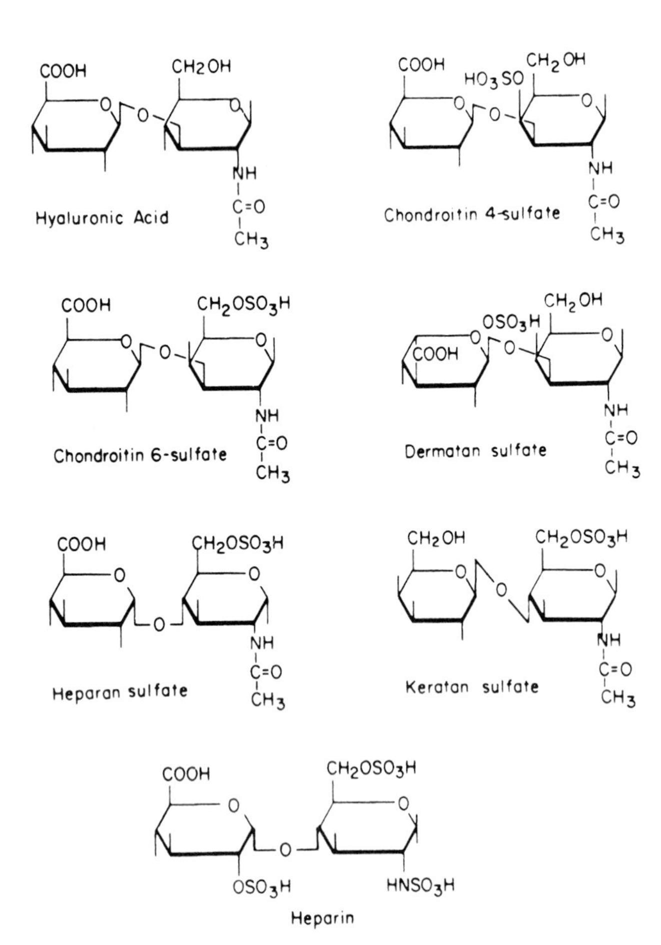

FIGURE 1 Chemical structure of the repeating disaccharide units of the glycosaminoglycans.

one type of glycosaminoglycan predominates on a single core protein, giving rise to four major families: chondroitin sulfate proteoglycan (CS-PG), dermatan sulfate proteoglycan (DS-PG), heparan sulfate proteoglycan (HS-PG), and keratan sulfate proteoglycan (KS-PG). More than one type of glycosaminoglycan can, however, be inserted on the same core, such as is found in cartilage CS/KS and in some epithelial-derived PGs (HS/CS). The presence of sulfate and/or carboxyl groups on each of the disaccharide units makes the chains strong polyanions and this characteristic contributes significantly to their physical properties as well as governs many of their interactions with other molecules. Hyaluronic acid differs from the other glycosaminoglycans in that it is nonsulfated, and current available evidence indicates that hyaluronic acid is not covalently linked to protein.

Chondroitin sulfate has a disaccharide repeat pattern similar to hyaluronic acid, but it contains galactosamine instead of glucosamine, and the galactosamine usually has a sulfate ester attached at the 4 or 6 position. The degree of sulfation can vary within a single preparation and from one tissue to another. The isomeric chondroitin 4- and 6-sulfates occur in tissues independently of each other, although in some instances, chondroitin 4-sulfate and chondroitin 6-sulfate disaccharide units may be present in the same chain.

Dermatan sulfate is an isomer of the chondroitin sulfates in which some (varies between 10% to 80%) of the D-glucuronic acid is replaced by L-iduronic acid; however, not all the glucuronic acid is replaced in this polysaccharide and, therefore, dermatan sulfate may be viewed as a copolymer of disaccharides containing both iduronic and glucuronic acid. The formation of iduronic acid residues occurs by the conversion of glucuronic acid already incorporated into the growing polymer by an epimerization reaction that is tightly coupled to the sulfation process.

Keratan sulfate has a repeating disaccharide residue containing *N*-acetylglucosamine and galactose but no uronic acid. It is usually of relatively low molecular weight and has limited distribution, being present in cornea, cartilage, and nucleus pulposus. Keratan sulfate differs from other glycosaminoglycans not only by the fact that it lacks uronic acid, but it also is not formed on the typical xylose–serine linkage between the core protein and the GAG chain, which is characteristic of all the other glycosaminoglycans.

Heparin and heparan sulfate are glycosaminoglycans of closely related structure. Heparin has a repeating disaccharide unit composed of glucosamine and either L-iduronic acid or D-glucuronic acid. A large portion of the glucosamine residues contain *N*-sulfate groups instead of *N*-acetyl groups, although a small proportion of the glucosamine residues are N-acetylated. Most glucosamine residues also carry a sulfate ester group in the C-6 position, and in addition most of the iduronic acids are sulfated at C-2. Heparin exists in tissue as single polysaccharide chains with molecular weights ranging from 5000 to 15,000 but it is synthesized as a proteoglycan with a molecular weight of 1,000,000 containing a small core protein and several long (~70-kDa) chains.

The structure of heparan sulfate is based on the same disaccharide repeating unit as that of heparin but it differs markedly in its sulfate content. It contains as much *N*-acetyl as *N*-sulfate substitution and a lower degree of O-sulfation than heparin such that on the average there is only one sulfate group per disaccharide unit. The chain appears to contain a block structure in which some regions contain no *N*-sulfate and very little *O*-sulfate. In addition, glucuronic acid is the predominant uronic acid component in heparan sulfate rather than iduronic acid, as is the case for heparin. The L-iduronic acid: D-glucuronic acid ratio generally increases with increasing sulfate content. In broad terms, low-sulfated, D-glucuronic acid-rich polysaccharides are classified as heparan sulfate, whereas high-sulfated, L-iduronic acid-rich species are designated as heparin. Since samples of each class do possess intermediate properties from a structural standpoint, it is most likely appropriate to consider heparin, heparan sulfate, and the intermediates as members of the same family of heparinlike polysaccharides.

Proteoglycans from cartilage have been the most widely studied primarily because cartilage contains large amounts of proteoglycans. The principal PG in cartilage is a large macromolecule (~2000 kDa) that contains both CS and KS chains. The core protein is approximately 250 kDa and about 300 nm in length. The protein portion constitutes only 5–10% of the entire molecule, and the rest of the molecule consists of complex carbohydrates. For example, there are approximately 80–100 CS chains (~20 kDa) on a typical cartilage CS-PG. The sulfated glycosaminoglycan chains are involved in strong intramolecular electrostatic repulsion, which give the PG molecule a very large solvent volume and contribute to the ability of PGs to absorb compressive loads. Each chain is attached to the core protein through a glycosidic bond between the hydroxyl group of serine and a xylose residue at the reducing end of the chain. Two galactose residues and one glucuronic acid residue complete the specialized linkage region. This linkage region is typical for all other PGs with the exception of KS-PG. Keratan sulfate is linked to protein via an *N*-acetylgalactosamine residue linked to either serine or threonine (in cartilage) or via an *N*-acetylglucosamine residue attached directly to asparagine (in cornea). In addition, both *N*- and O-linked oligosaccharides are covalently attached to the core protein of the cartilage CS-PG. Some of the serines and xyloses appear to be phosphorylated. A diagram showing the complex carbohydrate substitutions on the core protein is presented in Fig. 2. [*See* CARTILAGE.]

The large CS/KS proteoglycan in cartilage also possesses the capacity to noncovalently interact with hyaluronic acid and form large multimolecular aggregates. A specific region located in the N-terminal portion of the core protein (approximately 65 kDa) is responsible for this binding activity and is localized to a globular domain (G1) of the core protein. This globular domain contains a tandemly repeated double-loop structure enriched in cysteine residues. This portion of the molecule is termed the hyaluronic acid binding region. A set of one to three small molecules (~45 kDa), called link glycoproteins (LP), interacts with both hyaluronic acid and the core protein to stabilize the aggregate structure. Interestingly, the LPs also have a double-loop structure near the C-terminal end that is highly homologous to the same structure in the core protein. Both the hyaluronic acid binding region and the link glycoprotein–hyaluronic acid binding site require five consecutive repeating hyaluronic acid disaccharides for tight binding.

Other domains within the CS/KS core protein have been recognized using monospecific probes such as monoclonal antibodies and by examining amino acid sequences derived from cDNA clones to the core protein. For example, a specific region of the molecule (approximately one-half of the protein core) contains 117 Ser–Gly sequences and these serines are the sites for the attachment of the bulk of the CS chains. The distribution of Ser–Gly in this domain leads to clustering of the CS chains. A specific region of the core protein near the N terminus is the site for the attachment of KS chains. A region at the C-terminal end of the molecule is globular and

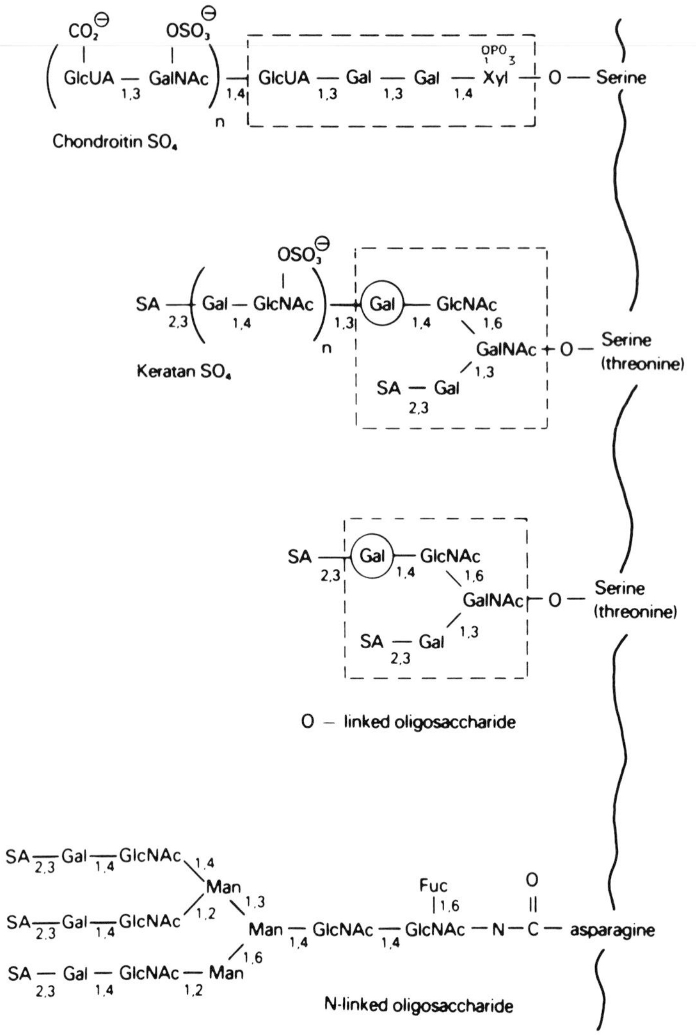

FIGURE 2 Structures of the complex carbohydrates attached to the cartilage CS/KS proteoglycan core protein. Xyl, xylose; Gal, galactose; GlcUA, glucuronic acid; GalNAc, N-acetylgalactosamine; GlcNAc, N-acetylglucosamine; Man, mannose; SA, sialic acid. Reproduced with permission from V. C. Hascall and G. K. Hascall, *in* "Cell Biology of Extracellular Matrix" (E. Hay, ed.), p. 139, Plenum, New York, 1981.

deduced amino acid sequences derived from cDNA clones show that this region has a high homology with a hepatic lectin and other vertebrate carbohydrate binding proteins. Thus, this region may have an organizational function involving interaction with other molecules in the extracellular matrix. A diagram of the different domains of a PG molecule is shown in Fig. 3.

The strategic placement of the component complex carbohydrates and protein domains within the cartilage CS/KS proteoglycan confers an important biological property to this class of molecule. The high charge density achieved when PGs aggregate with hyaluronic acid permits these molecules to attract large volumes of water. These molecules can occupy a solution volume of 30–50 times their dry weight. In cartilage, these large hydrodynamic volumes are limited to much smaller volumes by the network of collagen fibrils. The resulting swelling pressure provides cartilage with its resiliency and its ability to absorb compressive forces with minimal deformation. Any alteration in the charge density of this molecule severely compromises its ability to absorb compressive loads.

II. Distribution and Function

Large aggregating PGs are not restricted to cartilage but are found in other tissues that are subjected

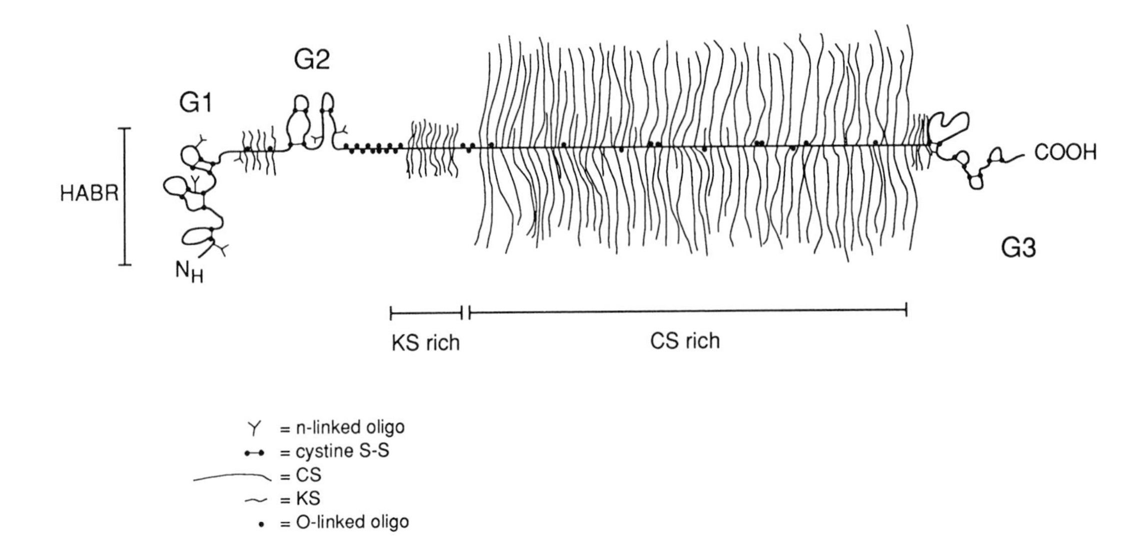

FIGURE 3 Schematic model of the core protein of the large CS/KS proteoglycan from cartilage. The molecule is divided into a number of domains. G1, G2, and G3 indicate the globular domains. The hyaluronic acid binding region (HABR) is followed by a keratan sulfate-rich region. The bulk of the protein core has CS chains attached, forming the CS-rich region. Adapted from V. C. Hascall, *ISI Atlas of Science: Biochemistry* **1,** 189, 1988.

to compressive forces. For example, large aggregating CS-PGs have been found in aorta and tendon and they appear similar but not identical to the CS/KS proteoglycan from cartilage. Some tissues such as skin contain *large nonaggregating CS-PGs*. The amino acid composition of skin CS-PG is similar to that of the large aggregating cartilage PG but it appears to lack cysteine and methionine. These two amino acids are present in the hyaluronic acid binding region of the cartilage molecule, where cysteine disulfide linkages are essential for the functional integrity of this region of the core protein.

There is a class of *small interstitial proteoglycans* (~100–300 kDa) that contain either CS or DS chains that are present in the ECM of most tissues. Two major forms of DS-PGs have been described that vary in size and iduronic acid content. One form has an overall size of 90 to 140 kDa with a core protein of ~40 kDa. It is thought to contain only one DS chain. This molecule has been named *decorin* because it appears to "decorate" collagen fibrils in a regular array. The iduronate content of the chains varies from 45 to 90%. Peptide mapping and cloning studies reveal similarities in the core protein of this DS-PG class among various tissues, such as skin, cartilage, sclera, aorta, and tendon. Deduced amino acid sequences from cDNA clones reveal that decorin contains a repeat of a 24-amino-acid unit that has an arrangement of conserved leucine sequences, characteristics of many proteins that bind to other proteins. This small DS-PG binds to type I collagen through its core protein and is capable of influencing collagen fibrillogenesis and may regulate the organization of collagen in some tissues. In addition to DS-PG, a small KS-PG binds to collagen in some tissues such as cornea. [*See* Collagen, Structure and Function.]

A larger form of DS-PG (~160–240 kDa) has been isolated and characterized from bone and cartilage, although it appears to be present in other tissues. This form, called *biglycan*, has (nearly) an identical core protein size to that of *decorin* but the two core proteins are only 55% homologous. The precise location and function of biglycan are not yet known. It may be that these two DS-PG forms arose as a result of gene duplication from a short gene that may have been used many times in diverse organisms for generating protein domains with the capacity to bind to other proteins.

Other PGs are confined to specialized structures of the ECM such as *basement membranes*. Basement membranes, which lie beneath epithelial and endothelial cells, consist of a sheet of ECM containing a number of different proteins that provide a framework for the attachment of these cells as well as a filtration barrier for the exchange of oxygen and plasma constituents. Although basement membranes from different sources contain a number of different proteins, HS-PG as well as type IV collagen and laminin is believed to occur in all basement membranes. The basement membrane HS-PGs appear to be quite heterogeneous but the major one is large, containing a core glycoprotein that varies between 200 and 400 kDa depending on the tissue source. This large HS-PG may act as a precursor, generating smaller HS-PGs that have been identified in some basement membranes. This particular PG confers a selective charge filtration barrier to basement membranes by virtue of the strategic placement of charged glycosaminoglycan chains attached to the core protein. Disruption of this charge barrier leads to marked alterations in vascular permeability, which occurs in some diseases such as diabetes and glomerular nephritis. In addition, HS-PGs interact with other basement membrane components such as laminin, type IV collagen, and fibronectin through specific binding sites within these molecules. Thus, these PGs contribute to the structural stability of this ECM structure. A specialized form of the basement membrane is found at the nerve muscle synapse, where both HS-PGs and CS-PGs are concentrated. These PGs may function in the transport of synaptic proteins, regulate diffusion of ions, and/or provide proper spacing for the pre- and postsynaptic membranes.

III. Plasma Membrane Proteoglycans

Proteoglycans appear to be associated with the plasma membrane of cells in at least three ways: (1) intercalated within the membrane, (2) associated with the membrane via a phosphatidylinositol linkage, and (3) bound to particular membrane receptors. Heparan sulfate PGs, which appear distinct from those present in basement membrane, are common as part of the surfaces of most cells. There are a number of different HS-PGs associated with cell membranes. For example, an 80-kDa species that is lipophilic is believed to be intercalated in the plasma membrane of rat liver hepatocytes. Another similarly sized, nonlipophilic species is associated with the membrane via a putative receptor. A larger species (~350 kDa) is present on the surface of a variety of cells. For example, a number of different HS-PGs have been identified on the surface of cultured human fibroblasts. One of them has two core proteins of about 90 kDa linked together by disulfide bonds. This molecule closely resembles the transferrin receptor and may play a role in cellular growth control.

Other HS-containing PGs are associated with endothelial cell surfaces, such as the HS-PG that binds antithrombin III, facilitating the inactivation of thrombin at the endothelial cell surface. An HS-PG that specifically binds lipoprotein lipase has also been identified on the endothelial cell surface and therefore participates in the metabolism of lipids at that site. One of the better-known HS-containing plasma membrane PGs is a 250-kDa PG containing both HS and CS chains. The molecule is present and restricted to the surface epithelia in a variety of tissues. This PG interacts with a variety of ECM components such as collagen types I, III, and V, fibronectin, and thrombospondin and therefore is thought to be one form of an ECM receptor. Recent success at achieving cDNA clones to this molecule reveals specific domains within the 33-kDa protein core. The sequence shows discrete cytoplasmic, transmembrane, and NH_2-terminal ECM domains. The amino-terminal domain contains about 235 amino acids and the Ser–Gly glycosaminoglycan attachment sites while the transmembrane domain contains a hydrophobic stretch of about 25 amino acids. The cytoplasmic domain is short, containing only 34 amino acids. This molecule has been named syndecan (from the Greek *syndein*, "to bind together").

Although HS-PGs appear to be associated with most cell surfaces, CS-PGs are also associated with plasma membranes. For example, a large CS-PG (~400 kDa) has been identified on the surface of cultured cells from human malignant melanoma tissue. This PG contains a core protein of 250 kDa and two to three large (60-kDa) CS chains. Although the core protein has not been sequenced, this PG fulfills most of the criteria for an intercalated plasma membrane component. The PG is hydrophobic, it can be inserted into liposomes, and it can be localized to specific microspike structures as part of the cell surface. Functional studies suggest that this PG is important in cell adhesion and may be a factor in regulating the growth of these malignant cells. A specific ligand for this membrane-associated PG has not yet been identified.

There are some PGs that are present within the membrane that bind to a specific external protein in a highly selective manner. One example is the high-molecular-weight component of the TGF β receptor. This is a 200- to 350-kDa PG (core protein ~100–200 kDa) that binds to transforming growth factors $\beta1$ and $\beta2$ in some cells. TGF β is part of a large family of factors that influence differentiation and growth of a variety of cell types.

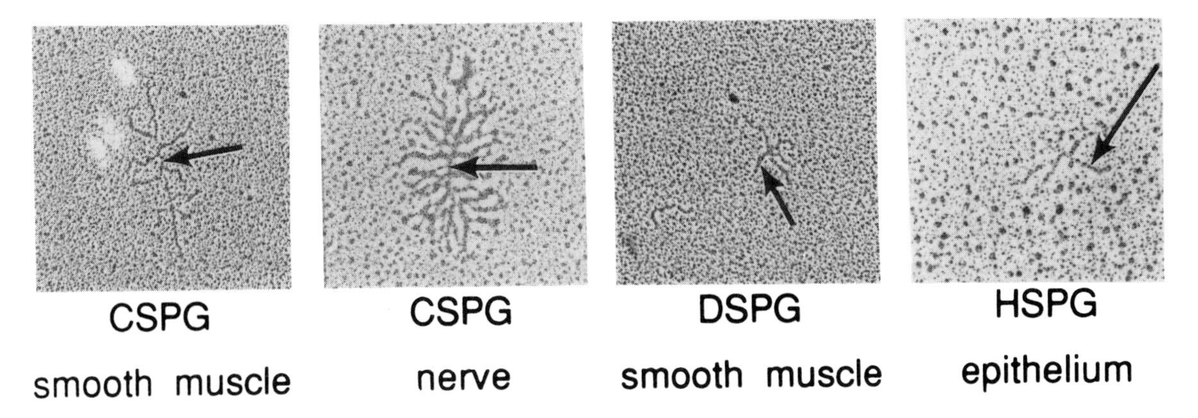

FIGURE 4 Electron micrographs of purified PGs prepared from different sources. Micrographs are printed at the same magnification (×84,750). Arrows indicate the core protein with varying numbers of glycosaminoglycans attached. Reproduced with permission from T. N. Wight, *Arteriosclerosis* **9**, 1, 1989.

IV. Intracellular Proteoglycans

A number of cells that participate in immune and inflammatory reactions, including mucosal mast cells, basophils, monocytes, eosinophils, neutrophils, and natural killer cells, contain PGs within their storage granules. Perhaps the best example of an intracellular PG is the heparin PG, which is synthesized and stored within mast cells as 1000-kDa PG. The heparin PG consists of a small core glycoprotein (~20 kDa) containing multiple Ser–Gly repeats that are attachment sites for heparin chains of 80–100 kDa. These chains are subsequently cleaved by a mast cell-derived endoglycosidase to yield partially degraded chains of 7–25 kDa. These chains represent active heparin, which is secreted by mast cells upon activation. Heparin is a powerful anticoagulant in that it potentiates the inactivation of thrombin by antithrombin III. Heparin binds to lysyl residues in antithrombin and accelerates thrombin–antithrombin formation. This accelerating activity is believed to reside in a specific pentasaccharide sequence within the heparin molecule. Some HS-PGs contain HS chains with "heparinlike" sequences. Other cells that participate in the immune response also synthesize and store similar but not identical PGs. For example, basophils synthesize and store a highly sulfated CS-PG (~150 kDa). This PG differs from other mammalian CS-containing PGs in that it is highly resistant to proteolysis and thought to be important in stabilizing, concentrating, and regulating the activity of specific storage granule enzymes such as proteases important in immune and inflammatory reactions.

Other cell types contain PGs within their secretory vesicles. For example, a pituitary tumor cell line At-T-20 stores a CS-PG in its secretory vesicle and secretes it coordinately with pituitary adrenocorticotropic hormone. Proteoglycans are also present in cholinergic synaptic vesicles and may facilitate delivery of certain components of the synapse such as acetylcholinesterase, which is known to interact with a HS-PG. Examples of the structural diversity of different PGs are illustrated in Fig. 4, and the structural characteristics of some of the well-characterized PG types is presented in Table I.

V. Biosynthesis

The synthesis of PGs occurs in a number of well-defined steps. The synthesis of the core protein and the assembly of N-linked high-mannose oligosaccharides take place in the rough endoplasmic reticulum of the cell. The majority of posttranslational modifications occur when the core protein (sometimes referred to as precursor protein) is processed in the Golgi complex. This processing includes: (1) stepwise addition of monosaccharides to form the GAG chains on appropriate serine residues, (2) addition of O-linked oligosaccharides onto appropriate serine and threonine residues, (3) conversion of high-mannose N-linked oligosaccharides to complex forms, (4) possible processing of protein to remove a portion of the peptide, and (5) O-sulfation, N-sulfation, and glucuronic to iduronic acid epimerization. In some cases, phosphate may be added onto serine residues in the core protein and/or onto many of the xylose residues that link the GAG chains to the core protein. Sulfation of the GAGs is

TABLE I Diversity of Proteoglycan Families

Type (suggested names)	Average size[a] (kDa)	No. of GAG chains[b]	GAG (kDa)	Core protein[c] (kDa)	Source
CS/KS-PG (Aggrecan)	2500	100 CS 30–60 KS	20 6	250	Cartilage
CS-PG	1000–2000	12–20	40–60	300–400	Aorta
CS-PG	1000	50	20	50	Skin
CS-PG (Tap-1) (PG-1000)	1000	20	40	200	Nerve muscle synapse
CS-PG (Versican)	1000	ND[d]	ND	350	Skin fibroblasts
CS-PG	400-500	4	60	250	Melanoma surface
CS-PG (PG-19) (Serglycin)	500	14	40	20	Parietal yolk sac
DS-PG	2000-3000	20	56	500	Ovarian granulosa cells
DS-PG (DS-PG-1) (Biglycan)	160–285	2	20	40	Bone, cartilage
DS-PG (DS-PG-II) (Decorin)	70–120	1	20	40	Tendon, skin, aorta, cartilage
Heparin-PG	750–1000	10–15	60–100	4	Mast cells
HS-PG	750	3–4	70	350–400	Basement membrane tumor
HS-PG	350	8–12	20	2 × 90	Fibroblast cell surface
HS/CS-PG (Syndecan)	200–260	4 HS 2 CS	36 17	56	Epithelial cell surface
HS-PG	130–200	4	14–30	18–250	Kidney glomerular basement membrane
HS-PG	80	4	15	30	Hepatocyte cell surface
KS-PG	80	1–3	5–7	40–55	Cornea

[a] Most of these values are estimates based on molecular sieve properties of intact molecules and therefore should be considered as approximations.
[b] GAG = glycosaminoglycans.
[c] Most of these values are derived from size estimates after removing the GAG chains. Since all of these proteins also contain N- and O-linked oligosaccharides that are not removed by the enzymes, the size estimates will include these carbohydrates.
[d] ND = no data.

mediated by 3′-phosphoadenosine-5′- phosphosulfate, either simultaneously with chain polymerization or as a final step after chain completion.

Once synthesized, the PGs can have a variety of fates. They may enter a storage granule such as observed for the heparin PG in mast cells and the HS-PG in the synaptic vesicles. They may insert into the membrane of secretory vesicles and be deposited on the surface as an intercalated, integral membrane component like the CS-PG of human melanoma cells and hybrid CS/HS proteoglycans of mouse mammary epithelial cells, or they may be packaged into secretory vesicles and secreted into the ECM as a structural component such as the large CS-PG from both cartilage and blood vessels and the small DS-PGs in cartilage skin and tendon.

VI. Involvement in Development and Disease

Formation of a precisely organized, functional tissue or organ is the culmination of a complex series of specific cellular events. These events involve several common types of cell behavior, notably movement, proliferation, shape change, recognition, and adhesion. Molecules present in the ECM and associated with the surface of cells are important for providing the proper environment for these events to take place and for contributing appropriate signals to the cells involved. There are a number of developing systems where changes in PGs and hyaluronic acid have been correlated with different stages of morphogenesis. For example, in the developing cornea, hyaluronic acid is prominent at early epithelial stages. This environment is conducive for mesenchymal migration into the hydrated cornea stroma. This phase is followed by removal of hyaluronic acid by a secreted form of the hyaluronidase enzyme, which causes the tissue to lose water and condense. At this time, mesenchymal cells differentiate into corneal keratinocytes and synthesize CS-PGs, KS-PGs, and collagen to form the differentiated tissue.

Some other examples of PG changes during development include changes during hematopoiesis, branching morphogenesis in kidney, lung, and salivary gland, and the development of the vascular system. The manner by which PGs influence development in all of these systems appear to be mediated through their ability to modify cellular events such as cell adhesion, proliferation, and migration. For example, HS-PGs have been found in points of cell contact with their ECM and it is thought that these molecules stabilize adhesion by interacting with other adhesion proteins. While HS-PGs appear to promote cell adhesion, both CS-PG and DS-PG inhibit the attachment of cells to a variety of substrata and therefore may facilitate cell detachment and movement. In fact, increased amounts of CS-PG have been observed during cell movement in a number of systems. Certain PGs also appear to influence cell proliferation. The glycosaminoglycan, heparin, and some forms of HS-PG inhibit the proliferation of some types of cells such as vascular smooth muscle cells. The precise mechanism by which these molecules influence growth is not understood. Proteoglycans can interact with growth factors. For example, HS-PGs have been shown to bind growth factors such as fibroblast growth factor, preventing its proteolysis and possibly making the growth factor available for maximal activity within tissue.

Proteoglycans have also been implicated in a number of human diseases. Usually, any change in the structure and/or content of PGs within tissue creates an abnormal ECM that severely compromises the normal functioning of the tissue. For example, PGs that are undersulfated in embryonic cartilage contribute to a reduction in the size of the forming growth plate, which consequently results in stunted and shortened growth—a condition known as *chondrodystrophy*. A group of genetic diseases, the *mucopolysaccharidoses*, is characterized by the abnormal accumulation of specific types of glycosaminoglycans in lysosomes of different cell types. The accumulation is due to a deficiency in specific lysosomal enzymes involved in the degradation of glycosaminoglycans. These diseases often lead to severe joint and corneal abnormalities and early death.

Proteoglycans also accumulate in the intimal layer of blood vessels during the early phases of *atherosclerosis* and this accumulation is thought to predispose the blood vessels to further complications of atherosclerosis, such as lipid accumulation, calcification, and thrombosis, by virtue of their ability to interact with component molecules involved in these processes. In *cancer*, changes in PG in the stroma surrounding malignant cells are often observed and thought to contribute to the proliferative and invasive properties of the malignant cells themselves. In some diseases of the nervous system, such as *Alzheimer's disease*, PGs have been observed as components of amyloid deposits and may play a role in amyloid deposition in this disease. [*See* ALZHEIMER'S DISEASE; ATHEROSCLEROSIS.]

Bibliography

Evered, D., and Whelan, J. (eds.) (1986). "Functions of Proteoglycans" (CIBA Symposium No. 124). Wiley, New York.

Gallagher, J. T., Lyon, M., and Steward, W. P. (1986). Structure and function of heparan sulfate proteoglycans. *Biochem J.* **236,** 313.

Hascall, V. C. (1989). Proteoglycans: The chondroitin sulfate/keratan sulfate proteoglycan of cartilage. *ISI Atlas of Science: Biochemistry* **1,** 189.

Hay, E. (ed.) (1981). "Cell Biology of Extracellular Matrix." Plenum, New York.

Höök, M., Woods, A., Johanson, K., Kjellen, L., and Couchman, J. R. (1986). Function of proteoglycans at the cell surface. *In* "Functions of Proteoglycans" (CIBA Symposium No. 124) (D. Evered and J. Whelan, eds.). p. 145. Wiley, New York.

Muir, H. (1983). Proteoglycans as organizers of the intercellular matrix. *Biochem. Trans.* **11,** 613.

Poole, A. R. (1986). Proteoglycans in health and disease: Structure and functions. *Biochem. J.* **236,** 1.

Ruoslahti, E. (1988). Structure and biology of proteoglycans. *Annu. Rev. Cell Biol.* **4,** 229.

Toole, B. P. (1981). Glycosaminoglycans in morphogenesis. *In* "Cell Biology of Extracellular Matrix" (E. Hay, ed.). Plenum, New York.

Wight, T. N. (1989). Cell biology of arterial proteoglycans. *Arteriosclerosis* **9,** 1.

Wight, T. N., and Mecham, R. (eds.) (1987). Biology of proteoglycans. *In* "Cell Biology of Extracellular Matrix" (E. Hay, ed.). Academic Press, New York.

Protozoal Infections, Human

J. EL-ON, *Ben Gurion University of the Negev, Beer Sheva, Israel*

Glossary

Commensalism Two organisms living together without causing injury to each other

Cyst Nonactive, resting form that is responsible for disease transmission

Prophylaxis Treatment given to people at risk for protection against disease

Protozoan parasite Microorganism that lives within or on another organism, the host

Pseudopods, cilia, flagella, undulating membrane Locomotion organelles

Reservoir Animal harboring a disease in addition to and while humans are not available

Trophozoite Active, mobile form that is generally responsible for pathological manifestations

Vector Arthropod that transmits parasitic disease from one host to another

Zoonosis Disease affecting both animals and humans

PROTOZOAN PARASITES are microscopic unicellular organisms with worldwide distribution affecting both animals and man. These parasites generally live either within or on the host in one of two ways: as commensals, causing no injury to the host, or as parasites responsible for various damages to host tissues that may lead to severe disease or even death. In the past, high prevalence of tropical diseases has been limited to certain areas, especially tropical and subtropical climates. In recent decades, these diseases have expanded their geographical distribution, mainly as a result of the increased frequency of travelers to the tropics. This article was compiled as a source of information covering the important protozoal diseases of man.

I. Introduction

Among the infectious diseases affecting man, those caused by protozoa are still highly important. These unicellular, eukaryotic cells are microscopic and are capable of performing all the vital functions within a single cell. They have a worldwide distribution and cause morbidity and mortality in millions of people, particularly in temperate and tropical areas. The distribution and prevalence of these infectious diseases are considerably affected by socioeconomic conditions, malnutrition, and genetic factors. Acquired and natural immune responses are also significantly important and can regulate infections and affect their expression. Protozoan parasites have generally complex life cycles that may involve vertebrate and invertebrate hosts and might be associated with morphological, biological, and biochemical changes. They replicate within the host, either asexually or sexually, to produce overwhelming infections. The development of resistant protozoa to available drugs and the complicated mechanisms that they have developed to invade the host immune response have raised serious problems regarding their control and eradication.

II. Amoebiasis

The amoebas are cosmopolitan parasites with variable infective rates and are highly common in populations with poor hygienic conditions. Apart from *Entamoeba gingivalis,* which habitates the mouth, all the others—*Entamoeba histolytica, E. hartmani, E. coli, E. polecki, Endolimax nana, Iodoamoeba butchlii,* and *Dientamoeba fragilis*—are intestinal parasites. The amoebas are variable in size (6–50 μm), they move by protoplasmic extensions called pseudopodia, and they multiply by binary fission. Amoebas appear as a motile active trophozoite or as a nonmotile round cyst with a thick wall. The trophozoite form is involved in the pathogenicity, while the cyst transfers the disease. Only two amoebas are pathogenic to man, *E. histolytica* and *D. fragilis,* of which the former may cause fatal infection. Invasive amoebiasis is a major health and social problem in Africa, Southeast Asia, China, and Latin America and represents one of the fatal parasitic intestinal diseases. Improving environmental sanitation may reduce infectivity and mortality.

A. Entamoeba histolytica

Entamoeba histolytica is a parasite generally limited to adults, although occasionally young children up to 2 years old may also be infected. It infects almost 10% of the world's population and results in about 40,000 deaths annually (Walsh). [*See* AMOEBIASIS, INFECTION WITH *Entamoeba histolytica*.]

Man becomes infected by ingesting the cyst in contaminated food or water. The cyst (see Color Plate 2), 8–20 μm in diameter, may have up to four nuclei and a bar-shaped condensed chromatoid material containing ribosomes and glycogen. Excystation occurs in the ileocecal region of the intestine, forming one to four trophozoites from each cyst. The trophozoite is 15–30 μm in diameter with one nucleus containing a central karyosome (see Color Plate 3). Most infections are asymptomatic. Invasion of the intestinal mucosa of the large intestine generally starts 10 days to several weeks after infection, leading to flask-shaped lesions accompanied by diarrhea or amoebic dysentery. The rectosigmoid area is generally less affected with minimal tissue reaction. In severe cases, the disease is accompanied by bloody diarrhea, sharp abdominal pain, vomiting, constipation, fever, nausea, and leukocytosis.

Amoebic granuloma (amoeboma), an inflammatory lesion resembling carcinoma, may develop in the intestinal wall mainly due to mixed bacterial and amoebic infection. The amoeboma is generally loaded with eosinophils, lymphocytes, and fibrous tissue. It usually occurs in the cecum and disappears within several weeks after successful treatment.

Penetration into circulation through the intestinal wall may lead to extra-intestinal infection. The upper part of the right lobe of the liver is the most commonly infected region, although the parasite may spread to the lung, the brain, or other organs. The development of liver abscess may be associated with fever, chills, sweating, and low erythrocyte sedimentation rate. Pain in the right side, the right lower chest, and the stomach area may also develop. Approximately 50% of the patients with amoebic liver abscess have no history of intestinal infection. It is interesting to note that the amoebic liver abscess is generally not associated with inflammatory reaction of the surrounding tissue, therefore, no scar tissue forms after successful treatment.

For diagnostic purposes, *E. histolytica* should be distinguished from other nonpathogenic amoebas harboring in the intestine. Amoebic intestinal infection is confirmed by detecting the trophozoite or the cyst in fresh stools. The presence of red blood cells in the *Amoeba* cytoplasm is highly indicative of invasive amoebiasis. The presence of Charcol–Leyden crystals (coalescence of eosinophil granules in the stool) may also suggest amoebic infection. A selective medium supplemented with starch is generally used for cultivating the parasite. Negative examination results of three stools obtained on alternate days is highly indicative for lack of amoebic infection. Sigmoidoscopy may reveal typical amoebic ulcers in about 85% of moderate cases. Tests for circulating antibodies (including indirect hemagglutination (IHA) test, indirect fluorescent antibody test (IFA), enzyme-linked immunosorbent assay (ELISA), and gel diffusion) are indicative in severe intestinal (60%) and extraintestinal (95%) amoebiasis. In cases of liver abscesses, amoebic cysts may be found in approximately 15% of the patients. Computerized tomography scanning of the liver, using X-rays will demonstrate an altered area, and ultrasound examination is important for both diagnosis and abscess aspiration.

Since nonpathogenic *E. histolytica* may be transformed into a virulent invasive parasite, a treatment should be given for all cases of amoebiasis, both symptomatic and asymptomatic. The latter is treated with either diiodohydroxyquin (Diodoquin), diloxanide furoate (Furamide), or paromomycin sulfate (Humatin). Mild intestinal disease is treated with a combination of metronidazole (Flagyl) or tinidazole (Fasigyn) and diiodohydroxyquin or paromomycin sulfate, and severe intestinal amoebiasis is treated with a combination of metronidazole or dehydroemetine plus diiodohydroxyquin. Hepatic abscess is treated with metronidazole plus diiodohydroxyquin or dehydroemetine followed by chloroquine phosphate (Aralen) plus diiodohydroxyquin. Dehydroemetine might be highly toxic. It reduces dysentery but does not terminate the infection if given alone. Occasionally, aspiration of the abscess is required in addition to chemotherapy.

It was recently shown (Mirelman *et al.*) that carbohydrate recognition plays an important role in the interaction of the amoeba with both bacterial and mammalian cells. The recognition is based on interaction of specific sugars on the cells' surface and sugar-recognizing proteins (lectins) on the surface of the parasite. The interaction induces the conversion of nonpathogenic parasites to the virulent state, characterized by changes of amoebic enzymes recognizable by their electrophoretic profile (zymodeme).

Virulency of invasive amoebiasis after contact with the target cells is initiated by the release of proteins that create holes in the host cell membrane (amoebapore) and toxins with protease activity. Patients with invasive amoebiasis may suffer from immune dysfunction, caused by deficiencies of macrophages and other accessory cells (Denis and Chadee).

Dientamoeba fragilis is an additional pathogenic intestinal amoeba. This parasite occurs only as a trophozoite having either one (40%) or two (60%) nuclei with a karyosome composed of four to eight granules. It is limited to the intestine, in which it may cause abdominal pain associated with diarrhea. Either diiodohydroxyquin, furamide, or tetracycline is considered an effective drug against the disease.

III. Trypanosomiasis

The trypanosomes are elongated hemoflagellates, 15–40 μm long, belonging to the family Trypanosomatidae. There are two major diseases caused by the genus *Trypanosoma:* sleeping sickness in tropical Africa and Chagas' disease in Central and South America. Parasites appear in the blood of the vertebrate host as a trypomastigote—an elongated flagellate with lateral undulating membrane that emerge as free flagellum with a single nucleus and an additional rod like organelle, the kinetoplast, containing mitochondrial DNA. Beneath the parasite's outer membrane, there are a number of microtubules. When blood forms are ingested by the invertebrate host, they transform into nonvirulent, elongated, procyclic forms with a single, well-developed mitochondrion, which runs the whole length of the parasite. They multiply in the midgut of the host and after several multiplication cycles they develop into metacyclic infective forms in either the salivary gland (African trypanosomes) or the rectal area (American trypanosomes) of the insect. Once they have reached the mammalian host by either inoculation during the insect feeding time or contamination with insect feces containing trypanosomes, they initiate infection. [*See* TRYPANOSOMIASIS.]

A. African Trypanosomiasis

African trypanosomes are the causative agents of sleeping sickness in humans and animals in Africa. The disease is considered endemic in tropical Africa between the latitudes of 20° north and south of the equator. The distribution of the disease corresponds to that of the tsetse fly (genus *Glossina*), which transmits it during the feeding process. Although there are several species of trypanosomes, only two affect man: *T. gambiense* causes chronic disease in West and Central Africa and *T. rhodesiense* causes acute disease in East Africa with a 100% mortality rate in untreated patients. Other species, for example, *T. brucei* (see Color Plate 4), *T. evansi*, and *T. congolense*, are animal parasites. *Trypanosoma gambiense* is considered mainly a human parasite, while *T. rhodesiense* is a human and game animal parasite. Both parasites are transmitted by several species of *Glossina* flies, which are either highly associated with water (i.e., *G. palpasis* and *G. tachinoides*, which transmit *T. gambiense*) or associated with drier areas (i.e., *G. morsitans*, which transmits *T. rhodesiense*).

An estimated 50 million people living in Africa are at risk; there are about 20,000 new cases annually. Inoculation of the parasites into the skin by the tsetse fly may lead to variable induration and swelling at the site of inoculation, which may then develop into a trypanosomal chancre within 2–3 days postinfection. After several additional days, the parasites spread into the blood circulation via the lymph nodes, where they undergo extensive multiplication by binary fission. Their invasion of the circulation is associated with pathological manifestations, including lymphadenopathy, hepatosplenomegaly, hypochromic anemia, increased erythrocyte sedimentation rate, and a high elevation of IgM titer. Headaches, joint pains, cramps, weakness, and erythematous rash also develop. In advanced infection, the nervous system may be involved, leading to mental disturbances, apathy, and paralysis, followed by diffuse meningoencephalitis associated with coma, convulsions, and death. The invasion of the nervous system is more common in *T. gambiense* infections, while *T. rhodesiense* generally kills the host before brain damage occurs (de Raadt).

African trypanosomes in the blood have wave appearances with 1- to 8-day intervals. The survival of trypanosomes within the circulation is due to their ability to vary their surface membrane glycoproteins, a phenomenon known as "antigenic variation." Most of the parasites are killed by the antibodies produced by the host, but several resist the killing as a result of the expression of new variant surface glycoproteins. Each trypanosome might show more than 100 variable antigens during infection, and the variability causes fatal infection by avoiding host defenses. Antigenic variation, which was shown to occur in either the presence or the absence of antibody response, represents a major obstacle regarding the development of an effective vaccination.

Several methods have been used for the detection of trypanosomal infection. Active motile trypanosomes may be detected in either the lymph node aspirate or by a microscopic examination of a wet drop or stained smear of blood. Negative daily examinations for 12 consecutive days are required to exclude infection. Cerebral spinal fluid aspirate should also be examined to confirm dissemination to the nervous system. Invasion of the nervous system may be associated with raised white blood cells and up to five times the protein concentration, particularly IgM in the spinal fluid. Parasites may also be cultivated in special culture media and inoculated into rats, but this technique is useful only in *T. rhodesiense* infection. Since the African trypanosomes are morphologically similar, electrophoretic enzyme variants (isoenzymes) and DNA profile analysis have to be done for species identification. Pentamidine isethionate (Lomidine) is the drug of choice for *T. gambiense* and sodium suramin (Naphuride, Antrypol) for *T. rhodesiense* infections. Arsenic compounds, for example, melarsen oxide (Mel B, Arsobal), are used in the meningoencephalitic stages when the nervous system is involved. Pentamidine has a marked prophylactic activity. All these drugs are toxic. Both pentamidine and suramine are nephrotoxic, and pentamidine may also cause a form of reversible diabetes. Melarsen oxide occasionally causes encephalopathy. Recently, α-difluoromethylornithine (DFMO), an inhibitor of polyamine synthesis, has shown antitrypanosomal activity in experimental animals. This drug, which is highly tolerable in humans, is presently being clinically examined.

B. American Trypanosomiasis

American trypanosomiasis (Chagas' disease) is a zoonotic disease caused by the hemoflagellate *T. cruzi*. This disease is transmitted to man by domestic species of blood-sucking triatomine bugs that have adapted to life in the walls of shanty-type houses and feed on human blood. The parasite develops in the insect gut as an epimastigote, which transforms into the metacyclic infective form in the posterior part of the gut within 10 to 14 days after infection. Humans become infected when metacyclic trypanosomes are deposited with the insect feces at the time of blood feeding, and the feces are rubbed into the bite wound or mucous membranes. In the mammalian host, the parasite generally grows and multiplies as intracellular amastigotes of 1.5–4 μm length in various cells (Fig. 1), particularly those of the reticuloendothelial system and heart muscle. In heavy infection, they also reach the peripheral blood, in which they appear as elongated trypomastigotes, 20 μm long with a large posterior, or kinetoplast (see Color Plate 5). The amastigote forms are responsible for the pathological manifestations, while the trypomastigotes transfer and spread the disease (Marsden).

Parasites in the skin multiply locally to form induration and swellings called chagomas. If the conjunctiva is the site of trypanosome entry, a unilat-

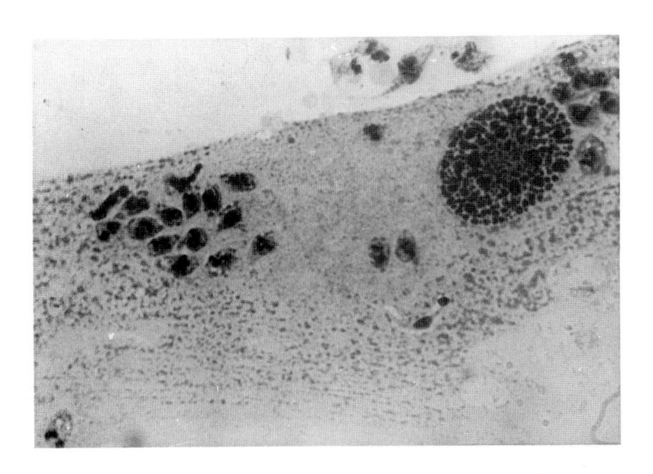

FIGURE 1 *Trypanosoma cruzi* amastigotes in a fibroblast.

eral painless periorbital edema (Romanas' sign) appears. From the skin, the parasites are disseminated to the internal organs, where they invade the cells of many tissues, particularly phagocytic cells and heart muscle fibers.

Trypanosoma cruzi infection is characterized by an early acute phase during which the parasites multiply intracellularly, but they may also appear in the blood. Infected patients, particularly children, may die during the acute phase, and the remaining survivors generally develop a chronic infection associated with mild manifestations. The disease may relapse some years later, causing cardiac and gastrointestinal manifestations with a high mortality rate.

Patients suffering from Chagas' disease develop lymphadenitis and hepatosplenomegaly accompanied by irregular fever. The release of toxins and the development of autoimmune response mediated by the parasite generally cause local and systemic reactions. It is believed that in advanced cases the parasite causes damage to parasympathetic ganglia leading to cardiomyopathy. Also, invasion of the nerve plexuses in the alimentary tract wall lead to dilatation and discoordination of muscular movement and the development of megacolon and megaesophagus.

In the mammalian host, *T. cruzi* develops mainly as intracellular amastigotes. The parasite presumably avoids being killed by macrophages through a mechanism involving escape from the phagolysosomal vacuole into the cytoplasm. It was also suggested that the parasite shares common antigens with the host's tissue, leading to severe autoimmune pathology that causes considerable difficulties in regard to immunization and vaccine development. However, further evidence is required to demonstrate that autoimmunity is responsible for cardiac damage. Also, it is not clear whether autoimmunity is elicited by the parasite or by altered tissue antigens resulting from damage caused by the parasite (Kierzenbaum and Hudson).

Typical trypomastigotes with big kinetoplasts could be detected in the blood within 1–2 weeks after infection. The disease can also be confirmed serologically using a complement fixation (CF) test. Xenodiagnosis is an additional test for *T. cruzi*. Four to six uninfected bugs are fed on a suspected patient, and within 7 to 10 days, the bugs and their feces are examined for the presence of parasites.

Therapy is limited. Nifurtimox (Lampit), benznidazole, and primaquine phosphate, the drugs available against this disease, are only partially effective. It is also advisable, particularly in endemic areas, to add a 1% concentration of gentian violet to blood used for transfusion to kill the trypanosomes and to prevent the infection of the blood recipient.

IV. Leishmaniasis

Leishmaniasis is a disorder produced by a protozoan of the genus *Leishmania*. Depending on the parasite species, the disease may be either generalized (systemic), destructive, affecting skin and internal mucus as (mucocutaneous), or chronic, being purely cutaneous. All forms are transmitted by several species of sand flies belonging to the genus *Phlebotomus* (Old World) or the genus *Lutzomia* (New World). The disease is widely distributed in many parts of the New and Old World, including the Far East, the Middle East, India, southern Russia, the Mediterranean, Africa, and Central and South America. So far, no accurate estimate of the number of people suffering from leishmaniasis is available. A figure of 12 million has been suggested, with 5000 deaths and 400,000 new cases annually. Leishmaniasis is a zoonotic disease with various animals, including rodents, small animals, and canine species serving as reservoirs. This disease is of clinical importance owing to its chronicity, local destruction, resistance to chemotherapy, and potentially high mortality. The leishmanias are obligate intracellular parasites of the mononuclear phagocyte, in which they grow and multiply within the digestive vacuole. It appears that *Leishmania* amastigotes avoid digestion by interfering with the activity of lysosomal enzymes and the oxidative burst activity, probably through secreted glycoproteins and glycolipids. [*See* LEISHMANIASIS.]

In the sand fly, the parasite is developed into an elongated (20 μm) free-living flagellate—the promastigote (Fig. 2)—with a single nucleus and a rod-like kinetoplast containing mitochondrial enzymes. After 5–7 days of several biological cycles in the midgut of the insect, it travels to the insect's mouth, from which it is transferred to the mammalian host through bites. In the mammalian host, including man, the parasite is engulfed by the phagocytic cells, the macrophages, in which it transforms into a small (2–5 μm) form—the amastigote (Leishman Donovan body) (Fig. 3). All *Leishmania* are morphologically similar and classification is generally based on serological and biochemical (isoenzyme profiles and DNA pattern) characterization.

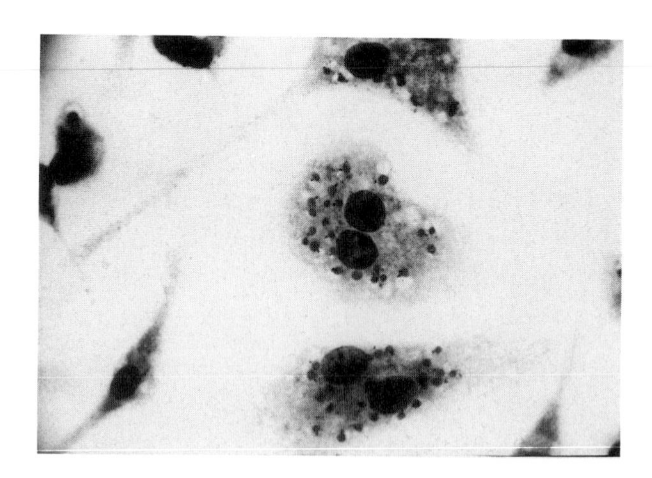

FIGURE 3 *Leishmania major* amastigotes in mononuclear phagocytes.

A. Cutaneous Leishmaniasis

Cutaneous leishmaniasis (CL) is one of the most important cases of chronic ulcerative skin lesions (Fig. 4). The disease is endemic to several tropical and subtropical parts of the world and occurs in several clinical forms: acute CL, chronic CL, recurrent CL, and diffuse CL. The introduction of the parasite into the skin is followed by the development of a papular lesion at the site of the fly bite, which may enlarge, ulcerate, and persist for up to 18 months. The ulcer tends to be painless unless secondarily infected. Spontaneous resolution of a large lesion often leaves an unsightly scar. The acute form of CL, caused by *Leishmania major,* can develop into chronic CL, which follows a course of years, and that caused by *L. tropica* may develop into recurrent CL or leishmaniasis recidivans. The diffuse form of the disease (DCL), which occurs in a small number of cases infected with either *L. aethiopica, L. mexicana amazonensis,* or *L. m. pifanoi,* is characterized by multiple lesions in the form of nodules and plaques widely disseminated on the entire body surface. In humans, a self-healing lesion is followed by a long-lasting immunity to reinfection.

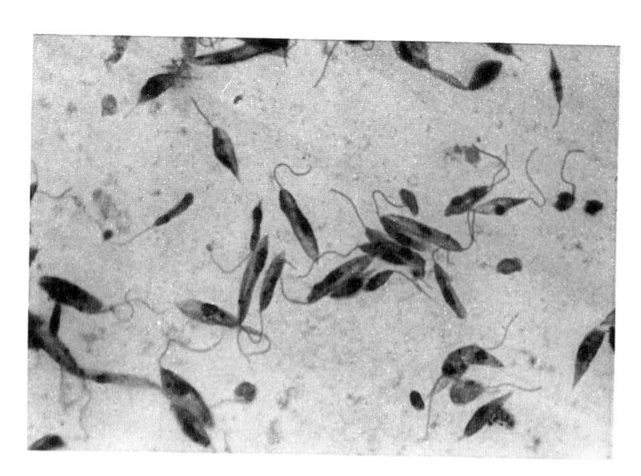

FIGURE 2 *Leishmania major* promastigotes.

B. Mucocutaneous Leishmaniasis

The most destructive lesion of CL is the mucocutaneous form (MCL) caused by *L. brazilienses braziliensis* and *L. b. panamensis*. The disease starts as a small nodule that ulcerates within several weeks and tends to metastasize to the mucocutaneous junction either spontaneously or years after the original lesion has healed, causing destruction of the adjacent tissue over a period of years.

C. Visceral Leishmaniasis

Visceral leishmaniasis (kala azar), caused by *L. donovani*, is characterized by systemic infection of the reticuloendothelial system. This disease is associated with enlargement of the spleen and the liver, accompanied by massive involvement of the bone marrow and the lymph nodes. Irregular fever, anemia, leukopenia, thrombocytopenia, and an elevated IgM level may also develop. Advanced stages of the disease are generally associated with generalized immunosuppression, leading to secondary bacterial and viral infections. The visceral disease is

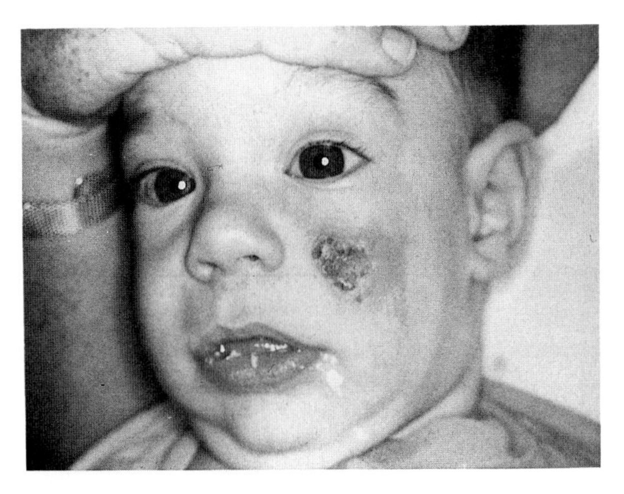

FIGURE 4 Ulcerative skin lesion caused by *Leishmania major*.

the most lethal, producing almost 100% mortality in untreated cases.

Diagnosis of all forms of leishmaniasis is made by demonstrating the amastigotes in aspirate, made from either the edge of the lesion (CL and MCL) or lymph nodes (VL). The parasite is detected microscopically after Giemsa staining. The parasite could also be easily cultivated in blood agar medium at 28°C, in which free-living promastigotes develop. Serological tests, including the IFA test, hemagglutination, the CF test, and ELISA, are also available and are mainly used for the diagnosis of visceral leishmaniasis. The leishmanin test is a skin test that is particularly suited for the cutaneous and mucocutaneous disease. This test measures delayed-type hypersensitivity reaction and becomes positive during the acute phase of the foregoing diseases, but remains negative in the diffuse and visceral forms. The pentavalent antimonials, pentostam and glucantime, are the drugs of choice against all the clinical forms of the disease. Infections resistant to antimonial are treated with either pentamidine or amphotericin B. These drugs are highly toxic and are generally associated with severe side effects. Recently, rifampicin and ketoconazole, given orally, were found to be effective against cutaneous and mucocutaneous forms. Topical treatment using paromomycin sulfate and methylbenzethonium chloride in soft white paraffin are also available against the cutaneous forms.

Immunization against CL with living, fully virulent parasites has been used in Iran, Israel, and the USSR (Greenblatt). This is the only protozoan "vaccine" presently available for humans. However, it entails the inoculation of living promastigotes, leading to normal infection, but at a selected site on the body and at a suitable time. The disadvantage of this vaccination procedure is the possible development over many years, of microbial contamination, allergic reactions, and chronic disease, in addition to lesion formation. This vaccination process that mimics long-term natural infection is limited to CL and cannot be applied against MCL or VL.

V. Trichomoniasis

Trichomoniasis is caused by flagellated protozoan parasites in animals and humans. The parasites inhabit the gastrointestinal and the urogenital systems, living on bacteria and cell debris. The trophozoite, the only form that exists, is pear shaped, measures 5–12 μm, and has three to five anterior flagella, an axostyle, and an undulating membrane. The parasite multiplies by simple binary cell fission and is transmitted from person to person through intimate contact.

Three *Trichomonas* species infesting man: *T. tenax* (6–10 μm) inhabits the mouth and is transmitted through mouth-to-mouth contact; *T. hominis* (8–12 μm) inhabits the intestine and is transmitted by the fecal–oral route and *T. vaginalis* (7–10 μm) a parasite of the urogenital tract, is generally transmitted during sexual intercourse. *Trichomonas tenax* and *T. hominis* are considered commensal, harmless flagellates and are highly prevalent in populations with low hygienic conditions. *Trichomonas vaginalis* is the only virulent parasite of the three and can cause either asymptomatic or symptomatic infection with severe inflammation.

Trichomonas vaginalis is currently the most common pathogen in the human urogenital tract. In the female, it inhabits the vagina, urethra, paraurethral glands, and the bladder. In the male, the urethra and prostate gland are the most affected organs, although the bladder, the paraurethral gland, seminal vesicles, testes, and epidydimis might also be infected. The disease is generally transmitted during sexual intercourse, although nonvenereal routes of transmission through contaminated towels cannot be excluded. The frequency of urogenital trichomoniasis is relatively lower in men than in women, and thus it is generally agreed that the infected female is the reservoir of the parasite and that the male is usually the vector of transmission.

Trichomoniasis is an international health problem, and the distribution of the disease in Western countries is relatively high as a result of more liberal attitudes and practices. Reports of infection rates in women vary widely. According to various sources, the prevalence of trichomoniasis in the United States is 3–15% among healthy women, 10% in private gynecological practices, 30% in gynecological clinics, and 38–56% in women attending venereal disease clinics. Data from the 1978 survey of the clinics in the United Kingdom showed a prevalence of about 5%. Similar findings have been reported recently in other countries. Trichomoniasis in women is generally associated with vaginal, vulva, and cervix inflammation, accompanied by a discharge that ranges from scanty and diluted to malodorous and cream colored. Vaginal tenderness, vulval pruritis, and burning are the most common symptoms. Men harboring the parasites are generally asymptomatic, although urethral and other urological manifestations might develop.

Trichomoniasis elicits cellular and humoral responses. Serological tests, such as hemagglutination, complement fixation, immunofluorescence, radioimmunoassay, and ELISA, have been used to measure antibodies for the disease. In patients infected with *T. vaginalis*, there is an antitrichomonal response with IgG, IgM, and IgA antibodies. Sera from infected patients have also been shown to exhibit a complement-mediated lytic activity on *Trichomonas* in culture and to protect rodents against the parasites' infections. Antibodies of the IgG and IgA classes were also demonstrated in vaginal secretions of infected women. However, very little correlation has been detected between the antibody level and the severity of the vaginal inflammation.

The most accurate diagnostic method is the detection of the mobile trophozoite in fresh vaginal or urethroprostatic secretions under a light microscope. Because of the rapid death of the parasite outside the body, the microscopic examination of the sample should be done as soon as possible after it has been obtained. The parasite may also be cultivated in a variety of liquid media at pH 5.5–6.0 at 37°C. Cultivation of vaginal exudate is a more sensitive diagnostic procedure that may detect twice as many *Trichomonas* infections.

Metronidazole (Flagyl) and other 5-nitroimidazoles (tinidazole and secnidazole) are the drugs of choice against the disease. Alternative treatments include a mixture containing diiodohydroxyquinoline, dextrose, lactose, and boric acid (Floraquin) or a combination of furazolidone and nifuroxime (Tricofuron). Both partners should be treated simultaneously.

VI. Giardiasis

Giardia lamblia occurs worldwide and is a flagellated parasite of man, particularly children, and probably domestic animals. Up to 30% of the world population is infected (Benenson). Adults generally develop resistance to the local strain, but exposure to a new strain in a new geographical area may initiate new infections. Infection occurs through food or water contaminated by fecal cysts. About 100 cysts per person are required to produce infection. Person-to-person contact may serve as an additional route of transmission, particularly among children in nursery school, mentally handicapped people, and homosexuals. Most infections are endemic, but contamination of water supplies with the parasite may cause epidemics.

The parasite has a simple asexual life involving motile trophozoites that divide and develop into infective cysts that are shed in the stool. The trophozoite, the form causing the disease in humans, is pear shaped with bilateral symmetry (see Color Plate 6). It is 10–18 μm long with four pairs of flagella, two nuclei, and two central karyosomes. It has an axostyle, parabasal bodies, and a sucking disc that occupies about half of the ventral surface of the parasite and serves to attach the parasite to the intestinal wall. The cysts (Fig. 5), 10–14 μm long with two to four nuclei, are produced when conditions are unfavorable. They are protected by a thick wall and are hardly affected by the normal level of water chlorination. The cysts remain infective even after several weeks in fresh water.

Giardiasis is asymptomatic in most cases, despite the presence of cysts in the feces. In invasive cases, the parasite may cause irritation to duodenal and jejunal mucosa associated with acute or chronic diarrhea and impaired absorption of fat (steatorrhea), D-xylose, and vitamin B-12. In severe cases, anorexia, watery diarrhea with yellow stools, abdominal cramps, urticaria, weakness, and general malaise may develop. Symptom development is due partly to host characteristics and partly to parasite strain differences. Most of the pathology mediated by *Giardia* is due to mechanical injury of the epithelial cells of the intestinal villi to which the parasite is attached via its sucking disc. Several other compo-

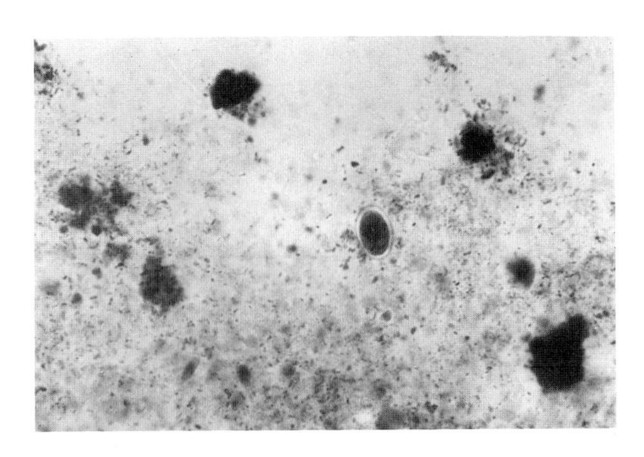

FIGURE 5 *Giardia lamblia* cyst in stool.

nents, for example, parasite toxin, and other microbial pathogens may also be responsible for the pathological effect.

The role of immunological response in giardiasis was recently reviewed by den Hollander *et al.* The fact that adults develop resistance to infection, whereas immunologically compromised individuals, particularly those with hypogammaglobulinemia, show increased incidence of infection, suggests that the immunological status of the host may play an important role in the outcome. However, although circulating antibodies against *Giardia* have been demonstrated in 18–77% of infected individuals, very little is known about their function in the disease. It was also suggested that prolonged infection may segregate the human leukocyte antigens (HLA) markers. No correlation with ABO blood groups was demonstrated.

Diagnosis is based on microscopial detection of the trophozoite and/or the cyst in the stool. The formol-ether concentration technique may increase the chances of finding the cyst. In cases suspected for giardiasis but with negative stool examinations, the hairy string test (Enterotest) is required to detect the parasite. This test uses a small capsule containing a coiled thread. The capsule is swallowed by the patient while the free end remains attached to the patient's cheek. Two hours later, the released thread is withdrawn and examined for attached trophozoites. Cultivation of the parasite in artificial medium is possible, although this procedure has not yet been applied for routine examinations. Recently, an ELISA was developed for detecting parasite antigens in the stool. Several negative stool examinations over a period of 1 to 2 weeks are required to exclude infection.

The nitroimidazole compounds metronidazole (Flagyl) and tinidazole (Fasigen), as well as quinacrine and chloroquine, are the drugs of choice against this disease. Flagyl and Fasigen may be associated with nausea and a metallic taste in the mouth. Flagyl was found to be mutagenic *in vitro* and teratogenic in experimental animals and therefore should be avoided during pregnancy.

VII. Balantidiasis

Balantidium coli is a ciliated protozoan of man, primates, pigs, and swines of worldwide distribution. The parasite, which has a cyst and trophozoite form is the largest protozoan of man. The trophozoite is pear shaped, 30–150 μm long, with an elongated, kidney-shaped macronucleus, a single micronucleus, and a subterminal cytostome. The body is covered with fine cilia that are arranged in rows and are responsible for the parasite's movement. The cyst, 40–60 μm, has a spherical appearance and is involved in the disease transmission. The parasite is reproduced by either simple binary fission or conjugation. *Balantidium coli* is a parasite of the large intestine and in most cases it is commensal, living on starch, bacteria, and cell debris in the lumen of the large intestine. Invasive balantidiasis is associated with formation of flask-shaped ulcers, similar to those caused by *E. histolytica,* accompanied by diarrhea and dysentery. In severe cases, urethritis, cystitis, and myeloneplevitis may also develop. Penetration of the parasite into the peritoneum through intestinal perforation is highly dangerous, causing about 30% mortality.

Microscopical detection of either the cyst or the trophozoite in the stool is the most accurate diagnostic method available. The parasite may also be cultivated in various media, although a large inoculum is required for successful results. Tetracycline, diiodohydroxyquine, and metronidazole are the drugs of choice against the disease.

VIII. Malaria

Malaria, a sporozoan infection, is still one of the most important protozoal diseases affecting many millions of people in tropical and subtropical areas. The disease has a global distribution and kills about 1 million children in Africa every year. It is a threat to 200 million additional people all over the world.

The disease is caused by intracellular parasites of the genus *Plasmodium,* and four species affect humans. *Plasmodium falciparum* and *P. vivax* cause malignant and benign tertian malaria, respectively, and account for more than 95% of all malarial infections. The other two species, *P. ovale* and *P. malariae,* are of regional importance, causing benign tertian and benign quartan malaria (Gilles). [*See* MALARIA.]

The disease is initiated by the bite of an infected female anopheline mosquito, which releases small (2–3 μm) spindle-shaped sporozoites into the circulation. The inoculated sporozoites rapidly penetrate the liver parenchymal cells in which they multiply asexually by binary fission to produce preerythrocytic schizonts (see Color Plate 7). During this stage, the patient remains free of symptoms. Within several days, the infected hepatocytes rupture and mature merozoites are released into circulation, where they invade the red blood cells and start the erythrocytic phase. *Plasmodium falciparum* and *P. malariae* are known to have only one preerythrocytic cycle, after which they are passed from the liver cells into circulation, while *P. vivax* and *P. ovale* are reproduced simultaneously in both erythrocytes and liver cells to form the exoerythrocytic (EE) phase over a period of years. This may account for the relapses of the disease observed long after the original blood forms have been cleared by chemotherapy. In addition, merozoites that have emerged from ruptured erythrocytes are incapable of infecting hepatocytes. Therefore, patients infected by blood transfusions will suffer only from blood-phase parasites no mater which species it is and should be treated with anti-blood form drugs only.

In the erythrocyte, the merozoite grows into a trophozoite (see Color Plate 8) and multiplies by asexual schizogony, forming a blood form schizont (see Color Plate 9). After 48 to 72 hours, depending on the *Plasmodium* species, the erythrocyte is ruptured and the releasing merozoites invade new red blood cells and start the cycle again. Within several cycles, a synchronous maturation and liberation of parasites and their metabolites into circulation is established, leading to typical malaria fever attacks every 48 hours (*P. falciparum, P. vivax,* and *P. ovale*) or 72 hours (*P. malariae*). During this period, the trophozoites may develop into sexual forms, the gametocytes (see Color Plate 10). These forms are able to further their development only in the insect gut, where fertilization occurs to form the ookinete, which penetrates the intestinal walls and develops into an oocyst. The mature oocyst is then ruptured and thousands of sporozoites are released. These initiate a new infection following the bite of an infective mosquito. Malaria infection may also be acquired congenitally during pregnancy, by blood transfusions, and by contaminated syringes among drug users.

The clinical manifestations of malaria are caused by the erythrocytic phase of the parasite. Merozoite invasions of red blood cells and rupture of the mature schizont-infected erythrocyte are followed by new invasions of other red blood cells by the released merozoites. Once inside the erythrocyte, *Plasmodium* ingests 75% of the hemoglobin content of the cell. Consequently, toxic metabolites and malaria pigment are accumulated in the red blood cells and are released into the circulation when the cell ruptures, leading to a malarial fever attack. The attack generally lasts for 8 hours and includes a cold stage (0.3–1 hour) and a hot stage (3–8 hours), followed by sweating (2–3 hours), and is the main symptom of malaria. As a result of parasite development, many erythrocytes are destroyed and anemia develops. The severity of anemia is affected not only by the level of the parasitemia but also by immunological mechanisms and depression of the hematopoietic system. Clinical signs include headaches, malaise, nausea, vomiting, abdominal discomfort, aching muscles, and joint pain. Hepatosplenomegaly is commonly observed in malaria patients, and jaundice as well as renal failure associated with edema and hemoglobinuria may also occur.

In the *P. falciparum* infection, severe symptoms that could be fatal may develop suddenly. It was found that erythrocytes infected with late trophozoites and schizont presented electron-dense excrescences (knobs) on their surface membrane. These knobs used to be attached to receptors on venous endothelium, thus leading to massive sequestration of parasites in deep capillaries. As a result, brain capillaries are blocked and damaged and cerebral malaria develops and is associated with convulsions and comas. In addition, damage to the fetus with high mortality rate may be caused from the sequestration of infected cells in the placenta.

Conditioned protective immunity (premunition) to malaria develops very slowly in residents living in endemic areas. Immunological studies clearly indicate roles for both cellular and humoral responses in the defense mechanism against malaria. However, malarial infection may also induce immunosuppression, leading to increased frequency and high susceptibility to viral and bacterial secondary infection.

Genetic and epidemiological studies indicate that patients with certain red blood cell abnormalities, that is, glucose-6-phosphate dehydrogenase (G-6-PD) deficiency, sickle-cell anemia, and β-thalassemia, are protected from lethal infections of *P. falciparum*. Oxidative damage was shown to limit the parasite multiplication within these cells, allowing a better immunological response and recovery from infection.

Malaria is diagnosed by detecting the parasite in blood smears and thick blood drops. Several blood films taken every 6 hours over a period of 2 to 3 days should be examined in order to exclude malaria. Serological examinations, including IFA, ELISA, and RIA, are also available. Recently, parasite DNA has been used as a target molecule for the development of diagnostic tests based on hybridization technology. This technique was shown to be highly sensitive and could detect 5–25 pg of parasite DNA and parasitemia level of 0.001%.

Chloroquine (Aralen) and other 4-amino quinolines, that is, amodiaquin and camoquin, are the drugs of choice against the disease. These compounds are effective only against blood forms. For a radical cure of *P. vivax* and *P. ovale* infections, an additional treatment with primaquine is required. The drugs are usually well tolerated with only minor side effects of nausea and dysphoria. In chronic administration, myopathy and retinopathy may develop. Drugs useful for resistant malaria are quinine, quinidine, or mefloquine, given either alone or in combination with sulfa-antifolate. Quinine treatment may be associated with massive intravascular hemolysis leading to dark urine (black water fever) and renal failure. Chemoprophylaxis includes chloroquin or other 4-amino quinolines, either alone or in combination with Fansidar (pyrimethamine and sulfadoxine), in areas having chloroquine-resistant parasites. Recently, proguanil has been suggested as a substitute for Fansidar, since the combination of chloroquine and Fansidar could be fatal. Prophylaxis should start 1 week before exposure and be continued for 4 weeks after exposure.

IX. Babesiosis

Babesiosis is a tick-borne, highly fatal, economically important disease of animals and livestock. The causative agents are several members of the genus *Babesia* (i.e., *B. microti* and *B. divergens*) that also infect man through hard ticks of the genus *Ixodes*. The parasite invades the reproductive organs of the tick, and from them it is transmitted to the tick's eggs. The developing tick embryo becomes infected, and once the parasite reaches the salivary gland, the tick becomes infective and may transmit the disease to the vertebrate host during the feeding process. Adult ticks generally do not transmit the disease.

In the vertebrate host the 3-μm-long parasite invades the erythrocyte, in which it grows and multiplies by either binary fission or schizogony, causing infection resembling that of *P. falciparum* malaria. Hemolytic anemia and hemoglobinuria are the major pathological manifestations, accompanied by fevers, chills, sweating, and malaise. In splenectomized patients, the disease may be fatal from damage caused to the liver and the kidneys. Microscopic examination of Giemsa-stained blood smears is used to detect the parasites. An IFA test is also available, although it is generally used for epidemiological studies. Inoculation of the patient's blood into splenectomized hamsters might be useful in cases with negative smear results. Chloroquine phosphate, although not curative, is the drug of choice against the disease. Pentamidine isethionate is used mainly in severe infections.

X. Toxoplasmosis

Toxoplasmosis is a common sporozoan disease of animals and man. The disease is economically important in veterinary medicine, as it is a major cause of ovine abortion. *Toxoplasma*, the causative agent of toxoplasmosis, is an obligatory intracellular parasite, 3–5 μm long, with a crescent shape, and generally appears in pairs or groups in various nucleated cells. Very limited information is available regarding the timing of initial infection in either humans or animals.

Man becomes infected either by ingesting the oocyst that is shed in a cat's feces or by eating raw meat contaminated with *Toxoplasma* cysts. Infection may also be acquired through organ transplantation and blood transfusion. Parasites released from either an ingested oocyst or cyst-contaminated food invade the epithelial intestinal cells, reproduce by endodyogeny, and destroy the cells. The released parasites may invade additional intestinal cells or alternatively reach the internal organs through the lymphatic and blood vessels. The parasite invades various nucleated cells, particularly those of the reticuloendothelial system and epithelial cells. Within these cells, the parasite reproduces very fast, producing pseudocysts containing many tachyzoites. The mature pseudocyst is then ruptured and the released parasites invade new cells. Most of the parasites are killed by the host's immune response 8–10 days after infection, leaving only a few parasites (bradyzoites) that multiply very slowly to form 5- to 100-μm cysts in various organs, in striated muscles, and particularly in the brain. The cysts, which may contain more than 100 bradyzoites, remain in an active resting state with no inflammatory reaction over a period of years and may be long-lived. Both humoral and cellular responses are involved in this conditioned immunity (premunition), and immunosuppression of the host may cause a relapse of the disease followed by invasive infection.

In the cat family, the parasite invades only the intestinal epithelial cells. After ingesting oocysts or meat containing cysts, parasites are released in the intestine and invade the epithelial cell, in which they reproduce asexually. After several cycles of reproduction, microgametes (male) and macrogametes (female) are formed to start the sexual cycle, producing zygotes that develop into oocysts containing two spores, each with four sporozoites. The oocysts are shed in the cat's feces and become infective within several days. The oocysts may remain infective in the ground over a period of a year.

In man, toxoplasmosis is generally asymptomatic. In symptomatic infants the disease is considered neurotropic and in adults it is viscerotropic. Lymphadenopathy is the major clinical manifestation of acute, acquired toxoplasmosis. Malaise, fever, headache, sore throat, and hepatosplenomegaly with or without chorioretinitis may develop. In severe infection, pathological effects may also appear in the heart, kidneys, lungs, and other organs. An immunosuppressed host may develop fatal infection if not treated. The fetus carried by an infected mother may be infected through the placenta; such congenital infection may cause severe eye injury, hydrocephalus, and mental disturbances associated with convulsions in the newborn. In the central nervous system of these infants, the parasites are present in the neurological cells, the capillary endothelium, and the mononuclear phagocyte, causing small lesions and abscesses.

In the mononuclear phagocyte, the parasite, once enclosed in an endocytic vesicle, was found to resist lysosomal digestion by avoiding phagosome–lysosome fusion by producing a protein that is released into the host cell and sticks to the surface of the vesicle.

Several serological tests, including hemagglutination, the IFA test, and ELISA, are available to detect infection. The Sabin–Feldman dye test, which measures IgG antibodies, is considered highly specific and sensitive. This test, introduced in 1948, is based on the lytic effects of live parasites mediated by serum containing anti-*Toxoplasma* antibodies. These antibodies are detectable 1 to 2 weeks after infection and reach a titer of $\geq 1:1000$ within 6 to 8 weeks. A low titer of 1 : 4 to 1 : 64 commonly persists for the patient's entire lifetime. Approximately 5 to 30% of young individuals and 10 to 67% of individuals over 50 years old may have positive serological reactions. The detection of IgM anti-*Toxoplasma* antibodies is important for the diagnosis of acute and infant infections. IgM antibodies appear as early as 5 days after infection, rise rapidly to 1 : 80–1 : 1000, and fall to 1 : 10–1 : 20 within several weeks.

A combination of pyrimethamine and sulfonamide in conjunction with folinic acid is used for treating *Toxoplasma*. These drugs should be avoided during pregnancy because of the possible teratogenic effect of pyrimethamine. Spiramycin is an alternative, safer drug.

XI. Sarcosporidiosis

Sarcosporidiosis is caused by several members of the genus *Sarcocystis*, which includes a group of parasites of various domestic and wild animals with worldwide distribution. Most species of *Sarcocystis* are nonpathogenic, but several of them may cause severe symptomatic infection, particularly in animals. The parasites are mainly located in the fibers of striated muscles, in which they appear as elongated cyst-like bodies—sarcocysts—from a few microns to 5 cm long and containing many uninucleated crescentlike spores, of 10 to 15 μm in length.

Several species, for example, *S. bovihominis, S. suihominis,* and *S. linemani,* have been found to infect man. The disease is acquired by ingesting uncooked meat contaminated with sarcocysts. In the intestine, the released parasites invade the intestinal epithelial cells, within which they reproduce by binary fission. They are then carried out by the lymph and blood vessels to the muscles, in which they reproduce and form sarcocysts. Infection in humans is very rare and generally asymptomatic (Bunyaratvej *et al.*), causing neither damage nor inflammatory reactions. Fever, diarrhea, abdominal pain, and weight loss occasionally develop. The disease is generally detected at histological examinations. No treatment is available.

XII. Cryptosporidiosis

Cryptosporidium is a relatively new recognized coccidian protozoan parasite of man. Since 1976, when *Cryptosporidium* was first demonstrated as a human pathogen, a considerable number of biological and epidemiological studies have described the zoonotic characteristics of this disease, which consists of gastrointestinal manifestations in both humans and livestock (Tzipori *et al.*). So far, very little information is available regarding host specificity, although more than 30 host species have been described as susceptible to infection. The disease may be transmitted directly from human to human or animal to human, as well as by waterborne routes. The parasite appears to be extremely widespread organism of global distribution. People of all ages are susceptible to infection, although young children are considered the most susceptible. The parasite is now accepted to be a major cause (7%) of diarrhea among children in Third World countries.

Cryptosporidium is a small protozoan with a life cycle that includes six major stages: release of infective sporozoites (excystation), asexual reproduction (merogony), gamete formation (gametogony), fertilization, oocyst production, and sporozoite formation (sporogony) (Current *et al.*). The parasite is considered an obligatory intracellular parasite of epithelial cells confined to the microvillus region of the intestine. The oocyst, up to 6 μm in size, sporulates within the host cell and is infective when passed in the feces.

Man generally becomes infected by ingesting food and water contaminated with oocysts. The disease in man is usually limited to between 10 and 20 days and is associated with watery diarrhea. In malnourished children and in immunodeficient hosts, such as patients with AIDS, the disease may cause a prolonged life-threatening, choleralike syndrome associated with abdominal discomfort, anorexia, fever, nausea, and weight loss. Staining of fecal smears with carbol fuchsin and methylene blue is generally used for detecting the parasite oocyst by light microscopy (see Color Plate 11). If needed, the formaline–ether technique might be used to concentrate the oocysts. A commercial, indirect fluorescent antibody test for stool examination is also available.

There is almost no data available regarding the role of the immune response in human cryptosporidiosis, although specific antiparasite IgM and IgG antibodies are detected in the sera of most patients. The major therapy in humans is supportive dehydration care. Spiramycin, a macrolide antibiotic, the only drug available against the disease, is only occasionally effective.

XIII. Isospora Infection

Isospora is a common, worldwide intestinal coccidian protozoan causing intestinal manifestations in animals and man. Although most of the *Isospora* species are animal parasites, three—*I. hominis, I. belli,* and *I. natalensis*—may also infect man, particularly children in warm climates. Similar to other coccidian parasites, *Isospora* is an obligatory intracellular parasite of the epithelial cells of the small intestine, which reproduces both sexually and asexually within the same host. Man may acquire the disease by ingesting undercooked meat containing oocysts. Each oocyst (20–33 μm) contains two sporocysts (12–14 μm), each of which contains four sporozoites. In the intestine, the crescent-shaped sporozoites are liberated from the ingested oocysts and penetrate the epithelial cells of the mucosa. After extensive reproduction, the cells are ruptured, releasing merozoites that invade adjacent epithelial cells. After several asexual cycles, gametocytes are formed, followed by fertilization and oocyst formation. The oocyst may be mature by the time it passes into the feces (*I. hominis*) or it may require 1 (*I. natalensis*) to 4 days (*I. belli*) outside the body for complete maturation. The disease in humans is generally asymptomatic, causing no harm to the patient. In a few cases, fever, headache, and abdominal cramps associated with mild diarrhea, lasting a few days, may develop within about a week after infection. In immunocompromised and immunosuppressed patients, severe disease accompanied by water diarrhea, anorexia, and weight loss may develop over a long period.

Diagnosis of the disease is made by the microscopic demonstration of the oocyst in the stool. The formalin–ether concentration technique may also be employed. No effective treatment is available and only symptomatic treatment is possible. *Isospora belli* is partially sensitive to a combination of either pyrimethamine/sulfonamide or trimethoprime/sulfamethoxazole.

XIV. Pneumocystosis

Pneumocystis carinii is a cosmopolitan organism found in the lungs of malnourished children and immunocompromised hosts. Both sexes and all ages are susceptible to infection. There is not yet a final agreement regarding the classification of this parasite and some authors have classified it as a yeast. Four stages of the parasite have been described in the infected lungs of humans and animals: trophozoites, intermediate immature cysts, mature cysts, and intracystic bodies. The trophozoites, 2–8 μm long, develop from the ruptured cyst and multiply by binary fission, budding, or endodyogeny. The trophozoite may appear either as a single cell with one or two nuclei or as clumps. The cyst is spherical, 4–6 μm, and generally contains eight crescent-shaped intracystic bodies, 1–1.5 μm each, which multiply by sporogony. All forms occur extracellularly in the alveoli, and although intracellular forms may be detected, it is not clear whether they can survive intracellularly (Matsumoto and Yoshida).

Both the cyst and the intracystic bodies are involved in the transmission of the disease. Since more than 75% of all children present anti-*Pneumocystis* antibodies by the age of 4 years, it was suggested that infection with *Pneumocystis* occurs in very early childhood, probably through inhalation of contaminated air. Intrauterine infection has also been reported. Most initial infections are asymptomatic, leading to a latent phase that may flare up in immunocompromised patients. Children with congenital immune deficiency, organ transplant recipients receiving immunosuppressive agents, patients with malignancies receiving antineoplastic therapy, and those with AIDS are considered highly susceptible to infection, with 100% mortality if not treated.

The parasite lives in the alveolar lining layer attached to the alveolar type I epithelial cells (Yoshida *et al.*). It causes an acute, diffuse alveolar pneumonitis with marked impairment of the pulmonary functions. The lungs fill up with thick and tenacious material containing masses of plasma cells and parasites that act as a mechanical barrier to oxygen. Hypoxia, cyanosis, dyspnea, and tachypnea associated with a nonproductive cough may develop.

Diagnosis of the disease is made by detecting the trophozoite and/or the cyst in either sputum, bronchopulmonary lavage, or lung biopsy. Samples are examined microscopically after staining with Giemsa or Gomori's methenamine silver stain. Serological tests for detecting both anti-*Pneumocystis* antibodies and parasite antigen are available, but have only minor use, since antigens and antibodies are also demonstrated in asymptomatic patients.

Trimethoprim/sulfamethoxazole or pyrimethamine/sulfadizine are the drugs of choice against the disease. In resistant cases, pentamidine is introduced. This drug is highly toxic and may cause hyper- or hypoglycemia, hypotension, renal damage, and sterile abscesses at the site of injection. Trimethoprim/sulfamethoxazole is also used as prophylactic therapy to individuals at high risk, that is, before transplantation or prior to toxic therapy administration.

Bibliography

Benenson, A. S. (1985). "Control of Communicable Diseases in Man." 14th ed. Amer. Public Health Assoc., Washington, D.C.

Bunyaratvej, S., Bunyawongwiroj, P., and Nitiyanant, P. (1982). Human intestinal sarcosporidiosis: Report of six cases. *Amer. J. Trop. Med. Hyg.* **31,** 36.

Current, W. L., Reise, N. C., Errest, J. V., Baily, W. S., Heyman, M. B., and de Raadt, P. (1985). African trypanosomiasis. *Med. Internat.* **4,** 146.

den Hollander, N., Riley, D., and Befus, D. (1988). Immunology of giardiasis. *Parasitol. Today* **4,** 124.

Denis, M., and Chadee, K. (1988). Immunopathology of *Entamoeba histolytica* infections. *Parasitol. Today* **48,** 247.

Gilles, H. M. (1985). Malaria. *Med. Internat.* **4,** 141.

Greenblatt, C. L. (1980). The present and future of vaccination for cutaneous leishmaniasis. *In* "New Developments with Human and Veterinary Vaccines." (A. Mizrahi and I. Hertman, eds.). Alan R. Liss Inc. New York. pp. 259–285.

Kierzenbaum, F., and Hudson, L. (1985). Autoimmunity in Chagas disease: Cause or symptoms? *Parasitol. Today* **1,** 4.

Marsden, P. D. (1985). Chagas' disease. American trypanosomiasis. *Med. Internat.* **4,** 151.

Matsumoto, Y., and Yoshida, Y. (1986). Advances in *Pneumocystis* biology. *Parasitol. Today* **3,** 137.

Mirelman, D., Feingold, C., Wexler, A., and Bracha, R.

(1983). Interactions between *Entamoeba histolytica,* bacteria and intestinal cells. *In* "Cytopathology of Parasitic Disease" (Ciba Foundation Symposium No. 99, pp. 2–18). Pitman, Bath, England.

Tzipori, S., Smith, M., Brich, C., Barnes, G., and Bishop, R. (1983). Cryptosporidiosis in hospital patients with gastroenteritis. *Amer. J. Trop. Med. Hyg.* **32,** 931.

Walsh, J. A. (1983). Problems in recognition and diagnosis of amoebiasis: Estimation of the global magnitude of morbidity and mortality. *Rev. Infectious Diseases* **8,** 228.

Yoshida, Y., Matsumoto, Y., Yamada, M., Okabayashi, K., Yoshikawa, H., and Nakazawa, M. (1984). *Pneumonocystis carinii:* Electron microscopic investigation on the interaction of trophozoite and alveolar living cell. *Abl. Bakt. Hyg. A* **256,** 390.

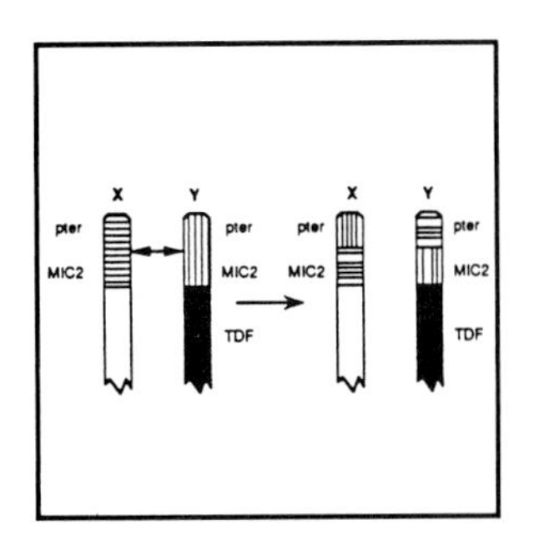

Pseudoautosomal Region of the Sex Chromosomes

FRANÇOIS ROUYER, *INSERM, Institut Pasteur*

Glossary

Alu sequences Repetitive deoxyribonucleic acid (DNA) sequence family; about 3×10^5 Alu sequences are interspersed in the human genome

Crossing-over Reciprocal exchange of DNA segments between homologous chromatids during meiosis

Genetic linkage Cosegregation of two genetic markers at meiosis according to their proximity along the chromosome

Positive interference Effect that decreases the probability of a second crossing-over when a first crossing-over has occurred on the same chromatid

Pseudoautosomal Genetic behavior of a locus located on the sex chromosomes but displaying partial or no genetic linkage to the sex locus

Recombination fraction (or recombination frequency) Summed frequency of recombinant chromatids among the total number; varies from 0 to 0.5 (0–50%) and can be expressed as a genetic distance in centiMorgans (0–50 cM) between the two markers

Restriction fragment-length polymorphism DNA polymorphism that alters the length of a restriction fragment; it can be generated by mutations suppressing or creating restriction sites and by deletion or insertion events within the fragment

Synaptonemal complex Structure observed between paired homologous chromosomes at meiosis, bearing some nodules supposed to be the sites of crossing-over and named recombination nodules

Telomere End of linear chromosome; human chromosomes are divided into short arms (p) and long arms (q), separated by the centromere; a distal position means toward the telomere, and a proximal one toward the centromere

IN MAMMALS, SEX DETERMINATION is directed by a pair of heterologous chromosomes, X and Y, and the presence of a Y chromosome triggers male differentiation of the organism. This pair of sex chromosomes is thought to derive evolutionarily from a common ancestor. The divergence between the X and Y chromosomes during evolution would have started with the acquisition by the Y chromosome of a male differentiating function, the testis-determining factor (TDF). Nevertheless, the X and Y chromosomes of mammals retain a common region in which there is X-Y crossing-over, possibly ensuring their proper segregation at male meiosis. Because it is not strictly sex-linked, this portion of the sex chromosomes is named the pseudoautosomal region. In humans, the occurrence of an X-Y genetic interchange at each male meiosis confers to the pseudoautosomal segment some properties unique in the genome. This makes it an interesting model for studying normal and abnormal meiotic recombination.

I. Definition

A. Meiotic Pairing Between the X and Y Chromosomes

In 1934, Koller and Darlington observed that the sex chromosomes of *Rattus norvegicus* paired at male meiosis. Since then, this observation has been

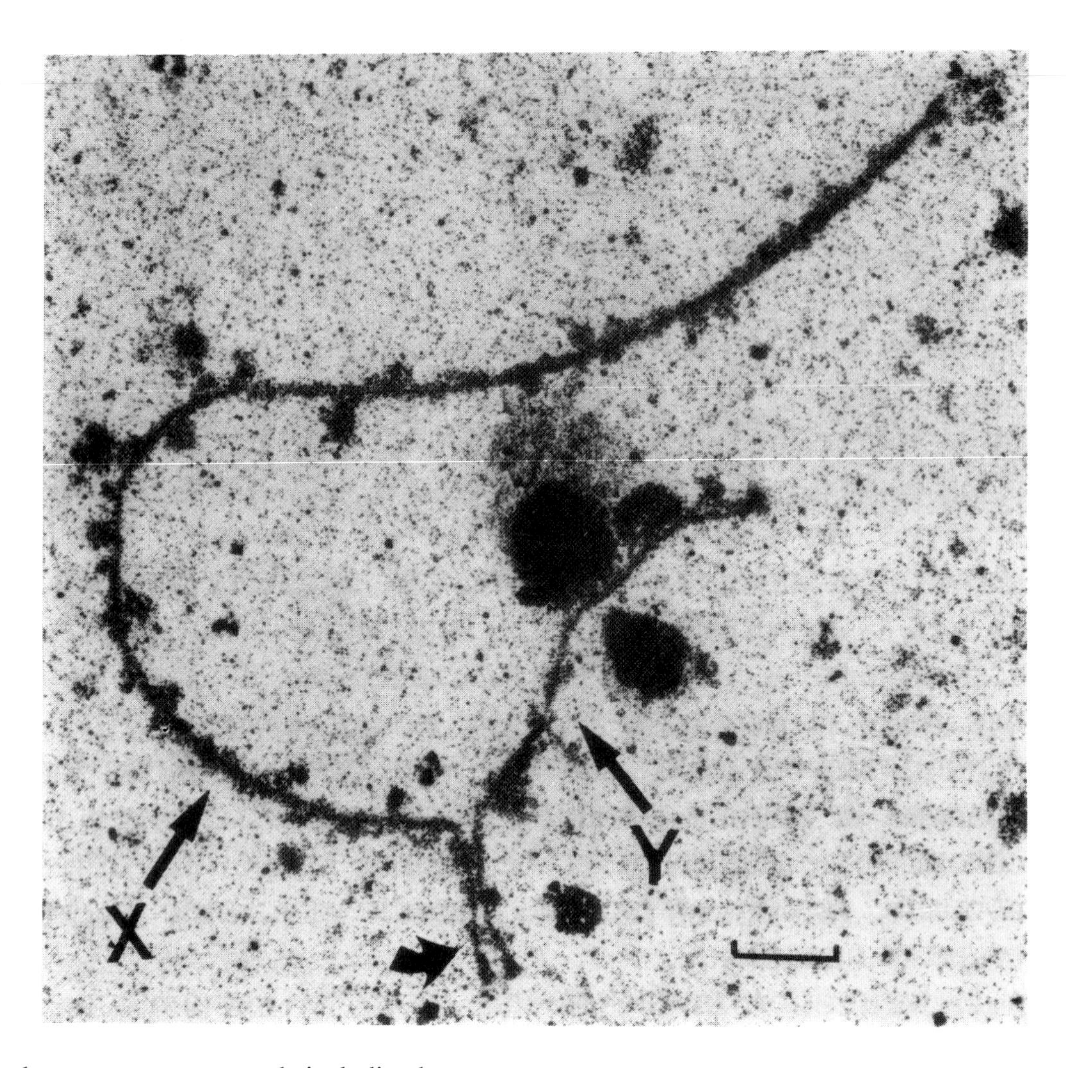

FIGURE 1 The X–Y pair. Electron microscopy microspread preparation showing the X–Y pairing at the pachytene stage of meiosis. Arrow indicates the location of the synaptonemal complex between the short arms of the chromosomes. Bar 1 μm. [Reproduced, with permission, from A. C. Chandley (1984). On the nature and extent of X–Y pairing at meiotic prophase in man. *Cytogenet. Cell Genet.* **38,** 241–247, S. Karger AG, Basel.]

enlarged to numerous mammals including humans. Pairing is a well-known behavior of homologous autosomes at meiotic prophase and is supposed to allow the exchange of genetic material and proper segregation of homologues to the haploid gametes. The molecular basis of such a mechanism is unknown but is supposed to be directed by DNA sequence homology. In the case of the human sex chromosomes, the synapsis encompasses the entire short arm of the Y chromosome and the tip of the short arm of the X chromosome (see Fig. 1). However, the short arm of the Y chromosome also contains Y-specific DNA sequences as well as sequences homologous to the long arm of the X chromosome. Therefore, the pairing segment of the short arms of the X and Y chromosomes must be divided into a homologous part and a nonhomologous part (see Fig. 2). [*See* CHROMOSOMES; MEIOSIS.]

B. Pseudoautosomal Sequences at the Tip of the Short Arms of the Sex Chromosomes

Figure 2 shows a model that assumes that the mammalian X and Y chromosomes undergo a meiotic crossing-over in their pairing region and that this crossing-over is limited to a particular region at the tip of their short arms, which is strictly homologous between X and Y. Such DNA sequences show par-

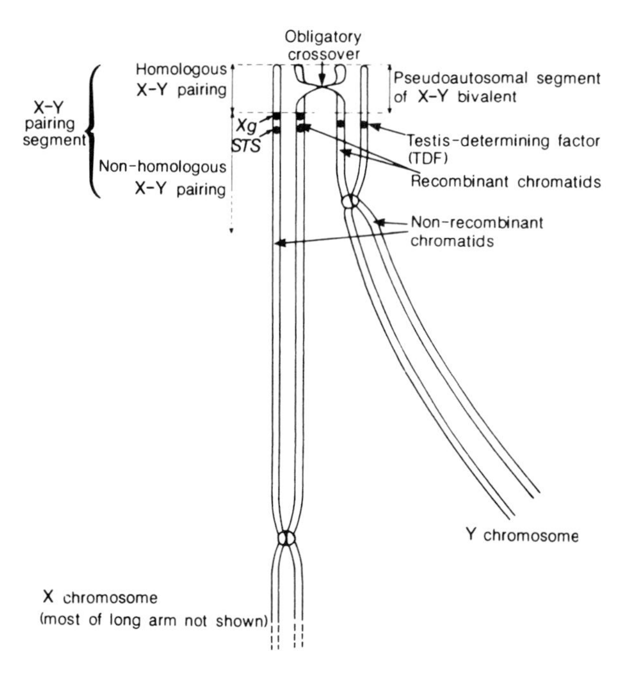

FIGURE 2 X–Y crossing-over. X and Y chromosomes both are divided into two chromatids. The two internal chromatids undergo crossing-over. The pseudoautosomal segment is only the distal part of the pairing region, which therefore comprises a proximal portion of nonhomologous pairing. Centromeres are represented by the joining of the chromatids. Black circles indicate X- or Y-linked genes. [Reprinted by permission from, P. S. Burgoyne (1986). Mammalian X and Y crossover. *Nature* **319**, 258–259. Copyright © 1986 Macmillan Magazines Ltd.]

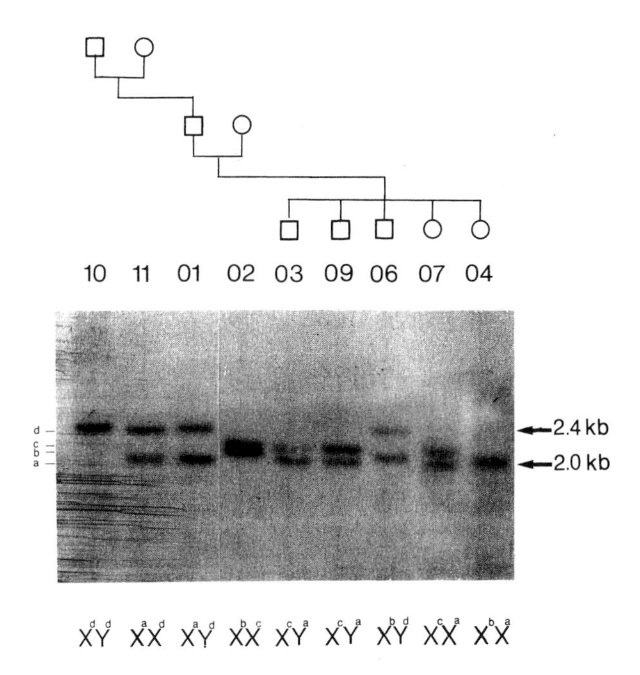

FIGURE 3 Inheritance of the pseudoautosomal locus DXYS15 in a three-generation family. Individuals are two paternal grandparents (10 and 11), two parents (01 and 02), and five children (3, 9, 6, 7, and 4). Squares and circles represent male and female individuals, respectively. DNA samples are digested with the restriction enzyme TaqI, electrophoresed on agarose gel and transferred to a nylon membrane by the Southern blot technique. The membrane is hybridized to the radiolabeled probe, washed, and then autoradiographed. Four different allelic DNA fragments (a, b, c, and d) with sizes ranging from 2 to 2.4 kb are observed in this family. The genotypes of each individual are shown at the bottom and indicate the alleles of the DXYS15 locus borne by the sex chromosomes. The X-linked allele (a) of the father, 01 (inherited from the grandmother, 11), segregates with the Y chromosome in the two sons 03 and 09. Therefore, two X-Y recombination events can be detected between DXYS15 and TDF among the five paternal meioses analyzed in this progeny. [Reprinted by permission from M. C. Simmler, F. Rouyer, G. Vergnaud, M. Nyström-Lahti, K. Y. Ngo, A. de la Chapelle and J. Weissenbach (1985). Pseudoautosomal DNA sequences in the pairing region of the human sex chromosomes. *Nature* **317**, 692–697.]

tial or no sex linkage and are thus pseudoautosomal sequences.

In humans, the existence of the pseudoautosomal region of the sex chromosomes has been demonstrated by the finding of DNA fragments, which recognize loci unlinked to the sex on the Y chromosome. Probes detecting restriction fragment-length polymorphisms (RFLPs) in the DNA represent useful genetic markers along the chromosome. The allelic forms of such polymorphic markers are represented by DNA fragments of different sizes, as observed on a Southern blot. Figure 3 shows the segregation of a pseudoautosomal locus in a five-child family and demonstrates the inheritance of the X-linked allele of the father by two of his sons. This is interpreted as an X-Y interchange, which occurred in these two meioses, between the locus detected by the probe and the sex-determining locus TDF. Results obtained from numerous family studies can be cumulated to determine the recombination fraction between two such loci.

II. Genetic Properties

A. A Gradient of Sex Linkage in Male Meiosis

The use of several DNA probes allowed a complete analysis of genetic recombination in the pseudoautosomal region. The finding of a DNA probe located near the telomere of the short arm of the X and Y chromosomes has been particularly helpful. This telomeric marker is not linked to the sex-determining locus but recombines with TDF at a rate of

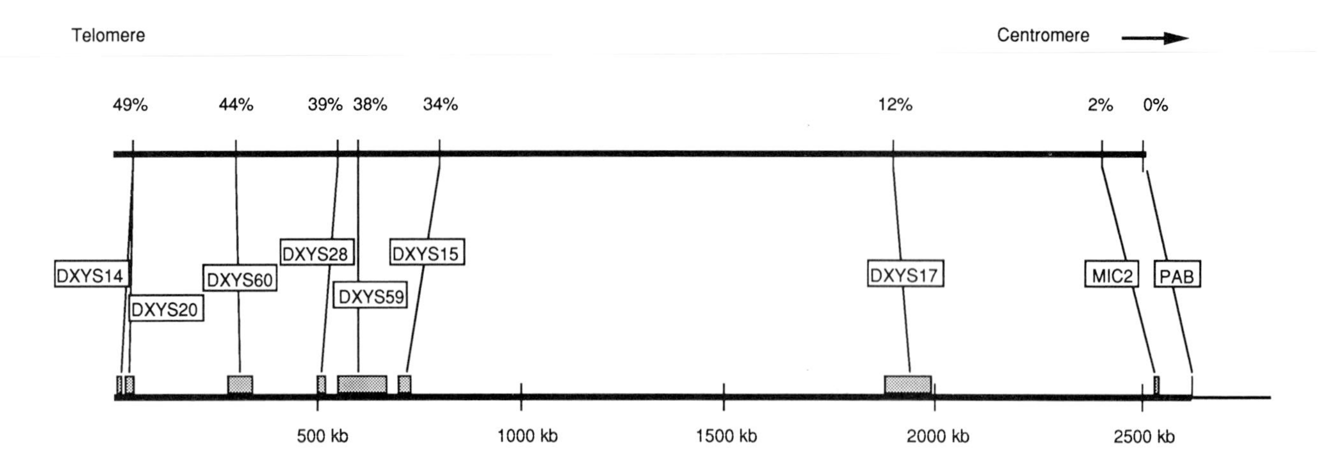

FIGURE 4 Genetic and physical maps of the pseudoautosomal region. Top: The complete genetic map stretches over 50% recombination or 50 cM in male meiosis. Sex recombination frequencies of eight loci, including the pseudoautosomal boundary (PAB), are indicated. Bottom: Physical map of the Y pseudoautosomal region. Shaded boxes above the map reflect the localization intervals of the corresponding loci as defined by pulse field gel electrophoresis. Distances (in kilobases) are measured from the telomere.

50%. The absence of sex linkage for the telomere indicates that an apparently single obligatory X-Y crossing-over occurs between TDF and the end of the short arm of the Y chromosome at male meiosis; however, this crossing-over can take place at varying positions in the region, and the different pseudoautosomal loci show a hierarchy of partial sex linkage, which reflects their order on the map (see Fig. 4). For example, the proximal locus MIC2 is exchanged between X and Y chromosomes in only 2% of meioses, whereas the pseudoautosomal boundary (locus PAB) is totally sex-linked and displays a 0% sex recombination rate. In other words, the pseudoautosomal loci define a gradient of sex linkage from the unlinked telomere to the proximal part of the map tightly linked to TDF. [*See* TELOMERES, HUMAN.]

No double recombination events have been observed in the pseudoautosomal region for the 150 meiosis analyzed thus far. In agreement with this, the sum of the recombination frequencies of two or more adjacent intervals equals the frequency measured between the outermost loci (see Table I, XY meiosis). Therefore, the rules of additivity of recombination fractions appears to apply to the whole pseudoautosomal region, which stretches >50 cM. This represents additional, strong evidence in favor of a single crossing-over per male meiosis between the human sex chromosomes. The human pseudoautosomal region is thus characterized by an almost complete positive interference.

B. Comparison of Male and Female Recombination Frequencies

The recombination rates measured between pseudoautosomal loci in female meiosis strongly contrast with the rates in male meiosis, as shown in Table I. For the same interval, comprising almost the entire pseudoautosomal region, 2–3% recombination is observed in X-X meiosis compared with nearly 50% recombination in X-Y meiosis. This considerable excess of male recombination can be interpreted as a consequence of the short length of the pseudoautosomal segment compared with the size of the X chromosome. In male meiosis, the X-Y exchange is limited to the pseudoautosomal region, whereas the X-X crossing-over can occur along the entire X chromosome in female meiosis.

C. Comparison of Physical and Genetic Maps

A physical map of the human pseudoautosomal region has been established by pulse field gel electrophoresis. The region spans >2,600 kb and is delimited on its distal side by the telomere and on its proximal side by sequences specific to the X or Y chromosomes, which define the pseudoautosomal boundary. This physical map can be compared to the genetic maps obtained in male and female meioses and is shown in Figure 4. Assuming a genetic distance of 50 cM for the entire pseudoautosomal region, 1 cM corresponds to 50–60 kb in male meiosis. This represents a recombination excess of about 18-fold compared to both the mean value estimated

TABLE I Recombination among Pseudoautosomal Loci in Male and Female Meiosis[a]

	XX meiosis			
	MIC2	DXYS17	DXYS15	DXYS20–DXYS14
MIC2	*	0	0.03	0.02
DXYS17	0.11	*	0.03	0.02
DXYS15	0.35	0.24	*	0
DXYS20–DXYS14	0.48	0.37	0.13	*
	XY meiosis			

[a] Recombination fractions measured between four pseudoautosomal loci in male (top) and female (bottom) meiosis. Only loci for which a sufficient number of informative meioses was available have been considered. Data for the very closed terminal markers DXYS20 and DXYS14 are pooled. The paucity of recombination events explains the null values observed in female meiosis for intervals (MIC2–DXYS17 and DXYS15–DXYS20–DXYS14).

for the human genome (1 cM for 1,000 kb) and that measured in the pseudoautosomal region in female meiosis.

If crossing-over is necessary to ensure a proper segregation of the X-Y pair, the recombination excess observed in male meiosis could be a direct consequence of an interchange having to take place between X and Y chromosomes in the short pseudoautosomal segment. This very high rate of recombination measured in the pseudoautosomal segment could simply be due to elimination of the nonrecombined X-Y pairs during meiosis by a selection mechanism (this represents 90–95% meioses according to the recombination excess in the male). Alternatively, a particular recombination process could act in the pseudoautosomal region at male meiosis to ensure one crossing-over in each X-Y pair. Electron microscopy studies show that a recombination nodule is present on each X-Y pair at meiosis, as it is for other chromosomes. This suggests the existence of a control mechanism that could promote the occurrence of crossing-overs in chromosome pairs including X and Y, whatever the extent of their pairing.

III. Abnormal Interchanges

Two major types of sex reversal syndromes are known in the human: XX males and XY females. XX males (about 1/20,000 newborn males) have a 46,XX karyotype and show a sterile male phenotype, whereas XY females (about 1/50,000 newborn girls) have a 46,XY karyotype and are sterile females. The majority of XX males and a small proportion of XY females are due to abnormal genetic interchanges between X and Y chromosomes during paternal meiosis.

At each meiosis, a normal crossing-over takes place between the X and Y chromosomes within the pseudoautosomal region. Therefore, this portion of the Y chromosome constitutes a hot spot of recombination located just distal to the sex-determining region. Exceptionally, some accidental events occur and promote X-Y recombination outside the pseudoautosomal segment. A normal interchange between the pseudoautosomal regions of the X and Y chromosomes is shown in Fig. 5A. The crossing-over can take place proximally to the TDF locus on the Y chromosome, as shown in Fig. 5B and 5C. In such cases, the sex-determining region is transferred from the Y to the X chromosome, and the region distal to the X breakpoint is transferred to the Y chromosome. The spermatozoa bearing such abnormal paternal sex chromosomes will give birth to XX males if the rearranged X is inherited (chromosomes I and III in Fig. 5) or XY females if the rearranged Y is inherited (chromosomes II and IV in Fig. 5).

Notably, the majority of XY females show the presence of the sex-determining region on their Y chromosome and, therefore, do not result from this etiology. Mutations in TDF or related autosomal genes could account for the female phenotype of such patients. Conversely, abnormal X-Y interchange represents the major cause of XX maleness and has essentially been studied through this disorder. The position of the recombination breakpoint on the short arm of the Y chromosome is variable among XX males but obviously proximal to the TDF

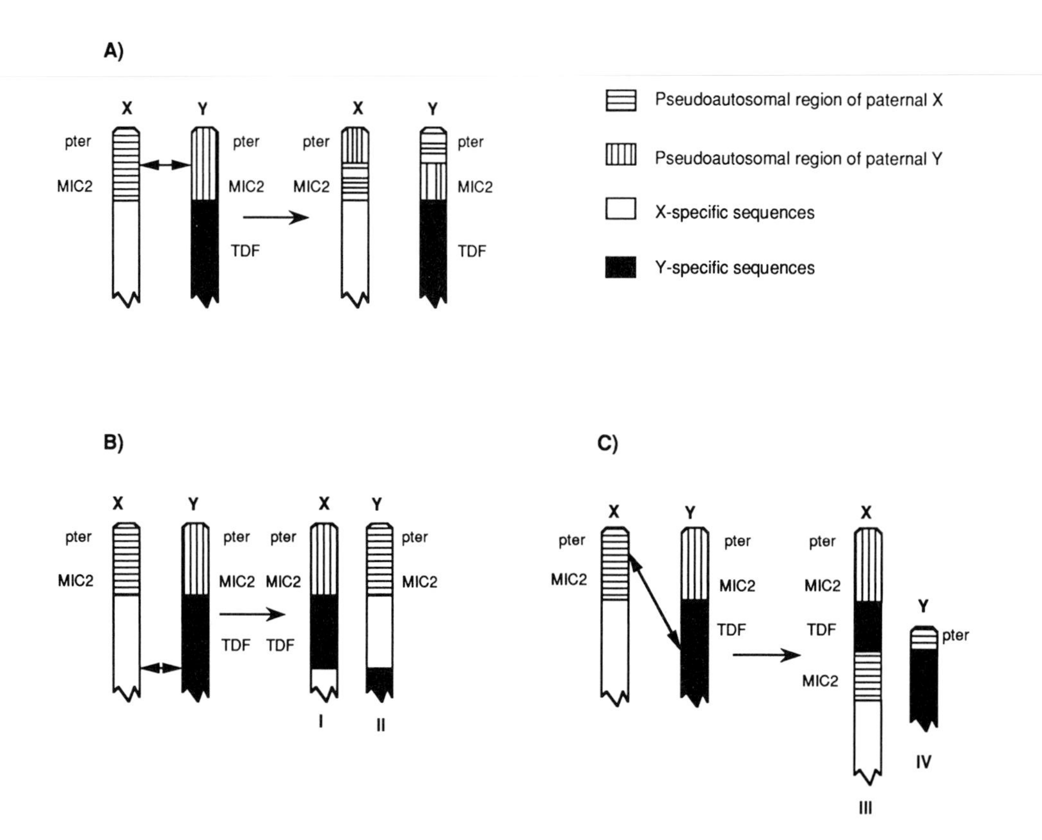

FIGURE 5 Normal and abnormal X-Y interchanges at male meiosis. Paternal chromosomes and the reciprocal meiotic products are drawn on the left and right side of the single-head arrows, respectively. Double-head arrows indicate the localization of the recombination sites. The terminal part of the chromosomes is indicated by pter. Roman numerals below the chromosome drawings refer to the text. A. Normal crossing-over within the pseudoautosomal region. B. Abnormal crossing-over with a breakpoint located in X-specific sequences. C. Abnormal crossing-over with a breakpoint located in the pseudoautosomal region of the X-chromosome.

locus. On the short arm of the X chromosome, the breakage occurs principally within X-specific sequences (chromosome I in Fig. 5B), but in some cases it is located inside the pseudoautosomal region (chromosome III in Fig. 5C). These patients exhibit a partial trisomy of the pseudoautosomal region but do not seem to differ phenotypically from the other XX males. Analysis of DNA sequences surrounding the breakpoints of rearranged chromosomes provides clues about the mechanism involved in these X-Y interchanges. The breakpoint was analyzed at the molecular level in a single case of XX maleness, with an X chromosome of type III according to Fig. 5C. This study revealed that the abnormal X-Y interchange had been promoted by homologous recombination between two Alu repetitive sequences from two nonhomologous loci of the X and Y chromosomes.

IV. Expression

A. Pseudoautosomal Genes

At present the MIC2 gene is the only pseudoautosomal gene cloned in man. The MIC2 gene encodes a ubiquitous cell-surface antigen—12E7, which is also known as the E2 antigen, a glycoprotein involved in T-cell adhesion processes. MIC2 occurs in two allelic forms: high and low expression, restricted to red blood cells and related to Xg blood groups. The Xg blood group system is defined by the red blood cell Xg^a antigen, encoded by the Xg gene. The Xg locus is located in the X-specific region proximal to the pseudoautosomal segment. In females, Xg(a+) individuals show high levels of 12E7 antigen expression, whereas Xg(a−) individuals are low-level expressors of 12E7, indicating that Xg^a could be an activator of the MIC2 gene. In males, Xg(a+) individuals are high-level expressors of 12E7, whereas Xg(a−) individuals can be either high- or low-level expressors. These genetic data

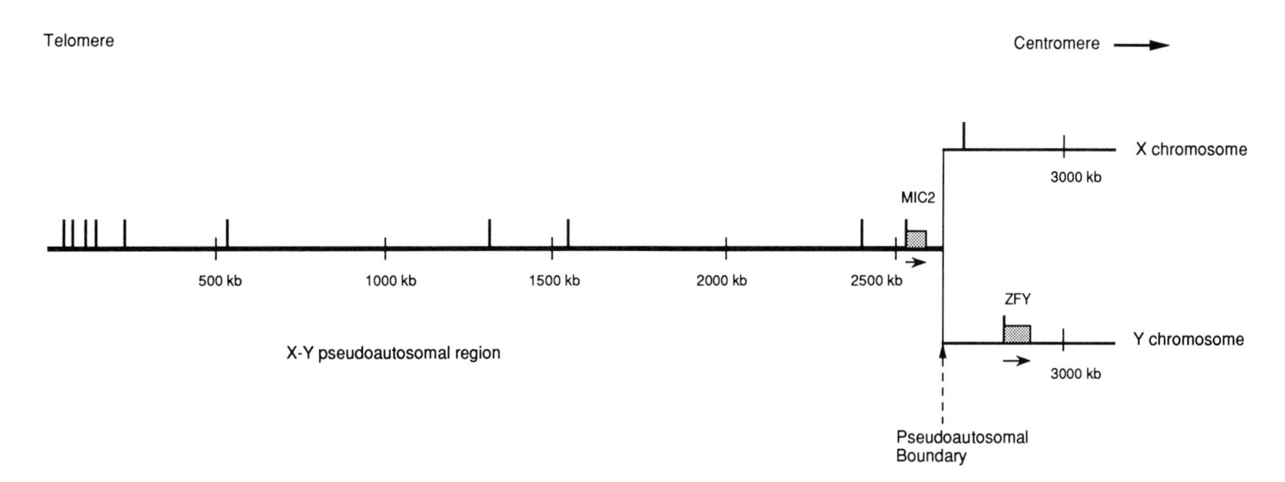

FIGURE 6 Map of the genes and HTF islands of the pseudoautosomal region. HTF islands are indicated by thick vertical bars. The pseudoautosomal region is represented by a thick line, and X- or Y-specific sequences by thin lines. Distances (in kilobases) are measured from the telomere. Genes are represented by shaded boxes according to their respective size on the chromosome, and the arrows below give the transcription direction.

led to proposal of the existence of an Xg-regulating locus (XGR), cis-acting on both Xg and MIC2 genes. The finding of rare recombination events between XGR and TDF suggests that XGR could be pseudoautosomal.

The MIC2 locus is located in the proximal part of the pseudoautosomal region and is exchanged between X and Y chromosomes at a frequency of 2–3%. As shown in Fig. 6, the gene stands about 50 kb of genomic DNA, and its 3′ end is located about 40 kb distal to the pseudoautosomal boundary. An Hpa II tiny fragment (HTF) island is observed at the 5′ end of the MIC2 gene. HTF or CpG islands are small DNA regions with a high G/C nucleotide content and numerous unmethylated CpG dinucleotides. They are associated with the promoter region of numerous genes in vertebrates, especially genes whose function is vital for the cell (''housekeeping'' genes). Therefore, HTF islands can be considered as gene markers of the vertebrate genome. Several other HTF islands scattered in the pseudoautosomal region, with particular abundance in the terminal part, are shown in Fig. 6. Some of these HTF islands could be associated with further pseudoautosomal genes. Proximal to the pseudoautosomal boundary, HTF islands are located on both the X- and Y-specific regions. On the X chromosome, an HTF island is found at about 50 kb proximal to the boundary and could correspond to the Xg^a-encoding gene; on the Y chromosome, an HTF island occurs 150 kb from the boundary. Apparently, this HTF island corresponds to the one located in the 5′ region of the zinc finger Y-encoded (ZFY) gene.

Based on the sex distribution of afflicted individuals within families, a recent genetic study suggests the existence of a pseudoautosomal gene predisposing to schizophrenia. [*See* SCHIZOPHRENIC DISORDERS.]

B. Pseudoautosomal Sequences Escape X-Inactivation

In mammals, one of the two X chromosomes is inactivated in females. The inactivation process allows the dosage compensation of X-linked genes between females with two X chromosomes and males with a single X chromosome; therefore, genes shared by both sex chromosomes, such as pseudoautosomal genes, should be noninactivated. Studies on the chromatin structure or DNA replication suggest that the tip of the short arms of the X and Y chromosomes escape X-inactivation. At the molecular level, only the MIC2 gene has been investigated and is clearly noninactivated, as shown by its expression in human–rodent hybrid cells containing an inactive human X chromosome.

In vertebrates, methylation of cytosines within CpG doublets is associated with extinction of gene activity. HTF islands are present in the 5′ region of several genes from the inactivated part of the X chromosome, namely the genes encoding glucose-6-phosphate dehydrogenase, phosphoglycerate kinase (PGK), and hypoxanthine phosphoribosyl transferase (HPRT). In an XX cell, HTF islands are methylated on the inactive X chromosome and un-

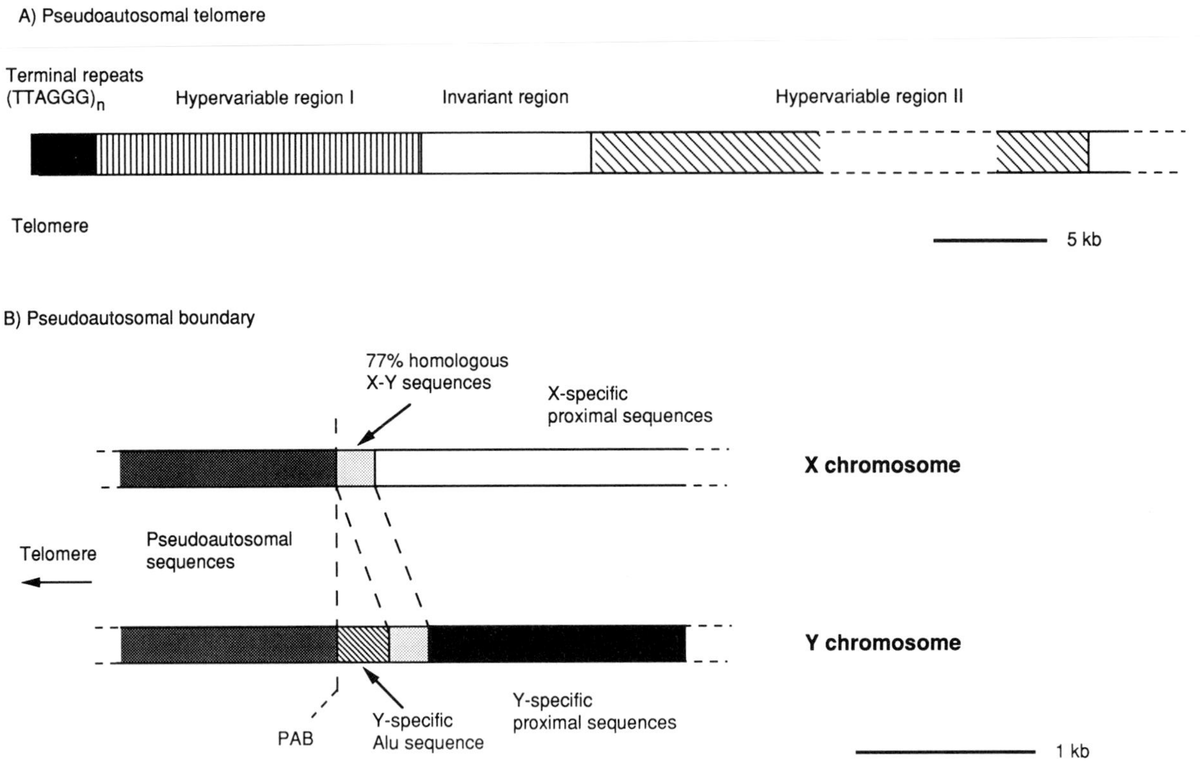

FIGURE 7 Boundaries of the pseudoautosomal region. A. Telomere. The length of the terminal repeats at the end of the chromosome is arbitrarily defined, as it is variable according to the cell type. The size of the hypervariable region II is in the range of a few tens of kilobases. B. Boundary region. The Y-specific Alu sequence and the partially X-Y homologous region are 300 and 225 bp long, respectively.

methylated on the active X chromosome. Conversely, the MIC2-associated HTF island remains unmethylated on the inactive X as well as on the active X and Y chromosomes.

The molecular bases of X-inactivation are unknown, but studies with X-autosome translocations indicate that inactivation can spread over several million base pairs within autosomal sequences, although it is not as stable as it is in X-linked sequences. The noninactivation of the pseudoautosomal segment could, therefore, be promoted by specific sequences of this region. Interestingly, the noninactivated region at the tip of the short arms of the X and Y chromosomes is not limited to the pseudoautosomal region and spreads over proximal X-specific sequences. The Xg and steroid sulfatase (STS) genes, located in this proximal X-specific region, escape X-inactivation, although only partially in the case of the STS gene.

V. Boundaries

The distal and proximal limits of the pseudoautosomal region display some original features (see Fig. 7). At the distal end (Fig. 7A), the telomere is constituted of tandem arrays of the sequence TTAGGG, which has been identified as the component of all vertebrate telomeres as well as telomeres of some lower eucaryotes. A telomeric restriction fragment containing the last kilobases of the pseudoautosomal region can be used as a functional telomere in the yeast *Saccharomyces cerevisiae*, showing the very high conservation of these structures during evolution. In humans, the $(TTAGGG)_n$ array is variable in length according to the cell lineage but occupies a few kilobases at the end of the chromosome, and has a particularly large size in the germ line. Several kilobases proximal to the telomere, two stretches of hypervariable sequences occur (hypervariable regions I and II in Fig. 7A), separated by an invariant region. Hypervariable sequences are due to copy number variations of small repeated nucleotide (or minisatellite) sequences. They provide very polymorphic markers (i.e., highly variable in different individuals, but inherited in a Mendelian fashion) for the end of the pseudoautosomal region, which are particularly useful for genetic analysis. Moreover, other minisatellite sequences are present in other parts of the pseudoautosomal segment.

At the proximal end (Fig. 7B), a discrete boundary separates pseudoautosomal sequences regularly exchanged between sex chromosomes, and proximal X- or Y-specific unrelated sequences. These X- or Y-specific DNA sequences could have been brought into the vicinity of the pseudoautosomal region by a major rearrangement on the Y chromosome during evolution (see below). On the Y chromosome, but not on the X, an Alu repetitive sequence has been inserted 225 bp distal to the former pseudoautosomal border, defining a new boundary. The accumulation of mutations in this short 225-bp interval led to a sequence divergence of 23% between X and Y chromosomes, showing that the insertion of the Alu element on the Y chromosome has prevented subsequent proximal X–Y recombination. Among 150 sex chromosomes tested for the occurrence of this Alu element, neither gain of the sequence by an X chromosome nor its loss by a Y chromosome has been observed. The presence of the Alu sequence on the Y chromosome of great apes and its absence in Old and New World monkeys suggest this new boundary to be about 50 million years (Myr) old.

The presence of 150 kb of Y-specific sequences separating the pseudoautosomal border from the proximal ZFY gene raises the question of rare abnormal X–Y interchanges occurring in this interval. If the products of these interchanges do not confer sterility to their carriers, they should be propagated in the population. This would tend to an extension of the pseudoautosomal region up to the most distal Y-specific gene, which is at present ZFY. Alternatively, the occurrence of a gene involved in male sex functions, distal to ZFY on the Y chromosome could prevent X-Y exchanges proximal to the pseudoautosomal boundary.

VI. Evolutionary Considerations

A. Conservation of Pseudoautosomal Sequences in Mammals

In the absence of polymorphic DNA markers, the occurrence of pseudoautosomal sequences has not been established in most species except for human, chimpanzee, and mouse. Comparison of genetic properties of the pseudoautosomal region of mouse and humans shows differences in the fact that, unlike human species, double interchanges seem to occur frequently between X and Y chromosomes in the mouse. Furthermore, the STS gene is pseudoautosomal in mouse, whereas it is X-specific in humans. Interestingly, the human gene is only partially inactivated, as would be expected for a gene in the process of becoming an X-specific gene. The finding of STS-related sequences on the long arm of the human Y chromosome strongly supports the occurrence of a pericentric inversion of the Y chromosome during evolution (type 2 homology in Fig. 8). This would have excluded the STS locus from the pseudoautosomal region, preventing X-Y recombination. As shown in Fig. 8, other regions homologous to the Y chromosome occur on the X chromosome. This complex patchwork of DNA segments probably accounts for several rearrangements of the Y chromosome, as it has diverged from the X.

Another feature of the human pseudoautosomal segment is the presence of a repeated sequence called subtelomeric interspersed repeat (STIR), scattered in the region. Some STIR sequences are also found in the X-specific region proximal to the pseudoautosomal segment, suggesting an ancient pseudoautosomal origin of this region. The STIR elements appear to be very good markers of pseudoautosomal sequences, their presence being observed in the early replicating segment located at the tip of the sex chromosomes of all species of primates tested thus far. Some STIR-related sequences are also present in the genomes of several other mammalian species with the notable exception of rodents. In chimpanzee, this region is located on the Y chromosome at the tip of the long arm, and crossing-over has been demonstrated by using conserved human pseudoautosomal DNA probes.

B. Pseudoautosomal Sequences: Remnants of an Ancestral Homologous Chromosome Pair

The X and Y chromosomes of mammals are thought to derive from a common ancestor. The numerous DNA sequences shared by both sex chromosomes, especially the pseudoautosomal sequences, attest to this common origin. During evolution, a male sex-determining segment would have been differentiated, defining the so-called Y chromosome. Accompanying the divergence between the X and Y chromosomes, inactivation would have progres-

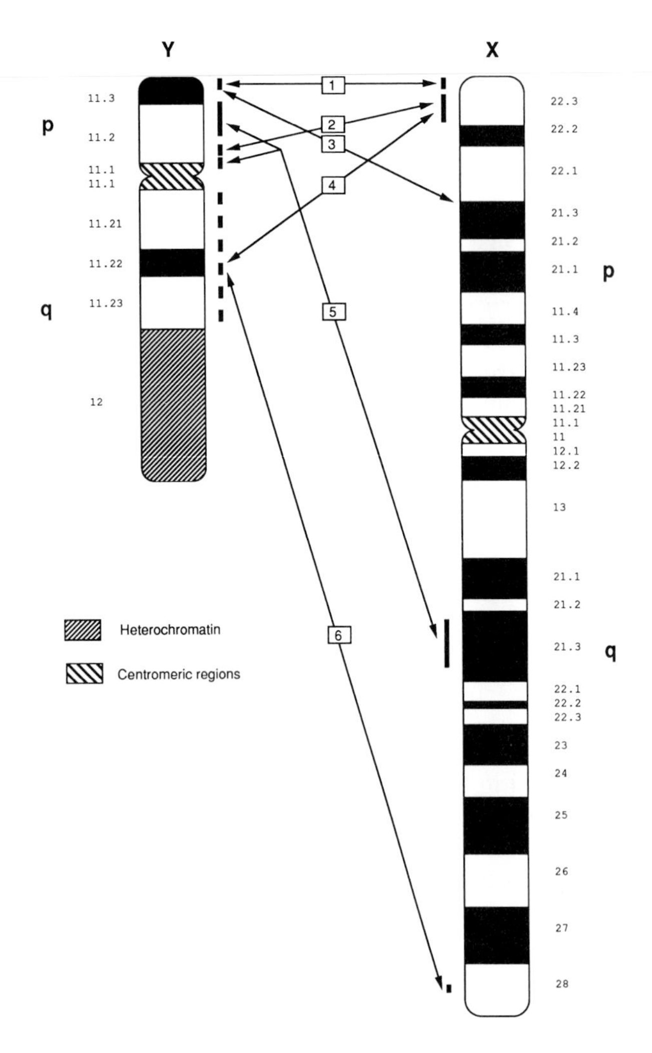

FIGURE 8 DNA sequence homologies between X and Y chromosomes. Chromosomes are drawn according to the conventional representation of their respective banding patterns, including numbering of the bands. Vertical bars joined by arrows show regions displaying DNA sequence homology between X and Y chromosomes. 1. Pseudoautosomal region. 2, 4. Sequences shared by the distal short arm of the X and the proximal short arm or long arm of the Y chromosome (contain STS sequences on Xp and Yq). 3. ZFY gene and its X-linked counterpart ZFX, which escapes to X inactivation. 5. X-Y homologous sequences recently transposed from the X onto the Y chromosome (absent in the Y chromosomes of great apes). 6. Highly homologous sequences between Yq and Xqter.

sively spread over loci on the X chromosome, after their loss by the Y chromosome. According to this view, the pseudoautosomal region of eutherian mammals, including humans, represents the remnants of the ancestral chromosome. The requirement of a pairing segment for proper meiosis would dictate the need of pseudoautosomal sequences submitted to homologous recombination. Fortunately, the range of species encountered in nature provides us with some clues about this evolutionary process. Monotremes, who diverged from other mammals 150–200 Myr ago, have sex chromosomes whose long arms display the same G-banding pattern and undergo homologous pairing at meiosis. Strikingly, sex chromosomes of marsupials, whose divergence from other mammals (eutherians) is dated at 130–150 Myr ago, lack of common region and do not pair at meiosis. These latter could have developed a different system to ensure the proper segregation of their sex chromosomes.

Bibliography

Burgoyne, P. S. (1982). Genetic homology and crossing over in the X and Y chromosomes of mammals. *Hum. Genet.* **61,** 85.

Cooke, H. J., Brown, W. R. A., and Rappold, G. A. (1985). Hypervariable telomeric sequences from the human sex chromosomes are pseudoautosomal. *Nature* **317,** 687.

Ellis, N. A., Goodfellow, P. J., Pym, B., Smith, M., Palmer, M., Frischauf, A., and Goodfellow, P. N. (1989). The pseudoautosomal boundary in man is defined by an Alu repeat sequence inserted on the Y chromosome. *Nature* **337,** 81.

Goodfellow, P. J., Darling, S. M., Thomas, N. S., and Goodfellow, P. N. (1986). A pseudoautosomal gene in man. *Science* **243,** 740.

Page, D. C., Bieker, K., Brown, L. G., Hinton, S., Leppert, M., Lalouel, J. M., Lathrop, M., Nyström-Lahti, M., de la Chapelle, A., and White, R. (1987). Linkage, physical mapping, and DNA sequence analysis of pseudoautosomal loci on the human X and Y chromosomes. *Genomics* **1,** 243.

Petit, C., Levilliers, J., and Weissenbach, J. (1988). Physical mapping of the human pseudoautosomal region; comparison with genetic linkage map. *EMBO J.* **7,** 2369.

Rouyer, F., Simmler, M.-C., Johnsson, C., Vergnaud, G., Cooke, H. J., and Weissenbach, J. (1986). A gradient of sex linkage in the pseudoautosomal region of the human sex chromosomes. *Nature* **319,** 291.

Simmler, M. C., Rouyer, F., Vergnaud, G., Nyström-Lahti, M., Ngo, K. Y., de la Chapelle, A., and Weissenbach, J. (1985). Pseudoautosomal DNA sequences in the pairing region of the human sex chromosomes. *Nature* **317,** 692.

Weissenbach, J., Levilliers, J., Petit, C., Rouyer, F., and Simmler, M. C. (1987). Normal and abnormal interchanges between the human X and Y chromosomes. *Development* Suppl. **101,** 67.

Psychoanalytic Psychotherapies

ROBERT S. WALLERSTEIN, *University of California, San Francisco*

Glossary

Expressive psychotherapy Psychoanalytically based and informed psychotherapy based on expressing or uncovering or analyzing unconscious conflicts (albeit less comprehensively than in psychoanalysis proper) by methods primarily of interpretation leading to ego reintegrations and enhanced insightful mastery

Psychoanalysis (as a therapy) The clinical techniques and interventions as innovated by Freud, based centrally on interpretation and "working through" leading to insight and mastery

Psychoanalytic psychotherapies are the broad spectrum of dynamic psychotherapies based on the theoretical understandings of psychoanalysis as a theory of the mind, ranging from the most supportive to the most expressive psychotherapy including psychoanalysis proper as a specific therapy

Supportive psychotherapy Psychoanalytically based and informed psychotherapy directed toward maintaining and strengthening the ego against neurotic conflicts and symptoms by varieties of techniques aimed at enhancing the ego's capacity to defend against inner and outer disequilibrating pressures

PSYCHOANALYSIS, in a famous statement by Sigmund Freud, its creator, is three things:

It is a *theory* of the mind, a network of psychological concepts to explain the normal and abnormal functioning of the mind in its developmental unfolding over the human life span. Many of its central conceptions—the unconscious, psychic determinism, infantile sexuality and the theory of drives, the nature of ambivalence, of anxiety, the defense mechanisms, psychic conflict, etc.—are by now part of our scientific culture and, to a considerable extent, our shared intellectual heritage. As a theory of the mind, it was summarized in a well-known aphoristic statement that psychoanalysis is "*nothing but* human behavior considered from the standpoint of conflict"—i.e., the picture of the mind divided against itself with attendant anxiety, with adaptive and maladaptive defensive and coping strategies, and with symptomatic behaviors when the defenses fail.

Psychoanalysis is secondly an investigative *method,* the well-known method of free association by the analysand. This makes it possible for the analytic observer to obtain access to the overt and covert rational and irrational data of mental functioning, which then the theory of psychoanalysis renders intelligible and coherent. It is the phenomenological data of psychoanalysis that may be irrational; the method and the theory are rational.

And thirdly, psychoanalysis is a specific psychological *therapy* using the method of free association to obtain the psychic data, thoughts, feelings, memories, fantasies, dreams, etc. that are arranged and comprehended within the theoretical understandings of the theory. Through interpretation of these data leading to insight and "working through" the treatment process is carried progressively forward. It is upon this aspect of psychoanalysis, as a specific therapeutic modality, with its range of indications within its specific reach and limitations, that this article will focus.

Dynamic psychotherapy (or psychoanalytic psychotherapy) is an intensive psychological therapy that is grounded within the framework of psychoanalytic *theory* which, however, uses a different method of inquiry (not free association) and deploys different techniques of intervention (not just interpretation leading to insight) in order to treat a whole other range of patients not suitable for psychoanalysis, again with specific reach and limitations.

Both psychoanalysis and the psychoanalytic psychotherapies are long-term and can go on for many years. A better way to describe this is "open-ended" in the sense that they start as compacts between two people to explore together in an open-ended way the psychological difficulties of the patient, for as long as this proves worthwhile, and with no predetermined duration or necessary end point. This is in contrast to brief psychotherapy which should be better called time-limited therapy; the predetermined limit is usually 12 or 20 once- or twice-weekly sessions of 50 minutes each. The central defining difference is not only the differing duration, though that is of course a major difference (between 10 weeks, say, and several years), but the fact that the time-limited therapy is from the very first within the shadow of the preordained termination date, so that everything that is to be done must be done within that preset time, with all the attendant temptations to withhold various mentally painful areas from psychological scrutiny. By contrast, in open-ended or long-term therapy (whether psychoanalysis *or* dynamic psychotherapy) whatever is not talked about now, can (and will) come out later, since the treatment will not draw to its natural conclusion until all the relevant issues are explored and resolved to whatever extent they can be, however long that takes—given that the external conditions of sufficient available time and resources are maintained. Long-term (open-ended) therapy is therefore very different in its psychological conditions from short-term (time-limited) therapy; it is not just the same thing for more hours.

I. Psychoanalysis as Therapy

To turn to the nature of psychoanalysis as a specific psychological treatment: It was developed by Freud in its major essentials just before the turn of the twentieth century as a purified product out of the congeries of therapeutic approaches in vogue in his time or experimentally introduced by him and his coworker, Breuer—electric stimulations, rest cure, hypnotic suggestion, forced associations on command, etc.—and soon became the first scientific psychology and scientific psychotherapy. It rests on the continuity of focus on the internal workings of the mind on a day-to-day basis, so that each day one can pick up from and develop upon the leaving-off point of the preceding day. Ideally, this would be seven days each week for an hour each day. (This became the "fifty-minute hour" in order to allow the analyst ten minutes between patients to order his/her thoughts and make whatever written notations he/she wished about the patient who had just left and then to set his/her mental focus upon the patient about to come.) However, because of the personal life needs of the analyst, the work week is confined within the traditional five working days, and Freud complained of the "Monday crust," the sealing over the open mental surfaces by the intervening weekend, with the Monday task including that of reopening the continuity of daily mental exploration. Because of limited treatment resources and the need to accommodate more patients, in many cities psychoanalyses are now conducted four days a week. Less than that most analysts do not consider proper psychoanalysis, because the vital element of day-to-day continuity is lost.

Within this time frame, the classical treatment is conducted with its well-known features: the patient's recumbent position on the couch; the analyst sitting behind and out of the line of direct vision; the minimization of stimuli, of any external intrusion; the agreement by the patient to *try* to say whatever comes to mind no matter how seemingly remote, irrelevant, trivial, repugnant, shameful, or anxiety provoking (the so-called "fundamental rule"); the renunciation of motor activity with the agreement to try to put all mental content into expressed verbal form; the understanding that the analyst will choose when and how to intervene verbally and not necessarily in usual conversational dialogue; the analyst's constant endeavor to focus exclusively on the understanding of the shifting contents of the patient's mind, excluding as much as humanly possible the analyst's own concerns, predilections, values, and judgments. The purpose of all this is to promote the fullest untrammeled and continuous access to the contents of the patient's mind, and on to its remotest reaches into its past and its unconscious depths. In this process one thing leads to another while the patient (apparently randomly) free associates, which the analyst gradually apprehends by a counterpart process of "free floating attention," i.e., without preconceptions as to what is important, what leads to, connects with, or explains what.

It is within this "regressive" analytic process that the patient's mental contents gradually emerge around the figure of the analyst. Past and long-forgotten dispositions, interaction patterns, affects, traumata, and various defensive or adaptive efforts at their mastery all eventually reemerge in the interactions with the analyst, in the "here and now" of the immediate analytic situation (the transference). That is, the psychic past is reenacted in the analytic present; it is recognized and interpreted via the inappropriateness of the present (transference) reactions of the patient to the reality of the ongoing interaction with the analyst. This full revival of the past in the present is called the "regressive transference neurosis." It is through the systematic interpretation of these transference phenomena in all their endlessly ramifying complexity that past unresolved problems are reworked, new more adaptive solutions found, and concomitantly the prior maladaptive, neurotic solutions are more or less thoroughly abandoned. In this sense psychoanalysis is the fullest possible journey of exploration of the human mind, in the course of which the patient has "rewritten" his/her autobiography, and has along the way, and as a by-product of this process, shed as fully as possible the neurotic problems that brought him/her to treatment in the first place.

In bringing this treatment to a reasonably successful conclusion, psychoanalysis relies centrally on the technique of interpretation leading to enlarging insights (that is, enlarging understandings of the mental scheme of things, how meanings hang together in ever more comprehensive and plausible configurations that "make sense") which then have to be "worked through" again and again as they reemerge around ever-widening new mental contents. Edward Bibring described five essential therapeutic techniques—suggestion, abreaction (catharsis), manipulation, clarification, and interpretation—that in varying combinations, characterize the different psychoanalytically based psychotherapies, psychoanalysis included. What distinguishes psychoanalysis as a specific therapy within this spectrum is that interpretation is the supreme agent in the hierarchy of techniques characteristic of analysis, and that all the other techniques are subordinate to it; that is, all the others are employed only in the service of making proper interpretation more possible and more effective.

Who are the patients for whom this most intensive, time-consuming, and often stressful of therapies is specifically indicated? It has been stated quite succinctly that psychoanalysis is the treatment of choice for that narrow middle band of patients who are sick enough to need it, but also well enough to tolerate it. I will explain. There are many psychiatric patients (in fact the largest number) whose symptoms and problems in living can be satisfactorily resolved with dynamic psychotherapies that are less intensive and/or less prolonged. Such patients who are deemed too well for psychoanalysis and don't need the thoroughgoing life and character reconstruction that it provides, include those with acute reactive illnesses, posttraumatic stress syndromes, various kinds of situational maladjustments, *circumscribed* neurotic illnesses, including many symptom-neuroses (reactive depressions, anxiety and phobic states, some obsessive and compulsive states), and many character neuroses (like varieties of passive-aggressive, dependent, compliant, compulsive, even some masochistic character difficulties).

There are also many patients who come to psychotherapy, often with concomitant psychoactive drug management, whose illnesses are severe and who cannot tolerate the anxiety-provoking stresses of psychoanalysis, i.e., the intensive regression-inducing of a treatment contrived to pull for ever more elaboration of a hidden internal world of primitive passions and archaic (infantile) constellations, with the concomitant minimization of external correction by reality testing—the recumbent position out of eye contact with the analyst, the often one-sided dialogue with a minimum of corrective feedback and responsive cuing, the shielding from reality intrusion, and the ban (except in emergencies) on contact between the analyst and the patient's outer world. For patients with vulnerable "ego strength," an effort at psychoanalysis can become a psychologically disorganizing process with an ever-present danger of regressive psychotic swings, of severe acting out and flight from treatment, of suicidal pressures, etc. Such patients who are deemed too ill for psychoanalysis and need to be treated by other dynamic psychotherapies comprise patients who are designated borderline personalities, narcissistic characters, character disorders (gamblers, kleptomaniacs, delinquent and antisocial characters, etc.), addictive disorders (alcohol and/or drugs), severe sexual disorders, severe character neuroses (especially rigidly compulsive or paranoid or masochistic or hypochondriacal characters), and even some few very severe symptom neuroses.

Who then are left for psychoanalysis of those who come for psychiatric treatment? A narrow middle band: Assessments of the size of this segment vary in relation to differing technical perspectives within the field; this is the controversy between those who advocate the "narrowing" versus those who advocate the "widening" scope of the indications for psychoanalysis. My own perspective is that of a "narrowing scope" and I would count about 5% of those who come to psychiatric evaluation as being suitable for psychoanalysis per se. These are patients with classical symptom neuroses and also moderate character neuroses set within a strong "ego organization." These patients need and are amenable to psychoanalysis and are also able to tolerate it.

Given this perspective on the limited place of psychoanalysis as a specific therapy for neurotic disorder, where then does its social value as well as its scientific importance lie? This is in three realms: research, education, and treatment. As a research instrument, psychoanalysis affords the most far-reaching and unparalleled access to the innermost workings of the mind, to knowledge of developmental unfolding, character formation, and normal and abnormal mental functioning. It is where the knowledge of psychoanalysis as a theory of the mind has evolved and out of which the specific therapeutic applications of psychoanalysis *and* also of the range of psychoanalytically based psychotherapies has been created. This need for continued knowledge of the mind as it manages inner conflicts and the outer tasks of life is a most important social and scientific value of psychoanalysis.

As an educational tool, the personal psychoanalysis of the therapist, required for all those who seek training as practitioners of psychoanalysis, and often sought by those who seek enhanced effectiveness as dynamic psychotherapists, plays an indispensable role in creating successive generations of clinicians capable of deploying both psychoanalysis and the whole range of dynamic psychotherapies with ever-increasing effectiveness. And as a specific treatment, for that number for whom psychoanalysis is specifically indicated, it offers a possibility for thoroughgoing resolution of neurotic problems and for fundamental character reconstruction not offered by any other treatment modality.

There are however others, as I have indicated, who hold to a "widening scope" of indications for psychoanalysis, who feel it to be a superb therapeutic tool to be extended and made available as broadly as possible across the spectrum of nosologic entities. This is linked to the value assumption that, because it is a treatment that offers a more thoroughgoing reconstructive approach to personality organization, it is therefore a better treatment, achieving results, when it succeeds, that are more complete, more stable, more enduring, and more proof against subsequent adverse environmental vicissitude. Psychoanalysis has therefore over the years been extended—usually with more or less modification to render it applicable to the particular situation—to varying categories of people, types of problems, and treatment situations beyond those with whom it was generated. This has included children, and later, adolescents, groups (in of course very modified form), delinquents, psychosomatic patients, even the overtly psychotic (albeit in extremely modified form), and very currently, narcissistic characters, and borderline personality organizations. This whole movement attempting to extend analysis towards the nosologic periphery was reviewed in a most balanced way by Leo Stone in a widely remarked article on "the widening scope" of psychoanalysis; it was in Anna Freud's discussion of this paper that she undertook to spearhead the opposed trend towards narrowing the indications of analysis. She had herself been most instrumental in creating the field of child and adolescent analysis, which has been the one fully successful extension of psychoanalysis as a therapeutic modality beyond the original patients around whom it was devised, psychoneurotic adults.

This same possibility for a widening scope of psychoanalysis was presented in another influential article by Edward Glover on the indications for psychoanalysis, where he divided the potential patients into three categories, which he designated the ideally suitable, the moderately suitable, and those for whom the prospects were dubious but for whom no other treatment approach was seen as offering any hope for substantial help. These patients in the third category (borderline, narcissistic, addictive, etc.) became the patients taken into psychoanalysis on the basis of so-called "heroic indications" where expectations would be limited, and failure would be seen as "honorable failure." It is the concept of psychoanalytic treatment for these much sicker patients than those seen in usual outpatient psychoanalytic practice that has been a major rationale for the psychoanalytic sanatorium, where intensive psychoanalytic treatment with just such patients could be carried on within a protective sanatorium milieu with its total life management.

The accounting in my book, Forty-two Lives in Treatment, of a research program studying the patients at such a sanatorium, The Menninger Foundation, speaks to the failure of fulfillment of reasonable treatment goals with these sicker patients treated by psychoanalysis on the basis of "heroic indications." Other workers in the field still currently feel otherwise. Basically, the controversy can be captured aphoristically as the difference between the perspective of, as much as possible, fitting the treatment to the patient (the "narrowing scope") and the perspective of, as much as possible, fitting the patient to the treatment ("the widening scope").

II. Psychotherapy vis-à-vis Psychoanalysis

To turn now to the description of all the other psychoanalytically based psychotherapies as specific psychological treatments. This is the domain of the varyingly expressive and varyingly supportive psychodynamic psychotherapies available for that much larger realm of potential patients who are either too well for psychoanalysis or too sick for it. As distinct from its psychoanalytic parentage, which originated in Freud's late-nineteenth-century Vienna, in a world in which no other scientific psychotherapy was anywhere available, psychodynamic psychotherapy is a peculiarly American creation (though now practiced worldwide). It was brought into being in the interregnum between World Wars I and II and crystallized into a coherent body of theory and technique in the post–World War II decade, when its parent psychoanalytic theory became the dominant psychological perspective within American psychiatry. The dynamic psychotherapies, expressive and supportive, arose as a pragmatic response to the treatment needs of the vast majority of patients coming to psychiatric clinics in universities and hospitals, as well as to the offices of private practitioners, who were clearly not appropriate for psychoanalysis proper.

These psychoanalytically based psychotherapies are divided conceptually into two theoretically distinguishable modes. On one side are treatment approaches whose aim is expressive, i.e., to uncover (or make conscious) psychological conflict through analyzing the "defenses" and the "resistances," and thereby to resolve conflict through interpretation, insight, and insight-driven change (just as with psychoanalysis proper). On the other side are seemingly polar opposite treatment approaches whose aim is supportive, i.e., to diminish the force of external or internal pressures through whatever available "ego-strengthening" technique, so as to strengthen the individual's capacity to suppress mentally painful emergent conflict and its dysphoric or symptomatic expression, and thereby to produce change through means other than interpretation and insight.

Conceptually, this dichotomization makes psychoanalysis proper a more thoroughgoingly uncompromising form of expressive psychotherapy. In terms of the entire spectrum of psychoanalytic psychotherapies ranging from psychoanalysis proper through expressive psychotherapy and supportive psychotherapy, it becomes conceptually more important to subdivide them between the expressive psychotherapeutic modalities (psychoanalysis included) and the supportive psychotherapeutic modalities rather than between psychoanalysis and all the other dynamic therapeutic approaches.

Yet useful as this expressive-supportive dichotomization is for heuristic, prescriptive, and prognostic purposes, it also can be misleadingly oversimplifying. For clearly, in a larger sense, all psychiatric treatment that is experienced as helpful must be truly supportive even when most uncompromisingly expressive. After all, what can be more supportive than an open-ended treatment like psychoanalysis, occurring daily, and going on for as many years as need be, in which the patient is enabled to express any verbal content, no matter how repugnant, embarrassing, inconsequential, etc., without fear or hindrance, without moral censure or behavioral pressure, and where the entire enterprise consists of two people, focused, as exclusively as human capacity permit, on the concerns of the one. On the other hand, any treatment, no matter how single-mindedly supportive in the sense of strengthening the defenses, suppressing unwanted conflict and symptom expressions, nonetheless must in some way also be expressive of some aspect of the patient's concerns. The issue, seen in this way, is not expressive vs. supportive, but rather expressive of what, and when, and how, in regard to the patient's mental and emotional life, and concomitantly supportive of what, and when, and how, in regard to that same mental and emotional life.

How do these two opposed treatment approaches differ in techniques, and who are the patients for whom they are differentially indicated? In outer form and structural arrangements, dynamic psychotherapies, however expressive *or* supportive, are much alike, while differing sharply from psychoanalysis. The patient sits up vis-à-vis the therapist, with direct eye contact and the expectation of feedback and reciprocal exchange. In that sense, unlike psychoanalysis, where the burden is on the patient to keep saying whatever comes to mind with the analyst choosing when and how to intervene, in psychotherapy the format is more like a conversational exchange. Nor has the patient undertaken a commitment to try to say everything that comes to mind without censorship; the patient has only agreed to bring problems for joint consideration as he/she feels able and willing to (and there is no "fundamental rule" that is being violated by decisions to withhold particular mental content).

Frequency is more flexible in the psychotherapies, ranging (in both expressive and supportive treatments) anywhere from once weekly to three or four times. Unless a variety of time-limited therapy is selected, the duration is open-ended as with psychoanalysis; though in practice this usually means less years in duration, it can be just as long, and even on occasion (unlike psychoanalysis) lifelong. Such patients, who have been called "therapeutic lifers" are in an unending supportive relationship akin to the lifelong medical maintenance regimens of diabetics or patients with cardiac disability. Overall, dynamic psychotherapies are carried out at lesser frequency than psychoanalysis (once or twice weekly) and for fewer years (one to two years as against four to six years); thus they run for between 50 and 200 therapeutic hours as against roughly 500 to 900 hours in analysis. The length of the treatment is again usually 50 minutes but there are exceptions; in some long sustaining, supportive psychotherapies, especially with schizoid individuals, fearful of interpersonal intimacy, sessions are in some instances curtailed to no more than 30 minutes; less frequently in both expressive and supportive therapies, at times of acute emergency, sessions are extended as long as necessary up to two hours or more (this would be excessively rare in psychoanalysis, by virtue of the kinds of patients deemed appropriate for it in the first place). And occasionally in the psychotherapies there are emergency weekend or evening hours, and—at the other end of the frequency spectrum—there are supportive treatments less frequent than once a week. When the frequency is low, say only once a month, the sessions become a follow-up reporting, rather than a continuing psychotherapy with a therapeutic process in motion.

And again, in contrast to psychoanalysis, where such events are rare occurrences to deal with unanticipated extreme situations, in the psychotherapies there is a greater range of external impingements upon the therapeutic process. They can include adjuvant psychoactive drug management, coordination with general health care by the family physician, telephone, or other extra-session contacts, involvements with third parties, whether members of the patient's family (in patients in dependent circumstances, like youngsters or the elderly), other treating physicians, occasionally the employer, school, court, lawyer, etc. All of these kinds of extra activities are more frequent, the more supportive the particular psychotherapy is intended to be.

III. Expressive Psychotherapy

Within this overall common structure, how do the central technical interventions differ between the more expressive and the more supportive psychotherapies? Essentially, in expressive psychotherapy, within the vis-à-vis situation, with the patient free (but not required) to bring up problems and anxieties in whatever way and at whatever pace he/she desires, the therapeutic endeavor is intended to be as interpretive and insight-aiming as possible, and intent to bring about as much change as possible by intrapsychic conflict resolution based on the uncovering of unconscious conflict and the consequent enlargement of insightful mastery. The "techniques" of psychoanalysis—though not true free association, or dream analysis, or reconstructive interpretation of infantile emotional constellations—are applied, however, within different structural arrangements, the vis-à-vis conversational mode, which allows a more constant reality contact, and minimizes the regressive pull towards the more archaic, buried fantasies, and memories.

Expressive psychotherapy is then the treatment of choice for those persons with the requisite ego-strength, intelligence, tolerance for anxiety, etc. who have serious enough, but nonetheless circumscribed neurotic conflicts. These are individuals who don't *need* full psychoanalysis, because they are not that deeply embedded in their neurotic characters and problems. If such patients will assume responsibility for their character traits and their problems in living, and are willing to look introspectively at the irrational aspects of their relationships, significant change can be effected without the full-scale reconstructive effort to uncover the infantile roots of the neurotic personality development. Measured along a temporal parameter, the question is, can the marital problems be resolved without the need to recreate and to work through insightfully the earlier prototype, the infantile conflicts with the mother, repressed behind the childhood amnesia? In this sense, the distinction is made between psychoanalysis as the therapy that *analyzes* the conflicts back to their infantile roots, and analytically oriented expressive psychotherapy which *recognizes* (and only partially "analyzes") those same conflicts, and rationally utilizes this recognition in the therapy. Increasing insight is a concomitant of this kind of therapy, but is only equivalent to the "depth" of the problem dealt with. It clearly stops at the unconscious infantile genesis of the original conflicts. Appropriate here (and also for psychoanalysis) are patients who fall into the categories of the classical symptom neuroses (the depressed, anxious, phobic, obsessional, and compulsive states) and the character neuroses (aggressive, passive-aggressive, passive-dependent, depressive, masochistic, compulsive, even some paranoid character formations, etc.).

The distinction here between those who would seem to *need* psychoanalysis and those who can be (and therefore should be) treated by a less intense therapy can be illustrated by the example of a traumatic neurosis. The therapeutic work would be limited to a defined sector of the individual's life and problems, anchored to the stresses precipitating the immediate breakdown, and enough of their roots to make it possible to resolve the current stress. Thus, with the survivor of an accident, grief-stricken and guilty at his companion's death when he himself luckily survived, the events of the death, the complex of ambivalent feelings about the companion, and maybe even the manner in which this current friendship had recreated the conflictual sibling relationships of childhood, might all come within the scope of the expressive therapeutic work—though not the still earlier roots in the infantile relationship with the parents. Or rather than a treatment confined to a specific sector of difficulty, an expressive psychotherapy could involve a thorough concern with characterological difficulties and their maladaptive roles in the individual's life functioning, staying, however, at the "level" of the individual's willingness (and capacity) to assume responsibility for their existence and their modification in the present, without the (need for the) concomitant uncovering of their infantile roots. Such treatment can be long term and can range widely over the whole of the patient's life adjustments.

The possibility for adequate conflict resolution and relief from distressing symptoms in such expressive psychotherapies rests of course on the relative "autonomy" of the conflict in the present (with the wife) from the earlier infantile prototype (with the mother), though clearly a developmental linkage can be traced from present-day neurotic configuration to the original pathogenic template. Actually, this relative autonomy is often enough so, as to provide the large patient population for the large body of psychiatrists, trained in dynamic psychotherapy (who outnumber probably by a factor of ten those who are further trained as psychoanalysts). With the value judgments (already mentioned) rife in the field, that the more expressive-interpretive treatments are "better" because they promise changes that are more stable, enduring, and proof against subsequent environmental pressures, the tendency is fostered among practitioners of dynamic psychotherapy, to (in the words of a popular training aphorism) "be as expressive as you can be, and as supportive as you have to be."

IV. Supportive Psychotherapy

The techniques of expressive psychotherapy are relatively clear but the terrain of the supportive therapeutic mode is the much more ambiguous. Expressive psychotherapy is a foreshortened version of its psychoanalytic progenitor (which has become over the years substantially codified and well articulated), whereas supportive psychotherapy is a melange of all else that goes on in psychotherapy. In the early days of psychoanalysis, the distinction was constantly made between psychoanalysis as the first scientific psychotherapy, contrasted with all other preexisting psychotherapeutic activities. These were oversimplistically seen as only varieties of suggestion, and therefore inherently unpredictable and unstable. Hypnosis was looked at as the prototype of such suggestive therapies. As employed by nonanalytically trained practitioners these other (supportive) psychotherapies often carried heavy doses of reassurances based on common sense; and this too came to be considered a hallmark of the supportive approach. But this is misleading and oversimplified. Actually, explicit reassurance is very seldom reassuring to the patients whose problems are of sufficient significance to have them in the therapist's consulting room. In such instances the effort to reassure often backfires; it only convinces the patient that the therapist who offers the reassurance simply does not understand the nature of the difficulty and/or is unwilling to hear about it, and seeks to inhibit its expression.

What then does supportive psychotherapy actually consist of? Within the psychodynamic literature one of the earliest efforts to delineate its complex nature was that of Merton Gill, who identified three kinds of technical interventions that he felt "strengthened defenses." This was against the opposite expressive approaches that undertook to uncover and interpret the defenses as a step towards an eventual integration. These explicitly supportive interventions are (1) the consistent encouragement of adaptive combinations of impulse and defense with the simultaneous active discouragement of maladaptive combinations; (2) the deliberate refraining from interpretive focus on those character configurations, no matter how rigid or maladaptive, that are adjudged vital to the patient's stability of functioning; and (3) the partial uncovering of some aspect of neurotic conflict (say within a troubled marriage or work situation) in order to reduce the inner pressures towards unwanted symptoms or dysphoric affects, thereby altering the balance of psychic forces and rendering the repression of the core of the neurotic conflict easier to accomplish.

Within the framework of the five therapeutic techniques adumbrated by Bibring that I have earlier outlined—suggestion, abreaction, manipulation, clarification, and interpretation—psychoanalysis could be described as resting centrally on interpretation (leading to insight and insight-based change) as the vital therapeutic vehicle, with all other techniques employed *only* when necessary to render interpretation more possible and more effective. Expressive psychotherapy could be described as resting importantly on interpretation, but also considerably on clarification and with some admixture of the other techniques. Supportive psychotherapy could be described as that variet*ies* of psychotherapy that deploy all five of these techniques in whatever mix is most indicated by the specific therapeutic needs of the particular patient, interpretation itself included.

The most comprehensive effort to systematize the technical ingredients that constitute the complex array of therapeutic approaches that are differently supportive is the 30-year-long Psychotherapy Research Project of The Menninger Foundation. The central common supportive mechanism described in that study is the evocation and firm establishment of a positive dependent emotional attachment to the therapist (the meanings of which are not interpreted or "analyzed") within which varieties of conflicted emotional needs achieve varying degrees of overt or covert symbolic gratification. This dependent emotional attachment seems in turn to be an essential precondition to the operation of the various other supportive mechanisms that are then varyingly admixed with it. It is the basis of the so-called "transference cure," the willingness and capacity of the patient to reach therapeutic goals, changes in behaviors, symptoms, and modes of living, as something being done "for the therapist," as the quid pro quo for the emotional gratifications received within the benevolent dependent attachment. It is, in effect, a psychic trade; "I do it—make the agreed-upon and desired changes—for you, the therapist, in order thereby to earn and/or maintain your support, your esteem, your love."

Upon this base, a further array of supportive maneuvers can be erected as indicated by the differing needs of the particular patients. Where the need for the continuing emotional gratifications within the benevolent dependent attachment cannot be otherwise transferred, or attenuated, or developed into a stable set of internal mental representations of the helpful therapist, it can be maintained within a non-ending therapeutic relationship. These become those maintenance supportive psychotherapies (analogized earlier as akin to the maintenance of the diabetic or cardiac patient) required by some vulnerable sicker patients whose hold on reality-adapted psychic functioning is tenous. On the other hand, those patients with comparable dependent tendencies but with stronger psychological resources—a greater capacity to develop identifications with the helping, problem-solving therapist—are often able to terminate treatments, but do so through a kind of planned tapering and "weaning" process. These are patients who can identify successfully with the therapist's way of approach towards and mastery of conflict pressures and can thus come to carry on, on their own.

Intermediate between those capable of being helped to psychological autonomy via such identifications, and those for whom continued therapeutic contact, possibly lifelong, is necessary to maintain a level of adequate functioning, are those patients with whom the attachments and the gratifications can be "transferred" within the patient's now improved life situation, a transfer usually to the spouse. How successful this is depends not only on the effectiveness of the therapeutic work within the ongoing treatment, but also on the capacity and willingness of the other (the spouse) to carry this transferred emotional burden indefinitely. And indeed, the vulnerability of this kind of resolution, is precisely its dependence on a continuing fully supportive life context, in which some patients are more fortunate than others.

Another change mechanism within the supportive therapeutic mode is that of fostering the displacement of the neurotic dispositions into the treatment situation with the alleviation of the manifestations of the impaired functioning and/or symptoms in the outside life. Typical here is the conversion of an unduly dependent and submissive individual into more satisfying and assertive behaviors outside the treatment, on the basis of a (covert) new submissiveness (within the treatment) to the therapist. The patient thus experiences the altered external behaviors as the price of the continuation of the dependent gratifications within the treatment. Again, the ultimate success of such a maneuver depends upon life circumstance, reinforcing positive feedback and enhanced self-esteem, when the new behaviors bring not neurotically feared disaster but real reward and gratification.

Quite opposite to the so-called "transference cure" is the mechanism of change by what I have called the "antitransference cure." These are changes not "for the therapist" but "against the therapist," in which the patient changes on the basis of perceived contrary expectations by the therapist, usually as an act of triumph over the therapist in the treatment struggle. Such "cures" of course have to be buttressed in some way agains their potential instability, either through reinforcement from a stabilizing life situation, or from positive environmental feedback in response to enhanced life functioning.

A particular concept that has had considerable currency is the "corrective emotional experience," at times invoked as the construct used to explain the supportive therapeutic approaches, in almost the counterpart role to interpretation leading to insight as the central mechanism in the expressive therapeutic approaches. The concept can of course be used thus broadly to cover the whole range of supportive therapeutic maneuvers, since by its nature everything that goes on in a therapeutic context is indeed intended in one sense or another to be a corrective emotional experience. More properly, I feel the concept should be used in a more restricted way in relation to treatments that provide—as a central mechanism—a kindly, understanding, reality-oriented figure (the therapist) who is able steadfastly to meet the patient's pressures and importunities with a (benevolent) neutrality that does not fall victim to the interacting entanglements by which the patient has managed to maintain his/her neurotic suffering in the pretreatment life experience.

Only subtly different from such corrective emotional experience are the operative mechanism, reality-testing and reeducation. Here again, reality-testing and reeducational activities can be conceived broadly as part of every psychotherapy, whether supportive therapy at one end of the spectrum or psychoanalysis at the other. However, again, I am using a narrower construction of educational activities as part of the content focus in a majorly supportive treatment. In it the therapist plays a directly educational role in the transmission of advice, information, and education to society's normative standards and expectations. Again, here at issue is the therapist's capacity to play this role in a way that the patient perceives as nonjudgmental and, if at all coercive, as guided solely by the patient's best interests. Clearly, the distinction between such activities and the steady provision of a corrective emotional experience cannot be a clear one. With both strategies the patient is supported and educated towards reality-oriented problem-solving and reality-corrected emotional responses on the basis of the "borrowed strength" that comes from the identification with the therapist. And again, the maintenance of changes achieved on this basis is dependent on feedback reinforcement from a fulfilling environment admixed with some transfer of the attachments to significant life others.

To shift in supportive mechanisms under consideration, there is the kind of life-manipulation involved with the sicker patients in supportive psychotherapies who often come to hospital, residential care, and day hospital settings (the alcoholic, the addicted, the acting out, the suicidal, etc.), the planned disengagement (temporarily or permanently) from unfavorable and noxious life situations. And of course the counterpart position needs to be kept in mind, that for other patients, it is critical to their best chance for treatment success that their therapy be conducted against the backdrop of an ongoing interaction *within* their usual life situation, and that if this cannot be properly maintained, for whatever reason, their chances for an optimal result diminish.

Still another and different helping mechanism can be called the "collusive bargain," quite common in tacit and minor ways, though at times quite explicit, and playing a major treatment role. At issue here is the agreement, however tacitly or explicitly arrived at, to exclude particular areas of personality functioning or symptoms from therapeutic scrutiny—leaving more or less consequential islands of maintained psychopathology—in return for the patient's willingness to make changes in (all) other areas of personality functioning or symptoms. This is actually akin to the basis of the "transference cure," since the patient is here making changes "for the therapist" in return for a specific reward within the therapeutic relationship, that is, the shielding from therapeutic probing of a particularly tenacious or rewarding symptom or behavior. The success of such a maneuver depends on the meaning and value of such a symptom or behavior to the patient—it can be a homosexual lifestyle—as well as the capacity of the patient sufficiently to detach the symptom or behavior at issue from the other aspects of life functioning, which the patient and therapist can then commit themselves to trying to change. Since the symptom or behavior "allowed" to the patient in this compromise-solution is experienced as in some ways rewarding or gratifying, these particular therapeutic outcomes have a built in stability.

It should be clear from all this that there are a great variety of specifiable ways in which psychotherapies can support improved functioning, and that ways can additionally be built in to maintain such improvement in stable and enduring fashion. Clearly, these various mechanisms interact in distinctive configurations in particular therapeutic courses, with varying combinations of meeting emotional demands and expectations, of agreed-upon therapeutic "trades," of specific substitutions of more adaptive impulse-defense configurations for less adaptive ones, of specific agreements about what to talk about and what not to talk about, of various manipulations in relation to the patient's life situation and various kinds of transfer of the emotional attachments, of specific engagements with and disengagements from ongoing life context, and of managing the positive feedback reinforcements that result in enhanced self-esteem from behaviors and relationships altered in desired directions.

Given this great variety of approaches to the implementation of the supportive therapeutic goals that have to be differently fashioned for each patient in relation to that patient's individual problems, two corollaries follow directly. The first is the high degree of skill, experience, and psychodynamic understanding that goes into the appropriate fashioning and execution of a supportive psychotherapy; this is contrary to the common misconception that in contrast to the skill in psychodynamics required by expressive psychotherapy, that supportive therapy consists of no more than common sense, goodwill, and kindly reassurance. Actually, neither kind of psychotherapy involves any less knowledge or skill. This links to the second corollary, that supportive psychotherapy, comprising a more diverse array of techniques, necessitates a greater flexibility (albeit disciplined via a sound framework of theoretical understanding) and permits (and at times requires) a wider deployment of such "extras" as adjuvant psychoactive drug management, contacts with third parties, telephone or other extra-session contact, etc.

Within this overall conceptual framework it should be clear that supportive psychotherapy is the treatment of choice for a more diverse range of patients than is expressive psychotherapy; in fact, supportive psychotherapy is indicated both for some of the patients among those "not sick enough" for analysis, and, even more so, for the great preponderance of the patients in the "too sick" category, who are usually too sick for analysis or *any* intensive expressive approach. The first category of basically healthier individuals comprises many of those caught up in disruptive responses (anxiety, depression, rage) to intense traumatic or disturbing situations; these include some grief reactions, acute anxiety states, situational maladjustments, etc. Although, depending on the individual circumstance, expressive-interpretive activity is often also indicated in such situations, there is also and just as often the need to slow up, to take stock, to consider the situation, and to reintegrate, over time, to the best of one's coping and mastery potential. Such therapies are usually shorter rather than longer; they usually comprise admixtures of supportive and expressive elements.

The other larger and more important category of patients for supportive psychotherapies are those much sicker individuals who require often long-time sustaining therapeutic relationships (at times lifelong) and who at best exhibit slow rates of improvement. Stability of psychological functioning at the best achievable level is often the therapeutic goal; at times the goal is more curative. To be placed here are most of the borderline (and sicker) patients, some of the narcissistic characters, some of the other very severe and refractory character neuroses (very rigid paranoid or masochistic or hypochondriacal characters), the most severe addictions, alcoholisms, sexual disorders, delinquent and sociopathic characters, etc. The caveat should of course be added that with almost all of these some degree of expressive therapeutic work can also usually be done (remember the aphorism already quoted, "Be as expressive as you can be, and as supportive as you have to be"). The problem for expressive work with these patients is that they have poor impulse control, poor anxiety tolerance, and a ready vulnerability to regressive (psychotic or suicidal) swings in psychological functioning. The eruption of a florid psychotic state is thus often a present danger. It is among just such patients, incidentally, that therapeutic extensions have been made into more expressive approaches, even at times psychoanalysis, on the basis of attempts to "widen the scope" leading ultimately to the concept of "heroic indications" as already discussed. Though there is major controversy about this, the preponderant view is that such extensions of therapeutic ambition have worked poorly.

This concludes this overview of psychoanalysis and the long-term dynamic psychotherapies (expressive *and* supportive) from the point of view of their overall nature and scope, the varying techniques that characterize them, their reach and their limitations, and their differential indications for them in relation to the range of psychopathology.

Bibliography

Bibring, E. (1954). Psychoanalysis and the dynamic psychotherapies, *J. Amer. Psychonal. Assn.* **2,** 745–70.

Brenner, C. (1955). "An Elementary Textbook of Psychoanalysis." International Universities Press, New York.

Freud, S. (1916). "Introductory Lectures on Psycho-Analysis," standard ed. Vols. 15 & 16, Hogarth Press, London (1963).

Gill, M. M. (1951). Ego psychology and psychotherapy, *Psychoanal. Quart.* **20,** 62–71.

Glover, E. (1954). The indications for psychoanalysis, *J. Mental Science* **100,** 393–401.

Stone, L. (1954). The widening scope of indications for psychoanalysis, *J. Amer. Psychoanal. Assn.* **2,** 567–94.

Wallerstein, R. S. (1986). "Forty-Two Lives in Treatment: A Study of Psychoanalysis and Psychotherapy." Guilford Press, New York.

Psychoanalytic Theory

JACOB A. ARLOW, ARNOLD D. RICHARDS, *New York University College of Medicine*

Glossary

Borderline state Descriptive term referring to a group of conditions that manifest both neurotic and psychotic phenomena without fitting unequivocally into either diagnostic category. Some ego functions are fairly well preserved but other ego functions show impairment, resulting in reduced flexibility, adaptability, and interference with the overall evaluation of reality

Identification Psychological process whereby the subject assimilates an aspect, property, or attribute of the other and is transformed wholly or partially after the model the other provides. It is by means of a series of identifications that personality and character of the individual are constituted

Narcissism Concentration of psychological interest upon the self. This may range from healthy self-esteem and pride and pleasure in one's own body and mind, relationships, and achievements to pathological forms of brooding, painful self-consciousness, and an increased propensity for shame

Psychic apparatus Hypothetical division of the mind into various systems, agencies, or groups of functions to aid in the understanding of the psychological development, experience, and behavior of human beings. It does not imply specific anatomical or neurophysiological nervous system structures

Transference Displacement of patterns of feelings and behavior originally experienced with significant figures of one's childhood to individuals in one's current relationships

PSYCHOANALYSIS HAS A threefold character. It is a method for studying the function of the human mind, a means of treating certain psychological disorders, and a body of knowledge derived from psychoanalytic investigation. Mental activity reflects the function of the brain. It constitutes the result of the dynamic interaction of conflicting psychological forces operating within and beyond the scope of consciousness. The nature of an individual's intrapsychic conflicts and the various compromise formations instituted in an attempt to resolve conflicts are deeply influenced by the vicissitudes of individual development.

I. Psychoanalysis as a Biological Science

Psychoanalysis takes its place among the biological sciences as a naturalistic discipline deriving conclusions from observations within a standard setting. Psychoanalysis views the functioning of the mind as a direct expression of the activity of the brain. This activity reflects the experience of the total organism and operates according to certain inherent biological principles. A primary principle is the tendency of the human organism to seek pleasure and to avoid pain or unpleasure. Clearly, this principle must have had survival value in the course of evolution since painful sensations are likely to be noxious in nature (i.e., threatening the integrity of the organism) while pleasurable sensations are usually associated with gratification of biological needs, with security, and safety.

This last point is especially pertinent because of the helpless, immature state of the human infant at birth. The newborn child is totally incapable of

Encyclopedia of Human Biology, Volume 6.

fending for itself. Without the nurturing and protective care of adults, it would soon perish. Furthermore, this state of dependency continues for a long time, longer than in the case of other mammals. The consequences of this fact for the development of the human psyche are profound. It underlies the importance of the "others" upon whom the individual depends. Perforce, man is destined to become a social animal, keenly aware of the distinction between himself and others, as well as his relations to others. Recent studies of infant development have demonstrated certain features of preadaptation to socialization in the form of inherent patterns of behavior, patterns that serve to stimulate pleasurable reactions in the mother and dispose the infant in turn to respond to the mother's expressions. Human communication begins in the context of the *pleasure–unpleasure principle*.

At the beginning of life, it would appear that the quest for complete and instantaneous gratification is the paramount principle of mental activity. The exigencies of human existence, however, are such that attaining endless, unalloyed pleasure is impossible. Pleasurable sensations disappear as new needs arise. Tensions rise when needs are not immediately met and new experiences do not correspond fully with memories of lost pleasures. Thus, in pursuit of pleasure, gratification of needs must be postponed and some activity relating to the external world must be instituted. Help must be enlisted and substitute gratifications accepted.

The capacity for pleasure is biologically rooted. It is related to the physiology of the body and ordinarily seems to follow a consistent course of maturation and development. During the earliest months of life, the pleasurable sensations connected with feeding seem to be the most important ones. These include not only the alleviation of hunger but also the concomitant sensations associated with the experience—bodily contact, warmth, the mother's gaze, the sound of her voice. The central need of the child is to be nourished and the mother is the object that satisfies the need. Wishes emanating from this phase are called the *oral phase* and remain active throughout life in one form or another, even though augmented and modified by subsequent wishes and events. During the second year of life, other interests come to the fore. The child becomes aware of himself as an independent entity. He begins to assert his independence from his mother and he begins to appreciate developing capacities and a sense of mastery. The biological sources of pleasure now center about digestive and excretory functions, the mastery of one's body and its content. The child tests his ability to manipulate his body contents in fact and in fantasy and observes how these activities influence others. This period, roughly covering the age span from two to four, is referred to as the *anal phase*.

Perhaps the most crucial period for psychological development begins about the age of three to three-and-a-half and culminates between the ages of five-and-a-half to six. During this period the child is intensely preoccupied with the activity of his genitals and their potential for pleasure. He begins to contemplate such fundamental issues as the differences between the sexes, the mysteries of conception and birth, the powers and privileges of the adult, and the puzzle of death. Sexual urges become very strong and become manifest in speech and play or covertly in fantasies and dreams. Since, in both sexes, the possession of a phallus becomes a central issue and because of the fact that, at the same time, sexual wishes are often directed towards the parents, this phase has been called the *phallic-oedipal period*. Sexual wishes appear in the content of fantasies accompanying masturbation, a practice common during this phase.

From the study of sexual perversions, observations of children, the psychology of dreams, and the structure of neurotic symptoms, as well as many aspects of normal sexual experience, Freud concluded that these biologically based pleasure-seeking activities of childhood represented constituent elements of the sexual impulse, which attains its final genitally dominated character only towards the end of adolescence.

II. Fundamentals of Psychoanalytic Psychology

Several other principles are fundamental to psychoanalytic theory. Foremost among these is the concept of *determinism*. Mental life is not random or chaotic. Psychological experiences demonstrate the persistent effects of significant antecedent events in the life of the individual. What happens in mental life is part of an ongoing historical process that gives form and meaning to what the individual thinks and feels. Both nature and nurture contribute to developmental sequences of mental functioning, resulting in a patterning of experience that is clearly motivated and in which events are causally related.

Psychoanalysis, furthermore, is a *dynamic* psychology. It conceptualizes mental functioning in terms of an apparatus that performs work, an apparatus that is propelled into motion, driven to action by inner urges consciously experienced as wishes. Some of these urges, as indicated above, are clearly biological in nature. They stem from the physiological functioning of the organs of the body. This is clearly true in the case of the sexual drives. Other dynamic forces in the mind are less clearly linked to specific physical zones or bodily functions, but they are equally important. Foremost among these are the propelling drives toward aggression, hatred, and destructiveness. Unlike the sexual drive, evidence suggesting the operation of a persistent urge towards aggression has no clear base in biology. The evidence for this concept is psychological in nature. Psychoanalysts differ as to whether other motivational forces, such as safety, security, self-esteem, integrity, and mastery should be considered primary drives or should be subsumed in some way under the broader categories of the *sexual and aggressive drives*. [*See* AGGRESSION.]

A note on the history of terminology must be inserted here. Freud used the German world "trieb" to indicate the dynamic, driving forces acting upon the mind. Unfortunately, the term was mistranslated into English as "instinct" and it is in this form that the term for the concept has persisted in the literature, sometimes expressed as instinctual drive. The term "drive" conveys the precise meaning of the concept.

The drive concept, of course, is an abstraction. In practice, what one observes are mental representations of the drives. They take the form of specific, concrete wishes and are designated drive derivatives. As the individual develops and matures, the manifest forms of the *drive derivatives* undergo change and transformation. This occurs in keeping with the pleasure principle, with the tendency to avoid or mitigate the potential experiencing of unpleasant or painful affects.

Certain specific fears play a leading role in the transformation of the drive derivatives. The first is the danger that the organismic distress the infant experiences in situations of unfulfilled needs might assume devastating proportions. At a later stage of cognitive functioning, with a beginning appreciation of the perception that the mother's face signals impending relief, the failure of the mother to reappear melds into a fear of separation from her. This combination of events becomes a danger situation, fraught with the possible evolution of intense unpleasure or pain. At a still more advanced level of interaction with the mother and a fuller appreciation of her as an independent object, the child comes to face another set of potential dangers. He feels threatened by situations that may lead to the loss of the mother's love. Somewhat later, the child perceives another set of threatening situations, the all-pervasive fear of punishment, specifically by means of physical mutilation of the genitals. Subsequently, as the individual conscience begins to develop, a new source of danger comes into being, namely, the painful affects connected with self-condemnation, the sense of guilt, and the need for punishment. The latter constitute an example of self-directed aggression. It is important to note at this juncture that the drives, sexual and aggressive, operate towards the self and one's own body just as they do towards other objects. An individual may feel love and hate for himself even at the same time, very much as he may experience these feelings towards others. The urge to punish one's self evokes unpleasant affects, which the individual perceives as a signal of impending unpleasure or pain.

These considerations lead us to another fundamental principle of psychoanalysis. Psychoanalysis is a psychology of *conflict,* a psychology of dynamic forces in opposition to each other. Intrapsychic conflict is ubiquitous and never-ending. The forces in conflict are multiple and diverse. They bespeak the contradiction among wishes and fears, threats and warnings, hopes and anticipations for the future, regrets of the past. Freud formulated his final theory of the structure of the psychic apparatus according to the role that each mental element played in intrapsychic conflict. Thus, the persistent wishes of the past, operating as continuous stimuli to the mind and giving rise to innumerable, repetitive, relatively predictable patterns of mental representations, collectively constitute a structure of the mind, a system of the psychic apparatus designated as the *id*. The term *id* derives from the Latin word for *it,* and the choice of the term reflects how alien and unacceptable some of its derivative manifestations are when presented to consciousness. The id is the vast reservoir of motivational dynamic. It consists of sexual and aggressive wishes, primitive in nature, self-centered and often antisocial.

Another source of motivational dynamic consists of the ideal aspirations, the moral and behavioral imperatives, the judgment of right and wrong. This group of relatively stable functions of the mind

Freud called the *superego,* in recognition of the fact, as he thought, that it developed later in the life of the mind, but also from its function as an observer and critic, seeming to stand above and beyond the self, passing judgment on it. Frequently, but by no means always, impulses emanating from the superego oppose the demands of the id. Moral considerations are repetitive and relatively predictable but, like the id, the superego is full of internal contradictions.

The third structural component of the psychic apparatus is made up of those functions that serve to integrate and to mediate the complementary or contradictory aims of the other agencies of the mind. At the same time it takes into account the nature of the objective, realistic situation in which the individual finds himself. Freud called this set of functions the *ego.* It comprises activities that identify it as the executant for all the agencies of the mind. It is the mediator between the internal and the external world, between the world of thoughts and feelings on one hand, and the world of perception and objects on the other.

The concept of psychic structure should not be taken too rigidly. The essential criterion that applies to any mental representation is the role it plays in intrapsychic conflict. Id, ego, and superego constitute only abstract conceptualizations of the patterning of the forces in conflict within the mind. In the clinical setting, situations of intense conflict delineate most clearly the boundaries of the constituent structures of the psychic apparatus. Under more harmonious psychological circumstances, the contributions of the component systems tend to fuse. It is the function of the ego to integrate and to resolve intrapsychic conflicts in order to avoid the danger of unpleasant affects, particularly depressive affect and anxiety. The end products of the ego's efforts, so to speak, represent compromise formations, which is to say that all the various participants in the internal conflict find at least some representation of their dynamism in the final mental product. The compromise formations effected by the ego may be successful ones in the sense that they ward off or circumvent the appearance of pain or unpleasure or they may fail in the sense that the final product of the process is fraught with a greater or lesser component of pain or brings the individual into conflict with the environment and actual danger to his person. Various affects, anxiety in particular, serve as signals, warning that a danger situation of one of the several types mentioned earlier may be developing. The potential danger may be actual or imagined, real or fantasy. The functioning of the mind is an endless exercise in adaptation, reconciliation, and integration. Needless to say, this effort is not always successful.

The ego has at its disposal many different methods for dealing with danger situations. For example, certain mental processes signaling danger, from whatever source, may be rendered nonexistent, which is to say they are promptly forgotten and cannot be recalled to mind. They no longer are available to consciousness. The individual is not aware of them and, as far as he is concerned, they never happened. Such mental processes are said to have been *repressed.* In a definitive act of repression, the repressed element leaves no traces, which is to say it ceases to function as a driving force stimulating the psychic apparatus into action. Psychically it has become nonexistent. It is only when the process of repression is incomplete, when the element excluded from consciousness continues to exert some dynamic impetus, as evidenced by the appearance of derivative representations, that one may infer the presence of a repressed element, of an unconscious mental process. Unconscious mental processes are not apprehended directly. They are inferred from an examination of the data of observation, utilizing criteria of interpretation applicable to any form of communication.

The ubiquitous influence of unconscious processes on conscious mental functioning is a major principle of psychoanalytic theory. The role of unconscious processes in mental functioning follows inexorably from the dynamic principle in psychoanalysis, from the theory of forces activating the mental apparatus. Deriving from the metaphorical use of surface for conscious and depth for unconscious, psychoanalysis has come to be considered a "depth psychology." While the concept of surfaces or layering is hardly appropriate, it has nonetheless served as the basis for a term applied to this aspect of psychoanalytic theory, namely, the *topographic principle,* the relationship between conscious and unconscious mental processes.

One of the empirical findings of psychoanalysis is the persistent and powerful influence of early childhood experience. Although the events and wishes of the years before the age of six seem to fade from memory and very little of them can be recalled in later years, these events nonetheless affect psychological development and personality structure in the most profound ways. For ages, educators have appreciated this principle intuitively. Freud reached the conclusion empirically. Ethologists have confirmed the principle experimentally.

The human mind develops as the interplay between the maturation of inherent, biologically determined capacities and the vicissitudes of experience. In the transformation of the newborn infant into a mature human being, every stage of development presents the individual with a specific set of problems, with fresh goals to be achieved. How problems are solved at one stage will influence the ability of the individual to negotiate the next set of developmental challenges. The specific needs, achievements, and conflicts of one phase are superseded but not displaced or eliminated in the next phase. As a rule, but not always, the successful resolution of developmental challenges of earlier phases seems to facilitate successful resolution of later developmental challenges. On the other hand, accidents of fate, such as severe illness, inborn physical or psychological deficits, cruel treatment, negligent care, or abandonment may all have adverse effects upon individual development, causing pathological development immediately or rendering the individual incapable of mastering subsequent developmental challenges. Such events or influences constitute psychological trauma when they overwhelm the ego's capacity to master the terrifying and painful dangers that may occur at all periods of life. This is the beginning of the process of pathogenesis, which may lead to inhibitions, symptoms, perversions, and character deformations. Emphasis on the importance of childhood events for normal and pathological development constitutes the *genetic principle* in psychoanalysis. It is an empirical finding, not a derivative hypothesis.

Before considering the form that the persistent wishes of childhood take, it is necessary to appreciate the nature of children's thinking. The child's wishes are urgent, imperious, and uncompromising. His interests are self-centered. The distinction between objective perception and inner wishful thinking is not firmly established. Fantasies of the magical power of thought, of omnipotence and destructiveness are taken very seriously. The child's grasp of reality and causality take a long time to develop. In many ways, infantile notions of magic persist in the minds of many adults. Evidence for this fact is readily available in the prevalence of superstitious beliefs, etc.

It is against this background that the child tries to formulate answers to fundamental existential problems. He must confront the inevitability of frustrations of his needs and wants, the limitations of his control over his own body, and the discrepancy between himself and adults. Issues of procreation, life and death, sex and violence, the different roles of men and women—the challenges are universal, true for children all over the world, no matter the cultural level of the environment in which they are being raised. Each child attempts to answer these questions with the limited intellectual resources at his command. The assistance that he gets from grownups is not always cognitively useful or affectively satisfying.

The child creates his own fantasy solutions to these problems and these fantasies serve as vehicles for the powerful driving wishes, fears, and self-punitive notions typical for that period of development. Originally, it may be presumed, such fantasies are exclusively or primarily imagined representations of wishes fulfilled but, as the child becomes aware of the dangers connected with the emergence or the expression of such wishes, even their fantasy expressions are modified by the process of compromise formation. Because the more primitive expressions of these wishes prove to be dangerous, they are repressed and forgotten. As an adult, the individual is not at all aware that he ever harbored such notions. Such repressions, however, are rarely definitive. Although the wishes remain unconscious, they continue to exert a dynamic role, making their effects discernible in derivative forms. Some of these derivative forms emerge as fantasies, watered-down versions, symbolically altered, less threatening representations of the original wishes. Derivative, acted-out forms of the same wishes may take the shape of habits, character traits, special interests, choice of profession, and a wide range of psychopathological formations. In summary, the unacceptable wishes of childhood take the form of persistent unconscious fantasies, exerting a continuous stimulus to the mind, eventuating in compromise formations. Some of these compromise formations are adaptive and are considered normal. Others are maladaptive—"abnormal."

Just how early in life the ability to create fantasy emerges is difficult to say. The process is certainly facilitated by the acquisition of language, but the ability to fantasize seems to antedate the appearance of language to some degree. This occurs probably in the second year of life. The complex fantasy life that certain observers believe to characterize the psychological activity of infants even as far back as the first six months of life does not seem to be a tenable proposition. Fantasy thinking is metaphoric in nature and makes extensive use of symbolism. This is an inevitable outcome of the fact that human thought is inherently metaphoric. Fundamentally, perceptual experience is processed according to the criteria of pleasant or unpleasant, familiar or unfamiliar. Memories of perceptual experiences are stored, organized, and patterned in keeping when these principles. The facile transfer of meaning from one mental element to another on the basis of similarity or difference, of association in memory or experience leads quite directly to metaphoric thinking and symbolism.

While the nature of the unconscious fantasies remains constant, their derivative manifestations evolve and are transformed in the course of time. They change with the advancing cognitive capabilities, with the appreciation of the real environment, and with the consolidation of moral values. The fundamental plot of the fantasies, however, remains the same. The characters and the settings change and "grow up" with time. There is good evidence to suppose that unconscious fantasying goes on all the time we are awake and a good deal of the time we are asleep. Every individual harbors a set of unconscious fantasies typical for him. They represent the special way in which that individual integrated the major experiences and relationships, the important traumata and drive conflicts of his childhood years. The persistent unconscious fantasies serve as a mental set against which the sensory data are perceived, interpreted, and responded to. Furthermore, specific perceptions of events in the external world may resonate with elements in the individual's unconscious fantasy system and may evoke conscious representations of unconscious fantasy wishes. The derivatives consist of compromise formations that may be adaptive or maladaptive. The emergence of maladaptive compromise formations of unconscious fantasy wishes marks the beginning of the process of pathogenesis.

III. Psychoanalytic Methodology and Therapy

The fundamental operational principles of psychoanalysis—determinism, dynamic conflict, and the role of unconscious mental elements—all enter into the organization of the standard mode of psychoanalytic investigation. This is known as the psychoanalytic situation. The patient reclines on the couch, looking away from the analyst. He is asked, as far as possible, to report with complete candor whatever thoughts or feelings present themselves to consciousness. In effect, he is asked to function as a nonjudgmental reporter of his own mental functioning. No consideration justifies the exclusion of any element that occurs to the analysand's mind. This technique of reporting is called *free association*. Its aim is to obtain a dynamic record of the analysand's mode of mental functioning, reflecting an endogenously determined flow of the individual's thought. External influences are reduced to a minimum. When external influences do intrude upon the analysand's awareness, they are examined from the point of view of the dynamic, evocative power they exert upon the stream of the analysand's associations, on his mode of mental functioning. During the analytic session, external intrusions may take many forms. The siren of a passing fire engine, the perception of a change in the decor or furniture arrangement in the analyst's office, the odor of flowers in the room, and, most important of all, whatever the analyst does or says. In any event, the approach to each of these perceptual experiences remains the same. What the analyst studies is the dynamic, evocative effect of the perceptual experience upon the nature of the analysand's thoughts.

The analyst has a dual role in this procedure. He is an observer of how the patient's mind works. At the same time, however, through the things that he says and does, he becomes a participant in the process. Functioning as a participant observer, the analyst pays special attention to the effects his interventions produce in the stream of the analysand's associations. It is for this reason that the analyst must pursue rigorously a stance of nonjudgmental neutrality regarding the analysand's realistic decisions, moral dilemmas, or partisan conflicts. The fundamental concern of the analyst is to observe and to understand how the patient's mind works. The analyst's opinions and prejudices are irrelevant to this work and, in fact, introducing them may be counterproductive. Such efforts would be suggestive and educational in nature, subtly directing the analysand away from trends in his own thinking that he would consider to be counter to the analyst's point of view or interests. The technical aim of the analyst is to supply understanding and insight, not to furnish a set of directives or to act as a model for the analysand to emulate. This conjoint investigation of the workings of a person's mind has no parallel in any other form of human communication.

From the point of view of theory, the results of this inquiry may be examined from several hierarchically related levels of abstraction. From the fundamental, experiential level of *clinical observation,* the analyst gets to know things not available to other observers. More than that, he becomes aware of the form and content in which thoughts appear, the patterning and configurations they assume, the repetition of certain themes, the irrelevant or unexpected intrusions of ideas and actions, the struggle of the patient to hide some elements or to minimize or repudiate their significance.

The next level is that of *clinical interpretation.* As in any form of communication or dialogue, more meaning is conveyed than is contained in the explicit verbal or motor expressions alone. The context in which ideas occur, the relationship to contiguous elements, the position of the idea in a sequential series of thoughts, the similarity to antecedent mental presentations or their persistence and repetition—all of these, augmented, to be sure, by the nature of the quality of the analysand's speech, the affective mood that is projected, and the motor concomitants of communication, enable the analyst to make connections that are not immediately apparent in the manifest text of the analysand's productions. In this respect, he is aided by the analysand's use of figurative speech, especially metaphor and symbol. Their use enhances and extends the communicative significance of the analysand's productions. At this level of the analytic work, what the analyst does is to make connections among the analysand's thoughts, connections that the analysand has been unable or unwilling to acknowledge on his own. From the stream of the analysand's associations, the analyst infers meanings and motives unknown to the analysand, which he communicates to him.

How is it possible for the analyst to learn something about another person's thinking, something of which the latter is himself not at all aware? In the genesis of interpretation, a number of important processes take place in the mind of the analyst. First is the experience of *empathy.* The analyst identifies with the patient, that is, he puts himself in the patient's position psychologically but is aware at the same time that the moods and thoughts that occur to him represent his reflections and reactions to what the patient has been telling him. The second process is *intuition,* according to which the analyst organizes the patient's productions and integrates them with what he has learned about the analysand previously, but it is a process that takes place outside the scope of the analyst's awareness. The end result of this integrative working over of the patient's productions into a meaningful hypothesis presents itself to consciousness through the process of *introspection.* The thought thus formed may be incomplete, incorrect, or usually only a step in the direction of the proper apprehension of the meaning of the analysand's associations. A more accurate interpretation comes about when the results of this intrapsychic communication to the analyst are consciously and cognitively examined in the light of the criteria already mentioned. Meaning derives from context, contiguity, sequence, similarity, figurative language, especially metaphor and symbolism, and other elements which, in general, hold sway in communication.

An interpretation is actually an hypothesis offered to the analysand. Interpretations vary in the amount of data they attempt to comprehend. There are interpretations concerning minute sequences of mental processes, ranging to comprehensive formulations concerning the meaning, origin, and purpose of lifelong patterns of behavior, thought, and feeling. Like any hypothesis, an interpretation must be consistent with the data of observation and it must be coherent. Furthermore, a psychoanalytic interpretation is not proferred in the course of analytic investigation merely as a summary of the relationship among observable data. The dynamic impact of a particular intervention called interpretation is what is important. An immediate acceptance or rejection of the interpretation by the analysand can be totally misleading and is actually beside the point. The effect of the intervention on the subject's flow of thought, the new material that it brings to light—these are of paramount significance. In effect, each interpretation the analyst makes constitutes a sort of experimental intervention.

From the experience of clinical observation and interpretation, certain *clinical generalizations* become possible. Such generalizations may apply to the meaning of repetitive, diverse, but related patterns of mental activity or syndromes articulating similar compromise formations. They may be observed repetitively in the individual; they may be recapitulated from patient to patient. In a compulsion neurosis, for example, the compulsive symptom generally protects the individual against unpleasant affects resulting from conflicts over murderous impulses. For the fetishist, the presence of the fetish is an essential condition for making sexual pleasure possible by presenting the individual with an actualization of an unconscious fantasy of a female phallus, a concept necessary in the case of the fetishist, to deny the possibility of genital mutilation.

From clinical interpretations and clinical generalizations, it becomes possible to formulate certain theoretical concepts that flow logically from the interpretations and to which the interpretations may lead. Most of the basic, operational, theoretical concepts of psychoanalysis belong to this category. Among these, for example, are the concepts of repression, defense, unconscious fantasy, compromise formation, etc. This is the level of *clinical theory*.

Finally, there are those abstract theoretical concepts not directly derived from clinical observational experience. Accordingly, this level of abstraction is referred to as *metapsychology,* i.e., beyond psychology. Metapsychology concerns such issues as the compulsion to repeat, the nature and origin of mental energy, the relationship of quantitative changes in drive energy to the experience of pleasure and pain, or whether repression takes place because a mental element is divested of its drive energy or is opposed by countervailing drive force supplied by the ego. Most of these concepts Freud borrowed from physics and biology. Waelder, who delineated the levels of psychoanalytic propositions just mentioned, said that these levels are not of equal importance for psychoanalysis. The data of observation and clinical interpretation are entirely indispensable, not only for the practice of psychoanalysis but for an appreciation of the empirical basis upon which psychoanalytic propositions are founded. Clinical generalizations and clinical theory are necessary too, though perhaps not to the same degree. Studying the record of free associations in context, a person may be able to understand a situation, a symptom, or a dream with little knowledge of clinical theory. Metapsychological abstractions, however, bear little relevance to the interpretations of the observational data and to the generalizations drawn from them. In actual therapeutic experience, metapsychological theories play hardly any role in the formulation of psychoanalytic conclusions. [*See* PSYCHOANALYTIC PSYCHOTHERAPYS.]

IV. Psychic Structure

Psychoanalytic theory is not static nor unchanging. It has been the subject of continuing revision, sometimes radical in nature, in keeping with new insights and fresh discoveries. The current division of mental organization according to the specific function that each psychological element plays in intrapsychic conflict has been designated the *structural hypothesis*. Persistent patterns of functioning, repetitive in nature, more or less predictable in function, are grouped together as a component structure of the mind. This theory replaces an earlier concept of mental organization that Freud enunciated, the topographic theory. According to that theory, accessibility to consciousness was the paramount criterion of mental organization. Repressed elements collectively constituted the system *unconscious,* usually designated *ucs,* and the elements it contained shared certain characteristic modes of functioning. They were primarily instinctual wishes, driving impulsively towards gratification. The wishes were basically primitive in nature and operated in complete disregard for logic or the fixed categories of mental concepts, functioning indifferently to realistic considerations. Freud abandoned this theory when his observational data indicated that forces opposed to the drives, fixed mental concepts, defensive activities, and adaptive mechanisms also operated outside of the individual's awareness. The system ucs contained more than representations of repressed drives. According to structural theory, portions of the ego, the id, and the superego, that is to say, parts of all of the psychic structures, function outside of consciousness and exert a significant influence on conscious mental activity.

The delimitations of the structural components of the psyche are not as sharp as one might think. In a manner reminiscent of behavior in political organizations, there are shifting alliances among the component psychic systems and there are even contradictions within the component parts of each system. Severe intrapsychic conflict lays bare most clearly the outlines of the different psychic agencies. Typically, in the hysterias, the ego must deal with a sharp conflict between the wishes of the id for forbidden gratifications and the countervailing condemnation of the superego. In severe depressions, the id and superego impulses join forces in a murderous assault upon the self. In certain forms of psychopathy, an alliance is made between the id and the ego. In asceticism, superego and ego combine against the pleasure-seeking impulses of the id. To be sure, these formulations represent extreme simplifications but they serve to illustrate the clinical conceptualization of intrapsychic conflict within the structural theory.

Pursuing its function of attempting to reconcile the conflicting demands made upon it, the ego has available to it a wide range of mental mechanisms. The mechanism of repression has already been mentioned, but there are many others in addition. A drive impulse may be diverted from its primary object onto other objects, objects less important and less threatening. Or the impulse may be transformed into or displaced by a different impulse, even an opposite one, e.g., hatred into love, or vice versa. The existence of the disturbing impulse may be acknowledged, but it may be mistakenly perceived as being present in someone else. These mechanisms rarely function in isolation in regard to a particular conflict. Because they were first described as modes of ego functioning, serving the common purpose of fending off an impending danger, these operations were referred to as mechanisms of defense. Closer examination, however, reveals that the ego makes use of such operations for many purposes other than defense. Fundamentally, these are mechanisms of the mind, serving the process of adaptation in the broadest sense of the word.

The unconscious defensive displacement of fantasy wishes from the original or primary object onto others is a constant feature of psychological life. Unconsciously, the individual may transfer these persistent wishes onto other objects who in some way become associated with the original object of the individual's drives. It is as if the individual foists a preconceived scenario onto people and events, a process that endows a unifying pattern upon the course of the individual's life. This process of transferring fantasy wishes onto other persons, called *transference,* plays an important role in psychoanalytic treatment. In contrast to other people in the course of everyday life, the analyst does not respond to the transferred wishes the patient directs towards him. Accordingly, derivatives of the patient's fantasy life and experiences of the past emerge with special clarity in the course of the analytic relationship. Analyzing the transference enables the analyst to demonstrate to the patient how the unconscious wishes of childhood have persisted in the patient's life, causing him to confuse fantasy with reality, past and present. Analysis of the transference is a particularly effective instrument to demonstrate to the analysand how the past is embedded in the present.

In addition to the formative role played by gratification and/or frustration of the drives, there are other aspects of the interrelationship with the "others" that play a crucial role in developing psychic structure. The term used for the interaction with other individuals is *object relations*. The term distinguishes between self and object, but it probably owes its origin to earlier concepts in the history of psychoanalysis. Freud used the term originally to apply to the mental representation that was the object of an instinctual drive. Strictly speaking, the term "object" could refer to the self, to a portion of one's body or to representations of other individuals. In actual practice, it has come to apply almost exclusively to other persons. The term *narcissism* applies when the self is taken as the object of the erotic drive.

From his interaction with objects in his immediate world, the individual acquires a wide catalog of methods for coping with difficult situations, preferred solutions to conflicts, skills to master, ideals to aspire to, etc. Basically, these acquisitions result from an identification that the individual effects with certain objects. The individual remodels himself after the object and takes over some aspect of the personality of the other. Identification, however, is never complete; one person is never a psychological clone of another. Nor is it possible to predict in advance with which aspect of an object an individual will identify. Affective considerations play an important role in this process. Nor is it essential for one to have direct experience with an object in order to identify with him. Fantasy objects, individuals from literature, history, or religious teaching may become models from whom the individual acquires modes of thinking and acting. While the personality of the individual is shaped greatly by the identifications effected with primary objects during childhood, it is a fact that identification is possible later in life, especially during adolescence. In many instances, such late identifications help to give the final stamp of character upon the personality.

Identifications the individual makes in childhood tend to be primitive, highly idealized, and invested with grandiose illusions. Furthermore they are self-centered, that is to say, narcissistic. Part of the process of attaining maturity consists of replacing the exaggerated, grandiose, ideal aspirations of childhood with more realistic, obtainable goals, as well as developing a more objective evaluation of the primary object and of one's objects and of one's self. Some residues of early ideal formation persist in everyone and they serve as a standard against which the individual measures his self-worth. The feeling of self-esteem depends, to a large extent, upon how he judges himself in terms of the distance between his actual situation and the ideals and standards he has set for himself. According to the conclusions reached, he will either like or dislike himself. Some individuals, however, continue to harbor the grandiose ego ideals of childhood. They are impelled towards impossible goals and exaggerated expectations, all of which are doomed to failure and disillusionment. Such issues become central in the psychology of narcissistic persons. They require endless supplies of narcissistic gratification in order to feel worthwhile. When successful in the pursuit of their goals, they are elated; when unsuccessful, they become depressed and even suicidal.

V. Compromise Formation

The turbulent conflicts of childhood come to a head during the phallic-oedipal phase. Powerful though his wishes are, the child ultimately recognizes that they can never be fulfilled. He dare not risk the loss of his parents nor can he master the fear of retaliation. In a sense, his own wishes come to represent a source of great danger to him. Either he renounces his wishes or masters them by modifying them. Injury to the genitals, the so-called castration anxiety, is the danger typical for this period. How the child resolves the conflicts of this period has fateful consequences for the development of his personality and the course of his life. The manner in which such conflicts are resolved varies from individual to individual. The inherent drive endowment of the individual, the nature of the relationships with his parents and other significant figures, the specific events in his life and other experiential and maturational factors all play a role in shaping the final outcome of the conflicts of the oedipal phase. Certain forms of the wishes may be renounced entirely; others may be repressed but only incompletely. Taking the form of persistent unconscious fantasies, such wishes may exert a continuing influence upon mental functioning for the rest of the individual's life.

The turmoil of the oedipal phase is followed by a period of relative quiescence. This is known as the latency period and it lasts until the onset of adolescence. The intensity of conflicts abates and the individual becomes socialized and educable. But, while the conflicts have become relatively quiescent, they are hardly extinct. The turmoil of adolescence represents a secondary, more sophisticated attempt to resolve the conflicts of the oedipal phase at a higher level. A certain degree of restructuralization of the psychic apparatus occurs, out of which emerge the individual's sexual identity, his social and professional roles, and his inner moral commitments. Whether the secondary recapitulation of the conflicts of the oedipal phase during adolescence is primarily the result of biological or sociological factors is impossible to ascertain.

How the individual resolves the inevitable conflicts of life decides the issues of normality or pathology, of health or illness. Largely, this is a question of degree, since it is impossible for all compromise formations instituted by the ego to be equally effective under all circumstances. Some individuals, for example, seem never to have been able to master their childhood fears. They continue to suffer from irrational inhibitions, fears, and compulsions throughout latency and adolescence and into adult life. In other cases, pathology originates when some event in life upsets the balance of compromises that the ego has managed to maintain. This situation activates a latent conflict from childhood, and the individual responds to fantasied dangers articulating the effects of persistent unconscious childhood wishes. The reactivation of such latent unconscious conflicts is called *regression*. The responses to the inherent dangers are automatic, that is to say, outside of the individual's control. In addition, they are maladaptive and may be painful. The goal of psychoanalytic therapy is to give the patient insight into the nature of his fears and to demonstrate to him the automatic, inappropriate, largely unconscious measures he has undertaken to cope with his irrational anxieties. As the patient comes to recognize how the past is operative in his functioning in the here and now, he becomes able to organize more effective, less conflictual compromise formations to manage the effects of his persistent conflicts.

VI. Social Applications

By applying some of the principles derived from the knowledge of the psychology of individuals, psychoanalysis has been able to afford some measure of insight into social and group phenomena. Certain group phenomena appear to replicate *en masse* some of the themes and psychological mechanisms observable in individuals. Themes and modes of expression common to myths, fairy tales, literary works, and religious traditions and rituals often repeat, sometimes in minute detail, the derivative expressions of unconscious conflicts observed in individuals. They prove appealing to the individual because, like fantasies and dreams, they represent disguised representations of repressed wishes. In large measure, the fantasy life of each individual represents a secret rebellion against the need to grow up and to renounce the gratification of his drives. In a very specific sense, it constitutes rebellion against society and certain of the strictures that civilization imposes upon the individual. In the process of mythopoesis and literary creation, the poet and the mythmaker transform their private daydreams into creations compatible with the ideals of the community, capable of giving pleasure and conveying at the same time disguised, transformed expressions of forbidden wishes that other members of the community share in common with them.

In the myth or the work of art, an individual may find in the external world a projected representation of his own unformed and unexpressed wishes. By identifying with the principal characters in a work of fiction or with the hero of the historical or religious myth, the individual attains some measure of gratification of his repressed, unconscious wishes, impulses that are ordinarily forbidden. Those who participate in this process, the audience or the members of a cult, constitute a group by virtue of sharing with each other disguised indulgence in forbidden wishes. In doing so, they exculpate each other by the knowledge that they have all participated, however transiently, in the pleasure of an unconscious fantasy shared in common. As Sachs said, everybody's guilt is nobody's guilt.

An unconscious fantasy shared in common serves as the bridge from individual conflict to mass participation. The capacity of the political or religious leader to evoke unconscious fantasies shared by large numbers of the population has played a significant role in many of the mass movements of this century. Mass movements are often organized around a central charismatic figure, who is intuitively perceived as the hero of a latent or manifest group mythology and who comes to represent the ego ideal of the members of the group. Because of the inherent methodological problems, however, validation of psychoanalytic insights into mass phenomena remains problematic.

VII. More Recent Theoretical Formulations

Many new developments have occurred in psychoanalytic theory during the past few decades. Some may be considered extensions or elaborations of the basic concepts of the drive-conflict-compromise formulation with renewed emphasis on certain vicissitudes of development. Other approaches are clearly revisionary in nature. Foremost among these are object relations theory, attachment theory, and self-psychology. Some of these have coalesced into identifiable orientations or schools of thought, but none so far has achieved preeminence. Whether these newer theoretical formulations expand the internal consistency or explanatory power of the more traditional Freudian theory remains to be seen.

Modern object relations theories challenge the motivational importance of biologically grounded drives in human experience, including the role of the drives in establishing the infant–mother bond. Instead, they emphasize that drive-related strivings gain expression and are really understandable only in terms of the early experiential interactions with gratifying or nongratifying objects. The early patterns of self–object relations are internalized and repressed but they exercise a dynamic power upon the mind, leading the individual to repeat earlier patterns of interaction with significant objects. Thus, from these perspectives, it is not the derivatives of conflictual childhood wishes *per se* that are repudiated but the direct relational constellations through which these wishes came into being. It is the tendency to repeat earlier pathological patterns of interaction with significant objects that brings about pathology. Proponents of this approach believe that object relations formulations are especially relevant to the understanding of more disturbed patients, patients often referred to as borderline or narcissistic. Characteristic of such patients is a tendency to split the mental representation of the primary objects into two parts, one entirely good and the other entirely bad. This primitive defense of splitting, it is maintained, is more germane to the psychopathology of such patients and their regressive reactivation of unconscious fantasy wishes typical for neurotics.

A variant of the object relations approach, one that is somewhat closer to the conflict concept in psychoanalytic theory, is the separation-individuation paradigm proposed by Margaret Mahler. Based upon observations of infants and toddlers, she conceptualized a series of stages through which the individual achieves psychological as well as physical separation from the mother. Successful separation culminates in subjective feelings of autonomy and the emergence of the individuality that become the foundations of the self, of psychological personhood. From this standpoint, pathology results from an inability to negotiate successfully the psychological challenges of the separation-individuation process. In such instances, the individual finds it difficult to engage and resolve the conflicts of the oedipal period. Psychotic pathology reflects the failure to emerge from the state of symbiotic fusion with the mother; neurotic pathology would point to an inability to reach that stage of autonomy in which the object is firmly conceptualized as an independent and constant one. While Mahler's research concentrated on the neonate's dependence on the mother, more recent studies by infant researchers have stressed the infant's inborn perceptual preferences, experiences of perceptual unity, and programmed capacity to interact with the environment. Evidence of the infant's ability to order and differentiate a variety of stimuli and, by inference, to experience the process of emerging organization has led some researchers to postulate a series of different senses of selfhood present from birth on. The findings of such investigations present a direct challenge to some of Freud's earlier metapsychological theories, such as the notion of "primary narcissism," a theory which assumed that, at the beginning of life, all pleasurable drive tendencies are vested in the self.

Theories that delineate successive senses of self-experience during the first two years of life, based upon direct observations of neonates, parallel certain psychological theories of the self introduced by other psychoanalysts in recent decades. These theories elevate the concept of the self to a superordinate position in explaining mental development, psychopathology, and psychoanalytic treatment. Such theories hope to replace explanations framed in terms of Freud's structural hypothesis and the concepts of intrapsychic conflict. Theories of the self claim to address two types of deficiencies they discern in traditional psychoanalytic therapy. They assert their explanations of clinical phenomena to be closer to actual experience than those explanations framed in terms of the traditional model, which they regard as mechanistic, experience-distant, and wed to a metapsychology based on dated neurobiological and energic concepts. Second, they claim superior insight into the more disturbed patients, particularly those with severe character pathology. According to them, the pathology of the more disturbed patients is based not on the conflicts associated with the oedipal phase of development but on what they call "archaic," "preoedipal," "narcissistic," or "self object" transferences that characterize the object relations of such patients. Conflicts concerning mothering and nurturing become central in treatment. The classic conflicts of the oedipal phase, conflicts concerning envy, rivalry, hostility, and fear, are held to be of secondary importance. Self psychologists see their patients relating to them in primitive ways that correspond to the modes in which infants and young children use and depend upon their parents, especially their mothers. For such analysands, the analyst becomes an object who promotes their development by mirroring and fostering emergent feelings of adequacy and self worth. The analyst becomes an object for idealization, someone who calms and soothes the patient, thus aiding the patient in the regulation of his inner tensions.

Unlike most object relations theories, these theories of the self tend to understand early development in terms of the programmed unfolding of a constellation of functions that collectively constitute the self. The self in its regulatory, mediating, integrating, and initiative-taking activities thus supplants the ego in Freud's structural theory. Whether in the guise of a "self schema," a "self organization" or a "bipolar nuclear self," such theories understand mental health as the adequate realization of certain maturational potentials inherent in the self. Correlatively, pathology is viewed as a derailment of this biopsychological program at some point in early life. Thus, some self theories, construing early development in terms of successive stages or phases to be transversed, are led to view psychopathology from the standpoint of a deficit psychology, whereby pathology is seen to intrude when an individual fails to negotiate particular developmental challenges. Proponents of these theories trace psychological difficulties back to developmental arrests, nodal points in early life from which the self was unable to proceed with its programmed agenda. This position stands in contrast to the conflict psychology through which psychoanalysis has traditionally approached psychopathology and psychotherapy. The relationship between deficit and conflict psychologies, as is also the question of whether the self has explanatory or heuristic power greater than that of the Freudian ego, is a subject of ongoing debate within contemporary psychoanalysis. According to some self psychologists, pathology is a consequence of a failure to achieve a vital, cohesive, "nuclear" self, a failure generally rooted in unempathic, unresponsive mothering that has not mirrored or reinforced the early feelings of grandiosity that, under conditions nearer optimal, can blossom into healthy feelings of vitality and self-worth. While self psychology is offered as a rival theory to the conflict-danger-compromise approach, it may be that actually it only adds a further danger situation to the four typical dangers mentioned previously, namely, the danger of loss of self.

Among recent theoretical developments, biological formulations of psychoanalytic concepts deserve special mention, but with a proviso. Descriptions of neurobiological mechanisms have not as yet been shown capable, and indeed may never supplant psychological propositions. The most productive work in neurobiology involves work on the neuroanatomical structures and the neurophysiological processes that subtend psychoanalytic concepts. Rather than modifying its understanding of mental functioning on the basis of neurobiological findings, psychoanalysis has pointed neurobiology in the direction of particularly promising conceptual and experimental endeavors. Thus we now have a body of literature that addresses topics of analytic concern: neurophysiological demonstrations of unconscious mental processes, e.g., between right brain and left brain, that relate to Freud's notions of primary and secondary processes; research on the neurophysiological pathways of affective expression; and research on the impact of perceptual environmental experiences on neurological development. Notions of "neuroplasticity" (the notion of critical periods during which certain experiences are necessary for optimal brain development) and of a "neurorepresentational system" have been proposed as bridge concepts linking neurobiological functioning to mental activity, including unconscious mental processes. A recent study equates the Freudian unconscious with biogenetically ancient mechanisms that involve REM sleep and are located in the prefrontal cortex and associated structures. Research that seeks to elaborate a neurobiological substrate to psychoanalytic concepts provides the same useful function as research into the psychological processes associated with neurobiological events. In the case of neurobiology and psychoanalysis, benefit is derived from reporting research in coordinate rather than in casual terms so that, in the words of Smith and Ballinger, "For the neurobiologist, psychological events are markers of neurobiological processes and for the psychologist, neurobiological events as markers of psychological processes."

Bibliography

Blum, H., Kramer, Y., Richards, A. K., and Richards, A. D., eds. (1988). "Fantasy, Myth, and Reality: Essays in Honor of Jacob A. Arlow." International Universities Press, Madison, Conn.

Eagle, M. N. (1984). "Recent Developments in Psychoanalysis: A Critical Evaluation." McGraw-Hill, New York.

Edelson, M. (1988). "Psychoanalysis: A Theory in Crisis." University of Chicago Press, Chicago and London.

Kernberg, O. F. (1984). "Severe Personality Disorders: Psychotherapeutic Strategies" Yale University Press, New Haven, Conn.

Kohut, H. (1984). "How Does Analysis Cure?" University of Chicago Press, Chicago.

Pine, F. (1985). "Developmental Theory and Clinical Process." Yale University Press, New Haven.

Reiser, M. F. (1984). "Mind, Brain, Body: Toward a Convergence of Psychoanalysis and Neurobiology. Basic Books, New York.

Richards, A. D., and Willick, M., eds. (1986). "Psychoanalysis, the Science of Mental Conflict: Essays in Honor of Charles Brenner." The Analytic Press, Hillsdale, N.J.

Smith, J. H., and Ballinger, J. C. (1981). Psychology and neurobiology, *Psychoanalysis and Contemporary Thought* **4(3),** 407–21.

Stern, D. N. (1985). "The Interpersonal World of the Infant." Basic Books, New York.

Wallerstein, R. S. (1986). "Forty-Two Lives in Treatment: A Study of Psychoanalysis and Psychotherapy." The Guilford Press, New York.

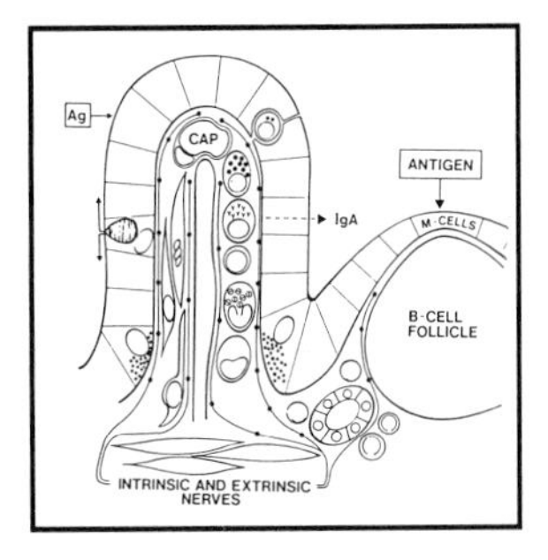

Psychoneuroimmunology

ROBERT ADER AND NICHOLAS COHEN, *University of Rochester School of Medicine and Dentistry*

Glossary

Autoimmunity Immune reactions generated against an individual's own tissue or cellular antigens that can lead to diseases such as systemic lupus erythematosus and rheumatoid arthritis

Cellular immunity Immunity mediated by antigen-stimulated, thymus-derived effector T lymphocytes (e.g., cytotoxic T cells) that does not involve antibody

Conditioned response Response to a previously neutral (conditioned) stimulus after the neutral stimulus has been paired with a stimulus that unconditionally elicits the particular response being studied

Humoral immunity Effector mechanism or immunity mediated by circulating antibodies produced by bone marrow–derived B lymphocytes. For many antigens, B cells produce antibody only after the occurrence of complex interactions among helper and suppressor T lymphocytes, certain accessory cells, and B lymphocytes

PSYCHONEUROIMMUNOLOGY is perhaps the most recent convergence of disciplines (i.e., the behavioral sciences, the neurosciences, and immunology) that has evolved to achieve a more complete understanding of how the interactions among systems serve the equilibrium of the body and influence health and disease. The hypothesis that the immune system constitutes an underlying mechanism mediating the effects of psychosocial factors on the susceptibility to and/or the progression of some disease processes is tenable, however, only if it can be shown that the brain is capable of exerting some regulation or modulation of immune responses. This subject is now under intense investigation. Psychoneuroimmunology, then, refers to the study of the neural and endocrine mediation of the effects of behavior on immune function—and vice versa. Or, more simply, it is the study of nervous system–immune system interactions. The present synopsis concentrates on the behavioral component of psychoneuroimmunologic research.

I. Introduction

In practice, research in the field of immunology has proceeded on the implicit assumption that the immune system is autonomous—an agency of defense that operates independently of other psychophysiological processes. Indeed, the self-regulating capacities of the immune system are remarkable and the interactions among subpopulations of lymphocytes occur and can be studied *in vitro*. The same can be said for other physiological processes. The endocrine system, too, was once thought to be autonomously regulated, and it is only within modern times that neural influences were identified. It may well be that immune responses can be made to occur in a test tube, devoid (presumably) of the (confounding or modulating) influences of the natural environment, but it is the immune system functioning in the natural environment that is of ultimate concern. We cannot deny that the immune system is capable of considerable self-regulation. At the same time, converging data from the behavioral and brain sciences suggest a critical role for the nervous system in the modulation of immunity.

It has been known for some time that manipulation of brain functions can influence immune function. Lesions or electrical stimulation of the hypothalamus, for example, can alter humoral and cell-mediated immune responses. Conversely, elicitation of an immune response (i.e., exposing an animal to an antigen) can influence hypothalamic activity. After stimulation with different antigens, the firing rate of neurons within the ventromedial hypothalamus increases, and the change in the firing rate of these neurons occurs at the time of the peak antibody response. Some of the most recent data documenting the potential for brain–immune system interactions comes from neuroanatomic studies of the innervation of lymphoid tissues (Fig. 1), which provides a foundation for functional relations between these systems.

The several observations that neurochemical and endocrine signals influence immune responses are reinforced by the identification of receptors on lymphocytes for a variety of hormones and neurotransmitters. Moreover, data suggest that lymphocytes themselves are capable of producing neuropeptides and hormones. Again, not only do variations of hormonal state or neurotransmitter levels influence immune responses, but the immune response to antigenic stimulation induces neuroendocrine changes. At the neural and endocrine levels, then, abundant evidence indicates the potential for interactions between the brain and the immune system. [*See* ENDOCRINE SYSTEM; LYMPHOCYTES.]

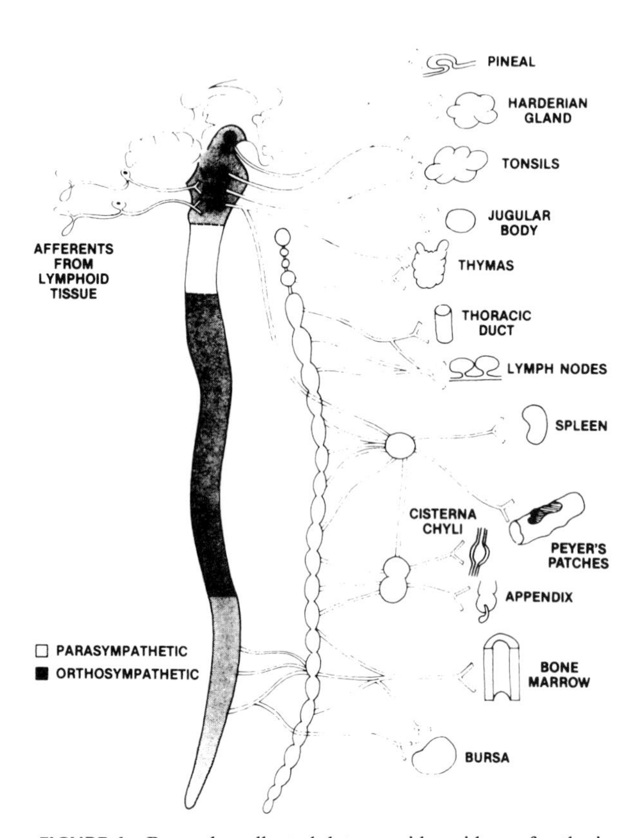

FIGURE 1 Recently collected data provide evidence for the innervation of lymphoid organs. [Reproduced from Martin, J. B. (1988). *Prog. Neuroendocrinimmunol.* **1,** 5–8, with permission.]

II. Stress and Immune Function

A. Studies in Humans

Data suggesting a link between behavior and immune function go back to the earliest observations of a relation between psychosocial factors, including "stress," and susceptibility to those disease processes that we now recognize as involving immunologic mechanisms. There are, now, abundant clinical data documenting an association between psychosocial factors and disease.

The death of a family member, for example, is rated highly on scales of stressful life events and has been associated with depression and an increased morbidity and mortality in the case of a variety of diseases, many that are presumed to involve immune defense mechanisms. Bereavement and/or depression has also been associated with changes in some features of immunologic reactivity, such as a reduced lymphoproliferative response to mitogenic stimulation and impaired natural killer cell activity. Other studies have documented changes in immune reactivity associated with the affective responses to other "losses" such as marital separation and divorce. These are provocative and important observations. It should be emphasized, however, that the association between the response to "losses" and increased morbidity and the association between the response to "losses" and alterations in immune function do not, in themselves, establish a causal link between psychosocial factors, immune function, and health or disease. [*See* DEPRESSION.]

The death of a spouse is intuitively stressful, but changes in immune function can be observed in humans using intuitively less severe, but nonetheless effective, naturally occurring "stressful" circumstances. The level of distress during examination periods, for example, is invariably greater than during control periods, and in a series of experiments conducted at Ohio State University, transient impairments in several parameters of immune function were observed in medical students at such times. Relative to a nonstressful baseline measurement, examination periods are associated with a decrease in mitogen responsiveness, natural killer cell activity, percentage of helper T lymphocytes, and interferon production by stimulated lymphocytes. In students that are seropositive for Epstein-Barr virus (EBV), EBV titers are elevated, interpreted as a poorer cellular immune response control over the latent virus during examination than control periods. The incidence of self-reported symptoms of infectious illness is also increased during examination periods. Personality tests have not yielded differences between the student volunteers and their classmates, and other life changes that could have influenced immune function during examination periods were minor and were not related to changes in immunologic reactivity.

An old (and a continuing) experimental and clinical literature suggests that immune function can be altered by psychological means (e.g., hypnosis), and an increasing number of studies are attempting to relate different personality characteristics and affective states with alterations in immune function. These correlational studies must be considered preliminary at this time. They are difficult studies to implement, so it is hardly surprising that unequivocal evidence for causal relations does not yet exist.

B. Studies in Animals

Most of the evidence for stress-induced alterations in immunity comes from basic research on animals. For example, a variety of stressors (i.e., avoidance conditioning, restraint, noise) can, under appropriate experimental circumstances, influence the susceptibility of mice to a variety of infectious diseases (i.e., Coxsackie B, herpes simplex, polyoma, and vesicular stomatitis viruses). It thus appears that stressful circumstances can alter the host's defense mechanisms, allowing an otherwise inconsequential exposure to a pathogen to develop into clinical disease. Adult mice are generally resistant to Coxsackie virus and show no manifestations of disease when exposed to either virus or stress alone. Symptoms of disease are observed, however, when mice are inoculated with virus *and* exposed to stressful environmental conditions. These results parallel clinical observations that the presence of pollen alone may not be sufficient to elicit symptoms in a subject with hay fever, but the combination of pollen and a threatening life situation is.

Separation experiences (i.e., "losses") have also been studied in animals. Periodic interruptions of mother–litter interactions and/or early weaning decrease lymphocyte proliferation to mitogenic stimulation and reduce the immunologic response to subsequent challenge with sheep red blood cells (SRBC) in rodents. Monkey infants and their mothers respond to separation with a transient depression of *in vitro* mitogen responsiveness. Separation of squirrel monkeys from their mothers results in several changes in immunologic reactivity including a decline in complement protein levels (an effector mechanism in humoral immune responses), macrophage function, and IgG antibody responses to immunization with a benign bacteriophage, the magnitude of one or another of the effects being a function of the psychosocial environment in which the animals were housed after separation.

In adult animals, a variety of behavioral manipulations interpreted as being stressful to the organism are capable of influencing immune responses. Heat, cold, and restraint (each of which is commonly thought to elicit a representative stress response) have different effects on the same immune response, and the same stressor (e.g., restraint) can have different effects on different immune responses. The intensity and "chronicity" of the stressor are also likely to influence the immune response. The initial response to auditory stimulation, for example, is a depression of mitogenic responsivity. Repeated exposure to the loud noise (i.e., chronic stress), however, is associated with an enhancement of this same response. Moreover, the initial association of elevated adrenocortical steroid levels with decreased immunologic reactivity is not necessarily observed under conditions of chronic stress. In another series of experiments, there was a graded suppression of mitogen responses corresponding to increasing intensities of electric shock stimulation, a relation that could not be caused by adrenocortical steroids, because it persisted in adrenalectomized animals. As has been observed for other psychophysiological responses, it may be the organism's capacity to cope with stressful environmental circumstances that determines the extent to which immune processes are affected.

Neuroendocrine states provide the internal milieu within which immune responses occur. Stimulating animals during prenatal and early life, varying social interactions among adult animals, and exposing animals to environmental circumstances over which they have no control induce neuroendocrine changes that are now implicated in the modulation of immune responses. Our knowledge of interactions between neuroendocrine and immune function under normal and stressful conditions, however, is incomplete. Glucocorticoids, for example, are, in general, immunosuppressive. It is generally assumed, therefore, that adrenocortical steroid elevations, the most common manifestation of stress, are responsible for the frequently observed suppression of immune function that is associated with stress. This assumption is supported by numerous examples of stress-induced, adrenocortically mediated alterations of immune responses, particularly, *in vitro*; however, numerous other observations of stress-induced alterations in immune function are independent of adrenocortical activation. The subtleties that can characterize the involvement of hormones and neuropeptides in the mediation of stress-induced alterations of immunity are illustrated by the results of different regimens of electric shock stimulation. Intermittent and continuous schedules of inescapable electric footshocks result in an analgesia to subsequent footshock. Only the intermittent shock, however, was found to be an opioid-mediated analgesia and the only one resulting in a suppression of natural killer cell activity. These results apparently reflect the immunomodulating potential of endogenously released opioids. Thus, even a cursory review of the literature makes it evident that the *in vivo* immunologic consequences of stress involve extremely complex neural, endocrine, and immune response interactions. Considering that immune responses are themselves capable of altering levels of circulating hormones and neurotransmitters, these interactions are likely to include complex feedback and feedforward mechanisms, as well. [*See* NEUROENDOCRINOLOGY.]

At this time, the results of studies of stress-induced alterations in immune function yield an inconsistent picture of the direction, magnitude, and duration of the effects. At the very least, the effects of stress appear to be determined by (1) the quality and quantity of stressful stimulation; (2) the capacity of the organism to cope effectively with stressful circumstances; (3) the quality and quantity of immunogenic stimulation used to elicit an immune response; (4) the temporal relation between stressful stimulation and immunogenic stimulation; (5) the parameters of immune function and the times chosen for measurement; (6) the experiential history of the organism and the prevailing social and environmental conditions on which stressful stimulation and immunogenic stimulation are superimposed; (7) a variety of host factors such as species, strain, age, sex, and nutritional state; and (8) interactions among these several variables. Although we are far from a definitive analysis of the effects of stress and the means by which perceived events are translated into altered physiological states capable of modulating immune functions, the available data do provide a body of evidence that the immune system is sensitive to the modulating effects of psychobiological processes ultimately regulated by the brain.

III. Effects of Conditioning on Immune Responses

A. Background

Behaviorally conditioned alterations in immune function provide one of the more dramatic lines of research that implicate the brain in the modulation of immune responses. These studies derived from serendipitous observations of mortality among animals in which a saccharin-flavored drinking solution had been paired with an injection of cyclophosphamide (CY), a powerful immunosuppressive drug. Pairing a novel taste stimulus, a conditioned stimulus (CS), with an agent that induces temporary gastrointestinal upset, the unconditioned stimulus (UCS), will, after a single pairing, result in an aversion to that taste stimulus. Repeated re-exposures to the CS in the absence of the UCS will result in extinction of the avoidance response. During the course of repeated extinction trials, conditioned animals began to die, and more critically, mortality rate, like the magnitude of the avoidance response, varied directly with the *volume* of saccharin consumed on the single conditioning trial. In an attempt to explain these results, it was hypothesized that at the same time that a behavioral response was conditioned, an immunosuppressive response was conditioned and was being elicited in response to re-exposure to the CS (Fig. 2). If so, this might have increased the susceptibility of these animals to any latent pathogens in the laboratory environment.

B. Conditioned Changes in Immune Function

To test this hypothesis, a study was designed in which rats were conditioned by a single pairing of saccharin-flavored water and an injection of CY. When the animals were subsequently immunized

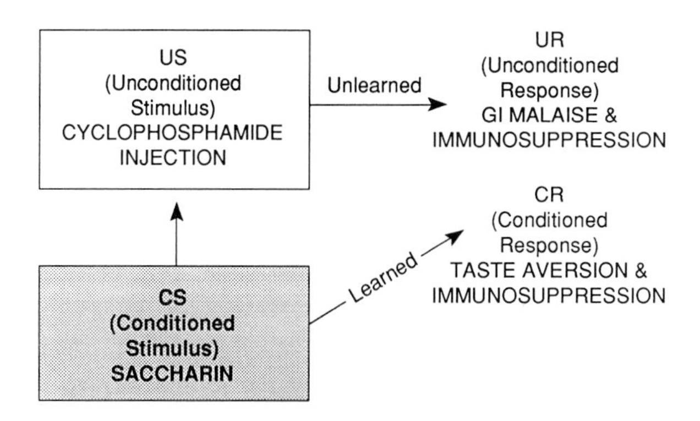

FIGURE 2 Schematic representation of the relation between conditioned and unconditioned stimuli and conditioned and unconditioned responses. [Reproduced from Ader, R. (1987). *Immunopathol. Immunother. Lett.* **2,** 6–7, with permission.]

with SRBC, one subgroup of conditioned animals remained untreated (to assess the effects of conditioning *per se*); another subgroup of conditioned animals was injected with CY (to establish the unconditioned immunosuppressive effects of the drug on the response to antigen); and an experimental subgroup was re-exposed to the saccharin-flavored water (the CS). A nonconditioned group was also provided with saccharin-flavored water, which had not previously been paired with their injection of CY. As hypothesized, conditioned animals re-exposed to the CS at the time of antigenic stimulation showed an attenuated antibody response to SRBC in relation to nonconditioned animals and conditioned animals that were not re-exposed to the CS. These results were taken as evidence of behaviorally conditioned immunosuppression and have subsequently been independently verified and extended.

The magnitude of the effects of conditioning in altering immunologic reactivity has not been large. The effects have, however, been quite consistent under a variety of experimental conditions. The magnitude and/or the kinetics of the unconditioned response vary as a function of the dose of CY, but we can still observe conditioned changes in immunologic reactivity using different doses of CY. Also, in addition to the effects of conditioning on antibody responses described above, conditioning is capable of influencing different parameters of cell-mediated immunity as well as a variety of nonspecific host defense responses. Release of histamine by sensitized animals has been conditioned, as well as an increase in a specific mediator of mucosal mast cell function.

Immunomodulating agents other than CY (as the UCS) and gustatory stimuli other than saccharin (as the CS) in the taste aversion paradigm also result in conditioned alterations in immunologic reactivity. Moreover, the phenomenon is not confined to conditioned immunopharmacologic effects. In several experiments, the immunomodulating effects of stress have been conditioned.

The prevalent use of aversive stimuli may be a limitation of these experiments. However, there is no evidence that conditioned immune responses are inextricably linked with conditioned avoidance responses or that there is some direct relation between conditioned behavioral responses and conditioned immunologic responses. Taste aversions can be expressed without concomitant changes in immune function, and conditioned changes in immune function can be obtained without observable conditioned avoidance responses. Consistent with the relation between conditioned behavioral and autonomic or endocrine responses, the available data suggest that different (multiple) conditioning processes and mechanisms are involved in the conditioning of behavioral responses and in the conditioning of different immune responses.

In other studies both the acquisition and extinction of the conditioned *enhancement* of immunologic reactivity have been observed using an antigen rather than pharmacologic agents as the UCS. In one study, CBA mice were anesthetized, shaved, grafted with skin from C57BL/6J mice, and left bandaged for 9 days. In response to the grafting of allogeneic tissue, there was an increase in the number of precursors of cytotoxic T lymphocytes (CTLp) that could react against alloantigens of the foreign tissue. The conditioning manipulations were repeated three times at 40-day intervals since it required 40 days for CTLp numbers to return to baseline levels. On the fourth trial, the procedures were repeated again, except that the experimental animals did not receive the tissue graft but a sham graft instead. Approximately half the animals, however, showed an increase in CTLp in response to the procedures. When these "responders" were divided into groups that were exposed to additional conditioning trials (including the grafting of C57BL/6J shin) or extinction trials (sham grafting), all those that experienced additional conditioning trials showed a conditioned increase in CTLp whereas none of the previous responders that were given unreinforced (extinction) trials showed a conditioned response.

The physiologic mediation of conditioned alterations in immune function are not yet known. Some investigators, however, have explicitly or implicitly assumed that the conditioned suppression of immunologic reactivity is the direct result of stress-induced responses. Because glucocorticoid elevations, taken as an index of stress, frequently suppress some immune responses, the hypothesis is attractive because it provides a ready "explanation" of a phenomenon for which no other explanation yet exists. The existing data, however, provide no support for stress-induced elevations in "stress hormones," notably adrenocortical steroids, as the mediator of conditioned alterations in immune function. In fact, most of the data stand in direct contradiction to such a hypothesis.

Although there are problems in interpreting extirpation experiments, it should be noted that in one study, immunosuppression was not observed in adrenalectomized mice. However, the "stress mediation" hypothesis is unable to account for several other observations. For example, although lithium chloride (LiCl) is an effective UCS for inducing taste aversions, no conditioned suppression of antibody production is observed when a drug, lithium chloride (LiCl) that is immunologically neutral under defined experimental conditions, is used as the UCS or when steroid levels are elevated by injections of LiCl or corticosterone at the time of immunization. Conditioned suppression and/or enhancement of antibody- and/or cell-mediated responses occur in the presumed absence of or with equivalent elevations in corticosterone and, presumably, other stress hormones. In a two-bottle preference testing procedure, fluid consumption is equal in experimental and control groups, and in contrast to the presentation of a single bottle containing the CS solution, the thirsty animal is not faced with the conflict of choosing between drinking and the noxious effects that are associated with the CS solution. Under these experimental circumstances, conditioned immunosuppression is still observed in the antibody response to T-dependent and T-independent antigens, a graft-versus-host reaction, and in the white blood cell response to CY. Finally, in a discriminative conditioning paradigm, both the stimulus (bovine serum albumin) that signaled presentation of the UCS (the CS+) and a signal that was not associated with the UCS (the CS−) induce an elevation in adrenocortical steroid levels but only the CS+ induces a conditioned release of histamine in previously sensitized animals. It is reasonable to hypothesize that conditioned alterations of immunologic reactivity may be mediated by *conditioned* neuroendocrine responses, but the data collected thus far are inconsistent with the hypothesis that such effects are mediated simply by stress-induced changes in hormone levels.

Although the mechanisms underlying conditioned suppression and enhancement of immunologic reactivity are not known, there is no shortage of potential mediators of such effects. Multiple processes are probably involved. Conditioned immunosuppressive responses, for example, occur when conditioned animals are re-exposed to the CS before as well as after immunization. This observation could imply that the mechanisms do not involve antigen-induced immunologic or neuroendocrine changes; they could also indicate that different mechanisms are involved when conditioning is superimposed on a resting or on an antigen-activated system. Also, different immunomodulating agents have different sites of action, and the same immunomodulating drug may have different effects on an activated or nonactivated lymphocyte. We now know that the immune system is innervated, that leukocytes and neurons share certain neuropeptide/neurotransmitter receptors, that lymphocytes can produce several neuroendocrine factors, and that cells of the immune system and the nervous system can produce and respond to the same cytokine (interleukin-1). Thus, conditioned changes in neural and/or endocrine activity that can be recognized by activated lymphocytes or, conversely, the effects of conditioning on the release of immune products capable of being recognized by the nervous system constitute potential pathways for the conditioned modulation of immune functions.

Current research provides compelling evidence for the acquisition and extinction of conditioned suppression and enhancement of immunologic reactivity. These studies dramatically illustrate the role of behavior in the modulation of immune responses, a modulation that is, presumably, ultimately regulated by the brain.

C. Clinical Implications of Conditioning Studies

Some strains of mice spontaneously develop an autoimmune disease that is strikingly similar to systemic lupus erythematosis in humans. In this disorder, a suppression of immunologic reactivity is in the biological interests of the organism. The (NZBxNZW)F1 female mouse, for example, develops a lethal glomerulonephritis that can be delayed by weekly injections of CY. Therefore, in an effort to determine the biologic significance of conditioned alterations in immunologic reactivity, the effects of conditioning were assessed in this animal model of autoimmune disease. If, as indicated above, the immunosuppressive effects of CY can be conditioned, can conditioned immunopharmacologic effects be applied to a pharmacotherapeutic regimen, i.e., can CSs be substituted for some proportion of the active drug treatments received by these animals to delay the development of disease?

One group of mice (Group C100), treated under a traditional pharmacotherapeutic regimen, was given a saccharin-flavored solution to drink, and after each exposure to saccharin, the animals were injected with CY. As expected, this protocol delayed the development of proteinuria and mortality. An experimental (conditioned) group (Group C50) was injected with CY after saccharin on only half the weekly occasions when the CS was presented. A nonconditioned control group (Group NC50) received the same number of saccharin and CY presentations as Group C50, but for this group, saccharin and CY were never paired.

Nonconditioned and placebo-treated animals did not differ. This indicated that half the total dose of CY administered to mice that were treated under the standard pharmacotherapeutic protocol was ineffective in modifying the course of autoimmune disease. Group C50 was also treated with half the cumulative dose of CY given to Group C100, but these animals developed proteinuria significantly more slowly than placebo-treated mice and significantly more slowly than nonconditioned mice treated with the same cumulative amount of drug.

The mortality data yielded the same results. There was no difference between nonconditioned animals and untreated controls, but conditioned animals treated with the same amount of CY as nonconditioned animals survived significantly longer than untreated controls *and* nonconditioned animals. The mortality rate of conditioned mice did not differ significantly from animals in Group C100 that received twice as much drug. These results indicate that within the context of a pharmacotherapeutic regimen, conditioning effects were capable of influencing the onset of autoimmune disease using a cumulative dose of active drug that was not by itself sufficient to alter the course of disease.

After the period of pharmacotherapy, groups were divided into thirds that (1) continued to receive saccharin and CY on whatever schedule existed during therapy, (2) continued to receive saccharin and intraperitoneal injections of saline but no CY, or (3) received neither saccharin nor CY. Consistent with the interpretation that the effects on the development of lupus were conditioning effects, unreinforced presentations of the CS influenced the development of autoimmune disease in conditioned animals but not in nonconditioned animals. Among the animals conditioned under a traditional or continuous schedule of reinforcement (Group C100), mice that continued to receive CS exposures after the termination of active drug therapy survived significantly longer than similarly treated mice that were deprived of both the CS and the drug. In fact, animals that continued to be exposed to saccharin plus intraperitoneal injections of saline did not differ from animals that continued to be treated with active drug. These data, too, are consistent with the conditioned immunopharmacologic effects described above. These results are confirmed by studies in which repeated exposures to a CS previously associated with CY accelerated tumor growth and mortality in response to a transplanted syngeneic plasmacytoma and attenuated an experimentally induced arthritic inflammation. These findings document the biologic impact of conditioned immunopharmacologic responses. They also suggest that there may be some heuristic value in conceptualizing pharmacotherapeutic protocols as a conditioning (learning) process.

IV. Immunologic Effects on Behavior

In the same way that there are reciprocal relations between neural and immune functions and endocrine and immune functions, data are accumulating to suggest that there are immunologic influences on behavior in addition to behavioral influences on immune function. Several investigators have described the behavioral effects of (early) viral infections, cognitive and emotional sequelae of autoimmune diseases, and behavioral differences between normal mice and those with a genetic susceptibility to autoimmune disease. [*See* AUTOIMMUNE DISEASE.]

Recent data further suggest that behavioral changes associated with immunologic dysfunctions may actually be adaptive with respect to the maintenance or restoration of homeostasis within the immune system. Lupus-prone (NZBxNZW)F1 mice do not acquire conditioned taste aversions in response to immunosuppressive doses of CY that are effective in inducing conditioned avoidance responses in healthy control (C57BL/6) mice. Also, when tested after the development of signs of autoimmune disease (lymphadenopathy and elevated autoantibody titers), Mrl-lpr/lpr mice, another strain of animals that spontaneously develop a lupus-like disorder, do not avoid flavored solutions paired with doses of CY that are effective in inducing taste aversions in congenic (Mrl+/+) control mice. These differences in behavior do not result from a learning deficit in the lupus-prone mice because there are no substrain differences before the development of symptoms of disease and there are no substrain differences when a nonimmunosuppressive drug is used as the UCS. Phenomenologically, it would appear that lupus-prone mice "recognize" the existence of their immunological deficit and/or the ameliorating effects of the immunosuppressive drug, despite its noxious gastrointestinal effects.

Mrl-lpr/lpr mice with symptoms of autoimmune disease also voluntarily consume more of a flavored drinking solution containing CY than asymptomatic controls. Moreover, they drink sufficient amounts of the CY-laced solution to attenuate lymphadenopathy and autoantibody titers. Although not previously described with respect to the immune system, these data are consistent with a large literature indicating that behavioral responses are a primary means by which animals maintain and regulate some physiological states. Whether, in the case of a dysregulated immune system, the animal is responding to nonspecific, immunologically induced pathophysiological changes in one or another target organ or, consistent with the bidirectional pathways that link the CNS and immune system, the brain is capable of receiving and processing information emanating from the (dysregulated) immune system directly, remains to be determined. To the extent that the brain is capable of acting on information provided by the immune system, it would appear that behavioral processes have the potential to serve an *in vivo* immunoregulatory function.

V. Summary

The observations and research described in this article derive from a nontraditional view of the "immune system." It has become abundantly clear that there are probably no organ systems or homeostatic defense mechanisms that are not, *in vivo*, subject to the influence of interactions between behavioral and physiological events. The complex mechanisms underlying these interactions and their relation to health and illness, however, are imperfectly understood. The most imperfectly understood, perhaps, are the interrelations among brain, behavior, and immune processes.

Without attempting to be exhaustive, we have pointed out, using stress effects and conditioning phenomena as illustrations, that behavior is capable of influencing immune function. We have also noted that the immune system is capable of receiving and responding to neural and endocrine signals. Conversely, it would seem that behavioral, neural, and endocrine responses are influenced by an activated immune system. Thus, a traditional view of immune function that is confined to cellular interactions occurring within lymphoid tissues is insufficient to account for changes in immunity observed in subhuman animals and humans under conditions that prevail in the real world. The clinical significance of these interactions will not be fully appreciated until we understand more completely the extent of the interrelations among brain, behavior, and immune functions. Behavioral research represents a new dimension in the study of immunity and immunopharmacology, but it has already yielded basic data that suggest new integrative approaches to an analysis of clinically relevant issues.

Bibliography

Ader, R., and Cohen, N. (1982). Behaviorally conditioned immunosuppression and murine systemic lupus erythematosus. *Science* **214,** 1534.

Ader, R., and Cohen, N. (1985). CNS-immune system interactions: Conditioning phenomena. *Behav. Brain Sci.* **8,** 379–426.

Ader, R., Cohen, N., and Felten, D. L., eds. (1990). "Psychoneuroimmunology," second edition. Academic Press, New York.

Berczi, I., and Kovacs, K., eds. (1987). "Hormones and Immunity." MTP Press, Lancaster, England.

Cooper, E. L., ed. (1984). "Stress, Immunity, and Aging." Marcel Dekker, New York.

Felten, D. L., Felten, S. Y., Bellinger, D. L., Carlson, K. D., Ackerman, K. D., Madden, K. S., Olschowka, J. A., and Livnat, S. (1987). Noradrenergic sympathetic neural interactions with the immune system: Structure and function. *Immunol. Rev.* **100,** 225–260.

Goetzl, E. J., ed. (1985). Neuroimmunomodulation of immunity and hypersensitivity. *J. Immunol.* **135** (Suppl. 2).

Jankovic, B. D., Markovic, B. B., and Spector, N. H. eds. (1987). Neuroimmune interactions: Proceedings of the Second International Workshop on Neuroimmunomodulation. *Ann. N.Y. Acad. Sci.* **496.**

Locke, S., Ader, R., Besedovsky, H., Hall, N., Solomon, G., and Strom, T., eds. (1985). "Foundations of Psychoneuroimmunology." Aldine Press, New York.

Psychophysiological Disorders

LAWRENCE WARWICK-EVANS, *University of Southampton, United Kingdom*

Glossary

Approach–avoidance Situation in which the competing drives to approach for a reward and to avoid for fear of punishment are simultaneously present

Asthma Recurrent respiratory disorder involving reversible narrowing of the airways due to autonomic and/or immune responses

Atherosclerosis Relatively permanent reduction in the cross-sectional area of (particularly coronary) arteries due to thickening of the arterial wall and deposition of mainly fatty substances

Biofeedback Procedure whereby physiological activity (heart rate, muscle tension, eccrine sweat gland activity, etc.) is recorded and, with humans, is continuously displayed to the subject while he or she attempts to control the activity; with animals, reward and/or punishment is contingent on increasing or decreasing the activity

Conditioning There are two types of conditioning: (1) Pavlovian or classical conditioning in which a neutral stimulus (e.g., a bell) is presented at the same time as an adequate stimulus (e.g., food) for a response (e.g., salivation); after one or more pairings the previously neutral stimulus evokes the response; (2) Operant or instrumental conditioning in which any behavior followed by reinforcement tends to be repeated, while nonreinforced behavior becomes less frequent

Coronary heart disease Occlusion of the coronary arteries that is serious enough to produce symptoms ranging from minor electrical abnormalities to infarction and death

Hypertension Resting blood pressure that is elevated with respect to a particular criterion such as 120 mm hg systolic or 80 diastolic. The degree of elevation is relatively arbitrary: it may be either systolic or diastolic or both and is often age related, e.g., over 160 systolic or over 90 diastolic at age less than 40 years

Neuropeptides Molecules of two or more amino acids linked by peptide bonds and capable of either facilitating or inhibiting neural transmission

Raynaud's syndrome Intermittent but excessive vasoconstriction of the peripheral circulation, precipitated by causes such as a drop in temperature, smoking, or psychological factors

Repression Hypothesized mental process in which disturbing ideas, feelings, or memories are kept out of conscious awareness, but may be brought into consciousness via techniques of word association or dream analysis

Risk factors Characteristics which indicate an increased probability of the presence or future occurrence of a disease (sometimes but not always causal)

Stress Pressure defined subjectively in terms of the psychological demands (real or imaginary) on a person relative to their coping resources and generally accompanied by emotional arousal

Ulcers Local lesions of the skin or mucous membranes exposing deeper unprotected tissue

IT IS CONSPICUOUSLY CLEAR that the mind (psyche) can influence many aspects of bodily (somatic) activity, as when fear quickens the pulse, or even that the mind can completely control bodily functions, such as the skilled sensorimotor coordination of reading aloud or driving an automobile. Moreover it is widely believed that this influence or control may adversely effect a wide range of physiological functions. The resulting dysfunctions are popularly known as psychosomatic disorders, but this term has been so extensively abused, in particular to refer to conditions in which patients believe that they are ill even in the absence of any physiological or organic pathology, that an alternative description became necessary. Consequently the American Psychiatric Association introduced into its Diagnostic and Statistical Manual (DSM) of Mental Disorders (1952) the category of "Psychophysiologic Autonomic and Visceral Disorders." These were defined as disorders due to disturbance of innervation or of psychic control and were subdivided into the following ten categories of reaction: musculoskeletal, respiratory, cardiovascular, gastrointestinal, genitourinary, nervous system, organs of special sense, endocrine, skin, and hemic and lymphatic. In the second edition of the DSM (1968) and in the eighth edition of the World Health Organization's (WHO) International Classification of Disease, the description "Psychophysiologic Disorders" was retained but redefined in terms of "physical symptoms that are caused by emotional factors and involve a single organ system, usually under the control of the autonomic nervous system."

However, a radically new diagnostic structure was introduced in the third edition of the DSM (1980) and retained in its revised version (1987). The innovation was the introduction of five independent axes of diagnostic classification. The third of these axes involves all physical symptoms or states of the patient. A new and much broader category of "psychological factors affecting physical condition" was introduced and defined as follows: "[Symptoms] can apply to any physical condition to which psychological factors are judged to be contributory. It can be used to describe disorders that in the past have been referred to as either 'psychosomatic' or 'psychophysiological.' " They then cite the following as common but not exhaustive examples: obesity, tension headache, migraine headache, angina pectoris, painful menstruation, sacroiliac pain, neurodermatitis, acne, rheumatoid arthritis, asthma, tachycardia, arrhythmia, gastric ulcer, duodenal ulcer, cardiospasm, pylorospasm, nausea and vomiting, regional enteritis, ulcerative colitis, and frequency of micturation. No reason was given for this change of emphasis but it may well have resulted from the following two considerations. First, there was a growing appreciation that psychiatric illness was not a necessary prerequisite for the development of a psychophysiological disorder and, second, that the "normal" behavior of a patient was a contributory cause to most illnesses. Nevertheless, a contemporary redefinition of the original term is still required, and the following is proposed: "Psychophysiological disorders are real, not imaginary, disorders of physiological systems, the onset, severity or duration of whose dysfunction is at least partly due to psychological factors."

Traditionally, psychological factors have been interpreted to refer only to psychodynamic processes, conditioning (both classical and operant), emotional experience, and other aspects of private mental life. If however, psychological factors are understood to include an organism's behavior, then the implications broaden. Since very little pathophysiology is of purely genetic origin it must follow that environmental factors are the major influence. But since the organism has the choice of an infinitude of alternative behaviors ranging from those with the highest risk (e.g., substance abuse) to those with the lowest risk (e.g., the choice of a healthy diet and adequate exercise), it must be behavior that mainly determines the extent of the environmental contribution to physiological dysfunction. On this broader interpretation of psychological factors it is clear that not just a limited set of dysfunctions invites the description psychophysiological but that the description applies to the majority of illness and disease states. An appreciation of the inevitability of this conclusion is essential for the prevention and management of illness. This article, however, will be restricted to psychophysiological disorders as defined by the traditional interpretation of "psychological factors."

I. The Concepts of a Physiological System and of Causation

A. The Systems Analysis of the Human Organism

It is an almost universal convention in the teaching of medicine and physiology to subdivide the body into a set of systems; to a great extent the practice of and specialization within medicine follow a similar convention and, to a lesser extent, research tends to focus within a particular system. Thus people learn about, specialize in, or research into the cardiovascular, respiratory, gastrointestinal, genitourinary, musculoskeletal, or endocrine or nervous system much as if they were *separate* systems, which they are not. They are, of course, *separable* systems in theory and to treat each as a closed system has many advantages. For example, one's area of interest or responsibility is thereby precisely delimited; quantifiable analytic concepts such as negative or positive feedback, system variable, set point, gain, delay, transfer function, damping, oscillation, and homeostasis may be used to model the system's behavior; and above all we are provided with a feeling of potential completeness of understanding. But the cost of these advantages is that we can lose sight of the more complex realities that (1) the dividing line between systems is to some extent arbitrary and (2) that there is a hierarchy of systems with each system open for dynamic interaction with other systems at the same level in the hierarchy, open to control from above, and also open to disturbance from the external world.

The first point may be illustrated by the following example: a major function of the cardiovascular system is to perfuse the tissues with oxygen, but this is also a function of the respiratory system, so any understanding of how the body achieves this necessitates the idea of a cardiorespiratory system. The second point, which is the more fundamental with respect to psychophysiological disorders, merits further expansion. The fundamental point is that it is the *brain* that has the capacity to regulate or disrupt the efficient functioning of all other systems and that has the potential for control of each of the systems in which psychophysiological disorders may arise. This is not to deny that a system is capable of autonomy or self-regulation or that a peripheral system does not exercise feedback control (e.g., via neuropeptides) on the brain itself. Dynamic interaction between systems is well illustrated by the interaction of visual, vestibular, other proprioceptive systems, the striate musculature, and the autonomic nervous system in the normal control of balance and movement and in the production of motion sickness. Finally, the direct influence of the external world over each system should not be forgotten: the immune system reacts dramatically to bacteria, foreign protein, and viruses (and catastrophically to HIV), gastrointestinal efficiency depends on the quality and quantity of food, and effective sexual functioning depends on the presence of appropriate stimulation. In conclusion, the systems-analytic approach not only offers a useful classificatory framework for psychophysiological disorders but because of the overarching influence of the brain this approach suggests how psychological factors might exercise an influence on most body systems.

B. Causation

Whereas many illnesses are defined in terms of a description of their symptoms, psychophysiological disorders are defined in terms of their causes. This immediately requires us to consider what is meant by a cause and what criteria must be satisfied before something can be accepted as a cause. But contemporary medicine has tended to relinquish the concept of cause in favor of the idea of a risk factor. This refers to any characteristic (genetic, biochemical, physiological, psychological, or sociological) the degree of whose presence increases the probability that a person has or will develop a disease. Epidemiological research, particularly prospective studies, has established that for each disease there are many risk factors. But the relationship between causes and risk factors is not straightforward, since any definition of cause must acknowledge two ideas: (1) that causes precede their effects and (2) to the extent that a cause is present then the probability of its effects will be increased. Additionally, if we can describe, preferably quantitatively, the intervening processes whereby a hypothesized cause produces its effect then we feel more justified in accepting that there is a causal relation between the two events. But since this third requirement merely reflects what happens at a particular moment to be the state of our knowledge, it cannot be used as a necessary criterion of causality. Just as there are many risk factors for each disease, the current consensus is that causation is usually multifactorial.

The main conclusion from these considerations is that while all causes of diseases must be risk factors, it is not necessarily the case that all risk factors have a causal relationship to a disease. For example, a risk factor may be associated with a disease either because the risk factor causes the disease or because the disease caused the risk factor or because a third variable was responsible for both the disease and the risk factor. In theory these three criteria would only be fully met by the following type of research study. Human subjects would be randomly allocated into either a control or experimental group. The latter would be exposed to severe and protracted psychological stress in a standardized laboratory environment (which should be both objectively described and subjectively rated), while continuous invasive measurements were made across an extensive range of variables, and the experiment would continue until severe physiological disturbances arose, at which point the stress would be discontinued but the subjects would continue to receive intensive and invasive study so as to establish whether the effects were acute and reversible or chronic and irreversible. Even if severe pathology did result, one could conclude only that psychological factors are *capable* of causing the disorder, not that whenever a disorder arises it is due to psychological factors. But since such an experiment is both impractical and unethical, the demands for conclusive proof that psychological factors cause physiological disorders have been set impossibly high. Nevertheless, there is an abundance of alternative lines of evidence in support of the hypothesis that psychophysiological disorders are both widespread and severe in contemporary societies. This evidence is presented and critically reviewed in Sections II and III.

II. Evidence for the Prevalence of Psychophysiological Disorders

A. Theoretical

The central requirement of all theories is that they attempt to explain how mental activity can exercise an adverse effect on physiological functions. In the case of disorders of the striate muscles such as tics, torticollis, tension headache, and muscle spasms, this presents no problem. The cortex (the locus of our essential humanness, of personality, of learning, and of psychodynamic conflicts) is hardwired via motor neurons through to the relevant muscle beds and can produce its effect directly, though not necessarily voluntarily. A similar argument can be deployed that the cortex via the limbic system, hypothalamus, and pre- and postganglionic fibers can dysregulate all effectors of the autonomic nervous system (ANS). Similarly, cortical efference to the hypothalamus and thence to the pituitary can effect hormonal activity, and therefore to some extent could disrupt the neurohormonal environment that is required for the optimal functioning of the immune system.

The main theoretical approaches to explain the occurrence of psychophysiological disorders are psychodynamic, evolutionary, conditioning, and personality or dispositional. Each will be described in turn. The evolutionary approach emphasizes the very protracted time scale of natural selection of those best adapted to survive and reproduce in a competitive world where predators abounded and prey was scarce. The chances of survival were greatly increased by the development of bodily capacities such as the fight and flight reactions for dealing with immediate demands for dynamic activity and a variety of catabolic processes that mobilize energy resources on a longer time scale. This emphasis on the adaptation of early man the hunter and the hunted has been invoked to explain both the metabolically inappropriate high cardiac output often seen in early hypertension and the excess free fatty acids and cholesterol associated with atherosclerosis. But the evolutionary approach also emphasizes the steady exercise requirements of man the nomadic gatherer, the amount of time spent in leisure and relaxation after basic needs for survival have been satisfied, and an individual's interdependence on other members of a relatively small group. By contrast, contemporary man takes little exercise, enjoys less leisure, and tends to live relatively independently of others in very large cities. Essentially the argument is that humans evolved via natural selection to cope successfully with the demands of a world that has virtually ceased to exist, and those very reflexes and dispositions that once served so well have become maladaptive and counterproductive in terms of physical (and mental) health.

Psychodynamic explanations derive from Freudian conceptions of the development and structure of the mind (ego, superego, id, subconscious, and unconscious), the dynamic power of repressed intrapsychic conflicts to be converted into physical symptoms, and the importance of symbolism. More specifically, a psychosomatic disorder was believed to result from a combination of the following three characteristics: (1) a personality with an important intrapsychic conflict pattern that evoked a defense mechanism such as repression, (2) an emotionally important precipitating life crisis, and (3) a constitutional weakness in an organ system. The conflict was said to be resolved or alleviated by being converted into bodily symptoms (hysterical conversion reaction). The main themes of this analysis have received so little experimental support and achieved so few therapeutic successes that they have been largely abandoned, but the idea of a constitutional (possibly genetic) predisposition to develop a psychophysiological disorder in a particular system may be incorporated into other theoretical approaches. [*See* PSYCHOANALYTIC THEORY.]

Not all people are equally susceptible to psychophysiological disorders and there have been alternatives to the psychodynamic approach in explaining this differential susceptibility. It has been proposed that some people have a genetically determined tendency to overreact to psychologically disturbing situations in a particular way, for example, by oversecretion of digestive acids or by excessive vasoconstriction. (This is clearly similar to the organ weakness theory of the preceding paragraph.) Although there is little evidence that this tendency is either widespread or strong in humans, laboratory animals may be selectively bred for a predisposition to develop ulcers in response to psychological stress. An alternative explanation of individual differences in susceptibility proposes that, particularly with increasing age, some people develop a fixed or "stereotyped" tendency to react in one specific system to a wide range of emotional experiences such as fear, anger, or anxiety, each of which would in most people evoke different patterns of physiological responses. This repeated disturbance will then, it is argued, lead to a breakdown of normal homeostatic restraints and ultimately to illness. There is little or no systematic evidence that this occurs.

Personality theory also attempts to explain not only why certain people succumb to psychophysiological disorders but also which particular disorders they will develop. It has been suggested that some people have personalities characterized by the frequent or protrated presence of particular attitudes or emotions. Each attitude or emotion is said to be accompanied by a disturbance of a particular physiological system, for example, the experience of helpless suffering and frustration is accompanied by hives or eczema, while the feeling of being in a state of constant readiness to deal with psychological demands is associated with hypertension. There is little evidence for the truth of these suggestions except that feelings of anger, hostility, and preparedness to meet real or imagined demands have quite frequently been reported in people with mildly elevated blood pressure.

The final theoretical veiwpoint asserts that psychophysiological disorders, since by definition they are not purely genetic, are acquired or learned at the hands of the environment. This learning may take the form of either classical (Pavlovian) conditioning or instrumental (operant) conditioning. In classical conditioning a previously neutral or unconditional stimulus (e.g., the sound of a bell) is presented along with a potent stimulus (e.g., food) that evokes a response, usually within the ANS (e.g., salivation). Usually, after several pairings the conditional stimulus when presented *without* the unconditional stimulus evokes the response, although the effect soon disappears. Sometimes, particularly if fear or avoidance of an aversive situation is involved, only a single pairing of the two stimuli is required for the association to be learned, moreover the conditional stimulus may retain its power to evoke the response for thousands of test trials or over many years. By contrast, in instrumental conditioning a response (e.g., bar pressing) that is rewarded or reinforced (e.g., by food or avoidance of shock) comes to be repeated with a probability that is a function of the frequency and intensity of the reinforcement. As with classical conditioning, the response usually disappears soon after the reinforcement is discontinued, but if the original reinforcer was avoidance of an aversive stimulus then the response may remain for years. Learning theorists have certainly demonstrated classical conditioning in which asthmatic attacks occur in response to a previously neutral visual stimulus. However, the phenomenon appears to occur only in a very few individual animals or humans. More surprisingly by pairing sodium saccharine (conditional stimulus) with cyclophosphamide (unconditional stimulus that depresses immune function) it has recently been shown that saccharine subsequently acquires the capacity to depress both humoral and cell-mediated immunity. This example of a classically acquired response is relatively easy to establish in several species, though there is as yet no evidence that the effect is long-lasting. By contrast, several animal studies have successfully used the instrumental paradigm of shock avoidance either alone or with food as the reward for the response of increases in blood pressure. The majority of animals learned to elevate their mean blood pressures by 30–60 mm Hg, but when the reinforcers were removed, the blood pressure returned to baseline. [*See* CONDITIONING.]

B. Empirical Research Paradigms

In theory, relevant research may be classified according to the following schema: human versus animal experiments (each subdivided into acute or chronic) and experimental versus epidemiological studies (the latter subdivided into retrospective or prospective).

1. Human Studies

Human experiments are limited by ethical considerations to brief and relatively mild manipulations of stressful stimuli (e.g., white noise, IQ tests, competitive reaction time tasks, mental arithmetic, penetrating interviews about anxiety-provoking incidents in the subject's life, or threat of electric shock), while the effects are recorded over a wide range of physiological and biochemical variables (e.g., blood pressure, cardiac output, regional blood flow, muscle tension, airway resistance, digestive acids, blood or urinary catecholamines, or corticosteroids). Even if only slight and temporary disturbances are found, it is usually implied that if the stimulation had been more severe and protracted then the disturbance would have been correspondingly more serious and chronic. Epidemiological studies of the retrospective type select patients with functional or even organic disorders (e.g., asthma, hypertension, myocardial infarction, dermatitis, migraine, impotence, or ulceration of the digestive tract), then using interviews or questionnaires they establish a link with possible psychological aspects of the patient's personality or history. These include intensive single-patient studies and surveys extending to many thousands of subjects. A large proportion of retrospective studies have concentrated not merely on stressful experiences but on any events that require major psychological changes in life-style.

The original inventory was the Social Readjustment Rating Scale, which included not only negative items such as death of a spouse, divorce, or jail term but also more positive events such as marriage, pregnancy, and vacations. Each event counted on a scale of life change units (LCUs) and a high overall score was expected to be associated with a high incidence of illness. Modest negative correlations have often been found between high scores and general measures of well-being (depression, anxiety, and minor infections), but the correlation between LCUs and real physiological disorders is often not significantly different from zero. In a series of attempts to improve these results the rating scale has undergone a number of revisions and amendments, as in the Psychiatric Epidemiology Research Interview (PERI). Also it is suggested that the effect of stress and change on health are buffered in some people by the presence of moderating variables such as physical fitness, social support, or characteristics like hardiness of personality. Even after statistical allowance is made for these factors, most correlations remain in the region of $r = 0.1$ or less. And even if clinically and statistically significant correlations were found they would still be open to the criticisms (1) that there was bias on the part of the interviewer or the respondent to the questionnaire and (2) that the correlation did not indicate causality but was due to the influence of confounding variables such as diet, alcohol, smoking, poverty, or many others. The first of these criticisms may be wholly and the latter partly countered by the use of prospective studies. In these, initially healthy subjects are recruited; they are assessed along many dimensions of biochemical, physiological, psychological, and sociological characteristics; these assessments are repeated over many years; and the records of those who eventually become ill are then searched for the characteristics that distinguish them from those who remain healthy. Such studies are rare.

2. Animal Studies

With respect to animal studies there is always the question of whether they provide a realistic physiological model, but in connection with psychosomatic disorders one must consider also whether their mental processes are sufficiently comparable to our own to justify extrapolating to humans. Epidemiological studies of free-living animals report that the incidence of any of the illnesses mentioned so far as possibly being psychologically caused has an extremely low prevalence.

In acute animal studies a hypothesized and often very intense stressor is continuously or repeatedly applied, usually for just sufficient time for a physiological effect to be found. In better experiments careful control is maintained over confounding physical variables like fighting and nutrition. For example, in the yoked control design, pairs of animals are restrained and wired electrically in series; each animal therefore receives the same number of shocks of the same intensity at the same time; each animal has access to an identical bar or lever; and one of these is connected to the circuitry and, if pressed at the correct time, will delay or cancel the shock, whereas the other is not connected. The only differences between the two conditions are the psychological experience of being in control rather than helpless or being able or not able to predict whether a shock will occur. Significantly more stomach ulcers are reliably found in the helpless animals for whom the shock is unpredictable.

In other designs a wide variety of psychosocial stimuli (exposure to the sight, sound, and smell of a predator, overcrowding, or disruption of established social networks) have been employed and the effects are frequently severe and extensive (elevated catecholamine and corticosteroid levels, increased blood pressure, reduced estrous and spermatogenesis, ulceration, increased rates of infection and cancer, adrenocortical enlargement, thymic involution, atherosclerosis, and premature death). Chronic experiments in which all animals are studied until their deaths are rarely reported; this is a serious gap in our knowledge since without them we cannot be certain whether some of the physiological effects (notably hypertension) are reversible when the stressor is removed. In short, animal work has established incontrovertibly that psychological factors can cause not only functional physiological disorders but eventually organic damage. What is more contentious is whether animals provide a reasonable model for human psychophysiological disorders.

If research over the last few decades had been systematically coordinated, then we would have from each of these paradigms evidence for or against the validity of each of the preceding theories with respect to each of the types of disorder discussed in the next section. In fact, research has been disproportionately concentrated on acute animal experiments, on retrospective epidemiological studies, and on only a few disorders.

III. Currently Available Evidence for Particular Disorders

In the 1950s there would have been no difficulty about which disorders to include in this section: they would have been the seven that were then the focus of attention for psychoanalysts, that is peptic ulcer, bronchial asthma, essential hypertension, ulcerative colitis, thyrotoxicosis, rheumatoid arthritis, and neurodermatitis. By the 1970s the list would have been classified according to the body system analysis proposed in DSM II and would have excluded thyrotoxicosis and rheumatoid arthritis, but would have been extended to include most disorders of each system, for example, musculoskeletal—tension headache, tics, torticollis, and muscle spasms; cardiovascular—essential hypertension, tachycardia, vasovagal fainting, arrhythmias, migraine headache, and Raynaud's syndrome; respiratory—psychogenic breathlessness. But with a growing belief that psychological factors (even if behavior is excluded and in the absence of psychiatric disorder) can precipitate, exacerbate, or prolong a wide range of disorders, the list is greatly extended so that only a very small proportion can be selected for discussion. The following criteria are proposed for the selection process: (1) the disorder must be serious, (2) it must have a high prevalence, and (3) evidence of a causal role for psychological factors must come from at least two of the paradigms described in Section II,B. The following three illustrative areas meet all these criteria—ulcers of the alimentary tract, cardiovascular disorders, and asthma.

A. Ulcers of the Alimentary Tract

Ulcers form a heterogeneous set of disorders, even if they are subdivided by locus or hypothesized cause. Mechanisms may involve over- or undersecretion of acid or digestive hormones, alterations in smooth muscle motility or regional blood supply, reduced mucus production, or impaired neutralization. As usual, both genetic and environmental factors influence these functions. But mere introspection, unfashionable though this is, reveals how susceptible the delicate interrelationships of the system are to psychological upsets. Invasive studies of patients with gastric fistulae confirm their functional vulnerability to psychological events such as anger, frustration, and distressing interveiws. Epidemiological work is inconclusive since, although sustained and/or severe exposure to stress frequently precedes ulceration, this may be mediated by behavioral responses to the stress such as increased smoking or alcohol intake. But the most consistent and dramatic evidence that psychological stress causes ulcers comes from laboratory work on rats and monkeys performed by a variety of experimenters over the last three decades. Stresses have included physical restraint, approach–avoidance conflict, and the use of controllable versus uncontrollable and predictable versus unpredictable electric shock; experimental durations have varied from 6 hours to 30 days.

Experimental designs have become increasingly sophisticated and typically employ multiple genetically matched control groups, often in a yoked control paradigm. By sacrificing animals at different times during and after exposure to the stressful situation it is possible to study the development of the pathology. Perhaps the most interesting result is that the ulcers seem to result not so much for sympathetic overarousal but from poststress parasympathetic rebound, although corticosteroid activity also appears to be implicated. One frequently overlooked aspect of these studies is that there is always disruption of the animals' normal eating patterns. Animals are either food-deprived for 6 to 72 hours before the experiment or the experimental procedures themselves disrupt normal patterns of eating. This disruption per se is not usually enough to cause ulceration since the control animals experience equal disruption without adverse effects. The essential point is that the *combination* of the psychological stress and food deprivation is required to provoke the ulcers—the effect is interactive.

B. Disorders of the Cardiovascular System

Cardiovascular disorders range from minor dysfunctions of vasoconstriction and vasodilation such as Raynaud's syndrome or migraine through arrhythmias and hypertension to coronary heart disease. The latter two conditions each meet the criteria of seriousness, prevalence, and evidence, and will be considered separately.

All definitions of hypertension are arbitrary, but all refer to a relatively permanent elevation of systolic and/or diastolic blood pressure with respect to a theoretical "normal" level that is often age related (e.g., current North American practice but not the WHO recommendation). Secondary hypertension is defined as being due to an organic cause like coarction of the aorta or pheochromocytoma and therefore cannot be of psychological origin. But primary or essential hypertension is defined in terms of the absence of organic cause and thereby becomes a candidate for a psychophysiological disorder. Estimates of the prevalence of the latter vary from 10 to 30% in Europe and the United States. It is serious both in its own right as a cause of death but more because of its causal role for stroke and coronary heart disease. Consequently its etiology has been the subject of extensive research for many decades. Nevertheless our understanding of its causes is still very limited. Some form of polygenic inheritance is estimated to account for 20 to 65% of the population variance in blood pressure, but even the genetic effect may act only by increasing one's susceptibility to as yet unconfirmed environmental causes. Suggested risk factors include dietary aspects (sodium or sodium/potassium intake, obesity, low fiber, high protein, excess trace elements, and high alcohol intake), smoking, caffeine, and psychological factors such as personality and stress. Evidence for a causal role of the latter two will be reviewed next. [*See* HYPERTENSION.]

Acute experimental studies of animals and humans are quite conclusive; psychological factors such as classical conditioning, operant conditioning, and stress reliably evoke large increases in blood pressure of the order of 10 to 50% of pretreatment levels. In explanation of this effect there are numerous regulatory mechanisms that may mediate between the cortically based psychological processes and the response of elevated blood pressure, for example, cardiac and vagal regulation of heart rate and stroke volume, vasoconstriction, renal blood flow, or hormonal changes in catecholamines, renin, or aldosterone. But the central and hitherto unresolved issue is whether psychological factors can produce relatively permanent (of the order of years) effects.

Chronic human experimental work does not exist at all and very few animal studies follow the animals for more than a few weeks after the end of the stress sessions. The general picture is that animals recover back to baseline levels either overnight or a few weeks or months after the last stress sessions. However, a few individual animals do develop left ventricular hypertrophy or interstitial nephritis, and this may sustain irreversible hypertension. Cross-sectional epidemiological surveys frequently report an association between personality variables (anger or anxiety) and level of pressure, or between stressful employment (air traffic control, bus driving), stressful environments (inner city housing), and elevated blood pressure, but the effect is usually slight (<10 mm Hg).

True prospective studies over more than a decade and focusing on psychological factors do not appear to have been carried out, although people with large short-term blood pressure increases to a variety of stimuli have been found to have a significantly higher incidence of hypertension at follow-ups of 5–45 years. Nevertheless, the role of psychological factors in provoking acute blood pressure increases is indisputable and, in theory, repeated or protracted stress could lead from elevated cardiac output to elevated peripheral resistance via baroreceptor resetting, hypertrophy of the ventricular or arterial wall, or whole-body autoregulation—but this is currently unproven.

There are three arguments why coronary heart disease (CHD) could be regarded as psychophysiological disorder. First, hypertension is a contributory cause of CHD; therefore, to the extent that hypertension is psychogenic so also is CHD. Second, large-scale prospective studies have shown that individuals with a type A personality (time urgent, competitive, and ambitious) have double the normal probability of developing CHD over the subsequent 8 years, though not over the subsequent 24 years. Finally, numerous animal experiments and human epidemiological studies have shown that laboratory and real life stresses are associated with increased atherosclerosis, although there have also been some negative findings. But it should be added that the effects of stress, even when present, are relatively small compared with major risk factors such as age, smoking, diet, and serum cholesterol. By contrast, the acute effects of stress in people with preexisting CHD may be calamitous, involving angina, arrhythmias, or myocardial infarctions. [*See* ATHEROSCLEROSIS.]

C. The Respiratory System

The major psychologically influenced disturbance of the respiratory system is asthma, which may be defined as a reversible, recurrent breathing disorder associated with hyperreactivity of the bronchial tree. It has a prevalence of roughly 10% and is potentially lethal if an attack cannot be controlled. There are two different types of reactivity; the first involves the parasympathetic branch of the ANS, while the second comprises an allergic/immune response. The relationship between these two types of response is very poorly understood and it is easiest to illustrate psychological influences on the ANS component. The initial events involve constriction of the smooth muscles of the bronchioles and an increase in the rate and viscosity of neurosecretion, both being under parasympathetic control and possibly leading to local edema and hyperemia. There is little animal evidence for the relevance of psychological factors, but epidemiological work confirms the relevance of emotional disturbance; and acute experimental work clearly demonstrates that the onset and course of an attack may be influenced by the expectation and beliefs of the sufferer and suggests that conditioning may be involved. [*See* PULMONARY PATHOPHYSIOLOGY.]

Numerous experimenters have explored the effects of suggestion on asthmatics and controls who are either healthy or suffer from other lung diseases. It has been frequently shown that if it is suggested to an asthmatic that he or she is inhaling a substance that will precipitate an attack then the respiratory efficiency will decline. Measures of breathing efficiency have ranged from self-report to whole-body plethysmography and forced oscillation techniques. The strength of the effect has varied from slight but statistically significant increases in airway resistance to full-blown dyspnea and wheezing. Such experiments also tend to confirm the direct dependency of the bodily response on the mental attitude because the effect can be reversed by providing a therapeutic placebo such as nebulized saline along with suitable instructions. These effects cannot always be obtained and in partial explanation of this it is suggested that asthmatics can be divided into three groups. One group's asthma is said to derive essentially from severe infectious disease of the lung, another group is said to suffer from an allergic form of asthma, and only in the last group are psychological factors a cause, and it is these sufferers who show the dramatic effects of suggestion. Case histories and surveys provide some support for this classification, and there is a relatively consistent subgroup who often identify emotional disturbance as the precipitating factor for their asthmatic episodes. However, it should be noted that some subjects show evidence of all three aspects.

IV. Concluding Comments

Although the idea that the mind can produce adverse effects on the body is still central to any conception of psychosomatic/psychophysiological disorders, there have been several changes of emphasis. First, such effects are no longer seen as necessarily symptomatic of mental illness. Second, whatever arbitrary analysis of the body systems is used, it is clear that psychological factors can either directly or indirectly influence them. Third, although a disorder may at first be "merely" functional it may develop into structural pathology; consequently it may not necessarily be reversible. Finally, because of the multifactorial causation of illness, it is clearly a logical error to classify certain groups of disorders as psychological in origin and others are due to physical causes. The point is that on one occasion an ulcer or asthmatic episode may be due to excess smoking or animal danders while on another occasion the same event in the same patient may be due to psychological factors.

The logic of the relationship between etiology and treatment is frequently misunderstood, and nowhere more so than with respect to psychophysiological disorders. The logic of this error is as follows: "A entails B," therefore "not-A entails not-B." Stripped to its logical bones, this argument lacks all conviction, but embedded in the form "stress causes psychophysiological disorders so suitable stress management or avoidance will cure the disorder," its appeal is more insidious. The conclusion about cure may well be correct but it is not logically entailed by the premise, particularly if the disorder has progressed to an organic form. In this case it may require surgical or pharmacological intervention. Nevertheless, psychotherapy, including psychoanalysis, meditation, relaxation, behavior modification, biofeedback, or family therapy, may assist the process of recovery from psychosomatic disorders.

Bibliography

Christie, M. J., and Mellet, P. G. (eds.). (1986). "Psychosomatic Approach: Contemporary Practice of Whole-Person Care." Wiley, Chichester.

Dutevall, G. (1985). "Stress and Common Gastrointestinal Disorders." Praeger, New York.

Fava, G. A, and Wise, T. N. (eds.). (1987). "Research Paradigms in Psychosomatic Medicine." Karger, Basel.

Kasl, S. V., and Cooper, C. L. (1987). "Stress and Health: Issues in Research Methodology." Whiley, Chichester.

Steptoe, A. (1983). Psychological aspects of bronchial asthma. *In* "Contributions to Medical Psychology" (S. Rachman, ed.). Pergamon, Oxford.

Puberty

ZVI LARON, *Beilinson Medical Center, Petah Tikva*

Glossary

Bone age Important maturational index determined with the aid of an Atlas using the number, size, and shape of the small bones of the hand and epiphyseal centers of the finger phalanges

Estrogens Female sex hormones

GnRH Gonadotropin-releasing hormone, a hormone secreted in pulsatile manner (every 90 min) by the hypothalamus and stimulating the synthesis and secretion of the gonadotropins: follicle-stimulating hormone (FSH) also developing the male sperm cells, and luteotrophic hormone (LH) also stimulating the testosterone-secreting cells in the male testicles

Gonadostat Hypothetical center of the central nervous system that governs the secretion of GnRH, the gonadotrophic secretion–stimulating hormone in the hypothalamus

Gonads Two sex glands: the ovaries in the female, which produce the female sex hormone (estrogen) and ova. The testicles are male sex glands producing testosterone and spermatozoa

Hypothalamus Part of the central nervous system that manufactures the stimulatory and inhibitory hormones of the pituitary

Menarche First menstrual bleeding

Neuroendocrine Interrelation between nervous and endocrine tissues

Pituitary Hormone-secreting gland situated in the bony part of the skull called *sella turcica,* the anterior part of which, originating from the pharynx, secretes a series of trophic hormones including the gonadotropins

Pubertal growth spurt Accelerated growth velocity caused by an increased secretion of sex hormones during puberty

Puberty Maturational stage of childhood that comprises the stages of sexual development

Testosterone Male sex hormone

PUBERTY IS a maturational stage between childhood and adulthood. It comprises the stages of sexual development, the end of which is the capacity to reproduce. It is a complex process governed by neuroendocrine factors modulated by hereditary traits.

During this period, boys and girls undergo profound physical changes (e.g., the appearance and progression of the secondary sexual characteristics and a growth spurt) (Fig. 1). In addition, profound psychological changes take place. The onset of puberty is governed by a hypothetical center in the hypothalamus, the "gonadostat," which is latent in the prepubertal period and, for reasons so far not understood, is activated at around age 8 years in girls and 10 years in boys. The relatively dormant period of the gonadostat is paralleled by a relative state of "lowered sensitivity" of the gonads and their target tissues. An imbalance of either gonadostat or the sensitivity of the peripheral tissues may cause an abnormal sequence of puberty.

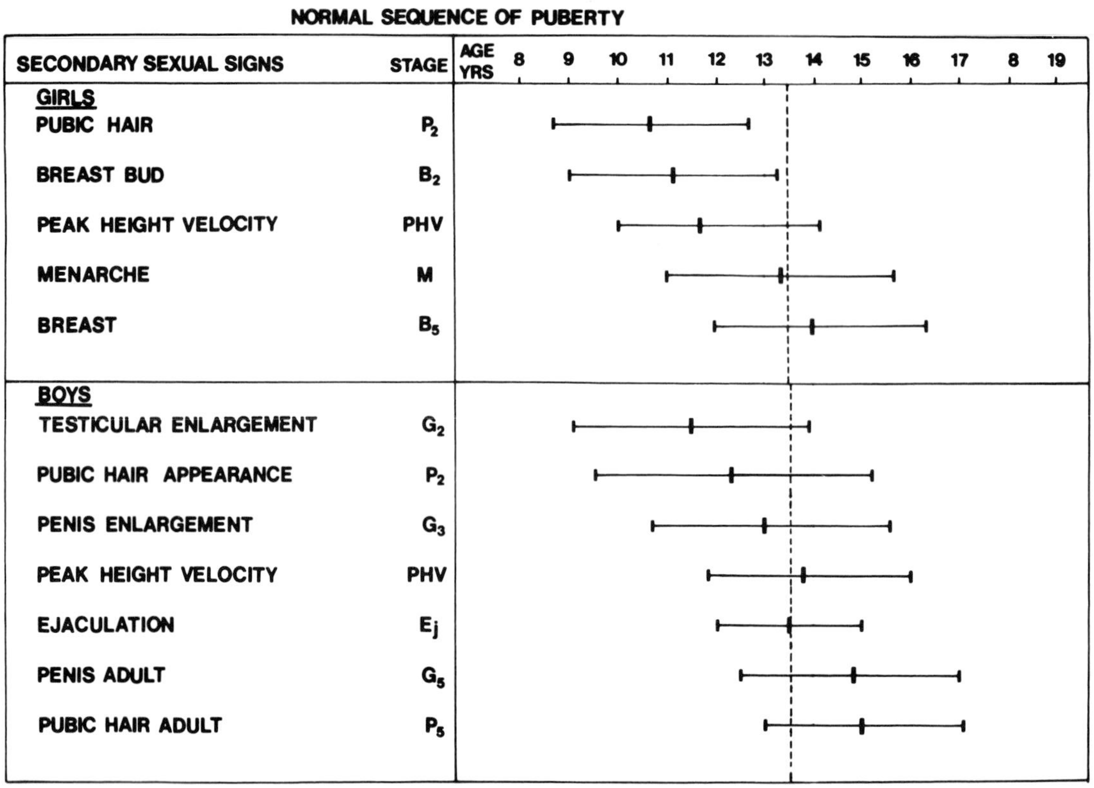

NORMAL SEQUENCE OF PUBERTY

SECONDARY SEXUAL SIGNS	STAGE
GIRLS	
PUBIC HAIR	P_2
BREAST BUD	B_2
PEAK HEIGHT VELOCITY	PHV
MENARCHE	M
BREAST	B_5
BOYS	
TESTICULAR ENLARGEMENT	G_2
PUBIC HAIR APPEARANCE	P_2
PENIS ENLARGEMENT	G_3
PEAK HEIGHT VELOCITY	PHV
EJACULATION	Ej
PENIS ADULT	G_5
PUBIC HAIR ADULT	P_5

AGE YRS 8 9 10 11 12 13 14 15 16 17 8 19

FIGURE 1 Sequence of pubertal development.

I. Nomenclature of Pubertal Grading

The usual rating of prepuberty and pubertal stages is done according to Tanner who graded the prepubertal stage P1 and rates the pubertal stages from P2 to P5 (the adult stage). Although four ratings are insufficient to grade the developmental stages, Tanner's staging is the most frequently employed, and we shall refer to it in this article. It is also advised for means of accuracy to rate each sign separately [e.g., breasts (B_{1-5}), axillary hair (A_{1-5}), pubic hair (P_{1-5}), genitalia (G_{1-5})].

II. Evaluation of the Progression of Puberty

Ascertainment of normal progression of puberty (i.e., the age at onset) and normal rate necessitates periodic complete physical examinations, including standing height (and/or supine length), ascertainment of appearance of secondary sexual characteristics such as preacne or acne, moustache, beard, axillary and pubic hair, and a change in the voice in boys, and, in females, development of breast tissue and redness and presence of white discharge from the vagina. In addition, the size of the testes and penis in boys and small and large labia in girls should be evaluated. An important indicator of the degree of maturation is the bone (skeletal) age, determined from a hand and wrist X-ray.

III. Sequence of Pubertal Development in Girls

A relation exists between the age of onset of breast development, height spurt, and age at menarche (Fig. 1). The first pubertal event may be the onset of the growth spurt and/or pubic hair. The growth spurt may precede the onset of breast development by as much as 1 year. Peak height velocity coincides with midpuberty (i.e., Tanner P_3). Menarche occurs after the time of peak height velocity has passed, at a bone age of 13–13.5 years. Growth after menarche ranges between 5 to 9 cm. In most girls, menarche occurs within 4–5 years after the onset of breast growth.

IV. Sequence of Pubertal Development in Boys

The initial manifestation is enlargement of the testicular volume and appearance of creases (rugae) of the scrotum to be followed by the appearance of pubic hair (Fig. 1). The penis growth parallels usually that of pubic hair. The onset of the pubertal growth spurt occurs approximately 1 year after the start of increase in gonadal growth. The peak height velocity occurs late in puberty, approximately at P_4. At the stage, there occurs also a change in voice and an increase in muscle mass. The milestone of male puberty is the age at first conscious ejaculation, which occurs at a bone age of 13–13.5 years. The growth after the first ejaculation to final height ranges between 16 to 22 cm.

V. Environmental Factors Influencing Puberty

Normal, early, or delayed puberty is influenced by hereditary characteristics. The normal age of onset of puberty in girls is around age 8 and in boys 10 years, but slight variations occur according to geographical areas, populations in warm climates have often earlier onset of puberty than nordic people. It has also been claimed by Frisch that there is an invariant mean weight of approximately 48 kg associated with the age of menarche, but this axioma is not universally accepted. Malnutrition or insufficient nutrition in absolute term (e.g., in anorexia) or in relative terms (e.g., in youngsters participating in competitive sports) may delay onset or progress in puberty. Growth hormone deficiency may also delay puberty. Obesity, however, may enhance the appearance of puberty.

The average onset of puberty during the past 200 years or more shows a secular declining trend. This has been explained to result from a deterioration in socio-economic, nutritional, and health conditions in the Middle Ages and in improvement in these conditions during the past two centuries. In developing countries, this trend seems to have stopped.

VI. Development of Secondary Sexual Signs

In many children, true gonadal-induced puberty is preceded by an arousal of the adrenal and sex hormones (adrenarche), the androgens playing the major role. These hormones account for the appearance of pimples (preacne), above or on the side of the nose, and axillary or pubic hair (mainly in females).

Whereas in females pubic hair seems to be mainly under the control of adrenal androgens, in males both the gonadal and adrenal androgens appear to play a role. The appearance of pubic hair is gradual from soft "fuzzy" to harder "curly" hair. The initial hair growth is at the base of the scrotum or along the labia majora and extends toward the pubis. In the adult stage (P_5), the female hair is distributed in an inverse triangular fashion, whereas in men there may be further spread up the pubic area along the linea alba.

Breast growth in females is controlled by the ovarian estrogen secretion. The first stage is characterized by appearance of subareolar tissue growth "breast buds," to be followed by areolar growth and bulging, nipple growth, and subsequent breast enlargement to full size. The size of the breasts at stage B_5 is variable among individuals and is influenced by hereditary characteristics. The growth is best determined by repeated measurements of the half circumference and grading of the nipples, areolae, and whole breasts.

During full-fledged puberty in males, the massive testosterone to estrogen conversion leads to swelling (sometimes painful) of the breasts. This is called *pubertal gynecomastia*. In most instances, this is transitory.

The initial manifestation of the male pubertal development is the enlargement of the testes (Fig. 2). The changing size of the testes can be appraised by the comparative palpation using an orchidometer or by measurement with a caliper. As there are individual variations, the early onset of testicular growth can be determined only if prepubertal measurements have been performed. Testicular growth is followed by enlargement and coarsening of the scrotal skin, which also reddens and becomes more pigmented.

The phallic growth starts in stage G_2 and is progressive both in length and width. It can be measured with a caliper, applying light stretching. Knowing that mean penis length at birth is 3.5 ± 0.5 cm and that mean prepubertal length is 5.5 ± 0.5 cm, early pubertal growth can be determined by sequential measurements (Fig. 3).

The first observed signs in females are redness and whitish discharge from the vagina and increase

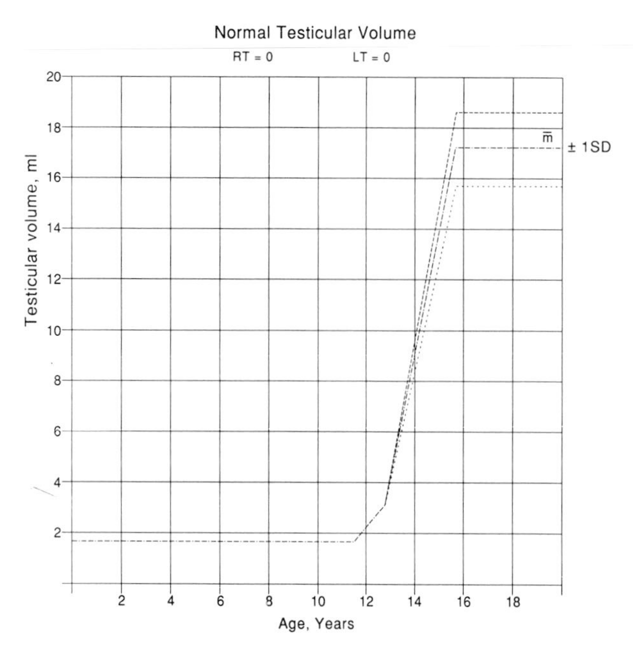

FIGURE 2 Normal testicular development. [From Zilka, E., and Laron, Z. (1969). *Harefuah* **77**, 511–513, with permission.]

in size of the labia minora and majora. Following is a progressive growth until adult dimensions.

Pubertal milestones are, in the female, the "menarche," and in the male, the "first conscious ejaculation." Both occur at a bone age of 13.5 years, which concurs with pubertal stage of P_{3-4}.

VII. Hormonal Changes at Puberty

There are marked changes of gonadotropins and gonadal and adrenal sex hormones before and during puberty; the progressive changes are clearly evident when we examine large groups, but not always when following single individuals.

After a period of increased secretion postnatally and during infancy, the levels of basal gonadotropins (LH and FSH), as well as the sex hormones, decrease to constant prepubertal levels.

During the peripubertal period FSH levels rise in girls, whereas the LH rise in boys is later; however, there is a progressively increasing response of LH in boys with each pubertal stage and of FSH in girls. However, in some girls the FSH response may be strongest in peripuberty.

As a consequence of the rise of the gonadotropins, the sex hormones [estrogens (mainly estradiol) in females (Table I) and testosterone in boys (Table II)] rise progressively.

There is also a rise in the adrenal androgens delta 4 androstenedione and dehydroepiandrosterone sulfate (DHEA-S), which may precede the rise of gonadal steroids. [*See* STEROIDS.]

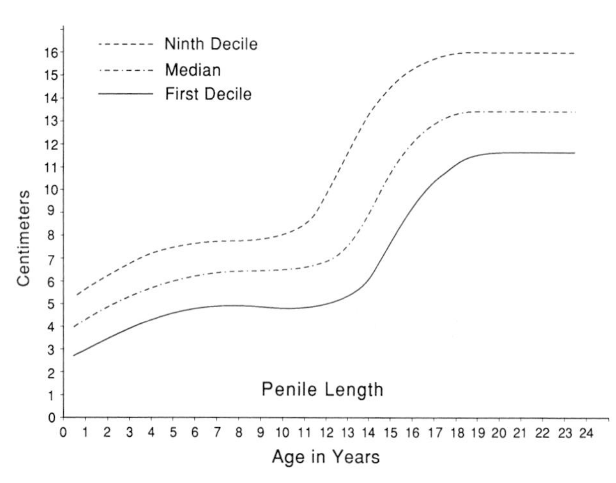

FIGURE 3 Normal penile growth.

VIII. Induction of Puberty

The increased secretion of gonadotropins before onset of puberty is a consequence of a change in neural and hormonal modulation of synthesis and secretion of GnRH. GnRH secretion changes from a tonic to a pulsatile form—the type of stimulation needed to induce and maintain LH and FSH release

TABLE I Normal Values for Plasma Estrone (E1) and Estradiol (E2) During Sexual Development (pg/ml) in Females[a,b]

Age (yrs)	*n*	E1	E2
Prepubertal (>7 years)	32	14	10
		7–29	<7–20
Stage 2	30	16	14
		<7–37	<7–35
Stage 3	24	29	30
		8–53	7–60
Stage 4	31	38	45
		10–77	12–93
Stage 5	36	53	63
		12–142	12–250
Adults	47	69	111
		20–182	17–290

[a] Based upon data from Bidlingmaier, F., and Knorr, D. eds. (1978). "Oestrogens, Pediatric Adolescent Endocrinology," vol. 4. S. Karger, Basel, with permission.
[b] E1 and E2 given as median and range. *n* = number of determinations.

TABLE II Mean (±SD) Plasma Concentration of Testosterone in Males According to Pubertal Stages[a]

Pubertal stage	Age (yrs)	Testosterone (nM/liter)
P_1	9–11	0.6 ± 0.3
P_2	11.5	1.8 ± 1.0
P_3	12.5	3.8 ± 2.0
P_3	13.5	7.3 ± 5.5
P_4	14.5	12.8 ± 4.3
P_5	15.7	15.2 ± 4.1

[a] Based upon data from Forest, M. G., ed. (1989). "Androgens in Childhood, Pediatric Adolescent Endocrinology," Vol. **19**. S. Karger, Basel.

and subsequently the female menstrual cycle. The area in the hypothalamus responsible for the change to pulsatility is called the *gonadostat*. Hereditary differences in the period of maturation of the gonadostat account for individual variations in the onset of puberty. The limits of what is considered within physiological range and abnormal are not clear and not uniformly accepted.

IX. Delayed Puberty

The upper limit to distinguish between constitutional delay and abnormal delay seems to be age 13 in girls and 15 in boys (e.g., when no signs of puberty are present at those ages an abnormality should be suspected and investigations done). However, it is practical to perform these even earlier, especially when there is no familial history of delayed puberty. Constitutional delay of puberty is more frequent in boys, and these are more concerned about the accompanying slowing in the growth rate than the slow sexual maturation. Differential diagnosis of constitutional delay are true hormonal deficiency, which includes isolated gonadotropin deficiency [normal prepubertal growth sometimes associated with anosmia (lack of smell), i.e., Kallman syndrome, Prader Willi syndrome (marked obesity, short stature, and mental retardation), and multiple pituitary hormone deficiencies including gonadotropins (show signs of other hormone deficiencies)]. Possible other diagnoses are primary gonadal failure (high basal or stimulated LH and/or FSH), such as in gonadal dysgenesis, Turner syndrome, anorchia, or acquired damage to the ovaries or testes (e.g., torsion, mumps, accidents).

Hyperprolactinemia is a frequent cause of delayed puberty in girls but rarely also in boys. It also causes secondary amenorrhea in girls.

Constitutionally delayed puberty can be treated by either short courses of small doses of androgens or anabolic steroids or by pulsatile GnRH infusions. The latter form is more physiologic, the former simpler and more acceptable by the individuals. As longer prepubertal height gain leads to a higher final height, it is advised to treat delayed puberty only if psycho-social factors demand it. Sex hormone replacement therapy in states of permanent hormone deficiency should be initiated as late as possible for the same reasons. Hyperprolactinemia is treated medically in cases of pituitary microadenoma or surgically in macroadenoma and selective cases of microadenoma.

X. Precocious Puberty

Precocious puberty is defined as the appearance of sexual characteristics before age 8 in girls and age 10 in boys. When it is due to a premature activation of the gonadostat it is defined as "true" or "central" in contradistinction to precocious "pseudo" puberty when the pubertal signs are caused by sex hormones not stimulated by the pituitary (e.g., congenital adrenal hyperplasia, gonadal or adrenal tumors, or exogenous sex hormones). There is also a form of "incomplete" sexual precocity [e.g., precocious thelarche (breast development) or precocious adrenarche (pubic and/or axillary hair development)]. These forms may be transitory (thelarche only), stationary, or develop into true precocious puberty.

True precocious puberty is four to five times more frequent in girls than in boys. In girls, in most instances, the cause is unidentified. In approximately half of the affected boys, an organic lesion or anomaly of the central nervous system [e.g., brain tumors (hamartoma, pinealoma) or hydrocephalus] is found. Precocious puberty should be suspected when an acceleration of the growth spurt and an advancement of bone age occur before the expected age.

Organic lesions are treated whenever possible by surgery. Idiopathic true precocious puberty is treated at present successfully with slow-release long-acting depot preparations of GnRH analogues.

XI. Conclusions

Puberty is a stormy period in the life of children. The sexual maturation leads to marked external and internal changes of the body and a spurt in linear growth and profoundly affects the behavior and emotions.

Any change in the orchestration of these complex processes may have long-lasting effects. Thus growth and development during prepuberty should be carefully followed and explored, if indicated, to ensure a normal pubertal process.

Bibliography

Bidlingmaier, F., and Knorr, D., eds. (1978). "Oestrogens, Physiological and Clinical Aspects, Pediatric Adolescent Endocrinology," vol. 4. S. Karger, Basel.

Bronson, P. H., and Rissman, E. F. (1986). The biology of puberty. *Biol. Rev.* **61,** 157–195.

Forest, M. G., ed. (1989). "Androgens in Childhood, Pediatric Adolescent Endocrinology," vol. 19. S. Karger, Basel.

Frisch, R. E. (1985). Fatness, menarche and female fertility. *Perspect. Biol. Med.* **28,** 611–633.

Laron, Z., Arad, J., Gurewitz, R., Grunebaum, M., and Dickerman, Z. (1980). Age at first conscious ejaculation: A milestone in male puberty. *Helv. Paediatr. Acta* **35,** 13–30.

Laron, Z., Dickerman, Z., and Kauli, R. (1987). Treatment of premature and delayed puberty with LH-RH and its analogs. *In* "Seminars in Reproductive Endocrinology" (R. H. Asch, ed.), pp. 421–429. Thieme Med. Publ., New York.

Stanhope, R., Abdulwahid, N. A., Adams, J., Jacobs, H. S., and Brook, C. G. D. (1985). Problems in the use of pulsatile gonadotropin releasing hormone for the induction of puberty. *Hormone Res.* **22,** 74–77.

Tanner, J. M. (1962). "Growth at Adolescence," 2nd ed. Blackwell Scientific Publ., Oxford.

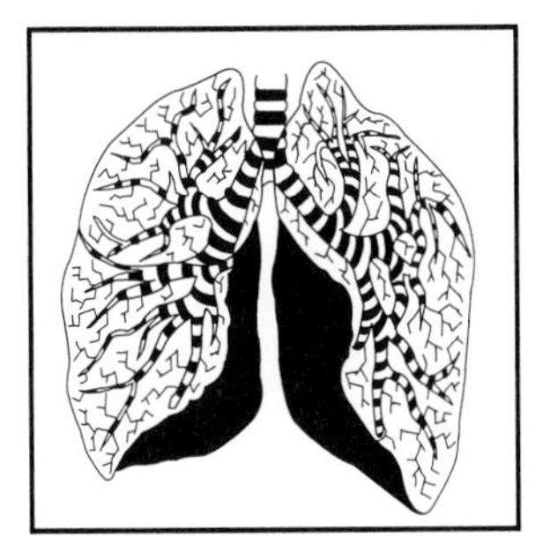

Pulmonary Circulation, Pharmacology

RAJAMMA MATHEW, *New York Medical College*

BURTON M. ALTURA, *State University of New York and Health Science Center at Brooklyn*

Glossary

Bleomycin Cytotoxic–antitumor antibiotic
CNS Central nervous system
EC Endothelial cells
EDRF Endothelium-derived relaxing factor. A labile relaxing agent produced by endothelium in response to certain stimuli
Eicosanoids 20 carbon fatty acids, which include prostaglandins, leukotrienes, and epoxy-, mono-, and dihydroeicosanoic acids
Oxygen radicals Partially reduced species of O_2 free radicals, which are reactive and cytotoxic byproducts of normal aerobic metabolism
Pulmonary hypertension Mean pulmonary artery pressure of more than 18 mm Hg
Vascular tone Partial contractile state of the vessel

MOST OF OUR understanding of human pulmonary circulation is based on animal experiments. The pulmonary circulation is a low-resistance system and is influenced by ventricular compliance and pressure and thoracic pressure. Not only does it serve the vital function of gaseous exchange, but it is also important in numerous metabolic functions that affect pulmonary arterial and bronchial smooth muscle cells, pulmonary microvascular permeability, and the systemic circulation. The pulmonary vasculature will react to external stimuli such as an increase in pulmonary artery pressure and flow secondary to cardiopulmonary defects as well as to various drugs. It is also influenced by local changes (e.g., oxygen tension, acid–base balance, injury, and the maturation of the system). The pulmonary vasculature has a large surface area lined by endothelial cells (EC), which provides a dynamic barrier between the underlying smooth muscle cells and the various vasoactive substances in the circulating blood. Various vasoactive substances are synthesized and metabolized in the lungs, and the end products are distributed throughout the systemic circulation to act on peripheral tissues. The responses to various stimuli depend on the existing vascular tone. Various types of studies such as *in vivo*, isolated vessel, and isolated perfused lung studies, as well as isolation and culture of endothelial cells, have been designed to understand the physiology and pathophysiology of the pulmonary circulation. In this chapter an attempt will be made to review some of these factors.

I. Maturational Aspects

During fetal life, lungs receive only 8–10% of cardiac output. The rest is diverted to the systemic circulation via a patent ductus arteriosus. The fetal lungs do not participate in gas exchange but are involved in various metabolic functions that, in fact, prepare the fetus to adapt to an independent life at birth. Phospholipid, glucocorticoid, and antioxidant synthesis increase as the gestational age advances. Prostaglandins (PGs) of the E and F series are present in both fetal and maternal circulation.

The functional activity of endocrine cells and peptide (i.e., bombesin-like) immunoreactivity increases during late gestation and the neonatal period, which may indicate that they have a role in regulation of the neonatal circulation; however, their exact roles are not understood.

The fetal circulation has high pulmonary vascular resistance and tone, and it also has a greater vascular reactivity compared with the adult pulmonary circulation. The reactivity increases with gestational age. In the fetal circulation, small doses of epinephrine or norepinephrine cause intense vasoconstriction. Acetylcholine and bradykinin, which have little effect on normal adult pulmonary circulation, produce dramatic drops in pulmonary arterial pressure and resistance. The profound effects that are seen in fetal pulmonary circulation are reduced after the pulmonary vascular resistance has fallen. Thus, the effects of these compounds depend on the initial vascular tone in the fetus as well as the adult. Bradykinin, a potent vasoactive peptide, is thought to have a physiological role in the perinatal circulation. It produces vasodilatation by PG-dependent and independent mechanisms in fetal lambs. The magnitude of the acetylcholine response progressively increases with gestational age. This is thought to be caused by maturation of the effector system rather than the vascular receptors. In rabbit pulmonary artery, from day 1 to 5 months, there is a progressive increase in beta-adrenergic receptor–dependent relaxation response to isoproterenol, and subsequently there is reduction in the relaxation response to isoproterenol with advancing age. However, relaxation to sodium nitrite, which is unrelated to beta-adrenergic receptors, is independent of age. The fetal carotid arteries contract in response to norepinephrine and serotonin earlier than they do to a neurogenic stimulus, suggesting that the alpha-adrenergic receptors on smooth muscle cells develop before the adrenergic nerves function. Whether it holds true for the pulmonary arteries or arterioles is not known. In piglets, the receptors for norepinephrine appear within 1–2 days after birth in intrapulmonary arteries, but receptors for serotonin, angiotensin II, and vasopressin are not present, although these receptors are present in intrapulmonary veins. Beta-adrenergic-related relaxation is seen within the first day of birth, but histamine H_2 receptors are not present even after 10 days of birth. Thus, there is a wide variation in development of various receptors. The contractile response to histamine increases with age in dogs as well as in lambs. It is thought to be caused by an increase in quantity of H_1 receptors and also to an increase in intrinsic contractile responsiveness.

Some of the PGs, too, show age-related changes in their action on pulmonary vasculature. PGD_2 causes marked decreases in pulmonary artery pressure and resistance in the first 3 days of life, and after 15 days, it acts as a vasoconstrictor. Likewise PGH_2 (endoperoxide, a metabolite of arachidonic acid) causes decreases in pulmonary arterial pressure in the unventilated fetal lamb, but increased pulmonary arterial pressure in ventilated fetal lambs. The production of an endothelium-derived relaxing factor (EDRF) in the fetus has not been established at the time of this writing. Some of the metabolic functions appear late in fetal life and do not reach the adult level at birth. The changes related to maturation also consist of changes in the ratio of subtypes of receptor populations. Thus, there is a distinct difference in response to vasoactive substances between the neonatal and adult animals. These differences will have effects on drug metabolism, clearance, and treatment of pulmonary disease processes.

II. Neurohumoral Aspects

In fetal lungs, the nervous tissue and neuroepithelial bodies appear early. Lungs are innervated from the anterior and posterior pulmonary plexi. The latter are formed by fibers from the sympathetic trunk and the vagi. From this network, the major pulmonary arteries receive a plexus of large nerve trunks coursing within the adventitial layer (myelinated and nonmyelinated). The terminal twigs pass to the region of smooth muscle cells (SMC) in both large and small elastic, as well as muscular, arteries down to 30-μm diameter. Fewer fibers are seen in veins. In small arteries, the fibers remain external to the medial coat. In arteries, less than 30 μm in diameter, the fibers are not present. Both sympathetic and parasympathetic fibers are found in close association with pulmonary vessels. In adult dogs, the major effect of sympathetic activation is to increase stiffness of large arteries. Adrenergic receptors in the pulmonary vessels are involved in local responses to hypoxia and hypercapnea. In the cat pulmonary vascular bed, both subtypes of postjunctional alpha adrenoceptors have been demonstrated.

Forebrain stimulation of the cat causes an active

change in pulmonary vascular resistance, suggesting the presence of a descending pathway that, when stimulated, causes vasodilation. It has been suggested that adrenal-medullary mechanisms may play a role in cardiovascular responses elicited by stimulation of CNS structures. The role of the hypothalamus in regulation of vasoconstrictor tone in the pulmonary vascular bed still remains to be established. Electrical stimulation of the left vagus causes vasodilation of the ipsilateral unexpanded lung, an effect abolished by atropine. The pulmonary vasculature of unexpanded fetal lung is sensitive to both parasympathetic and sympathetic stimulation. It is not yet clear if neurotransmitters are released in the unstimulated state. Thus, it is not certain if the CNS has any role in modulating pulmonary circulation under physiological conditions.

Pulmonary smooth muscle is equipped with functioning alpha- and beta-adrenergic systems. Alpha receptors predominate over beta receptors. Thus, a vasoconstriction response is favored. These receptors are functionally independent of the available autonomic nerve endings. Propranolol, a mixed beta-adrenergic blocker (acting on β_1 and β_2 receptors), increases the vasoconstrictor response to epinephrine and norepinephrine in perfused lungs and on pulmonary vascular smooth muscle (VSM). Histamine is contained in relatively large concentrations in the lungs of rats and dogs as well as other mammals. Two subtypes of histamine receptors exist, H_1 receptors, which mediate vasoconstriction, and H_2, which subserve vasodilation. A significant population of H_2-inhibitory receptors exist, which may play a role in hypoxic pulmonary vasoconstriction. Thus, histamine and adrenergic receptors may have a role in modulating vascular tone and may play a role in mediating, partly, hypoxic vasoconstriction.

Serotonin, a circulating hormone, is a powerful vasoconstrictor in the pulmonary circulation. It is contained in mast cells and platelets; it is metabolized by pulmonary ECs. In pulmonary arteries, infusion of serotonin causes pulmonary vasoconstriction but no platelet aggregation. Catecholamines (e.g., norepinephrine and epinephrine) are potent vasoconstrictors and are endogenous in origin. A large number of peptides have been identified in the lungs. Many of these peptides influence pulmonary smooth muscle tone, permeability, and inflammatory responses. Angiotensin II, spasmogenic lung peptide, substance P, and bombesin are vasoconstrictors. Substance P can affect pulmonary vessels directly, but it is also thought to be involved in pulmonary reflexes. Vasoactive intestinal peptide (VIP) is the only peptide so far shown to relax pulmonary vessels, and it is more potent than prostacyclin (PGI_2). Activated complement C_5a increases pulmonary vascular permeability as do the peptides containing lipids such as leukotrienes (LTC_4, LTD_4). Bradykinin, together with hypoxia or PGE_2, can induce moderate degrees of a high permeability type of pulmonary edema. A physiological role for these peptides, however, is not established as yet.

III. Factors Influencing Vasomotor Tone

A number of physiological, pharmacological, and experimental conditions can affect vasomotor tone. VSM differ in their ability to respond to various vasoactive substances depending on their location, structure, sensitivity and density, local temperature, pH, integrity of ECs, cell metabolism, and tissue injury to name a few. There are also species- and strain-related changes. The role of aging and neurohumoral factors have been covered elsewhere in this article. Improper handling of the tissue has been found to alter ionic equilibrium, Ca^{2+} and K^+ are lost, and the Na^+ level is intracellularly increased, which can alter vascular responsiveness to various stimuli. Some of the amine and zwitterion buffers used in various *in vitro* studies (e.g., Tris, HEPES, and MOPS) exert significant inhibitory effects on exchangeability and transmembrane movement of Ca^{2+} in VSM. General anesthetics commonly used can attenuate contractile responses to agonists in a dose-dependent fashion, which are not inhibited by known pharmacological antagonists or PG synthetase inhibitors. These anesthetic agents, in a dose-dependent manner, prevent uptake of radiolabeled Ca^{2+} in VSM.

For excitation–contraction coupling in VSM, Ca^{2+} is required. The process of contraction can involve influx of extracellular Ca^{2+} or release of Ca^{2+} from intracellular stores. Depending on the agonists used and the type of blood vessel, VSMs may vary in their mechanism for accumulation and recruitment of Ca^{2+}. It has been shown that calcium channel blockers (e.g., nitrendipine and verapamil) attenuate K^+-induced contractions but are ineffective in attenuating vasoactive agonists such as phenylephrine, norepinephrine, serotonin, angiotensin II, and prostanoids. These results are consistent with the concept that K^+-induced contractions of pulmonary VSM are primarily mediated via influx of extracellular Ca^{2+}, whereas most vasoactive agonist recruit Ca^{2+} from intracellular stores. Recently, two vasoactive peptides (VIP and substance P) have been identified within the nerve fibers and nerve terminals innervating pulmonary and systemic vessels. It is possible that these neuropeptides have a physiological influence on VSM tone.

IV. Isolated Vessel Studies

Isolated vessel studies can be performed with relative ease. Therefore, this method has been used extensively with various vessel segments from various sites and species to study the effects of different vasoactive substances. The pulmonary arteries have low resting tone. Therefore, in all such *in vitro* studies, basal tone has to be increased with a vasoconstrictor agent such as phenylephrine, 5-HT, PGF_{2alpha}, or their analogues. The vascular response varies with the species and the segment of the vessel studied. An obligatory role for endothelium was described for acetylcholine-induced vascular relaxation. It was thought to be secondary to a synthesis and release of EDRF, which was subsequently confirmed in isolated canine pulmonary arteries. The work done with isolated pulmonary arteries is too numerous to summarize here, and it is beyond the scope of this chapter. The segment of a vessel can be used either in a ring form or spirally cut to a narrow strip. Care is taken not to injure ECs, and in some cases endothelium is removed by gently rubbing the luminal surface with a pared wooden stick or filter paper. The presence or absence of endothelium is confirmed by the presence or absence of a relaxation response to acetylcholine. The arterial segment is mounted in a tissue bath containing physiological buffer solution, kept at 37°C and at pH 7.4, with an oxygen–carbon dioxide gas mixture. The contractile responses to various vasoactive agents are usually monitored isometrically by a force displacement transducer. The resting tone required to obtain the optimal response varies with origin of the vessel and the species.

It has been suggested that pathological destruction of endothelium can transform important endogenous circulating dilators to vasoconstrictors. In atherosclerosis and pulmonary artery hypertension, EC damage has been demonstrated, and this method provides an opportunity to study EC cell function in experimentally induced diseases. The effects of hypoxia have been extensively studied. In all mammalian species studied so far, contractile responses of pulmonary arteries to hypoxia have been demonstrated, but as yet, the mechanism of hypoxic vasoconstriction has not been elucidated. It has been suggested that during hypoxia endothelium may release a contractile agent(s) or that during hypoxia there may be inhibition of a cyclic 3′5′-guanosine monophosphate (cGMP)–associated relaxation mechanism; the role of Ca^{2+} has also been evoked.

V. Isolated Lung Perfusion Studies

Isolated perfused lung preparations have been used to evaluate the effects of various noxious substances such as hypoxia and other chemicals (e.g., bleomycin, alpha-naphthyl thiourea) and also to evaluate the uptake and metabolism of various amines and drugs and how they are affected in states of injury. The metabolism of various amines reflect the functional state of ECs. The response to pharmacological agents would vary, depending on the perfusate used. In physiological salt-perfused lungs, more PGs are released compared with blood-perfused preparations. The vascular tone of the preparation is important as the elevated tone would magnify the effects of vasodilators. Although the effects of hypoxia have been extensively studied, the precise mechanism has not been established. It has been shown that blood-perfused isolated rat lungs develop a large hypoxic pressor response compared with physiological salt-perfused lungs, but this difference might be dependent on differences in glucocorticoid activity in the two preparations. Effects of commonly used anesthetic agents on lung metabolism of various biogenic amines have been studied. Ketamine, halothane, and nitrous oxide inhibit 5-HT metabolism, whereas fentanyl, an opiod, does not have any effect. Thus isolated perfusion studies of lung allow us to examine the function of endothelium in normal and abnormal states as well as in the presence of various pharmacological agents. However, as explained elsewhere, not all the results obtained can be extrapolated to the living organism.

VI. Pharmacology of Endothelial Cells

Tremendous progress has been made in the understanding of ECs primarily because of the advent of the electron microscope and techniques for the isolation and culture of ECs. The pulmonary EC lining comes in contact with the entire cardiac output, blood-borne hormones, vasoactive substances, and formed elements of the blood itself. The ECs not only provide a physical barrier between the blood and VSM but are actively involved in synthesis, uptake, and metabolism of various vasoactive substances, antithrombolytic activities, and respiratory function. ECs from different sites differ in morphology and biochemical functions.

A. Barrier Function

The EC lining provides a physical barrier between the circulating blood elements and the underlying smooth muscle cells. Various substances such as lipoproteins and nucleotides are degraded on the luminal surface. Thus, these substances are prevented from reaching the extravascular space. Other compounds (e.g., adenosine) are taken up intracellularly and metabolized.

B. Modulation of Vascular Tone

Pulmonary ECs modulate pulmonary vascular tone. The cells release vasoactive substances in response to numerous stimuli.

1. Endothelium-Derived Relaxing Factor

Endothelium-derived relaxing factor was discovered in 1980 by showing that acetylcholine-induced relaxation was EC-dependent. Since then other agents have been shown to cause EC-dependent relaxation (e.g., norepinephrine, histamine, bradykinin, and thrombin). These responses are associated with increased intracellular Ca^{2+}. Local loss of ECs can lead to contraction. EDRF can be released from luminal and abluminal surfaces. There is evidence for spontaneously released EDRF from cascade bioassay experiments. The spontaneously released EDRF is similar to that released by acetylcholine. Basal release of EDRF has a major modulating influence on the response to contractile agents. The presence of basally released EDRF is manifest more in agonist-constricted than K^+-constricted or resting preparations. The increase in tone that results from inhibition of the basal release of EDRF varies with arteries studied and is Mg^{2+}-dependent. Stimulated release of EDRF critically depends on mitochondrial ATP synthesis. In rabbit aorta, stimuli such as flow-induced increases in shear stress, pulsatile stretching, or a low level of pO_2 can cause EDRF release but not anoxia. Calcium is necessary for EDRF production, but it is not inhibited by calcium channel blockers such as verapamil in all vessels studied. Both direct and indirect experimental evidence indicate that an increase in free cytosolic Ca^{2+} concentration in ECs is an initial step in synthesis or release of EDRF. EDRF has subsequently been shown to be nitric oxide–generated from endogenous L-arginine. In immunostimulated macrophages, the formation of nitric oxide is dependent on NADPH and is enhanced by Mg^{2+}. Whether this is true for pulmonary endothelium is not established at present. The pulmonary vascular relaxation by EDRF is produced by activation of guanylate cyclase and production of cGMP. Acetylcholine produces a time- and dose-dependent accumulation of cGMP but not cAMP. Atropine and methylene blue (inhibitor of soluble guanylate cyclase), carbonyl group reagents, K^+-borohydride, diethiothreitol, phenylhydrazine, phenidone, and nordihydroguaretic acid all inhibit relaxation as well as cGMP accumulation, where as the relaxation is potentiated by the phosphodiesterase inhibitor MB 22948. It appears that cGMP mediates muscarinic receptor-linked relaxation of intrapulmonary arterial smooth muscle. In bovine coronary arteries, guanylate cyclase is sensitive to activation by nitric oxide and nitrosoguanidine in the presence of Mg^{2+}, and these activations markedly increase tissue cGMP, whereas Ca^{2+} markedly inhibits activation of guanylate cyclase in the presence of Mg^{2+} but not Mn^{2+}. This inhibition is dependent on the Ca^{2+}/Mg^{2+} ratio.

2. Endothelium-Dependent Contracting Factor

Several laboratories have reported the existence of endothelium-dependent contracting factor (EDCF). Under appropriate experimental conditions, EDCF is expressed by the intact EC in contact with the underlying smooth muscle. The time course of contraction is accompanied by a similar time-dependent influx of extracellular Ca^{2+}. The contractile responses are abolished by the removal of Ca^{2+} and by Ca^{2+} antagonists. Endothelin, a recently isolated 21-amino-acid peptide, is a vasoconstrictor (more potent for the veins than arteries), but it also releases potent vasodilators such as PGI_2 and EDRF and is inactivated by lung. Thus, the contractile activity of circulating endothelin is limited by the production of vasodilators and its inactivation. It is not thought to be released in response to acute hypoxia.

C. Metabolic Activity

Pulmonary EC surfaces are important sites of pharmacokinetic functions. These include removal, biosynthesis, and release of vasoactive substances that affect cardiovascular regulation in health and disease states. Biogenic amines (e.g., norepinephrine and 5-HT are removed by a carrier-mediated and drug-sensitive transport process and are degraded intracellularly by catechol-*O*-methyl transferase and monoamine oxidase, respectively. A substantial amount of these amines is removed during a single passage through the pulmonary circulation. However, other biogenic amines (e.g., histamine, dopamine, and epinephrine) generally are not removed. Altered biogenic amine metabolism may thus be related to alteration in cardiovascular regulation. PGs of the E and F series are removed by the lungs via carrier-mediated and energy-requiring processes, and also rapid enzymatic degradation occurs. However, PGI_2 and thromboxane are not removed by the lungs. In response to shear stress and increased flow, the ECs increase the release of PGI_2, thus affecting lung function.

Angiotensin-converting enzyme (ACE) hydrolyzes peptidyl dipeptide bonds from the terminal group carboxyl of bradykinin and angiotensin I, resulting in inactivation of the former and the conversion of the latter to a circulating vasoactive hormone, angiotensin II. Several biologically active peptides (e.g., VIP, bombesin, and oxytocin) escape degradation. Substance P is hydrolyzed by cultured EC but not by intact lung. Other substances (e.g., insulin and natriuretic atrial factor) are also removed from pulmonary circulation. ECs are also thought to play a role in modifying interstitial concentrations of adenosine, which is an important link between tissue metabolism and VSM activity.

A number of drugs are metabolized and cleared by lung. Several pharmacological agents (e.g., antihistamines, antimalarials, morphine-like analgesics, anorectics, tricyclic antidepressants, and anesthetics) are concentrated in the lung as basic amines. The mechanisms of accumulation of several xenobiotic amines have been extensively studied by the isolated lung perfusion method. Persistence of some of the amines in lung is important in drug-induced pulmonary phospholipidosis. Phospholipidosis decreases uptake and metabolism of serotonin, but it enhances uptake and accumulation of chlorpromazine, chlorphentermine, and imipramine, which may limit their access to metabolic sites. Cocaine inhibits serotonin and norepinephrine clearance in rabbit perfused lungs. However, in *in vivo* studies norepinephrine clearance is not affected. Therefore, not all the results obtained from isolated perfused lung can be extrapolated to the entire organism.

D. Hemostasis and Antithrombogenic Function

Normal ECs do not activate platelets, leukocytes, blood coagulation, or fibinolytic factors. Endothelial proteoglycans (mainly heparin sulfate) provide a passive nonthrombogenic surface. The ECs shield the blood elements and coagulation factors from the underlying basement membrane, which is extremely toxic. Active antithrombogenic mechanisms involve synthesis and release of prostacyclin, secretion of plasminogen activators, degradation of proaggregatory ADP by membrane-associated ADPase, uptake and degradation of vaspoactive amines, uptake, inactivation, and clearance of thrombin, and contribution of thrombomodulin in thrombin-dependent activation of protein C. The thrombin activity is limited to the site of injury. Hemostasis is a complex interaction among the endothelial cells, platelets, and coagulation factors to form a mechanical seal, which subsequently gets removed by fibrinolysis.

E. Immune System Responses

Endothelial cells when stimulated with appropriate agents can be induced to express class II histocompatability antigens, express Fc receptors , engage in phagocytosis, and produce toxic oxygen radicals. In this respect, ECs resemble macrophages.

F. Respiratory Function

The ECs also participate in CO_2 release from the lungs. Carboanhydrase is a cytosolic enzyme and a membrane-bound ecto-enzyme and adds significantly to the conversion of CO_2 from plasma bicarbonate.

VII. Role of Eicosanoids

The term *eicosanoids* refers to oxygenated 20 carbon fatty acids including PGs, leukotrienes, epoxy-, mono-, and dihydroxy eicosanoic acids. Eicosanoids are released by enzymatic oxygenation of fatty acids [e.g., 8,11,14-eicosatrienoic acid (dihomo-gamma-linoleic acid), 5,8,11,14-eicosate-

COLOR PLATE 1 Photograph of a model of a globular protein, α-lactalbumin from the baboon, determined crystallographically. The atoms of the polypeptide backbone are white, those of the side chains are various colors. The spheres corresponding to the individual atoms have only one-third of their van der Waals radii. (The model was constructed by Jonathan Ewbank.)

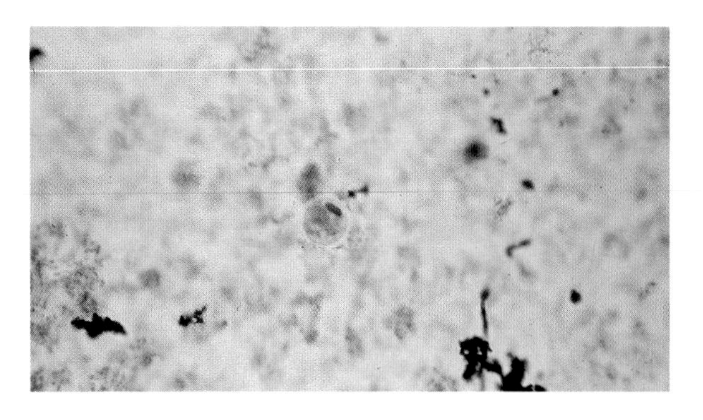

COLOR PLATE 2 *Entamoeba histolytica* cyst in human stool.

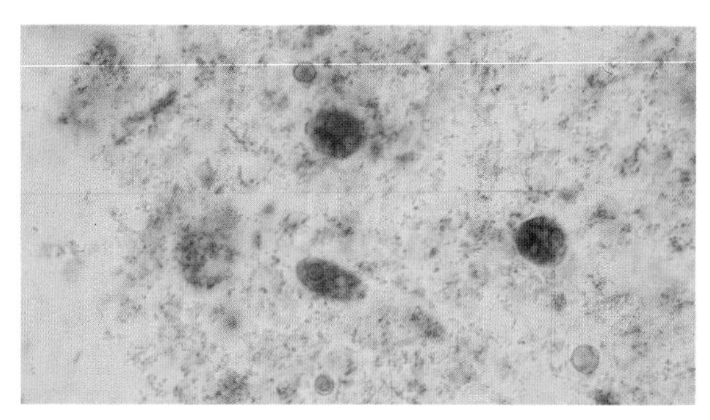

COLOR PLATE 3 *Entamoeba histolytica* trophozoite grown in culture medium at 37°C.

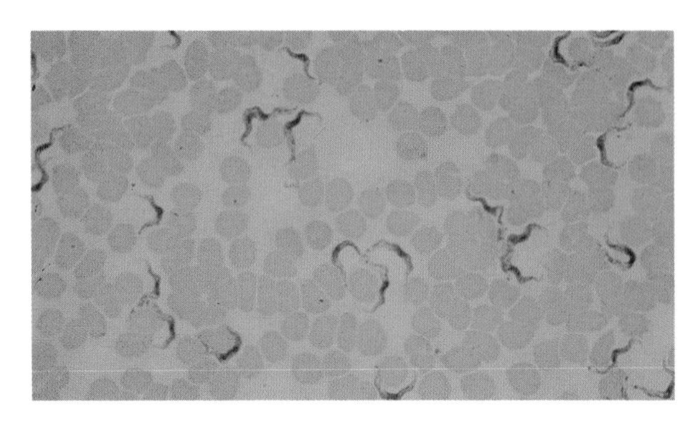

COLOR PLATE 4 *Trypanosoma brucei* trypomastigotes in rat's blood.

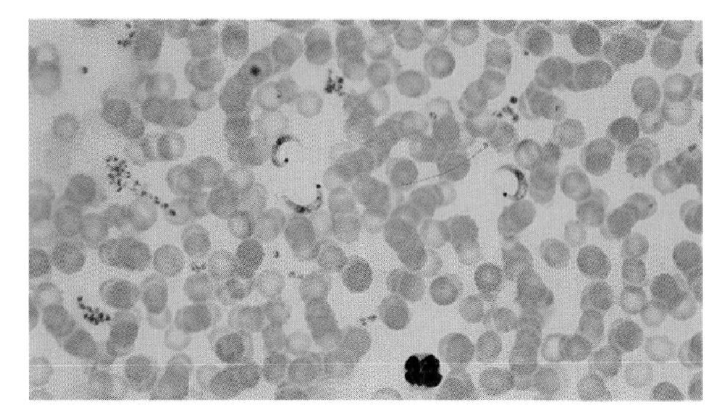

COLOR PLATE 5 *Trypanosoma cruzi* trypomastigotes in human blood.

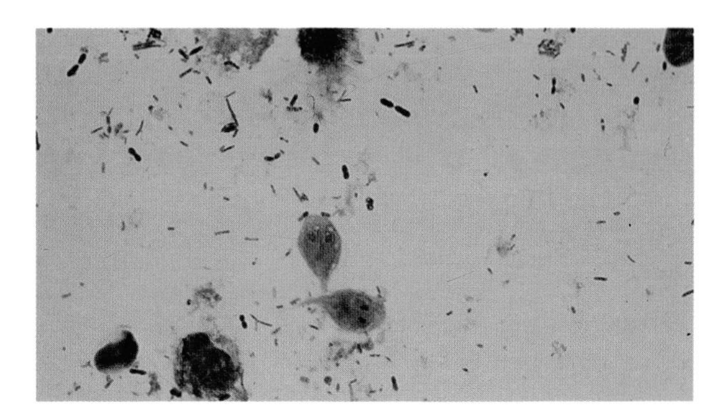

COLOR PLATE 6 *Giardia lamblia* trophozoite in stool.

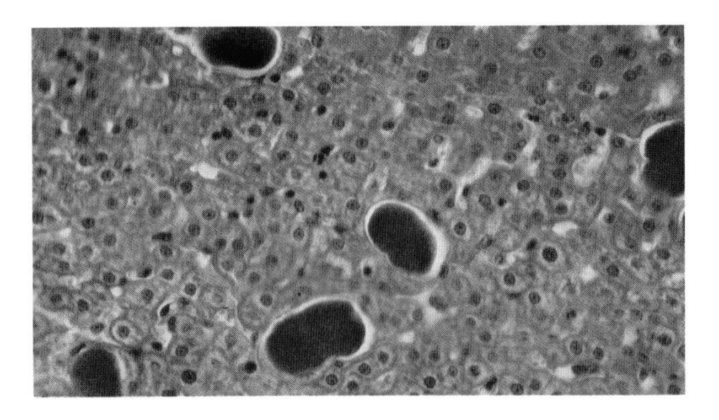

COLOR PLATE 7 *Plasmodium* spp. pre-erythrocytic schizonts in hepatocytes.

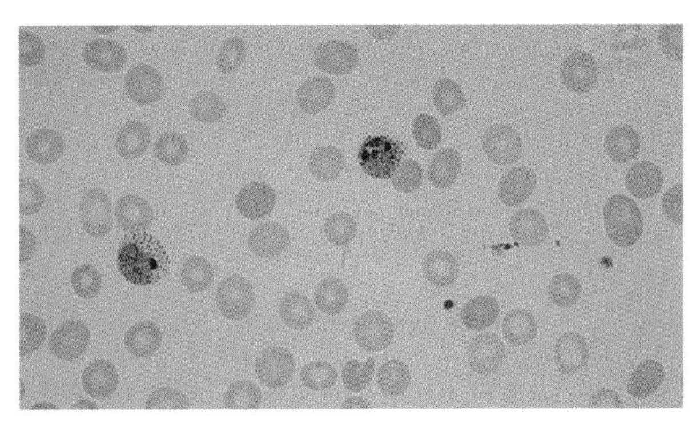

COLOR PLATE 8 *Plasmodium vivax* trophozoite in human blood.

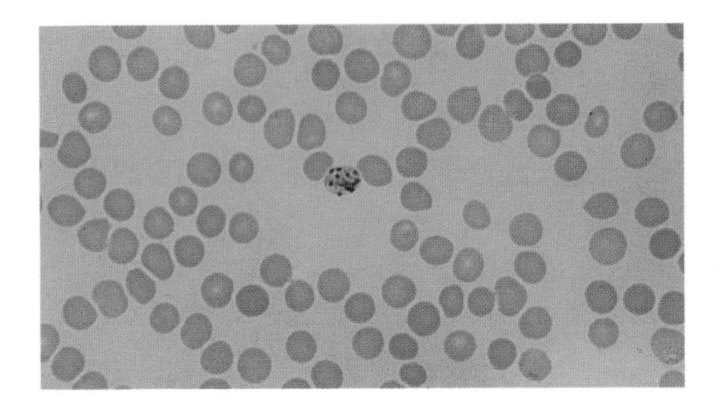

COLOR PLATE 9 *Plasmodium malariae* schizont in human blood.

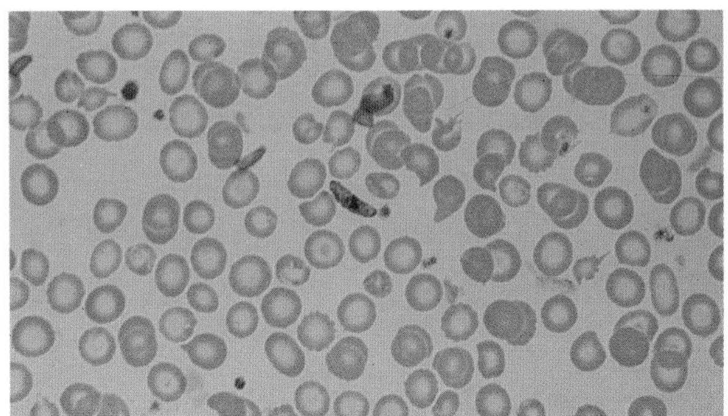

COLOR PLATE 10 *Plasmodium falciparum* gametocyte in human blood.

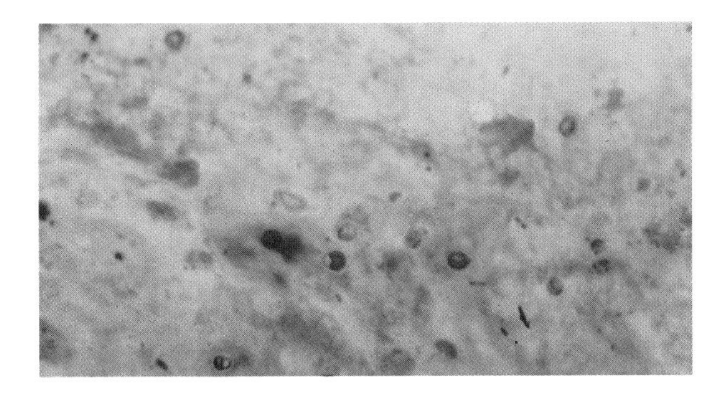

COLOR PLATE 11 *Cryptosporidium* oocysts in stool.

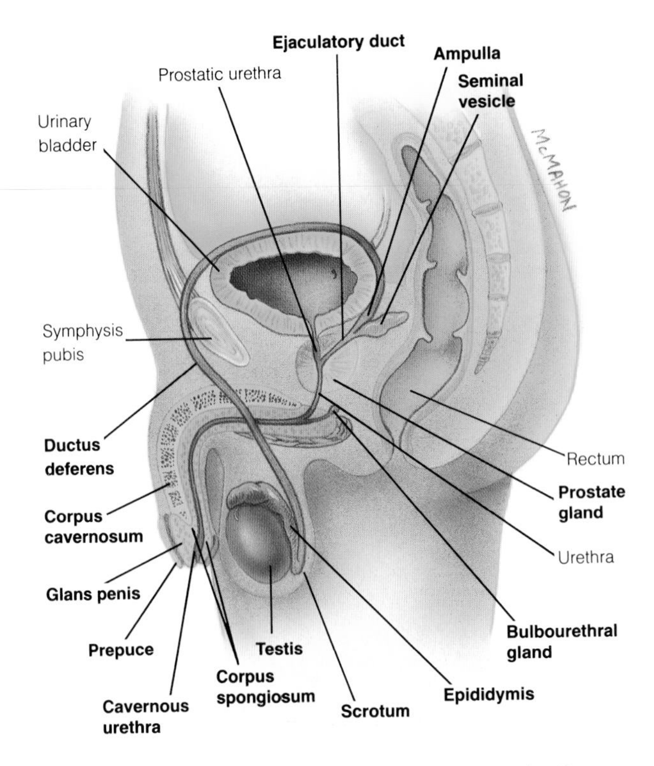

COLOR PLATE 12 Reproductive system of an adult male. [Source: Gaudin, A. J., and Jones, K. C. (1989). "Human Anatomy and Physiology." Harcourt Brace Jovanovich, San Diego. p. 699. Reproduced with permission.]

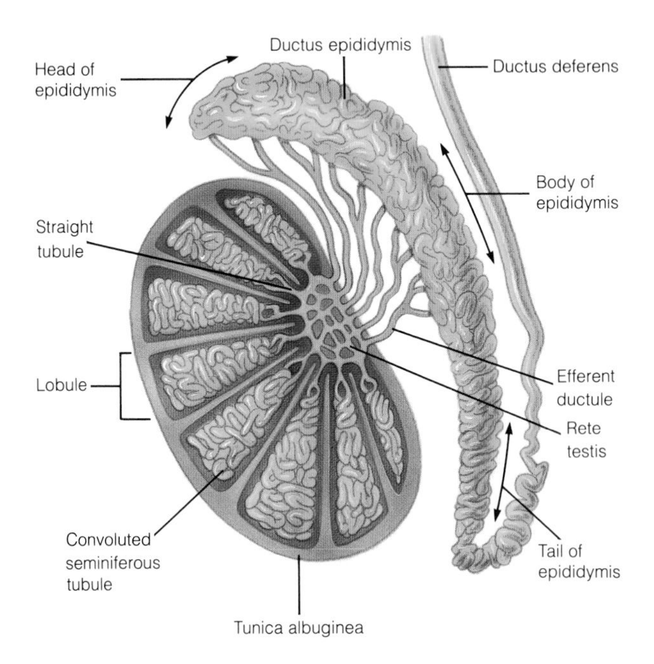

COLOR PLATE 13 The testis and associated ducts. [Source: Gaudin, A. J., and Jones, K. C. (1989). "Human Anatomy and Physiology." Harcourt Brace Jovanovich, San Diego. p. 702. Reproduced with permission.]

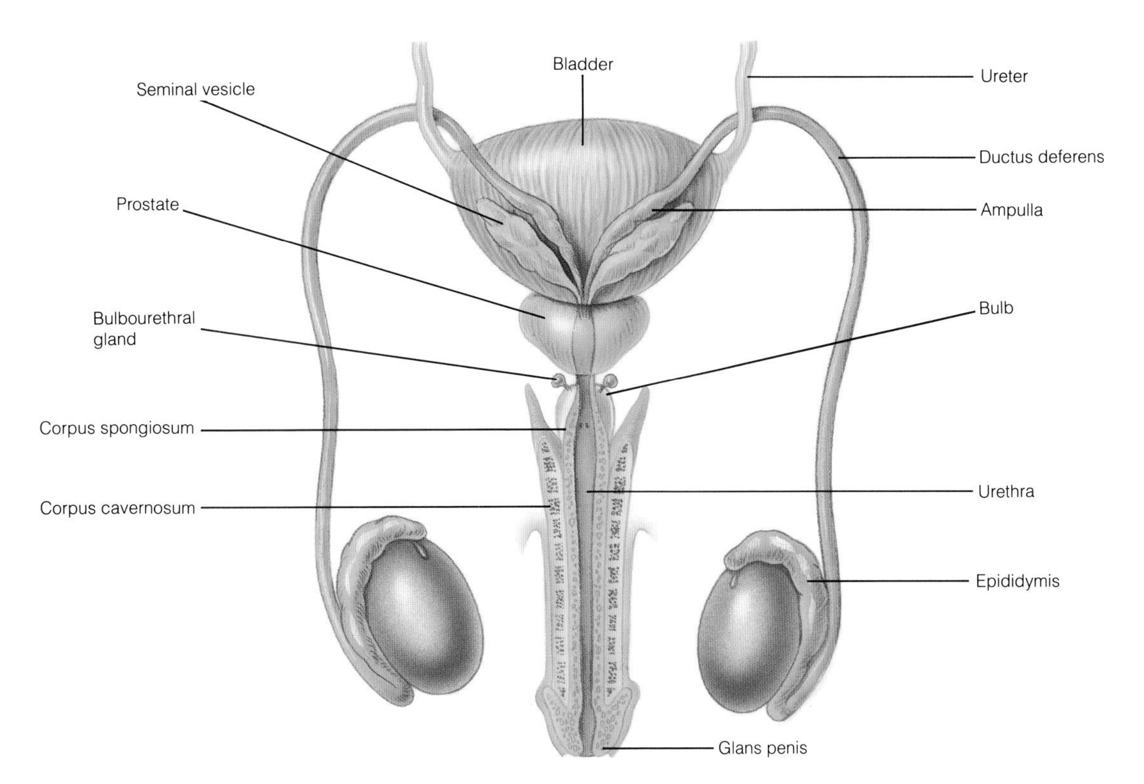

COLOR PLATE 14 Posterior view of the male reproductive system. [Source: Gaudin, A. J., and Jones, K. C. (1989). "Human Anatomy and Physiology." Harcourt Brace Jovanovich, San Diego. p. 703. Reproduced with permission.]

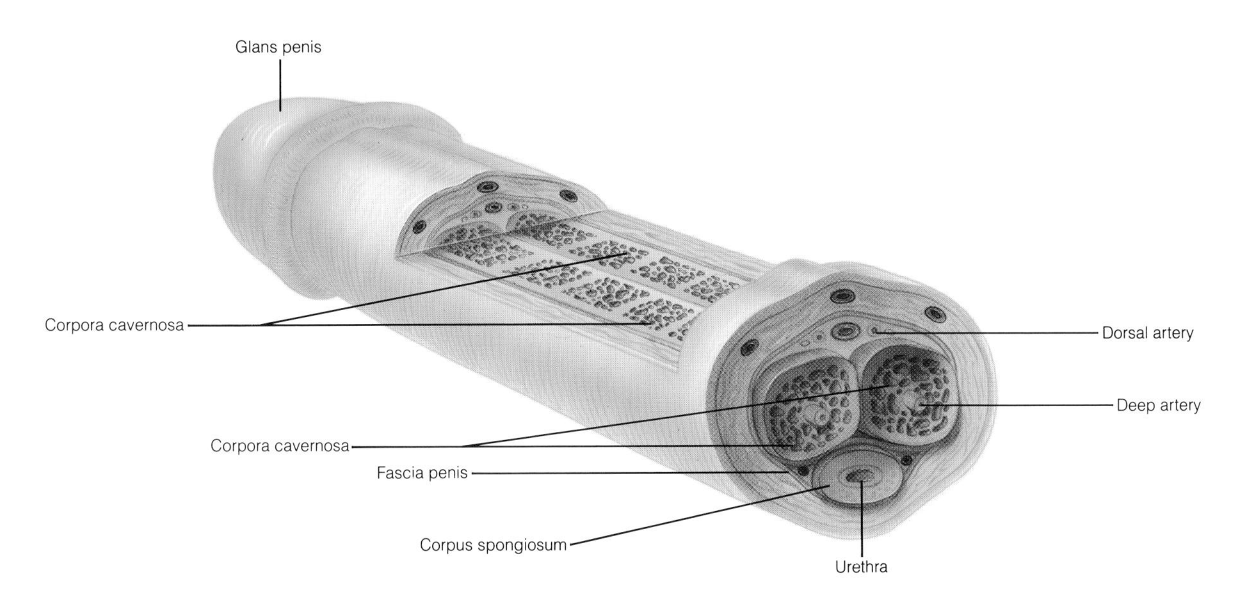

COLOR PLATE 15 Anatomy of a penis. [Source: Gaudin, A. J., and Jones, K. C. (1989). "Human Anatomy and Physiology." Harcourt Brace Jovanovich, San Diego. p. 707. Reproduced with permission.]

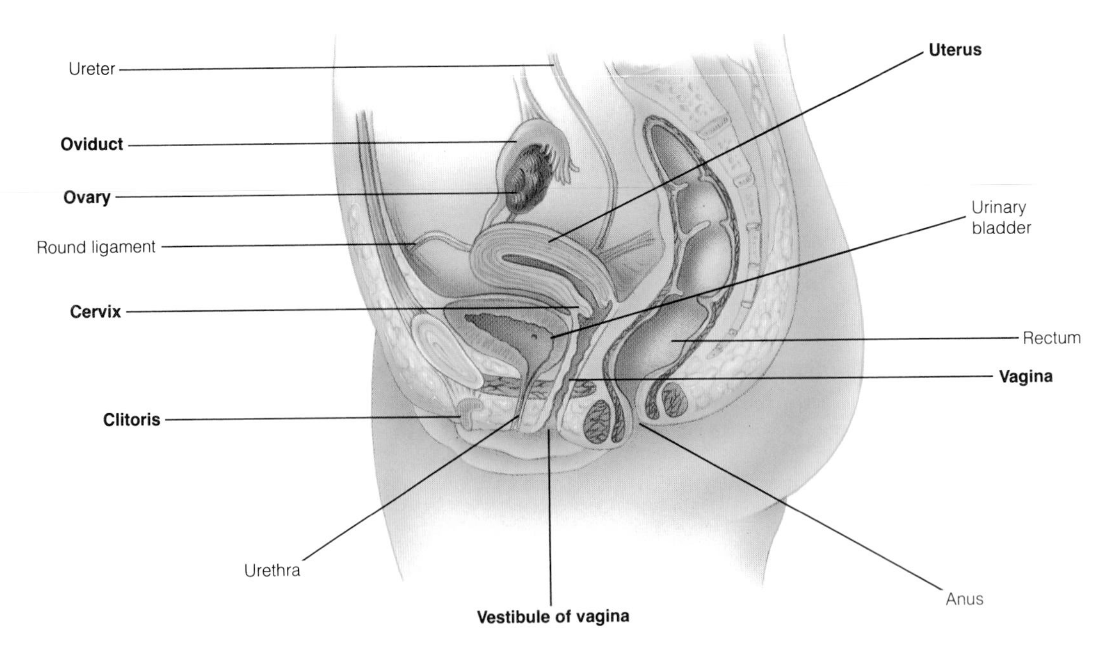

COLOR PLATE 16 Anatomy of the female reproductive tract. [Source: Gaudin, A. J., and Jones, K. C. (1989). "Human Anatomy and Physiology." Harcourt Brace Jovanovich, San Diego. p. 709. Reproduced with permission.]

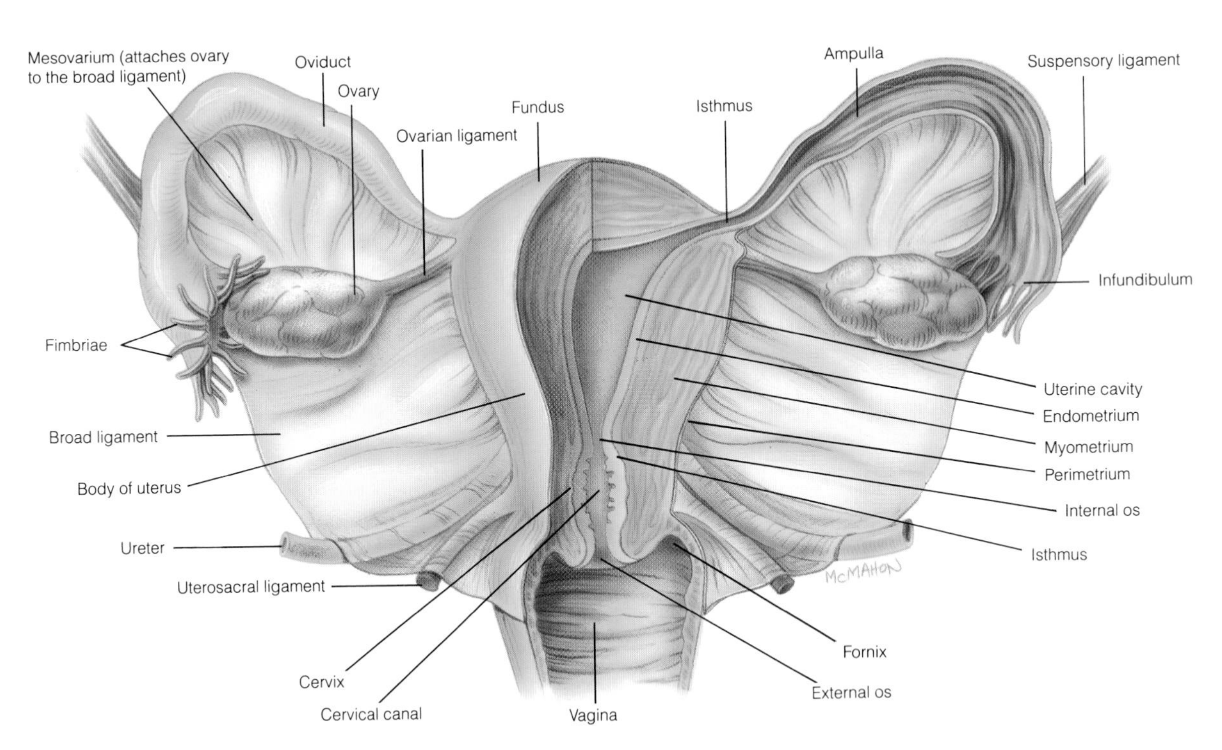

COLOR PLATE 17 Anterior view of the female reproductive tract. [Source: Gaudin, A. J., and Jones, K. C. (1989). "Human Anatomy and Physiology." Harcourt Brace Jovanovich, San Diego. p. 712. Reproduced with permission.]

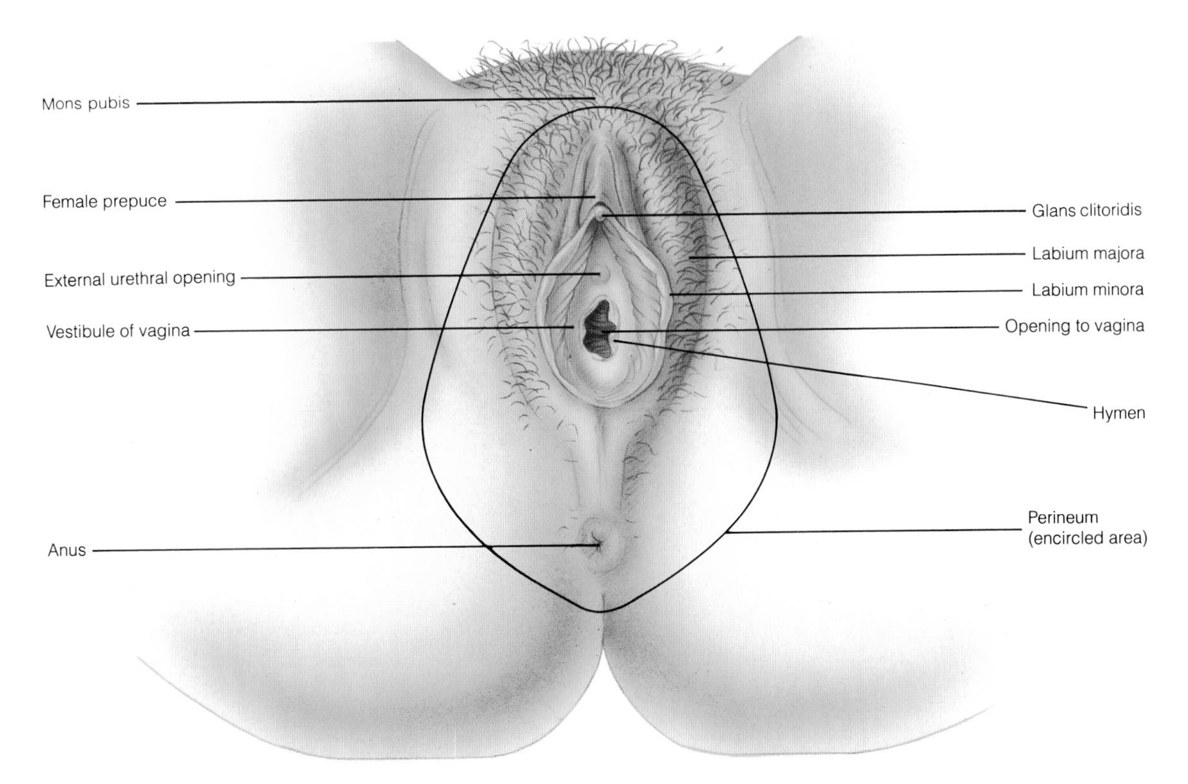

COLOR PLATE 18 External female genitalia (vulva). [Source: Gaudin, A. J., and Jones, K. C. (1989). "Human Anatomy and Physiology." Harcourt Brace Jovanovich, San Diego. p. 715. Reproduced with permission.]

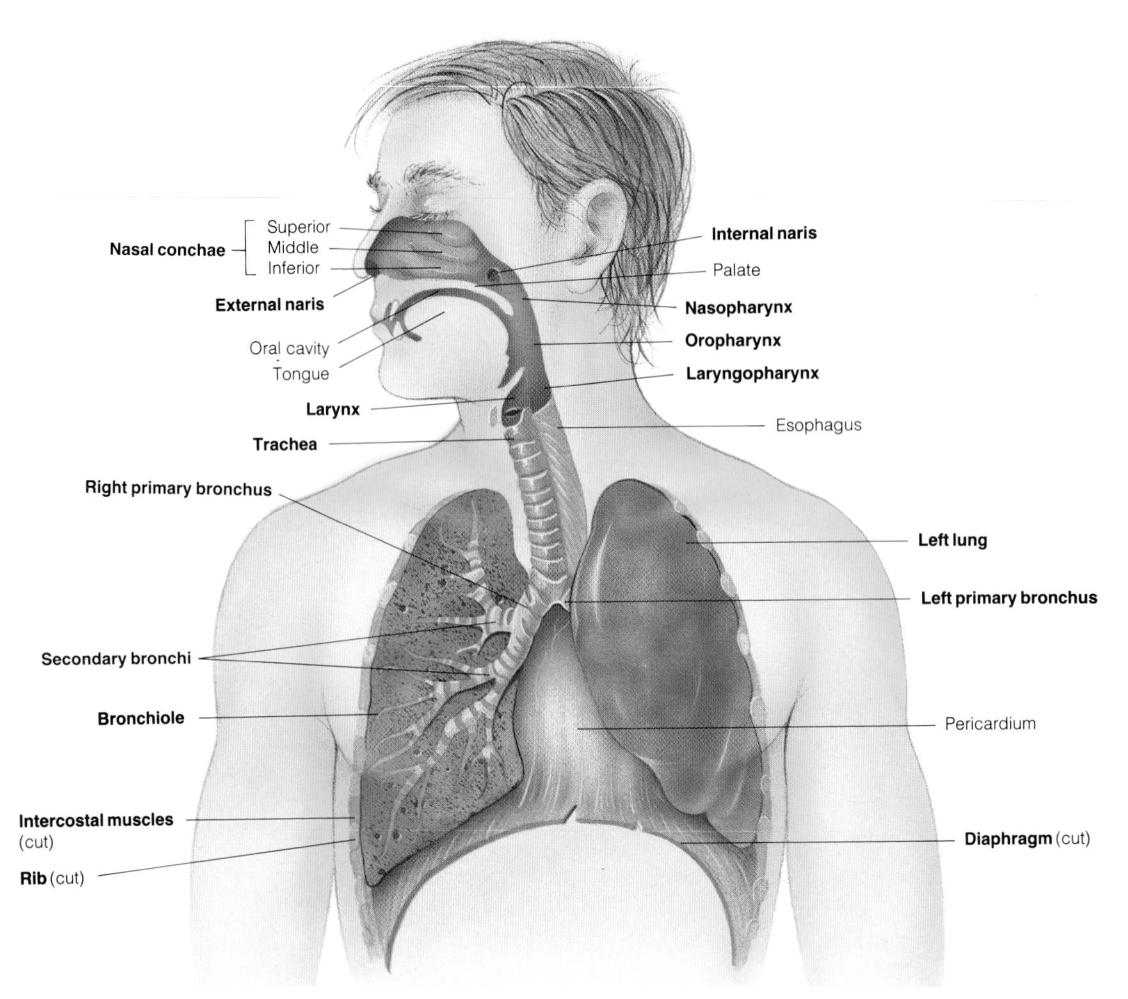

COLOR PLATE 19 The respiratory tract. [Source: Gaudin, A. J., and Jones, K. C. (1989). "Human Anatomy and Physiology." Harcourt Brace Jovanovich, San Diego. p. 505. Reproduced with permission.]

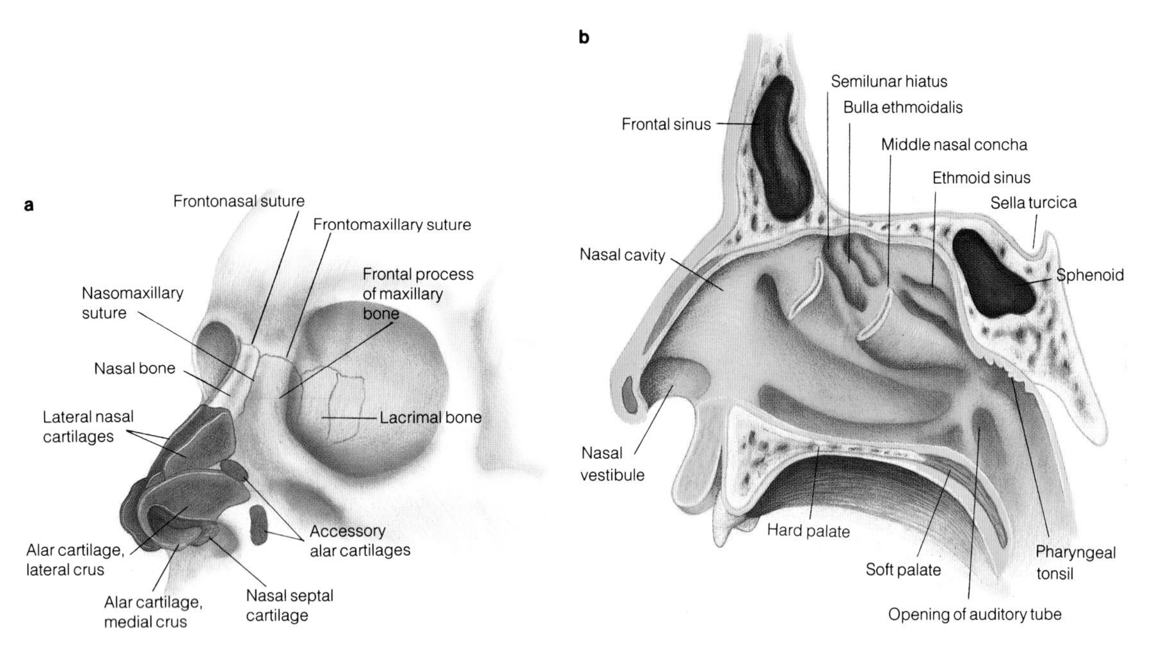

COLOR PLATE 20 The nose. (a) The external nose. (b) The nasal cavity. [Source: Gaudin, A. J., and Jones, K. C. (1989). "Human Anatomy and Physiology." Harcourt Brace Jovanovich, San Diego. p. 507. Reproduced with permission.]

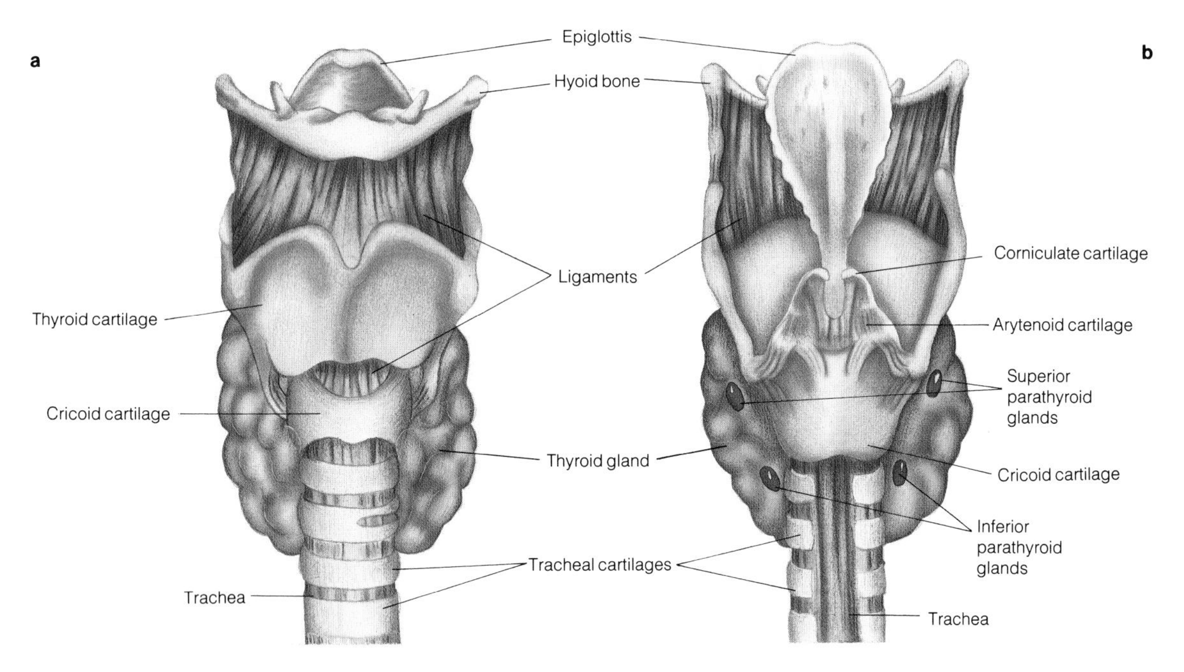

COLOR PLATE 21 The larynx. (a) External anterior view. The isthmus of the thyroid gland has been omitted to show the cricoid cartilage. (b) External posterior view. [Source: Gaudin, A. J., and Jones, K. C. (1989). "Human Anatomy and Physiology." Harcourt Brace Jovanovich, San Diego. p. 510. Reproduced with permission.]

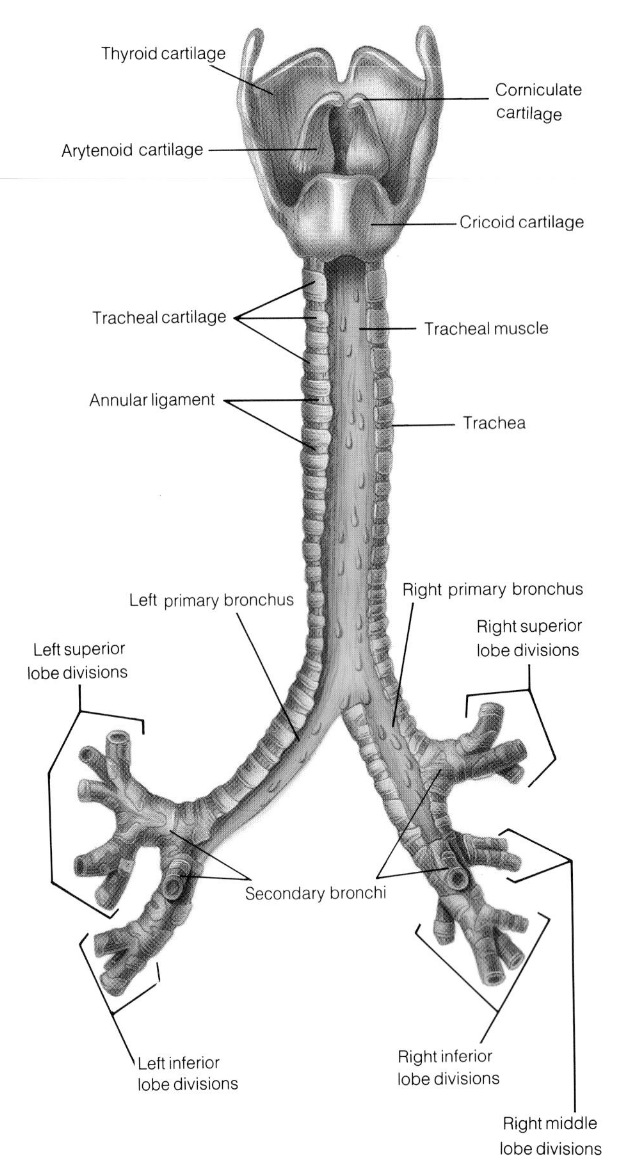

COLOR PLATE 22 The trachea and the bronchi in a posterior view. [Source: Gaudin, A. J., and Jones, K. C. (1989). "Human Anatomy and Physiology." Harcourt Brace Jovanovich, San Diego. p. 511. Reproduced with permission.]

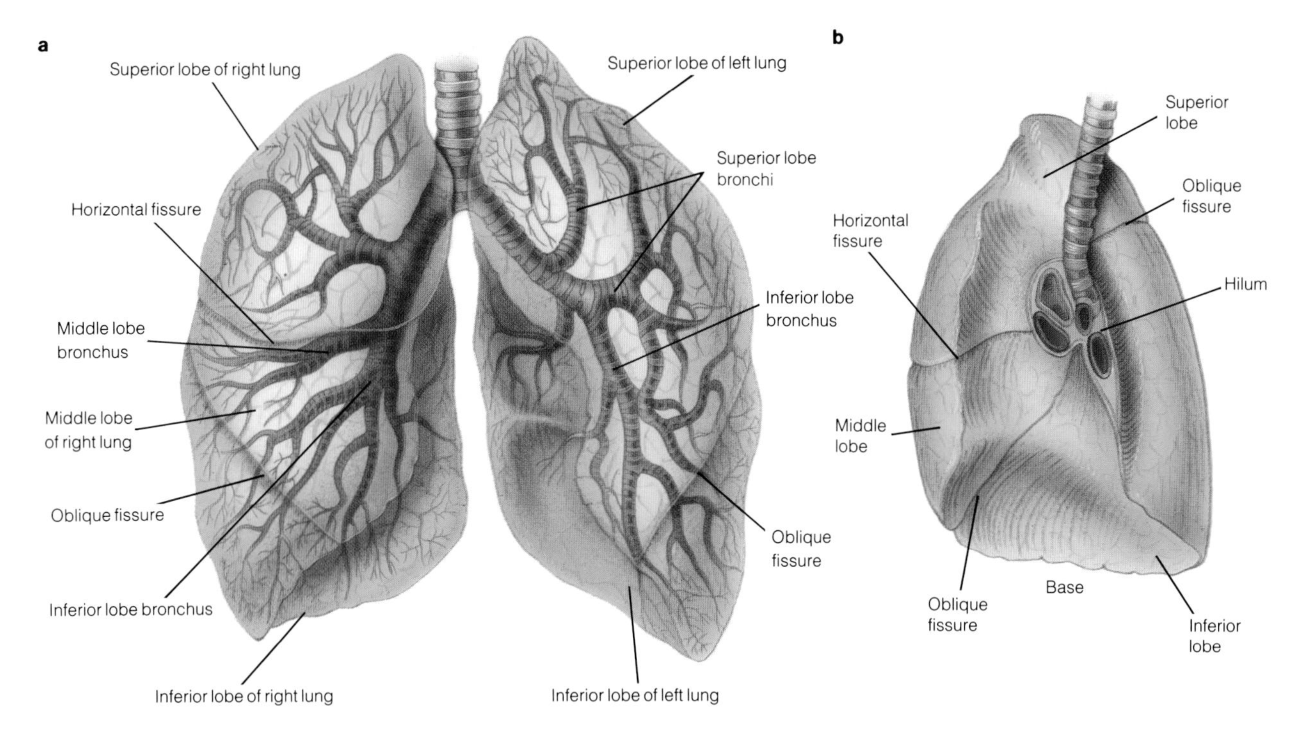

COLOR PLATE 23 (a) Anterior view of both lungs. (b) Medial view of the right lung, showing the hilum. [Source: Gaudin, A. J., and Jones, K. C. (1989). "Human Anatomy and Physiology." Harcourt Brace Jovanovich, San Diego. p. 512. Reproduced with permission.]

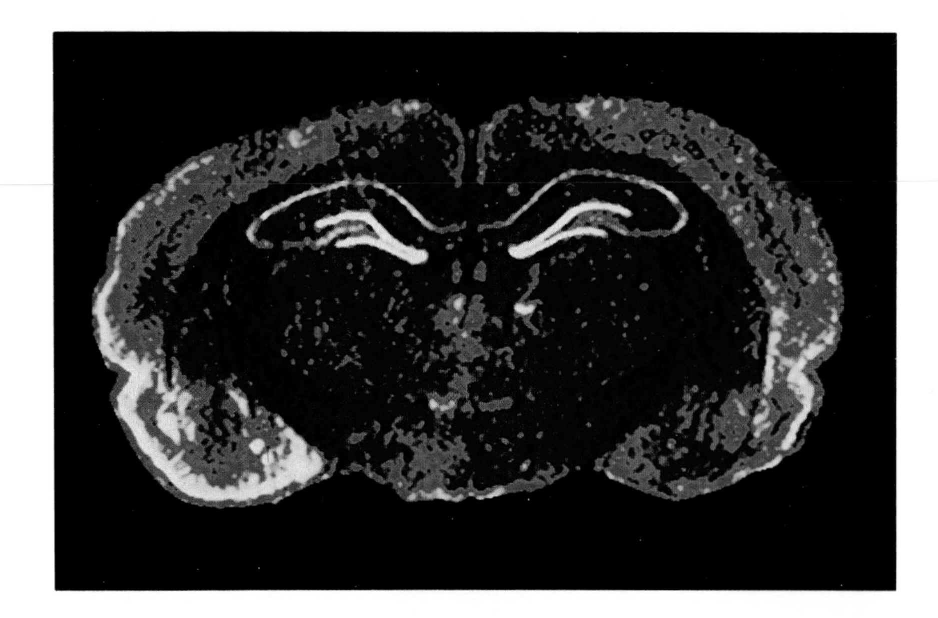

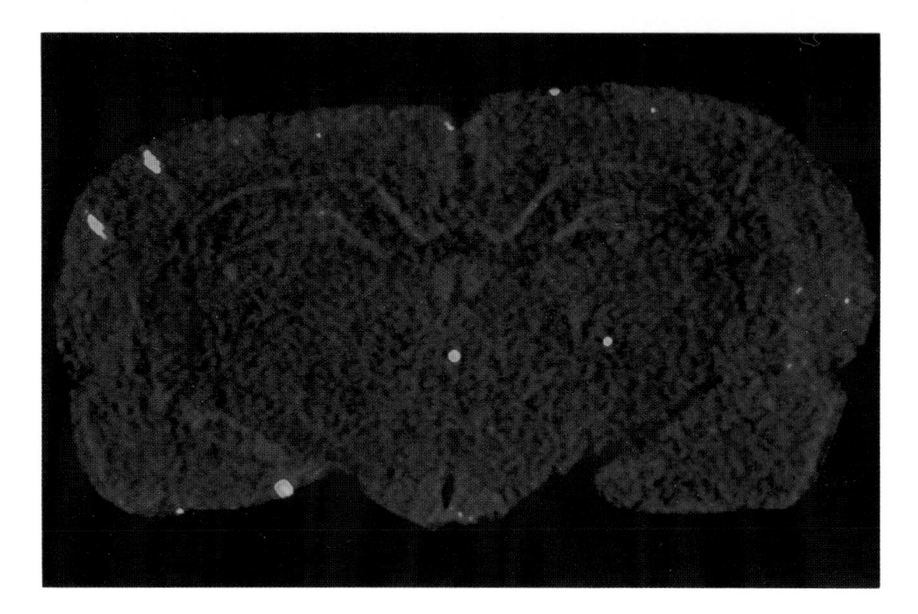

COLOR PLATE 24 Effect of seizures (evoked from area tempestas) on c-*fos* gene expression in various brain regions. Thin coronal sections of brain were exposed to a radioactively labeled probe that selectively recognizes c-*fos* messenger RNA. Brain regions with the highest density of binding to the probe are represented by the color white in the photograph; this reflects the highest c-*fos* expression. The color strip in the center shows the range of colors from lowest (blue) to highest (white) c-*fos* expression. Brain section from a rat that had a convulsive seizure triggered by placing a GABA antagonist in the left area tempestas (top). Similar brain section from a rat that did not have a seizure (bottom).

traenoic acid (arachidonic acid), and 5,8,11,14,17-eicosapentaenoic acid]. The lungs are rich in arachidonic acid. Other cells (e.g., leukocytes and platelets) are also capable of generating arachidonate products. Arachidonic acid is found as an esterified complex in the phospholipid component of the cell membrane. PGs are not stored in the cells but are synthesized and released on demand or stimuli. These stimuli include mechanical stimulation, embolization with air, endotoxin, and chemical stimulation with agents (e.g., acetylcholine, bradykinin, histamine, angiotensin II, serotonin, leukotrienes, and antigen). They have effects locally and, by in large, are not circulating hormones. Most of the eicosanoids are vasoactive substances. Thus, they are capable of modulating vascular tone and membrane permeability. They are also involved in events that occur concomitantly with lung injury.

Free fatty acids from membrane lipids are released by oxygenation, which is preceded by activation of phospholipase A_2 or diacylglyceride lipase. Once arachidonic acid is released, it is metabolized by different pathways.

A. Cyclo-oxygenase Pathway

A clyco-oxygenase enzyme, PG endoperoxide synthetase, converts arachidonic acid to biologically active and unstable PG endoperoxides (PGG_2 and PGH_2). These endoperoxides are further converted to PGE_2, prostacyclin (PGI_2), PGD_2, and thromboxane A_2 (TxA_2). PGI_2 and TxA_2 are unstable products, and they are converted to 6-keto-PGF_{1alpha} by nonenzymatic degradation and TxB_2 by spontaneous hydrolysis, respectively. The presence of these stable compounds is taken as an indicator of the parent compounds. PGs of the E and F series both are released during anaphylactic shock, and they play significant roles in fetal pulmonary circulation and childbirth. PGD_2 is a potent vasoconstrictor (a vasodilator during first 3 days of life), and it is released during anaphylaxis. TxA_2 is a potent vasoconstrictor and a powerful platelet-aggregating agent. It comprises one-third of the cyclo-oxygenase products released from human lung cells challenged with IgE or a calcium ionophore. There is species variation in TxA_2 production. Histamine acting via the H_1 receptor, bradykinin, slow-reacting substance of anaphylaxis (SRS-A) or mechanical stimulation all can release TxA_2. There is experimental evidence that TxA_2 plays a major role in mediating bronchoconstriction. A large amount of PGI_2 is released by the lungs. ECs produce much more PGI_2 compared with smooth muscle cells. It is a potent vasodilator and has platelet antiaggregatory properties. In large vessels, the ability to produce PGI_2 is of physiological importance for the prevention of platelet aggregation and injury to the EC surface.

B. Lipoxygenase Pathway

Arachidonic acid metabolized via the lipoxygenase pathway results in peptidoleukotrienes, which make up the slow-reacting substances of anaphylaxis. Peroxidation of arachidonic acid results in formation of unstable hydroperoxyeicosatetraenoic acids (HPETEs), which are then reduced to respective hydroxyacids (HETEs) and an unstable intermediate LTA_4, which is converted to LTB_4, LTC_4, LTD_4, and LTE_4. Peptidyltransferases in the tissue cleave the terminal amino acids from LTC_4 to yield LTD_4 and LTE_4. The main effects of leukotrienes are on the immunologic system and smooth muscle in the respiratory, cardiovascular, and gastrointestinal systems. LTB_4 is mainly produced by leukocytes, and it has chemotactic and cytokinetic properties. It induces adhesion of leukocytes to ECs. It has a weak contractile property. Thus, it is mainly involved in inflammation. LTC_4, LTD_4, and LTE_4 are potent vasoconstrictors, and they can induce increased mucus secretion and bronchoconstriction in various species. Leukotriene release has been shown to be increased in various pathologic states.

C. Cytochrome P450 Pathway

This is a relatively recently described pathway. The products from the cytochrome P450 pathway have effects on sodium transport and the vasculature. These products have been identified by their oxygenated peak formation (P_1 and P_2) in the presence of NADPH. Doses of P_1 relax isolated rabbit pulmonary arterial rings. A major component of P_1 appears to be a 5,6 epoxyeicosatrienoic acid derivative of arachidonic acid. P_2 has no effect on the vasculature.

VIII. Influence of Injury

Endothelial cells play an important role in lung injury. The abnormal EC function occurs concomitant with enhanced pulmonary vascular reactivity.

A. Endotoxin Injury/Sepsis

Lung injury often is the first symptom of sepsis in one-third of all trauma patients. In burn patients, sepsis-induced adult respiratory distress syndrome (ARDS) has a high mortality. Alterations in pulmonary circulation can result in pulmonary arterial hypertension (PAH) and increased pulmonary microvascular permeability. In addition, these patients develop ventilatory abnormalities such as decreased lung compliance and ventilation perfusion mismatch. These abnormalities have been duplicated in animal models by administering endotoxin or bacteria. Thromboxane A_2 is considered to be responsible for early PAH and bronchoconstriction seen after bacterial infusion, and thromboxane inhibitors can markedly attenuate these symptoms. Furthermore, an increase in plasma TxA_2 has been demonstrated in humans during sepsis-induced ARDS. However, TXA_2 is not responsible for the later stage of PAH. Lipoxygenase products (e.g., LTC_4 and LTD_4) are released during inflammation, and they are thought to play a role in accentuating the lung injury. However, neutrophils play an important role in increasing lung vascular permeability in sepsis and oxygen toxicity. The activated neurtrophils release proteases and oxygen radicals, which cause lung injury. Arachidonate products may modulate endotoxin-induced injury indirectly through interactions with the complement system and platelet or leukocyte functions. Endotoxemia releases both constrictor and dilator cyclo-oxygenase products as well as lipoxygenase products. Arachidonate products of leukocyte origin may affect airways. However, arachidonate products of platelet origin are not considered important in the pathogenesis of lung injury.

Antibody binding with ECs can result in platelet-activating factor (PAF) release. The released PAF will recruit polymorphonuclear cells (PMN) and induce their aggregation and degranulation leading to occlusion of small vessels, tissue injury, and pulmonary edema. PAF-stimulated neutrophils can induce *de novo* synthesis of LTB_4, which will contribute to, amplify, and perpetuate the inflammatory reaction. The activated neutrophils can release H_2O_2 and O_2^-, which have cytotoxic effects. It is suggested that H_2O_2 is converted to $OH^\cdot$, which may lead to lipid peroxidation injury.

B. Oxygen Toxicity

Prolonged exposure to high oxygen invariably leads to oxygen-mediated lung injury. The mechanism of oxygen toxicity is believed to be related to the intracellular production of free radicals, which are cytotoxic byproducts of a variety of normal processes of aerobic tissue metabolism. These free radicals are superoxide radical ($O_2^\cdot$), hydrogen peroxide (H_2O_2), hydroxyl radical ($OH^\cdot$), singlet oxygen ($^1O_2$), and peroxide radical ($ROO^\cdot$). Under normal circumstances, they are detoxified by enzymatic antioxidants (e.g., superoxide dismutase, catalase, and glutathione peroxidase). There are nonenzymatic antioxidants such as vitamin E, beta-carotene, ascorbate, glutathione cysteamine, nonessential polyunsaturated fatty acids, and other thiols. ECs are more susceptible to oxygen toxicity than are epithelial cells. Within 48 hr of hyperoxic exposure, the biochemical and morphological changes in the ECs can be detected, which lead to microvascular permeability. Normally in ECs the rate of cyanide-resistant O_2 consumption is high, which increases markedly in hyperoxia, thus creating intracellular imbalance and making ECs more vulnerable to hyperoxic insult. Bleomycin and radiation therapy injury are also related to the generation of oxygen radicals. If the capillary leak induced by oxygen toxicity is not restored, the lung damage becomes progressive and ventilation–perfusion mismatch develops, eventually leading to low respiratory reserve. A classical example of oxygen toxicity is seen in premature babies with hyaline membrane disease who go on to develop bronchopulmonary dysplasia after oxygen therapy. Increased O_2 concentration inhibits DNA and protein synthesis. Therefore, lung maturation is also affected in these newborns, because the maturation continues postnatally.

C. Pulmonary Arterial Hypertension

Acute and subacute lung injury often leads to PAH with its effects on pulmonary function. A variety of cardiopulmonary diseases are known to lead to pulmonary artery hypertension and pathological changes in the pulmonary vasculature. But the pathological mechanisms that lead to PAH are not understood. With left to right cardiac shunts and pulmonary embolism, the EC injury occurs, which accompanies arteriolar medial wall thickening. How lung parenchymal diseases such as chronic bronchitis, interstitial pneumonitis, etc., lead to vascular EC injury is not understood. EC injury is considered to be the key event in the pathogenesis of PAH. The injured ECs can activate cellular and humoral inflammatory pathways, arachidonic acid products, and coagulation factors, which in turn can potentiate EC injury. As has been stated above, the injured endothelium can activate leukocytes resulting in generation of oxygen radicals. Injured ECs can increase the release of procoagulant factor VIII and reduce plasminogen activator and prostacyclin, thus favoring thrombosis (blood clotting). EC dysfunction also leads to decreased metabolism of vasoactive peptides. Recently, it has been recognized that these cells can release various growth factors, which could then act on underlying smooth muscle. Histological changes that occur in the pulmonary vasculature are swelling, hypertrophy, and hyperplasia of the cells in the intima and the medial coat of the pulmonary arteries, leading to narrowing and occlusion of the lumen. There is also extension of the smooth muscle cells into nonmuscular pulmonary arteries. In some small arterial lumens, platelet or leukocyte aggregates or thrombosis can be seen, which again would lead to occlusion. These changes lead to PAH and right ventricular hypertrophy and, in late stages, right heart failure. There is a great deal of interest in various experimental models, because some of the changes seen in these models are similar to that seen in various clinical forms of PAH. Persistent PAH in newborn infants has a high morbidity and mortality. It is believed to be due to a lack of regression of high pulmonary vascular resistance of the fetal lung at the onset of air breathing at birth. Some of the important features of this illness are hypoxemia, PAH, pulmonary edema, and bronchoconstriction. High levels of leukotrienes have been demonstrated in the bronchial lavage of these infants, which is not the case in infants with other respiratory diseases. However, the exact role of leukotrienes in the development of PAH in this disease is not clear, but it may contribute to pulmonary edema and bronchoconstriction.

Acute exposure to hypoxia leads to increased PAH, wich is reversible. However, subacute and chronic hypoxia lead to persistent vasoconstriction, PAH, and structural remodeling. Within the first day of exposure to hypoxia, structural remodeling can be demonstrated. On return to room air, these hemodynamic and histological changes start to regress.

Monocrotaline (MCT) is a pyrrolizidine alkaloid of plant origin that causes PAH and associated cardiovascular changes in rats and other mammals by mechanisms not yet clearly understood. MCT is metabolized in the liver to an active pyrrole form, which causes pulmonary EC damage. The pulmonary ECs become swollen and damaged, leading to pulmonary edema. There is a time lag between the administration of MCT and the development of PAH. In the intervening period, an increased vascular response to vasoactive constrictor agents (e.g., angiotensin II and serotonin) and hypoxia has been demonstrated. EC dysfunction (e.g., decreased uptake of serotonin and norepinephrine) and diminished ACE levels are present. Platelets, in part, contribute to the structural modeling. MCT initiates the structural remodeling, but it is the extent of EC injury and later activation of biological activities that determine its degree and nature.

Acute and chronic injury to lungs lead to various biochemical and histological changes, which result in altered responsiveness of the pulmonary vasculature to pharmacological and physiological stimuli. The metabolism of various vasoactive substances are affected, which, in turn, worsen the disease process. In late stages, these alterations in functions become irreversible.

IX. Potential Clinical Application

Knowledge of the pharmacology of eicosanoids has already introduced a greater understanding of perinatal pulmonary circulation and has become useful in clinical practice. In a number of premature infants, the patent ductus arteriosus can be successfully closed by administering cyclo-oxygenase inhibitors (e.g., indomethacin). In infants with ductus arteriosus–dependent cardiac defects, the ductus can be kept patent temporarily by administering PGE_1. PGI_2 has been used in various cardiac catherization laboratories to assess pulmonary vascular reactivity and, to a limited extent, to predict the outcome of vasodilator therapy in cases of primary PAH. The role of the arachidonic acid cascade and oxygen radicals is well known in lung injury, and thus manipulation of PGs and administration of antioxidants will become a part of therapy.

Recently, considerable interest has been generated in the possible important role of Mg^{2+} in modulating vascular tone and integrity of excitable cells. Mg^{2+} deficiency has been shown to be associated with a high incidence of sudden cardiac death, ischemic heart disease, stroke, and preeclampsia. Dietary Mg^{2+} deficiency is known to cause systemic hypertension, structural changes in the peripheral and cerebral microvasculature, and pulmonary lesions that resemble those found in adult respiratory disease. Loss of cellular Mg^{2+} results in loss of critically important phosphagens (e.g., MgATP and creatine phosphate). Therefore, under certain pathophysiologic conditions (e.g., hypoxia, anoxia, ischemia, and cellular injury) in which cellular Mg^{2+} is depleted, the NA^+/K^+ pump, phosphagen stores, and membrane structure will be compromised, leading to alterations in resting membrane potentials. Increased synthesis of some eicosanoids may be linked to enhanced influx and translocation of Ca^{2+}. In Mg^{2+} deficiency, the lowered cyclic AMP can permit a high cyclo-oxygenase activity and a drastic increase in thromboxane. Thus, Mg^{2+} deficiency can affect the arachidonic acid cascade, which has been considered extremely important in pathogenesis of lung injury. There is evidence that Mg^{2+} acts primarily at the VSM cell membrane on specific Mg^{2+}/Ca^{2+} and Na^{2+}/Ca^{2+} exchange sites to regulate finely the entry of Ca^{2+} in smooth muscle cells, vascular tone, blood pressure, and local flow. During stress, Mg^{2+} leaves the cell, thus making it more vulnerable to toxicity of Ca^{2+}. External Mg^{2+} plays an important role in modulating PG-induced contractions and relaxations. It also enhances prostacyclin synthesis in cultured cells. It is a powerful inhibitor of blood coagulation and is a requirement for intracellular cement and endothelial–endothelial cell junctions. There have been few reports of Mg^{2+} treatment in PAH. In acute hypoxia-induced PAH in dogs, pulmonary artery pressure can be successfully lowered with infusion of $MgCl_2$. Mg^{2+} aspartate HCL has been demonstrated to attenuate MCT-induced pulmonary hypertensive changes in rates.

Although magnesium deficiency *per se* is not implicated as an etiologic factor of any of the diseases mentioned above, the deficiency certainly can make the cellular injury and response much worse, through various mechanisms mentioned. Thus, by increasing intracellular Mg^{2+} by dietary or therapeutic means, it is possible that the cells might be able to withstand injury much better.

X. Summary

Our understanding of pulmonary circulation has undergone a considerable expansion during the past decade. Recent studies have shed a great deal of light on the delicate balance between the vasoconstrictors and vasodilators and the multifaceted activities of ECs needed to maintain homeostasis. Further insight into these areas will lead to a better understanding of the pathophysiology of various disease states of the lung and the prevention and treatment of pulmonary diseases.

Bibliography

Altura, B. M. (1987). Pharmacology of the pulmonary circulation. *In* "The Pulmonary Circulation in Health and Disease" (J.A. Will, C. A. Dawson, K. E. Weir, and C. K. Buckner, eds.), pp. 79–95. Academic Press, New York.

Altura, B. M., Altura, B. T. (1989). Role of magnesium in pathogenesis of high blood pressure: Relationship to its actions on cardiac and vascular smooth muscle. *In* "Hypertension: Pathophysiology, Diagnosis and Management" (J. H. Laragh and B. M. Brenner, eds.), pp. 1003–1025. Raven Press, New York.

Carrol, M. A., Schwartzman, M., Baba, M., Abraham, N. G., and McGiff, J. C. (1987). Formation of biologically active cytochrome *P*450- arachidonate metabolites in renomedullary cells. *In* "Advances in Prostaglandin, Thromboxane and Leukotrienes Research," vol. 17. (B. Samuelssomn, R. Paoletti, and P. W. Ramwell, eds.), pp. 714–718. Raven Press, New York.

Chand, N., and Altura, B. M. (1980). Occurrence of inhibitory histamine H_2-receptors in isolated pulmonary blood vessels of dogs and rats. *Experientia* **36,** 1186–1187.

Chand, N., and Altura, B. M. (1981). Acetylcholine and bradykinin relax intrapulmonary arteries by acting on endothelial cells: Role in vascular diseases. *Science* **213,** 1376–1379.

de Nucci, G., Thomas, R., D'Orleans-Juste, P., et al. (1988). Pressor effects of circulating endothelin are limited by its removal in the pulmonary circulation and by the release of prostacyclin and endothelium derived relaxing factor. *Proc. Natl. Acad. Sci.* **85,** 9797–9800.

Fishman, A. P. (1985). Pulmonary circulation. *In* "Handbook of Physiology. Section 3: Respiratory System" (A. P. Fishman and A. B. Fisher, eds.), pp. 93–165. American Physiological Society, Baltimore, Maryland.

Gillis, N. C. (1986). Pharmacological aspects of metabolic processes in the pulmonary microcirculation. *Annu. Rev. Pharmacol. Toxicol.* **26,** 183–200.

Harker, L. A. (1988). Endothelium and hemostasis. *In* "Endothelial Cells," vol. 1. (U. S. Ryan, ed.), pp. 167–177. CRC Press, Boca Raton, Florida.

Highsmith, R. F., Aichholz, D., Fitzgerald, O., Paul, R., Rubanyi, G. M., and Hickey, K. (1988). Endothelial cells in culture and production of endothelium-derived constricting factors. *In* "Relaxing and Contracting Factors" (P. M. Vanhoutte, ed.), pp. 137–158. Humana Press, Mt. Kisco, New York.

Hyman, A. L., Lippton, H. L., and Kadowitz, P. J. (1988). Neurohumoral regulation of the pulmonary circulation. *In* "Vasodilators, Vascular Smooth Muscle, Peptides, Autonomic Nerves and Endothelium" (P. M. Vanhoutte, ed.), pp. 311–319. Raven Press, New York.

Martin, D. C., Carr, A. M., Livingston, R. R., and Watkins, C. A. (1989). Effects of ketamine and fentanyl on lung metabolism in perfused rat lungs. *Am. J. Physiol.* **257,** E379–E384.

Martin, W. (1988). Basal release of endothelium derived relaxing factor. *In* "Relaxing and Contracting Factors" (P. M. Vanhoutte, ed.), pp. 159–178. Humana Press, Mt. Kisco, New York.

Mathew, R. (1991). Fetal and neonatal circulation. II. Metabolic aspects. *In* "Neonatal and Fetal Medicine (Physiology and Pathophysiology)" (R. A. Polin and W. W. Cox, eds.). Grune and Stratton, New York.

Mathew, R., and Altura, B. M. (1988). Magnesium and the lungs. *Magnesium* **7,** 173–187.

Nowak, J. (1984). Eicosanoids and the lungs. *Ann. Clin. Res.* **16,** 269–286.

Plamer, R. M. J., Ashton, D. S., and Moncada, S. (1988). Vascular endothelial cells synthesize nitric oxide from L-arginine. *Nature* **333,** 664–666.

Ryan, U. S. (1988). Phagocytic properties of endothelial cells. *In* "Endothelial Cells," vol. 3. (U. S. Ryan, ed.), pp. 33–49. CRC Press, Boca Raton, Florida.

Said, S. I., ed. (1985). "The Pulmonary Circulation and Acute Lung Injury." Futura Publishing Co. Inc., Mount Kisco, New York.

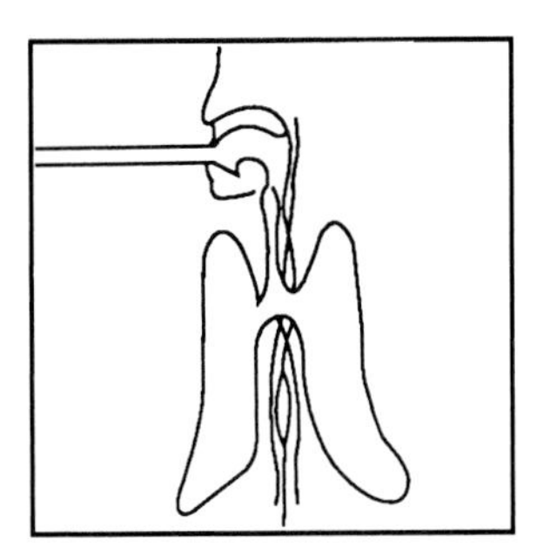

Pulmonary Pathophysiology

A. VERSPRILLE, *Erasmus University, Rotterdam*

Glossary

Airway obstruction Increased air flow resistance in the airways
Compliance Change in lung volume for a given change in lung recoil pressure, measured as intraesophageal pressure
Equal pressure point Sites in the bronchial tree where, during expiration, the pressure in the bronchi is equal to the pressure around the bronchi
FEV_1 Forced expiratory volume in 1 sec, i.e., the maximal lung volume to be expired in 1 sec after maximal inspiration
FIV_1 Forced inspiratory volume in 1 sec, i.e., the maximal lung volume to the inspired in 1 sec after maximal expiration
Functional residual capacity (FRC) Lung volume at the end of a normal expiration
Hyperinflation Increase in FRC
Hyperpnea Increase in ventilation corresponding to an increase in metabolic rate, e.g., during exercise, when the normal arterial P_{CO_2} is maintained
Hyperventilation Ventilation which is too large for a certain metabolic state, leading to a decrease in the CO_2 tension of the arterial blood below 30 mmHg
Hypoventilation Insufficient ventilation to maintain a normal arterial P_{CO_2}, which increases above 49 mmHg
Hypoxemia Decrease in oxygen saturation of the arterial blood below 90%
$MEF_{75,50,\text{or } 25}$ Maximal expiratory flow when 75, 50, and 25% of the vital capacity is in the lungs, respectively
MVV_{30} Maximal voluntary volume which can be breathed in and out in 30 breaths/min
Peak expiratory flow rate (PEFR) Maximal air flow shortly after the start of a forced expiration
Pneumotachometer Tube with a resistance to measure air flow based on the measurement of a pressure difference
Pulmonary edema Condition in which an increased amount of fluid is present in the lung interstitium and in the alveoli
Restriction Decrease in alveolar volume at TLC (total lung capacity), FRC, and RV (residual volume) levels
Residual volume (RV) Lung volume after a maximal expiration
Specific compliance Lung compliance per liter of lung volume
Spirometer Container submerged in a water bath, used to measure lung volume and volume displacement during breathing maneuvers
Spirometry Recording of lung volumes with use of a spirometer
Total lung capacity (TLC) Lung volume after a maximal inspiration

PATHOPHYSIOLOGY is a field in medical biology in which fundamental mechanisms of disorders in physiological functions are studied. The main goal is directed at defining the character of the relationships between the disordered processes. These relationships are expressed in quantitative terms to assess the gravity of an illness. The description of disorders in founded theories, i.e., the explanation of empirical knowledge in logic models, will contribute to medical science in two ways: (1) by detecting the essential variables for use as indices in an adequate diagnostic procedure, and (2) by predicting the results of medical interventions at a higher level of confidence than before.

In pulmonary pathophysiology the functional disorders of airways, alveoli, and gas exchange are the subjects of study.

A logical consequence of fundamental knowledge in this field is the application of physiological measurements in clinical routine. This type of applied physiology can be called clinical physiology, which deals especially with the methodology and interpretation of functional tests in patients.

In this chapter the pulmonary disorders will be considered in three categories:

- disorders of the airways which increase airflow resistance
- disorders of the alveolar volume which impair lung expansion
- disorders of gas exchange between alveolar capillary blood and ambient air

Usually the functional disorders will have a structural basis.

I. Airway Obstruction

A. The Nose

Three important functional structures in the nose must be considered.

1. The ciliated, pseudostratified columnar epithelium contains mucus cells, i.e., goblet cells, and mucus glands, which produce a fluid mixture of water, salts, and mucus. The water evaporates to saturate the inspiratory air with water vapor at 37°C. The mucus collects particles from the inspiratory air, in order to clean the air before it enters the lungs.

A second important structural part of the epithelium is formed by the cilia, which move the mucus and dust particles to the pharynx where it is accumulated and swallowed. The cilia have a fast beat in the direction of the nasal pharynx, and a much slower movement back. During the fast beat they are stiffened, and the viscous mucus moves in the direction of the pharynx. During the backward movement the cilia relax, more or less, and bend and pass through the mucus more easily without moving it backward. The movement of the cilia is coordinated in such a way that a wave moves from the front of the nose to the nasal pharynx. The mechanisms of coordination are unknown.

The cleaning process is usually supported by blowing one's nose. This forceful action causes a high pressure in the nose, which might press some mucus and dust, including bacteria, into the *paranasal sinuses,* causing infections. Sniffing repeatedly, or *sniffling,* to prevent mucus from running out of the nose, is also used as a cleaning process. During sniffling a negative pressure develops in the nasal cavity.

The motility of the cilia is affected by tobacco smoke, other chemical substances, and infectious materials.

2. Below the epithelium a rich plexus of blood vessels is present, supplying heat for both warming the air and evaporating water.

3. The nasal conchae, shell-like bony projections into the nasal cavity, covered with epithelium, enlarge the surface area of the nose, and intensify all the processes mentioned above. These conchae decrease the width of the nasal passage, which allows obstruction to occur more easily when epithelial swelling occurs.

B. The Nasal and Oral Pharynx

The nasal pharynx and the oral pharynx are conducting parts of the airway without special function. At this level the alimentary tract and the respiratory tract cross each other. During swallowing the larynx is moved upward and the epiglottic cartilage is pushed against the root of the tongue and bent over the entrance of the larynx. This closure function of the epiglottis seems redundant: after removal of the epiglottis the entrance of the larynx is directly pushed against the base of the tongue and sufficiently closed.

C. The Larynx

The main structures in the larynx cavity are the vocal folds, between which is a cleft, the rima glottidis. The vocal folds are a double layer of membraneous and mucosal tissue around the vocal ligaments. These ligaments are fixed in front to the thyroid cartilage and dorsally to the arytenoid cartilage. Movement of the latter can stretch the ligaments. [*See* LARYNX.]

Coughing is a function necessary to evacuate mucus mixed with foreign material from the large bronchi and trachea into the pharynx. The material will be either expectorated or swallowed; if the latter occurs, bacteria are destroyed by gastric acid in the stomach. Coughing is a deep inspiration followed by a forced expiration, initially with a closed vocal cleft, resulting in an elevated subglottic pressure which opens the vocal cleft shortly and repeatedly. Closure occurs each time by the pressure drop when air escapes. During such a short opening of the vocal cleft the forceful expiratory flow blows out the mucus. It even carries mucus away from the large bronchi all along the trachea. Coughing is elicited by mechanical or chemical stimulation of the tracheal and bronchial epithelium.

Reflexive closing off of the entrance of the trachea occurs in the presence of irritating and noxious gases, which stimulate the nasal mucosa. Water also stimulates such a reflex, causing a spasm of the larynx muscles, which stretch the vocal ligaments to close off the trachea. This phenomenon explains death by "dry drowning" during immersion.

Pressing is a voluntary action during defecation, micturition, and during labor in women, which is achieved by an expiratory action against closed vocal folds. The increased pressure in the lungs supports the diaphragm, when forceful contractions of the abdominal muscles are exerted to increase intraabdominal pressure.

The speech function of the vocal folds occurs in a coordinated action of the laryngeal muscles with the respiratory muscles, and the muscles of the pharynx and mouth. The respiratory muscles deliver the energy for the air flow which causes the vocal folds to vibrate. The laryngeal muscles stretch the vocal cords, the degree of which more or less determines the pitch of a sound. The muscles of the throat, tongue, and mouth determine the shape of the sounding board which acts as a filter on the higher harmonics of a tone. The higher harmonics make the tone of a sound. This accounts for the vowels. The consonants are produced in the mouth by closed or narrowed passages. However, they can be extended with a sound containing vowel, e.g., when the alphabet is said.

Paralysis of the larynx muscle, e.g., after thyroid surgery during which the recurrent nerves are injured, causes an extreme increase in airway resistance, especially during inspiration. The vocal ligaments are not stretched and the folds behave like half-moon-shaped sacs, which bulge out downward during inspiration by the lower pressure below the sacs. The ligaments are moved in a medial direction and cause a narrowing of the vocal cleft, causing a serious obstruction. During expiration the sacs are pressed in a lateral direction and the cleft is less narrowed. The obstruction can be so serious that formerly extirpation of the vocal folds was necessary and speech function had to be sacrificed. At present the vocal folds are fixed laterally in the larynx, giving enough cleft for air passage and maintaining the possibility of speech, although in a hoarse, throaty voice.

D. The Trachea and Bronchial Tree

1. Structure

The trachea, or windpipe, which extends from the larynx into the thorax, consists of open, horseshoe-shaped rings of cartilage. The open part of the horseshoe is located dorsally. The ends of the cartilage are connected to each other with fibrous tissue. This shape is important for the passage of food through the esophagus, which is located between the trachea and the cervical vertebrae. During the passage of food the esophagus bulges into the trachea.

The cartilage rings are joined by connective tissue. This structure of cartilage with connective tissue in between containing elastic fibers makes the trachea rigid transversely and flexible longitudinally. It keeps the trachea open for air passage, and allows the trachea to bend with neck movements. The elastic fibers recover the length of the trachea after stretch during both breathing and swallowing.

The pseudocolumnar ciliated epithelium, located in the trachea, functions like its counterpart in the nose. The cilia, however, beat in the opposite direction and move the mucus upward to the larynx. The same occurs in the bronchi. This cleaning function is inhibited by the same agents affecting the cilia of the nose.

Two main bronchi are formed at the bifurcation of the trachea. Their angle forms inside a keel, the carina, in the direction of the trachea. By further branching a bronchial tree is formed. The main function of the trachea and bronchial tree is conduction of air, but the functions mentioned for the nose are also still exhibited: warming of air, humidifying air by evaporation of water, and cleaning the air from dust. However, this cleaning process often fails. When the respiratory air is full of dust, these particles enter the alveoli, causing in the long run alveolitis, followed by fibrosis. A common cold and, more seriously, a flu are signs of insufficient protection of the epithelium by the mucus and cilia, and by our immune system as a secondary barrier.

The bronchial tree is subdivided into generations. The trachea is called generation zero. The two main bronchi are generation 1, etc. The bronchial tree is composed of 16 generations.

The main bronchi also contain horseshoe cartilage in their walls. In generations 2 to 11 the cartilage is irregularly spiralized around the bronchi. These generations are surrounded by connective tissue, and run in parallel with the pulmonary arteries. In these connective tissue layers the pressure is equal to pleural pressure. From generation 12 up to 16 the very small bronchi are called bronchioles. Bands of smooth muscle cells are formed into spirals around these bronchioles. The inner layer of the bronchioles consists of cubic epithelium without cilia. Connective tissue fibers run radially from the bronchioles into the septa of the surrounding alveoli, like the guy ropes of a tent. Contraction of the smooth muscle cells constricts the bronchioles and stretches the radial fibers. Relaxation of the muscles allows the fibers to dilate the bronchioles.

During inspiration airway resistance decreases by two mechanisms: (1) the more negative pleural and intrathoracic pressure, which dilates the larger bronchi, and (2) the radial fibers of the bronchioles, which stretch more and exert more traction on the bronchioles to dilate them. This latter mechanism is supported by the more negative intrathoracic pressure, which also surrounds the bronchioles.

The smooth muscles of the bronchioles have an important control function on distribution of the respiratory air in the alveoli. We will consider the effects of this control function when smooth muscles contract in one part of the respiratory tract and relax in another part. Next, we will analyze the effects of a general bronchiolar constriction, as could occur in asthmatic patients.

2. Airway Resistance and Airway Obstruction

Air flow needs a pressure difference (P) between the alveoli and the mouth. During normal inspiration alveolar pressure is one to two mmHg lower than ambient air pressure at the entrance of the nose. During expiration alveolar pressure is higher than ambient. When no turbulence should occur the relationship between the pressure difference and air flow obeys the aerodynamic equivalent of Ohm's law: $P = V'R$, where V' is volume flow of air (in ml/sec) and R is flow resistance in mmHg/(ml sec). R is a variable by definition, and cannot be measured independently from V' and P; it is the proportionality constant between the two. When turbulence occurs the linear relationship between V' and P is lost. Then, the relationship is described by $P = k_1V' + k_2V'^2$. Due to turbulence a much larger P value is needed for the same volume flow.

During normal breathing turbulence occurs at the bifurcations of the bronchi. During the forced breathing action turbulence also occurs in the larger bronchi themselves. Therefore, flow resistance is usually higher than can be predicted from the geometry of the bronchi assuming a nonturbulent flow. Such a prediction can be made from the equation experimentally found by the French physician J. L. M. Poiseuille: $R = 8\eta l/\pi r_4$ where η is the viscosity of the respiratory air, and l and r are the length and radius of each generation. When using this formula for R and the geometric data of the trachea and bronchial tree as published by Weibel (1964) at the lung volume level after normal expiration, the resistance of generations 0 to 11 is about 87% of total airway resistance. Thus, the remaining part of flow resistance in generations 12–16, including generations 17–23 (or 26) is about 13%. The approximation of airway resistance is given in Fig. 1. Due to turbulence the calculated value of R in the larger bronchi is slightly underestimated. For our

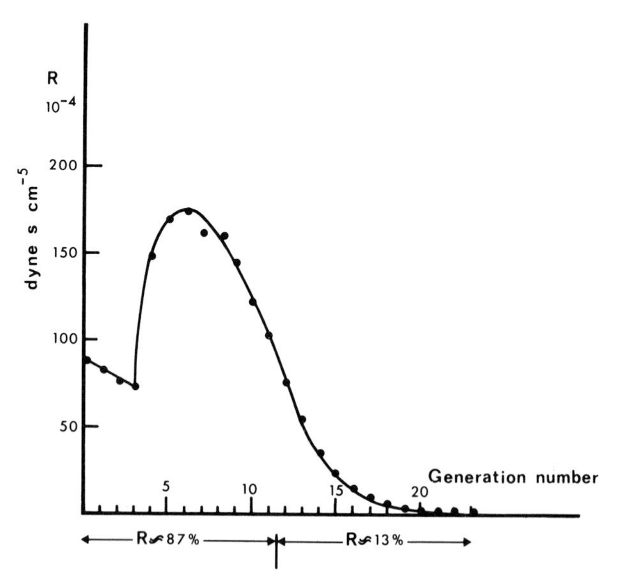

FIGURE 1 Airway resistance per generation. The calculation of airway resistance per generation is based on the flow resistance (R_i) of an individual average bronchus of a generation, which is determined according to the Poiseuille equation (see text) and the mean data of r (radius) and l (length) of each generation. The effective flow resistance (R_g) in a generation is found from $1/R_g = n(1/R_i)$, where n is the number of bronchi per generation.

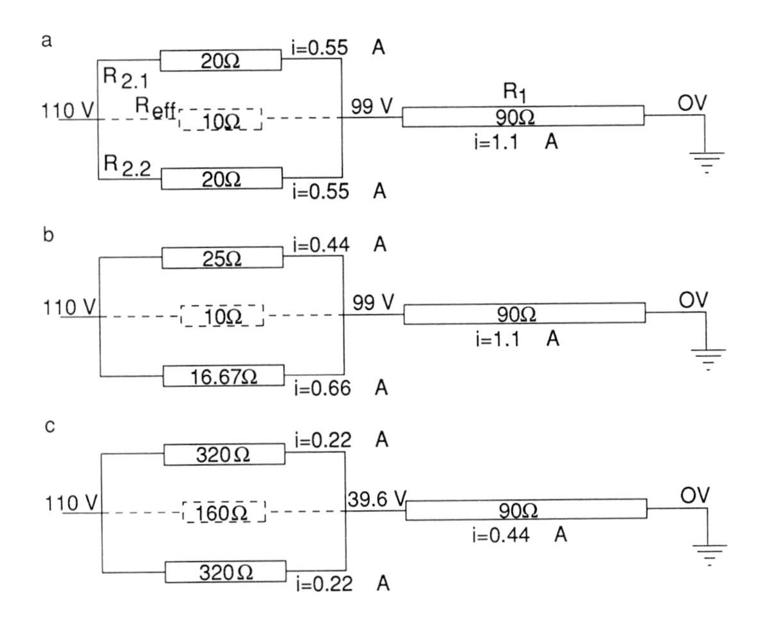

FIGURE 2 Effects of changes in bronchiolar flow resistance. The baseline resistances of each of the two bronchiolar compartments is 20 Ω, giving an effective resistance (R_{ef}) of 10 Ω. In between the bronchiolar resistances at the right side and the resistance of the large bronchi at the left a voltage of 11 V exists in (a) and (b). The figures in the model give the changes as discussed in the test. In (c) the voltage is 39.6 V between the "bronchioles" and the "bronchi." For further explanation see text.

modeling we therefore assume a ratio of 90 : 10 between generations 0–11 and 12–23, respectively. In the electric analogy this ratio is used to demonstrate the effects of distribution and extensive bronchial constriction in an asthmatic attack.

In Fig. 2a the normal model is presented. Total airway resistance (100%) has been set at 100 Ω. Then, trachea and bronchi, i.e., generations 0–11, are represented by resistance R_1 of 90 Ω and the bronchioles and alveolar ducts by 10 Ω. The bronchioles are subdivided into two compartments to demonstrate the effects of mutual changes in flow resistance on the distribution of air flow, using electric current as its equivalent. Both parts have a resistance of 20 Ω, and exhibit an effective resistance of 10 Ω according to the formula $1/R_{eff} = 1/R_{2.1} + 1/R_{2.2}$. A charge of these resistances with 110 V results in a current of 1.1 A. A current of 0.55 A runs through both $R_{2.1}$ and $R_{2.2}$.

In Fig. 2b $R_{2.1}$ = 25 Ω and $R_{2.2}$ = 16.67 Ω, giving again an effective resistance R_{eff} = 10 Ω and a current of 1.1 A. However, the current is differently distributed between $R_{2.1}$ and $R_{2.2}$. The voltage on $R_{2.1}$ and $R_{2.2}$ is 10% of 110 V. Thus, $R_{2.1}$ gets a current of 11 V/25 Ω = 0.44 A, and through $R_{2.2}$ runs a current of 11 V/16.67 Ω = 0.66 A. To increase flow resistance of a tube with 25% (from 20 to 25 Ω) a diameter decrease of approximately 6% according to the Poiseuille equation is necessary. Such a decrease in diameter causes a decrease in flow of 20% (from 0.55 to 0.44 A). The decrease in resistance of $R_{2.2}$ is 16.67%, corresponding to a diameter increase of approximately 4.5%, causing an increase in flow of 20%. This example of modeling demonstrates a large shift in air flow by a small change in diameter of the bronchioles.

In Fig. 2c we modeled a general constriction of the bronchioles elicited during an asthmatic attack. We assume a constriction to one-half the normal diameter. That implies a 2^4 or 16-fold increase in resistance. Total airway resistance then is 90 + 160 = 250 Ω. The decrease in current is from 1.1 to 0.44 A, which is 60%. To compensate for this decrease in airflow an effort 2.5 times larger than normal is needed during breathing. A constriction to one-quarter of the bronchiolar diameter increases bronchial resistance 256 times and total flow resistance 26.5 times, which means very large efforts are needed to breath.

The ability of the bronchioles to constrict and dilate by means of their smooth muscle cells layer is a favorable function to control the distribution of air. However, this ability is paid for by asthmatic patients who easily develop high flow resistances in the bronchioles by a variety of stimuli.

3. Control of Bronchiolar Diameter

The bronchial muscular tone is dependent on the activity of the autonomic system. Sympathetic activity is increased during physical work, causing a dilatation in favor of better ventilation without too much effort. The sympathomimetic hormones norepinephrine and epinephrine support this bronchial dilatory effect.

Isoprenaline, a sympathomimetic drug, is usually used to test the origin of increased airway resistance. When airway resistance diminishes after administration of this drug with use of a nebulizer during inhalation, increased muscular tone is assumed to be the reason for bronchial obstruction. Nowadays another sympathomimetic drug, terbutaline, with less cardiac side effects, is used for this test.

In normal subjects a dilating effect of isoprenaline and terbutaline is undetectable. This can be easily understood from Fig. 2a, when we assume a double diameter of the bronchioles in the normal model. The resistance will decrease 16 times ($R = 8nl/\pi r^4$), which is a fall to less than 1 Ω in the electric analogy. Then the total flow resistance is decreased to 90%, which results in a 10% increase in air flow. This difference, which is the maximal one, is barely significant.

The vagus nerve and the parasympathomimetic agents acetylcholine, pilocarpine, and carbachol, have a bronchial dilatory effect. The effect of the vagus is small.

An important stimulus is the alveolar CO_2 tension (P_{A,CO_2}). A decrease in P_{A,CO_2} causes a bronchial constriction. In some way the vagus nerve is involved in this reaction, because after the nerve is blocked by atropine the constrictor effect of a decrease in P_{A,CO_2} is smaller. A decrease in P_{A,CO_2} can be caused by either hyperventilation or by a decrease in perfusion of blood to some pulmonary area. By this mechanism regional ventilation is adapted to regional perfusion. Bronchial constriction occurs where blood perfusion is small, causing a smaller ventilation, matched to the perfusion.

4. Pathologic Stimuli of the Bronchial Smooth Muscle

During exercise spasmodic contractions of the bronchial smooth muscles can occur, causing a serious asthmatic attack. The hypothesis of a decrease in P_{A,CO_2} is not very popular anymore. Another hypothesis to explain the effect is a stimulation of the bronchial mucosa with cold air or cooling of the epithelium by evaporation of water.

Allergens are external substances which, when inhaled, provoke an asthmatic attack of a type called extrinsic asthma. When an asthmatic attack cannot be traced to such external allergens, by use of a skin test, the spasmodic smooth muscular contractions are called intrinsic asthma. [*See* ALLERGY.]

Psychological factors by themselves will not be sufficient to cause a spasmodic contraction of the bronchial muscles. Emotions might facilitate an attack of asthma in patients who are constitutionally predisposed to hyperreactivity of the bronchial smooth muscles.

5. Other Types of Airway Obstruction

Other reasons for airway obstruction can be swelling of the bronchial walls, as in bronchitis, edema, caused by many disorders, and loss of elasticity of the connective tissue, as in emphysema.

In emphysema two main mechanisms cause an increased obstruction of the airways, especially during expiration. Due to the loss of elasticity two changes occur: (1) negative thoracic pressure is less negative than normal, and (2) the walls of the bronchioles are insufficiently connected by connective tissue fibers to the alveolar interstitial tissue. The effect of both mechanisms is to decrease the diameters of the airways, especially during expiration.

During inspiration this effect does not have much influence, because the walls of the bronchi are flaccid. During inspiration a negative alveolar pressure is needed to cause the flow of ambient air (which pressure is reset to zero) to the alveoli. This negative pressure causes a more negative intrathoracic pressure. Then a positive pressure exists between pressure in the airways and pressure surrounding the airways. The flaccid airways will be kept open easily and air flow will hardly be obstructed.

During expiration a positive alveolar pressure is needed with respect to ambient air pressure. As a consequence intrathoracic pressure surrounding the airways will be increased above ambient air pressure. As considered for normal breathing this increased intrathoracic pressure causes a larger airway resistance during expiration than during inspiration. In emphysema patients this mechanism is worsened due to the higher intrathoracic pressure and the less firm bronchi. Emphysema patients can inspire well, but have great difficulties in expiration.

During expiration a pressure drop occurs from the alveoli through the airways to the nose or mouth due to the flow resistance of the bronchi (Fig. 3a

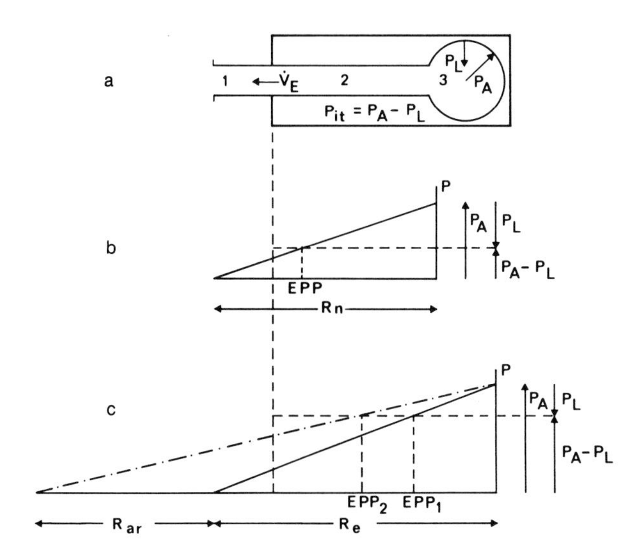

FIGURE 3 Pressure fall during expiration in the intra- and extrathoracic airways. (a) A schematic conceptual model of extra- (1) and intrathoracic (2) airways as a tube of constant resistance between alveoli (3) and mouth or nostrils. This implies a nonlinear projection of anatomical length of the airways on this tube. Parts with a relatively large resistance as the upper airways will take a longer part of the tube than peripheral bronchioli where the resistance is low (see Fig. 1). P_A is alveolar pressure during expiration; V'_E is expiratory flow; P_L is recoil pressure; P_{it} (intrathoracic pressure) = $P_A - P_L$; P_{it} surrounds the airways. (b) Pressure fall from alveoli to nostrils over a normal airway resistance (R_n). At the site where pressure in the airways is decreased to $P_A - P_L$, transmural pressure is zero because airway pressure is equal to P_{it}. This is the equal pressure point (EPP). Downstream (toward the left) the pressure outside the airways is higher than the intralumenal pressure, causing some compression of the airways. (c) Conceptual diagram for a patient with emphysema. Airway resistance is increased, therefore the abscissa is lengthened to the right (R_e) with respect to R_n. Extrathoracic airway resistance is constant. A higher P_A is needed for expiration. Due to loss of elasticity P_L is decreased. Therefore, the difference between P_{it} and P_A is smaller than in normal conditions. The point of zero transmural pressure (EPP_1) is shifted to the periphery with respect to normals. An additional external resistance (P_{ar}) decreases the pressure gradient, giving a shift of EPP_1 to proximal airways (EPP_2) and a smaller positive pressure on the airways downstream of EPP_2.

and b). This pressure gradient will have a value at a certain point in the airways equal to intrathoracic pressure: this is the equal pressure point. Downstream in the upper airways pressure in the lumen of the airways is even smaller than the surrounding intrathoracic pressure, resulting in a positive pressure difference from outside to inside. The bronchi will be easily compressed, resulting in a large airway obstruction.

In an advanced state of expiratory airway resistance many patients expire with pursed lips. Pursed lip breathing increases airway resistance even more (Fig. 3c). But the additional resistance is not within, but downstream from the intrapulmonary airways. As a consequence of the increased total expiratory resistance air flow is decreased and, therefore, the pressure gradient from the alveoli through the airways is decreased. This results in a shift of the equal pressure point to the upper airways, which are more rigid than the smaller bronchi. This shift diminishes the possibility of airway collapse. These emphysematous patients try to control a more gradual expiration.

E. Tests of Airway Obstruction

1. Spirometry

Airway resistance has been defined above as a variable calculated from the pressure difference between alveoli and the ambient air divided by the expiratory and inspiratory air flow, respectively. Alveolar pressure cannot be measured directly. Therefore, several indirect measures have been applied as substitutes for airway resistance. The first substitutes were based on the capability to breath out or to breath in a maximal volume of air in a limited time, e.g., a second. Such a volume is called a dynamic lung volume. FEV_1 is the forced expiratory volume which can be maximally expired in 1 sec after maximal inspiration. FIV_1 is the forced inspiratory volume, which can be maximally inspired in 1 sec after maximal expiration. MVV_{30} is the maximal voluntary volume which can be breathed in and out at a rate of 30/min. The rate is given by a metronome. These volumes can be obtained by means of spirometry, which is presented in Fig. 4 and explained in its legend. In Fig. 5 a recording of FEV_1, FIV_1, and MVV_{30} of a normal person is shown. These values are diminished when airway obstruction is present.

2. Pneumotachography

Another method is pneumotachography, which is the recording of the forced expiratory and inspiratory flow with use of a pneumotachometer (Fig. 6). The flow signal is not recorded as a function of time but plotted against the volume which is breathed out or in, respectively (Fig. 7). This volume is obtained from the integration of the flow signal. From this flow–volume curve several indices of obstruction are obtained.

Peak expiratory flow rate (PEFR) is the maximal

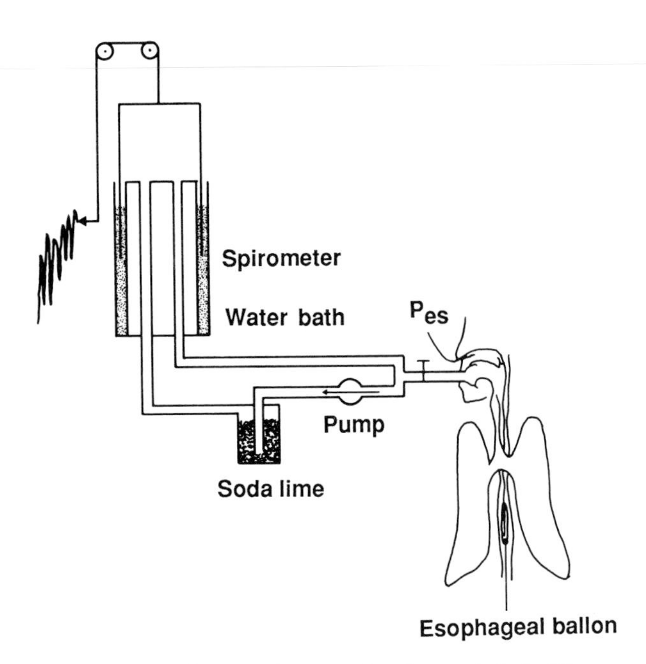

FIGURE 4 A spirometer is a container in a waterbath. The patient's lung volume changes are reflected in this container by volume displacement. CO_2 production is trapped in soda lime. O_2 from the spirometer is used. Therefore, the volume of the spirometer decreases gradually, as seen by the gradual increase of the level on the recording paper. With use of an esophageal balloon the change in pleural, i.e., intrathoracic pressure can be measured during normal breathing (P_{es}).

flow obtained shortly after the start of forced expiration. The maneuver is performed after maximal inspiration. This variable is often used in follow-up studies of a patient at home. The device is simple and is based on a measurement which indicates only the maximal flow expired. Several methods of measurement have been developed. This variable is not an accurate measure of airway obstruction because the value is very sensitive to the efforts of the patient. It needs maximal cooperation.

MEF_{75} is the maximal expiratory flow when 75% of the volume which will be expired during this maneuver is not yet expired. MEF_{50} and MEF_{25} are the corresponding flows when 50 and 25%, respectively, are not yet expired.

These indices have relatively large standard variations when compared in a group of normal volunteers to be used as a reference group for comparison with a patient's value.

The shape of the curve can also give us important information. In obstruction of the peripheral airways, i.e., the smaller bronchi and bronchioles, usually the down-slope of the flow–volume curve from PEFR to the end is concave.

II. Abnormal Changes in Volume

Changes in volume are mainly due to changes in alveolar volume. The main features of the alveoli will be considered first to better understand their functional disorders.

A. The Primary Lobule

Generation 16, the terminal bronchioles, gives entrance to the alveolar region of the lungs. Each tubule of generation 17 with all its branches is called a primary lobule. It also could be called a functional unit, because it represents the principal function of the lung, as the nephron does in the kidney. The first three generations, 17–19, of the primary lobule, are partly conducting tubules and partly gas-exchanging alveoli, because alveoli bulge from the walls of the tubules. In Fig. 8 a schematic drawing of generations 17–23 is given. Three generations, 20–22, of alveolar ductules have walls of alveoli. The last generation, 23, contains alveolar saccules. In Table I data are given about the number of alveoli in a primary lobule. The total number of alveoli in the lungs is $2277 \times 2^{17} \approx 300{,}000{,}000$. Other studies of the total number of alveoli estimated the number to be between 200×10^6 and 600×10^6.

The diameter of the alveoli is approximately 0.2 mm (200 mμ). However, the diameter is different for different lung regions. At the top of the lungs the diameter is larger and at the bottom it is smaller. At the bottom the lung tissue is less stretched than at the top, because the top is stretched by the weight of the lung. The specific gravity of the lung tissue is approximately 0.3. This vertical gradient from top to bottom changes in a vertical gradient from sternum to spine when the vertical position is changed for the supine position.

B. The Structure of the Alveoli

The alveoli contain different types of cells which have different functions.

1. Pneumocytes Type I

These simple squamous cells compose the thin epithelial layer of the alveoli. The diffusion membrane between alveolar air and blood is composed of this epithelium, its basement membrane, a thin layer of connective tissue, and the endothelial layer

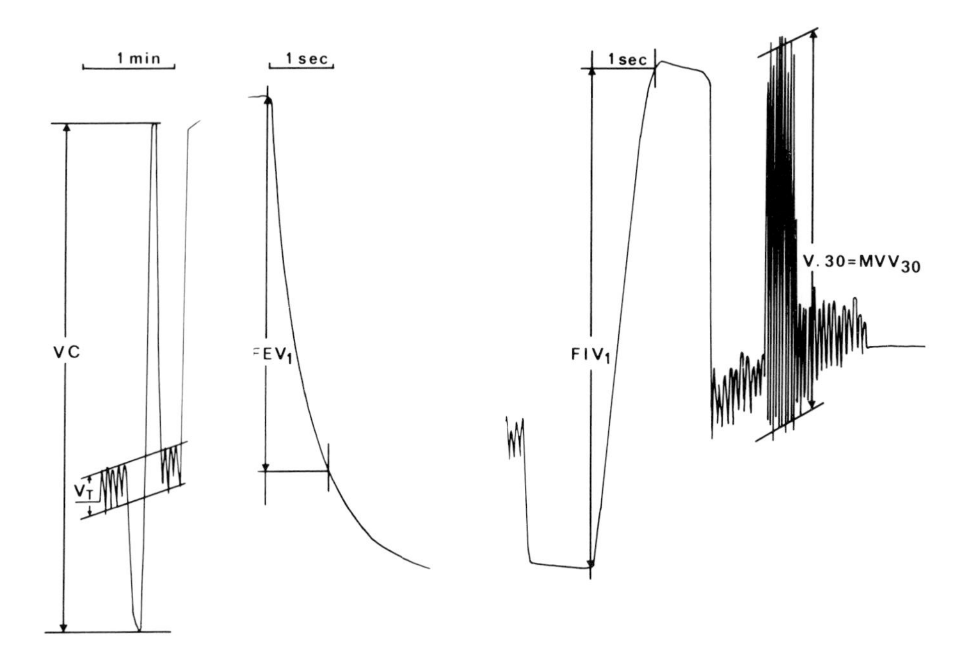

FIGURE 5 Spirometric recording. The procedures used to determine several static and dynamic volumes as defined in the text. *Note:* During the recording of the dynamic volumes the recording paper is transported faster.

plus its basement membrane of the blood capillaries. On the epithelium a thin fluid layer is present, containing at its surface substances which decrease the surface tension of the fluid–air interface.

2. Pneumocytes Type II

These secretory cells are connected to the squamous cells and are located at the surface of the alveolar tissue. They contain lamellar bodies in which the secretions are collected. These secretions contain the surfactants which decrease the surface tension. Phospholipids (68%), especially dipalmitoyl lecithin (47%), are the main constituents. Proteins (9%) and cholesterol (8%) are other components of the surfactant.

Surface tension is the result of the mutual attractive forces exerted by the molecules in fluid phase compared with the mutual attractive forces of the molecules in gas phase. A drop of water, without molecular attraction, would immediately fall apart into molecules. The attractive forces keep the molecules together in a drop and cause lateral tension at the surface: surface tension. When the surface of the fluid–air interface is curved, as in a water drop, the surface tension exerts a pressure on the fluid. In the negative image of the drop the water bubble pressure is exerted on the air in the bubble. This pressure (P) depends on the surface tension (γ) and the radius (r) of the bubble according to Laplace's law: $P = 2\gamma/r$. But when the pressure is kept at zero in the bubble the surface tension exerts a negative tension on the surrounding water.

In pure water $\gamma = 72$ mN/m. Lung surfactant decreases this value to $\gamma = 20$ mN/m. The alveoli with a diameter of 0.2 mm will exert a negative pressure on the surrounding tissue of $-P = -2.20$ mN $m^{-1}/0.0001$ m $= -400$ $Nm^{-2} = -0.4$ kPa ≈ -4 cm H_2O. This is a mean value based on the assumption that the alveoli are balloon shaped. Actually the alveoli are irregular polyhedra, with flat interalveolar membranes and strongly curved parts at the edges of the membranes, where two or more alveoli are connected. Therefore we must assume different values of negative pressure in the alveolar intersti-

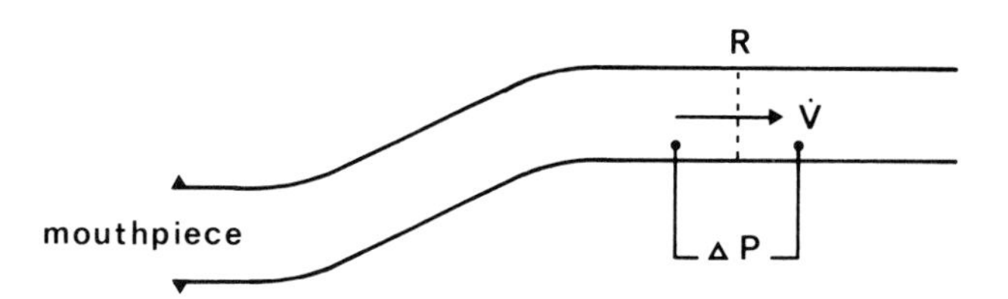

FIGURE 6 Pneumotachometer. This design gives only the principles of the pneumotachometer and is not a longitudinal section of an industrial type. In a tube a grid causes a slight flow resistance (R), which has a mathematical relationship to P and V', according to the aerodynamic law $P = RV'$. The tube near the mouth is bent to trap the sputum droplets.

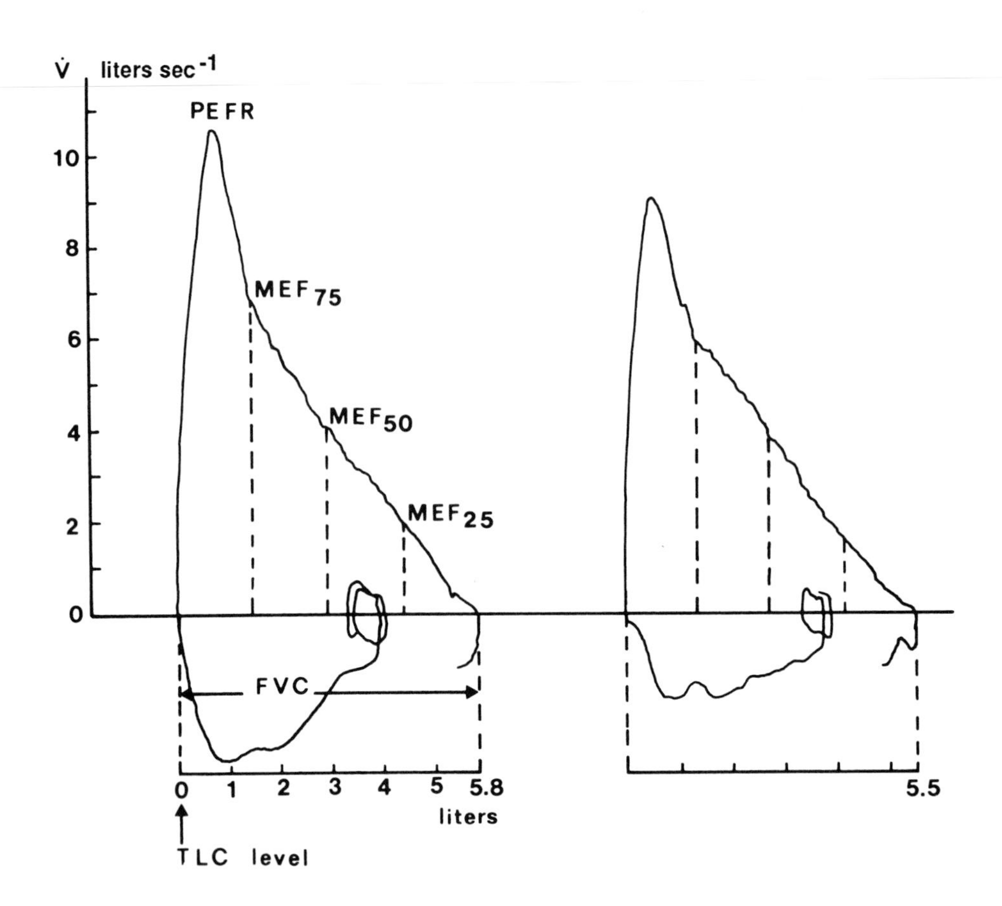

tium. Overall, the surface tension of the alveolar lining fluid contributes for about half the value to the negative pleural pressure.

FIGURE 7 Two flow–volume curves. These two curves represent flow velocity, plotted on the *y* axis, against expired volume, plotted on the *x* axis, during a forced and maximal expiration from maximal inspiratory level (TLC). In normal lungs this forced vital capacity (FVC) is equal to the inspiratory VC (Fig. 5). In airway obstruction FVC is usually smaller than inspiratory VC. These two curves were made consecutively. They demonstrate the much larger variation in PEFR than in MEF_{50} and MEF_{25}. These latter two values are almost independent of the amount of the voluntary muscle forces on the expiratory air. The reason is that a larger pressure inside the alveoli increases intrathoracic pressure with the same amount, causing a higher flow resistance which compensates for the higher alveolar expiratory driving pressure.

3. Alveolar Macrophages

These phagocytes originate from monocytes in the blood. They are present in the interstitium and in the alveolar lumen. Their phagocytic function depends on the destructive enzymes from the lysosomes. One of these enzymes is elastase. These enzymes are set free in the surrounding tissue when macrophages die. Then autodigestion of the alveolar septa is a real danger. Under normal conditions two mechanisms protect the lung against this threat: (1) the elastase is transported with mucus to the larynx to be coughed out, and (2) the elastase is inactivated by the formation of a conjugate with the plasma protein α_1-antitrypsin. In the blood the conjugate is changed to α_2-macroglobulin and next destroyed in the liver. Smoking has a negative effect on both protection mechanisms. It is generally accepted that a deficiency in α_1-antitrypsin is one of the conditions leading to the development of emphysema, although not a *sine qua non*. [*See* MACROPHAGES.]

C. The Interstitium

The interstitium is in between the basement membranes of the alveolar epithelium and the vascular endothelium. On one side of the alveolar septum it is a very thin layer. There the blood–air barrier is about 0.4 μm. At this side, the active side, the respiratory gases are exchanged. At the other side, the service side, the connective tissue layer is much

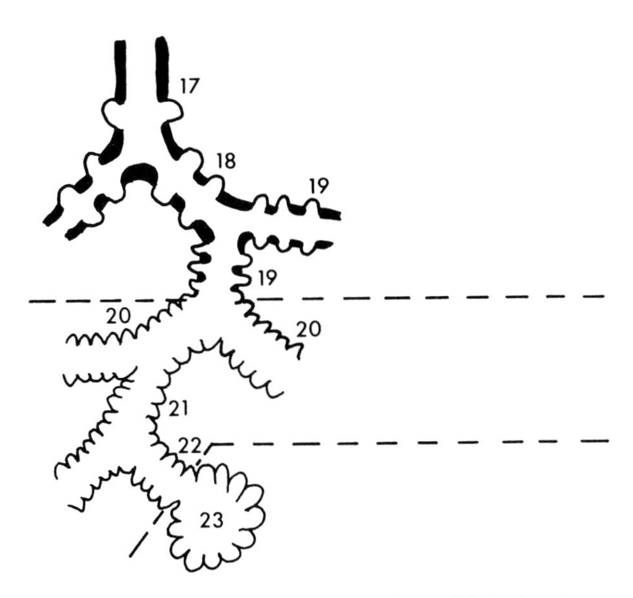

FIGURE 8 Primary lobule. A schematic model of the primary lobule and the distribution of alveoli. More detailed information is given in the text and Table I.

TABLE I The Number of Alveoli in a Primary Lobule[a]

Generation number	Number of alveoli per unit	Units per generation	Number of alveoli
17	5	1	5
18	8	2	16
19	12	4	48
20	20	8	160
21	20	16	320
22	20	32	640
23	17	64	1088
		Total number of alveoli:	2277

[a] Data according to Weibel, 1964.

thicker. This layer contains collagen, elastic and reticular fibers. The fibers provide the lungs with a fragile framework. Active and service sides alternate to both sides of the alveolar septa, and the fibers are more or less plaited with the blood vessels.

The framework can be subdivided into two parts: (1) a basic rete of thick elastic fibers, in a relatively small amount, which are the main constituents of the framework, and (2) a more delicate network of reticular and collagenic fibers, which supports the capillaries and the epithelium. The fibers are also connected to the muscular bronchioles.

In the elastic fibers recoil forces are present due to the stretch of the lung tissue. Together with the surface tension these recoil forces cause the negative pleural, and thus intrathoracic, pressure. When pulmonary volume increases, the surface area of the alveoli increases. As a consequence the concentration of the surfactant at the surface of the alveolar lining decreases, which increases the surface tension γ. Moreover, the elastic fibers are more stretched, which increases their recoil forces. Thus, during inspiration pleural and intrathoracic pressure will become more negative by both mechanisms.

The distinction between active and service side could make sense from a pathophysiological point of view. When pulmonary edema develops, which is an excess of extravascular fluid in the interstitium and secondarily in the alveoli, the fluid will first accumulate at the service side before entering between the basement membranes at the active side. Therefore, pulmonary edema does not coincide necessarily with a decrease in the diffusion of oxygen from alveoli to blood.

In patients suffering from pulmonary fibrosis, which is an excessive accumulation of connective tissue, some are characterized by a decreased diffusion capacity whereas others are not. We might conclude from such observations that different types of fibrosis will exist with respect to the distribution of the increased amount of connective tissue. When diffusion capacity is not changed the amount of connective tissue will not (yet?) be located at the active side.

D. Stability and Atelectasis

According to Laplace's law, $P = 2\gamma/r$, the negative pressure developed by a small alveolus on the surrounding tissue is larger than that by a large alveolus, when both have the same surface tension in their alveolar lining fluid. This could imply that in a normal lung, where alveoli with different diameters exist, the smaller alveoli empty into the larger alveoli. This does not occur because of two self-controlling mechanisms: (1) when the alveoli enlarge the elastic fibers are more stretched and a larger counter force is developed, and (2) the concentration of surfactant in the superficial layer decreases per unit of surface area and γ increases; when the alveoli become smaller the opposite happens.

When insufficient surfactant is available, as in premature infants, this control of stability fails more or less. Small alveoli empty into the larger ones and collapse, which is called atelectasis. The adult respiratory distress syndrome (ARDS) is among other features characterized by a deficiency of surfactant, caused by edema and destruction of the type II pneumocytes. Theoretically γ could increase maximally to 72 mN/m. For the averaged alveolus this should need a pressure of $(72 - 20) \times 2/0.0001 \approx 10$ cm H_2O. To keep open the smaller alveoli this pressure must be even higher. Moreover, the stability mechanism of the surfactant is lost and the smaller alveoli will empty into the larger ones, causing often excessive atelectasis.

E. Alveolar Volumes and Lung Volumes

Alveolar volume is estimated to obtain specific information of pulmonary pathology. Usually the volume of the airways, i.e., anatomical dead space of about 150 ml, is included in the value, giving lung volume. Lung volume is routinely determined at three specific levels.

1. Functional residual capacity (FRC) is the volume of alveoli and airways after a normal expiration (Fig. 9).
2. Total lung capacity (TLC) is the volume in the lungs and airways after maximal inspiration.
3. Residual volume (RV) is the lung volume after maximal expiration.

The difference between TLC and RV is the vital capacity (VC), which is the maximal volume a person can inspire after maximal expiration. A normal person can expire as much as he/she can inspire in one maneuver. Emphysema patients cannot easily expire VC after maximal inspiration. They mostly interrupt expiration by an inspiration before the level of RV is attained. Therefore, routinely inspiratory VC (Figure 5) must be determined.

The FRC is determined by use of the spirometer. First an amount, e.g., z ml, of pure helium is injected in the spirometer. This causes a helium concentration of $y\%$ after equilibration, giving the volume of the spirometer and additional tubes according to $V_1 = z \times 100/y$. The patient is breathing ambient air through a three-way stopcock of the spirometer by way of a mouthpiece and a clamp on the nose. The patient is connected to the spirometer at the end of a normal expiration, when the patient's lung volume is FRC, by turning the three-way stopcock. The patient must breathe quietly for a few minutes to distribute the helium concentration equally in the spirometer and his lungs. It is not

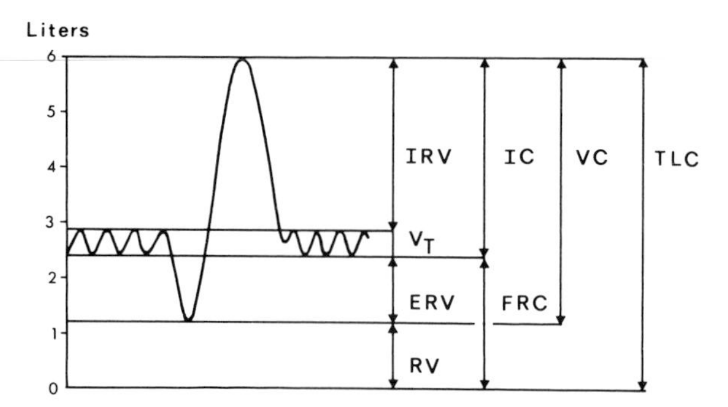

FIGURE 9 An idealized diagram giving all static volumes. The RV is determined by the helium-dilution method. Inspiratory reserve volume (IRV) is the maximal volume to be inspired after a normal inspiration. Expiratory reserve volume (ERV) is the maximal volume that can be expired after a normal expiration, and inspiratory capacity, or IC, is that volume which can be maximally inspired from the normal expiratory lung volume level, FRC. All other symbols are given in text.

necessary to maintain V_1 in the spirometer during this test. Thus, when oxygen consumption decreases the volume and carbon dioxide production does not compensate for it (because it is trapped by the soda lime), the spirometer volume will be decreased. The recording will indicate accurately the volume decrease. When the helium is distributed the test is stopped again after a normal expiration. Then the total volume of spirometer and lungs is V_2 + FRC. The helium concentration is $x\%$ now, thus the following mass balance can be formulated: $V_1 y = (V_2 + \text{FRC})x$, from which FRC can be solved easily.

When FRC is known RV can be found easily by maximal expiration from the FRC level. Total lung capacity (TLC) is found when VC is added to RV.

The volumes TLC, FRC, RV, and VC are called static volumes, because the patient can take his time for their determination, in contrast to the dynamic volumes FEV_1, FIV_1, and the flow–volume indices.

Another volume indicated in Figure 5 but not yet mentioned is V_T, tidal volume, which is the normal volume breathed in and out with each ventilation. Expiratory volume is smaller than inspiratory volume, because oxygen uptake from the alveolar air into the blood is larger than carbon dioxide output from the blood into the alveolar air. An accurate indication of the differences is given by the higher N_2 concentration in the expiratory air with respect to the inspiratory air, whereas no N_2 exchange between blood and alveolar air will occur.

F. Pathophysiologic Aspects of Lung Volume Changes

1. Obstruction

Lung volume changes in patients are usually due to changes in alveolar volume. They can be typical for certain diseases. In obstructive diseases such as emphysema, where extensive expiratory obstruction coincides with hardly any increase in inspiratory obstruction, patients accumulate alveolar volume and breathe at a much larger lung volume. The FRC is increased, which is called hyperinflation. The TLC is also increased. Due to the decreased elasticity the lung recoil forces are diminished and inspiration can reach a higher level. The RV is increased because of a collapse of the flaccid bronchi during maximal expiration. Alveolar air remains trapped behind these collapsed bronchi. The increase in RV is usually larger than the increase in TLC, giving a decrease of VC. A decrease in VC will also decrease FEV_1. However, in such an obstructive disease FEV_1 is much more decreased than VC, which is indicated with the ratio FEV_1/VC as an additional but important indication for obstruction (see Section III,A,2).

These volume changes are characteristic for an airway obstruction due to a loss in elasticity. However, they can also be seen in chronic obstructive diseases by swellings of mucosa and mucus accumulations in the bronchi and bronchioles. Thus, the pattern of volume changes is characteristic but not specific for emphysema.

It happens regularly that patients with an extensive obstruction do not show any increase in these static volumes. This can be due to the fact that part of the alveolar regions is trapped behind closed bronchi during normal breathing when FRC is determined by the helium dilution method as described above. These closed regions will not be involved in the dilution of helium. Thus, the FRC value is observed too small, as are the RV and TLC values.

Using body plethysmography, a description of which is beyond the scope of this article, total thoracic gas volume can be determined, including the trapped air. An additional test with this device will reveal the differences when trapped air is present.

In asthma, when an expiratory obstruction coincides with an increased inspiratory obstruction, the static volumes are not necessarily increased.

2. Restriction

Restriction of lung volume means a decrease in alveolar volume. Pneumonectomy and lobectomy, i.e., resection of one lung and one of the lung lobes, respectively, are obvious reasons. Total lung capacity and VC will be decreased, and therefore also FEV_1, but FEV_1/VC will not be decreased. Thus, the decrease in FEV_1 is not indicative of an obstruction; only when FEV_1/VC is decreased do we suspect the patient to have an obstructive disorder.

Restriction is found in a variety of interstitial diseases. These disorders cause restriction via two mechanisms: (1) a replacement of air volume by tissue expansion, and (2) a decrease in alveolar volume by increased elasticity (stiffness) of the lung tissue.

In both types of disorder TLC, RV, FRC, and VC will be decreased. FEV_1 will also be smaller, but FEV_1/VC is usually normal, or even higher than normal. When the lung tissue is more elastic, i.e., when it develops larger recoil forces than normal at the same lung volume level, negative intrathoracic pressure is more negative and the bronchioles are more stretched by the radial fibers running into the alveolar interstitium. Then, the airway resistance is decreased and expiration is facilitated. Although FEV_1 is decreased due to the decrease in VC, it is increased with respect to VC and thus FEV_1/VC is larger than normal. An increase in this ratio suggests increased elasticity or stiffness of the pulmonary tissue. By itself it is not a measure for it.

G. Compliance

As a measure for lung elasticity compliance or, even better, specific compliance is determined. Compliance is the reverse of elasticity: it indicates how much the lungs yield when a certain pressure is established. During normal breathing pressure in the lungs is zero at end-inspiration and at end-expiration. Only during these actions is alveolar pressure slightly negative and positive, respectively. The established pressure is the pleural pressure outside the lungs, which is negative at end-expiration and becomes more negative at end-inspiration. These changes in negative pleural pressure can be measured in the esophagus by use of an air-filled balloon about 0.5–1 cm in diameter and about 10 cm long, fixed on a sufficiently rigid but flexible tube.

Lung volume changes are measured by use of the spirometer. The changes in lung volume and intrathoracic pressure are plotted in an x–y diagram (Fig. 10). Then, compliance $C = V_T/(P_{it,e} - P_{it,i})$, where $P_{it,e}$ and $P_{it,i}$ are expiratory and inspiratory pressures, respectively. C is the slope of the straight line in Figure 10. For one-half the lung volume the same change in intrathoracic pressure (P) is caused by only $\frac{1}{2}V_T$. Then a compliance value will be found that is one-half the value of the total lung, whereas no change in elasticity occurred. Thus, compliance depends on lung volume, which is not a good indication of the elasticity of its tissue. Therefore, compliance is expressed per liter of lung volume, giving specific compliance (C_{sp}). Specific compliance is calculated according to $C_{sp} = V_T P(\text{FRC}+\frac{1}{2}V_T)$.

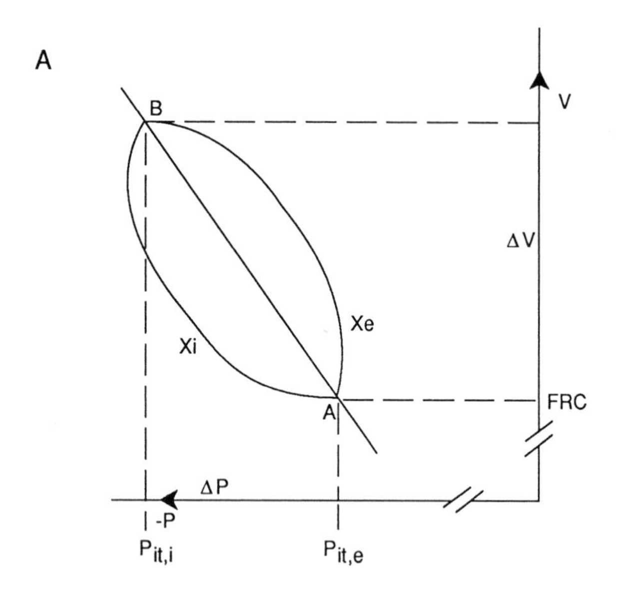

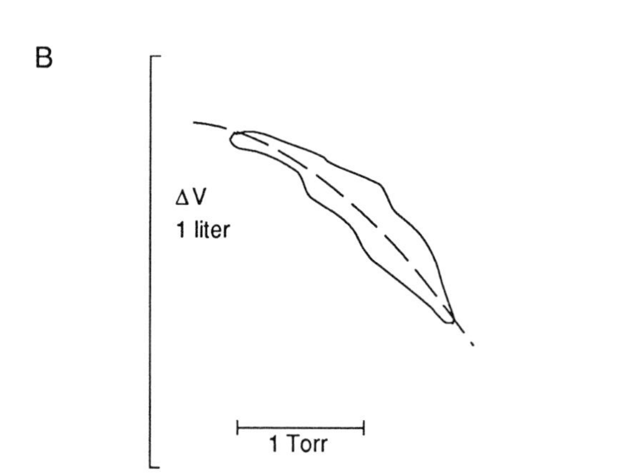

FIGURE 10 Volume–pressure diagram. (A) Lung volume (V) is plotted on the x axis against esophageal pressure on the y axis ($-P$). $P_{it,i}$ and $P_{it,e}$ are the esophageal pressures as a substitute for intrathoracic pressure at inspiratory and expiratory volume levels, respectively. Starting inspiration at point A an extra negative intrathoracic pressure is developed, following curve X_i due to the negative alveolar pressure. The reverse occurs during expiration, following curve X_e starting at point B. The slope of the straight line from point A to point B is the value of compliance. (B) A curve of a patient suffering from fibrosis. The slope is smaller, indicating a lower compliance. Moreover, the slope is not straight, because the curve has a typical banana shape. The waves in the curve are due to cardiac oscillations.

C and C_{sp} are decreased in restrictive diseases dependent on an increased elasticity of the pulmonary tissue. Because TLC and VC are decreased, too, C is additionally decreased by the loss of lung volume.

III. Disorders in Gas Transfer

A. Ventilation–Perfusion Ratio

A schematic model of ventilation and perfusion is given in Fig. 11 for the total lung and pulmonary circulation. V_T is tidal volume, which is the volume breathed in and out. Because inspiratory volume (V_I) is larger than expiratory volume (V_E), due to a higher oxygen uptake from the alveoli into the blood than carbon dioxide output from the blood into the alveoli, V_T is replaced by the two mentioned volumes: $V'_E = fV_E$ and $V'_I = fV_I$, where f is the ventilatory rate per minute. V_D is airway dead space, i.e., anatomical dead space. During each breath a part of the tidal volume is left behind in V_D, which is (per minute) $V'_D = fV_D$. Thus, alveolar ventilation V'_A is either equal to $V'_I - V'_D$ for inspiration or equal to $V'_E - V'_D$ for expiration.

Normal ventilation is about 7 liters/min. When ventilatory rate is 14 per min and the volume of the airways is about 150 ml, $V'_D \approx 2$ liter per min. Thus, total alveolar volume exchange is 5 liters/min of fresh air and 2 liters/min of alveolar air, returning from the airways. V'_A is used only for ventilation with fresh air.

Normal alveolar perfusion, i.e., cardiac output, is about 6 liters/minute. Thus, the overall ratio between alveolar ventilation (with fresh air) and perfusion (V'/Q' ratio) is $5/6 = 0.8$. Normal values are usually given between 0.8 and 1.

B. Disorders in CO_2- and O_2-Exchange

Perfusion delivers CO_2 from the tissues into the alveoli and takes up O_2 from the alveoli into the blood. Ventilation transports CO_2 from the alveoli

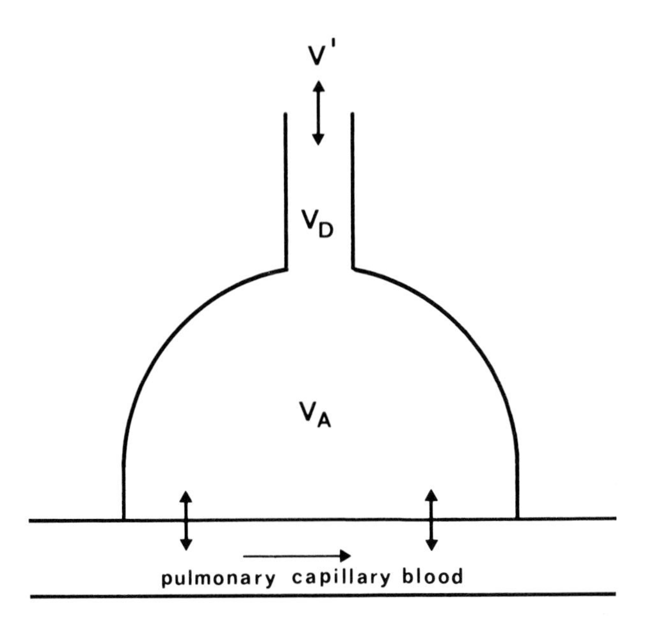

FIGURE 11 Three-compartment model of airways, alveoli, and pulmonary capillaries. V_D is the volume of the airways (anatomical dead space), V_A is the volume of the alveoli. V, lung ventilation.

to the ambient air and refreshes the alveoli with ambient air to maintain alveolar O_2 concentration. Therefore, alveolar CO_2 and O_2 concentrations depend on both ventilation and perfusion of the alveoli.

1. Carbon Dioxide

CO_2 output (V'_{CO_2}) in both lungs is equal to the product of alveolar ventilation and CO_2 concentration in the alveolar air, according to

$$V'_{CO_2} = (V'_E - V'_D)F_{A,CO_2} - (V'_I - V'_D)F_{I,CO_2} \quad (1)$$

where F_{A,CO_2} and F_{I,CO_2} are the CO_2 fractions of alveolar air and inspiratory air, respectively. For accurate calculations all volumes must be corrected to standard temperature (273 K) and pressure (760 Torr), which we have neglected here for reasons of simplicity. F_{I,CO_2} is very low and can be neglected for the same reason. Then, Eq. (1) can be rewritten as

$$V'_{CO_2} = V'_A F_{A,CO_2} \quad (2)$$

$$F_{A,CO_2} = P_{A,CO_2}/(P_B - P_{s,H_2O}) \quad (3)$$

where P_B is ambient air pressure, and P_{s,H_2O} is the saturated water vapor tension in the alveolar air. Thus, the fraction of CO_2 is part of the dry alveolar air.

From Eqs. (2) and (3) follows

$$V'_{CO_2} = V'_A P_{A,CO_2}/(P_B - P_{s,H_2O}) \quad (4)$$

When perfusion decreases to an alveolar area, where ventilation is normal, total CO_2 output (V'_{CO_2}) from the blood will be decreased, but CO_2 output per milliliter of blood will be increased. As a consequence F_{A,CO_2} will be decreased. Thus, in circumstances of high V'/Q' ratios a decrease in F_{A,CO_2} and thus in P_{A,CO_2} will occur. When P_{A,CO_2} decreases arterial P_{CO_2} (P_{a,CO_2}) will also decrease.

Under normal V'/Q' conditions P_{a,CO_2} is about 40 Torr. When CO_2 production in the tissues is also normal CO_2 output into the lungs and from the lungs into the ambient air will be normal, otherwise an accumulation or a depletion of CO_2 should occur.

A depletion will occur when V'_A increases. As a consequence P_{A,CO_2} and thus P_{a,CO_2} will decrease until a new equilibrium in Eq. (4) is established. When P_{a,CO_2} falls below 30 Torr the condition is called hyperventilation.

An accumulation occurs when V'_A decreases. P_{a,CO_2} will rise until the multiplication between P_{A,CO_2} and V'_A in Eq. (4) is equal to V'_{CO_2}. When P_{a,CO_2} rises above 49 Torr the condition is called hypoventilation.

In lung diseases accompanied with V'/Q' disorders regions with a variety of V'/Q' ratios will exist, from high V'/Q' ratios to low V'/Q' ratios. Because breathing is controlled, among other factors, by P_{a,CO_2}, overall ventilation will be controlled as long as possible to maintain the normal P_{a,CO_2}. Then the effects on P_{a,CO_2} by the regions with low V'/Q' ratios will be compensated by the effects on P_{a,CO_2} in the regions with the high V'/Q' ratios.

When the production of CO_2 is increased during exercise and the ventilation is also increased to maintain normal P_{a,CO_2} at 40 Torr, we call this type of ventilation hyperpnea.

When considering ventilation at high altitude Eq. (4) can be easily misinterpreted. At high altitude P_B is lowered, thus we could be tempted to conclude that P_{A,CO_2} will be decreased to maintain the same V'_{CO_2}. However, that is a hasty conclusion. At high altitude the pressure in the ventilatory volume is decreased also. However, according to Eq. (3) F_{A,CO_2} will be increased at the same ratio as $P_B - P_{s,H_2O}$ is decreased, when P_{A,CO_2} remains the same. This latter condition will be true when V'_A and V'_{CO_2} remain the same [see Eq. (2)]. When a hypoxic drive of the ventilation is not yet present, there is no reason to assume a change in V'_A. [*See* RESPIRA-

TORY SYSTEM, PHYSIOLOGY AND BIOCHEMISTRY.]

2. Oxygen

Arterial oxygen tension (P_{a,O_2}) is approximately equal to alveolar oxygen tension (P_{A,O_2}) under normal circumstances. The difference between both variables ($D_{(A-a)O_2}$) is a few Torr. Nevertheless, in normal conditions our considerations of alveolar P_{O_2} count also for aterial P_{O_2}.

Oxygen consumption V'_{O_2} is equal to oxygen entering the lungs during inspiration $[f(V_I - V_D)F_{I,O_2}]$ minus oxygen leaving the lungs during expiration $[f(V_E - V_D)F_{E,O_2}]$.

$$V'_{O_2} = f(V_I - V_D)F_{I,O_2} - f(V_E - V_D)F_{A,O_2} \quad (5)$$

Although we are aware of the fact that V_I is slightly larger than V_E, they are assumed equal for the benefit of showing the effects on alveolar and arterial P_{O_2}. Then these equations follow:

$$V'_{O_2} = V'_A(F_{I,O_2} - F_{A,O_2}) \quad (6)$$

$$F_{A,O_2} = F_{I,O_2} - V'_{O_2}/V'_A \quad (7)$$

$$P_{A,O_2} = F_{A,O_2}(P_B - P_{s,H_2O}) \quad (8)$$

From Eqs. (7) and (8) come the following:

$$P_{A,O_2} = F_{I,O_2}(P_B - P_{s,H_2O}) - (V'_{O_2}/V'_A)(P_B - P_{s,H_2O}) \quad (9)$$

$$P_{A,O_2} = [F_{I,O_2} - (V'_{O_2}/V'_A)]\,(P_B - P_{s,H_2O}) \quad (10)$$

When oxygen consumption is normal the reasons for a fall in P_{A,O_2}, and thus in P_{a,O_2}, which is called hypoxemia, can be easily deduced from Eq. (9): (1) a decrease in F_{I,O_2} by oxygen consumption in an enclosed room, e.g., a bathroom with a gas water heater, (2) a decrease in P_B at high altitude in the mountains, and (3) a decrease in V'_A.

When we regard F_{I,O_2} in the term $F_{I,O_2}(P_B - P_{s,H_2O})$ of Eq. (9) as the fraction of the air coming into the alveoli after having passed the airways we can rewrite Eq. (9):

$$P_{A,O_2} = P_{I,O_2} - (V'_{O_2}/V'_A)(P_B - P_{s,H_2O})$$

Three other reasons leading to arterial hypoxemia will be considered.

(4) Impaired diffusion: When the active side of the alveolar membrane between air and blood is thickened, e.g., as in fibrosis, diffusion is inhibited and $D_{(A-a)O_2}$ can increase. This does not usually occur in resting conditions, but as soon as some work is done the equilibrium between alveolar air and capillary blood is no longer attained and arterial P_{O_2} is decreased. Therefore, an exercise test could be useful to detect diffusion disorders. However, a direct measurement of diffusion with carbon monoxide as indicator gas is more specific.

(5) Right-to-left shunt: When blood passes atelectatic regions it will not be oxygenated. In this part of the blood P_{O_2} and oxygen saturation of hemoglobin (S_{O_2}) will remain at the level of the P_{O_2} and S_{O_2} of the mixed venous blood. This blood mixes with normal arterialized blood, causing a lower S_{a,O_2} and also P_{a,O_2}. The phenomenon of no saturation of a part of the blood during lung passage is called right-to-left shunting. Such a shunt occurs in the nonventilated regions. The phenomenon of admixture of mixed venous blood with normal oxygenated blood is called venous admixture. A right-to-left shunt also causes an arterial hypoxemia. The amount of shunt or venous admixture can be calculated from the amount of oxygen transported in the shunt blood, which is $Q'_{sh}C_{v,O_2}$, and the amount transported in the ideal saturated blood, which is $(Q'_p - Q'_{sh})C_{pc,O_2}$, where Q'_p is the total lung perfusion (= cardiac output) and C_{pc,O_2} the oxygen content, in the pulmonary capillary blood entering the veins. The sum of the oxygen transport in the shunt blood and that in the blood effectively perfusing the lungs is equal to the amount transported in the arterial blood, which is Q'_pC_{a,O_2}. Thus

$$Q'_{sh}C_{v,O_2} + (Q'_p - Q'_{sh})C_{pc,O_2} = Q'_pC_{a,O_2}$$

which can be resolved into

$$Q'_{sh}/Q'_p = (C_{pc,O_2} - C_{a,O_2})/(C_{pc,O_2} - C_{v,O_2})$$

Q'_{sh}/Q'_p is called the shunt fraction of total cardiac output. To determine the shunt (in ml/min) cardiac output also must be determined.

(6) V'/Q' disorders: When pulmonary disorders are characterized by disorders in the V'/Q' ratios of the different regions hypoxemia will occur also. The regions with high V'/Q' will be characterized by a high ventilation rate with respect to perfusion, i.e., oxygen uptake (local V'_{O_2}). In the blood perfusing this region P_{A,O_2} and P_{a,O_2} will increase according to Eq. (10). However, because of the almost full saturation of the hemoglobin in this blood hardly any extra oxygen will be taken up.

The regions with low V'/Q' will cause a decrease of P_{A,O_2} and P_{a,O_2} [see Eq. (10)]. When P_{a,O_2} decreases S_{a,O_2} also decreases. The arterial blood is a mixture of blood with a normal oxygen content and blood with a decreased oxygen content. It will be characterized by hypoxemia.

V'/Q' disorders and a shunt have the same effect on arterial carbon dioxide content. When the regions with low V'/Q' or no V' at all in the shunting are compensated by some extra ventilation in the normal regions CO_2 output will be normal and P_{a,CO_2} will be normal (see Section III,B,1).

A differentiation between the low V'/Q' and a shunt is possible by application of oxygen, i.e., by an increase of F_{I,O_2}. An increase in F_{I,O_2} in the regions with low V'/Q' will cause an increase in P_{A,O_2} according to Eq. (10). We assume V'/Q' is half its normal value, which implies a relatively large uptake of oxygen (V'_{O_2}) with respect to V'_A [see second term at the right side of Eq. (10)]. P_{A,O_2} will be decreased. When F_{I,O_2} is increased P_{A,O_2} in that region becomes normal, and P_{a,O_2} in the blood leaving that region will be normal again.

In case of a shunt an increase in F_{I,O_2} is not effective, because when blood passes an atelectatic region without any ventilation no exchange with oxygen is possible.

3. Pulmonary Disorders and Gas Exchange

In obstructive disorders the amount of obstruction will usually not be distributed homogeneously in both lungs. Consequently a disturbance in the distribution of the inspired air will occur. The alveolar regions, which are drained by bronchi with a low flow resistance, will be emptied easier and earlier during expiration than the regions with a high flow resistance. The same is assumed to occur during inspiration. The alveolar regions of bronchi with a high flow resistance will be ventilated less than the other regions. There will be a high P_{A,CO_2} in the regions with low V'/Q' and a low P_{A,CO_2} in the regions with high V'/Q'. When we register the expiratory CO_2 concentration an indication of the existence of V'/Q' disorders can be obtained. In Fig. 12a a normal curve of the expiratory CO_2 concentration is given. Phase I is the phase without CO_2, when fresh air from the airways passes the CO_2 cuvette. Phase II is a transient phase when air passes coming partly from the alveoli and partly from the most distant airways. It is a mixture of fresh air and alveolar air in which the contribution of the first one decreases and that of the second increases. This transient phase also depends on the gradient in the CO_2 concentration from alveolar concentration to fresh air concentration in the region of the respiratory bronchioles. Phase III is the alveolar plateau, which increases slightly. In emphysema (Fig. 12b) phase II and phase III are one continuous curve, often ending at a CO_2 concentration which has a corresponding partial pressure above 49 Torr. This indicates regions with hypoventilation, or low V'/Q' ratios.

In adult respiratory distress syndrome (ARDS) V'/Q' disorders and extensive shunting due to edema causing atelectic regions are often present. Oxygen therapy does not help satisfactorily. The best therapy is to reopen alveoli. This needs mechanical ventilation with an increased (positive) end-expiratory pressure and superimposed insufflations.

In fibrosis the active side of the alveolar membrane can be thickened, which will impair gas exchange. The measurement of O_2 transfer is not possible in clinical routine. Diffusion depends linearly on the difference in gas tension between both sides of the membrane. The determination of the difference in oxygen tension between the alveolar air and the pulmonary blood not only needs invasive techniques, such as catheterization of the pulmonary artery, it also demands complicated calculations, because the oxygen tension in the capillary blood increases nonlinearly during passage through the capillaries, and the oxygen tension in the alveoli changes continuously during breathing. Therefore, carbon monoxide (CO) is used. The diffusion capacity (D_{CO}) is defined as the amount of CO that passes the diffusion membrane per second and per CO pressure difference in Torr.

A well known volume of air, approximately equal to VC, containing a low CO and helium concentration is inspired after maximal expiration. The dilution of the CO fraction in the alveolar air is calculated from the dilution of the helium, assuming that no helium disappeared from the alveolar air between the moment of inspiration and that of expiration, giving the CO fraction in the alveolar air immediately after inspiration. The air is kept in the lungs for about 10 sec. This time of breath holding is accurately recorded. Then the air is breathed out fast and CO and helium concentrations are measured. The difference in CO concentration between the start and the end of the breath-holding period gives (1) the amount which diffused into the blood, and (2) the two principle values of the monoexponential decay during that period. D_{CO} can be calculated from these data. D_{CO} depends on the surface area and thickness of the alveolar–capillary membrane, and the amount of blood, specifically hemoglobin,

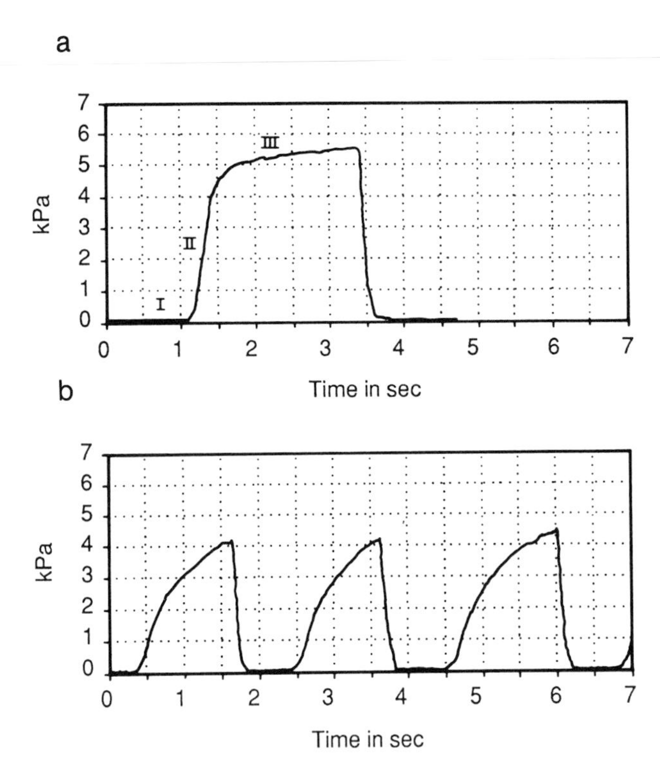

FIGURE 12 A normal and a pathologic capnogram. (a) A capnogram is the expiratory CO_2 concentration (%) or CO_2 tension (kPa as in this figure, 1 kPa = 7.5 mmHg) plotted against time (sec as in this figure). A normal capnogram is composed of three phases (see text). (b) Phases II and III coincide in the curve of a patient suffering from emphysema.

in the pulmonary capillaries able to take up the CO. In large lungs the diffusion area is larger than in small lungs, coinciding with a larger D_{CO}. To avoid the effect of volume on D_{CO} total D_{CO} is corrected for alveolar volume and expressed per liter lung volume from D_{CO}/TLC.

In fibrosis usually a restriction of alveolar volume is present. Diffusion capacity (D_{CO}) will be impaired not only by a thicker air–blood barrier but also by the decrease in alveolar volume (i.e., the decrease in alveolar surface area). Therefore D_{CO}/TLC is determined. A decrease in this value is not only caused by a thick membrane at the active side, it also results from a decrease in alveolar capillary blood. It is not possible to discriminate between both factors.

IV. Pulmonary Circulation

A. Pressure Fall

The pressure gradient in the pulmonary circulation from about 12 mmHg at the arterial side to about 2 mmHg at the venous side is not only smaller than that in the systemic circulation (100–0 mmHg), it is also different in percentages of the total pressure fall in corresponding vessels compared with the systemic circulation. In the systemic circulation about 60% of the pressure fall is located in the arterioles, and 10% in the capillaries. In the pulmonary circulation these figures are approximately 10 and 50%, respectively. Although about half of the pressure gradient is lost in the pulmonary capillaries, the total fall from 10 to 5 mmHg is smaller than that in the capillaries of the systemic circulation, from 30 to 20 mmHg. These figures indicate that the level of pressure in the lung capillaries is also lower than that in the capillaries of the systemic circulation.

We will see in Section IV,D that low pressure in the pulmonary circulation is important to avoid filtration of fluid from the blood vessels into the pulmonary interstitium and alveoli.

B. Pulmonary Vascular Flow Resistance

The pulmonary circulation is more complicated with respect to pressure fall and perfusion than might be concluded from the general data presented above. The pulmonary vessels, especially the capillaries, are collapsible tubes. When the pressure at the venous side of the capillaries is about 5 mmHg (≈6.8 cm H_2O), measured in the horizontal plane at the level of the entrance of the pulmonary artery into the lungs, this capillary pressure will be negative at a height more than 6.8 cm above the pulmonary artery, implying a collapse of the capillaries at the venous side by the alveolar pressure. Analogies of this type of perfusion are the Starling resistor and the waterfall concept, which are illustrated in Fig. 13a and b.

As in aerodynamics (Section I,D,2), flow resistance in the circulation obeys the Poiseuille resistance when flow is not turbulent, blood viscosity is constant, and the vessels are continuously open. Then the hemodynamic law $\Delta P = Q'R$, where ΔP is the pressure fall from pulmonary artery to left atrium, can be applied. However, when Starling resistors exist in part of the branched circuit of the pulmonary circulation this hemodynamic equation, equivalent to Ohm's law, cannot be applied, because in those parts the capillaries are not continuously open. As is illustrated by the waterfall model (Fig. 13b), in these vessels no continuous pressure gradient exists, and a sudden pressure drop occurs after the rim (in the Starling resistor after the collapse in the capillaries). Then the downstream pres-

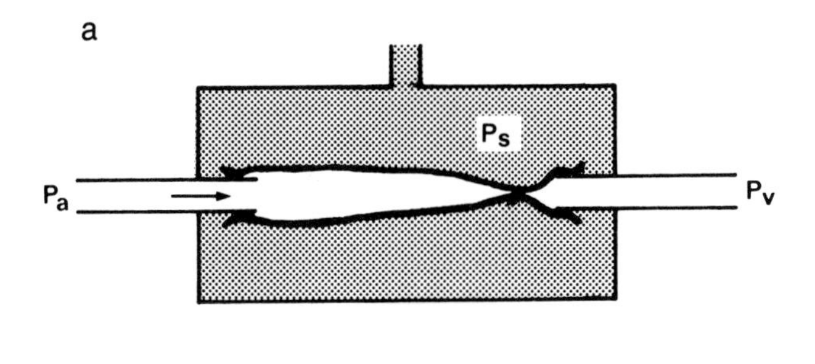

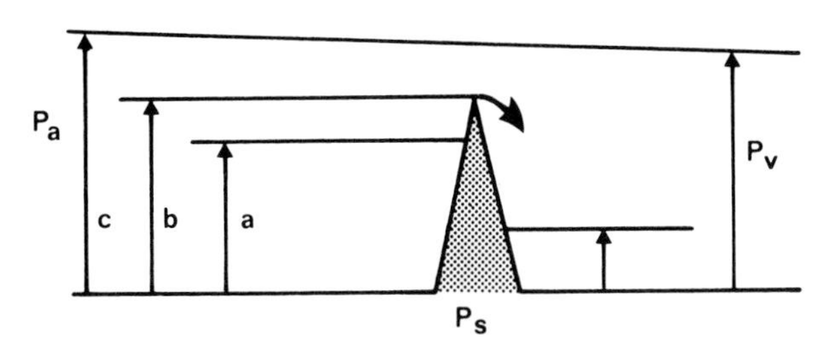

FIGURE 13 Starling resistor and waterfall model. (a) A *Starling resistor* composed of a latex tube in a closed chamber. When a flow from the left meets a closed Starling resistor due to P_s and flow is kept constant, pressure before the resistor will increase until it exceeds P_s. Then the "valve" will open and fluid passes the resistor. The pressure before the resistor will be maintained at the level of P_s independent of the amount of flow and independent of the pressure behind the resistor. (b) The *waterfall concept* represents the same mechanism. When the water before the rim of the fall is lower (situation a) than the height of the rim, no water will pass the fall. When water is continuously supplied before the rim, the water level will rise until it reaches the rim and starts to fall down, maintaining the level before the rim at the same height (situation b) independent of the amount of flow. The level downstream of the rim does not affect the flow over the rim. Only when the downstream level is higher than the rim (situation c) will we see a normal streaming river. The same is true for the Starling resistor; when the downstream pressure exceeds P_s the latex tube remains fully open and a continuous flow dependent on the upstream-to-downstream pressure difference will be established.

sure has no effect and flow is not governed by the arterial-to-venous pressure difference. Under these circumstances flow depends on the arterial-to-capillary (i.e., alveolar) pressure difference. As a consequence calculations of the pulmonary blood flow resistance (R_p), based on the hemodynamic law mentioned above, must lead to unreliable data. This certainly occurs when alveolar pressure is increased, as during mechanical ventilation.

C. Control of Pulmonary Perfusion

1. Sympathetic Control

When the activity of the sympathetic part of autonomic nervous system is increased the small muscular arteries are dilated. This dilation increases pulmonary flow for the same driving pressure. A corresponding effect is produced by the sympathomimetic hormones. During exercise sympathetic activity is increased, which supports a large flow through the lungs without an increase in pulmonary artery pressure. This effect is beneficial for the right ventricle, when it delivers a large cardiac output during exercise.

2. Alveolar O_2 Concentration

Another important stimulus on the small muscular arteries in the lungs is the alveolar O_2 concentration. This stimulus is important for the maintenance of normal V'/Q' ratios in the different lung regions.

When in an alveolar region the O_2 concentration decreases, the muscular arteries which supply this region with blood will constrict, causing a smaller flow through this region. This is a useful reaction to avoid an insufficient oxygen transfer to this part of the blood. When for some reason, e.g., a clot of mucus in a bronchus, ventilation of the corresponding alveolar area is decreased, the O_2 concentration in this area will decrease, as explained in Section III,B [Eq. (10)]. Then perfusion of the region is adapted by the hypoxic vasoconstriction and the V'/Q' ratio is restored. At the end of Section I, we considered the adaptation of ventilation to alveolar CO_2 concentration. Also this stimulus is a control mechanism to maintain normal V'/Q' ratios in the different lung regions. To maintain normal V'/Q' ratios it is much more efficient to control V' as well as Q' than to adapt only one to the other.

Another important function of the hypoxic vasoconstriction is its effect on the pulmonary circulation before birth. A high flow resistance in the pulmonary arteries allows only a small blood flow through the lungs. Therefore, the majority of the blood from the inferior vena cava is shunted to the left side of the heart through the foramen ovale. Thus, the oxygenated blood coming from the placenta is transmitted primarily to the aortic arch, supplying the brain and heart.

The hypoxic vasoconstriction does not only have beneficial functions. Under pathologic conditions, coinciding with atelectasis and low V'/Q' ratios, as in ARDS and emphysema, the flow resistance in the pulmonary muscular arteries can be increased to such an extent that a considerable increase in pulmonary arterial pressure results. This leads to a hypertrophy of the right ventricle, *cor pulmonale*.

At high altitude the decrease in P_B will lead to a decrease in P_{A,O_2} [see Eq. (10)], causing hypoxic vasoconstriction. Under these circumstances edema could develop, which will contribute to a further development of vasoconstriction, causing a severe illness called mountain sickness. The mechanism leading to edema under these high altitude conditions is not well understood.

D. Extravascular Lung Water and Pulmonary Edema

The extravascular fluid in the lung or extravascular lung water (EVLW) is continuously produced from the blood capillaries and is removed by the system of lymph vessels. The lymph vessels drain the lungs in the septa along the bronchi up to the respiratory bronchioles, but are not present in the alveolar septa. The amount of lymph transported by the lymph vessels in humans is unknown. About 5 ml/hr has been reported for dogs and sheep. The transport of lymph is insufficiently increased when relatively large amounts of EVLW develop in the interstitium. This is a reason for the supposition of insufficiently developed lymph valves in these lymph vessels.

Production of lymph from the blood capillaries is governed by the interaction of several forces, which are combined into the Starling equation:

$$Q'_f = K_f[(P_c - P_t) - \sigma(\pi_p - \pi_t)]$$

where

Q'_f is the amount of fluid (ml/sec) which is filtered through the vascular wall of the pulmonary capillaries,

K_f is the fluid filtration coefficient. This term indicates the amount of fluid filtration through a membrane for a given pressure difference. The term between the brackets implies the effective pressure gradient over the vascular wall. This term can be simplified to ΔP, giving the equation $Q'_f = K_f \Delta P$. Compared with the hemodynamic law $\Delta P = Q'R$ we conclude that $K_f \approx 1/R$. K_f is a reciprocal value of flow resistance, which is conductance. Thus, K_f indicates the fluid permeability of the vascular wall.

P_c is the hydrostatic pressure in the lung capillaries and perhaps also in the lung arterioles and venules.

P_t is the hydrostatic pressure of the interstitial fluid.

π_p is the oncotic pressure of the plasma proteins.

π_t is the oncotic pressure of the interstitial fluid.

σ is the osmotic reflection coefficient, which corrects the effectivity of the proteins, because the capillary wall is not perfectly impermeable for these proteins. When the wall should be perfectly impermeable $\sigma = 1$. For substances which pass the wall freely, as the small molecular minerals in the plasma, $\sigma = 0$.

Most of these factors are not very well known. π_p is the only one which can be measured, because it is easy to obtain blood plasma from a patient. P_c can only be approximated by measuring the pulmonary capillary wedge pressure, P_{pcw}, with use of a Swan–Ganz catheter in the pulmonary artery. When the balloon at its tip is filled up with air the catheter will close the artery, and no flow will exist in this artery and its branches. In these vessels the pressure will become equal to the pressure at the venous side of the system where the veins coming from the branches of the closed artery connect with vessels where flow was maintained.

The other terms are unknown.

P_t will be different in different parts of the alveolar septa, as explained in Section III,B,2. P_t will be increased when surfactant is insufficiently active. Edema decreases surfactant activity and causes a higher surface tension, which causes a more negative interstitial pressure, leading to a vicious circle of EVLW formation.

Destruction of the capillary wall by noxious substances will decrease σ and k_f. The decrease in σ will diminish $\pi_p - \pi_t$ and the decrease in K_f will increase the passage of fluid. Both will contribute to the production of edema.

An increase in P_c by left ventricular failure or by mitral valve stenosis will cause a cardiogenic pulmonary edema.

An increase in EVLW has two counteracting effects on its own production, suggesting some self-limiting mechanism. An increase in EVLW increases P_t and decreases π_t. These two changes have a negative effect on Q'_f, according to the Starling equation.

Bibliography

Clark, J. T. H., and Godfrey, S. (1977). "Asthma." Chapman and Hall, London.

Clausen, J. L. (ed.) (1984). "Pulmonary Function Testing. Guidelines and Controversies." Grune and Stratton, New York.

Cotes, J. E. (1979). "Lung Function. Assessment and Application in Medicine," 4th Ed. Blackwell Scientific Publications, Oxford.

Harris, P., and Heath, D. (1986). "The Human Pulmonary Circulation," 3rd Ed. Churchill Livingstone, Edinburgh.

Staub, N. C. (ed.) (1978). "Lung Water and Solute Exchange." Marcel Dekker, New York.

Weibel, E. R. (1964). Morphometrics of the Lung. *In* "Handbook of Physiology, Section 3: Respiration" (W. O. Fenn and H. Rahn, eds.), Vol. 1. American Physiological Society, Washington, D.C.

Weir, E. K., and Reeves, J. T. (eds.) (1989). "Pulmonary Vascular Physiology and Pathophysiology." Marcel Dekker, New York.

West, J. B. (ed.) (1977). "Regional Differences in the Lung." Academic Press, New York.

Wilson, A. F. (ed.) (1985). "Pulmonary Function Testing. Indications and Interpretations." Grune and Stratton, New York.

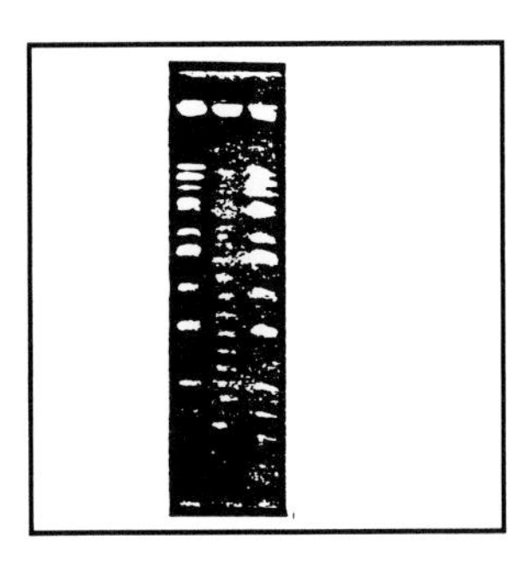

Pulsed Field Gel Electrophoresis

BRUCE BIRREN, *California Institute of Technology*

ERIC LAI, *University of North Carolina, Chapel Hill*

Glossary

Field inversion electrophoresis Electrophoresis in which the reorientation angle is 180°
Gene linkage Occurrence of different genes on the same chromosome, such that they are inherited together
Reorientation angle Angle between the alternating electric fields with which the DNA must realign
Restriction enzymes Proteins that recognize a specific nucleotide sequence, usually four to eight base pairs long, and cleave DNA at that site
Sizing ladders Series of size markers generated by end-to-end joining of a unit-length molecule such that each step reflects a constant increment in length
Switch interval (switch time) Frequency with which the alternating electric fields are activated

PULSED FIELD GEL ELECTROPHORESIS is the process whereby large DNA molecules are separated according to size as they migrate through a matrix under the influence of periodically reorienting electric fields. Conventional agarose electrophoresis, which makes use of a constant electric field, has a maximum size limit for DNA resolution of about 50,000 base pairs. In contrast, pulsed field agarose gels can resolve DNA that is longer than 10 million base pairs. Methods for preparing and manipulating DNA have been developed that minimize physical breakage, thus permitting analysis of extremely large, intact DNA molecules. The ability to separate such large DNA fragments makes pulsed field gel electrophoresis a radical improvement in the technology of gene mapping and analysis.

I. The Need for Pulsed Field Gel Electrophoresis

Pulsed field gel electrophoresis bridges the gap in resolution between the techniques of traditional genetics, which measures on a gross scale of chromosomal distances, and the much finer scale of molecular biology. Human recombinational mapping (carried out by studying the inheritance of well-characterized genes in families or in cell cultures) and cytogenetics (the analysis of chromosomes) can detect deletions, duplications, and rearrangements involving large portions of a chromosome. Such large regions would be expected to contain many genes. These methods are incapable of resolving loci (the sites of genes) that are physically separated by less than 1–3 million base pairs along the chromosome. The techniques of molecular biology offer a more detailed description of the structure and organization of genes by exploiting enzymes that recognize specific sites in the DNA sequence. However, technical limitations of the methods used in isolating and characterizing DNA fragments have limited their application to molecules typically smaller than several thousand bases (kilobases, or kb). Pulsed field gel electrophoresis now permits the techniques of molecular biology to be applied to much larger DNA fragments. It can be used preparatively to isolate large fragments for cloning as well as analytically to construct megabase-scale restriction maps around specific regions by using Southern hybridization. This is accomplished by cleaving chromosomal DNA with a restriction enzyme and separating the resulting fragments in an agarose gel.

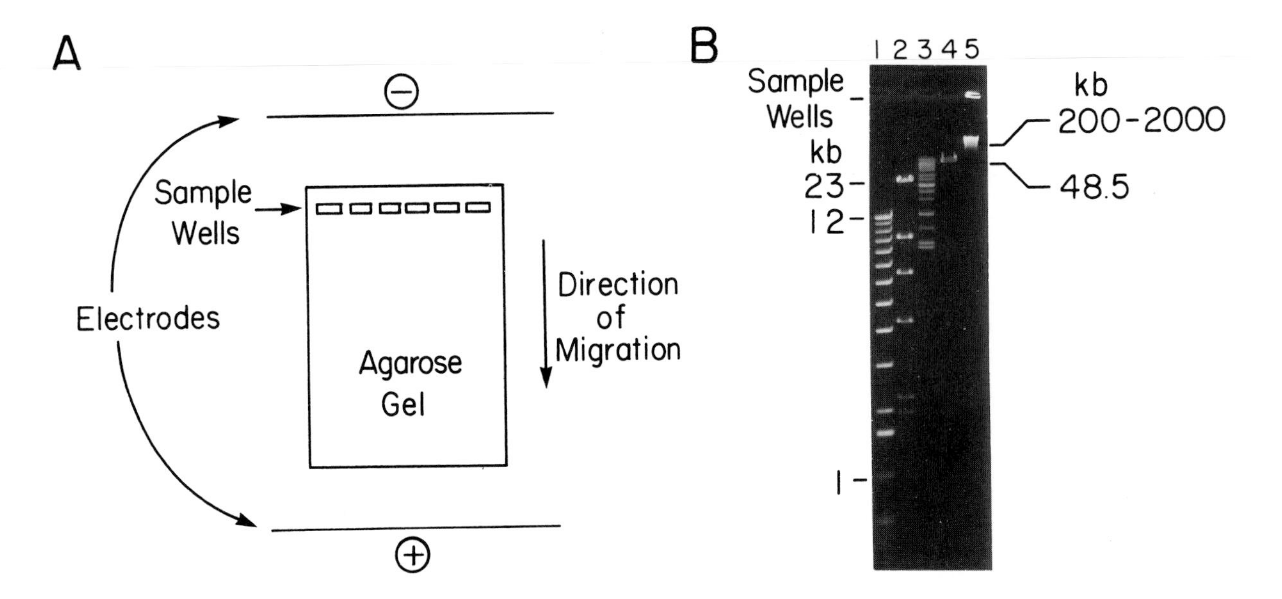

FIGURE 1 Conventional agarose electrophoresis. (A) The principle of separation of DNA molecules in a conventional agarose gel. DNA loaded in the sample wells will migrate in the direction of a constantly applied electric field at a rate reflecting the size of the fragments. (B) Photograph demonstrating the limitation of conventional gels in separating DNA fragments above 40 kb. DNA samples were loaded in the sample wells and run for 16 hours at room temperature in a 0.6% agarose gel with a voltage gradient of 1 volt/cm. Lane 1: 1-kb ladder from BRL (1- to 12-kb fragments); lane 2; lambda DNA digested with *Hin*dIII (2- to 23-kb fragments); lane 3; high-molecular-weight marker from BRL (8- to 48.5-kb fragments); lane 4; linear lambda DNA (48.5 kb); lane 5; yeast chromosomes (200–2000 kb).

The separated fragments are then transferred from the gel to a membrane. The position of specific sequences can be detected by allowing labeled probe DNA to bind (hybridize) to homologous sequences on the membrane. Thus, the size of the original restriction fragments containing the region of interest can be established. The ability to work with such large pieces of DNA reduces the number of steps and specific DNA probes required to create physical maps of human DNA and permits rapid movement from one genetic marker to the next. [*See* GENETIC MAPS.]

II. The Effect of Alternating Fields

A. Conventional Electrophoresis

In conventional electrophoresis, molecules become aligned and then move through a matrix under the influence of a static electric field. Sieving by the matrix produces size-based separation, with molecules migrating at a rate proportional to the log of their size. Figure 1A shows a schematic diagram of a conventional agarose gel in which DNA samples are loaded into sample wells at the top of the gel and migrate in the direction of the applied electric field. Figure 1B shows a photograph of such a gel in which samples of a range of different sizes were loaded at the top of the gel and stained for visualization after separation. Note that the spacing between the different sample bands is greater for the small fragments that have run farthest at the bottom of the gel and gradually decreases with increasing fragment size. The resolution of molecules based on their size breaks down for DNA fragments larger than 40 kb in conventional gels, as all molecules above this size limit migrate with the same mobility. Thus, in Fig. 1B there is no difference between the mobility of the lambda DNA (50 kb) and the chromosomes of the yeast *Saccharomyces cerevisiae* (approximately 200 to 2000 kb).

B. Pulsed Field Electrophoresis

As indicated in Figs. 2A and 2B, pulsed field gel electrophoresis makes use of alternating electric fields that differ in their orientation with respect to the gel. DNA molecules must realign with each reorientation of the alternating fields prior to migrating in the new direction. By activating each field for an equal length of time, the molecules will migrate down the gel in a series of short "zigzag" steps. The key to pulsed field gel electrophoresis is that the time required by DNA fragments to reorient when each new field is applied is directly proportional to the size of the molecules. Small fragments

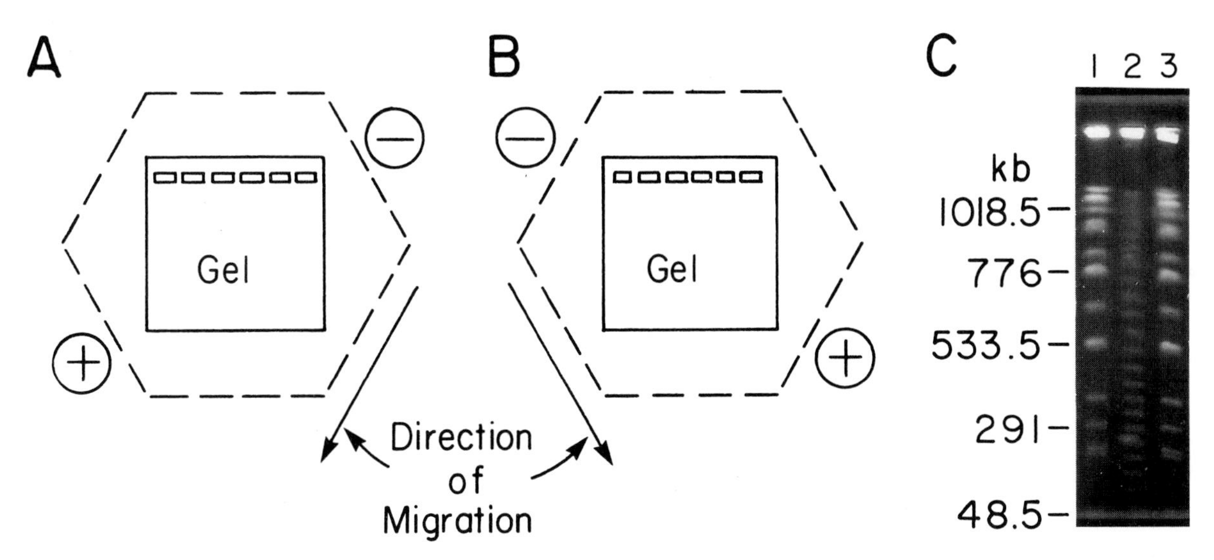

FIGURE 2 Pulsed field gel electrophoresis. (A) The gel and the orientation of the "A" electric field. After a period of time (the switch interval), the orientation of the electric field is changed to that shown in (B). The "A" and "B" electric fields alternate throughout the gel run and are of the same duration. (C) Photograph showing the stained bands of DNA separated by a pulsed field gel run. Samples were run in a 1% gel for 36 hours at 14°C with a voltage gradient of 6 volts/cm. The switch interval was linearly ramped from 45 seconds at the beginning to 90 seconds at the conclusion of the run. Lanes 1 and 3 contain chromosomes of the yeast *S. cerevisiae*, strain D273-10B, and lane 2 contains a ladder of phage lambda DNA (monomeric length of 48.5 kb).

reorient quickly and will therefore begin to migrate soon after the field is switched. Larger molecules spend a greater proportion of each cycle realigning and will not have as much time to migrate before the field is switched again. Thus, periodically reorienting the electric fields permits DNA molecules too large to be fractionated by conventional electrophoresis to be separated according to their size. Figure 2C is a photograph of a pulsed field gel in which molecules from 50 kb to greater than 1000 kb have been separated. Unlike in the conventional gel shown in Fig. 1B, the distance traveled by fragments in the pulsed field gel is a linear function of size so that the spacing between molecules of a common difference in size is relatively constant throughout the gel.

III. The Process of Pulsed Field Gel Separations

A. Equipment

Pulsed field gel equipment has evolved with a greater understanding of the basis for the separation. The earliest device made use of electric fields that varied both in intensity and in the angle of intersection (reorientation angle) across the gel (hence the early term "pulsed field gradient gel" electrophoresis). These properties lead the samples to migrate in a characteristic "bent" trajectory down the gel, a pattern accentuated toward the sides of the gel. More recent apparatus designs make use of homogeneous fields and constant reorientation angles and give rise to straight-line sample migration. Pulsed field electrophoresis requires reorientation of the electric field with respect to the gel. This can be achieved mechanically either by discontinuously rotating the gel through an unchanging electric field or by rotating the electrodes around the gel. Field orientation can be also accomplished by alternately activating different sets of electrodes fixed around the gel, either electronically or by using relays. One of the simplest forms of pulsed field gel electrophoresis uses a conventional gelbox and periodically reverses the polarity of the electric field. Such field inversion electrophoresis employs either a higher voltage or a longer duration for one of the two electric fields and thereby induces net migration of samples down the gel.

B. Field Switching

Pulsed field gels work because the reorientation of DNA molecules after field switching is dependent on size. A pulsed field gel will only be able to resolve those molecules that can reorient within the time provided before the field again changes direction. All molecules that are larger than this size limit will migrate with a common mobility. By increasing the switch interval, increasingly long DNA molecules can be resolved. Switch intervals of fractions of a second to longer than one hour have been used to separate different size ranges of DNA. Figure 3 shows photographs of three different pulsed field gels using different switch intervals. Each gel contains ladders of phage lambda DNA and yeast chromosomes. At the shortest switch interval of 30 seconds, only 4 bands in the sample of yeast chromosomes have been resolved (maximum size 450 kb) and the rest of the yeast chromosomes are seen to comigrate in an intensely staining band. As the switch interval is increased to 60 seconds, separation extends to 830 kb and 10 bands of yeast chromosomes can be clearly separated. With a switch interval of 90 seconds, 16 different bands have been separated and the resolution extends beyond 1000 kb. However, as the size range increases, resolution for the smaller fragments is diminished (note the decrease in the spacing between the first two yeast chromosomes as the switch time is increased). Thus, the user must select the frequency of field switching to balance obtaining the highest possible resolution while still separating the complete size range desired. The voltage chosen is also a key factor in the success of pulsed field gels, as the separation of molecules larger than 2 Mb requires the use of low-voltage gradients.

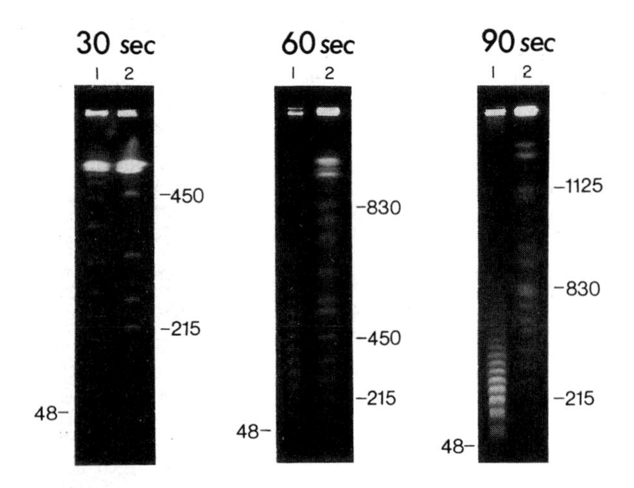

FIGURE 3 Effect of switch interval on pulsed field separations. Lambda concatamers (lane 1) and chromosomes of the yeast *S. cerevisiae* (lane 2) strain YNN295 were separated using identical conditions (1% agarose, 14°C, 6 volts/cm for 30 hours) at three different switch intervals. The switch interval is indicated above the photograph of each gel, and the sizes of the DNA fragments are indicated to the side of each photograph.

C. Sample Preparation and Handling

The development of pulsed field gel electrophoresis required the codevelopment of techniques for preparing and handling large DNA fragments without physical breakage. Traditional procedures for preparing DNA introduce shear forces through pipetting or mixing solutions of DNA, which causes fragmentation of long DNA molecules. However, embedding intact cells in agarose prior to digestion of the cell wall, the protein, and the lipid components provides physical protection to the molecules. The digestion products diffuse out of the agarose, leaving nucleic acids. Endogenous nucleases can be inhibited during these digestions, yielding DNA molecules from human cells that are longer than 10 million base pairs (Mb). Restriction enzymes and other DNA-modifying enzymes can diffuse into the agarose and act on the DNA trapped there, permitting all the standard manipulations of molecular biology to be carried out while the DNA is protected in the solid block of agarose.

D. Size Markers

The initial challenge of pulsed field gel electrophoresis was to accurately determine the sizes of the migrating molecules, which were larger than any previously characterized. The linear genomes of several bacteriophages had been estimated by physical methods and therefore provided some assistance for a few specific sizes but were not useful for a broad size range. The genome of bacteriophage lambda exists as a linear DNA molecule whose 48.5-kb length is known precisely. Each end of the molecule has short single-stranded regions capable of specifically and stably binding the ends of other such molecules. Through such end-to-end joining, one can generate molecules of various lengths, each differing by the number of 48.5-kb monomeric lengths added. In practice, when lambda DNA is embedded in agarose and allowed to polymerize, a collection of molecules representing the addition of random numbers of monomers is obtained. When these ladders of lambda molecules are separated by pulsed gel electrophoresis, sizes from 48.5 kb to over 1000 kb can be distinguished.

Using lambda ladders for comparison, the sizes of chromosomes from several microorganisms have been determined and are frequently used as size standards. The most useful have been those of yeasts *Saccharomyces cerevisiae* and *Schizosac-*

charomyces pombe. *Saccharomyces cerevisiae* chromosomes span a size range from about 200 to 2000 kb. The three *S. pombe* chromosomes are much larger, being 3.5, 4.7, and 5.7 Mb. Each of the chromosomes of these organisms has been mapped using restriction enzymes and pulsed field gels, so that the size is known with a high degree of accuracy.

IV. Applications

A. Principle

A primary goal of genome analysis is to establish the normal location of genetic markers and structural elements on specific chromosomes. This involves determining both the order of markers on the chromosome and the distances between them. In addition, rearrangements such as deletions and inversions can alter gene expression from a region and produce disease when vital functions are disrupted. Pulsed field gel electrophoresis has proven valuable in both of these situations, that is, by establishing linkage between markers and demonstrating large genome rearrangements. Markers that are too close to be easily distinguished or ordered using genetic recombination can be readily distinguished by gel analysis. In addition, because pulsed field gels can resolve fragments up to 5 Mb, linkage of more distantly separated markers can be demonstrated by hybridization of the separate probes to the same fragments. Because such long fragments can be analyzed, fewer probes are required to map a large region. The high degree of accuracy associated with the sizing of pulsed field gel electrophoresis facilitates detection of genome rearrangements.

B. Choice of Restriction Enzymes for Pulsed Field Gels

The first step in the application of pulsed field gel electrophoresis to the analysis of the human genome is the cleavage of chromosome-sized molecules to specific DNA fragments that are small enough to be separated in a gel but large enough to allow linkage of distant markers. The common restriction enzymes, such as *Eco*RI or *Bam*HI, have six base pair recognition sequences and produce fragments that are too small to allow linking for all but the closest markers (average fragment size 1 to 50 kb). In general, the frequency of occurance of a restriction site is related to the length of the sequence being recognized. Enzymes that recognize octameric sequences such as Not I and Sfi I cleave less frequently and are very useful in generating large mammalian DNA fragments. These enzymes have produced specific fragments as large as 5 Mb upon digestion of human DNA. Other restriction enzymes have been shown to generate large DNA fragments from human DNA even though they only recognize a hexameric sequence. This is due to the fact that the recognition sequences for these enzymes include the dinucleotide CpG, which is greatly underrepresented in the human genome. Enzymes that contain two CpG dinucleotides in their recognition sequence are considered to be "rare cutters" for human DNA.

C. Identifying Genes by Southern Hybridization with Pulsed Field Gels

After DNA has been cleaved with rare-cutting restriction enzymes and separated in a pulsed field

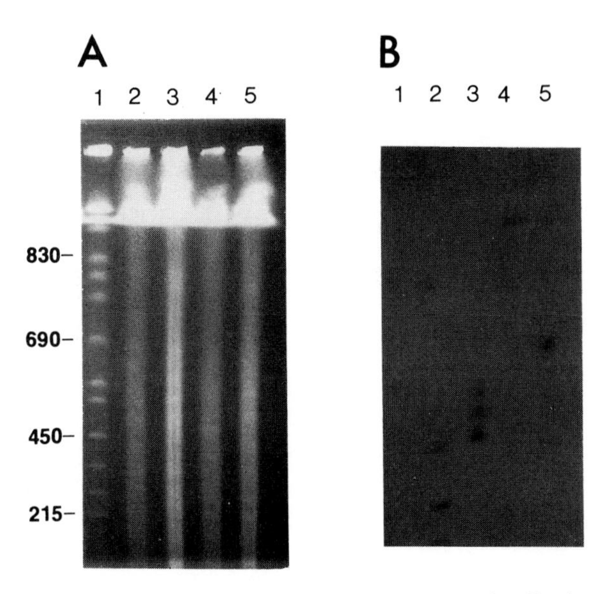

FIGURE 4 Detection of large DNA fragments by Southern transfer of pulsed field gels. Human DNA was digested with restriction enzymes and separated by pulsed field gel electrophoresis. The separated DNA stained with ethidium bromide is shown in (A). Lane 1 contains yeast chromosomes as size markers; lane 2 contains DNA digested with Mlu I; and lane 3 contains DNA digested with Nru I. The sizes of the yeast chromosome markers are indicated. After transferring the separated DNA to nylon, those fragments homologous to the β subunit of transducin were detected by hybridization with a radioactive probe. The autoradiograph is shown in (B) and reveals the presence of multiple bands with each enzyme.

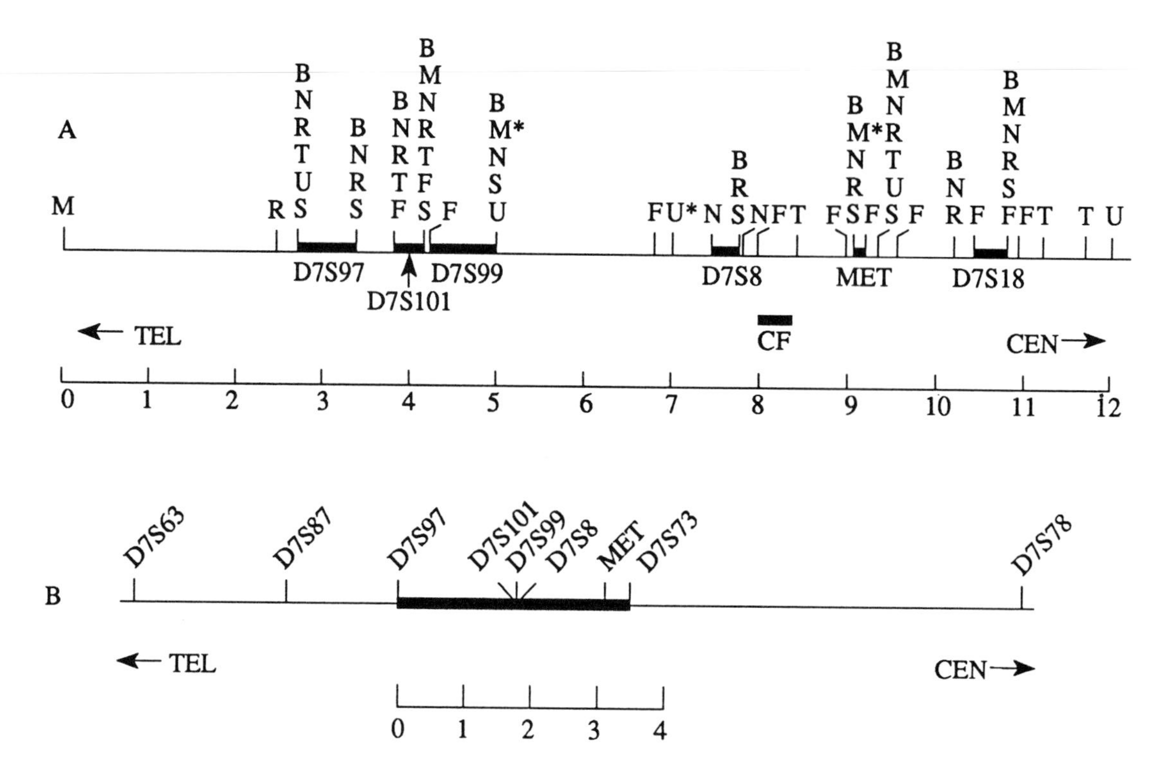

FIGURE 5 Genetic and long-range physical maps of the cystic fibrosis locus. (A) Multipoint genetic linkage map. The positions of the probes are indicated above the line. The region that has been mapped by pulsed field gel electrophoresis is indicated by the bar. The scale is in Kosambi centiMorgans. (B) Long-range restriction map. The restriction sites are shown above the line. The restriction site abbreviations are as follows: B=BssH II, N=Nae I, R=Nar I, T=Not I, U=Nru I, F=Sfi I, S=Sst II. The scale is in megabase pairs. The approximate location of the cystic fibrosis locus is shown as CF. CEN=centromere and TEL=telomere. Reprinted with permission from Fulton T.R. *et al., Nucl. Acids Res.* **17,** 271–284, 1989.

gel, a smear of all the different length products is obtained. Figure 4A shows a photograph of stained DNA after digestion with two different restriction enzymes and separation on a pulsed field gel. The human DNA has been purified separately from two different sources, blood and a cultured cell line. The yeast chromosomes provide size standards for the human DNA. Transferring the DNA from the gel to a membrane permits detection of specific fragments by hybridization (specific association) with probes (DNA fragments recognizable for their radioactivity or fluorescence). Figure 4B shows the DNA fragments identified by hybridization with a β subunit of human transducin. Note that different-sized bands are produced by each of the two different restriction enzymes. In addition, the sizes of the bands differ when DNA from different sources is compared. For example, the Mlu I digest of DNA from the cell line (Fig. 4B, lane 2) produces three bands with sizes from 175 to 400 kb. The Mlu I digest of DNA from blood (lane 3) also generates three bands, but in this case the fragments range from 400 to 550 kb. Nru I (lanes 4 and 5) also generates multiple bands that vary in size for the two separate DNA sources. Differences between the fragments identified when DNA from different sources is digested are most often due to differences in the degree of methylation as discussed in the last section of this article. To establish that another marker is linked to this gene, hybridization with the second probe would have to identify the same-sized fragments when a number of different restriction enzymes are used.

D. Pulsed Field Mapping of the Cystic Fibrosis Region

Cystic fibrosis (CF) is an autosomal recessive genetic disorder that has a frequency of about 1 : 2000 in the Caucasian population. The molecular basis for the disease is still unknown. The CF locus was initially mapped to human chromosome 7 by genetic linkage analysis. A number of DNA probes and markers were found to be close to the CF locus and were assumed to be within 1 Mb of the CF gene though physical distances and an exact order could not be unambiguously determined. More detailed mapping and localization of the CF gene within this region by genetic linkage analysis would be both difficult and extremely time-consuming. On the other hand, molecular cloning with current approaches is not feasible for such a large region. However, by using the linked probes, a long-range restriction map was constructed around the CF locus. This map is shown in Fig. 5.

Although no probe existed for the central (CF gene-containing) region, hybridization of the flanking probes to overlapping restriction fragments defined the distances between the existing probes and therefore established limits for the boundaries of the CF gene. Genetic linkage studies placed the CF gene in between the D7S8 and MET markers. The long-range pulsed field map shown in Fig. 5 defined the minimum distance between these loci to be approximately 1.4 Mb. Note that the sites for the rare-cutting restriction enzymes are not evenly distributed along the DNA but occur grouped in clusters. This reflects the nonrandom distribution of the CpG dinucleotide in the human genome. These clusters of sites for the rare-cutting enzymes frequently are found associated with genes and thus their presence can be used to direct the search for new genes in uncharacterized regions of the genome. In addition, with such a map, newly isolated probes can rapidly be placed with respect to the CF gene by hybridization to similar blots. The CF gene has recently been isolated by using a combination of techniques that rely on pulsed field gel electrophoresis. [*See* CYSTIC FIBROSIS, MOLECULAR GENETICS.]

E. Pulsed Field Gel Electrophoresis Characterization of the Duchenne Muscular Dystrophy Defect

Duchenne Muscular Dystrophy (DMD) is an X-linked recessive neuromuscular disease affecting about 1 in 4000 males. It has the highest known spontaneous mutation rate of around 1 in 10,000. In addition, the DMD phenotype had been observed to coincide with other chromosomal abnormalities such as translocation, deletions, and other mutations involving the X chromosome. These observations lead to the belief that the DMD locus is distributed over a large chromosomal region. While a number of markers for the region had been obtained, their exact position relative to each other and the distance separating them were not known. Pulsed field gels were used to create a physical map of the region and unambiguously place these markers. This work established that the DMD gene spanned at least greater than one million bases on the chromosome. Figure 6 shows this long-range physical map around human Xp21. The entire region shown between the markers CX5.4 and B24 represents over 4.5 Mb. In addition, pulsed field gel electrophoresis was vital in establishing the nature of the DMD defect. When DNA from DMD patients was analyzed it was apparent that large portions of this region were often deleted. In addition, the end points of the deletions defined using the pulsed field gels were unique for each patient. These deletions were not large enough to be seen using conventional chromosome analysis. Thus pulsed field gel analysis contributed to the view that the gene for DMD was extremely large and prone to delete at many different sites, each giving rise to DMD. In addition, a number of other diseases that frequently appeared in DMD patients had also been mapped to the same locus. The long-range restriction map established that specific areas of the region were associated with these different diseases. [*See* MUSCULAR DYSTROPHY, MOLECULAR GENETICS.]

F. Interpreting Pulsed Field Data

Several aspects of pulsed field gel electrophoresis require caution in the interpretation of data. First, the mobilities of DNA fragments are extremely sensitive to the amount of DNA loaded onto a gel lane. Thus, apparently different sizes can be observed for

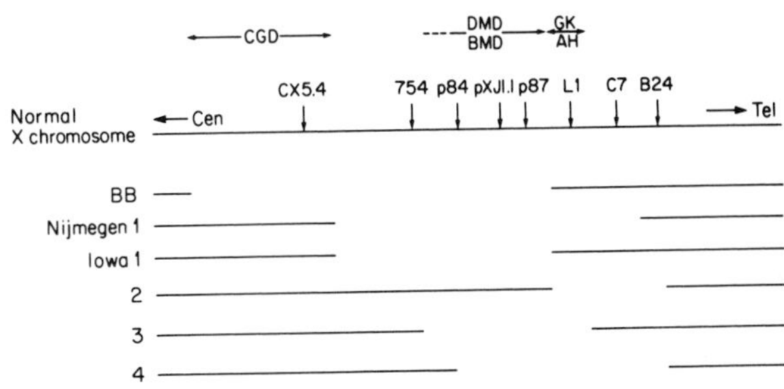

FIGURE 6 Deletions in the Duchenne muscular dystrophy region identified by pulsed field gel mapping. The positions of the DNA probes used to generate the long-range restriction map are shown above the normal X chromosome. The approximate positions of the chromic granulomatous disease (CGD) gene, Duchenne muscular dystrophy (DMD) gene, Becker muscular dystrophy (BMD) gene, glycerol kinase (GK) gene, and the adrenal hypoplasia locus (AH) are indicated. Deletions in the DMD locus identified in the DNA of six patients are shown below the normal chromosome. Modified with permission from van Ommen G.-J. *et al., Cell* **47,** 499–504, 1986. Copyright by Cell Press.

the same fragment run with different DNA concentrations, or conversely, identical sizes can be observed for different amounts of fragments of different lengths. Second, different DNA samples may vary in the extent to which the DNA is methylated. Since most of the enzymes used for pulsed field gel mapping are sensitive to methylation of the restriction site, differences in the size of restriction fragments could indicate differences in methylation rather than differences in physical distance. Often methylation differences arise when comparing DNA obtained from different tissue sources. Third, ambiguity is introduced by allelic polymorphisms. Identification of multiple fragments by hybridization with a single copy probe can reflect allelic differences between the chromosomes.

Finally, a feature common to all pulsed field gel systems is the existence of regions of the gel in which the relationship between size and mobility is not linear. It is possible for a 2000-kb fragment to migrate ahead of a 1500-kb fragment under certain conditions. This behavior is easily detected in the separation of distinct bands such as the yeast chromosomes, but it is not obvious in digests of mammalian DNA, which generate a smear. Thus, it is crucial to use appropriate markers with each gel to confirm that optimal separation conditions are being used for the size range of interest.

Bibliography

Barlow, D. P., and Lehrach, H. (1987). Genetics by gel electrophoresis: The impact of pulsed field gel electrophoresis on mammalian genetics. *Trends in Genetics* **3**, 167.

Olson, M. V. (1989). Pulsed field gel electrophoresis. *In* "Genetic Engineering" (J. K. Setlow, ed.). Vol. 11. Plenum, New York.

Smith, C. L., Klco, S. R., and Cantor, C. R. (1988). Pulsed field gel electrophoresis and the technology of large DNA molecules. *In* "Genome Analysis, a Practical Approach" (K. E. Davies, ed.). IRL Press, Oxford.

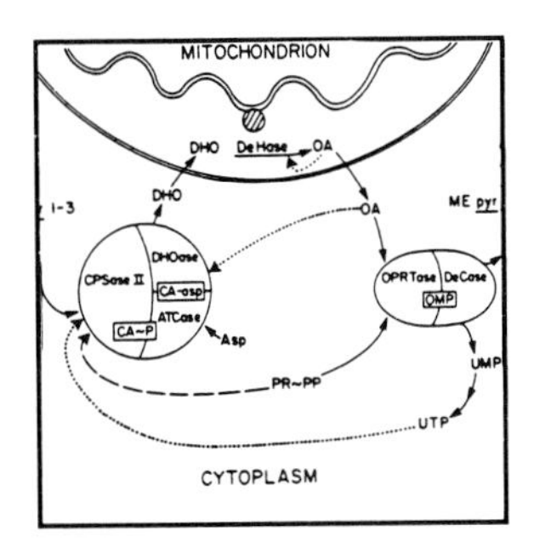

Purine and Pyrimidine Metabolism

THOMAS D. PALELLA, IRVING H. FOX, *University of Michigan*

Glossary

Base Generic term for the heterocyclic (i.e., containing two kinds of atoms—carbon and nitrogen) ring structures of purines and pyrimidines

***De novo* Biosynthesis** Metabolic pathways through which nonpurine and nonpyrimidine precursors are combined to form purines and pyrimidines, respectively

Nucleoside Compound of a sugar (e.g., ribose or deoxyribose) with a purine or pyrimidine base by way of an N-glycosyl link

Nucleotide Combination of a purine or pyrimidine base, sugar, and phosphate group or groups comprising the fundamental units of nucleic acids; the base sugar linkage is N-glycosyl; the sugar–phosphate combination is via an ester linkage, usually to the 5′ hydroxyl group; mononucleotides contain a single phosphate group, dinucleotides contain two phosphate groups linked by a phosphodiester bond, and trinucleotides contain three phosphate groups

Purine Heterocyclic nine-member ring ($C_5H_4N_4$) that is the parent compound of adenine, guanine, and hypoxanthine

Pyrimidine Heterocyclic six-member ring ($C_4H_4N_2$) that is the parent compound of thymine, cytidine, and uracil

PURINES AND PYRIMIDINES are present in all forms of plant and animal life. They constitute the building blocks of deoxyribonucleic acid (DNA) and ribonucleic acid (RNA) and, thus, are fundamental to reproduction and inheritance. In addition, purine nucleotides serve as molecular energy sources. Both purine and pyrimidine nucleotides are donors of high-energy phosphate in a plethora of enzymic reactions. Finally, they also serve as coenzymes, neurotransmitters, and second messengers. In humans, several important diseases result from disordered purine or pyrimidine metabolism. For all these reasons, purine and pyrimidine metabolism have been extensively studied. This chapter summarizes only the most important aspects of these expansive topics.

I. Purines

The overall scheme of human purine metabolism is depicted in Fig. 1. This representation is conveniently divided into interfacing compartments: (1) *de novo* purine synthesis, (2) salvage of preformed purine bases, (3) nucleotide interconversions, and (4) degradation. The parent compound, the purine ring, is the starting point. Because the final degradative product in humans is uric acid, disorders of purine metabolism are often reflected in abnormal concentrations of uric acid or urate in the blood and other body fluids.

A. Metabolic Pathways

1. *De Novo* Purine Synthesis

The purine nucleus, consisting of fused pyrimidine and imidazole rings, is the parent compound for all purines. The origins of its individual atoms have been defined in a variety of prokaryotic and eukaryotic systems. The biosynthetic pathway through which nonpurine precursors are combined to form the purine ring is referred to as *de novo* purine synthesis.

The first reaction in this synthesis is the formation of an important regulatory intermediate, 5-

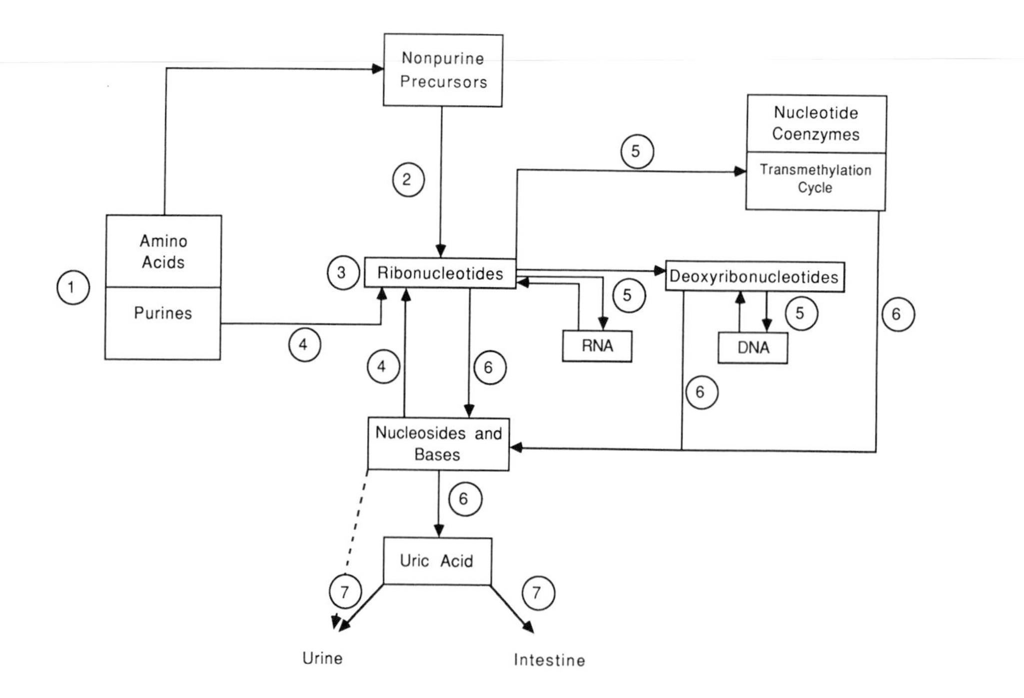

FIGURE 1 Overview of human purine metabolism. Human purine metabolism is focused on the synthesis of purine ribonucleotides by two pathways. In the *de novo* pathway, dietary intake of amino acids (1) provides nonpurine precursors for purine biosynthesis (2) leading to IMP formation. IMP is converted to other purine ribonucleotides by the interconversion pathways (3). Alternatively, preformed purine bases and nucleosides may be reutilized by conversion to ribonucleotides by the salvage pathways (4). The substrates for the salvage reactions derive from dietary purines (1) or breakdown of nucleotides (6). Purine ribonucleotides are converted from the monophosphate form to diphosphate and triphosphate forms, which serve as essential substrates for a variety of pathways (5). Purine nucleoside monophosphates are also the main substrates for the degradative pathways of purine metabolism (6). The final step in the degradation of purine nucleotides in humans is the formation of uric acid. Uric acid is the major excretory product of human purine metabolism although small amounts of nucleosides and bases are excreted into the urine (7). Uric acid is excreted primarily in urine but smaller amounts are excreted into the intestine in which bacterial uricolysis occurs (7). [Modified, with permission, from an article written by I. H. Fox, "Clinical Medicine" (J. A. Spittell, Jr., ed.), Harper & Row, 1982.]

phosphoribosyl-α-1-pyrophosphate (PP-ribose-P). It has two main roles: it condenses with L-glutamine in the first committed step of the *de novo* pathway, and it serves as the phosphoribosyl donor in salvage reactions whereby purine bases are converted into ribonucleotides.

In a reaction catalyzed by PP-ribose-P synthetase, in the presence of magnesium (Mg^{2+}) and inorganic phosphate (P_i), the terminal pyrophosphate (PP_i) of adenosine-5′-triphosphate (ATP) is transferred to carbon 1 of ribose-5-phosphate yielding adenosine-5′-monophosphate (AMP). This allosteric enzyme is regulated by negative feedback inhibition by purine and pyrimidine nucleotides. [*See* ADENOSINE TRIPHOSPHATE (ATP).]

$$\text{ribose-5-phosphate} + \text{ATP} \xrightarrow{Mg^{2+}} \alpha\text{-PP-ribose-P} + \text{AMP} \qquad \text{Reaction (1)}$$

The first committed step in *de novo* synthesis generates 5-β-phosphoribosyl-1-amine. In this irreversible reaction, which is catalyzed by the enzyme amidophosphoribosyltransferase, the amide group of glutamine displaces pyrophosphate from PP-ribose-P. The activity of this enzyme is regulated by PP-ribose-P and nucleotide concentrations. In the process, the substituents are inverted such that a β-linkage is generated.

$$\alpha\text{-PP-ribose-P} + \text{glutamine} + H_2O \xrightarrow{Mg^{2+}} 5\text{-}\beta\text{-phosphoribosylamine} + \text{glutamic acid} + PP_i \qquad \text{Reaction (2)}$$

Subsequent reactions in the *de novo* pathway form the purine ring and lead to inosine-5′-monophosphate (IMP) formation (Fig. 2).

2. Nucleotide Interconversions and Catabolism

All purine ribonucleotides and deoxyribonucleotides derive from IMP. These reactions are summarized in Fig. 3 (reactions 12–15).

Mononucleotides are converted to nucleosides by

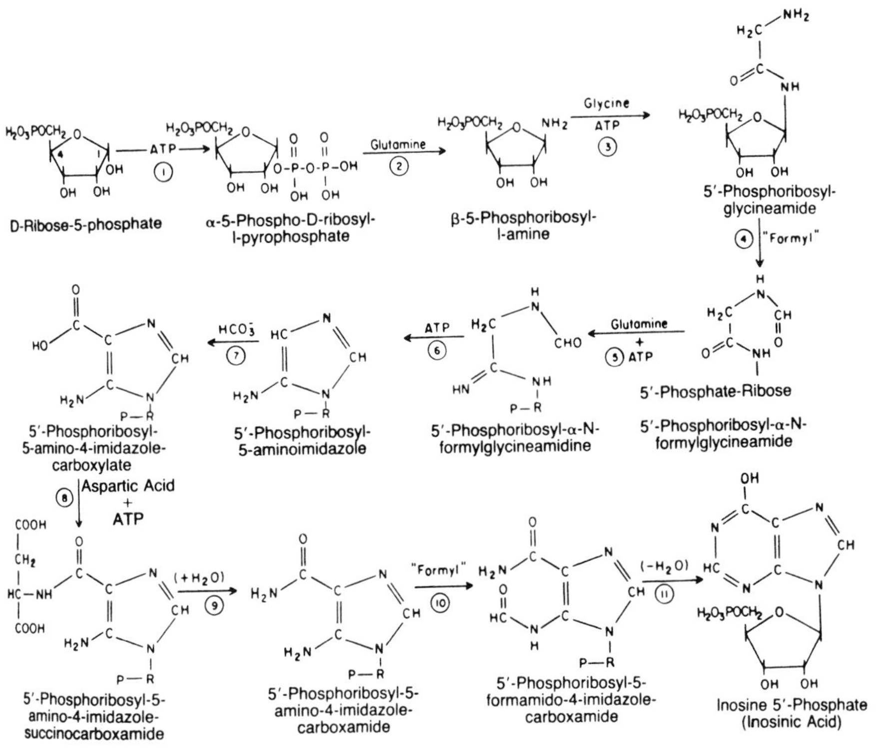

FIGURE 2 Biosynthesis of the purine ring. 1. PP-ribose-P synthetase. 2. Amidophosphoribosyltransferase. 3. Phosphoribosylglycineamide synthetase. 4. Phosphoglycineamideformyltransferase. 5. Phosphoribosylformylglycineamidine synthetase. 6. Phosphoribosylformylglycineamidine cycloligase. 7. Phosphoribosylaminoimidazolecarboxylase. 8. Phosphoribosylaminoimidazole succinocarboxamide synthetase. 9. Adenylosuccinate lyase. 10. Phosphoribosylaminoimidazolecarboxamideformyltransferase. 11. Inosinate cyclohydrolase. [Modified, with permission, from J. B. Wyngaarden and W. N. Kelley, 1976, "Gout and Hyperuricemia," Grune & Stratton, Orlando.]

a variety of enzymes falling into two broad classes: (1) nonspecific phosphatases and (2) 5′-nucleotidases (Reaction 16). Purine nucleosides may be further catabolized by phosphorolysis via purine nucleoside phosphorylase (Reaction 17). AMP deaminase converts AMP to IMP with the liberation of ammonia (Reaction 18).

Adenosine is converted to inosine via adenosine deaminase (Reaction 19), whereas adenosine kinase converts adenosine to AMP, consuming ATP in the process (Reaction 20).

3. Salvage Pathways

De novo purine synthesis is metabolically expensive. A minimum of six ATP molecules are consumed per purine nucleotide formed. Thus, salvage of purine moieties is energetically economical. Catabolized nucleotides are salvaged in two distinct ways. As noted above, purine nucleoside phosphorylase is a freely reversible reaction in which nucleosides are converted to the corresponding base and ribose-1-phosphate. The characteristic of this reaction favors the synthetic direction.

Furthermore, the purine nucleus is salvaged via the actions of phosphoribosyltransferases. In these reactions, the free bases condense with PP-ribose-P, forming ribonucleotides in a single step. Phosphoribosyltransferase reactions have the following general form:

$$\text{purine base} + \text{PP-ribose-P} \xrightarrow{Mg^{2+}} \text{purine mononucleotide} + PP_i \quad \text{Reaction (21)}$$

In humans, two different purine phosphoribosyltransferases exist. Adenine phosphoribosyltransferase (APRT) converts adenine to AMP. Hypoxanthine–guanine phosphoribosyltransferase (HPRT) converts hypoxanthine and guanine to IMP and guanosine-5′-monophosphate (GMP), respectively. Physiologically, the phosphoribosyltransferase pathways are the most active in the salvage of preformed purine bases. The origin of these bases is from endogenous catabolism of purines, nucleic acid breakdown, and the exogenous intake or administration of purine compounds.

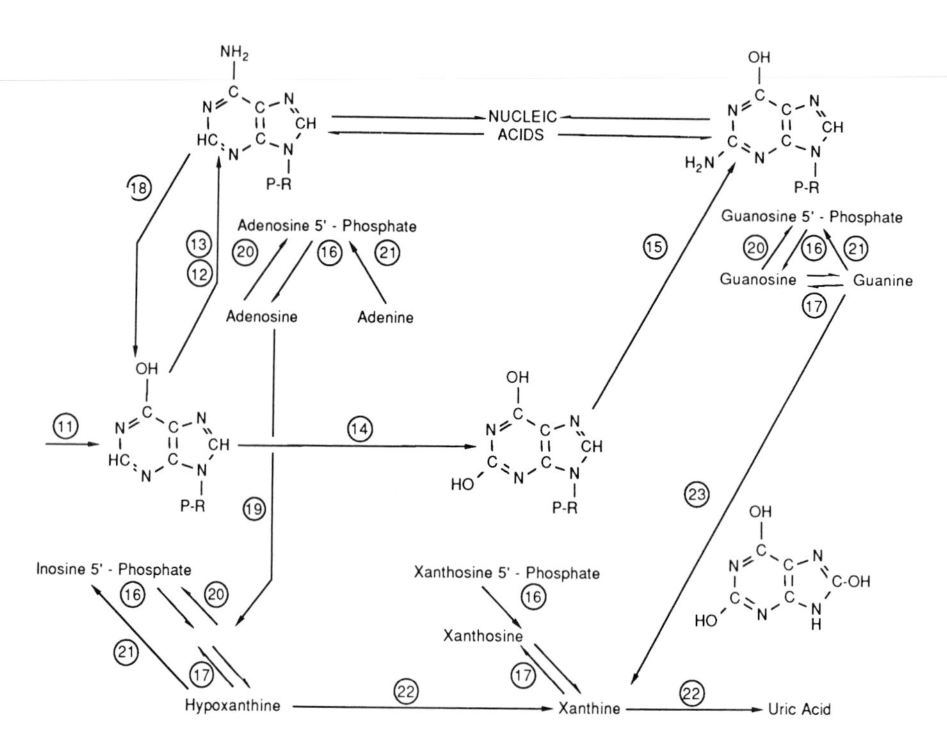

FIGURE 3 Biosynthesis and degradation of purine ribonucleotides, ribonucleosides, and bases. 12. Adenylosuccinate synthetase. 13. Adenylosuccinate lyase. 14. IMP oxidase. 15. XMP-deaminase. 16. 5′-Nucleotidase. 17. Purine nucleoside phosphorylase. 18. AMP deaminase. 19. Adenosine deaminase. 20. Adenosine kinase. 21. Adenine phosphoribosyltransferase and hypoxanthine–guanine phosphoribosyltransferase. 22. Xanthine oxidase. [Modified, with permission, from J. B. Wyngaarden and W. N. Kelley, 1976, "Gout and Hyperuricemia," Grune & Stratton, Orlando.]

4. Synthesis of Uric Acid

Xanthine oxidase is a flavoprotein containing both iron and molybdenum, which oxidizes a wide variety of purines and pteridines. A soluble form of the enzyme having dehydrogenase activity (xanthine dehydrogenase, or D-form) has been isolated from a variety of sources and is likely the normal form of the enzyme. The D-form is converted to the oxidase form (O-form) by the oxidation of thiol groups in the protein. The O-form uses oxygen directly as a substrate. This enzyme converts hypoxanthine to xanthine and xanthine to uric acid:

$$\text{hypoxanthine} + H_2O + 2O_2 \longrightarrow \text{xanthine} + 2O_2^- + 2H^+ \quad \text{Reaction (22a)}$$

$$\text{hypoxanthine} + H_2O + O_2 \longrightarrow \text{xanthine} + H_2O_2 \quad \text{Reaction (22b)}$$

$$\text{xanthine} + H_2O + 2O_2 \longrightarrow \text{uric acid} + 2O_2^- + 2H^+ \quad \text{Reaction (22c)}$$

or

$$\text{xanthine} + H_2O + O_2 \longrightarrow \text{uric acid} + H_2O_2 \quad \text{Reaction (22b)}$$

Superoxide anion or hydrogen peroxide are thus generated. Hydrogen peroxide may then be converted to free hydroxyl radicals:

$$Fe^{2+} + H_2O_2 \longrightarrow Fe^{3+} + OH^- + OH^{\cdot}$$

These compounds (O_2^-, H_2O_2, OH^-) are important mediators of inflammation and tissue destruction. Xanthine oxidase may thus have a significant pathophysiologic role when the flow of blood is blocked (ischemia) or there is tissue injury, which both lead to accentuated adenine nucleotide breakdown.

B. Regulation of Purine Biosynthesis

1. Synthesis of Phosphoribosylamine

The first committed step in purine biosynthesis is condensation of L-glutamine and P-ribose-PP to form phosphoribosylamine. Phosphoribosylamine is the first specific precursor of purines through which *de novo* synthesis proceeds in a direct manner to formation of inosinic acid, from which all other purines are subsequently derived. The enzyme catalyzing this initial reaction, amidophosphoribosyltransferase, is subject to regulation by the end products of purine metabolism, i.e., purine ribonucleotides. The rate at which purine biosynthesis proceeds is controlled at this step by PP-ribose-P: High intracelluar concentrations of PP-ribose-P accelerate purine biosynthesis, whereas depletion of this substrate slows the rate of *de novo* synthesis. Glutamine is rate-limiting for purine synthesis only under very restricted conditions.

2. Purine Ribonucleotide Interconversions

The next major site of regulation of purine ribonucleotide synthesis is at the level of interconversions. As noted above, a variety of reactions occur by which the various nucleotides are interconverted. Each of these reactions is regulated by its nucleotide end product, which typically inhibits the first enzyme in the pathway. The conversion of IMP to AMP is inhibited by the products of the reaction, AMP and guanosine-5′-diphosphate (GDP). Correspondingly, the synthesis of GMP from IMP is inhibited by guanosine-5′-triphosphate (GTP) in a two-step process, which requires ATP as an energy donor. Thus, availability of GTP and ATP, each of which controls the synthesis of the other nucleotide, as well as inhibition by products of the reactions with respect to their own biosynthesis, regulates the steady-state levels of adenyl and guanyl nucleotides.

3. Ribonucleotide Degradation

In vivo and *in vitro* evidence indicates that nucleotide breakdown is regulated in a complex manner. When cultured ascites tumor cells are incubated with 2-deoxyglucose or glucose, these compounds are rapidly phosphorylated by ATP. The abrupt decrease of ATP concentrations from its rapid utilization results in elevations of AMP and IMP. These nucleotides are dephosphorylated to purine nucleosides by 5′-nucleotidases and are subsequently either catabolized or reutilized. An important product of 5′-nucleotidase activity of AMP is adenosine, which is a biologically active compound. Adenosine acts by stimulating its own cell-surface receptors.

Regulation of nucleotide degradation is critically controlled by AMP deaminase (Reaction 18) and 5′-nucleotidases (Reaction 16). AMP deaminase is an allosteric enzyme that is activated by ATP and adenosine-5′-diphosphate (ADP) and inhibited by GTP and P_i. Release of inhibition of AMP deaminase results in accelerated production of uric acid. Regulation at the level of dephosphorylation is more complex and is the focus of intensive investigation. At least three soluble 5′-nucleotidase activities have been described. One form requires very low concentrations of both AMP and IMP for optimal activity and is inhibited by nucleotides. A second form, isolated from a wide variety of sources, hydrolyzes IMP and GMP preferentially but works with much higher concentrations of these substrates. The "high K_m" form is activated by ATP and ADP and inhibited by P_i. Finally, nonspecific phosphatases cleave AMP and also require high concentrations of substrate. Evidence indicates that these three forms of enzymes are located in the same cells.

C. Production and Excretion of Uric Acid

The major determinants of purine synthesis are (1) rate of *de novo* purine synthesis, (2) rates of purine salvage, and (3) rate of exogenous supply of purines. The elimination of purines is controlled by the rates of (1) nucleotide breakdown, (2) synthesis of uric acid, and (3) purine excretion in urine and by extrarenal routes. Thus, the overall rate of purine synthesis reflects the difference between purine intake and *de novo* synthesis and excretion of purine metabolic products, particularly uric acid. In practice, these variables are complex and interrelated, making it very difficult to asses synthetic rates in intact organisms. Estimates of the minimal level of purine production have been obtained by severe dietary restriction of purine intake in humans. A very wide distribution of these rates in normal adults is seen. Correspondingly, urinary uric acid excretion rates also vary widely, because they are a reflection of overall purine production. Values for urinary uric acid excretion in 24 hr are defined on a statistical basis under controlled conditions of dietary intake on the basis of the mean, plus or minus two standard deviations. On isocaloric purine free diets, the value in normal adult men was 418 ± 70 mg in one study and 426 ± 81 mg in another.

Sustained overexcretion of uric acid indicates increased synthesis of purines *de novo*. Normal excretion of uric acid in a gouty subject does not exclude the possibility of overproduction, however, because renal impairment may alter the evident excretor status of the patient. In such subjects, extrarenal disposal of urate is increased and may account for the majority of urate turnover. Probably $<10\%$ of subjects with primary gout overproduce urate.

1. Renal Handling of Uric Acid

Uric acid is fully excreted by the glomeruli. The relative contributions of reabsorption and secretion of urate in the human kidney are difficult to estimate. A four-component model of the renal handling of urate has been proposed, which agrees closely with a variety of observations regarding the renal handling of uric acid. In this model, there are two sites for urate reabsorption in the proximal nephron, separated by the urate secretory site. This model represents the best explanation to date for the accumulated observations on renal urate handling in humans. The relative fluxes of urate through each component remain undetermined and authentication of their existence awaits direct demonstration.

2. Extrarenal Excretion of Uric Acid

When ^{15}N-uric acid is injected into normal subjects, only 75% of the radiolabel is recovered in the urine. Approximately 25% ^{15}N is recovered in urinary allantoin, urea and ammonia, and fecal nitrogen. These compounds originate from bacterial uricolysis within the intestines.

D. Important Disorders of Purine Metabolism

Many acquired and inherited disorders are associated with the purine metabolic pathways (Table I). In addition, these pathways are the target for drugs that have anticancer, antiviral or immunosuppressive properties. Hyperuricemia and gout are the most common abnormalities.

1. Hyperuricemia and Gout

"Gout may be defined as a heterogenous group of diseases found exclusively in man, which in their full development are manifest by (a) an increase in the serum urate concentration; (b) recurrent attacks of a characteristic type of acute arthritis, in which crystals of monosodium urate monohydrate are demonstrable in leukocytes of synovial fluid; (c) aggregated deposits of monosodium urate monohydrate (tophi) occurring chiefly in and around the joints of the extremities and sometimes leading to severe crippling and deformity; (d) renal disease involving glomerular, tubular, and interstitial tissues and blood vessels; and (e) uric acid urolithiasis (kidney stones). These manifestations can occur in different combinations" (Wyngaarden and Kelley, Gout and Hyperuricemia, 1976).

Hyperuricemia needs to be distinguished from gout. Although only a minority of hyperuricemic patients ever become gouty, all patients with gout have hyperuricemia at some stage in their clinical course. In one study, only 12% of 200 hyperuricemic patients had gout.

Using the uricase differential spectrophotometric method for measuring urate valves, the serum urate value is theoretically elevated when it is >7.0 mg/100 ml, the limit of solubility of monosodium urate in serum at 37°C. An elevated serum urate concentration is defined as a value exceeding the upper limit of the mean serum urate value plus two standard deviations in a sex- and age-matched healthy population. In most epidemiological studies, the upper limit has been rounded off at 7.0 mg/100 ml in men and 6.0 mg/100 ml in women. Finally, a serum urate value >7.0 mg/100 ml begins to carry an increased risk of gouty arthritis or renal stones. When methods are used for measuring uric acid that are not specific for urate, the upper limit of normal will be >7.0 mg/100 ml.

a. Clinical Features Symptomatic hyperuricemia includes acute gouty arthritis, chronic gout, and renal disease. These complications of hyperuricemia result from the precipitation of uric acid in body fluids and tissues. The occurrence of this phenomenon depends on the physical properties of uric acid. It is a weak acid that is ionized at position 9 with pK of 5.75. The pH determines the form of uric acid that is deposited. In normal body tissues at physiologic pH, sodium urate is deposited, whereas in urine, which generally has a lower pH, uric acid is precipitated.

The most common complication of hyperuricemia is the development of gouty arthritis. Gout usually passes through four states: asymptomatic hyperuricemia, acute gouty arthritis, intercritical gout, and chronic tophaceous gout. The phase of asymptomatic hyperuricemia ends with the first attack of gouty arthritis or of urolithiasis. While only a minority of patients with hyperuricemia will ever develop gout, in most instances when gout does develop, it is after 20–30 yr of sustained hyperuricemia.

After gouty arthritis, renal disease appears to be the most frequent complication of hyperuricemia. Several types of renal disease have been associated with hyperuricemia. The first type, urate nephropathy, is attributed to the deposition of monosodium urate crystals in the renal interstitial tissue and is thought to be associated with chronic hyperurice-

TABLE I Disorders Associated with Altered Purine Metabolism

Disorder	Basis for Disorder
Hyperuricemia and gout	Idiopathic
	Inherited
	Hypoxanthine–guanine phosphoribosyltransferase deficiency
	PP-ribose-P synthetase overactivity
	Glucose-6-phosphatase deficiency
	Decrease renal urate clearance
	Secondary to drug or disease
Renal calculi	
Uric acid	Hyperuricemia and gout
Xanthine	Xanthine oxidase deficiency
	Inherited
	Acquired
	Allopurinol therapy
2,8-Dehydroxyadenine	Complete adenine phosphoribosyltransferase deficiency
Oxypurinol	Allopurinol therapy
Immune deficiency	Adenosine deaminase deficiency
	Inherited
	Acquired
	Purine nucleoside phosphorylase deficiency
	Ecto-5′-nucleotidase deficiency
	Inherited
	Acquired
Anemia	
Hemolytic	Adenylate kinase deficiency
	Adenosine deaminase increased activity
	Purine nucleoside phosphorylase deficiency
	Acquired adenosine deaminase deficiency
Megaloblastic	Hypoxanthine-guanine phosphoribosyltransferase deficiency
	Purine nucleoside phosphorylase deficiency
	PP-ribose-P synthetase deficiency
Central nervous system disease	Hypoxanthine-guanine phosphoribosyltransferase deficiency
	PP-ribose-P synthetase overactivity
	Adenylosuccinate lyase deficiency
	Guanine deaminase deficiency
	Purine nucleoside phosphorylase deficiency
Myopathy	Myoadenylate deaminase deficiency
	Xanthine oxidase deficiency
	Metabolic myopathies
No disease	Partial deficiency of adenine phosphoribosyltransferase

mia. In contrast, uric acid nephropathy is related to the formation of uric acid crystals in the collecting tubules, pelvis, or ureter, with subsequent impairment of urine flow. This disorder is caused by elevated concentrations of uric acid in the urine and can appear as either acute uric acid nephropathy or uric acid calculi.

Hyperuricemia and gout are well-recognized complications of chronic lead intoxication, but its occurrence varies considerably. The basis for this association may be related to the ability of lead urate to undergo nucleation (nucleus around which a crystal forms) at a lower concentration than is required for the nucleation of monosodium urate.

Acute renal failure can result from the precipitation of uric acid crystals in the collecting ducts and ureters. This complication most commonly occurs in patients with leukemias and lymphomas as a result of rapid malignant cell turnover, often during chemotherapy.

b. Treatment The natural history of hyperuricemia and gout may be modified by lowering the serum urate levels. This is accomplished by drug

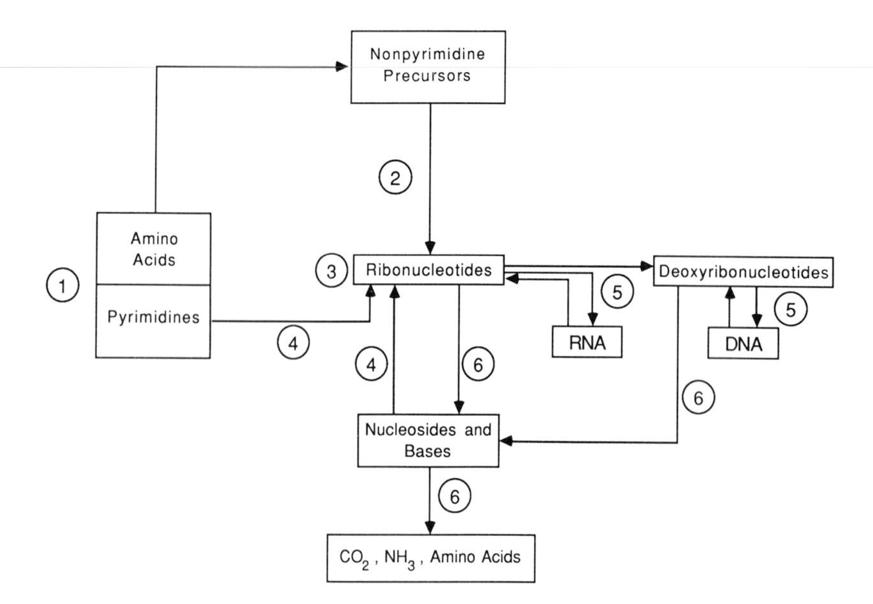

FIGURE 4 Overview of pyrimidine metabolism. Human pyrimidine metabolism is focused on the synthesis of the pyrimidine ribonucleotides in a manner analogous to purines (Fig. 1). Dietary intake of amino acids (1) provides nonpurine precursors for *de novo* biosynthesis (2) leading to UMP formation. UMP is converted to other pyrimidine nucleotides by interconversion pathways (3). Preformed pyrimidine nucleosides and bases from the diet (1) and nucleotide catabolism (5) may be reutilized via salvage pathways (4). Pyrimidine nucleoside monophosphates are converted to diphosphate and triphosphate forms (3), which are used in the synthesis of nucleic acids (5). Degradation proceeds through the ribonucleotides to nucleosides and bases (6). Free pyrimidine bases are further degraded to amino acids, CO_2, and NH_3, the final products of pyrimidine catabolism. [From T. D. Palella and I. H. Fox *in* "The Molecular & Metabolic Basis of Acquired Disease" (R. D. Cohen, ed.), Balliere Tindall, in press 1989. Used with permission.]

therapy, which either inhibits uric acid synthesis at xanthine oxidase (allopurinol) or increases the renal excretion of uric acid (sulfinpyrazone or probenecid). The acute attacks of gout are treated by nonsteroidal anti-inflammatory drugs, such as indomethacin.

II. Pyrimidines

Pyrimidine metabolism is analogous to purine metabolism (Fig. 4). In pyrimidine metabolism, all pyrimidine nucleotides derive from uridine-5′-monophosphate (UMP). Pyrimidine nucleotides may be synthesized *de novo* or by reutilization of preformed bases and nucleosides. Interconversion of pyrimidines occurs at the monophosphate level. Degradation of pyrimidines results in formation of an amino acid, ammonia, and carbon dioxide. The overall economy of pyrimidine metabolism is regulated to some extent at each level.

A. Metabolic Pathways

1. Pyrimidine Biosynthesis

The *de novo* synthesis of the parent pyrimidine compound UMP is depicted in Fig. 5. The initial step is the formation of the unstable high-energy compound carbamylphosphate (CAP). Although two different enzymatic activities catalyzing the formation of CAP have been described in eukaryotes, CAP-synthetase II (CPS II) appears to be the activity associated with *de novo* pyrimidine metabolism.

CPS II is a cytoplasmic enzyme that exists in mammalian tissues as part of a three-enzyme complex. The other two enzymes, aspartate transcarbamylase and dihydroorotase, catalyze the second and third steps of *de novo* pyrimidine biosynthesis, respectively. This complex is designated *pyr* 1–3 (Fig. 5). The product of these reactions, dihydroorotic acid, is reversibly oxidized to orotic acid to form the pyrimidine ring in a reaction catalyzed by dihydroorotic acid dehydrogenase. This enzyme is associated with the outer surface of the inner mitochondrial membrane in contrast to the cytosolic location of the other five enzymes of the pyrimidine *de novo* pathway.

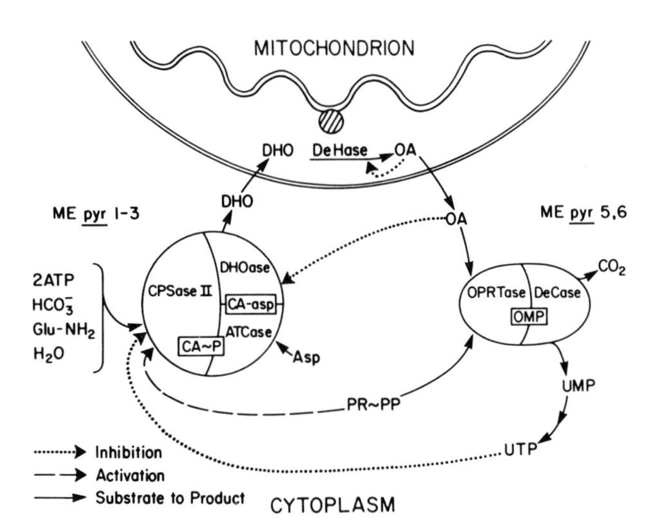

FIGURE 5 *De novo* pyrimidine biosynthesis. Multienzyme (ME) *pyr* 1–3 and *pyr* 5,6 are cytosolic proteins. ME *pyr* 1–3 has the enzyme centers for carbamylphosphate synthetase (CPSase II), aspartate transcarbamylase (ATCase), and dihydro-orotase (DHOase). ME *pyr* 5,6 has the active centers for orotate phosphoribosyltransferase (OPRTase) and orotidylate decarboxylase (OMP DeCase). Dihydro-orotate dehydrogenase (DHO DeHase) is on the outer surface of the inner membrane of the mitochondrion. The substrates are ATP, L-glutamine (Glu-NH_2), HCO_3^-, L-aspartate (Asp), and 5-phosphoribosyl-1-pyrophosphate (PR-PP). The intermediates of the pathway are carbamyl phosphate (CA-P), carbamyl-L-aspartate (CA-asp), dihydro-orotate (DHO), orotate (OA), orotidylate (OMP), and uridylate (UMP). Solid lines with arrows indicate substrate-to-product reactions. CA-P, CA-asp, and OMPase were placed in boxes because these intermediates are not readily released from the multienzymic proteins into the cytoplasmic solution. [From an article by M. E. Jones. Reproduced, with permission, from the Annual Review of Biochemistry, Volume 49, © 1980 by Annual Reviews Inc.]

In the penultimate steps of the *de novo* pathway (*pyr* 5,6), orotidine-5′-monophosphate (OMP) is formed by orotate phosphoribosyltransferase (OPRT), a reaction similar to the salvage reactions of purine metabolism catalyzed by HPRT and APRT. OPRT activity is closely associated with the catalytic activity of the next enzyme in the pathway, orotidine-5′-monophosphate decarboxylase (ODC). Whether OPRT and ODC activities are associated as a single multifunctional protein or as a complex whose aggregation is dependent on the presence of two enzymes is not clear.

Pyrimidine nucleotides may also be formed by reutilization of preformed bases and nucleosides from dietary sources or nucleotide catabolism (Fig. 6). Uracil may be salvaged by two mechanisms. Uracil is directly converted to UMP by OPRT (Reaction 1) or is converted to uridine by the ubiquitous enzyme uridine phosphorylase (Reaction 2), which has specificity only for uracil and its analogues. Uridine is then converted to UMP by the action of uridine kinase (Reaction 3), which phosphorylates uridine and cytidine.

Uracil and thymine are ultimately converted to deoxyribonucleosides by a distinct enzyme, deoxythymidine phosphorylase, which requires deoxyribose-1-phosphate as a substrate. A separate reaction catalyzed by deoxythymidine synthetase (Reaction 15) converts thymine to deoxythymidine. Thymidine (i.e., 2-deoxythymidine) is converted to its monophosphate form (dTMP) by thymidine kinase (TK):

$$\text{deoxythymidine} + \text{ATP} \rightarrow \text{dTMP} + \text{ADP} \quad \text{(Reaction 14)}$$

TK is widely distributed in human tissues. Like uridine kinase, TK is present in fetal and adult liver in distinct forms. Deoxycytidine is converted to 2′-deoxycytidine-5′-monophosphate (dCMP) by a different enzyme, deoxycytidine kinase (Reaction 17), which also phosphorylates deoxyguanosine and adenosine.

2. Interconversions

Pyrimidine interconversions are also summarized in Figure 6. Ribonucleoside monophosphates are phosphorylated by kinases to di- and triphosphate forms. One kinase subserves phosphorylation of cytidine-5′-monophosphate (CMP), UMP, and dCMP (Reaction 11). A second activity phosphorylates deoxyadenosine-5′-triphosphate (dTMP) and 2′-deoxyuridine-5′-monophosphate (dUMP) (Reaction 16). Either ATP or 2′-dTMP (dATP) can donate the phosphate moiety. Deoxycytidylate deaminase (Reaction 6) catalyzes the deamination of dCMP to dUMP. TMP (dTMP) synthesis from dUMP is controlled by the availability of folate derivatives by a reaction catalyzed by thymidylate synthetase:

$$\text{dUMP} + \text{5,20-methylene-5,6,7,8-tetrahydrofolate} \rightarrow \text{dTMP} + \text{7,8-dihydrofolate} \quad \text{(Reaction 10)}$$

Pyrimidine nucleoside diphosphates are converted to triphosphates via nucleoside diphosphokinase. The reaction has the following general form:

$$\text{NDP} + \text{N}_1\text{TP} \rightarrow \text{NTP} + \text{N}_1\text{DP}$$

and

$$\text{dNDP} + \text{N}_1\text{TP} \rightarrow \text{dNTP} + \text{N}_1\text{DP} \quad \text{(Reaction 12)}$$

In this reaction, N and N_1 may be any purine or pyrimidine base. (Deoxynucleotides are indicated as dNDP and dNTP.)

Nucleoside diphosphates are reduced to deoxy

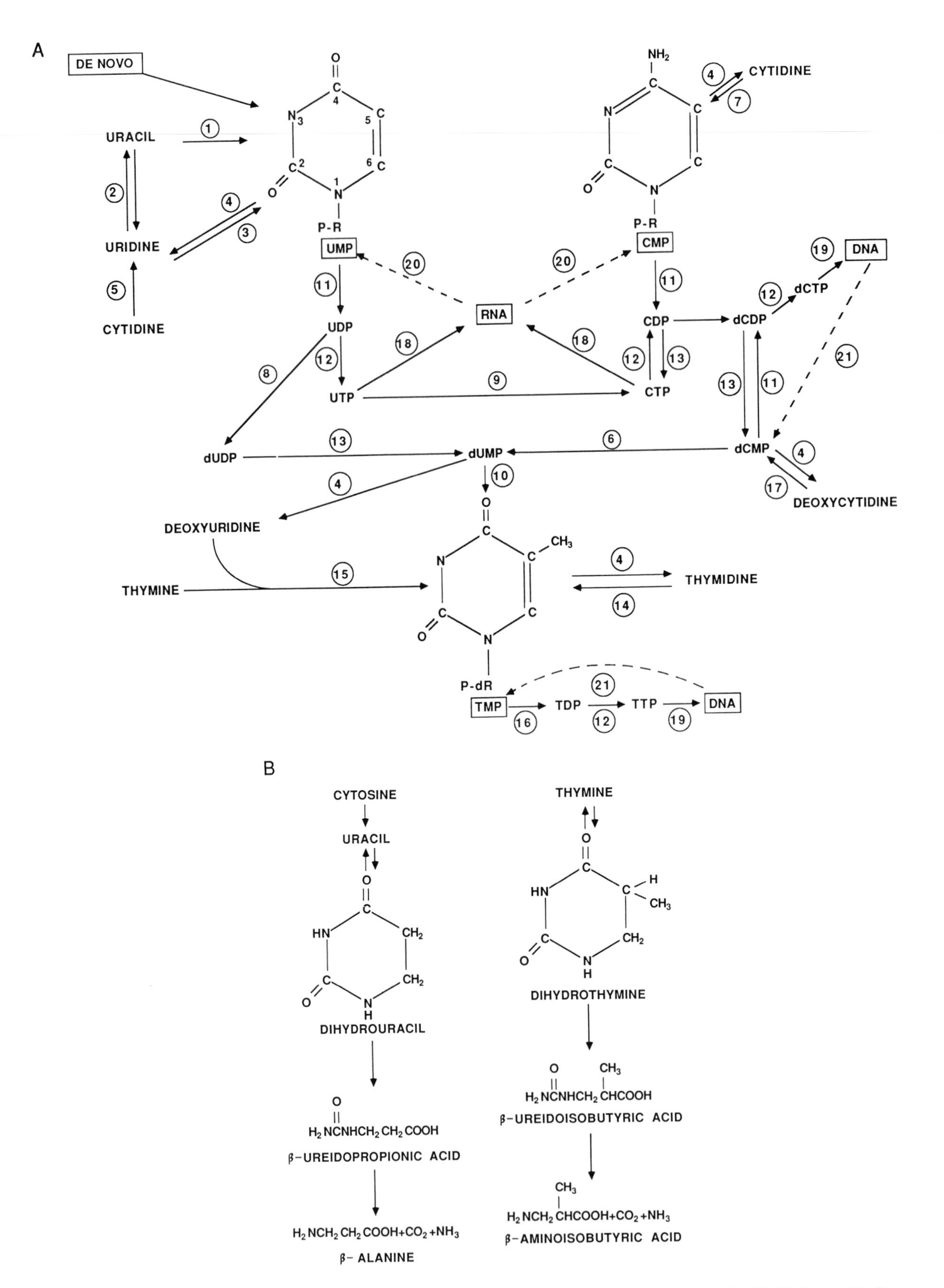

FIGURE 6 A. Biosynthesis of pyrimidine nucleosides, nucleotides, and bases. 1. Orotate phosphoribosyltransferase. 2. Uridine phosphorylase. 3. Uridine kinase. 4. 5′-Nucleotidase. 5. Cytidine deaminase. 6. Cytidylate deaminase. 7. Cytidine kinase. 8. Ribonucleotide reductase. 9. CTP synthetase. 10. Thymidylate synthetase. 11. CMP–UMP–dCMP kinase. 12. Nucleoside diphosphate kinase. 13. Phosphatase. 14. Thymidine kinase. 15. Thymidine synthetase. 16. TMP–dUMP kinase. 17. Deoxycytidine kinase. 18. RNA polymerase. 19. DNA polymerase. 20. Ribonucleases. 21. Deoxynucleases. B. Degradation of pyrimidines. [From T. D. Palella and I. H. Fox *in* "The Molecular & Metabolic Basis of Acquired Disease" (R. D. Cohen, ed.), Balliere Tindall, in press 1989. Used with permission.]

forms by ribonucleotide reductase:

$$\text{NDP} + \text{thioredoxin-(SH)}_2 + \xrightarrow{Mg^{2+}} \text{dNDP} + \text{thioredoxin} - S_2 \quad \text{(Reaction 8)}$$

Thioredoxin-$(SH)_2$ is regenerated by the action of thioredoxin reductase. Ribonucleotide reductase is subject to complex feedback regulation.

3. Degradation

Pyrimidine mononucleotides are catabolized to nucleosides by the action of both nonspecific phosphatases and 5′-nucleotidase (Reaction 4). Cytidine is deaminated to uridine by cytidine deaminase (Reaction 5). Uridine and thymidine are catabolized to uracil and thymine by pyrimidine nucleoside phosphorylase (Reaction 2). Uracil and thymine are converted to dihydro forms by specific reactions catalyzed by dihydrouracil dehydrogenase and dihydrothymine dehydrogenase, respectively (Fig. 6). The resultant dihydropyrimidines are hydrolyzed to their respective carbamyl-β-amino acids by dihydropyrimidinase. Products of these reactions are degraded to corresponding amino acids, CO_2, and ammonia by specific enzymes (e.g., β-ureidoproionase). Thus, the end products of pyrimidine metabolism are common metabolites that are not unique to a single metabolic pathway.

B. Regulation of Pyrimidine Metabolism

The first reaction in *de novo* pyrimidine biosynthesis, which is catalyzed by CPS II, is regulated by several mechanisms. CPS II is saturated with respect to HCO_3^- and glutamine but not to ATP. CPS II activity increases with elevated PP-ribose-P or ATP concentrations as well as with reduced UTP concentrations.

The association of CPS II, aspartate transcarbamylase, and dihydroorotase activities in the multifunctional protein *pyr* 1–3 results in coordinate regulation of the initial three steps of *de novo* pyrimidine synthesis. This reflects changes in the amount of enzyme protein, which has been demonstrated in a variety of systems including rapidly growing tissues, fetal tissues, neoplastic tissues, and blast-transformed lymphocytes. In one experimental model, gene amplification resulting in increased amounts of a single mRNA and higher production of the enzyme species has been suggested. Evidence also suggests that PP-ribose-P may increase *pyr* 1–3 levels by direct action on the gene encoding the multifunctional complex.

Other enzymes may also serve as regulatory points within the pyrimidine synthetic pathway. Evidence suggests that OPRT may be rate-limiting for the entire pathway. Because OPRT works at 10% saturation with PP-ribose-P, in cells, levels of this substrate may regulate the rate of the entire pathway through this enzyme. Finally, the rate of pyrimidine catabolism may influence the synthetic rate through modulation of uridine-5′-triphosphate (UTP) concentrations. The terminal enzyme in pyrimidine catabolism, N-carbamyl-β-alanine (NCβA) aminohydrolase is regulated in opposing fashion by its substrate, NCβA, and its product, β-alanine. This effect is mediated by ligand-induced changes in enzyme polymerization.

In summary, the weight of evidence suggests that CPS II predominates as the major regulatory point in pyrimidine biosynthesis. When ATP concentrations are increased and both PP-ribose-P and uridine nucleotide levels are low, OPRT may become rate-limiting and thus an important regulatory site.

C. Interrelationship of Purine and Pyrimidine Metabolism

PP-ribose-P is an essential substrate and important regulatory intermediate in both purine and pyrimidine metabolism. In purine biosynthesis, PP-ribose-P is rate-limiting in three reactions—i.e., those catalyzed by amidophosphoribosyltransferase, hypoxanthine-guanine phosphoribosyltransferase, and adenine phosphoribosyltransferase. In pyrimidine metabolism, PP-ribose-P activates CPS II, may induce *pyr* 1–3, and is a rate-limiting substrate for OPRT. Thus, increases in intracellular PP-ribose-P levels accelerate both purine and pyrimidine biosynthesis, whereas decreases in PP-ribose-P levels decrease the synthetic rates. Therefore, a potential mechanism for the coordinated synthesis of nucleotides for nucleic acid synthesis exists.

D. Important Disorders of Pyrimidine Metabolism

Only a few clinical disorders exist in the pyrimidine pathway. Orotic aciduria refers to the excessive excretion of orotic acid. It may be a rare inherited disorder in babies with severe anemia or occur secondarily to other diseases or drug intake.

Like the purine pathways, the pyrimidine pathways are the target for antiviral and anticancer drugs. A relatively common defect is pyrimidine 5′-nucleotidase deficiency, which is described below.

Pyrimidine 5′-nucleotidase deficiency has been observed in the erythrocytes of patients with an inherited hemolytic anemia or an acquired disorder related to lead intoxication.

Patients with lead intoxication have pyrimidine 5′-nucleotidase deficiency that worsens with increasing lead concentrations in the blood, a mild anemia, and large numbers of red cells with basophilic stippling. The inherited disorder associated with erythrocyte pyrimidine 5′-nucleotidase deficiency is characterized by congenital hemolytic anemia, enlargement of the spleen, and conspicuous basophilic stippling of peripheral erythrocytes. The hemoglobin ranges from 6 to 12 g/dl and the hematocrit (red cell volume) from 28 to 34%. There is a prominent reticulocytosis (presence of red cell precursors in circulations) with values ranging as high as 50%, poikilocytosis (red cell shape irregularities), moderate polychromasia, and mild anisocytosis; there frequently is hyperbilirubinemia.

The diagnosis of hemolytic anemia due to 5′-nucleotidase deficiency may be proven by demonstrating the enzyme deficiency in peripheral circulating erythrocytes using either UMP or CMP as substrate:

$$\text{UMP} + H_2O \rightarrow P_i + \text{uridine}$$
$$\text{CMP} + H_2O \rightarrow P_i + \text{cytidine}$$

The enzyme deficiency is a sensitive indicator of lead intoxication and is found to be decreased even when most other biologic indicators of lead intoxication are negative.

The enzyme deficiency leads to an inability to degrade pyrimidine nucleotides causing elevated cytidine nucleotide diphosphodiesters. Cytidine diphosphate choline and cytidine diphosphate ethanolamine account for 55% of the abnormal red cell pyrimidine nucleotides accumulated in this disorder. This leads ultimately to the basophilic reticulum of normal reticulocytes, and basophilic granules or stippling in erythrocytes are composed of RNA. The granules result from the retarded RNA degradation secondary to inability to dephosphorylate and render diffusible pyrimidine degradation products of RNA. The hemolysis results from reversible inhibition of the pentose phosphate shunt by elevated levels of lead and the pyrimidine 5′-nucleotides.

Treatment of lead-induced 5′-nucleotidase deficiency can be performed with calcium disodium edetate intramuscularly in addition to stopping toxic lead intake. Blood lead levels showed a linear decrease while erythrocyte pyrimidine 5′-nucleotidase levels show a linear increase in value.

BIBLIOGRAPHY

Dieppe, P. A., Doherty, M., and MacFarlane, D. (1983). Crystal-related arthropathies. *Ann. Rheumatic Dis.* **42,** 1.

Fox, I. H. (1981). Metabolic basis for disorders of purine nucleotide degradation. *Metabolism* **30,** 616.

Fox, I. H. (1982). Disorders of purine and pyrimidine metabolism. *In* "Clinical Medicine" (J. A. Spittel, ed.). Harper & Row, Philadelphia.

Kelley, W. N., Fox, I. H. and Palella, T. D. (1989). Gout and related disorders of purine metabolism. *In* "Textbook of Rheumatology" (W. N. Kelley, E. D. Harris, Jr., S. Ruddy, and C. Sledge, eds.). W. B. Saunders, Philadelphia.

Palella, T. D., and Fox, I. H. (1989). Disorders of purine and pyrimidine metabolism. *In* "The Metabolic and Molecular Basis of Acquired Disease" (R. D. Cohen, K. G. M. M. Alberti, B. Lewis, and A. M. Denman, eds.). Bailliere Tindall, London. In press.

Palella, T. D., and Fox, I. H. (1989). Hyperuricemia and gout. *In* "The Metabolic Basis of Inherited Disease" (C. R. Scriver, A. L. Beaudet, W. S. Sly, and D. Valle, eds.). p 965–1006. McGraw-Hill Book Co., New York.

Wyngaarden, J. B., and Kelley, W. N. (1976). "Gout and Hyperuricemia." Grune & Stratton, New York.

Quinolones

VINCENT T. ANDRIOLE, *Yale University School of Medicine*

Glossary

Cinnolones Subgroup of the quinolone class of antimicrobial agents, identified by an additional nitrogen in the 2-position of the nucleus, and also referred to as 2-aza-4-quinolones

DNA gyrase Bacterial enzyme that nicks double-stranded chromosomal DNA, introduces supercoils, and then seals the nicked DNA. Also referred to as DNA topoisomerase and nicking–closing enzyme

Naphthyridines Subgroup of the quinolone class of antimicrobial agents, identified by an additional nitrogen in the 8-position of the nucleus, and also referred to as 8-aza-4-quinolones

Pyrido-pyrimidines Subgroup of the quinolone class of antimicrobial agents, identified by additional nitrogens in the 6- and 8-positions of the nucleus, and also referred to as 6,8-diaza-4-quinolones

Quinolones Completely synthetic class of highly effective antimicrobial agents

QUINOLONES ARE a class of antimicrobial agents that are completely synthetic and highly effective in the treatment of many different types of infectious diseases, primarily those caused by bacteria. These agents also have variable activity against *Mycoplasmas, Chlamydiae, Rickettsiae,* and some *Plasmodium* species, but are not known to be active against viruses and fungi. The quinolone antimicrobial agents are analogs of nalidixic acid, which was developed and introduced in 1962. The older analogs, pipemidic acid, oxolinic acid, and cinoxacin, were developed later. Shortly thereafter the development of the newer quinolones progressed rapidly and was spearheaded by the insertion of a fluorine at the 6-position in the basic nucleus. This chemical modification was observed to enhance and broaden the antibacterial activity of these agents and led to the discovery of newer 4-quinolones with antibacterial activities 1000 times that of nalidixic acid. More than 1000 quinolones and their analogs have been synthesized and evaluated for their activity against bacteria. The clinical importance of the newer quinolones is based on their extremely broad antibacterial spectrum, unique mechanism of action, good absorption from the gastrointestinal tract after oral administration, excellent tissue distribution, and low incidence of adverse reactions.

I. Chemistry and Classification

The quinolone antibacterial agents are all structurally similar compounds. Yet there are some differences in the basic nucleus so that they can be divided into four general groups, i.e., naphthyridines, cinnolines, pyrido-pyrimidines, and quinolines (Fig. 1). A common skeleton, 4-oxo-1,4-dihydroquinolone, more commonly called 4-quinolone, is produced by the addition of an oxygen at the 4-position in the basic nucleus. The naphthyridines (nalidixic acid and enoxacin), with an additional nitrogen in the 8-position, are 8-aza-4-quinolones. The cinnoline cinoxacin, with an additional nitrogen

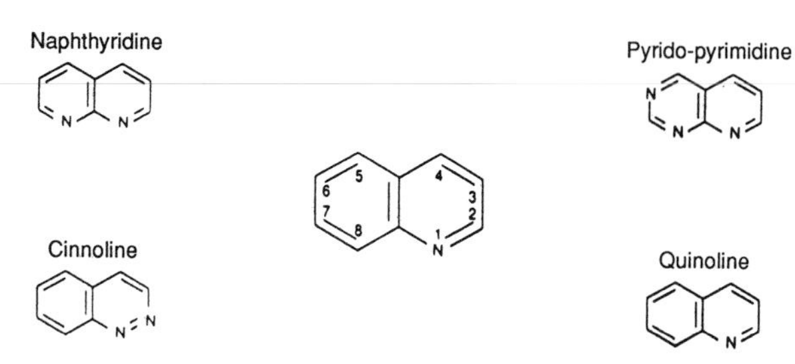

FIGURE 1 Chemical structure of the four general groups of the 4-quinolones and its system of ring numbering. Reprinted with permission from V. T. Andriole (1989). Quinolones. *In* "Principles and Practice of Infectious, Diseases" (G. L. Mandell, R. G. Douglas, and J. E. Bennett, eds.), 3rd Ed. Churchill Livingstone, New York.

in the 2-position, is a 2-aza-4-quinolone. The pyrido-pyrimidines (pipemidic and piromidic acids), with additional nitrogens in the 6- and 8-positions, are 6,8-diaza-4-quinolones (Fig. 2). All of the other highly active agents [acrosoxacin (rosoxacin), amifloxacin, ciprofloxacin, difloxacin, fleroxacin, flumequine, irloxacin (pirfloxacin), lomefloxacin, norfloxacin, ofloxacin, oxolinic acid, pefloxacin, piroxacin, and temafloxacin] are classified as 4-quinolones (Fig. 3). Numerous additional compounds have been synthesized and are undergoing development.

The 4-quinolones have certain common structural features which include a carboxyl group at the 3-position and a piperazine ring (except flumequine and oxolinic acid) at the 7-position of the quinolone nucleus. Substitutions at the 1- and 8-positions of the quinolone or the naphthyridine nucleus and the para position of the piperazine ring plus the introduction of a fluorine at the 6-position are responsible for differences in the *in vitro* activity and pharmacologic properties of the newer 4-quinolone antimicrobial agents (Figs. 2 and 3).

FIGURE 3 Chemical structure of the 4-quinolones that are derivatives of the quinolone nucleus. Reprinted with permission from V. T. Andriole (1989). Quinolones. *In* "Principles and Practice of Infectious, Diseases" (G. L. Mandell, R. G. Douglas, and J. E. Bennett, eds.), 3rd Ed. Churchill Livingstone, New York. →

Nalidixic acid (1-ethyl-7-methyl-1,8-naphthyridine-4-one-3-carboxylic acid) is one of a series of 1,8-naphthyridine derivatives. This compound is only slightly soluble in water but is soluble in dilute alkali and is stable in urine. It is marketed as Negram (Winthrop, New York, 1964). Oxolinic acid (5-ethyl-5,8-dihydro-8-oxo-1,3-dioxolo[4,5-*g*]quinoline-7-carboxylic acid), a quinolone derivative, is a crystalline substance that is a weak organic acid. It is marketed as Utibid (Warner-Lambert, Morris Plains, New Jersey). Cinoxacin (1-ethyl-1,4-dihy-

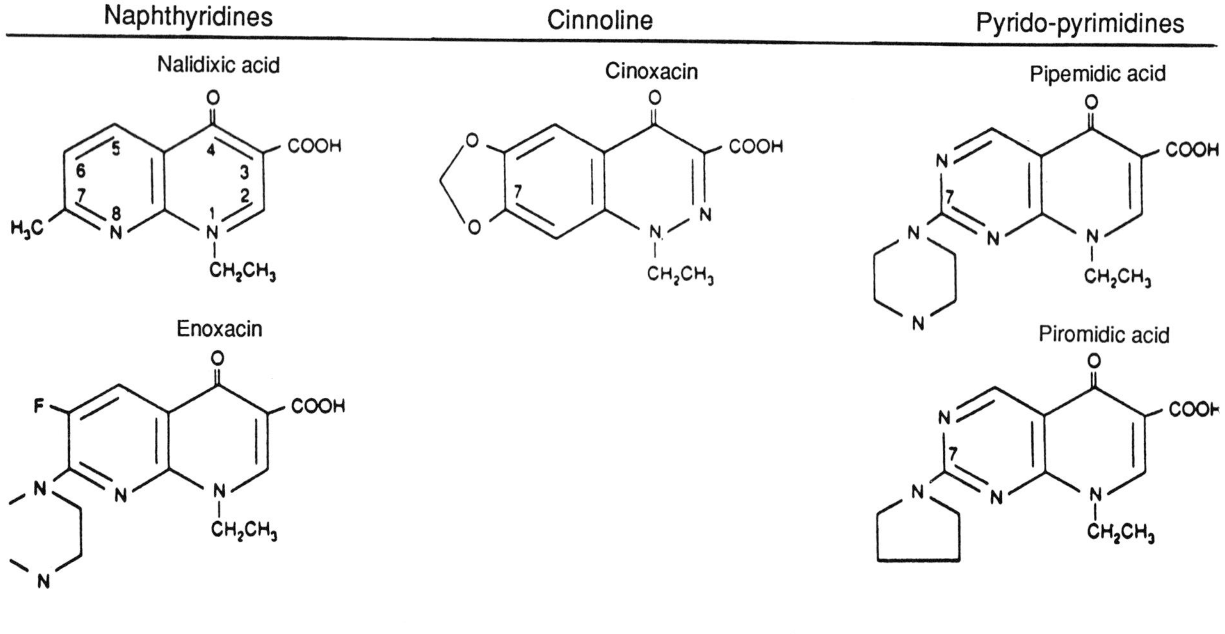

FIGURE 2 Structures of napthyridine, cinnoline, and pyrido-pyrimidine derivatives. Reprinted with permission from V. T. Andriole (1989). Quinolones. *In* "Principles and Practice of Infectious, Diseases" (G. L. Mandell, R. G. Douglas, and J. E. Bennett, eds.), 3rd Ed. Churchill Livingstone, New York.

Naphthyridines
Nalidixic acid
Enoxacin

Cinnoline
Cinoxacin

Pyrido-pyrimidines
Pipemidic acid
Piromidic acid

Acrosoxacin

Amifloxacin

Ciprofloxacin

Difloxacin

Fleroxacin

Flumequine

Irloxacin

Lomefloxacin

Norfloxacin

Ofloxacin

Oxolinic acid

Pefloxacin

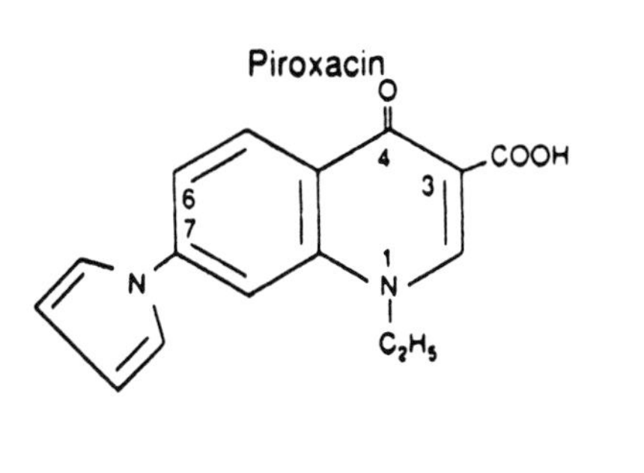

Temafloxacin

dro-4-oxo-1,3-dioxolo[4,5-*g*]cinnoline-3-carboxylic acid) is a yellow–white crystalline solid with a pK_a of 4.7. It is insoluble in water, poorly soluble in lipids, but is soluble in alkaline solution. It is marketed as Cinobac (Dista/Lilly, Indianapolis, Indiana). Norfloxacin [1-ethyl-6-fluoro-1,4 dihydro-4-oxo-7-(1-piperazinyl)-3-quinolone carboxylic acid] is a yellow–white crystalline solid only slightly soluble in water. Norfloxacin is marketed as Noroxin (Merck, West Point, Pennsylvania). Ciprofloxacin [1-cyclopropyl-6-fluoro-1, 4-dihydro-4-oxo-7-(1-piperazinyl)-3-quinolone carboxylic acid hydrochloride] is a light yellow crystalline substance slightly soluble in water. Ciprofloxacin is marketed as Cipro (Miles/Bayer, West Haven, Connecticut). Other newer quinolones which are currently being tested and which may be marketed soon include the following: enoxacin [1-ethyl-6-fluoro-1,4-dihydro-4-oxo-7-(1-piperazinyl)-1,8-naphthyridine-3-carboxylic acid] (Warner-Lambert, Morris Plains, New Jersey), ofloxacin [9-fluoro-2,3-dihydro-3-methyl-10-(4-methyl-1-piperazinyl)-7-oxo-7*H*-pyrido[1,2,3-*de*]-1,4-benzoxacine-6-carboxylic acid] (Ortho, Raritan, New Jersey), fleroxacin [6,8-difluoro-1-(2-fluoro-ethyl)-1,4-dihydro-7-(4-methyl-1-piperazinyl)-4-oxo-3-quinoline carboxylic acid] (Roche, Nutley, New Jersey), lomefloxacin [1-ethyl-6,8-difluoro-1,4-dihydro-7-(3-methyl-1-piperazinyl)-4-oxo-3-quinoline carboxylic acid] (Searle, Skokie, Illinois), and pefloxacin [1-ethyl-6-fluoro-1,4-dihydro-7-(4-methyl-1-piperazinyl)-4-oxo-3-quinoline carboxylic acid] (Bellon/Dianippon).

Many other newer quinolone compounds are currently under various stages of development. Some of these include temafloxacin [6-fluoro-1-(2,4-difluoro-phenyl)-1,4-dihydro-7-(4-methyl-1-piperazinyl)-4-oxo-3-quinolone carboxylic acid] (Abbott, Chicago, Illinois), difloxacin [6-fluoro-1-(4-fluoro-phenyl-1,4-dihydro-7-(4-methyl-1-piperazinyl)-4-oxo-3-quinolone carboxylic acid] (Abbott), acrosoxacin (rosoxicin) [1-ethyl-1,4-dihydro-4-oxo-7-(4-pyridyl)-3-quinolone carboxylic acid] (Sterling), amifloxacin [6-fluoro-1,4-dihydro-1-methylamino-7-(4-methyl-1-piperazinyl)-4-oxo-3-quinolone carboxylic acid] (Sterling), flumequine [9-fluoro-6,7-dihydro-5-methyl-1-oxo-1H,5H-benzo[ij]-2-quinolicine carboxylic acid] (Riker), irloxacin (pirfloxacin) [1-ethyl-6-fluoro-1,4-dihydro-4-oxo-7-(1-pyrrolyl)-3-quinolone carboxylic acid] (Esteve), and piroxacin [1-ethyl-1,4-dihydro-4-oxo-7-(1-pyrrolyl)-3-quinolone carboxylic acid] (Esteve).

II. Mechanism of Action

The molecular basis for the potent antibacterial effects of the newer quinolones has not yet been determined definitively. However, nalidixic acid has been shown to rapidly inhibit DNA synthesis in susceptible bacterial cells but not in mammalian cells, whereas protein and RNA synthesis are not inhibited. However, bacterial cells exposed to nalidixic acid will resume growth when placed in drug-free media. Thus, inhibition of DNA synthesis by nalidixic acid is reversible. Although nalidixic acid does not bind to purified DNA, recent work suggests that some quinolones may bind to DNA, and that the bactericidal activity of nalidixic acid as well as other newer 4-quinolones is reduced significantly if RNA or protein synthesis is inhibited. All 4-quinolones are bactericidal. However, these drugs have a single concentration that is most bactericidal, so that higher or lower concentrations produce less bacterial death. This paradoxical effect of decreased killing at greater concentrations is most likely caused by a dose-dependent inhibition of RNA synthesis.

Other more recent studies indicate that the mechanism of action of nalidixic acid, as well as the newer quinolones, is by inhibition of DNA topoisomerases (gyrases), of which four subunits (two A monomers and two B monomers) have been defined. The topoisomerases, which have been found in every organism examined, supercoil strands of bacterial DNA in the bacterial cell. Each chromosomal domain is transiently nicked during supercoiling, which results in single-stranded DNA. When supercoiling is completed, the single-stranded DNA state is abolished by an enzyme that seals the nicked DNA. The sealing action of this enzyme is inhibited specifically by nalidixic acid. Thus the enzyme, termed DNA gyrase or topoisomerase II (nicking–closing enzyme), nicks double-stranded chromosomal DNA, introduces supercoils, and seals the nicked DNA. The A subunits are thought to introduce the nicks, the B subunits to supercoil, and then the A subunits seal the nick they produced initially. Nalidixic acid prevents the A subunits from sealing the nicks in chromosomal DNA. However, the newer 4-quinolones may act slightly differently and may affect both the A and B subunits of DNA gyrase since mutations that affect the B subunit change the bacterial sensitivity to the 4-quinolones. The identification of DNA gyrase has provided the opportunity to develop new quinolone

TABLE I *In Vitro* Activity of Selected 4-Quinolones[a]

Organism	Nalidixic acid	Ciprofloxacin	Enoxacin	Norfloxacin	Ofloxacin	Pefloxacin
Gram-negative aerobes						
Escherichia coli	8 (4–128)	0.03 (0.015–0.06)	0.5 (0.25–1)	0.125 (0.06–0.5)	0.125 (0.06–0.25)	0.125 (0.125–0.25)
Klebsiella pneumoniae	8 (1–128)	0.125 (0.06–0.25)	0.5	0.25 (0.125–1)	0.25	0.5
Enterobacter spp.	32 (4–128)	0.125 (0.03–0.5)	0.5 (0.25–4)	0.5 (0.125–2)	0.5 (0.125–1)	0.5 (0.25–1)
Citrobacter spp.	8 (4–>100)	0.03 (0.03–0.06)	0.5	0.25 (0.125–0.5)	0.5	0.5
Serratia marcescens	>128 (16–>256)	1 (0.25–2)	2 (0.5–4)	1 (0.5–8)	1 (0.25–2)	1 (0.25–2)
Shigella spp.	4	0.03 (0.015–0.06)	0.125	0.06 (0.03–0.125)	0.125 (0.06–0.125)	0.125
Salmonella spp.	8 (4–8)	0.015 (≤0.015–0.03)	0.25 (0.125–0.25)	0.125 (0.06–0.125)	0.125 (0.06–0.125)	0.125 (0.06–0.25)
Proteus mirabilis	16 (4–32)	0.06 (0.03–0.125)	0.5 (0.25–1)	0.25 (0.125–0.5)	0.25 (0.25–0.5)	0.5 (0.25–1)
Proteus spp. (indole positive)	8 (4–16)	0.06	0.25 (0.25–0.5)	0.125 (0.06–0.125)	0.25	0.25
Morganella morganii	8 (2–8)	0.015 (0.015–0.03)	0.25 (0.25–0.5)	0.125 (0.03–0.25)	0.125 (0.125–0.25)	0.25 (0.25–0.5)
Pseudomonas aeuruginosa	≥128	0.5 (0.25–1)	4 (2–8)	2 (0.06–8)	4 (2–4)	4 (2–8)
Hemophilus influenzae	1 (1–2)	0.015 (0.015–0.03)	0.125 (0.06–0.25)	0.06 (0.03–0.125)	0.03 (0.03–0.06)	0.06 (0.03–0.06)
Legionella pneumophila	NA	(0.03–0.125)	NA	(0.125–0.5)	NA	NA
Neisseria gonorrhoeae	1 (1–2)	≤0.015	0.03 (0.015–0.06)	0.06 (0.015–0.125)	0.03 (0.015–0.06)	0.06 (0.03–0.06)
Neisseria meningitidis	2	0.004	0.06	0.03	0.015	0.03
Gram-negative anaerobes						
Bacteroides fragilis	128 (64–256)	8 (4–32)	32 (16–128)	64 (16–>128)	4(4–8)	16 (8–16)
Other *Bacteroides* spp.	256	16 (16–32)	32 (32–64)	128 (128–256)	NA	NA
Gram-positive aerobes						
Staphylococcus aureus (MS)	≥128 (32–>128)	0.5 (0.25–1)	2 (1–4)	2 (1–4)	0.5 (0.25–1)	0.5 (0.125–1)
Staphylococcus aureus (MR)	>64 (32–128)	0.5 (0.5–1)	2	2	0.5 (0.25–0.5)	1 (0.5–1)
Staphylococcus epidermidis	>64 (64–128)	0.25 (0.125–0.5)	1	2 (1–4)	0.5 (0.25–1)	1 (0.5–2)
Streptococcus pneumoniae	≥128 (64–≥256)	1 (0.5–2)	16	16 (4–16)	2 (1–2)	8 (8–16)
Streptococcus pyogenes	≥128	1 (0.5–2)	8 (8–16)	16 (8–32)	4	8
Streptococcus agalactiae	>128 (>128–512)	1 (0.5–2)	16 (16–32)	16 (8–32)	2 (1–4)	16
Enterococcus	>128 (64–>128)	2 (0.5–2)	8 (8–16)	8 (4–32)	2 (2–4)	4
Gram-positive anaerobes						
Peptococcus	256	2	8	8	4	NA
Peptostreptococcus	≥64	1 (0.5–8)	8	4 (2–4)	2	NA
Clostridium spp.	≥256	16 (8–32)	32 (32–64)	64 (32–128)	16 (8–16)	NA

[a] MIC_{90} – mcg/ml; mean value and range. NA, Not available; MS, methicillin sensitive; MR, methicillin resistant; MIC_{90}, minimum inhibitory concentration that inhibits 90% of bacteria studied; Mcg, micrograms.

Reprinted with permission from V. T. Andriole (1989). Quinolones. *In* "Principles and Practice of Infectious Diseases" (G. L. Mandell, R. G. Douglas, and J. E. Bennett, eds.) 3rd Ed. Churchill Livingstone, New York.

compounds that may have increased activity against DNA gyrase.

III. Antimicrobial Activity

Nalidixic acid has greater antimicrobial activity against gram-negative rods than against gram-positive bacteria (Table I). It is active against most Enterobacteriaceae, including strains of *Escherichia coli, Proteus mirabilis,* other *Proteus* species, *Klebsiella* species, *Enterobacter* species, and other coliform bacteria at concentrations (16 μg/ml or lower) that are easily achieved in the urine. Some strains of *Salmonella, Shigella,* and *Brucella* may be sensitive. *Pseudomonas* and *Serratia* species are resistant. Bacterial resistance to nalidixic acid may develop by exposure to increasing concentrations of the drug *in vitro*, and in patients during treatment of bacteriuria. Gram-positive bacteria (e.g., *Staphylococcus aureus, Streptococcus pneumoniae,* and *Streptococcus faecalis*) are resistant to nalidixic acid.

Oxolinic acid has an antibacterial spectrum of activity similar to that of nalidixic acid but is significantly more active than nalidixic acid *in vitro* and *in vivo*. Oxolinic acid has no significant activity against other gram-positive bacteria or fungi, except for strains of *S. aureus* that are inhibited at concentrations of 6.25 μg/ml. Bacterial resistance to oxolinic acid can develop *in vitro* and during treatment of patients with bacteriuria similar to that observed with nalidixic acid. Cross-resistance between these two agents has also been demonstrated.

Cinoxacin has an antibacterial spectrum of activity similar to that of nalidixic acid so that it is also active against most strains of gram-negative bacteria that cause urinary tract infections, but has negligible activity against *Pseudomonas aeruginosa* and gram-positive cocci such as *S. aureus, Staphylococcus saphrophyticus, Streptococci* species and enterococci. Cinoxacin also has *in vitro* activity against *Alcaligenes, Acinetobacter, Moraxella,* and *Hemophilus* species as well as against *Clostridium perfringens, Clostridium tetani, Neisseria meningitidis,* and *Pseudomonas pseudomallei.* Cross-resistance among bacterial isolates develops against cinoxacin, nalidixic acid, and oxolinic acid.

Norfloxacin is 100 times more active than nalidixic acid, with a spectrum that includes strains of *E. coli, Klebsiella* species, *Salmonella* and *Shigella* species, *Citrobacter* species, *Enterobacter cloacae, aerogenes,* and *agglomerans, Proteus mirabilis, vulgaris, and rettgeri, Morganella* species, *Providencia* species, *Pseudomonas aeruginosa,* including gentamicin-resistant strains, *Pseudomonas maltophilia, Serratia marcescens, Acinetobacter calcoaceticus,* and *Yersinia, Arizona, Aeromonas,* and *Campylobacter* species. Norfloxacin is also active against *Hemophilus influenzae, Neisseria gonorrhoeae,* regardless of β-lactamase activity, and *Branhamella catarrhalis.* Norfloxacin is less active against *Staphylococcus aureus,* methicillin-resistant strains of *S. aureus, S. saprophyticus, Staphylococcus epidermidis, Streptococcus pyogenes, Streptococcus agalactiae, Enterococcus faecalis, Streptococcus pneumoniae,* and *Listeria* species. *Bacteroides fragilis* and most other anaerobes are relatively resistant. Norfloxacin has some activity against *Gardnerella vaginalis* and *Ureaplasma urealyticum.*

Ciprofloxacin is even more potent than norfloxacin and is active against most strains of gram-negative and gram-positive bacteria that cause infections at concentrations that are easily attained in most tissues and body fluids. Ciprofloxacin has excellent activity against strains of *E. coli, Klebsiella* species, *Salmonella* and *Shigella* species, *Citrobacter diversus* and *freundii, Enterobacter cloacae* and *aerogenes, Proteus mirabilis* and *vulgaris, Morganella morganii, Providencia stuartii, Pseudomonas aeruginosa, maltophilia,* and *cepacia, Serratia marcescens, Acinetobacter calcoaceticus, Yersinia enterocolitica, Campylobacter jejuni, Helicobacter pylori, Aeromonas hydrophilia,* and *Pasteurella multocida,* and *H. influenzae, Branhamella catarrhalis, Gardnerella vaginalis,* and *Neisseria gonorrhoeae,* regardless of β-lactamase activity. In contrast to norfloxacin, ciprofloxacin is more active against *Bacteroides fragilis, melaninogenicus/oralis* group, and *urealyticus, Fusobacteria* species, *Mobiluncus* species, *Peptococcus* species, *Peptostreptococcus* species, and *Clostridia* species. Similarly, ciprofloxacin is more active than norfloxacin against *Staphylococcus aureus,* including methicillin-resistant strains, *Staphylococcus epidermidis, Streptococcus pyogenes, Streptococcus agalactiae, Enterococcus faecalis, Streptococcus pneumoniae,* viridans streptococci, *Legionella pneumophila, Mycobacterium tuberculosis* (atypical mycobacteria are less susceptible), and *Listeria monocytogenes.* Compared to norfloxacin, ciprofloxacin has increased activity against all bacterial species studied, and also has excellent activity against *Chlamydia trachomatis* and genital mycoplasmas.

The antibacterial activities of newer 4-quinolones under development are similar to that of ciprofloxacin, which, with few exceptions, is the most potent of the newer 4-quinolones. [*See* ANTIMICROBIAL DRUGS.]

IV. Mechanisms of Bacterial Resistance

Serial exposure of both gram-positive or gram-negative bacteria to subinhibitory concentrations of the quinolones *in vitro* leads to the selection of bacterial variants with reduced susceptibility to the drug. The resulting strains may exhibit cross-resistance to other quinolones. The mechanism of resistance usually involves either (1) mutations in the gene coding for DNA gyrase so that there is reduced quinolone affinity for the A or B subunit, or (2) mutations which change the outer membrane porins. Relative resistance to antibiotics unrelated to the quinolones has been observed when reduced susceptibility to the quinolones is caused by reduced outer membrane porin F activity.

Although the exact mechanism for the development of bacterial resistance to the quinolones has not been determined, recent work suggests that high-level resistance to the quinolones occurs when serine in the 83-position of subunit A is replaced by tryptophan.

Since quinolones interfere with DNA gyrase activity, which is necessary for plasmid replication, quinolones were expected to promote loss of plasmids and to inhibit transfer of R factor-mediated resistance. However, recent work suggests that plasmid-mediated resistance may be possible, though rare.

V. Pharmacology

All of the quinolones are well absorbed from the gastrointestinal after oral administration and most are excreted by the kidney into the urine. Some are metabolized in the liver.

After oral administration and absorption from the gastrointestinal tract, nalidixic acid is rapidly metabolized in the liver to biologically active hydroxynalidixic acid, and to inactive monoglucuronide conjugates which, along with the parent compound, are rapidly excreted by the kidney into the urine. Plasma levels of 20–50 μg/ml may be attained 2 hr after a single oral dose of 1 g of the drug. Nalidixic acid does not accumulate in tissues even after prolonged administration, and the kidney is the only organ in which tissue concentrations may exceed plasma levels. The drug does not diffuse into prostatic fluid, but does appear in mother's milk and may be harmful to the newborn. Excretion is almost completely via the kidney into the urine in concentrations of active drug, after a 0.5- to 1-g oral dose in adults, in the range of 25–250 μg/ml. Bactericidal levels of the drug are also attained in the urine of patients with moderate or advanced renal failure. Although increased toxicity has not been observed in these patients, nalidixic acid should be used cautiously in patients with advanced renal failure and in patients with liver disease, since conjugation of the drug may be impaired in this latter group.

Oxolinic acid, after oral administration, produces low plasma levels but effective bactericidal concentrations for most susceptible microorganisms in the urine. Oxolinic acid should be used cautiously in patients with severely impaired renal function and, since the drug is excreted in human milk, it is contraindicated in nursing mothers.

Cinoxacin is rapidly and almost completely absorbed from the gastrointestinal tract after oral administration and has a serum half-life of approximately 1 hr, which may increase threefold in patients with creatinine clearances of less than 30 ml/min/1.73 m^2. Peak plasma levels occur in 2–3 hr and range from <4 to 14.8 μg/ml and from 2.8 to 28 μg/ml after an oral dose of 250 and 500 mg, respectively. Urine concentrations are high and bactericidal for susceptible microorganisms but are decreased in patients with impaired renal function. Cinoxacin concentrations in human prostatic tissue range from 0.6 to 6.3 μg/g, and in renal tissue exceed those in serum. Cinoxacin is excreted by both glomerular filtration and tubular secretion and probenecid inhibits cinoxacin excretion by the kidney. Although it is not known whether cinoxacin is excreted into human milk, the drug should not be used by nursing women because of the potential for serious adverse reactions in nursing infants.

Norfloxacin, after oral administration, is also well absorbed from the gastrointestinal tract, producing low plasma levels and high concentrations in the urine. Norfloxacin crystals are occasionally observed during microscopic examination of freshly voided urine collected after large doses of 1200 and 1600 mg, but crystalluria is not encountered at lower doses. Norfloxacin is not excreted into human milk in concentrations detectable by bioassay.

Ciprofloxacin, after oral administration, is also rapidly absorbed from the gastrointestinal tract, and is the most thoroughly studied quinolone in all respects. Peak plasma levels occur in 1–1.5 hr and are approximately 2–3 μg/ml after an oral dose of 500 mg. Its extravascular penetration into tissues and other body compartments is better than or comparable to other, newer quinolones. Ciprofloxacin penetrates blister fluid well, with 57% of the serum concentration recoverable. It is not known whether ciprofloxacin is excreted into human milk. A parenteral preparation of ciprofloxacin is also available.

The pharmacokinetic properties of some of the newer quinolones are summarized in Table II. The newer quinolones, in general, exhibit linear pharmacokinetics. Peak serum concentrations occur 1–3 hr after oral administration. Food and histamine H_2-receptor antagonists (ranitidine) delay absorption so that serum peaks appear later and are moderately lower. Absorption is also reduced by concomitant administration of magnesium or aluminum hydroxide antacids, and by other drugs that decrease peristalsis or delay gastric emptying time. The newer quinolones are not extensively bound to serum proteins. Their long serum half-life allows dosing once or twice daily. The newer quinolones undergo renal and hepatic metabolism. Renal elimination is by glomerular filtration and active tubular secretion which is blocked by probenicid (except fleroxacin). The antibacterial activity of the quinolones is reduced at lower urinary pH values (5.5–6.0 vs 7.4). Pefloxacin and difloxacin undergo extensive hepatic metabolism, followed by enoxacin and, to a lesser degree, norfloxacin, ciprofloxacin, and fleroxacin. Hepatic metabolism is least with lomefloxacin and ofloxacin. Biliary concentrations of ciprofloxacin, enoxacin, ofloxacin, and pefloxacin are two to eight times the simultaneous serum con-

TABLE II Pharmacokinetic Properties of Selected Newer Quinolones[a]

Drug	Dose (mg)	C_{max}[b] (mg/liter)	Half-life (hr)	Protein-binding (%)	Bioavailability (%)	VD[c] (liters)	Urinary excretion (%)	
							Unchanged	Metabolites
Ciprofloxacin	500	2–3	3–4.5	35	85	250	30–60	10
Norfloxacin	400	1.5	3–4.5	15	80	225	20–40	20
Ofloxacin	400	3.5–5.0	5–6	8–30	85–95	100	70–90	5–10
Enoxacin	400	2–3	4–6	43	90	190	50–55	15
Pefloxacin	400	4–5	10–11	25	90	110	5–15	55
Fleroxacin	400	4–6	10	23	96	100	60–70	10
Lomefloxacin	400	3	8	NA[d]	NA	190	70	10
Difloxcin	400	4–5	26	42	NA	140	10	20

[a] Reprinted with permission from V. T. Andriole (1989). Quinolones. *In* "Principles and Practice of Infectious Diseases" (G. L. Mandell, R. G. Douglas, and J. E. Bennett, eds.), 3rd Ed. Churchill Livingstone, New York.
[b] C_{max}, Peak serum concentration.
[c] VD, Volume of distribution.
[d] NA, Data not available.

centrations. Ciprofloxacin and norfloxacin have high concentrations inside human neutrophils, whereas pefloxacin penetrates poorly into neutrophils and alveolar macrophages. The tissue penetration of some of the newer quinolones is summarized in Table III. [*See* PHARMACOKINETICS.]

VI. Toxicity and Adverse Reactions

Oral nalidixic acid, oxolinic acid, and cinoxacin are usually well tolerated. Adverse reactions include nausea, vomiting, diarrhea, and abdominal pain (gastrointestinal); pruritis, nonspecific rashes, urticaria associated with eosinophilia, and edema of the extremities (dermatologic); photosensitivity reactions involving skin surfaces exposed to sunlight manifested as a sunburn or rarely as a bullous eruption (nalidixic acid); blurred vision, diplopia, photophobia, abnormal accommodation, and changes in color perception, all of which disappear with cessation of therapy (opthalmologic); headaches, drowsiness, asthenia, giddiness, vertigo, syncope, restlessness, insomnia, tinnitus, sensory changes,

TABLE III Penetration of Selected Quinolones into Body Fluids and Tissues[a]

Fluid or tissue	Ciprofloxacin	Norfloxacin	Ofloxacin	Enoxacin	Pefloxacin	Fleroxacin
Blister fluid	++++[b]	++++	++++	++++	+++	++++
Saliva	++	++	+++	+++	+++	+++
Bronchial secretions	++	—	+++	++++	++++	—
Pleural fluid	+++	—	—	—	—	—
Nasal secretions	+++	+++	++++	++++	+++	++++
Tears	++	++	+++	++	+++	+++
Sweat	+	+	++	++	++	++
Cerebrospinal fluid	+	—	++	—	+++	—
Prostatic fluid	+++	++	++++	++	—	++
Ejaculate	+++++	—	++++	++++	—	++++
Lung	++++	—	++++	+++++	++++	—
Kidney	+++++	+++++	+++++	++++	—	—
Bone	++++	—	++	++	++	—
Skin	++++	—	—	++++	—	—
Muscle	++++	—	—	++++	—	—
Fat	++++	—	—	+++	—	—

[a] Reprinted with permission from V. T. Andriole (1989). Quinolones. *In* "Principles and Practice of Infectious Diseases" (G. L. Mandell, R. G. Douglas, and J. E. Bennett, eds.), 3rd Ed. Churchill Livingstone, New York.
[b] +, AUC (area under curve) ratios or concentration ratios <0.1; ++, AUC ratios or concentration ratios 0.1–0.5; +++, AUC ratios or concentration ratios 0.5–1; ++++, AUC ratios or concentration ratios 1–4; +++++, AUC ratios or concentration ratios >4.

grand mal seizures, and acute reversible toxic psychosis as well as pseudotumor cerebri with intracranial hypertension, papilledema, and bulging fontanelles in infants and young children which reverses after cessation of therapy (neurologic). Rarely, abnormal liver function tests, renal function values, and reduced hematocrit, hemoglobin, and leukocyte counts have been observed. Nalidixic acid has been associated, rarely, with cholestatic jaundice and with hemolytic anemia that sometimes is associated with glucose-6-phosphate dehydrogenase (G6PD)-deficient red blood cells. The use of nalidixic acid, oxolinic acid, and cinoxacin in prepubertal children and during pregnancy is not recommended.

The newer fluoro-quinolones, norfloxacin, ciprofloxacin, ofloxacin, enoxacin, and pefloxacin are considered relatively safe agents. They have very similar side effects but in low incidences. Gastrointestinal side effects are the most frequent (0.8–6.8% of patients) and include nausea, vomiting, dyspepsia, epigastric/abdominal pain, anorexia, diarrhea, flatulence, and dry mouth. Antibiotic-associated colitis has been seen rarely.

Central nervous system side effects occur in 0.9–1.8% of patients. Mild reactions include headache, dizziness, tiredness, insomnia, faintness, agitation, listlessness, restlessness, abnormal vision, and bad dreams. Hallucinations, depressions, psychotic reactions, and grand mal convulsions (severe reactions) are rare and disappear when therapy is discontinued.

Skin and allergic reactions occur in 0.6–2.4% of patients, and include erythema, urticaria, rash, pruritis, and photosensitivity reactions of skin surfaces exposed to sunlight. Hypotension, tachycardia, nephrotoxicity, thrombocytopenia, leukopenia, anemia, and transient elevations in liver enzymes have been observed rarely. Anthropathy, gait abnormalities, and articular cartilage lesions in weight-bearing joints in juvenile animals have been observed. Thus, the newer quinolones have not been approved for use in pediatric patients in the United States.

VII. Drug Interactions

Nalidixic acid–glucuronide conjugates may produce a false-positive reaction for urine glucose when tested with Benedict's solution, but not with glucose oxidase test strips. Oxolinic acid may enhance the effects of the oral anticoagulants bishydroxycoumarin and warfarin by displacing these drugs from serum albumin binding sites. Nitrofurantoin interferes with the therapeutic action of nalidixic and oxolinic acid.

Some of the newer quinolones increase theophylline plasma concentrations, e.g., enoxacin (111%), ciprofloxacin (23%), pefloxacin (20%), and ofloxacin (12%). The 4-oxo metabolite of the piperazine ring is thought to complete with theophylline for hepatic enzymes and interfere with theophylline clearance. Theophylline doses probably should be halved in patients receiving enoxacin. No routine reduction in theophylline dose is recommended for patients receiving ciprofloxacin, ofloxacin, or pefloxacin, but theophylline levels should be monitored.

Caffeine clearance is interfered with by the newer quinolones. Enoxacin increases the plasma concentration of caffeine by 41% and reduces the clearance by 78%. Ciprofloxacin increases the half-life of caffeine only modestly (15%), and ofloxacin only minimally.

VIII. Clinical Uses

The newer quinolones have proved to be effective therapies for infection of the urinary tract, respiratory tree, gastrointestinal tract, skin, soft tissue and bone, and for sexually transmitted bacterial diseases.

A. Urinary Tract Infections

Nalidixic acid, oxolinic acid, cinoxacin, norfloxacin, and ciprofloxacin have established roles in treating urinary tract infections. For acute and recurrent uncomplicated urinary infections due to susceptible organisms, doses are as follow:

1. Nalidixic acid: Adults, 1 g qid for 1–2 weeks, then 0.5 g qid (four times daily) if needed; children, 55 mg/kg/day in four divided doses for 1–2 weeks, then 33 mg/kg/day if needed.
2. Oxolinic acid: Adults, 750 mg bid (twice daily) for 2 weeks; children, not recommended.
3. Cinoxacin: Adults, 250 mg qid or 500 mg bid for 1–2 weeks; children, not recommended.

Long-term therapy with nalidixic acid for frequently recurrent bacteriuria has resulted in poor follow-up cure rates and resistance has commonly emerged during treatment.

The newer quinolones are as effective as other well-established agents for the treatment of uncomplicated urinary infections. Single doses of norfloxacin (800 mg), ciprofloxacin (100 or 250 mg), and ofloxacin (200 mg) are highly effective in women with simple cystitis caused by Enterobacteriaceae, but may be less effective against *Staphylococcus saprophyticus*. Also, norfloxacin, ciprofloxacin, ofloxacin, or enoxacin given for 3–10 days has resulted in excellent bacteriologic cure rates in uncomplicated urinary infections.

Norfloxacin, ciprofloxacin, ofloxacin, and enoxacin, given for 5–10 days to patients with nosocomial or complicated urinary infections, have resulted in excellent cure rates.

Ciprofloxacin (1000 mg/day), ofloxacin (300–600 mg/day), pefloxacin (800 mg/day), and norfloxacin (800 mg/day), given to patients with either acute or chronic prostatitis for 28 (range 5–84) days cured 63–92% of patients.

B. Respiratory Tract Infections

Patients with purulent bronchitis, acute exacerbations of chronic bronchitis, or pneumonia who were treated for 10 (range 7–15) days with ciprofloxacin, ofloxacin, enoxacin, or pefloxacin experienced clinical cure or improvement (76–91%) and bacteriologic cure (68–83%). However, in 49% of patients with *Pseudomonas aeruginosa* infections, 39% with *Streptococcus pneumoniae,* and 33% with *Staphylococcus aureus* infections, bacteriologic persistence, relapse, or treatment failure occurred. Most physicians are reluctant to use the newer quinolones to treat either community-acquired or aspiration pneumonia because of their reduced activity against *Str. pneumoniae* and against those microaerophilic and anaerobic bacteria associated with aspiration pneumonia. However, the newer quinolones may have value in hospital-acquired pneumonia caused by aerobic gram-negative bacteria. Ciprofloxacin (750 mg twice daily) has been effective in cystic fibrosis patients with exacerbations of acute pulmonary infections, although resistant organisms may emerge. Malignant external otitis caused by *Pseudomonas aeruginosa* may also respond to ciprofloxicin therapy.

The newer quinolones should not be used for acute sinusitis because of the possible presence of pneumococci and anaerobic streptococci, but may be useful in specific cases of chronic sinusitis caused by susceptible aerobic gram-negative bacteria. These agents should not be used for otitis media in pediatric patients. Norfloxacin has not been approved and should not be used for any type of respiratory tract infection.

C. Gastrointestinal Infections

The newer quinolones are highly active against those bacterial pathogens causing diarrheal disease, including toxigenic *E. coli, Salmonella, Shigella, Campylobacter* and *Vibrio* species. The quinolones provide high drug concentrations in the lumen of the gut and the mucosa, which contribute to the eradication of these pathogens from the intestine within 48 hr of initiating therapy. Ciprofloxacin, 500 mg bid, and norfloxacin, 400 mg bid, for 5 days cure greater than 90% of patients with either acute bacterial diarrhea or acute traveler's diarrhea and are comparable to trimethoprim–sulfamethoxazole. Bacterial resistance may develop more rapidly with the indiscriminate use of the newer quinolones. Thus, these newer agents should not be used as prophylactic agents to prevent acute traveler's diarrhea because this disease responds promptly to treatment once symptoms develop. All patients with typhoid fever treated with ciprofloxacin, 500 mg twice daily for 2–15 (mean 13) days or ofloxacin, 200 mg twice daily for 6–30 days, were cured. None of these patients relapsed or became chronic carriers. Also, the chronic salmonella carrier state was eliminated in 86% of patients treated with ciprofloxacin, 500–750 mg twice daily for 4 weeks, and followed for 10–12 months.

Helicobacter pylori, which has been associated with antral gastritis, is inhibited by the newer quinolones. However, these agents have not been effective in the treatment of *H. pylori*-associated gastritis. Although a preliminary report suggests that ciprofloxacin may have some value in *Clostridium difficile* enterocolitis, the newer quinolones have also been associated with this disease. Some relapses have been reported in patients with *Brucella* infections who have been treated with the newer quinolones.

D. Skin and Soft Tissue Infections

Ciprofloxacin, ofloxacin, and enoxacin, given orally, effectively treat a variety of bacterial skin and skin structure infections in patients with cellulitis, subcutaneous abscesses, wound infections, and infected ulcers in diabetic patients. Clinical cure or improvement was observed in 95% of patients treated with oral ciprofloxacin (750 mg twice daily for 14 days). Bacteriologic cures were lower in patients infected with gram-positive organisms than were observed for infections caused by gram-negative aerobic bacteria. Quinolone therapy failed in 25% of anaerobic infections. Colonization with methicillin-resistant *Staphylococcus aureus* (MRSA) was eradicated in 50–79% of evaluable patients treated orally with ciprofloxacin (750 mg twice daily for 7–28 days). When rifampin was combined with ciprofloxacin, the eradication rate was 100%, when the isolates were susceptible to both agents. These patients remained free of MRSA for at least 1 month. Although ciprofloxacin may eradicate MRSA colonization, or cure MRSA infection, the development of ciprofloxacin-resistant strains may occur.

E. Osteomyelitis

The newer oral quinolones, primarily ciprofloxacin, have been effective as monotherapy for osteomyelitis, particularly when caused by gram-negative aerobic pathogens. Oral ciprofloxacin in a dose of 750 mg twice daily for 8 weeks (range 4 days to 6 months) has been used in most patients with either acute or chronic osteomyelitis, in either native bone or complicated by a foreign body. Clinical cure or improvement occurred in approximately 80% of patients with follow-up of at least 6 months to more than 1 year. Treatment failures occurred in 15% and recurrent infection occurred in a few patients. The development of resistant strains occurred in only 0.5% of patients, primarily those with *Pseudomonas aeruginosa* infections. Thus, ciprofloxacin, and possibly other newer quinolones, has an established efficacy in the treatment of osteomyelitis.

F. Sexually Transmitted Diseases

The newer quinolones have been used to treat a variety of sexually transmitted diseases. Ciprofloxacin, and other newer quinolones, are extremely active *in vitro* against *Neisseria gonorrhoeae,* including penicillinase-producing strains (PPNG), and against *Hemophilus ducreyi. Chlamydia trachomatis* isolates are most susceptible to ciprofloxacin, ofloxacin, and difloxacin but are resistant to enoxacin and norfloxacin. Although *Gardnerella vaginalis* and *Ureaplasma urealyticum* are relatively resistant to these agents, ciprofloxacin is active against the latter about 50% of the time. [*See* SEXUALLY TRANSMITTED DISEASES (PUBLIC HEALTH).]

1. Gonococcal Infections

Single oral doses of either 500, 250, or 100 mg of ciprofloxacin, 600, 400, or 200 mg of enoxacin, 400 mg of pefloxacin, 800, 600, 400, or 200 mg of ofloxacin, or 800 mg of norfloxacin cured 95–100% of uncomplicated gonococcal infections in both men and women, including patients infected with PPNG. Thus, the lowest effective oral single dose of the newer quinolones (100 mg of ciprofloxacin) has cured almost 100% of patients with urethral as well as rectal gonorrhea, and is probably effective for pharyngeal gonococcal infections. However, there is little experience with these newer agents in the treatment of *disseminated gonococcal* infections.

2. Chlamydia Urethritis, Postgonococcal Urethritis, Nongonococcal Urethritis

The current quinolones are not effective as single-dose therapy for *C. trachomatis* urethritis, nor do they prevent postgonococcal urethritis (PGU) when used as single-dose therapy in gonococcal infections. However, ciprofloxacin in a dose of 750 mg orally twice daily for 4 days, eradicated *C. trachomatis* in 60% of coinfected patients, and reduced the incidence of PGU from 35 to 13%. Most patients with nongonococcal urethritis (NGU) caused by *C. trachomatis* have been cured when treated with ciprofloxacin (500 mg three times daily) or ofloxacin (100 mg three times daily) for 14 days, whereas norfloxacin is not effective. However, the newer quinolones are less effective than doxycycline in NGU patients with chlamydial infections alone. Further studies are needed to accurately define the efficacy of the newer quinolones in NGU.

3. Chancroid

A single oral dose of 500 mg of ciprofloxacin or 500 mg of ciprofloxacin twice daily for 3 days has cured 95 and 100% of patients with chancroid and *H. ducreyi* infections, respectively.

4. Nonspecific Vaginitis

Clinical and bacteriologic cures occurred in 73% of women with nonspecific vaginitis caused by *Corynebacterium* species, *Bacteroides* species, and *Gardnerella vaginalis* when they were treated with 500 mg of ciprofloxacin orally twice daily for 7 days. Clinical improvement without bacteriologic eradication occurred in an additional 18%, and 9% failed to respond. Vaginal colonization with *Candida albicans* occurred in 32% of these patients but without clinical signs or symptoms of yeast infection.

G. Other Infections

1. Immunocompromised Host

In preliminary studies, ciprofloxacin and norfloxacin have been successful as prophylactic agents in granulocytopenic patients. Ciprofloxacin prevented, whereas norfloxacin reduced but did not prevent, colonization by gram-negative bacilli, nor did norfloxacin influence the incidence of gram-positive bacteremia. Limited clinical experience with ciprofloxacin, ofloxacin, and pefloxacin in the treatment of severe infections in immunocompromised patients suggests a potential role for these agents, at least in febrile neutropenic patients, but further trials are needed to establish their value.

2. Central Nervous System Infections

Ciprofloxacin, ofloxacin, and pefloxacin do penetrate into the cerebrospinal fluid and brain tissue but there is little clinical experience with these newer therapeutic agents for central nervous system bacterial infections. The newer quinolones should be reserved for special cases caused by multiantibiotic-resistant aerobic gram-negative bacteria.

The newer quinolones, particularly ciprofloxacin, have been shown to be effective in eradicating the meningococcal nasopharyngeal carrier state. Ciprofloxacin and ofloxacin have inhibitory activity against *M. tuberculosis* and *Mycobacterium avium-intracellulare* and may be useful in drug-resistant mycobacterial infections with these organisms, but clinical studies are needed to define their effectiveness. Ciprofloxacin has been used successfully to treat patients with Mediterranean spotted fever caused by *Rickettsia conorii,* and has also been used to treat patients with recalcitrant bacterial endocarditis. The newer quinolones may also be of value in the treatment of patients with malaria.

In the future, further modifications of the chemical structure of the newer 4-quinolones should lead to newer agents which may have the potential to treat infections in addition to those mentioned above.

Bibliography

Andriole, V. T. (ed.) (1988). "The Quinolones." Academic Press, New York.

Andriole, V. T. (1989). The quinolones. *In* "Principles and Practices of Infectious Diseases" (G. L. Mandell, R. G. Douglas, and J. E. Bennett, eds.), 3rd ed. Churchill Livingstone, New York.

Andriole, V. T. (1990). Quinolones. *In* "Infectious Diseases in Medicine and Surgery" (S. L. Gorbach, J. G. Bartlett, and N. R. Blacklow, eds.). Saunders, Philadelphia, Pennsylvania.

Chu, D. T. W., and Fernandes, P. B. (1989). Structure-activity relationships of the fluoroquinolones. *Antimicrob. Ag. Chemother.* **33,** 131.

Hooper, D. C., and Wolfson, J. S. (1985). The fluoroquinolones: Pharmacology, clinical uses, and toxicities in humans. *Antimicrob. Ag. Chemother.* **28,** 716.

Sorgel, F., Jaehde, U., Naber, K., and Stephan, U. (1989). Pharmacokinetic disposition of quinolones in human body fluids and tissues. Clin. Pharmacokin. **16S,** 5.

Wolfson, J. S., and Hooper, D. C. (1985). The fluoroquinolones: Structures, mechanisms of action and resistance, and spectra of activity *in vitro. Antimicrob. Ag. Chemother.* **28,** 581.

Wolfson, J. S., and Hooper, D. C. (eds.) (1989). "Quinolone Antimicrobial Agents." American Society for Microbiology, Washington, D. C.

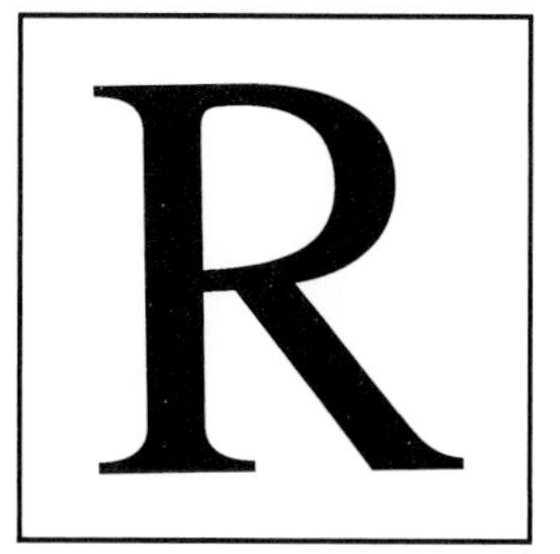

Radiation, Biological Effects

JOSEPH H. GAINER, *Food and Drug Administration*

Glossary

Antibody affinity Binding energy of antibody sites for specific antigen epitopes. It determines the stability of antigen–antibody complexes, influencing the ability of antibody to neutralize microbial agents and toxins

Aplastic anemia Disease in which there is a persistent reduction in the number of circulating erythrocytes caused by a failure of adequate stem cell production

Gray (Gy) SI unit for the absorbed dose of ionizing radiation. 1 Gy = 1 joule of absorbed energy per kg of irradiated material. 1cGy = 10^{-2} Gy

Inflammation Response of a tissue or organ to a physical, biological, or chemical irritant with swelling (edema), redness (erythema), and pain, accompanied by cellular infiltration

Phagocytosis Engulfment of bacteria, fungi, and other foreign particles by white blood cell elements

Rad Absorption of 100 ergs per gram of tissue. One rad is approximately one roentgen (R)

Sievert (Sv) SI unit of dose equivalent. This is calculated by multiplying the dose in Gy by a quality factor, which adjusts for differences in relative biological effectivenesses of different radiations. 1cSv = 10^{-2} Sv

RADIATION SICKNESS is the illness produced after exposure to ionizing radiation. With heavy exposure it is characterized by malaise, nausea, emesis, diarrhea, and leukopenia. The mean survival time of mammals including humans is inversely a function of radiation dose. At doses to 10,000 R or greater, the exposed individual dies in a matter of minutes to a couple of days, with symptoms consistent with pathology of the central nervous system. From 1,000 R to several thousand R, there is a constant mean survival time. Below 1,000 R, individuals die with symptoms of bone marrow pathology.

The radiation dose producing an LD_{50}, lethal dose 50% (at which 50% exposed subjects are killed), varies among animal species and humans, computed during a 30–42-day period. Sheep and burros have LD_{50} of 155 R; swine 195; monkey 395; marmoset, goat, dog, and humans 300–400; rabbit, mouse, rat, hamster 840–900; gerbil 1,059; and wild mice 1100–1,200.

Radiation sickness was reported soon after the discovery of the X-ray. Bequerel and Curie suffered from acute radiation dermatitis. The earliest known case of radiation-induced cancer was reported in 1902. Mm. Curie herself died of aplastic anemia, probably caused by prolonged exposure to radiation. Nine deaths caused by bone cancer were recorded in 1924 among watch industry workers who painted dials with radium, which emits radiation. Leukemia was induced in mice by a single whole-body exposure of 400 R or by a closely spaced, fractionated dose of 900 R. A high incidence of skin cancer and leukemia was observed among radiologists exposed to X-rays during work. Human data on the effects of ionizing radiation come primarily from the people of Hiroshima and Nagasaki exposed at that time. The explosions at Three Mile Island and at Chernobyl, with injuries and deaths at Chernobyl, and the potential for increased incidence of cancer and leukemia have heightened the concerns over the safety of nuclear power.

Living with radiation from "natural exposure" presents the following risks of cancer. Residing within 20 miles of a nuclear power plant for 1 year gives a dose of 0.02 mrem/yr or less than one cancer if all the U.S. population were exposed. One round-trip transcontinental flight gives 5 mrem/yr and 135 cancers if the whole U.S. population were exposed. Radioactive potassium normally in the body gives a dose of 30 mrem/yr or 1,000 cancers; cosmic radiation at sea level and medical X-rays each give 40 mrem/yr or 1,100 cancers. Finally, radon gas (1.5 pCu/L of air) gives 500 mrem/yr, resulting in 13,500 cancers if all the U.S. population were exposed. Radon then presents substantial radiation exposure. Being a user of tobacco products, which themselves contain radioactive products, adds significantly more exposure to any of these events. Debate among radiation experts ensues as to how significant the radiation exposure is to the induction of lung cancer from radon. In 1985, according to the Surgeon General, smoker deaths from lung cancer were 106,000 but only 3,800 in nonsmokers.

I. Theories on the Biological Effects of Radiation

French scientists in 1906 proposed that less-differentiated cells are more radioresistant than highly differentiated ones and that proliferating tissues are more radiosensitive than nonproliferating ones. An exception to this is that lymphocytes and oocytes are sensitive, although they neither differentiate nor divide. Target theory predicts that the proportion of unaltered biological molecules decreases as a negative exponential function of dose; however, it was found inadequate to explain cellular radiation injuries. The indirect action of radiation was then proposed; it was stated that biological molecules in aqueous solution are inactivated by free radicals formed when radiation interacts with water. Oxygenated tissues are more sensitive to radiation than anoxic ones, owing to formation of oxygen free radicals. Another hypothesis was that radiosensitivity of a cell is directly proportional to its interphase chromosomal volume. This is consistent with observations on plant tissues, but for mammalian species its validity is not established.

The concept of RBE (relative biological effectiveness) evolved because different types of radiation which produce different degrees of damage with the same dose. This is due to the fact that the linear energy transfer (LET)—that is the amount of energy released per unit length along the radiation path—for each type of radiation is different. For the same total dose, the radiation of high LET (i.e., alpha particles, protons, or neutrons) produces greater damage than that of low LET (X-ray and gamma ray). Furthermore, the oxygen effect, so marked with low LET radiation, is negligible with high LET.

The discovery of several radioprotective and radiosensitizing agents increased the knowledge of radiation injuries. A therapeutic regime for exposed individuals involves a "functional replacement therapy" requiring transfusions of fresh platelets and whole blood and antibiotic administration. Spleen, splenic cells, and bone marrow transplantation are beneficial. Current research in immunology, including studies on radiation and its effects on immunity, and the development of products such as biological response modifiers including interferons, interleukins, cytokines, and monokines provide additional chemotherapeutic treatment regimes for radiation sickness.

II. Effects of Radiation on Organ Systems

A. Central Nervous System

Neuroblasts and neuroglia are radiosensitive, whereas neurons are radioresistant. Blood vessels are radiosensitive. The acute response of the central nervous system to radiation is an acute inflammation. Oligodendrocytes show swelling and degenerative changes soon after acute inflammation. Early demylination occurs. In the subacute period, the patient may completely recover or may have convulsions, ataxia, and incoordination. In the chronic period, delayed necrosis and demyelination are observed. Malignant intracranial neoplasms after radiation therapy have been observed, as also has spinal cord myelitis. Peripheral nerves are highly radioresistant.

Learning behavior in animals is not adversely affected by ionizing radiation, but nonintellectual behavior is affected. Depression of total body activity in monkeys and depression of running-wheel activity in rats have been observed. Decreased mating activity in boars and rats has been seen after fetal radiation; learning capacity of animals radiated *in utero* markedly decreases.

Radiation injury to the central nervous system is a rare event in clinical medicine, but it is catastrophic for the patient in whom it occurs. By the mid-1960s, 57 patients with documented clinical necrosis had been reported. Cerebral necrosis can be diagnosed with the computerized tomogram scan, and it does not repair itself. Necrosis has been reported as less than 1% for doses less than 5,200 rads, as high as 4% for doses up to 6,000 rads, and greater than 6,000 rads an incidence of 16%.

1. Mental Retardation and Radiation

Analysis of the occurrence of mental retardation and (or) intelligence test scores on individuals exposed *in utero* to the atom bombs in Japan revealed the following. (1) The data did not suggest a commingling of different distributions, such as might arise from inclusion of a qualitatively different group of individuals (i.e., unrecognized cases of mental retardation). The cumulative distribution suggests a general phenomenon, a shift in the distribution of scores describing mental intellection with exposure. (2) There is no evidence of a radiation-related effect on mental retardation or intellection generally for those individuals exposed in the first 8 weeks of life. (3) The mean test scores are consistently significantly heterogeneous among exposure categories for those individuals exposed at 8–15 weeks after conception and less heterogeneous for those groups exposed at 16–25 or 26 or more weeks of gestational age. (4) Among the latter two groups of individuals there is a significant decrease in intelligence test score with increasing exposure. (5) Within the most sensitive group, individuals exposed 8–15 weeks after conception, and with the better fitting linear-quadratic model, the diminution in intelligence score is 21–26 points per Gy of exposure, or about 0.2 points/cGy.

B. Bone Marrow

Erythrocytes immediately decline after 200 R with a severe depression of red cell precursors; regeneration occurs early. After 400–800 R, red cell count falls slowly during 7–8 days and then more abruptly. After 700 R, the red cell precursors are depleted by day 1 and are totally absent by day 3.

There is no change in hematopoietic elements with less than 100 R. With 175 R, granulocytes decrease the first 8 days but not below 200/mm^3. They continue at this level to 40 days, then show a slow increase. Lymphocytes markedly fall initially. At from 200 to 500 R, there is first a granulocytosis, then a fall, with recovery beginning at day 36. Lymphocytes are low at 3–4 days, continuing the same for 5 weeks. Platelets fall to a minimum in 8–10 days, returning to normal in 7 weeks. [*See* HEMOPOIETIC SYSTEM.]

C. Skin and Mucous Membranes

The degree of radiation response of the skin and mucous membranes depends on (1) radiation dose, (2) quality of radiation, (3) time over which the dose is administered, (4) size of the field, and (5) anatomical location. The extent of damage may range from minimal degenerative changes to germinal cells to total necrosis. Dermal changes are an initial erythema, desquamation, and then necrosis. Hyperpigmentation or depigmentation may occur, and neoplasms of the skin may develop. Grafted skin radiosensitivity depends on the age of the graft; a 3-month-old graft shows greater sensitivity than normal skin, a 3–12-month graft shows responses similar to normal skin, and an old graft, longer than 1 year, shows little reaction to radiation. [*See* SKIN, EFFECTS OF UTRAVIOLET RADIATION.]

The mucous membrane effect is radiation mucositis; it initially appears patchy and later spreads over the entire irradiated area. Varying degrees of squamous cell metaplasia and hyperplasia occur, and neoplasms of the laryngopharynx have been seen. [*See* RADIOSENSITIVITY OF THE INTEGUMENTARY SYSTEM.]

D. Urogenital System

The kidneys are only moderately sensitive to radiation. Acute radiation nephritis occurs 6–12 months after radiation therapy. Five adults died of acute radiation nephritis 5–6 months after 2,300 R to the kidneys.

Concerning the genital system, the ovary and the testis are radiosensitive. Spermatogonia are most sensitive to radiation, whereas spermatids and spermatozoa are radioresistant. In the male, radiation-induced sterility is never immediate because of the high resistance of the spermatids and the spermatozoa. The male accessory organs are highly radioresistant. In animals, the lower the dose rate (within a certain range), the higher the sterility in the male.

Short-time fractionated doses produce a higher incidence of sterility in males than does a single dose.

In the female, radiation not only destroys the radiosensitive gametogenic epithelium but also much of the production of the sex hormones. In growing follicles, postradiation damage of the granulosa cells is seen before changes in the oocytes. The female is more sensitive than the male; sterility in the female is immediate if the dose level is sufficiently intense. The uterus is radioresistant, and the vaginal mucosa is similar to that of other mucous membranes. The mammary glands, vulva, labia, and clitoris are relatively radiosensitive.

Sensivity of oocytes to radiation varies as a function of the stage of development. In mice the radiation of oocytes varies with age, species, strain of mouse, stage of growth of the follicles about the oocytes, and chromosomal configuration.

E. Human Fetus

Recent data suggest that a dose of 1–5 rads may be harmful to the human fetus. The gestation period between 18 and 38 days is highly radiosensitive for the fetus and gives a greater number of organ anomalies. Nervous tissue and optic tissue are especially sensitive. Diagnostic radiation (i.e., 0.7–5 rads) of young pregnant women increases the probability of a Down's syndrome baby by a factor of 10.

F. Liver

The liver is a highly radioresistant organ based on morphologic changes. Functionally hepatic enzyme changes have been reported after radiation.

G. The Cardiovascular System

1. Heart

Myocardial cells are very radioresistant based on morphological changes, while small vessels of the heart are moderately radioresistant. Direct irradiation on the heart involves primarily the fine vasculature, with secondary effects on the connective tissue and indirect effects on the myocardium.

Radiation induces swelling of the myocardial fiber, loss of striations, homogenization of the sarcoplasm, granulation of the sarcoplasm, disappearance of protoplasm, and persistence of hollow sarcolemma. Nuclei show pyknosis or hyperchromasia around the nuclear membrane, fragmentation, and lysis. Arterioles show thickened intima and degeneration and hyalinization of the media. The structural changes in the myocardium are observed only after high doses of radiation (i.e., 3,000–10,000 R), however, damage to the pericardium is produced at lower doses.

Electrocardiographic changes have been reported after irradiation of the thorax. Myocardial infarction, although rare, may occur within 6 mo of radiation therapy. Progressive deleterious changes in the fine vasculature of the pericardium or myocardium may lead to secondary focal degeneration of the pericardial or cardiac tissue and focal replacement fibrosis many months or even years after the irradiation. The degree and the extent of the damage depend upon the dose and the contribution of other factors. Pericarditis, pericardial effusions, pericardial adhesions, and pericardial fibrosis also occur.

Similarly, in animals, high doses of radiation are required to produce heart damage. In the dog, myocardial necrosis and hyaline degeneration have been observed after 38,000 R given to the heart over 8 wk. Myocardial hemorrhages with epicardial and endocardial hemorrhage were reported in irradiated monkeys receiving 700 or 800 R whole-body irradiation.

2. Aorta and Major Vessels

Hemorrhage is the main manifestation of damage to the major vessels after radiation therapy. Doses in the range of 5000 R in 41–47 d have produced rupture of the aorta. The intima was thickened and granular. In some areas, fibrin covered the intimal surface, and elsewhere there were mural thrombi.

Medium-sized muscular arteries show degeneration similar to those in large arteries, but they are not as severely damaged as are the arterioles and capillaries. The large and medium-sized veins show similar changes, as do the lymphatics. Repair of acute injury to the blood vessel walls by fibrosis causes advanced sclerosis, sometimes completely blocking the vessel lumen.

Thus irradiation can induce such vascular changes as degeneration, necrosis, and proliferation of the endothelium, with edema, fibrosis, thickening, and infiltration of the vascular wall. Subsequent thrombosis and/or occlusion may occur in the severely damaged blood vessels. Vascular alterations can produce marked changes in tissues dependent on them for their blood supply.

H. The Respiratory System

The richly ramified vascular system and lymphatic tissues of the respiratory system are radiosensitive, but cartilage is radioresistant. Infection renders lung more radiosensitive. In humans, radiation pneumonitis has been observed 4–6 mo after a fractionated exposure of 3,000–5,000 R.

The accumulation of fibrin-rich exudate within alveoli and the thickening of alveolar septa by fibrillar material, cellular proliferation, or fibrous tissue are seen following therapeutic irradiation of the thorax. Outpouring of alveolar edema fluid rich in fibrin, associated with the initial congestion and edema, in the absence of cellular inflammatory changes, suggests that injury to the fine vasculature plays a primary role in the development of the characteristic picture of radiation pneumonitis. Fibrin condenses at the alveolar walls to produce the so-called "hyaline membrane." Fibrin membranes were found in 41% of all irradiated lungs, most frequently and most prominently 6 mo to 2 yr after doses greater than 2,000 R. Increased deposition of fibrillar connective tissue in the alveolar septa was frequent after radiation doses greater than 3,000 R and after postirradiation intervals longer than 6 mo. The increased accumulation of histiocytes and fibroblasts in the alveoli was seen most frequently at dose levels between 2,000 and 5,000 R.

Residual radiation damage in lung, in the form of sclerosis and reduction of fine vasculature and increased parenchymal fibrosis, may not be progressive with time. However, these changes may be additive to similar changes occurring as a result of other insults or normal aging. These combined changes constitute a progressive sclerotic deterioration of the lung parenchyma, with a gradual reduction in the functional reserve capacity and in the tolerance of additional infection, insult, or stress. Therefore the clinical problem of pulmonary fibrosis may appear years after irradiation.

The dose rate appears to affect the severity of the radiation pneumonitis. At a dose rate of 1,000 rads/wk, acute reactions are fairly common; when the dose rate is reduced to 700 rads/wk, this reaction becomes of little importance.

III. Radiation Carcinogenesis

Carcinogenesis is likely the single largest concern of radiation exposure. X-irradiation induces several types of neoplasms. Data from Hiroshima and Nagasaki indicate that the threshold of a gamma-ray dose for leukemia is about 50 rads, whereas no threshold dose exists in the case of mixed radiation of gamma rays and neutrons. The threshold dose for osteosarcoma induction in children appears to be 1,000 rads; in adults, 3,000 rads. The threshold dose for thyroid tumor is about 50 rads, and 50 rads or more induce a high incidence of breast cancer. In earlier analyses, a high dose (i.e., 3,000–4,000 rads) produced a significant increase in lung cancer, but recent data from Hiroshima and Nagasaki show that a dose of larger than 128 rads increases the incidence by a factor of 2.

A. Risk of Breast Cancer After Irradiation of the Thymus in Infancy

Exposure of the female breast to ionizing radiation in infancy increases the risk of breast cancer later in life. In an average follow-up of 36 years, there were 22 breast cancers among 1,201 women X-rayed for enlarged thymus in infancy: there were 12 cancers in 2,469 nonirradiated controls. This gives an adjusted rate ratio of 3.6. The estimated mean absorbed dose to the breast was 0.69 Gy. There was a dose-response linear relation with a relative risk of 3.48 for 1 Gy of radiation and an additive excess risk of $5.7/10^4$ person years/Gy.

B. Breast Cancer Risk from Mammography

The risk of breast cancer associated with radiation decreases sharply with increasing age at exposure. Even a small benefit to women of screening mammography outweighs any possible risk of radiation-induced breast cancer.

C. Mechanism of Radiation-Induced Cancer

The mechanism is not completely understood. Two major hypotheses have been suggested: (1) Radiation causes somatic mutation, which is responsible for the malignant transformation of normal cells; and (2) radiation makes the cellular environment compatible for viral replication and viral-induced malignant transformation. The latter hypothesis applies especially to the animal species; the former one applies more generally. There is no evidence that radiation produces tumors of the ovary, uterus, pituitary gland, adrenal, brain, liver, or alimentary tract in humans; however, radiation produces these tumors in animals. Skin cancer is high among persons with chronic radiation dermatitis.

Recent developments in the molecular biology of DNA are leading to new ideas concerning how DNA alterations might be involved in radiation carcinogenesis. Alteration of certain proto-oncogenes generates oncogenes which initiate the development of malignancy. A chromosome break often occurs at or near the location of a specific oncogene in Burkitt's lymphoma. Such breaks could represent initial lesions in a translocation process that activates the oncogene by inserting it at a new location (e.g., near an active promoter). DNA breakage is one of the principal ways that ionizing radiation affects mammalian cells. It might be involved as an initial event in carcinogenesis.

D. Diagnostic and Therapeutic Radiation Effects

Human exposure to ionizing radiations is ubiquitous. Current estimates of the carcinogenic risks associated with exposure are that radiations are responsible for only a relatively small proportion of all cancers. It is estimated that exposure of 1 million persons to 1 rad of ionizing radiation will induce about 20 leukemias and 100 fatal cancers. On this basis, radiation might well result in about 150 (5%) of the 3,000 leukemia deaths a year in England and Wales and 750 (0.6%) of the 120,000 deaths from other malignant disease. Less than 3% of cancers in the United States may be attributed to radiation.

1. Aplastic Anemia

Aplastic anemia has been observed after radiation. Deaths caused by aplastic anemia in patients receiving 112–3,000 R for the treatment of ankylosing spondylitis were about 30 times higher than expected. The number of observed deaths caused by aplastic anemia was 17 times greater than expected among American radiologists.

2. In Utero Exposures

In utero exposure to diagnostic radiation might lead to the development of childhood leukemia and other cancers. A study was made in 1962 in the United States of over 700,000 children born between 1947 and 1954 in 37 maternity hospitals. Deaths from cancer among children in the period 1947 to 1960 were traced, and the frequency of prenatal exposure to radiation among those who had died of cancer was contrasted with that of a 1% sample of all children who had been born in the same 37 hospitals. Irradiated children had a mortality rate from both leukemia and other cancers about 45% higher than that of nonirradiated children. Cancer risk was directly related to the number of times that X-ray exposure had occurred in the pregnancy. A steady declining risk was found with year of birth.

3. Radiologists

Radiologists have been studied for longer than any other defined population to assess the late effects of exposure to ionizing radiations received as a consequence of their occupation. Among them there is evidence of a substantial excess of cancer: for leukemia, four deaths against 0.65 expected, and for cancers of the skin, six deaths against 0.77 expected.

Three studies on patients treated with radiotherapy give seemingly contradictory results in terms of carcinogenesis from radiation.

4. Radiation-Induced Menopause

From the 1930s onward, X-irradiation of the ovaries was commonly used to induce artificial menopause among women with benign menopausal bleeding. About 2,000 women who had had a radiation-induced menopause at three Scottish radiotherapy centers between 1940 and 1960 were studied. By 1970, 25% of the women had died, and there had been seven deaths from leukemia, against 2.7 expected ($p < 0.03$). There was a significant excess of deaths from cancers of the pelvic sites but no excess of other cancers. The excess cancers in the radiated sites became apparent 5–9 years after treatment, and an excess cancer risk persisted beyond 20 years after treatment.

5. Radiation Treatment of Cervical Cancer

Cervical cancer is usually treated by the insertion of radium or by high doses of X-rays directed at the cancer or by both methods. No excess of leukemia cases was found in a study of more than 70,000 women treated in this way (16 observed against 14 expected). This finding was confirmed in another study of more than 80,000 women treated with radiotherapy for cervical cancer. There were 77 cases of leukemia against about 66 cases expected.

6. X-Ray Treatment of Ankylosing Spondylitis

The study of patients given radiotherapy for ankylosing spondylitis was one of the first to be set up to assess the carcinogenic hazards of radiation exposure. It has been among the most informative for several reasons: (1) A large group of patients, 14,000, was followed; (2) there was a long period of follow-up of the irradiated population; and (3) reasonably high radiation doses were used in the treatment of this condition. There was clearly an increased risk of leukemia and aplastic anemia (67 deaths against 6.0 expected) and of cancer (285 deaths against 194.5 expected). The excess cancers were largely confined to those sites likely to have been directly in the radiation beam.

7. Induction Periods in Radiation Carcinogenesis

The way in which the risk of a radiation-induced cancer varies with time since exposure is easiest to look at in populations in which only one exposure of short duration has occurred. Two groups in this respect are the atomic bomb survivors and patients with ankylosing spondylitis after a single treatment. Neither group, however, is well-suited to examine the risk in the period immediately after exposure because the A-bomb cohort was not defined until 1950. It thus does not include deaths from cancers other than leukemia occurring in the first 5 years after the bombings.

Leukemias appeared in both the spondylitic population and the A-bomb survivors before other cancers. In spondylitics, the leukemia risk was greatest 3–5 years after exposure, subsequently declining, such that by 20 years there was no evidence of an excess risk. Among the A-bomb survivors, the greatest measured risk was 5–9 years after exposure. Subsequently, the excess risk declined, and there was only a small excess mortality 25 or more years after exposure.

8. Dose-Response Relations

Most public health interest centers on the likely effect of low doses of radiation, and it is to these that a substantial proportion of the population may be exposed. The spondylitics were, in general, exposed to large doses of radiation, as were many of the A-bomb survivors, although most of the latter group received doses of less than 10 rads.

The simplest assumption to make is that effects are linearly related to dose and extrapolate backward on this basis. This has been the general approach adopted for many of the analyses of data from studies of the A-bomb survivors. For leukemia—the cause of death for which the radiation-induced excess is most apparent—this form of dose-response relation fits the data well, although it is a better fit for the Hiroshima data than for the Nagasaki data, and also the leukemia risk, for a specified dose, seems to be higher in Hiroshima than Nagasaki. If the data from the two cities are combined and linear dose-response curves are fitted to leukemia excess mortality, the estimate of the induction rate is 1.9 leukemia deaths/million person years at risk/rad, and for other cancers the rate is 2.2/million person years at risk/rad. Table I provides leukemia incidence from the atomic bombings in Japan, illustrating the higher risk with exposure at an older age and at a younger age.

The greatest risk of leukemia induction is in patients with a mean marrow dose of 100–200 rads. At higher doses the risk appears to be reduced. The data are not well-fitted by a linear dose-response curve, and the assumption that the excess risk is unrelated to dose fits the data better.

It appears that a simple linear dose-response relationship for the induction of leukemia may be incorrect. Little or no excess risk of leukemia among women irradiated for the treatment of cervical cancer occurred. Women given a radiation-induced menopause dose received a substantially smaller dose to the bone marrow and yet showed a statistically significant excess leukemia. It was suggested that patients treated for cervical cancer may not be at greatly increased risk of leukemia because the radiation treatment is given in such a way that some of the bone marrow receives a very high dose of radiation (enough to sterilize the marrow) while the dose to the marrow falls off rapidly with distance from the cervix, so that the mean dose to surviving cells may be small. Thus the "effective" dose for leukemia induction in cervical cancer patients under radiation therapy is small.

In ankylosing spondylitis, usually only the spine and sacroiliac joints were irradiated, so the dose to cells directly in the radiation field may be higher than the mean dose by a factor of 2 or more, and many cells may have received a killing dose.

TABLE I Leukemia Incidence Since the Atomic Bombs[a]

Age at exposure (years)	Years after exposure 5	10	20	30	40
15	100	40	3	2	2
15–29	76	40	6	4	2
30–44	36	40	35	20	17
45+	10	40	60	75	78

[a] Data are from Castellani, A. (1984). "Epidemiology and Quantitation of Environmental Risk in Humans from Radiation and Other Agents." Plenum Press, New York, with permission.
[b] Approximate excess risk of leukemia per 100,000 by years after the Hiroshima and Nagasaki atomic bomb exposures.

9. Age at Exposure

Concerning *in utero* radiation exposure, it is suggested that some fetuses may be many times more susceptible than others to radiation carcinogenesis. Evidence to support this has been unconvincing. There is good evidence, however, that age at exposure is related to radiation risk; the fetus seems more sensitive to radiation than the adult. With spondylitics, there is a steep and highly significant increase as the age at exposure increases. The risk for leukemia among children is similar to that among adults up to the age of 50 years, rising slightly after this. It is suggested that radiation may be interacting in a multiplicative way with other factors in inducing leukemia.

For cancers other than leukemia, the situation appears to be similar. Excess risk increases regularly with age at exposure among adults, although the risk of radiation-induced breast cancer was highest among those exposed at 10–19 years of age.

Other authors have studied the risks from the atomic bombings in Japan, and they propose a number of points important for risk estimation after low-level radiation exposure. First is the issue of linear versus nonlinear models. The purely quadratic model (i.e. risk increasing with the square of the dose) gives a reduced risk we estimate when compared with a linear model or a linear-quadratic model. At a dose of 1 rad, the difference in the risk estimate is of one or two orders of magnitude. The second point for which the epidemiological data are lacking is whether there is a reduced risk for protracted or low dose-rate exposures. Information for risk assessment purposes must come from experimental systems to account properly for dose-rate effects. From the data available, there are large differences in the risk estimates, depending on how time is treated. Also there is the critical issue of possibly increased effects for individuals exposed at younger ages. Differences of upward of a factor of 20 were observed in the risk estimates for epithelial tumors at the youngest category of 0–9 years. Comparisons were made with BEIR estimates because any differences would be primarily due to differences in modeling approaches. For acute leukemia incidences, the estimates were similar, assuming linearity, but for exposure at the lower ages the ankylosing spondylitis data suggest that the risk may be underestimated. Some mix of linear and quadratic seemed to be the model of choice.

For epithelial cancers, the time-linear model appeared more suitable and suggested a somewhat higher risk than given in the BEIR study. If, however, the relative risk model is the correct model, the risks for the youngest age-at-exposure category would have been underestimated. The epithelial cancers were adequately described by a linear dose-response function.

IV. Chemical Protection to Radiation

Thiols, such as glutathione (GSH), limit oxidative stress, thereby protecting cells against radiation. Elevation of GSH to 20 mM will protect cells at intermediate oxygen tension.

Vitamin E, an antioxidant, is an efficient inhibitor of free radical–based reactions including lipid peroxidation. Drugs producing hypoxia through reducing blood flow should be effective in reducing radiosensitivity.

Hypothermia-lowered body temperature reduces metabolic activity, allowing for more complete and efficient repair of radiation damage. Radiation-induced hyperthermia elevates body temperature, by 1–15 Gy in animals and is mediated via PGE2 and another PG, possibly PGD2. Body temperature is lowered by 20–200 Gy from a direct action of radiation on the brain. Central administration of WR-2721, *S*-(3-aminopropylamino)-ethylphosphorothioic acid, attenuates radiation-induced hypothermia, this effect being correlated with the inhibition of oxygen uptake.

There is a close correlation in rats between skin pigmentation/coat color and susceptibility to radiation injury. The different properties of different strains of rats on exposure to thermal injury or to radiation injury may provide an opportunity to separate them further by selectively breeding for the genetic factors determining host responses to these types of injury.

V. Genetic Effects of Radiation

That radiation induces gene mutation in many organisms dramatizes the hazards of radiation. Evaluation of the genetic risk to humans of exposure to ionizing radiation rests on a variety of experiences. These include: (1) exposure to diagnostic and therapeutic doses of X-ray and radioactive materials; (2) occupationally incurred exposures (e.g., in uranium mining or in the maintenance of nuclear reactors); (3) geographic areas with high natural or man made background radiation levels; and (4) the atomic bombings of Hiroshima and Nagasaki. Various approaches have been employed to determine the occurrence of mutations; these include a search for changes in the frequency of (a) certain population characteristics, e.g., the occurrence of major congenital defects and premature death; (b) sentinel phenotypes; (c) chromosomal abnormalities; and (d) altered proteins.

Genetic surveillance of children born in Hiroshima and Nagasaki began in 1948. The indicators studied include sex ratio, weight, viability, presence of gross malformation, death during the first month of life, and physical development at age 8–10 months. Analysis of data through 1953 revealed marginal findings on the sex ratio and survival of liveborn infants; this prompted a continued collection of data on these two variables, using cohorts matched in age, city, and sex. One study involved all infants liveborn in the two cities between May 1946 and December 1958 one or both of whose parents were within 2 km of the hypocenter at the time of the bombing. The second study used was a cohort for which one parent was exposed at 2,500 m or beyond and the other similarly exposed or not exposed at all. The third study used was a cohort in which neither parent was present in the cities at the time of the atomic bombing.

A. Clinical Findings

In 1948 through 1953, 76,617 pregnancy terminations were studied. The findings identified those pregnancies that terminated with a child with a major congenital defect, who was stillborn, or died during the first week of life. The increase in frequency of an untoward outcome per Sievert of gonadal exposure, as measured by regression coefficient, was 0.001824 (standard error was 0.003).

The observed frequencies among all parents of untoward pregnancies were to define a series of "maternal age–paternal age parity"-specific probabilities of an untoward outcome. These values were used to calculate the expected number of such pregnancies in five dose categories (i.e., less than 1, 1–9, 10–49, 50–99, and 100 and above cGy) based on the combined parental exposures. The differences between the observed and expected numbers in each category, the "excess cases," were then regressed on average dose. The data gave a regression coefficient of 63 untoward pregnancy outcomes per million pregnancies per cGy. To compare this value with the absolute risk of leukemia or cancers other than leukemia, the genetic risk must be related to time. Because the cancer risks are on an annual basis and because the genetic one represents the time from the bombing to the average year of birth—6 years, as a first approximation—the pregnancy risk per year per cGy is about 10. This differs less than a factor of 2 from the excess cancer incidence cases per million person year per rad.

Among the 63,817 persons whose parents' exposures are known, there have been 3,786 deaths. Distance from the epicenter is known, but the shielding is not known. Through December 1971, 3,231 deaths occurred; 321 more deaths occurred from 1972 to 1980. The impact of parental exposure to one Sievert is only one-fifth to one-eighth that associated with a year's difference in the time of birth during these years. Seventy-two of these 3,552 deaths were ascribed to malignancy, the most common being 35 leukemia cases. The frequency of deaths per 1,000 at risk varies from 0.36, both parents exposed, to 1.18, only father exposed. Cancer of the stomach is common in Japan; there was no trend of its occurrence with atomic bomb exposure.

B. Sentinel Phenotypes

Sentinel phenotypes are those that have a high probability of being caused by dominant mutation. Among these phenotypes are aniridia, chondrodystrophy, epiloia, neurofibromatosis, and retinoblastoma. No relation between maternal radiation and Down's syndrome could be demonstrated.

The frequency of sex-chromosome aneuploids in the children of parents exposed beyond 2,499 m was 13/5,058 or 0.00257. The frequency in the children of parents exposed within 2,000 m, with an average combined exposure of 0.87 Sv, was 16/5,762 or 0.00278; the difference is not significant.

C. Biochemical Variants

Children born to parents exposed in Hiroshima and Nagasaki were examined in 1976 for rare electrophoretic variants of 28 proteins of the blood and erythrocytes. Since 1979 a subset of the children was further studied for deficiency variants of 11 of the erythrocyte enzymes. Another author had information on 642,004 locus tests in 13,052 children whose parents had an average combined gonadal dose of approximately 0.59 Sv. Three probable mutations were described: (1) a slow migrating variant of glutamate pyruvate transaminase, (2) a slow migrating variant of phosphoglucomutase-2, and (3) one involving nucleophosphorylase. Three probable mutations have also been seen in 478,803 locus tests on 10,609 children whose parents, one or both, were exposed beyond 2,400 m (i.e., received less than 1 cSv). These variants involve the haptoglobin, 6-phosphogluconate dehydrogenase, and adenosine deaminase systems. The two estimates of the rate of mutation were 0.47 and 0.63×10^{-5}, respectively. Patently, neither these data nor those on structured variants are sufficiently extensive to warrant strong inferences.

D. Estimates of Risk

Two parameters were used to characterize the risk of radiation-related mutation. One is the probability of a mutation at a specific locus per unit of exposure and the other the so-called doubling dose, the dose at which the ratio of the spontaneous mutation rate to the slope of the dose-response curve is precisely 2. Estimates of the doubling dose on untoward pregnancy outcome, F_1 mortality, and sex chromosome aneuploids, respectively, were made. As rough approximations to the three standard deviations of the doubling dose estimates, 0.93, 3.88, and 24.26 were obtained for untoward pregnancy outcome, F_1 mortality, and sex chromosome aneuploids. The weighted average of the estimates is 1.4 Sv, with a standard deviation of 1.6 Sv. An analysis that yields an estimate of the risk of an untoward pregnancy outcome with exposure was 63 per million pregnancies per rad. Of a million pregnancies terminating under the circumstances that prevailed in these years, exposure to 1 cGy would have increased the number of 47,500 "naturally ended" pregnancies by some 63 cases. The relative risk would have increased by 1.001. This pregnancy risk is about twice that of cancer among the survivors. If half the pregnancy risk is attributable to radiation-related mutations, a conservative estimate presumably, the genetic and cancer risks would be approximately equal.

VI. Immunologic Effects of Radiation

Ionizing radiation was shown to be immunosuppressive in the early 1900s. Ionizng radiation exposure increases the risk of infectious diseases as a result of tissue invasion by microorganisms, through radiation damage to hemopoietic and intestinal tissues leading to leukopenia and increased permeability of the intestinal mucosa. Alterations of nonspecific defense mechanisms play important roles in decreasing the resistance to infections.

The radiation-induced impairment of specific immune responsiveness is also a relevant component of the pathogenesis of infectious diseases. Studies were begun in the early 1950s pointing out the immunological effects of radiation. This area of research continues actively to define further the direct effects of radiation on immunity as well as to expand basic understanding of immunity itself.

Radiation strongly influences antibody response, reflecting radiation damage mainly of the inductive rather than the productive phase. Antibody response is depressed when antigen is injected shortly before or immediately after radiation in doses of 200–700 rad. Recovery of the response after sublethal doses starts after about a week and may be complete in 2 months. When antigen is given a week or a month after 25–100 rad, prolonged production and transiently higher antibody peak titer may be observed in irradiated as compared with unirradiated animals.

The antibody molecule itself is indeed radioresistant, in that whole body radiation with 600 rad failed to change the rate of degradation of passively transferred antibodies; kilorads of X-rays failed to affect the antigen-binding capacity of the antibody molecules. Enhancement of the antibody titer occurs either as a true stimulation if antigen is injected before relatively low doses of radiation, 25–300 rad, or by the formation of increased amounts of antibody produced at a lower rate if antigen is injected before relatively large doses of radiation, 500–700 rad. Cellular mechanisms may account for this (e.g., disproportionate repopulation of the depleted lymphoid tissues by rapidly dividing antigen-activated cells, the adjuvant effect of bacteria or endotoxins entering the blood stream from the radiation-damaged intestinal mucosa, or preferential inactivation of suppressor T cells). The secondary response which takes place after a second exposure to the same antigen is more radioresistant than is the primary antibody response.

The affinity of antibody for the antigen was found substantially higher in mice irradiated with 450 rads than in unirradiated controls although the titer of antibody was lower than in unirradiated controls.

Antibody affinity, however, was up to 20-fold higher in irradiated than in control mice.

Changes in T- and B-cell populations were studied by determining the *in vitro* mitogenic responses of spleen cells from unimmunized mice exposed to 450 rads at various times before culture. B cells are relatively more sensitive than T cells. The recovery of T- and B-cell mitogenic responsiveness was found to be incomplete 8 weeks after irradiation, thus providing an explanation for the lower antibody concentration in radiated than in control mice. The high sensitivity of antibody affinity to radiation-induced alterations of the immune system persisted with even smaller doses of radiation, 25 rad 5 days after immunization. The *in vitro* mitotic responses of spleen cells from these immunized and irradiated mice were measured and again indicated a shift in the recovering lymphocyte population in favor of B cells and a relative lack of suppressor T cells. Enhancement of anti-DNP antibody affinity has also been observed in radiation chimeras. Enhanced affinity may play a role in the protection from radiation-induced immuno-deficiencies; it may overcome quantitative defects of the antibody response.

A. Immunological Effects of Radiation at the Cellular Level

Dysfunctions of the antibody response in radiated animals reflect cellular alterations of the immune system.

1. Auxiliary Cells

Auxiliary cells or antigen-presenting cells are polymorphonuclear leukocytes, reticular cells, and macrophages, all involved in antigen handling after immunization. Effects of radiation on B cells will be discussed later.

Macrophage migration and phagocytosis are not affected by radiation even in the kilorad range. Intracellular catabolism of ingested antigens is variably reduced depending on the antigen used, but only from high radiation, in the kilorad range. However, macrophage intracellular levels of lysosomal enzymes are variably increased after irradiation. Thus radiation-induced dysfunctions of the antibody response are likely to depend mainly on lymphocyte radiosensitivity. [*See* MACROPHAGES.]

2. Lymphocytes

The radiosensitivity of lymphocytes is marked; the peripheral lymphocyte count is significantly depressed by 25 rad and reduced to 25% of control by 100 rad. A small number of thoracic duct lymphocytes can still be collected from mice 4 days after 800 rad. Most of these lymphocytes surviving interphase death are T cells, which, however, are unable to survive after entering mitosis on *in vivo* or *in vitro* antigenic stimulation. Immunofluorescence of T cells has shown the existence of two cell subpopulations, one (8%) of which is extremely radioresistant. B cells are more radiosensitive than T cells. Functional heterogeneity of B cells seems to be associated to some extent with different radiosensitivities. On antigen or mitogen stimulation, both B and T cells acquire some degree of radioresistance, presumably as a result of activation of repair mechanisms preventing interphase death. Antigenically stimulated B cells synthesizing antibodies appear to be much more radioresistant than unstimulated B cells. Exposure of spleen cells in a diffusion chamber to 10,000 rad during the secondary response to an antigen induce a relative increase in the proportion of plasma cells, the cells that release antibodies into the circulation, but no change in the rate they synthesize antibody and no changes in cytoplasmic structures. Estimates of the radiosensitivity of mouse antibody–forming cells provided survival curve Do value of 6,000 rad for IgM-PFC, plaque forming cells, and Do value of 1,500 rad for IgG-PFC. A threshold of inactivation was also evident.

Recovery of the lymphocyte populations after sublethal irradiation (5, 50, or 500 rad) starts during the first week and seems to be faster for T than for B cells during the subsequent 3 weeks. However, from 4 to 8 weeks after 450 rad, B cells recover faster and to a greater extent than the T cells. At 8–9 weeks after radiation, the two cell populations reach the same degree of recovery, but neither one attain the control value of unirradiated mice. [*See* LYMPHOCYTES.]

Gamma radiation of B cells damages the plasma membrane via highly reactive free radicals. There is a rapid increase in plasma membrane permeability and swelling of the cells; it may play a major role in causing interphase death. Five hundred to 1,000 rad caused resting B cells to enlarge slightly; 3,000 rad resulted in cell doubling in size within 3–4 hours. Sensitivity to the membrane damaging effects of gamma radiation was in the order of resting B cells > resting T cells > a long-term $L3T4^{+}$ T-cell clone > a B-cell lymphoma. The radiation effects could be ameliorated by excluding oxygen at the time of irradiation or by adding the free radical scavenger agent cysteamine.

Mice given 200 rad have normal splenic PFC responses to T-independent type 1 antigen but reduced responses to thymus-independent type 2 antigen. Single-parameter FACS analyses demonstrated a diminution in both B-cell number and the heterogeneity of membrane antigen expression within the surviving B-cell pool. Multiple-parameter FACS analyses indicate that B cells with the sIgM $\gg$ sIgD phenotype are more radiosensitive than B cells of the sIgM $\ll$ sIgD phenotype. Enhanced radiosensitivity of marginal zone B cells is also observed.

The major effects of radiation to a resting B-cell antigen presenting cell (APC) are a reduction in the effective display of antigen plus class II molecules and a loss in the ability to provide APC-derived costimulatory signals.

3. Selective Effects of Radiation on T Cells

The immune system is a complex network of interacting cellular and soluble elements under genetic control. The antibody response is induced in B cells by macrophage-processed antigen and modulated by signals passed among different types of immunoregulatory T cells. Cellular members of this network express characteristic membrane structures associated with genetically programmed functions. Some cell types in addition bear membrane receptors specific for antigen and/or for self-structures encoded by MHC, Igh-V, and Igh-C genes. Assessment of the radiation effects on helper (T_h) and suppressor (T_s) cell activities represents a major issue for understanding radiation-induced immunological dysfunctions.

a. Helper T Cells Radiation has selective effects on the T_h cell population leading to inactivation of some but not other cell subpopulations. This selectivity gives rise to abnormal T-B cell cooperation, which apparently selects B cells producing antibodies of higher affinity for the antigen.

b. Suppressor T Cells Suppressor T cells are one of the major cellular components of the immune network regulating the antibody response. T_s cells are induced during antigenic stimulation and exert their functions by reducing the amplification signals provided by T_h cells or by acting directly on effector B cells.

Radiosensitivity of T_s cells has not been thoroughly investigated, but it is a general consensus that they are more radiosensitive than T_h cells. T_s cell activity was found radio-sensitive at early but not at later times after induction. A decrease in number of T_s cells may account for the increase in antibody affinity observed in mice irradiated after immunization. The relative radiosensitivity of antigen-unstimulated precursors of T_s cells compared with precursors of T_h cells may provide an explanation for the increase in affinity observed in mice immunized after irradiation.

4. Recent Radiation Effect Findings on Immunity

a. T-Cell Receptor Genes Radiation is being used to define T-cell receptor genes. Clonal selection of T-cell receptor V beta repertoire, irrespective of positive or negative selection, appears to occur at the early stage of T-cell differentiation (i.e., on the blast-like $CD4^+$ $CD8^+$ thymocytes).

The sequential differentiation patterns of thymocyte were observed with cell surface phenotypes and the expression of T-cell antigen receptor in 800 rad irradiated mice. Intrathymic radioresistant stem cells for T thymocytes seem to proliferate and differentiate after irradiation with the same pattern as was seen in the fetal thymus development.

b. Cytokines and Their Antiradiation Effects

i. IL-1 Alpha, TNF Alpha, G-CSF, GM-CSF Interleukin-1 alpha (IL-1 alpha), tumor necrosis factor alpha (TNF alpha), granulocyte colony–stimulating factor (G-CSF), and granulocyte-macrophage colony–stimulating factor (GM-CSF) are molecularly distinct cytokines acting on separate receptors. Administered alone, human recombinant (hr) IL-1 alpha and hrTNF alpha protect lethally irradiated mice from death, whereas murine recombinant GM-CSF and hrG-CSF did not. On a dose basis, IL-1 alpha was a more efficient radioprotector than TNF alpha. The relative effectiveness of TNF alpha and IL-1 alpha probably depends on the genetic makeup of the host. The two cytokines together result in additive radioprotection, suggesting that they act through different radioprotective pathways. Suboptimal, nonradioprotective doses of IL-1 alpha also synergize with GM-CSF or G-CSF to confer optimal radioprotection. [*See* CYTOKINES IN THE IMMUNE RESPONSE.]

ii. TNF on Hemopoietic Reconstitution After Sublethal Irradiation Pretreatment of mice with murine rTNF alpha enhanced hematopoietic reconstitution after sublethal irradiation of mice with 7.5 Gy whole body, suggesting a possible therapeutic potential for this agent in the treatment of radiation-induced myelosuppression.

iii. IL-1-Enhanced Survival from Radiation IL-1 enhanced the survival of lethally irradiated mice, treated with isogenic or allogenic bone marrow cells. It may prove clinically useful in patients undergoing bone marrow transplantation.

iv. Immune Modulators Immune modulators [glucan-F (GF), glucan-P (GP), krestin (K), lentinan (L), pecibanil (P), azimexone (A), ciamexone (C), MVE-2 (M), picolinic acid (PA), and poly ICLC (PI)] were studied in female C3H/HeN mice being given 20 hours before irradiation. Hemopoietic stem cells were measured by endogenous spleen colony assay after 6.5 Gy. Survival was observed after 9.0 Gy. Except for A, C, PA, and PI, all immune modulators increased spleen colony assays in a dose-dependent manner ($P > M > GP > K > GF > L$). Only P, GP, and GF significantly enhanced survival. These studies suggested that immunomodulated-mediated postirradiation hemopoietic and survival enhancement can be dissociated.

c. T Cells Inactivation of naturally occurring radiosensitive suppressor T cells is associated with increased production of IL-2. The increased levels of IL-2 could account for the radiation-induced enhancement of a host's immune system with a consequent effect on tumor regression.

In fact tumor-activated T cells, but not resting T cells, survive irradiation. They are free to expand in number to cause tumor regression in the absence of radiosensitive suppressor T cells.

T helper cells cooperate with radioresistant T effector cells to produce autoimmune thyroiditis in chickens.

Mice given sublethal irradiation to both the thymus and the peripheral lymphoid tissues have major transient and some persistent disruptions in thymic architecture and in thymic stromal components. This radiation regimen induced both transient and long-term effects, for at least 4 months after total lymphoid irradiation.

Irradiation of the thymic microenvironment during marrow ablative preparative regimens may be in part responsible for some of the immune alterations observed in marrow transplant recipients. It may provide a valuable tool for studying the roles of the thymus on T-cell maturation.

B. Miscellaneous Immunological Effects of Radiation

Delayed-type hypersensitivity (DTH) was not inhibited by irradiation at 200 R but was slightly diminished by 900 R irradiation of the host. These results imply that immunosuppressive agents, which inhibit cell division, do not inhibit the initiation of DTH but can affect the amplification phase contributed by the host.

Using DNA content as a quantitative measure of cellular infiltration, it was shown that selective irradiation does not prevent the development of an inflammatory response. This response does not require T lymphocytes.

Host resistance to alloengraftment develops between 8 and 15 days after irradiation and T-cell-depleted syngeneic bone marrow administration. The mechanism of this resistance is unknown but could involve the level of host engraftment and/or the return of host alloreactivity by the time of allogeneic bone marrow transplantation.

ATP-$MgCl_2$ (60 umol/kg IV) during radiation therapy of 4,500 cGy of external pelvic exposure over 4 weeks in random bred pigs decreases colorectal seromuscular ischemia, skin injury in radiation therapy portals, and bowel inflammation. ATP-$MgCl_2$ thus appears to offer a significant cytoprotection during preoperative radiation therapy.

Copper (II)(3,5-diisopropylsalicylate) (Cu-DIPS) was shown to have radioprotective activity, in C57BL/6 mice, injected subcutaneously with 80 mg/kg Cu-DIPS and exposed to 8.0 Gy 3 hours later. Cu-DIPS increased the survival rate, 30 day, from 40% to 86% without overt toxicity.

The number of white blood cells in unirradiated mice is $10 \times 10^3/mm^3$ blood, and the reserve leukocytes mobilized by dextran sulfate increase this to $30 \times 10^3/mm^3$. The number of reserve white blood cells is thus $20 \times 10^3/mm^3$. One day after 1 or 3 Gy of radiation the number of reserve cells available falls to 11 or $3 \times 10^3/mm^3$, respectively. Leukocyte reserves then can serve as a good biological indicator for net impairment of hemopoiesis (i.e., total body burden).

VII. Miscellaneous Recent Findings on Radiation Effects

A. Endothelial Cells and Radiation Injury

Radiation injury in cultured endothelial cells is accompanied by delayed progression through S-phase, blocked progression at G2/M-phase, and blocked progression 4 hours before the end of G1-phase.

Radiation-induced cell loss was not evident, but increased solute flux as tested by albumin and sucrose levels was observed. Both albumin and sucrose flux was dose-dependent over the radiation range of 1, 5, and 10 Gy, being 69% more than controls at 4 hours after 5 Gy and 115% more than controls after 10 Gy. The increase occurs without the presence of inflammatory cell or addition of mediators.

DMFO (alpha difluoromethyl ornithine) or low doses of P (putrescine) reversed radiation-induced permeability changes. Greater than 10 mM P did not protect the cells, but addition of 25 mM or more P to unexposed cells results in morphologic evidence of injury. These data suggest that polyamines may be important in oxidant-induced pulmonary endothelial cell injury.

B. Cyclooxygenase Inhibitors and Radiation

Decreased aortic responsiveness to U 46619, a thromboxane A2 mimic, after irradiation of the whole body after 20 Gy may be related to a radiation-induced increase in cyclooxygenase product release. It is concluded that cyclooxygenase inhibitors may be important in attenuating vascular injury seen with radiotherapy.

WR 2721, a radioprotectant, was tested on the radiation-induced decrease in vascular activity to U 46619. WR 2721 attenuates the radiation-induced decrease in vascular activity to U 46619 in rat aorta, exposed to 20 Gy whole-body irradiation.

C. Vascular Endothelium and Smooth Muscle

Vascular endothelium (EC) and smooth muscle cells (SMC) respond differently to ionization radiation. SMC cultures show no change in cell number or size after large radiation doses (i.e., 3,000 rads). In contrast, after radiation with doses comparable with clinical radiotherapy (i.e., 200–400 rads), EC cultures show dose-dependent cell loss, and subsequent hypertrophy of the remaining cells. A two to four time increase in protein content 48 hours after radiation is proportional to both cell loss and cell size. FACS showed no change in the proportion of cells in $G_0/_1$, S, and G_2.

D. Cardiac Effects of Radiation

In rats irradiation with 5 Gy causes a change in the relative amount of cardiac myosin enzymes v_i, v_2, and v_3, but there are no significant increases in serum levels of the enzymes creatine kinase and lactic dehydrogenase which typically reveal cardiac damage.

E. Gastric Effects of Radiation

Sublethal doses of ionizing radiation, 300 rads of total body in 100 rad/minute, inhibited gastric acid secretion in intact animals by a direct action on the stomach, because abdominal shielding reduced the radiation injury. The radiation damage is not likely to be caused by direct effect on the parietal cell receptors for histamine or gastrin because radiation did not affect the *in vitro* response of the gastric glands.

F. Radiation-Induced Lung Tumors

Six lung tumors at necropsy from Beagle dogs that inhaled particles of either $^{239}PuO_2$ or ^{90}Sr as young adults were examined for the aberrant expression of 22 of the known oncogenes. Sequences similar to the *N*-ras and myc oncogenes appeared to be constitutively expressed in both the tumor and control lung tissues. In one tumor, obtained from a $^{239}PuO_2$ exposed dog (1145T), the H-ras, v-abl, erb-B, and met hybridizations may be due to cross-hybridization between homologous tyrosine kinase domains, suggesting that the 1145T tumor was expressing a protein tyrosine kinase not observed in the other five tumors examined or in the control tissues.

G. Residual Hemopoietic Damage After Fractionated Gamma-Irradiation

Residual damage in hemopoietic progenitor cells, spleen, and granulocyte-macrophage colony–forming cells (CFU-S and GM-CFC) was detected in mice after 15 daily fractions as low as 0.1 Gy. The injury was dose-dependent, and after higher total fractionated doses of 7.5–10 Gy, the CFU-S cells recovered to about 50% of control in 2–12 months. Residual damage was also detected in the stroma, in the form of reduced numbers of fibroblastoid colony–forming cells and of CFU-S in ossicles under the kidney capsule. The response to a second course of 15 fractions, 3 weeks after the end of the first course, was similar and additive to the response to the first course in the short term; however, in the long term, recovery levels were similar after either one or two courses.

H. Radiation-Induced Lung Damage: Dose-Time-Fractionation Considerations

An isoeffect formula has been specifically developed for radiation-induced lung damage, a formula based on a linear-quadratic model including a factor for overall treatment time, allowing for the simultaneous derivation of an alpha/beta ratio and a gamma/beta time factor. From published animal data, the derived alpha/beta and gamma/beta ratios for acute lung damage are 5.0 ± 1.0 Gy and 2.7 ± 1.4 Gy/day, respectively, whereas for late damage the suggested values are 2.0 Gy and 0.0 Gy^2/day. Data from two clinical studies, one prospective and the other retrospective, were also analyzed and ratios determined. For the prospective clinical study, the resultant alpha/beta and gamma/beta ratios were 0.9 ± 2.6 Gy and 2.6 ± 2.5 Gy^2/day. Combining the retrospective and prospective data yielded alpha/beta and gamma/beta ratios of 3.3 ± 1.5 Gy and 2.4 ± 1.5 Gy^2/day, respectively. This isoeffect formula might be applied to both acute and late lung damage.

VIII. Summary

A discussion is presented of the acute, subacute, and chronic effects of ionizing radiation on humans and experimental animals at the whole-body, organ, cellular, biochemical, and immunological level. Cancer and leukemia risk estimates are presented from data from the atomic bomb exposures in Japan and from other specific human radiation exposures such as radiologists, women with menopausal problems, and patients with ankylosing spondylitis. Concerns about mental retardation effects and genetic effects of ionizing radiation also arose from the atom bomb exposures; studies have been ongoing now for many years to make these assessments. Radon carcinogenesis is widely discussed currently among radiation experts; but there is no consensus yet on its significance.

Anticarcinogenic states and radioprotection have been considered. The rapidly expanding science of the biological response modifiers provides new approaches for the treatment of radiation sickness. Especially important is radiation injury in the behavior of the immune system.

Bibliography

BEIR Committee. (1980). "The Effects on Populations of Exposure to Low Levels of Ionizing Radiations." National Academy of Sciences/National Research Council, Washington, D.C.

Burns, F. J., Upton, A. C., and Silini, G. (1984). "Radiation Carcinogenesis and DNA Alterations." Plenum Press, New York.

Castellani, A. (1984). "Epidemiology and Quantitation of Environmental Risk in Humans from Radiation and Other Agents." Plenum Press, New York.

Cerutti, P. A., Najjard, O. F., and Simic, M. G. (1987). "Anticarcinogenesis and Radiation Protection." Plenum Press, New York.

Diethelm, L., Heuck, F., Olsson, O., Strnad, F., Vieten, H., and Zeippinger, A. (1985). Radiation exposure and radiation protection *In* "Handbuck der Medizenischen Radiologie (Encyclopedia of Medical Radiology)" (F. Heuck and E. Scherere, eds.). Springer-Verlag, New York.

Doria, G., Agarossi, G., and Adorini, L. (1982). Selective effects of ionizing radiations on immunoregulatory cells. *Immunol. Rev.* **65,** 24.

FASEB *Journal*. (1987, 1988, 1989). "Abstracts on Radiations Effects on Immunology," vols. 1–3.

Journal of Immunology. (1987, 1988, 1989). "Papers on Ionization Radiation Effects on Immunology," vols. 138–143.

Livesey, J. C., Reed, D. J., and Adamson, L. F. (1985). "Protective Drugs and Their Reaction Mechanisms." Noyes Publications, Park Ridge, New Jersey.

Prasad, K. N. (1974). "Human Radiation Biology." Harper and Row, New York.

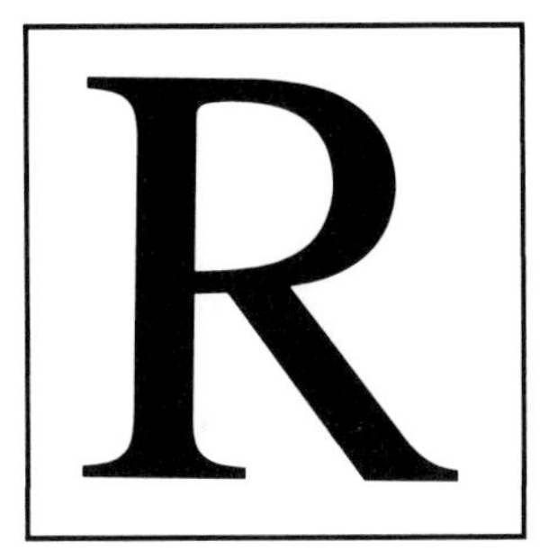

Radiation in Space

R. J. M. FRY, *Oak Ridge National Laboratory*

Glossary

Absorbed dose Energy imparted to matter by ionizing radiation per unit mass of irradiated material at the point of interest; unit of absorbed dose has been the rad and now, in System of International Units (SI), is the gray (Gy) (100 rad = 1 Gy)
Alpha particles Nuclei of helium atoms consisting of two protons and two neutrons in close association; they have a net positive charge of +2 and therefore can be accelerated in large electrical devices similar to those used for protons, and they are also emitted during the decay of some radioactive isotopes
Bremsstrahlung Secondary photon radiation produced by deceleration of charged particles
Dose equivalent Quantity that expresses the biological effect of interest in radiation protection for all kinds of radiation on a common scale for calculating the effective absorbed dose; defined as the product of the absorbed dose in rad, or Gy, and the quality factor (Q) for the particular radiation based on its relative biological effectiveness. The unit of dose equivalent has been the rem and is now the Sievert (Sv) in SI units (100 rem = 1 Sv)
Extravehicular activity Any activity undertaken by the crew outside a space vehicle
Fluence Number of particles divided by the cross-sectional area of a sphere that the particles enter
Flux Number of particles or photons per unit of time passing through a surface
Gray (Gy) New SI unit of absorbed dose of radiation (1 Gy = 1 J kg^{-1} = 100 rad)
Heavy ions Nuclei of elements such as nitrogen, carbon, oxygen, neon, argon, or iron that are positively charged due to some or all of the planetary electrons having been stripped from them
Inclination of orbit Acute angle that the orbit's trajectory makes with the Earth's equator
Linear energy transfer Average amount of energy lost per unit of particle track length and expressed in kilo-electron volt (keV) μm^{-1}
Proton Positively charged nucleus of the hydrogen atom
Relative biological effectiveness Factor used to compare the biological effectiveness of absorbed radiation doses from 250 kilo Voltage peak (kVp) X rays relative to other different types of ionizing radiation; more specifically, the experimentally determined ratio of an absorbed dose of a radiation in question to the absorbed dose of a reference radiation (e.g., 250-kVp X rays) required to produce an identical biological effect in a particular experimental organism or tissue; if 10 mGy of fast neutrons equaled in lethality 20 mGy of 250-kVp X rays, the relative biological effectiveness of the fast neutrons would be 2
Sievert (Sv) SI quantity of radiation dose equivalent equal to dose in gray times a quality factor that depends on linear energy transfer (1 Sv = 100 rem)
Z Atomic number; the number of protons in an atomic nucleus or the positive charges in the nucleus

RADIATION IN SPACE[1] is conventionally divided into trapped particle radiation, galactic cosmic radi-

1. Research sponsored by the Office of Health and Environmental Research, U.S. Department of Energy, under contract DE-AC05-840R21400 with the Martin Marietta Energy Systems, Inc. The submitted manuscript has been authored by a contractor of the U.S. Government under contract No. DE-AC05-840R21400. Accordingly, the U.S. Government retains a nonexclusive, royalty-free license to publish or reproduce the published form of this contribution, or allow others to do so, for U.S. Government purposes.

ation, and solar particle radiation. The exposures incurred during space missions are influenced by altitude, inclination of the orbit, shielding, and the duration of the mission.

Missions within the Earth's geomagnetic field at low altitudes and inclinations are protected, largely, from galactic cosmic radiation and solar particle event radiation.

In geosynchronous orbit, the dose from radiation is primarily from electrons of the outer radiation belt, which exhibit marked temporal variations in intensity. Galactic cosmic radiation and solar particle events also contribute to the dose.

The major source of radiation outside the magnetosphere is galactic cosmic rays. Although protons are the predominant radiation of galactic cosmic rays, there are helium ions and heavier ions called High-Z and High-Energy (HZE) particles. These particles are of particular interest radiobiologically and are of concern for risk estimates of late effects of radiation. The dense ionization along a particle track of the heavy ions that traverses many cells is in contrast to gamma or X rays.

In deep space, solar particle events occur unpredictably. When these events are large, the proton dose rate may rise rapidly to levels that can cause acute radiation damage. The design of shielding to prevent these effects is an important feature in the planning of deep space missions.

The risks from exposure to protons and electrons can be estimated with some confidence; this is not the case for HZE particles because of the density of the ionization and the length of the particle track. Fortunately, the HZE component of cosmic rays is small.

I. Sources of Space Radiation

There are three sources of radiation in space: (1) trapped particle radiation, (2) galactic cosmic radiation, and (3) solar particle radiation.

A. Radiation Belts

Within the magnetosphere above the densest part of the Earth's atmosphere, there is a complex system of magnetic fields, which are populated by the trapped electrons, protons and some low-energy heavy ions. These particles are reflected back and forth between regions of magnetic field strength or mirror points in the Northern and Southern hemispheres. The movement of the particles is also complex and, because of the different charges, the electrons drift eastward and the protons westward. Thirty-one years ago, cosmic ray detectors were put on Explorer 1 and Explorer 3. Initially, about 600 miles above the Earth's surface, the cosmic ray flux appeared to decrease dramatically. However, Van Allen and colleagues deduced and determined that their detectors were saturated by a huge and unexpected flux of trapped electrons and protons, now known as the Van Allen radiation belts (Fig. 1). There are two regions of trapped electrons, designated inner and outer, although they are not completely distinct. The maximum intensities and energies of the electrons in the outer zone are much higher than those in the inner zone. The inner zone extends out to about 2.8 earth radii (R_e) (about 18,000 km), and the outer zone from about 2.8 to 12 R_e (Fig. 2). In the outer zone, the trapped electrons are affected by the changes in the geomagnetic field caused by solar activity. At the altitude of geosynchronous orbits, about 36,000 km, the intensity of electrons shows a diurnal cycle caused by the interaction between the solar wind and the magnetosphere. These interactions produce changes in the magnetic field locally and therefore the trapping strength; as a result, the intensities of the electrons may vary by a factor of 10 in amplitude. Smaller variations occur with a periodicity of the solar cycle, and magnetic storms cause sporadic fluctua-

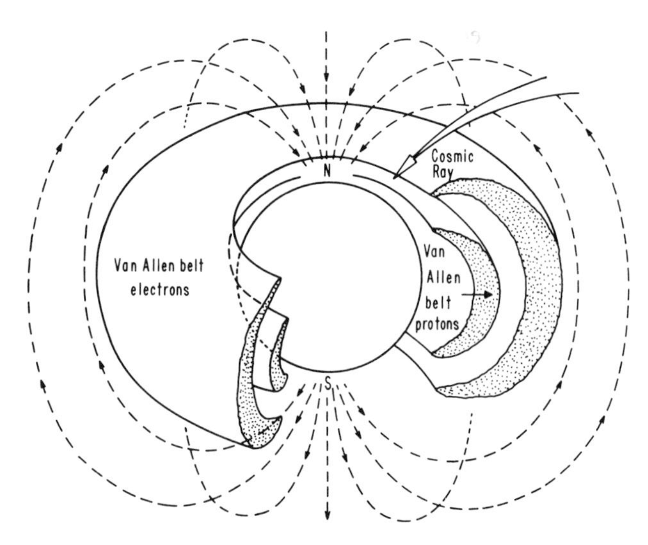

FIGURE 1 Schematic of the magnetosphere and the radiation belts. The inner and outer Van Allen belts are shown as distinct and separate for illustrative purposes, but the separation is not complete.

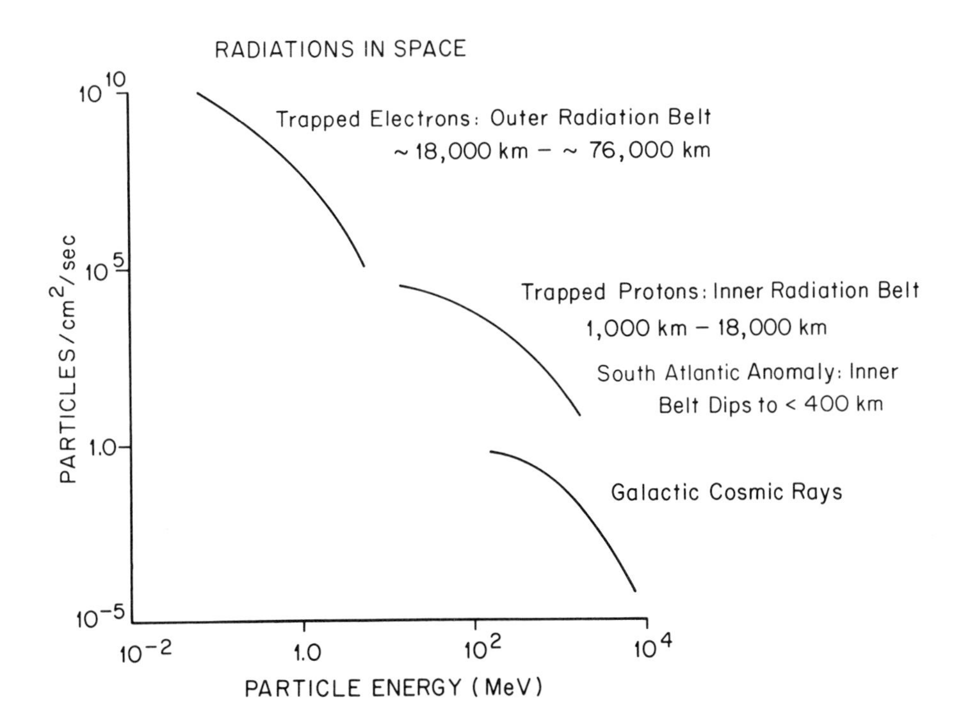

FIGURE 2 Schematic to illustrate the ranges in flux and energy of the particles.

tions. The flux of electrons with energies >45 keV may increase rapidly (in a matter of minutes) by an order of magnitude. So-called substorms occur frequently and result in the injection of electrons with energies between 50 and 150 keV from the magnetosphere tail.

The flux of the higher-energy electrons, >200 keV, remains relatively constant. This degree of detail in the description of the radiation environment is important, because short-term missions to altitudes of geosynchronous orbit that may involve extravehicular activity are likely to be required in the future. Good planning of such activities based on the knowledge of the wide swings in radiation intensity and their periodicity could limit the exposure to radiation of the astronauts markedly. The flux of electrons in the inner zone is less, by about a factor of 10, than the peak flux in the outer zone and contributes little to the radiation exposure of space vehicles and crews in low Earth orbits.

B. South Atlantic Anomaly

The trapped protons, which predominate in the inner radiation belt, extend out to about 3.8 R_e. The higher-energy protons are trapped in a much smaller volume than the electrons but in the range of altitudes that are of importance for low Earth orbits. In the region between Africa and South America, the spiraling protons reach down closer to the Earth than in other regions. This anomalous distribution of the trapped protons is due to a displacement of the magnetic dipole from the Earth's center that alters the geomagnetic field.

The exposure to radiation in the South Atlantic Anomaly of the inner belt is important for space missions in low Earth orbits. The altitude and the orbital inclination of the spacecraft determines the rate of exposure to radiation in the South Atlantic Anomaly. Figure 3 shows that the dose rate in the radiation environments of low Earth orbits is influenced by the altitude of the orbit. All the data shown in Figure 3 came from missions of the shuttle vehicles, all of which had an orbital inclination of 28.5°. The choice of orbit is influenced by a number of factors including the site of the launch. Most of the Soviet missions have been at higher orbital inclinations such as 62.8° and also usually at lower average altitudes. Both of these features of the missions reduce the exposure to the radiation encountered in the South Atlantic Anomaly. The dose rate increases with altitude as the orbit involves increasing radiation exposure in the South Atlantic Anomaly. The solar activity also influences the trapped protons, particularly those with lower altitudes in the

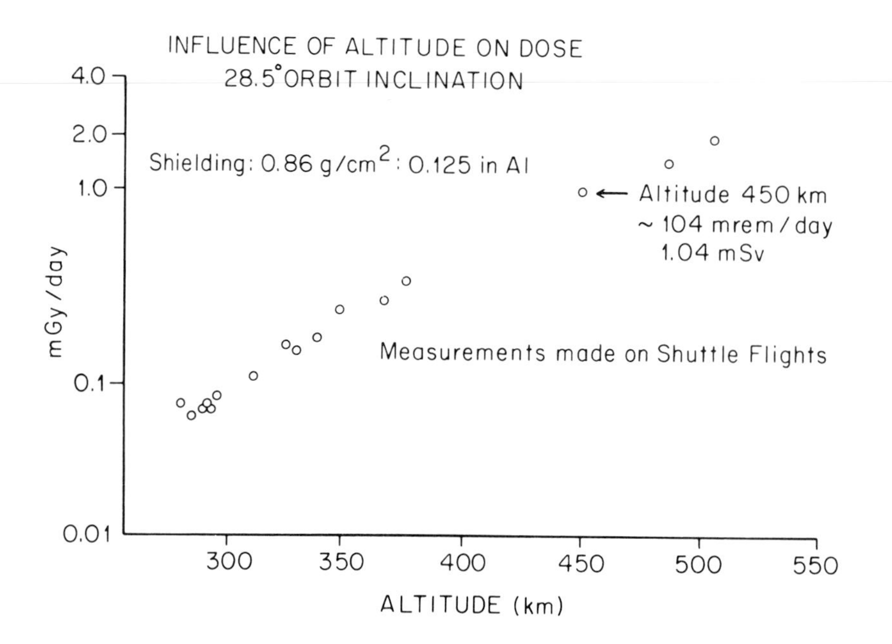

FIGURE 3 Dose rate of radiation as a function of altitude of the space shuttle in an orbit of 28.5°. Note that dose rate is plotted on a log scale indicating the marked influence of altitude. The dose equivalent at 450 km is given in the old (rem) and new (mSv) SI units, which take into account the different effects of the radiations involved. The figure is based on data reported in (1988). *Health Physics* **55:** 159–164, with permission.

radiation belt. In the active part of the solar cycle, the proton intensities are reduced, in contrast to the electrons in the outer region of the radiation belt. There is also a small flux of low-energy trapped heavy ions. The energy of these heavy particles appears to be too low to affect either humans or electronic components of satellites.

On missions to deep space, the space vehicle must pass through the radiation belts. Although the flux of both protons in the inner region and electrons in the outer region is high, the time to traverse the radiation belts is short, and this limits the total dose. The dose equivalent incurred in the low Earth orbits and traversal through the Van Allen radiation belts to reach deep space has been estimated to be 20 mSv. This dose equivalent is about the same as or less than the dose incurred with a whole-body computerized axial tomography (CAT) scan used for medical diagnoses.

C. Galactic Cosmic Rays

In the early part of this century, instruments were sent up in balloons to measure the Earth's natural background radiation at high altitudes. The results were contrary to expectation. It had been assumed that as the instruments got farther and farther from Earth the background radiation from Earth would decrease correspondingly. Surprisingly, the radiation levels increased with altitude. The measured levels rose because of radiation streaming in from the cosmos. The cosmic rays that were the source of the radiation detected by the instruments in the balloons were secondary cosmic rays. The primary cosmic rays are altered in energy and composition by nuclear reactions with atoms in the atmosphere and by the magnetosphere. The cosmic rays reaching the Earth contribute about the same amount of radiation to our background radiation on Earth as the terrestrial radiation that comes from rocks and soil (excluding radon).

In deep space, the radiation consists of galactic cosmic rays and solar particle radiation. Cosmic rays are thought to be distributed uniformly outside the solar system. This uniformity results in an equal flux of charged particles that approach the solar system from all directions.

About 98% of the primary galactic cosmic rays are protons and heavier ions. The flux rate is greatest for particles in the 100 MeV to 10 GeV per nucleus energy range, which is 87% constituted by protons, 12% by helium ions (alpha particles), and 1% by heavier ions. The ions heavier than helium, which have been described as HZE particles, are of particular interest to the radiobiologist. The abundance of the various ions varies. Because the capacity for ionization is proportional to Z^2, it is the High-Z particles that are of concern. The fluences of the ions above atomic number of iron (26) are very low; thus, iron with a High Z and a significant fluence is of particular importance.

The precise source of the galactic cosmic rays is still a matter of discussion. The discovery of *Cygnus X-3*, a neutron star in the Cygnus constellation, has identified a possible source of at least some of the cosmic rays in the emission by *Cygnus X-3* of high-energy gamma rays that could be by-products of the production of the cosmic rays.

As noted above, the cosmic rays detected on Earth were largely secondary cosmic rays, because the magnetosphere and the Earth's atmosphere act as a giant shield. The only gaps in that shield are the polar regions (Fig. 1), where cosmic rays can enter the Earth's atmosphere with less restriction. Thus, in low Earth orbits at 90° orbital inclination, so-called polar orbits, the flux of galactic cosmic rays is greater than at lower inclinations. The contribution from galactic cosmic rays to the total dose in a polar orbit is about three times that in a comparable low Earth orbit.

The energies of the galactic cosmic rays may reach as high as 10^{20} eV, compared with the maximum energy of about 10^7 eV of radioactive substances. The sun has a major influence on the galactic cosmic rays. The solar wind formed from highly ionized gas emitted by the sun carries magnetic fields that point radially away from the sun. These magnetic fields reduce the intensity of the lower-energy particles, especially during the active part of the solar cycle, which is called solar maximum. The fluence rate, which is highest during the less active part of the solar cycle, called solar minimum, is about 4/cm^2/sec.

D. Solar Flares

The variation in the number and size of the dark areas in the photosphere of the sun, called sunspots, have been an interest for astronomers for centuries. The observation that the activity on the sun's surface varied led to the identification of the 11-year solar cycle. The number of sunspots shows this approximate 11-year periodicity, but the number also varies among different cycles.

There is not a constant emission of solar particles because the solar particle radiation is associated with solar flares. Solar particle events are large emissions of the charged particles, protons, helium, and heavier ions that occur at the Earth in association with about 12.5% of the solar flares. Solar particle events occur intermittently during the active part of the solar cycle, and their number and size vary from cycle to cycle. Solar particle events have been classified into small, intermediate, and very large (Table I). The very large events occur infrequently, about one to three per solar cycle: three in the 19th cycle, one in the 20th, and none in the 21st.

The amount of energy released in the very large particle events can be extraordinarily large. In 1972, a series of flares and a so-called anomalously large solar particle event released 10^{25} or 10^{17} kilowatt-hours of energy. That amount of energy is hundreds of thousands of times the current annual energy consumption in the United States. When the particles hit the Earth's magnetosphere, the resulting magnetic storms interfere with satellites and power supplies in the northern latitudes, and the aurora borealis is spectacular. Another effect of the sudden injection of energy into the Earth's magnetosphere and upper atmosphere is the heating and expansion of the atmosphere, to an extent sufficient to alter the drag on space vehicles. Changes in atmosphere caused by solar flares altered the paths of the space vehicles Columbia and Skylab, bringing them down to orbits at lower altitudes. Despite an increase of several orders of magnitude in the flux of protons during solar particle events, the increase in the radiation does not present a danger to missions in the low-inclination, low-altitude orbits. In the case of polar orbits, the rise in radiation levels is significant and could prohibit extravehicular activities.

The major concern about very large particle events is related to missions to the moon or Mars. These events result in a shower of protons varying in energy. The dose rate rises rapidly and within hours reaches a rate sufficient to cause acute radiation effects to crews on missions in deep space. Because of the preponderance of low-energy protons in these very large particle events, the dose in the radiosensitive deep organs, such as the bone marrow and gut, fortunately, is likely to be much less than in the more radioresistant skin. There has been a concerted effort to develop methods of predicting the occurrence of the events. The difficulty is in deciding which active region on the sun's surface will develop into solar flares and the fact that very few of these become large particle events. Predicting the eventual size of the solar flare by monitoring the development of the flare and how the

TABLE I Characteristics of Solar Particle Events[a]

	Type of event		
	Small	Intermediate	Very large
Frequency (solar cycle dependent)	5–20/yr	3–6/yr	1–3/cycle
Peak flux (p cm^{-2} sec^{-1})(>10 MeV)	$10–10^2$	$10^3–10^4$	10^5
Fluence (p cm^{-2})(>10 MeV)	$10^5–10^7$	$10^8–10^9$	$>10^9$

[a] G. R. Heckman, personal communication.

proton dose rate builds up has been more successful. The time taken by energetic protons from a solar flare to reach Earth varies depending on the position of the origin on the sun and other factors, from <1 hr to several hours. Geosynchronous orbit Earth satellites provide continuous proton monitoring and transmission of data. Hopefully, an X-ray imager can be put in space to detect flares that arise behind the limbs of the sun. With this knowledge, even in deep space, sufficient warning would allow astronauts to seek a shielded area before the dose rate reached levels that could cause radiation effects. Possibly, when a return to the moon or a visit to Mars is planned, the trip will be planned to take place in the less active period of the solar cycle, called solar minimum. If not, an adequate shield storm shelter must be provided.

II. Radiation and Space Missions

A. Low Earth Orbits

In low Earth orbits, the radiation exposure rate is considerably greater than on earth. Increases in radiation exposure are of concern because of the assumption that there is no threshold for the induction of cancer or genetic effects, the so-called stochastic effects of radiation. In the case of acute radiation effects, such as damage to skin or blood-forming tissues, clinically significant effects do not occur until the threshold doses for the specific effects are exceeded. Threshold doses for noncancer or nonstochastic effects will not be exceeded on any space mission within the magnetosphere. A most important feature of radiation in space, as well as on Earth, is a low dose rate. All the effects of radiation are reduced by decreasing the dose rate. In the case of nonstochastic effects, the low dose rate allows time for repair of the damage to DNA, and, in tissues that undergo cell renewal, time to replace damaged cells. The effect of dose rate on cancer induction in humans is not known adequately, but experimental animal studies indicate that lowering dose rate reduces the carcinogenic effect of radiation.

The doses incurred by the crews of spacecraft have been low. The highest doses received by U.S. astronauts occurred in the Skylab missions, which had the longest duration of any of the U.S. space missions. The average total bone marrow dose equivalent was about 77 mGy for Skylab 4, which orbited for 84 days at an altitude of 435 km and an orbital inclination of 50°.

In space missions that are within the magnetosphere, the radiation environments can be predicted with considerable confidence. Exposures to crews can be controlled by limiting the duration of the mission and providing appropriate shielding. The current flight plan for the U.S. Space Station has an altitude of about 450 km and an orbital inclination of 28.5°. Assuming a shielding of 1 g cm^{-2} Al, the daily dose equivalent from worst-case solar minimum conditions is estimated to be about 1.1 mSv. Almost all of this dose is contributed by protons in the South Atlantic Anomaly; only 0.08 mSv is attributed to galactic cosmic rays. Under those flight conditions, a dose equivalent equal to the maximum permissible annual dose for a radiation worker on Earth would be incurred in about 50 days. Modifications of the shielding or lowering the altitude during low-drag solar minimum should increase the possible sojourn time considerably. The long duration on the Soviet Station Mir has been possible because the level of radiation at the altitude of the Mir mission is considerably less than at the altitude proposed for the U.S. Space Station.

B. Deep Space Missions

In deep space, low dose-rate proton radiation predominates, and the risks are relatively well understood. However, there is also the small heavy ion component in galactic cosmic rays, and the risks that these heavy-charged particles pose is not well understood. Tobias of the Lawrence Berkeley Laboratory predicted that astronauts would see infrequent flashes of light in deep space because the heavy ions would be capable of stimulating the retina; the astronauts did see flashes. The concern is whether or not the high linear energy transfer (LET) radiations with long particle tracks of dense ionizations and accompanying electron tracks that traverse a number of cells can cause damage that is more severe than that induced by low-LET radiations such as X rays. In other words, is the relative radiobiological effectiveness (RBE) of the heavy ions much greater than X rays for both cancer and noncancer late effects of radiation. The RBEs for a number of effects such as cell killing, mutation, cancer induction, and cataractogenesis increase with increasing LET, up to about 200 keV/μm^{-1}. Information about the risks of the late effects of radiation from HZE particles is under study. For example, if the energy deposited by an iron nuclei is 676 times (26^2) greater than that deposited by a cosmic ray proton, how much greater are the biological effects? Without such information, the risks of missions in deep space, such as to Mars, will be difficult to estimate.

The doses received by the crews on the Apollo missions were low because of the short duration of the missions; the longest mission, Apollo 15, was about $12\frac{1}{2}$ days. There was a certain amount of luck that the doses were so low in the case of Apollo 15 and 16 because they were undertaken within months of the largest recorded solar particle event. If either mission had coincided with the solar particle event, the dose to the skin could have been as high as 14–15 Sv and about 1 Sv to the bone marrow. Such dose levels would have caused serious radiation effects. The highest dose received by any of the crews of the Apollo missions was about 11 mGy on Apollo 14. This dose is of the same order as that a patient would receive from a whole-body CAT scan and poses no significant risk for late effects from the radiation exposure.

As discussed earlier, large solar particle events that occur during solar maximum are a hazard for deep space missions, and appropriate shielding, which can be effective against proton radiation, must be incorporated.

Although the radiation environments in space are much less benign than on Earth, there is good reason to believe that humans can travel and work in space with acceptable levels of risk of radiation effects. The design of space vehicles and careful mission planning can provide adequate conditions of radiation safety for missions that involve other considerable hazards.

Bibliography

Blakely, E. A., Ngo, F. Q. H., Curtis, S. B., and Tobias, C. A. (1984). Heavy-ion radiobiology: Cellular studies. *Adv. Radiat. Biol.* **11,** 295.

Eirich, F. R., Bücker, H., Horneck, G., Cox, A. B., and Fry, R. J. M. (eds.) (1986). "Advances in Space Research. Life Sciences and Space Research XXII(1)," Vol. 6. Pergamon Press, Oxford.

Leith, J. T., Ainsworth, E. J., and Alpen, E. L. (1983). Heavy-ion radiobiology: Normal tissue studies. *Adv. Radiat. Biol.* **10,** 191.

McCormack, P. D., Swenberg, C. E., and Bücker, H. (1988). "Terrestrial Space Radiation and Its Biological Effects." NATO AS1 Series, Series A: Life Sciences, Vol. 156. Plenum Press, New York.

National Academy of Sciences/National Research Council (NAS/NRC) (1967). "Radiobiological Factors in Manned Space Flight," Report of Space Radiation Study Panel of the Life Sciences Committee (W. H. Langham, ed.). National Academy Press, Washington, D.C.

National Council on Radiation Protection and Measurements. (NCRP) (1989). Guidance on Radiation Received in Space Activities. NCRP Report 98. National Council on Radiation Protection and Measurements, Bethesda, Maryland.

Rust, D. M. (1982). Solar flares, proton showers and the space shuttle. *Science* **216,** 939.

Simpson, J. A. (1983). Introduction to the galactic cosmic radiation. *In* "Composition and Origin of Cosmic Rays" (M. M. Shapiro, ed.), p. 1. Reidel Publishing, Dorcht, The Netherlands.

Radiation Interaction Properties of Body Tissues

DAVID R. WHITE, *St. Bartholomew's Hospital, London*

Glossary

Absorbed dose (D) Mean energy ($d\bar{\varepsilon}$) imparted by ionizing radiation to a given mass (dm) of material ($D = d\bar{\varepsilon}/dm$) [Units: Gray (Gy); 1 Gy = 1 J kg^{-1}]

Compton scattering Interaction of a photon (X ray or γ ray) with an electron of a target atom in which the photon is scattered (deviated from its path) with a reduction in its energy and the electron recoils with additional energy. Compton (recoil) electrons contribute to the absorbed dose in irradiated body tissue

Electron stopping power (S/ρ) The energy lost (dE) by an electron in traversing a given thickness (dx) of material of mass density, ρ [$S/\rho = (1/\rho)(dE/dx)$]. (Units: MeV m^2 kg^{-1})

Exponential attenuation Narrow, monoenergetic beam of photons passing through a material of thickness, x, is reduced in number according to the exponential relationship, $N = N_0e^{-\mu x}$, where N is the number of transmitted photons, N_0 the number of incident photons, and μ the linear attenuation coefficient

Mass attenuation coefficient (μ/ρ) The fraction of photons (dN/N) that experience interactions in traversing a given thickness (dx) of material of mass density, ρ [$\mu/\rho = (1/\rho N)(dN/dx)$]. (Units: m^2 kg^{-1})

Pair production An interaction of a high-energy photon ($\geq$1.02 MeV) as it passes near to the nucleus of a target atom. The photon disappears and an electron–positron pair of particles is produced (A positron is a "positive" electron)

Photoelectric absorption Interaction of a photon with an electron of a target material in which all the energy of the photon is transferred to the electron. The electron is ejected from the target atom. These photoelectrons contribute to the absorbed dose in irradiated body tissue

Radiation energy (electron-volt, eV) Kinetic energy gained by an electron when accelerated through a potential difference of 1 V. (1 eV = 1.6 × 10^{-19} J; 10^3 eV = 1 keV; 10^6 eV = 1 MeV)

Tissue substitute Material used to simulate a particular body tissue with respect to a given set of physical characteristics. The physical characteristics may be the radiation interactions in the body tissue or the absorbed dose at a point of interest in the tissue. A volume of a tissue substitute is called a phantom

RADIATION INTERACTIONS that occur in irradiated body tissues are dependent on the type and energy of the radiations being used and the elemental composition and mass density of the tissues. Photons (X rays and γ rays) interact with tissue by an energy-dependent combination of photoelectric absorption, Compton scattering, and pair production. Electrons passing through tissue lose energy by electronic collisions and, to a lesser extent, by the production of bremsstrahlung photons. The magnitude of the radiation interactions is influenced by the compositions of the irradiated tissues. These compositions, which fall into two categories, the soft tissues (adipose tissue, muscle, etc.) and the hard skeletal tissues (cortical bone), vary with the age, nutrition, state of health, and physical activity of the individual.

Encyclopedia of Human Biology, Volume 6.

I. The Composition of Body Tissues

A. Introduction

The composition of body tissues depends on the age, nutrition, state of health, and physical activity of the individual. Changes may be very large as in severe malnutrition or in fatty infiltration of specific organs such as the heart or liver. On the other hand, cortical bone remains constant with respect to elemental composition throughout adult life, although its total mass in the body decreases with the aging process.

In order to determine elemental composition, the basic components of a tissue specimen are first evaluated by chemical and physical separations. Once the water, lipid, protein, carbohydrate, and mineral contents are established, the overall elemental composition of the tissue can be calculated from the accepted compositions of the components. Mass density can be derived from mass and volume estimations on the fresh specimen. As these measurements are sparse in the literature, mass densities must frequently be estimated by calculation, knowing the densities of the components and their mass proportions in the specimen.

The discussion here is devoted to *human* body tissue. The composition and radiation interaction data given will be predominantly for healthy tissues, with only an outline of the probable variations in composition due to over- and undernutrition, disease, and physical activity.

B. Age Dependence

The chemical composition of the body's organs vary before and after birth. Changes in composition continue until the organs have matured, which may be soon after birth for some (e.g., liver and muscle) or much later for others (e.g., cortical bone).

These changes with age are illustrated in Table I. Adipose tissue, often incorrectly referred to as "fat," is composed of a protein matrix supporting cells (adipocytes) highly specialized for the storage of lipid, a mixture of triglycerides of various long-chain fatty acids. It is the most variable tissue in the body, in particular its contribution to body mass, typically making up approximately 18 and 29%, by mass, of the healthy adult male and female body, respectively. The lipid in the adipose tissue of a full-term newborn baby (34.7%, by mass) will more than double by adulthood (74.1%, by mass). [*See* ADIPOSE CELL.]

Muscle (skeletal) comprises the connective tissue, blood vessels, blood lymph, etc., usually associated with striated muscle and accounts for some 40 and 30%, by mass, of the adult male and female body, respectively. Fetal muscle has a very high water content (90.4%, by mass). During development the percentage of water decreases and the concentration of protein increases. Fatty infiltration also increases with age. [*See* CONNECTIVE TISSUE; SKELETAL MUSCLE.]

Cortical bone is made up of a protein matrix supporting minerals containing phosphorus and calcium, the major compound being hydroxyapatite, $Ca_{10}(PO_4)_6(OH)_2$. The bone mineral steadily increases with age at the expense of the water content. A rapid increase in bone mineral occurs from the twentieth week of gestation (33.4% by mass) to full term (49.6%, by pass). This is followed by a slower increase, reaching 58.0%, by mass, by 18 years of age and staying at this level during most of adult life. Whole bones will have average compositions that are significantly different from that of cortical bone due to the "diluting" effect of other skeletal components present, such as red and yellow marrow.

The resulting elemental composition of adipose tissue, muscle (skeletal), and cortical bone for the same age groups considered in Table I have been listed in Table II. The importance of the hydrogen, carbon, nitrogen, and oxygen elemental groups in the soft tissues and, in addition, phosphorus and calcium in cortical bone is evident from the tabulation.

The trends indicated for muscle (skeletal) in Tables I and II are repeated in the other soft tissues, with time scales being dictated by the rate of maturation of the specific tissues. Once they have reached their mature state, soft tissues have similar compositions, with water contents typically in the range of 60 to 80%, by mass (Fig. 1). Higher water contents are found in body fluids such as cerebrospinal fluid and urine, while reduced amounts occur in adipose tissue and yellow marrow.

C. Other Influencing Factors

Marked changes in body tissue composition can be caused by both over- and undernutrition. In overnutrition, adipose tissue may be changed considerably, with both the amount of adipose tissue in the

TABLE I The Components of Adipose Tissue, Muscle (Skeletal), and Cortical Bone (Fetus to Adult)

Body tissue and Age group	Percentage by mass				
	Water	Lipid	Protein	Carbohydrate	Minerals[a]
Adipose tissue					
Newborn	59.7	34.7	5.4	—	0.2
Infant (2 days–10 months)	47.6	47.2	5.0	—	0.2
Child (1–18 years)	41.1	55.0	3.7	—	0.2
Adult	21.2	74.1	4.4	—	0.3
Muscle (skeletal)					
Fetus (15 weeks)	90.4	0.9	7.9	—	0.8
Newborn	80.6	2.0	14.0	2.8	0.6
Infant (3 months)	79.2	2.0	17.2	1.0	0.6
Infant/child (6 months–18 years)	(As adult, but lipid increasing 2.1 to 4.2%)				
Adult	74.1	4.2	19.8	1.0	0.9
Cortical bone					
Fetus (20 weeks)	41.2	—	20.4	(5)[b]	33.4
Newborn	21.1	—	24.3	(5)[b]	49.6
Infant (3 months)	23.2	—	24.8	(5)[b]	47.0
Child (1 year)	20.7	—	26.1	(5)[b]	48.2
Child (10 years)	16.9	—	25.8	(5)[b]	52.3
Adult	12.2	—	24.6	5.2	58.0

[a] Includes remaining elements with atomic numbers above 8.
[b] Estimated, not measured values.

body *and* the lipid concentration within the tissue increasing sharply. An obese individual may have adipose tissue with a lipid content of over 85%, by mass. Conversely, a lean person may have less than 50% lipid, by mass. Severe malnutrition has a large effect on soft tissues such as muscle. Protein is lost and the water content increases, producing elevated concentrations of sodium and chlorine and suppressed concentrations of potassium and phosphorus. In children, undernutrition may restrict

TABLE II The Elemental Composition and Mass Density of Adipose Tissue, Muscle (Skeletal), and Cortical Bone

Body tissue and age group	Percentage by mass											Mass density
	H	C	N	O	Na	Mg	P	S	Cl	K	Ca	($kg\ m^{-3}$)
Adipose tissue												
Newborn	11.1	29.7	0.9	58.0	0.1	—	—	0.1	0.1	—	—	990
Infant (2 days–10 months)	11.2	39.2	0.9	48.4	0.1	—	—	0.1	0.1	—	—	970
Child (1–18 years)	11.3	44.5	0.6	43.3	0.1	—	—	0.1	0.1	—	—	960
Adult	11.4	59.8	0.7	27.8	0.1	—	—	0.1	0.1	—	—	950
Muscle (skeletal)												
Fetus (15 weeks)	10.8	4.9	1.3	82.1	0.2	—	0.1	0.1	0.3	0.2	—	1030
Newborn	10.4	10.3	2.4	76.2	0.1	—	0.1	0.1	0.2	0.2	—	1050
Infant (3 months)	10.3	11.2	2.9	74.8	0.1	—	0.1	0.2	0.2	0.2	—	1050
Adult	10.2	14.3	3.4	71.0	0.1	—	0.2	0.3	0.1	0.4	—	1050
Cortical bone												
Fetus (20 weeks)	6.4	12.8	3.5	57.1	0.1	0.1	6.1	0.2	0.2	—	13.5	1430
Newborn	4.4	15.3	4.1	47.7	0.1	0.2	8.5	0.2	0.1	—	19.4	1720
Infant (3 months)	4.7	15.4	4.2	48.5	0.1	0.2	8.2	0.2	0.1	—	18.4	1680
Child (1 year)	4.5	15.9	4.4	46.7	0.1	0.2	8.7	0.3	0.1	—	19.1	1710
Child (10 years)	4.0	15.9	4.4	45.0	0.1	0.2	9.6	0.3	0.1	—	20.4	1790
Adult	3.4	15.5	4.2	43.5	0.1	0.2	10.3	0.3	—	—	22.5	1920

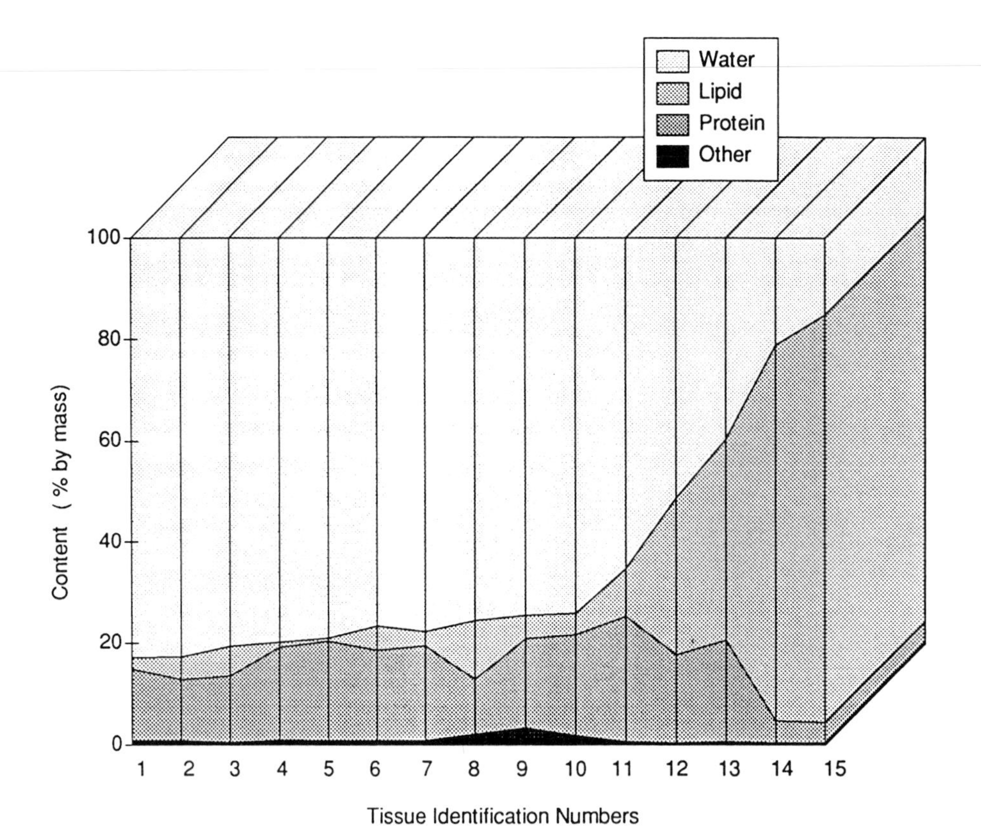

FIGURE 1 The water, lipid, and protein components of adult soft tissues, plotted in order of decreasing water content. Tissue identification: (1) Ovary, (2) testis, (3) GI tract (intestine), (4) lung, (5) blood (whole), (6) kidney, (7) heart, (8) brain (whole), (9) liver, (10) muscle (skeletal), (11) skin, (12) mammary gland, (13) red marrow, (14) adipose tissue, (15) yellow marrow.

growth and the expected changes in composition with development are curtailed or slowed. [*See* MALNUTRITION; OBESITY.]

Significant increases in lipid content at the expense of water and to a somewhat lesser extent protein, can occur in certain disorders. Cirrhosis of the liver due to chronic alcoholism may cause lipid concentrations in the liver to increase from approximately 5 to 19%, by mass. Fatty infiltration can occur in and around organs such as the heart. A 12-fold increase in the lipid content of the muscle (skeletal) in a young male suffering from muscular dystrophy has been reported. [*See* LIPIDS.]

Vigorous exercise over long periods of time may also alter body tissue composition. Total body adipose tissue masses of one-third normal values have been recorded for athletes compared to sedentary people.

II. Photon and Electron Interaction Properties

A. Basic Radiation Interactions

The amount of radiation energy absorbed by an irradiated mass of body tissue (the absorbed dose) depends upon the type and magnitude of the radiation interactions that occur. For both photons and electrons, the types of interactions depend upon the energy of the incident radiation. The magnitude of the interactions is influenced by the elemental composition and mass density of the tissue.

In the energy interval 10 keV to 10 MeV, photons may undergo a combination of photoelectric absorption, Compton scattering, pair production, and other minor effects. At 10 keV photoelectric absorption, which is strongly dependent on the elemental atomic numbers (Z) present in the tissue, accounts for 93% of the total interactions occurring in muscle (skeletal) at this energy. In cortical bone with its higher Z elements, this proportion rises to 98%. Photon scattering becomes more important as the energy increases. For 1-MeV photons, over 99% of the total interactions in both muscle (skeletal) and cortical bone are due to the Z-independent Compton scattering process. At higher energies the

TABLE III Radiation Interaction and Mass Density Data for Adult Body Tissues

	Mass attenuation coefficient, μ/ρ ($m^2\ kg^{-1}$)				Electron mass stopping power, S/ρ (MeV $m^2\ kg^{-1}$)				
	Photon energy				Electron energy				
Body tissue	10 keV ($\times 10^{-1}$)	100 keV ($\times 10^{-2}$)	1 MeV ($\times 10^{-3}$)	10 MeV ($\times 10^{-3}$)	10 keV	100 keV ($\times 10^{-1}$)	1 MeV ($\times 10^{-1}$)	10 MeV ($\times 10^{-1}$)	Mass density ($kg\ m^{-3}$)
Adipose tissue	3.27	1.69	7.08	2.14	2.33	4.21	1.88	2.14	950
Brain[a]	5.41	1.70	7.04	2.20	2.25	4.11	1.85	2.14	1040
Cortical bone	2.85	1.86	6.57	2.31	1.93	3.61	1.65	2.02	1920
Heart[b]	5.48	1.70	7.01	2.20	2.24	4.09	1.84	2.13	1060
Liver	5.40	1.69	7.01	2.19	2.24	4.08	1.84	2.13	1060
Lung	5.46	1.70	7.01	2.20	2.24	4.08	1.87	2.22	260 (Inflated)
Mammary gland	4.30	1.69	7.03	2.17	2.27	4.13	1.86	2.13	1020
Muscle (skeletal)	5.36	1.69	7.01	2.19	2.24	4.09	1.84	2.13	1050
Red marrow	4.35	1.69	7.02	2.16	2.28	4.14	1.85	2.13	1030
Skin	4.95	1.69	6.99	2.18	2.24	4.08	1.84	2.12	1090
Yellow marrow	3.10	1.69	7.08	2.14	2.33	4.22	1.88	2.14	980
Water[c]	5.33	1.71	7.07	2.22	2.26	4.12	1.86	2.15	1000

[a] Contains 50% gray matter and 50% white matter, by mass.
[b] Blood filled.
[c] Water is included for reference purposes.

importance of Compton scattering decreases and pair production processes increase. For 10-MeV photons, 22.7 and 31.5% of the total interactions in muscle (skeletal) and cortical bone are due to the weakly Z-dependent pair production process, with Compton scattering contribution the remainder.

In the energy interval 10 keV to 10 MeV, electrons lose energy when traversing body tissues by electronic collisions and the production of bremsstrahlung photons or "braking radiation." Up to 500 keV, electronic collisions are the predominant effect. At 10 MeV, bremsstrahlung production becomes more important, contributing 8.4 and 12.4% of the total mass stopping power for muscle and cortical bone, respectively.

B. Photon and Electron Interaction Data

Photon and electron interaction data for a selection of adult body tissues are given in Table III. Data are listed for 10-keV, 100-keV, 1-MeV, and 10-MeV photons and electrons. Total mass attenuation coefficients (μ/ρ), relating to the fractional reduction in the numbers of transmitted photons, are tabulated. Total electron mass stopping powers (S/ρ), relating to the energy lost as an electron traverses a tissue, are given.

Table III also shows that mass densities range from 950 kg m^{-3} for the fatty tissues such as adipose tissue to 1050 kg m^{-3} for a soft tissue such as muscle (skeletal). Cortical bone has a much higher density of 1920 kg m^{-3}. Larger differences in attenuation (i.e., absorption and scattering) occur in body tissues at low photon energies (e.g., 10 keV) where photoelectric absorption predominates compared to high photon energies. At 100 keV and above, equal masses of all soft tissues absorb photons similarly, which is also the case for electrons. Water, included for reference purposes, has similar photon and electron interaction properties to soft tissues.

III. Some Practical Implications

A. Medical X-Ray Imaging

Conventional X-ray imaging relies on the differential attenuation that occurs when a beam of X rays passes through a body section. If adjacent tissues within the section attenuate the X rays diffently, the transmitted X-ray distribution reflects the tissue pattern which may be recorded on a photographic film (static image) or a fluorescent screen (dynamic image). By carefully selecting the incident X-ray energies, different parts of the body may be examined in this way.

A particularly challenging application of X-ray imaging is mammography, the radiography of the breast. For a given breast thickness the requirements are for adequate transmission through the breast to the film, maximum differentiation of the soft tissues present, while keeping the absorbed dose to the breast as low as reasonably practicable. X rays generated around 30 kV with suitable filtration maximize the photoelectric interaction processes in the soft tissues to yield the necessary image contrast. Similar methods are adopted to image bone detail in thin body sections.

As the thickness of the body section increases, higher energy X rays are required to penetrate the tissues. A subsequent decrease in photoelectric processes results and the ability to differentiate soft tissues of similar composition and density is diminished. Computerized techniques are now employed to accentuate these small differences in photon attenuation. Computed tomography (CT) is widely used to provide detailed, enhanced images of thin body slices.

B. Radiotherapy

Radiotherapy uses X-ray techniques to sterilize the reproductive ability of tumor cells, while sparing adjacent normal cells. Megavoltage radiotherapy uses photon energies of 1 MeV or above. The importance of Compton scattering at these energies leads to a more uniform absorbed dose distribution even in thick body sections. In addition, the fact that the maximum absorbed dose does not occur at the skin surface but at an energy-dependent depth in the tissues produces an important ''skin-sparing'' effect.

As the major portion of an electron beam's energy is deposited in a limited depth of tissue, high absorbed doses may be given to superficial tumors, while sparing underlying tissues. Changing the electron energy affects the depth of tissue treated.

C. Tissue Substitutes

In order to investigate experimentally the radiation interactions in body tissues, substitute materials are necessary. Such materials should ideally have absorption and scattering properties which are the same as the body tissues being considered. In practice this requirement is approximated, although tissue substitutes are now available that give absorbed doses that differ by no more than 1% from those in the body tissues being simulated. Typical materials that have been used are listed below:

1. Adipose tissue: Polyethylene, ethoxyethanol
2. Muscle (skeletal): Polystyrene, water
3. Cortical bone: Aluminum, saturated solution of dipotassium hydrogen orthophosphate

A number of resin and polymer-based materials containing corrective powdered fillers, simulating a range of body tissues, are also available.

Volumes of tissue substitutes, or phantoms, are used for the measurement of absorbed dose, the assessment of image quality, and for the calibration of radiation detectors. At one extreme they may be simple blocks, stacked sheets, or cuboid tanks of water. Alternatively they may be elaborate representations of body sections or whole bodies with embedded skeletons (natural and artificial) and replicated organs.

Bibliography

Forbes, G. B. (1987). ''Human Body Composition. Growth, Aging, Nutrition and Activity.'' Springer-Verlag, New York.

Hubbell, J. H. (1982). Photon mass attenuation and energy-absorption coefficients from 1 keV to 20 MeV. *Int. J. Appl. Radiat. Isotopes* **33,** 1269.

ICRP (International Commission on Radiological Protection) (1975). ''Report of the Task Group on Reference Man,'' ICRP Publication 23. Pergamon Press, Oxford.

ICRU (International Commission on Radiation Units and Measurements) (1984). ''Stopping Powers for Electrons and Positrons,'' ICRU Report 37. ICRU, Bethesda, Maryland.

ICRU (International Commission on Radiation Units and Measurements) (1989). ''Tissue Substitutes in Radiation Dosimetry and Measurement,'' ICRU Report 44. ICRU, Bethesda, Maryland.

ICRU (International Commission on Radiation Units and Measurements) (1991). ''Photon, Electron, Proton and Neutron Interaction Data for Body Tissues,'' ICRU Report (in press). ICRU, Bethesda, Maryland.

Woodard, H. Q., and White, D. R. (1986). The composition of body tissues. *Br. J. Radiol.* **59,** 1209.

White, D. R., and Woodard, H. Q. (1988). The effects of adult human tissue composition on the dosimetry of photons and electrons. *Health Phys.* **55,** 653.

White, D. R., Widdowson, E. M., Woodard, H. Q., and Dickerson, J. W. T. (1991). The composition of body tissues (II) fetus to young adult. *Br. J. Radiol.*, (in press).

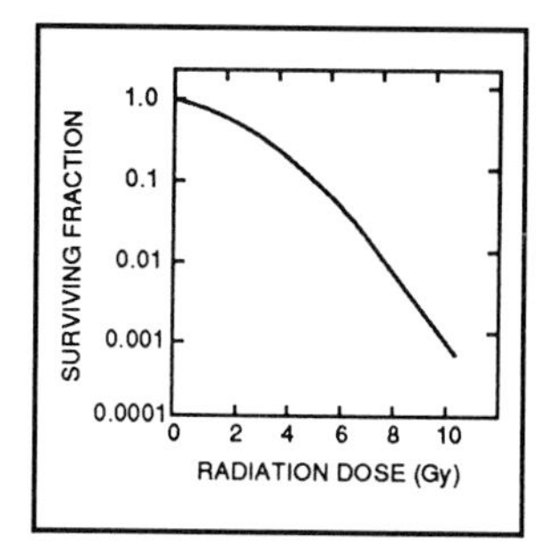

Radiobiology

SARA ROCKWELL, *Yale University School of Medicine*

Glossary

Gray International System of Units unit for the absorbed dose of ionizing radiation (abbreviated Gy); 1 Gy = 1 joule of absorbed energy per kilogram of irradiated material

Ionization Removal of one or more orbital electrons from an atom or molecule. "Ionizing" radiations produce ionization in the material that absorbs them.

Linear energy transfer Factor describing the average rate at which a charged particle transfers its energy to the medium (abbreviated LET); generally given in kilo electron volts per micron of unit density material

Relative biological effectiveness Quantity used to compare the biological effects of different radiations; abbreviated RBE. RBE = D_x/D_r, where D_r is the dose of the experimental radiation needed to produce a specified biological effect and D_x is the dose of the reference radiation (usually 250-kV X-rays) needed to produce the same effect.

Sievert International System of Units unit of dose equivalent, calculated by multiplying the dose (in Gray) by a quality factor, which adjusts for differences in the relative biological effectiveness of different radiations (abbreviated Sv); 1 mSv = 10^{-3} Sv

RADIOBIOLOGY IS the study of the effects of radiation on biological systems. In its broadest sense radiobiology encompasses many radiations, ranging from radiowaves through visible and ultraviolet (UV) light to ionizing electromagnetic and particulate radiations; these different radiations have widely differing biological effects. The biological systems considered in radiobiology range from biological molecules and subcellular structures to simple organisms (e.g., viruses and bacteria), higher plants and animals, and humans. This article focuses on the effects of ionizing radiations on humans.

I. Sources and Kinds of Radiation

Radiation is and always has been an inescapable part of the human environment. Natural background radiation comes from several sources (Table I). Naturally occurring radioactive elements are found in the soil and in rocks. The dose from these radiations varies considerably at different locations; regional averages vary from 0.25 to 0.9 mSv/yr in the United States. In addition, people are exposed to radiation from natural radioactive materials which have been taken into the body in food or inhaled with the air.

One such element, radon, has recently received considerable attention. This radioactive gas is produced by the decay of radium, a radioactive nuclide which is a natural component of the rocks in many areas of the country. Levels of radon in the air inside buildings vary greatly, depending on such factors as the amount of radium in the strata below the building, the characteristics of the water supply, the construction of the building, and the amount and the nature of ventilation in the building. It now appears that radon could be the major source of background radiation in many areas of the country, contributing approximately 2 mSv to the average annual radiation dose equivalent. Because radon

TABLE I Estimated Annual Effective Dose Equivalents to the U.S. Population in 1980–1982[a]

Source	Average annual effective dose equivalent (mSv)
Natural sources	
Radon	2.00
Cosmic rays	0.27
Radionuclides in body	0.39
Terrestrial and atmospheric sources	0.29
Rounded total	3.00
Medical exposures	
Diagnostic X-rays	0.39
Nuclear medicine	0.14
Total	0.53
Occupational exposures	0.009
Nuclear energy	0.0005
Miscellaneous environmental sources	0.0006
Consumer products	
Tobacco	b
Other	0.05–0.13
Rounded total (nonsmokers)[b]	3.6

[a] Data are summarized and condensed from the 1987 report of the National Council on Radiation Protection and Measurement (see Bibliography). The doses are averaged over the entire population of exposed and unexposed persons.
[b] The calculation of the effective dose equivalent from tobacco products is exceedingly problematic, as described in the National Council on Radiation Protection and Measurement report. If the estimated value of ~13 mSv for an average smoker is correct, tobacco represents the largest source of radiation exposure for smokers and raises their total annual effective dose equivalent from 3.6 to ~16 mSv.

and its radioactive daughters emit α particles, which have short ranges, only the cells lining the airways of the lungs are irradiated.

Cosmic rays also contribute to our radiation dose. Cosmic rays are energetic positively charged particles (i.e., atomic nuclei) emitted from the sun and other celestial sources. Although some of these particles are deflected by the Earth's magnetic field and others are absorbed by the atmosphere, some reach the surface of the Earth; the intensity of cosmic radiation increases near the poles and at high altitudes (doubling with every 2000 m, from about 0.26 mSv/yr at sea level).

The dose of background radiation that people receive from natural sources varies considerably, as it depends on such factors as the altitude, the local composition of the Earth's crust, and the design and ventilation of one's home. People's activities also influence their exposure to natural radiations; for example, a single high-altitude transatlantic flight results in an extra 0.05 mSv of exposure from cosmic radiation.

Medical radiations dominate our exposure from man-made radiation, with diagnostic and dental X-rays and diagnostic nuclear medicine procedures contributing most of this dose. Occupational exposure for those working with radioactive materials contributes slightly to the average dose for the total population. Of approximately 1.3 million people in the United States engaged in radiation work in 1980, only one-half actually received measurable exposures. For those exposed, the average dose equivalent received was 2.3 mSv. The nuclear fuel cycle and miscellaneous environmental sources (including fallout from past nuclear weapons testing, emissions from certain nonpower reactors, and the release of naturally occurring radionuclides from aluminum, copper, zinc, and lead processing facilities) contribute small doses to the average population exposure.

The dose received from consumer products is variable and difficult to estimate. The doses received from manufactured products which deliberately use radioactive isotopes (e.g., smoke detectors and luminous watches) or produce ionizing radiation (e.g., airport inspection systems and television receivers) are small, estimated at 0.0015–0.012 mSv. Greater radiation exposures result from

TABLE II Electromagnetic Radiations and Their Effects[a]

Radiation	ν	E	Effects
Radiowaves, radar waves microwaves	$<3 \times 10^{11}$	<0.00124 eV	Molecular vibrations; heating
Infrared	3×10^{11}–4.3×10^{14}	0.00124–1.77 eV	Molecular vibrations; heating
Visible light	4.3–7.5×10^{14}	1.77–3.1 eV	Absorption by specific molecules, causing excitation, possibly leading to chemical reactions
Ultraviolet light	7.5×10^{14}–3×10^{16}	3.1–124 eV	Absorption by specific molecules, causing excitation, leading to chemical reactions
X-Rays, γ-rays	$>3 \times 10^{16}$	>124 eV	Ionization, nonspecific absorption of energy; nonspecific damage

[a] ν, Frequency of the radiation (in cycles per second). The energy (E) contained in a single photon of radiation is related to the frequency, by the relationship $E = h\nu$, where h is the Planck's constant.

the release of naturally occurring radionuclides during the combustion of coal or natural gas (0.003–0.006 mSv), and from exposure to radionuclides present naturally in building materials (0.036 mSv), domestic water supplies (0.01–0.06 mSv), and mining and agricultural products (less than 0.01 mSv).

Tobacco products probably contribute the highest radiation dose of all consumer products, although it is difficult to estimate. For an average smoker the decay of ^{210}Po and other radionuclides which are concentrated naturally by tobacco leaves has been calculated to produce average doses to a small region of the bronchial epithelium of ~160 mSv, and a calculated effective (weighted) dose equivalent of 13 mSv. This radiation dose is thought to contribute somewhat to the excess risk of lung cancer in smokers; however, the larger portion of the risk to smokers probably reflects the chemical and physical effects of tobacco products. [*See* TOBACCO SMOKING, IMPACT ON HEALTH.]

Electromagnetic and particulate radiations are both important in radiobiology. The electromagnetic radiations with enough energy to produce ionization, X-rays, and γ-rays (Table II) are identical in their characteristics, but are produced in different ways. γ-Rays are emitted when the nuclei of certain radioactive elements decay. In contrast, X-rays are emitted when orbital electrons make transitions between high- and low-energy states, and they carry the energy released in these transitions. X-Rays are often produced by accelerating electrons, using strong electric fields, so that they carry large amounts of kinetic energy, then stopping these electrons suddenly by bombarding a metal target; the resulting reactions convert the kinetic energy of the electrons into X-rays. Particles accelerated to fast velocities (i.e., high energies) also carry enough kinetic energy to produce ionization along their paths. Electrons, protons, neutrons, and a variety of heavier particles can produce ionization when accelerated (Table III).

II. Nature of Radiation Damage

The biological effects produced by electromagnetic radiations vary dramatically with the photon energy and, therefore, with the frequency of the radiation

TABLE III Some Particles of Interest in Radiobiology

Particle	Symbol	Mass (Atomic Mass Units)	Charge
Proton	p	1	+1
Neutron	n	1	0
Electron	e, e^-, β^-	0.00055	−1
α Particle	α	4	+2
Heavy charged particles, cosmic rays	None	Variable	Variable

(Table II). Radiowaves and microwaves, for example, carry only very small amounts of energy in each photon; when a molecule absorbs such a photon, it acquires a quantum of energy which is insufficient to alter its electronic configuration; this energy is dissipated as vibrational energy, or heat. Visible and UV light are absorbed selectively by specific biological molecules having electron orbital structures such that absorption of a photon of the correct energy causes the transition of an electron from a lower-energy state to a higher-energy state. The excited molecule then loses its excess energy through relatively specific and predictable chemical reactions. For example, when DNA is irradiated with UV light ~260 nm in wave length, the light is absorbed primarily by the rings of the pyrimidine nucleotides, thymidine and cytidine. The major damage to UV-irradiated DNA results from reactions between the excited pyrimidines and other nearby pyrimidines, producing intrastrand pyrimidine–pyrimidine dimers; these distort the DNA helix and interfere with normal DNA synthesis, and could be mutagenic or cytotoxic.

The biological effects of these "nonionizing" radiations are different from those of "ionizing" radiations; this reflects differences in the localization of the energy absorbed and in the nature of the resulting damage. The energy carried by a single photon or particle of ionizing radiation exceeds the binding energy of the inner-shell electrons surrounding atomic nuclei. This energy is imparted in random reactions, producing highly localized areas of ionization and intensive damage.

The mechanisms by which electromagnetic radiations interact with matter vary with photon energy and with the composition of the absorbing material. All three absorption processes—photoelectric effect, Compton scattering, and pair production—result in the release of lower-energy photons and energetic "fast" electrons. As these fast electrons interact with atoms along their path, they strip electrons from the shells of the atoms, leaving a track of ion pairs, each of which consists of a negatively charged relatively low-energy electron, plus a positively charged atom. The chemical reactions of these exceedingly reactive ions with nearby molecules lead to the production of large numbers of ions, free radicals, excited molecules, and other chemically reactive species. The chemical reactions between these moieties and critical biological molecules eventually lead to biologically measurable damage.

Fast charged particles (Table III) produce their biological damage through analogous mechanisms. The initial event is an interaction between the moving charged particle and the electron shell of a nearby atom, which produces an ion pair. A fast charged particle therefore leaves a track of ion pairs along its path which react further, as described above.

The physics of the absorption of neutrons is more complex. These uncharged particles interact with atomic nuclei in the absorbing material, through scattering reactions which produce recoil protons, α particles, and heavier nuclear fragments. The interactions of these moving charged particles with nearby atoms then lead to the production of radiation damage.

Two things should be noted about the damage produced by ionizing radiation. First, the interactions between the radiation and the irradiated material are random; unlike visible or UV light, ionizing radiations are not absorbed by specific molecules. Second, many different kinds of biological lesions are produced by ionizing radiation, because most of the damage reflects nonspecific chemical reactions occurring after the initial ionization event. A mammalian cell is approximately 70% water; much of the damage in a irradiated cell therefore reflects the chemical reactions of radiation-damaged water molecules. Molecules of many kinds are damaged during irradiation; most can be replaced with no lasting effects on the cell. However, a few lesions, or even a single, critical lesion, in DNA can alter the function of the cell and can even be lethal. As a result damage to DNA is the critical factor producing the cytotoxic effects of ionizing radiation.

Irradiation produces many different lesions in DNA and in the chromosomes that provide the structure organizing DNA in human cells (Table IV). These lesions lead to a variety of genetic changes, including simple point mutations, deletions of small or large amounts of genetic informa-

TABLE IV Some Lesions Produced in the DNA and Chromosomes of Irradiated Cells

Damage to DNA	Damage to Chromosomes
Base changes	Deletions
Deletions	Translocations
Single-strand breaks	Inversions
Double-strand breaks	Acentric fragments
Crosslinks	Dicentric chromosomes
Adducts	Ring chromosomes

tion, and DNA or chromosomal rearrangements which alter the expression of genetic information. The genetic changes seen in irradiated cells are nonspecific and are qualitatively similar to those associated with aging or with exposure to nonspecific oxidative and chemical damage. As a result the spectrum of mutations observed in irradiated cells or in the offspring of irradiated individuals resembles that of spontaneous mutations observed in a natural population. No novel mutations have been observed in large populations of irradiated *Drosophila,* mice, or people.

III. Cellular Radiobiology

A. Cellular Lethality—Delayed Cell Death

The nature of the damage in irradiated cells underlies both the unique therapeutic uses and the unique hazards of ionizing radiation. Because lethally damaged cells die as a result of DNA damage, they do not die immediately, or even rapidly. Rather, lethally irradiated cells function relatively normally and even divide once or a few times, producing "abortive clones" of daughter cells, all of which eventually die as the genetic injury becomes manifest. Cells which never attempt to divide might never express their lethal injury; such "radiation-sterilized" cells might function normally for years. This fact provides one of the biological bases for the treatment of cancer with radiation: Radiation sterilizes the malignant cells, preventing them from indefinite growth, while the quiescent cells of irradiated normal tissues continue to function relatively normally for long periods. The hazards of high doses of radiation result primarily from the death of dividing cells. As described in detail in Section V.B.1, large doses of radiation are generally required to deplete normal tissues sufficiently to produce overt injury.

In contrast, the hazards of low doses of radiation generally result from the production of DNA damage which is not lethal to the cell, but which could become manifest under appropriate circumstances months, or even years later. For example, damage to the DNA of a resting hematological stem cell could lie dormant for years before being expressed in leukemic transformation. Similarly, damage to the DNA of a spermatogonium might not kill the developing sperm and might not interfere with fertilization, but could be deleterious to the embryo resulting from conception by a sperm carrying the radiation-induced genetic charges.

B. Cell Survival

The effects of radiation on mammalian cells have been studied extensively using primary cell cultures initiated from normal and malignant tissues or established cell lines. A typical survival curve for cells irradiated by X-rays or γ-rays is shown in Fig. 1. Because radiation-sterilized cells do not die immediately, cell survival is measured by testing the ability of individual cells to grow into colonies composed of dozens or hundreds of cells. The survival of the cells decreases continuously as the radiation dose increases. Different human cell lines have slightly different survival curves, but the differences are generally relatively subtle.

1. Repair of Radiation Damage

A number of factors modulate the response of cells to radiation. The first of these is the ability of the cell to repair radiation-induced lesions in DNA. Different cells have different intrinsic repair capacities. For example, cells of hematological origin, both normal and malignant, generally are less able to repair DNA damage than are cells from other tissues. Cells from persons with certain genetic diseases associated with defects in DNA repair (e.g., ataxia–telangiectasia), are unusually sensitive to ionizing radiation; people with these diseases might be correspondingly hypersensitive to radiation in-

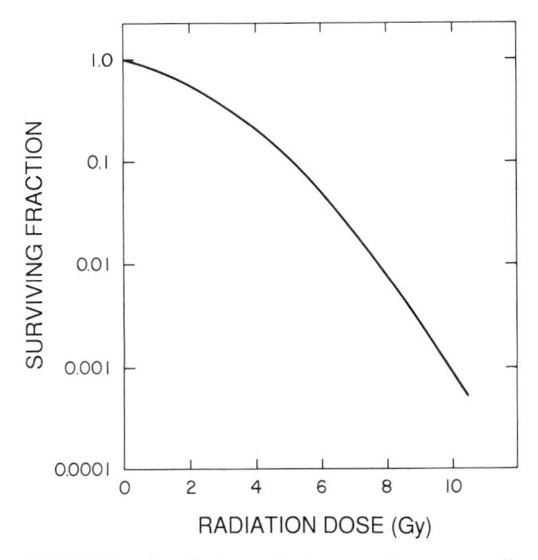

FIGURE 1 Typical survival curve for mammalian cells *in vitro*. Surviving fraction represents the fraction of cells in the population which remain capable of prolonged proliferation.

jury, just as people with xeroderma pigmentosum, a genetic disease which results from a defect in the repair of UV-induced DNA damage, are unusually sensitive to the toxic and carcinogenic effects of the UV radiation present in sunlight. Many factors modulate the ability of the cell to repair radiation damage, including whether the cell is proliferative or quiescent, the position of the cell in the cell cycle, and the environment of the cell (e.g., the extracellular pH and the adequacy of the supplies of nutrients, energy sources, and oxygen). [*See* REPAIR OF DAMAGED DNA.]

2. Radiation Sensitizers and Radiation Protectors

Because ionizing electromagnetic radiations produce DNA damage indirectly, through a variety of nonspecific chemical reactions, the production of DNA damage can be altered by chemicals that interact with the reactive intermediates. Radiosensitizers increase the amount of damage produced by a given dose of radiation. One of the most important radiosensitizers for radiobiology is molecular oxygen (O_2). When O_2 is present, this extremely reactive electron-affinic molecule participates in, multiplies, and modifies the reactions of irradiated water and biological molecules. As a result cells which are aerobic at the time of irradiation are approximately three times more sensitive to the effects of radiation than are severely hypoxic cells. The O_2 concentrations in most human tissues are sufficient to produce maximal radiosensitization by O_2; the chemistry of the radiation reactions in most human tissues, except those with compromised vasculatures, is therefore characteristic of well-oxygenated cells. Other radiosensitizers of interest in medically oriented radiobiology are the thymidine analogs bromodeoxyuridine and iododeoxyuridine. When incorporated into the DNA of proliferating cells, these halogenated pyrmidines sensitize cells to radiation.

Radioprotectors act by scavenging the relatively long-lived free radicals, thereby reducing the amount of damage produced in biological macromolecules. A variety of radical scavengers, including alcohols and sulfhydryl-containing compounds, are effective radioprotectors. Naturally occurring endogenous intracellular nonprotein sulfhydryls, including glutathione, are thought to play an important role in minimizing the production of DNA damage by radiation and other radical-producing processes, including normal oxidative metabolism. A number of chemical radioprotectors have been tested as possible agents for protecting individuals who are about to be exposed to dangerous levels of radiation; however, the high intracellular concentrations required for radioprotection and the toxicities of the compounds limit their potential usefulness.

3. Effect of Radiation Dose Rate

The response of mammalian cells to radiation is affected by the rate at which the radiation is delivered. The above discussions have assumed that radiation is delivered at the relatively high dose rates characteristic of single therapeutic and diagnostic radiology treatments (i.e., several Grays per minute). However, when a large dose of radiation is fractionated into many smaller treatments, or when the cells are irradiated continuously over a long period at a low dose rate, the radiation is less effective in producing biological damage. As the duration of irradiation increases from a few minutes to a few hours for a given total dose, repair of the damage occurs during irradiation; because of the curvilinear shape of the acute radiation dose–response curve (Fig. 1), this repair decreases the overall efficacy of the radiation. Moreover, with longer irradiation times the proliferation of the surviving cells increases the size of the cell population during irradiation, thereby decreasing the damage received by any particular cell. In the limit of very low dose rate irradiations, the growth, viability, and characteristics of an irradiated cell population will be indistinguishable from those of control cultures.

4. Effect of Radiation Quality

The preceding discussion focuses on the radiobiology of electromagnetic radiations. These sparsely ionizing radiations deposit energy randomly and relatively uniformly throughout the cell. The effects of other sparsely ionizing radiations (e.g., high-energy protons and electrons) are similar. Some radiations, however, deposit energy very densely along their paths; these radiations are said to have a high linear energy transfer (LET). High-LET radiations have somewhat different biological effects than sparsely ionizing low-LET radiations, because the ionizations they produce are closely clustered in time and space along the tracks of the particles (Figs. 2 and 3). As a result some submicroscopic areas are intensely damaged by multiple ionization events, while other areas are essentially unirradiated. High-LET radiations, therefore, produce intense local

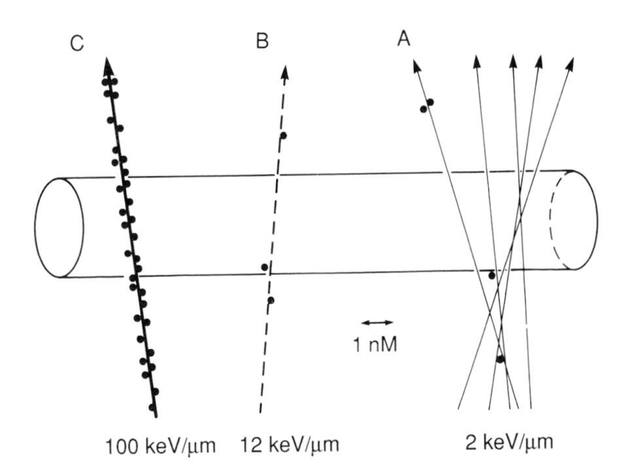

FIGURE 2 Ionization patterns resulting from (A) a low-LET radiation (electrons from 250 kV X-rays, LET ~2 keV/μm; (B) a medium-LET radiation (a 15-MeV neutron, track average LET ~12 keV/μm; and (C) a high-LET radiation (2.5 MeV α particle, track average LET ~100 keV/μm). The cylinder represents a double helix of DNA. Each line represents the path of a single photon or particle of radiation. Each dot represents one ion pair.

DNA damage, which is difficult to repair and has a high likelihood of being lethal, while low-LET radiations often result in single, simple, reparable lesions. As a result as the LET of the radiation increases, its relative biological effectiveness increases.

Moreover, the indirect damage (i.e., the damage produced through chemical reactions) becomes less important, because the intense local ionization produces sufficient direct damage to the critical structures to render the indirectly produced lesions largely superfluous. As a result the modulating effects of sensitizers, protectors, cell age, repair processes, fractionation, dose rate, etc., decrease as the LET of the radiation increases, until, at track average LETs of approximately 200 keV/μm, they become negligible.

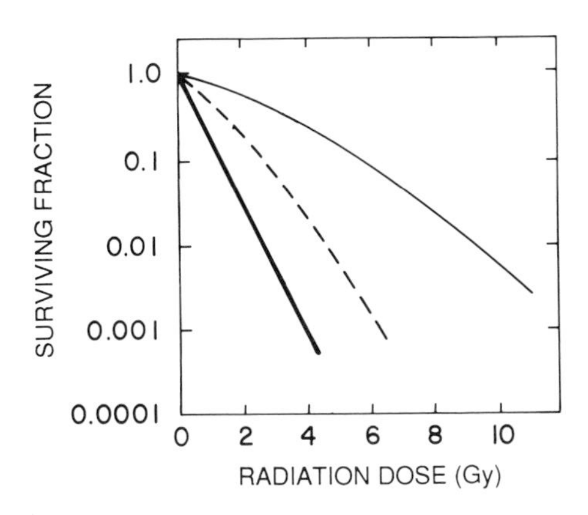

FIGURE 3 Survival curves for aerobic cells irradiated with the low (thin solid line)-, medium (dashed line)-, and high (thick solid line)-LET radiations in Fig. 2. As the LET of the radiation increases, the survival curve becomes straighter and steeper. (These changes are more fully discussed by E. J. Hall, "Radiobiology for the Radiologist," 3rd ed. Lippincott, Philadelphia, Pennsylvania, 1988.)

IV. Medical Uses of Radiation

Radiation is widely used in the diagnosis and treatment of human disease. Diagnostic radiology uses low-energy X-rays to map differences in tissue density. The basis of their use is the fact that the absorption of low-energy X-rays occurs primarily through photoelectric absorption, which is dependent on the atomic number (Z) increasing approximately as Z^3. Small differences in the atomic composition of tissues (e.g., the different calcium levels in bone and muscle) therefore produce large differences in the absorption of the radiation. The uses and the techniques of diagnostic radiology are discussed elsewhere in this encyclopedia. [*See* DIAGNOSTIC RADIOLOGY.]

Nuclear medicine uses injected radionuclides and radiolabeled molecules to diagnose and treat diseases. The tracers used in diagnostic nuclear medicine have been developed and chosen because they localize in specific areas or participate in specific biological reactions. The use of radioactive iodine to study thyroid function and of radioactive gallium or radiolabeled antibodies to locate tumors are examples of routine diagnostic nuclear medicine techniques. Therapeutic applications of nuclear medicine [e.g., the treatment of thyroid cancer or hyperthyroidism with radioactive iodine or the (still experimental) attempts to treat certain cancers with radiolabeled antibodies] are likewise based on selective localization of the radiation; in these therapeutic applications the large doses of high-specific-activity radionuclides yield local radiation doses large enough to produce cytotoxic effects.

Radiation therapy delivers high doses of localized radiation to treat cancer and certain other life-threatening hyperproliferative diseases. The biological basis of radiation therapy is the fact that radiation sterilizes the malignant cells (as described above), and thereby destroys their capacity to cause the tumor to grow, invade nearby tissues, and initiate metastases. Radiation, of course, can also injure normal tissues. Radiotherapy, therefore, must be carefully planned and delivered, so as to maximize antineoplastic effects and to minimize normal tissue injury.

The choice of radiation source is important. Deep-seated tumors are generally treated using high-energy linear accelerators delivering 4–28 MeV X-rays. Because these radiations are penetrating and their absorption is relatively unaffected by tissue density, uniform radiation doses can be delivered to tumors deep within the body. At the other extreme superficial cancers can be treated with low-energy electrons, which penetrate only a few centimeters and therefore spare normal tissues beneath the tumor. Tumors are also treated from several different angles, so that the tumor is always irradiated, but different areas of normal tissue are exposed during each treatment; each area is therefore held to low doses, producing tolerable damage.

The delivery of radiation therapy is carefully planned for each individual patient, based on detailed considerations of the location and pathology of the tumor and of the relative radiosensitivities of the tissues surrounding and involved by the tumor. In addition, the radiation treatment is delivered in many small fractions, generally given each day over ~6 weeks; this allows the normal tissues to recover from radiation injury between each treatment. The differential effect of radiation on tumors and normal tissues is also aided by the differential effect of radiation on proliferating and quiescent cells (described in Section III,A).

Radiation can also be delivered as brachytherapy, the application to or implantation into the tumor of sealed sources of intensely radioactive isotopes, which irradiate the tumor with highly localized radiation continuously over a period of hours, days, or weeks. Brachytherapy is another approach to achieving the localization of radiation dose and the protraction of irradiation that is used in external beam radiotherapy.

Radiation therapy is a valuable modality in the treatment of neoplastic disease. Radiation is used either alone or in combination with surgery and/or chemotherapy in the curative treatment of many early localized cancers. In some cancers these treatments are extremely effective, allowing over 95% of the patients to be cured and to lead a full normal life. In incurable disease radiation therapy can also play a valuable role in the palliation of local symptoms and the amelioration of pain. Currently, over 60% of all cancer patients receive radiotherapy at some point during the treatment of their disease. Major research efforts are directed toward the improvement of radiation therapy and the development of improved regimens for combining radiation with other modalities of cancer treatment.

V. Hazards of Radiation

A. Acute Radiation Injury

1. Injury to Specific Tissues

Acute radiation injuries occur only after large doses of radiation. In general, these reactions reflect the toxicity of radiation to the rapidly proliferating progenitor cells in the tissues. The nature of these acute radiation reactions is well illustrated by the processes occurring in the irradiated intestine. Radiation sterilizes the proliferating cells in the gut crypts, but the quiescent cells of the villi are unaffected and continue to perform their differentiated functions. However, these differentiated cells have a relatively short life span, normally being shed into the lumen of the intestine after 5–8 days. As the differentiated cells are shed, the death of the precursor cells becomes manifest as a deficit in differentiating cells. The result after low doses of radiation is a shortening of the villi and the crypts and a resulting impairment in gut function. Large doses of radiation result in more serious deficits in gut cellularity and more serious impairments in function. Very large doses of radiation can lead to an actual breakdown of the intestinal lining, resulting in serious or life-threatening injury. The biological basis of early reactions in other rapidly renewing tissues (e.g., skin, bone marrow, and hair follicles) is analogous). These reactions occur only after doses thousands of times higher than those used in diagnostic radiology. During radiotherapy their severity is minimized by fractionating the radiation dose into many small fractions, allowing these rapidly renewing tissues to recover between treatments. [*See* RADIOSENSITIVITY OF THE INTEGUMENTARY SYSTEM; RADIOSENSITIVITY OF THE SMALL AND LARGE INTESTINES.]

2. Acute Effects of Whole-Body Irradiation

The acute effects of whole-body irradiation result primarily from the cytotoxic effects of radiation on rapidly renewing tissues (described in Section V,A,I). Our understanding of the mechanism of these effects is derived primarily from animal studies and from extrapolations based on observations of locally irradiated tissues in patients receiving radiotherapy. The effects of total-body exposures on people are known primarily from the medical records of cancer patients and of people receiving dangerous doses of radiation in nuclear accidents (e.g., Chernobyl) and from the fragmentary medical records available in Hiroshima and Nagasaki.

The symptoms of radiation damage can be divided into two phases: a "prodromal syndrome," which occurs minutes or hours after exposure, and the delayed symptoms of the potentially lethal injuries, which develop days to weeks later. The prodromal syndrome is mediated through the response of the autonomic nervous system, and includes a variety of gastrointestinal and neuromuscular symptoms. Higher radiation doses produce a more rapid onset, a longer duration, and more severe symptoms. Prodromal symptoms seldom appear at doses below 1 Gy. At doses of 1–3 Gy, symptoms include anorexia, nausea, vomiting, diarrhea, and fatigue. At higher doses these symptoms increase in severity, and other symptoms (e.g, apathy and headache) also occur. At supralethal doses hypotension, fever, and shock can also occur rapidly. The prodromal symptoms then subside, and the patient enters a relatively symptom-free latent period before the onset of the delayed symptoms, which are manifestations of cytotoxic injury.

The hematopoietic system is the most sensitive tissue to radiation. Doses in excess of 3 Gy result in severe depletion of the hematological progenitor cells. Immunosuppression occurs rapidly. After a latent period of ~3 weeks, decreases in the levels of circulating granulocytes and platelets occur, leading to hemorrhage and secondary anemia. These symptoms resolve as the progenitor cells that survived irradiation proliferate and repopulate the bone marrow and the hematopoietic system recovers. At doses of 4–5 Gy, treatment with whole-blood or platelet transfusions, antibiotics, and supportive care during the hematological crisis could be critical. At doses of 5–8 Gy, survival becomes increasingly problematic, even with intensive medical support, because the severe persistent depletion of the hematological progenitor cells renders recovery of the bone marrow uncertain. The role of bone marrow transplants remains controversial, especially in view of the limited success of this procedure in aiding victims of the Chernobyl accident. [*See* HEMOPOIETIC SYSTEM.]

At radiation doses of ~10 Gy, radiation damage to the gut becomes manifest ~3–8 days after radiation, reflecting damage to the progenitor cells (described in Section V,A,1). Death from mucosal damage, dehydration, electrolyte loss, and infection may result ~1 week after exposure. Victims surviving gastrointestinal syndrome inevitably develop the hematopoietic syndrome.

At radiation doses of at least 100 Gy, cerebrovascular syndrome results in death within two days of exposure. This probably results from acute vascular permeability changes, histamine release, and chemical damage to critical cellular molecules, resulting in overwhelming central nervous system (CNS) injury. The symptoms of the prodromal syndrome merge with those of lethal CNS injury.

B. Late Radiation Hazards

Late radiation injuries can be divided into two kinds: degenerative effects, which result from the killing of large numbers of cells, and stochastic effects, which result from heritable sublethal damage in one or a few cells. Degenerative effects occur only after large doses of radiation, which produce significant tissue damage, such as the doses received by cancer patients treated with intensive radiation therapy or by people involved in very serious radiation accidents. The severity of these injuries increases with the radiation dose. Stochastic effects, in contrast, can result from a single sublethal lesion in a single cell; such effects therefore occur at a certain probability among individuals in a population of people exposed to small doses of radiation. Stochastic effects are probabilistic: They occur with a probability that depends on the radiation dose, but the biological effects per se are the same whenever they occur, regardless of the radiation dose. Stochastic effects of ionizing radiation include carcinogenesis and the induction of mutations.

1. Late Degenerative Injuries

Late radiation damage can develop months to years after irradiation in tissues receiving a large radiation doses (e.g., in tissues irradiated with ~40–60 Gy during the treatment of cancer). Injury can result either from depletion of the parenchymal cells (i.e., the cells that perform the unique tissue-specific functions) or from damage to the stromal elements (i.e., the blood vessels and the connective tissue that nourish and support the parenchyma of the tissue). The nature of the injury in a given tissue and the time at which injury develops depend on the relative sensitivity of the parenchymal and stromal cell populations and on the proliferation patterns of these cells.

At one extreme are tissues such as mature muscle and brain, in which the parenchymal cells are totally quiescent and extremely radioresistant; in such tissues late radiation injury reflects injury to the stromal elements. The other extreme is exemplified by the ovaries and the testes, in which diminution in fertility reflects the depletion of certain radiosensitive precursor cell populations. The different radiation responses in the male and female gonads reflect the intrinsic differences in the patterns of cellular proliferation in the ovaries and the testes.

Radiation-induced cataracts result from the death of parenchymal cells within the lens of the eye. The lens has an extremely unusual pattern of cell renewal. New cells are produced by mitotic activity in the germinal zone of the epithelium. These cells differentiate into lens fibers and migrate toward the equator. There is, however, no mechanism for the removal of dead cells; they simply accumulate at the posterior pole. If many cells are killed by radiation or by other agents, the accumulation of dead fibers becomes large enough to produce an opacity visible upon ophthalmic examination; greater cytotoxicity produces larger opacities, which could interfere with vision.

Because degenerative late reactions result from the death of large numbers of cells, they occur only in sites receiving large doses of radiation. There is a threshold dose, below which the reactions never occur. In general, the severity of late degenerative reactions increases with the radiation dose and the time after irradiation.

2. Genetic Effects of Radiation

Science fiction books and horror movies have popularized the impression that radiation induces bizarre monstrous mutations. This is not the case. Rather, as described in Section II, radiation produces DNA damage which is similar to that occurring naturally; the effect of irradiation is therefore to increase the frequency of the same mutations which occur spontaneously. The fact that it is impossible to distinguish between radiation-induced and spontaneous natural mutations complicates studies of the genetic effects of radiation. The effect of radiation is seen as an increase in the rate of occurrence of genetic diseases and the rate of changes in specific genetic markers; at low doses of radiation, the effect of irradiation becomes impossible to detect, because it is lost in the background of the spontaneous mutations. [*See* MUTATION RATES.]

Studies with human populations exposed to radiation have proved especially difficult, because of the genetic heterogeneity of the populations and also because social, environmental, and physiological factors (e.g., parental age, inbreeding, and exposure to other genotoxic agents) can influence the rates of genetic disease and congenital malformations. None of the epidemiological studies performed on humans has found a statistically significant increase in genetic disease in the offspring of irradiated persons. The most extensive and careful of these studies is that performed on the offspring of the survivors of the atomic bombs at Hiroshima and Nagasaki, where four genetic indicators are being monitored: (1) untoward pregnancy outcome (e.g., stillbirths, major congenital defects, and perinatal deaths); (2) death before age 17; (3) sex chromosome abnormalities; and (4) mutations in blood proteins (i.e., electrophoretic variants in 28 plasma and erythrocyte proteins; activity variants in eight erythrocyte enzymes). The frequency of these four indicators has not increased significantly in either the low (0.01–0.09 Gy)- or high (over 1 Gy)-dose populations.

Our knowledge of the genetic effects of radiation, therefore, comes from the extensive studies performed on plants (especially *Tradescantia*), insects (especially the fruit fly, *Drosophila melanogaster*), and mice; the mouse appears to be the best model for predicting genetic effects in humans. These studies show that most mutations, whether spontaneous or induced, are deleterious. Studies with mice suggest that there is repair of radiation damage by the gonadal precursor cells and show that delaying conception for ~6 months after radiation exposure decreases the risk of genetic anomalies in the offspring. Because of the differences in the radiation responses of the ovaries and the testes, almost all of the heritable radiation damage is carried by the male, rather than the female. Laboratory studies support the concept that there is no threshold for genetic effects: Any radiation dose, however small, poses a finite statistical risk of producing genetic damage. In fact, background radiation probably produces ~1–6% of the spontaneous mutations occurring in people. The number of mutations increases with dose. The dose required to double the spontaneous mutation frequency in a population of exposed people has been estimated to be approximately 1 Sv.

3. Carcinogenesis

Radiation induces cancer in people and animals. The mechanism by which this occurs is unclear, and it is possible that several mechanisms contribute to radiation carcinogenesis. First, genetic changes (i.e., mutations or chromosomal rearrangements) induced in somatic cells could start a normal cell down the path of evolution toward malignancy. Second, radiation could induce oncogenes or inactivate oncogene suppressors, thereby causing malignant transformation. In one strain of laboratory mice, induction of an endogenous oncogenic virus by radiation is responsible for radiation-induced leukemia, raising the possibility of a similar mechanism in humans. At very high radiation doses local chronic tissue damage could contribute to the carcinogenic process; skin cancers in early radiation workers developed primarily on the edges of unhealing radiation ulcers.

At low radiation doses radiation carcinogenesis appears to have the characteristics expected of a stochastic process: There is no threshold; the radiation-induced cancers cannot be distinguished from spontaneous cancers; and the frequency of cancer increases with dose, but the severity of the induced cancers is independent of dose and indistinguishable from that of spontaneous cancers.

Carcinogenesis has been studied extensively in people exposed to radiation. These epidemiological studies are subject to all of the problems inherent in defining changes in the rates of relatively common diseases which are also influenced by many other biological, genetic, and life-style factors. Nevertheless, these data are extensive enough that human data provide the framework for predicting the carcinogenic risk of radiation. Populations that have been or are being studied include (1) early radiation scientists exposed to large doses of X-rays or natural radioactive materials; (2) patients receiving large doses of localized radiation for the treatment of cancer or benign diseases (including ankylosing spondylosis, ringworm, and postpartum mastitis); (3) survivors of the bombings of Hiroshima and Nagasaki; (4) populations exposed to radiation during atomic bomb tests (e.g., Marshall Islanders); (5) persons exposed to radiation during accidents; (6) women who ingested radium while painting luminous clock dials; (7) patients injected with large doses of radioactive material for the diagnosis or treatment of disease; and (8) patients receiving multiple high-dose fluoroscopies to monitor the efficacy of treatment for tuberculosis. Many of these populations were exposed to very large doses of radiation in the early 1900s, before the delayed hazards of radiation were appreciated and before radiation protection practices were implemented in the workplace or in medical practice.

The conclusions from these epidemiological studies can be summarized as follows: There is a long latent period between exposure to radiation and the development of cancer. Excess cases of leukemias begin to appear ~2 years after irradiation, peak at ~10 years, and essentially disappear by 25 years. The latent period for solid tumors can be much longer and is known to be at least 20–40 years. Because most large cohorts of irradiated people being monitored are still living, it is not yet possible to say whether the excess risk of solid tumors decreases, remains constant, or increases at longer times after exposure.

Many kinds of benign and malignant tumors have been shown to be induced by radiation. Acute and chronic myeloid leukemias account for the excess leukemia incidence observed in irradiated adults, while acute lymphocytic leukemia is induced in children. Thyroid carcinoma, breast cancer, lung cancer, stomach cancer, bone cancer, skin cancer, and other malignancies have been documented convincingly in various exposed populations. In general, persons exposed to highly localized radiation (from external sources or from internally deposited radionuclides) seem to be at excess risk only for tumors arising in the irradiated tissues, but not for tumors arising in unirradiated sites.

There is considerable controversy concerning the magnitude of the risk of carcinogenesis, especially at low radiation doses. This results from the nature of the data. Although many of the studies have followed relatively large numbers of people, the actual numbers of observed excess cancers are not large. When the data are subdivided by radiation dose to assess the dose–effect curve for carcinogenesis, the number of cases in each class is small and the error limits on the points are large. As a result the slopes (and even the shapes) of the individual dose–response curves are uncertain and vary from one study to another. In addition, it is difficult to estimate accurately the radiation doses received by the exposed individuals. The identification and follow-up of the appropriate exposed populations also raise serious logistical problems, because the studies must span several decades. Moreover, the natural variation in the spontaneous cancer incidence makes it difficult to establish the appropriate control incidence, while the long latent periods before the cancers appear mean that the total cancer incidence is still evolving in most populations being monitored.

Radiation appears to increase the relative risk of specific cancers; the risk of a radiation-induced neoplasm therefore varies with the incidence of that cancer among the unexposed members of the same population. For example, radiogenic stomach cancer is frequent among irradiated populations in Japan, reflecting the high background incidence of stomach cancer in these populations, but is infrequent in irradiated populations in Europe where stomach cancer is a rare disease. Evidence that relative risks for the induction of some cancers vary with age, sex, and other factors further complicates the development of risk estimates.

One of the best current evaluations of the risk of radiation-induced cancer is found in the 1990 report of the National Research Council Committee on the Biological Effects of Ionizing Radiation (see Bibliography.) A simplified summary of the committee's major risk estimates is shown in Table V. The uncertainty in these estimates is large because of the complicating factors described above. Moreover, risk estimates for continuously irradiated populations also rely upon some exceedingly tenous assumptions to extrapolate from the effects observed after a large dose of radiation received in a short period of time and predict the effects of small radiation doses accumulating continuously over many decades.

C. Effects of Prenatal Irradiation

The effects of prenatal irradiation are of special concern, because the developing embryo is extremely sensitive to injury from many toxic agents. As there are few data from humans on this critical subject, our knowledge of the effects of prenatal irradiation comes primarily from studies with rodents deliberately irradiated at known times after conception.

The effects of irradiation and the sensitivity of the embryo vary with the stage of development. In the mouse and the rat preimplantation embryos are exquisitely sensitive to the lethal effects of radiation. A dose as small as 0.05 Gy can kill fertilized mouse eggs. However, the embryos that survive develop normally both before and after birth. The effect of irradiation is therefore seen as a decrease in litter size and an increase in embryonic resorption. Extrapolation to humans suggests that the major effect of radiation during this period (i.e., the first 10 days postconception) would be embryonic death, probably before the mother knew she was pregnant. The limited available data from humans support this concept.

Irradiation of rodents with ~2 Gy during the period of rapid organogenesis results in the production of a variety of structural abnormalities. Specific organs are most sensitive during the beginning of their differentiation, when they contain small numbers of rapidly proliferating cells. Later, during the fetal period, rodents become more resistant to radiation, and the primary effects appear to be generalized growth retardation and developmental abnormalities in late-developing tissues (e.g., the gonads and the CNS). CNS and behavioral damage has been noted in rodents at doses as low as 0.1 Gy.

TABLE V Estimates of Excess Cancer Mortality in Irradiated Populations (Excess Mortality per 100,000 Exposed Persons)

Irradiation conditions	Males		Females	
	Leukemia	Non-leukemia	Leukemia	Non-leukemia
Single acute exposure to 0.1 Sv	110	660	80	730
Continuous lifetime exposure 1 mSv/yr	70	450	60	540
Continuous exposure ages 18–65 10 mSv/yr	400	2480	310	2760

Values are summarized from the 1990 Report of The National Research Council Committee on the Biological Effects of Ionizing Radiations (BEIR V) (see Bibliography).

Extrapolation of these data to humans suggests that irradiation during weeks 4–11 postconception might produce severe developmental abnormalities, while irradiation between weeks 11 and 16 might lead to a few structural abnormalities in late-developing organs and possibly to stunted growth, microcephaly, and retardation. Because the CNS and the skeleton continue to develop throughout the fetal period and into infancy, irradiation during these periods could also cause growth and mental retardation. The limited data available on humans support these general extrapolations. There are reported instances of structural abnormalities and mental retardation in children whose mothers received pelvic radiotherapy during the first trimester of pregnancy. Children exposed *in utero* to the atomic bombs have excess incidences of microcephaly and mental retardation (judged by intelligence tests and school performance.) The incidence of severe retardation increased with dose; significant effects were observed at doses as low as 0.1–0.2 Gy. The effect is highly dependent on gestational age; maximal retardation was seen after irradiation at 8–15 weeks of gestation, smaller effects were seen for 16–26 weeks of gestation, and no significant effects were observed in persons irradiated either earlier or later in gestation. Growth retardation, mental retardation, and learning disabilities have been observed in some children receiving radiotherapy as infants.

These data illustrate the need for caution in the treatment of pregnant and potentially pregnant women with diagnostic or therapeutic radiation. An especially troubling aspect of radiation teratogenesis is the fact that the embryo is most sensitive during the first few weeks, before the pregnancy is obvious or even definitively diagnosed. The doses required for fetal injury are much larger than those received by most occupationally exposed women, but are small enough to mandate special precautions to protect unborn children of radiation workers. Such precautions have been implemented.

Bibliography

Committee on the Biological Effects of Ionizing Radiations (BEIR V). (1990). ''Health effects of exposure to low levels of ionizing radiation.'' National Academy Press, Washington, D. C.

Hall, E. J. (1988). ''Radiobiology for the Radiologist,'' 3rd ed. Lippincott, Philadelphia, Pennsylvania.

Johns, H. E., and Cunningham, J. R. L. (1983). ''The Physics of Radiology,'' 4th ed. Thomas, Springfield, Illinois.

National Council on Radiation Protection and Measurements (NCRP). (1987). ''Ionizing Radiation Exposure of the Population of the United States,'' NCRP Rep. 93. NCRP, Bethesda, Maryland.

United Nations Scientific Committee on the Effects of Atomic Radiation (UNSCEAR). (1988). ''Sources, effects, and risks of ionizing radiation.'' United Nations, New York.

Radioimmunoassay

ROSALYN S. YALOW, *Veterans Administration Medical Center and The Mount Sinai School of Medicine*

SOLOMON A. BERSON, *The Mount Sinai School of Medicine*

Glossary

Double-antibody method Precipitation of antigen-antibody complex with a second antibody directed against the antibody in the complex

Hapten Substance which when coupled to a carrier protein elicits immune response

Immunogen Substance that stimulates antibody response

Iodination Addition of iodine to a molecule

Prohormone Precursor for biologically active hormone

THE MEASUREMENT OF INSULIN in human plasma using radioimmunoassay (RIA) methodology was first reported in 1959. Immediately, this methodology was appreciated as having the potential for widespread applicability in endocrinology because, by that time, advances in peptide chemistry had made available highly purified preparations of animal and human peptide hormones. The potential sensitivity and specificity of RIA made possible physiologic studies of dynamic changes in concentrations of circulating peptide hormones that would have been impossible with the bioassays available at that time. By the 1970s, RIA methodology had spread beyond endocrinology into many areas of biologic interest including, among others, pharmacology and toxicology, infectious diseases, oncology, hematology, and neurosciences.

RIA is now employed in thousands of laboratories around the world, even in the less-developed nations, to measure hundreds of substances of biologic interest. At present, commercial kits are available for many substances routinely measured in clinical or nuclear medicine laboratories; however, "in-house" RIAs remain a powerful tool in the research laboratory. Therefore, the value of the underlying RIA principle as well as its problems and pitfalls is appreciable.

I. The Radioimmunoassay Principle

RIA is based on a simple principle (Fig. 1). The concentration of an unknown substance is determined by comparing its inhibition of binding of labeled antigen to specific antibody with the inhibitions observed with a set of known standards. A typical RIA is performed by the simultaneous preparation of a series of tubes containing standards and unknown samples to which fixed amounts of antibody and labeled antigen are added. The fraction of the labeled antigen, which binds to the fixed amount of antibody, is inversely related to the concentration of the unlabeled antigen in standards or in unknown samples. Hence, after an appropriate time, usually ranging from hours to days, the labeled antigen that is bound to antibody (B) is separated from that which is free (F), and the radioactivity associated with each fraction is measured. The ratio of B/F for each standard tube is then plotted as a function of the concentration in that tube. The unknown concentration is determined by comparing its observed B/F ratio with the standard curve (Fig. 2).

A variety of other modes of plotting the data have been used for the standard curve. These include, among others, the percentage of the tracer bound in the absence of unlabeled hormone or even simply the counting rate of the bound fraction as a function of hormone concentration. Linearity of the standard curve is sometimes approached by use of logarithmic or semilogarithmic plotting.

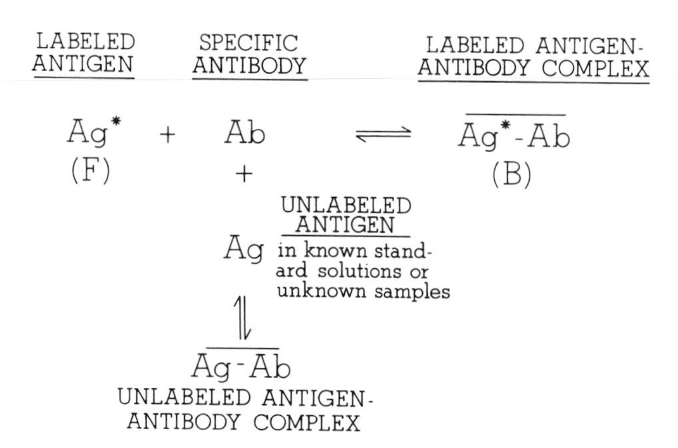

FIGURE 1 Competing reactions that form the basis of RIA. [Reproduced from R. S. Yalow, 1978, Radioimmunoassay: A probe for the fine structure of biologic systems, *in* "Les Prix Nobel en 1977," pp. 243–264, Nobel Foundation, Stockholm.]

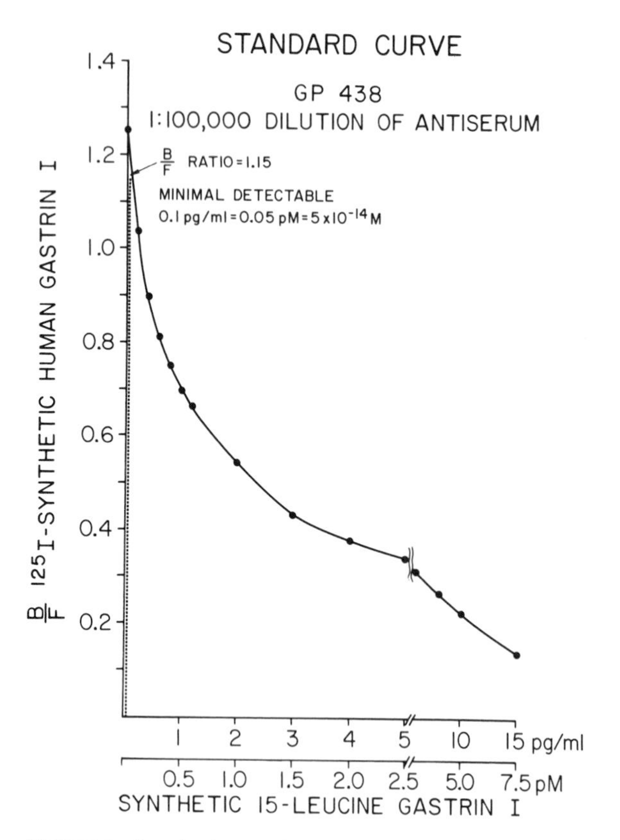

FIGURE 2 Standard curve for the detection of gastrin, a digestive hormone, by RIA. Note that as little as 0.1 pg of gastrin per milliliter of incubation mixture (<0.05 picomolar) is readily detectable. [Reproduced from R. S. Yalow, 1978, Radioimmunoassay: A probe for the fine structure of biology systems, *in* "Les Prix Nobel en 1977," pp. 243–264, Nobel Foundation, Stockholm.]

The RIA principle is not limited to immune systems but can be extended to systems where, in place of the specific antibody, there is a specific reactor (or binding substance). This reactor could be a specific binding plasma protein (e.g., thyroxine-binding globulin or cortisol-binding globulin) or a tissue-receptor site. In place of the radioisotopic labels, other markers such as enzymes or fluorescent dyes have been used. The general term ligand assay has been applied to these assay systems.

RIA differs from traditional bioassay in that it is an immunochemical method in which the measurement depends on only the interaction of chemical reagents in accordance with the law of mass action. Immunochemical activity may or may not be identical with or even reflect biologic activity. The validation of an immunoassay procedure simply requires that the concentration in the unknown sample be independent of the dilution at which it is assayed. However, the relationship between RIA and bioassay quantification often requires additional information because in both systems specific and nonspecific factors often play a role.

II. Preparation of Reagents

A. Sensitivity of RIA

The sensitivity in a classic RIA system is dependent on the equilibrium constant of the reaction between antigen and specific antibody. Furthermore, as evident from Figure 2, the amount of labeled antigen employed in the assay should generally be no larger than the minimal detectable antigen concentration because the initial portion of the standard curve is the region of greatest sensitivity. Thus, the crucial factors limiting the sensitivity of a RIA procedure are the availability of a suitable antiserum and the specific activity of the labeled antigen.

B. Preparation of Antisera

The availability of a suitable antibody is the *sine qua non* of RIA. Generally, polypeptides of molecular weight >2,000 are sufficiently immunogenic when administered as an emulsion in Freund's adjuvant. Lower molecular weight substances (e.g., thyroidal hormones, steroids, prostaglandins, drugs) may be rendered antigenic by coupling to a protein or a large peptide by a variety of methods. Commonly used methods include coupling through carbodiimide to bovine serum albumin or through glutaraldehyde to thyroglobulin. Although a variety of other coupling agents or proteins have been employed, there is no uniform agreement as to which protein or which coupling agent might be optimal for any hapten.

For the preparation of antisera for peptide hormones, the guinea pig has proven to be most satisfactory for a number of reasons. The antigenicity of the immunogen is related to its "foreignness," and many of the guinea pig peptides differ uniquely from those of other mammalian species. Randomly bred guinea pigs have an advantage over inbred animals in that a genetic diversity is more likely, which increases the probability that one of the immunized animals will be a hyper-responder to the immunogen. Obviously, the probability of obtaining a satisfactory antiserum increases with the number of animals immunized, and several guinea pigs can be housed in the same space as a single rabbit. Space considerations are, of course, important for a small laboratory. If the immunogen is a large protein or a hapten coupled to a large protein, a rabbit may be the animal of choice because it is simpler to obtain larger volumes of blood from rabbits than from guinea pigs.

Because the probability of obtaining a satisfactory antiserum increases with the number of animals immunized and the presence of other immunological reactions does not interfere with the reaction of labeled antigen with specific antibody, each animal is immunized simultaneously with several unrelated antigens. This permits reduction of the number of animals to be maintained and bled by a factor equal to the number of immunogens employed.

Animals are injected subcutaneously three times with the antigen(s) homogenized in complete Freund's adjuvant at about 2-wk intervals and are tested about 2 mo after the first dose. Animals without useful antibody titers at this time are generally abandoned. Additional immunizations at monthly intervals may result in further increases in titer. When the titer appears to have stabilized, immunization is discontinued and the animal is bled regularly. When the titer falls to about one-half, the animal is then reimmunized.

It has been our experience that undiluted antisera can be stored at −20°C for at least three decades without loss of potency. Working stock solutions can be prepared at 1 : 100–1 : 2,000 dilution in normal saline containing a bacteriostatic agent and can be fortified with 1–2% nonimmune plasma from the same species. These solutions are generally stable for several years when stored at 4°C but they cannot be frozen.

Monoclonal antibodies, which provide a virtually unlimited amount of homogeneous antibodies against a specific antigenic site, have proven useful in RIA procedures. However, for research laboratories requiring only limited amounts of antisera, the simplicity of antibody production in animals compared with the increased effort required for production of monoclonal antibodies must be considered. [*See* MONOCLONAL ANTIBODY TECHNOLOGY.]

C. Labeled Antigen

RIA was first applied to the measurement of peptide hormones, most of which contain a tyrosyl or histidyl residue that can readily be radioiodinated. The chloramine-T technique for the oxidation of the radioiodine is commonly employed. This method has the advantage over other methods in that the solutions employed are maintained alkaline, and the radioiodine does not volatilize, an important feature from the perspective of radiation protection.

For more than a quarter-century, ^{125}I ($T_{1/2}$ = 60 days) has been the isotope of choice for labeling. If there is no more than one ^{125}I atom per molecule, the preparation is likely to be as stable as the uniodinated preparation. Appreciably, iodination at an average of one radioiodine atom per molecule does not mean that all or even the major fraction of the radioactivity is incorporated into molecules containing one radioactive atom. For instance, iodination of insulin in aqueous solution results in the same distribution of iodine atoms among the tyrosyl residues, which is independent of experimental methods and depends only on the average iodine number. Assuming that there is an average of less than one iodine atom per molecule and that iodination occurs only at the two A-chain tyrosyl residues, with an equal probability of iodinating each residue, a Monte Carlo simulation was used to calculate the theoretical distribution of iodine atoms in labeled insulin preparations. The theoretical analysis indicated that even at an average of only 0.8 radioiodine atom per molecule, approximately half of the radioactivity would be in other than monoiodoinsulin. Thus, purification methods that separate the monoiodinated substance from the uniodinated and multiply iodinated forms are required to assure maximally stable products with high specific activity.

For the assay of steroids, drugs, and other substances, which are generally present at quite high concentrations in plasma, labeled tracers of high specific activity are not necessary. Therefore, ^{3}H-labeled tracers were initially employed for many of these assays. In principle, because ^{3}H has a 12-yr half-life, the labeled antigen has excellent long-time stability. Currently, however, most commercial kits employ ^{125}I-coupled tracers to avoid the need for liquid scintillation counting and the problems associated with disposal of organic scintillation fluids.

D. Separation Methods

The classic immunologic method for separation of antibody-bound from free antigen was based on spontaneous precipitation of antigen–antibody complexes. However, RIA is generally used because of its high potential sensitivity. This requires that the molar concentration of the reagents be so low that spontaneous precipitation does not occur and the antigen–antibody complexes remain soluble. Therefore, a wide variety of other methods have been used to effect separation of antibody-bound and free antigen.

The methods in most common use include (1) precipitation of antigen–antibody complexes with a second antibody directed against the antibody complex (double antibody), (2) the use of organic solvents or salting out to precipitate complexes, (3) adsorption or complexing of antibody to solid-phase material, and (4) adsorption of free antigen to solid-phase material such as cellulose, charcoal, silicates, or ion-exchange resins.

The double-antibody method is generally the method of choice in developing new RIA procedures. However, the cost of the second antibody may make this prohibitively expensive when thousands of samples are to be analyzed. The use of aqueous polyethylene glycol for precipitation of antigen–antibody complexes is often employed after an assay has been validated with double-antibody methodology. The adsorption or complexing of antibody to solid-phase material has the advantage of being a generally applicable method and is frequently employed in commercial kits; however, it has the disadvantage that, because of chemical alterations in the antibody molecule introduced by the coupling procedure or because of steric hindrance, the assays using solid-phase techniques may be somewhat less sensitive than assays employing the same antiserum when it is not complexed.

For routine procedures, a method dependent on adsorption of free antigen to solid-phase material has been widely used, because it is quite inexpensive. Certain common principles apply to all antigen-adsorbent techniques. A given mass of adsorbent is generally more effective if its total surface area is increased, i.e., if the adsorbing particles are made smaller. However, trace amounts of antibody as well as of free antigen may be adsorbed to materials such as charcoal, cellulose, and silicate, unless the concentrations of plasma or other proteins in the incubation mixture are sufficiently high to saturate the binding sites for gamma-globulin. The antigen-adsorption methods have proven to be most satisfactory for small antigens, those with molecular weights of 25,000 or less. The greater affinity of the absorbent for the low-molecular weight substances in the presence of plasma proteins permits their near-total adsorption even in the presence of virtually undiluted plasma or high concentrations of other proteins.

The fact that a large number of different methods of separation of antibody-bound from free-labeled hormone have been employed is a consequence of the variety of chemical properties of the now hundreds of substances for which RIA has been employed as well as of the experimental predilections of the many independent laboratories that have developed such procedures.

III. Problems and Pitfalls

A. Nonspecific Interference in the Immune Reaction

In general, antigen–antibody reactions are optimal in a pH range from 6.5 to 8.5. However, for some antisera to basic peptides such as secretin, the maximum binding of the tracer occurs at a pH of 5, and its binding at pH 4 may be quite comparable to that at pH 7. Furthermore, although it has been shown that the reaction of labeled insulin with many antisera is independent of buffer and pH in the 6.5–8.5 range, the reaction with other antisera may be strongly dependent on both buffer and pH. In general, but not always, an increase in the ionic strength of the buffer decreases the binding of antigen to antibody. Apparently, one cannot predict *a priori* in RIA the optimal pH or buffer to be employed, and these conditions must be determined for each substance and each antiserum.

The binding of antigen to antibody can be inhibited by a variety of substances including proteins, anticoagulants such as heparin, bacteriostatic agents such as merthiolate (ethylmercurithiosalicylate), and enzyme inhibitors such as Trasylol. Not every substance interferes in all immune reactions to the same extent; therefore, the possible effect of each substance must be tested or, alternatively, it must be assured that the milieu of the unknown sample is identical with that of the known standards. Furthermore, blood and other biologic fluids may contain proteolytic enzymes that can damage labeled antigen or antibody. These nonspecific effects decrease the binding of antigen to antibody and result in falsely high antigen levels.

The factors that interfere with the chemical reaction in a nonspecific fashion must be appreciated but can and should be avoided.

B. Heterogeneity of Antigen

It is now commonly appreciated that many, if not most, peptide hormones are found in more than one form in plasma and in glandular and other tissue extracts. These forms may or may not represent precursor(s) or metabolic product(s) of the well-known, well-characterized, biologically active hormone.

Fortunately, the first RIA was described for insulin. The 6,000-dalton peptide with full biologic activity is the predominant form in the circulation of virtually all subjects in the stimulated state. Only in patients with an insulin-secreting tumor or in those with a rare genetic abnormality that prevents cleavage of the C-peptide does the prohormone appear to predominate. However, there are assays of other hormones in which the usual biologically active form is not predominant or in which more than one biologically active form exists. Some examples of these problems and how they may or may not be dealt with are considered below.

Immunochemical heterogeneity was first demonstrated for parathyroid hormone (PTH) when it was observed that a factor, which could be used to superimpose a plasma dilution curve on a curve of standards obtained from dilution of an extract of a normal parathyroid gland for two antisera, failed to effect superposability when another antiserum was employed and that the rate of disappearance of immunoreactive PTH after parathyroidectomy depended on the antiserum employed. Subsequently, it was shown that this was a consequence of the presence in plasma of a metabolic COOH-terminal fragment of PTH, which crossreacted with some, but not all, antisera, and which disappeared more slowly than intact hormone. This fragment is removed from the circulation primarily by the kidney, and its turnover time can be markedly reduced in the presence of impaired renal function. Thus, although in spite of this problem the PTH RIA could be clinically useful, the results in different laboratories were not quantitatively identical because different antisera with different immunochemical specificities were generally employed. The problem appears to have been solved recently with the development of a two-site immunoradiometric assay, which appears to be specific for intact PTH (1–84). The assay depends on the preparation of heterogenous antisera to PTH and subsequent purification of two different antisera on affinity columns: one antiserum is directed to PTH (1–34) and the other to PTH (39–84). To perform the assay, the anti-PTH (39–84) is immobilized on plastic beads. The unknowns or the PTH (1–84) standards in hormone-free plasma are then added. This is followed by addition of ^{125}I-anti-PTH (1–34). The assay thus depends on the binding of intact hormone to the COOH-terminal immobilized antiserum and its capturing the ^{125}I-antibody directed against NH_2-terminal PTH. This assay appears to have sufficient sensitivity to distinguish among plasmas from subjects with primary hyperparathyroidism, normal function, and the hypercalcemia of malignancy and seems to have solved the problem of the heterogeneity of PTH in plasma. [*See* PARATHYROID GLAND AND HORMONE.]

The RIA for gastrin has been complicated by the presence of two biologically active forms of gastrin in plasma. It has been shown that the predominant form of gastrin in plasma collected in the fasted state from normal subjects or from hypersecretors such as patients with Zollinger–Ellison syndrome or pernicious anemia (PA) is generally, but not always, a 34-amino acid precursor peptide (G34), not the 17-amino acid peptide (G17) initially extracted and purified from the antrum by Gregory and colleagues. Both hormonal forms are stimulated by feeding in normal subjects and PA patients. The infusion of equimolar amounts of G34 and G17 results in about the same acid response in the dog. Thus, on the basis of administered dose and resultant biologic effect, G34 and G17 can be claimed equally biopotent. However, the plasma concentrations of G34 during such infusions are about four times higher than those of G17 because of a fourfold slower turnover time. Similar differences in turnover times of G17 and G34 are also found in humans. Thus, a plasma concentration of 50 pg G17/ml is equivalent to an *in vivo* biologic potency equal to a plasma concentration of 200 pg G34/ml. Obviously, however, an occasional patient with marked hyperacidity might present with a tumor that secretes primarily G17 rather than G34, but their plasma gastrin levels are well within what is believed to be the normal range. Therefore, plasma concentrations per se may not be sufficient for diagnostic differentiation between normal subjects and those with a gastrin-secreting tumor. Fractionation of plasma immunoreactive gastrin on Sephadex columns has been used to differentiate between the molecular forms of gastrin. However, the determination of the clinical reason for hypergastrinemia may require additional studies using appropriate tests for stimulation of gastrin release.

Problems in RIA also arise when different peptides share common amino acid sequences that might result in immunologic and/or biologic cross-reactivity. Such systems include, among others, gastrin and cholecystokinin, which share the same COOH-terminal pentapeptide, adrenocorticotropin and melanocyte-stimulating hormone, which have similar NH_2-terminal sequences but different biologic activities, and lipotropin, which contains within it the complete structure of β-MSH.

The application of RIA in pharmacology also has problems relating to the specificity of antisera. Structurally related compounds or metabolites may have significant immunoreactivity with some antisera but not with others and may or may not constitute a problem, depending on the purpose of the assay. For instance, if the clinical problem relates to the toxicity of a particular drug, then the question as to whether or not the assay measures only the biologically active form is relevant. If the question relates simply to whether or not a drug had been taken surreptitiously, then the reactivity of metabolites or variation of the immunoreactivity with the exact form of the drug may be irrelevant.

Perhaps the most widely used RIA in therapeutic drug monitoring is that for digoxin. However, recently a series of reports has suggested the presence of endogenous digoxinlike factors in various clinical conditions as well as the presence in the circulation of metabolites of high immunologic activity but low cardioactivity. These problems with the digoxin assay have not as yet been adequately resolved. Similar problems are likely to occur in assays for other drugs, but such problems have not as yet been fully identified.

IV. Conclusion

Generally, the purpose of RIA is to determine quantitatively the presence of biologically active substances. However, evidently the presence of heterogeneous forms of a variety of substances, both hormonal and nonhormonal, may introduce problems of quantification. Similar problems are present in bioassay as well.

For example, stimulation of amylase release from dispersed acini from guinea pig pancreas can be effected by both cholecystokinin and vasoactive intestinal peptide. If both of these peptides were present in a biologic fluid at equimolar concentrations than the apparent biologic activity (stimulated amylase release) to immunologic activity of either peptide would be greater than one. It must be appreciated that biologic activity *in vivo* can be enhanced by the presence of substances other than the one presumably being assayed. For instance, notably several peptides enhance and others suppress pancreatic secretion of water and bicarbonate. Thus, in any bioassay system in which multiple peptides interact, it is essential that the possible presence of the other substances be considered. Whenever the ratio of biologic and immunologic activity is less than or greater than one, the existence of factors other than the specific molecular form of the peptide to be assayed must be evaluated.

RIA is now used in thousands of laboratories around the world to measure multiple classes of substances of biologic interest, both peptidal and nonpeptidal in nature. Evidently, if there is a need to measure an organic substance of biologic interest and there is no simple method to do so, some imaginative investigator will develop an appropriate RIA procedure. Although new assays are continuously being described and their usefulness documented, it is important to appreciate the problems and pitfalls that limit the precise quantification of substances measured by RIA.

Bibliography

Yalow, R. S. (1983). "Radioimmunoassay" (R. S. Yalow, ed.). Hutchinson Ross Publishing Co., Stroudsberg, Pennsylvania.

Yalow, R. S. (1984). Radioimmunoassay in oncology. *Cancer* **53,** 1426.

Yalow, R. S. (1985). Radioactivity in the service of humanity. *Interdiscip. Sci. Rev.* **10,** 56.

Yalow, R. S. (1985). Radioimmunoassay of hormones. *In* "Williams Textbook of Endocrinology" (J. D. Wilson and D. W. Foster, eds.). W. B. Saunders Co., Philadelphia.

Yalow, R. S. (1987). Radioimmunoassay—A historical perspective. *J. Clin. Immunoassay* **10,** 13.

Radiosensitivity of the Integumentary System

FREDERICK D. MALKINSON, *Rush-Presbyterian–St. Luke's Medical Center*

Glossary

Collagen Main protein substance (matrix) in connective tissue of skin and certain other structures
D_0 Radiation dose required to reduce the fraction of surviving cells to 0.37 of their previous number
Fibrosis Increased collagen (scar) formation
Gray (Gy) One Gray = 100 rads
Keratinocytes Epidermal cells
Keratoses Superficial scaling, red, premalignant lesions
Orthovoltage In X-ray therapy, voltage in the range of 140–400 kV
Telangiectasia Permanent dilatation of superficial blood vessels

BIOLOGICAL, PHYSICAL, and chemical events in irradiated skin are best interpreted from the general responses of all cells and tissues to ionizing radiation. This article is a brief summary of the major radiobiological effects occurring in skin and hair.

I. Radiobiology of the Skin: Some General Considerations

Our knowledge of the responses of skin to ionizing radiation is derived from two sources: the clinical observations of radiotherapists and experimental animal data. The latter source is of critical importance, since certain data for human subjects can be provided only by incidents of accidental exposure to ionizing radiation. In its response to single and fractionated radiation doses, pig skin reacts similarly to human skin qualitatively and quantitatively. Mice and rats have also been widely used to study radiation effects in skin.

Exposure of human skin to high doses of fractionated ionizing radiation (2.0 Gy daily orthovoltage radiation over 4 weeks, for a total dose of 40.0 Gy) induces an acute reaction. Erythema, epilation, moist desquamation, and erosions are ultimately followed by healing. Somewhat lower doses produce less severe changes of scaling and hyperpigmentation. Over the next few months or years atrophy, telangiectasia, hypopigmentation, fibrosis, keratoses, ulcerations, and carcinoma may develop. After a *single* exposure to 8.0 Gy, an early erythema develops in 1–3 days followed by increased erythema in the second to fourth weeks; hyperpigmentation appears after the third week. In general, acute epidermal changes result primarily from damage to germinative cells; later chronic changes reflect injury to structures deeper in the dermis, particularly collagen and blood vessels.

In vitro experiments revealed D_0 values of 0.92 Gy for human keratinocytes and 1.39 Gy for mouse keratinocytes. These values correlated well with those obtained in previous *in vivo* studies. [*See* KERATINOCYTE TRANSFORMATION, HUMAN.]

Encyclopedia of Human Biology, Volume 6.

The character and magnitude of cutaneous responses to radiation depend on many factors, including total dose, dose fractionation, dose rate, radiation quality, area or volume of tissue irradiated, anatomic site, vascular supply, and age. For single exposures to ionizing radiation, the larger the dose, the greater the skin injury. For multiple exposures over time, however, higher dosages are required to produce the same degree of injury when they are divided into two or more fractions. For example, to obtain a certain severity of acute reaction in pig skin following X irradiation (8 MV) requires a total dose of 20.2 Gy given as one dose in 1 day, 32.0 Gy given as three equal fractions in 3 days, and 38.0 Gy given as five equal fractions in 5 days. Increased tolerance with fractionation results from the repair of sublethal damage and cell repopulation. Repair of sublethal damage occurs in less than 24 hr or in one cell cycle after irradiation; repopulation takes days and depends on the tissue's regenerative capacity. [*See* SKIN; SKIN, EFFECTS OF ULTRAVIOLET RADIATION.]

The rate at which radiation is administered also affects cutaneous response. Lower dose rates produce less injury, particularly for late radiation effects. Radiation quality also influences the degree of tissue damage. More energetic radiation is more penetrating, increasing the injury to deeper tissues.

In regard to skin surface area, acute reactions to fixed single doses in pig skin were reduced when progressively smaller fields were irradiated. But for irradiated fields larger than 2 cm in diameter, such area effects were lost. These differences appeared to be related to migration rates of epidermal cells in the healing process.

Concerning age factors, repair rates for radiation-induced DNA damage in proliferating rat epidermal cells are reduced in older animals.

II. Epidermal Cell Survival: Effects of Single and Multiple Radiation Exposures

Studies with an *in vivo* epidermal cell-cloning technique in mice revealed a linear relationship between X-ray dose and epidermal cell survival. The D_0 was 1.34 Gy in air, 1.12 Gy in hyperbaric oxygen (4 atm), and 3.5 Gy under anoxic conditions. With this same technique two-dose-fraction irradiation studies revealed that fractionation intervals beyond 24 hr up to 6 days yielded progressive increases in cell survival. In two human subjects receiving fractionated radiation therapy, a D_0 of 4.9 Gy was estimated for epidermal cells by this method.

The higher rate of cell survival, or the diminished intensity of gross epidermal reaction, that occurs when a single dose of radiation is divided into two fractions yielding the same total dose in Gy, reflects the repair of sublethal radiation damage. The extent of this repair is related to the magnitude of the shoulder of the cell survival curve. For short time intervals (hours), however, the effects of sublethal damage repair may be modified by induction of a partially synchronous cell population from the first radiation dose, thereby inducing a cell cycle phase-dependent "radiosensitizing" or, conversely, "radioprotective" effect.

III. Radiosensitizers

The skin is a readily available site for testing augmented radiation responses to radiosensitizers. Studies of skin injury in patients breathing air or exposed to hyperbaric oxygen revealed oxygen enhancement ratios of 1.3 or greater, increasing with higher radiation dosages.

Electron affinic agents (nitroimidazoles) are effective radiosensitizers of hypoxic mammalian cells *in vitro* and of rodent and human skin cells *in vivo*, yielding enhancement ratios of 1.25–2.22, compared with an oxygen enhancement ratio of about 2.6. No radiosensitization occurs in well-oxygenated cells. The older nitroimidazoles are cytotoxic, especially neurotoxic, however, and to date they have not produced significant benefits in tumor therapy.

Severe cutaneous reactions to ionizing radiation (erythema, moist desquamation) have been reported in patients receiving certain chemotherapeutic agents during courses of radiotherapy for a variety of malignancies. These agents may also induce moist desquamation well beyond the time that radiation treatment courses have been completed ("recall" phenomenon). Such drugs include adriamycin, actinomycin D, methotrexate, daunomycin, vinblastine sulfate, and hydroxyurea. Similar effects occur in other tissues or organs. Reduced cutaneous reactions occur following reduced radiation dosages or increased time intervals between radiation courses and drug administration. With some of these agents (actinomycin D, vinblastine sulfate), true radiosensitization may occur, as shown in drug–radiation studies of hair. For other cytotoxic compounds, however, enhanced radiation effects may only reflect additive cell toxicity of concomitant or sequential administration.

Where it does occur, the effects of various chemotherapeutic agents in producing radiosensitization (or, occasionally, radioprotection) can be attributed in part to the ability of these compounds to position proliferating cells in cycle phases conferring relative sensitization (G_2, M) or protection (S) at the time of subsequent irradiation.

IV. Radioprotectors

Cutaneous injury from radiation delivered to deeper tissues and organs has stimulated investigations of radioprotective agents, especially those that are active topically. Thiol and disulfide compounds presumably act largely as free radical scavengers. They may also promote the repair of radiation-damaged molecules before these molecules can react with oxygen. Topical applications of cysteamine in several species of laboratory animals was radioprotective for acute skin reactions with dose-reduction factors of 1.4–1.5. Studies with WR-2721, administered systemically, revealed dose reduction factors of 2.0–2.4 for cutaneous ulceration and desquamation. For single or fractionated irradiation in mice, optimal protection required drug injection 30–60 min before irradiation.

V. Hyperthermia

Heating skin before, during, or after irradiation enhances the cutaneous response. For skin, thermal enhancement ratios range from 1.1 at 40°C to 2.0 at 43°C.

When heat was applied to mouse skin for intervals of 7.5 min to 4 hr, combined with orthovoltage radiation, an increase in temperature of 0.5°C was equivalent to increasing the dose of radiation 10–15%. Each rise of 1°C reduced by one-third the time required to enhance the radiation response. For human skin, thermal enhancement ratios of 1.7–2.0 were found in temperature ranges of 43–46°C.

Since the thermal enhancement of radiation damage tends to spare normal tissues but preferentially injures hypoxic cells, which are radioresistant, hyperthermia may play an important role in cancer therapy. For example, laboratory studies of human and mouse melanoma cells have revealed radiosensitization by concurrent or postradiation hyperthermia. [*See* HYPERTHERMIA AND CANCER.]

As might be expected, cooling of the skin induces some *radioprotection*, probably resulting from lowered oxygen tension in vasoconstricted sites.

VI. Early Vascular Effects

Postradiation erythema results from changes occurring within hours to days after irradiation. These evolve over months and years and, in laboratory animals, can be roughly quantitated by dye leakage, which is most intense 2.5 weeks after irradiation. A dose of several Gy induces dye leakage, and the intensity of extravasation and its time of appearance are dose dependent.

Studies with the Sandison–Clark rabbit ear chamber after single dose irradiation revealed serial intimal swellings and proliferation appearing along blood vessel walls 3–4 days postradiation. These repair processes followed endothelial nuclear swelling, enlargement, and pyknosis. Dose-dependent reductions in capillary density presumably impeded oxygen delivery to the skin. Such early events may have significance for the development of telangiectasia and other long-term vascular changes after irradiation.

In vitro studies of endothelial cells in radiation-induced vascular reactions have revealed D_0 values of 1.68–2.40 Gy.

VII. Late Effects

The magnitude of the early acute skin reaction is not a good indicator of the severity of late changes. β-Irradiation of pig skin, for example, in doses producing minimal erythema without desquamation, may result in reduced dermal thickness (atrophy) of up to 50% 2 years later. Late radiation changes may be more dose limiting than changes seen in the first few weeks after exposure, perhaps reflecting much slower cell turnover rates and recovery processes in the dermis than in the epidermis. Extensive clinical data suggest that late effects of radiation are usually independent of overall treatment times, unlike early reactions, and that they are more dependent on the size of the daily fraction.

Radiation injury to the vasculature may be the most important factor in late radiation damage in the skin. Endothelial cell damage and complement-mediated inflammation, present for several months postradiation, reduce circulation. These changes may also contribute to the development of progressive fibrosis.

Studies of single dose exposures in mouse skin demonstrated that increased postradiation collagen synthesis is an early (1 week) change and not only a late radiation effect. By 12 weeks postradiation, collagen synthesis levels were enhanced, in dose-dependent fashion, up to 50%, and this effect was sustained for at least 1 year. Increased collagen content of skin and microscopic fibrosis in the lower dermis were associated findings. Subcultures of fibroblasts isolated from irradiated skin 1–48 weeks postradiation were threefold overproducers of collagen. [*See* COLLAGEN, STRUCTURE AND FUNCTION.]

VIII. Radiobiology of Hair

Throughout the mammalian life span hair matrix cells undergo alternate periods of exceedingly active cell proliferation (anagen) and total reproductive inactivity (telogen). The most useful of numerous indices used to evaluate radiation effects on matrix cells are quantitative assessments of hair loss and growth rates, the incidence of microscopic morphologic changes, and kinetics studies in irradiated hair matrix cell populations with [^{3}H]thymidine techniques.

Anagen hairs are 2.5 times more radiosensitive than telogen hairs, but repair radiation damage more quickly. Telogen matrix cells sustain and "store" radiation damage, which is then expressed in succeeding cell generations following a mitotic stimulus (plucking). Marked differences in vascular supply between anagen and telogen matrices may well contribute an oxygen effect to their different radiosensitivities. In mice, large radiation doses (20.0–25.0 Gy) permanently alter anagen–telogen cycle times and hair growth rates, and reduce hair matrix cell mitotic indices by 50%. The resultant partial permanent alopecia presumably reflects radiation destruction of some entire hair matrix cell populations.

Radioprotective drug–radiation effects were first noted when systemic cysteine prevented radiation-induced epilation in dose-dependent fashion in guinea pigs receiving doses of 7.0–9.5 Gy. No protection occurred at doses above 12.0 Gy. Glutathione, cysteamine, or WR-2721 yields dose-reduction factors ranging to 1.7 for radiation-induced anagen alopecia. Up to threefold greater radioprotection is seen with certain systemically administered prostaglandins, and these compounds are active topically as well. Conversely, radiosensitizing drug–radiation effects inducing increased alopecia have been noted with actinomycin D, colchicine, hydroxyurea, bleomycin, and other compounds. The time of administration of these agents prior to irradiation is critical, as they act largely by influencing progression of hair matrix cells through the cell proliferation cycle. Their "positioning" of matrix cells at the time of subsequent radiation results in increased or, occasionally, decreased hair loss depending on whether the matrix cells are irradiated in relatively radiosensitive or radioresistant phases of the cell cycle.

The use of hair as a biological indicator system for combined drug–radiation effects has obvious implications for studies in other cell systems and for investigations of therapeutic approaches to experimental and clinical malignancies. Considerations of the various factors involved in combined treatment regimens are complex, however, and several problems have been encountered in the applications of investigative findings to tumor therapy. [*See* HAIR.]

IX. Radiobiology of Pigmentation

Postradiation hyperpigmentation occurs in skin, partly due to enhanced tyrosinase activity in melanocytes and later increases in the numbers of these cells. Hyperpigmentation is directly related to radiation intensity and to total accumulated dose, and is characterized by increased melanin deposition throughout the epidermis. Although low-dose or chronic irradiation produces increased melanin synthesis, higher doses of radiation destroy melanocytes and result in epidermal depigmentation or in graying of hair. Follicular melanocytes are destroyed by radiation much more readily than epidermal melanocytes.

In the mouse, hair graying is dose dependent and varies with the growth cycle of the hair. In contrast to radiation effects on hair matrix epithelial cells, melanocytes in telogen hairs are far more susceptible to radiation destruction than those in anagen matrices. Melanocytes in resting follicles have a D_0 value of 1.50–2.0 Gy for single dose radiation. For fractionated radiation D_0 values range from 2.15 Gy (two fractions) to 4.15 Gy (eight fractions).

Both melanocyte survival in hair follicle squashes and depigmentation of hair in laboratory animals have been used to assess quantitatively combined drug and radiation effects.

X. Radiobiological Effects on Langerhans Cells

Confirming earlier studies, electron microscopic investigations revealed that single doses of 20.0 Gy in mice reduced Langerhans cell (LC) numbers to 18% of controls within 10 days. Cell repopulation was rapid by day 16 and cell numbers were essentially normal by day 30. LC loss postradiation is dose dependent. When whole-body irradiation was given to mice receiving local skin irradiation in addition, repopulation of the epidermis by LCs was delayed by another 3 weeks. The latter findings largely reflect the bone marrow origin of LC precursors, since mitoses in LCs postradiation were not observed in the epidermis.

In regard to late radiation effects, full recovery in LC numbers in mice was followed by substantial reductions at 19 months of age, with a drop to 60% of normal at 24 months of age. Late changes of radiation-induced fibrosis and impaired circulatory function in skin may impede replacement of LCs from bone marrow precursors. Postradiation LC loss may have implications for the subsequent development of skin tumors and reduced susceptibility to induced contact hypersensitivity.

Bibliography

Cole, S., Humm, S. A., James, D. R., and Townsend, K. M. S. (1986). Langerhans cells: Quantitative indications of X-ray damage in mouse skin? *Br. J. Cancer* **53,** Suppl. VII, 75.

Hamlet, R., Heryet, J. C., Hopewell, J. W., Wells, J., and Charles, M. W. (1986). Late changes in pig skin after irradiation from beta-omitting sources of different energy. *Br. J. Radiol.* Suppl. 19, 51.

Hanson, W. R., Pelka, A. E., Nelson, A. K., and Malkinson, F. D. (1988). 16,16 dm prostaglandin E_2 protects from acute radiation-induced alopecia in mice (abstract). *Clin. Res.* **36,** 906A.

Malkinson, F. D., and Keane, J. T. (1978). Hair matrix cell kinetics: A selective review. *Int. J. Dermatol.* **17,** 536.

Malkinson, F. D., and Hanson, W. R. (1991). Radiobiology of the skin. *In* "Biochemistry and Physiology of the Skin" (L. Goldsmith, ed.), 2nd Ed. Oxford University Press, New York.

Panizzon, R. G., Hanson, W. R., Schwartz, D. E., and Malkinson, F. D. (1988). Ionizing radiation induces early, sustained increases in collagen biosynthesis: A 48-week study in mouse skin and skin fibroblast cultures. *Radiat. Res.* **116,** 145.

Peel, D. M., Hopewell, J. W., Wells, J., and Charles, M. W. (1984). Non-stochastic effects of different energy beta emitters on pig skin. *Radiat. Res.* **99,** 372.

Vegesna, V., Withers, H. R., and Taylor, J. M. G. (1987). The effect on depigmentation after multifractionated irradiation of mouse resting hair follicles. *Radiat. Res.* **111,** 464.

Wambersie, A., and Dutreix, J. (1986). Cell survival curves derived from early and late skin reactions in patients. *Br. J. Radiol.* Suppl. 19, 31.

Withers, H. R. (1967). Dose survival relationship for irradiation of epithelial cells of mouse skin. *Br. J. Radiol.* **40,** 187.

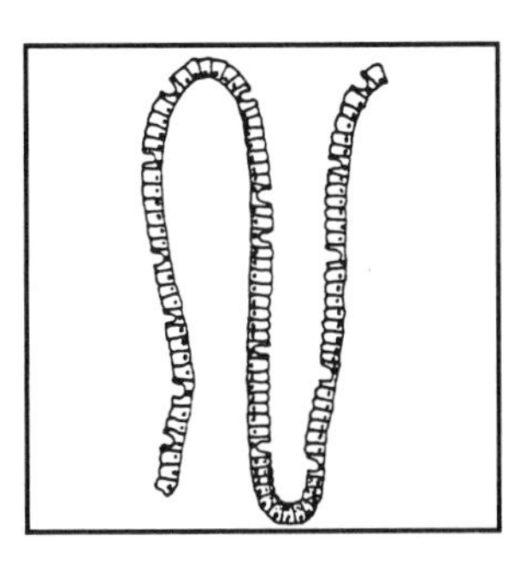

Radiosensitivity of the Small and Large Intestines

ALDO BECCIOLINI, *University of Florence, Italy*

Glossary

Adventitia or tunica adventitia Part of the connective tissue covering the blood vessels

Cell cycle time (T_c) Time it takes a cell to move through the different cycle phases from one mitosis to another

Circadian pattern Biologic parameter which has a rhythmic variability occurring about every 24 hr

Crypt (of Lieberkühn) Localized in the lower part of the villus, it corresponds to the proliferative compartment of the intestinal epithelium

Duodeum First part of the small intestine, about 25 cm long

Glycocalyx Structure made up of mucopolysaccharides which appears as a filamentous excrescence on the microvillous tip of differentiated epithelial cells

Gray (Gy) SI unit of radiation absorbed dose. 1 Gy = 1 J/kg = 100 rad

Hypovolemic shock Shock caused by reduced blood volume which may be due to loss of blood or plasma

Ilcum Distal portion of the small intestine localized in the lower abdomen

Jejunum Proximal part of the mesenteric small intestine, occupying the upper portion of the abdominal cavity

Linear energy transfer (LET) Rate at which energy is deposited when ionizing radiation travels through matter. It is expressed in kiloelectron volts per micrometer

Microvilli or striated border or brush border Closely packed, parallel, cylindrical processes visible with the electron microscope

Mucosa Lumenal surface of the small intestine, greatly increased by grossly circular folds (plicae) and by microscopic finger-like structures (the intestinal villi)

Muscolaris Constituted by an external and an internal layer of the muscolaris coat that covers the intestines

Pinocytosis Mechanism of engulfing particles or dissolved materials by a process of vesciculation. Lipid droplets are absorbed in this way by the epithelial cells of the small intestine

Valvulae conniventes (or valve of Kerckring) Highly visible folds that increase the lumenal surface of the small intestine

Villus Finger-like processes that increase the absorptive surface of the small intestine

GASTROINTESTINAL EPITHELIUM presents a high radiosensitivity because of its high proliferative activity. The irradiation of the small and large intestines with an acute exposure produces an early appearance of injury due to the fast cell turnover.

When a high-enough dose is used, specific clinical conditions defined as gastrointestinal radiation syndrome are observed. The syndrome is characterized by nausea, vomiting, and diarrhea that could cause death within about 10 days.

When sublethal doses are used this syndrome is less severe and lasting, in fact, the high proliferative activity of the epithelium allows a rapid replacement of dead cells.

The experiments on laboratory animals showed that the intestine is a good model to study radiation effects on the mechanisms of proliferation, repair, and cell differentiation.

In humans the gastrointestinal tract is included in the irradiated volume of patients affected by neoplasias localized in the abdominopelvic region and treated by radiotherapy. The total dose administered is quite high, but being divided into daily fractions (2 Gy/day, 5 days/week), it causes damage that can be recovered from within a few weeks after treatment. However the induced modifications are so severe as to produce temporaneous morphologic alterations and a malabsorption of many molecules in the diet.

The irradiation of the alimentary canal can induce later, even severe, damage such as sclerosis, intestinal occlusion, and carcinogenesis.

I. Introduction

The energy given off by ionizing radiation produces chemical bond breakage and, therefore, modifications in atom organization which can be seen at a molecular level. This series of events takes place in from 10^{-18} sec to a few seconds. It leads to the alteration of macromolecular structures which is later manifested by morphological and functional cell modifications and their consequent biological effects. If these effects are greater than the capacity of the repair mechanisms, clinical damage can be observed both in early and in late phases depending on dose, linear energy transfer (LET), irradiation schedule, and, especially, on the proliferative activity of the cell lines that form the tissue. DNA and the enzymes necessary for its replication and repair are thought to be the principal targets for radiation damage. This implies a difference of effect between proliferative cells and differentiated ones.

Therefore a cell system with a high proliferative activity will be profoundly altered by radiation doses which will bring about only modest lesions in cell systems where the mitoses are rare or absent. The time necessary for the lesion to appear depends on the turnover of the irradiated cell system as well as on the time necessary for its return to normality.

In a tissue, composed of several types of cells, acute injury is concentrated mainly in that cell component having the highest proliferative activity. Damage to the component with the least mitotic activity or the slowest cell turnover appears only much later. The same occurs in cell lines such as intestinal epithelium when the dose is high enough to sterilize the proliferative compartment, the epithelium disappears and ulceration is produced. Depending on the extent of ulceration and on connective tissue damage, cicatrization may or may not occur with time. When the dose is sublethal the mitotic activity is blocked at first, but later, depending on cell turnover and damage, which is dose dependent, the epithelium returns to normal conditions. Proliferative activity of the epithelium in the gastrointestinal tract is high and therefore the tissue should be considered highly radiosensitive. The small intestine is more radiosensitive than the stomach and large intestine, the duodenum is more than the jejunum and ileum, and the colon more than the rectum.

Irradiation in man can occur accidentally or purposely as in the treatment of neoplasia: here a part of the gastrointestinal tract is included in the irradiated area and therefore receives elevated doses. In other conditions, for example following radiodiagnostic examinations of the digestive tract or genitourinary tract or for nuclear medicine examinations, the radiation doses absorbed by the intestine are much lower, and unable to produce acute injury.

II. Alimentary Tract

The alimentary tract includes the esophagus, stomach, small and large intestines, and rectum. The mucosa consists of (1) a single layer of epithelium resting on a lamina of connective tissue, (2) a layer of smooth muscle (submucosa), and (3) two layers of muscle tissue, one placed transversally, the other longitudinally.

The structure of the mucosa in the different parts of the alimentary tract is similar, being made up of glands whose proliferative compartment is located in the intermediate region in the stomach and at the base of the crypts in the small and large intestines and in the rectum (Fig. 1). The size of the proliferative compartment is proportionally greater in the colon and rectum whereas it is limited to the lower third of the crypts in the small intestine. In this part of the intestine the proliferation rate is higher.

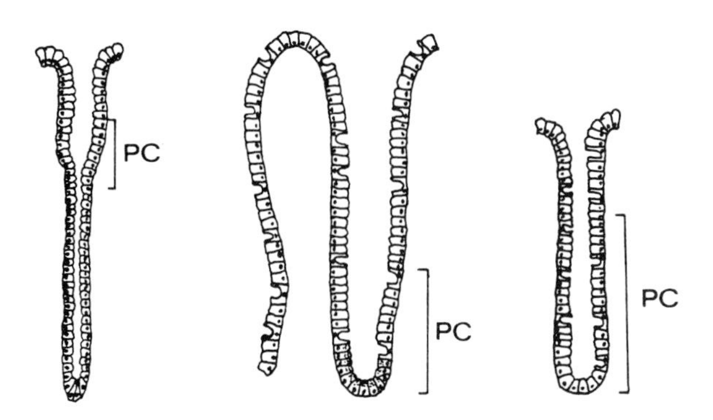

FIGURE 1 Schematic representation of different types of epithelium lining the gastrointestinal tract: (a) stomach, (b) small intestine, (c) colon and rectum. PC, Proliferative compartment.

The proliferative cells go through the different phases of the cycle during their migration toward the villus. Once they reach a certain level of the proliferative compartment, they lose their capacity to divide, change their morphology, and begin to synthesize the molecules which characterize mature cells.

The renewal process of the gastrointestinal epithelium involves proliferation, migration, differentiation, and loss or death of the cells. Studies of these cells is performed by using [^{3}H]thymidine. Information obtained by directly injecting the labeled nucleoside into patients with preterminal illnesses showed that the cell cycle time in the human intestine lasts 1–2 days, depending on the part of the alimentary tract; the S phase (of DNA synthesis) is about 60% of the cell cycle time (T_c). Turnover time (the average life of a cell) in the small intestine epithelium is longer in the proximal tract (about 5 days), where the epithelium is formed by a greater number of cells, than in the distal ileum. In the human large intestine and rectum turnover time is about 6 days. From a physiological point of view, the mucosa cells actively participate in the digestive process by synthesizing enzymes and by absorbing the hydrolyzed molecules coming from the diet. Furthermore, the epithelium layer acts as a protective barrier against acidity, proteolytic enzymes, and the bacterial flora and their toxins. These agents, together with the continuous microtraumas provoked by the passage of food, necessitate continuous substitution of the epithelial cells. For this reason, elevated proliferative activity and rapid turnover exist along the entire gastrointestinal tract and these factors explain why the alimentary canal is highly sensitive to ionizing radiations.

III. Acute Gastrointestinal Radiation Syndrome

When a mammal is exposed to sufficiently high doses of ionizing radiations over its entire body, it manifests a series of symptoms referred to as radiation syndrome. Bone marrow syndrome, gastrointestinal syndrome, and central nervous system (CNS) syndrome appear as the dose increases and therefore within progressively shorter time periods. [*See* Radiation, Biological Effects.]

Whereas the CNS syndrome in man is poorly defined because of the high dose necessary for its induction, the other two syndromes have been widely studied both in accidently exposed individuals and in patients with tumors who have undergone radiotherapy. More than 2 Gy is needed to produce, in a few hours, the initial symptoms of the gastrointestinal (GI) syndrome: these are similar to air- or seasickness and are influenced by psychological factors and individual susceptibility. Nausea and vomiting follow a state of loss of appetite and apathy, and these symptoms progressively worsen on the second and third day. Then, if a higher dose has been absorbed, all the symptoms characterizing the GI syndrome appear: anorexia, nausea, vomiting, and high fever. A week after exposure, severe dehydration, reduction of plasma volume, circulatory collapse, as well as profound alterations of the intestinal mucosa, can lead to death. About 5 Gy is sufficient to induce a quite serious GI syndrome and, with whole-body irradiation, there is a marked bone marrow syndrome characterized by lymphopenia, neutropenia, and platelet reduction. Doses above 10 Gy in man and in laboratory animals induce death in 3–8 days, depending on the dose.

Numerous studies of the GI syndrome have been performed but they have not shown whether death is due to a single factor or to a series of factors. The profound alterations of the epithelial cells throughout the entire gastrointestinal tract cause several functional alterations. These include water and electrolyte loss, which can lead to hypovolemic shock, and loss of the mucus layer, the latter permitting greater mucosa sensitivity to microtraumas and autodigestion by gastric and pancreatic enzymes as well as possible entrance of the intestinal microflora into the organism. Studies with germfree mice demonstrate the role of infection: if exposed to a lethal dose, their survival time is about two times greater than that of normal mice because the turnover time in the former is nearly double that found in the latter.

IV. Small Intestine

This part of the alimentary tract is devoted to the absorption process. For this reason the small intestine is organized in valvulae conniventes, crypt–villus system, and microvilli in the cells or the villus. This complex of structures increases the absorption surface by more than 600 times. The proliferative compartment is localized in the crypt, which is in the lower third of the mucosa (Figure 1).

The epithelium is a single layer containing four types of cells: columnar cells (more than 90%), mucus cells (6%), very few enteroendocrine cells, and Paneth cells, located at the base of the crypt. Some recent theories favor the hypothesis of a common origin for all epithelial cells, the totipotent stem cells being located in the first positions at the base of the crypt. These cells are normally included in the nonproliferative compartment in the G_0 phase of the cell cycle. However, when necessary, they become active and produce partially differentiated proliferative cells which subsequently divide and allow the repopulation of the different types of cells that make up the intestinal epithelium: this occurs when the mucosa is damaged by physical or chemical agents.

The proliferative columnar cells undergo more than one division during their migration along the crypt. In laboratory animals and in man cell cycle time is approximately 14 hr, with differences depending on the position of the proliferative cells in the crypt. Turnover time in man is about 5 days whereas in mice and rats turnover is about 3 days. The use of [^{3}H]thymidine, a labeled nucleoside which is taken up by cells in the DNA synthesis phase, has made it possible to demonstrate the irregular distribution of these cells along the crypt. One hour after the injection of the labeling agent the frequency of labeled cells is very low at the base of the crypt but rapidly increases, showing that more than 50% of the proliferative cells are in the DNA replication phase in the lower third portion. In the superior positions the frequency of S phase cells decreases rapidly so that in the upper third there are no labeled cells. Since the cells go through the S phase and then the G_2 and M phases, mitotic figures can be observed in positions above the labeled cells.

In the upper third of the crypt, the columnar cells lose their proliferative capacity and synthesize enzyme molecules which characterize the function of the mature cells. Other molecules, such as thymidine kinase and other enzymes, which are involved in DNA synthesis, are, on the contrary, no longer produced.

The main function of the small intestine is the digestion of food and the absorption of the various nutrients that are a part the diet. A partial hydrolysis of starches and proteins takes place in the stomach. Then the pancreatic and biliary secretions present in the small intestine continue the digestive process. The simplest molecules are absorbed through active or passive transport mechanisms or through pinocytosis as occurs for fats. Partially broken down molecules must be further hydrolyzed in order to be absorbed. This process takes place in the microvilli of the columnar cells by enzymes, such as tri- and dipeptidases and tri- and disaccharases, located in the brush border and is referred to as *membrane digestion*. The monomers (amino acids, sugars) produced by this hydrolysis are absorbed by the intestinal mucosa. This process of terminal digestion takes place within the glycocalyx under sterile conditions. The brush border enzymes are synthesized during the migration of the columnar cells along the villus. Brush border enzymes are absent in the stomach, gradually increase from the duodenum to the jejunum, then decrease to reach very low levels in the terminal part of the ileum. They are absent in the colon and in the rectum. These enzymes have a fundamental importance in the digestive process because their congenital or acquired deficiency leads to intolerance and malabsorption, which are particularly severe in childhood and in some intestinal pathologies.

In mammals brush border enzyme activity is strictly correlated with food intake and shows marked modifications during the daily light : dark period. In nocturnal animals (e.g., rat, mouse) the activity during the night is twice that during the early afternoon whereas in man brush border enzyme activity is higher during the light period.

Intracellular enzymes, which have no specific dietary function as do the lysosomal enzymes, do not show variations during the light : dark cycle.

A. Experimental Studies

More experimental studies on the effects of ionizing radiations have been performed on the small intestine than on the stomach and on the large intestine.

There are many morphological and functional similarities between the small intestine of experimental animals and that of man. For this reason the small intestine has been widely used as a model in radiobiological studies to obtain information which can be applied to man.

Total body irradiation or irradiation only of the abdomen brings about alterations of the alimentary canal which have already been described in the discussion of gastrointestinal syndrome above. The evolution of radiation injury can be analyzed by considering the various levels of organization, as follows:

- Cell level: the proliferative cells die.
- Tissue level: the height of the crypt–villus system is reduced because of the lack of new cells and the physiological loss of cells at the villi apex.
- Organ level: loss of water and electrolytes, malabsorption, disepithelialization.
- Organism level: the preceding conditions lead to a state of physical debilitation, which, for doses greater than 10 Gy, is aggravated by ulcerations, entrance of germs into the body, and a complex of causes that lead to death.

The evolution of intestinal damage in man is similar, if we consider a certain delay in the sequence of effects caused by the slower turnover.

Cell damage begins after a latent period of about 1–2 hr, because it occurs in the lower portion of the proliferative compartment. With time, degenerative phenomena (nuclear and cytoplasmic alterations) and cell death involve a progressively increasing part of the crypt and after 1–2 days also the villus (Fig. 2). The reduction in the number of cells increases as the dose increases and reaches a maximum in 48–72 hr with sublethal doses. But already less than 1 day after exposure the proliferative activity begins again and the percentage of cells that synthesize DNA and that are in mitosis rapidly exceeds the preirradiation values. The size and the length of the reductions in cell number, as well as the successive phase of repopulation, are directly dose dependent.

After approximately 20 hr an inflammatory infiltration is observed in the stroma, which increases progressively during the entire acute phase of damage. Later the infiltrating cells progressively decrease and the epithelium returns to normal conditions.

On an ultrastructural level, the cell modifications concord with those observed using the optic microscope. A few hours after irradiation with a sublethal dose, a large number of free or aggregated ribosomes appear in the rough endoplasmic reticulum of the cells in the upper part of the crypt. Phagocytic cells, expansion of the Golgi complex, and enlarged mitochondria are observed. Later marked alterations are present in the entire crypt–villus system. After a few days, the epithelium recovers and the morphology of the columnar cells seems to return to normal.

The study of irradiated intestinal epithelium has permitted the formulation of interesting hypotheses with possible clinical applications. As observed with plant tissue and with some tumoral cell lines, the results demonstrate that the ionizing radiations are able to induce an early differentiation, which is very evident 24–36 hr after exposure. This happens when the proliferative cells are damaged sublethally and lose their capacity to divide. They begin to synthesize molecules, for example the brush border enzymes, which characterize the differentiated state, before the time of their normal differentiation. Altered differentiation is confirmed by a significant increase in the number of goblet cells.

Another interesting aspect is related to the phase of repopulation: at first it appears to involve almost exclusively columnar cells; the goblet cells reappear only later. The late return to a mucosa with normally functioning capacities is confirmed by the activity of the brush border enzymes (Fig. 3).

B. Modifications after Fractionated Irradiation

Radiobiological studies have demonstrated that when a dose of radiation is divided into fractions the resulting effect is inferior to that induced by a single dose of the same size. This phenomenon demonstrates the capability of cells to repair sublethal injury. In fact, in the interval between fractions, endocellular repair occurs which, if the time period is sufficient, will completely eliminate the acute damage.

The use of dose fractionation in tumor radiotherapy is based on time differences and efficacy differences in sublethal damage repair and recovery between healthy and neoplastic tissues exposed to low LET radiations. The small intestine has been used to evaluate tissue tolerance with this type of irradiation and as a model to study the effects of repeated doses on mechanisms regulating proliferative activity and cell differentiation. The effects of dose fractionation on epithelial cell number in rat jejunal

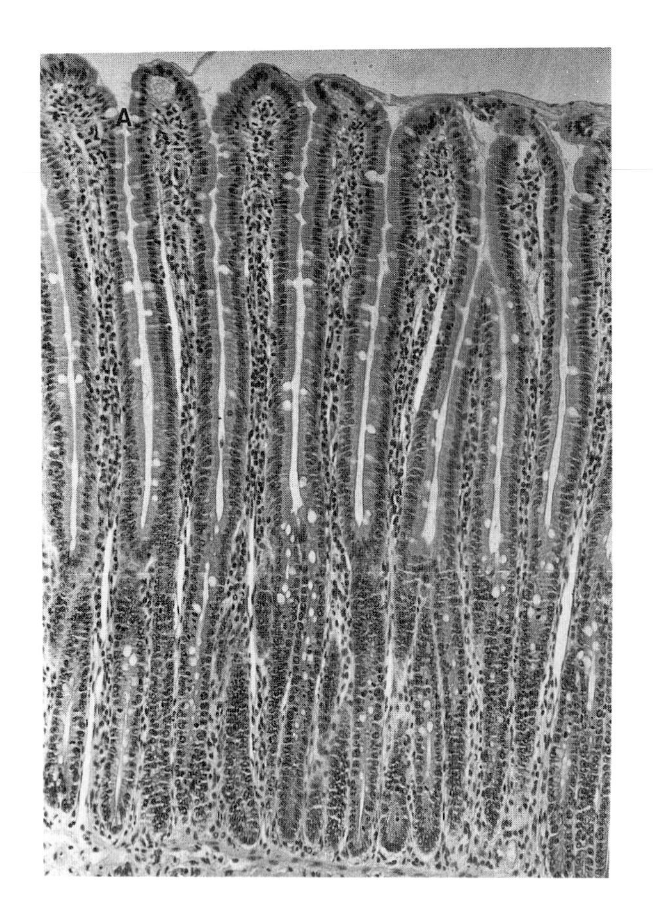
A

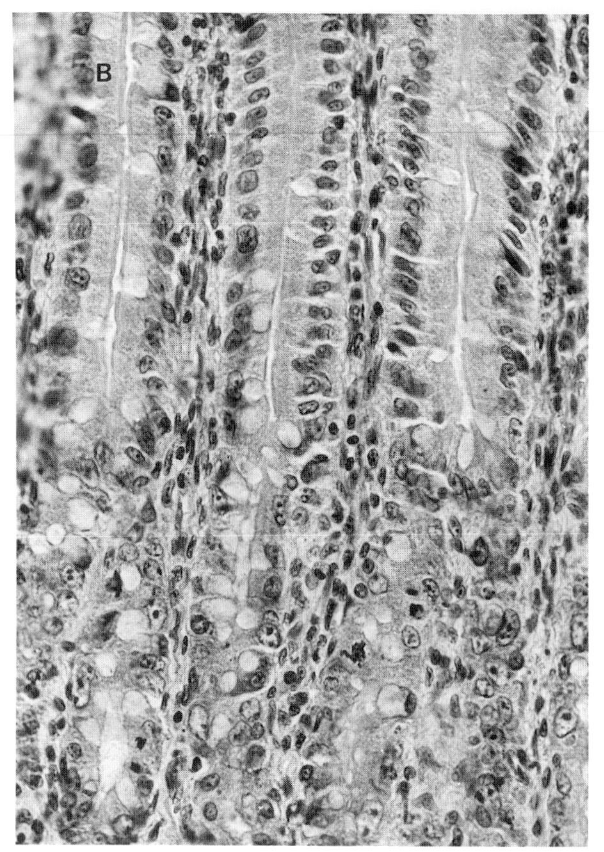
B

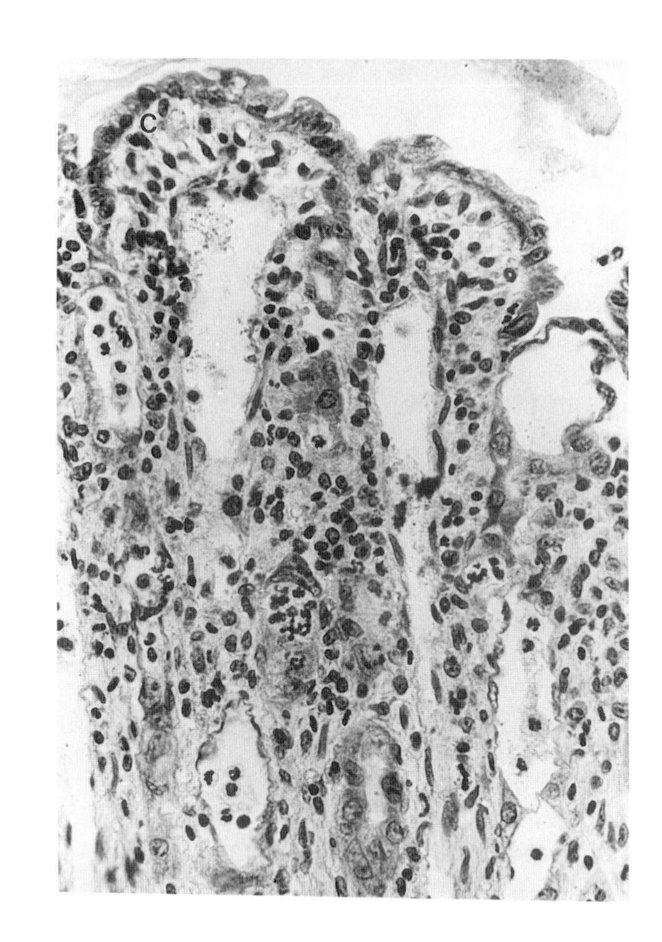
C

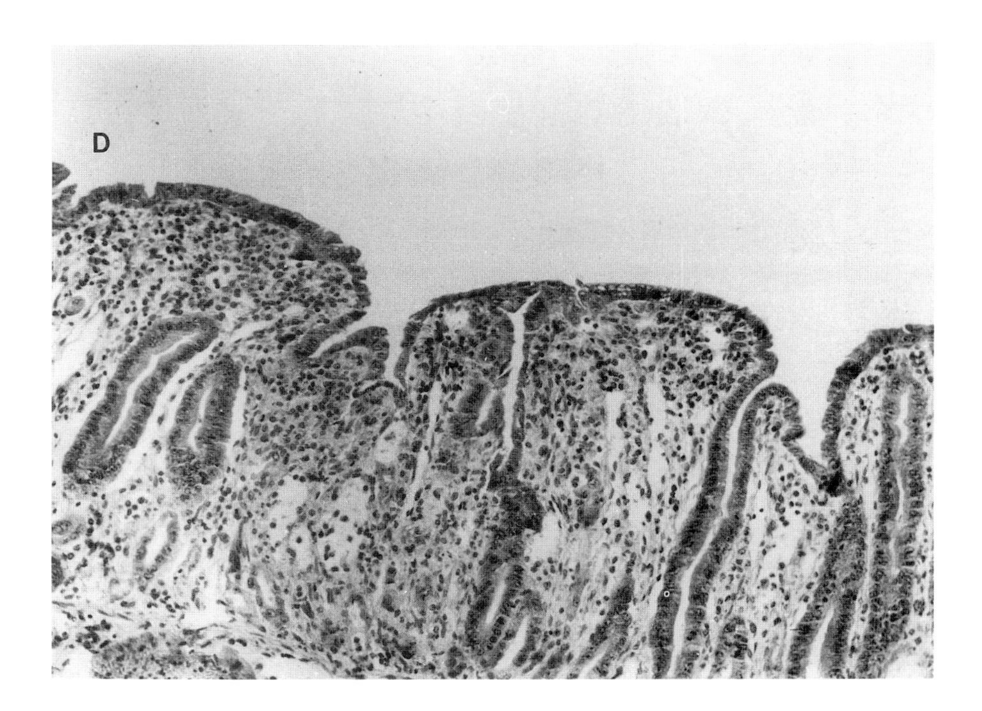

FIGURE 2 The progression of radiation injury in the rat jejunum at different times after 12 Gy abdominal irradiation from a telecobalt unit. The dose is high enough to produce 35% death for intestinal radiation syndrome. (A) Control mucosa: morphology of normal small intestine epithelium. (B) Twenty-four hours after irradiation. Epithelial cells lining the whole crypt and the bottom of the villus are very irregular, misaligned, have enlarged nuclei, and are reduced in number. The epithelial cells of the upper part of the villus are still regular in appearance. (C) Seventy-two hours after irradiation: Severe morphologic alterations are present in the entire epithelium; the villi have lost their individuality and the number of cells is markedly reduced. The vessels are dilated and the lamina propria is hemorrhagic. (D) One hundred and twenty hours after irradiation: The epithelium is flattened and severe morphologic alterations are still evident. Marked proliferative activity is observed in some crypts.

crypt are represented in Fig. 4. The behavior after 2 fractions of 3 Gy and 4 fractions of 3 Gy, both with a 12-hr interval between fractions, is compared with a 3- and 8-Gy single doses.

The other morphological and biochemical parameters show the same kind of difference. In general, when low LET radiations are used, the time between each fraction is particularly important in the manifestation of the effect.

When the interval is short (a few hours) the sequence described above is induced after a single dose. If the second fraction is administered during the phase of acute injury, this phase lasts longer because of lesion overlapping. If the fractions are separated by a few days, the initial damage is repaired and the second appears as a function of the size of the dose of the second fraction. The ultrastructural modifications, the determination of brush border enzyme activity, and the lysosomal enzyme activity confirm the observations made after a single dose.

The morphological and biochemical–functional alterations produced by irradiations in a single or fractionated dose lead to a reduction in the weight of the intestine, a highly radiosensitive tissue, which reaches a minimum in laboratory animals after 2–3 days. Body weight also decreases but to a much lesser degree. During the acute phase of injury there is a notable decrease in DNA content while protein synthesis and protein content in the small intestine decrease to a lesser degree and more slowly. It should be observed that the longer the interval between fractions, or the less the dose per fraction, the less the effect due to the greater efficacy of the repair and repopulation mechanisms.

C. Clinical Studies

Doses received due to accidental exposure are only rarely high enough to produce the gastrointestinal syndrome. However, more or less extensive intestinal irradiation is quite frequent in treatment of gynecological tumors, testicular seminomas, etc.,

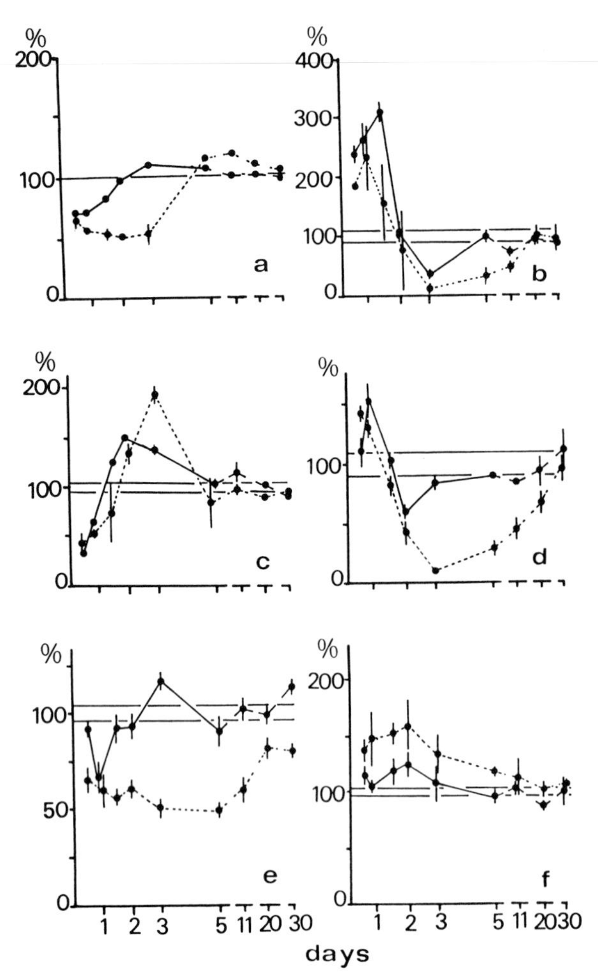

FIGURE 3 Effects of two different sublethal doses on some morphological and biochemical parameters of the small intestine. Rats were exposed to a 3-Gy (continuous line) or 8-Gy (dotted line) dose with a γ-ray source (60 Co) and sacrificed at different time intervals (on the abscissa) after irradiation. The animals were caged under a constant light/darkness cycle (6:30–18:30). Each point represents the mean value ±SEM of every parameter obtained from five or six rats and the values are expressed as a percentage of controls sacrificed at the same time of the day. The first 3 parameters [(a) number of epithelial cells, (b) goblet cell index, and (c) labeling index], are related to at least 50 side well-aligned jejunal crypts; the other three [(d) brush border maltase activity, (e) DNA content, and (f) lysosomal β-glucuronidase activity], are referred to the whole small intestine. Three phases are well evident: a few hours after irradiation a reduction in cell number and proliferative activity is observed. During this period goblet cell index and maltase activity increase, supporting the hypothesis of an early differentiation process. The second phase, acute morphologic damage, is characterized by high proliferative activity in the crypts whereas brush border enzyme activity and goblet cell index are significantly lower than control value. During the last part of this phase repopulation in the whole epithelium occurs but cells of regular morphology appear not completely differentiated. In fact, brush border enzyme activities are significantly lower than controls and goblet cells are very few. The third phase is characterized by a return to a normal functional ability. Lysosomal enzyme activities increase during acute injury. The higher dose produces higher effects and a delayed return to normal conditions.

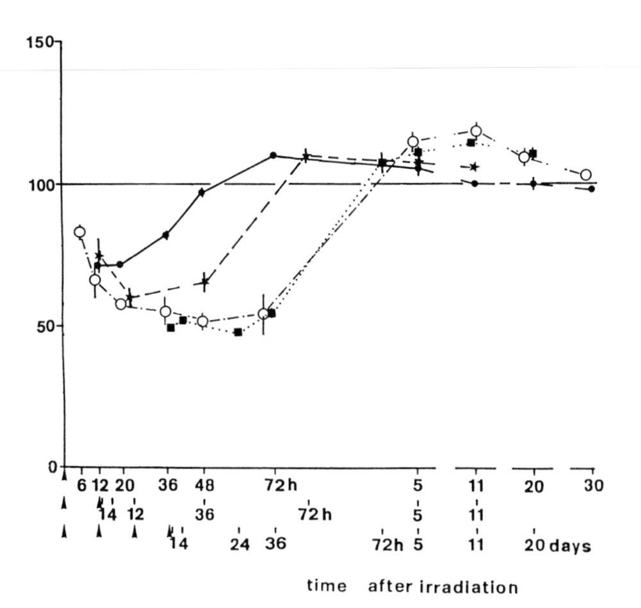

FIGURE 4 Effect of dose fractionation on the number of epithelial cells in the jejunal crypt. The values are represented as percentage ±SEM of control animals. The time of exposure at the 3-Gy fractions is indicated by the arrowheads. The continuous line represents the effect of a 3-Gy single dose; the dashed line represents those of 2 fractions of 3 Gy every 12 hr and the dotted line those of 4 fractions of 3 Gy every 12 hr. The dot–dash line represents the effect of an 8-Gy single dose. The behavior is similar to that observed in Fig. 3.

and especially in total body irradiation prior to bone marrow transplant.

Studies conducted between the end of the 1960s and the beginning of the 1970s pointed out that several substances were malabsorbed during the acute phase of damage. However, few studies were adequately performed because the pretreatment levels of the morphological and functional parameters were often abnormal and the tests were only sometimes repeated in the same patient after increasing doses.

The conventional treatment for gynecological or testicular tumors uses fractions of 2 Gy/day, 5 days/week for up to a maximum total dose of 45-55 Gy. The irradiated volume can include the abdomino-pelvic region, the lower hemiabdomen, or the epigastric region. In these patients nausea and vomiting generally appear after the first sessions. These symptoms vary greatly in severity and last about 1 week. They are more frequent and more intense when the upper half of the abdomen or the epigastric region have been irradiated. Diarrhea generally appears during the third week (halfway through treatment, 25–30 Gy), especially in women receiving lower abdomen irradiation. Drug administration limits symptom severity.

To evaluate intestinal conditions during treatment, a series of tests were performed. Morphological damage, evaluated by biopsy, shows the same modifications as observed in laboratory animals. When the biopsies are repeated during treatment the mitotic index progressively decreases until about 40 Gy. Then the values remain constant until the end of treatment. During this phase the cells in the crypt–villus structure show the previously described alterations and decrease in number. The mucosa returns to normal within the first months after the end of the treatment. Morphological modifications are also observed in the intestinal loops outside the irradiated field.

Radiological investigations with barium sulfate show motility modifications which are directly dependent on the exposed volume. When the region of the lower hemiabdomen is irradiated, acceleration of the transit is very frequently observed, whereas in other cases the transit time changes are modest.

Patients with diarrhea present the most striking alterations. An altered motility modifies the contact time between molecules present in the food and the mucosa, and this can therefore influence absorption.

Oral administration of molecules mixed with a test meal and their assay in blood, urine, and feces permits evaluation of changes in absorption during abdomen irradiation. A state of malabsorption is present in most patients, but this is highly dependent on individual susceptibility. Malabsorption is generally very noticeable halfway through treatment and can worsen when therapy has been completed. It is more serious and frequent in patients irradiated in the abdominopelvic region. Results may differ depending on the substances used in the absorption tests.

Carbohydrate absorption, measured with glucose, which uses active transport as its absorption mechanism, is significantly reduced halfway through and at the end of treatment. Saccharose, instead, must be previously hydrolyzed by the brush border enzymes and is more markedly reduced, whereas xylose, a portion of which is thought to be absorbed by passive transport, shows an evident reduction. The reduction at the end of treatment appears to be, respectively, 32% for glucose, 40% for saccharose, and 33% for xylose.

Reduction in absorption is much less and of a similar size when proteins and amino acids were administered orally (less than 10%). The reduction of fat absorption is evident but differs for oleic acid and triolein (12 vs 28%, respectively, at the end of treatment). Other substances show a marked malabsorption. At the end of the treatment the remaining absorption rate is improved and the reduction is only of 30%. Vitamin B_{12} absorption is reduced by 35 and 30%, respectively. There is loss of water and electrolytes but it does not produce important modifications since diet and suitable drugs can easily compensate. The modest variations observed in the plasma and blood volumes and in the total body water confirm this result.

Changes in biliary and pancreatic secretions are modest and generally return to the normal range of physiologic variations. With epigastric irradiation, a reduction of gastric acidity is observed. Bile salts play an important role in inducing diarrhea. These molecules are reabsorbed in the terminal ileum; irradiation of this segment, by provoking a reduced absorption, can be one cause of diarrhea.

An estimate of the intestinal damage by irradiation can be obtained by evaluating the daily quantity of feces. The initial quantity increases twofold during the second week of treatment and remains high until the end of therapy.

Studies of the intestinal microflora demonstrate that in irradiated patients there is a variation in the number of the preexisting microorganisms, rather than in the appearance of new species. Malabsorption increases molecule permanence in the intestinal lumen and therefore produces an osmotic effect. The molecules can be utilized by the microflora, with the possible production of toxic compounds.

In accidental exposure, the administration of water, electrolytes, and antibiotics increases the survival probability. Some studies have demonstrated that administration of an elemental liquid diet immediately before and during radiotherapy of the abdomen and pelvis leads to better alimentation and therefore facilitates absorption of elemental nutrients even during acute intestinal radiation syndrome. The diet must be milk free, gluten free, low fat, and low residue to limit the possibility of diarrhea.

V. Large Intestine and rectum

The main features are (1) the epithelium in the large intestine lacks villi; (2) the crypts decrease in length, going from the colon to the rectum; and (3) there are more goblet cells than in the small intestine. The cell cycle time is about 24 hr and the turnover is about 6 days; the lesser proliferative activity explains the greater tolerance of this segment. The main function of the large intestine is linked to the reabsorption of water and electrolytes and to the elimination of residues.

Radiation injury has the same features as in the small intestine; that is, cell death and inhibition of mitosis a few hours after irradiation. After several days these alterations involve the entire epithelium, and edema and vascular changes in the submucosal and serosal layers are observed. Hypermotility and tenesmus can be observed and the rectal mucosa appears hyperemic and edematous. A few weeks are necessary for repair. With elevated doses ulcerations throughout the intestine can be observed.

Between 2 and 12 months following the end of treatment subacute damage can be observed: the mucosa regenerates but alterations of the endothelial cells remain, the small arteries and capillary vessels are still swollen, and fibrin plugs can cause thrombosis. Lack of circulation is progressive and contemporarily there is the beginning of a chronic collagen degeneration.

The submucosa becomes thicker and fibrotic. These mechanisms can lead to the appearance of atrophy in the glandular mucosa with consequent malabsorption and the formation of ulcers. The progressive degeneration of the connective tissue can lead to obstruction of the canal. For elevated doses holes through the wall (fistulas) can form. Late intestinal injury can appear even more than 20 years after radiotherapy, but 5 years is considered to be the period of risk.

Radiosensitivity can be expressed on the basis of the *tolerance dose* (TD), that is, the dose administered according to a conventional fractionation schedule (2 Gy/day, 5 days/week) which produces the injury in a certain percentage of cases after a fixed time. The TD 5/5 is the dose that causes severe injury in 1–5% of patients 5 years after radiotherapy. The TD 50/5 is the dose that produces severe injury in 50% of patients in 5 years. The TD 5/5 and the TD 50/5 have been calculated in the alimentary canal by assessing the incidence of ulcers and strictures and are as follow: for esophagus, 60 and 75 Gy; for stomach, 45 and 50 Gy; for small intestine and colon, 45 and 65 Gy; and for rectum, 55 and 80 Gy. The small intestine is the most radiosensitive but significant injuries are usually avoided because of its mobility.

Evident alterations are observed during intracavity or external irradiation for carcinoma of the uterus, and they are particularly severe when adhesions are present. Late reactions or complications include inflammation of the rectum, perforation, and vesicovaginal or rectovaginal fistulas. Damage is much more severe when large doses are used for each session. With external radiotherapy of the pelvis for tumors of the cervix, uterus, or bladder, about 70% of the patients have strictures and more than 50% show inflammation of rectum and colon.

Rectal bleeding and modifications in bowel habits are observed when there are large intestine complications. Sometimes these symptoms are initially attributed to a recurrence because diagnosis is difficult owing to ulcers and necrotic areas. Surgical intervention is difficult in these patients because of the presence of serious fibrosis in the irradiated tissue which can induce necrosis and fistulas even several days after surgery. The presence of hardened vessels in the area near the irradiated field complicates the possibility of reestablishing normal conditions following surgery.

The rectosigmoid tract of the bowel is less mobile than the rest of the alimentary canal. Therefore, though it is more radioresistant than the small intestine, early and late alterations are similar to those in the other segments: that is, fibrosis, strictures, ulcerations, fistulas, and perforations. In tumors of the uterus and of the prostate gland the rectum is also irradiated, and injury occurs in this tissue as well as in the urethra and the bladder. Here, too, damage is more probable, after laparotomies and previous surgical procedures, due to immobilization of the intestinal loops. The results obtained demonstrate that treatment with ionizing radiations on areas including the intestines must be very carefully conducted. All data confirm the scarce tolerance of the bowel and that early damage cannot be considered a prognostic factor for late injury.

VI. Predisposing Factors in the Production of Radiation Injury

Besides total dose, time, irradiation modality, and irradiated volume, other risks factors are gender (risk is higher in women than in men), age, hypertension, vascular disease, pelvic chronic inflammation, leanness, and diabetes mellitus. Patients who have undergone previous surgery have adhesions;

for this reason the intestine is less mobile, and therefore during the treatment the possibility of repeated irradiation of the same intestinal loops is greater.

Abdominal irradiation of children and infants with tumors produces more severe alterations than in adults, even if lower doses (15–40 Gy) are used. Note that generally the irradiated volume is greater in children than in adults. In 70% of the cases immediate tolerance is poor and vomiting and diarrhea are present. Surgery and concomitant chemotherapy are also unfavorable factors in these cases. Few observations exist concerning chronic and late effects due to a short rate of survival.

VII. High Linear Energy Transfer Radiation

High linear energy transfer (LET) radiations (α particles, neutrons, protons) are available for radiotherapy only in a few centers. The relative biological effectiveness of neutrons is much greater than that of X rays, γ rays, and accelerated electrons, because they release large quantities of energy per unit track (μm) of the material they traverse. For equal physical doses absorbed, the higher density of interactions leads to much greater cell damage and to less possibility of repair. The effect of high LET radiations is scarcely influenced by the presence of oxygen or by cell parameters, properties that could make the utilization of this type of radiation valid in the therapy of tumors.

The quality of acute and late lesions induced in the intestinal epithelium is similar to those caused by low LET irradiations, but the injury is greater. In a high percentage of cases doses greater than 20 Gy in patients with advanced pelvic tumors lead to serious complications, including fibrosis, obstructions, necrosis, and bowel perforations. However, in about 50% of the cases, thought to be uncurable, complete tumor regression is obtained.

VIII. Radiation Carcinogenesis

The capacity of ionizing radiations to induce tumors is well documented and it is probable that a second tumor will eventually appear in patients undergoing radiotherapy as late injury. However, it is impossible to distinguish between a second spontaneous tumor and a radioinduced tumor. A second tumor in the irradiated volume can appear from 5 to 30 years after exposure, and this makes accurate controls even more necessary in the posttreatment period.

Usually the induced tumors are adenocarcinomas, and their incidence depends on the dose administered. Some studies have demonstrated that irradiation of the abdomen and of the pelvis for primary cervical or ovarian tumors can induce leukemia. The risk of acute leukemia is increased by a factor of 2.1 and that of acute nonlymphocytic leukemia by 4.0 or more. A lower incidence has been observed in women affected by tumors of the cervix or uterus compared to cases irradiated for benign gynecologic disorders: in the first case the dose to the bone marrow was higher and resulted in killing bone marrow cells rather than inducing their transformation.

IX. Conclusions

The modifications produced by ionizing radiations on the alimentary canal have been the object of many studies, but increased understanding of molecular and cellular biology makes continuous research updating necessary.

The intestine is a tissue that has always been widely used in radiobiological studies, and it is considered a valid model. In fact, its morphological and functional characteristics permit easy recognition of the proliferative compartment and, especially in the small intestine, of the differentiated compartment which is characterized by the synthesis of specific, easily assayed molecules. It is possible, therefore, to obtain accurate analysis of the effects of chemical or physical agents on different cellular activities.

Rapid cell turnover permits manifestation of damage within a brief time period, and the elevated proliferative activity explains the high tolerance of this tissue, which even after high doses of radiations has a rapid repopulation. The induction of an early differentiation process and the presence of cells with a regular morphology during recovery, but with incomplete acquisition of their functional capacities, will be an interesting base of study to understand the mechanisms that lead to cell damage and those that intervene in the passage from proliferative to differentiated cells.

Different modalities of dose administration and

association with other physical and chemical agents will contribute further to a better understanding of the modifications induced by antineoplastic treatments. The intestinal tract can also be useful for analyzing alterations that take place in the connective tissue and in the vasal endothelium, which are the principle causes of severe late injury.

The knowledge of the cellular mechanisms that lead to the radiation injury and to its recovery is expected to provide information for better effectiveness of therapy for human tumor, and/or to reduce the damage to healthy tissues.

Bibliography

Becciolini, A. (1987). Relative radiosensitivity of the small and large intestine. *Adv. Radiat. Biol.* **12,** 83–128.

Casarett, G. W. (1980). "Radiation histopathology." CRC Press, Boca Raton, Florida.

Fajardo, L. F. (1982). "Pathology of Radiation Injury." Masson, New York.

Quastler, H. (1956). The nature of intestinal death. *Radiat. Res.* **4,** 303–320.

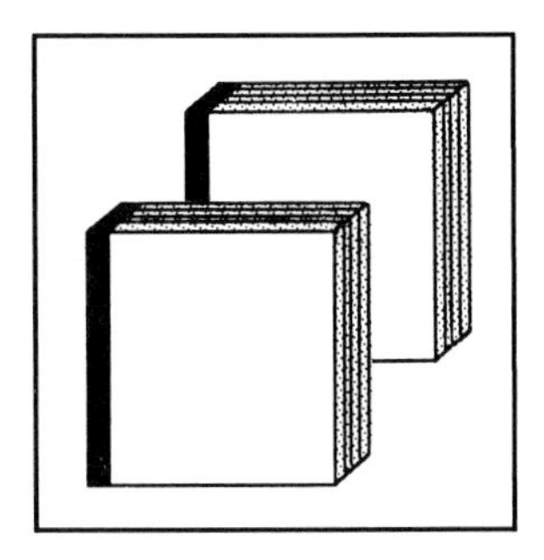

Reading Processes and Comprehension

KEITH RAYNER AND ALEXANDER POLLATSEK, *University of Massachusetts*

Glossary

Anaphora When a word in text refers to a person or thing mentioned earlier

Garden-path effects When the early part of a sentence leads a reader to expect one interpretation of the sentence, but later information indicates a different interpretation is correct

Inner speech Activity in the speech tract (subvocalization) and the inner voice during silent reading (phonological coding)

Phonological coding The inner voice during silent reading

Saccades Movements of the eyes during reading and other perceptual activities

Subvocalization Activity in the speech tract during silent reading

Visual acuity The ability to accurately discriminate letters or words when they are not in the center of vision

HOW DO WE READ? That is, what are the mental processes involved when we are actually engaged in reading and what do we remember later about what we read? These two questions have been the focus of a great deal of research by experimental and educational psychologists. The reasons why so much research on reading has been undertaken are clear. Nearly a hundred years ago, a pioneer in reading research, Edmund B. Huey, argued that to understand what the mind does during reading would be the "acme of a psychologist's achievements, since it would be to describe very many of the most intricate workings of the human mind, as well as to unravel the tangled story of the most remarkable specific performance that civilization has learned." In addition, it is clear that understanding the process of skilled reading is necessary in order to make generalizations about how reading should be taught to children and how remediation techniques should be developed to help those who do not read well.

In this article the focus will be first on the process of skilled silent reading and then on comprehension. The view that will be advocated is that the *products* of reading comprehension cannot be understood apart from the *processes* that occur during the act of reading.

I. The Process of Reading

There are three important components of the reading process which we will describe in relation to the basic processes that occur during reading. We will discuss (1) eye movements, (2) word recognition, and (3) inner speech. We will first describe some basic facts about each of these components and then discuss each of them in more detail.

When we read, it seems as though our eyes move smoothly across the page of text. However, this is largely an illusion since our eyes make a series of jumps (called saccades) that last about 20–35 msec. Between these saccades, our eyes are relatively still for periods of about 200–250 msec called fixations. Information is encoded from the text during the fixations, since the eyes are moving so quickly during the saccades as to preclude obtaining useful information. (We are unaware of the "blur" during the saccade, however.) The pattern of information extraction during reading is thus a bit like seeing a slide show. You see a "slide" for about a quarter of a second, there is a brief "off time," and then a new "slide" of a different view of the text appears for about a quarter of a second.

The second way in which our subjective impression is an illusion is that our eyes do not move forward as relentlessly as we think. While most of our saccades in reading move forward, about 10 to 15% move backward in the text and are termed regressive saccades (or regressions for short). Think of regressions this way: since we make about four to five saccades in a second, we make a regression about once every 2 sec. Thus, we are largely unaware of most regressions. While some regressions reflect major confusion requiring us to go back a considerable distance in the text to straighten things out, the majority are quite short, only going back a few letters.

The reason why eye movements are made so frequently in reading becomes obvious when we consider some basic facts about visual acuity. Relative to our fixation point, a line of text can be divided into three regions: foveal, parafoveal, and peripheral regions. The *fovea* (extending out from the fixation point about 1° of visual angle) is the area in the center of vision where the ability to discriminate fine detail is greatest. Acuity drops off markedly outside of the fovea. Thus, in parafoveal vision (which extends to about 5° of visual angle on either side of fixation), it is much more difficult to discriminate the letters making up words; in peripheral vision, it is virtually impossible. We therefore make frequent eye movements to bring the next words in the text that have not been processed into foveal vision for detailed processing. [*See* EYE MOVEMENTS.]

While the process by which words are identified is complex, it is very fast and automatic. By this we mean that skilled readers can recognize words in text very quickly. Our estimate is that the time needed to identify a word may be as little as 50 msec and is usually less than about 200 msec. Skilled readers identify words in isolation almost as quickly as when the words are preceded by a relevant context. In contrast, beginning and unskilled readers derive much more benefit from a relevant prose context than do skilled readers. The reason that skilled readers do not benefit much from context in identifying words is that word recognition is such a rapid and effortless process that they rely on it rather than context. Unskilled readers, on the other hand, often rely heavily on context and thus make many errors of word identification when reading.

When we read, we experience an inner voice saying the words that our eyes are falling on. The role of this inner speech in reading has been controversial, but its function is now becoming clearer. It is not absolutely essential to identifying words, although current research indicates that it plays a more important role than previously thought. However, inner speech appears to be critically important after a word has been identified as an aid for comprehending larger units of text. Again, there seems to be a difference between skilled and unskilled readers in the extent to which they are able to use sound codes as aids to comprehension. Having presented this brief sketch of the process of reading, we now turn to a slightly more detailed consideration of each of these three topics.

A. The Work of the Eyes

Figure 1 shows part of a page of text with a record of a reader's eye movements superimposed on the text. On this particular page for this reader, the average saccade length was 8.5 characters, but the range was from 1 character to 18 characters. The average fixation duration was 218 msec, but the range was 66 to 302 msec. Notice that, for the most part, words are fixated only once. However, the word *enough* is fixated twice while *pain* and *those* are not fixated at all. Since a fixation lands on or near almost all words, it is clear (as argued above) that the major purpose of eye movements is to bring all words close to the fovea. But what is causing the variability? Why are some words not fixated while others are fixated twice? Is this just miscalculation of the eye movement or does it reflect something deeper?

While there is clearly some noise associated with the motor programming of the saccadic eye movement system, much of the variability associated with fixation times and saccade lengths is due to processes involved in comprehending text. That is, the ease or difficulty of processing a fixated word strongly influences how long the reader looks at that word. This observation is most obvious from the fact that word frequency has a powerful influence on fixation time on a word; on the average, a low-frequency word is fixated for about 50 msec longer than a high-frequency word. Not only are fixation times sensitive to the difficulty of identifying indi-

```
Roadside joggers endure sweat, pain and angry drivers in the name of
   *        *       *      *               *     *   *        *
   1        2       3      4               5     6   7        8
  286      221     246    277             256   233 216      188

fitness.  A healthy body may seem reward enough for most people.  However,
    *      *  *       *       *      *     *  *      *     *        *
    9     10 12      11      13     14    15 16     17    18       19
   301   177 196    175     244    302   112 177    266   188      199

for all those who question the payoff, some recent research on physical
      *       *     *      *     *           *         *         *
     21      20    22     23    24          25        26        27
     216     212   179    109   266         245       188       205

activity and creativity has provided some surprisingly good news.
 *     *       *   *            *        *      *          *
29    28      30  31           32       33     34         35
201   66      201 188          203      220    217        288
```

FIGURE 1 An excerpt from a passage of text with fixation sequence and fixation durations indicated. The asterisk represents the fixation location.

vidual words, they are also sensitive to the various processes that take the meanings of individual words to form a more global understanding of the text (such as processes involved in constructing syntactic structures in individual sentences and processes involved in linking the meanings of individual sentences into a coherent overall mental structure). When a word (or a group of words) is difficult to process, the eyes remain fixated on the region for a longer period of time than when that region is easy to process.

While comprehension processes associated with understanding words, phrases, sentences, and larger units all influence how long the eyes remain fixated on each word, the primary influence on how far the eyes move is the length of the words in the text. Short words (three characters or less) are often not directly fixated by the eyes but processed on the fixation prior to which they are skipped. Most of the time, however, readers are trying to move their eyes to a place in the next word that will be maximally efficient for word identification processes.

This brings us to the interesting question of how much information we are able to utilize from each individual fixation. Phenomenologically, it seems as though we can see the entire line of text and identify words from a large region around our fixation point. However, this subjective impression again does not coincide with the facts. Remember that we only move our eyes about 1.5 words on average with each saccade. If we could identify many more words than that, why would we make so many eye fixations? The reason again has to do with decreased visual acuity in the parafoveal and peripheral regions.

A great deal of research has been done to determine the size of the perceptual span (or region of useful vision) during an eye fixation. The best technique for this purpose has been one which limits the amount of information that is available to the reader on each fixation. In this moving window paradigm, readers' eye movements are monitored by a very accurate eye tracking system that is interfaced with a computer, which in turn is interfaced with a cathode ray tube (CRT) on which text is presented. Wherever the reader looks, normal text is exposed while outside of the window region containing text the letters are all mutilated in some way. The window either coincides with word boundaries or is determined by a set number of letters. Figure 2 shows examples of both kinds of windows. The basic logic of the method is as follows: if the presence of a window of a certain size has no effect on reading (i.e., reading comprehension and reading speed are undiminished), then we can conclude that readers are extracting no useful information outside this window region. Variations on this technique employ more local display changes that enable more precise conclusions to be drawn about the nature of visual information extraction during reading (see the bottom of Fig. 2 for examples).

The central finding of this research is that the

```
Roadside joggers endure sweat, pain, and angry drivers (NORMAL TEXT)
           *
_____________________________________________________

XXXXXide joggers eXXXXXXXXXXXXXXXXXXXXXXXXXXXXXXXXXXX   (13-LETTER WINDOW)
           *
XXXXXXXXXXXXXers endure swXXXXXXXXXXXXXXXXXXXXXXXXXXX
                   *
_____________________________________________________

XXXXXXXX joggers endure XXXXXXXXXXXXXXXXXXXXXXXXXXXXX   (2-WORD WINDOW)
           *
XXXXXXXXXXXXXXXX endure sweat, XXXXXXXXXXXXXXXXXXXXXX
                   *
_____________________________________________________

Roadside joggers cxlosr sweat, pain, and angry drivers (BEFORE BOUNDARY)
           *   |

Roadside joggers endure sweat, pain, and angry drivers (AFTER BOUNDARY)
               |   *
_____________________________________________________
```

FIGURE 2 Examples of the moving window paradigm. The first line shows normal text (with fixation location indicated by the asterisk). The next two lines show two successive fixations in the moving window situation with a 13-letter window. The next two lines show two successive fixations with the window defined by word boundaries. The final two lines show a variation of the technique. When the reader is fixated on the word *joggers* a nonword occupies the next position. However, when the reader's eye crosses over an invisible boundary location, the nonword is replaced by the word *endure*. Because the change takes place during the saccade, readers do not see the change.

perceptual span is small: readers obtain information from a region extending 3–4 letter spaces to the left of fixation (but no more than the beginning of the currently fixated word) to about 15 character spaces to the right of the current fixation point. That is, they appear to be processing the fixated word, the word to the right, and perhaps another word if the words are all small. Note that the perceptual span is asymmetric. This asymmetry turns out to be related to attentional factors and the direction the eyes move, since the span is asymmetric to the left of fixation for readers of Hebrew (which goes from right to left across the line rather than from left to right as is the case with English). Other research indicates that certain kinds of information can be extracted further from fixation than others. For example, it appears that information about the location of word boundaries can be extracted out to 15 characters while letter information is extracted from a somewhat smaller region.

B. Word Identification

Most of what we know about how words are identified comes from experiments in which skilled readers process isolated words (rather than words embedded in coherent discourse), although there is a resurgence of research with children and in trying to understand the influence of textual context on word identification. Clearly, the attraction of studying skilled readers and isolated words is that the experiments are simpler and the data are less variable. Fortunately, the newer research indicates that most of the findings in the more controlled settings generalize well to more natural situations. The discussion that follows will rely on experiments using words in isolation, although the story is pretty much the same with words in coherent discourse.

The psychological literature has not made much progress on the ultimate question of pattern recognition: how the pattern of excitation on the retina of the eye is classified into meaningful patterns. In particular, we still have only the vaguest idea of how readers in fact recognize the letters of the alphabet. Instead, psychologists have made most headway with questions about how processes at various levels relate to each other. We first consider the relation of letters to words, and then move to questions of how encoding of the sound of a word relates to encoding its meaning.

One proposal for how letter identification relates to word identification is that words are identified by a serial scan of the component letters (probably from left to right). Several lines of research indicate that letters are not serially scanned at even a high rate. One of the most important methods employs words presented briefly enough so that there are errors in identification. In these experiments, when subjects are asked to identify a preselected letter (e.g., the last letter in *WORK*), they are more accurate in identifying it than in identifying a letter in isolation (e.g., a *K* in the same location as the *K* in *WORK*). This finding has been dubbed the word superiority effect. The word superiority effect is inconsistent with a serial scanning hypothesis, since several letters, no matter how meaningful, would take longer to process than a single letter.

A second theory is that words are unified visual patterns or "templates" and thus that word identification in skilled readers is a process that occurs independently of identification of the component letters. While this theory seems unlikely—since the logic of alphabetic languages is that words are made up of component letters—proponents argue that any attempt to use the component letters to recognize a word slows down word identification. There are several findings that rule out the word as template theory. The first is that letters in "pseudowords" (i.e., pronounceable nonwords) such as *MARD* are identified better than letters in isolation and about as well as letters in words. The second is that text (or isolated words) written in aLtErNaTiNg CaSe LiKe ThIs is just about as easy to read as normal text if you have even minimal practice in reading it. Thus, in either case, there is no striking difference between whole visual patterns actually experienced and those being seen for the first time.

The bulk of the word identification research indicates that words are identified through parallel processing of the component letters. That is, it appears that visual information about letters in words (and nonwords) excites component letter detectors in parallel, which in turn feed excitation and inhibition into a set of word detectors (e.g., *W* in the first position excites the "word," "work," "wood," and other detectors, while inhibiting the "cork," "cord," "care," and other detectors). The word detectors also feed back excitation and inhibition to component letters (e.g., the "work" detector excites the "w" detector in the first position but inhibits the "b" detector). When sufficient visual information has accrued, only the correct word and letters will be excited because all the others will be inhibited, but with insufficient visual information, errors of identification may occur. In fact, several versions of such a model have accounted well for a wide variety of data on word identification (including the word superiority and "pseudoword superiority" effects described above). Thus, such a model appears to be a good working hypothesis if not revealed truth.

In most current theories, word detectors are viewed as being entries in a "lexicon" or mental dictionary. Other information, such as the pronunciation and meaning of a word, are accessed from this lexical entry (pretty much the way you would find them in a standard dictionary). Of interest is whether the information flow is this simple in skilled reading (i.e., the lexical entry for the visual word is accessed which in turn independently accesses the sound and meaning of the word). An alternative theory (the *dual access* theory) is that there are two independent pathways to meaning. One is the *direct* lexical access of meaning sketched above, with the second being an *indirect* route whereby the letters activate a sound code through some constructive process (possibly using the "spelling rules" of English), and this sound code in turn is used to help activate the meaning of the word. (The latter process is presumably much the same one as when a spoken word accesses the lexicon.)

There are several lines of support for this dual access theory. One of the most important comes from experiments with people with brain damage in which either one or the other route seems to be selectively impaired. Thus, when asked to pronounce words or pseudowords, one group (called surface dyslexics) can pronounce both pseudowords and words, but tend to regularize irregular words (e.g., pronounce *ISLAND* as /iz-land/). On the other hand, phonemic dyslexics cannot pronounce pseudowords but can pronounce most words (both regular and irregular) correctly. Thus, it appears that the surface dyslexics have an impaired direct route, while the phonemic dyslexics have an impaired indirect route. One of the more dramatic demonstrations of indirect access to meaning in normal readers is that words or pseudowords that are homophones of another word are often mistaken for the real thing. For example, under time pressure, subjects will falsely classify *MEET* as a type of food or *SUTE* as an article of clothing much more of the time than visually similar controls such as *MELT* and *SUTH*. Logically, of course, the indirect route cannot be the only access to meaning, since we can tell homophones apart.

Originally, the indirect route had been thought to work through abstract spelling-to-sound "rules." However, in languages like English with fairly complex and chaotic spelling, it is quite hard to think of rules that would work for a majority of words and how these rules could be employed to help access meaning within a quarter of a second or less. Accordingly, the current view is that the sound code in the indirect route is "constructed" from stored sound representations. For example, *MARD* may activate sound representations of words that are visually similar (such as /hard/, /mark/, /mare/, /ward/, and /card/) and the sound code of *MARD* is constructed from the sound codes of its "visual neighbors." (A similar process would also work for words.)

C. Words in Context

We argued above that identification of words is not that different for skilled readers when they are in context than when they are out of context. There are two primary arenas where context would be expected to influence word identification. First, if a word is predictable from the prior text, one might expect the identification of words to be facilitated. Second, prior context would be expected to disambiguate words such as *bank* that have more than one meaning. We will discuss these in turn.

The evidence is clear that a word is identified more quickly when prior prose context makes a word highly predictable. When the predictable word is fixated (which for nouns, verbs, and adjectives would occur more than half the time), fixation time on the word is speeded only by about 20–30 msec. However, a highly predictable word is more often skipped over by the eyes than a less predictable one, affording significantly greater reductions in processing time. A caveat is in order here, however. Just because a word is not fixated by the eyes does not mean that the reader guessed the word in the absence of visual information. While such guessing may occasionally happen, the available evidence is that the reader is identifying the skipped word on the basis of both the visual information in the parafovea and the prior context. It makes sense that context has a greater effect on word identification in the parafovea than on words actually fixated since the quality of the visual information in the parafovea is poorer and thus in more need of help from context.

Readers make use of prior context to disambiguate words. Thus, when you read "After spreading some over his garden, he put the remaining straw in the box.", you undoubtedly conclude that *straw* refers to dried grass rather than to a drinking implement and are probably unaware of any conscious decision about which meaning of *straw* was intended. However, some very clever research indicates that even in some of these circumstances where prior context makes totally clear which meaning is intended, both meanings of the word are in fact accessed. However, the unintended meaning is inhibited within about 200 msec, explaining why you are not conscious of having accessed the "wrong" meaning as well as the "right" one. A similar finding occurs with ambiguous words such as *boxer* where one meaning is dominant (i.e., much more frequent than the other). In these cases, it is often the case that both meanings are initially accessed but the less dominant meaning is quickly suppressed.

Thus, a lot of the resolution of lexical ambiguity goes on below the level of consciousness and in fact does not appear to disrupt the reading process. However, when there is conflict—either when neither meaning is dominant and prior context is neutral or when one meaning is dominant but prior context indicates the other meaning was intended—reading is disrupted (and the reader is probably aware of the ambiguity).

D. Inner Speech in Reading

We have already noted that when we read silently, we hear our voice saying the words that our eyes are going over. Sometimes, we are even able to hear the voice of someone we know well when we read something (such as a letter) that they have written to us. Why do we hear this voice? It has frequently been argued that we learn to read orally before we read silently, and that the inner voice we hear is the result of bad habits formed in oral reading. Thus, according to this point of view, inner speech is simply a drag on reading and primarily serves to slow us down. This issue will come up again when we discuss "speedreading." For now, we will simply assert that there is more to inner speech than this view suggests. Indeed, the consensus view is that inner speech is centrally involved in the process of comprehending text.

Actually, there are two forms of inner speech. One is subvocalization (covert activity in the speech tract). It is clear that subvocalization generally occurs during silent reading, since when surface electrodes are placed on any part of your speech tract during silent reading, there is greater activity than when you sit quietly. However, there is a second form of inner speech, an inner voice in your head (sometimes referred to as phonological coding). You can do a simple experiment that shows that the two phenomena are not the same, and that the inner voice can be quite independent of the speech apparatus. Try reading while at the same time saying aloud the letters of the alphabet. After a bit of practice, you will be able to do both tasks at the same time and you will be able to hear your inner voice saying the words while your mouth is also saying the letters.

While it is agreed that subvocalization generally occurs during silent reading, there has been a good deal of controversy about whether it is necessary. The data come from experiments where subvocalization is suppressed either by a biofeedback technique or by forcing the reader to do irrelevant vocalizations as in our example above. The basic findings are that if the text is easy, suppressing subvocalization has little effect on comprehension, while if the text is difficult, subjects are unable to comprehend and remember much of what they read. Thus, it appears that subvocalization, while not absolutely necessary for comprehending text, plays a part when things get difficult.

The demonstration we suggested above indicates that phonological coding may occur when subvocalization is suppressed and may even play a more central role in reading. One piece of evidence for this is that tongue twisters and similar material that is difficult to say are also difficult to read. Since the reader cannot turn off phonological coding even when it is interfering, it appears that phonological coding during silent reading is not optional. Indeed, several lines of evidence indicate that phonological coding plays an important function in keeping material that you have just read in short-term memory, so that material earlier in a sentence can be more easily integrated with material later in a sentence or with material in succeeding sentences. Another role that inner speech may play in silent reading is to add stress and prosodic cues to the written message to aid in understanding.

The exact relation between phonological coding and subvocalization is still not clear. Several lines of evidence indicate that suppression of subvocalization has little effect on the formation of phonological codes (that is, the look-up of the sound of a word by either the direct or indirect route). In contrast, suppression of subvocalization appears to interfere with (but not eliminate) maintenance of these phonological codes in short-term memory and thus to interfere with comprehension processes that depend on integration of current material with what has just been read (these processes are discussed in the next sections). Since difficult text usually requires more reliance on such comprehension processes, it makes sense that suppression of subvocalization interferes more with comprehension of more difficult text.

In the remainder of this article, when we talk about inner speech, we will be referring to phonological coding and not to activity in the speech musculature. We should make clear, however, that phonological coding is not always an inner voice of which you are conscious. That is, when you introspect while reading, you are bringing the inner voice to consciousness. However, the production and maintenance of phonological codes need not be conscious and probably often is not during silent reading. In fact, it is likely to be this bringing of phonological codes to consciousness, rather than phonological coding per se, that slows down the reading process.

II. Comprehension Processes

The primary goal when we read is to be able to understand the information (or message) that the author is trying to communicate. Earlier, we made a distinction between the mental process of comprehension when reading and the later memory product of that comprehension process. While we believe that the product of reading comprehension is very important, we will focus primarily on the process because the product is not unique to reading. That is, when we listen to speech we also try to comprehend and store away in memory what we are listening to. Likewise, when we watch television or a movie, we store away in memory certain aspects of the plot, the main characters, and generally certain details that seem important in one way or another.

Is there any reason to think that comprehension processes and the product of our mental activity associated with reading, listening, and watching television differ? At a certain level, we suspect that comprehension processes for these (and other) activities are quite similar. But the way in which the information is initially processed is somewhat different. For example, consider the case of listening to a lecture. While the sounds go by fairly rapidly (although not as rapidly as we read; the rate of most speech is around 200 words/min, whereas most skilled readers read at rates around 300 words/min), they occur sequentially. With reading, on the other hand, the letters in the words are processed in parallel and we sometimes identify more than one word during an eye fixation. In addition, with listening, once the words have gone by, we do not have the opportunity to have them replayed if we did not understand something (unless, of course, we have a tape recorder or we ask the lecturer to repeat what he or she just said). In reading, on the other hand, we can always look back at something in the text that we did not understand. Also, because speech is slower than reading, we may rely more heavily on contextual information to anticipate what will come next than we do in reading.

The point that we have been trying to make is that the form in which text or discourse is initially encoded in listening and reading is quite different and this fact may have implications for how comprehension processes work in the two situations. However, it is undoubtedly the case that the processes become quite similar once we get to the level where we know the meanings of the words and sentences that we listening to or reading. In this section, we will discuss a few other places where there appear to be differences between the exact characteristics of what is remembered during listening and reading. Though the general comprehension product in the two situations is quite similar, people generally have somewhat better comprehension of material from reading than listening.

As an end to this preamble, we wish to stress again that we will focus primarily on the process of reading comprehension, although we will also discuss the product of reading comprehension. Again, the reason for so doing is that the study of the product of reading comprehension is really the study of memory processes and such processes are largely beyond the scope of this article. In the remainder of this section, we will discuss (1) sentence parsing, (2) what we remember from what we read, and (3) specific comprehension processes.

A. Sentence Parsing

In order to understand text, we must parse different words into their appropriate syntactic constituents or units. For example, consider the sentence "The horse raced past the barn fell." Most readers have immense difficulty understanding this sentence when it is first presented to them. Your initial reaction might be that the sentence does not make sense and that it is ungrammatical. In reality, the sentence does make sense and it is not ungrammatical. The sentence can be understood in the context of two horses, one of which is trotting along the trail and the other is being raced past a barn. Thus, the translation of this very difficult-to-comprehend sentence is really something like "The horse which was raced past the barn fell down"—it is the horse that is falling down and not the barn. People have trouble with this type of sentence construction (called a *reduced relative* sentence) because they want to parse "the barn" as the direct object of the sentence and do not know what to do with "fell," which seems to be attached to "barn."

Consider some less extreme examples. The sentence "Mary knew the answer was correct." is generally read more slowly than the sentence "Mary knew the answer to the question.", and the sentence "While Mary was mending the sock fell off her lap." generally causes readers to pause for a long time when they encounter the word *fell*. Readers are able to recover and make the appropriate analysis by consulting the inner speech sound track or by moving their eyes back to where they went wrong in the sentence and then starting over. (Often the recovery process is fluent enough that they are unaware of having had any difficulty.)

In all of these examples, there is the same tendency to want to parse the sentence in one way, only to find that the parse is wrong and recomputation is needed. For the two latter examples that we have presented, you may be thinking that normal text is generally better constructed than these examples so that the overt complementizer *that* would be inserted in the Mary example ("Mary knew that the answer was correct") or a comma would be inserted after *mending* in the second example to prevent you from being "led down the garden path." However, both the overt complementizer and the comma are optional in these examples (that is, they are not required by any laws of grammar) and they often are not present in text. One particularly obvious place where we often have trouble parsing sentences is with newspaper headlines that are designed to be short, but which often have double meanings that are hard to discern.

Something else that you may be thinking is that surely we would never be "led down the garden path" in normal text because the context would always make it clear how we should parse the sentence. Somewhat surprisingly, it seems to be the case that context does not always help us out and we can still temporarily be "garden pathed" even when the context is trying to help us. Such findings have led to the interesting observation that there may be special language processing modules that do not always interact with each other as well as they should (perhaps due to processing or capacity limitations).

B. What Do We Remember from What We Read?

When we read, our primary goal is to comprehend and remember the information from the text. It is fairly clear from a number of experiments examining the *product* of comprehension processing that what we remember about the text is primarily the *gist* of what we have read. That is, for the most part we remember the important points, the main ideas the author is trying to get across, and some of the important details. However, it is also the case that we quickly lose the surface form in which the information was presented. We generally do not remember the exact typeface in which the text was presented and the syntax is quickly lost.

This latter point is nicely demonstrated in an experiment in which subjects listened to short passages of text and were then tested for their recognition memory for the specific forms in which sentences had originally been presented. So, for example, in a story about Lippershey (the man who invented the telescope), the original text included the sentence: "He sent a letter about it to Galileo, the great Italian scientist." At various intervals after that sentence had been presented, subjects had to judge whether or not a test sentence such as "He sent Galileo, the great Italian scientist, a letter about it." was identical to the first sentence. Unless the test was immediately after the target sentence, subjects were as likely to respond "yes" to the altered sentence as to the original one that they actually had heard. While their ability to detect changes from active to passive voice (or vice versa) and changes in wording (i.e., substitution of a synonym) was little better than chance, subjects detected a meaning change (e.g., reversal of subject and object) a high proportion of the time.

The study that we have just described was originally done with listening rather than reading. Subsequent experiments in which people read the text (rather than listened to it) revealed essentially the same results. However, readers do tend to remember more of the surface form of the text than do listeners. Indeed, if you are asked to read text presented in unusual formats (e.g., in mirror-image form or reversed so that the words and letters are presented going from right to left), you will have a surprisingly good memory for whether you read a particular sentence in one of these unusual formats (as opposed to normal text). In addition, readers seem to have a reasonably good spatial memory of text they have read. If you were tested on the material in this text and asked where on the page certain information was presented, you will do surprisingly well on such a test. And when you misinterpret a sentence in the text because it is syntactically ambiguous, your eyes tend to be quite accurate in moving back to the place in the text where you started to go wrong.

While readers do have some memory for the surface form of the text, the major finding that still emerges is that readers primarily tend to remember the gist of what they read.

C. Text Understanding

The primary activities involved in understanding printed text consist of understanding (1) individual words, (2) sentences, and (3) the higher level representations beyond individual words and sentences. If readers do not know the meanings of the words making up the sentence, they will obviously not understand the text. Likewise, if they cannot correctly parse the words in the sentences into their appropriate constituent units, such as phrases and clauses, they will have problems understanding the meaning of the text. Since we have already discussed word processing and sentence processing, we will not dwell on those activities here. Rather, we will focus on comprehension as reflected in the higher level representations that readers arrive at in processing text. Largely, what we will focus on here is the extent to which readers must make constructions upon the text in order to comprehend its meaning.

Perhaps the simplest construction process is the comprehension of *anaphora*, where a word in the text refers to a person or thing mentioned earlier. While anaphora often involves the use of pronouns, earlier concepts are also referred to by nouns, either by repetition or by using a synonym. A deep question about text comprehension is how linking of a noun or a pronoun to its antecedent is accomplished. For pronouns, the usual problem is to decide which antecedent noun the pronoun is referring to, while for nouns, an important problem is deciding whether there is an antecedent or not. For example, if you read "A bus came roaring round the corner. A pedestrian was nearly killed by the vehicle.", *vehicle* refers to *bus* in the earlier sentence, while if the second sentence is replaced by "It nearly hit a horse-drawn vehicle.", *vehicle* does not refer to *bus*.

Solving the problem of whether there is an antecedent in the text for a word often involves a search of memory for possible antecedents and an evaluation of whether what is retrieved is in fact an antecedent (although these processes need not be conscious). Obviously, the interval between the anaphor and the antecedent influences the chances that the search for the antecedent will be successful (since we are careful not to use pronouns to refer to antecedents far back in the text). Recent research indicates, however, that the distance between the anaphor and the antecedent is not the important factor; instead, the most important factor determining ease of search is whether the antecedent is the current topic or focus of discourse, which it often is if it has been mentioned recently.

The "vehicle" example above indicates that the determination of whether there is an antecedent for a word is not the result of applying simple grammatical rules, but rather of understanding the meaning of the text. It appears that when the concept itself is clearly different (e.g., "horse-drawn vehicle" is not likely a bus), no search for antecedents is undertaken, but when only pragmatic information is needed to decide that a word has no antecedent, search is required (as when the second sentence is "It nearly smashed a vehicle.").

A whole host of cues is used in guiding this linkage process, including those mentioned above, standard cues (such as gender and number), and more subtle cues relating to discourse conventions (such as parallelism of construction). Current research has focused on the question of how and when these various sources of information are used, employing eye movements and other techniques for studying "on-line" processes. Not surprisingly, the clearer cues, such as using gender to assign the appropriate antecedent to a pronoun, may sometimes actually be completed before the eye has actually left the pronoun. In most cases, however, the process of assigning the appropriate antecedent continues while the rest of the sentence is being read.

Assigning antecedents to anaphors is probably the simplest linking problem facing the reader. Many connections must be constructed through a variety of inferential processes. For example, in the pair of sentences, "The boy cleared the snow from the stairs. The shovel was heavy.", the reader must make the inference that the shovel was the instrument used for clearing the stairs. A major question regarding such inferences is when they are made. Earlier research tended to suggest that readers typically made such inferences at the earliest possible moment (e.g., that the inference would be made about the time of reading "cleared the snow from the stairs"). However, the bulk of the findings indicate that readers tend to make such inferences reluctantly and generally make them only when they need to. One piece of evidence is that in sentences like the above, the reader is generally slowed down when reading the word "shovel" relative to when it was explicitly mentioned in the first sentence (indicating that at least part of the inferential process waits until then). Similar findings have been obtained for other types of inferences, such as causal inferences.

The ultimate question about the comprehension process is what constitutes understanding of the text; that is, what mental structure is created from the words on the printed page. A complete answer to that question would clearly necessitate almost a complete understanding of the human mind, as the earlier quote from Huey indicated. Current research on this topic as been involved with trying to characterize what the units of such a structure would be and some general principles for how they are organized. In many of the current models, the units of meanings are "propositions," which are linked pairs or triples of *arguments* (such as subject–verb–object arguments or modifier noun pairs), together with rules for linking these propositions together through overlapping arguments (i.e., the anaphoric and inferential links we discussed earlier).

These theories have had some success in predicting aspects of reading. For example, the difficulty of reading text can be predicted by the number of propositions or the structure needed to join the propositions together. Moreover, what is remembered can be predicted to some extent by such analyses. However, there is fairly universal consensus that such analyses can be only a small part of the story. First of all, our knowledge of the text cannot be captured merely by achieving links between abstract "arguments," but must be captured by a richer system that includes our knowledge of the world: aspects such as visual and other sensory representations, knowledge of causation, and knowledge about motivation of actors. At present, research has just begun to scratch the surface of understanding these aspects of reading.

III. Speedreading

One of the most controversial topics related to reading processes and comprehension is the issue of speedreading. Speedreading proponents, now usually represented by some type of profit-making concern, have made some remarkable claims. Let us examine these claims and then discuss what the evidence suggests.

Speedreading proponents claim that you should be able to increase your reading speed from about 300 words/min (the average rate for skilled adult readers) to 2000 words/min or even faster. There have been speedreaders who purport to have rates in excess of 10,000 words/min. One of the major claims of speedreading proponents is the idea that the brain is rather lazy and only effectively processes a small proportion of what it is capable of doing. In particular, speedreading proponents argue that reading speed can be increased by taking in more information per eye fixation and by eliminating inner speech, which is seen as a drag on reading speed. By processing more on a fixation, the brain is presumably forced to operate closer to maximum capacity, and supposedly there will be no loss of comprehension. In order to process more per fixation, speedreaders are often taught to move a finger rapidly across a line. The eyes are supposed to keep up with the finger so that the speed of visual processing is increased. Eventually, the finger is used as a pointer as it zigzags down the left-hand side of the page (and up the right hand-side so that the reader is actually taking in information in a different order than the author intended!) with the reader presumably taking in large chunks of information. The central claim of speedreading proponents is thus that reading speed can be dramatically increased without any loss of comprehension.

While the claims of speedreading proponents are certainly interesting and provocative, the evidence suggests that the claims are not valid. First, readers are not able to process entire lines of text at once. Indeed, as we argued earlier, the perceptual span is relatively small, and fast readers do not have much larger spans than slower readers. Second, the evidence is pretty clear that if inner speech is eliminated or interfered with, comprehension suffers.

When speedreaders are carefully studied some interesting results emerge. Speedreaders are about as good as normal skilled readers at answering general comprehension questions or questions about the gist of a passage. However, if tested on specific or detailed information, speedreaders cannot answer the question correctly unless they had fixated on the region where the answer was located. In addition, there is evidence that many of the general comprehension questions used in these tests can be answered even without reading the passages.

In general, the results of research on speedreaders' eye movements suggest that they are skimming the text and not really reading it in the sense of reading each word. They are probably doing a lot of filling in on the basis of what they already know about the topic being read or what they can surmise from those portions of the text that they have actually read. It is also interesting to note that when normal, skilled readers are asked to skim text, both their eye movement patterns and their comprehension performance are similar to that of so-called speedreaders.

IV. Beginning Reading and Dyslexia

Our primary goal in this chapter has been to describe skilled reading. However, at times we did mention dyslexic readers and children beginning the reading process. It is clear that skilled reading differs from beginning reading and what is a relatively automatic process in skilled readers is often a laborious and effortful process in beginning readers. As beginning readers become more proficient, many components of reading skill develop: for example, more complex syntactic forms can be understood, there is a greater knowledge base, and the capacity for making inferences about causation and motivation improves. However, the major skill that beginning readers must acquire is the ability to decode and comprehend individual words. Thus, the primary breakthrough for beginning readers comes when they realize that there is a systematic relationship between printed letters and sounds and that they can apply analytic strategies in determining how to decode new and unfamiliar words. Instruction which points out the alphabetic principle (that there is a relationship between letters and sounds) aids beginning readers in acquiring this skill. By the time children are in fourth grade, they have acquired the basic mechanics of reading skill; they differ from skilled readers in that they do not have nearly as extensive vocabularies or world knowledge to bring to bear on comprehending text.

Though some types of developmental dyslexia (problems associated with learning to read) can be traced to perceptual problems, the major difficulty seems to be a language processing deficit. In fact, for most developmental dyslexics the major difficulty is at the word decoding level. Likewise, in acquired dyslexia (inability to read due to brain damage), formerly proficient readers have lost the ability to do so and much of the problem seems to reside at the word processing level.

V. Summary

In this article, we have discussed the basic processes involved in reading and how readers comprehend text. As progress is made in understanding reading, it is clear that there are many complex processing activities that are finely orchestrated as readers process and comprehend text. It is also clear that in order to understand reading, we need to know about the basic processes involved and how each contributes to that whole complex we call "understanding." Indeed, as Huey pointed out nearly 100 years ago, to understand what we do when we read is tantamount to undertanding the processing of the mind.

Bibliography

Balota, D. A., Flores d'Arcais, G. B., and Rayner, K. (1990). "Comprehension Processes in Reading." Erlbaum, Hillsdale, New Jersey.

Besner, D., Waller, T. G., and MacKinnon, G. E. (1985). "Reading Research: Advances in Theory and Practice," Vol. 5. Academic Press, San Diego.

Daneman, M., MacKinnon, G. E., and Waller, T. G. (1988). "Reading Research: Advances in Theory and Practice," Vol. 6. Academic Press, San Diego.

Huey, E. B. (1908). "The Psychology and Pedagogy of Reading." McMillan, New York. (Republished by MIT Press, Cambridge, Massachusetts, 1968.)

Just, M. A., and Carpenter, P. A. (1987). "The Psychology of Reading and Language Comprehension." Allyn and Bacon, Boston, Massachusetts.

Perfetti, C. A. (1985). "Reading Ability." Oxford Press, New York.

Rayner, K., and Pollatsek, A. (1989). "The Psychology of Reading." Prentice Hall, Englewood Cliffs, New Jersey.

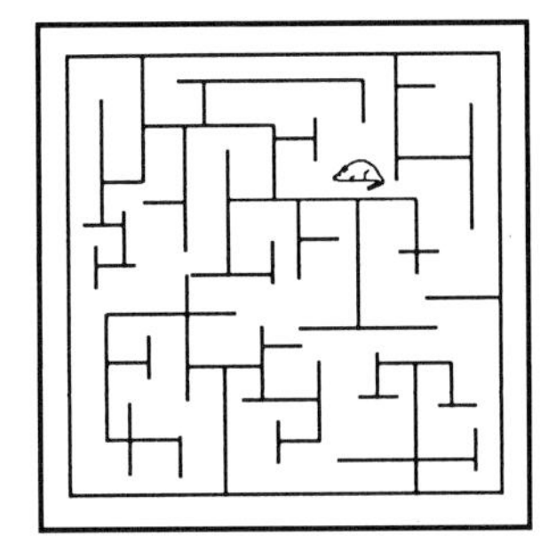

Reasoning and Natural Selection

LEDA COSMIDES, JOHN TOOBY, *Center for Advanced Study in the Behavioral Sciences*

Glossary

Adaptation Aspect of an organism that was created by the process of natural selection because it served an adaptive function

Adaptive Contributing to the eventual reproduction of an organism or its relatives

Bayes's theorem Specifies the probability that a hypothesis is true, given new data; $P(H|D) = P(H)P(D|H)/P(D)$, where H is the hypothesis and D is the new data

Cognitive psychology Study of how humans and other animals process information

Natural selection Evolutionary process responsible for constructing, over successive generations, the complex functional organization found in organisms, through the recurring cycle of mutation and subsequent increased reproduction of the better design

Normative theory Theory specifying a standard for how something *ought* to be done (as opposed to how it actually *is* done)

Valid argument Argument that is logically derived from premises; a conclusion may be valid, yet false, if it is logically derived from false premises

THE STUDY OF REASONING is an important component of the study of the biology of behavior. To survive and reproduce, animals must use data to make decisions, and these decisions are controlled, in part, by processes that psychologists label "inference" or "reasoning." To avoid predators, for example, a monkey must infer from a rustle in the grass and a glimpse of fur that a leopard is nearby and use information about its proximity to decide whether to take evasive action or continue eating. Because almost all action requires inferences to regulate it, the mechanisms controlling reasoning participate in almost every kind of behavior that humans, or other animals, engage in. Human reasoning has traditionally been studied without asking what kind of reasoning procedures our ancestors would have needed to survive and reproduce in the environment in which they evolved. In recent years, however, an increasing number of researchers have been using an evolutionary framework.

I. What Is Reasoning and How Is It Studied?

When psychologists study how humans reason, they are trying to discover what rules people use to make inferences about the world. They investigate whether there are general principles that can describe what people will conclude from a set of data.

One way of studying reasoning is to ask "If one were trying to write a computer program that could simulate human reasoning, what kind of program would have to be written? What kind of information-processing procedures (rules or algorithms) would the programmer have to give this program, and what kind of data structures (representations) would those procedures operate on?"

Of course, the human brain was not designed by an engineer with foresight and purposes; it was "designed" by the process of natural selection. Natural selection is the only natural process known that is capable of creating complex and organized biological structures, such as the human brain. Contrary to widespread belief, natural selection is not "chance"; it is a powerful positive feedback process fueled by differential reproduction. If a change in an organism's design allows it to outreproduce other members of its species, that design change will become more common in the population—it will be *selected for*. Over many generations that design change will spread through the population until all members of the species have it. Design changes that enhance reproduction can be selected for; those that hinder reproduction are selected against.

When evolutionary biologists study how humans reason, they are asking, "What kind of cognitive programs was natural selection likely to have designed, and is there any evidence that humans have such programs?"

A. Mind Versus Brain

At present, researchers find it useful to study the brain on different descriptive and explanatory levels. Neuroscientists describe the brain on a physiological level—as the interaction of neurons, hormones, neurotransmitters, and other organic aspects. Cognitive psychologists, on the other hand, study the brain as an information-processing system—that is, as a collection of programs that process information—without worrying about exactly how neurophysiological processes perform these tasks. The study of cognition is the study of how humans and other animals process information.

For example, ethologists have traditionally studied very simple cognitive programs: A newborn herring gull, for instance, has a cognitive program that defines a red dot on the end of a beak as salient information from the environment, and that causes the chick to peck at the red dot upon perceiving it. Its mother has a cognitive program that defines pecking at her red dot as salient information from her environment, and that causes her to regurgitate food into the newborn's mouth when she perceives its pecks. This simple program adaptively regulates how the herring gull feeds its offspring. (If there is a flaw anywhere in the program—if the mother or chick fails to recognize the signal or to respond appropriately—the chick starves. If the flaw has a genetic basis, it will not be passed on to future generations. Thus natural selection controls the design of cognitive programs.)

These descriptions of the herring gull's cognitive programs are entirely in terms of the functional relationships among different pieces of information; they describe two simple information-processing systems. Of course, these programs are embodied in the herring gull's neurological "hardware." Knowledge of this hardware, however, would add little to our understanding of these programs as information-processing systems. Presumably, one could build a silicon-based robot, using hardware completely different from what is present in the gull's brain, that would produce the same behavioral output (pecking at red dot) in response to the same informational input (seeing red dot). The robot's cognitive programs would maintain the same functional relationships among pieces of information and would therefore be, in an important sense, identical to the cognitive programs of the herring gull. But the robot's neural hardware would be totally different.

The specification of a cognitive program constitutes a complete description of an important level of causation, independent of any knowledge of the physiological hardware the program runs on. Cognitive psychologists call this position "functionalism," and they use it because it provides a precise language for describing complex information-processing architectures, without being limited to studying those few processes that neurophysiologists presently understand. (Eventually, of course, one wants to understand the neurophysiological processes that give rise to a cognitive program as well.) Cognitive scientists use the term "mind" solely to refer to an information-processing description of the functioning of the brain, and not in any colloquial sense.

II. The Mind as Scientist: General-Purpose Theories of Human Reasoning

Traditionally, cognitive psychologists have acknowledged that the mind (i.e., the information-processing structure of the brain) is the product of evolution, but their research framework was more strongly shaped by a different premise: that the mind was a general-purpose computer. They thought the function of this computer was self-evident: to discover the truth about whatever situation or problem it encountered. In other words, they started from the reasonable assumption that the procedures that governed human reasoning were there because they functioned to produce valid knowledge in nearly any context a person was likely to encounter.

They reasoned that if the function of the human mind is to discover truth, then the reasoning procedures of the human mind should reflect the methods by which truth can be discovered. Because science is the attempt to discover valid knowledge about the world, psychologists turned to the philosophy of science for *normative theories*—i.e., for theories specifying how one *ought* to reason if one is to produce valid knowledge. Their approach was to use the normative theories of what constitutes good scientific reasoning as a standard against which to compare actual human reasoning performance. The premise was that humans should be reasoning like idealized scientists about whatever situation they encountered, and the research question became: To what extent is the typical person's reasoning like an ideal scientist's?

The normative theories of how scientists—and hence the human mind—should be reasoning fall broadly into two categories: inductive reasoning and deductive reasoning. Inductive reasoning is reasoning from specific observations to general principles; deductive reasoning is reasoning from general principles to specific conclusions.

Ever since Hume, induction has carried a heavy load in psychology while taking a sound philosophical beating. In psychology, it has been the learning theory of choice since the British Empiricists argued that the experience of spatially and temporally contiguous events is what allows us to jump from the particular to the general, from sensations to objects, from objects to concepts. Many strands of psychology, including Pavlovian reflexology, Watsonian and Skinnerian behaviorism, and the sensory-motor parts of Piagetian structuralism, have been elaborations on the inductive psychology of the British Empiricists. Yet when Hume, a proponent of inductive inference as a psychological learning theory, donned his philosopher's hat, he demonstrated that induction could never justify a universal statement. To use a familiar example, no matter how many white swans you might see, you could never be justified in concluding "All swans are white," because it is always possible that the next swan you see will be black. Thus Hume argued that the inductive process whereby people were presumed to learn about the world could not ensure that the generalizations it produced would be valid.

Only recently, with the publication in 1935 of Karl Popper's *The Logic of Scientific Discovery*, has a logical foundation for psychology's favorite learning theory been provided. Popper argued that although a universal statement of science can never be proved true, it deductively implies particular assertions about the world, called hypotheses, and particular assertions can be proved false. Although no number of observed white swans can prove that "All swans are white" is true, just one black swan can prove it false. Generalizations cannot be confirmed, but they can be falsified, so inductions tested via deductions coupled with observations are on firmer philosophical ground than knowledge produced through induction alone.

This view had broad consequences for psychologists interested in learning. Psychologists who assumed that the purpose of human learning is to produce valid generalizations about the world reasoned that learning must be some form of Popperian hypothesis testing. Inductive reasoning must be used to generate hypotheses, and deductive reasoning coupled with observation must be used to try to falsify them. Furthermore, these reasoning procedures should be general-purpose: They should be able to yield valid inferences about any subject that one is interested in.

A broad array of cognitive psychologists such as Jean Piaget, Jerome Bruner, and Peter Wason, adopted a version of hypothesis testing—often an explicitly Popperian version—as their model of human learning. They used it to set the agenda of cognitive psychology in the 1950s and 1960s, and this view remains popular today. Some psychologists investigated inductive reasoning, by seeing whether people reason in accordance with the normative theories of inferential statistics; others investigated deductive reasoning, by seeing whether people reason in accordance with the rules of inference of the propositional calculus (formal propositional logic.

A. Deductive Reasoning

Psychologists became interested in whether the human mind included a "deductive component": mental rules that are the same as the rules of inference of the propositional calculus. They performed a wide variety of experiments to see whether people were able (1) to *recognize* the difference between a valid deductive inference and an invalid one, or (2) to *generate* valid conclusions from a set of premises. If people have a "deductive component," then they should be good at tasks like these. For example, in reasoning about conditional statements, one can make two valid inferences and two invalid inferences (see Fig. 1).

One of the most systematic bodies of work exploring the idea that people have reasoning procedures that embody the rules of inference of the propositional calculus was produced by Peter Wason and P. N. Johnson-Laird, together with their students and colleagues. Their research provides strong evidence that people do not reason according to the canons of formal propositional logic. For example:

(1) *Recognition of an argument as valid.* To see whether people are good at recognizing an argument as valid, psychologists gave them arguments like the ones in Fig. 1; for example, a subject might be asked to judge the validity of the the following argument: "If the object is rectangular, then it is blue; the object is rectangular; therefore the object is blue." In some of the experiments, unfamiliar conditionals were used; in others, familiar ones were used. These experiments indicated that people are good at recognizing the validity of a *modus ponens* inference, but they frequently think *modus tollens* is an invalid inference and that the two invalid inferences in Figure 1 are valid. Furthermore, they frequently view logically distinct conditionals as implying each other, and they have a pronounced tendency to judge an inference valid when they agree with the conclusion and invalid when they do not agree with the conclusion, regardless of its true validity.

(2) *Generating valid conclusions from a set of premises.* In other experiments, psychologists gave people sets of premises and asked them to draw conclusions from them. Many of the problems requiring the use of *modus ponens* were done incorrectly, and most of those requiring the use of *modus tollens* were done incorrectly. It may at first seem puzzling that people who are good at recognizing a *modus ponens* argument as valid would have trouble using *modus ponens* to generate a conclusion from premises. However, analogous experiences are common in everyday life: sometimes one cannot recall a person's name but can recognize it on a list. If humans all had rules of reasoning that mapped on to *modus ponens*, then they would be able both to generate a valid *modus ponens* inference and to recognize one. The fact that people cannot do both indicates that they lack this rule of reasoning. They may simply be able to recognize a contradiction when they see one, even though they cannot reliably generate valid inferences.

Valid inferences		Invalid inferences	
Modus ponens	Modus tollens	Affirming the Consequent	Denying the Antecedent
If P then Q P	If P then Q not-Q	If P then Q Q	If P then Q not-P
Therefore Q	Therefore not-P	Therefore P	Therefore not-Q

FIGURE 1 "P" and "Q" can stand for any proposition; for example, if "P" stands for "it rained" and "Q" stands for "the grass is wet," then the *modus ponens* inference above would read "If it rained, then the grass is wet; it rained, therefore the grass is wet." *Affirming the consequent* and *denying the antecedent* are invalid because the conditional "If P then Q" does not claim that P is the only possible antecedent of Q. If it did *not* rain, the grass could still be wet—the lawn could have been watered with a sprinkler, for example.

Perhaps the most intriguing and widely used experimental paradigm for exploring deductive reasoning has been the Wason selection task (see Fig. 2a). Peter Wason was interested in Popper's view that the structure of science was hypothetico-deductive. He wondered if learning were really hypothesis testing—i.e., the search for evidence that contradicts a hypothesis. Wason devised his selection task because he wanted to see whether people really do test a hypothesis by looking for evidence that could potentially falsify it. In the Wason selection task, a subject is asked to see whether a conditional hypothesis of the form "If P then Q" has been violated by any one of four instances, represented by cards.

A hypothesis of the form "If P then Q" is violated only when "P" is true but "Q" is false—the rule in Figure 2a, for example, can be violated only by a card that has a D on one side and a number other than 3 on the other side. Thus, one would have to turn over the "P" card (to see if it has a "not-Q" on the back) and the "not-Q" card (to see if it has a "P" on the back)—i.e., D and 7 for the rule in Figure 1a. The logically correct response, then, is always "P and not-Q."

Wason expected that people would be good at this. Nevertheless, he and many other psychologists have found that few people actually give this logically correct answer (<25% for rules expressing unfamiliar relations). Most people choose either the "P" card alone or "P and Q." Few people choose the "not-Q" card, even though a "P" on the other side of it would falsify the rule.

A wide variety of conditional rules that describe some aspect of the world ("descriptive" rules) have been tested; some of these have expressed relatively familiar relations, such as "If a person goes *a.* **Abstract Problem (AP)**

Part of your new clerical job at the local high school is to make sure that student documents have been processed correctly. Your job is to make sure the documents conform to the following alphanumeric rule:

"If a person has a 'D' rating, then his documents must be marked code '3'."
(If P then Q)*

You suspect the secretary you replaced did not categorize the students' documents correctly. The cards below have information about the documents of four people who are enrolled at this high school. Each card represents one person. One side of a card tells a person's letter rating and the other side of the card tells that person's number code.

Indicate only those card(s) you definitely need to turn over to see if the documents of any of these people violate this rule.

D	F	3	7
(P)	(not-P)	(Q)	(not-Q)

b. **Drinking Age Problem (DAP; adapted from Griggs & Cox, 1982)**

In its crackdown against drunk drivers, Massachusetts law enforcement officials are revoking liquor licenses left and right. You are a bouncer in a Boston bar, and you'll lose your job unless you enforce the following law:

"If a person is drinking beer, then he must be over 20 years old."
(If P then Q)

The cards below have information about four people sitting at a table in your bar. Each card represents one person. One side of a card tells what a person is drinking and the other side of the card tells that person's age.

Indicate only those card(s) you definitely need to turn over to see if any of these people are breaking this law.

drinking beer	drinking coke	25 years old	16 years old
(P)	(not-P)	(Q)	(not-Q)

c. **Structure of Social Contract (SC) Problems**

It is your job to enforce the following law:

Rule 1 — Standard Social Contract (STD-SC): "If you take the benefit, then you pay the cost."
(If P then Q)

Rule 2 — Switched Social Contract (SWC-SC): "If you pay the cost, then you take the benefit."
(If P then Q)

The cards below have information about four people. Each card represents one person. One side of a card tells whether a person accepted the benefit and the other side of the card tells whether that person paid the cost.

Indicate only those card(s) you definitely need to turn over to see if any of these people are breaking this law.

	Benefit Accepted	Benefit NOT Accepted	Cost Paid	Cost NOT Paid
Rule 1 — STD-SC:	(P)	(not-P)	(Q)	(not-Q)
Rule 2 — SWC-SC:	(Q)	(not-Q)	(P)	(not-P)

FIGURE 2 Content effects on the Wason selection task. The logical structures of these three Wason selection tasks are identical; they differ only in propositional content. Regardless of content, the logical solution to all three problems is the same: To see if the rule has been violated, choose the "P" card (to see if it has a "not-Q" on the back) and choose the "not-Q" card (to see if it has a "P" on the back). Fewer than 25% of college students choose "P & not-Q" for the Abstract Problem (*a*), whereas about 75% choose both these cards for the Drinking Age Problem (*b*)—a familiar social contract. *c* shows the abstract structure of a social contract problem. A "look for cheaters" procedure would cause one to choose the "benefit accepted" card and the "cost NOT paid" card, regardless of which logical categories they represent. For Rule 1, these cards represent the values "P & not-Q," but for Rule 2 they represent the values "Q & not-P." Consequently, a person who was looking for cheaters would appear to be reasoning logically in response to Rule 1 but illogically in response to Rule 2. * The logical categories (P and Q) marked on the rules and cards here are only for the reader's benefit; they never appear on problems given to subjects in experiments.

to Boston, then he takes the subway" or "If a person eats hot chili peppers, then he will drink a cold beer." Others have expressed unfamiliar relations, such as "If you eat duiker meat, then you have found an ostrich eggshell" or "If there is an 'A' on one side of a card, then there is a '3' on the other side." In many experiments, performance on familiar descriptive rules is just as low as it is on unfamiliar ones; some familiar rules, however, do elicit a higher percentage of logically correct responses than unfamiliar ones. Even so, familiar descriptive rules typically elicit the logically correct response in fewer than half of the people tested. Recently, rules expressing causal relations have been tested; the pattern of results is essentially the same as for descriptive rules.

It is particularly significant that performance on the Wason selection task is so poor when the descriptive or causal rule tested is unfamiliar. If the function of our reasoning procedures is to allow us to discover new things about the world, then they must be able to function in novel—i.e., unfamiliar—situations. If they cannot be used in unfamiliar situations, then they cannot be used to learn anything new. Thus, the view that the purpose of human reasoning is to learn about the world is particularly undermined by the finding that people are not good at looking for violations of descriptive and causal rules, especially when they are unfamiliar.

B. Inductive Reasoning

The hypotheses that scientists test do not appear from thin air. Some of them are derived from theories; others come from observations of the world. For example, although no number of observations of white swans can prove that all swans are white, a person who has seen hundreds of white swans and no black ones may be more likely to think this hypothesis is worthy of investigation than a person who has seen only one white swan. The process of inferring hypotheses from observations is called inductive inference.

Using probability theory, mathematicians have developed a number of different normative theories of inductive inference, such as Bayes's theorem, null hypothesis testing, and Neyman–Pearsonian decision theory. These theories specify how scientists should make inferences from data to hypotheses. They are collectively known as inferential statistics.

A number of psychologists have studied the extent to which people's inductive reasoning conforms to the normative theories of inferential statistics and probability theory. One of the most extensive research efforts of this kind was spearheaded by Amos Tversky and Daniel Kahneman, along with their students and colleagues. They tested people's inductive reasoning by giving them problems in which they were asked to judge the probability of uncertain events. For example, a subject might be asked to reason about a diagnostic medical test: "If a test to detect a disease whose prevalence is 1/1000 has a false positive rate of 5%, what is the chance that a person found to have a positive result actually has the disease, assuming you know nothing about the person's symptoms or signs?" If the subject's answer is different from what a theory of statistical inference says it should be, then the experimenters conclude that our inductive-reasoning procedures do not embody the rules of that normative statistical theory.

The consensus among many psychologists is that this body of research demonstrates that (1) the human mind does not calculate the probability of events in accordance with normative probability theories, and (2) the human mind does not include information-processing procedures that embody the normative theories of inferential statistics. In other words, they conclude that the human mind is not innately equipped to do college-level statistics. Instead, these psychologists believe that people make inductive inferences using heuristics—cognitive shortcuts or rules of thumb. These heuristics frequently lead to the correct answer but can also lead to error precisely because they do not embody the formulas and calculational procedures of the appropriate normative theory. These psychologists also believe that humans suffer from systematic biases in their reasoning, which consistently lead to errors in inference.

Recently, however, a powerful critique by Gerd Gigerenzer, David Murray, and their colleagues has called this consensus into serious doubt. Their critique is both theoretical and empirical. Gigerenzer and Murray point out that the Tversky and Kahneman research program is based on the assumption that a statistical problem has only one correct answer; when the subject's response deviates from that answer, the experimenter infers that the subject is not reasoning in accordance with a normative statistical theory. However, Gigerenzer and Murray show that the problems subjects are typically asked to solve do not have only one correct answer. There are several reasons why this is true.

1. Statistics Does Not Speak With One Voice

There are a number of different statistical theories, and not all of them give the same answer to a problem. For example, although subjects' answers to certain problems have been claimed to be incorrect from the point of view of Bayes's theorem (but see below), their answers can be shown to be correct from the point of view of Neyman–Pearsonian decision theory. These subjects may be very good "intuitive statisticians," but simply applying a different normative theory than the experimenter is.

2. Concepts Must Match Exactly

For a particular statistical theory to be applicable, the concepts of the theory must match up precisely with the concepts in the problem. Suppose, for example, that you have some notion of how likely it is that a green cab or a blue cab would be involved in a hit-and-run accident at night. You are then told that there was a hit-and-run accident last night, and that a witness who is correct 80% of the time reported that it was a green cab. Bayes's theorem allows you to revise your prior probability estimate when you receive new information, in this case, the witness's testimony.

But what should your prior probability estimate (i.e., the estimate that you would make if you did not have the witness's testimony) be based on? It could, for example, be based on (1) the relative number of green and blue cabs in the city, (2) the relative number of reckless driving arrests for green versus blue cab drivers, (3) the relative number of drivers who have alcohol problems, or (4) the relative number of hit-and-run accidents they get into at night.

There is no normative theory for deciding which of these four kinds of information is the most relevant. Yet Bayes's theorem will generate different answers, depending on which you use. If subjects and experimenter differ in which kind of information they believe is most relevant, they will give different answers, even if each is correctly applying Bayes's theorem. Indeed, experimental data suggest that this happens. If one assumes that the subjects in these experiments were making certain very reasonable assumptions, then they *were* answering these questions correctly.

3. Structural Assumptions of the Theory Must Hold for the Problem

Assume that nature had selected for statistical rules; then it also should have selected for an assumption-checking program. For a particular statistical theory to be applicable, the assumptions of the theory must hold for the problem. For instance, a frequent assumption for applying Bayes's theorem is that a sample was randomly drawn. But in the real world there are many situations in which events are not randomly sampled: Diagnostic medical tests, for example, are rarely given to a random sample of people—instead, they are given only to those who already have symptoms of the disease. By their content, certain problems tested invited the inference that the random sampling assumption was violated; given this assumption, the "incorrect" answers subjects were giving were, in fact, correct. Indeed, in an elegant series of experiments, Gigerenzer and his colleagues showed that if one makes the random sampling assumption explicit to subjects, they do appear to reason in accordance with Bayes's theorem.

These experiments and theoretical critiques cast serious doubt on the conclusion that people are not good "intuitive statisticians." Evidence suggests that people are very good at statistical reasoning if the problem is about a real-world situation in which the structural assumptions of the theory hold. Their apparent errors may be because they are making assumptions about the problem that are different from the experimenters', or because they are consistently applying one set of statistical principles in one context and other sets in different contexts. What is clear from the research on inductive reasoning, however, is that the content of the problem matters, a theme we will return to in Section III.

C. Did We Evolve to Be Good Intuitive Scientists?

Good design is the hallmark of adaptation: To demonstrate that human reasoning evolved to fulfill a particular function, one must show that our reasoning procedures are well designed to fulfill that function. If the human mind was designed by natural selection to generate logically valid, scientifically justifiable knowledge about the world, then we ought to be good at drawing correct inductive and deductive inferences. Moreover, this ability ought to be context-independent, to allow us to learn about new, unfamiliar domains. After all, everything is initially unfamiliar.

But the data on deductive reasoning indicate that our minds do not include rules of inference that conform to the canons of deductive logic. The data on inductive reasoning indicate that we do not have inductive-reasoning procedures that operate independently of content and context. We may have inductive-reasoning procedures that conform to normative theories of statistical inference, but if we do, their application in any particular instance is extremely context-dependent, as the issues of conceptual and structural matching show. The evidence therefore suggests that we do not have formal, content-independent reasoning procedures. This indicates that the hypothesis that the adaptive function of human reasoning is to generate logically valid knowledge about the world is false.

III. The Mind as a Collection of Adaptations: Evolutionary Approaches to Human Reasoning

Differential reproduction is the engine that drives natural selection: If having a particular mental structure, such as a rule of inference, allows an animal to outreproduce other members of its species, then that mental structure will be selected for. Over many generations it will spread through the population until it becomes a universal, species-typical trait.

Consequently, alternative phenotypic traits are selected for not because they allow the organism to more perfectly apprehend universal truths, but because they allow the organism to outreproduce others of its species. Truth-seeking can be selected for only to the extent that it promotes reproduction. Although it might seem paradoxical to think that reasoning procedures that sometimes produce logically incorrect inferences might be more adaptive than reasoning that always leads to the truth, this will frequently be the case. Among other reasons, organisms usually must act before they have enough information to make valid inferences. In evolutionary terms, the design of an organism is like a system of betting: What matters is not each individual outcome, but the statistical average of outcomes over many generations. A reasoning procedure that sometimes leads to error, but that usually allows one to come to an adaptive conclusion (even when there is not enough information to justify it logically), may perform better than one that waits until it has sufficient information to derive a valid truth without error. Therefore, factors such as the cost of acquiring new information, asymmetries in the payoffs of alternative decisions (believing that a predator is in the shadow when it is not versus believing a predator is not in the shadow when it is), and trade-offs in the allocation of limited attention may lead to the evolution of reasoning procedures whose design is sharply at variance with scientific and logical methods for discovering truth.

Although organisms do not need to discover universal truths or scientifically valid generalizations to reproduce successfully, they do need to be very good at reasoning about important adaptive problems and at acquiring the kinds of information that will allow them to make adaptive choices in their natural environment. Natural selection favors mental rules that will enhance an animal's reproduction, whether they lead to truth or not. For example, rules of inference that posit features of the world that are usually (but not always) true may provide an adequate basis for adaptive decision-making. Some of these rules may be general-purpose: For example, the heuristics and biases proposed by Tversky and Kahneman are rules of thumb that will get the job done under the most commonly encountered circumstances. Their availability heuristic, for instance, is general-purpose insofar as it is thought to operate across domains: One uses it whether one is judging the frequency of murders in one's town or of words in the English language beginning with the letter "k." However, there are powerful reasons for thinking that many of these evolved rules will be special-purpose.

Traditionally, cognitive psychologists have assumed that the human mind includes only general-purpose rules of reasoning and that these rules are few in number. But natural selection is also likely to produce many mental rules that are specialized for reasoning about various evolutionarily important domains, such as cooperation, aggressive threat, parenting, disease avoidance, predator avoidance, and the colors, shapes, and trajectories of objects. This is because different adaptive problems frequently have different optimal solutions. For example, vervet monkeys have three major predators: leopards, eagles, and snakes. Each of these predators requires different evasive action: climbing a tree (leopard), looking up in the air or diving straight into the bushes (eagle), or standing on hind legs and looking into the grass (snake). Accordingly, vervets have a different alarm call for each of these three predators. A single, general-purpose alarm call would be less efficient because the monkeys would not know which of the three different evasive actions to take.

When two adaptive problems have different optimal solutions, a single general solution will be inferior to two specialized solutions. In such cases, a jack of all trades is necessarily a master of none, because generality can be achieved only by sacrificing efficiency.

The same principle applies to adaptive problems that require reasoning: There are cases where the rules for reasoning adaptively about one domain will lead one into serious error if applied to a different domain. Such problems cannot, in principle, be solved by a single, general-purpose reasoning procedure. They are best solved by different, special-purpose reasoning procedures. We will consider some examples of this below.

A. Internalized Knowledge and Implicit Theories

Certain facts about the world have been true for all of our species' evolutionary history and are critical to our ability to function in the world: The sun rises every 24 hours; space is locally three-dimensional; rigid objects thrown through space obey certain laws of kinematic geometry. Roger Shephard has argued that a human who had to learn these facts through the slow process of "trial and possibly fatal error" would be at a severe selective disadvantage compared to a human whose perceptual and cognitive system was designed in such a way that it already assumed that such facts were true. In an elegant series of experiments, Shepard showed that our perceptual–cognitive system has indeed internalized laws of kinematic geometry, which specify the ways in which objects move in three-dimensional Euclidean space. Our perceptual system seems to expect objects to move in the curvilinear paths of kinematic geometry so strongly that we see these paths even when they do not exist, as in the phenomenon of visual apparent motion. This powerful form of inference is specific to the motion of objects; it would not, for example, help you to infer whether a friend is likely to help you when you are in trouble.

Learning a relation via an inductive process that is truly general-purpose is not only slow, it is impossible in principle. There are an infinite number of dimensions along which one can categorize the world, and therefore an infinite number of possible hypotheses to test ("If my elbow itches, then the sun will rise tomorrow", "If a blade of grass grows in the flower pot, then a man will walk in the door"; i.e., "If P then Q," "If R then Q," "If S then Q" *ad infinitum*). The best a truly unconstrained inductive machine could do would be to randomly generate each of an infinite number of inductive hypotheses and deductively test each in turn.

Those who have considered the issue recognize that an organism could learn nothing this way. If any learning is to occur, then one cannot entertain all possible hypotheses. There must be constraints on which hypotheses one entertains, so that one entertains only those that are most likely to be true. This insight led Susan Carey and a number of other developmental psychologists to suggest that children are innately endowed with mental models of various evolutionarily important domains. Carey and her colleagues call these mental models *implicit theories*, to reflect their belief that all children start out with the same set of theories about the world, embodied in their thought processes.

These implicit theories specify how the world works in a given domain; they lead the child to test hypotheses that are consistent with the implicit theory, and therefore likely to be true (or at least useful). Implicit theories constrain the hypothesis space so that it is no longer infinite, while still allowing the child to acquire new information about a domain. Implicit theories are thought to be domain-specific because what is true of one domain is not necessarily true of another. For example, an implicit theory that allows one to predict a person's behavior if one knows that person's beliefs and desires will not allow one to predict the behavior of falling rocks, which have no beliefs and desires. The implicit-theory researchers have begun to study children's implicit theories about the properties of organisms, the properties of physical objects and motion, the use of tools, and the minds of others.

B. Reasoning about Prescriptive Social Conduct

The reasoning procedures discussed so far function to help people figure out what the world is like and how it works. They allow one to acquire knowledge that specifies what kind of situation one is facing from one moment to the next. For example, a rule such as "If it rained last night, then the grass will be wet this morning" purports to describe the way the world is. Accordingly, it has a truth value: A descriptive rule can be either true or false. In contrast, a rule such as "If a person is drinking beer, then that person must be over 21 years old" does not describe the way things are. It does not even describe the way existing people behave. It *prescribes*: It communicates the way some people want other people to behave. One cannot assign a truth value to it.

From an evolutionary perspective, knowledge

about the world is just a means to an end, and that end is behaving adaptively. Once an organism knows what situation it is in, it has to know how to act, so reasoning about the facts of the world should be paired with reasoning about appropriate conduct. For this reason, the mind should have evolved rules of reasoning that specify what one ought to do in various situations—rules that prescribe behavior. Because different kinds of situations call for different kinds of behavior, these rules should be situation-specific. For example, the rules for reasoning about cooperation should differ from those for reasoning about aggressive threat, and both should differ from the rules for reasoning about the physical world. Recent research by Cosmides & Tooby, Manktelow & Over, and others has explored such rules.

Social exchange, for example, is cooperation between two or more people for mutual benefit, such as the exchange of favors between friends. Humans in all cultures engage in social exchange, and the paleoanthropological record indicates that such cooperation has probably been a part of human evolutionary history for almost 2 million years. Game-theoretic analyses by researchers such as Robert Trivers, Robert Axelrod, and W. D. Hamilton have shown that cooperation cannot evolve unless people are good at detecting "cheaters" (people who accept favors or benefits without reciprocating). Given a social contract of the form "If you take the benefit, then you pay the cost," a cheater is someone who took the benefit but did not pay the required cost (see Fig. 2c). Detecting cheaters is an important adaptive problem: A person who was consistently cheated would be incurring reproductive costs, but receiving no compensating benefits. Such individuals would dwindle in number, and eventually be selected out of the population.

Rules for reasoning about descriptive relations would lead one into serious error if applied to social contract relations. In the previous discussion of the Wason selection task, we saw that the logically correct answer to a descriptive rule is "P and not-Q," no matter what "P" and "Q" stand for (i.e., no matter what the rule is about). But this definition of violation differs from the definition of cheating on a social contract. A social contract rule has been violated whenever a person has taken the benefit without paying the required cost, *no matter what logical category these actions correspond to*. For the social contract expressed in Rule 1 of Figure 2c, a person who was looking for cheaters would, by coincidence, produce the logically correct answer. This is because the "benefit accepted" card and the "cost NOT paid" card correspond to the logical values "P" and "not-Q," respectively, for Rule 1. But for the social contract expressed in Rule 2, these two cards correspond to the logical values "Q" and "not-P"—a logically incorrect answer. The logically correct answer to Rule 2 is to choose the "cost paid" card, "P," and the "benefit NOT accepted" card, "not-Q." Yet a person who has paid the cost cannot possibly have cheated, nor can a person who has not accepted the benefit.

Thus, for the social contract in Rule 2, the adaptively correct answer is logically incorrect, and the logically correct answer is adaptively incorrect. If the only reasoning procedures that our minds contained were the general-purpose rules of inference of the propositional calculus, then we could not, in principle, reliably detect cheating on social contracts. This adaptive problem can be solved only by inferential procedures that are specialized for reasoning about social exchange.

The Wason selection task research discussed previously showed that we have no general-purpose ability to detect violations of conditional rules—unfamiliar descriptive and causal rules elicit the logically correct response from <25% of subjects. But when a conditional rule expresses a social contract, people are very good at detecting cheaters. Approximately 75% of subjects choose the "benefit accepted" card and the "cost NOT paid" card, regardless of which logical category they correspond to and regardless of how unfamiliar the social contract rule is. This research indicates that the human mind contains reasoning procedures that are specialized for detecting cheaters on social contracts. Recently, the same experimental procedures have been used to investigate reasoning about aggressive threat. Although there is only one way to violate the terms of a social contract, there are two ways of violating the terms of a threat: Either the person making the threat can be bluffing (i.e., he does not carry out the threat, even though the victim refuses to comply), or he can be planning to double-cross the person he is threatening (i.e., the victim complies with his demand, but the threatener punishes him anyway). The evidence indicates that people are good at detecting both bluffing and double-crossing. Similarly, two British researchers, Kenenth Manktelow and David Over, have found that people are very good at detecting violations of "precaution rules." Precaution rules specify what precautions should be taken to avoid danger in hazardous situations.

Situations involving social contracts, threats, and precaution rules have recurred throughout human evolutionary history, and coping with them successfully constituted powerful selection pressures. An individual who cannot cooperate, cannot avoid danger, or cannot understand a threat is at a powerful selective disadvantage in comparison to those who can. More important, what counts as a violation differs for a social contract rule, a threat, and a precaution rule. Because of this difference, the same reasoning procedure cannot be successfully applied to all three situations. As a result, there cannot be a general-purpose reasoning procedure that works for all of them. If these problems are to be solved at all, they must be solved by specialized reasoning procedures. Significantly, humans do reason successfully about these problems, suggesting that natural selection has equipped the human mind with a battery of functionally specialized reasoning procedures, designed to solve specific, recurrent adaptive problems.

IV. Summary

Reasoning procedures are an important part of how organisms adapt. Adaptive behavior depends on adaptive inferences to regulate decisions. Although initial approaches within psychology to the study of human reasoning uncovered many interesting phenomena, the search for a few, general rules of reasoning that would account for human-reasoning performance and explain how humans cope with the world was largely unsuccessful. The recent emergence of an evolutionary perspective within cognitive psychology has led to a different view of how inference in the human mind is organized. Instead of viewing the mind as a general-purpose computer, employing a few general principles that are applied uniformly in all contexts, an evolutionary perspective suggests that the mind consists of a larger collection of functionally specialized mechanisms, each consisting of a set of reasoning procedures designed to efficiently solve particular families of important adaptive problems. In the last decade, a growing body of research results has validated this approach, indicating that humans have specialized procedures for reasoning about such things as the motion of objects, the properties of living things, cooperation, threat, and avoiding danger.

Bibliography

Carey, S. (1985). "Conceptual Change in Childood." MIT Press, Cambridge, Massachusetts.

Cosmides, L. (1989). The logic of social exchange: Has natural selection shaped how humans reason? Studies with the Wason selection task. *Cognition* **31,** 187–276.

Cosmides, L., and Tooby, J. (1989). Evolutionary psychology and the generation of culture, part II. Case study: A computational theory of social exchange. *Ethol. Sociobiol.* **10,** 51–97.

Dawkins, R. (1976). "The Selfish Gene." Oxford University Press, Oxford.

Dawkins, R. (1986). "The Blind Watchmaker." Norton, New York.

Evans, J. St. B. T. (ed.) "Thinking and Reasoning." Routledge and Kegan Paul, London.

Gigerenzer, G., Hell, W., and Blank, H. (1988). Presentation and content: The use of base rates as a continuous variable. *J. Exp. Psych.: Hum. Percep. Perform.* **14,** 513–525.

Gigerenzer, G., and Murray, D. (1987). "Cognition as Intuitive Statistics." LEA, Hillsdale, New Jersey.

Johnson-Laird, P. N. (1982). Thinking as a skill. *Q. J. Exp. Psych.* **34A,** 1–29.

Kahneman, D., Slovic, P., and Tversky, A. (1982). "Judgment Under Uncertainty: Heuristics and Biases." Cambridge University Press, Cambridge.

Keil, F. (1989). "Concepts, Kinds, and Cognitive Development." MIT Press, Cambridge, Massachusetts.

Manktelow, K. I., and Over, D. E. (1987). Reasoning and rationality. *Mind Lang.* **2,** 199–219.

Manktelow, K. I., and Over, D. E. (1990). Deontic thought and the selection task. *In* "Lines of Thinking," Vol. 1, (K. J. Gilhooly, M. T. G. Keane, R. H. Logie, and G. Erdos, eds.). Wiley, New York.

Shepard, R. N. (1984). Ecological constraints on internal representation: Resonant kinematics of perceiving, imagining, thinking, and dreaming. *Psychol. Rev.* **91,** 417–447.

Wason, P., and Johnson-Laird, P. N. (1972). "Psychology of Reasoning: Structure and Content." Harvard University Press, Cambridge, Massachusetts.

Williams, G. C. (1966). "Adaptation and Natural Selection." Princeton University Press, Princeton.

Receptor Molecules, Fertilization

PAUL M. WASSARMAN, *Roche Institute of Molecular Biology*

Glossary

Acrosome Lysosome-like vesicle located in the head of mammalian and nonmammalian sperm. In mammals, the acrosome contains enzymes which permit sperm to penetrate the zona pellucida

Acrosome reaction Fusion of sperm outer acrosomal membrane with sperm plasma membrane that results in exposure of the egg extracellular coat to acrosomal enzymes and other acrosomal components

Cortical granules Lysosome-like vesicles located in the cortex of mammalian and nonmammalian eggs. Cortical granules contain enzymes and other components that modify the egg extracellular coat and, thereby, establish a block to polyspermy following fertilization

Cortical reaction Fusion of cortical granule membrane with egg plasma membrane that results in exposure of the egg extracellular coat to cortical granule components following fertilization

Egg-binding protein Molecule located on the sperm head that, together with the sperm receptor, is responsible for species-specific binding of sperm to eggs

Fertilization Union of male and female gametes, sperm and egg, respectively, to form a zygote, or one-cell embryo

Fertilization envelope Remodeled form of the vitelline envelope, altered by cortical granule and jelly coat components, that establishes a block to polyspermy following fertilization

Jelly coat Thick, gelatinous-like, extracellular coat that encompasses the vitelline envelope of many nonmammalian eggs and is the site of the acrosome reaction inducer

Polyspermy Fertilization of an egg by more than one sperm; commonly a lethal condition

Sperm receptor Molecule located on the egg extracellular coat that, together with the egg-binding protein, is responsible for species-specific binding of sperm to eggs

Vitelline envelope Relatively thin extracellular coat that encompasses many nonmammalian eggs and is the site of species-specific sperm receptors

Zona pellucida Relatively thick extracellular coat that encompasses mammalian eggs and is the site of species-specific sperm receptors

Zona reaction Changes in the zona pellucida, introduced by cortical granule components, that establish a block to polyspermy following fertilization

FERTILIZATION OF EGGS by sperm, the means by which sexual reproduction takes place in nearly all multicellular organisms, is fundamental to maintenance of life. In order to maintain speciation, fertilization in both plants and animals exhibits a relatively high degree of species specificity which is attributable to receptor molecules located on the surface of gametes. Among animals, species-specific receptors for sperm are located on extracellular coats that encompass both mammalian and nonmammalian eggs and restrict access to egg plasma membrane. Only a sperm that bears compatible binding molecules, complementary to sperm receptors on an egg, can successfully bind to and penetrate the extracellular coat and, ultimately, fertilize an egg by fusing with egg plasma membrane. Following fertilization of the egg by a single sperm, receptors are inactivated in order to prevent fertilization by additional sperm which would jeopardize

Encyclopedia of Human Biology, Volume 6.

normal embryonic development. Thus, sperm receptors regulate interactions between gametes that can lead to formation of a zygote.

I. Fertilization and Sperm Receptors

Fertilization in animals involves the union of male (sperm) and female (egg) gametes to produce a single cell, the zygote, which is capable of giving rise to a new organism that expresses and maintains characteristics of the species. Various mechanisms have been devised in nature to ensure that, in general, fertilization of eggs by sperm takes place only when gametes are derived from the same (homologous) species. Occasionally, when fertilization of eggs from one species by sperm from a different (heterologous) species does occur (heterospecific fertilization), in most instances either no progeny are produced or the progeny are infertile. [*See* FERTILIZATION.]

It is well established that among animals a major barrier to heterospecific fertilization exists at the level of gamete recognition. Complementary molecules located at the surface of eggs and sperm from the same species recognize one another in much the same way that antibodies recognize specific antigens. As a result, gametes from homologous species interact with one another and fertilization ensues. On the other hand, molecules located at the surface of eggs and sperm from different species do not complement one another and, in most cases, the gametes fail to interact properly and fertilization does not occur. The interaction between eggs and sperm can be likened to that between viruses and their host cells, which exhibit a relatively high degree of species specificity that is attributable to cell surface molecules.

Historically, for both mammalian and nonmammalian organisms, egg surface molecules that are specifically recognized by complementary sperm surface molecules have been called sperm receptors. The term sperm receptor is used in much the same way that the term "virus receptor," a molecule located on the surface of host cells, is used in virology. Complementary molecules on the surface of the sperm head have been called egg-binding proteins, as well as other names. Although other factors play a role, the interaction between sperm receptors and egg-binding proteins is largely responsible for species-specific fertilization in animals, as well as plants.

In general, sperm receptors are associated with envelopes that surround animal eggs and restrict access to the egg plasma membrane. For example, many nonmammalian eggs, such as echinoderm and amphibian eggs, are surrounded by a relatively thin vitelline envelope and thick jelly coat, while mammalian eggs are surrounded by a single, relatively thick coat, the zona pellucida (Fig. 1). Both the vitelline envelope and zona pellucida contain the sperm receptors that restrict heterospecific fertilization. Removal of the zona pellucida from mammalian eggs, thereby exposing egg plasma membrane directly to sperm, largely eliminates the barrier to heterospecific fertilization *in vitro*. Thus, the receptor is essential for species-specific fertilization. Following fertilization (i.e., fusion of sperm and egg), sperm receptors are inactivated during either conversion of the vitelline envelope into a fertilization envelope (nonmammalian organisms) or completion of the zona reaction (mammalian organisms) (Fig. 2). Such inactivation of sperm receptors assists in the prevention of polyspermy, that is, fertilization by more than one sperm (the so-called slow block to polyspermy). [*See* SPERM.]

II. Pathways to Fertilization

A. Fertilization Pathway in Nonmammals

Based primarily on studies with sea urchin gametes, the fertilization pathway for many nonmammalian organisms consists of several steps that occur in a compulsory order. Briefly, these steps include the following:

1. Binding of sperm with an intact acrosome to the jelly coat: Peptides (speract and resact) associated with the jelly coat stimulate sperm respiration and motility in a species-specific manner, and may also serve as species-specific chemoattractants.
2. Completion of the acrosome reaction by sperm bound to the jelly coat: A fucan sulfate polysaccharide component of the jelly coat is the acrosome reaction inducer and an M_r 210,000 glycoprotein may be the sperm surface molecule that binds to the jelly coat polysaccharide.
3. Formation of an acrosomal process, at the tip of the sperm head, that penetrates the jelly coat: Acrosomal process formation is associated with extensive polymerization of actin into microfilaments.
4. Species-specific binding of sperm, via their acrosomal process, to the vitelline envelope: Binding

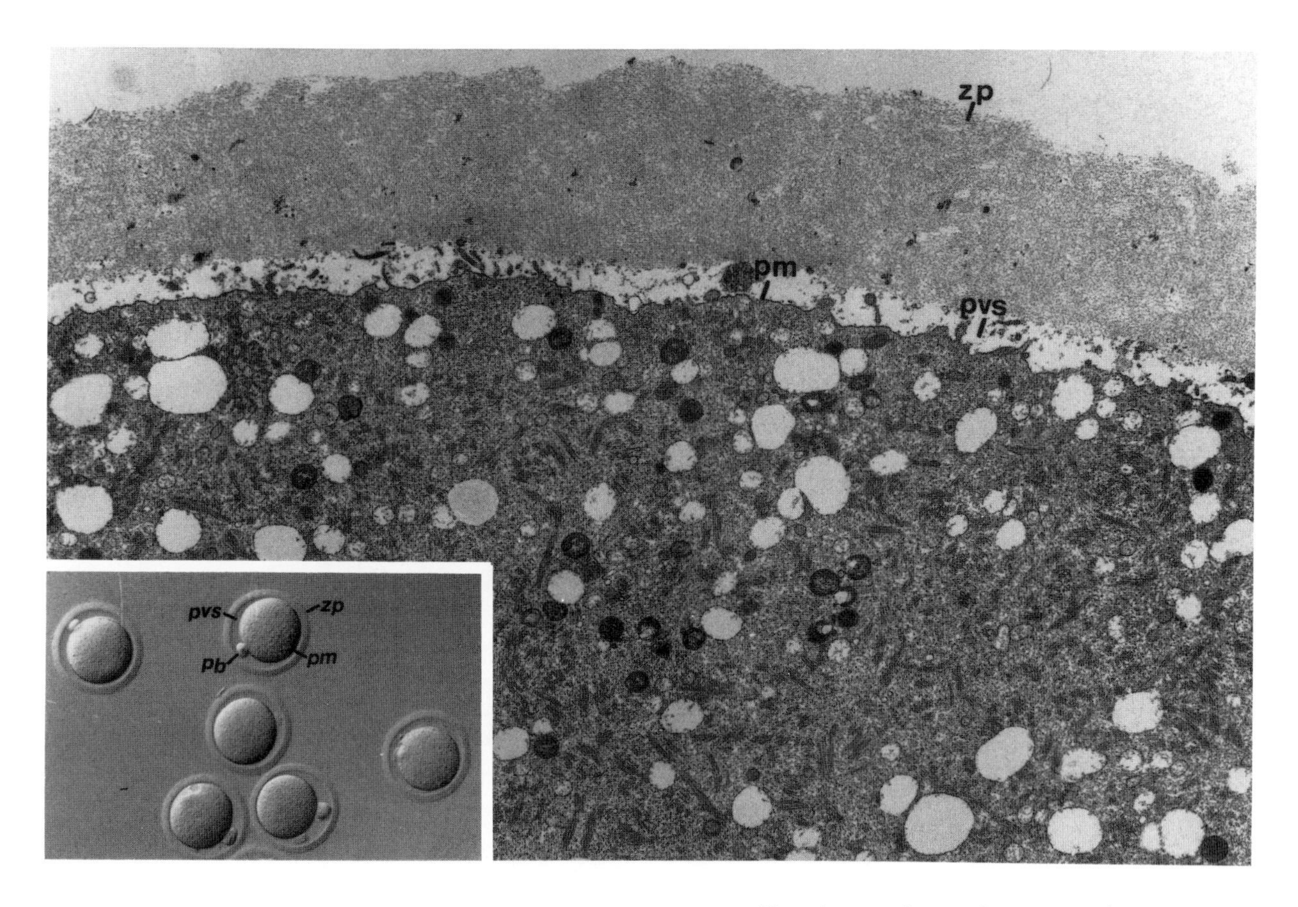

FIGURE 1 Light and electron micrographs of unfertilized mouse eggs. Shown are a transmission electron micrograph of a thin section taken through an unfertilized mouse egg and a light micrograph (*inset*) of some intact unfertilized mouse eggs. The light micrograph was taken using Nomarski differential interference contrast (DIC) optics. zp, Zona pellucida; pm, plasma membrane; pvs, perivitelline space (between the zp and pm); pb, polar body.

is supported by interactions between sperm receptors and egg-binding proteins.
5. Penetration of the vitelline envelope by acrosome-reacted sperm.
6. Fusion of a single sperm with the egg.
7. Establishment of a fast block to polyspermy, at the egg plasma membrane, triggered by sperm–egg fusion: This block is probably due to a change in electrical potential caused, in turn, by a change in the permeability of the plasma membrane to certain ions.
8. Completion of the cortical reaction (i.e., fusion of cortical granule and plasma membranes with exocytosis of cortical granule contents) triggered by sperm–egg fusion.
9. Completion of fertilization envelope formation triggered by cortical granule components.

B. Fertilization Pathway in Mammals

Based primarily on studies with mouse gametes, the fertilization pathway for many mammalian organisms consists of several steps that occur in a compulsory order. Briefly, these steps include the following:

1. Species-specific binding of acrosome-intact sperm to the zona pellucida (Fig. 3): Binding is supported by interactions between sperm receptors and egg-binding proteins.
2. Completion of the acrosome reaction by sperm bound to the zona pellucida: The mammalian acrosome reaction is not accompanied by emission of an acrosomal process. Rather, it consists of fusion of outer acrosomal and plasma membrane at the anterior region of the sperm head, with formation of hybrid membrane vesicles and exposure of inner acrosomal membrane.
3. Penetration of the zona pellucida by acrosome-reacted sperm.
4. Fusion of a single sperm with the egg.
5. Establishment of a fast block to polyspermy, at the egg plasma membrane, in response to sperm–egg fusion.

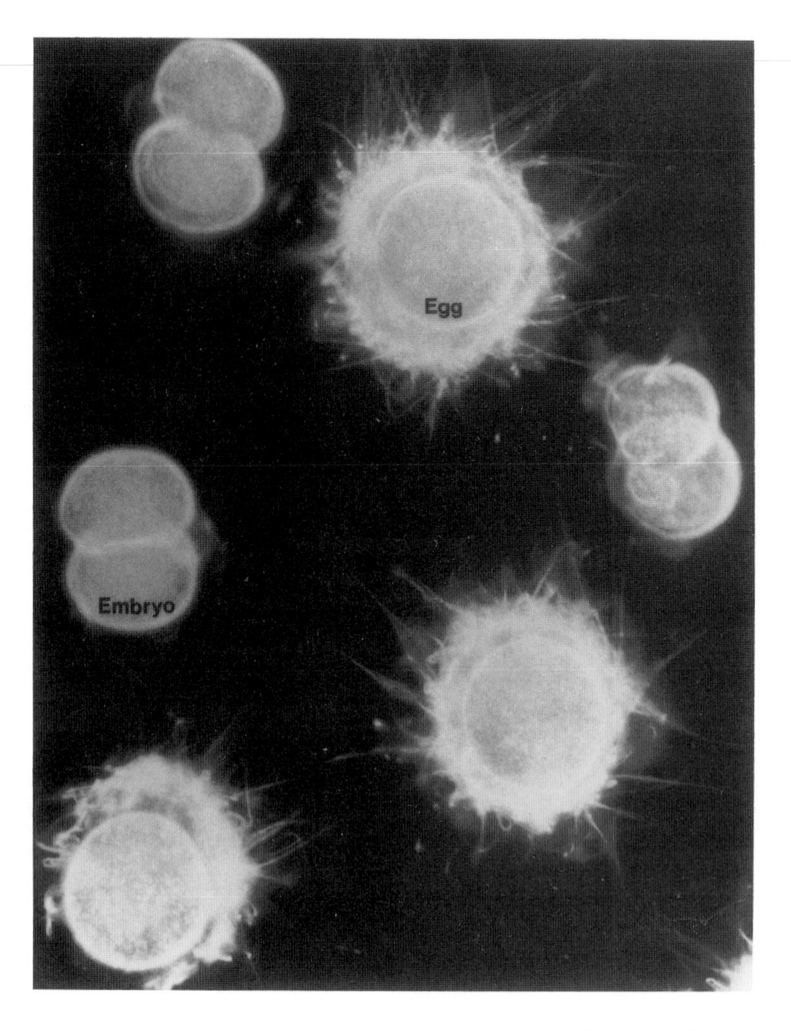

FIGURE 2 Light micrograph of mouse sperm bound to unfertilized mouse eggs in the presence of two-cell mouse embryos *in vitro*. Sperm bind to the zona pellucida of unfertilized mouse eggs, but not to the zona pellucida of two-cell embryos due to inactivation of sperm receptors following fertilization. The micrograph was taken using dark-field optics.

6. Completion of the cortical reaction triggered by sperm–egg fusion.
7. Completion of the zona reaction triggered by cortical granule components.

Fertilization of mouse eggs *in vitro* follows an approximate timetable. Binding of some sperm to eggs occurs as early as 1 to 2 min after combining gametes, with maximum binding observed within 10 to 20 min. Once bound to the zona pellucida, some sperm undergo the acrosome reaction within another 10 to 20 min. It takes at least 15 to 20 min for a bound, acrosome-reacted sperm to penetrate the zona pellucida and reach the egg plasma membrane. The sperm head is incorporated into egg cytoplasm (sperm–egg fusion; fertilization) within another 10 min or so. Thus, sperm enter egg cytoplasm as early as 1 to 2 hr after combining mouse gametes *in vitro*. Mouse eggs fertilized *in vitro* give rise to viable offspring following transfer of fertilized eggs to oviducts of foster mothers.

C. Features Common to Mammalian and Nonmammalian Fertilization Pathways

The fertilization pathways for mammalian and nonmammalian organisms have certain features in common. These include the following:

1. Species-specific binding of sperm to receptors located in the egg extracellular coat (either zona pellucida or vitelline envelope).
2. Induction of the acrosome reaction by a component of the egg extracellular coat (either zona pellucida or jelly coat).

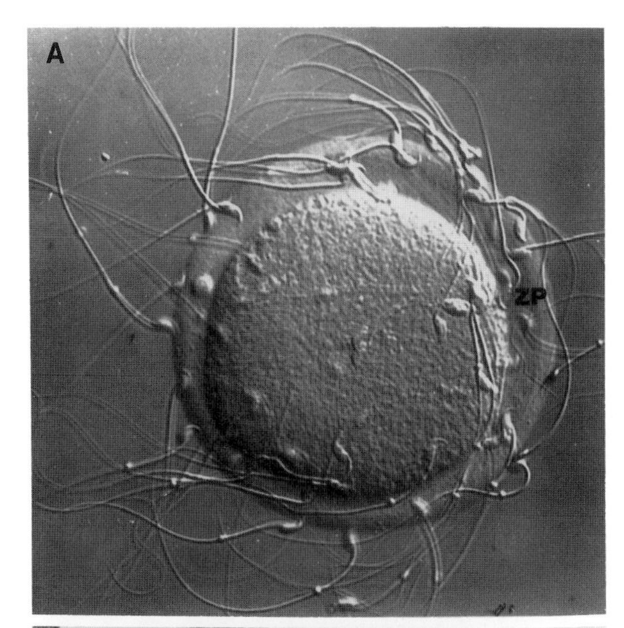

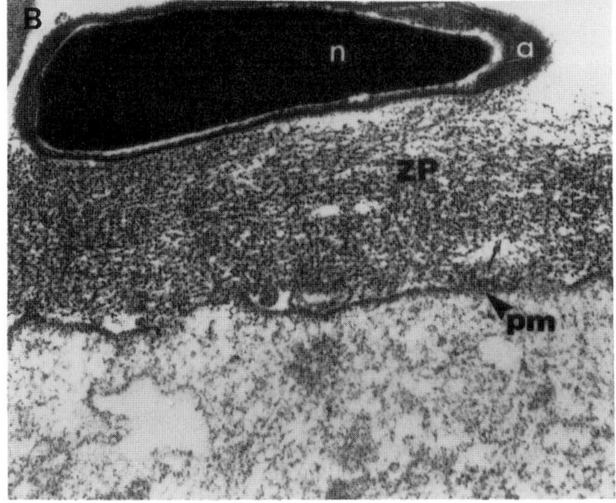

FIGURE 3 Light and electron micrographs of mouse sperm bound to unfertilized mouse eggs *in vitro*. (A) A light micrograph of multiple sperm bound to an unfertilized egg. (B) A transmission electron micrograph of a thin section taken through a sperm bound to an unfertilized egg. Binding occurs between sperm receptors located on the zona pellucida and egg-binding proteins located on the plasma membrane overlying heads of acrosome-intact sperm. zp, Zona pellucida; pm, plasma membrane; n, nucleus; a, acrosome.

3. Establishment of a fast block to polyspermy, at the egg plasma membrane, in response to sperm–egg fusion.
4. Completion of the cortical reaction triggered by sperm–egg fusion.
5. Modification of the egg extracellular coat (either zona reaction or fertilization envelope formation), which includes inactivation of sperm receptors, following fertilization.

It is noteworthy that each pathway includes three events that involve fusion of gamete membranes; the acrosome reaction (sperm), union of sperm and egg, and cortical reaction (egg). Thus, despite the evolution from external to internal fertilization, estimated to have occurred over a period of 100 million years or so, many of the cellular and molecular mechanisms involved remain the same.

III. Nature of Sperm Receptors

Relatively recent studies of the mouse sperm receptor have revealed a great deal about its structure and multiple functions. The lessons learned from such studies in mice apparently apply to other mammals as well, including humans. Furthermore, from studies of the sea urchin sperm receptor, it is clear that mammalian and nonmammalian sperm receptors have several important features in common.

A. Biochemical Characteristics of the Mouse Sperm Receptor

The mouse sperm receptor is a glycoprotein, called ZP3, that is found in more than a billion copies in the zone pellucida. ZP3 consists of a single polypeptide chain (402 amino acids; M_r 43,900) to which sugars are attached at certain asparagine, serine, and threonine residues. ZP3 contains three or four asparagine-linked oligosaccharides. On gel electrophoresis ZP3 migrates as a relatively broad band having an average molecular weight of 83,000. The appearance of ZP3 on gels is due to the heterogeneity of oligosaccharides attached to the polypeptide chain. The acidic nature of ZP3 is also due to its oligosaccharides, not to the polypeptide chain.

B. Structural Characteristics of the Mouse Sperm Receptor

ZP3 has a unique polypeptide chain, unrelated to any other known proteins. Compared to the average vertebrate protein, ZP3 contains an unusually high number of serine (43), threonine (29), and proline (29) residues. The polypeptide chain has six potential asparagine-linked oligosaccharide acceptor sites. Secondary structure prediction methods indicate that ZP3 polypeptide chain consists of stretches of 5–29 residues in an extended chain conformation, interrupted by stretches of 4–10 residues in reverse turn or coil conformations. Analyses of the local hydrophobicity along the ZP3 polypeptide chain indicate that it is neither strongly hydrophobic nor hydrophilic.

C. Molecular Genetics of the Mouse Sperm Receptor

The mouse ZP3 polypeptide chain is encoded by a single-copy gene located on chromosome 5. The transcription unit, approximately 8.5 kb, consists of eight exons and seven introns. The 8 exons provide a messenger RNA with 1272 nucleotides of protein-coding sequence and 45 nucleotides of noncoding sequence. Fully processed ZP3 messenger RNA, approximately 1.5 kb, is polyadenylated (200–300 residues) at its 3′ terminus. The messenger RNA encodes a 424-amino acid polypeptide chain, which includes a 22-amino acid signal sequence that is cleaved from the amino terminus during processing of the ZP3 polypeptide chain. The ZP3 gene is expressed, and ZP3 is synthesized and secreted, only during oogenesis by growing mouse oocytes, not by any other cell type. Therefore, ZP3 is an example of sex-specific gene expression. Molecular probes, constructed on the basis of mouse ZP3 gene sequence, recognize the analogous gene in a variety of other mammals, including humans. The sperm receptor polypeptide chain is approximately the same size in several mammals, although the extent of glycosylation of the sperm receptor apparently varies considerably among different mammals.

D. Functional Characteristics of the Mouse Sperm Receptor

ZP3 has two known functions during fertilization. First as a sperm receptor during species-specific binding of acrosome-intact sperm to eggs and then as an acrosome reaction inducer following the binding of sperm to eggs. Each acrosome-intact sperm is capable of binding by its head to tens of thousands of ZP3 molecules located at the surface of the zona pellucida. ZP3 is located every 15 nm or so along the filaments that make up the zona pellucida. Interactions between egg-binding proteins, located in plasma membrane overlying the sperm head, and ZP3 in the zona pellucida, are sufficient to support gamete adhesion and to trigger the acrosome reaction. Induction of the acrosome reaction is dependent on the presence of external calcium and probably requires multivalent interactions between receptors and binding proteins. Acrosome-reacted sperm remain bound to the egg by interacting with another zona pellucida glycoprotein, called ZP2, that is present as part of a ZP2–ZP3 dimer. Binding to ZP2 may be mediated by a sperm proteinase, perhaps acrosin. Thus, ZP3 and ZP2 serve as primary and secondary sperm receptors, respectively. Only bound, acrosome-reacted sperm can penetrate the zona pellucida, probably by using the proteinase acrosin to digest a path through the extracellular coat, and fuse with egg plasma membrane. Following fertilization of eggs, both ZP3 and ZP2 are modified by cortical granule enzymes as a result of the cortical and zona reactions. The modified zona pellucida glycoproteins are rendered unable to perform their normal functions and free-swimming sperm are unable to bind to fertilized eggs.

IV. Mechanism of Action of Sperm Receptors

A. Involvement of Carbohydrate in Sperm Receptor Function

The sperm receptor function of ZP3 is attributable to the glycoprotein's oligosaccharides and, in particular, to a specific class of serine/threonine-linked oligosaccharides (O-linked oligosaccharides). These oligosaccharides have a galactose residue at their nonreducing terminus, in α-linkage with the penultimate sugar, that is essential for sperm receptor function. The role of carbohydrates in the initial steps of the mammalian fertilization pathway suggests that the relative species specificity of gamete adhesion may be due to variations in the structure of oligosaccharides (e.g., composition, sequence, conformation) present on sperm receptors of different mammals. It also suggests that inactivation of sperm receptors following fertilization is due to modification of the oligosaccharides by (a) cortical granule glycosidase(s).

Like ZP3, the sea urchin sperm receptor, located in the egg vitelline envelope, is a glycoprotein. It has a very high molecular weight (M_4 $>10^7$) and is extremely insoluble. It is approximately 85% protein by weight and contains highly sulfated glycosaminoglycan-like chains composed of iduronic acid, galactosamine, and fucose. Also like ZP3, sea urchin sperm receptor function is dependent on the glycoprotein's carbohydrate chains. Similarly, sperm receptor function in ascidians, amphibians, and plants is also dependent on carbohydrates. Thus, gamete adhesion in a variety of organisms, from plants to animals, is supported by carbohydrate (egg)–protein (sperm) interactions.

B. Characteristics of a Nonmammalian Egg-Binding Protein

The egg-binding protein in sea urchins, called bindin, has a molecular weight of approximately 24,000 and coats the surface of the sperm acrosomal process. Bindin is an extremely hydrophobic protein that recognizes and binds to specific sequences of sugar residues. For example, bindin recognizes jelly coat fucans (polymers of fucose sulfate esters), fucoidan (fucose sulfate ester), and a variety of other sulfated polysaccharides. Apparently, the sulfate moiety is extremely important for binding. Differences in bindin primary structure have been found for different sea urchin species and, together with differences in sperm receptor structure, probably account for the observed species specificity of fertilization in echinoderms. Bindin may also play a role in the fusion of sperm and egg.

C. Candidates for a Mammalian Egg-Binding Protein

The identity of the mammalian egg-binding protein located on sperm is quite problematical. Several candidates, most of which interact with carbohydrates, have been suggested. Among others, these candidates include galactosyl and fucosyl transferases, proacrosin (binds to carbohydrates), fucose- and galactose-binding proteins, α-fucosidase and α-mannosidase, an M_r 56,000 protein that recognizes ZP3, and an M_r 95,000 protein that recognizes ZP3 and is a tyrosine kinase substrate. This aspect of receptor-mediated fertilization in mammals remains to be resolved.

V. Concluding Remarks

Sperm receptors are specific glycoconjugates (protein and carbohydrate) located on egg extracellular coats, either zona pellucida (mammals) or vitelline envelope (nonmammals), that support species-specific binding of sperm to eggs and, consequently, restrict heterospecific fertilization. The protein and carbohydrate constituents both play roles in sperm receptor function, with carbohydrate playing a primary role in binding of sperm to the receptor. In some nonmammalian organisms the egg-binding protein, which interacts with sperm receptors, is a carbohydrate-binding molecule associated with inner acrosomal membrane of the sperm head. The precise nature of the mammalian egg-binding protein is unclear at present, although nearly all candidates for this function bind to carbohydrates and are associated with plasma membrane overlying the sperm head. In many mammals, binding to receptors induces sperm to undergo the acrosome reaction, which is required for penetration of the extracellular coat and gamete fusion. Sperm receptors are inactivated following fertilization and egg extracellular coats undergo structural rearrangements, thus assisting in the prevention of polyspermy.

Bibliography

Garbers, D. L. (1989). Molecular basis of fertilization. *Annu. Rev. Biochem.* **58,** 719.

Gwatkin, R. B. L. (1977). "Fertilization Mechanisms in Man and Mammals." Plenum, New York.

Hartmann, J. F. (ed.) (1983). "Mechanism and Control of Animal Fertilization." Academic Press, New York.

Hedrick, J. L. (ed.) (1986). "The Molecular and Cellular Biology of Fertilization." Plenum, New York.

Longo, F. J. (1987). "Fertilization." Chapman and Hall, London.

Metz, C. B., and Monroy, A. (eds.) (1985). "Biology of Fertilization," Vols. 1–3. Academic Press, New York.

Schatten, H., and Schatten, G. (eds.) (1989). "The Cell Biology of Fertilization." Academic Press, New York.

Schatten, H., and Schatten, G. (eds.) (1989). "The Molecular Biology of Fertilization." Academic Press, New York.

Shapiro, B. M., Schackmann, R. W., and Gabel, C. A. (1981). Molecular approaches to the study of fertilization. *Annu. Rev. Biochem.* **50,** 815.

Trimmer, S., and Vacquier, V. D. (1986). Activation of sea urchin gametes. *Annu. Rev. Cell Biol.* **2,** 1.

Wassarman, P. M. (1987). Early events in mammalian fertilization. *Annu. Rev. Cell. Biol.* **3,** 109.

Wassarman, P. M. (1987). The biology and chemistry of fertilization. *Science* **235,** 553.

Wassarman, P. M. (1988). Fertilization in mammals. *Sci. Amer.* **256** (December), 78.

Wassarman, P. M. (1990). Profile of a mammalian sperm receptor. *Development* **108,** 1.

Wassarman, P. M. (ed.) (1990). "Elements of Mammalian Fertilization." CRC Press, Boca Raton, Florida.

Yanagimachi, R. (1988). Mammalian fertilization. *In* "The Physiology of Reproduction" (E. Knobil and J. D. Neill, eds.). Raven, New York.

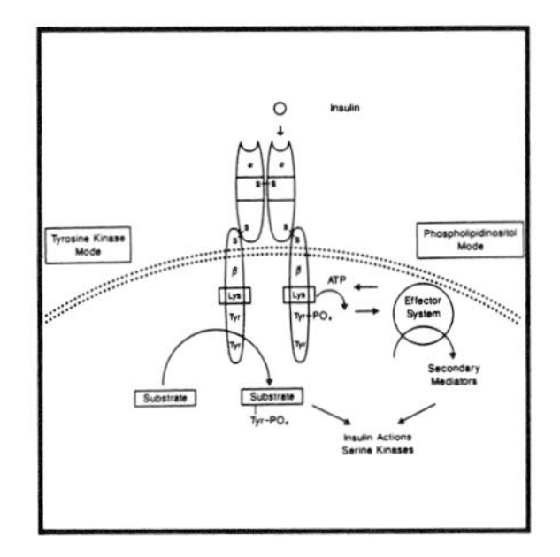

Receptors, Biochemistry

JOHN J. MARCHALONIS, *University of Arizona*

Glossary

Affinity Strength of binding between a ligand and the receptor combining site; it is usually represented by the association constant for the monovalent binding of one ligand to one site

Agonist Drug that mimics the stimulatory properties of the endogenous component that activates a receptor

Antagonist Drug that mimics the inhibitory capacity of endogenous inhibitors of receptor function and blocks the action of an agonist

Antibody Induced, circulating proteins that bind specifically to foreign antigens and brings about their removal and destruction

Antigen Molecule that is recognized as nonself or foreign by specific lymphocytes of the immune system

Dissociation constant (Kd) Constant that describes the equilibrium state of the reversible reaction between two molecular species A and B such that $A + B \rightleftharpoons AB$. The dissociation constant is $\frac{(A)(B)}{(AB)}$ where (A) and (B) are the concentrations at equilibrium of the free reactants and (AB) is that of the complex. The units of Kd are mol/l. It is the reciprocal of the association constant.

Effector Second attribute of a receptor; following specific binding of ligand, the receptor carries out an effector function such as protein kinase activity or ion transport, which alters cell function

Immunoglobulins Family of proteins to which antibodies belong; the antigen-specific receptors of bone marrow-derived lymphocytes and thymus-derived lymphocytes are also special types of immunoglobulins

Kinase Enzyme that catalyzes the phosphorylation of proteins

Lectins Proteins that are generally of unknown function but are characterized by their capacity to bind certain sugars

Ligand Molecule that is bound specifically by a receptor; this can be a hormone, antigen, or any number of molecules that show specific binding

Receptor Cell-associated molecule that binds ligands specifically; this binding initiates effector mechanisms leading to cell stimulation or regulation

A PROCESS INVOLVING a cellular receptor is one in which specific binding of a ligand to this molecule initiates a physiological response. The ligand that binds the receptor can be either of endogenous (e.g., hormones, neurotransmitters) or exogenous (e.g., antigen) origin and may consist of low-molecular weight organic molecules (e.g., acetylcholine, adrenaline) or larger molecules such as the polypeptides insulin or growth hormone. Although there are examples of specific cell activation by binding of lectins or microbial toxins to plasma membrane glycolipids, the usual situation is one in which the receptor is a protein that shows a high degree of specificity for the particular ligand and binds it with a relatively high-binding affinity. The high binding affinity corresponds to the fact that relatively low concentrations (nanomolar or less) of ligand such as hormones are required for cell activation. The first attribute of a receptor is its specific binding capacity; the second key attribute is the effector function carried out by the receptor after it has bound the ligand. The binding is generally reversible and can be inhibited, in the case of trace amounts of labeled ligand, by excess native ligand. There is an extensive literature on the pharmacology of drug receptors, which was developed based on demonstration of the ability of particular receptors to be stimulated by a variety of defined agonists and inhibited by a variety of defined antagonists. This chapter reviews pertinent data on the types, cellular distribution, and biochemical activation of major classes of receptors for neurotransmitters, hormones, and foreign antigens. Although the receptors considered here are often described in the context of their relationship to a particular cell or physiological function, they are found on many types of cells throughout the body. However, differences usually are profound in number of binding affinity in corresponding receptors expressed on various cell types.

I. Types of Receptors

Figure 1 illustrates the cellular location and models for activation of six characterized types of receptors. Intracellular receptors for steroids (e.g., thyroid hormone, estrogen, and some hormonal lipids such as vitamins A and D) are located within the cell nucleus (A). The hormone binds specifically to the receptor (R_1), and the complex then induces the synthesis of specific proteins by binding to nuclear chromatin and enhancing the transcription of the appropriate genes. The other receptors considered here occur on the plasma membrane of cells. Receptors for several neurotransmitters such as acetylcholine γ-aminobutyric acid and the amino acid glycine are either ion channels themselves or are associated with ion channels in the cell membrane. These ion channels open in response to the physiological neurotransmitters (and to its agonists), with the receptors then controlling cell membrane potential and the internal ionic composition. Receptors for several polypeptide growth factors including insulin (C) are themselves ligand-stimulated, plasma membrane-bound protein kinases. The physiological substrates for these kinases are just being identified, but their initial phosphorylation is localized to tyrosyl rather than to seryl or threonyl residues. The latter is the usual case with subsequent phosphorylation dependent on cyclic adenosine monophosphate (cAMP)-dependent kinases. Numerous receptors (D) such as β- and α-adrenergic receptors act to either stimulate or inhibit adenylate

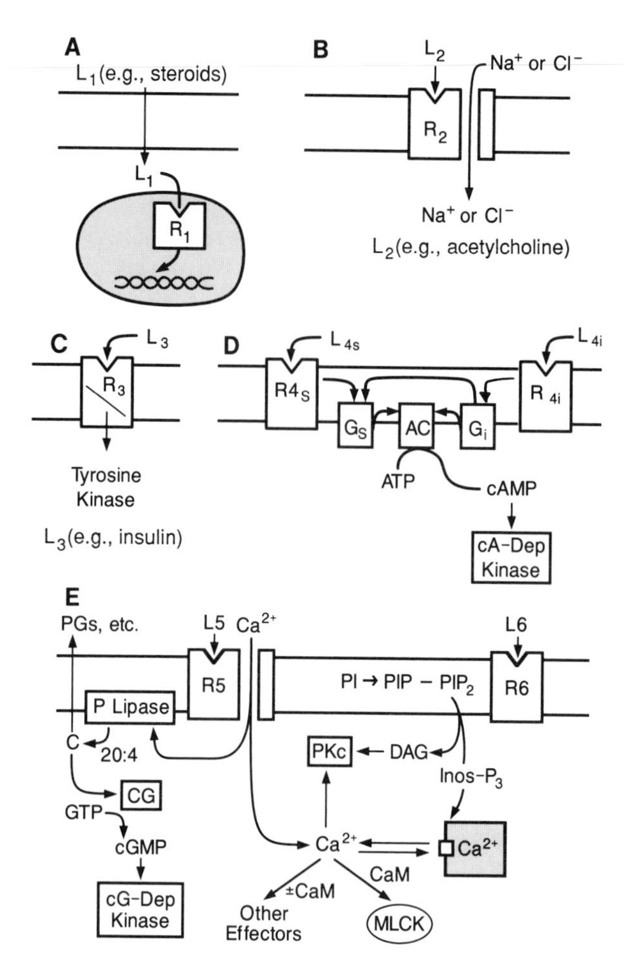

FIGURE 1 Schematic diagram of major types of receptor-mediated regulatory mechanisms. The types of receptors are designated R_1–R_6 and their corresponding ligands L_1–L_6. All of these classes of receptors can coexist on certain individual cells. The abbreviations and identification of receptors are given in the text.

cyclase via a mechanism involving distinct guanosine triphosphate (GTP)-binding regulatory proteins (G_s or G_i, respectively). The cumulative effect of the function of these receptors is either an increase or a decrease in intracellular concentration of cAMP. Following binding of a stimulatory ligand to $R4_S$, G_s is activated to a molecular form that can stimulate the production of adenylate cyclase by binding GTP. This binding is accelerated by the appropriate agonist–receptor ($R4_S$), complex. Deactivation occurs by the hydrolysis of GTP to guanosine diphosphate. In analogous fashion, G_i can be activated to a form that inhibits the activation of G_s and most probably inhibits the function or activation of adenylate cyclase. The cAMP nucleotide acts within the cell to stimulate cAMP-dependent protein kinases that catalyze the phosphorylation of various enzymes and other proteins on the seryl residues.

The calcium ion (Ca^{2+}) localized to the cell cytoplasm regulates some functions directly and initiates others only when it is bound to the intracellularly Ca^{2+}-dependent regulatory protein calmodulin (CaM). The calmodulin–calcium complex directly controls the functions of some protein and effects others by activating a distinct group of protein kinases, of which myosin light-chain kinase (MLCK) is an important example. Calcium and cAMP are major intracellular second messengers, but receptors for many hormones and neurotransmitters (E) apparently can cause the accumulation of multiple second messengers. For example, a fundamental event in some receptor-mediated systems (e.g., lymphocyte receptors for antigen) appears to be the specifically stimulated formation of an inositol-1,4,5-triphosphate ($InosP_3$) and diacylgylcerol (DAG) as the consequence of the hydrolysis of a membrane phospholipid, phosphatidylinositol-4,5 by phosphate (PIP_2). The formation of INOS-P_3 is closely linked to the release of Ca^{2+} from intracellular depots. In addition to the key functions of calcium described above, this ion in the presence of DAG also activates a distinct protein kinase (kinase C). Other secondary events that occur in these activations and regulatory cascade involve the hydrolysis of arachidonic acid ($C_{20:4}$) from membrane phospholipids by calcium-activated phospholipases with the subsequent generation of prostaglandins, prostacycline, leukotrines, and eicosanoids. The formation of these modified active lipids are oxidative events that lead further to the activation of guanylate cyclase, which results in the elevation of the intracellular concentration of guanosine 3′,5′ monophosphate (cyclic GMP [cGMP]). This cyclic nucleotide is the activator of yet another class of protein kinases, the cG-dependent (cG-DEP) protein kinases.

A number of these receptors can occur on a single cell. Antigen-specific lymphocytes, for example, can express uniquely specific receptors for an individual antigen, receptors for insulin, adrenergic receptors, and receptors for interleukins (polypeptide molecules regulating cell activation and differentiation) simultaneously (Table I). Complex cascades of biochemical reactions can be involved in activation of single receptors or receptor systems. Moreover, in the cases where a cell has several receptors that utilize a single effector mechanism, multiple extracellular signals may be integrated to yield a cumulative intracellular signal. For example, the submaximal stimulation of two individual receptors that activate adenylate cyclase and of one receptor that inhibits this enzyme will be finally expressed as a unique rate of synthesis of cAMP. Moreover, receptors that act by different primary mechanisms may be coordinated at other levels (i.e., release of intracellular calcium and the activation of adenylate cyclase can lead to phosphorylation and activation of the same metabolic enzymes or of distinct enzymes with opposing or synergistic functions).

TABLE I Example of Many Types of Receptors Found on a Single Cell Type (Lymphocytes)

Type of Receptor
Receptors for antigen
B cells
Antigen-binding receptor (membrane Ig)
T cells
Antigen-binding T-cell receptor
MHC-restricted T-cell receptors (α/β heterodimers); T3/Ti complex
Triggering
B cells
B-cell growth factor
B-cell differentiation factor
T cells
IL-2
Mitogens
Supportive
Hormones
Insulin
Growth hormone
Steroids
Carrier proteins
Transferrin
Low-density lipoproteins
α_2-HS glycoprotein
Transcobalamin
Dihydroxycholecaliciferol [1,25$(OH)_2D_3$]
G_c (vitamin D-binding protein)
Miscellaneous
Fc portion of IgG, IgM, IgA, and IgE
Complement component C3
Interferon
β-adrenergic
Histamine
(?) Acetylcholine

Furthermore, receptors are not the only determinants of acute regulation of physiological and biochemical function, but are themselves subject to regulatory and homeostatic control. The continued stimulation of cells with agonists generally results in a state of desensitization, also known as refractoriness or downregulation, such that subsequent exposure of the cell to the same concentration of ligand results in a markedly diminished response. A documented example is the observation that repeated use of β-adrenergic broncodilators such as isoproterenol (see Section III.A, for the treatment of asthma requires increasing dosages of this agonist). Conversely, hyper-reactivity or supersensitivity to receptor agonists is also frequently observed to follow reduction in the chronic level of stimulation of receptors by a particular ligand. Analogous situations of either increased or diminished sensitivity occurs following multiple stimulation of lymphocytes via their antigen-specific receptors. Depending on the manner of presentation of antigen and factor such as dosage, it is possible to observe either an enhanced "secondary stimulation" or diminished reactivity termed tolerance or immune paralysis.

II. Intracellular Receptors

A variety of steroid hormones exist. The estrogens, progesterones and androgens, are produced by the male and female sex organs and affect function and development of the sex organs. Adrenocorticotropic hormone stimulates the adrenal cortex, which itself generates a number of steroids that regulate either electrolyte balance (mineralocortcoids) or carbohydrate metabolism (glucocorticoids). All of the steroid hormones are thought to act by controlling the rate of synthesis of proteins by a process that is initiated by the binding of the steroids to receptor proteins in the cytoplasm of sensitive cells, where they form a steroid–receptor complex. Steroids can have profound effects on many cell types; for example, not only are cells of the gonads, placenta, and adrenal tissues sensitive to steroid hormones but peripheral tissues such as liver, fat, skeletal muscles, and hair follicles can form steroids and are affected by them. Some lymphocytes are exquisitely sensitive to steroids and this is a commonly used immunosuppressive regime. [*See* STEROIDS.]

Following binding of the steroid to its intracellular receptor, the complex undergoes a structural modification and moves into the cell nucleus where it binds to chromatin. This binding results in the transcription of mRNA corresponding to specific sets of enzymes and other proteins.

III. Cell Membrane Receptors For Endogenous Ligands

A. Adrenergic and Cholinergic Receptors

1. Adrenergic Receptors

Adrenergic receptors function in the physiologic control of circulation and in the response to drugs that act via the sympathetic nervous system and on the vasculature. Two general types of adrenergic receptors, α and β, each of which contains subtypes (α_1, α_2, β_1, β_2) are activated by epinephrine (adrenalin). The α and β-receptors have opposite effects (inhibitory or stimulatory) on the adenylate cyclase system. Analysis of gene and derived protein sequence has established that the two receptors are homologous to one another, to rhodopsin, and to other receptors that are coupled to the guaninine nucleotide regulatory proteins (G proteins). The molecules contain seven hydrophobic domains, which most probably represent transmembrane-spanning segments. A schematic diagram of the adrenoceptor system is shown in Figure 1D. Figure 2 depicts the likely extracellular, transmembrane helical and intracellular portions of α_2- and β_2-adrenergic receptors. The specificity for coupling to the stimulatory G protein lies within the region extending from the amino terminus of the fifth hydrophobic domain to the carboxyl terminus of the sixth. The major determinants of the α_2- and β_2-adrenergic receptor agonist and antagonist ligand binding are restricted to the seventh membrane-spanning domain. [*See* ADRENERGIC AND RELATED G PROTEIN-COUPLED RECEPTORS.]

Adrenergic receptors are found on neurons and on muscle tissue including skeletal and heart muscle

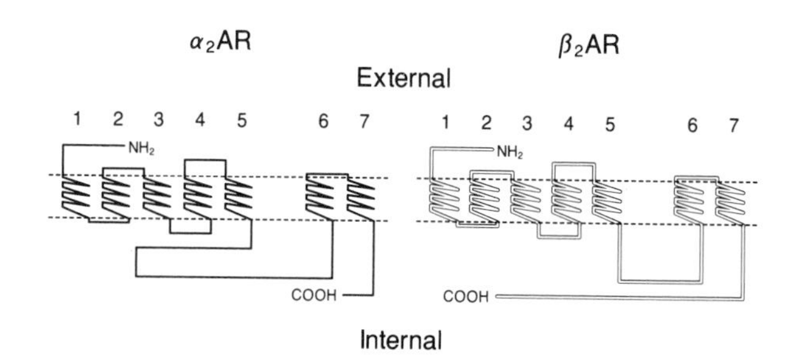

FIGURE 2 Diagram of the α_2-adrenergic receptor (α_2AR) and the β_2-receptor (β_2AR). The hydrophobic domains are depicted as helices that span the plasma membrane. The adrenergic receptors, the muscarinic, cholinergic receptors, and the opsins (e.g., rhodospins) have a similar overall organization and most probably represent homologous molecules.

cells and on smooth muscle. Activation of β-adrenergic receptors brings about an increase in adenylate cyclase. The β_1- and β_2-receptors can be generally activated by agonists such as isoproterenol or by selective agonists such as norepinepherine, which stimulates β_1-adrenoceptors, and procaterol, which selectively activates the β_2-receptor. Propranolol serves as an antagonist to β-adrenoceptors. In addition to the association of these receptors with the central and autonomic nervous system, they have been found on other types of cells including circulating lymphocytes.

2. Cholinergic Receptors

Cholinergic receptors were first detected on the basis of their sensitivity to the neurotransmitter acetylcholine. These receptors are prominent in the central nervous system, including motor neurons, hippocampus, and cerebral cortex and in the spinal cord motor neurons and autonomic ganglia. Such receptors have also been found on cells that are neither nerve nor muscle, such as lymphocytes. These receptors are associated with ion transport channels in the plasma membrane that open in response to agonists and, thereby, control cellular membrane potential and ionic composition. Two general types of cholinergic receptors were classified on the basis of their reactivity with known agonists and antagonists. There is a set of acetylcholine receptors where the effect of acetylcholine is mimicked by the alkaloid muscarine and is selectively antagonized by atropine. This subset is termed muscarinic receptors. The second broad subset of acetylcholine receptors is that in which the stimulatory effects of acetylcholine are mimicked by the agonist nicotine but are not antagonized by atropine. However, the stimulatory effect of nicotine is selectively blocked by other agents such as tubocurarine or α-bungarotoxin. This subset of cholinergic receptors is termed the nicotinic receptors. Adrenergic, muscarinic, and cholinergic receptors are known to be derived from different genes, but a number of similarities exist between these proteins in terms of overall structure and function. Muscarinic cholinergic and adrenergic receptors as well as others involved in modulating cellular function via guanine nucleotide proteins (G proteins) are apparently members of a multigene family of membrane proteins that has shown considerable conservation in evolution. For example, muscarinic receptors from insects to humans show a high degree of conservation based on immunological and biochemical comparisons. The receptor monomer units consist of approximately 500 amino acids but have larger apparent masses (approximately 80 kDa) by acrylamide gel electrophoresis because of glycosylation and conformational properties. As of this time, five different genes for the muscarinic receptors m, m_2, m_3, m_4, and m_5 have been described.

B. Receptors for Insulin and Insulinlike Growth Factors

Insulin is a polypeptide with a mass of approximately 6,000 daltons that is made of two chains of amino acids joined by disulfide bonds. This hormone is synthesized in the pancreas. It functions as the primary regulator of blood glucose levels. Insulin acts on cells to stimulate glucose, protein, and lipid metabolism as well as to regulate RNA and DNA synthesis by modifying the activity of a variety of enzymes in transport processes. The understanding of the molecular pathways of insulin action is essential to unravelling the pathogenesis of insulin-dependent diabetes (type I), and noninsulin-dependent (type II) diabetes mellitus. This knowledge is also important to understanding other insulin-resistant states including obesity, uremia, and glucocorticoid and growth hormone excess as well as a variety of rare genetic disorders such as leprechaunism and lipotropic diabetes. Insulin-dependent cell activation and differentiation are initiated by the specific binding of insulin to its plasma membrane receptor. The insulin receptor is present on virtually all mammalian cells with the density varying from as few as 40 receptors per cell to circulating red blood cells to more than 200,000 receptors per cell on fat cells and liver cells. [*See* INSULIN AND GLUCAGON.]

The insulin receptor is a heterodimeric glycoprotein comprised of two α-subunits of approximate mass 135 kDa and two β-subunits of approximate mass 95 kDa linked by disulfide bonds to give a β-α-α-β structure. The α-subunit is completely extracellular and contains the insulin-binding site. Approximately one-third of the β-subunit is external to the cell. The β-subunit contains a membrane-spanning helical region of approximately 24 amino acid residues and an intracellular domain that is a tyrosine kinase. Binding of insulin to the external α-subunit initiates an allosteric change that stimulates tyrosine phosphorylation of the β-subunit of the insulin receptor. The insulin receptor is a member of a fam-

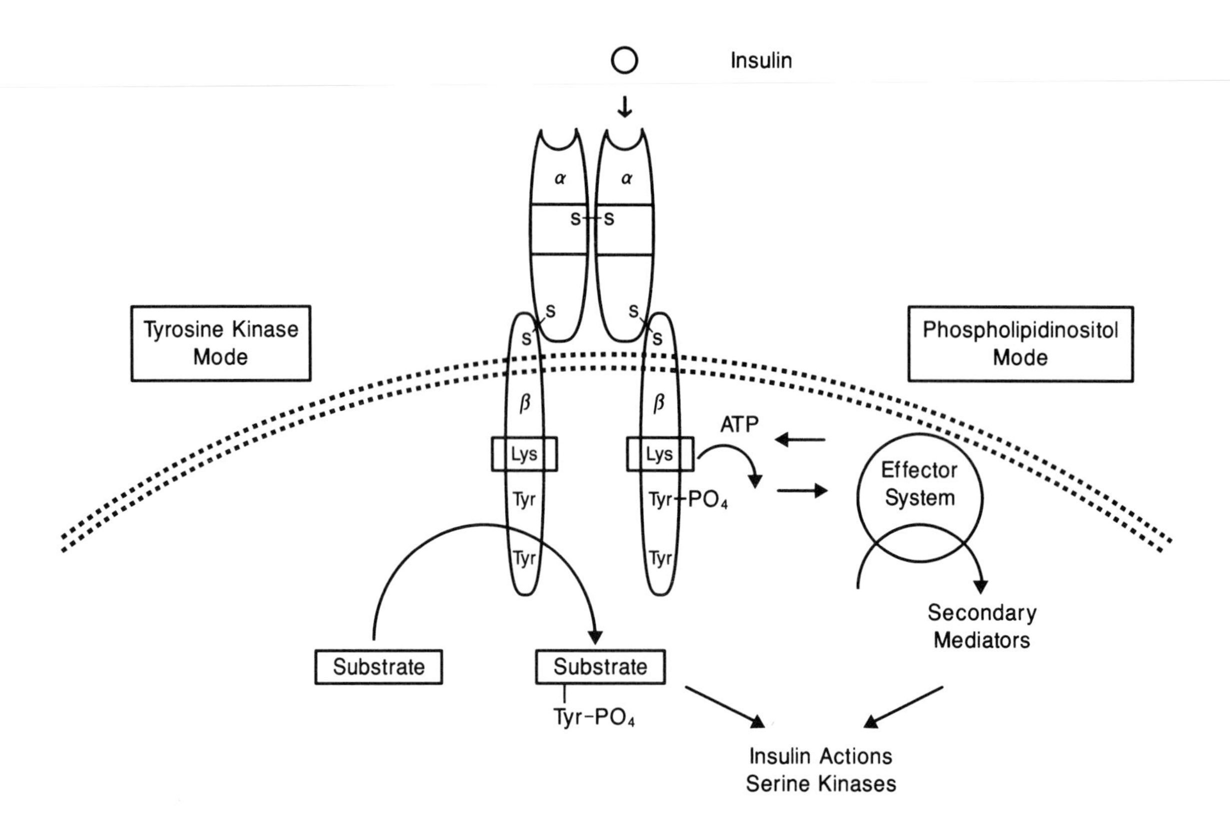

FIGURE 3 Schematic representation of the insulin receptor, its association with the cell membrane, and two general modes of its biochemical function. The binding of insulin is carried out by the α-subunits. The β-subunits are linked covalently to the α-subunits via disulfide bonds. The major portion of the β-subunit is intracellular and is tyrosine (Tyr) kinase. The first mode of activation shown is that involving the tyrosine kinase mechanism. The second mode implicates a phospholipid inositol-glycan head group, which is released in response to insulin that acts via an allosteric mechanism to activate one or more serine kinases.

ily of protein kinases that phosphorylate their substrates on tyrosine residues. Two groups of such kinases are related to insulin. The first group includes receptors for growth factors such as epidermal growth factor (EGF), platelet-derived growth factor (PDGF), insulinlike growth factor I, and colony-stimulating factor I. The second group of kinases homologous to the insulin kinase subunit consists of viral oncogene products and their related cellular homologues. These include V-ERB B, V-FMS, V-ABL, V-FES/FPS, V-SRC, V-ROS, and V-YES. The sequence homologies and functional similarities between the two groups of tyrosine kinases suggest that there is a structural similarity between the catalytic domains of growth factor receptors and oncogene products. Moreover, there has been considerable conservation of insulin receptor structure in evolution because the complementary DNA for insulin receptors of humans and the fruit fly *Drosophila* are extremely similar to one another. Figure 3 illustrates the cascade of biochemical mechanisms initiated by the specific binding of insulin to the insulin receptor.

C. Interleukins

Interleukins or cytokines are polypeptide hormones that regulate the activation and growth of cells involved in the immune system such as macrophages and lymphocytes but can also interact with a variety of nonlymphoid cells. As of this time, more than nine cytokines have been cloned and characterized molecularly, but we will focus on one here, interleukin 2 (IL-2), because the molecule has been structurally and functionally characterized as has its specific receptor. IL-2 is a small single-chain protein of mass 15.5 kDa that stimulates thymus-derived lymphocytes (T cells) to proliferate following antigenic stimulation. In the case of lymphocytes, the binding of the ligand antigen generally does not by itself stimulate cell division, and interleukins are required to initiate this process. The density of receptors for IL-2 on resting or nonactivated lymphocytes are generally too low to be detectable. However, following activation of T cells by mitogens, for example, IL-2 receptors are demonstrable within 4–8 hr and reach a peak in the range of 30,000–60,000 receptors per cell 48–96 hr following activation. IL-2 receptors can be detected using a monoclonal antibody (anti-TAC) that recognizes the 55-kDa receptor subunit. The majority of resting T cells, bone marrow-derived lymphocytes (B cells), or monocytes do not bind the monoclonal antibody. However, following activation, the receptor becomes detectable. [*See* CYTOKINES IN THE IMMUNE RESPONSE; INTERLEUKIN-2 AND THE IL-2 RECEPTOR.]

High-affinity (Kd of approximately 10 picomolar) and low-affinity (Kd of approximately 10 nanomolar) forms of the human IL-2 receptor have been identified. The receptor consists of two IL-2-binding peptides: a 55-kDa peptide that reacts with the anti-TAC monoclonal antibody and a 75-kDa IL-2-binding peptide that does not react with the anti-TAC antibody. The functional receptor probably exists as a dimer composed of one 55-kDa subunit and one 75-kDa molecule, which interact noncovalently in a cooperative manner. Both subunits have an external portion involved in the specific binding of the polypeptide hormone, a transmembrane helical portion and an intracellular domain. There is currently considerable interest investigating the pharmacology of IL-2 receptors because they are essential for the proliferation of antigen-specific activated T cells in immunity and also in malfunctions of the immune system such as T-cell leukemia, where continued stimulation of some leukemic cells is maintained by an autocrine mechanism in which the cells produce both IL-2 molecules and IL-2 receptors in large quantities. As an example of potential therapeutic use, two effective immunosuppressive pharmaceuticals, glucocorticoids and cyclosporin, both mediate their effects by inhibiting the production of IL-2 antigen-activated T cells.

D. Surface Carbohydrates as Receptors for Lectins

Both proteins and lipids associated with the outerface of cell plasma membranes are often glycoslated. The various glycosyl moieties can serve as targets or "receptors" for lectins, which are carbohydrate specific binding proteins found in both the serum or hemolymph and on the cells of many vertebrates and invertebrates and are also common constituent plant seeds. The binding of the lectins to cell-surface glycoproteins or glycolipids is usually or relatively low affinity (e.g., 1–6 micromolars for binding of concanavalin-A to murine lymphocytes), but the binding of certain lectins often initiates cellular activation and differentiation mimicking that shown by interaction of specific ligand with its receptor. Insulin receptors, for example, are glycoproteins that can be activated by lectins reacting with the glycosyl moieties on the chains. A number of characterized surface proteins of human cells are glycosylated and have been shown to bind lectins, particularly concanavalin-A, lentil lectin, or phytohemagglutinin to cite some of the more frequently used lectins. The concanavalin-A and lentil lectins are of plant origin and have specificities for a α-D-mannose and α-D-glucose. The carbohydrate specificity of phytohemagglutinin has not been established, but this lectin will activate T cells and initiate the cascade of events normally associated with binding of IL-2 by its receptor. Biologically important membrane surface components known to bind lectins include the following: membrane receptor immunoglobulins (IgM_m, IgD_m) of resting B cells, major histocompatibility complex (MHC) products, class I and class II; the α/β heterodimeric antigen receptor of T cells and the insulin receptor. In addition to the glycoproteins, some glycolipids serve as targets for the binding of lectins or of specific toxins such as cholera toxin. The capacity of glycoproteins or glycolipids to bind lectins is not a specific receptor property, as described above, but merely reflects the fact that glycosylated membrane molecules may have the right sugar composition and sequence to combine with the lectins. The lectin molecules are generally multivalent so cross-linking of the membrane molecules occurs with a variety of biochemical consequences, comparable with that shown for specific binding of insulin to its receptor or antigen to its receptor on T cells. Approximately 10% of plasma membrane proteins of lymphocytes are glycosylated and capable of reacting specifically with a major lectin such as concanavalin-A. The capacity of cells of various types to bind lectins has provided useful schemes for the fractionation of distinct types of cells and also has facilitated studies of the biochemistry of activation.

IV. Lymphocyte Membrane Receptors For Antigen

The receptors considered above are found on a number of cell types, and, with the exception of the binding of lectins, function normally in response to endogenous molecular signals. By contrast, the antigen-specific receptors of lymphocytes are restricted to immunologically committed lymphocytes and are adapted to respond to foreign molecular configurations termed antigens. This unique association only with committed lymphocytes occurs because at least two gene rearrangements involving variable (V), joining (J), and constant (C) gene segments are required for expression. [*See* LYMPHOCYTES.]

There are two major classes of lymphocytes that express antigen-specific surface receptors. One class, termed B cells, expresses membrane associated immunoglobulin as its surface receptor for antigen. Following stimulation and differentiation, these cells and their progeny (plasma cells) eventually produce large quantities of immunoglobulin (antibody) that is secreted into the serum. [*See* B-CELL ACTIVATION, IMMUNOLOGY.]

The other major class of lymphocytes, termed thymus-derived lymphocytes or T cells, does not itself produce antibodies but recognizes antigen and serves a central regulatory role in the generation of antibodies by either helping B cells produce antibodies (helper or amplifier T cells) or suppressing the formation of antibodies (suppressor T cells). T cells usually interact with accessory cells such as macrophages in their regulatory role. This interaction requires compatibility between the cells in the major histocompatibility complex (MHC). T cells also carry out antigen-specific cellular immune reactions such as delayed type hypersensitivity, rejection of allografts, destruction of tumors, and elimination of virally infected cells. The capacity of T cells to recognize antigen is considered to be "MHC-restricted or -dependent," and it is suggested that T cells can recognize antigen only in the context of particular determinants of the MHC. MHC products possess a surface groove that allows them to present processes peptide antigens to the low-affinity α/β-receptors on T cells.

An antigen is a molecular structure that is normally foreign ("non/self") and is recognized by antigen-specific receptors of B cells (immunoglobulins) or T cells (α/β or γ/δ heterodimers). Although large structures including cells and viruses are often considered to be antigens, only a relatively small portion of their molecular architecture comprises the actual determinants recognized. Antigens can be protein, carbohydrate, or lipid, and the size of the bound determinant (epitope) corresponds to approximately six amino acids or six monosaccharide units. This molecular size corresponds to the size of the combining site of the antibody or T-cell receptor for antigen. In addition, the antigen receptors of T and B cells react to small organic molecules such as 2,4-dinitrophenol, a small molecule that is termed a hapten. Although the affinity of antibody following boosting or secondary stimulation for antigens such as haptens can be high (nanomolar range), the antigen receptors on T cells and on unstimulated B cells have low affinities for antigens (micromolar range). The low affinity of the T-cell antigen receptor is most probably compensated for by the presentation of certain antigens such as peptides to helper T cells by the MHC. The number of membrane immunoglobulin antigen receptors on B (IgM_m or IgD_{2m}) or T cells (α/β or γ/δ heterodimers) is in the range of 10,000–100,000 molecules/cell. [*See* T-CELL RECEPTORS.]

A simplified diagram illustrating the properties of antigen receptors of B- and T-lymphocytes is given in Figure 4. As depicted here, the surface receptor of B cells is essentially an immunoglobulin that has a combining site and light polypeptide chains identical to those of the secreted antibody. In this particular case, the B-cell receptor is membrane IgM immunoglobulin and the secreted antibody is IgG. IgG molecules appear on the surface of memory B cells (those that are the progeny of cells previously stim-

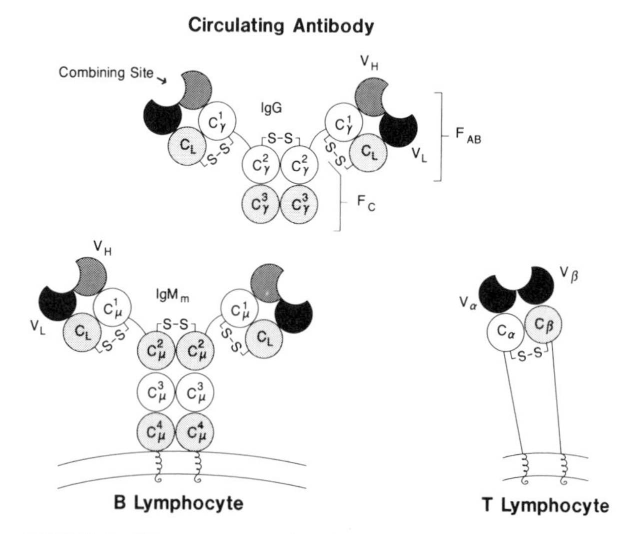

FIGURE 4 Diagram comparing the antigen-specific receptors of B- and T-lymphocytes with the major class of circulating antibody in humans. The combining site for antigen in all cases is formed via the interaction of the variable (V) domains of heavy and light chains in the case of antibody and B-lymphocyte receptors and V_α and V_β in the case of the α/β T-lymphocyte heterodimeric receptor. The α/β heterodimer depicted here does not include the accessory T3 complex with which is it invariably associated. F_{AB}, antigen binding fragments; F_C, common or crystalizable fragments.

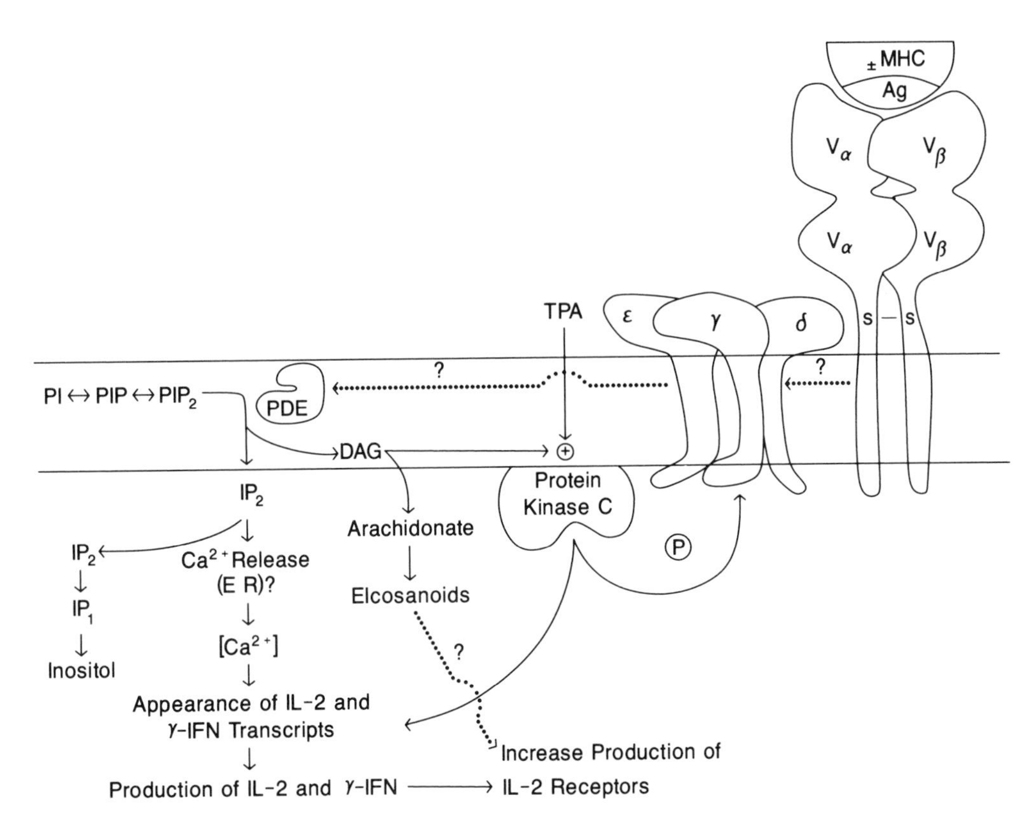

FIGURE 5 Membrane expression of the disulfide-bonded α/β heterodimer existing in association with the T3 complex (δ, γ, ε). This scheme illustrates the membrane association and the processes of transmembrane signalling initiated by binding of either antigen plus MHC to the combining site of the α/β heterodimer (V_α/V_β) or by antibodies against the T3 complex. A detailed description is given in the text. Although the biochemical mechanisms elicited by binding of antigen to the T-cell receptor complex are similar to those associated with the insulin receptor or with the scheme given for receptor 6 in Figure 1 above, the rearranging α/β- or γ/δ-receptors are found only in T cells.

ulated by antigen and expressing the same specificity) and is the major class of the circulating antibody. Membrane IgM is one of the two major antigen receptor classes on B cells that have not been previously stimulated by antigen. The short transmembrane piece found on the membrane form of immunoglobulin is lacking on the secreted molecules. The antigen receptor on T-lymphocytes is a member of the immunoglobulin family and contains a combining site for antigen that is formed of the interaction of variable regions of the α- and β- or γ- and δ-receptors. The α/β or γ/δ T-cell receptors are disulfide bond linked heterodimers that occur in association with a complex of molecules termed the T3 complex, which is itself not antigen-specific and appears to be necessary for transmission of a signal to the cell interior. The affinity for binding a foreign antigen by the IgM, IgD, or TCR receptors is generally quite low (micromolecular range), but the affinity of IgG type receptors on memory B cells and circulating antibodies can be substantially higher. In the case of T-cell recognition of antigen, this is usually considered to be restricted by the MHC because MHC molecules can bind peptide determinants and present these determinants to the low-affinity TCR α/β-receptor.

Binding of antigen by the surface immunoglobulin receptors of B cells is in itself not sufficient to activate the B cells. Generally, helper T-cell activity is required in the stimulation and differentiation of B cells. At least two types of mediators known as B-cell stimulating factor and B-cell differentiation factor are required for activation and differentiation to IgG secreting plasma cells (see Table I). These factors are interleukins resembling the IL-2 molecule described above. Likewise, for T-lymphocytes an antigen-specific response is initiated by the binding of antigen by the T-cell receptor. This process can also be initiated by the binding of molecules that mimic the bonafide ligand such as antibodies to T-cell receptor determinants or mitogenic lectins. Furthermore, monoclonal antibodies directed against the T3 complex can activate T cells in a manner comparable with that observed for antigen or anti-idiotypic antibodies that react directly with the combining site region of the receptor.

Upon binding of ligand to the T-cell receptor, second messengers (Fig. 5) are generated by the phosphatidylinositol (PI) pathway that is initiated by the phosphodiesteratic cleavage by phospholipase C of three phosphoinositides associated with the plasma membrane: PI, diphosphoinositide (PIP), and triphosphoinositide (PIP_2). A cleavage product of each of these phosphoinositides, diacygycerol (DAG), activates a protein kinase (C kinase) that is distinct from the cAMP- or cGMP-activated protein kinases. In addition to its requirement for DAG, protein kinase C requires phospholipids and Ca^{2+} for maximum activity, an increase in cytosolic Ca^{2+} is the result of mobilization of the ion from endoplasmic reticulum as mediated by inositol-1,4,5-triphosphate (IP_3), which is an immediate breakdown produce of PIP_2. The DAG branch and IP branch of the phosphoinositide signal cascade appeared to act synergistically on kinase C to phosphorylate proteins that control the activation process. The protein kinase C phosphorylates the γ-subunit of the T3 complex, but no functional attribute of this phosphorylation has been determined other than it serving as a prerequisite for activation.

The physiological consequences of T-cell activation are blastogenesis, occurring within 48 hr, in which the small resting T-lymphocyte is transformed into an immature blast form, the secretion of lymphokines including IL-2 and γ-interferon and increased production of receptors for IL-2. The overexpression of IL-1 receptors coupled with IL-2 production by the same cell enables it to grow in an "autocrine" manner and allows cell division to take place, resulting in a clonal expansion of the original cell type.

Bibliography

Ciardelli, T., and Smith, K. A. (1989). Interleukin-2: Prototype for a new generation of immunoactive pharmaceuticals. *Trends Pharmacol. Sci.* **10,** 239–243.

Frielle, T., Kobilka, A. B., Lefkowitz, R. J., and Caron, M. G. (1988). Human β-1 and β-2 adrenergic receptors: Structurally and functionally related receptors derived from distinct genes. *Trends Neurosci.* **11,** 321–324.

Kahn, C. R., and White, M. F. (1988). The insulin receptor and the molecular mechanism of insulin action. *J. Clin. Investigation* **82,** 115–116.

Kerlavage, A. R., Fraser, C. M., Chung, F. Z., and Venter, J. C. (1986). Molecular structure and evolution of adrenergic and cholinergic receptors. *Proteins: Struc. Func. Genet.* **1,** 287–301.

Kobilka, B. K., Kobilka, T. S., Daniel, K., Regan, J. W., Caron, M. G., and Lefkowitz, R. J. (1988). Chimeric α-2, β-2 adrenergic receptors: Delineation of domains involved in effector coupling and ligand binding specificity. *Science* **240,** 1310–1316.

Marchalonis, J. J., and Schluter, S. F. (1989). Evolution of variable and constant domains and joining segments of rearranging immunoglobulins. *FASEB J.* **3,** 2469–2479.

Marchalonis, J. J. (1988). "The Lymphocyte. Structure and Function," 2nd ed. Marcel Dekker, Inc., New York.

Ross, E. M., and Gilman, A. G. (1985). Pharmacodynamics: Mechanisms of drug action and the relationship between drug concentration and effect. *In* "The Pharmacological Basis of Therapeutics" (Goodman, Gilman *et al.*, eds.), pp. 35–48. Macmillan Press Inc., New York.

Waldmann, T. A. (1989). Multichain interleukin-2 receptor: A target for immunotherapy in lymphoma. *J. Nat. Cancer Inst.* **81,** 914–923.

Zick, Y. (1989). The insulin receptor: Structure and function. *Crit. Rev. Biochem. Mol. Biol.* **24,** 217–269.

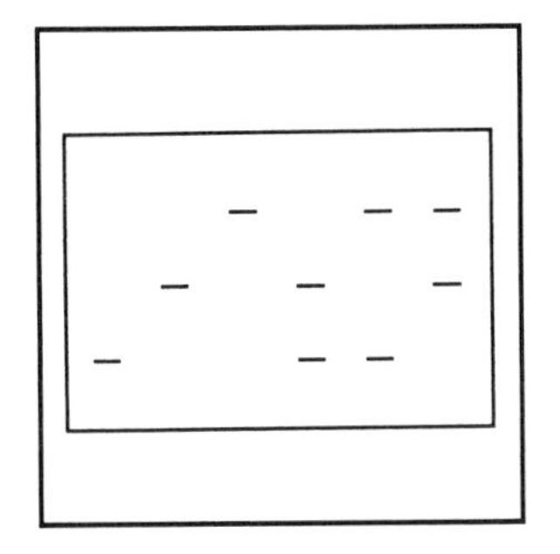

Recombinant DNA Technology in Disease Diagnosis

ALFREDO PONTECORVI, RENATO MARIANI-COSTANTINI, *Universitá G. D'Annunzio, Chieti*

PAOLO ROSSI, *Universitá "Tor Vergata," Rome*

MARCO CRESCENZI, LUIGI FRATI, *Universitá "La Sapienza," Rome*

Glossary

Alleles Multiple forms of the same gene

DNA mutation Any structural alteration of the sequence of DNA that may range from single deoxyribonucleotide substitutions (point mutations) to extensive deletions or rearrangements

Gene probes Molecularly cloned DNA sequences, labeled with radioactive or nonradioactive tracers, which hybridize to complementary DNA regions and allow their visualization

Haplotype Combination of multiple restriction fragment-length polymorphisms that identify specific alleles

Hybridization Duplex formation between complementary nucleic acid sequences

Oligonucleotides Single-stranded DNA sequences that can be synthesized *in vitro*

Oncogenes Cellular or viral genes that contribute to tumorigenesis

Polymerase chain reaction Procedure that allows the *in vitro* amplification of DNA sequences between sites on opposite strands complementary to synthetic oligodeoxynucleotides

Proto-oncogenes Cellular genes that can be converted into oncogenes by structural or functional alterations

Restriction enzymes Enzymes that cleave DNA only at specific deoxyribonucleotide sequences

Tumor suppressor genes Class of genes that can affect tumor incidence by acting as negative regulators of tumor growth in opposition to oncogenes

IN RECENT YEARS, the progress in the field of molecular biology has produced the development of an entirely new technology that involves recombinant DNA methods and reverse genetic strategies for the isolation, characterization, and manipulation of genes. This powerful and continuously improving technology, as well as the accumulating information on the structure and function of many genes, is beginning to see practical application in the fields of agriculture, veterinary medicine, drug production, and human clinical medicine. In disease diagnosis, recombinant DNA techniques are now successfully applied to the prenatal identification and carrier detection of genetic disorders, to the diagnostic and prognostic evaluation of cancer, and to the identification of pathogens in infectious diseases. This chapter describes the diagnostic use of recombinant DNA techniques in the clinical investigation of diseases at the molecular level.

I. Diagnosis of Genetic Diseases

Prior to the introduction of diagnostic methods based on recombinant DNA techniques, phenotypic analyses of limited sensitivity could be performed only in a restricted number of inherited diseases. The development of numerous and highly specific genetic probes and the precise identification of molecular DNA defects responsible for genetic disorders enabled genotypic analysis in the majority of these diseases. Because inherited genomic alterations are present in each cell of the same individual,

Encyclopedia of Human Biology, Volume 6.

it is sufficient to analyze DNA extracted from easily available cells, such as peripheral blood lymphocytes or skin fibroblasts for genetic diagnosis in adults and fetal amniocytes or trophoblasts from chorionic villi for prenatal diagnosis.

According to whether or not the molecular basis of the disease is known, two different strategies can be followed in the diagnosis of genetic disorders. Indirect diagnostic analyses are usually performed when affected genes have been identified and molecularly cloned but the specific DNA alterations are unknown. The same approach is applicable to genetic diseases that have not yet been assigned to cloned genes. Direct diagnostic analyses are available when both the affected genes and its specific alterations (mutations, deletions, etc.) are known. In this case, the highest level of diagnostic sensitivity and specificity can be achieved.

A. Indirect Diagnostic Analysis of Genetic Diseases

The genomes of two individuals are estimated to differ in 1 out of 100–300 base pairs (bp). This variability represents the main source of DNA polymorphism in the human species and accounts for the multitude of gene forms (alleles) known for the same chromosomal locus. Fortunately, most DNA mutations are silent, due to the prevalence of noncoding DNA regions (introns), the degeneration of the genetic code in which several triplets of bases code for the same amino acid, and the possible structural and functional equivalence of substituting amino acids in the polypeptide chain. Only in rare cases are DNA mutations responsible for severe impairment of the structural and biochemical properties of the encoded proteins, which may result in disease. By modifying the primary structure of DNA, mutations frequently produce the loss or gain of specific sequences recognized by distinct restriction endonucleases, thus generating DNA fragments of different lengths, which can be easily resolved by Southern blot analysis. This technical procedure is based on the analysis of electrophoretically size-fractionated restriction fragments, which are transferred to nucleic acid-binding membranes, hybridized with labeled genetic probes, and visualized by autoradiography. The restriction fragment-length polymorphism (RFLP) of DNA is commonly used for the detection of carriers and for prenatal diagnosis of most genetic disorders; however, it should be emphasized that mutations generating RFLPs are usually not responsible for disease. [*See* POLYMORPHISM, GENES.]

1. Indirect Diagnostic Analysis with Probes for Affected Genes

Genetic diseases assigned to molecularly cloned genes can be investigated by Southern blot using the specific gene probes. This procedure allows the study of the pattern of coinheritance of the disease with specific RFLPs (Fig. 1). This strategy was first applied to the diagnosis of sickle-cell anemia. A RFLP was found associated at high frequency with the β-globin sickle-cell alleles but not with the normal alleles. The RFLP analysis of genetic diseases is applicable only in families with member(s) affected by the disease. The RFLP patterns generated using different restriction endonucleases (haplotype), which provide data for the proper identifica-

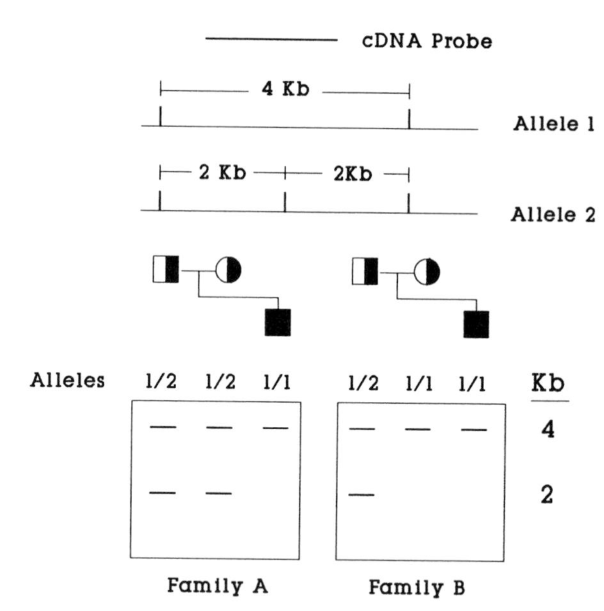

FIGURE 1 RFLP analysis of an inherited recessive disease. In this schematic example, the presence of an additional restriction endonuclease site on allele 2 allows the identification of each allelic form of the gene in all family members. The analysis of family A is fully informative and can be easily interpreted because each parent, heterozygous for the disease, carries both alleles (genotype 1/2), whereas the affected child is homozygous for allele 1 (genotype 1/1). In family B, the analysis is only partially informative because the mother is homozygous for allele 1 (genotype 1/1), thus preventing the identity of the maternal mutant allele. Upon analysis of embryonic DNA in future pregnancies, the finding of the 2-Kb band (identifying the normal allele 2) will predict the absence of phenotypic manifestations of the disease.

tion of the normal and diseased alleles, are frequently necessary for the usefulness of the analysis. [*See* DNA MARKERS AS DIAGNOSTIC TOOLS.]

A classical example of haplotype analysis is provided by the study of families "at risk" for phenylketonuria (PKU). PKU, the most frequently inherited disorder of amino acid metabolism, is an autosomal recessive disease with an estimated frequency of 1/10,000 newborns (Table I). It is caused by the deficiency of phenilalanine hydroxylase (PAH) and, if not treated with a phenylalanine-restricted diet from the first days of life, results in severe mental retardation. The cloning of the human PAH gene and the identification of several haplotypes associated with the disease permit prenatal diagnosis and carrier detection of PKU. Recent studies demonstrated the association between some haplotypes and PAH gene mutations responsible for PKU. In particular, four haplotypes at the PAH locus occur in about 90% of diseased alleles. A similar association between the mutated Z allele and an RFLP generated after digestion with the restriction endonuclease *Ava* II has been reported in α-1-antitrypsin deficiency. [*See* PHENYLKETONURIA, MOLECULAR GENETICS.]

TABLE I Frequency of Common Genetic Diseases

Genetic diseases	Frequency in newborns[a]
Autosomal dominant	
Familial hypercholesterolemia	1/500
Polycystic kidney disease	1/1,250
Huntington's chorea	1/2,500
Hereditary spherocytosis	1/5,000
von Willenbrand's disease	1/8,000
Marfan syndrome	1/20,000
Autosomal recessive	
Sickle-cell anemia	1/655
Cystic fibrosis	1/2,500
Tay–Sachs disease	1/3,000
α1-antitrypsin (ZZ type)	1/3,500
Phenylketonuria	1/10,000
Mucopolysaccharidoses (all types)	1/25,000
Glycogen storage diseases (all types)	1/50,000
X-linked	
Thyroxine-binding globulin (TBG) deficiency	1/5,000
Duchenne muscular dystrophy	1/7,000 males
Hemophilia A	1/10,000 males
Fragile X mental retardation	1/2,000 males

[a] Disease frequency may vary among ethnic groups. Data modified from Scriver *et al.* (1989).

2. Indirect Diagnostic Analysis without Probes for Affected Genes

The analysis of RFLP is also applied to trace the pattern of inheritance of diseases that have not yet been assigned to cloned genes. In these cases, genetic markers that are in strict linkage with disease transmission must be used. In fact, independent segregation of distant genetic markers to different gametes frequently occurs at meiosis by crossing-over, a process that results in the exchange of regions among homologous chromosomes. However, closely located genetic markers have a high probability to segregate to the same gamete and to be transmitted in linkage to the progeny. The identification of such genetic markers is a difficult laboratory task, but the availability of an ever-increasing number of genetic markers allows identification of carriers and prenatal diagnosis of many diseases of unknown genetic origin. However, a variable degree of uncertainty in predicting the disease-transmission pattern is recognized when using these genetic markers.

For example, prior to the cloning of the dystrophin gene and the identification of its DNA sequence, the prenatal diagnosis of Duchenne muscular dystrophy has been performed with high accuracy by linkage analysis despite the incomplete knowledge of the genomic organization of the affected gene. Linkage analysis has also been successfully applied to the prenatal diagnosis of cystic fibrosis, a disease whose gene has only recently been cloned and which represents one of the most common recessive inherited disorders among Caucasians (Table I). [*See* CYSTIC FIBROSIS, MOLECULAR GENETICS; MUSCULAR DYSTROPHY, MOLECULAR GENETICS.]

In human DNA, another type of polymorphism can be identified. It is generated by the variation in the number of tandem nucleic acid repeats (VNTRs), noncoding DNA sequences associated with several chromosomal loci that are variably repeated in different genotypes. The presence of VNTRs in regions explored by genetic probes confers a high degree of polymorphism to the genotypic analysis, thus improving its diagnostic accuracy (Fig. 2). For example, in adult polycystic renal disease (Table I), responsible for progressive renal failure in adults, a genetic marker identifies a VNTR's

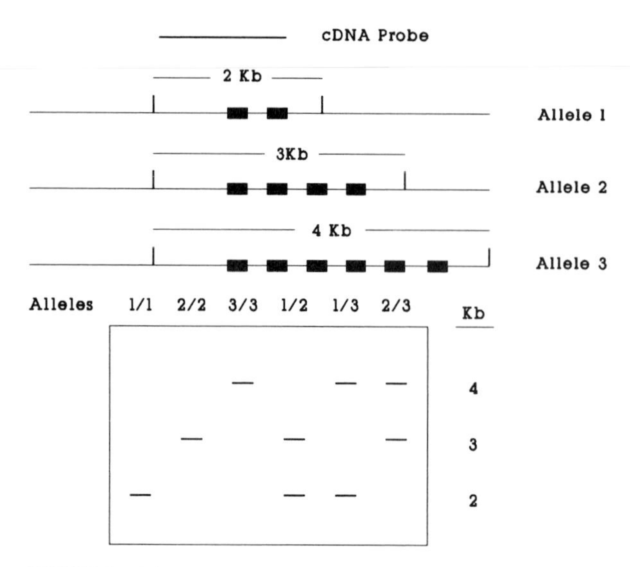

FIGURE 2 Advantages offered by VNTRs in the RFLP analysis of genomic DNA. The presence of VNTRs within restriction fragments generated by endonuclease digestion produces additional differences in the size of DNA fragments, conferring to the analysis a superior polymorphic character, which can be useful in identifying multiple allelic forms of a gene.

region on chromosome 16, which is in linkage with the disease locus and also with the α-globin gene.

B. Direct Diagnostic Analysis of Genetic Diseases

Genetic diseases assigned to molecularly cloned genes can be diagnosed using techniques that allow the direct detection of causal mutational events. Different technical approaches should be followed according to whether DNA mutations have been already characterized or unknown mutational events have to be identified *ex novo*.

The recent development of an *in vitro* DNA amplification technique, the polymerase chain reaction (PCR), has dramatically improved the sensitivity of direct tests for the analysis of DNA mutations. PCR allows the selective amplification of distinct DNA sequences by highly sensitive and specific, fast, and reproducible means. PCR has rapidly found wide application in the investigation of the molecular pathogenesis of diseases as well as in diagnostic medicine; with this technique, only a minimal amount of starting genetic material needs to be extracted from a small number of cells (even a single one in extreme cases). Two oligodeoxynucleotide primers, flanking the DNA sequence of interest and hybridizing to the opposite DNA strands, are used in combination with a thermostable DNA polymerase (*Taq polymerase* isolated from the bacterium *Thermus aquaticus*) to synthesize, in an exponential fashion, the DNA fragment of interest with a yield of about a millionfold amplification. The amplified DNA fragment can be subsequently analyzed by different methods, and mutations and polymorphisms of DNA sequences can be directly and rapidly identified without gene cloning. However, because of its recent introduction, diagnostic applications of PCR are not yet validated, but its spread in diagnostic services is an easy prediction.

1. Direct Detection of Known Mutations

a. Restriction Enzymes Restriction endonucleases can be directly used for carrier detection and prenatal diagnosis of diseases due to DNA mutations that cause the loss or gain of the specific sequences recognized by the enzymes. For example, in sickle-cell anemia, the mutation associated with the beta^s allele produces the loss of a *Mst* II restriction site, thus generating a 1.35-Kb instead of a 1.15-Kb fragment. The use of PCR-amplified DNA fragments allows the utilization of restriction endonucleases, which, by recognizing four-nucleotide sequences, cut DNA more frequently and are, therefore, less suitable for the analysis of genomic DNA. An additional advantage is that PCR-amplified fragments may be directly identified after electrophoresis and ethidium bromide staining of the DNA without radioactively labeled probes.

b. Allele-Specific Oligodeoxynucleotides An accurate detection of known mutations can be obtained by differential hybridization using allele-specific oligodeoxynucleotides (ASO) that can be easily prepared in large amounts by DNA synthesizers. The diagnostic method uses oligodeoxynucleotide probes, labeled with radioactive or nonradioactive tracers. These probes under stringent hybridization and washing conditions, which stabilize only perfectly matched hybrids, recognize either the mutant or the normal DNA sequence. This highly specific and accurate technique allows the identification of any known DNA mutation in both homozygotes and heterozygotes. The introduction of PCR has greatly improved the sensitivity and specificity of ASO. The combined PCR–ASO technique was first applied in the diagnosis of sickle-cell anemia. In this case, a 110-bp fragment of the β-

globin gene, encompassing the mutation sites of the beta^{s} and beta^{c} alleles, was amplified and subjected to differential ASO hybridization with normal and mutant oligodeoxynucleotide probes, which allowed the rapid screening for the diseased alleles.

2. Direct Detection of Novel DNA Mutations

a. Ribonuclease A Cleavage This technique is based on the ability of ribonuclease A (RNase A) to cleave single-stranded RNA at the 5′-phosphate linkage between a pyrimidine and any adjacent nucleotide. The enzyme does not attack double-stranded RNA–DNA hybrids unless a base-pair mismatch occurs. In the diagnostic application of this method, radioactive RNA probes, obtained by *in vitro* transcription of normal DNA sequences (riboprobes), are hybridized to genomic or PCR-amplified DNA. The hybrids are subsequently digested with RNAse A, separated by gel electrophoresis, and analyzed by autoradiography. In the presence of a DNA mutation, the RNA probe will be cut in two smaller fragments at the level of the base mismatch(es), while a single RNA band, corresponding to the intact probe, will be apparent when the riboprobe hybridizes to normal DNA sequences (Fig. 3). In a clinical setting, this procedure has been successfully applied to the diagnosis of mutations that cause the Lesch–Nyan syndrome (hypoxanthine phosphoribosyltransferase deficiency) or, in conjunction with PCR, to the analysis of defects of the gene coding for the urea cycle enzyme ornithine transcarbamylase.

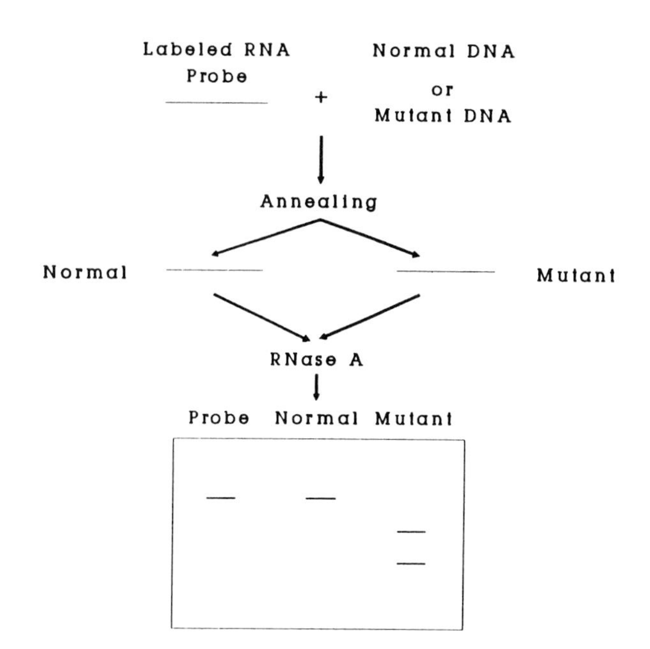

FIGURE 3 RNase A cleavage method for the direct detection of unknown DNA mutations. The presence of a base mismatch between sample DNA and the RNA-labeled probe, transcribed *in vitro* from the normal gene (left lane), allows RNase A cleavage of the riboprobe that, after gel electrophoresis and autoradiography, will appear as two bands of lower molecular weight (right lane), instead of the unique band (uncut probe) generated in the presence of normal DNA (middle lane).

b. Denaturing Gradient Gel Electrophoresis Denaturing gradient gel electrophoresis (DGGE) allows the separation of DNA–DNA hybrids differing for as little as a single-base mismatch. The technique is based on the differential melting property (Tm) of DNA fragments of different lengths (melting domains) exposed to increasing concentration gradients of denaturants (usually formamide). The nucleotide sequence of the DNA fragments is the main determinant of the Tm of each melting domain: Single-base mismatches, by decreasing the Tm, result in a lowered stability of the double-stranded DNA fragment relative to the fully matched hybrid. Probe DNA–sample DNA duplexes, migrating through linear gradients of increasing denaturant concentration in polyacrylamide gels, melt upon reaching their Tm, thus generating branched structures that exhibit retarded electrophoretic mobility. Mismatched duplexes that melt at a lower Tm than perfectly matched hybrids show a shorter migration within the gel (Fig. 4). The technique offers the advantage of detecting base-pair mismatches, which may be present at any level of the nucleotide sequence.

c. DNA Sequencing Cloned genes or PCR-amplified DNA fragments can be resolved in their primary structure by DNA sequencing techniques,

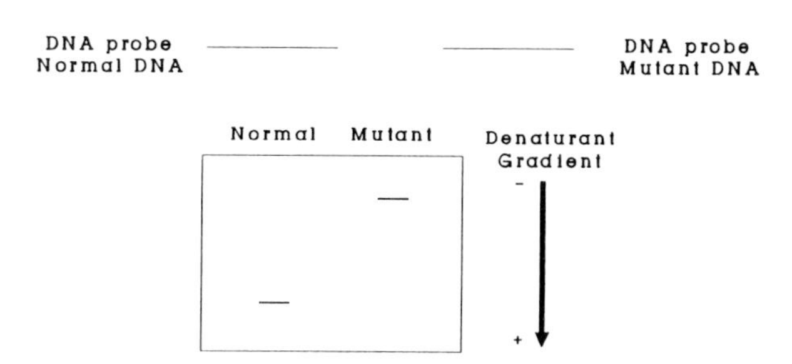

FIGURE 4 DGGE analysis for the direct detection of unknown DNA mutations. A mismatched sample DNA–probe DNA hybrid will be denaturated at a lower denaturant concentration in DGGE, resulting in its retarded electrophoretic migration (right lane) relative to the perfectly matched hybrid (left lane).

which allow the comparison of their nucleotide sequence with that of the prototype normal gene. DNA sequencing is very useful in the investigation of the molecular basis of unknown diseases and in the detection of novel mutations but cannot yet be routinely applied in extensive diagnostic programs because it is costly and cumbersome. [*See* DNA, RNA, AND PROTEIN SEQUENCES ANALYSIS BY COMPUTER.]

II. Diagnosis of Cancer

Alterations of gene structure and function play a key role in the mechanisms leading to the development of malignancy. Damage of dominantly acting proto-oncogenes may be responsible for the origin and evolution of sporadic tumors, whereas recessive germ-line lesions of putative tumor suppressor genes may help to explain the inheritance of cancer predisposition. Although a full understanding of the molecular pathogenesis of cancer is still distant, the present knowledge of the genetic bases of neoplasia has already produced relevant clinical applications, which are progressively increasing the sensitivity and specificity of cancer diagnosis. The detection of genetic alterations is in fact an important element for the diagnosis, staging, and prediction of the clinical evolution of several neoplasms. [*See* ONCOGENE AMPLIFICATION IN HUMAN CANCER; TUMOR SUPPRESSOR GENES.]

A. Diagnostic Analysis of Cancer-Associated DNA Mutations

Chromosomal translocations, gene rearrangements, and DNA mutations responsible for the activation of proto-oncogenes may generate modifications in the restriction endonuclease cleavage pattern of the involved genes.

In the case of gross gene rearrangements, DNA alterations can be easily revealed when present in a sizable fraction of the tumor-cell population from which genomic DNA is extracted. Southern blot analysis, using radiolabeled, molecularly cloned sequences of the gene of interest, may detect the presence of novel hybridization bands or show the comigration of previously unrelated DNA fragments, thus indicating the presence of alterations of the gene structure (Fig. 5). However, this type of analysis may not be sensitive enough to identify genetic lesions present in a restricted number of tumor cells. In these cases, the PCR techniques greatly help in selectively amplifying the altered DNA sequences, thus increasing the diagnostic potential. Chromosomal translocations and gene rearrangements, which juxtapose previously separated DNA sequences, can be revealed using primers complementary to each genomic region involved in the phenomenon. In these cases, amplification will be possible only when the position of the two primers is brought within the range of PCR performance by gene rearrangements (Fig. 6). For the detection of DNA mutations, the amplified PCR product may be analyzed by restriction endonuclease cleavage or differential ASO hybridization, when the mutated sequence is known, and by DGGE or RNase A cleavage, to screen for the presence of unknown modifications in the nucleotide sequence.

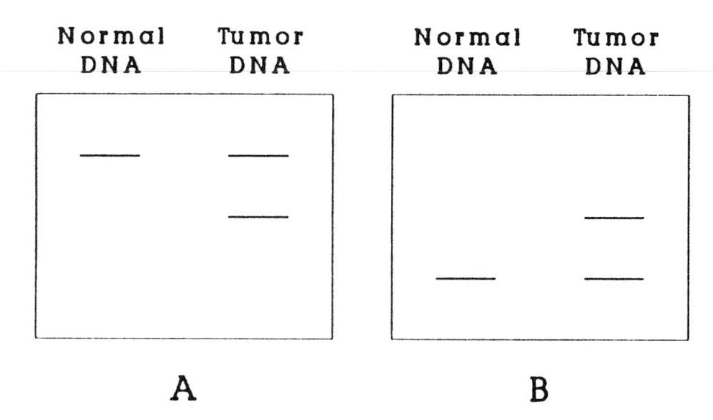

FIGURE 5 Southern blot analysis of gene rearrangement in cancer. Two different probes (A and B), specific for genes involved in reciprocal rearrangements, are sequentially hybridized, after probe-stripping, with the same filter containing DNA extracted from normal and tumor cells digested with the same restriction endonuclease. The presence in tumor DNA of a band hybridizing to both probes indicates the comigration of the two genes in the same restriction fragment, which is diagnostic for gene rearrangements.

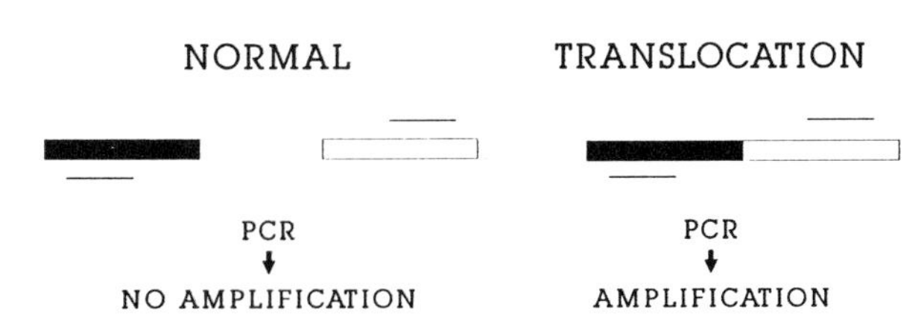

FIGURE 6 Use of PCR in the identification of chromosomal translocations in cancer. Primers complementary to opposite DNA strands of two distant genes will yield an amplification product only when the primer target sequences are brought in close proximity and within the amplification range of *Taq* I DNA polymerase, following the occurrence of reciprocal translocation in tumor DNA.

1. Analysis and Staging of Leukemia and Lymphoma

The impact of recombinant DNA technology is particularly perceived in the diagnosis of tumors of the hemopoietic system. Malignancies of white blood cells, either circulating in the blood stream (leukemias) or growing as a solid mass (lymphomas), supply larger numbers of easily available tumor cells than most solid cancers. This is reflected in the vast knowledge that is available on the molecular basis of tumorigenesis and particularly on the genetic abnormalities associated with neoplasms of the hemopoietic system. The search for cells harboring tumor-associated genetic lesions may be important not only in performing a correct initial diagnosis but also for the detection of residual tumor cells after therapy. [*See* LEUKEMIA; LYMPHOMA.]

Two classic examples of chromosomal translocations are characteristic of Burkitt's lymphoma and of chronic myelogenous leukemia. In the former, the c-*myc* proto-oncogene, located on the long arm of chromosome 8, is involved in reciprocal translocations with the immunoglobulin heavy-chain gene on the long arm of chromosome 14. In chronic myelogenous leukemias, the breakpoint cluster region on the long arm of chromosome 22 is the site of reciprocal translocations with the c-*abl* proto-oncogene located on chromosome 9, which result in an abnormally truncated chromosome 22 (Philadelphia chromosome) detected in >90% of the cases. This genetic abnormality is also present in approximately 20% of acute lymphoblastic leukemias.

Point mutations may be responsible for the malignant conversion of proto-oncogenes. This is well documented for the *ras* family of proto-oncogenes (c-H-*ras*, c-K-*ras*, c-N-*ras*), where point mutations at codons 12, 13, and 61 confer transforming activity to the encoded protein product. While the diagnostic value of the detection of *ras* gene mutations is still unclear in solid tumors, in chronic myelogenous leukemia the presence of neoplastic cells carrying a *ras* point mutation is indicative of a poor prognosis. Mutations of *ras* proto-oncogenes also appear to be a predictive marker of the clinical evolution of myelodysplasia, a disorder of hemopoiesis characterized by the alteration of blood-cell differentiation, which progresses to acute myelogenous leukemia in 10–30% of cases. Therefore, the detection of mutated *ras* proto-oncogenes may help in identifying myelodysplastic syndromes susceptible of a malignant evolution.

The diagnosis and typing of lymphoid neoplasms is facilitated by the analysis of antigen receptor genes, i.e., immunoglobulin genes in B lymphocytes and T-cell receptor genes in T lymphocytes. In the absence of phenotypic markers, the characterization of the stage of the somatic recombination of antigen receptor genes allows the assignment of neoplasms to the B- or to the T-cell lineage of lymphoid neoplasms. In addition, the level of differentiation or the stage of maturational arrest of the neoplastic cells can be precisely defined on the basis of the stage of antigen receptor gene rearrangements observed. The characterization of neoplastic cell populations provides information that is useful in disease diagnosis as well as in monitoring the persistence or recurrence of the disease after therapy.

2. Analysis and Staging of Solid Tumors

The genetic amplification of specific proto-oncogenes, particularly of those coding for growth factors or growth-factor receptors and of gene-conferring multidrug resistance (*mdr* genes), may have a prognostic value in defining tumors that have high biological aggressiveness and metastatic potential or are resistant to chemotherapy. This is the case of the c-*erb*-B2 proto-oncogene (also known as HER-2, or *neu*), which encodes a cell-membrane receptorlike protein related to the epidermal growth-factor receptor (EGFr). A significant increase in HER-2 copy number (more than 5 copies per diploid genome) has reportedly been associated with an aggressive clinical behavior, characterized by a decrease of disease-free and overall survival, in carcinomas of the breast and ovary. The genetic amplification of the c-*erb*-B2 proto-oncogene may identify those breast cancers that, despite the absence of apparent metastases, will have poor clinical evolution and require chemotherapy to prevent disease relapse after surgery. In analogy, the genetic amplification of the N-*myc* proto-oncogene in neuroblastomas and of the c-*erb*-B1 (EGFr gene) in gliomas appears to be highly correlated with rapid tumor progression, even in patients at early stages of the disease. [*See* BREAST CANCER BIOLOGY.]

The presence and extent of a proto-oncogene amplification is usually determined on autoradiograms of Southern blots or dot blots hybridized with the specific cDNA-labeled probe. The amplification level is evaluated by comparing the hybridization signal of the amplified gene with that of another gene that is not affected by cancer-associated alterations. The comparison should be made using appropriate radioanalytical imagery systems, such as soft laser densitometry. As an alternative to genomic analysis by Southern blotting, PCR may be used to estimate the level of amplification of the gene of interest in comparison to a single-copy gene of reference. The degree of genetic amplification is reflected in the relative ratio between the levels of the two PCR products (which is determined by electrophoresis and autoradiography) after incorporation of a radiolabeled deoxyribonucleotide during PCR or ethidium bromide staining. As a caveat, the detection of gene amplification is influenced by the ratio between tumor and normal cells in the biopsy sample from which DNA is extracted. A significant contamination by normal cell DNA may in fact dilute the signal of the amplified target gene.

3. Analysis of Cancer Predisposition

Hypothetically, the familial predisposition to cancer may be associated with recessive germ-line mutations of putative tumor-suppressor genes. The inheritance of a mutated allele from one parent followed by a somatically acquired mutation on the other allele may account for the high incidence of cancer in at-risk families. An inherited cancer-predisposing mutation at a specific chromosomal locus can be suspected when tumor DNA becomes homozygous for markers located on that locus. This tumor-associated loss of heterozygosity has been documented in retinoblastoma, a malignant tumor of the retinal cells of the eye, and in osteogenic sarcoma, a malignant bone tumor, both of which share a genetic predisposition mapping to a gene (RB-1) located on the long arm of chromosome 13. Several other types of cancer predisposition have also been mapped to specific chromosomal regions. In perspective, the molecular cloning of these genes may allow the identification of recessive genotypes in at-risk families, which require an adequate follow-up for the early recognition of the insurgence of neoplasias. At present, these developments are speculative, but conceivably preventive diagnostic approaches to cancer, analogous to those already applied for genetic disorders, will become available in the future. [*See* CHROMOSOME PATTERNS IN HUMAN CANCER AND LEUKEMIA.]

III. Diagnosis of Infectious Diseases

The diagnosis of infectious diseases by recombinant DNA techniques is independent from antigenic properties of pathogens and from the immune response of patients. In addition, molecular hybridization methods do not require the viability of the infectious agent and allow the precise identification of a small number of pathogen DNA sequences within the much more abundant human DNA. Following the molecular characterization of the genomic structure of a variety of infectious agents, methods based on nucleic acid hybridization have been successfully applied to the detection of bacteria, parasites, and viruses in cells and body fluids. For example, this is the case of pathogen detection in the stools where the fast and more standardized recombinant DNA techniques have proven to be superior to traditional methods requiring lengthy and complex cultural conditions.

The molecular identification of pathogens is usually performed either after chemical extraction of nucleic acids, which are subsequently assayed by dot blot or Southern blot hybridization analyses, or by *in situ* hybridization techniques, which allow the direct localization of pathogens at the cellular level. The specificity of *in situ* hybridization techniques, for example, permits the distinction of viral types that are associated with a different clinical evolution of the disease. This is the case of human papilloma viruses, where types 16 and 18 are frequently detected in preneoplastic lesions of the uterine cervix, whereas types 3, 6, 10, and 11 are associated with lesions that have a benign course. Recombinant DNA methods may also recognize antibiotic-resistant strains, as shown in the case of *Neisseria gonorrhoeae*. [*See* PAPILLOMAVIRUSES AND NEOPLASTIC TRANSFORMATION.]

In vitro amplification of pathogen DNA by PCR may be of great help particularly when infectious agents are present in limited numbers. However, caution is needed in the application of PCR to the diagnosis of common infections because of the high probability of occasional contaminations and because previous infections, unrelated to the present pathology, may lead to incorrect diagnostic associations. In the field of viral diseases, PCR-based diagnostic tests have been developed for the human immunodeficiency virus (HIV), the hepatitis viruses, the cytomegalovirus, the human papilloma viruses, and several others. An important area of PCR application concerns the screening of subjects at risk for those viral infections that may escape the detection by less sensitive techniques. This is particularly the case of HIV-1, the etiologic agent of acquired immunodeficiency syndrome (or AIDS), whose identi-

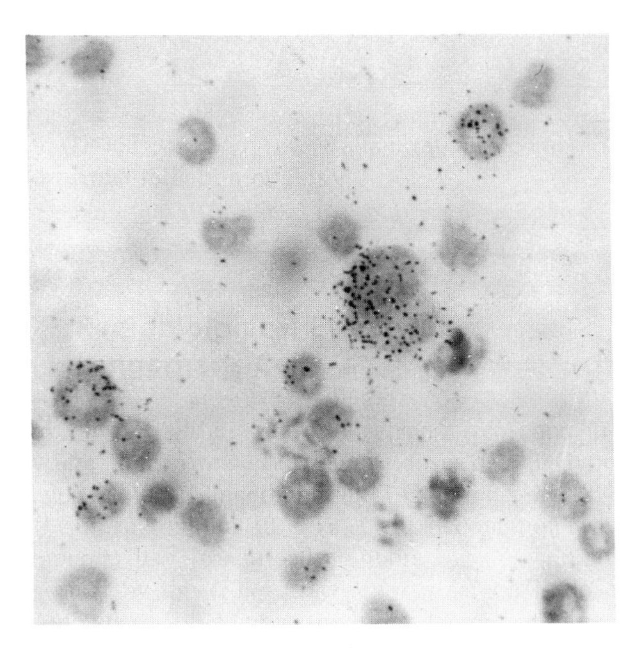

FIGURE 7 Use of *in situ* hybridization for the identification of peripheral blood lymphocytes expressing HIV-1 RNA in a patient, which resulted seronegative by immunoenzymatic and Western blot analyses but positive for the amplification of HIV-1 proviral DNA by PCR. The clustering of autoradiographic grains identifies cells containing viral RNA hybridizing to the complementary labeled riboprobe. (×900).

fication in seronegative carriers and blood products used for transfusions is essential to establish the appropriate clinical follow-up and to prevent the spread of the disease (Fig. 7).

IV. Summary

Recombinant DNA technology provides powerful tools that result in an extraordinary improvement in our capability to diagnose and prevent human diseases. By introducing higher levels of sensitivity and specificity in the prenatal diagnosis and carrier detection of genetic disorders, in the diagnosis and staging of cancer, and in the identification of infectious agents, these new techniques, so far restricted to the investigation of selected cases in research laboratories, will soon become the backbone of diagnostic services for the general population.

Bibliography

Berger, S. L., Kimmel, A. R. (eds.) (1987). Guide to molecular cloning techniques. *In* "Methods in Enzymology," Vol. 152. Academic Press, New York.

Bishop, J. M. (1988). The molecular genetics of cancer: 1988. *Leukemia* **2,** 199.

Caskey, T. C. (1987). Disease diagnosis by recombinant DNA methods. *Science* **236,** 1223.

Crescenzi, M., Seto, M., Herzig, G. P., Weiss, P. D., Griffith, R. C., and Korsmeyer, S. J. (1988). Thermostable DNA polymerase chain amplification of t(14;18) chromosome breakpoints and detection of minimal residual disease. *Proc. Natl. Acad. Sci. USA* **85,** 4869.

Croce, C. M. (1987). Role of chromosome translocation in human neoplasia. *Cell* **49,** 155.

Orkin, S. H. (1987). Genetic diagnosis by DNA analysis. *N. Engl J. Med.* **317,** 1023.

Saiki, R. K., Scharf, S., Faloona, F., Mullis, K. B., Horn, G. T., Erlich, H. A., and Arnheim, N. (1985). Enzymatic amplification of beta-globin genomic sequences and restriction site analysis for diagnosis of sickle cell anemia. *Science* **230,** 1350.

Schochetman, G. S., Ou, C.-Y., and Jones, W. K. (1988). Polymerase chain reaction. *J. Infect. Dis.* **158,** 1154.

Scriver, C. R., Beaudet, A. L., Sly, W. S., and Valle, D. (eds.) (1989). "The Metabolic Basis of Inherited Disease," 6th ed. McGraw-Hill, New York.

Slamon, D. J., Clark, G. M., Wang, S. G., Levin, W. J., *et al.* (1987). Human breast cancer: Correlation of relapse and survival with amplification of the HER-2/*neu* oncogene. *Science* **235,** 177.

Various authors (1989). Nucleic acid probes: Definition and diagnosis of human disease. *Clin. Chem.* **35,** 7(B).

$[Ca^{2+}]$

Red Cell Calcium

ROBERT M. BOOKCHIN, *Yeshiva University*
VIRGILIO L. LEW, *University of Cambridge*

Glossary

$[Ca^{2+}]_i$ Intracellular (cytoplasmic) concentration of Ca^{2+}, the free ionized (unbound) form of calcium

MAMMALIAN RED CELLS have the lowest total calcium content of any cell in the body. After loosely bound calcium is washed from their outer membrane surface, human red cells contain less than 5 μmol of calcium/L cells. This low total calcium content reflects the absence of the calcium-accumulating compartments and calcium binding structures found within other cells, whose total calcium content is about 100 to 500 times that of mature mammalian red cells.

I. States of Calcium

Calcium is found in a variety of physical states within organisms. It can be found precipitated with phosphate-rich compounds in all sorts of hard structures (e.g., bones, teeth, and shells); it may be tightly or loosely bound to a variety of organic molecules, which are either freely soluble or part of cell or intercellular structures; or it may exist in free ionized form (Ca^{2+}) within intracellular or extracellular compartments. There is a dynamic equilibrium among these various forms of calcium, with all transfers between the different physical states occurring through the water-soluble, ionized form. The chemical versatility of calcium, whereby it readily forms insoluble salts or reversible high-affinity bonds with organic molecules, allows it to mediate such diverse biological functions as forming skeletons or transmitting transient signals to intracellular targets. It seems a bit ironic that red cells, with no known Ca^{2+}-mediated signaling, exhibit many of the functional features crucial to the signaling process, so that studies of these features in the readily available red cells have provided clues to help understand the messenger functions.

The strategy by which Ca^{2+} transmits messages to intracellular targets involves three distinct processes: (1) the maintenance (and eventual restoration) of large Ca^{2+} gradients across a membrane (this, figuratively, puts the explosive powder in the gun); (2) the opening and closing of Ca^{2+} channels by primary chemical or electrical signals, which causes transient increases in $[Ca^{2+}]_i$ (equivalent to pulling the trigger and converting the potential chemical energy of the gun powder into kinetic energy of the bullet); and (3) the reaction of Ca^{2+} with a target molecule (the bullet hits the target). It is on the first of these processes that work with red cells provided fundamental information.

II. Passive and Active Ca^{2+} Transport in Red Cells

A. Calcium Distribution and Permeability

Red cell calcium might seem at first to be a rather dull subject, in view of the minute total calcium content of these cells, and particularly after noting that nearly all the 4 to 5 μmol/L cells of total red cell calcium is bound to unknown membrane components, leaving less than 1% in free ionized form in the cytoplasm. The concentration of Ca^{2+} in the cytoplasm of normal human red cells, $[Ca^{2+}]_i$, was found to be about 10 to 30 nM. Yet, it is precisely this minute cytoplasmic Ca^{2+} concentration (a feature shared by all cells, with minor variations) that is particularly interesting; it is the combination of this feature, together with the low total calcium content, and the lack of "complicating" organelles that makes the red cell such a useful experimental model. The low cytoplasmic Ca^{2+} level is interesting because the cells are surrounded by fluids containing at least a hundred thousand-fold higher Ca^{2+} concentrations ($\approx$1–1.2 mM in mammals). The true inward Ca^{2+} gradients are even steeper than suggested by the concentration differences alone, because all cells maintain large electrical potential differences across their membranes, with the inside of the cell negative in relation to outer surface. The combined electrical and chemical inward Ca^{2+} gradient therefore represents a formidable force. If the cells were permeable to Ca^{2+}, this force would induce a large Ca^{2+} inflow, rapidly dissipating the gradient.

Experiments with calcium tracers (^{45}Ca) and additional experiments demonstrating a requirement of external Ca^{2+} for contractile and secretory cell functions showed that all cells have a finite permeability to calcium. The red cell is among the least permeable, with an inward passive calcium flow of about 50 μmol/L cells/hr under physiological conditions. Because human red cells have a normal lifespan of about 120 days in the circulation, such a calcium influx, if unopposed, would raise $[Ca^{2+}]_i$ to over several thousand times the normal value in less than a day; but measurements show little variation in $[Ca^{2+}]_i$ throughout the cells' life.

B. The Ca^{2+} Pump

To sustain such a huge Ca^{2+} gradient throughout the red cell lifespan, when the normal "leaks" would tend to dissipate the gradient in a fraction of that time, human red cells use a powerful Ca^{2+} pump, which derives energy from the hydrolysis of adenosine triphosphate (ATP) to extrude Ca^{2+}. This pump, which is present in the plasma membrane of most animal cells, is part of a family of ATP-fueled pumps, which transport sodium, potassium, protons, and calcium ions across biological membranes against electrochemical gradients. The plasma membrane Ca^{2+} pump is an integral membrane protein with a molecular weight of about 140 kDa. It transports one calcium ion out per molecule of ATP hydrolyzed, and its activity may be regulated by calmodulin, a ubiquitous calcium-binding cytoplasmic polypeptide. [*See* ADENOSINE TRIPHOSPHATE (ATP).]

In human red cells, the Ca^{2+} pump is the sole active transporter of calcium ions against their gradient. It is interesting to note, however, that one other such calcium transporter is a sodium : calcium exchanger, which uses the energy stored in the normally large inward sodium gradient to transport Ca^{2+} out of the cell. This transporter is found only in the plasma membrane of excitable animal cells, with one exception thus far: In dog red cells, a sodium : calcium exchange is present, which operates as a Na pump, because the inward Ca^{2+} gradient maintained by the ATP-fueled Ca^{2+} pump is steeper than the sodium gradient. All cells have a tendency to swell, which results from excess impermeant solutes relative to their environment. Unlike plant cells, whose cell walls can sustain hydrostatic pressures, soft-walled animal cells require a mechanism to pump out sodium—a job generally performed by an ATP-fueled sodium pump. In dog red cells, which lack the sodium pump entirely, the sodium : calcium exchange thus represents an interesting alternative sodium extrusion strategy compared with that of other animal cells.

Experimentally, divalent cation-selective ionophores can be used to increase the Ca^{2+} permeability of intact red cells in a controlled manner, to load red cells with Ca^{2+} and study the properties of the Ca^{2+} pump and the Ca^{2+}-binding components of the cytoplasm. Of the total calcium thus loaded into the cells, 15% to 30% remains in free ionized form, while the remainder is bound to the cytoplasmic calcium buffers. The calcium pump is activated by Ca^{2+} with a high affinity ($K_{\frac{1}{2}}$, the Ca^{2+} concentration needed for half-maximal activation of the pump, $\leq$1 μM), and when saturated, it can extrude Ca^{2+} at rates that vary widely between 5 and 25 mmol/L cells/hr.

This raises the important question as to why red cells, with a minute passive inward calcium leak of less than 50 μmol/L cells/hr, have a pump capable of extruding Ca^{2+} at least a hundred times faster. By contrast, the sodium pump, for instance, operates at about 25% to 30% of its capacity in physiological conditions. Alternative explanations have been offered for the enormous spare capacity of the calcium pump, but more evidence will be needed to establish its purpose.

One suggestion is that as red cells squeeze rapidly through the capillaries, shear forces acting on their membranes rub Ca^{2+} channels open, transiently increasing Ca^{2+} permeability to values much higher than those measured in unstressed experimental conditions. During such transient permeabilizations, the powerful Ca^{2+} pump would secure rapid restoration of normal $[Ca^{2+}]_i$. Another suggestion is that if there is an upper chemical limit to the affinity with which a pump can bind Ca^{2+}, the only way to prevent $[Ca^{2+}]_i$ from rising above the low physiological levels is by increasing the overall pumping capacity.

Implicit in all such arguments is the notion that high $[Ca^{2+}]_i$ is "bad" for the cells. The answers as to whether this is so and the effects of sustained elevated Ca^{2+} levels differ from one type of cell to another, depending on which of a large variety of specific functions are controlled by $[Ca^{2+}]_i$ in that cell type. The abnormalities that may result from elevated $[Ca^{2+}]_i$ in human red cells are discussed in Section III. The catalog of claimed disasters, including the induction of normal cell senescence, is large indeed, but specific evidence on how each instance of abnormally elevated $[Ca^{2+}]_i$ may affect cells in the living organism is sparse. Because of the functional importance of $[Ca^{2+}]_i$, considerable effort and ingenuity has been devoted to its accurate measurement in various cells. All such measurements indicate that $[Ca^{2+}]_i$ is less than 100 nM in resting cells, and it is the prevailing belief that this low resting level is a universal, primitive feature of all cells. Experimental studies with red cells have been instrumental in elucidating the mechanisms by which a low $[Ca^{2+}]_i$ is maintained.

III. Effects of Elevated Red Cell $[Ca^{2+}]_i$

A. General Effects

In view of the low $[Ca^{2+}]_i$ level in normal human red cells, which appears to be rather constant, questions arise about the consequences of any increases in this level, either in special physiological circumstances or in pathological states. Most *in vitro* studies have examined the effects of raising red cell $[Ca^{2+}]_i$ by permeabilizing the cells with the divalent cation ionophore A23187 at various levels of $[Ca^{2+}]_o$. These effects include (1) reduction or depletion of ATP caused by increased activation of the $[Ca^{2+} + Mg^{2+}]$ATPase, with accumulation of inosine monophosphate (IMP) in calcium-equilibrated cells, associated with irreversible ATP depletion; (2) K^+ permeabilization caused by activation of the Ca^{2+}-dependent K channel, when $[Ca^{2+}]_i$ exceeds about 40 nM; in plasma-like low-K media, this is accompanied by net KCl and water efflux and red cell dehydration; (3) inhibition of the Na pump, by more than 10 μM $[Ca^{2+}]_i$, and (4) activation of a red cell transglutaminase capable of cross-linking membrane proteins, by $[Ca^{2+}]_i$ levels greater than 50 μM.

B. Increased Red Cell Calcium in Sickle Cell Anemia

An increased total calcium content has been described in several types of red cell disorders associated with abnormal hemoglobins. Among these, the most extensively studied is sickle cell anemia, which occurs in persons homozygous for the mutant gene directing synthesis of the abnormal hemoglobin S. This example, which illustrates some of the complexities of the distribution and effects of intracellular calcium, even in such a structurally simple-appearing, enucleate cell as the mature human red cell, will be discussed here in some detail.

The fundamental abnormality in sickle cell anemia (SS) red cells is the predominance of hemoglobin S. When the SS cells are deoxygenated, the hemoglobin S polymerizes into bundles of long rod-like chains, which produce rigid projections in the cells and often endow them with curved "sickle" shapes. As a result, the cells lose their normal deformability, become mechanically fragile, and tend to be rapidly destroyed in the circulation and to occlude small blood vessels. On reoxygenation in the circulation, hemoglobin S depolymerizes and resumes normal solubility, and the cells "unsickle." A portion of the SS red cells tends to become dehydrated, however, which increases their rigidity and their propensity to sickle. In many of these dense, dehydrated SS cells the cell membrane becomes fixed in a relatively rigid, elongated shape, even when the cells are oxygenated and the hemoglobin is not polymerized; these are the "irreversibly sickled cells" (ISCs). [*See* SICKLE CELL HEMOGLOBIN.]

Sickle cell anemia cells were found to have an increased calcium content, averaging about 50 μmol/L of red cells, with as much as 200 μmol/L of red cells or more in the dense, ISC-rich fraction, and to show a temporary increase in calcium permeability (of about four- or fivefold) during deoxygenation-induced sickling. These findings initially prompted the view that all the effects described above in red cells loaded with micromolar levels of Ca^{2+} *in vitro*, using the ionophore, might occur in SS cells, thereby accounting directly for the features of ISCs. Detailed investigations of SS cells, however, revealed none of the expected effects of a continuously high $[Ca^{2+}]_i$. Using recently developed methods to measure free cytoplasmic Ca^{2+} accurately at low nanomolar levels, it was shown that $[Ca^{2+}]_i$ in oxygenated SS red cells was normal or only minimally increased. Other studies demonstrated that virtually all the excess (micromolar) calcium in SS cells is compartmentalized in intracellular membrane vesicles with inwardly directed calcium pumps ("inside-out vesicles," or IOVs), which are largest and most numerous in the ISCs. These vesicles would tend to accumulate calcium during periods of increased calcium influx associated with deoxygenation and sickling, but accumulated vesicular calcium is not accessible to the pump or the cytoplasmic side of the cell membrane.

Among the effects of experimentally increased $[Ca^{2+}]_i$ described above, only the K channel is activated by submicromolar levels of Ca^{2+}. Accordingly, it appears that the small transient increases in $[Ca^{2+}]_i$ that must occur during sickling-induced calcium permeabilization are sufficient to activate the Ca^{2+}-dependent K channel in some SS red cells, contributing to their dehydration.

C. Increased Red Cell Calcium in Nonsickling Disorders

An increased total calcium content has been reported in certain other congenitally abnormal red cells, including homozygous hemoglobin C disease (another structurally abnormal hemoglobin that tends to crystallize in the red cells), and beta-thalassemia (in which there is deficient synthesis of one of the two polypeptide chains making up the hemoglobin tetramers, the beta chains, leaving an excess of alpha-hemoglobin chains, which are unstable when not combined with the beta chains). In those two disorders, red cell $[Ca^{2+}]_i$ was measured, and as in SS cells, it was found to be normal or minimally increased in the nanomolar range. The origin of the increased total calcium and any possible role it may play in the red cell pathology of these disorders has not yet been established.

A word of caution is in order about interpreting findings of increased mean red cell calcium in severe hemolytic disorders or after membrane damage induced *in vitro*. Because of the huge inward Ca^{2+} gradient, a small percentage of severely damaged or "prehemolytic" cells that become permeabilized and equilibrate with the millimolar calcium in the extracellular medium can raise the mean level of red cell calcium well into the high micromolar range. Simple measurements of mean red cell calcium cannot distinguish this pattern from one of a more general increase in calcium affecting most of the cells. In interpreting the possible effects of increased red cell calcium on the overall population of cells, it is therefore important to have additional information about the heterogeneity of calcium distribution among the cells.

Bibliography

Bookchin, R. M., and Lew, V. L. (1983). Red cell membrane abnormalities in sickle cell anemia. *Prog. Hematol.* **13,** 1–23.

Bookchin, R. M., Ortiz, O. E., and Lew, V. L. (1986). Red cell calcium transport and mechanisms of dehydration in sickle cell anemia. *In* "Approaches to the Therapy of Sickle Cell Anaemia" (Y. Beuzard, S. Charache, and F. Galacteros, eds.). Colloque INSERM Vol. 141, Paris.

Lew, V. L., and Garcia-Sancho, J. (1987). Measurement and control of intracellular calcium in intact red cells. *In* "Methods in Enzymology, Biomembranes, Part M, Biological Transport. Eukariotic, Cellular and Subcellular Transport" (S. Fleischer and B. Fleischer, eds.). Academic Press, New York.

Lew, V. L., Hockaday, A., Sepulveda, M. I., Somlyo, A. P., Somlyo, A. V., Ortiz, O. E., and Bookchin, R. M. (1985). Compartmentalization of sickle cell calcium in endocytic inside-out vesicles. *Nature* **315,** 586–589.

Lew, V. L., Tsien, R. Y., Miner, C., and Bookchin, R. M. (1982). The physiological (Ca^{2+}) level and pump-leak turnover in intact red cells measured with the use of an incorporated Ca chelator. *Nature* **298,** 478–481.

Rega, A. F., and Garrahan, P. J. (1986). "The Ca^{2+} Pump of Plasma Membranes." CRC Press, Boca Raton, Florida.

Schatzmann, H. J. (1982). The Plasma-membrane calcium pump of erythrocytes and other animal cells. *In* "Membrane Calcium Transport" (E. Carafoli, ed.). Academic Press, London.

Red Cell Membrane

STEPHEN B. SHOHET, *University of California at San Francisco*

Glossary

Band 2.1 (ankyrin or syndein) Globular protein which serves to tie the spectrin network to the bilayer through a high-affinity binding reaction with the intrinsic membrane protein, band 3

Band 3 Important intrinsic membrane protein which spans the bilayer. It contains at least one of the major ion channels (that for anions) and probably has permeability functions in the membrane. Its cytoplasmic segment is connected to the membrane skeleton via band 2.1 and is one of the two major sites for the skeleton

Band 4.1 Globular protein which serves to connect spectrin and red cell actin in the membrane skeleton, and which by a second binding reaction with the intrinsic membrane protein, glycophorin, which is a second site, serves to tie the skeleton to bilayer

Cholesterol Neutral lipid, based on the polycyclic benzanthracene nucleus which is a major structural component of the membrane bilayer. At the molecular level it fits closely with phosphatide fatty acids in the bilayer and probably provides much of the membrane aqueous barrier function

Intrinsic membrane proteins Proteins which are difficult to extract with detergents. They interact comparatively strongly with the phosphatides and cholesterol of the lipid bilayer. They all have hydrophobic regions and hence, at least partially, are found in the hydrophobic core of the membrane bilayer

Membrane bilayer Double array of structurally bifunctional lipid molecules typically found in the plasma membranes of cells. Comparatively water-soluble portions of the lipids distribute to the aqueous interfaces on the outside and inside of the cell, while the water-insoluble fatty acid tails and polycyclic hydrocarbons distribute to a central hydrophobic core in this structure. The bilayer lipids primarily provide a barrier function for the cell. Hydrophobic proteins which are often also inserted in the bilayer provide other specialized transport, enzymatic, and recognition functions for the membrane

Membrane skeleton A horizontally extended anastomotic network of extrinsic membrane proteins which lies directly underneath the membrane bilayer and is coupled to its intrinsic proteins. It is essential for bilayer support and membrane integrity. It also probably has important roles in membrane deformability and the regulation of cell shape

Peripheral membrane proteins Proteins which are comparatively easily extracted from the membrane and solubilized with nonionic detergents. They are predominantly found in the membrane skeleton

Phosphatide Lipid made up of esterified fatty acids and a phosphus-linked base on a glycerol backbone. Together with cholesterol, phosphatides are the major lipid components of the membrane

Spectrin Predominant protein component of membrane skeleton. This elongated, high-molecular-weight, fibrillar protein serves as the major structural element in the membrane skeleton network. Its imaginative name reflects its origin from the red cell ghost or specter

RED CELL MEMBRANE is the structure composed of lipids and proteins which separates the interior of the cell from the plasma in which it is suspended. Inherent in its basic role of confining the interior contents of the cell, the membrane is responsible for the structural integrity and many of the physical properties of the cell as well as its shape. In addition, the membrane has important ancillary functions in terms of transporting molecules in and out of the cell and as the site of immunologic interactions with other cells and the plasma.

I. Introduction

Ever since Leeuwenhoek's original observations in the seventeenth century it has been known that the hemoglobin which circulates in the blood is contained in small discrete particles which he called "globules." Subsequent observations rapidly established that these more or less uniform packages of hemoglobin were surrounded by a "delicate pellicle," or skin, which served to prevent the random diffusion of the hemoglobin, thus providing the essential characteristic of cellularity to these "globules." This pellicle, of course, is what is now called the cell membrane. In addition to its primordial function of packaging the hemoglobin into small, highly concentrated and easily transported elements, this membrane has several important secondary functions and characteristics.

First, the membrane serves as a semipermeable barrier which, while effectively containing the hemoglobin, still allows the passage of smaller electrolytes, carbohydrates, and peptides to provide nutritional and excretory pathways for the cell. Complementary to this, the membrane also contains various "pumps" and enzyme systems designed to facilitate the accumulation or discharge of certain smaller molecules against concentration gradients.

Second, the membrane provides the cell with a unique biconcave configuration which maximizes the efficiency of oxygen transport in this small circulating bag of concentrated hemoglobin. Perhaps derivative to this function, the membrane also contains enzyme systems to repair and remodel itself, and to modulate the asymmetry of its own components.

Third, and in support of its cardinal role in dividing the hemoglobin mass into stable cells, the membrane is wholly responsible for the cell's extraordinary strength and much of its remarkable deformability. These two characteristics are essential to enable the cell to circulate for 120 days and to traverse over 175 miles of capillaries during its lifetime. In contrast to many other cell types there is no internal structural apparatus beyond the membrane in the red cell to assist in maintaining these characteristics.

Finally, certain proteins contained within the membrane have portions exposed on its external surface which identify the cell immunologically and hence have a role in making the cell a target for various pathologic immunological conditions. These membrane proteins may also have an analogous role in directing the normal senescent turnover of red cells by possible immunological recognition mechanisms.

Again starting with Leeuwenhoek, generations of investigators have recognized that the red cell is the most easily biopsied tissue in man, and that its isolation from the other elements in the blood is fortuitously straightforward. Further, hypoosmotic treatment of red cells causes them to swell and then burst, releasing their hemoglobin; hence, virtually pure membrane can be isolated by simple washing and centrifugation. The ease of obtaining red cells from living subjects, together with the ease of isolating membranes from those cells, has made this membrane extraordinarily valuable as an archetypal model for mammalian cells in general. Indeed, a great deal of what has been learned about membrane structure and function in the red cell has been promptly and directly applied to the plasma membranes of much more complicated cells such as those from kidney, brain, heart, and fat among many other tissues.

The remainder of this article will be devoted to a brief discussion of the composition and biochemical "anatomy" of the red cell membrane and some selective comments about some preliminary observations of a few apparently typical abnormalities of the membrane in various disease states. When possible an effort will be made to indicate how the composition and disposition of the membrane components may be related to the various membrane functions which have just been mentioned, and the analogous dysfunctions which occur in disease.

II. Membrane Lipids

All the lipids in the mature cell are contained within the membrane and are partially responsible for many of its physical characteristics. For example, both the passive cation permeability and the mechanical flexibility of the red cell can be influenced by modifying the lipid composition of its membrane. [*See* LIPIDS.]

A. Lipid Composition

Approximately one-half of the mass of the human red cell membrane consists of lipids, largely arranged as a bilayer.

Phospholipids and nonesterified cholesterol account for more than 95% of the total lipids. Small amounts of glycolipids, glycerides, and free fatty acids are also present. On a molar basis, the phospholipids and cholesterol are present in nearly equal amounts, and there is evidence that considerable interaction may occur between these major lipid classes within the membrane (e.g., cholesterol "condenses" and stabilizes bimolecular phosphatide leaflets).

B. Phospholipids

The phospholipids are divided into subclasses distinguished, with the exception of sphingomyelin, by the base group which is in phosphodiester linkage to the third carbon atom of their glycerol backbone (Fig. 1). Usually, there are two esterified fatty acids on the 1- and 2-positions of the glycerol backbone, although vinyl ether linkages occur to a substantial extent on the 2-position in some acidic phosphatides, such as phosphatidylethanolamine. The major phospholipids, which are usually named after their bases, and their approximate concentration in human erythrocytes are as follow: phosphatidylcholine (PC), 30%; phosphatidylethanolamine (PE), 28%; phosphatidylserine (PS), 14%; and sphingomyelin (SM), 25%.

Sphingomyelin is distinct structurally and probably dynamically from the rest of the group and may

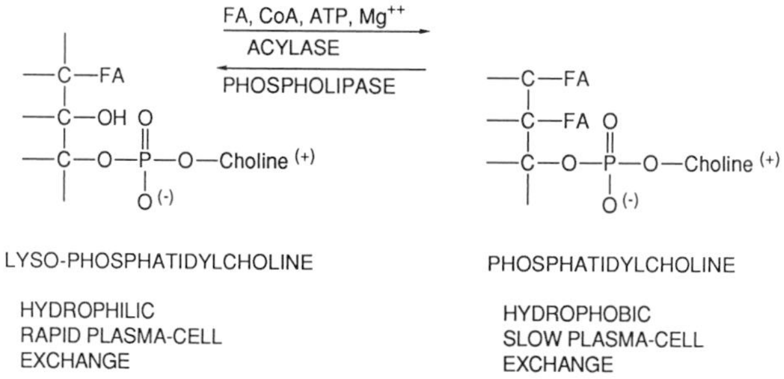

FA, CoA, ATP, Mg^{++}
ACYLASE
PHOSPHOLIPASE

—C—FA
—C—OH O
—C—O—P—O—Choline (+)
O (-)

—C—FA
—C—FA O
—C—O—P—O—Choline (+)
O (-)

LYSO-PHOSPHATIDYLCHOLINE
HYDROPHILIC
RAPID PLASMA-CELL EXCHANGE

PHOSPHATIDYLCHOLINE
HYDROPHOBIC
SLOW PLASMA-CELL EXCHANGE

FIGURE 1 The acylation reaction basic to red cell membrane lipid renewal. A lysophosphatide is esterified with a free fatty acid to produce a complex phosphatide with profoundly different physical properties. FA, Fatty acid; CoA, coenzyme A.

have a more structural role in the membrane. Small, but perhaps physiologically significant, amounts of phosphatidylinositides, phosphatidic acid, and polyglycerol phosphatides are also present in red cells. There are characteristic patterns of esterified fatty acids within each phospholipid class, and these also serve to distinguish subgroups.

Phospholipids containing only one fatty acid are known as lysophosphatides (see Fig. 1). The absence of one of the acyl (fatty acid) groups profoundly influences the physical characteristics of the phospholipid. Although the phospholipids with two fatty acids are highly lipophilic, the lysophosphatide compounds are nearly balanced in terms of lipophilic and hydrophilic characteristics. They tend to concentrate, therefore, at lipid–aqueous phase interfaces. This change in relative solubility increases both their detergent qualities and their rate of exchange between the cell membrane and the plasma. Because of these properties, lysophosphatides in low concentrations (2×10^{-4} *M*) can cause the lysis of red cell membranes—hence their trivial name. In even smaller concentrations, they can produce profound, eventually irreversible, shape changes ("echinocytogenesis") in the membrane (Fig. 2a–d). These morphologic distortions may be due to small changes in the relative packing density of the phospholipids in the inner and outer leaflets, which operate as a "bilayer couple" with consequent binding of the membrane, analogous to the bending of the bimetallic strip in a thermostat (see Fig. 3). Red cell membranes usually contain only small amounts of lysophosphatides and the cell has reacylation and dismutation enzyme systems to convert lysophosphatides back to phosphatides if their concentration begins to rise excessively. These reactions, together with the slow, but steady, exchange of several important membrane lipids with those of the plasma, also serve to maintain and renew much of the lipid of the membrane. This may account for the remarkable durability of this anuclear, but much traveled cell. These lipid renewal reactions are schematically summarized in Fig. 4.

C. Lipid Disposition

Although many structural relationships have been proposed, the precise anatomic localization of all of the lipids within the membranes is still unknown. There is, however, little doubt that a large percentage of the lipid is arrayed in the form of a bimolecular leaflet. In this bilayer disposition, the polar head

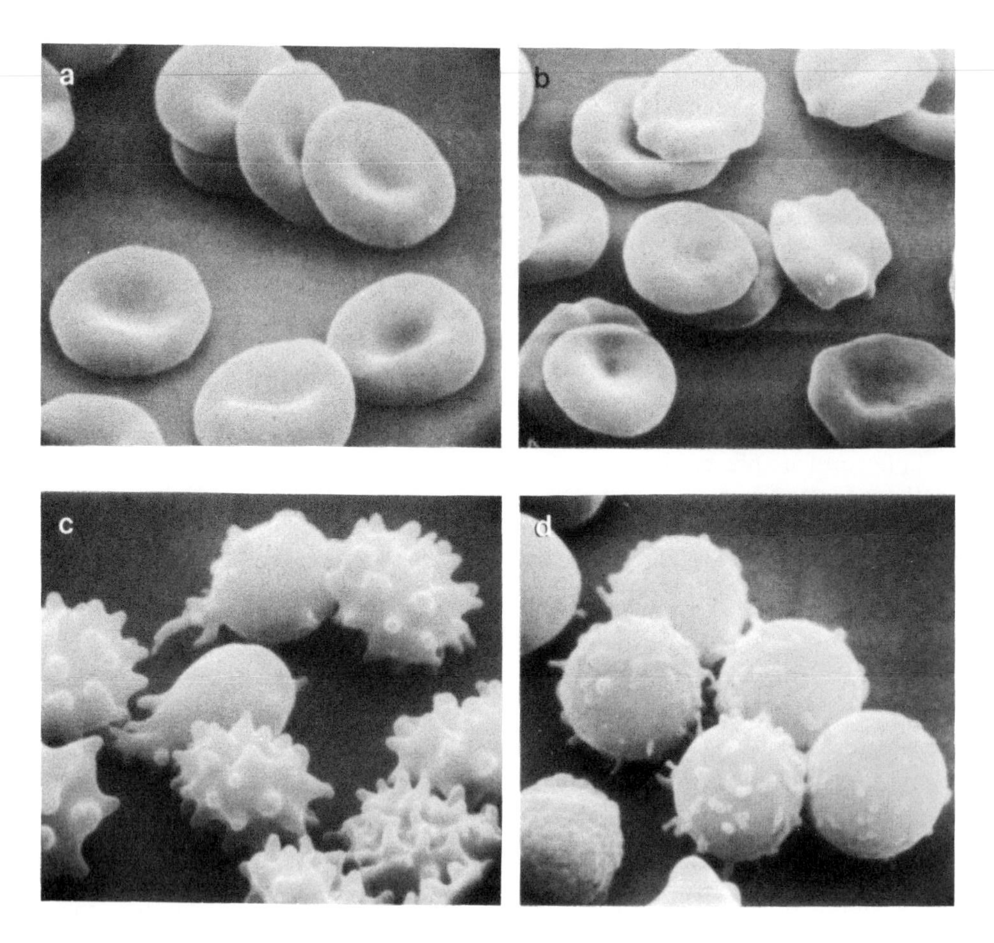

FIGURE 2 Scanning electron micrographs of erythrocytes subjected to increasing concentrations of membrane lysophosphatidylcholine (LPC). (a) Membrane LPC, 0.12 μmol/cm^3 cells; (b) membrane LPC, 0.15 μmol/cm^3 cells; (c) membrane LPC, 0.30 μmol/cm^3 cells; (d) membrane LPC, 0.50 μmol/cm^3 cells. The bulk of the changes in (a) to (c) could be reversed by washing the cells with defatted albumin. The changes seen in the red cells in (d) are irreversible and represent, in part, membrane loss due to microvesiculation induced by the high LPC concentration.

groups of each lipid layer face away from the center of the membrane into the hydrophilic environments of the cytoplasm and the plasma, while the long acyl tails of the lipids form a central hydrophobic core for the membrane. This hydrophobic core is in a liquid–crystalline state at normal temperatures and may facilitate the physiologically essential flexibility and deformability of the red cell membrane. It has been proposed that many of the protein elements of the membrane are inserted into this lipid matrix in much the same way as icebergs float in the ocean (Fig. 5). In this model, some of the proteins and glycoproteins are confined to one leaflet, while others, especially those assumed to have transport and shape-mediating roles, span the entire membrane. In addition, both the inserted proteins and the lipids are relatively free to move laterally within the plane of the membrane at comparatively rapid rates. In contrast, motions across the bilayer from one leaflet to the other are much more restricted. Since it is reasonable to assume that the lipid composition partially determines membrane viscosity, any significant changes in lipid composition which affect the membrane's internal microviscosity (e.g., an increase in cholesterol) might be expected to have some effect on the flexibility of the whole cell. Indeed, in some variants of liver disease, where a distortion of the normal exchange pathway markedly elevates red cell membrane cholesterol, stiff and spiculated "spur" cells are produced which circulate poorly and anemia may occur. However, usually the protein constituents of the membrane are predominant in regulating cell deformability, and calculations of the physical forces involved suggest that lipid–protein interactions must be much more important than pure lipid effects. Further, membrane flexibility is not the only, or even the

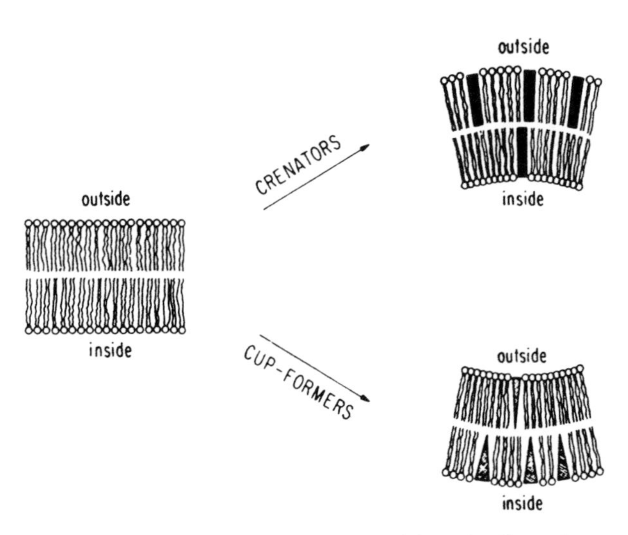

FIGURE 3 The hypothetical treatment of the red cell membrane as a "bilayer couple" is presented schematically here. Amphipathic compounds, including various drugs and phospholipids, can intercalate preferentially into either the inner or outer leaflet of the membrane because of individual characteristics (e.g., net charge at a given pH). Those that preferentially inserted into the outer leaflet crowd that leaflet, creating a bending moment which tends to induce evagination of the membrane, or crenation; while conversely, those that concentrate in the inner leaflet induce invagination of the membrane, or cup formation.

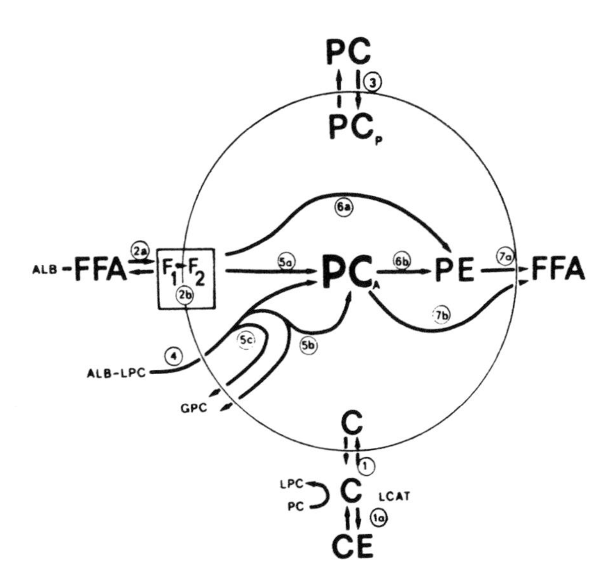

FIGURE 4 Schema of the major exchange and metabolism pathways for lipids in the mature erythrocyte. Alb-FFA, Albumin-bound free fatty acid; F_1, membrane surface pool of fatty acid which is freely exchangeable with plasma FFA; F_2, "deeper" membrane pool of free fatty acid used as a source of acyl groups for phosphatides within the membrane; PC_A, phosphatidylcholine actively synthesized within the membrane; PC_P, phosphatidylcholine passively acquired by exchange with the plasma by the membrane; PE, phosphatidylethanolamine; Alb-LPC, albumin-bound lysophosphatidylcholine, which, together with Alb-FFA, serves as the precursor for PC_A, GPC, glycerolphosphorylcholine; C, cholesterol which exchanges with the membrane; CE, cholesterol ester which does not exchange with the membrane; LCAT, lecithin cholesterol acyltransferase which catalyzes the formation of cholesterol ester and hence modulates the rate of the cholesterol exchange.

predominant, requirement for whole-cell deformability. The cell must avoid reductions in surface area and increases in intracellular viscosity to maintain optimal deformability. In the latter case, it is likely that the physical characteristics of the membrane are subordinate to its permeability and water regulation characteristics in maintaining cell deformability.

D. Lipid Asymmetry

The lipids are not symmetrically distributed between the inner and outer leaflets of the membrane. It appears that the majority of phosphatidylethanolamine and phosphatidylserine is contained within the inner or cytoplasmic leaflet of the membrane, while the majority of phosphatidylcholine and the sphingomyelin is contained in the leaflet facing the plasma. The biochemical basis of this asymmetry may be the combined result of some site specificity of the phospholipid exchange and renewal reactions and the sluggish lipid exchange rates between the inner and outer membrane leaflets. In addition, the action of a recently described phospholipid "flipase," which preferentially translocates acid phosphatides from the outer to the inner leaflet of the bilayer, is likely to be involved in maintaining this asymmetry. Although the full extent of the physiologic consequences of this asymmetry is not known, at the very least considerable transmembrane charge potential is induced by the excess of positive charges on the inner leaflet. This asymmetry is apparently disturbed in some abnormal states (e.g., sickled and irreversibly sickled cells), and the speculation has been made that such externalization of acidic red cell phospholipids may serve to activate the coagulation system and pathologic thromboses. The asymmetric distribution of the typical intramembranous particles seen on freeze-fracture electron microscopy of the cleaved membrane may also be a consequence of lipid asymmetry.

Another potential consequence of lipid asymmetry may involve coupling of the membrane bilayer to an important and novel structure, the membrane skeleton. This structure, which will be discussed in detail below with the membrane proteins, consists primarily of spectrin, actin, and proteins 2.1 (ankyrin) and 4.1. It appears to serve as a scaffolding for the lipid bilayer and to have a dominating role in influencing membrane stability and deformability. Although elegant studies have established specific protein interactions between this membrane "skeleton" and protein components embedded in the lipid bilayer, such as band 3 and glycophorin, spectrin : phosphatidylethanolamine and spectrin : phosphatidylserine interactions have also been observed, as well as band 4.1 : phosphatidylserine interactions. However, it is not clear if the membrane phospholipid asymmetry which places these acidic phospholipids in intimate contact with spectrim at the interface between the inner leaflet of the lipid bilayer and the membrane skeleton is a cause or a consequence of such interactions.

Finally, the predominant location of small quantities of variously phosphorylated phosphatidylinositides in the inner leaflet may be important for the regulation of the affinity of the skeletal protein band 4.1 in red cells, and for their role in the generation of "second messengers" in the membranes of other cells.

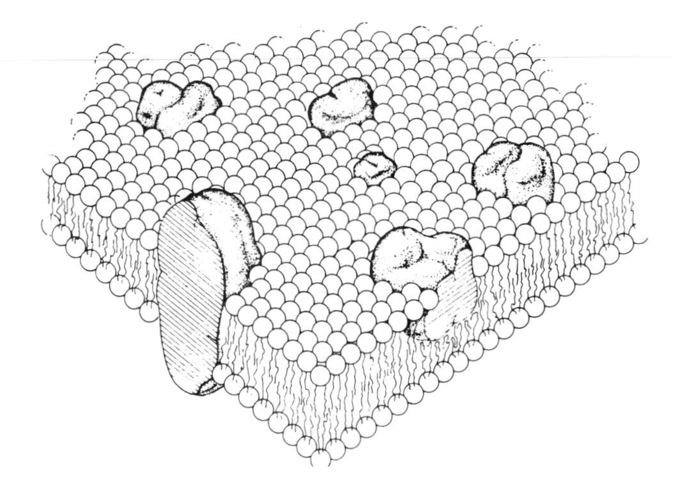

FIGURE 5 A schematic representation of the Singer–Nicholson fluid mosaic model for the structure of cell membranes. Irregular proteins are seen penetrating both into and through the bimolecular leaflet composed of regular arrays of phospholipid molecules. The head groups of the phospholipids face the cytoplasmic and plasma environments, while their acyl tails are enmeshed to form the lipophilic membrane core.

III. Membrane Proteins and the Membrane Skeleton

A. Analysis

Study of red cell membrane proteins was initially difficult because of their insolubility in aqueous media of physiologic ionic strength. Indeed, any membrane proteins which did not have such properties would be lost as membranes were washed in the course of preparing hemoglobin-free ghosts. It is possible, however, to dissolve red cell membranes completely in sodium dodecyl sulfate and to accurately analyze protein subunits varying in size from the larger subunit of spectrin (M_r 240,000) to traces of globin monomers (M_r 16,000) with great sensitivity. The study of red cell membrane proteins has been appreciably advanced by this technique and by the uniform numbering of such polypeptides which most investigators now use (Fig. 6a).

A variety of stratagems have been employed to deduce the relative positions of the various protein subunits in the membrane. For example, it has been possible to determine whether or not a human protein is exposed at the outer or the inner surface of the membrane by labeling membrane preparations on the inside or the outside with radioactive iodine using lactoperoxidase. Also, cross-linking studies have been employed to determine which proteins may be regarded as "neighbors" in the intact membrane. The fact that spectrin and actin (bands 1, 2, and 5) and band 4.1 play a major role in maintaining the shape of the erythrocyte may be deduced from the fact that ghosts of various poikilocytes (abnormally shaped cells) extracted with nonionic detergents such as Triton X-100 maintain their initial poikilocytic shape. These components of the membrane are sometimes designated as the "extrinsic" membrane proteins because they can be released from the membrane by treatment with very low ionic strength, slightly alkaline solutions without disrupting the lipid bilayer. In contrast, the "intrinsic" proteins of the membrane, those which are imbedded in the lipid bilayer, are removed only by detergent treatment.

B. Protein Disposition and the Membrane Skeleton

Through these types of studies, the concept of a "membrane skeleton" of proteins which may have a role in modulating cell shape and deformability has emerged. This structure is composed of the ex-

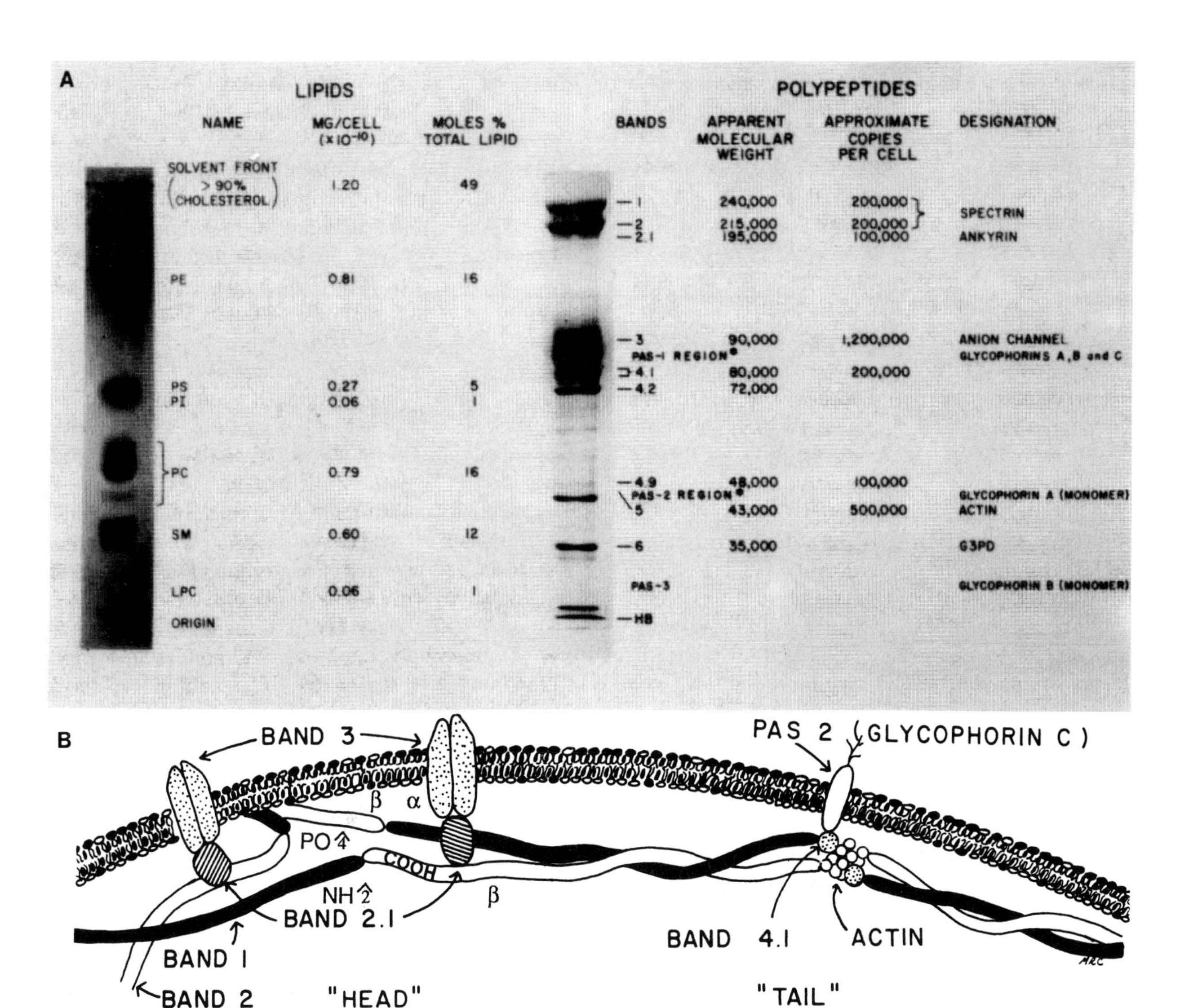

FIGURE 6 Current concepts of red cell membrane composition and organization. (A) Thin-layer chromatogram of red cell membrane lipids (left) and a Coomassie Blue-stained SDS polyacrylamide gel electrophoretogram of red cell membrane proteins (right). (B) Diagrammatic cross-section of membrane bilayer and supporting "skeleton." The predominant protein of the membrane, spectrin, occurs as a heterodimer (bands 1 and 2) linked together into a fibrous network. The linkage between the "tail" ends of the dimers appears to be mediated by actin (band 5) and band 4.1. Linkage between the "head" ends of the dimers occurs by direct contact between complementary strands of the heterodimer (the carboxy terminus of the β chain and the amino terminus of the α chain). Attachment of the skeleton to the membrane is produced by a specific association between band 2 of spectrin and band 3 in the lipid bilayer via the spectrin-binding protein, band 2.1 (ankyrin), near the head end of the spectrin dimer. An additional association of the skeletal complex with the lipid bilayer may be provided by a connection between spectrin and another bilayer protein, glycophorin C or A, via band 4.1. Bands 2, 2.1, 3, and 4.1 can be phosphorylated, and some of that phosphorylation is cyclic AMP dependent. The phosphorylation sites of spectrin band 2, which are not shown in this diagram, appear to be close to the carboxy end of the dimer. The outer leaflet of the bilayer is composed predominantly of choline-containing phospholipids (indicated by black head groups), and the inner leaflet is predominantly composed of acidic phospholipids, such as phosphatidylethanolamine and phosphatidylserine (indicated by white head groups). Cholesterol (indicated by black ovals) is shown embedded symmetrically in each leaflet among the fatty acid side groups of the phospholipid, although this has not yet been experimentally verified. The PAS1 and PAS2 bands have been better defined by additional biochemical studies, including gel electrophoresis techniques, with improved resolution. It is now known that the PAS1 region contains the dimer of the sialoglycopeptide glycophorin A and that the PAS2 region resolves into three bands: the dimer of glycophorin B (M_r 47,000), the monomer of glycophorin A (M_r 38,000), and glycophorin C, also called glycoconnectin (M_r 35,000). Unfortunately, there is not, as of yet, a universally accepted nomenclature for these PAS-staining sialoglycopeptides. PE, Phosphatidylethanolamine; PS, phosphatidylserine; PI, phosphatidylinositol; PC, phosphatidylcholine; SM, sphingomyelin; LPC, lysophosphatidylcholine; G3PD, glucose-3-phosphate dehydrogenase.

trinsic proteins which are coupled to the bilayer via the intrinsic proteins. In electron micrographs, this skeleton appears to be a moderately dense meshwork of intimately interconnected proteins immediately subjacent to the lipid bilayer. Recent elegant preparations of osmotically stretched or partially extracted membranes show an underlying pattern to this structure, which probably depicts the actual spectrin–actin–band 4.1 connections in a hexagonally arrayed network (see Fig. 7). Of course, it should be recognized that these preparations are not native cells, and the treatments which were required for the clarity of these remarkable micrographs may themselves have modified the arrangement of this structure. Nevertheless, for the present, they offer an effective insight into the nature of the skeleton. Whatever the exact biochemical anatomy of this structure, it seems to be clear that it is immediately beneath the bilayer, that it is anchored to the bilayer by at least one and probably two integral membrane proteins (band 3 and glycophorin C), and that it supports the bilayer and acts as an essential scaffolding for that otherwise weak and ephemeral structure.

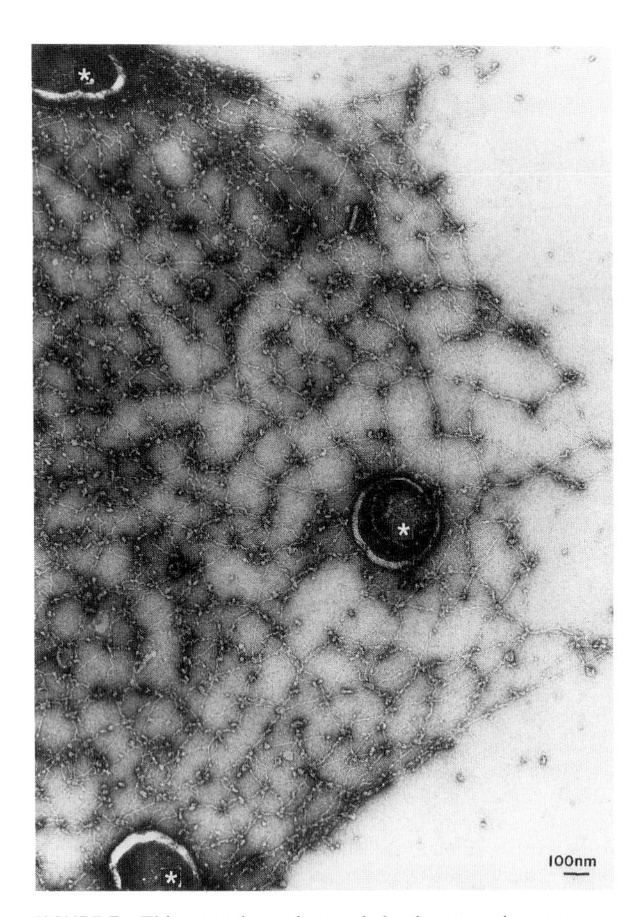

FIGURE 7 This spread membrane skeletal preparation was prepared by Triton extraction of ghosts which were expanded by hypotonic treatment. This negative-stained electron micrograph shows the distinctive hexagonal lattice typical of the membrane skeleton in these preparations. The long filaments consist predominantly of spectrin tetramers, whereas the central junctional complexes are most likely actin and band 4.1. The small globular areas decorating the middle of many of the spectrin strands probably represent ankyrin bound near the head groups of the spectrin dimers.

C. Spectrin and Its Interactions

The structural backbone of this skeleton appears to be composed primarily of the protein, spectrin, a long fiber-like molecule ideally suited to be the major structural element in a network. Substantial characterization of the basic organization of this very large molecule has been achieved by careful proteolytic digestion studies. As shown in Fig. 8, each chain consists of a considerable number of homologous repeating triple-helical segments which are linked by short, flexible, nonhelical regions. This catenary organization is very well designed to produce a strong but flexible elongated structural element. Fastidious amino acid analyses and early efforts to obtain cDNA probes for both α and β spectrin began to unravel the intimate molecular details of this crucial component of the membrane, and both α and β spectrin have now been completely cloned and sequenced.

Spectrin occurs as heterodimers (made up of two unequal subunits) in solution, but the heterodimers self-associate to form predominantly tetramers (four subunits) and higher oligomeric forms in the membrane skeleton by the "head-to-head" associations indicated in Figure 6b. The tetramers, in turn, are laterally ramified and bound together into an anastomotic network by interactions with band 4.1 and actin at the opposite (tail) ends of the basic dimeric unit. Importantly, this entire network is then connected to the membrane bilayer by at least one additional linking protein, band 2.1. This protein, also called ankyrin or syndein, has high-affinity binding sites for both spectrin and the integral bilayer membrane protein, band 3. Hence, it serves to "anchor" the membrane skeleton to the bilayer, forming the complete membrane unit. A second anchorage site, which connects band 4.1 to another integral membrane protein, glycophorin C, is also likely. The net effect of this highly orchestrated series of spectrin : spectrin interactions and specific

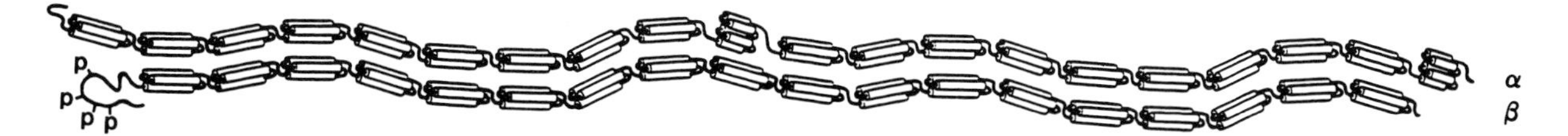

FIGURE 8 The spectrin dimer model: α and β subunits composed of multiple triple helical segments are connected by short nonhelical regions. Each segment contains 106 amino acids and, with the exception of the tenth, and perhaps the twentieth, subunit in the α chain, they are all homologous in basic structure, though not identical in amino acid composition. The two chains are antiparallel with the NH_2 terminus of the α subunit to the left. The carboxy terminus of the β chain is multiply phosphorylated.

bilayer binding interactions is to produce a strong, flexible, skeletal network which is closely applied to, and coupled with, the undersurface of the bilayer, and which is capable of supporting it and providing it with remarkable physical integrity.

D. Defects of the Membrane Skeleton

As might be expected, a large number of red cell morphologic abnormalities and hemolytic anemias with premature destruction of the circulating cells are associated with abnormalities in both the constituents of this newly defined structure and in their interactions. These abnormalities include quantitative deficiencies of particular proteins, such as spectrin in both recessive and dominant hereditary spherocytosis (hemolytic anemias with reduced surface area red cells) and perhaps in pyropoikilocytosis (severe hemolytic anemia with heat sensitive fragile red cells) as well as in murine models of spherocytosis. Deficiency of band 4.1 has also been found in cases of severe, possibly homozygous, hereditary elliptocytosis. Also, deficiency of band 2.1 (ankyrin) has been found in a rare heat-sensitive fragmentation hemolytic anemia and in some severe variants of heeditary spherocytosis.

Qualitative abnormalities in the elements of this skeletal structure and in their interactions include defects in spectrin : 4.1 binding in other cases of hereditary spherocytosis, and an extensive series of discrete abnormalities in spectrin : spectrin interactions in hereditary pyropoikilocytosis. In addition, qualitative defects in interactions between membrane skeletal elements may be acquired through damage that occurs to red cells in prolonged blood bank storage or during various oxidative stresses. Finally, both short and long variants of band 4.1, and a short variant of spectrin, have been reported in patients with poikilocytic hemolytic anemias, and it is reasonable to propose that some consequent qualitative membrane skeletal malfunction is responsible for the hemolysis in these conditions.

It seems probable that many more abnormalities in the components and interactions of the membrane skeleton will eventually be found when hemolytic disorders of the membrane are exhaustively examined. Indeed, such studies are likely to show that the limited morphologic variations which we see in the clinical blood film in hemolytic anemias represent a final common pathway of a much more diverse variety of biochemical abnormalities in this unusual and important membrane structure.

Bibliography

Bennett, V. (1985). The membrane skeleton of human erythrocytes and its implication for more complex cells. *Ann. Rev. Biochem.* **54,** 273.

Conboy, J., Kan, Y. W., Shohet, S. B., and Mohandas, N. (1986). Molecular cloning of protein 4.1, a major structural element of the human erythrocyte membrane skeleton. *Proc. Natl. Acad. Sci. U.S.A.* **83,** 9512.

Fairbanks, G., Steck, T. L., and Wallach, D. F. H. (1971). Electrophoretic analysis of the major polypeptides of the human erythrocyte membrane. *Biochemistry* **10,** 2606.

Knowles, W., Marchesi, S. L., and Marchesi, V. T. (1983). Spectrin: Structure, function and abnormalities. *Semin. Hematol.* **20,** 159.

Kopito, R., and Lodish, H. F. (1985). Primary structure and transmembrane orientation of the murine anion exchange protein. *Nature (London)* **319,** 234.

Lux, S. E. (1979). Dissecting the red cell membrane skeleton. *Nature (London)* **281,** 426.

Mohandas, N., Clark, M. R., Jacobs, M. S., and Shohet, S. B. (1980). Analysis of factors regulating erythrocyte deformability. *J. Clin. Invest.* **66,** 563.

Palek, J., and Lambert, S. (1990). Genetics of the red cell membrane's skeleton. *Semin. Hematol.* (in press).

Sheetz, M. P., and Singer, S. J. (1974). Biological membrane as bilayer couples. A molecular mechanism of drug–erythrocyte interactions. *Proc. Natl. Acad. Sci. U.S.A.* **71,** 4457.

Shen, B. W., Josephs, R., and Steck, T. L. (1984). Ultrastructure of unit fragments of the skeleton of the human erythrocyte membrane. *J. Cell Biol.* **99,** 810.

Shohet, S. B. (1972). Hemolysis and changes in erythrocyte membrane lipids. *N. Engl. J. Med.* **286,** 577, 638.

Singer, S. J., and Nicolson, G. L. (1972). The fluid mosaic model of the structure of cell membranes. *Science* **175,** 720.

Speicher, D. W., and Marchesi, V. T. (1984). Erythrocyte spectrin is comprised of many homologous triple helical segments. *Nature (London)* **311,** 177.

Tchernia, G., Mohandas, N., and Shohet, S. B. (1981). Deficiency of skeletal protein band 4.1 in homozygous elliptocytosis: Implications for membrane stability. *J. Clin. Invest.* **68,** 454.

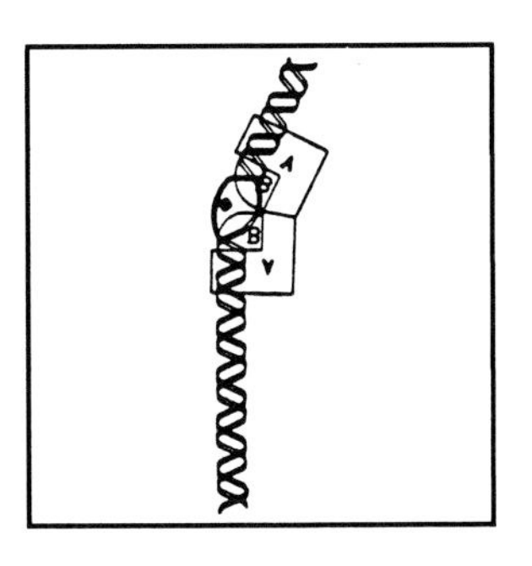

Repair of Damaged DNA

LAWRENCE GROSSMAN, *The Johns Hopkins University*

Glossary

Endonuclease Nuclease that hydrolyzes internal phosphodiester bonds

Excision Removal of damaged nucleotides from incised nucleic acids

Exonuclease Nuclease that hydrolyzes terminal phosphodiester bonds

Glycosylase Enzymes that hydrolyze N-glycosyl bonds linking purines and pyrimidines to carbohydrate components of nucleic acids

Incision Endonucleolytic break in damaged nucleic acids

Ligation Phosphodiester bond formation as the final stage in repair

Nuclease Enzyme that hydrolyzes the internucleotide phosphodiester bonds in nucleic acids

Resynthesis Polymerization of nucleotides into excised regions of damaged nucleic acids

THE ABILITY OF CELLS to survive hostile environments is due in part to surveillance systems that recognize damaged sites in DNA and are capable of either reversing the damage or removing damaged bases or nucleotides, generating sites that lead to a cascade of events restoring DNA to its original structural and biological integrity.

Both endogenous and exogenous environmental agents can damage DNA. A number of repair systems are regulated by the stressful effects of such damage, affecting the levels of responsible enzymes, or by modifying their specificity. Repair enzymes appear to be the most highly conserved proteins showing their important role throughout evolution.

Either the enzyme systems can directly reverse the damage to form the normal purine or pyrimidine bases or the modified bases can be removed together with surrounding bases through a succession of events involving nucleases, DNA polymerizing enzymes, and polynucleotide ligases, which assist in restoring the biological and genetic integrity to DNA.

I. Damage

As a target for damage, DNA possesses a multitude of sites that differ in their receptiveness to modification. On a stereochemical level, nucleotides in the major groove are more receptive to modification than those in the minor groove, the termini of DNA chains expose reactive groups, and some atoms of a purine or pyrimidine are more susceptible than others. As a consequence, the structure of DNA represents a heterogenous target in which certain nucleotide sequences also contribute to the susceptibility of DNA to genotoxic agents. [*See* DNA AND GENE TRANSCRIPTION.]

A. Endogenous Damage

Even at physiological pHs and temperatures in the absence of extraneous agents, the primary structure of DNA undergoes alterations. A number of specific reactions directly influence the informational content as well as the integrity of DNA. Although the rate constants for many reactions are inherently low, because of the enormous size of DNA and its persistence in cellular life cycles, the accumulation of these changes can have significant long-term effects.

1. Deamination

The hydrolytic conversion of adenine to hypoxanthine (Fig. 1), guanine to xanthine, and cytosine to uracil-containing nucleotides is of sufficient magnitude to affect the informational content of DNA.

2. Depurination

The glycosylic bonds linking guanine in nucleotides are more sensitive to hydrolysis than the adenine and pyrimidine glycosylic links. The resulting apurinic (AP), or apyrimidinic, site is recognized by surveillance systems and, as a consequence, is repaired.

3. Mismatched Bases

During the course of DNA replication, there are those noncomplementary nucleotides that are incorrectly incorporated into DNA and manage to escape the editing functions of the DNA polymerases. The proper strand as well as the mismatched base is recognized and repaired.

4. Metabolic Damage

When thymine incorporation into DNA is limited either through restricted precursor deoxyuridine triphosphate (dUTP) availability or inhibition of the thymidylate synthetase system, dUTP is utilized as a substitute for thymidine triphosphate. The presence of uracil is identified as a damaged site and acted upon by repair processes.

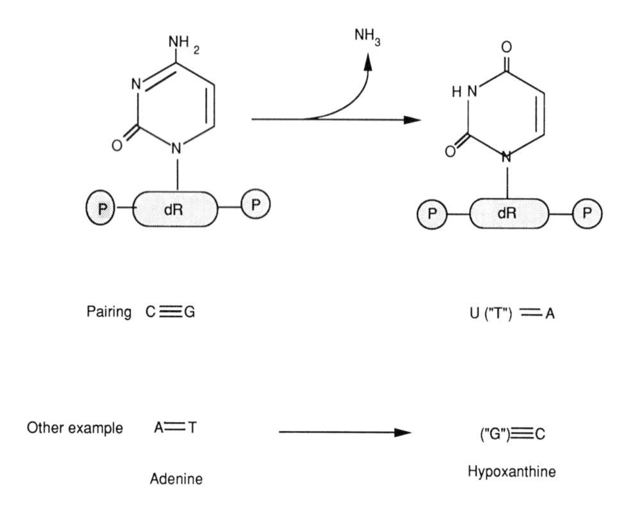

FIGURE 1 Deamination reactions have mutagenic consequences because the deaminated bases cause false recognition.

5. Oxygen Damage

The production of oxygen radicals (superoxide or hydroxyl radicals) as a metabolic consequence as well as at inflammatory sites causes sugar destruction, which eventually leads to strand breakage.

B. Exogenous Damage

The concept of DNA repair in biological systems arose from studies by photobiologists and radiobiologists studying the viability and mutagenicity in biological systems exposed to either ionizing or ultraviolet irradiation. Target theories, derived from the random statistical nature of photon bombardment, led to the identification of DNA as the primary target for the cytotoxicity and mutagenicity of ultraviolet light. The photoproducts in DNA responsible for the effects of such irradiation have been attributed to 5,6-pyrimidine cyclobutane *cis,syn* dimers, 6,4-pyrimidine-pyrimidone dimers, and 5,6-water-addition products of cytosine (hydrates). In addition, most of the structural and regulatory genes controlling DNA repair in *Escherichia coli* were identified facilitating the molecular characterization of the relevant enzymes.

1. Ionizing radiation

The primary cellular effect of ionizing radiation is the radiolysis of water, which mainly generates hydroxyl radicals (HO·). HO· is capable of abstracting protons from the C-4′ position of the deoxyribose moiety of DNA, thereby labilizing the phosphodiester bonds and generating single- and double-strand breaks. The pyrimidine bases are also subject to HO· addition reactions.

2. Ultraviolet irradiation

Most ultraviolet photoproducts are chemically stable; their recognition provides direct biochemical evidence for DNA repair. The major photoproducts are 5,6-cyclobutane dimers of neighboring pyrimidines (intrastrand dimers), 6,4-pyrimidine-pyrimidone dimers (6–4 adducts), and 5,6-water-addition products of cytosine (cytosine hydrates).

3. Alkylation

This modification takes place on purine-ring nitrogens (cytotoxic adducts), the O^6 position on guanine, the O^4 positions of the pyrimidines (mutagenic lesions), and the oxygen residues of the phosphodiester bonds of the DNA backbone (biologically silent). Alkylating agents are environmentally

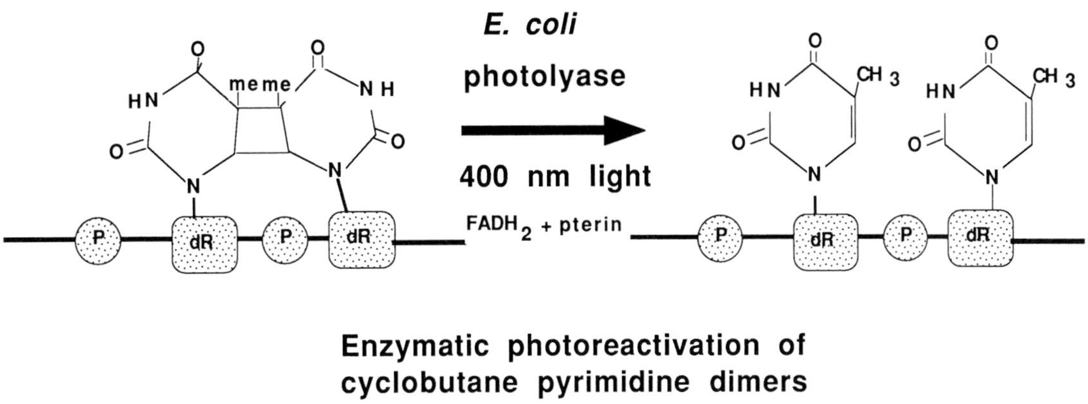

E. coli photolyase
400 nm light
$FADH_2$ + pterin

Enzymatic photoreactivation of cyclobutane pyrimidine dimers

FIGURE 2 The direct photoreversal of pyrimidine dimers in the presence of visible light. $FADH_2$, reduced flavin adenine dinucleotide.

pervasive arising indirectly from many foodstuffs and from automobile exhaust, in which internal combustion of atmospheric nitrogen results in the formation of nitrate and nitrites.

4. Bulky adducts

Large, bulky, polycyclic, aromatic hydrocarbon modification occurs primarily on the N-2 and C-8 position of guanines invariably from the metabolic activation of these large hydrophobic, uncharged macromolecules to their epoxide analogues. The major source of these substances is from the combustion of tobacco, petroleum products, and foodstuffs.

II. Direct Removal Mechanisms

The simplest repair mechanisms involve the direct photoreversal of pyrimidine dimers to their normal homologues and the removal of O-alkyl groups from the O^6-methylguanine and from the phosphotriester backbone as a consequence of alkylation damage to DNA.

A. Photolyases (Photoreversal)

The direct reversal of pyrimidine dimers to the monomeric pyrimidines is the simplest mechanism (Fig. 2), and parenthetically it is chronologically the first mechanism described for the repair of photochemically damaged DNA. It is a unique mechanism characterized by a requirement for visible light as the sole source of energy for breaking two carbon-carbon bonds.

The enzyme protein has two associated light-absorbing molecules (chromophores), which can form an active light-dependent enzyme. One of the chromophores is reduced flavin adenine dinucleotide ($FADH_2$), and the other is either a pterin or a deazaflavin, capable of absorbing the 365–400 nm wavelengths required for photoreactivation of pyrimidine dimers. It is suggested that photoreversal involves energy transfer from the pterin molecule to $FADH_2$ with electron transfer to the pyrimidine dimer resulting in nonsynchronous cleavage of the C5 and C6 cyclobutane bonds.

Enzymes that carry out photoreactivation have been identified in both prokaryotes and eukaryotes.

B. Alkyl Group Removal (Methyl Transferases)

Cells pretreated with less than cytotoxic or genotoxic levels of alkylating agents before a lethal mutagenic dose are more resistant. This is an adaptive phenomenon with antimutagenic and anticytotoxic significance. During this adaptive period, a 39-kDa *Ada protein* is synthesized, which specifically removes a methyl group from a phosphotriester bond and from an O^6-methyl group of guanine (or from O^4-methyl thymine) (Fig. 3). The O^6-methyl group of guanine is not liberated as free O^6-methyl guanine during this process but is transferred directly from the alkylated DNA to this protein; the ada protein (methyl transferase) and an unmodified guanine are simultaneously generated. These alkyl groups specifically methylate cysteine 69 and cysteine 321, respectively, in the protein.

The methyl transferase is used stoichiometrically in the process (i.e., does not turnover) and is permanently inactivated in the process. Nascent enzyme is, however, generated because the mono- or di-methylated transferase activates transcription of its own "regulon," which includes, in addition to

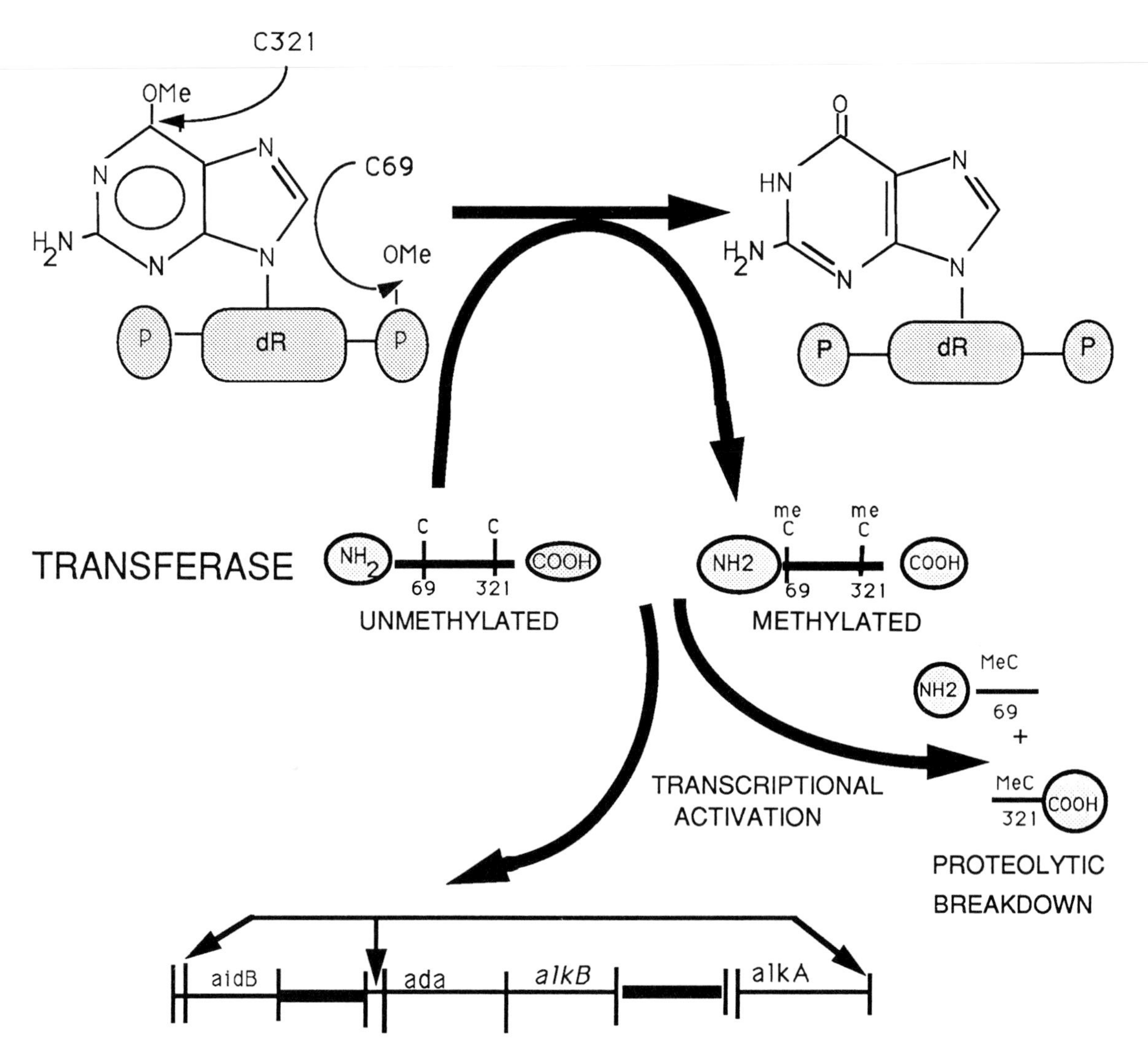

the *ada* gene, the *alk*B gene of undefined activity and the *alk*A gene (which controls a DNA glycosylase). The latter enzyme acts on 3-methyl adenine, 3-methyl guanine, O^2-methyl cytosine, and O^2-methyl thymine. The methylated Ada protein can specifically bind to the operator of the *ada* gene acting as a positive regulator. Down-regulation may be controlled by proteases acting at two hinge sites in the Ada protein.

FIGURE 3 The direct reversal of alkylation damage removes such groups from the DNA backbone and the O^6-position of guanine. Such alkyl groups are transferred directly to specific cysteine residues on the transferase, the levels of which are influenced adaptively by levels of the alkylating agents. The methylation of the transferase inactivates the enzyme, which is used up stoichiometrically in the reaction. The alkylated transferase acts as a positive transcriptive signal-turning on the synthesis of unique mRNA. Regulation of transferase levels may be influenced by a unique protease.

III. Base-Specific Responses

A. Base Excision Repair by Glycosylases and AP Endonucleases

Bases modified by deamination can be repaired by a group of enzymes called DNA glycosylases, which specifically break the N-glycosyl bond of that base and the deoxyribose of the DNA backbone generating an AP site (Fig. 4). These are rather small, highly specific enzymes, which require no cofactor for functioning. They are the most highly conserved proteins attesting to the evolutionary unity both structurally and mechanistically from bacteria to humans.

As a consequence of DNA glycosylase action, the AP sites generated in the DNA are acted upon

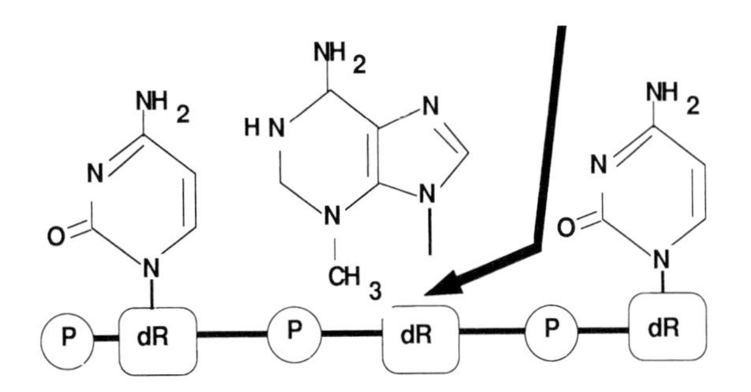

FIGURE 4 DNA glycosylases hydrolyze the N-glycosyl bond between damaged bases and deoxyribose generating an apurinic (AP), or apyrimidinic, site.

by a phosphodiesterase specific for such sites, which can nick the DNA either 5′- and/or 3′- to such damaged sites (Fig. 5). If there is a sequential action of a 5′-acting and 3′-acting AP endonuclease, the AP site is excised, generating a gap in the DNA strand.

B. Glycosylase-Associated AP Endonucleases

An enzyme from bacteria and phage-infected bacteria, encoded in the latter case by a single gene (*den*V), breaks the N-glycosyl bond of the 5′-thymine moiety of a pyrimidine dimer followed by hydrolysis of the phosphodiester bond between the two thymine residues of the dimer (Fig. 6). This enzyme, referred to as the pyrimidine dimer DNA glycosylase, is found in *Micrococcus luteus* and phage T4-infected *E. coli*. This small, uncomplicated enzyme does not require cofactors and is presumed to act by a series of linked beta-elimination reactions.

An enzyme behaving in a similar glycosylase-endonuclease fashion but acting on the radiolysis

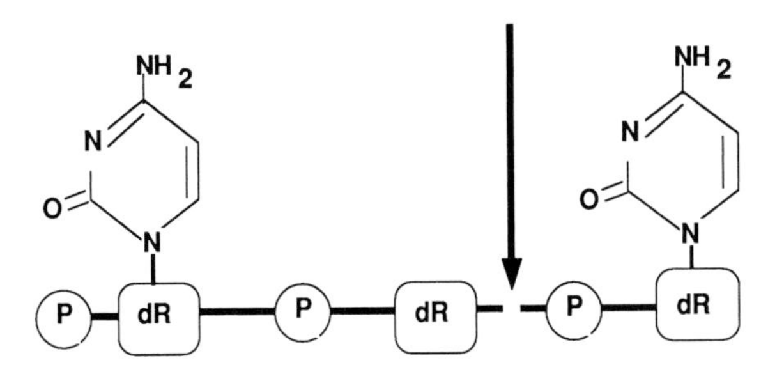

FIGURE 5 Endonucleases recognize AP sites and hydrolyze the phosphodiester bonds either 3′-, 5′- or both sides of the deoxyribose moiety in damaged DNA.

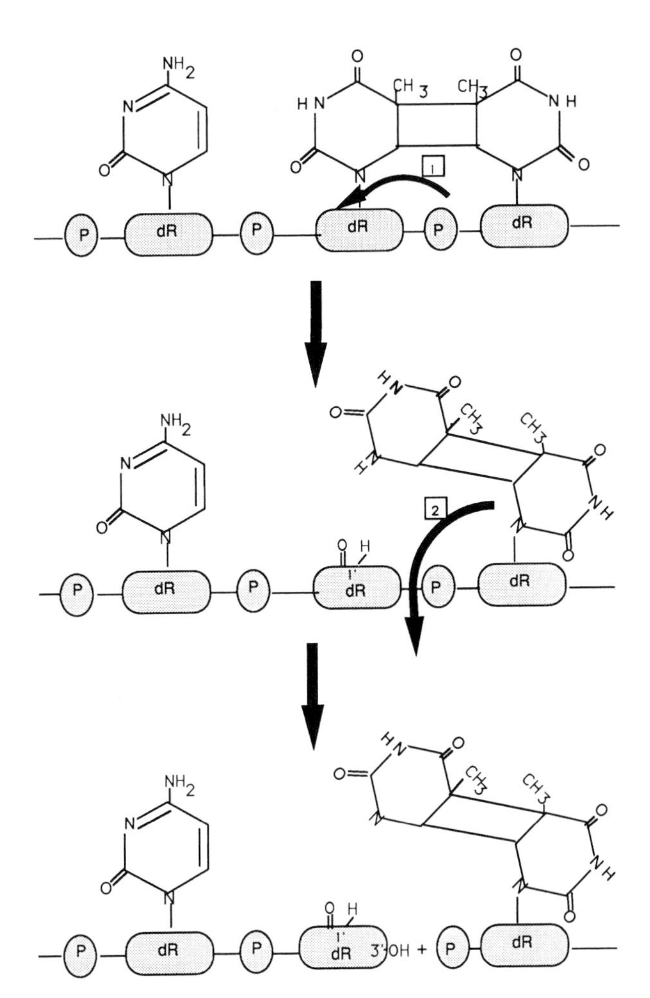

FIGURE 6 The same enzyme that can hydrolyze the N-glycosyl residue of a damaged nucleotide also hydrolyzes the phosphodiester bond linking the AP site generated in the first N-glycosylase reaction. UV, ultraviolet.

product of thymine, thymine glycol has been isolated from *E. coli* and is referred to as endonuclease III.

IV. Nucleotide Excision Repair

A. Prokaryotes

The ideal repair system is one that is somewhat indiscriminate and can respond to virtually any kind of damage. Such a repair system has been characterized in *E. coli,* where it consists of at least six gene products of the *uvr* system. These proteins consist of the UvrA protein, which binds as a dimer to DNA in the presence of adenosine triphosphate (ATP), followed by the UvrB protein, which cannot

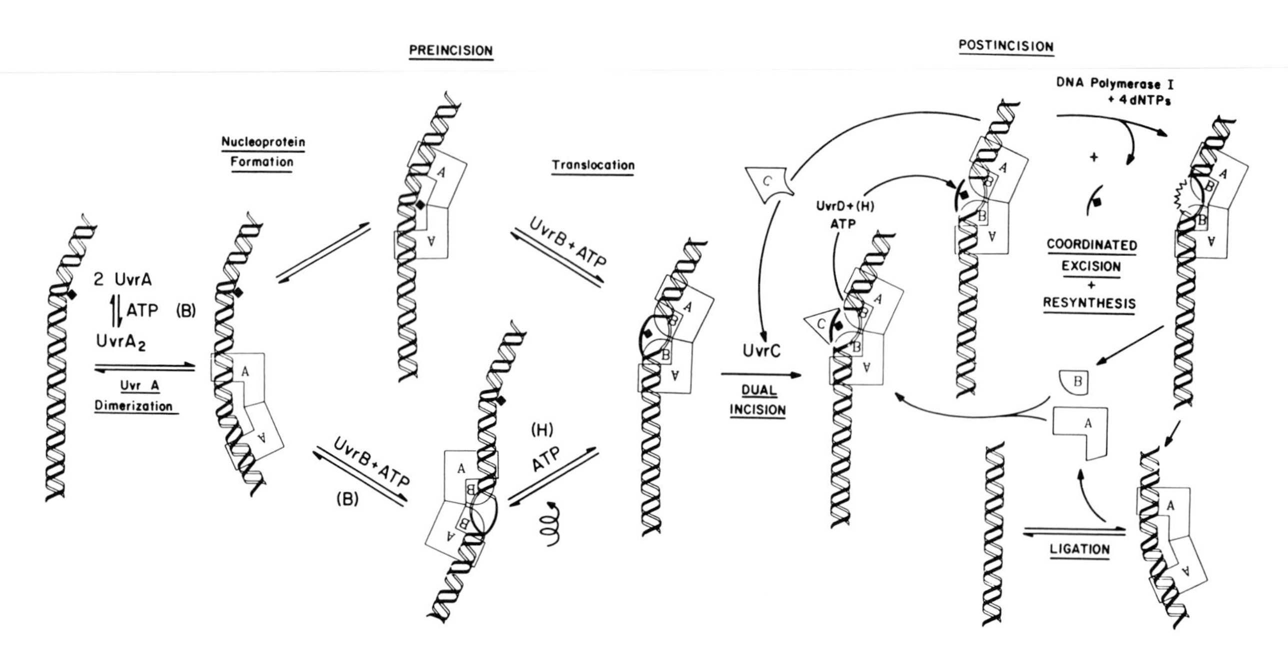

bind DNA by itself. Translocation of the UvrAB complex from initial undamaged DNA sites to damaged sites is driven by a cryptic ATPase associated with UvrB, which is activated by the formation of the UvrAB-undamaged DNA complex. This complex is now poised for endonucleolytic activity catalyzed by the interaction of the UvrAB-damaged DNA complex with UvrC to generate two nicks in the DNA, seven nucleotides 5′- to the damaged site and three to four nucleotides 3′- to the same site (Fig. 7). These sites of breakage are invariant regardless of the nature of the damage. In the presence of the UvrD (helicase III) and DNA polymerase I, the damaged fragment is released, accompanied by the turnover of the UvrA, UvrB, and UvrC proteins. The continuity of the DNA helix is maintained based on the sequence of the opposite strand. The final integrity of the interrupted strands is restored by the action of DNA polymerase I, which copies the other strand, and by polynucleotide ligase, which seals the gap.

The levels of the Uvr proteins are regulated in *E. coli* by an "SOS" regulon monitoring a large number of genes, which includes the *uvrA, uvrB,* possibly *uvrC,* and *uvrD* as part of the excision repair system; it also includes the regulators of the SOS system, the *lexA* and *recA* proteins; cell-division genes *sul A* and *sulB;* recombination genes *recA, recN, recQ, uvrD,* and *ruv;* mutagenic bypass mechanisms (*umuDC* and *recA*); damage-inducible genes; and the lysogenic phage lambda. The lexA protein negatively regulates these genes as a repressor by binding to unique operator regions. When the DNA is damaged, for example, by ultraviolet light, a signal in the form of a DNA repair intermediate induces the synthesis of the recA protein. When induced, the recA protein acts as a protease assisting the lexA protein to degrade itself activating its own synthesis and that of the recA protein as well as some 20 other different genes. These genes permit the survival of the cell in the face of life-threatening environmental damages. Upon repair of the damaged DNA, the level of the signal subsides, reducing the level of recA and stabilizing the integrity of the intact lexA protein and its repressive properties on all the other genes (Fig. 8); then the cell returns to its normal state.

FIGURE 7 Nucleotide excision reactions. In this multiprotein enzyme system, the UvrABC proteins catalyze a dual incision reaction seven nucleotides 5′ and three to four nucleotides 3′ to a damaged site. The UvrA protein, as a dimer, binds to undamaged sites initially and in the presence of UvrB, whose cryptic ATPase provides the energy, is able to translocate to a damaged site. This pre-incision complex interacts with UvrC leading to the dual incision reaction. The incised DNA–UvrABC and requires the coordinated participation of the UvrD and DNA polymerase reactions for damaged fragment release and turnover of the UvrABC proteins. Ligation, the final reaction, restores integrity to the DNA strands.

B. Eukaryotes

Although the molecular mechanisms of nucleotide excision repair in eukaryotes have not been docu-

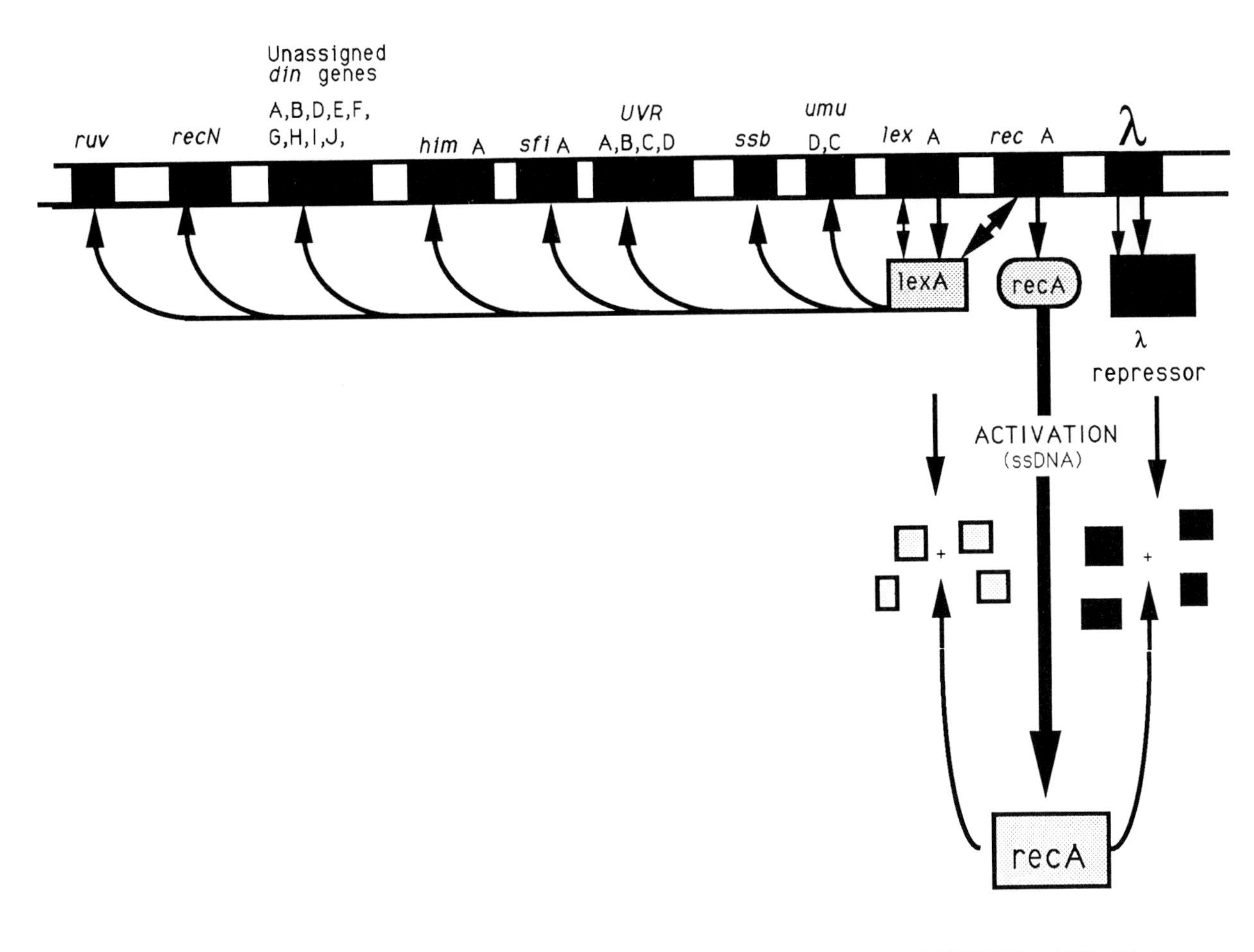

FIGURE 8 Regulation of the nucleotide excision pathway by the "SOS" system. The lexA and lambda repressors negatively control a multitude of genes, which are turned on when cells are damaged leading to the overproduction of the *recA* protein, which assists in the proteolysis of the *lexA* and lambda proteins, thereby derepressing the controlled gene systems. When DNA is fully repaired, the level of *recA* declines restoring the "SOS" system to negative control.

mented, it is assumed that there are evolutionary similarities to the prokaryotic ones for the following two reasons. Human cells with mutations affecting DNA repair show a sensitivity to the same spectrum of damaging agents as *E. coli uvr* mutants. As a consequence, the mechanism proposed for the *E. coli UVR* system is probably similar to the human DNA nucleotide excision repair system. Bacterial and human uracil DNA glycosylases have been shown to have as much as a 73% sequence homology and may be the most highly conserved group of proteins pointing to the evolutionary unity of repair mechanisms. However, inherent structural and physiological differences exist between bacterial and eukaryotic cells, which must be mechanistically accommodated. The chromosomal structure imposes not only a steric challenge to a complicated multiprotein repair complex but also to the highly regulated nature of the cell cycle. Apparently, the structural and biological specificity associated with transcriptional processes limit DNA repair to those damaged regions of the chromosome undergoing transcription, and that damage in those quiescent regions is persistent.

V. Mismatch Repair

A number of mechanisms recognize not damage, but rather mispairing errors that occur in all biological systems. In *E. coli,* mismatch correction is con-

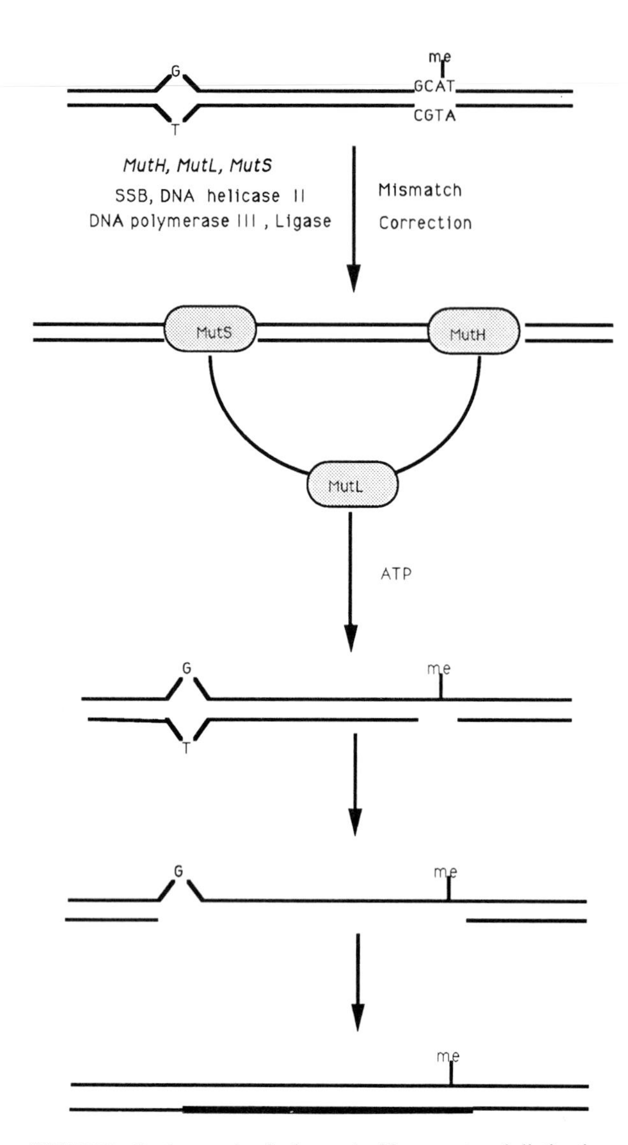

FIGURE 9 In the repair of mismatched bases, strand distinction can be achieved by the delay in adenine methylation during replication. It is the nascent, unmethylated strand that serves as a template for the incision reactions catalyzed by a number of proteins specifically engaged in mismatch repair processes.

trolled by seven mutator genes, *dam, mut*D, *mut*H, *mut*L, *mut*S, *mutU*, (*uvrD*), and *mut*Y. In mismatch correction, one of the two strands of the mismatches is corrected to conform with the other strand (Fig. 9). Strand selection is one of the intrinsic problems in mismatch repairs, and the selection is operated in bacterial systems by the adenine methylation, which takes place at d(GATC) sequences. Because such methylation occurs after DNA has replicated, only the template strand of the nascent duplex is methylated. In mismatch repair, only the unmethylated strand is repaired, thus retaining the original nucleotide sequence. The MutH, MutL, and MutS proteins appear to be involved in the incision reaction on this strand with the remainder of the proteins plus DNA polymerase III and polynucleotide ligase participating in the excision–resynthesis reactions.

Bibliography

Friedberg, E. C. (1984). "DNA Repair." Freeman, New York.

Grossman, L., Caron, P. R., Mazur, S. J., and Oh, E. Y. (1988). Repair of DNA containing pyrimidine dimers. *FASEB J* **2,** 2696–2701.

Lindahl, T., Sedgwick, B., Sekiguchi, M., and Nakabeppu, Y. (1988). Regulation and expression of the adaptive response to alkylating agents. *Ann. Rev. Biochem.* **57,** 133–157.

Modrich, P. (1989). Methyl-directed DNA mismatch correction. *J. Biol. Chem.* **264,** 6597–6600.

Sancar, A., and Sancar, G. B. (1988). DNA repair enzymes. *Ann. Rev. Biochem.* **57,** 29–67.

Walker, G. C. (1985). Inducible DNA repair systems. *Ann. Rev. Biochem.* **54,** 425–457.

Weiss, B., and Grossman, L. (1987). Phosphodiesterases involved in DNA repair. *Adv. Enzymol.* **60,** 1–34.

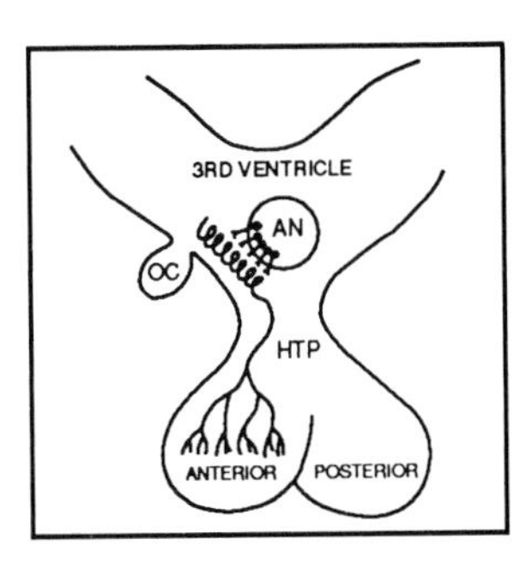

Reproductive Processes, Neurochemical Control

RICHARD E. BLACKWELL, *University of Alabama at Birmingham*

Glossary

Arcuate nucleus Hypothalamic center that regulates the secretion of gonadotropin-releasing hormone (GnRH)

Catecholamines Neurotransmitters that are involved in the regulation of GnRH secretion

Dominant follicle Of the 10 to 15 follicles that are recruited at the beginning of the cycle, the dominant follicle is the one that ultimately releases an ovum

Endorphins Body's own endogenous painkillers that are involved in the regulation of GnRN secretion

17-β-Estradiol Principal estrogen produced by the ovary, 95% of which comes from the dominant follicle

Follicle stimulating hormone (FSH) Glycoprotein secreted from the anterior pituitary gland which acts on the ovary to stimulate follicle growth

Gonadotropin releasing hormone (GnRH) Decapeptide produced in the base of the brain that stimulates the release of luteinizing hormone (LH) and follicle stimulating hormone (FSH) from the anterior pituitary gland

Inhibin Produced by the granulosa cells of the dominant follicle, it inhibits the secretion of FSH

Luteinizing hormone (LH) Glycoprotein that is secreted from the anterior pituitary gland which stimulates ovarian stroma to produce androgens. It ultimately brings about the formation of the corpus luteum

Serotonin Neurotransmitter that is involved in the regulation of GnRH secretion

I. History of the Neuroendocrinology of Reproduction

In 1932, Holweg and Junkman suggested that a sex center might exist in the brain which would regulate human reproduction. Subsequently, Harris, in 1937, electrically stimulated the median eminence of the brain and produced ovulation. In addition, Westman and Jacobson, in 1937, had demonstrated that a section of the pituitary stalk blocked ovulation. Markee, in 1946, showed that direct stimulation of the pituitary gland failed to duplicate this response. These studies strongly suggested that the brain produced some chemical or chemicals that were secreted into the hypothalamic portal system, traveled to the pituitary gland, and regulated the events of ovulation.

The hypothalamic portal system had been described by Popa and Fielding in 1930. Houssay, in 1935, had shown that blood flowed from the brain to the pituitary and in 1950 Harris sectioned the hypothalamic portal vessels and produced target end organ atrophy. It was not until 1955, however, that Guillemin and Rosenberg incubated fragments of the hypothalamus *in vitro* with pituitary tissue and were able to show an increased secretion of the hormone that stimulates the adrenal gland (adrenocorticotropic hormone, ACTH). It was postulated that ACTH secretion was controlled by a small polypeptide and it was thought that each of the classic pituitary hormones was controlled by one small protein, thus giving rise to the one peptide–one hormone hypothesis. This was confirmed for the reproductive hormones in 1971 and 1972 when Schally *et al.* and Guillemin *et al.* isolated and presented the structure of gonadotropin releasing hormone (GnRH). Concomitant with these studies, it was demonstrated by Bergland that blood flowed not only from the hypothalamus to the pituitary gland but in a retrograde manner. These studies opened the way for our current understanding of the neurochemical control of the reproductive cycle. [*See* NEUROENDOCRINOLOGY.]

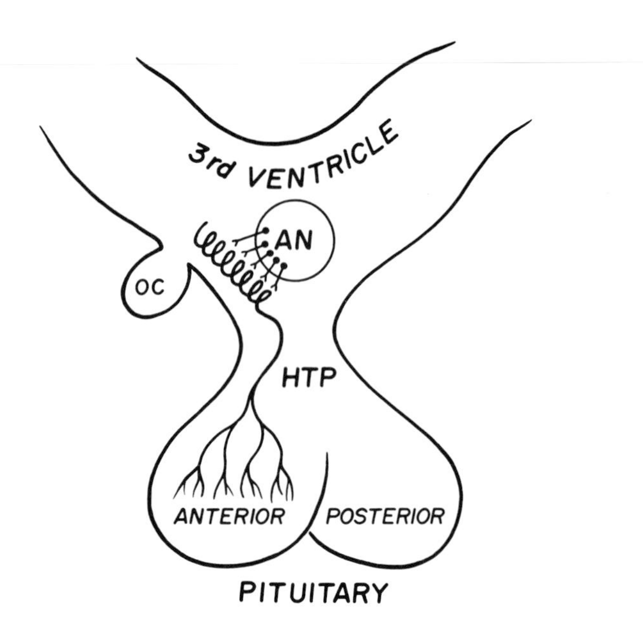

FIGURE 1 Diagram showing the relative positions of pituitary gland, hypothalamus, and brain, and the portal system connecting the median eminence of the hypothalamus with the anterior pituitary. AN, Arcuate nucleus; OC, optic chiasm; HTP, hypothalamic pituitary portal capillaries.

II. Anatomy of the Median Eminence and Anterior Pituitary Gland

The hypothalamus is phylogenetically old and found in mammals throughout evolution. It weighs approximately 1000 g and is located at the base of the brain just above the juncture of the optic nerves (optic chiasm). The arcuate nucleus is one of the medial hypothalamic nuclei. It lies just above the median eminence and adjacent to the third ventricle. The median eminence, which is in close contact with the arcuate nucleus, is the final common pathway for the neurohumoral control of anterior pituitary function. It receives peptidergic neurons which contain releasing and inhibiting hormones. The median eminence delivers these hormones to hypothalamic pituitary portal capillaries and these neurochemicals are subsequently transmitted to the anterior pituitary gland, where they act on the gonadotropes to release both luteinizing hormone (LH) and follicle stimulating hormone (FSH) (Fig. 1). [*See* HYPOTHALAMUS; PITUITARY.]

III. Neurochemicals Involved in Reproduction

At ovulation it was assumed that when increased amounts of LH and FSH were secreted, this was the result of an increased production of GnRH. This notion was dispelled when Knobil *et al.* demonstrated that when GnRH was delivered into the hypothalamic portal system of the rhesus monkey in a fixed amount at hourly intervals, ovulation occurred. It appears that GnRH secretion is controlled by a push/pull mechanism involving the neurotransmitters dopamine and norepinephrine (DA and NE, respectively, in Fig. 2). Infusion of dopamine produces an inhibition of LH secretion whereas norepinephrine has the opposite effect. Besides catecholamines, the endorphins (EOP) have a significant effect on gonadotropin secretion. These compounds have a suppressive effect on LH secretion and this is mediated through dopamine neurons. Serotonin, another neurotransmitter, may evoke inhibitory influences on GnRH neuronal activity (Fig. 2).

IV. Neurochemical Control of the Menstrual Cycle

As described above, the arcuate nucleus is the final integrator of the menstrual cycle. It receives exogenous and endogenous input and releases GnRH into the hypothalamic pituitary circulation which gives

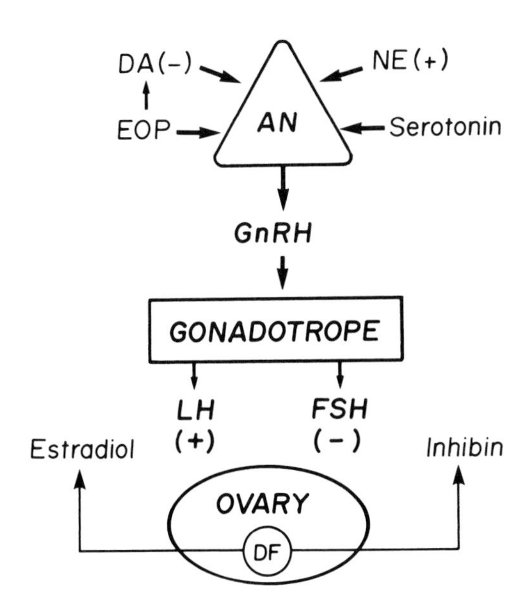

FIGURE 2 Sources and targets of neurochemicals controlling the menstrual cycle. DA, Dopamine; NE, norepinephrine; AN, arcuate nucleus; EOP, endorphins; GnRH, gonadotropin releasing hormone; LH, luteinizing hormone; FSH, follicle stimulating hormone; DF, dominant follicle.

rise to hourly pulses of both LH and FSH secretion. These hormones travel to the ovary where LH acts on the stroma to stimulate the production of androgens. Follicle stimulating hormone act on the granulosa cells of 10 to 15 primary oocytes to induce their maturation and growth. Follicle stimulating hormone increases the absolute number of granulosa cells, therefore the size of the follicle, induces an enzyme called aromatase which allows the follicle to metabolize androgens to estrogens, the latter being important in follicle growth, and it induces the formation of luteinizing hormone receptors which will ultimately be stimulated to give rise to the corpus luteum. Of the 10 to 15 follicles that are recruited, one becomes the dominant follicle (DF) (see Fig. 2). It begins to secrete significant amounts of 17-β-estradiol and inhibin; the latter compound inhibits the secretion of its parent hormone. Subsequently, the follicle begins to produce progesterone. When combined with estrogen, a synergy occurs that triggers the midcycle surge of luteinizing hormone. Following the surge of luteinizing hormone, a myriad of events occurs in the microenvironment of the follicle in preparation for ovulation. Once the follicle has been exposed to the stimulus of the luteinizing hormone for approximately 18–20 hr, the ovum is released. Luteinizing hormone secretion brings about the formation of the corpus luteum with its ever-increasing production of progesterone. Progesterone feeds back on the hypothalamic pituitary central axis and reduces the secretion of GnRH to one pulse every 4 hr. Progesterone and estrogen production increases until approximately the middle of the luteal phase (postovulatory day 8). If conception does not occur at this point these hormones begin to inhibit secretion from the central axis, leading to the ultimate demise of the corpus luteum and menstruation.

Bibliography

Blackwell, R., and Guillemin, R. (1973). Hypothalamic control of adenohypophyseal secretions. *Annu. Rev. Physiol.* **35,** 357–390.

Fritz, M., and Speroff, L. (1985). The endocrinology of the menstrual cycle: The interaction of folliculogenesis and neuroendocrine mechanisms. *In* "Modern Trends in Infertility and Conception Control" (E. E. Wallach and R. D. Kempers, eds.), Vol. 3, pp. 5–25. Year Book Medical Publishers, Chicago, Illinois.

Hodgen, G. (1985). The dominant ovarian follicle. *In* "Modern Trends in Infertility and Conception Control" (E. E. Wallach and R. D. Kempers, eds.), Vol. 3, pp. 26–45. Year Book Medical Publishers, Chicago, Illinois.

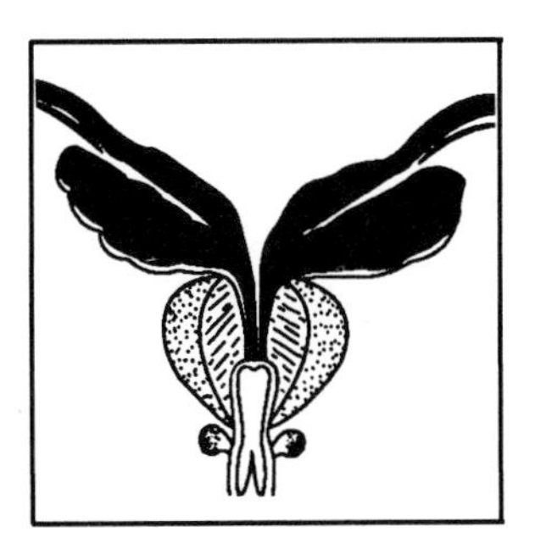

Reproductive System, Anatomy

ANTHONY J. GAUDIN, *California State University–Northridge*

KENNETH C. JONES, *California State University–Northridge*

Glossary

Atresia Degeneration and disappearance of follicles in the mammalian uterus
Bartholin's glands Two small mucous glands located on each side of the vaginal opening
Bulbourethral glands Two small mucous glands on each side of the prostate gland that secrete part of the seminal fluid
Corpora cavernosa Erectile tissue in both the penis and the clitoris
Corpus spongiosum Erectile tissue surrounding the urethra in the male penis
Cremasteric muscle Thin muscle that suspends and surrounds the testis and the spermatic cord
Endometrium Mucous membrane forming the inner lining of the uterus
Epididymis Convoluted tubule adjacent to the testis that contains mature and maturing sperm
External os The opening of the cervical canal leading into the vagina
Fimbriae Fingerlike projections surrounding the abdominal opening into the oviduct
Infundibulum Cavity formed by the fimbriae surrounding the opening into the oviduct
Internal os Opening leading from the cervical canal into the uterus
Isthmus Narrow end of the uterus leading into the cervix
Leydig cells Interstitial cells of the testis that secrete testosterone
Mesovarium A fold of peritoneal tissue that supports the ovary
Myometrium Muscular wall of the uterus
Oocyte Early or primitive ovum prior to development
Seminiferous tubules Numerous coiled tubules in the testis, where sperm are produced
Sertoli cells Cells in seminiferous tubules that support and nourish developing sperm
Tunica albuginea Coat of white fibrous tissue surrounding the testis
Tunica dartos Layer of muscle in the wall of the scrotum that permits shrinking and shriveling of the scrotal skin at cold temperatures
Tunica vaginalis Thin serous membrane surrounding the testis
Vaginal fornix Connection of the vagina to the fornix

UNLIKE OTHER SYSTEMS in the body, the reproductive systems differ in two sexes. Although these differences are the primary criteria by which one is identified as male or female, they are not the only ones. The reproductive systems have the related functions of manufacturing sex hormones responsible for male and female characteristics and producing sex cells used to generate offspring. Whereas muscles, nerves, blood, and digestive organs, for example, are all essential to life and the maintenance of healthy internal conditions, reproductive organs are not vital to life, nor are they generally involved in maintaining the internal equilibrium, without which life could not exist. Rather than being essential to the survival of an individual, the reproductive systems are essential to the survival of the species.

I. Male Reproductive System

The adult male reproductive system consists of four general regions (see Color Plate 12). The testes are paired organs in which sperm (i.e., male reproductive cells) develop. The testes are carried in a small sac, called the scrotum. Sperm leave the testes and are stored in a network of tubules (the second region), through which they pass when they are expelled from the penis. Accessory glands produce most of the fluid that carries the sperm and compose the third major region. The fourth major part of the male reproductive system is the penis, an organ through which sperm are delivered to the female reproductive system during sexual intercourse, of coitus. [*See* SPERM.]

A. Testes

Adult testes, or testicles, are egg-shaped structures. Each testis is about 25 × 50 mm and is divided into 200–300 lobules (see Color Plate 13). Lobular walls are extensions of a thick fibrous covering, called the tunica albuginea, which surrounds each testis. External to the tunica albuginea is a thinner membrane. the tunica vaginalis.

Each lobule in the testis consists of a collection of highly coiled seminiferous tubules. Located between the seminiferous tubules are Leydig cells, also called interstitial endocrinocytes, or interstitial cells of Leydig (Fig. 1). These cells secrete hormones important to sperm production. Microscopic examination of the inner wall of a seminiferous tubule reveals sperm in various stages of development. The periphery of each tubule is marked by a basement membrane. Sperm in the earliest stage lie just inside this membrane, with progressively more mature sperm lying nearer the lumen. Also embedded in the wall of a seminiferous tubule are Sertoli cells, also known as sustentacular, or nurse cells, which provide chemical assistance to developing sperm. [*See* REPRODUCTIVE SYSTEM, ANATOMY.]

Seminiferous tubules connect and fuse at one side of each testis to form another network of tubules, the rete testis. From the rete testis emerges a smaller collection of larger-diameter tubules, the efferent ductules. These empty into a single tubular structure, the epididymis, located outside of the testis, but within the scrotum.

B. Epididymis

The epididymis is formed by the fusion of efferent testicular ductules into the ductus epididymis, a tightly coiled tube (see Color Plate 13). An epididymis lies alongside each testis, curved in shape to conform to the testis (see Color Plate 14). The upper end is the head; the middle section, the body; the lower end, the tail. Although each epididymis is only about 4 cm long externally, the ductus epididymis included in it is nearly 6 m long. The duct is only 1 mm in diameter and is tightly coiled. The epididymis is where sperm undergo final maturation and where mature sperm are stored.

C. Scrotum

The scrotum, the sac that holds the testes and the epididymis, is divided into two halves exteriorly by a median ridge of tissue, the scrotal raphe. Internally, the scrotum is partitioned into two lateral compartments by a septum of connective tissue. Each compartment contains one testis. The scrotal skin covers subcutaneous tissue that contains a layer of smooth muscle, called the tunica dartos. The muscle extends into the septum that divides the scrotum and enables the wall of the scrotum to contract and thicken in response to low temperature. This provides an important mechanism for temperature control in the testes. Sperm development requires temperature lower than that of the body, which is achieved by suspension of the testes outside the body in the scrotum. When the external temperature is too cold for optimal sperm development, the tunica dartos contracts, thickening the scrotal wall and bringing the testes closer to the body, where the temperature is warmer. Another small muscle, the cremasteric muscle, assists in this process. It is located in the inguinal region, above the testes. In warmer temperatures these muscles relax, lowering the testes and allowing the scrotal wall to become thin. Richly endowed with sweat glands, the scrotum also secretes sweat, which cools the testes by evaporation.

D. Ductus Deferens and Ejaculatory Duct

The tail of each epididymis leads to the ductus deferens (or vas deferens), a tube about 45 cm long (see Color Plates 13 and 14). The ductus deferens from each testis passes out of the scrotum through the inguinal canal, a narrow opening in the abdominal wall, and then enters the pelvic cavity. Each duct is accompanied by testicular arteries, veins, nerves, and lymphatic vessels. The entire assemblage, the spermatic cord, is surrounded by the cremasteric muscle and connective tissue sheaths. Once inside the pelvic cavity, the ductus deferens loops over the ureter and passes behind the urinary

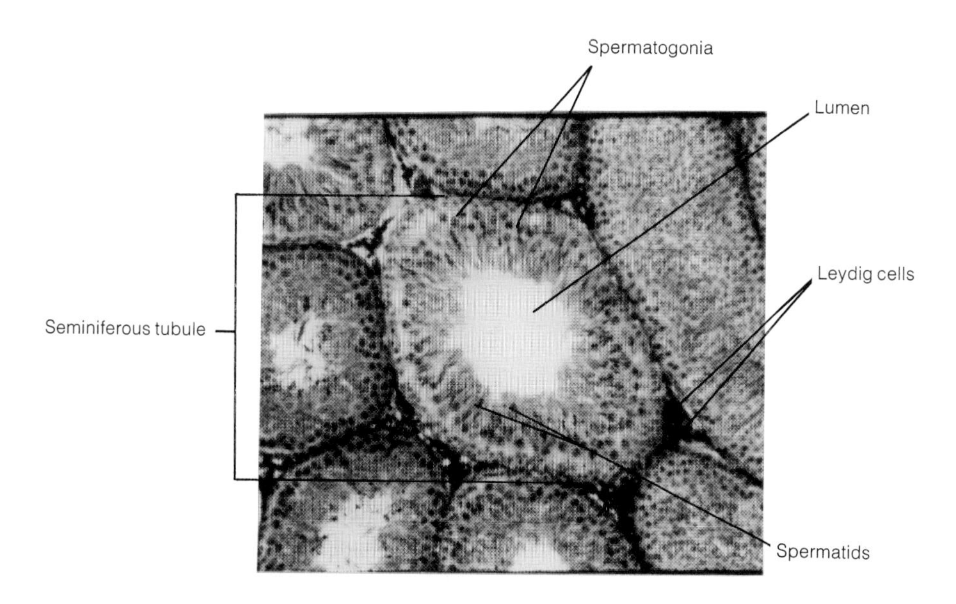

FIGURE 1 A seminiferous tubule in a testis. [From A. J. Gaudin and K. C. Jones (1989). "Human Anatomy and Physiology." Harcourt Brace Jovanovich, San Diego, California.]

bladder (see Color Plate 14). The terminal portion is enlarged in diameter, forming the ampulla.

The ductus deferens receives fluids from a seminal vesicle, an accessory gland. The union of the ductus deferens and a duct from the seminal vesicle produces the ejaculatory duct. This short duct, only about 2 cm long, conducts sperm into the urethra.

E. Urethra

The urethra is a single tube extending from the urinary bladder to the tip of the penis (see Color Plates 12 and 14). In males the urethra conducts urine from the bladder during urination and provides a path for sperm during sexual activity. The initial portion of the urethra is the prostatic urethra, a section about 3 cm long that passes through the prostate gland. Prostatic secretions enter the urethra in this section. At the base of the prostate gland, the membranous urethra passes through the urogenital membrane. This portion is less than 2 cm long and leads directly to the cavernous (or penile) urethra, which passes through the corpus spongiosum of the penis. The penile urethra, approximately 16 cm long, terminates at the urethral orifice.

F. Accessory Glands

In addition to fluid produced in the testes, fluid is added to the mixture of sperm by cells of sacs in the ampulla wall of the ductus deferens. Once sperm have passed through the ductus deferens, they are carried in fluid produced by accessory glands, which include the seminal vesicles, prostate gland, and bulbourethral glands.

The paired seminal vesicles are small pouches connected by short ducts to the junction of the ampulla and the ejaculatory duct. Seminal vesicles, each of which is 5–10 cm long, consist of a tightly twisted and convoluted tubule (Fig. 2). Each tubule is lined with secretory cells that release an alkaline fluid rich in fructose, a sugar which the sperm use for energy. The alkaline nature of the secretions helps neutralize the acidity of urine remaining in the urethra. The mixture of sperm and secretions is semen, or seminal fluid. Seminal vesicles contribute slightly more than one-half of the seminal fluid volume.

The prostate gland lies below the bladder, where it surrounds the urethra and the two ejaculatory ducts. Shaped like a cored apple, the prostate gland is divided into several branched compartments, each of which opens through a duct directly into the urethra. The compartments are separated from one another by walls of smooth muscle and connective tissue. The entire organ is surrounded by a jacket of connective tissue. The prostate gland produces 1–2 ml of fluid per day, which collects and is released into the urethra just prior to ejaculation.

The paired bulbourethral glands (or Cowper's glands) are located below the prostate gland. Each is about the size of a pea and empties into the cavernous urethra through ducts located close to the

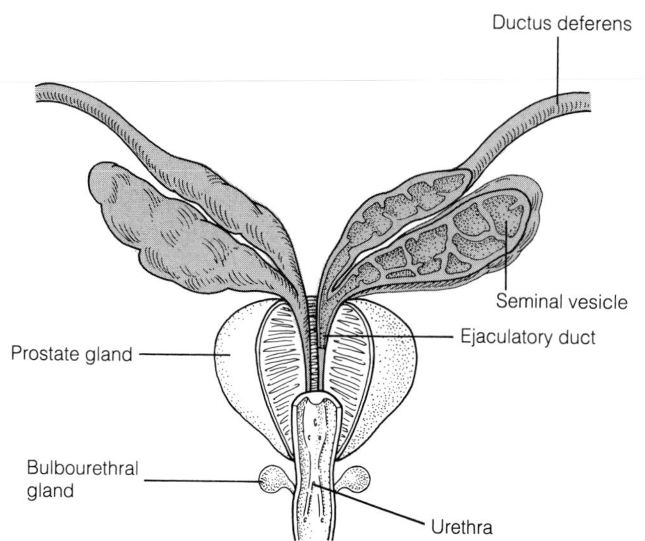

FIGURE 2 The seminal vesicles and the prostate gland. [From A. J. Gaudin and K. C. Jones (1989). "Human Anatomy and Physiology." Harcourt Brace Jovanovich, San Diego, California.]

base of the penis. Bulbourethral secretions, whose release is triggered by erotic activity, precede ejaculation and provide lubricants for sexual intercourse and the nutrients that sperm require for motility. They also help neutralize the acidity of any urine remaining in the urethra.

Numerous outpocketings lie in the walls of the cavernous urethra, some of which are branched chambers connected to the urethra by short ducts. These structures, called urethral glands, produce mucus that serves as a lubricant to semen.

G. Penis

The penis consists of three cylindrical bodies of spongy tissue, surrounded by sheaths of connective tissue and skin (see Color Plate 15). Two of the spongy bodies compose the corpora cavernosa. These paired structures lie parallel to one another in the superior portion of the penis. The third body, the corpus spongiosum, lies inferior to the corpus cavernosum, extending past the distal ends of the corpus cavernosum into an enlarged tip, the glans penis. The urethra enters the corpus spongiosum at its base and terminates at the glans penis in a slitlike opening, the urethral orifice. The glans penis is covered by a loose extension of the skin of the penis: the prepuce, or foreskin. The prepuce is removed by circumcision.

The two spongy cylinders of the corpora cavernosa are separated from one another for most of their length by a tough inelastic sheath, the tunica albuginea. The corpus spongiosum is also surrounded by a sheath, but one of somewhat more elastic structure. The three bodies, in turn, are surrounded by the fascia penis, a sheath that lies just below the skin. Blood is supplied to the penis by the internal pudendal arteries, which branch from the internal iliac arteries. Each pudendal artery divides further into a network of arteries that provide blood to superficial and deep tissues. [*See* CARDIOVASCULAR SYSTEM, ANATOMY.]

II. Female Reproductive System

The adult female reproductive system consists of five general regions (see Color Plate 16): two ovaries in which eggs, or ova, are produced; two tubular oviducts, through which eggs pass after release; a saclike uterus, in which an embryo develops; a vagina, which leads from the uterus to the exterior; and the vulva, a collective term for the external genitalia. Mammary glands used to provide milk for the newborn infant are also considered part of the female reproductive system.

A. Ovaries

Ovaries are almond shaped paired organs about 2–5 cm long, located on either side of the lower portion of the abdominal cavity, where they are held in place by the mesovarium. This ligament, in turn, connects to and supports other organs of the female reproductive tract. Medially, ovaries are connected to the uterus by the ovarian ligament; laterally, they are attached to the abdominal wall by the suspensory ligament.

Histologically, each ovary consists of an inner medulla (a cortex surrounding the medulla) and a single layer of cells (a modified mesothelium that covers the cortex) (Fig. 3). Nerves, lymph, and blood vessels lie within the medulla and extend into the cortical region. The cortex and the mesothelium are surrounded by a layer of connective tissue, the tunica albuginea. Embedded within the cortex are 300,000–400,000 undeveloped, but potential, egg cells. In the course of a woman's reproductive years, only about 400–500 of these cells will complete development and be released from the ovary. A mature egg cell, an ovum, is a female gamete capable of fertilization.

Several other structures are also visible in the ovary. Potential egg cells in the cortex of the ovary of a young woman are surrounded by a single layer

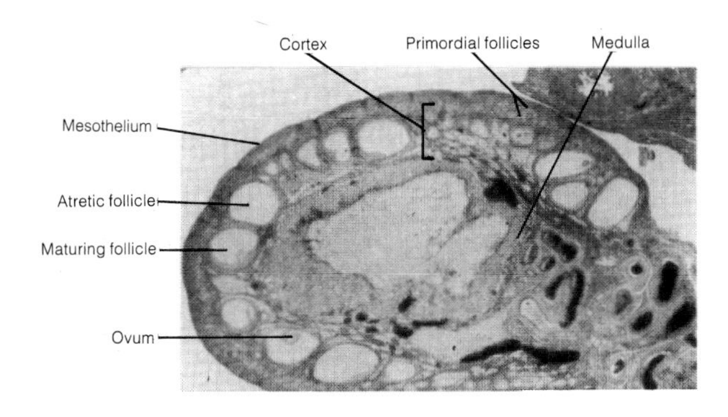

FIGURE 3 A longitudinal section through an ovary. [From A. J. Gaudin and K. C. Jones (1989). "Human Anatomy and Physiology." Harcourt Brace Jovanovich, San Diego, California.]

of cuboidal or columnar cells. The cell, at this stage, is a primary oocyte. The oocyte and the cells surrounding it form a primordial ovarian follicle. Numerous follicles in various stages of development are present, as well as some that failed to develop completely and have degenerated. Follicles that cease developing and subsequently degenerate are said to undergo atresia, and their remains are referred to as atretic follicles.

In addition to producing eggs, the ovaries are important sources of estrogens, steroid hormones responsible for developing the body form and other features characteristic of a female. [*See* PUBERTY.]

B. Oviducts

The oviducts are about 10–12 cm long and extend from the superior surface of the uterus (see Color Plate 17). The end of the oviduct distal to the uterus and adjacent to the ovary consists of a mass of highly convoluted fingerlike projections, called fimbriae. Collectively, fimbriae form a funnellike opening to the tube, the infundibulum. The infundibulum leads directly into the ampulla, an enlarged portion occupying about one-third to one-half the length of the oviduct. The interior wall of the ampulla is highly convoluted. It is in this region that fertilization normally occurs.

The remainder of the oviduct includes the isthmus, a portion that leads from the ampulla to the uterus, and an interstitial segment that penetrates the uterine wall. There is a progressive decrease in the degree of infolding of the wall and the percentage of ciliated cells from the infundibulum to the interstitial segment. At the end of the uterine tube closest to the uterus, the infoldings are reduced to several longitudinal ridges.

C. Uterus

The uterus is a hollow, pear-shaped, thick-walled sac that rests on the floor of the abdominopelvic cavity, located between the urinary bladder and the rectum (see Color Plates 16 and 17). It is held loosely in place by broad ligaments on either side, uterosacral ligaments that connect the cervix to the sacrum, lateral cervical ligaments that connect the cervix to the pelvic diaphragm, and round ligaments, which are anchored in the tissues beneath the labia majora. Some of the ligaments also carry uterine blood vessels and nerves. The bladder, rectum, and other adjacent organs also help position in the uterus, and because of its flexibility, the uterus can assume a number of different forms and orientations as the bladder and the rectum fill and empty.

The normal uterus in a nonpregnant woman is about 6 cm high, 4.5 cm wide, and 2.5 cm deep. The hollow interior of the uterus, the uterine cavity, connects to the outside through cervical and vaginal canals. Regions within the uterus include the fundus, the domelike cap on the body (or corpus) of the uterus; the isthmus, a region below the corpus where the uterus narrows; and the cervix, a narrow necklike extension that protrudes into the vagina. The junction of the uterine cavity and the cervical canal forms a narrow opening, called the internal os (see Color Plate 17). The opposite end of the cervical canal, where it opens into the vagina, is the external os. The uterus is usually bent in anteflexion, a position in which the body of the uterus projects anteriorly over the urinary bladder, and the cervix projects posteriorly, entering the upper end of the vagina at nearly a right angle.

The wall of the body of the uterus consists of an inner endometrium, a complex layer of epithelial cells, glands, and blood vessels; a middle layer, the myometrium, within which lie layers of smooth muscle, connective tissue, and many large blood and lymphatic vessels; and an outermost perimetrium. The perimetrium is continuous with the broad ligaments that support the uterus and the uterine tubes.

The endometrium consists of two functionally and structurally distinct regions: the stratum functionalis and the stratum basalis. Prior to menstruation these two regions can be as thick as 6–7 mm, but during menstruation the cells of the stratum functionalis die and slough off, reducing the endometrium essentially to the stratum basalis, a layer only about 1 mm thick. [*See* UTERUS AND UTERINE RESPONSE TO ESTROGEN.]

D. Cervix

Although the cervix is part of the uterus, the structure and function of its walls are quite distinct. The cervical wall is relatively thick, and its inner surface is covered by an elaborate network of ridges and valleys that secrete mucus, which fills the cervical canal, preventing the inward movement of microbes and external substances.

E. Vagina

The vagina, about 10–15 cm long, provides a passage from the uterus to the outside world. The vagina is largely fibromuscular tube and has a wall consisting of three tissue layers. Outermost lies a thick layer of connective tissue, covering a middle layer of muscle. Blood vessels and nerve bundles lie in this wall. The innermost layer of the vagina is a folded layer that secretes mucus.

The connection of the vagina to the cervix forms the vaginal fornix, an enlarged portion of the vagina, in which sperm collect during sexual intercourse. Deposited sperm are thus in close proximity to the external os, enhancing the likelihood of sperm's continuing into the cervix.

F. External Genitalia (Vulva)

The vagina leads inferiorly and anteriorly from the cervix and opens into the vulva, or pudendum, which is a collective term for the external genitalia (see Color Plate 18). This opening is surrounded by a membranous ring of tissue, the hymen, which consists of relatively fine and highly vascularized tissue. Normally, an opening exists through the hymen, but sometimes the hymen covers the opening of the vagina completely. In such cases the hymen must be opened surgically once menstruation begins, to permit menstrual flow to escape.

The external genitalia consist largely of modified skin and subcutaneous tissues. The mons pubis is a rather thick layer of fatty tissue that lies over the public symphysis, the junction of the two pubic bones that lie at the base of the abdominopelvic cavity. The mons pubis divides into two thick fatty pads that proceed posteriorly to join again in the space between the openings of the vagina and the anus. These two folds of fatty tissue are the labia majora, which enclose the remainder of the external genitalia in the pudendal cleft. Just inside the two labia major are two smaller folds of mucous membrane: the labia minora. These labia join at the anterior end of the pudendal cleft to form a hoodlike structure that covers the clitoris and surrounds the vestibule of the vagina, the cleft between the labia minora. As puberty is reached, the mons pubis and the labia majora become covered with a thick mat of coarse public hair. Hair does not develop on the labia minora, which are kept moist by glandular secretions. The entire region enclosed within a roughly diamond-shaped area from the clitoris anteriorly to the coccyx posteriorly and laterally to the region external to the ischial tuberosity of the coccygeal bone is the perineum. (Some authorities, however, restrict their definition of the perineum to only the small area between the anus and the vulva.) The triangular anterior half of the perineum forms the urogenital triangle; the posterior triangle of the perineum is the anal triangle.

From the developmental standpoint the clitoris and the penis are equivalent, because they are derived from the same embryonic structure. The clitoris, however, is much smaller and does not contain the urethra. Like the penis, it contains erectile tissue that becomes engorged with blood during periods of sexual stimulation, causing the clitoris to become erect. The spongy tissue responsible for erection of the clitoris is the corporus cavernosum. A corpus spongiosum is absent. The clitoris terminates in a glans clitoridis, which is nearly covered by the folds of the labia minora. The folds covering the glans compose the female prepuce.

The vestibule also contains the opening of the urethra and several mucus-secreting glands that provide lubrication for sexual intercourse. Paraurethral glands open into the area surrounding the urethra; a pair of small round Bartholin's glands, lies posterior to the base of each of the labia minora and opens through ducts on either side of the vaginal orifice. In addition to paraurethral and Bartholin's glands, the vestibule contains many smaller glands that also contribute mucus.

G. Mammary Glands

Mammary glands are considered part of the reproductive system because of the important role they play in nurturing an infant. Mammary glands begin to secrete milk 2–3 days after a baby is born. The milk is rich in nutrients and immunoglobulins that convey disease resistance to the suckling infant. It is a mixture consisting of about 88% water, 7% lactose, 4% fat, 1% protein, and minute amounts of other valuable nutrients.

A mature lactating mammary gland consists of 15–20 compartments, or lobes, separated by fatty and connective tissue. The lobes consist of subdivisions (i.e., lobules) made up of numerous grapelike clusters of alveoli, which produce the milk. A duct emerges from each cluster of alveoli and fuses with others to form lactiferous (or mammary) ducts, each one carrying the milk produced by the alveoli of a lobe. Each lactiferous duct has an enlarged region, an ampulla, which lies just beneath the nipple. Each ampulla leads into a short continuation of the lactiferous duct that opens at the nipple.

As milk is produced, it accumulates in each breast, collecting in ampullae and in the extensive duct system. This system opens externally in the nipple. The nipple is surrounded by a pigmented ring of tissue called the areola.

Bibliography

Burger, H. G., and Baker, H. W. G. (1987). The treatment of infertility. *Annu. Rev. med.* **38,** 29–40.

Cormack, D. H. (1987). "Ham's Histology," 9th ed. Lippincott, Philadelphia, Pennsylvania.

Fink, G. (1986). The endocrine control of ovulation. *Sci. Prog.* **70,** 403–423.

Gaudin, A. J., and Jones, K. C. (1989). "Human Anatomy and Physiology," Harcourt Brace Jovanovich, San Diego, California.

Longo, F. J. (1987). "Fertilization," Chapman & Hall, New York.

McMinn, R. M. H., and Hutchings, R. T. (1988). "Color Atlas of Human Anatomy," 2nd ed. New York Med. Pub. Chicago.

Money, J. (1981). The development of sexuality and eroticism in human kind. *Q. Rev. Biol.* **56,** 379–404.

Short, R. V. (1984). Breast feeding. *Sci. Am.* **250,** 35–41.

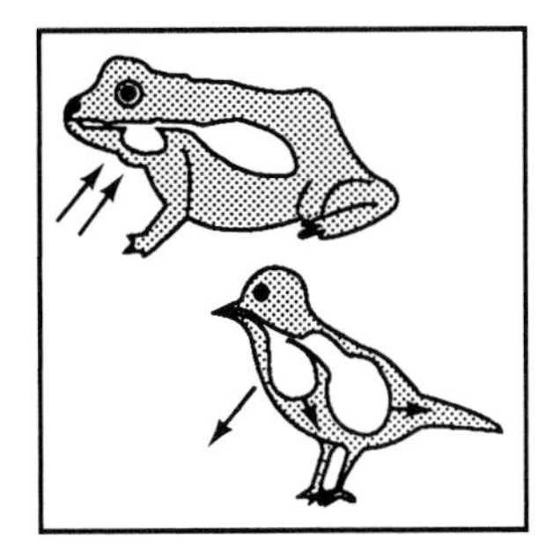

Respiration, Comparative Physiology

JACOPO P. MORTOLA, *McGill University*

Glossary

Allometric relationship Any relationship relating a variable to body weight; usually represented in the log-transformed version

Altricial species Species born at an early stage of development, therefore immature at birth

Chemoreceptors Receptors sensitive to chemical variations

Compliance Changes in volume per unitary changes in pressure

Dead space Total volume of the airways which do not participate in gas exchange

Functional residual capacity Amount of air left in the lungs at end expiration

Hematocrit Volume of blood cells expressed in percentage of total blood volume

Hering–Breuer inflation reflex Inhibition of inspiratory activity following lung inflation

Hypocapnia Decrease in CO_2

Hypoxia Decrease in O_2

Mechanoreceptors Receptors sensitive to mechanical stimuli, usually pressure or tension

Oxygen consumption Amount of oxygen used per unit time

Perfusion Amount of blood circulating per unit time

Resting volume Volume of the respiratory system when no forces are applied to it

Tidal volume Volume of air inhaled with each breath

Transpulmonary pressure Pressure across the lungs, i.e., the pressure difference between the airways and the pleural space

Ventilation Amount of air exchanged per unit time

Vital capacity Maximal amount of air which can be exhaled after a full inspiration

RESPIRATION IS gas exchange, specifically the exchange by a living organism of carbon dioxide (CO_2), a waste product formed during the oxidation of food molecules, for oxygen (O_2), which the organism needs to continue oxidizing its food. At the cellular level, gas exchange occurs by diffusion, according to the partial pressure gradient of the gas. In large cell aggregates and more complex organisms diffusion alone would not fulfill the minimal cellular metabolic requirements. Therefore, gas convection mechanisms are coupled to diffusion in what constitutes the respiratory apparatus. In this article the various mechanisms of convection in invertebrates and lower vertebrates are presented briefly. Mammals are then examined in more detail by addressing the question of how change in body size influences metabolic and ventilatory requirements, and how these requirements are met by modifications in structural design and functional properties of the respiratory system. Finally, the main modifications dictated by age (newborn) or by life in special conditions (diving, burrowing, and high altitude) are summarized.

I. The Dependence of Oxygen: Gas Diffusion and Convection

The respiratory system is designed to provide gas exchange, i.e., to fulfill the cellular necessity for O_2 and to eliminate the byproduct of cellular respiration, CO_2. Anaerobic mechanisms to generate energy are available but they are much less efficient and usually represent emergency or short-term routes of energy production; if anaerobic means are adopted, then aerobically produced energy must be spent to pay back the oxygen debt. Because the pressure of O_2 in the environment is much higher than in the cells (and in the cells the pressure of CO_2 is much higher than in the environment, where it is almost zero) we should expect a continuous diffusion of O_2 into, and of CO_2 out of, the cells. This is indeed the case, and diffusion represents the basic physical process governing respiration. However, the number of molecules that can diffuse in and out per unit time is inversely proportional to the distance between the organelles where cellular respiration takes place (mitochondria) and the environment. Hence, only in very simple organisms can diffusion alone provide the minimum amount of gas exchange per unit time (ml/min; conveniently expressed by O_2 consumption, $\dot{V}_{O_2}$, or CO_2 production, $\dot{V}_{CO_2}$) required for cell survival. In more complex aggregates gas exchange would take too long to fulfil the metabolic needs. Nature has circumvented this limitation by coupling the diffusion process to convection, whereby the environmental air is effectively brought into contact with all the cells of the organism via finely controlled structures. The way convection operates varies remarkably throughout the animal kingdom, and the design of the convection system is intimately related both to the metabolic needs of the organism and to other nonrespiratory functions of the respiratory structures. In vertebrates, two systems of pipes with their respective pumps convey gases to and from the cells, the respiratory system, which operates with a gaseous or liquid medium, and the cardiovascular system, in which blood represents the convection medium. They are coupled together at the pulmonary (or gill) gas exchange area where gas transfer depends on diffusion, similar to that which occurs at the cellular level (Fig. 1).

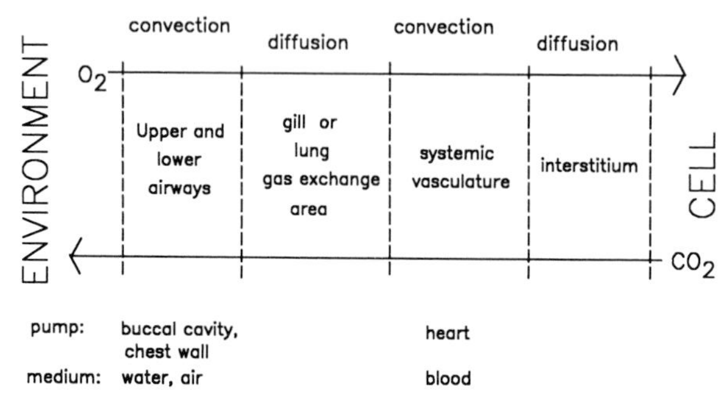

FIGURE 1 Schematic representation of the physical (top) and structural basis (center and bottom) for gas transport between environment and cell in vertebrates.

II. Mechanisms of Gas Convection in Vertebrates

A sudden drop in ambient O_2 could be accommodated by changing the environment in the search for better conditions. Indeed, an increase motility, in some cases with coordinated locomotive responses, represents a common reaction to hypoxia in the simplest living organisms. However, it is apparent that the strategy of "looking for a better place" is of limited value because motion is energetically demanding and because it does not solve problems of tissue hypoxia of internal origin. An alternative solution to an acute drop in O_2 availability would be a decrease in O_2 demand at the cellular level. In fact, decreasing $\dot{V}_{O_2}$ (O_2 conformity) is a very common phenomenon in hypoxic invertebrates and lower vertebrates, while it represents the last resort in adult mammals and birds.[1] O_2 conformity is rarely the only response to hypoxia, and some forms of acclimatization, whereby gas diffusion is implemented by convection, can be recognized even in the most primitive living organisms.

In insects, both diffusion and convection are enhanced by the presence of a complex system of tubes, the tracheoles, connecting the body surface to the innermost tissues. In fact, in some cases the tracheoles penetrate the cells, and blood circulation is therefore redundant, at least for the purposes of gas convection. Ventilation ($\dot{V}E$) is regulated via spiracles, according to control programs which compromise between the metabolic necessities of the insect and the problem of water loss which would accompany excessive exposure of inner body surfaces.

1. The drop in metabolic rate, and body temperature, during hypoxia is common in newborn mammals, which, in this respect, behave like heterothermic lower vertebrates.

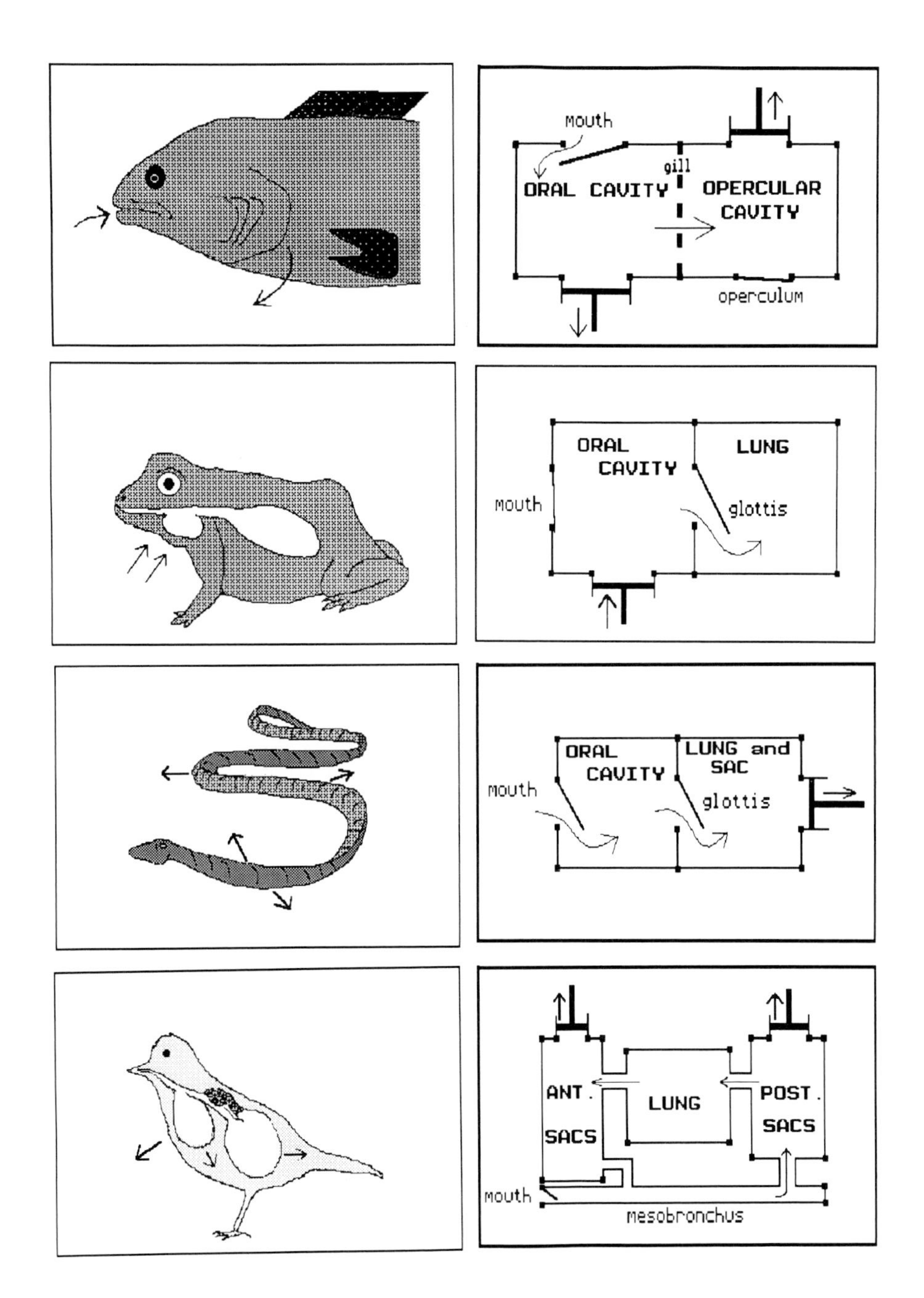

FIGURE 2 Schematic representation of the mechanisms of gas convection in some classes of vertebrates.

Enormous increases in gas exchange area can be achieved not only by inward foldings, as in the case of the insect tracheoles, but also by outward protrusions, as in the case of the multiple lamellae of fish gills. In fish, the large diffusion area of the gills is placed along the unidirectional stream of water entering the mouth and exiting via the operculum, behind the gill filaments. The unidirectional flow is usually helped by two aspiration–compression pumps (Fig. 2), in the oral cavity and in the opercular cavity, which act in coordination with the opening and closure of mouth and operculum. This design of unidirectional water flow across the gills from the front to the sides has the potential for providing sufficient water convection simply by swimming forward with the mouth open. Indeed, in some fishes, gill ventilation (i.e., external ventilation, $\dot{V}_E$; in ml/min) depends more on locomotion than on active pumping, and in these cases (e.g., tuna) continuous forward swimming becomes a fundamental necessity for $\dot{V}_E$. From an energetic viewpoint, one of the best solutions is that of *Remora remora*. This fish, by attaching itself to the body of a shark, gets not only a free ride and the leftovers of the shark's meals but, by keeping its mouth open, also a free $\dot{V}_E$. The inhibition of pump activity, in these cases, would be mediated by receptors functionally equivalent to the vagal stretch receptors of the mammalian lung. In addition, the unidirectional flow design offers an opportunity for extremely efficient gas exchange. In fact, in the lamellae, the anatomical arrangement is such that the direction of blood flow is opposite to that of the water (countercurrent pattern). Hence, the venous blood reaching the gill capillaries is meeting water with progressively higher P_{O_2} (the partial pressure of oxygen, in mmHg), and the arterialized blood leaving the gill capillaries can have a P_{O_2} higher than that of the water leaving the gas exchange area. This would never be the case with the back-and-forth design of the mammalian $\dot{V}_E$, whereby the arterial P_{O_2} can approach but never exceed alveolar P_{O_2}.

Gas convection in amphibians is a clear example of positive pressure ventilation. The upward lifting of the mouth floor generates a positive pressure which forces air into the lungs (Fig. 2). This mechanism is in many respects simpler than the double-pump arrangement of the fish, and is aided only by the valve action of the nares and the glottis. As far as lung inflation goes, it is certainly more efficient than the mammalian suction pump since chest expansion of the lung by positive pressure eliminates the inefficiency caused by chest distortion. On the other hand, the mouth $\dot{V}_E$ of the amphibians has obvious disadvantages. The inflation volume is limited by the size of the mouth, and $\dot{V}_E$ in amphibians must compromise with the nonrespiratory functions of the upper airways much more than it does in reptiles, birds, and mammals.

Turtles were first thought to breathe with a mechanism similar to that of the amphibians, but it is now clear that they operate by negative pressure ventilation as in other reptiles. Contraction of chest wall muscles, laterally located in the turtle, more circularly placed in lizards and snakes, lowers the pressure inside an air sac generating inspiratory flow (Fig. 2). The sac is larger than the functional lung, and in snakes it can be almost as long as the whole body although the gas exchange area is only in its proximal portion. The reptilian breathing cycle is characterized first by active expiration, then by active inspiration, terminating with closure of the glottis and relaxation, with positive airway pressure in between breaths. Inspired air is therefore sucked through the gas exchange region and retained for a period in the sac. Gas from the sac is forced through the same gas exchange region in the opposite direction during expiration.

The same idea in design, but more sophisticated and efficient, is found in the avian respiratory system. In birds, numerous large avascular sacs, some located anteriorly and others posteriorly, are the compliant structures expanding and compressing with the phases of the breathing cycle, while the lungs are small, almost rigid organs connected to the vertebral side of the ribs. Although both anterior and posterior sacs are connected to the lungs, the inhaled air heads almost entirely to the posterior sacs via the mesobronchus (Fig. 2). During inspiration the anterior sacs expand too, but they receive mostly air from the lungs. During the following expiration, the previously inhaled air moves into the lung, while the air stored in the anterior sacs is exhaled. Therefore, the bolus of air inspired takes two full cycles to be exhaled, going from the trachea and the mesobronchus to the posterior sacs during the first inspiration, to the gas exchange area in the first expiration, to the anterior sacs in the second inspiration, and finally out with the expiratory phase of the second cycle. This flow pattern, the direction of which is determined by regional pressure differences rather than anatomic valves, guarantees a continuous unidirectional flow through the lung irrespective of the phase of the breathing cycle. In addition, air and blood flows in the lung are arranged according to a cross-current pattern, which, although not as efficient as the true countercurrent arrangement, is still a better gas exchanger than the $\dot{V}_E$–perfusion matching of the mammalian lung. The result of this sophisticated convection system is that the blood can load more O_2 and does so from an exchange organ with a higher O_2 concentration than that present in the mammalian alveoli. The ultimate test of efficiency may be represented by the ability to survive and perform in hypoxia, and the results are unequivocal. When a mammal and a bird of comparable size, metabolic rates, and blood O_2 affinity are exposed to severe hypoxia, the former lays down panting while the latter is still flying. Some birds are known to migrate flying high above the Himalayas, where no mammal would be able to survive.

III. The Design of the Mammalian Respiratory System

Quite differently from the avian model, the mammalian solution to the gas exchange problem is a back-and-forth convection of air through the same system of conducting pipes, with the pulmonary gas exchange area located at the terminal end of the conducting airways. The mammalian tidal respiration implies that in the alveoli P_{O_2} and P_{CO_2} will be, respectively, lower and higher than in the environment. Because the alveoli are the terminal structures, the blood leaving the pulmonary capillaries cannot be as well arterialized as with the flowthrough system of the avian model. In addition, the common pathway for inspiration and expiration means that part of the inhaled air will not participate to gas exchange,[2] reducing the mechanical efficiency of $\dot{V}_E$ in comparison with the air and blood flow arrangements of fish and birds. On the other hand, that part of the body space which in reptiles and birds is occupied by the sac(s) in mammals is used for the lung itself, which results in a gas exchange area which is very large relative to body mass and fully protected inside the body.

As in reptiles and birds, the mammalian ventilatory pump operates by a negative pressure suction mechanism. The major difference is the separation of the visceral cavity into a thoracic and an abdominal compartment by a muscular layer, the diaphragm. Contraction of this muscle lowers pleural pressure, which results in lung expansion (caused by the rise in transpulmonary pressure) and the rib cage tending to collapse (caused by the negative transrib cage pressure). To what extent, during diaphragmatic contraction, the inward movement of the rib cage actually occurs depends on a number of factors which include the relative compliance of lung and chest wall and the action of the extradiaphragmatic muscles. Therefore, the respiratory pump in mammals is prone to a great deal of deformation, and the intercostal muscles have the important function of stabilizing the rib cage against the distorting action of the diaphragm.

On the one hand, having the main inspiratory muscle (the diaphragm) facing only one side of the thorax may seem poor mechanical design since a substantial fraction of diaphragmatic force is dissipated not in ventilating the lungs but in distorting the pump itself. On the other hand, the complete separation of the coelomic cavity into two compartments decreases the disturbing effects of postural changes, locomotive activities, and variations in abdominal pressure on the distribution of pulmonary $\dot{V}_E$. [*See* RESPIRATORY SYSTEM, ANATOMY; RESPIRATORY SYSTEM, PHYSIOLOGY AND BIOCHEMISTRY.]

2. The larger the volume of the conductive airways (also called anatomical dead space, V_D), the smaller the fraction of tidal volume (V_T) reaching the gas exchange area (also called alveolar volume, V_A). In most mammals during resting breathing V_D approximates one-third of V_T, hence, alveolar ventilation $\dot{V}_A$ = two-thirds of $\dot{V}_E$.

IV. Animal Size and Metabolic Requirements

About 4300 species of mammals are presently known to exist, covering an extremely large range in body size, from the shrew, weighing a few grams, to the African elephant, weighing several tons; marine mammals can have even larger weights,[3] and the blue whale can reach 100–120 tons. Because all mammals are homeotherms (neglecting some special conditions like hibernation, estivation, and torpor, and the newborns of some species) and their body temperature is maintained within a very narrow range (37–39°C), the huge differences in size suggest that either thermodispersion, or thermoproduction, or both, must vary among species. Thermodispersion is mostly determined by body surface. Heat production is determined by the metabolic activities of the cells and hence is proportional to body mass. If, for simplicity, we think of an animal as a sphere, because sphere surface is proportional to length^2 and sphere volume is proportional to length^3, it follows that the larger the animal body the smaller its surface-to-volume ratio. Hence, large animals should be much warmer than smaller animals; the fact that this is not so is because the metabolic activity of the cells of large animals is not as pronounced as it is in the smaller species. Indeed, it has long been recognized that, although the $\dot{V}_{O_2}$ of the whole organism increases with the size of the species, after normalization by the animal's weight ($\dot{V}_{O_2}$/kg) the opposite is true: $\dot{V}_{O_2}$/kg progressively decreases with the increase in animal's weight (Fig. 3). In other words, if we had

3. Mass and weight are used as synonyms throughout the text.

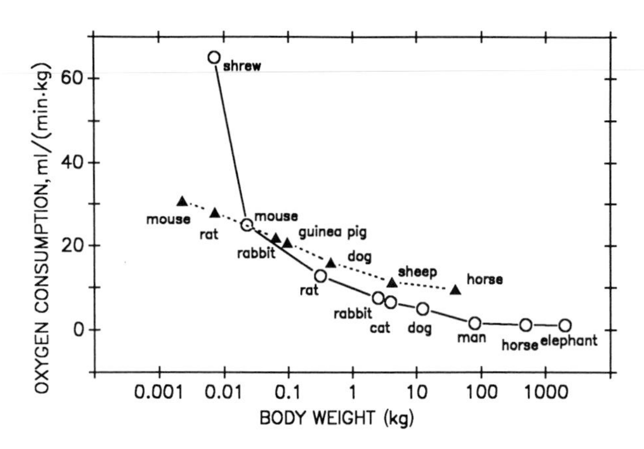

FIGURE 3 Oxygen consumption normalized by body weight in some adult (○) and newborn (▲) mammals. The $\dot{V}_{O_2}$/kg of the newborn is higher than that of the corresponding adult and usually also higher than that of same-weight adults of different species. Only in the smallest newborn species $\dot{V}_{O_2}$/kg is not as high as expected from the adult curve.

1-g samples of "average" flesh from a 100-g rat, a 20-kg dog, a 70-kg man, and a 5-ton elephant, we would find that $\dot{V}_{O_2}$ progressively decreases from the rat's sample, to the dog's, man's, and elephant's. A similar finding would occur if we were comparing samples from different-sized animals within the same species. In fact, it turns out that $\dot{V}_{O_2}$ *within* a species is proportional to the two-thirds power of body weight ($\propto BW^{0.66}$), which is the exponent to be expected according to the surface–volume argument presented above. $\dot{V}_{O_2}$ *among* species, on the other hand, scales to $BW^{0.75}$, which is clearly below unity, yet significantly above the expected 0.66. The reason for this discrepancy is not fully understood, and may relate to the existence of a systematic variation from the simplistic sphere analogy (possibly related to the influence of gravity in shaping the animal's body) and in the mechanisms of thermodispersion.

V. Respiratory Adaptation to Changes in Body Size

Given that $\dot{V}_{O_2}$/kg is not constant among different-sized animals, if the structure and function of the ventilatory apparatus were strictly proportional to size (and therefore not proportional to metabolic rate), major differences among species should be expected in the P_{O_2} and P_{CO_2} of the alveolar gas and blood. This is not the case. It would have been surprising if there were differences since mammalian functions have an optimal operational acid–base range, just as they have an optimal temperature range. Size-related modifications in the mammalian design of the respiratory system occur both at the structural and functional levels to accommodate the different metabolic needs, and will be examined systematically in the following sections.

A. Structure of the Lungs and Respiratory Pump

Comparison of adult mammalian species over a 10^4 range in size have indicated that the mass of the lung is a fixed proportion of the animal's total weight, approximately 1%. Also the amount of air present in the lung, at a prefixed transpulmonary pressure, it directly proportional to body and lung mass. This implies that lung compliance (change in volume per unitary change in transpulmonary pressure), whether per unit of lung tissue or per body weight, is an interspecies constant. Direct measurements of lung compliance show that this is the case; in addition, the value of the exponential constant of the lung deflation function, which describes the shape of the pressure–volume curve, varies very little among species.

The gross aspects of the internal architecture of the lung are similar among mammals. The volume of the conducting airways, which contribute to the anatomical dead space, is a constant fraction of total lung volume,[4] implying that the volume of the gas exchange region, at the lung periphery, is a fixed proportion of body and lung size. However, the subdivision of this peripheral volume varies because the radius of the alveoli is smaller in smaller species. It follows that, per unit of lung volume, small species have a larger number of alveoli and also a greater gas exchange surface area. Indeed, lung surface area is proportional not to body weight or lung volume but to $\dot{V}_{O_2}$. Therefore an important adaptation of the respiratory function to the changes in animal size is achieved structurally, by varying the internal compartmentalization of the lung. The larger alveoli of the bigger species require slightly longer diffusion times, but this is manageable in the larger species, which breathe at low rates. On the other hand, the abundance of surfac-

4. This is essentially true even in the giraffe, in which the trachea is very narrow with respect to its length.

tant on the alveolar wall maintains low surface tension even in the smallest alveoli, therefore avoiding the requirement of large inflating pressures.[5]

B. Passive Mechanical Properties

The fact that larger mammals have bigger lungs than smaller mammals yields the obvious expectations that compliance should increase with size, while resistance (the pressure difference per unitary change in flow) should decrease. The former is in fact mostly determined by the size of the organ, the latter is mostly determined by the length and diameter of the airways (i.e., the longer and narrower the conducting airways, the more resistance they offer).[6] Indeed, both lung compliance and respiratory system compliance are linearly proportional to the animal's body weight,[7] and the resistance of the respiratory system decreases in bigger animals according to the proportionality $BW^{-0.75}$.

The fact that the changes in resistance with animal weight are not proportional to those of compliance implies that the product of these two variables, which is the time constant of the respiratory system, varies among species, gradually increasing in larger animals.[8] This has important functional implications because the time constant of the respiratory system reflects the time of the ventilatory response to a pressure applied: the shorter the time constant, the faster the volume change following pressure generation by the respiratory muscles. Stated differently, for the same muscle pressure generated by a rat or by an elephant, the lungs will inflate more quickly in the rat, because, relatively to body weight, its resistance is not as great, and the time constant not as long, as in the elephant.

5. According to the Young–Laplace relationship, the recoil pressure P generated by an air–liquid spherical interface (as it occurs in the alveoli where the wet surface is in contact with air) is proportional to the product of surface tension T and the radius of curvature r of the interface. Hence, for a given T, the smaller the r the larger is the inflatory pressure required to overcome P.

6. In laminar flow conditions the driving pressure is directly proportional to flow (Poiseuille's Law), the proportionality constant being $R = (8\beta l)/(\pi r^4)$, where β is viscosity, and l and r represent, respectively, length and radius of the airways.

7. This neglects a trend for the largest and bulkiest species to have slightly stiffer chests.

8. Compliance of the respiratory system (C_{rs}) $\propto BW^1$ and resistance of the respiratory system (R_{rs}) $\propto BW^{-0.75}$; hence, the time constant of the respiratory system (τ_{rs}) $= C_{rs}R_{rs} \propto BW^{(1-0.75)} = BW^{0.25}$.

C. Ventilation and Dynamics of Breathing

The amount of air inhaled per unit time is commonly called ventliation ($\dot{V}_E$).[9] This is the product of the amount of air inhaled with each breath (tidal volume) and the number of breaths per minute (breathing frequency). Since, in mammals, pulmonary $\dot{V}_E$ represents the only means for the gas exchange of the venous blood, it is not surprising to find a close relationship between $\dot{V}_E$ and $\dot{V}_{O_2}$. If this was not so then animals with a high $\dot{V}_E/\dot{V}_{O_2}$ would have lower alveolar (and arterial) P_{CO_2} than animals with low $\dot{V}_E/\dot{V}_{O_2}$, contrary to the actual observation that alveolar and arterial blood gases are similar among species.

Because $\dot{V}_E \propto \dot{V}_{O_2}$ and $\dot{V}_{O_2} \propto BW^{0.75}$, $\dot{V}_E$ is also $\propto BW^{0.75}$, i.e., relative to animal size, small animals have higher $\dot{V}_E$ than larger animals do. For example, a rat, relative to its weight, ventilates 10–15 times more than an elephant. It is interesting that this difference is not due to differences in tidal volume per kilogram, which is approximately 8–10 ml/kg in all mammals, but to breathing frequency, which can be above 100 breaths/min in the rat, and less than 10/min in the elephant. Why differences in metabolic requirements, and in $\dot{V}_E$, are met entirely by differences in frequency rather than tidal volume is not immediately obvious, except for the consideration that lung mass is directly proportional to the animal's weight, and it is probably a good design to have the stroke of a pump proportional to its size. On the other hand, it is apparent that the very high values of breathing frequency in the smallest species, possibly above 600 breaths/min in the shrew, require rapid recruitment of the inspiratory motoneurons, and immediate muscle responses to quickly lower the pleural pressure and generate very high inspiratory flows per kilogram. It is interesting that these dynamic requirements are accommodated by some structural and functional properties of the respiratory pump. First, as mentioned earlier, resistance · kg is lower in small animals, and the time constant, whether measured in passive conditions or during contraction of the inspiratory muscles, is proportional to $BW^{0.25}$. This is numeri-

9. Inspired $\dot{V}_E$ is usually slightly larger than expired $\dot{V}_E$, because the ratio between CO_2 produced and O_2 used is less than 1, but the difference is very small. Because there is no evidence for a systematic change in $\dot{V}_{CO_2}/\dot{V}_{O_2}$ with animal size, whether we consider inspired $\dot{V}_E$ or expired $\dot{V}_E$ does not modify the reasoning.

cally the same exponent, although opposite in sign, of that relating breathing rate to body weight (frequency $\propto BW^{-0.25}$), implying that the product of these two parameters, frequency and time constant, is an interspecies constant. Second, although the mass of the respiratory muscles is directly proportional to the animal's mass, the diaphragm of smaller animals has a greater proportion of fast twitch fibers, higher activity of enzymes involved in muscle contraction, and faster rate of pressure development than in larger species.

D. Distribution of Ventilation

The distribution of tidal volume in the lung is not uniform, since it is subjected to the direction of the gravitational vector with respect to the thorax. In a standing subject breathing quietly from functional residual capacity most of the inspired air, per unit of lung tissue, is directed into the middle–lower lobes, and relatively less air reaches the top portions of the lungs. In supine position, the areas less ventilated would become those located more ventrally, while those relatively more ventilated are the gravity-dependent regions of the back.

The causes for these major differences in inspired air distribution are essentially two in number, the curvilinearity of the lung pressure–volume relationship, and the gravity-related variation in pleural pressure. In fact, in a standing subject, pleural pressure is more negative (i.e., more subatmospheric) around the upper lobes of the lung than in the dependent regions of the lower lobes, implying that the lung top is more distended than the bottom. Because, due to the shape of the pressure–volume curve, lung compliance decreases with increased distention, the overdistended top lung regions are less ventilated than the less inflated regions of the lower lobes. The magnitude of the unevenness in $\dot{V}_E$ distribution depends therefore on the curvilinearity of the pressure–volume curve and on the pleural pressure inequalities along the pleural space. The shape of the pressure–volume curve is relatively similar among mammals. On the other hand, thoracic dimensions clearly differ, and if the pleural pressure value at the lung surface was strictly determined by the gravitational field, one should expect the presence of huge top-to-bottom pleural pressure differences in the largest mammals and almost no inequalities among regional pleural pressure values in tiny animals like the shrew or the mouse. This is not the case, and large mammals do not have a distribution of the inspired air much worse than small mammals do because the pleural pressure gradient (which is the change in pleural pressure per unitary change in lung height) is smaller the bigger the animal. The precise reasons for why this is so are not clear, since it is not clear how gravity determines the regional differences in pleural pressure, but the end result is interesting; small and large mammals have similar *absolute* differences in pleural pressure among lung regions despite the enormous differences in size. Hence, one would expect that tidal volume is unevenly distributed within the lung in just about the same way in all species, and the measurements available support this prediction.

In the elephant, and possibly in some other very large mammals with very heavy lungs, some direct attachments between the two pleuras (thoracic and visceral) have been described. This limits the lung configurational freedom within the chest, but represents an additional mechanism for an adequate ventilation of all lung regions.

E. Transport of Gases

Although gases dissolve in blood, neither O_2 nor CO_2 are primarily transported in the dissolved form. O_2 is almost entirely carried by the hemoglobin and CO_2 is carried predominantly in the form of bicarbonate ions, HCO_3^-. Since the solubility of CO_2 in the blood by far exceeds that of O_2, the transport of O_2 from the lungs to the peripheral tissues is more critical than the elimination of CO_2 from the tissues.

Because red blood cells are very similar among mammals with respect to size and hemoglobin concentration, the amount of O_2 which can be loaded in the blood essentially depends upon two parameters: (1) the total mass of circulating blood and (2) the blood concentration of hemoglobin. In addition, because the amount of O_2 that binds to, and is released from, hemoglobin depends on the affinity of this molecule for the gas, it is important to consider also (3) the O_2–hemoglobin dissociation curve. The total mass of circulating blood per unit of animal weight and the hemoglobin concentration are almost constant in all mammals. Blood mass is about 60–70 ml/kg, and the value of hemoglobin concentration oscillates among species around the value of 15 g/100 ml of blood, which, fully saturated, corresponds to an O_2 concentration of about 20 ml/100 ml of blood; this implies that the amount of O_2 in the fully arterialized blood is a relatively fixed proportion of the animal's body weight, or about 13 ml O_2/kg. The relatively constant values in hemoglobin concentration and hematocrit probably reflect the optimal compromise between the advantages of a high O_2 concentration in the blood and the energetic disadvantages of an increased work load on the heart. A higher hematocrit would raise body viscosity, thus the increase in O_2 delivery would be accompanied by an increase in cardiac work. Only in circumstances of chronic low O_2 availability (e.g., at high altitude, in patients with cyanotic heart disease, or in shrews, some bats, and other very small mammals with high metabolic requirements) does the hematocrit increase break the equilibrium between O_2 content and cardiac energetics in favor of the O_2 transport.

Differences in metabolic rates appear to be matched mostly by the way O_2 is unloaded at the tissue level. In almost all healthy mammals, regardless of size, at an alveolar P_{O_2} of 100 Torr (as it occurs in resting conditions at sea level) the hemoglobin is 100% saturated with O_2. Conversely, the P_{O_2} at which the hemoglobin is 50% saturated (also called P_{50}) increases with decreasing species body weight. It is important for small species, because of their high metabolic requirements, to have an hemoglobin with low O_2 affinity (i.e., a high P_{50}), capable of easily unloading O_2 at the tissue level.

The affinity of hemoglobin for O_2 is influenced by a number of factors including the blood pH. With high $[H^+]$ (i.e., low pH) the hemoglobin dissociation curve shifts to the right, and the hemoglobin affinity for O_2 is decreased (Bohr effect). The $[H^+]$ at the tissue level depends largely on $\dot{V}_{CO_2}$, according to the reaction

$$CO_2 + H_2O \rightarrow H_2CO_3 \rightarrow H^+ + HCO_3^-$$

The first of these reactions (the formation of carbonic acid) is catalyzed by carbonic anhydrase, an enzyme present in the red blood cells where most of the HCO_3^- is formed. In the red cells of the smallest mammals carbonic anhydrase activity is higher than in the red cells of larger mammals; this further decreases the hemoglobin affinity for O_2 and helps unload O_2 in the tissue capillaries.

In summary, the differences in metabolic requirements among mammals are not met by differences in the O_2 content of the arterial blood, which is an almost fixed proportion of body weight, but by differences in the ability to unload O_2 in the tissue capillaries.

F. Aspects of the Regulation of the Breathing Pattern

Because of the differences in $\dot{V}_{O_2}$/kg and $\dot{V}_E$/kg among species, it is conceivable that the main regulatory feedback loops controlling the breathing pattern vary with body size, either as a result of the differences in $\dot{V}_E$, or the factors contributing to it. Very little information, however, is available for meaningful generalizations.

In mammals, because of the parallel mechanical arrangement of lungs and chest wall, either structure would be an appropriate anatomical location for receptors designed to sense changes in lung volume. Mechanoreceptors are indeed located both on the chest wall and in the airways, and both sets project to the respiratory center and influence the breathing pattern. The former group, however, is mostly concerned with the reflex control of the respiratory muscles and the integration of their respiratory activity with nonrespiratory functions, including postural control during locomotion. The mechanoreceptors in the airways, and specifically the slowly adapting subgroup, send information via the vagus, which is used to regulate the depth and frequency of breathing. As far as is known, all mammals have both sets of receptors with similar general properties. Whether or not their reflex contribution is also similar is difficult to conclude on the basis of the scattered data. The Hering–Breuer inflation reflex is often interpreted as an index of the reflex effect of the stimulation of the vagal slowly adapting pulmonary receptors. This reflex is particularly pronounced in small species, such as the mouse, rat, guinea pig, and rabbit, while it is weaker in cats, dogs, and humans. Because, as previously mentioned, both tidal volume per kilogram and the pleural pressure swing are similar among species, a more powerful Hering–Breuer reflex would cut off inspiratory activity at an earlier time, possibly contributing to the higher breathing rates of the smallest species.

The $\dot{V}_E$ responses to changes in inspired concentrations of O_2 or CO_2 have been studied in many species, but little is known about the relation between body weight and the various factors which take part in the response. Small mammals seem to respond to hypoxia much more briskly than larger mammals do. For example, rats, mice, and hamsters almost double their resting $\dot{V}_E$ when exposed to 12% of inspired O_2, while cows, men, and dogs only increase it by 10–20%. This could be interpreted as a difference in "gain" of the central respiratory regulator, similarly to what may occur during resting normoxic conditions, or as a difference in the sensitivity of the chemoreceptors to hypoxia, hypocapnia,[10] or both.

VI. Respiration in the Newborn Mammal

Because newborns are smaller than adults of the same species, one may expect that differences in the absolute values of structural or functional variables should relate to the above-discussed effects of size on the respiratory function (Section V). In addition, age-related differences occur because of developmental changes not necessarily related to size. One method of separating size-dependent factors from development-dependent factors, either or both of which may cause differences between newborns and adults, is to compare same-sized animals of different age and species, e.g., comparing a 250-g newborn puppy with a 250-g adult rat. More generally, one could compare the allometric[11] relationships of newborns and adults (or any other age); were the curves overlapping, the conclusion could be reached that size is the primary factor in the age-related differences of the parameter under consideration; on the other hand, differences between the allometric relationships indicate the contribution of maturational aspects not related strictly to body size.

Such an analysis has revealed that, in newborns of most species, $\dot{V}_{O_2}$/kg is higher than in adults of the same species not only because of the size difference (i.e., the surface–volume argument presented above) but also because of developmental factors related to the large anabolic requirements of the growing organism and incomplete thermoregulation (Fig. 3). However, in newborns smaller than 80–100 g, $\dot{V}_{O_2}$/kg can be equal to or even less than the value expected in same-sized adults of different species. A similar pattern occurs for $\dot{V}_E$/kg, since in newborns, as in adults, $\dot{V}_E$ is tightly coupled to the metabolic requirements. The reasons for this deviation at the lowermost end of the allometric scale are not clear. However, in order for the newborn mouse to have the same newborn–adult ratio as larger species,[12] it would need to breathe approximately 500 times/min (i.e., four times the adult rate of 120 breaths/min). Even assuming a massive recruitment of the inspiratory motoneurons and a very rapid generation of pleural pressure, the time constant of the respiratory system should be extremely short in order to inflate the lungs in the very brief inspiratory time available. In a newborn mouse the time constant is about 60 msec, i.e., much shorter than in newborns of larger species or adults, but this is still too long to permit ventilation at 500 breaths/min. Some species are born at a very immature stage, e.g., the opossum and other marsupials. In these newborns chest wall compliance is very high and their lungs are voluminous with respect to body size, with large peripheral units. Both aspects do not favor a small value of compliance of the respiratory system, and therefore present a structural limit to further shortening of the time constant. Thus, it is possible that in the smallest newborn species the mechanical characteristics of the respiratory system pose a limit to the ventilatory performance, and therefore to $\dot{V}_{O_2}$.

The characteristic large chest wall compliance of newborn mammals is an essential structural requirement at birth and favors the mechanics of delivery. After birth, the high ratio of chest wall-to-lung compliance contributes to important differences, with respect to the adult, in the dynamic properties of the respiratory system. The high ratio causes a small resting volume of the respiratory system, which is usually compensated for in the newborn by a dynamic elevation of the end-expiratory level. This is achieved via partial closure

10. As the hypoxic animal hyperventilates, alveolar and arterial P_{CO_2} decrease. Decreased P_{CO_2} tends to decrease $\dot{V}_E$ and thus tends to counteract the response to hypoxia by a factor which depends on the level of hyperventilation and the animal's sensitivity to changes in P_{CO_2}.

11. *Allometric analysis* is a special case of normalization by body weight, by which the effect of animal size (either within a species during growth or among same-age animals of different species) on a given structural or functional parameter Y is examined according to the function $Y = a\mathrm{BW}^b$. The log-transformed version of this function, $\log Y = \log a + b \log \mathrm{BW}$, is particularly useful; when the slope b, which is the exponent of the original exponential function, equals unity, the variable Y increases in direct proportion to animal body weight, the proportionality factor being a. Slopes higher or smaller than unity indicate that the variable under consideration increases, respectively, disproportionately more or less with the increase in body weight.

12. For example, during resting conditions a newborn infant breathes with approximately the same tidal volume per kilogram of the adult man (about 8 ml/kg), but its rate, at 50–60 breaths/min, is about four times higher.

of the glottis and activity of the inspiratory muscles during expiration, i.e., by prolonging the time of deflation with respect to the neural duration of expiration. The high chest wall-to-lung compliance ratio, and the incomplete maturation of the control of the intercostal muscles, make the newborn's chest wall more susceptible to deformation than the adult's. During diaphragmatic contraction and the parallel drop in pleural pressure, a paradoxical inward movement of the thorax during inspiration is a frequent observation in the young of many species, including the human infant.

As one may expect, many of the adult regulatory mechanisms are not completely developed at birth, particularly in the altricial species (i.e., the least mature; the human infant is one of these), with the result that the newborn mammal is more sensitive than the adult to environmental changes, and less capable of protecting its own internal environment. The $\dot{V}_E$ response to hypoxia has been extensively studied; newborns, like adults, hyperventilate in hypoxia, but they also drop the metabolic rate, i.e., they combine mechanisms of acclimatization with pure adaptation. On the one hand the adaptive response prolongs their living in hypoxic or asphyxic conditions, and indeed newborns are formidable survivors. On the other hand, the low metabolic rate reduces and alters body and organ growth with potentially dramatic short- and long-term consequences.

VII. Respiration in Mammals Adapted to Special Environments

A. Diving

For many land mammals, including humans, breath-holding time is of the order of a few minutes. The progressively evolving asphyxia eventually interrupts the $\dot{V}_E$ inhibition, via the stimulation of the chemoreceptors. Some diving mammals, on the other hand, can breath-hold for much longer times, and dives lasting more than 1 hr have been observed in some seals and whales.

The physiological basis for these achievements is not completely clear, but it is probably the result of many combined mechanisms. The ventilatory response to hypoxia and hypercapnia is usually decreased in adult diving mammals; this decrease is probably acquired with diving experience, since at birth newborn seals have a ventilatory response to hypoxia as brisk as that of other precocial species.

Seals free to swim in a covered tank with only one opening to air, when confronted with hypoxia or hypercapnia respond by shortening the dive duration and the time intervals between dives, rather than by increasing the depth and rate of breathing at the air hole. The suggestion was made that the long breath-holding time may also be favored by a blood buffering capacity higher than in most land mammals. Other, less studied mechanisms which could prolong the resistance to asphyxia include the ability to shut off the perfusion of some body districts, therefore avoiding or delaying systemic acidosis, and overall reduction in metabolic rate.

While some divers, like otters and seals, can spend a long time at the surface or even out of the water, others, like dolphins and whales, ventilate only in the short time intervals between dives. This implies the generation of very high inspiratory and expiratory flow rates. The problem of expiratory air flow limitation is reduced by having relatively stiff airways, with solid continuous cartilaginous rings in the larger airways, in place of the flexible horseshoe cartilages of most land mammals. During deep dives, the stiff conductive airways also accommodate the gas shifted from the periphery of the lung during chest compression by the hydrostatic pressure. The elimination of air from the gas exchange region into the dead space limits the amount of nitrogen dissolved in the blood and in the tissues, and is therefore an important precaution against potentially fatal embolic problems during the rapid decompression of surfacing. In this respect it is of interest that, at least in the seal, the chest wall opposes the recoil of the lung only at very small volumes, and therefore does not hinder lung compression by the hydrostatic pressure during a dive. Of course, this mechanical arrangement means that, when out of water, the resting lung volume of the seal is very low. However, the potential problems of a low resting volume on land are eliminated by adopting a peculiar breathing pattern, with occlusions of the airways at end inspiration or in the middle of expiration, and therefore maintaining the lung volume elevated between breaths. This pattern is rarely seen in adult land mammals, but is very much reminiscent of the strategy adopted by reptiles and by all newborn mammals during the first hours after birth, at a time when the resting lung volume is very low.

B. High Altitude

Many mammalian species manage to survive at altitudes above 3500 m, where barometric and O_2 pressures are below 60% of the sea level values. Some of them are actually permanent inhabitants of these low O_2 regions, e.g., some Camelidae of the Andes (llama, alpaca, vicuña, guanaco), the bovid mountain goat in the Rockies, and yak in the Himalayas. Some mammals are known to visit, at least temporarily, regions above 5000 m. Humans have settled high-altitude regions in the Andes and Tibet between 3500 and 4500 m. On the other hand, it is well known that lowlanders exposed to these altitudes often face a number of problems which, in some instances, are life threatening and resolved only by a rapid return to lower levels. The question of what major physiological features permit life at high altitude is of interest. The fact that no clear answer is as yet available probably indicates that there is no unique strategy among species for high-altitude adaptation.

High-altitude populations are characterized by short stature, relatively large thorax, and big lung volumes. At least some of their pulmonary features are probably not genetic traits but are acquired with the prolonged exposure to hypoxia, although the functional basis for these acquisitions, and whether or not they begin during the embryonic and fetal period, are not known. Pregnant women at high altitude give birth to small infants with slightly higher values of hematocrit, hemoglobin, and compliance. Highlanders typically have a lower $\dot{V}_E$ and a blunted $\dot{V}_E$ response to hypoxia when they are compared to lowlanders or lowlanders who moved to and lived at high altitude for long time. Very high values of hematocrit and hemoglobin concentration are other characteristics of the native highlanders, who also have a rightward shift of the hemoglobin dissociation curve (with high P_{50}[13]). Whether or not these features can be considered as the physiological prerequisite for high-altitude adaptation is not clear. Many high-altitude animals do not have the same acclimatization properties observed in humans. For example, llamas, yaks, and sheep respond to acute hypoxia with a hyperventilation similar to that of the corresponding low-altitude animals. Their values of hematocrit and hemoglobin concentrations are not particularly elevated; in addition, high-altitude Camelidae, rodents, and ruminants have a leftward, rather than rightward, shift of the hemoglobin–O_2 affinity, all characteristics which may reflect a better adaption to extremely hypoxic environments than that observed in humans.

13. P_{50} is the partial pressure of O_2 (in mmHg) at which 50% of the hemoglobin is saturated.

C. Burrowing

The strategy of hiding under ground against predators is common among animals, and many mammals, particularly rodents, marsupials, and insectivores, adopt it either permanently or intermittently. Values of O_2 and CO_2 concentrations in the burrow depend on many factors, including the design of the tunnels and the characteristics of the soil; the concentration of O_2 can be as low as 4–6% in the burrows of the marmot and pocket gopher, and that of CO_2 as high as 6–8% in the burrows of chipmunk, echidna, marmot, or ground squirrel. Hence, as opposed to the hypoxic conditions at high altitude, life in the burrow is characterized by an asphyxic environment, both hypoxic and hypercapnic.

In general, burrowers have a lower $\dot{V}_E$ than nonburrowing mammals, and whether their arterial P_{CO_2} is elevated or not depends on metabolic rate, which in some burrowers is less than in nonburrowing mammals of similar body size. The $\dot{V}_E$ response to hypercapnia is often described to be less brisk than in man or dog, and the few data available would also suggest the presence of a blunted $\dot{V}_E$ response to hypoxia. The O_2-carrying capacity, reflected by hematocrit and hemoglobin concentration, is increased in some but not in all the burrowing species. The finding of a reduced P_{50}, i.e., of a higher affinity of hemoglobin for O_2, seems to be more consistent. The blood-buffering capacity is increased in some burrowing species, while in others it is within the range found in nonburrowing mammals. In the first case blood pH is protected despite the hypoventilation, while in the latter case the acidosis must be more pronounced, but, via the Bohr effect (leftward shift of the hemoglobin dissociation curve), blood oxygenation is improved.

Bibliography

Andersen, H. T. (1966). Physiological adaptations in diving vertebrates. *Physiol. Rev.* **46,** 212–243.

Boggs, D. F., Kilgore, D. L., Jr., and Birchard, G. F. (1984). Respiratory physiology of burrowing mammals and birds. *Comp. Biochem. Physiol.* **77,** 1–7.

Dejours, P. (1975). "Principles of Comparative Respiratory Physiology." North-Holland, Amsterdam.

Leith, D. E. Comparative mammalian respiratory mechanics. *The Physiolog.* **19,** 485–510.

Mortola, J. P. (1987). Dynamics of breathing in newborn mammals. *Physiol. Rev.* **67,** 187–243.

Schmidt-Nielsen, K. (1986). "Scaling. Why is animal size so important?" Cambridge University Press, Cambridge.

Tenney, S. M., and Bartlett, D., Jr. (1981). Some comparative aspects of the control of breathing. *Lung Biol. Health Disease* **17,** 67–101.

Wood, S. C., and Lenfant, C. (1979). Evolution of respiratory processes. A comparative approach. *Lung Biol. Health Disease* **13.**

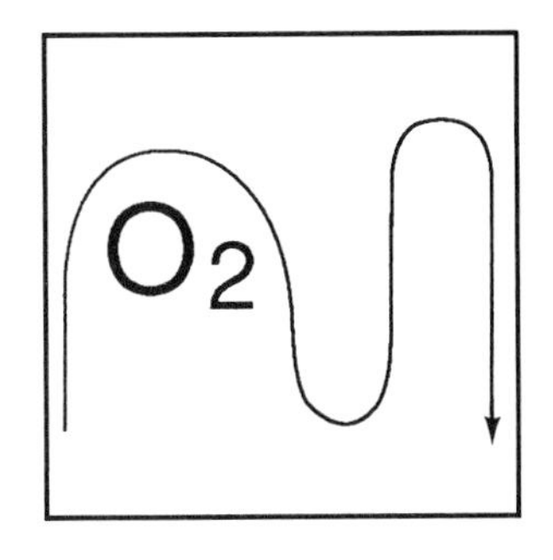

Respiratory Burst

ANTHONY J. SBARRA, CURTIS CETRULO, ROBERT KENNISON, MARY D'ALTON, JOSEPH KENNEDY, JR., FARID LOUIS, JOHANNES JONES, CHRISTO SHAKR, DALE REISNER,
St. Margaret's Hospital for Women; Tufts University School of Medicine

Glossary

Hexose monophosphate pathway Direct oxidative oxidation of glucose
Myeloperoxidase Heme enzyme located in the azurophilic granules of neutrophils
NADPH Oxidase Nicotinamide adenine dinucleotide phosphate reduced is a respiratory coenzyme
Phagocytosis Engulfment of particulate material (viable and/or nonviable) by a cell
ROS Reactive oxygen species. Electron reduced species of oxygen.
Respiratory burst Interaction of material (particulate or nonparticulate) with cells resulting in a burst of oxidative activity

WHEN PHAGOCYTES, MAINLY POLYMORPHONUCLEAR neutrophilic leukocytes, monocytes, eosinophils, and lung and peritoneal macrophages are exposed to a number of different stimuli, particulate and nonparticulate, they all undergo immediate and dramatic changes in the manner in which they process oxygen. Oxygen uptake undergoes a significant and rapid increase, frequently 50-fold over resting cells; direct oxidation of glucose by the hexose monophosphate pathway (HMP) is similarly dramatically increased. These increased oxidative changes noted in cells exposed to particulate material were initially known as the respiratory burst. The major functions of this oxidative burst were initially thought to generate powerful bactericidal agents by the partial reduction of oxygen. Clearly, however, this respiratory burst has much wider biological application and physiological function.

I. Introduction

It has been established that the respiratory burst, which all phagocytic cells undergo, is nonmitochondrial during phagocytosis and is due principally to activation of a unique membrane-bound nicotinamide adenine dinucleotide phosphate reduced (NADPH) oxidase. The physiological significance of this oxidative burst was highlighted when it was noted that the absence of the burst in phagocytes from children with chronic granulomatous disease (CGD) were found to lack the ability to kill phagocytized bacteria. This resulted in an increased susceptibility to infections in these children. By studying the leukocytes from children with this disease, it was shown that the primary product of the respiratory burst, the superoxide anion (O_2^-), was significantly reduced. At physiological pH and ionic concentration, the dismutation of superoxide to H_2O_2 and O_2 occurs. The H_2O_2 produced is not in sufficient quantity to kill phagocytized bacteria, but when reacted with myeloperoxidase (MPO) released from the azurophilic granules of the neutrophils a powerful antimicrobial system is formed, which is primarily responsible for the oxygen-dependent antimicrobial activity of the phagocyte. In the absence of oxygen, this antimicrobial system is significantly inhibited. [*See* PHAGOCYTES.]

Although initially the respiratory burst was thought to include only the immediate increase in oxygen consumption, flow of glucose through the HMP, and increased glycolysis during phagocytosis, obviously a more inclusive list of products would be identified and associated with it. The generation of O_2^-, H_2O_2, singlet oxygen, hydroxyl radicals, and, in addition, some MPO-related products, namely hypochlorous acid (HOCl) and chloromines, are such products.

II. Biochemical and Enzymatic Basis of the Respiratory Burst

A. Oxygen Consumption and the Respiratory Burst Oxidase

The respiratory burst oxidase was at first controversial, but it is now firmly established as NADPH oxidase and not nicotinamide adenine dinucleotide reduced (NADH) oxidase. This is consistent with the activity of the HMP as it is controlled by the generation of nicotinamide adenine dinucleotide phosphate (NADP) from NADPH. Studies with CGD leukocytes have substantiated this finding conclusively. In normal leukocytes, the burst in glucose oxidation by the HMP and in CO_2 production is regulated by NADPH oxidase.

The degree of oxygen stimulation is, to a considerable degree, dependent on the stimulating agent. The generation of superoxide O_2^- by neutrophils during phagocytosis is the result of the one-electron-reduced species of oxygen. The stoichiometry between O_2 consumption and O_2^- generation is controversial due to the methodology used for assay. Values of 4 : 1 to 2 : 1 between O_2^- generation and O_2 uptake have been reported. The generation of O_2^- is quickly followed by its dismutation to H_2O and O_2. Also, it has been suggested by some, and disputed by others, that the spontaneous dismutation of O_2^- results in generation of singlet oxygen $^1O_2$.

B. Nature and Molecular Structure of NADPH Oxidase

The properties of NADPH oxidase are well defined; its molecular structure, however, is not. Most evidence indicates that the O_2^- generating system of phagocytes is formed by different components forming an electron transport chain:

$$\text{NADPH} \rightarrow \text{flavoprotein} \rightarrow \text{cytochrome } b_{558} \rightarrow O_2$$

The precise structure of components comprising the NADPH oxidase is still uncertain. A major problem in its purification is its extreme lability on detergent extraction and on the usual purification manipulation methods. Despite these and other difficulties, molecular weights of some preparations reportedly are between 150,000 and 1,000,000. Electrophoresis of some of these preparations gave major components molecular weights of 87,000 and 32,000. By using different preparation methods, electrophoretic patterns of still different molecular weights were obtained. Apparently, the real nature of NADPH oxidase will not be known until the difficulties of purifying to homogeneity are solved.

C. Intracellular Killing of Microorganisms

MPO is a heme enzyme localized in the azurophilic granules of neutrophils. Release of MPO from these granules to phagolysosomes generates the reactive oxygen species (ROS), which are thought to be responsible for the bactericidal activity of the phagocyte. H_2O_2, at concentrations not normally antibacterial, when acting in concert with MPO and a halide, mainly chloride, form an effective antimicrobial and cytotoxic system. The major primary product produced by this system is HOCl. The primary product of the respiratory burst is O_2^-, and H_2O_2 is the secondary product generated from the dismutation of O_2^-. The hydroxyl radical (OH·) is also thought to be generated in an iron-catalyzed reaction in which O_2^- and H_2O_2 are reactants (Haber–Weiss reaction). The evidence for OH· involvement in the mediation of the respiratory burst-dependent processes is based on the use of radical scavengers such as mannitol. Oxygen consumed in the respiratory burst is ultimately recovered as H_2O_2. All of the reactive intermediates formed are relatively short-lived; however, the generation of relatively stable and reactive chloramine species are also noted. There are formed from the reaction of HOCl and peptides released from cells during the respiratory burst. The intracellular killing of microorganisms can be separated into phases: (1) microorganisms attach to the cell surface and are then engulfed by the pseudopods of the phagocyte; (2) the intracellular granules fuse with the phagocytic membrane; and (3) the NADPH oxidase is activated. Within the phagosome, O_2^- is dismutated to H_2O_2, which then reacts with a halide and MPO to form HOCl. All these end products are thought to have antimicrobial and cytotoxic activity. [*See* NEUTROPHILS.]

III. The Respiratory Burst and Different Cell Types

A. Eosinophils

Eosinophils can be stimulated by a number of different agents and respond with an increased oxygen uptake, O_2^- and H_2O_2 production, and stimulation of glucose through the hexosmonophosphate shunt. A major difference between the respiratory burst of neutrophils and eosinophils is that with eosinophils the burst is sensitive to sodium azide. Also, eosinophils reportedly exhibit a higher metabolic burst than neutrophils. Eosinophils have NADPH oxidase activity, and this activity is stimulated in phagocytizing cells. The cells are able to kill a number of different bacteria, including *Staphylococcus aureus* and *Escherichia coli*.

The inhibitory effect of azide on the respiratory burst of eosinophils and not on neutrophils is presently unexplained.

B. Lymphocytes

Generally, lymphocytes are believed to not experience a respiratory burst as do phagocytic cells. However, the respiratory burst and the resulting production of the different ROS by phagocytic cells can affect the lymphocyte. Because these cells are intimately involved in the immunological and inflammatory response, the effects could be highly significant. Presently, it is agreed that the functional capacity of blood T and B lymphocytes are impaired by oxidant damage resulting from the respiratory burst of phagocytizing cells. [*See* LYMPHOCYTES.]

IV. The Respiratory Burst and the Metabolism of Drugs

Theoretically, all ROS generated by the respiratory burst have the ability to metabolize drugs. A number of model systems have been developed and are being used to study the metabolism of drugs. These include the horseradish peroxidase–H_2O_2 system, the catalase–H_2O_2 system, the arachidonic acid metabolism to prostaglandins, and the myeloperoxidase–H_2O_2–halide system. Drugs such as hydroquinones, aromatic amines, hydroxylamines, and hydrazines have been shown to be metabolized by the myeloperoxidase–H_2O_2–halide system of neutrophils.

Recently, it has been shown that products of the respiratory burst, HOCl and taurine chloramine solutions produced drug-free radical intermediates from chlorpromazine, aminopyrine, and phenylhydrazine. On the basis of these findings, it has been postulated that the oxidation of certain compounds by neutrophils *in vivo* could generate damaging electrophilic free-radical forms. The MPO–H_2O_2–halide system could possibly serve as a unique metabolic pathway for the biotransformation of these drugs *in vivo* and in drug toxicity and chemical carcinogenesis. These observations and postulations provide some evidence that the generation of ROS by the MPO pathway presents a biochemical mechanism *in vivo* for the metabolism of these drugs. Also, the toxic compounds produced could result in tissue damage.

V. The Respiratory Burst and Its Effect on Tissue

Many studies have been carried out that have clearly demonstrated that the ROS generated by the respiratory burst are implicated in inflammation and tissue injury. The mechanism(s) responsible for

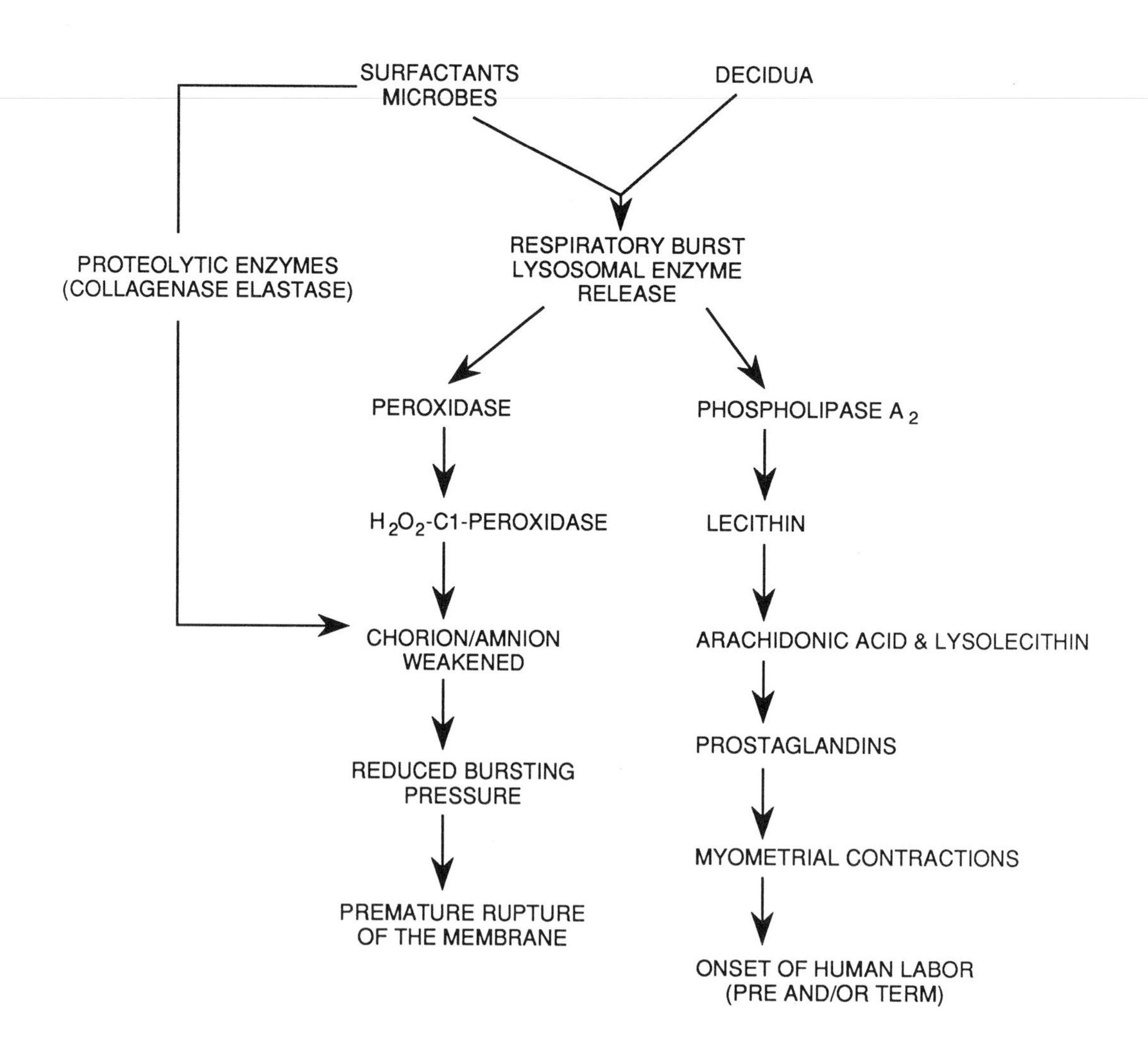

FIGURE 1 Postulated sequence of biochemical events relating the respiratory burst to premature rupture of the membranes and of preterm and term human labor.

these reactions are complex and not precisely defined. Changes in membrane function, brought about by lipid peroxidation, have been reported. Carbohydrates and proteins have also been shown to be modified by ROS. Specifically, O_2^-, H_2O_2, OH·, and HOCl have all been associated with tissue destruction and the inflammatory process. These products are considered to be highly cytotoxic. Cytotoxic activity exhibited by these products is dependent on different factors including target cell or area of attack and initiating stimulus. An example of tissue injury by ROS is the H_2O_2-mediated lung injury utilizing the glucose–glucose oxidase H_2O_2-generating system. The oxidase is directly added to tissue, and cytotoxicity is subsequently monitored. Interestingly, lung injury can be inhibited by catalase. [*See* INFLAMMATION.]

Overwhelming evidence indicates that ROS produced during the respiratory burst are involved with tissue injury; however, because of the complexity of the system(s) involved much remains to be learned.

VI. The Respiratory Burst and Its Effect on Different Physiological States

The respiratory burst has been shown to be involved in a number of different diseases and physiological states. Some of these diverse states include diabetes mellitus, atherosclerosis, aging, fertilization, psoriasis, and pregnancy.

Human pregnancy presents a unique example of

the involvement of the respiratory burst in another physiological process. It has been suggested that a common denominator exists among the onset of human labor, preterm labor, and premature rupture of the membranes. That common denominator is phagocytosis and the accompanying respiratory burst. A summary of the events leading to the onset of human labor, preterm labor, and PROM may be seen in Fig. 1. The phospholipase A_2 can be released by microorganisms as well as by lysosomal release as a result of surfactants reacting with the phagocyte. At term, surfactant arising from the fetus is at its maximal concentration. It has been postulated that this surfactant interacting with decidual cells will exhibit a respiratory burst and lysosomal phospholipase A_2 will be released. This enzyme can then initiate the events leading to the onset of normal human labor. Peroxidase can also be released from the cells. This enzyme will form a powerful cytotoxic system with metabolic H_2O_2 and a halide. It can attack the chorion amnion and weaken it. This has been observed by the target tissue's significantly reduced bursting pressure when compared with unexposed target tissue. By reducing the bursting pressure of the chorion-amnion, one can expect to see a premature rupturing of the membranes.

VII. Summary

As apparent from above, the respiratory burst is not limited, as was originally thought, to simply supplying reactive oxygen species (i.e., O_2^-, H_2O_2 and OH·) that are used to kill intracellular microorganisms. It has been shown to have a much wider scope. This is not surprising if phagocytosis and pinocytosis are the mechanism(s) that cells used to eat and drink. Its much wider biological and physiological applications have been presented.

Bibliography

Babior, B. M. (1984). Oxidants from phagocytes: Agents of defense and destruction. *Blood* **64,** 959.

Clark, R. A. (1983). Extracellular effects of the myeloperoxidase-hydrogen peroxide-halide system. *Adv. Inflam. Res.* **5,** 107.

Klebanoff, S. J., and Clark, R. A. (ed.) (1978). "The Neutrophil: Function and Clinical Disorders." Elsevier/North-Holland Biomedical, Amsterdam.

Rossi, F. (1986). The O_2^- forming NADPH oxidase of the phagocyte; nature, mechanisms of activation and function. *Biochim. Biophys. Acta* **853,** 65.

Sbarra, A. J., and Strauss, R. R. (ed.) (1988). "The Respiratory Burst." Plenum Press, New York.

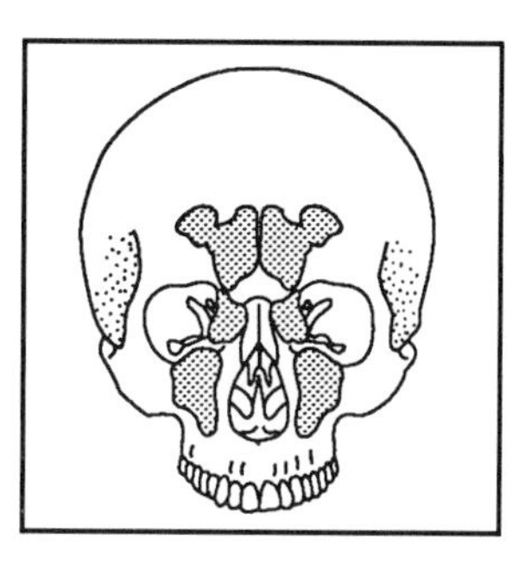

Respiratory System, Anatomy

ANTHONY J. GAUDIN, *California State University–Northridge*

KENNETH C. JONES, *California State University–Northridge*

Glossary

Alvelous Microscopic sac that provides a unicellular surface for the movement of oxygen between the blood and the interior of the lung
Epiglottis Cartilaginous flap of tissue that functions to close the glottis during swallowing
Glottis Opening from the pharynx into the larynx
Mediastinum Space between the lungs that contains the heart and several large blood vessels
Trachea Tube that extends from the larynx to the first branch of the primary bronchi

THE ACQUISITION of oxygen and the elimination of carbon dioxide is accomplished by the respiratory tract, a system of passageways that leads from the nose and the mouth to tiny chambers in the lungs. Oxygen diffuses into the blood from the gas in these chambers and is carried to the rest of the body as the blood is pumped through the circulatory system. As oxygen diffuses into the blood, carbon dioxide carried by the blood from active tissues diffuses into the chambers. The respiration accomplished by the lungs consists of pulmonary ventilation, used to move air into and out of the lungs; external respiration, the exchange of gases between the lungs and the blood; and internal respiration, the exchange of gases between the blood and the tissues of the body. Pulmonary ventilation is further divided into either inspiration (or inhalation), the mechanism that draws air into the lungs, or expiration (or exhalation), the mechanisms that forces air out of the lungs. A more common term for pulmonary ventilation is "breathing."

I. Organization of the Respiratory Tract

The respiratory tract consists of an upper and a lower respiratory tract (see Color Plate 19). The upper respiratory tract includes the external nose, and nasal cavity that lies posterior to the nose, a chamber behind the nasal cavity called the nasopharynx, and the larynx. The mouth can also be involved in inhalation and exhalation, but it is not normally considered part of the respiratory tract. The pharynx is a region common to both the respiratory and digestive tracts and is traversed both by food on its way to the stomach and by air on its way to the lungs.

The lower respiratory tract consists of a network of branching tubes and tubules that lead to a system

Encyclopedia of Human Biology, Volume 6.

of sacs, where gas exchange occurs. The first portion of the tubular system consists of the trachea, a tube that descends downward from the larynx. The trachea branches into bronchi, which, in turn, branch into successively smaller subdivisions called bronchioles. The bronchioles ultimately terminate as blind sacs, called alveoli. Collectively, the network of tubules and alveoli compose the lungs.

A. Upper Respiratory Tract

The nose provides the opening to the respiratory system. It is formed by cartilage and bone, covered by integument (see Color Plate 20). The two openings at its base are the nostrils, or the external nares. Each nostril leads into a vestibule, an enlarged region immediately behind the nostrils. The vestibule leads into the nasal cavity, which extends from the internal portion of the nose into the skull and terminates at the internal nares (or choanae), a pair of openings in the posterior portion of the nasal cavity. The external nares are rimmed by many relatively coarse hairs that filter large particles from inhaled air.

1. Nasal Cavity

The roof of the nasal cavity is formed by the cribriform plate of the ethmoid bone. This plate is perforated by olfactory sensory cells. The floor of the nasal cavity is formed by the superior surface of three structures: the maxilla, the palatine bone of the hard palate, and, posteriorly, the soft palate. The cavity is divided laterally into two nasal fossae by the vertical nasal septum, which extends from the floor of the cavity to the roof. The nasal septum consists of an anterior cartilaginous portion that attaches to the flat perpendicular plate of the ethmoid bone suspended from the roof of the cavity. The lower and posterior portion of the septum consists of another flat perpendicular bone that rests on the floor of the cavity formed by the vomer. Each fossa is also divided into passageways (i.e., meatus) by three bony projections of the lateral walls of the cavity called nasal conchae, or nasal turbinates. [*See* SKELETON.]

The wall of the nasal cavity is a mucous membrane consisting of epithelial tissue richly supplied with blood vessels and mucus-secreting glands. Blood carried in the vessels delivers heat to the inspired air, warming the air as it passes over the tissue. At the same time air is moistened by the evaporation of water from the mucus-covered surface, so air passing through the nasal cavity is warmed (or cooled, depending on the temperature outside), cleaned, and moistened. Mucus is sticky, so it traps much of the material that has passed through the hairs at the nostrils. As the air is moistened, the mucous surface becomes drier and forms semisolid material, in which the fine particulate matter is trapped. This material is carried toward the posterior portion of the nasal cavity by a current of fluid established by ciliated epithelial cells in the tissue lining the cavity. Rhythmic and coordinated motion of these cilia moves the mucus out the posterior exit of the nasal cavity and into the oropharynx. Once in the oropharynx, the material is either swallowed or spat out.

During development outcroppings from the nasal membrane delineate cavities surrounded by the bony tissue of the developing skull. After development is complete the cavities remain as the paranasal sinuses. Maxillary sinuses, the largest of the paranasal sinuses, are located in the body of the maxillary bones on either side of the nasal fossae. Frontal sinuses are located in the frontal bone above the eyes, the ethmoid sinuses are found in the ethmoid bone in the lateral wall of the nasal cavity (Fig. 1). Unlike other sinuses, ethmoid sinuses consist of numerous individual air cells, clustered in groups which collectively compose the sinuses. The sphenoid sinus is a single chamber in the sphenoid bone, just above and behind the junction of the ethmoid bone and the vomer in the posterior portion of the nasal septum (see Color Plate 20).

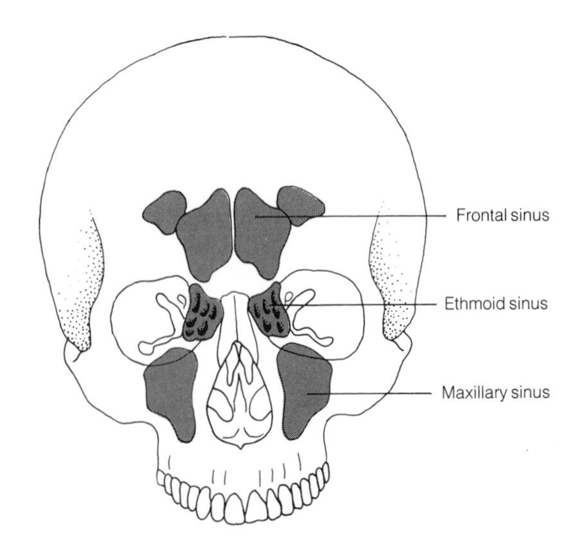

FIGURE 1 The sinuses. (From A. J. Gaudin and K. C. Jones, "Human Anatomy and Physiology," (1989). p. 508. Harcourt Brace Jovanovich, San Diego.)

Each sinus is lined with epithelial tissue similar to that of the nasal cavity itself. The tissues also produce a mucus that drains through ducts that pass from the sinus into the nasal cavity. These ducts are relatively narrow, and when they swell as a result of infection, drainage of the mucus is blocked. When this happens, the mucus and other fluids collect in the sinuses, causing a build-up of pressure that can be very painful.

2. Pharynx

The pharynx is a chamber about 12 cm long that connects the oral and nasal cavities with the esophagus and the larynx. As such, it provides a common pathway for food on its way to the digestive tract and air on its way to the respiratory tract (Fig. 2). Air passing through the nasal fossae exits the nasal cavity through the internal nares and enters the upper part of the pharynx in a region above the soft palate, called the nasopharynx.

The roof of the nasopharynx is formed by the sphenoid bone at the base of the skull, and its floor is formed by the soft palate. The lateral walls of the nasopharynx contain the openings of the auditory tubes, called the pharyngeal apertures. The posterior portion of the nasopharynx contains a broad flat mass of lymphoid tissue, called the pharyngeal tonsils (or adenoids). Infection can cause the adenoids to swell and interfere with the passage of air and mucus through the nasopharynx.

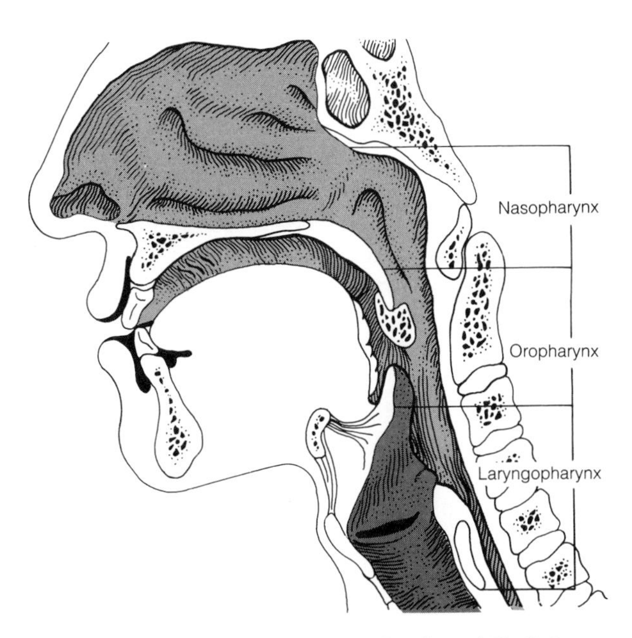

FIGURE 2 The pharynx. (From A. J. Gaudin and K. C. Jones, "Human Anatomy and Physiology," (1989). p. 509. Harcourt Brace Jovanovich, San Diego.)

At the posterior edge of the soft palate, the nasopharynx opens into the oropharynx. The oropharynx communicates anteriorly with the oral cavity, making it directly visible through the open mouth. The oral cavity communicates with oropharynx through the isthmus of the fauces, a narrow passageway formed by the base of the tongue, the soft palate, and two curved folds of tissue that lie anterior and posterior to the palatine tonsils. The anterior folds form the palatoglossal arch and mark the boundary between the oral cavity and the oropharynx. The oropharynx leads downward into the laryngopharynx, a short tube that lies posterior to the larynx.

3. Larynx

After passing from the nasal cavity into the nasopharynx, inhaled air passes through the oropharynx and enters an opening in the anterior wall just below the base of the tongue: the laryngeal aperture. Separating the aperture from the base of the tongue is a leaflike cartilaginous flap of mucous membrane-covered tissue, called the epiglottis. During swallowing, the larynx is elevated, forcing the epiglottis down over the opening into the larynx and preventing food or liquid from passing into the respiratory tract.

Air passing past the epiglottis and through the laryngeal aperature enters the larynx. The larynx acts as a valve that controls access to the tubular system that lies below it (see Color Plate 21).

The larynx consists of bony and cartilaginous structures held together by ligaments, muscles, and other tissues. The larynx lies below the horseshoe-shaped hyoid bone. Suspended from the hyoid bone by ligaments and membrane is the thyroid cartilage, the largest unit of the larynx. The thyroid cartilage produces a bulge in the front of the neck called the laryngeal prominance, better known as the Adam's apple.

Suspended from the thyroid cartilage by ligaments and muscle is the cricoid cartilage, a ringlike structure that nests at the base of the thyroid cartilage. The cartilages give the larynx its form and support the tissues used by the larynx in acting as a valve and a sound-producing organ.

The interior of the larynx is lined with mucous membrane that forms folds of tissue that extend into the passageway (Fig. 3). Just within the laryngeal aperture is an expanded region called the vestibule,

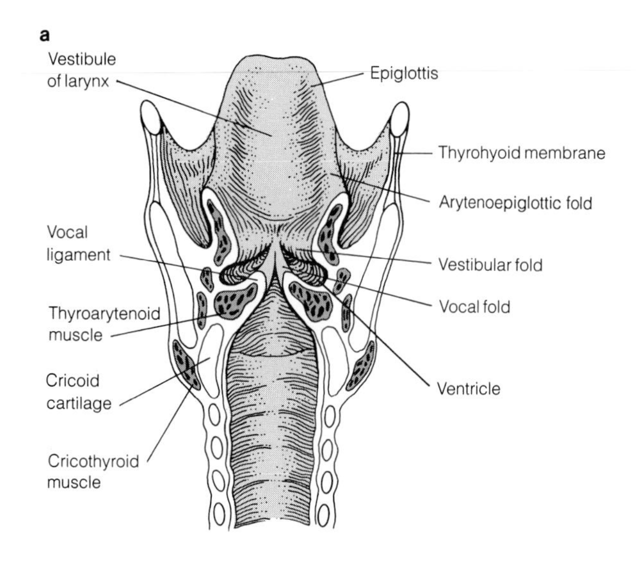

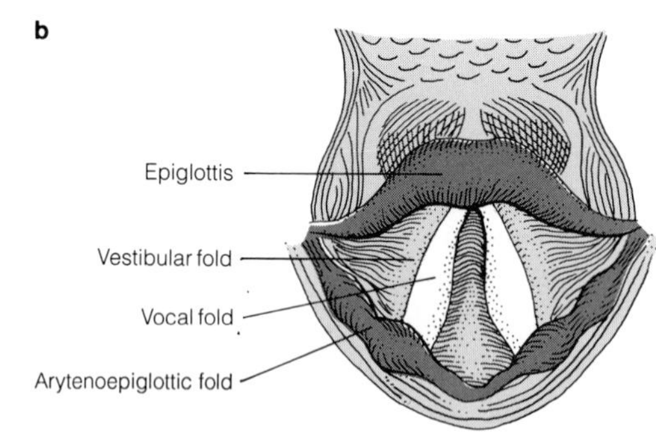

FIGURE 3 Interior organization of the larynx. (a) Posterior view. (b) View from the pharynx, with the glottis open. (From A. J. Gaudin and K. C. Jones (1989). "Human Anatomy and Physiology," p. 510. Harcourt Brace Jovanovich, San Diego.)

which is limited inferiorly by two such folds, called the vestibular folds, or "false vocal cords." Just below these folds there is another widening of the channel, followed by two more folds of the membrane. These are the vocal folds, or "true vocal cords," and are responsible for producing sound, as air passing over them from the lower respiratory tract causes them to vibrate. The space between the two vocal folds is known as the glottis. Inferior to the vocal folds, the laryngeal channel widens and continues into the trachea, the next organ in the tract.

The larynx contains several muscles that control the diameter of the passageway through it. With the exception of the posterior cricoarytenoid muscles, which lie on the anterior surface of the base of the larynx and dilate the passage, a spasm of the laryngeal muscles can close the folds and prevent air flow into the lungs.

4. Voice and Singing

The vestibular folds are controlled by muscle tissue in the wall that can be contracted to open and close the glottis. These muscles help close the larynx to the expulsion of air from the lungs when holding one's breath, lifting a weight, or otherwise straining.

The vocal folds are rimmed by vocal ligaments that border the glottis. Within the folds themselves are muscles that control the tension on the ligaments and control the size of the opening between the folds. Several other laryngeal muscles are also involved in determining the shape and orientation of the vocal folds, by adjusting the orientation of the various cartilages. (The quality of sound produced by the vibrating folds is determined in this way.) The muscles primarily involved in establishing tension in the vocal ligaments are the paired cricothyroid and thyroarytenoid muscles.

B. Lower Respiratory Tract

The trachea, more commonly called the windpipe, is a tube that extends downward from the base of the larynx and branches into two primary bronchi (see Color Plate 22). The trachea is about 12 cm long × about 2.5 cm in diameter in an adult and lies anterior to the esophagus.

Unlike the esophagus, which is soft and pressed flat when empty, the trachea is kept permanently open. This is accomplished by 16–20 horizontally oriented C-shaped bands of cartilage in the tracheal wall. The open portion of the "C" is spanned by trachealis muscle. The bands are connected to one another by intervening annular ligaments and are oriented so that their open portions are on the posterior side of the trachea, next to the esophagus.

The inner surface of the trachea is lined with ciliated mucous membrane laid on a supporting submucosa. This lining is similar in composition to the mucosal lining of the lower portion of the larynx, the nasal cavity, and the nasopharynx. It also helps to filter particulate matter from air passing through the trachea. Externally, the trachea is surrounded by a sheath of connective tissue.

II. Gross Morphology of the Lungs

The lungs are located in the chamber of the chest formed by the ribs laterally, the vertebral column posteriorly, the sternum anteriorly, and the diaphragm inferiorly. This region, the thoracic cavity, also contains several other structures, including the heart, the esophagus, the thoracic lymphatic duct, major nerves, and major arteries and veins that carry blood to and away from the lungs. These organs are located in the mediastinum, the medial portion of the thoracic cavity. The lungs themselves lie on either side of the mediastinum, enclosed in pleurae, a pair of double-walled sacs of serous membrane (see Color Plate 23). The pleura on each side of the thoracic cavity is a single "bag" that folds back on itself, as if one were to push a fist into a partially inflated balloon until the opposing inner surfaces were in contact.

Each pleura consists of a thin serous membrane that lies tightly pressed to the inner wall of the rib cage, folding back on itself in the region of the mediastinum and continuing back out and around the outer surface of the lung, to which it is also closely pressed. The parietal pleura consists of that portion of the serous bag that is not closely adherent to the lungs. The portion that adheres to the lung surfaces is the pulmonary, or visceral, pleura. The portion of the parietal pleura that adheres to the thoracic wall, the costal pleura, goes on to adhere to a portion of the diaphragm, where it forms the diaphragmatic pleura. The latter folds upward from the diaphragm and outlines the mediastinum. This portion is called the mediastinal pleura.

Returning to the analogy of the partially filled balloon with a fist in it, this fist corresponds to the lungs, and the inner surfaces of the balloon in contact with one another are analogous to the surfaces of the parietal and pulmonary pleurae. The portion of the balloon where the fist has entered and which would surround the wrist is equivalent to a region known as the pulmonary hilum. This is a roughly triangular hole, through which the primary bronchus, a pulmonary artery and vein, bronchial arteries and veins, nerves and lymphatic vessels pass into the lung (see Color Plate 23). Collectively, these structures compose the root of the lung.

The two lungs are not mirror images of one another, the left lung being somewhat smaller than the right. Each lung has deep fissures on its surface, which divide it into lobes. The right primary bronchus branches from the trachea and gives rise to three secondary bronchi, each of which enters a lobe. In contrast, the left primary bronchus gives rise to only two secondary bronchi, each of which enters a lobe. Consequently, the right lung has three lobes and the left lung has two. The pulmonary pleura follows the contours of the lobes, descending to the base of each fissure and out again, adhering closely to the surface of the lungs at all points except at the root (Fig. 4).

III. Pulmonary Tree

Inspired air passes into the lungs through the primary bronchi and is distributed throughout the lungs by an elaborate system of tubules which composes the pulmonary tree (Fig. 5). The pulmonary tree consists of two functionally distinct regions, one in which air passes and another in which gas exchange occurs. The first of these is the conducting division and the second, the respiratory division of the pulmonary tree.

A. Conducting Division

The two primary bronchi descend downward and laterally into the chest cavity for a few centimeters,

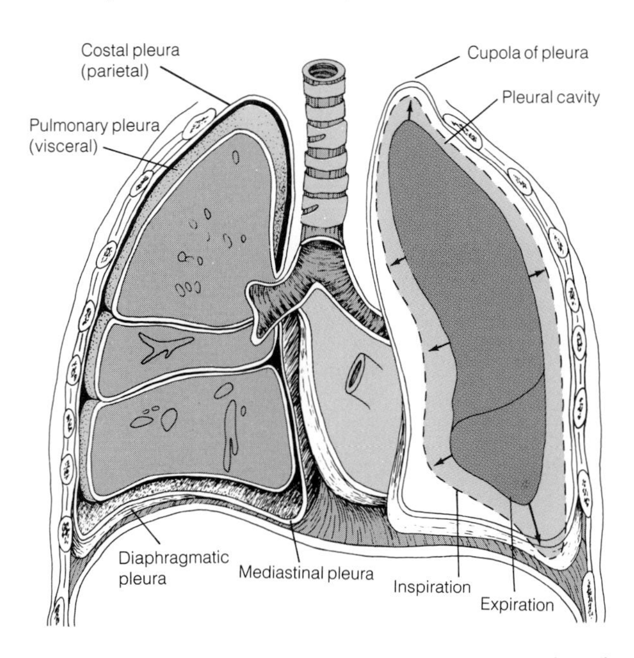

FIGURE 4 Organization of the pleurae. (From A. J. Gaudin and K. C. Jones, "Human Anatomy and Physiology," (1989). p. 513. Harcourt Brace Jovanovich, San Diego.)

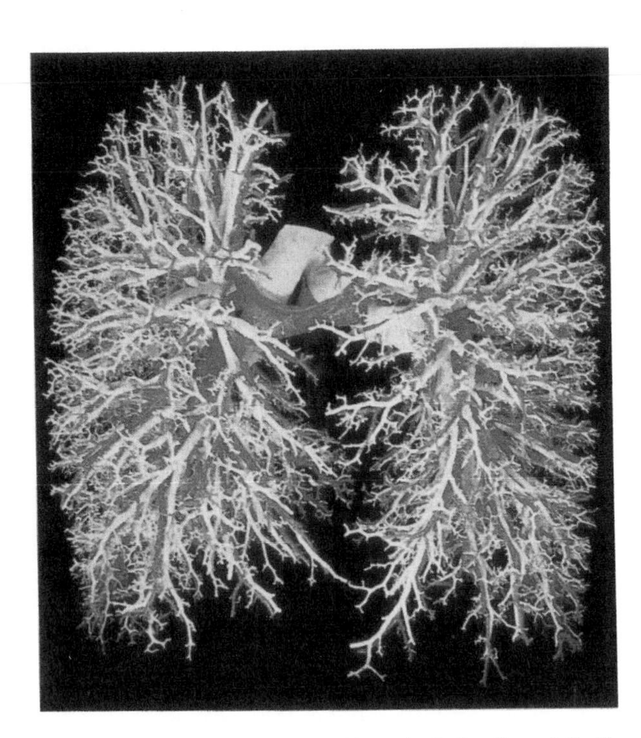

FIGURE 5 The pulmonary tree. (From A. J. Gaudin and K. C. Jones, "Human Anatomy and Physiology," (1989). p. 513. Harcourt Brace Jovanovich, San Diego.)

before passing through the hilum. The right branch extends a somewhat shorter distance than the left one and also has a somewhat larger diameter than the left branch and is oriented more vertically. As a result, if a foreign body succeeds in passing through the larynx, it tends to fall through and lodge in the right bronchus more readily than in the left.

The bronchial wall in this region has a cartilaginous skeleton similar to that of the trachea, although the bands are not so regular. In the region where the bronchus passes into the lung, the bands become even more irregular and are gradually replaced by cartilaginous plates of variable shape and size. Continued branching produces tubes with reduced cartilaginous support, and after a few branchings the cartilage is missing entirely. At that point the tubules are referred to as bronchioles. Bronchioles continue to branch smaller and smaller until they become terminal bronchioles. Normally, 16 branch points occur from the trachea to the terminal bronchioles, producing nearly 66,000 terminal bronchioles. These are distributed approximately equally between the right and left lungs.

Lacking a cartilaginous skeleton, bronchioles are softer and more flexible than bronchi. The bronchioles also contain smooth muscle, the constriction of which reduces their diameter and the amount of air that can pass through them.

B. Respiratory Division

Terminal bronchioles branch still further to produce a network of respiratory bronchioles. Typically, three levels of branching of respiratory bronchioles occur, with the final respiratory bronchiole in the sequence branching to produce a pair of alveolar ducts. Each alveolar duct can branch further, terminating in an alveolar sac. The walls of the passageways from the respiratory bronchioles to the alveolar ducts are perforated by openings that lead into small chambers, called alveoli. Alveoli are the functional units of the lung, where gas exchange occurs. A respiratory bronchiole and its subsequent divisions collectively compose a unit of the lung called a primary lobule (Fig. 6).

There are typically 23 levels of branching between the trachea and the alveoli, giving rise to about 150 million alveoli in each lung. The average diameter of an alveolus is about 300 μm, so that the entire functional surface of the lungs through which gas exchange occurs amounts to about 75 m^2, equal to a little over 800 square feet, or a square measuring more than 28 feet per side. [*See* RESPIRATORY SYSTEM, PHYSIOLOGY AND BIOCHEMISTRY.]

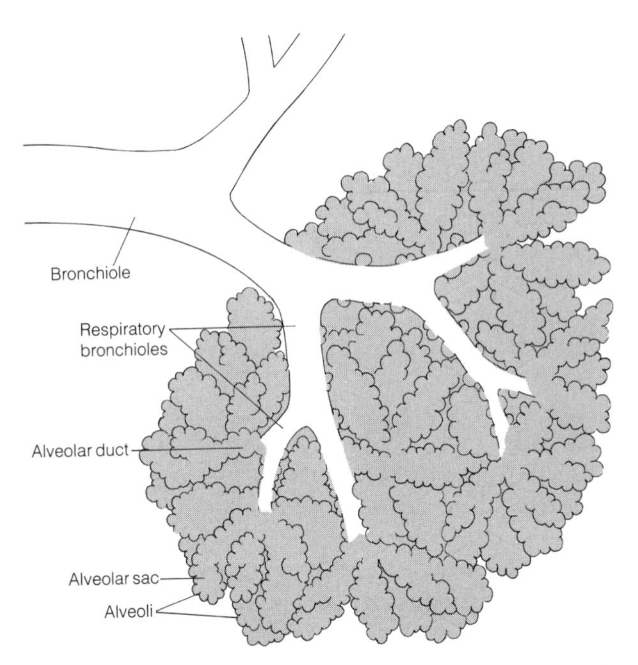

FIGURE 6 A primary lobule. (From A. J. Gaudin and K. C. Jones, "Human Anatomy and Physiology," (1989). p. 514. Harcourt Brace Jovanovich, San Diego.)

IV. Structure of the Alveolar Wall

The alveoli are supported by a network of fine bundles of smooth muscle fibers and fibrous protein that lie in the walls formed where alveoli come into contact with one another. These walls are called interalveolar septa. The septa also have the important function of carrying the capillaries in which blood flows as it absorbs oxygen from the alveolar air and releases carbon dioxide into it.

Figure 7 shows the organization of one of these extremely delicate septa. The surfaces on either side of the septum that face the interior of an alveolus are covered primarily by an epithelium that is so thin that much of it can only be visualized by the electron microscope. The only parts that are thick enough to see are the regions where the nuclei are located. The surface toward the septum rests on a basement membrane of glycoproteins and mucopolysaccharides secreted by the epithelial cell.

Sandwiched between the epithelial cells on either side of the septum are the capillaries. They are fine tubular vessels formed by flat endothelial cells wrapped in cylindrical form; blood passes through these cylinders. Like the epithelial cell of the alveolus, the endothelial cell of a capillary produces a thin basement membrane which, at some points, merges with that of the epithelium.

Also included in the septum are occasional septal cells that secrete a phospholipid material that adheres to the alveolar surface of the epithelial cells. This substance acts as a surfactant, which reduces the surface tension of the moisture on the cell surface. This, in turn, facilitates the diffusion of oxygen into the epithelium from the alveolus. Septal cells are roughly cuboidal and can be present on only one surface of a septum or can extend all the way through, providing surfactant for both sides of the wall.

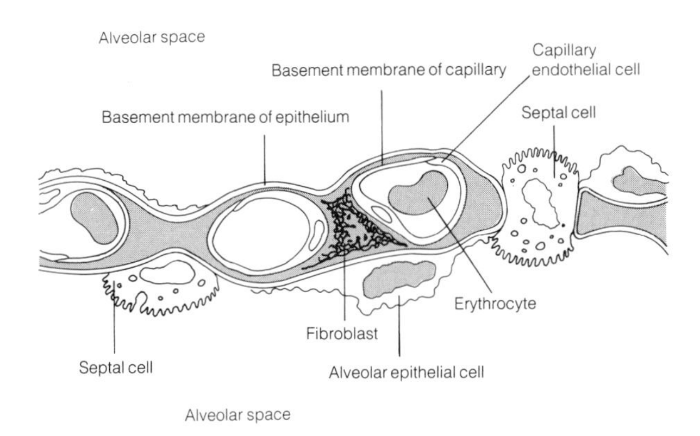

FIGURE 7 An interalveolar septum. (From A. J. Gaudin and K. C. Jones, "Human Anatomy and Physiology," (1989). p. 515. Harcourt Brace Jovanovich, San Diego.)

An alveolus contains macrophages which remove inhaled microorganisms and particulate matter. These macrophages are unusual, in that they spend considerable time in the alveolus, foraging over the epithelial surface, where they ingest dust and soot particles, as well as microbes. Once macrophages have ingested this foreign matter, they are carried out through the bronchi to the pharynx and swallowed. Digestive chemicals within the stomach destroy the macrophages and the materials they contain. [*See* MACROPHAGES.]

Bibliography

Ball, D. (1987). Black lungs and black walls. *New Sci.* **113,** 32.

Cormack, D. H. (1987). "Ham's Histology," 9th ed. Lippincott, Philadelphia, Pennsylvania.

Ganong, W. F. (1985). "Review of Medical Physiology," 12th ed. Lange, Los Altos, California.

Gaudin, A. J., and Jones, K. C. (1989). "Human Anatomy and Physiology." Harcourt Brace Jovanovich, San Diego, California.

Kelly, D. E., Wood, R. L., and Enders, A. C. (1984). "Bailey's Textbook of Microscopic Anatomy," 18th ed. William & Wilkins, Baltimore, Maryland.

Mines, A. H. (1986). "Respiratory Physiology," 2nd ed. Raven, New York.

Nadel, J. A., and Barnes, P. J. (1984). Autonomic regulation of the airways. *Annu. Rev. Med.* **35,** 451–468.

VanBolde, L. M. G., Batenburg, J. J., and Robertson, B. (1988). The pulmonary surfactant system: Biochemical aspects and functional significance. *Physiol. Rev.* **68,** 374–455.

Weisfeldt, M. L., and Chandra, N. (1981). Physiology of cardiopulmonary resuscitation. *Annu. Rev. Med.* **32,** 435–442.

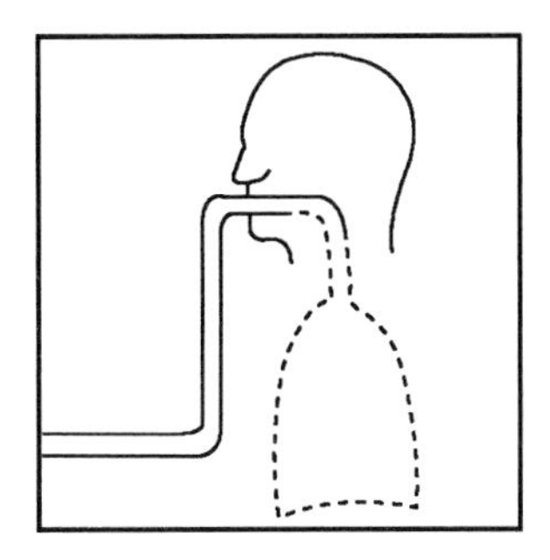

Respiratory System, Physiology and Biochemistry

JOHN B. WEST, *University of California–San Diego*

Glossary

Bohr effect Increase in the oxygen affinity of hemoglobin as a result of increasing the pH of blood
Chemoreceptor Sensor that responds to a change in the chemical composition of the blood or other fluid surrounding it
Dead space Volume of the lungs not participating in gas exchange
Fick's law of diffusion Principles determining the rate of diffusion of a gas through a thin tissue sheet
Fick principle Method for measuring cardiac output based on the fact that the amount of oxygen entering the lungs via the mouth is the same as the amount of oxygen leaving the lungs via the blood
Haldane effect Increase in carbon dioxide concentration of the blood, caused by a reduction in the hemoglobin oxygen saturation
Hypoxemia Condition in which the partial pressure of oxygen in the arterial blood is abnormally low
Partial pressure Proportion of the total gas pressure occupied by one component
Plethysmograph Airtight box like a telephone booth, in which a subject sits for measurements of lung volume and other respiratory variables
Shunt Admixture of poorly oxygenated and well-oxygenated blood, which results in hypoxemia

THE FUNCTION OF the respiratory system is to move oxygen from the air of the environment to the mitochondria of the body cells, where it is utilized, and to move carbon dioxide in the opposite direction. This is called gas exchange. The various links in the overall process include (1) pulmonary ventilation, or the moving of oxygen from the air into the alveoli in the depth of the lungs (and carbon dioxide in the opposite direction); (2) pulmonary blood flow, which moves oxygen out of the lungs after it has been taken up by the blood; (3) pulmonary gas exchange, or the movement of oxygen and carbon dioxide across the blood–gas barrier in the lungs; (4) blood gas transport, or the carriage of oxygen and carbon dioxide in the blood; (5) the mechanics of breathing, or the forces involved in supporting and moving the lungs and the chest wall; (6) the control of ventilation, or the mechanisms which regulate the gas exchange function of the lungs; and (7) peripheral gas exchange, or oxygen delivery to cells and intracellular respiration.

I. Pulmonary Ventilation

Pulmonary ventilation is the process of getting gas to and from the alveoli of the lungs. From a physiological standpoint the airways of the lungs can be divided into a conducting portion and a respiratory region (Fig. 1). The function of the conducting airways is to deliver inspired air to the alveoli of the respiratory region. The conducting airways do not contain alveoli, where gas exchange can occur, and therefore constitute the "anatomic dead space." The volume of dead space in human lungs is about 150 ml. Beyond the conducting airways increasing numbers of alveoli line the walls, and gas exchange

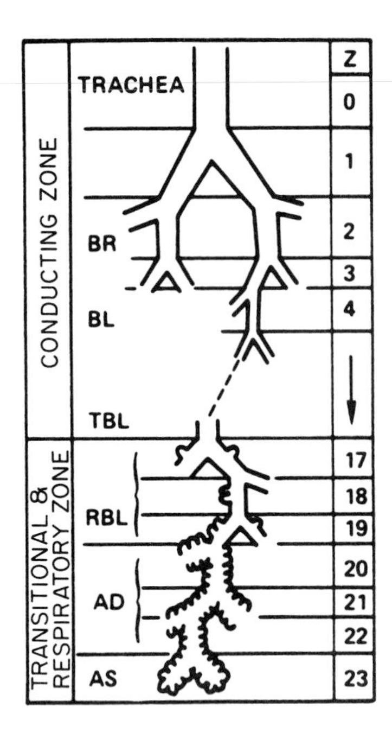

FIGURE 1 Idealization of the airways of the human lung. The first 16 generations (*z*) make up the conducting airways, and the last seven, the respiratory, or transitional and respiratory, zone. BR, bronchus; BL, bronchiole; TBL, terminal bronchiole; RBL, respiratory bronchiole; AD, alveolar duct; AS, alveolar sac. (From E. R. Weibel, "The Pathway for Oxygen." Harvard Univ. Press, Cambridge, Massachusetts, 1984.)

can therefore occur across the thin blood–gas barrier around the pulmonary capillaries (Fig. 2). The volume of the respiratory zone in the human lung is about 2.5–3.0 liters. [*See* RESPIRATORY SYSTEM, ANATOMY.]

A normal inspired volume of air (i.e., tidal volume) is about 500 ml, although there is considerable variation. Since 150 ml of this volume remains behind in the anatomic dead space, only 350 ml penetrates to the alveoli, where gas exchange can occur. The volume of fresh gas entering the alveoli (350 ml in this example) multiplied by the respiratory frequency is known as the alveolar ventilation. This is lower than the total ventilation, which is the tidal volume multiplied by the respiratory frequency.

Lung volumes can be measured with a spirometer, a light bell-shaped container immersed in a water tank (Fig. 3). As the subject exhales, the bell moves up and the pen moves down, marking the chart. In Fig. 3 normal breathing, giving the tidal volume, is followed by an inspiration to total lung capacity, which is followed by a maximal exhalation to residual volume. Not all of the air can be forcibly exhaled from the lungs. The maximal volume of air that can be exhaled from total lung capacity is called the vital capacity. The functional residual capacity is the volume of gas in the lungs at the end of a normal expiration.

Neither the functional residual capacity nor the residual volume can be measured with a simple spirometer. One way of obtaining these values is to connect the subject to a spirometer of known volume, containing a known concentration of the very insoluble gas helium. The subject then breathes in and out of the spirometer until the helium concentrations in the spirometer and the lungs are the same, and the volume of the lungs can then be derived. Another technique is to place the subject in a large airtight box, known as a plethysmograph, and

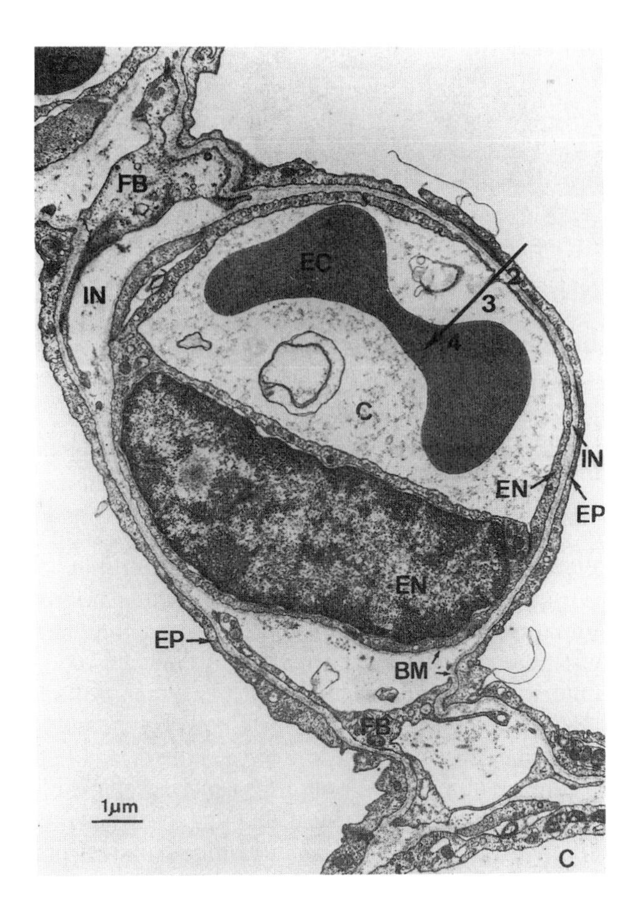

FIGURE 2 Electron micrograph showing a pulmonary capillary (C) in the alveolar wall. Note the extremely thin blood–gas barrier (less than 0.5 μm). The large arrow indicates the diffusion pathway from alveolar gas to the interior of the erythrocyte (EC) and includes the layer of surfactant (not shown), alveolar epithelium (EP), interstitium (IN), capillary endothelium (EN), and plasma. Also seen are parts of structural cells called fibroblasts (FB), basement membrane (BM), and a nucleus of an endothelial cell. [From E. R. Weibel, *Respir. Physiol.* **11**, 54–75 (1970).]

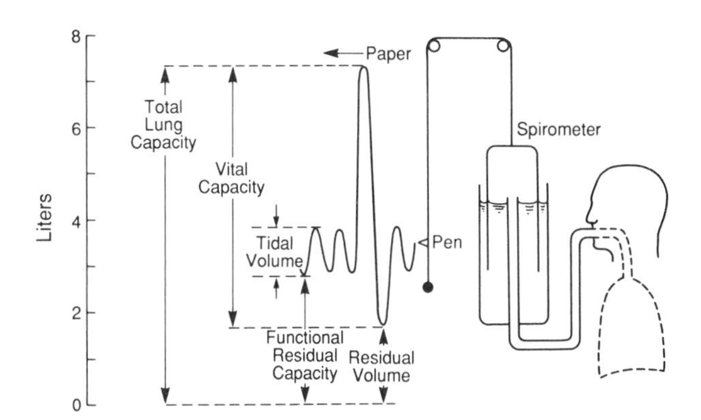

FIGURE 3 Lung volumes, as measured with a spirometer. Note that the functional residual capacity and residual volume cannot be measured without additional techniques. (From J. B. West, "Respiratory Physiology—The Essentials," 3rd ed. Williams & Wilkins, Baltimore, Maryland, 1985.)

to use Boyle's law (i.e., pressure multipled by volume as a constant at constant temperature) to derive the lung volume.

II. Pulmonary Blood Flow

Just as pulmonary ventilation brings oxygen to the blood–gas barrier, where gas exchange occurs, so the pulmonary circulation picks up the oxygen in the alveoli and delivers it to the left side of the heart, from which it is distributed to the rest of the body.

The pulmonary circulation begins at the main pulmonary artery, which receives blood pumped by the right ventricle of the heart. This artery then branches successively, following the system of airways (Fig. 1) ultimately feeding the capillaries. These form a dense network in the alveolar walls, giving an efficient arrangement for gas exchange. The oxygenated blood is then collected from the capillaries by small pulmonary veins, which eventually unite to form large veins, which drain into the left atrium of the heart.

The pulmonary circulation is characterized by the low pressures within it. For example, the mean pressure in the pulmonary artery in humans is only about 15 mm Hg, whereas the mean pressure in the aorta fed by the left heart is on the order of 100 mm Hg. A corollary is that the vascular resistance of the pulmonary circulation, defined as (pulmonary artery pressure − pulmonary venous pressure)/blood flow, is also very low.

The pulmonary circulation has a remarkable facility for accepting increases in cardiac output, with only small rises in pulmonary artery pressure. This is because some capillaries are normally closed or contain nonflowing blood, and these can open when capillary pressure rises, a phenomenon known as recruitment. In addition, the capillaries can distend when the pressure inside them rises. This is called distension.

Pulmonary blood flow, and therefore cardiac output, can be measured by the Fick principle. This states that the total amount of oxygen taken up by the lungs is equal to the cardiac output multiplied by the oxygen concentration difference between the mixed venous blood entering the pulmonary artery and the blood which leaves the lungs. This principle is simply a statement of the conservation of mass. In practice, systemic arterial blood is sampled with a needle to measure oxygen in the blood leaving the lungs, and a thin tube (i.e., a catheter) is passed into the pulmonary artery via a peripheral vein to sample mixed venous blood.

Because the pressures in the pulmonary circulation are so low, the hydrostatic differences between the top and the bottom of the lungs are significant and result in a larger blood flow through the bottom of the lungs than through the top. For example, in normal upright humans, the apical (i.e., uppermost) region of the lungs is only just perfused with blood. Since these regional differences of flow are determined by gravity, this difference disappears when a person lies flat. But then a difference in blood flow is seen between the uppermost and lowermost parts of the supine lungs. Upon exercise in the upright position, the topographical differences in blood flow become less obvious, because the pulmonary artery pressure increases.

An important active response of the pulmonary circulation is hypoxic pulmonary vasoconstriction. This refers to the fact that, when the alveolar gas is made hypoxic (e.g., when breathing a low-oxygen mixture or at high altitude), the small pulmonary arteries (arterioles) of the hypoxic region constrict. If the ventilation of a small region of the lungs is reduced, the vasoconstriction diverts blood flow from it. The mechanism of hypoxic vasoconstriction is not understood, but clearly does not depend on central nervous system connections, because it occurs in the lungs when they are isolated from the rest of the body.

Small amounts of fluid continually move across the walls of the pulmonary capillaries into the interstitial spaces of the lungs. If the pulmonary capillary pressure is increased, the rate of fluid movement rises, causing interstitial pulmonary edema. If the rate of fluid loss from the capillaries is greatly increased, fluid can spill over into the alveoli, causing alveolar edema. This is a potentially life-threatening situation and can occur, for example, as a result of a heart attack (i.e., myocardial infarction), when the left heart fails to pump blood as it should and the pressures in the pulmonary circulation consequently rise. Normally, the small amount of fluid draining from the pulmonary capillaries is carried away in the pulmonary lymphatics.

The pulmonary circulation has important metabolic functions, as well as its primary role in gas exchange. For example, the relatively inactive polypeptide angiotensin I is converted to the potent vasoconstrictor angiotensin II during the passage of blood through the lungs. This conversion is catalyzed by an enzyme (i.e., angiotensin-converting enzyme) which is located on the walls of the capillary endothelial cells. A number of vasoactive substances are completely or partially inactivated during passage through the lungs, including bradykinin, serotonin, and some prostaglandins.

III. Pulmonary Gas Exchange

The exchange of oxygen and carbon dioxide between the alveolar gas and the pulmonary capillary blood occurs across the thin blood–gas barrier shown in Fig. 2. Both gases pass across the tissue sheet by simple passive diffusion, and they obey Fick's law:

$$\dot{V}_{gas} = \frac{A}{T} D(P_1 - P_2)$$

where $\dot{V}_{gas}$ is the volume of gas per unit of time moving across the sheet, A is the area of the sheet, T is its thickness, D is a diffusion constant, and P_1 and P_2 are the partial pressures on either side of the sheet. Clearly, the properties of a sheet which would enhance diffusion are a large surface area and small thickness. The area of the blood–gas barrier in the human lungs is between 50 and 100 m^2, and the thickness is less than 0.5 μm in many places. Therefore, its geometry is ideally suited to rapid diffusion.

The diffusion constant (D) is proportional to the solubility of the gas and inversely proportional to the square root of its molecular weight. Carbon dioxide diffuses about 20 times faster than oxygen across a tissue sheet, owing to its much higher solubility and the small difference in molecular weights.

When mixed venous blood enters the pulmonary capillaries, its partial oxygen pressure (Po_2) is only about 40 mm Hg. The Po_2 of the alveolar gas is approximately 100, so there is a large pressure difference to promote diffusion. In fact, the diffusion of oxygen across the blood–gas barrier occurs so rapidly that the Po_2 of the capillary blood has almost reached that of the alveolar gas within about 0.25 second. Since the blood spends about 0.75 second in the capillary under resting conditions, there is ample time for diffusive equilibration. Upon exercise the time spent by the blood in the capillary is reduced, because of the greatly increased cardiac output. Even so, equilibration between the Po_2 of alveolar gas and end-capillary blood is believed to occur in the human lungs under all but the most exceptional conditions. The rate of combination of oxygen with hemoglobin (see next section) is also believed to delay the loading of oxygen onto the blood to some extent.

The diffusion characteristics of the human lungs can be measured by having humans inhale a low concentration of carbon monoxide and measuring the rate at which it is removed by the blood. It can be shown that the uptake of this gas is limited by the diffusion properties of the blood–gas barrier (i.e., thickness and area). Diseases in which the thickness of the blood–gas barrier is increased typically reduce the diffusing capacity of the lungs for carbon monoxide.

Various factors reduce the efficiency of the lungs for gas exchange. If the arterial blood leaving the chest has an abnormally low Po_2, the condition is known as hypoxemia. One cause is a reduced alveolar ventilation, because this results in a low alveolar Po_2 and a consequent depression of the arterial value. Causes include paralysis of the respiratory muscles and the depression of ventilation by the effects of drugs on the central nervous system. Hypoventilation, as this is called, also results in an increased arterial Pco_2, because carbon dioxide elimination is also affected.

Another cause of hypoxemia is shunt (i.e., the presence of channels that allow blood to bypass ventilated regions of the lungs). An example is a communication between the right and left sides of the heart, as in patients with cyanotic congenital heart disease ("blue" babies). This allows the admixture of poorly oxygenated venous blood with well-oxygenated arterial blood on the left side of the heart.

A common cause of inefficient pulmonary gas exchange is the mismatching of ventilation and blood flow in different regions of the lungs. It can be shown that the gas exchange which occurs in any lung unit depends on the ratio of its ventilation to blood flow. In the normal upright human lungs the differences in ventilation and blood flow of various levels result in a higher P_{O_2} at the apex of the lungs (e.g., compared with the base). In diseases such as chronic bronchitis and emphysema, the architecture of the lung is so disrupted that the normal matching of ventilation and blood flow is disturbed. This is known as ventilation–perfusion inequality. Arterial hypoxemia is then inevitable, and an increased arterial P_{CO_2} is also often seen.

If pulmonary gas exchange becomes inefficient, respiratory failure is said to occur. This is usually characterized by a low arterial P_{O_2} and an increased arterial P_{CO_2}. Patients with respiratory failure are often treated by raising the inspired oxygen concentration, and sometimes by connecting them to a mechanical ventilator.

IV. Blood Gas Transport

Both oxygen and carbon dioxide must be transported in the blood between the lungs and the peripheral tissues. Most oxygen is carried in the blood in combination with hemoglobin; only about 1% is carried in solution, owing to its low solubility (0.003 ml of O_2/100 ml of blood/mm Hg of P_{O_2}).

Hemoglobin consists of an iron–porphyrin compound, heme, joined to the protein globin, consisting of four polypeptide chains. The chains are of two types, α and β, and differences in their amino acid sequences give rise to various types of human hemoglobin. Normal adult hemoglobin is known as A. Hemoglobin F (fetal) has a high affinity for oxygen and is especially suited for transporting oxygen in the relatively hypoxic intrauterine environment. Hemoglobin S (sickle) is poorly soluble in its deoxygenated form, and patients with sickle cell disease are prone to clotting in their blood vessels.

Oxygen forms an easily reversible combination with hemoglobin to give oxyhemoglobin:

$$Hb + O_2 \rightleftharpoons HbO_2.$$

Figure 4 shows the relationship between the oxygen concentration in the blood and the P_{O_2}. The hemoglobin saturation is the proportion of the available binding sites of hemoglobin that are combined with oxygen. At a normal arterial P_{O_2} of 100 mm Hg, approximately 97% of the sites are bound to oxygen.

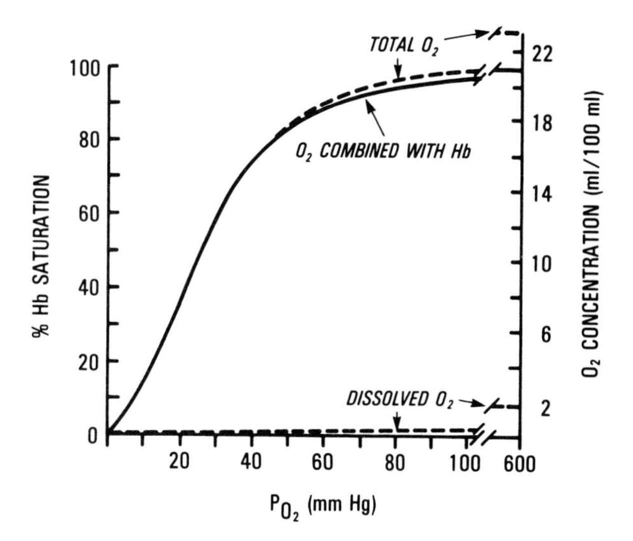

FIGURE 4 Typical oxygen dissociation curve for blood, showing dissolved oxygen and oxygen combined with hemoglobin (Hb). The total oxygen concentration is also shown for a hemoglobin concentration of 15 g/100 ml of blood. (From J. B. West, "Respiratory Physiology—The Essentials," 4th ed. Williams & Wilkins, Baltimore, Maryland, 1990.)

The curved shape of the oxygen dissociation curve (Fig. 4) has several physiological advantages. The flat upper portion means that even if the P_{O_2} of alveolar gas falls somewhat, loading of oxygen will be little affected. In addition, as the red blood cell takes up oxygen along the pulmonary capillary, a large partial pressure difference between the alveolar gas and blood continues to exist, even when most of the oxygen has been transferred. An advantage of the steep lower part of the dissociation curve is that peripheral tissues can withdraw large amounts of oxygen for only a small decrease in capillary P_{O_2}. This maintenance of capillary P_{O_2} assists the diffusion of oxygen into the tissue cells.

Several factors alter the affinity of hemoglobin for oxygen; that is, they shift the dissociation curve leftward or rightward. Increases in temperature P_{CO_2}, hydrogen ion concentration, and 2,3-diphosphoglycerate (2,3-DPG) all shift the curve to the right; that is, they reduce the oxygen affinity of the hemoglobin. The effect of hydrogen ion on the dissociation curve is known as the Bohr effect. The first three factors exert their effects in exercising muscle, which is relatively hot and has a high P_{CO_2}

and an increased hydrogen ion concentration. This reduced oxygen affinity assists the unloading of oxygen from peripheral capillaries.

2,3-DPG is an end product of red blood cell metabolism. An increase in its concentration occurs in several conditions, characterized by chronic hypoxia. Blood stored for transfusion shows a slow decrease in 2,3-DPG, with the result that after the transfusion of large quantities, the release of oxygen to tissues could be impaired. This change in 2,3-DPG in stored blood can be retarded by additives.

Carbon monoxide competes with oxygen for the same binding sites on hemoglobin. Furthermore, it has about 250 times the affinity of oxygen for hemoglobin and can therefore combine with the same amount of hemoglobin when the P_{CO} is 250 times lower than the P_{O_2}. For this reason small amounts of carbon monoxide can tie up large amounts of hemoglobin in the blood, making it unavailable for oxygen carriage. A heavy smoker can have 10% of his hemoglobin combined with carbon monoxide.

Carbon dioxide is carried by the blood in three forms: dissolved, as bicarbonate, and in combination with proteins. Since carbon dioxide is much more soluble than oxygen in blood, dissolved carbon dioxide plays a significant role in its carriage. However, most of the carbon dioxide in blood is carried as bicarbonate, which is formed according to the following reaction:

$$CO_2 + H_2O \overset{CA}{\rightleftharpoons} H_2CO_3 \rightleftharpoons H^+ + HCO_3^-$$

The first reaction is very slow in plasma, but fast within the red blood cell, which contains the enzyme carbonic anhydrase (CA). Carbonic acid can also combine with blood proteins, especially globin, forming carbamino compounds. Some 30% of the carbon dioxide released as the blood passes through the lung capillaries comes from carbamino compounds.

The relationship between carbon dioxide concentration and P_{CO_2} in blood is known as the carbon dioxide dissociation curve, which is considerably steeper than that for oxygen. One consequence of this is that the P_{CO_2} difference between mixed venous and arterial blood is only about 5–7 mm Hg, whereas it is about 60 mm Hg for oxygen. As blood loses oxygen in peripheral capillaries, it is better able to load carbon dioxide. This phenomenon is known as the Haldane effect.

The carriage of carbon dioxide by the blood has an important effect on the acid–base status of the body, because dissolved carbon dioxide forms carbonic acid. For example, a patient whose lungs are diseased, and therefore unable to properly excrete carbon dioxide, develops an increased hydrogen ion concentration in the blood, a condition known as respiratory acidosis. By contrast, if a normal subject hyperventilates and blows off carbon dioxide, he develops respiratory alkalosis. (This occurs in a newcomer to high altitude.) The level of arterial P_{CO_2} can be regulated rapidly by changes in ventilation, and therefore the lungs plays an important role in maintaining the correct acid–base status of the blood.

V. Mechanics of Breathing

In normal breathing at rest, inspiration is active, but expiration is passive. The most important muscle of inspiration is the diaphragm, a thin dome-shaped sheet of muscle supplied by two phrenic nerves originating from the spinal cord in the neck. When the diaphragm contracts, the abdominal contents are forced downward, and the vertical dimension of the chest cavity is increased. The rib cage moves out at the same time. The action of the diaphragm is assisted by external intercostal muscles, which connect adjacent ribs and slope downward and forward. When these muscles contract, the ribs are pulled upward, thus increasing both the lateral and anteroposterior diameters of the thorax. However, paralysis of the intercostal muscles alone does not seriously affect breathing, because the diaphragm is so effective. Accessory muscles of inspiration include neck muscles, which assist inspiration during vigorous exercise.

The most important muscles of expiration are those of the abdominal wall. When these contract, intraabdominal pressure is raised and the diaphragm is pushed upward. This action is apparently assisted by the internal intercostal muscles, the action of which is opposite that of the external intercostal muscles.

The lungs are elastic and are normally expanded by a reduction of pressure in the intrapleural space between the lungs and the chest wall. This space normally contains only a few milliliters of fluid to lubricate the surfaces of the two pleural membranes, but can enlarge if air, for example, enters it (i.e., a pneumothorax). The relationship between the intrapleural pressure around the lungs and its volume is shown in Fig. 5. Here, a lobe of the lungs

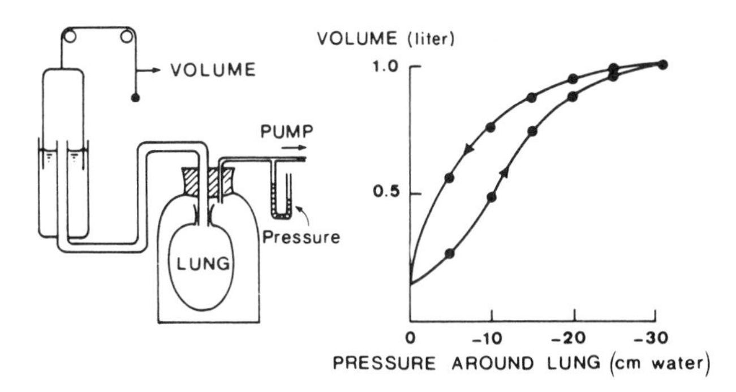

FIGURE 5 Measurement of the pressure–volume curve of a lung lobe. The lung is held at each pressure for a few seconds, while its volume is measured. Note the nonlinearity of the curve and that the inflation and deflation curves are not the same. (From J. B. West, "Respiratory Physiology—The Essentials," 4th ed. Williams & Wilkins, Baltimore, Maryland, 1990.)

is inflated by reducing the pressure around it, and its volume is measured with a spirometer. The pressure–volume curve is flatter at high states of lung inflation, and the lungs follow a different pathway during inflation compared with deflation. This is known as hysteresis. In this example the pressure inside the lungs remains atmospheric. The difference between alveolar and intrapleural pressures is sometimes called transpulmonary pressure.

The slope of the pressure–volume curve is known as the compliance, often measured over the normal working range of the lungs; for the total human lungs the value is about 200 ml per centimeter of water. In other words a normal tidal volume of about 500 ml requires a decrease in the intrapleural pressure of only about 2.5 cm of water. The compliance of the lungs is reduced in some diseases, for example, if fibrous tissue is deposited in the alveolar walls (i.e., pulmonary fibrosis). The compliance is increased by age and also by emphysema. In both instances an alteration in the elastic tissue of the lungs is probably responsible. The elastic behavior of the lungs can be partially attributed to elastic fibers in the lungs, including collagen and elastin. However, the extraordinary distensibility of the lungs has probably less to do with the simple elongation of these fibers than with changes in their geometrical arrangement. An analogy is a nylon stocking, which is very distensible because of its knitted makeup, although the individual nylon fibers are very difficult to stretch.

Another important factor in the pressure–volume behavior of the lungs is the surface tension of the liquid film lining the alveoli. Surface forces generate pressures in curved surfaces (e.g., bubbles), and these pressures are particularly large when the bubbles are small. This could be a serious problem for the lungs, because the alveoli have a diameter of only about 0.3 mm. Fortunately, some cells lining the alveoli secrete material which profoundly lowers the surface tension of the alveolar lining fluid. This surfactant includes the phospholipid dipalmitoyl phosphatidylcholine, and it is secreted by the type II alveolar cells.

The effects of this material on surface tension can be measured outside the lungs in a surface balance, and such studies show that the material can reduce the surface tension to extremely low values. This has the advantage of stabilizing the small alveoli, reducing the work required to expand the lungs and also reducing the tendency to pulmonary edema. Some premature babies are born with an immature surfactant system, because this develops relatively late in fetal life. This condition, known as the respiratory distress syndrome, is characterized by unstable stiff lungs. Recent work indicates that these babies can be treated by instilling artificial surfactant into their lungs after birth.

The lungs are contained within the chest wall, which is made up of the rib cage and the diaphragm. Like the lungs, the chest wall is elastic. At the end of a normal expiration (i.e., functional residual capacity), the tendency of the lungs to collapse because of their elastic recoil is balanced by the tendency of the chest wall to spring out. Indeed, the balance of these two forces is what determines the resting volume of the lungs and the chest wall. If one lung collapses, for example, as a result of a pneumothorax (i.e., air in the pleural space), the chest wall on that side springs out to some extent.

In order for inspiration to occur, the inspiratory muscles contract, intrapleural and alveolar pressures decrease, and air is drawn into the alveoli along the system of airways (Fig. 1). During quiet breathing expiration is passive. When the inspiratory muscles cease to contract, the lungs and the chest wall return to their resting positions. During expiration intrapleural pressure becomes less negative (Fig. 5), alveolar pressure rises slightly, and air moves from the alveoli to the mouth. Airway resistance can be calculated from the difference between alveolar and mouth pressures divided by flow.

At one time it was thought that the major site of airway resistance was in the very small airways. This was natural, because flow in the peripheral airways is laminar and Poiseuille's equation states that, under such conditions, resistance is inversely proportional to the fourth power of the radius. However, it is now known that because of the prodigious number of peripheral airways arranged in parallel, they actually contribute little to overall airway resistance, the main site of resistance being in the medium-sized bronchi. The fact that the peripheral airways contribute so little resistance is important in the detection of early airway disease. It is likely that the first abnormalities in chronic bronchitis, for example, occur in the small airways, but these changes are difficult to detect.

Airway resistance is sometimes increased by the contraction of muscle in the airway walls, as in asthma. The degree of contraction of the muscle is under the control of the autonomic nervous system. Drugs that stimulate the sympathetic nervous system reduce airway constriction. Airway resistance increases at low lung volumes and when gases of high density are breathed, as in deep sea diving.

During a maximal forced expiration some airways within the lungs are compressed by the high pressures developed by the expiratory muscles, and the expiratory flow rate is therefore limited. Under these conditions the expiratory flow rate cannot be raised by increasing the strength of contraction of the expiratory muscles; that is, flow is independent of effort. While this situation only occurs in healthy humans during forced expiration, it may limit the ventilation of patients with lung disease during even moderate exercise.

Work is required to move the lungs and the chest wall, but the oxygen cost of ventilation at rest is small. During heavy exercise the work of breathing might increase so much that the oxygen cost becomes a significant proportion of the total oxygen requirements of the body.

VI. Control of Ventilation

The level of ventilation is remarkably closely controlled, with the result that during normal activity, both at rest and during exercise, the arterial P_{CO_2} changes by only 2 or 3 mm Hg. This is in spite of the fact that, during heavy exercise, carbon dioxide production and oxygen uptake can increase 10-fold or more.

The three basic elements of the respiratory control system are the *sensors* that gather information, feeding it to the *central controller* in the brain. The controller coordinates the information and sends impulses to the *effectors* (i.e., respiratory muscles), which cause ventilation.

The central controller is located in the pons and the medulla of the brain. Part of the controller is the medullary respiratory center, which contains cells that generate the normal respiratory rhythm. This normal automatic process of breathing can be overridden voluntarily; for example, we can elect to hold our breath if we wish.

The sensors include the central and peripheral chemoreceptors. This term denotes a sensor which responds to a change in the chemical composition of the blood or other fluid around it. The central chemoreceptor is located near the ventral surface of the medulla and responds to changes in hydrogen ion concentration of the extracellular and cerebrospinal fluids in its vicinity. Carbon dioxide diffuses from cerebral blood vessels into the cerebrospinal fluid, releasing hydrogen ions, which stimulate the chemoreceptor.

There are also peripheral chemoreceptors located near the common carotid arteries and the arch of the aorta (Fig. 6). These receptors respond to decreases in arterial P_{O_2} and pH and increases in arterial P_{CO_2}. For example, they are responsible for the increase in ventilation that occurs at high altitude, when these receptors are stimulated by the low arterial P_{O_2}.

Other receptors have their nerve endings in the lungs and the upper respiratory tract. For example, the pulmonary stretch receptors are stimulated by inflation of the lungs and then tend to inhibit further

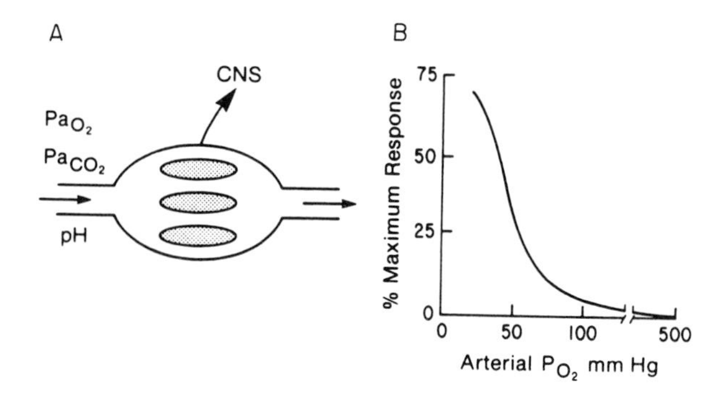

FIGURE 6 (A) A peripheral chemoreceptor (e.g., carotid body), which responds to changes of P_{O_2}, P_{CO_2}, and pH in arterial blood. Impulses travel to the central nervous system (CNS) through a small nerve. (B) The nerve impulse response of the chemoreceptor to arterial P_{O_2}. Note that the maximum response occurs below a P_{O_2} of 50 mm Hg. (From J. B. West, "Respiratory Physiology—The Essentials," 4th ed. Williams & Wilkins, Baltimore, Maryland, 1990.)

inspiratory muscle activity. Irritant receptors in the airway walls cause airway narrowing when stimulated by noxious gases or cigarette smoke. Juxtacapillary receptors are located in the alveolar walls and can be stimulated when fluid leaks out of the capillaries.

The integrated responses of these receptors result in increased ventilation in response to increased levels of carbon dioxide, low levels of inspired oxygen, or acidification of the blood (as in uncontrolled diabetes). However, the large increase in ventilation during exercise is not easily explained by these receptors and remains something of a mystery.

Sometimes breathing becomes unstable and follows a periodic pattern, with waxing and waning of respiration over several breaths, followed by a few seconds of breathholding. Such periodic breathing frequently occurs at high altitudes and is also seen in some types of lung and heart disease.

VII. Peripheral Gas Exchange

When oxygen reaches the peripheral capillaries of the body via the circulation, it diffuses to the mitochondria, where it is utilized. There is good evidence that the Po_2 at the site of utilization is low, probably less than 1 mm Hg. Therefore, the much larger Po_2 in the capillaries ensures an adequate diffusion head of pressure. Facilitated diffusion (which requires a chemical combination of oxygen with hemoglobin or myoglobin) can occur in skeletal muscle cells, which contain myoglobin; this helps to move the large amounts of oxygen consumed during heavy exercise.

Diffusion distances in tissues are typically much longer than in the lungs (Fig. 2). Capillaries can be on the order of 50 μm apart, resulting in very low Po_2 values in the center of the core of tissue supplied by adjacent capillaries, particularly near the venous end of the capillaries. This has been referred to as the "lethal corner."

The process of oxygen utilization within the mitochondria is known as oxidative phosphorylation. This series of reactions produces large amounts of energy, with 6 mol of ATP for every mole of oxygen consumed. [*See* ATP SYNTHESIS BY OXIDATIVE PHOSPHORYLATION IN MITOCHONDRIA.]

If oxygen is in short supply, some energy can be obtained by anaerobic glycolysis with the formation of lactate. However, only relatively small amounts of energy can be obtained from this reaction, and the lactic acid released perturbs the acid–base balance. This mechanism is chiefly reserved for short periods of heavy work, but it can also occur in abnormal states in which oxygen delivery is compromised (e.g., by a defective peripheral circulation).

Oxygen utilization by peripheral tissues can be hampered by a low Po_2 in the arterial blood (e.g., caused by lung disease), by a reduction in the amount of hemoglobin in the blood (i.e., anemia), by reduced local blood flow to the tissues (as in shock), and by tissue poisons (e.g., cyanide) which prevent the uptake of oxygen in the respiratory chain.

Bibliography

Murray, J. F. (1986). "The Normal Lung." Saunders, Philadelphia, Pennsylvania.

Weibel, E. R. (1970). Morphometric estimation of pulmonary diffusion capacity. *Respir. Physiol.* **11,** 54–75.

Weibel, E. R. (1984). "The Pathway for Oxygen." Harvard Univ. Press, Cambridge, Massachusetts.

West, J. B. (1989). "Respiratory Physiology—The Essentials," 4th ed. Williams & Wilkins, Baltimore, Maryland.

West, J. B. (1987). "Pulmonary Pathophysiology—The Essentials," 4th ed. Williams & Wilkins, Baltimore, Maryland.

RES

Reticuloendothelial System

MARIO R. ESCOBAR, *Virginia Commonwealth University*

Glossary

Dendritic cells Cells of macrophage lineage present in lymph nodes (follicular and interdigitating), spleen, other lymphoid organs, and skin (Langerhans cells) that bear class II MHC markers, have Fc receptors, and can process antigens for an immune response

Histiocytes Phagocytic cells of the macrophage series that are fixed in tissues

Interleukin-1 Macrophage-derived substance (MW 15,000) that has multiple biologic properties, including the ability to promote short-term growth of T cells (previously called leukocyte-activating factor)

Marginated monocytes Monocytes which adhere to endothelial cells when a chemoattractant is produced in intravascular sites

Opsonization Process of enhancing phagocytosis (e.g., by aggregated antibody or activated complement)

Phagolysosome Membrane-limited cytoplasmic vesicle formed by fusion of a phagosome and a lysosome

Phagosome Membrane-limited vesicle containing phagocytosed material

Professional phagocytes Phagocytic cells that possess specialized membrane receptors for IgG or C3.

Respiratory burst Metabolic function of phagocytic cells that provides the oxygen radicals employed by these cells for the destruction of susceptible bacteria

METCHNIKOFF ASCERTAINED IN 1892 that the physiologic role of phagocytic cells was to promote the resistance of the host. The term macrophage (so coined because these cells have the ability to ingest large particles) was introduced by him. Although since then many names (e.g., reticulohistiocytic system, lymphoreticular system, monocyte–macrophage system) have been considered for this group of highly phagocytic cells, the two most widely accepted are reticuloendothelial system (RES) and mononuclear phagocyte system (MPS). The term RES was first introduced by Aschoff in 1924 and was defined, according to functional and morphologic studies, as the multiorgan collection of wandering and sessile mononuclear phagocytes, in addition to a variety of lymphatic and sinusoidal cells. Fibroblasts and endothelial cells, which take up colloidal gold by endocytosis, were later added to the list of reticuloendothelial cells by other investigators. In the early 1970s, a number of scientists suggested that RES should be substituted by a more precise term. Accordingly, in 1975 van Furth, Langevoort, and Schaberg proposed the more restrictive term MPS, which consists only of bone marrow promonocytes, circulating blood monocytes, and both mobile and tissue macrophages. MPS as a conceptual framework for these cells excludes the vascular endothelium, reticulum cells, and dendritic cells of lymphoid germinal centers. One of the macrophage characteristics that some of these cells lacks was, for example, the presence of prominent phagolysosomes (e.g., reticulum cells). Recent studies, however, have revealed that many of the phenotypic differences among macrophage types and between macrophages and related cells are environmentally induced or due to variations between normal and experimental conditions. Hence, this chapter will maintain a broader perspective and focus on the RES concept.

I. Ontogeny, Organization, and Biological Role

A. Ontogeny

Mononuclear phagocytes arise in the bone marrow from a pluripotential stem cell common to all hematopoietic cells, including erythrocytes, megakaryocytes, granulocytes, and mononuclear phagocytes. As the stem cell becomes more committed through successive divisions, the mononuclear phagocytes and the cells of the granulocytic series continue to share a common committed stem cell. In culture, these bone marrow cells generate mixed colonies of granulocytes and macrophages under some conditions and monocytic colonies under others. The first progenitor cell, which can be recognized as part of the MPS, is the monoblast. This cell measures 10–12 μm in diameter and has a small rim of basophilic cytoplasm containing a few granules. Monoblasts are phagocytic and adhere to glass. They display Fc receptors as well as the esterase cytochemistry typical of the more mature progeny and are distinct from myeloblasts, which are the precursors in the granulocytic series. Each monoblast divides once, giving rise to the promonocyte with a cell-cycle time of approximately 12 hr. [*See* HEMOPOIETIC SYSTEM; PHAGOCYTES.]

The promonocytes measure about 15 μm in diameter and have an indented nucleus that occupies more than half of the cell. They share with the monoblast the typical features of mononuclear phagocytes, including prominent storage granules that strain azurophilic in smears, some of which are also positive for myeloperoxidase. The azurophil storage granules are synthesized only through this stage of maturation.

The promonocytes mature into monocytes, which have decreased numbers of peroxidase-positive granules and an increased ratio of cytoplasm to nucleus. In contrast to the neutrophils, the bone marrow reserve of preformed monocytes is small. In humans, they are released into the blood within 60 hr of their production, where they circulate with a half-life of about 8.5 hr, leaving randomly (i.e., unrelated to age) from the circulation to the extravascular pool. Daily monocyte turnover is approximately 7×10^6 cells/hr/kg body weight. The ratio of circulating to marginated monocytes in humans is approximately 1 : 3. During inflammation, the proliferation of monocytes is increased by the expansion of the promonocyte pool; their cell-cycle time is decreased, and they are released more rapidly into the circulation. [*See* NEUTROPHILS.]

Tissue macrophages arise by maturation of monocytes that have emigrated from the blood and by replication of immature macrophages in the resident macrophage population. This process is under the control of specific growth factors, termed colony-stimulating factors (CSF), which are produced by fibroblasts and lymphocytes. Of these, the best characterized is CSF-1, or M-CSF, which is lineage-specific for mononuclear phagocytes. Mononuclear phagocytes also proliferate and differentiate in response to CSFs that affect other hematopoietic lineages. These include GM-CSF and G-CSF, which control both myeloid and mononuclear phagocyte lineages, and interleukin-3, which also affects myeloid, erythroid, and lymphoid lineages. [*See* MACROPHAGES.]

Under normal conditions, >50% of the circulating monocytes probably settle in the liver as Kupffer cells, with another 15% settling in the pulmonary alveoli. The life span of the mature macrophage is estimated to be several months. During the inflammatory response, both the influx of blood monocytes and the local proliferation of tissue macrophages increase sharply, and in some granulomas the macrophage turnover may also be increased. During the process of inflammation, free-tissue macrophages may become activated, leading to structural and functional changes in response to mediators, such as gamma interferon released by antigen-stimulated lymphocytes and complement components. Multinucleated giant cells arise by either fusion of macrophages or failure of cytokinesis during mitosis. The epithelioid cell—another form of mature inflammatory mononuclear phagocyte—has decreased phagocytic and digestive capacities and increased endoplasmic reticulum, which may indicate that it has secretory roles. [*See* INFLAMMATION.]

Macrophages are seen early in the development of the lymphoid system and are known to play a role in tissue resorption associated with embryogenesis. Parallel with the maturation of the lymphoid system, the mononuclear phagocytes show increasing development during fetal and neonatal life, and a number of their functions are relatively immature at birth.

B. Organization

The RES or MPS is part of the so-called lymphoreticular system, which includes additionally granulocytes, platelets, and lymphocytes. The progenitors of these cells are pluripotential hematopoietic stem cells located within the bone marrow, fetal liver, and yolk sac of the fetus. Like the mononuclear phagocytic cells, lymphocytes are also mononuclear and interact closely with the cells of the RES but are nonphagocytic. Functionally, phagocytosis in humans is carried out as part of the nonspecific immune response primarily by cells of the RES as well as by neutrophils, and, to a lesser extent, by eosinophils. These types of cells have been referred to as professional phagocytes because their membranes possess specialized receptors for the Fc portion of IgG molecules (IgG1 and IgG3 subclasses) and for the activated component of complement C3. These receptors increase the efficiency of phagocytosis by assisting in the ingestion of microorganisms with IgG or activated C3 on their surfaces. On the other hand, nonprofessional, or facultative, phagocytes include endothelial cells, epithelial cells, fibroblasts, and other cells that will ingest microorganisms under specified conditions but do not possess specialized membrane receptors for IgG or C3. Mononuclear phagocytes, in contrast to the neutrophils or polymorphonuclear leukocytes, show much greater diversity in function and response. This diversity of structure and function is a result of the progressive maturation of these cells from their bone marrow precursors, their experiences with endocytosis, and their interaction with T lymphocytes. [*See* LYMPHOCYTES.]

C. Biologic Role

The primary role of the phagocytic cells in the body economy is the localization and removal of foreign substances, such as microorganisms. Several integrated functions may be required to achieve these goals. First, the phagocytic cells must reach the site of foreign configuration by a process called chemotaxis, which is a mechanism of unidirectional locomotion of the phagocytes toward an increasing gradient of a chemotactic stimulus or chemoattractant. The phagocytes must then ingest the foreign substance by phagocytosis. The process of phagocytosis is part of the nonspecific immune response and represents the host's initial encounter with non-self. Endocytosis is a more general term and includes both phagocytosis or the ingestion of particles and pinocytosis or the uptake of nonparticulates (e.g., fluid droplets). Both processes involve the engulfment and uptake of particles or fluid from the extracellular milieu. Finally, following one of a number of alternative mechanisms—each of which involves a series of biochemical events—the phagocytes must destroy the foreign substance(s) or inhibit the replication of the challenging microorganisms (i.e., microbial killing). The microbicidal activity of macrophages may vary depending on their source as well as on the type of parasite and its virulence. In general, bacteria of low virulence are not susceptible to killing. For example, in the facultative intracellular parasites, *Listeria monocytogenes* appears the most susceptible and mycobacteria the most resistant. The most microbicidal macrophages are the Kupffer cells of the liver, followed by the fixed macrophages of the spleen. Resident peritoneal macrophages and alveolar macrophages have much less activity, but even these cells can kill a substantial proportion of *L. monocytogenes* or opsonized *Salmonella typhimurium*. There is, however, no evidence that normal macrophages can kill pathogenic bacteria, even of such attenuated strains as *Mycobacterium bovis* bacillus Calmette-Guérin.

II. Organ-Specific Cellular Components and Their Individual Biologic Characteristics

A. Organ-Specific Cellular Components

Two general classes of mononuclear phagocytes of the RES are recognized: wandering and fixed. The wandering cells include the circulating monocytes in the peripheral blood and the free macrophages in the sinusoids of lymphoid organs or in the stroma of many other organs. In their normal state, these macrophages may be localized as temporary residents in connective tissue (histiocytes), lung (alveolar macrophages), and serous cavities (pleural and peritoneal macrophages). In contrast, the fixed cells are permanent residents in various tissue locations and constitute the majority of mononuclear phagocytes, which are distributed strategically throughout the body. They are found lining the microcirculation of the liver (Kupffer cells or sinusoid lining cells); spleen (sinusoid lining cells, reticular cells, dendritic macrophages); and lymph nodes, bone marrow, adrenals, and thymus (fixed tissue macrophages). In addition, they are present at other sites such as the intraglomerular mesangium of the kidney (mesangial macrophage), joints (synovial type A cells), bone (osteoclasts), and brain (microglia), or they are exposed to the external environment in the respiratory, gastrointestinal, and genitourinary tracts.

B. Individual Biologic Characteristics

1. Pulmonary Macrophages

Like the other cells of the MPS, the pulmonary or alveolar macrophages originate from the bone marrow. The differences between the majority of the tissue macrophages and the alveolar macrophages appear to be related to the distinctive functional milieu of the lung. In contrast to the phagocytes of the liver sinusoids and the peritoneum, the alveolar macrophages operate in an aerobic atmosphere. In addition, faced with a constant threat of an unexpectedly high load of inhaled particulates, the system must respond quickly with an enormous increase in cellular output. Apparently, these functions are supported by a labile population of interstitial macrophages, a cellular compartment between the circulating blood monocytes and the free alveolar cells. In this compartment, cells may undergo biochemical adaptation prior to emigration into the air sacs and, in response to demand, adaptive proliferation of interstitial cells may supplement the normal population of alveolar macrophages.

2. Peritoneal Macrophages

Present knowledge indicates that under normal steady-state conditions, peritoneal macrophages form a self-replicating population. The unstimulated peritoneal cavity may contain a number of monocytes, which can be considered to form a transient (traveling) population of cells different from that of the nontraveling and self-sustaining pool of resident peritoneal macrophages. When an inflammation is induced in the peritoneal cavity, there is an increased influx of bone marrow-derived monocytes from the peripheral blood into the peritoneal cavity, these monocytes differentiating to mature (exudate) macrophages at the site of the inflammation. Because the monocyte-derived macrophages differ in a number of respects (e.g., functionally) from the resident peritoneal macrophages, macrophages from a peritoneal exudate form a heterogeneous population of cells. Moreover, the way in which the peritoneal exudate is induced and the nature of the inducing agent itself may change the biologic properties of the macrophage population. For these reasons, doubt concerning the use of peritoneal exudates as a source of macrophages is unavoidable.

3. Liver Macrophages

These phagocytes, specifically known as Kupffer cells, are easily recognized on the basis of their fine structural characteristics and their distribution preferentially in the periphery of the liver lobule, where they maintain a close contact with endothelial cells. Although Kupffer cells can also be seen in direct contact with other sinusoidal elements such as fat-storing cells, pit cells, and reticulin fibers, contact between two Kupffer cells is seldomly observed. The three other types of liver sinusoidal cells differ from Kupffer cells with regard to morphology, reaction to experimental conditions, and endocytic capacity. Kupffer cells are endowed with specific functions such as their strong reaction to foreign particles, uptake of endotoxin and endocytosis of circulating tumor cells, bacteria, cellular debris, antigens, immune complexes, and fibrin degradation products.

4. Spleen Macrophages

In the spleen of an adult human, 3 million erythrocytes are destroyed per second, in addition to large numbers of neutrophils, eosinophils, and platelets. This reveals the extraordinary activity of the macrophages in the splenic red pulp, which alone are responsible for all this cytolysis. Contrary to earlier belief, most cells in the spleen (previously called reticulum cells) correspond to stages in the transformation from monocytes to macrophages. It would be too simplistic to attribute to all the red pulp macrophages a bone marrow origin as has been suggested for Kupffer cells or pulmonary macrophages. The spleen is a very active hematopoietic organ during the fetal period, and, therefore, it has been proposed that some of the macrophages seen in the cord at birth originate from monocytes produced locally. On the other hand, the macrophages seen in the germinal centers, which develop in the white pulp as a result of a humoral immune reaction, are different from those in the red pulp (Billroth cords). These white pulp macrophages have the unique property of phagocytizing lymphocytes. Lymphophagocytosis has only been observed in the germinal centers of the lymph nodes, spleen, thymus, tonsils, and appendix. Other cells, whose categorization as macrophages is still debatable, can be found at the junction of the red and white pulp, and in the white pulp, but are not associated with germinal centers. These are elongated and possibly stellate cells that take up and eventually completely degrade the antigenic material transported to the spleen by lymphocytes.

5. Lymph Node Macrophages

The concept of the RES does not distinguish clearly between reticulum cells and macrophages. In this classification, bone marrow-derived fixed macrophages are included in the cellular reticulum and, therefore, belong to the stationary elements or framework of the node in which lymphocytes reside for a certain time. However, the reticulum cells, which manufacture and sustain the intercellular skeleton, are essentially different from macrophages, which have the morphological characteristics of mononuclear phagocytes. According to the concept of the MPS, the cellular reticulum together with the intercellular skeleton forms the framework in which lymphoid cells and macrophages home and interact. The reticulum cells and macrophages in the various compartments are more or less specialized. The reticulum cells in the paracortex most resemble fibroblasts, forming a delicate reticulum throughout the area, except in the region of the germinal centers. In these structures, the antigen-retaining dendritic reticulum cells are highly specialized with a hypertrophic cell membrane, forming an intricate web in which antigen is localized. The macrophages in the outer cortex, or B-cell compartment, are strikingly different from those in the inner cortex, or T-cell compartment. In the outer cortex, the macrophages are large, actively phagocytic cells in various stages of development. They appear to migrate from the marginal sinus through the marginal zone and the follicles toward the base of the germinal centers in the region of the capillary venules. In the T-cell compartment, the characteristic type of macrophage is the interdigitating cell with a corresponding auxiliary role for T-cell function.

6. Bone Marrow Macrophages

In terms of structure, the macrophages or reticular cells of the bone marrow have many features in common with macrophages elsewhere. A distinctive feature is the intimate contact made by these cells with every other element in the bone marrow. Their processes partially invest the walls of sinusoids and, in places, penetrate the lumen; they have a central position in islands of erythroblasts and relate to developing leukocytes. The significance of these relationships lies in the functions of marrow macrophages. Phagocytosis and digestive functions are employed in the removal of effete red cells and particulate matter from the blood stream, in the disposal of extruded erythroid nuclei, and in the culling of defective cells during development. In association with their phagocytic and metabolic activities, they can act as storage cells, for example, in relation to inert particles, lipid, and iron; in certain circumstances, crystalloid material may be found in them. Their extensive relationships may enable the marrow macrophages to play an important role both in mediating hemopoiesis and in controlling the delivery of cells into the circulation. The phagocytic, digestive, and storage functions of normal marrow macrophages are highlighted by the changes in some pathological disorders involving the MPS.

III. Molecular Basis of Biologic Activities

The most prominent functional property of the macrophage is endocytosis. During phagocytosis, particles are bound to specific or nonspecific membrane receptors, then surrounded by the cell membrane, forming phagocytic vesicles. Receptors that bind the Fc portion of immunoglobulins and the C3 component of complement endow the macrophage with the ability to recognize opsonized particles.

There are at least four classes of Fc receptors. A proteinase-sensitive receptor that can bind monomers or complexes of IgG of certain subclasses (i.e., IgG1 and IgG3 in humans). Antibodies binding to this receptor are cytophilic; they bind to the macrophage before interacting with antigen. There is a proteinase-resistant Fc receptor that binds to and mediates endocytosis of antigen-antibody (immune) complexes or aggregates of IgG subclasses (IgG2 and IgG4 in humans). A third type of proteinase-resistant Fc receptor specific for IgG3 has been demonstrated in the mouse. Macrophages also have a receptor that binds IgE. During phagocytosis mediated by these Fc receptors, the receptors are cleared from the membrane, gradually returning over a subsequent period of 6–24 hr. The number of

Fc receptors may vary depending on the status of the host. For instance, it may increase four times during inflammation and in certain diseases.

The receptors for complement are independent of the Fc receptors. In unstimulated monocytes or macrophages, the C3 receptors are much more efficient at mediating binding than at mediating ingestion. It is likely that *in vivo,* the C3 and Fc receptors function synergistically. During macrophage activation, the complement receptors acquire the ability to mediate ingestion on their own. There at least three complement receptors on human mononuclear phagocytes. Complement receptor 1 is specific for C3b and complement receptor 2 for C3d. C5a is chemotactic for mononuclear phagocytes, and it is likely that this third complement receptor is expressed on these cells as well as on granulocytes.

Macrophages also have receptors for lymphokines, which are involved in macrophage activation, and for CSF, which regulate macrophage proliferation. Receptors for insulin have also been demonstrated on macrophages. In addition, macrophages have other receptors that recognize complex carbohydrates and fucosyl- and mannosyl-terminal glycoproteins. These receptors may be important in the clearance of glycoproteins and in the recognition of senescent cells, heterologous erythrocytes, yeasts and other fungi, bacteria, and parasites. Macrophages recognize α_2-macroglobulin-proteinase complexes, which may be important in *in vivo* clearance of enzymes such as thrombin, plasmin, kallikrein, and activated complement components. Receptors for proteins containing iron may play a role in the secretion of iron by macrophages. Receptors for fibrin–fibrinogen complexes may play an important role in the clearance of fibrin from the circulation or inflammatory sites. The receptor for fibronectin may aid in the adhesion of monocytes to areas containing breaches in the integrity of the endothelial lining of the vessels, and fibronectin may also act as an opsonin for certain particles. Finally, macrophages may also play an important role in the regulation of triglyceride and cholesterol metabolism through their receptors for normal and altered lipoproteins.

With regard to chemotaxis, macrophages contain on their surfaces and secrete proteolytic enzymes active at tissue pH that may be relevant to their capacity to migrate *in vivo*. Besides the chemotactic component of complement C5a or anaphylatoxin, there are other chemotactic substances such as bacterial products (e.g., N-formyl-methionyl peptides) and products from stimulated B and T lymphocytes that attract mononuclear phagocytes to sites of inflammation and delayed hypersensitivity reactions.

Factors produced by fibroblasts, fragments of collagen, elastin, and denatured proteins may help attract macrophages to sites of tissue injury. Incidentally, at least two classes of substances inhibit the random migration of macrophages and, thus, prevent migration away from sites of inflammation. Lymphokines (e.g., macrophage migration inhibitory factor, macrophage activation factor) and proteolytic enzymes produced during activation of complement (e.g., factor Bb) and of the fibrinolytic system (e.g., plasmin).

Although phagocytosis was recognized over a century ago, only in the last decade the importance of macrophages as secretory cells has been determined. Over 50 secretion products of macrophages have been identified. This secretory function of macrophages is under complex control and varies with their physiologic state. Some of these include lysozyme, complement components (C1–C5), arachidonic acid metabolites, acid hydrolases, neutral proteinases, plasminogen activator, elastases, collagenases, gelatinases, arginase, and lipoprotein lipase. Macrophages also secrete a variety of plasma proteins, many of which have previously been identified as secretion products of hepatocytes. These include α_2-microglobulin, α_1-proteinase inhibitor, tissue inhibitor of metalloproteinases, fibronectin, transcobalamin II, apolipoprotein E, tissue thromboplastin, and coagulation factors V, VII, IX, and X. Finally, among many other products produced by macrophages are those that serve to regulate the functions of other cells such as interleukin-1, angiogenesis factor, and interferon.

Although the biochemistry and metabolism of phagocytosis is beyond the scope of this chapter, worth noting is that, although the respiratory burst is intimately connected with phagocytosis, it is not essential. Recent evidence suggests that free-tissue macrophages and newly recruited monocytes—but not fixed-tissue macrophages—can respond to lymphokines and phagocytic stimuli by mounting a respiratory burst. The failure of fixed-tissue macrophages, such as Kupffer cells, to produce active metabolites of oxygen may be important in protecting tissues from damage during the scavenger functions of the macrophage. Many soluble agents, including antigen-antibody complexes, C5a, ionophores, and tissue promoters can trigger the respiratory burst without phagocytosis. The respiratory burst can also be triggered by opsonized particles or surfaces when phagocytosis is frustrated by the use of a drug such as cytochalasin B. Phagocytosis can also proceed without the respiratory burst. In particular, phagocytosis mediated by complement C3 or C3bi receptors does not trigger release of hydrogen peroxide or arachidonic acid metabolites. [*See* RESPIRATORY BURST.]

IV. Physiologic Interrelationships with Other Systems of Host Defense, Tissue Injury, and Homeostasis

Although macrophages were originally characterized by their phagocytic property, they are not unique in this regard because many other cell types are also phagocytic at one time or another during their development. For example, most endothelial cells are not phagocytic; however, many types of endothelial cells can manifest phagocytic behavior under special conditions, such as being stressed by small particulate challenges. Furthermore, in addition to this biologic activity, macrophages also play many other important roles: They contribute to host defense, tissue hygiene, wound healing, and general homeostatic mechanisms by interacting with many other cells and tissue fluids within the body. The unique versatility and easy adaptability of the macrophages to their tissue environments endow these cells with the ability to influence many processes negatively, positively, or in complex feedback loops. The invasion of a host by pathogenic microorganisms results in a pattern of cellular and humoral defense reactions. As will be explained later in this section, a pattern of systemic metabolic responses accompanies the mobilization of the immunological host-defense mechanisms.

As expected for a cell that (in evolutionary terms) preceded specific immunity, macrophages play a protean role in systemic host defense. As such, they are involved in resistance to infection, radiation injury, trauma, shock, and the metastatic spread of tumors. While the predominant role of neutrophils is the destruction of microorganisms, particularly extracellular pathogens that rely on the evasion of phagocytosis for survival, macrophages are concerned mainly with the control of those microorganisms that are able to survive intracellular residence and against which neutrophils are ineffective. In the host defense against infection with intracellular pathogens, the principal effector cells are monocytes and macrophages. Monocytes may serve as a backup system to neutrophils in acute infections, but they phagocytize less efficiently and lack many of the potent bactericidal systems of the neutrophil. Macrophages are much more important in chronic infections with intracellular pathogens, such as certain bacteria (e.g., *Mycobacterium leprae, Mycobacterium tuberculosis, L. monocytogenes, Salmonella typhi, Brucella abortus*), some fungi (e.g., *Candida albicans, Cryptococcus neoformans, Histoplasma capsulatum*), many protozoa (e.g., *Toxoplasma gondii,* malarial *Plasmodia, Leishmania donovani, Trypanosoma cruzi*), and most viruses. The growth of these microorganisms is not stopped by phagocytosis, and further steps are required to inhibit their multiplication and spread. These steps involve the accumulation of macrophages within infective foci and their activation. Once activation is achieved, the growth of the pathogen can be stopped. Sensitized lymphocytes play the central role in this process, thus enhancing the bactericidal activities of macrophages through direct cell-to-cell contact or by the intervention of soluble mediators, such as lymphokines, interleukins, and gamma interferon). Cellular immune reactions do not only occur against the microbes listed above but may also be generated against microbial products or against soluble proteins. The cellular landmark in some of these immune reactions involving cellular mechanisms is granuloma formation. The macrophages forming the granuloma are called epithelioid cells because they adhere closely to each other adopting an epithelioidlike characteristic. Some of the macrophages may fuse with each other forming multinucleated giant cells.

Apart from their historical and classical role in pathologic processes, such as inflammation and host resistance to microbial agents, macrophages represent one of the major—and, indeed, perhaps primary—effector mechanisms of the host in the surveillance and destruction of tumor cells. They have been identified as a cellular component of experimental as well as human tumors, constituting an appreciable portion of the total tumor mass (8–54%). Enhancement in the content of tumor macrophages has been interpreted as being associated with tumor regression and decreased incidence of metastases. Similarly, because macrophages possess the ability to exert an antiproliferative effect, the mitotic activity of tumor cells may also be influenced by the degree of infiltration of a tumor by macrophages. The validity of these findings with regard to human tumors can be supported by histologic observations where regression was characterized by a significant host-cell infiltration.

One of the most important contributions of macrophages to host defense is their central role in the initiation and regulation of the immune response, both *in vivo* and in culture. Macrophages may achieve this role in a number of ways with varying degrees of specificity. By a rather nonspecific mechanism, they can either improve the viability of lymphocytes or suppress their proliferation through thymidine, arginase, complement cleavage products, prostaglandin E, and interferon. Macrophages may also alter the function of lymphocytes more specifically by the interleukin-1 (formerly called leukocyte-activating factor) pathway. Yet, another more specific function of macrophages is their role in both humoral and cell-mediated immune responses. As stated previously, they are involved both in the initiation of responses as antigen-presenting cells and in the effector phase as inflammatory, tumoricidal, and microbicidal cells, in addition to their regulatory function. The uptake of antigens by macrophages is the first step in the processing of antigen leading to the production of circulating antibody. In such cases, antigen is not completely degraded by the macrophage but becomes bound to macrophage RNA or membrane. The macrophage is not the cell that recognizes antigen as foreign, but the macrophage nonspecifically processes the antigen so that it may be recognized by specific antigen-reactive lymphocytes. Processed antigens are expressed on the surface of antigen-presenting macrophages in conjunction with self-surface markers (class II major histocompatibility complex or MHC markers) that are recognized by T-cell receptors for antigen and for self-class II MHC. This function requires that T cells and macrophages display the same MHC-encoded class II determinants (HLA-DR antigens in humans). However, not all monocytes and macrophages express such determinants. The precise mechanism underlying this process at the molecular level is still not well understood.

Only in the last two or three decades, relevant methodological advances have been useful to the quantitative assessment of the biochemistry and physiology of mononuclear phagocytes. These advances have included sophisticated isotopic and nonisotopic tracer techniques, improved methods for the isolation and purification of macrophages, refinement of cell- and tissue-culture procedures, and the advent of powerful immunologic approaches. The production of monoclonal antibodies, the development of molecular biology, and the applications of scanning electron microscopy and flow cytometry as well as the use of sophisticated approaches in immunochemistry and immunohistology have expanded considerably our still relatively limited capabilities to perform quantitative and well-controlled studies on isolated cell populations.

Much of this progress in biotechnology and its application to studies on the RES have confirmed Aschoff's suggestion (made in his classic Janeway lecture in 1924) that the RES plays an important role in the metabolic functions of the host. In addition to the broad spectrum of activities displayed by macrophages, as described above, they also participate in a number of systemic responses involving carbohydrate, lipid, and protein metabolism. These systemic metabolic responses to infection and endotoxicosis are accompanied by either a generalized catabolic state with wasting of body tissue to supply amino acids and fatty acids as energy sources during the febrile state or, conversely, a selective anabolic state with protein synthesis of acute stress proteins and immunoglobulins.

V. Disorders of the RES

A detailed discussion of the pathobiology of the RES is beyond the scope of this chapter; the reader should refer to the bibliography section for further information. In brief, there are a number of disorders associated with either quantitative and/or functional abnormalities of monocytes and macrophages. For example, whereas an enlargement of the spleen and lymph nodes, which are rich in these cells, may be a result of the monocytosis or reactive hyperplasia produced in response to infection with *M. tuberculosis* in normal individuals, these cells may proliferate abnormally to extremely high levels in individuals with monocytic leukemia or other malignant histiocytic proliferative disorders.

One group of disorders of the RES develops as a result of the ingestion of a nondigestible substance, the overloading of iron as in hemosiderosis, or inborn errors of metabolism in which a specific genetic defect of macrophage enzyme function has occurred (e.g., Gaucher's disease, Hurler's disease).

These disorders fall under the category of storage diseases.

Another group of macrophage disorders includes certain genetic abnormalities, such as chronic granulomatous disease, in which both macrophages and polymorphonuclear leukocytes lack an enzyme important in the respiratory burst associated with phagocytosis. In this disease, the patient's phagocytes are unable to kill those pathogens that are susceptible to reactive metabolites of oxygen. Defects in synthesis and secretion of complement components can cause macrophage dysfunction because macrophages constitute a major source of certain complement components, and the products of complement activation are important for macrophage functions, such as phagocytosis and chemotaxis. In osteopetrosis, significantly decreased numbers of osteoclasts can be detected in bone, so that bone resorption is abnormal.

A number of iatrogenic or environmentally induced disorders also exist, such as the high concentration of glucocorticosteroids and ionizing radiation that may interfere with the macrophage defense system, including macrophage migration into tissues as well as macrophage proliferation. These acquired defects may lead to the onset of frequent opportunistic infections.

Although genetically determined differences in macrophage responses to lipopolysaccharides have been reported in animals, a similar heterogeneity in lipopolysaccharide responsiveness may likely occur in humans as well.

Bibliography

Bellanti, J. A. (1985). "Immunology III." W. B. Saunders, Philadelphia.

Friedman, H., Escobar, M. R., and Reichard, S. M. (gen. ed.) (1980–1988). "The Reticuloendothelial System—A Comprehensive Treatise," Vols. 1–10. Plenum Publishing Corporation, New York.

Jawetz, E., Melnick, J. L., and Adelberg, E. A. (ed.) (1987). "Review of Medical Microbiology," 17th ed. Appleton & Lange, Los Altos.

Roit, I. M., Brostoff, J., and Male, D. K. (1989). "Immunology," 2nd ed. C. V. Mosby Company, St. Louis.

Sell, S. (1987). "Immunology, Immunopathology and Immunity," 4th ed. Elsevier Science Publishing Company, New York.

Stites, D. P., Stobo, J. D., and Wells, J. V. (1987). "Basic & Clinical Immunology," 6th ed. Appleton & Lange, Los Altos.

Szentivanyi, A., and Friedman, H. (ed.) (1986). "Viruses, Immunity, and Immunodeficiency." Plenum Press, New York.

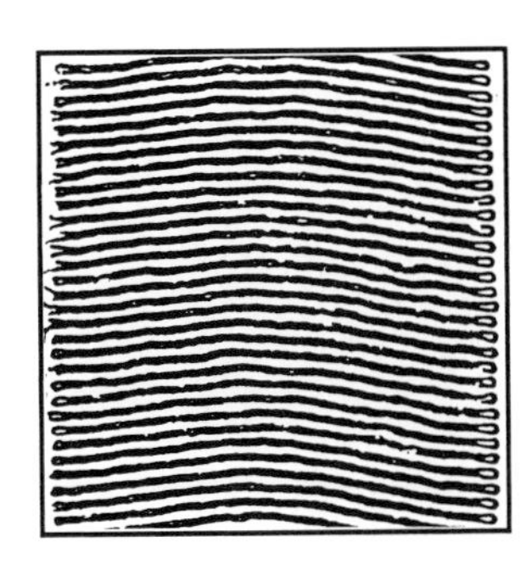

Retina

JOHN E. DOWLING, *Harvard University*

Glossary

Action potentials Transient, all-or-none potentials that usually serve to transmit information along nerve cell axons; also called impulses or spikes

Amacrine cells Axonless neurons whose processes are confined to the inner plexiform layer

Bipolar cells Output neurons of outer plexiform layer that carry information to inner plexiform layer

Fovea Specialized retinal region of highest visual resolution, containing only cones

Ganglion cells Third-order neurons in retina whose axons form the optic nerve and carry visual information from the eye to rest of brain

Graded potentials Local, sustained potentials whose amplitude is graded according to stimulus strength, generated mainly in nerve cell dendrites and sensory receptors

Horizontal cells Neurons whose cell bodies sit along the distal margin of the outer nuclear layer and that extend processes mainly in the outer plexiform layer

Interplexiform cells Neurons whose cell bodies reside in proximal part of the inner nuclear layer and that extend processes in both plexiform layers

Plexiform layers Regions consisting principally of neuronal processes, where synaptic interactions take place

Receptive field Area of retina that when illuminated influences the activity of a cell

Synapses Sites at which neurons make functional contact

Visual pigments Light-sensitive molecules in photoreceptors consisting of 11-*cis* retinal (vitamin A aldehyde) and protein (opsin)

THE RETINA is a thin layer of neural tissue that lines the back of the eye. It is a true part of the brain (central nervous system) displaced into the eye during development. In addition to the light-sensitive photoreceptor cells, the retina contains five basic classes of neurons and one principal type of glial cell, the Müller cell. The neurons are organized into three cellular (nuclear) layers, which are separated by two synaptic (plexiform) layers. Virtually all the junctions (synapses) between the retinal neurons are made in the two synaptic layers, and all visual information passes across at least two synapses, one in the outer plexiform layer and another in the inner plexiform layer, before it leaves the eye.

Processing of visual information occurs in both plexiform layers. The outer plexiform layer separates visual information into on- and off-channels and carries out a *spatial*-type analysis on the visual input. The output neurons of this layer, the on- and off-bipolar cells, demonstrate a center-surround antagonistic receptive field organization. The inner plexiform layer is concerned more with the *temporal* aspects of light stimuli. Many cells receiving input in this layer respond with transient responses and respond better to moving stimuli than to static spots of light. The output neurons of this layer, the ganglion cells, reflect either the processing of information in the outer plexiform layer (i.e., the cells respond in a sustained fashion to appropriately positioned stimuli) or the inner plexiform layer (i.e., the cells respond better to moving stimuli than to static ones).

Figure 1 is a light micrograph of a piece of human retina. The photoreceptors are located farthest from

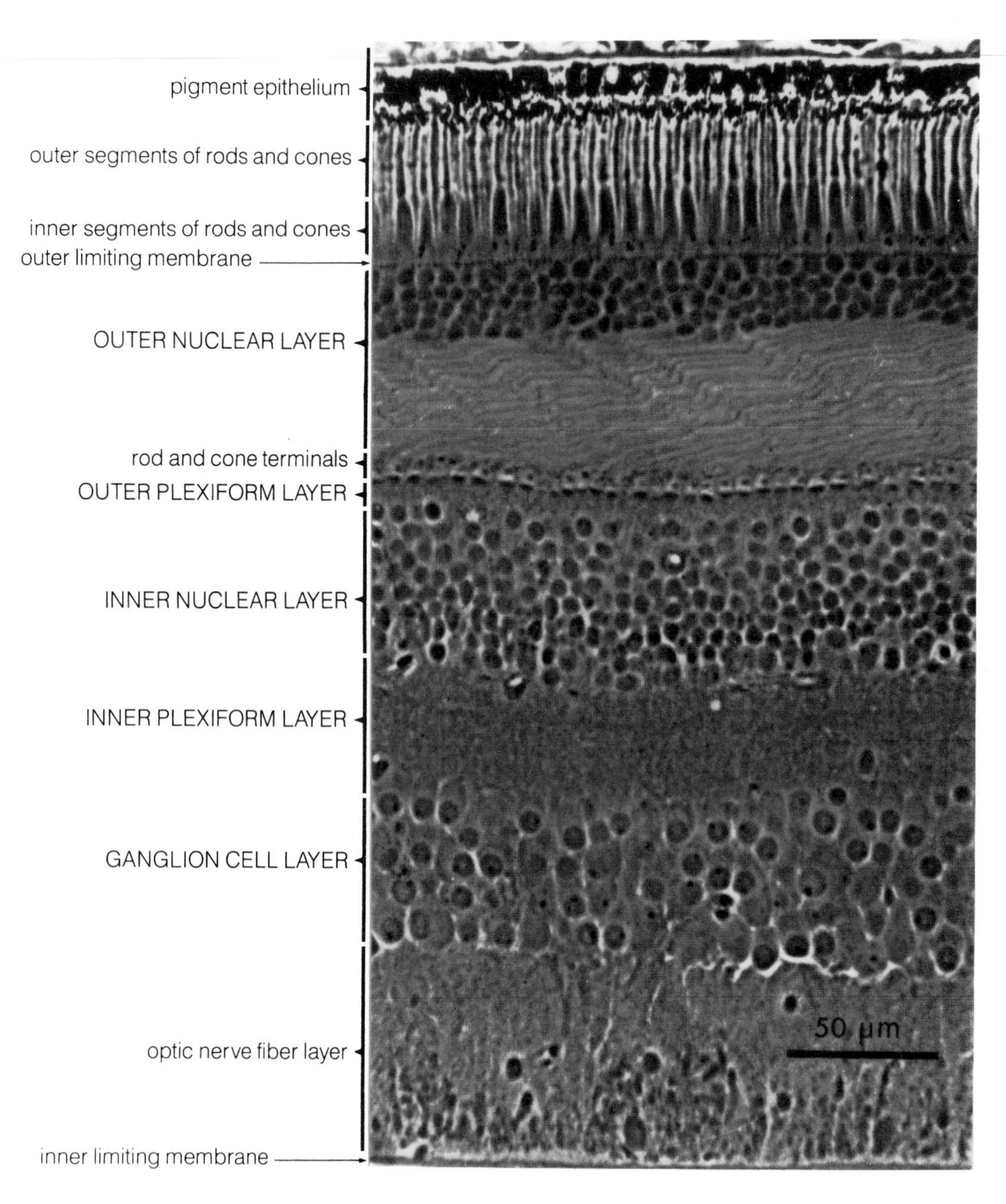

the front of the eye, at the top of the micrograph. Light, entering the eye, passes through the transparent retina and is captured by the pigment-containing outer segments of the photoreceptors. (Overlying the photoreceptors is the pigment epithelium, which serves to absorb stray light and to prevent backscatter of light into the retina.) The cell bodies of the photoreceptors are located in the outer nuclear layer, whereas the cell bodies of four of the basic classes of retinal neurons—horizontal, bipolar, amacrine, and interplexiform cells—are in the inner nuclear layer. The cell bodies of the ganglion cells make up the most proximal cellular layer. The outer and inner plexiform layers are interspersed, respectively, between the outer and inner nuclear layers and the inner nuclear and ganglion cell layers.

FIGURE 1 Vertical section through human retina. Micrograph shows an area about 1.25 mm from center of the fovea. In the foveal region of the retina, inner layers of the retina are pushed aside so that light can impinge more directly on receptors. Thus, around the fovea and for some distance away (as shown here), receptor terminals are displaced laterally from the rest of the photoreceptor cell.

In many primates, including humans, a small region of the retina is specialized for high-acuity vision. It is called the fovea and is centrally located (i.e., on the visual axis of the eye). The layers of the retina below the photoreceptor inner and outer segments are displaced aside from the fovea so that light can impinge directly on the photoreceptors. Only cones are present in this area, and the foveal cones are the thinnest and longest photoreceptors in the retina. The rod-free area is about 0.3 mm in diameter and contains approximately 10,000 cones. No blood vessels are found in the fovea, and furthermore, there are few blue-absorbing cones in the center of the fovea. These specializations serve to improve the visual resolution of the fovea. [*See* EYE, ANATOMY.]

I. Photoreceptors

Vertebrate retinas typically contain two types of photoreceptors, rods and cones, differentiated on the basis of their outer segment shape (Fig. 1). This criterion is not always reliable; in primates, for example, the outer segments of the cones found in the fovea show no significant taper (i.e., they are rod-shaped). Rods mediate dim-light vision, whereas cones function in bright light and are responsible for color vision.

A. Visual Pigments

Light sensitivity of the photoreceptors results from the presence of visual pigment molecules contained within their outer segments. One rod pigment, called *rhodopsin*, and three cone pigments are in the primate retina. Rhodopsin absorbs light maximally in the blue-green region of the spectrum (500 nm), whereas the primate cone visual pigments absorb maximally in the blue (440 nm), green (535 nm), and red-yellow (565 nm) regions of the spectrum. The cone pigments are segregated into separate classes of cones; thus blue, green, and red-yellow sensitive cones are in the primate retina. Color-blind individuals are missing or produce an altered visual pigment. Red-blind individuals (protanopes) are missing or possess an altered red-yellow pigment; green-blind individuals (deuteranopes) are missing or have altered the green pigment; blue-blind patients (tritanopes) are missing or have altered the blue visual pigment. In the primate eye, there are more red and green cones than blue cones, and as noted above, blue cones are extremely rare in the high acuity foveal region of the retina. [*See* COLOR VISION.]

The visual pigment molecules are concentrated to a high degree in the outer segments of the photoreceptors. The outer segments contain numerous transverse membranous discs (Fig. 2), and virtually all the visual pigment molecules are contained within the disc membranes. A typical outer segment may have as many as 2,000 transverse discs and may contain 10^9 visual pigment molecules.

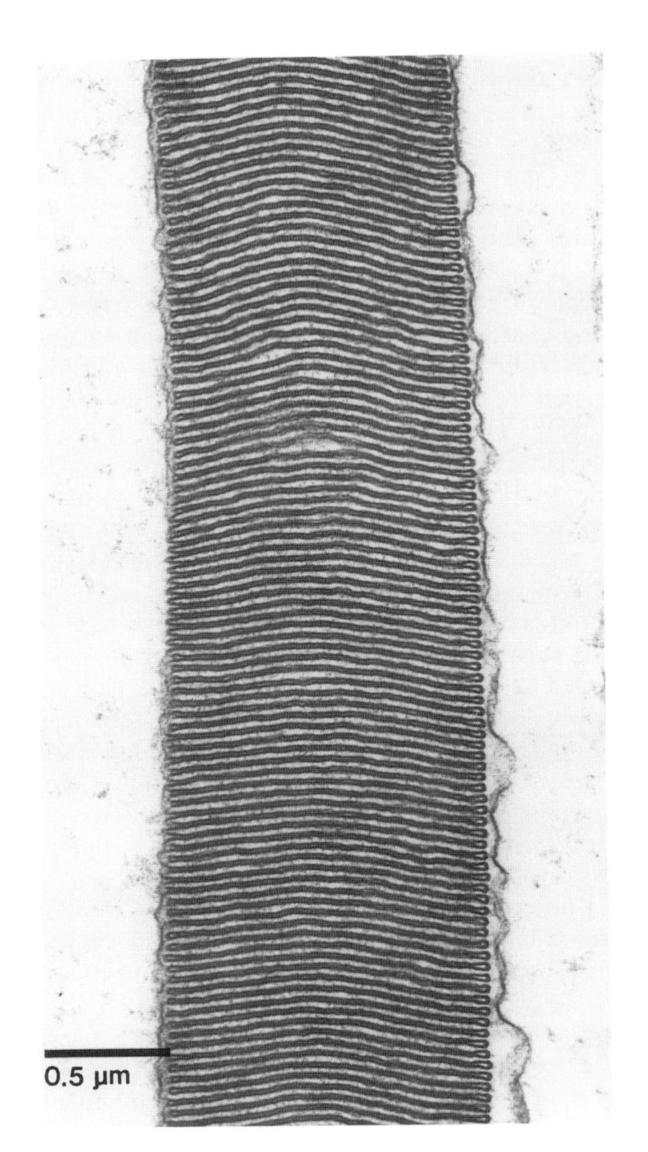

FIGURE 2 Electron micrograph of a portion of a cone outer segment. Contained within the structure are numerous transverse membranous discs.

1. Visual Pigment Chemistry

All visual pigments have a similar chemistry. They consist of two components, retinal (vitamin A aldehyde), termed a *chromophore*, bound to a protein called opsin. Different visual pigments have different opsins, and this accounts for the variations in their color sensitivity. The light sensitivity of the visual pigments is due to the retinal chromophore. When a visual pigment molecule absorbs a quantum of light, several molecular transformations occur, first in the chromophore and then in the protein (opsin) part of the molecule. These transformations lead to the excitation of the photoreceptor cell and also to the separation of the retinal chromophore from opsin. This latter process is called *bleaching* because it results in the loss of color of the visual pigment molecules and their ability to absorb visible light.

Retinal can exist in different shapes (i.e., several *cis-trans* isomers of the molecule are possible). All visual pigments require one particular isomer, the 11-*cis*, for their synthesis, and this form of the chromophore combines spontaneously with opsin to form visual pigment. When a visual pigment molecule absorbs a quantum of light, the first transformation is the isomerization of the chromophore from the 11-*cis* to the all-*trans* form. Indeed, this is the only action of light in the visual process to change the shape of the chromophore. The chromophore-shape change initiates a series of conformational changes in the opsin, and this leads to excitation of the photoreceptor cell and release of the chromophore from opsin.

A series of intermediates have been identified between the absorption of light by visual pigment molecules and the release of the chromophore from opsin. One of the intermediates, metarhodopsin II, appears responsible for excitation of the photoreceptor cell (i.e., it is the photoactive intermediate). Figure 3 shows a simplified scheme of the visual cycle for the rod visual pigment rhodopsin.

The retinal chromophore of rhodopsin derives from vitamin A, and it represents a slightly oxidized form of the vitamin. During the operation of the visual cycle, some retinal is lost and must be replaced from body stores of the vitamin. In vitamin A deficiency, this replenishment fails, a full complement of visual pigment can no longer be synthesized in the photoreceptors, and light sensitivity of the rods and cones is decreased. The loss of light sensitivity is more obvious in the dark, and hence the condition is known as night blindness. Refeeding of vitamin A to a vitamin A–deficient animal or patient usually restores visual sensitivity. [*See* VITAMIN A.]

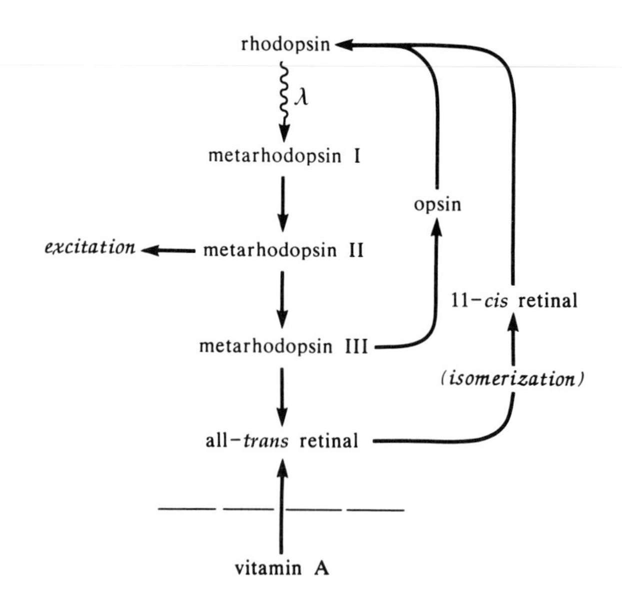

FIGURE 3 Scheme of the sequence of events that occurs after absorption of a quantum of light by the rod visual pigment, rhodopsin. Light initiates conversion of rhodopsin through a series of metarhodopsin intermediates. Metarhodopsin II is the active intermediate leading to excitation of the photoreceptor cell. Eventually, the chromophore of rhodopsin, retinal, separates from the protein opsin. For resynthesis of rhodopsin, the shape of retinal must be changed (isomerized), from all-*trans* to the 11-*cis* form. Retinal (vitamin A aldehyde) is replenished in the eye by vitamin A coming from the blood.

B. Photoreceptor Responses

Excitation of visual pigment molecules leads ultimately to a change of potential across the membrane surrounding the photoreceptor cell. Membrane potential controls the release of neurotransmitter molecules from synapses, and in this way light influences the exchange of information between photoreceptors and second-order neurons in the retina.

All vertebrate photoreceptors hyperpolarize in response to light (i.e., the membrane potential becomes more negative). Why photoreceptors hyperpolarize in the light is as follows: In darkness, the membrane of the outer segment is leaky to Na^+. Because Na^+ levels are higher outside the cell than inside, positive Na^+ ions enter the cell in darkness causing the cell to be partially depolarized (i.e., the membrane potential is more positive than is typically the case for neurons at rest). Light decreases the conductance (leakiness) of the outer segment membrane to Na^+, thereby decreasing the flow of positive ions into the cell and causing the cell to become more negative (i.e., to hyperpolarize).

The conductance of the outer segment membrane to Na^+ is controlled by a second-messenger molecule, cyclic GMP, which maintains channels in the membrane in an open state. In the light, levels of cyclic GMP fall because an enzyme (called *phosphodiesterase*), which breaks down cyclic GMP, is activated, and this causes the channels in the outer segment membrane to close. Phosphodiesterase in turn is activated by a protein called *transducin* (a so-called G protein), and transducin is activated by metarhodopsin II, the photoactive visual pigment intermediate. The cascade between the visual pigment molecule and the Na^+ channel in the outer segment membrane is shown in Fig. 4.

One reason for the cascade between the light activation of visual pigment molecules and the closing of Na^+ channels in the outer segment membrane is for amplification of the signal. That is, one photoactive visual pigment molecule can interact with many transducin molecules (as many as 500), and one phosphodiesterase molecule can break down about 2,000 cyclic GMP molecules per second. The cascade of reactions between photon absorption and cyclic GMP inactivation can result in an amplification of about 10^6.

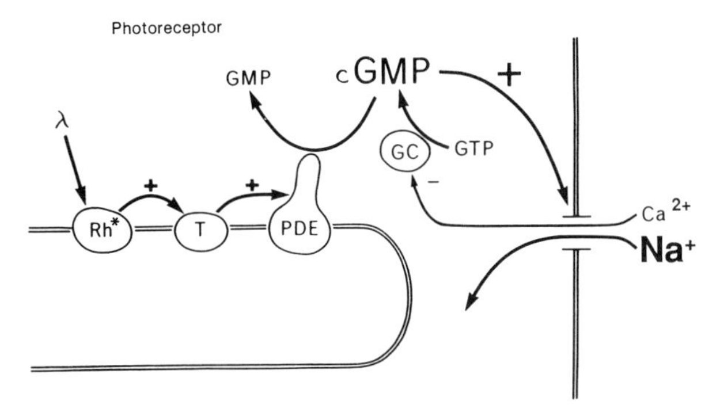

FIGURE 4 Summary diagram of interactions occurring in the rod outer segment during phototransduction. Light-activated rhodopsin (Rh*) activates transducin (T), which in turn activates the enzyme phosphodiesterase (PDE). These interactions occur in the disc membrane. Activation of PDE leads to breakdown of cyclic GMP (cGMP) to an inactive product (GMP). Cyclic GMP maintains channels in the outer segment membrane in an open configuration, thereby allowing both Na^+ and Ca^{2+} to enter the cell in the dark. With a fall in cyclic GMP levels in the light, channels in the outer segment membrane close. The resulting fall of Na^+ levels causes the cell to hyperpolarize. Decrease of Ca^{2+} levels enhances guanylate cyclase (GC) activity, an action that counters the effects of light and increases cyclic GMP levels in the outer segment.

The channels in the outer segment membrane controlled by cyclic GMP also allow some Ca^{2+} to enter the photoreceptor cell, and Ca^{2+} appears to play an important regulatory role in the phototransduction process. Ca^{2+} strongly inhibits an enzyme, guanylate cyclase, that promotes the synthesis of cyclic GMP. In the dark, the Ca^{2+} entering the cell inhibits this enzyme, and cyclic GMP synthesis is low. In the light, when the channels in the outer segment membrane are closed, Ca^{2+} entry into the cell decreases and intracellular Ca^{2+} levels fall. This results in an increased synthesis of cyclic GMP, which serves to counter the effect of light of lowering cyclic GMP levels. Thus, in continuous light, the photoreceptor response recovers partially to dark levels, enabling the photoreceptor to continue to respond even in bright light. This process is termed *adaptation*, and photoreceptor light and dark adaptation plays an important role in the ability of the visual system to respond over a wide range of ambient illumination.

II. Cellular and Synaptic Organization

A. Cellular Organization

Most of what is known about the classes of retinal cells has come from light microscopic studies of retinas processed by the silver-staining method of Golgi. This technique enables investigators to see the extent and distribution of the processes of the cells within the retina, to classify the cells, and to construct schemes of the cellular organization of the retina such as that shown in Fig. 5 for the primate retina. Although there are just five major classes of neurons in the retina, there are many morphological types and subtypes in each major cell class, and even today the total number of morphological types is not known. It is clear that some types are much more common than others, and the focus here will be on the major types of cells in the mammalian and primate retinas.

In the outer plexiform layer, two cell classes—horizontal and bipolar cells—receive input from the photoreceptors. The bipolar cells are the output neurons for the outer plexiform layer; all information passes from outer to inner plexiform layers via these neurons. Horizontal cells, however, extend processes widely in the outer plexiform layer, but

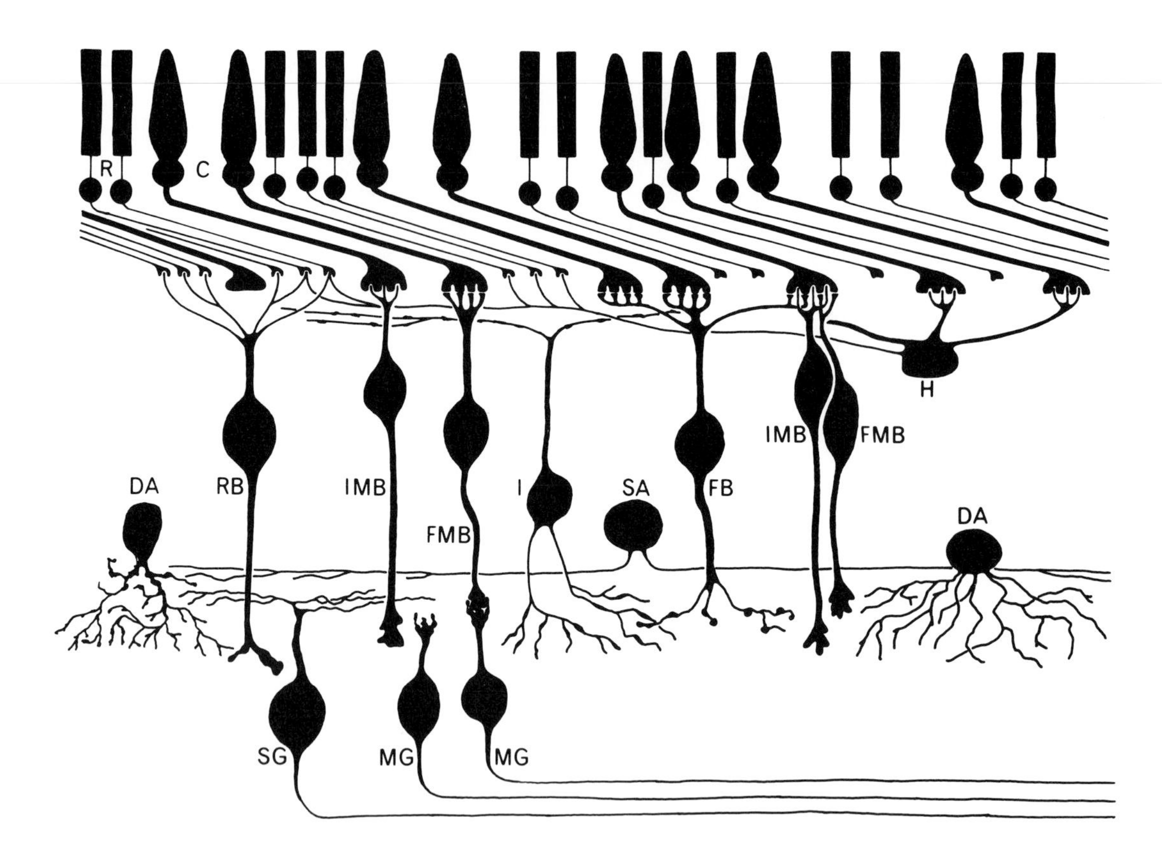

FIGURE 5 Major cell types found in primate retina as viewed in vertical sections of Golgi-stained retinas. See text for description of cells: R, rods; C, cones; RB, rod bipolar cell; IMB; invaginating midget bipolar cell; FMB, flat midget bipolar cell; FB, flat bipolar cell; H, horizontal cell; I, interplexiform cell; DA, diffuse amacrine cell; SA, stratified amacrine cell; SG, stratified ganglion cell; MG, midget ganglion cell.

their processes are confined to this layer. Their role is to mediate lateral interactions within this first synaptic zone.

Bipolar cell terminals provide the input to the inner plexiform layer, and three classes of cells—amacrine, interplexiform, and ganglion—are ultimately activated. Amacrine cells, like horizontal cells in the outer plexiform layer, spread processes widely in the inner plexiform layer, but their processes are confined to this layer. Interplexiform cells, however, extend processes in both plexiform layers. Ganglion cells are the output neurons for the retina; their axons run along the margin of the retina, collect at the optic disc to form the optic nerve, and carry all the visual information to higher visual centers.

1. Outer Plexiform Layer Neurons—Horizontal and Bipolar Cells

Most retinas contain two basic types of horizontal cells: a cell with a short axon that runs 400 μm or so before ending in a prominent terminal expansion and an axonless cell. The axonless cell has not been seen in the primate retina.

Horizontal cells of the cat have been studied in particular detail, and Fig. 6 shows examples observed by looking down on a flat mount of the whole retina. Electron microscopy has shown that the processes of the axonless cell (Fig. 6b) and the proximal dendritic processes of the short axon cell (Fig. 6a) connect exclusively with cones, whereas the axon terminal processes of the short axon cell end exclusively in the rod terminals. Why the input to the short axon cells is organized this way is not clear; an appealing suggestion is that this allows for segregation of rod and cone responses in different regions of the cells. In fishes, for example, there are separate rod and cone horizontal cells.

In primates, five principal types of bipolar cells are distinguished: one type is exclusively connected to rods and four types are exclusive to cones. The rod-related bipolars extend their dendrites into the rod synaptic terminals, and their axon terminals end deep in the inner plexiform layer. The rod bipolars contact as many as 30–50 rod terminals.

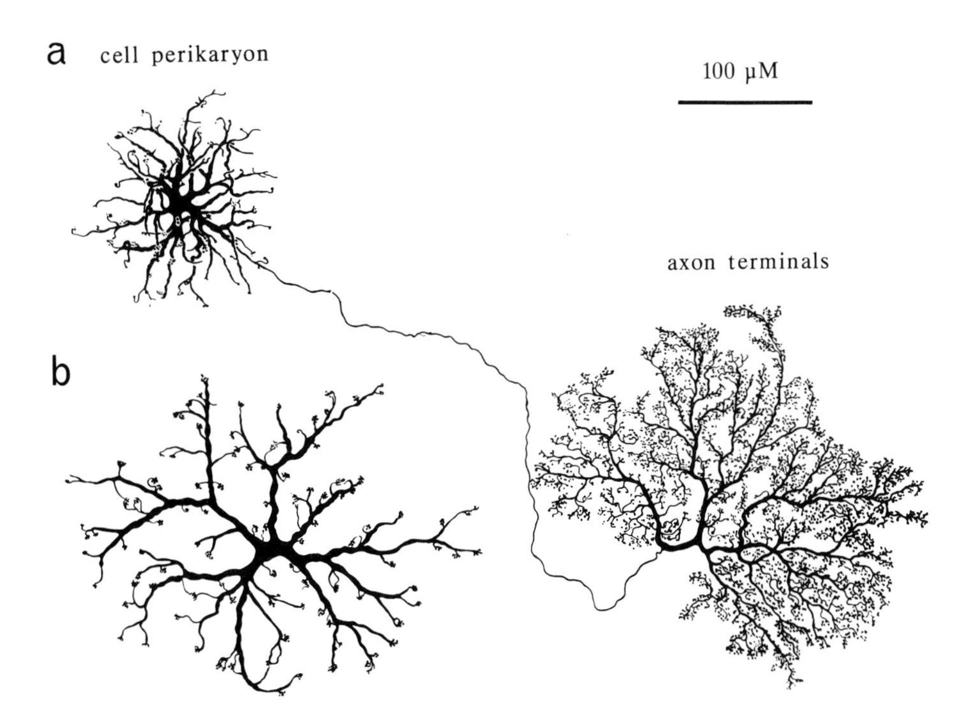

FIGURE 6 Drawings of Golgi-stained horizontal cells from the cat, viewed by looking down on a flat mount of retina. There are two types of horizontal cells in cat retina: an axonless cell (b) and a cell with a short axon (a).

Two of the cone-related bipolar cell types contact only a single cone terminal. These cells, called *midget bipolar cells*, make different kinds of synaptic contacts with the cone terminals—invaginating and flat contacts (see Section II,b). Their axon terminals end at different levels within the inner plexiform layer, and they appear to be related to the generation of either on- or off-responses to light in the retina. Every cone terminal in the primate retina probably makes connections with both kinds of midget bipolar cells (Fig. 5). In addition to the cone midget bipolar cells, there are also bipolar cells that contact several cones, probably as many as six or seven. They are called *flat bipolars* or *diffuse invaginating bipolars* based on the type of connection they make with the photoreceptors (see below).

2. Inner Plexiform Layer Neurons—Amacrine, Interplexiform, and Ganglion Cells

Amacrine cells have no axonal processes; all their processes usually look similar. Amacrine cells are diverse in terms of the extent and distribution of their processes, and it is possible to describe a large number of amacrine cell types in most species. Investigators typically classify amacrine cells into two major types: diffuse and stratified amacrine cells. Diffuse amacrine cells extend their processes throughout the thickness of the inner plexiform layer, whereas the stratified cells extend their processes on one or a few levels in the layer. This simple classification scheme can be expanded to include narrow- and wide-field diffuse or stratified amacrine cells, depending on how far their processes extend, and mono-, bi-, or multistratified cells, depending on whether their processes are confined to one, two, or several levels in the inner plexiform layers.

Interplexiform cells have been recognized as a separate class of retinal neuron only recently. Their perikarya (cell bodies) sit among the amacrine cells, but they send processes to both plexiform layers. Input to these cells is in the inner plexiform layer, whereas most of their output is in the outer plexiform layer. Hence they appear to be mainly a centrifugal type of neuron, carrying information from inner to outer plexiform layers.

Ganglion cells, like amacrine cells, are also diverse in their morphology. They too are classified into diffuse and stratified cells. In primates, particularly in the central region of the retina, ganglion cells are observed with limited dendritic fields. These cells, termed *midget ganglion cells*, receive input from one midget bipolar cell. The dendrites of these midget ganglion cells ramify either in the upper or lower parts of the inner plexiform layer, so that they receive input from one or the other of the two midget bipolar cells described above. This implies that one cone in the central (foveal) part of primate retina can send two separate messages to the rest of the brain via two midget ganglion cells. One cell is believed to signal increases in illumination of the cone (i.e., it is an on-cell); the other signals decreases in cone illumination (i.e., it is an off-cell).

3. Retinal Processing of Visual Information—Ganglion Cell Responses and Receptive Field Organization

As noted above, ganglion cells signal brightness and darkness information to the rest of the brain, but they also communicate much more than that. Indeed, two basic kinds of processing appear to be occurring in the retina; one carried out mainly in the outer plexiform layer and the other in the inner plexiform layer. Ganglion cells convey information to the rest of the brain that reflects these two stages of processing.

Ganglion cells typically respond to illumination of a restricted but relatively large region of the retina. This region is called the *receptive field* of the cell and is typically about 1 mm in diameter. The most common ganglion cell in the mammalian retina shows evidence of spatial processing of visual information in its responses. Ganglion cells of this type are often called *contrast-sensitive cells*, and they are subdivided into two mirror image classes; on-center, off-surround cells and off-center, on-surround cells. Each has a receptive field that is organized into two concentric zones that are antagonistic to each other (Fig. 7a).

On-center cells respond to an increase in the illumination of the receptive field center with a sustained burst of nerve impulses, whereas illumination of the surround inhibits the firing of nerve impulses by the cell for as long as the light is on. Off-center cells respond in the opposite way; illumination of the receptive field center inhibits the cell, whereas surround illumination provides a sustained excitation of the cell. In both cases the center and surround zones are antagonistic; if center and surround areas are simultaneously illuminated, the cell responds only with a weak response that usually reflects the central response. These ganglion cells mainly reflect the processing that occurs in the outer plexiform layer of the retina.

Other types of ganglion cells respond with more transient responses to retinal illumination, regard-

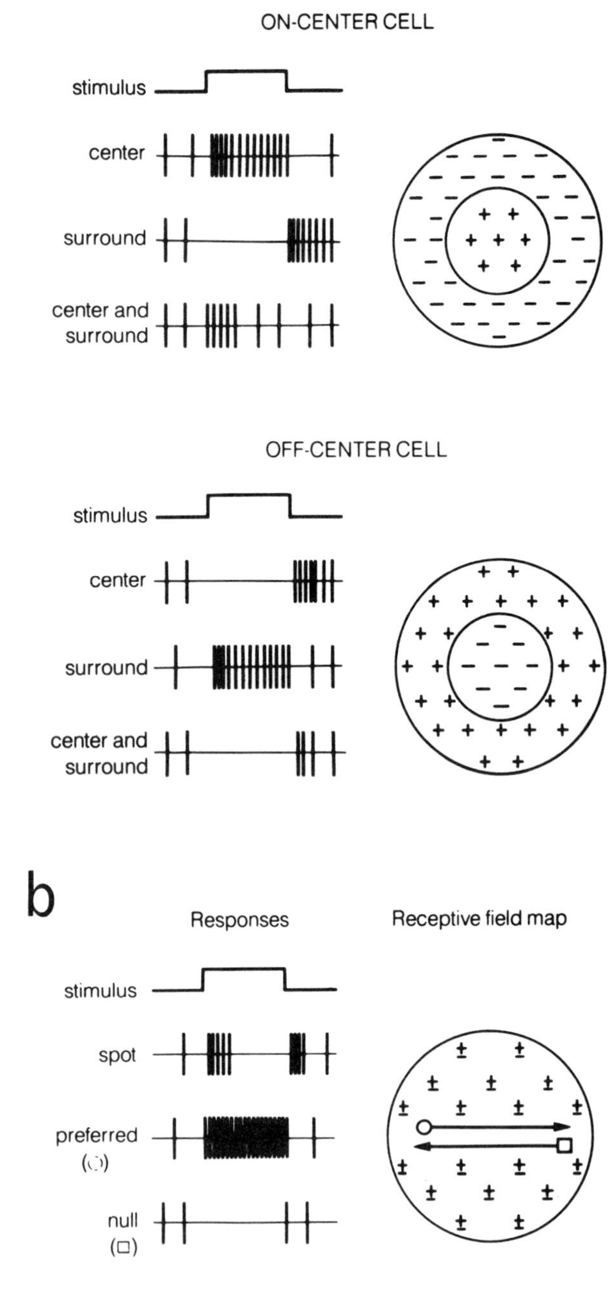

FIGURE 7 (a) Idealized responses and receptive field maps for on-center (*top*) and off-center (*bottom*) contrast-sensitive ganglion cells. Drawings on the *left* represent hypothetical responses to a spot of light presented in the center of the receptive field, in the surround of the receptive field, or in both center and surround regions of the receptive field. A + symbol on the receptive field map indicates an increase in firing rate of the cell (i.e., excitation); a − symbol indicates a decrease in firing rate (i.e., inhibition). (b) Idealized responses and a receptive field map for a direction-sensitive ganglion cell. Such cells respond with a burst of impulses at both onset and termination of a spot of light

less of the position of the illuminating spot in the receptive field, and they reflect more the responses of inner plexiform layer neurons. These cells typically respond with more vigorous responses to moving stimuli than to static spots of light. The receptive fields of some of these cells are organized into antagonistic center and surround regions like the sustained contrast–sensitive cells described above, but others give short on–off bursts of impulses to spots of light positioned anywhere in the receptive field. Some of the latter cells (particularly in nonmammalian species) show direction-selective properties (Fig. 7b). Movement of a light spot in one direction vigorously excites the cell whereas movement in the opposite direction inhibits the cell.

As noted above, midget ganglion cells receive input from a single midget bipolar cell, which in turn receives input from a single cone. The center of the receptive field corresponds to this direct pathway and is therefore cone-specific. The antagonistic surround response, however, probably reflects mainly horizontal cell activity, and the horizontal cells receive input from many cones. The center of the receptive field, therefore, is small, and it has spectral sensitivity different from the surround mechanism. Such ganglion cells are termed *color-opponent cells*.

B. Synaptic Organization

Two types of chemical synaptic contacts are observed in both plexiform layers. One of these is similar in morphology to known chemical synapses seen throughout the brain, and it is termed a *conventional synapse*. It is characterized by an aggregation of synaptic vesicles in the presynaptic terminal clustered close to the membrane. In the retina, conventional synapses are made by horizontal, amacrine, and interplexiform cells. The other type of synapse is characterized by an electron-dense ribbon or bar in the presynaptic process and is called a *ribbon synapse*. Photoreceptor and bipolar cells make ribbon synapses in the retina. Electrical (gap) junctions are also observed in both plexiform layers of the retina, and they are believed to mediate direct electrical interactions between certain retinal neurons.

Figure 8a shows a drawing of a bipolar ribbon synapse and an amacrine cell conventional synapse in the inner plexiform layer. Typically, there are two postsynaptic processes at the ribbon synapses of bipolar cells, whereas at conventional synapses, only one postsynaptic process is found. In this drawing, the amacrine cell synapse is made back onto the bipolar terminal that is making a synapse on it. Thus, a reciprocal or feedback synaptic arrangement is suggested, which is commonly seen between bipolar cell terminals and amacrine cell processes. Amacrine cell synapses are also made on the processes and cell bodies of ganglion, interplexiform, and other amacrine cells.

Photoreceptor cell synapses are particularly complex, and a summary drawing of a part of a cone synaptic terminal is shown in Fig. 8b. The synaptic ribbons are found above invaginations of the basal surface of the terminal. Processes from horizontal cells and invaginating midget and invaginating diffuse bipolar cells penetrate into the terminal invaginations. Horizontal cell processes lie lateral to the synaptic ribbons, whereas the dendrites of the invaginating bipolar cells are centrally positioned. In addition to the ribbon synapse, the cone photoreceptor terminals make a second, unusual synaptic contact with the flat bipolar cells, which is called a *flat or basal junction*. No ribbon or vesicle cluster is associated with this junction, but some specializations of the membranes on both sides of the synapse are seen. The flat midget bipolar processes are typically found immediately adjacent to the invaginating midget bipolar processes, whereas the processes of the other flat bipolar cells are positioned away from the invaginations. Rod photoreceptor terminals do not make basal junctions; all rod bipolar dendrites penetrate into invaginations of the rod terminal.

Figure 9 is a simplified summary diagram of the synaptic organization of the primate retina. Each cone in the primate makes connections with two midget bipolar cells, one that makes invaginating-type junctions and a second one that makes flat junctions (right side of figure). In the central region of the primate retina, these two bipolar cell types synapse on separate midget ganglion cells. In the periphery of the primate retina, midget bipolar cells

presented anywhere in the cell's receptive field. This response is indicated by ± symbols all over the map. Movement of a spot of light through the receptive field in the preferred direction (*open circles*) elicits firing from the cell that lasts for as long as the spot is within the field. Movement of a spot of light in the opposite (null) direction (*open squares*) causes inhibition of the cell's maintained activity for as long as the spot is within the receptive field.

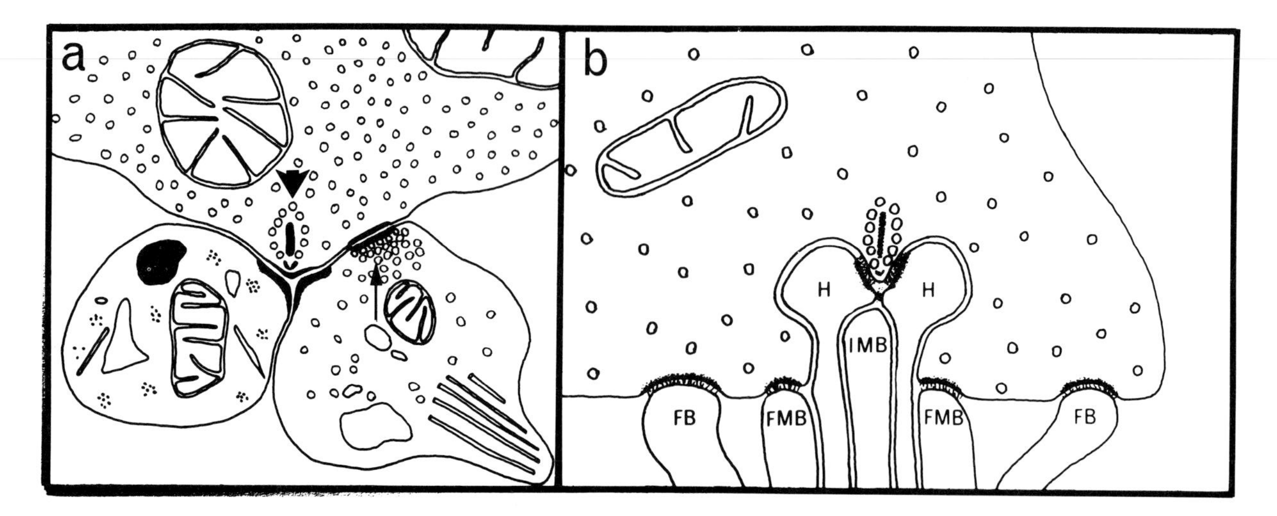

FIGURE 8 (a) Schematic drawing of a bipolar cell ribbon synapse (*arrowhead*) and a conventional amacrine cell synapse (*arrow*) back onto the bipolar cell terminal. One postsynaptic process at the ribbon synapse is an amacrine cell process (right); the other is a ganglion cell dendrite (left), a typical arrangement at cone bipolar cell terminals in primate. (b) Schematic drawing of synapses made by cone terminals in primates. See text. H, horizontal cell process; FB, flat bipolar cell dendrite; FMB, flat midget bipolar cell dendrite; IMB, invaginating bipolar cell dendrite.

synapse on ganglion cells that receive input from a few to many midget bipolar terminals.

All cone terminals in primates also make synapses with diffuse cone bipolars and the proximal (dendritic) processes of horizontal cells. The axonal processes of the horizontal cells extend to the rod terminals. Synapses made by the horizontal cells have been observed on bipolar cells in many species, but very rarely back onto the photoreceptor terminals. (There is, however, good physiological evidence for synapses from horizontal cells onto photoreceptors in many species.) Another interesting feature of the cone photoreceptor terminals is that they make junctions with each other and with adjacent rod terminals. Evidence in many species, including primate, suggests that these junctions are small electrical synapses.

All bipolar terminals make synapses onto amacrine cell processes, and these processes may make a reciprocal synapse back onto the terminal, as in Figure 8a. Amacrine cell processes also make junctions on ganglion cells, interplexiform cells, and other amacrine cells.

Rod bipolar terminals in the mammalian retina do not contact ganglion cells directly (see left side of Fig. 9). Rather amacrine cell processes are always postsynaptic at the ribbon synapses of the rod bipolar terminals. One of these makes a feedback synapse onto the terminal, whereas the other belongs to a special amacrine cell that makes both gap junctional (electrical) and conventional (chemical) synapses with cone bipolar terminals. These cone bipolar terminals then contact the ganglion cells. Thus all rod information in the mammalian retina passes through an amacrine cell before it is transmitted to the ganglion cells. Why this is so is not clear; it has been suggested that the amacrine cell serves to amplify the rod signal (see discussion of amacrine cell responses in Section III,d).

Finally, interplexiform cells receive their input from amacrine cells, and they make some synapses in the inner plexiform layer on amacrine and ganglion cells processes. Most of their synapses, however, are made in the outer plexiform layer on bipolar and horizontal cells.

It should be noted that Fig. 9 is highly simplified. It is unlikely that any one amacrine cell makes the variety of contacts shown for either of the amacrine cells drawn in the figure. For example, amacrine cells involved in the rod pathway make gap junctions only with diffuse invaginating bipolar terminals.

III. Neuronal Responses

In many nonmammalian species, intracellular recordings can be made routinely from most of the retinal cells. In mammalian retinas, intracellular recordings are much more difficult to make, and virtually none have been reported from the primate retina. The following discussion is based mainly on

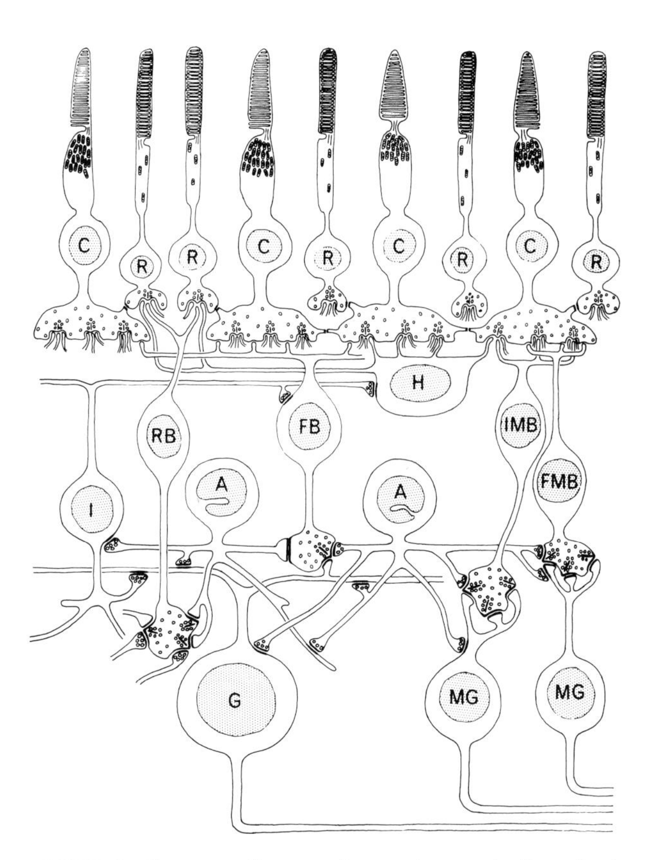

FIGURE 9 Summary diagram of synaptic organization of primate retina. See text for details. R, rod; C, cone; H, horizontal cell; RB, rod bipolar; FB, flat bipolar; IMB, invaginating midget bipolar; FMB, flat midget bipolar; I, interplexiform cell; A, amacrine cell; G, ganglion cell; MG, midget ganglion cell.

recordings from nonmammalian species, but the recordings made so far from mammalian retinal neurons, especially from rabbit cells, are similar.

The distal retinal neurons—receptors, horizontal cells, and bipolar cells—respond to light with sustained, graded membrane potential changes (Fig. 10). Unlike many neurons found elsewhere in the brain, they do not generate action potentials. This may be the case because these neurons have relatively short processes, and they do not need to transmit information over long distances; in other words, electrotonic spread of potential along the cell membrane is sufficient to transmit information from one end of the cell to the other. A second reason that the distal retinal cells may function with graded potentials is that such potentials are capable of discriminating a wider range of signals than can all-or-none events (i.e., action potentials).

Another unusual feature of the distal retinal neurons is that most of them respond to light with hyperpolarizing potentials; on illumination, the cell's membrane potential becomes more negative. Elsewhere in the nervous system hyperpolarizing potentials are usually associated with inhibition because they prevent neurons from generating action potentials. In the distal retina, the neurons do not generate action potentials; thus hyperpolarizing potentials can reflect excitation, and the photoreceptors are a good example of this.

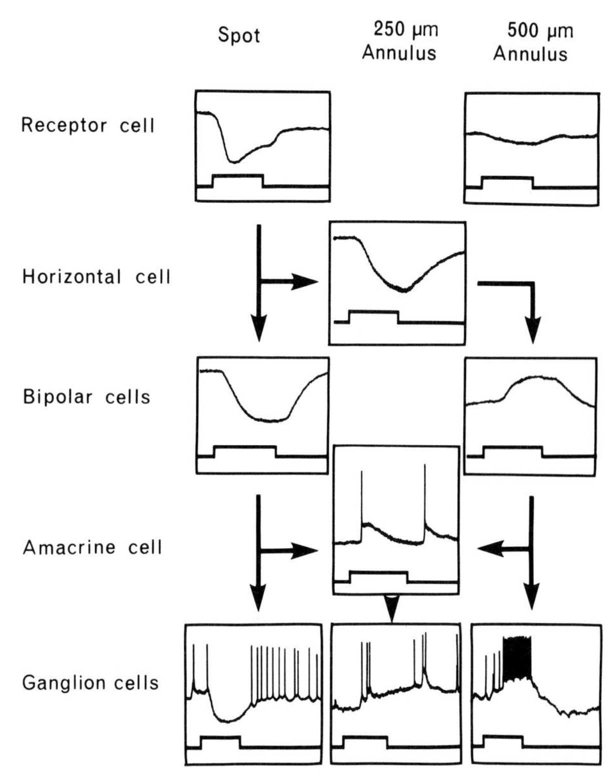

FIGURE 10 Intracellular responses from receptor, horizontal, bipolar, amacrine, and ganglion cells of mudpuppy retina. Distal retinal neurons (receptor, horizontal, and bipolar cells) respond to illumination with sustained graded potentials; proximal retinal neurons show both sustained and transient potentials and action potentials. Receptor, bipolar, and ganglion cells respond differently to center (*spot*) and (*annular*) surround illumination. Horizontal and amacrine cells usually respond similarly to spot and annular illumination; here responses to a small annulus (250 μm) are shown that stimulate both the center and surround of the receptive field. The bipolar cell illustrated is a center-hyperpolarizing cell, the amacrine cell shown is a transient amacrine cell, and the ganglion cell is an off-center cell. *Arrows* indicate in a general way how the responses are synaptically generated. That is, receptor cells directly drive horizontal and bipolar cells. Bipolar cell responses evoked with a large annulus (*right side*) are generated by horizontal cell activity. Bipolar cells provide the major input for amacrine cell responses and responses of the off- and on-center ganglion cells. The transient amacrine cells provide the major input for the on–off ganglion cells (center record).

A. Photoreceptors

Both rod and cone photoreceptors only hyperpolarize in response to illumination, but rods are more sensitive to light than are cones by about 100 times in the primate and other species. Why photoreceptors hyperpolarize in response to light and why in darkness the membrane potential of the cell is partially depolarized was discussed earlier. A typical resting value for the membrane potential of photoreceptors in the dark is −30 mV, and in response to bright illumination, the membrane potential goes to −60 mV, a typical resting potential for most neurons at rest. Thus, in the distal retina, photoreceptors and the other neurons behave as though darkness is the stimulus and light turns them off (i.e., it hyperpolarizes them). Why the system works this way is not known.

As noted earlier, there are electrical synapses between photoreceptors, but their role is not well understood. They may relate to the functioning of the photoreceptors rather than as a pathway for information flow. It has been proposed that electrical coupling between photoreceptors can reduce membrane noise, hence improve signal detection, and also that it may increase the amplification of signals transmitted from the photoreceptors at their synapses. Whatever the function of the electrical coupling between photoreceptors, it does increase the receptive field size of the cell somewhat. That is, the receptors respond to illumination over a wider area of the retina than that of a single receptor. Nevertheless, photoreceptors typically have the smallest receptive fields of any of the retinal neurons.

B. Horizontal Cells

Horizontal cells, like photoreceptors, have relatively low resting membrane potentials in the dark (−30 mV), and most often they only hyperpolarize in response to light (Fig. 10). Some horizontal cells respond with small depolarizing responses to certain wavelengths of light and with hyperpolarizing responses to other wavelengths. These cells appear to be involved in color processing and are termed *chromaticity (C) cells*. Horizontal cells that only hyperpolarize to illumination are called *luminosity (L) cells*.

The receptive fields of horizontal cells are characteristically large, often several millimeters in diameter. Thus, the receptive field size usually exceeds the dendritic spread of the cells. Many horizontal cells make extensive electrical junctions with each other, and the large receptive fields of these cells can be explained on the basis of the extensive electrical coupling between the cells.

C. Bipolar Cells

Bipolar cells, like receptors and horizontal cells, respond to light with sustained graded potentials (Fig. 10). However, two types of bipolar cells are found in all retinas: those that depolarize in response to central spot illumination and those that hyperpolarize to such stimuli. Furthermore, the bipolar cell receptive field is organized into antagonistic zones such that illumination of the surround antagonizes the response to spot illumination. Thus, bipolar cells show a center-surround receptive field organization, and there are separate on- (center depolarizing) and off- (center hyperpolarizing) bipolar cells in all species so far examined.

D. Amacrine Cells

In most retina two basic types of amacrine cell responses—transient and sustained—are observed. Transient amacrine cells usually give on- and off-depolarizing responses to illumination presented anywhere in their receptive field (Fig. 10), but there are also transient amacrine cells that respond only at the on or off of illumination. Transient amacrine cells typically generate action potentials, and these potentials are observed on the transient on- and off-depolarizations. Usually only one or two action potentials are observed on the transient depolarization, and thus it has been proposed that the action potentials may serve as a local amplifying mechanism for the amacrine cell potentials, rather than as the signal transmitted along the cell, as is the case for many neurons.

Sustained amacrine cell responses resemble horizontal cells responses. They may be either hyperpolarizing or depolarizing in polarity, and the amplitudes of the two types of responses are comparable. Usually sustained amacrine cells, like transient amacrine cells, give similar responses to spot illumination anywhere in their receptive field.

E. Interplexiform Cells

Only a few recordings from interplexiform cells have been reported, and they have all been from the retinas of nonmammalian species. The potentials are sustained with, in some cases, transient components. In other words, they are like amacrine cell responses.

F. Ganglion Cells

The receptive field properties of ganglion cells were discussed above, and it was pointed out that two basic types of ganglion cell responses are recorded in most retinas: sustained and transient responses. Intracellular recordings reveal the underlying potentials that give rise to these two response types. For example, Fig. 10 illustrates a sustained off-center, on-surround ganglion cell. With central spot illumination (left side of the figure), a sustained hyperpolarizing potential was evoked in the cell, and the cell was inhibited from discharging action potentials for the duration of the stimulus. With annular illumination (right side of the figure), a sustained depolarizing potential was produced, which elicited a steady discharge of action potentials from the cell for the duration of the stimulus.

Figure 10 also illustrates an on–off transient ganglion cell response (center response). With small annular illumination, the cell responded with transient depolarizing potentials at the onset and offset of the light. Each depolarization evoked a short burst of action potentials, but the durations of both the depolarizing potentials and action potential discharge were always shorter than the light stimulus. Such cells give similar responses to illumination anywhere in the receptive field.

G. Functional Organization of the Retina

Figure 10 shows in a simplified way how some of the potentials and certain of the receptive fields of the retinal neurons may be produced by the synaptic interactions occurring within the retina. The figure correlates the basic connections of the retinal cells with intracellular responses recorded from an amphibian (mudpuppy) with a spot of light, small annulus, or large annulus.

As noted above, receptors have small receptive fields. They respond well to spots of light centered over the receptor but poorly to surrounding annuli (i.e., to light that does not directly strike the cell). The anatomy of the retina indicates that bipolar and horizontal cells are both activated by the receptors and Fig. 10 shows that both cell types respond with sustained graded potentials that resemble in waveform the photoreceptor response. The figure further shows that the horizontal cells interact with bipolar cells and that this interaction is opposite in terms of sign from the receptor–bipolar interaction. In Fig. 10 the receptor causes the bipolar cell to hyperpolarize, whereas the horizontal cell causes the bipolar cell to depolarize. Horizontal cells have a much larger lateral extent than do bipolar cells; thus, a center-surround receptive field organization is observed in the bipolar cell response. The center response of the bipolar cell (left side of figure) is mediated by direct receptor–bipolar cell interaction, whereas the surround response (right side) reflects input to the bipolar cell from the horizontal cell.

As noted earlier, two types of synapses made by the cone terminals have been observed: invaginating and flat junctions. It appears that in many species the flat synapses result in the center-hyperpolarizing responses in bipolar cells, whereas the invaginating junctions result in the center-depolarizing bipolar cell responses. (Mammalian rod terminals make only invaginating contacts, thus all rod bipolars are center depolarizing.) The surround responses in all bipolar cells are provided by the horizontal cells, and these interactions may be mediated by direct horizontal-to-bipolar cell synapses or by inhibitory feedback synapses onto the photoreceptor terminals. [*See* CELL JUNCTIONS.]

The bipolar cell terminals carry the visual signal from the outer to inner plexiform layers. Depolarizing or on-bipolar cell terminals are found in the inner half of the plexiform layer, whereas hyperpolarizing bipolar cell terminals are in the outer half of the layer. Thus there is a division of the inner plexiform layer into on- and off-strata. On-responses of amacrine and ganglion cells are generated in the lower part of the layer, whereas off-responses are generated in the upper part. Thus, on-center ganglion cells have their processes in the lower or on-strata of the inner plexiform layer, off-center cells in the upper or off-layer, and on–off ganglion cells spread processes in both halves of the layer.

The responses of the two basic types of ganglion cells found in the mudpuppy (and other retinas) appear to be closely related to the responses of the input neurons to the ganglion cells (Fig. 10). The sustained on- or off-center ganglion cells appear to receive most of their synaptic input directly from the bipolar cells; their responses resemble those of the on- or off-bipolar cells, and their receptive fields primarily reflect the processing that occurs in the outer plexiform layer. The on–off transient ganglion cells, however, resemble in their properties the transient amacrine cells, and they appear to receive most of their input from amacrine cells. These cells reflect more the processing that occurs in the inner plexiform layer. As noted earlier, the transient ganglion cells respond better to moving than to static stimuli, and some of these cells show complex receptive field properties such as direction selectivity. Substantial anatomical evidence shows that complex ganglion cell receptive field properties, such as motion and direction sensitivity, are mediated in the inner plexiform layer, predominately as a result of interactions between amacrine cells and their processes. That is, in those species that have many motion and direction selective cells, there are more amacrine cell synapses per unit area of the inner plexiform layer than there are in species that have relatively few motion and direction selective ganglion cells.

In addition to the two basic types of ganglion cells described above, many ganglion cells have a mix of transient amacrine and bipolar cell characteristics. The Y-type ganglion cell of the cat is an example. The Y cells show a center-surround receptive organization like the bipolar cells, but their responses to both center and surround stimulation are quite transient and they are sensitive to moving stimuli. These cells appear to receive more of a mix of bipolar and amacrine cell synaptic input than the more sustained on-center or off-center ganglion cells.

It should be noted, finally, that all ganglion cells receive some amacrine cell input. The sustained ganglion cells, however, appear to receive less amacrine cell input than do the transient ganglion cells, and it seems likely that much of their amacrine cell input is from sustained amacrine cells. The above discussion also neglects the interplexiform cells and the role they may play in retinal function. This will be discussed in the next section.

IV. Pharmacology

The retina, like other regions of the brain, uses a large number of neuroactive substances. At the present time at least 15 substances are believed to be released from retinal neurons during retinal activity (Table I). These substances may be classified into two general categories: neurotransmitters and neuromodulators. Neurotransmitters act directly on retinal neurons, by altering membrane permeability to one or several ions. The ions move across the cell membrane causing a change of potential in either the depolarizing or hyperpolarizing direction. These changes in potential are rapid, and thus neurotransmitters are responsible for the fast excitatory and inhibitory pathways in the retina or brain. Neuromodulators, however, do not directly affect membrane permeability. Rather, they usually activate enzyme systems, and they modify neuronal activity biochemically. Neuromodulators do not usually initiate neural activity; rather they modify activity initiated by the neurotransmitters.

TABLE I Neuroactive Substances Found in the Retina

Amino acids
L-Aspartate
γ-Aminobutyric acid (GABA)
L-Glutamate
Glycine
Amines
Acetylcholine
Dopamine
Serotonin
Peptides
Cholecystokinin
Enkephalin
Glucagon
Neurotensin
Neuropeptide Y
Somatostatin
Substance P
Vasoactive intestinal peptide

Relatively few substances appear to serve as neurotransmitters in the retina. L-glutamate, acetylcholine, and perhaps L-aspartate mediate fast excitatory pathways in the retina, whereas GABA and glycine mediate fast inhibitory pathways. The bulk of the neuroactive substances released from retinal neurons (Table I) appears to be neuromodulatory in nature, although little is known about the action or role of most of these substances.

A. Amino Acids

Both photoreceptors and bipolar cells appear to employ L-glutamate as their transmitter. L-Glutamate depolarizes both horizontal cells and off-center (hyperpolarizing) bipolar cells, whereas it hyperpolarizes the on-center (depolarizing) bipolar cells. Neuroactive substances are released from neurons when they are depolarized, and because photoreceptors are maintained in a depolarized state in the dark, they release L-glutamate in the dark. When illuminated, photoreceptors hyperpolarize, and transmitter release is decreased. Thus, the light responses of the horizontal and bipolar cells reflect the withdrawal of transmitter from the cell. Horizontal cells are depolarized in the dark because of the dark release of L-glutamate from the photoreceptors; their light response is a hyperpolarization, reflecting the decrease in transmitter release from the photoreceptors in light. The same explanation holds for the off-center bipolar cells. These cells are depolarized in the dark; they hyperpolarize in light as transmitter release from the photoreceptor decreases. The on-center cell, however, is hyperpolarized in the dark, and it depolarizes in the light in response to the decreased release of L-glutamate from the photoreceptor.

Although the same substance, L-glutamate, mediates the photoreceptor input to horizontal cells and both types of bipolar cells, the receptor proteins with which the glutamate interacts differ somewhat in the three types of cells. Thus, it is possible to block the responses of one or another of the three cells with pharmacological agents while leaving the responses of the other cells intact. One substance, 2-amino-4-phosphonobutyric acid (APB), blocks specifically the on-center bipolar cells, and this results in the loss of all on-activity throughout the visual system. Monkeys treated with APB are unable to distinguish increases of illumination, although they can discriminate decreases of retinal illumination.

The amino acids GABA and glycine serve as inhibitory neurotransmitter agents in both the inner and outer plexiform layers. Many horizontal cells contain GABA, and approximately 80% of the amacrine cells in many retinas contain either glycine or GABA. Both GABA and glycine powerfully inhibit ganglion cells by opening channels in the cell membrane that cause hyperpolarization of the cell and inhibition of action potential generation.

GABA and glycine act on many retinal neurons, and they mediate specific inhibitory effects. For example, by blocking the effects of GABA in a retina with pharmacological agents, direction-sensitive ganglion cells lose their directional selectivity. Such cells now respond to spots of light moving in any direction across the retina. It is believed, therefore, that direction sensitivity is mediated by GABAergic amacrine cells in the inner plexiform layer.

B. Amines and Peptides

Most amines and neuropeptides appear to function as neuromodulatory agents in the retina, although little is known about the function of most of these agents, particularly the neuropeptides. The exception is acetylcholine, which functions in the retina as an excitatory neurotransmitter in the inner plexiform layer. Acetylcholine is found in amacrine cells, and some of these cells spread processes in the on-region of the inner plexiform layer and mediate transient excitatory responses in ganglion cells at the onset of illumination. Others extend processes in the off-region of the inner plexiform layer and mediate transient depolarizing responses in ganglion cells at the offset of illumination. The cell bodies of the on-acetylcholine-containing amacrine cells are found among the ganglion cells, whereas the cell bodies of the off-acetylcholine-containing amacrines are in the inner nuclear layer. The acetylcholine-releasing amacrine cells are believed to provide most of the excitatory input to the transient on–off ganglion cells, especially those that are direction sensitive.

The other amines, dopamines and serotonin, and the neuropeptides have also been principally localized to amacrine cells. These cells are, for the most part relatively scarce cells, accounting for no more than a few percent of the total number of amacrine cells. These cells usually spread their processes widely in the inner plexiform layer, and so they are capable of exerting wide effects. It is also the case that coexistence of neuroactive substances occurs in many amacrine cells. It has been reported that two peptides can coexist in the same amacrine cell, that a peptide and a monoamine coexist, or finally that a peptide or monoamine and an inhibitory amino acid are in the same cell. Some evidence has been provided that three or even more agents may be colocalized in the same neuron. As yet, the significance of the colocalization of two or more neuroactive agents in a single neuron is not understood.

C. Interplexiform Cells and Dopamine

Much of what we know about the action of neuromodulators in the retina has come from the study of dopamine in the teleost retina. In the teleost, dopamine is present in interplexiform cells and so these studies have also shed light on the role of these cells in retinal function. A brief summary of the findings are presented here as a model for the action of neuromodulators and interplexiform cells in the retina.

In teleosts, the synaptic output of the interplexiform cells is mainly on the cone-related horizontal cells. Two effects of dopamine on these cells have been observed; a loss of light responsiveness (i.e., light responses are reduced in amplitude after dopamine application to the retina), and a decrease in electrical coupling between horizontal cells. Dopamine does not exert these effects by acting directly on horizontal cell membrane channels; rather it interacts with a membrane receptor protein linked to the enzyme adenylate cyclase. This catalyzes the formation of the second-messenger molecule cyclic AMP (Fig. 11). Cyclic AMP, in turn, activates another set of enzymes called *kinases* that add phosphate groups to specific proteins. This process, called *phosphorylation*, serves to activate or inactivate cellular processes.

The kinases activated by cyclic AMP in horizontal cells appear to phosphorylate both the glutamate channels (i.e., the channels activated by the photoreceptor transmitter) and the gap junctional channels. The phosphorylation of these channels serves to alter their properties. In the case of the gap junctional channels, phosphorylation decreases the time that the channels remain open, thereby decreasing the flow of current that passes across the junction.

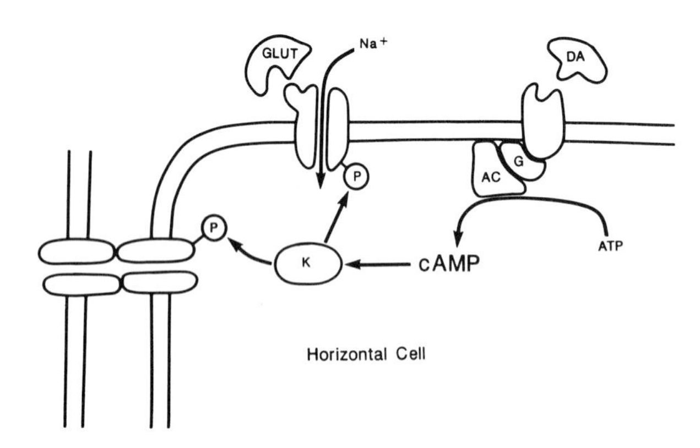

FIGURE 11 Summary scheme showing how dopamine (DA), acting via cyclic AMP, may influence responsiveness of horizontal cells to L-glutamate (the photoreceptor transmitter) and electrical coupling between horizontal cells. DA interacts with receptors that are linked to the enzyme adenylate cyclase (AC) via a G protein. Activation of adenylate cyclase results in conversion of ATP to cyclic AMP. Cyclic AMP interacts with kinases (K) that phosphorylate (P) glutamate (Glut) channels or the gap junction channels.

However, phosphorylation of the glutamate channels appears to modify the frequency of opening of the channels, which again modulates ion flow across the membrane. The action of dopamine on horizontal cells is not to initiate activity, but to modify the cell's response to the photoreceptor transmitter and the interactions between the cells. Furthermore, these effects are slow; they take many seconds to develop and they last for minutes.

Overall the effect of dopamine is to decrease the effectiveness of the horizontal cells in mediating lateral inhibitory effects in the outer plexiform layer. Decreasing light responsiveness of the cell and shrinking its receptive field size are effective ways of lessening its influence. As noted earlier, horizontal cells form the antagonistic surround response of bipolar cells, and thus a decrease in bipolar cell surround responses is observed after dopamine application to the retina.

What is the significance of the modulation of lateral inhibition and surround antagonism by dopamine and the interplexiform cells in the retina? It has long been known that after prolonged periods of time in the dark the antagonistic surround responses of ganglion cells are reduced in strength or even eliminated. An obvious speculation is that interplexiform cells and dopamine play such a role and regulate the strength of lateral inhibition and center-surround antagonism in the retina as a function of adaptive state. In fish, evidence in favor of this has been provided. After periods of prolonged darkness, horizontal cell receptive field size and light responsiveness are substantially decreased.

Bibliography

Baylor, D. A. (1987). Photoreceptor signals and vision. *Invest. Ophthalmol. Vis. Sci.* **28,** 34–49.

Boycott, B. B., and Dowling, J. E. (1969). Organization of the primate retina: Light microscopy. *Philos. Trans. R. Soc. Lond. B* **255,** 109–184.

Dowling, J. E. (1987). "The Retina: An Approachable Part of the Brain." Harvard University Press, Cambridge.

Ehinger, B., and Dowling, J. E. (1987). Retinal neurocircuitry and transmission. *In* "Handbook of Chemical Neuroanatomy, vol. 5. Integrated System of the CNS, Part I" (A. Bjorklund, T. Hokfelt, and L. W. Swanson, eds.), pp. 389–446. Elsevier, Amsterdam.

Kolb, H. (1970). Organization of the outer plexiform layer of the primate retina: Electron microscopy of

Golgi-impregnated cells. *Philos. Trans. R. Soc. Lond. B* **258,** 261–283.
Polyak, S. L. (1941). "The Retina." Chicago University Press, Chicago.
Rodieck, R. W. (1973). "The Vertebrate Retina: Principles of Structure and Function." W. H. Freeman, San Francisco.
Stryer, L. (1986). Cyclic GMP cascade of vision. *Annu. Rev. Neurosci.* **9,** 87–119.
Werblin, F. S., and Dowling, J. E. (1969). Organization of the retina of the mudpuppy, *Necturus maculosus.* II. Intracellular recording. *J. Neurophysiol.* **32,** 339–355.

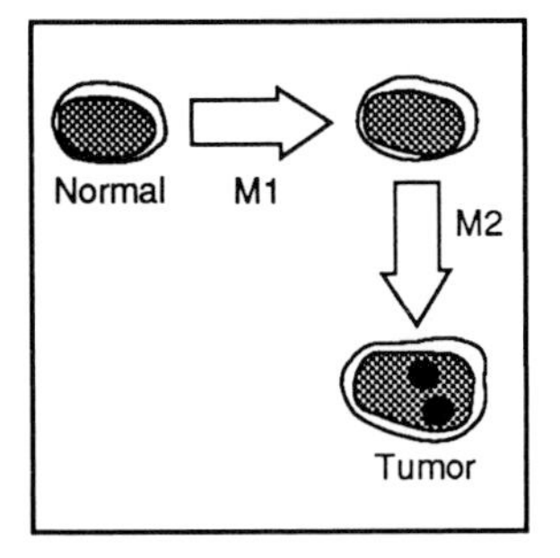

Retinoblastoma, Molecular Genetics

BRENDA L. GALLIE, *University of Toronto*

Glossary

Autosomal dominant inheritance Transmission of a hereditary disease that is manifest in most of the individuals, male or female, who inherit one mutant allele

Constitutional cells The normal, nonmalignant cells of an individual

Germ line mutation Mutation carried in the male or female germ cells that can contribute to the genetic makeup of the offspring

Loss of heterozygosity (LOH) Loss of part of one of a pair of homologous chromosomes, identified by absense of one copy of a polymorphic marker, that distinguished the two chromosomes in the constitutional cells

Recessive mutation A mutation that has no effect when a normal allele is present

Restriction fragment length polymorphism (RFLP) Unique DNA fragment showing variations in size between individual chromosomes due to differences in specific DNA sequences, revealed by digestion of the DNA with a specific restriction endonuclease

Retinoma A benign retinal tumor often observed in association with retinoblastoma

Somatic Nongerm line cells that will not contribute to the genetic inheritance of the offspring

RETINOBLASTOMA IS cancer of the infant retina, with clinical features that have led to the discovery of some fundamental mechanisms of cancer. Because the tumor in the eye could be directly seen (Fig. 1), early diagnosis and curative treatment were possible throughout the last century. Survivors demonstrated that retinoblastoma can be inherited as an autosomal dominant trait. In some rare patients, the hereditary predisposition to retinoblastoma is due to a microscopically detectable germ line chromosome 13q14 deletion. This suggested the genetic locus of the gene, *RB1*.

Application of molecular genetic technology to retinoblastoma has clarified the mechanism of tumor formation. Markers distinguishing the two chromosomes 13 of the individual revealed the recessive nature of *RB1*, and led to discovery of a previously unrecognized mechanism of human cancer initiation and progression, loss of heterozygosity (LOH). The *RB1* gene itself was identified because a chromosome 13 unique DNA sequence was completely missing from a few retinoblastoma tumors. The *RB1* gene codes for a phosphoprotein, located in the nucleus, that is important in control of the cell cycle. In addition to initiating retinoblastoma and probably osteosarcoma tumors, mutation of the *RB1* gene turns out to be a progressive event in malignancies not otherwise associated with retinoblastoma. The *RB1* mutations in many retinoblastoma tumors can be precisely identified and the germ line DNA of the patient and relatives examined for that specific mutation, achieving accurate detection of children at high risk for this cancer.

I. The Disease

A. Background

Retinoblastoma tumors are rare: only 1 in 20,000 newborn infants will develop retinoblastoma. This frequency is constant in all races and in all countries. In 40% of children, both eyes develop sepa-

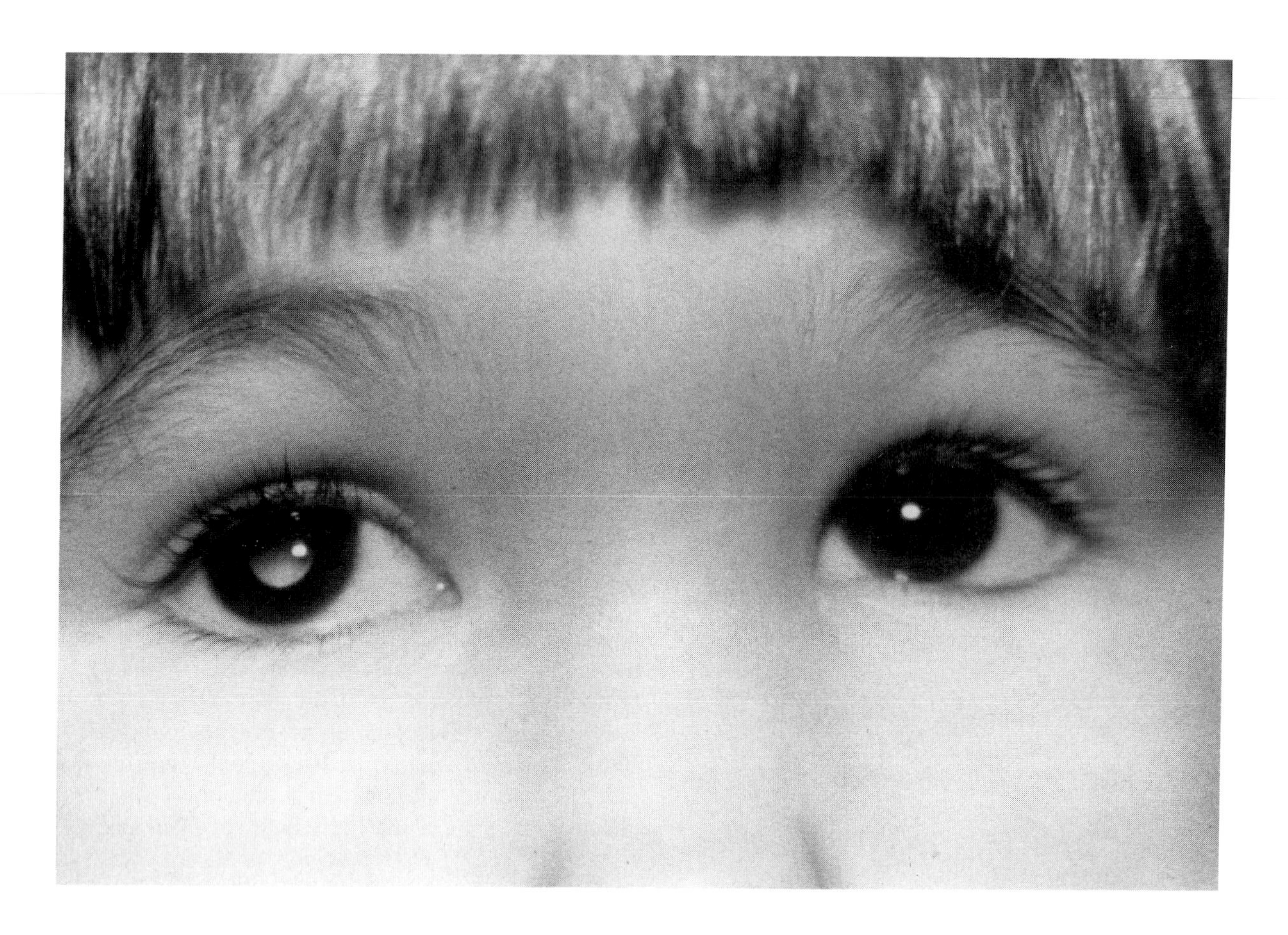

FIGURE 1 The right eye of this three-year old girl displays the "cat's eye reflex" commonly described by parents and indicating that retinoblastoma tumor is filling the eye.

rate retinoblastoma tumors. All these children have a germ line mutation of *RB1* in all of their constitutional cells. However, only 10% will have inherited the *RB1* mutation from a parent; the majority will have *new* germ line mutations.

Sixty percent of children develop only one retinoblastoma tumor. Of these unilaterally affected children, 85% do not have a germ line *RB1* mutation, but have suffered mutations of both *RB1* alleles in a single, somatic cell in the retina, which then developed into a single retinoblastoma tumor. However, 15% of the unilaterally affected patients do have germ line *RB1* mutations.

Approximately 2% of the retinal tumors do not behave as true cancers. A mass forms in the retina, similar in appearance to retinoblastoma tumor treated with radiation, but does not continue to grow. These tumors have been termed "retinoma," to distinguish them from malignant retinoblastoma tumors that continue to grow and spread. Many people with retinoma have been proved to have constitutional *RB1* mutations, because retinoblastoma tumors formed in their other eye or in the eyes of their relatives. [*See* RETINA.]

About 5% of children with retinoblastoma have multiple congenital abnormalities, such as abnormal fingers or toes, heart defects, and mental retardation. Examination of the chromosomes of such children show deletion of chromosome 13 band q14, large enough to be seen under a microscope. This observation suggested that 13q14 might contain the *RB1* gene. Studies of families with multiple members affected by retinoblastoma showed that the tendency to develop retinoblastoma tumors was inherited in conjunction with the alleles of a neighboring gene. Thus, the children with retinoblastoma because of chromosomal deletion indicated the genetic locus that predisposed other children to retinoblastoma tumors. However, most children with retinoblastoma have no other abnormalities, and are of normal intelligence. [*See* CHROMOSOME ANOMALIES.]

B. Clinical Management

Since most children with retinoblastoma have no family history of the disease, they present for medical attention only when the eye is full of retinoblastoma tumor. In countries with modern medical facilities, simple removal of the whole eye usually results in complete cure, and the cure rate for retinoblastoma is 94% in North America. Rarely, the tumor has extended beyond the eye, along the optic nerve to the brain, or through the blood to bone marrow. Metastatic retinoblastoma is rarely cured. People with a germ line *RB1* mutation are at an increased risk to develop other tumors, especially sarcomas, later in life.

If both eyes are affected, one eye usually contains smaller tumors. If an infant is at risk to develop tumors because a relative has retinoblastoma, the retina can be examined regularly after birth, searching for tumors. Small tumors can be treated with very little morbidity by direct local therapy: photocoagulation (heating by light) or cryotherapy (freezing through the wall of the eye). Eyes with medium-sized tumors, that still have useful vision, can be successfully treated with radiation therapy. Radiation causes growth abnormalities of the bone and tissues around the eye, leading to malformation of the face, and increases the risk of other types of cancer forming within the irradiated area in patients with germ line *RB1* mutations. Since the side effects and long-term consequences of radiation are severe, this treatment is carried out only when there is a good chance of keeping useful vision in the only seeing eye.

II. Tumor Suppressor Genes

A. Knudson's Hypothesis

It was noticed from the beginning of the century that children with retinoblastoma tumors in both eyes (who therefore must be predisposed to retinoblastoma by a germ line *RB1* mutation, even though they have no family history of retinoblastoma), were diagnosed to have tumors in their eyes at a younger age than the children with only one tumor in one eye (unilateral). Alfred Knudson examined this simple clinical data mathematically (Fig. 2a), and observed a straight line when the child's age at diagnosis was plotted against the logarithm of the fraction of cases not yet diagnosed. This indicated that a single event was rate limiting in tumor formation in these patients. A more complex curve was obtained for the unilateral patients, suggesting that two or more events were required to initiate retinoblastoma tumors in children with normal *RB1* alleles.

Knudson formulated the hypothesis that initiation of retinoblastoma tumors requires two mutations (Fig. 2b). In children with bilateral tumors, the

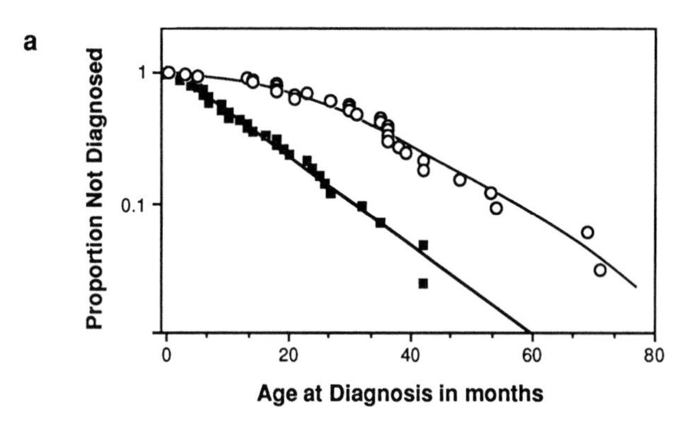

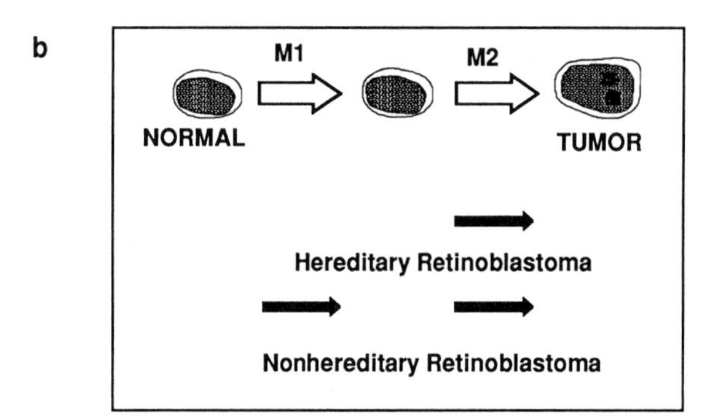

FIGURE 2 (a) The logarithm of the fraction of children not yet diagnosed was plotted against the age at diagnosis. The curve for children with hereditary disease (closed squares) is straight, suggesting a simple exponential with only one rate-limiting event initiating tumors. The curve for children with nonfamilial unilateral tumors (open circles) is more complex, suggesting two or more rate-limiting events. (b) Knudson's Hypothesis (now confirmed) was that all retinoblastoma tumors develop when two specific mutational events (M1 and M2) have occurred in developing retina. All cells of an individual have M1 in the hereditary retinoblastoma patients. Since M2 can happen more than once, multiple tumors usually form in the developing retina. Both M1 and M2 must occur in the same developing retinal cell to initiate a tumor in nonheritable retinoblastoma, and therefore these patients have only one tumor.

first mutation (M1) is present in each and every constitutional cell, involving the *RB1* gene. Retinoblastoma tumors form when a second mutation (M2) occurs in a developing retinal cell. On average, these predisposed children develop four separate tumors, usually involving both eyes. The simple exponential curve for age of tumor detection in bilateral patients reflects the rate of a single event, M2.

Knudson proposed that the nonhereditary (always unilateral) retinoblastoma tumors are caused by the two mutations, M1 and M2, occurring in a single developing retinal cell. The likelihood of two rare events happening in the same cell is so small that these children develop only one tumor.

B. Mutation of Both Alleles

When markers distinguishing the two chromosomes 13 of an individual inherited from each parent were developed, genetic changes around *RB1* were detected in retinoblastoma tumors. Enzyme and restriction fragment length polymorphism (RFLP) markers uncovered a previously unobserved mechanism for mutation in cancer cells, loss of heterozygosity (LOH). Tumors from individuals with heterozygous markers near *RB1* often showed loss of all or part of one chromosome 13. In familial cases, the chromosome inherited from the affected parent was retained in the retinoblastoma tumor, while the normal parent's chromosome was lost. This information suggested that both M1 and M2 involved *RB1*. Thus mutations of the *RB1* gene are recessive to the normal allele in individual cells, although the likelihood that a child with a germ line *RB1* mutation will develop retinoblastoma is so high that the predisposition to retinoblastoma is inherited as a dominant trait.

LOH was observed in 70% of retinoblastoma tumors in both bilateral (hereditary) and unilateral (usually nonhereditary) patients. Knudson's hypothesis was therefore supported: both M1 and M2 both involve mutation of *RB1*, in both hereditary and nonhereditary retinoblastoma.

C. Other Cancers

In order to locate similar tumor suppressor genes in other tumors, LOH was searched for in other cancers. Predisposition to Wilms tumor had been linked to chromosome 11p in children with chromosomal deletion, and those tumors showed LOH for chromosome 11p. Tumors in which genetic predisposition was not recognized, also showed LOH for tumor-specific genetic loci, such as chromosome 3p in small cell carcinoma of the lung, chromosome 22 in acoustic neuroma. The most systematic study, with markers on every chromosome arm, was carried out in colon cancer, and showed that LOH occurs for several different specific chromosomal regions, and can cause both cancer initiation and progression.

Survivors of bilateral retinoblastoma have an increased tendency to develop other tumors, particularly sarcomas, but do not appear to have a generalized cancer predisposition. However, many tumors, such as small cell carcinoma of the lung and breast carcinoma, show LOH for chromosome 13q14, and studies with the cloned *RB1* gene indicate that both alleles of *RB1* can be mutant in a wide variety of human cancers, particularly as those cancers advance and progress.

The human colon cancer survey of chromosomal regions undergoing LOH identified chromosome 17p as possibly containing a tumor suppressor gene. This included the locus of a gene that was cloned because its protein, p53, was superabundant in tumors. Study of colon cancer revealed that both p53 alleles were mutant, indicating that p53 is a human tumor suppressor gene. The overproduction of mutant p53 may result from mutations in regions of the gene that normally control levels of the protein. The mutant p53 can also bind and interfere with the function of the normal p53 protein, promoting tumor proliferation in the presence of a normal allele. Such a dominant mutation has not been observed in retinoblastoma.

III. The *RB1* Protein Product

A. Cell Cycle Regulator

The 110-kDa protein encoded by the *RB1* gene is present in all normal cells, where it is localized in the nucleus. Since absense of the normal protein is associated with malignant cellular proliferation, it is assumed that the active normal form of the RB1 protein blocks cell division. In resting, nondividing, normal cells, RB1 protein is partly unphosphorylated. As DNA synthesis starts, and the cells move toward cell division, additional sites on the RB1 protein become phosphorylated. This suggests that the unphosphorylated form of RB1 protein is active in inhibiting cell division.

Examination of the amino acid sequence of RB1 reveals little homology to known proteins. An imperfect "leucine zipper" motif, an α-helix with four or five leucine amino acids at regular intervals, which is involved in the dimerization of several DNA-binding proteins, is present in the RB1 protein. Specific DNA-binding by RB1 is not documented. At the present time, the normal function of the RB1 protein is not known, but must involve regulation of the cell cycle.

B. DNA Tumor Virus Interaction

DNA tumor viruses, such as simian virus 40 (SV40), adenovirus, and papilloma virus, induce proliferation of host mammalian cells by producing viral nuclear proteins that bind with cellular nuclear proteins. One of the cellular proteins bound by the "T/ E1A" viral protein family is the underphosphorylated RB1 protein. It is likely that the viral protein inactivates the RB1 protein, so that the cells can no longer "rest," but are driven to continual proliferation. In the process of cellular transformation, the p53 tumor suppressor gene is bound by a similar, but different, viral protein.

IV. The Mutations

A. Variety of *RB1* Tumorigenic Mutations

The mutations of *RB1* that result in retinoblastoma tumors include large and small deletions, duplications of parts of the sequence, insertions of unknown sequences, and point mutations. All regions of the 200-kbp gene can be affected. The largest deletions involve the whole *RB1* gene and neighboring genes, resulting in multiple congenital abnormalities if a germ line gene is involved. These deletions are evident by microscopic chromosomal studies. Submicroscopic deletions affecting large portions of the *RB1* gene were used to identify the gene clones, and are obvious when DNA from nontumor or tumor cells is examined by Southern blot electrophoresis. Only 20% of *RB1* mutations can be identified in this way.

More subtle mutations can be detected by examining the messenger RNA produced. Comparison of retinoblastoma tumor *RB1* messenger RNA with the normal messenger RNA, by hybridization followed by digestion of single-stranded RNA with RNase enzymes, reveals small deletions of 5 to 10 nucleotides, duplications, and point mutations that interfere with correct splicing, thereby producing an abnormal length messenger RNA. The *RB1* mutations of 50% of retinoblastoma tumors have been found in this way. For unknown reasons, some mutant messenger RNA molecules are detectable in retinoblastoma tumors that have no normal *RB1* gene, but are not detectable in the presence of a normal *RB1* allele, in the RNA of nontumor cells containing the same mutation.

The most subtle mutations, which change only a few amino acids or affect control of *RB1* messenger RNA transcription, are the most difficult to find. Direct sequencing, although time consuming and costly, can be applied to identify the mutation in tumors and patients, if the more efficient screen of RNA has not identified a mutation.

Mutations that alter specific functions can be useful in discovering the normal role of a protein. In a retinoblastoma tumor and a small cell lung tumor cell line, splice mutations have resulted in the absense of the 38 amino acids of exon 22 with the rest of the protein sequence normal, suggesting that this is an important region of the gene. However, most of the *RB1* mutations found in tumors result in no protein or a severely distorted protein, and therefore are not useful in identifying functional regions.

B. New Germ Line Mutations

The origin of each chromosome 13, from mother or father, can be determined by the use of RFLP markers. If LOH is the M2 event in a retinoblastoma tumor, the chromosome that is retained in the tumor is the chromosome that suffered the first mutation, M1. New germ line mutations of the *RB1* gene are for unknown reasons much more likely to involve the paternal chromosome than the maternal. Unilateral retinoblastoma tumors that have occurred because of somatic mutations only in the retinal cell that became malignant, show retention of either the maternal or paternal chromosome 13 when LOH occurs.

It is not easy to determine if new germ line mutations in other genes, leading to other diseases, involve the maternal or paternal chromosomes. Since the new germ line mutations of the *RB1* gene are random events, perhaps most human new germ line

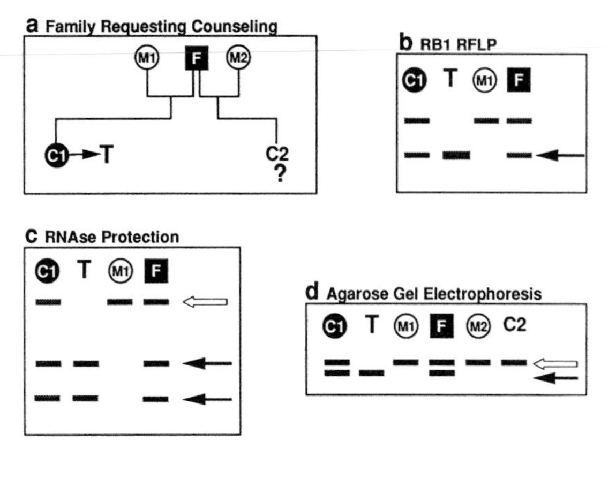

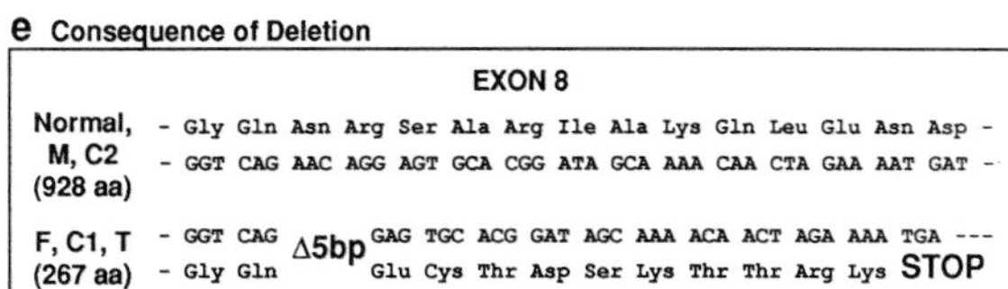

FIGURE 3 (a) The father (F) of this family had retinoblastoma as a child, treated by surgical removal of one eye, and radiation therapy of the other eye. His first child was not diagnosed to have retinoblastoma until age 1 year, at which time both eyes had to be removed. The father remarried, and when his second wife (M2) was pregnant, they requested information on the risk for the second child (C2) to be affected [see (b)]. (b) Southern blot analysis of DNA from the family and the retinoblastoma tumor of C1 using an RFLP within the RB1 gene. The tumor of C1 (T) had lost the normal allele from M1 and duplicated the mutant allele from the father (F) by the mechanism of LOH. (c) The position of the mutation in the RB1 gene in this family was located by RNase protection of messenger RNA. The father and C1 have both the normal, fully protected fragment (open arrow), and two smaller fragments (black arrows) in their normal cells. LOH in the tumor (T) is confirmed by the absense of the normal-sized protected fragment. (d) DNA from the family and the tumor (T) was amplified by polymerase chain reaction across the region of the deletion, and visualized on an agarose gel. The nontumor cells of the two affected individuals, F and C1, show both the deleted and the normal-sized DNA fragments. The tumor shows only the DNA that is deleted. DNA from the amniocentesis performed at 36 weeks gestation (C2) showed only the normal-sized fragment. If C2 had shown the deleted fragment, consideration would have been given to premature delivery, in order to treat retinoblastoma tumors earlier. (e) The DNA amplified by polymerase chain reaction was sequenced. The mutation was a 5-bp deletion in exon 8 of the RB1 gene. The deletion puts the nucleotide sequence out of frame, and leads to a downstream STOP codon. The mutant protein is expected to be only 267 amino acids long, in contrast to the normal protein of 928 amino acids.

mutations will involve the paternal chromosome. Whether this is due to the different biology of ovum and sperm development, different susceptibility of maternal and paternal chromosomes, or different environmental exposure of mothers and fathers is not clear at the present time.

C. Who Is at Risk?

Families with retinoblastoma are anxious to know which new baby will be at risk to develop tumors, and which family members carry an *RB1* mutation. If several children have had retinoblastoma, RFLP studies of the whole family can assign the *RB1* mutation to a particular chromosome 13. Anyone who inherits that chromosome is at risk. When only one member has retinoblastoma, this technique cannot be used.

Relatives of bilateral retinoblastoma patients, and individuals with one unilateral tumor, can be accurately assessed for a germ line *RB1* mutation if an *RB1* mutation can be identified in the retinoblastoma tumor, or in the nontumor cells of the patient. This information can be used as illustrated in the family shown in Fig. 3. The retinoblastoma tumor of child 1 developed because the normal chromosome of her unaffected mother was lost by LOH, while the affected father's mutant chromosome was retained in the tumor (Fig. 3b). The mutation of child 1 was located by RNase protection of messenger RNA from the retinoblastoma tumor (Fig. 3c), and shown to be a 5-bp deletion in exon 8 (Fig. 3e). The exon 8 DNA of any member of this family could then be examined to determine if the 5-bp deletion was present or not (Fig. 3d). Both the normal allele and the deleted allele are evident in the nontumor cells of child 1 and the father. The retinoblastoma tumor shows only the mutant allele. The father's second child had a 50% risk of having the father's germ line 5-bp deletion. The DNA studies of child 2 showed only normal alleles, indicating that this infant would be normal.

Since most retinoblastoma tumors arise because of new *RB1* mutations, each family has a different mutation. Accurate genetic counseling information can be achieved when the mutation of the family is identified.

Bibliography

Dunn, J. M., Phillips, R. A., Zhu, X., Becker, A. J., and Gallie, B. L. (1989). Mutations in the RB1 gene and their effects on transcription. *Mol. Cell. Biol.* **9,** 4594.
DeCaprio, J. A., Ludlow, J. W., Lynch, D., Furukawa,

Y., Griffin, J., Piwnica, W. H., Huang, C. M., and Livingston, D. M. (1989). The product of the retinoblastoma susceptibility gene has properties of a cell cycle regulatory element. *Cell* **58,** 1085.

Friend, S. H., Bernards, R., Rogelj, S., Weinberg, R. A., Rapaport, J. M., Albert, D. M., and Dryja, T. P. (1986). A human DNA segment with properties of the gene that predisposes to retinoblastoma and osteosarcoma. *Nature* (*London*) **323,** 643.

Gallie, B. L., Squire, J. A., Goddard, A., Dunn, J. M., Canton, M., Hinton, D., Zhu, X., and Phillips, R. A. (1990). Mechanism of oncogenesis in retinoblastoma, *Lab. Invest.* **62,** 394.

Knudson, A. G. J. (1971). Mutation and cancer: Statistical study of retinoblastoma. *Proc. Natl. Acad. Sci. U.S.A.* **68,** 820.

Sager, R. (1989). Tumor suppressor genes: The puzzle and the promise. *Science* **246,** 1406.

Vogelstein, B., Fearon, E. R., Kern, S. E., Hamilton, S. R., Preisinger, A. C., Nakamura, Y., and White, R. (1989). Allelotype of colorectal carcinomas. *Science* **244,** 207.

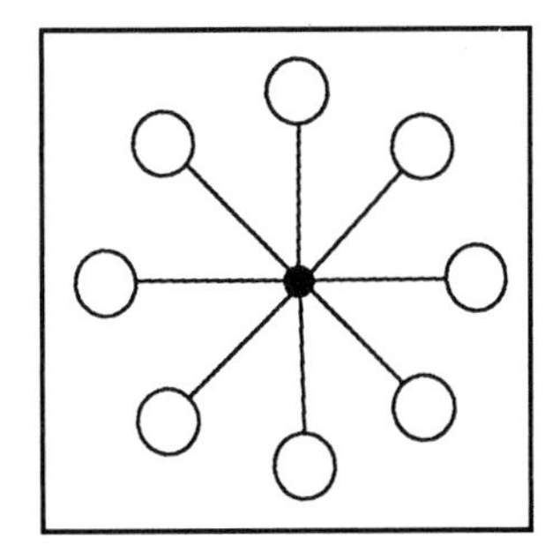

Retroviral Vaccines

DANI P. BOLOGNESI, *Duke University Medical Center*

Glossary

Adjuvants Substances which increase the potency of responses to immunogens by making them more easily recognizable to cells of the immune system

Antibodies Molecules produced by immune B lymphocytes which circulate in the blood and bind to foreign antigens with exquisite specificity

Antibody-dependent cell cytotoxicity (ADCC) An instance where antibodies associate with Fc receptors on cells which themselves have cytolytic properties. The antibodies provide the specificity of the reaction by binding to specific cell surface antigens and thereby direct the effector cells (usually macrophages or natural killer cells) to the target

Antigen Any substance which can evoke a humoral and/or cellular immune response

Cell-mediated immunity Cellular arm of the immune response is represented by a number of elements, among which are helper T cells, cytotoxic T cells, macrophages, natural killer cells, etc.

Cytotoxic T cells (CTL) T lymphocytes that become activated by helper T lymphocytes and develop the capacity to destroy cells which exhibit foreign antigens at their surface

Epitope Precise molecular configuration which defines a particular antigen

Fc receptor Structure found on the surface of cells to which antibodies are able to attach by their Fc portion.

Fusogenic event When membranes of different organelles coalesce and allow the contents of the respective structures to become unified. Many viruses use such a process to introduce their contents into cells in order to establish infection

Glycoprotein A protein which, in addition to a chain of amino acids, also consists of side chains of sugar (glyco) molecules

Helper T cells Cells of the immune system, which in response to appropriately presented antigens, send out chemical signals called lymphokines which help activate other lymphocytes and amplify the response to the antigen

Humoral immunity Humoral arm of the immune response represents that mediated by antibody molecules produced by B cells

Immunogen Substance which is able to provoke an immune response

Immunological memory Constitutes the generation of long-lived "memory" cells (T cells, B cells) of the immune system which are capable of recognizing an immunogen and stimulating a generalized response to it long after vaccination

Inflammatory reaction Refers to the use of adjuvants which provoke a strong response at the site of immunization which attracts many other components of the immune system to focus attention on the immunogen

Integration Insertion of a foreign DNA (e.g., a viral genome) into the DNA of the host cell

Interference The process by which cells infected by a given virus becomes resistant to second infection. This is usually but not solely related to competition for viral receptors on the surface of the susceptible cell

Neutralization Process by which antibodies directed against viral components, usually on the virion surface, are able to block infection of susceptible cells

Processing Refers to the proteolytic fragmentation of proteins through two independent pathways (see Figure 3) within the cell to generate peptides which can associate with major histocompatibility (MHC) molecules and thereby be recognized by helper and cytotoxic T cells

Retrovirus Belonging to a family of viruses whose genetic material is ribonucleic acid (RNA) but which replicate through a DNA intermediate. The process is catalyzed by a virus-encoded enzyme termed reverse transcriptase

Secretory antibodies Unique antibody molecules which are produced by cells of the secretory immune system which is associated with mucous membranes and is the first line of defense against invading organisms

Syncytium The end result when a number of cells fuse with one another

VACCINATION HAS historically proven to be the most simple, cost effective, safest, and efficacious means to prevent and thereby control the spread of disease in both animals and man. Among the most successful vaccines have been those prepared against viruses. The resounding triumphs of the smallpox and polio vaccines are of landmark proportions, but major advances have also been achieved in the control of diseases such as yellow fever, measles, mumps, and rubella through vaccination. More recently, impressive strides have been made in development of vaccines against more complex viruses such as hepatitis. On the horizon, one can envision eventual successes against certain members of the human herpes and papilloma virus families. This article will focus on vaccines against retroviruses, beginning with the experiences with animal retroviruses and ending with what is perhaps the most formidable and pressing challenge of all: a vaccine against the human immunodeficiency virus (HIV).

I. General Principles of Vaccine Development

The roots of vaccination lie in folk practices, dating back to before the seventeenth century in the far and middle east, which indicated that limited "inoculation" of the disease-causing organism itself by an "unnatural route" or exposure to it at a propitious age could lessen and even prevent disease. When Edward Jenner realized that exposure to the cowpox virus, which was similar to smallpox immunologically but was unable to cause disease in man, resulted in a comparable outcome, the platform for the science of vaccination and the principles of immunization were established. Indeed one can now name successful vaccines which represent derivations of a pathogenic organism which range from the organism itself, treated physically or chemically such that its ability to cause disease is severely limited, to actual subunits of the agent which are completely noninfectious and can even be produced by genetic engineering, as is the case for hepatitis B (Table I).

In order to be effective, these derivations must be capable of inducing protective immunity. It follows that those components of the disease organism that represent its critical targets for immune attack must be presented to the vaccinee in recognizable form.

TABLE I Current and Future Virus Vaccines[a]

Contemporary		Developmental and future	
Kind	Example	Kind	Virus targets
Live	Poliovirus	Subunits, peptides, recombinant, infectious vector, other (?)	Herpes (varicella zoster, simplex)
	Measles		Cytomegalovirus
	Mumps		Rotavirus
	Rubella		Dengue
	Yellow Fever		Human papilloma viruses
	Vaccinia		Human T cell leukemia viruses
	Adenovirus		HTLV-I, HTLV-II
Killed (whole virus)	Poliovirus		HIV
	Influenza		
	Rabies		
Subunit (natural and recombinant)	Hepatitis B		

[a] Courtesy of Maurice Hilleman, Merck, Sharpe and Dohme.

Otherwise, the disease organism will be invisible to the immune system. However, not all components of a pathogen represent beneficial immunological targets. In actuality, some parts of the organism may be undesirable elements because they elicit inappropriate immune responses which can sometimes mask those that are protective. Modern vaccine approaches thus take great pains to define what is effective and eliminate what may be deleterious.

While antibodies can directly recognize regions on native molecules, the cellular arm of the immune system requires prior processing of a given immunogen by what are known as antigen-presenting cells. These cells must internalize the antigen and modify it in a way such that it can associate with major histocompatibility (MHC) molecules and be recognized by antigen receptors on helper T lymphocytes. Antigen processing is also important for recognition at the surface of target cells by killer T lymphocytes, again in association with MHC.

Various means are used to enhance the recognition of an antigen by the immune system. In general, the more particulate the antigen, the more effective it is as an immunogen. By contrast, smaller subunits of a pathogen in the form of proteins or even peptides require formulation with "adjuvants" in order to achieve sufficient recognition. Complexing with adjuvants is largely empirical but formation of complex aggregates of antigen and adjuvants seems to be a guiding principle. The role of the adjuvant is that of a powerful immune stimulant in its own right; it establishes an inflammatory reaction at the site of inoculation, thereby attracting the attention of the immune system to the antigen. Today, various approaches are being devised to circumvent the "inflammatory response approach," which has been used most successfully in animals and for which a good counterpart does not exist in man because of serious adverse effects which have been experienced. This is being done by developing vehicles that will carry the antigen directly to antigen-presenting cells including nonpathogenic infectious vectors (usually innocuous bacteria or viruses) which would transmit the actual genes coding for a given antigen and allow their synthesis from within the antigen-presenting cells. As these methods are perfected they will have great bearing on development of HIV vaccines which are based on viral subunits.

Experience with both existing and certain experimental vaccines dictates that complete blockade against infection of the host by the pathogen is *not* a prerequisite for successful vaccination. Indeed, most vaccines protect against disease rather than infection per se. One can interpret this to signify that while viruses may replicate in a multitude of different cells in the body, the targets for disease are usually rather specific. For instance, in the case of polio, replication of the virus in epithelial cells of the gut is tolerable, whereas penetration of the virus within the central nervous system (CNS) spells disease. One can then envision at least two interdependent barriers against poliovirus infection. One line of defense would be mounted by the secretory immune system meeting the virus as it crosses mucous membranes. Although secretory antibodies do not prevent infection by poliovirus, they play the important role of limiting it to a level manageable by the systemic immune response. The latter is represented by both antibodies and immune lymphocytes found in the circulation and it is their role to prevent the virus from establishing a significant infection in the CNS. This is achieved by neutralization of virus infectivity and destruction of virus-infected cells, thereby clearing the infection before perceptible damage has occurred. A persistent infection in a sanctuary like the CNS, which is shielded from the immune system, generally results in serious consequences.

Finally, a successfully vaccinated individual possesses an immune system which is primed to respond rapidly upon encounter with the organism. A primed immune system consists of lymphocytes which retain the property of immunological memory for the lifetime of the individual. When such lymphocytes see the immunological target, they initiate a cascade of events that result in the generation of a full-blown effector phase of the immune response characterized by killer lymphocytes, neutralizing antibodies, and other protective elements which can effectively attack and clear the invading pathogen. [*See* LYMPHOCYTES.]

Turning to the issue of retroviruses, a question often asked is whether it is even possible to vaccinate against these agents. These viruses exhibit features that pose a number of new challenges for vaccinologists. First and foremost, retroviruses can integrate their genome into the cells of the host and thereby establish a permanent infection unless all of the infected cells are removed. A hallmark of these viruses is that they induce various forms of cancers in animals and a cell harboring an integrated retroviral genome can, in certain instances, become a malignant cell even in the absence of virus production. This is because certain regulatory elements in retroviruses can exert their effects not only on the viral genes but also on cellular genes. Thereby this added genetic information can disrupt normal cell growth and differentiation patterns which can lead to malignant transformation. This issue is of extreme importance for vaccines based on use of live attenuated or inactivated virus, particularly if they are destined for man. Human retroviruses not only possess the traditional gene regulatory elements but are replete with "accessory genes" that are able to impact even more extensively on the expression of normal cellular genes. Therefore, one must consider the use of vaccines containing retrovirus genetic material as extremely dangerous, unless this is done in individuals already infected as postexposure prophylaxis.

A successful vaccine against any retrovirus would thus represent an important milestone toward the development of a counterpart against disease causing human retroviruses. What has the experience been thus far? In short, not a great deal but a growing number of animal retrovirus models are receiving attention in this regard and such information is contributing to the evolving strategies toward vaccines for human use.

II. Vaccines against Leukemia Retroviruses of Mammals

Until recently there has been relatively little motivation to develop vaccines against animal retroviruses principally because the diseases they caused in natural circumstances have not been of sufficient impact to mount large-scale efforts in this regard. In one particular case, feline leukemia, vaccines have been approved for veterinary use but these have been largely ineffective. In the case of bovine leukemia, a disease of growing economic importance, an effective vaccine would prove of considerable value but other measures, such as eradicating infected and diseased animals, are proving successful for disease control.

Nevertheless, the efforts to develop experimental vaccines against such viruses are mounting rapidly, particularly because of their value as models for developing vaccines against human retroviruses.

A. Murine Retroviruses

The experimental studies with animal retrovirus vaccines date back to the early 1970s when the murine models received the greatest attention because of the extensive knowledge about the virion structure, its antigenic composition, and the practicality of a small and plentiful laboratory animal model.

The viral component responsible for the salient immunobiological features of a retrovirus is its major external glycoprotein (gp). First the glycoprotein is required for infection and mediates the attachment of the virus to the host cell. It is also the viral component that is directly involved in the phenomenon of interference, which reflects competition for viral receptors at the cell surface. Finally, the glycoprotein specifies the pattern of neutralization by antiviral antibodies. Consistent with these properties is its strategic location on the outer envelope of the virion (see Figure 1).

Another component found on the surface of the virion is a hydrophobic transmembrane protein (tmp) which noncovalently anchors the glycoprotein to the virion. The transmembrane protein can either contain or be devoid of carbohydrates, but in every case the degree of glycosylation is considerably less than in the exterior glycoprotein. The transmembrane protein can also be a target for neutralization by antibodies.

The viral glycoprotein transmembrane protein are also found in infected cells, concentrated at the sites of budding viruses but also present at other sites which are more uniformly distributed on the cell surface. As such, the envelope components also represent targets for destruction of infected cells.

Animals infected with retroviruses usually respond with easily detectable antibodies against the exterior glycoprotein. For the most part this immunity is specific for the infecting agent. Others represent determinants which are common to all viruses of a given species or may even extend to those found in widely different species. It is only in rare cases that animals respond to these latter domains under natural conditions. On the other hand, one can immunize animals with virus or purified glycoprotein and obtain antibody responses which are much broader in their reactivity than natural antibodies. Studies have also been done toward generating cytotoxic T cells (CTL) through use of recombinant vectors such as vaccinia virus, into which genes were incorporated coding for retroviral struc-

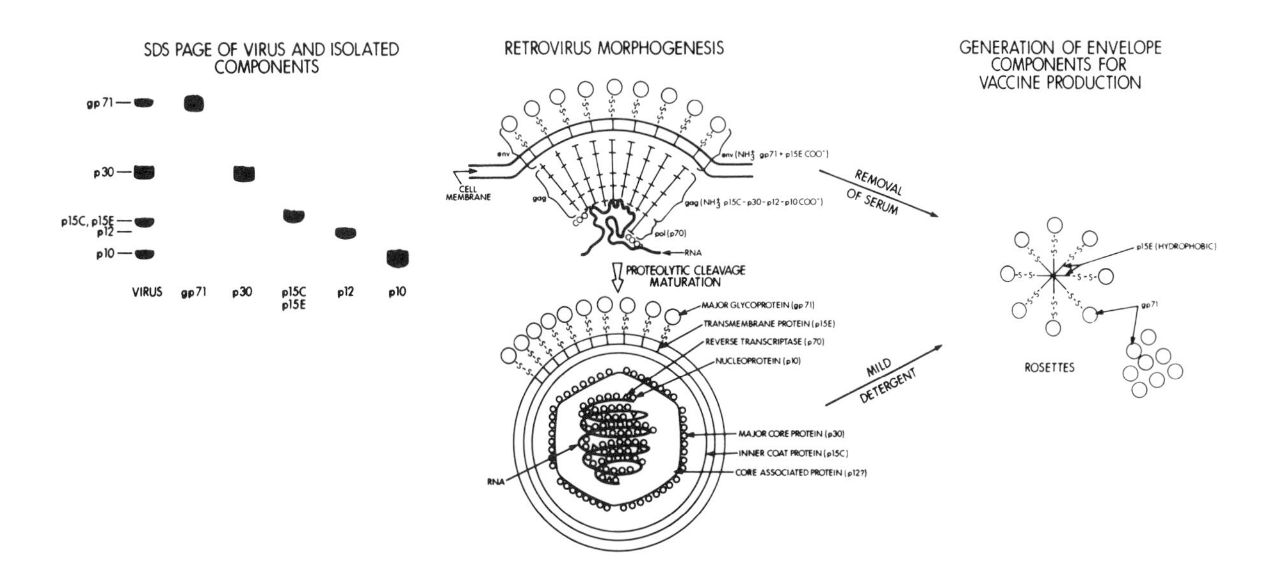

FIGURE 1 Morphogenesis, structure, and composition of a typical murine leukemia retrovirus. The surface components gp71 (gp) and p15E transmembrane protein (tmp) derived from the *env* gene of the virus can be recovered in homogeneous form either as multimers (rosettes) or monomers as a result of shedding from the cell surface or treatment with mild nonionic detergents.

tural antigens. As with other viral systems which have been studied, CTL could be generated against both external (envelope) and internal (core) components and these mirrored CTL arising during natural infection.

As noted above, immunization with glycoprotein elicits strong neutralizing antibodies as well as antibodies which are cytotoxic for infected cells. Mice immunized with purified glycoprotein can, indeed, resist substantial challenges of infectious leukemogenic virus. While monomeric glycoprotein is capable of inducing protective immunity, relatively large quantities of purified antigen were required to accomplish this reproducibly. On the other hand, the use of glycoprotein linked to the transmembrane protein so as to form multimeric aggregates resulted in a much more effective immunogen (see Figure 1).

While such experiments demonstrate that antibodies may be sufficient for protection it is important to note that many of these results were obtained under experimental conditions which would not be applicable for widespread vaccination. Moreover, experimental infection may be quite different than what occurs during natural transmission. To this end, it is also important to determine what other immune mechanisms might constitute a protective response. In this regard, there are a few but at the same time concrete examples that cellular immunity constituting both helper and cytotoxic lymphocytes can exert powerful protective immunity against infection and disease. While the approaches used to generate such responses remain to be formulated into a practical vaccine regimen, they emphasize the importance of harnessing the cell-mediated arm of the immune system.

B. Feline Retroviruses

Many of the principles emanating from studies with murine retroviruses were rapidly applied to the feline system with the intention of developing a practical vaccine for use against naturally occurring infections which were severely debilitating to household as well as free cat populations. The first licensed vaccine employed a mixture of viral antigens collected from infected cell cultures. While effective in experimental tests, it failed to generate significant protection when applied in veterinary clinics.

Experimental vaccines based on a technology whereby the virus envelope components (glycoprotein and transmembrane protein) are captured by glycoside lattices through hydrophobic interaction with transmembrane protein generate a multimeric matrix structure (ISCOM; immune stimulating complex) which is an unusually powerful immunogen. It may be that these complex structures are more easily recognized, presented, taken up, and processed by macrophages for antigen presentation to the immune system. Vaccines based on ISCOMs have been highly successful in field trials but remain to be developed in a commercial vaccine formulation.

A promising candidate, in this regard, is a vaccine based on a recombinant form of the exterior gp with a portion of the transmembrane protein produced in *Escherichia coli*. The antigen is devoid of carbohydrate and its sulfhydryl groups are blocked so that its secondary conformation is limited. The adjuvant used to elicit immunity is a purified saponin which itself is a constituent of ISCOMs but its activity in this setting appears to derive from a different mechanism. Efficacy trials in a limited number of animals approach 100% protection and emphasize once again the importance of neutralizing antibodies as a primary barrier against retrovirus infections. In these studies, both the immunization protocol and the amounts of antigen used would represent a practical vaccine and field trials are now in progress to determine if natural infection can be similarly affected.

Other studies have focused on the use of inactivated whole virus to protect animals against infection with a molecular clone of feline leukemia virus (FeLV) which specifically induces immunosuppression. This disorder precedes the leukemia/lymphoma malignancies and can result in death long before their onset. Successful vaccination against the immunosuppressive feline leukemia retrovirus offers hope that the knowledge gained can be applied to immunodeficiency viruses of simians and man.

C. Bovine Leukemia

The studies which have been conducted with the antigenic structure of the bovine leukemia virus (BLV) focus attention on its envelope components as the principal targets for vaccine design, much like the experience with murine and feline viruses. However, in contrast to the studies with subunit FeLV vaccines, which were deliberately presented in denatured form, the experience with BLV points to critical epitopes for neutralizing antibodies which depend on the native configuration of the envelope. This concept also derives from studies which demonstrated that natural antibodies to the envelope were protective against BLV transmission from mother to the offspring. Careful mapping of the various epitopes responsible for elicitation of neutralizing antibodies has been carried out. The outcome indicates that the regions responsible for induction of the strongest protective immunity may be involved in the process of viral entry through a fusogenic event with the cell membrane. BLV infection also results in cell/cell syncytium formation, which is one of the hallmarks of HIV infection.

III. Vaccines against Animal Lentiviruses

As with the animal retroviruses associated malignancies, vaccine efforts against various prominent members of the lentivirus family have been lagging, again despite their involvement in naturally transmitted chronic diseases which affect domestic animals such as sheep (Visna/Maedi), goats (caprine arthritis, encephalitis virus), and horses (equine infectious anemia virus). More recently they have been also implicated in immunodeficiency syndromes of subhuman primates. Indeed, a major impetus for further studying these models is that HIV is a member of this group of retroviruses and shares, to different degrees, certain of the unique properties of these agents. Among them are genomic organization, pathogenic processes, and, most poignantly, their high variability and propensity to escape from immune defenses.

A. Equine Infectious Anemia Virus (EIAV)

The natural mechanism of transmission of EIAV is via insect vectors. The disease is typified by progressive cycles of febrile illness with severe consequences to the affected animals, including death. The initial infection is met by a vigorous neutralizing antibody response, but variants emerge which are refractory to its effect. As the variant replicates, a second round of antibody is produced which is specific for the new virus; but again new variants emerge. By repeating this cycle of neutralization, mutation, and escape, the virus is able to complete its pathogenic mission. Antigenic variation and escape from neutralizing antibody also occur with Visna virus, but only very low and hardly measurable levels of neutralizing antibody are ever found with caprine arthritis encephalitic virus.

Using a formalin-inactivated whole-virus vaccine in the presence of a muramyl dipeptide adjuvant, it was possible to protect ponies against infection by a live challenge with avirulent EIAV. These animals apparently developed an immunity which was also highly protective against subsequent challenges with disease-causing viruses, including the highly virulent Wyoming strain. These results demonstrate that whole inactivated virus vaccines can be successful against both infection and disease induction by EIAV. Such studies set the stage for determining the parameters of protective immunity and identifying the elements of the virus particle which are responsible.

B. Simian Immune Deficiency Virus (SIV)

This group of viruses represents the closest relatives thus far known to the human immunodeficiency virus (HIV). Their ability to produce an acquired immunodeficiency syndrome (AIDS)-like illness in small subhuman primates (such as macaques, sooty monkeys, and mandrills) within months after inoculation provides a valuable animal model in which to exercise vaccine strategies.

The vaccine studies with SIV have only recently blossomed. Like those described above for EIAV they have chiefly focused on the use of whole inactivated virus preparations and have achieved comparable results. In what are now several independent examples, such preparations were able to delay the onset of disease for a significant period of time, in some cases more than 1 year and possibly permanently as the experiments may eventually demonstrate. Notably, some protection against disease could be achieved even in cases where actual infection by the virus was not prevented. The importance of this development is underscored because it suggests that the same general rules apply as have existed for other virus vaccines; namely that it may not be necessary to completely block infection in order to have a successful vaccine. If some degree of viral entry can indeed be tolerated a vaccine against viruses of this family is much more within reach. Nevertheless, in some of the studies, complete protection against infection was achieved, which is clearly the most favorable outcome.

IV. Vaccines against Human Retroviruses

Although animal studies are of extreme importance for providing concepts and strategies for vaccine development, none fully represents the host–virus relationships that exist between human retroviruses and man. In fact, it is the discovery of human retroviruses which has accelerated the studies of their animal counterparts.

A. Human T Cell Leukemia Viruses (HTLV)

Although, the HTLV viruses are important human pathogens it seemed, for some time, that they could be controlled by public health and education measures. Given the low attack rate of these viruses and the low incidence of disease following infection (1 in 100) the risk/benefit ratio of a vaccine would be relatively low. However, it is now clear that the distribution of these viruses is much wider and extends to individuals (i.e., drug users) where education methods and public health measures are ineffective. Moreover, these viruses are responsible for other diseases as well, such as tropical spastic paraparesis (TSP) and possibly other central nervous system disorders. Thus, it can be anticipated that a large number of HTLV carriers exists in the population which can further transmit the virus through various means, and that vaccination programs are therefore eminently justifiable to eradicate the infection.

Because of the morbidity and mortality risks associated with virus infection, approaches that would use whole inactivated virus have not been pursued. Instead, most of the attention has been focused on the envelope glycoproteins of the virus and the role of neutralizing antibodies and cytotoxic lymphocytes in the protective response. Various sites which represent important biologic and immunologic epitopes have been mapped. A vaccine consisting of an envelope subunit of HTLV produced in *E. coli* generated protective immunity in cynomolgus monkeys against primary infection by HTLV-I. Of interest is that this study employed HTLV-I-infected cells as the challenge vehicle, which is a step closer to natural transmission than virus itself. Protection correlated with the presence of neutralizing antibodies, indicating that humoral immunity can be an effective barrier against infection.

More intensive studies are currently being carried out in a rabbit model of HTLV-I infection. This approach will enable the optimization of candidate HTLV vaccines in terms of immunogenicity and efficacy.

B. HIV Vaccines

The barriers standing before the development of a vaccine against HIV are formidable and for some time have clouded the thinking and dampened the enthusiasm as to how such a task might be approached. However, as a result of progress along several fronts, there is cautious but growing optimism, which stems mainly from the demonstrations of efficacy of vaccines for both SIV and HIV in animal models. Complementing such efforts are major advances in understanding the variability of the virus and in how to design immunogens which can induce T and B cell memory to critical target epitopes of the virus. Finally, not only animals but humans are able to respond favorably to certain candidate immunogens.

1. Special Considerations for a Vaccine against HIV

A few key properties of HIV can be singled out in considering the development of a vaccine (Table II). First and foremost, like other members of the retrovirus family HIV is able to integrate its genetic information in the genome of its target cells. The survival of the virus is thus directly linked to the survival of the cell. Although HIV-1 can kill certain cells it infects, most notably T cells, other cells (e.g., monocyte/macrophages) can tolerate virus replication without succumbing. Thus, permanent reservoirs of infection can be established to serve as factories for production of virus which can then carry out its destructive effects on the host. [*See* ACQUIRED IMMUNODEFICIENCY SYNDROME (VIROLOGY).]

The virus can also establish latent infections, that is, the integration of its genome in that of the target cell without synthesis of virus or viral gene products. Latency is a well-known property of viruses of the herpes family, against which vaccines are still being sought. Latently infected cells are invisible to the immune system, unless signals are applied which are able to activate the expression of viral genes. For HIV these signals can be other viruses but also host factors, including elements that regulate immune function. From this alone, one can glean how intimately virus and target cell functions are intertwined, providing considerable selective advantage to the survival of the virus.

TABLE II Major Obstacles toward a Vaccine against HIV

1. Natural transmission includes free virus and infected cells
2. Virus can be transferred covertly and efficiently from cell to cell
3. Virus can establish latent infections in T cells and macrophages
4. Virus replicates as a swarm, generating numerous antigenic variants
5. Virus both impairs and destroys the immune system
6. Virus resides in the central nervous system
7. No HIV animal model exists for both infection and disease
8. No epidemiological evidence that antiviral immune response during natural infection is protective

Another feature of HIV-1 which bears some attention is its mode of transmission, which can occur either through free virus or virus-infected cells. Virus can also be transmitted covertly from cell to cell through the process of fusion and be invisible to immune defenses. HIV-infected monocyte/macrophages contain large concentrations of virus in intracellular vesicles. The virus is released within such structures rather than at the cell surface. Macrophages could disseminate the virus to other cells by a "trojan horse" fashion, as it occurs with other lentiviruses. Moreover, if macrophages carrying sacks filled with virus are ruptured, large quantities of HIV-1 would be liberated and could further propagate. Thus, immune attack on such cells may bear negative consequences.

HIV-1 is also notorious for its ability to escape immune attack through mutation within its envelope gene. In fact, large numbers of variants of this virus exist in the population. As is the case with EIAV, variants may be selected which are resistant to immune attack. This property is reminiscent of those of influenza viruses, but is much more extreme.

The very fact that macrophages can be infected by HIV-1 raises the question of the role of a special class of antibodies, the enhancing antibodies. Such antibodies would bind the virus and bring it to the macrophage surface by attachment to Fc receptors present on such cells, thereby enhancing the infectivity of the virus.

Thus, HIV is a formidable adversary. It attacks the immune system upon which a vaccine depends. It is able to hide from immune defenses by establishing latent infections and by developing covert mechanisms of transmission. It can infect sites forbidden to the immune system such as the central nervous system. The extensive variation of the virus allows it to escape immune defenses. These features have led to the speculation that a successful vaccine must be able to totally prevent infection. If this is indeed the case, it would place demands far above what has ever been required of vaccine for other pathogens, where complete blockade of infection is not a requirement for protection against disease.

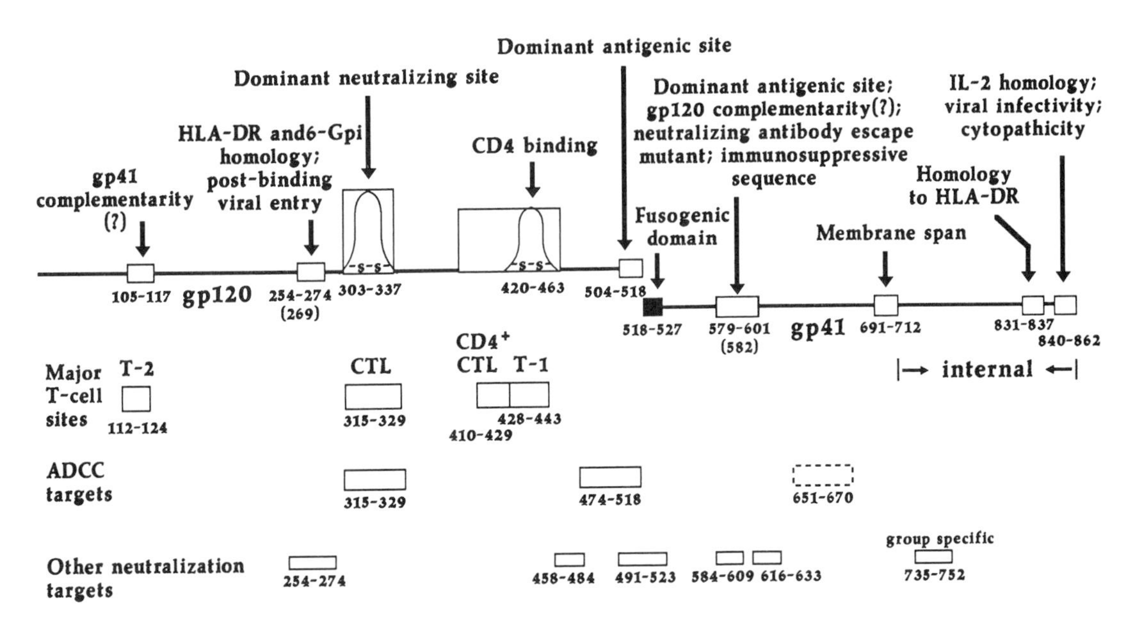

FIGURE 2 Composite of selected sites on the HIV envelope discussed in this article. Additional regions of the envelope that are critical for various viral functions have been identified by insertional deletion mutagenesis and are reviewed.

2. Progress in HIV Vaccine Research

In parallel with the promising results achieved with experimental vaccines against SIV, a few examples of vaccine success against HIV infection of chimpanzees have been reported using either whole inactivated virus preparations or viral subunits. In one instance, a whole virus vaccine administered to animals already infected with HIV apparently achieved the surprising result of clearing the infection within the bloodstream. Postexposure immunization with similar virus preparations has also been studied in man, but its effects do not compare to the results in chimps. Moreover, studies using various kinds of HIV subunit antigens have resulted in enhanced anti-viral immunity in patients with AIDS. Taken together, these results suggest that the immune system can be harnessed to protect man from HIV infection and perhaps improve on the course of disease.

In parallel with these studies, many laboratories are using a reductionist approach which seeks to define the essential components of the virus that would confer protection. A vast array of neutralizing epitopes, T cell epitopes, and targets for antibodies which mediate antibody-dependent cell cytotoxicity (ADCC) have been identified. Nearly 20 such sites have been mapped (Fig. 2).

The principal neutralizing epitope of HIV is situated within a highly variable region which forms a loop in the external envelope glycoprotein (gp120). When given with the virus, neutralizing antibodies to the loop can protect chimps against HIV infection. Moreover, when this segment of the envelope is used as a booster for chimps that were primed with other immunogens, high titers of virus-neutralizing antibodies were achieved, with at least temporary protection against HIV challenge. It has also been reported that antibodies to this region correlate with lack of HIV transmission from mother to infant.

Although these results are encouraging, the use of the loop in vaccine strategies is problematic because it undergoes extensive variation. However, a critical portion of this loop remains relatively constant, probably because it is essential for the process which triggers the fusion of the virus with the target cell. If this region can be exploited as a protective immunogen, it may be able to induce broad-spectrum immunity to a diversity of HIV isolates.

A number of unique T cell sites have also been defined on the virus envelope, its internal components as well as its regulatory elements. They may serve as targets for cytotoxic lymphocytes which are present in HIV-seropositive individuals. Along with the sites on the envelope which serve as targets for ADCC, a substantial arsenal is thus available with which to attack virus-infected cells. This is important for two reasons: (1) natural transmission of HIV occurs both through free virus and virus-infected cells and (2) some degree of infection of

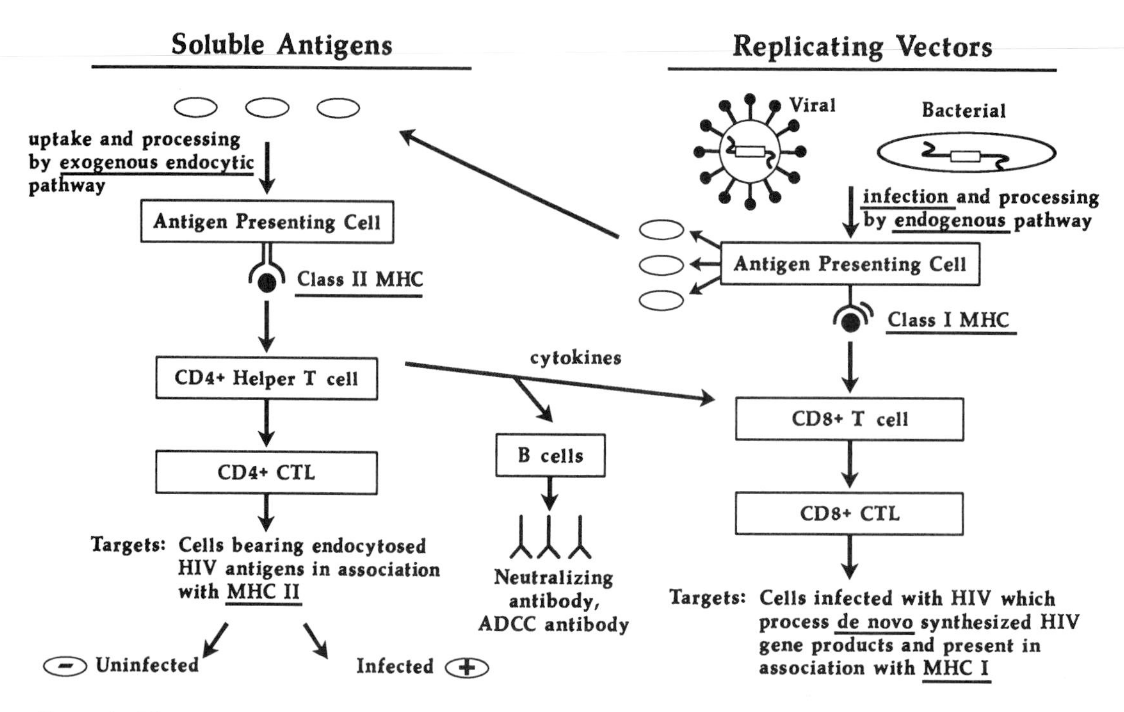

FIGURE 3 Generation of protective immune responses against HIV. Separate pathways are depicted, depending on whether the immunogen is self-replicating (right side) or inert (left side). The respective processing (endogenous vs endocytic) and presentation (class I MHC vs class II MHC) pathways are distinguishable. Presentation of a given antigen within the context of an innocuous replicating vector (viral, bacterial, or other) would appear to be the optimal route since it could encompass both pathways.

host cells probably cannot be avoided even with the ultimate vaccine strategy.

3. Conceptual Approaches for Vaccination against HIV-1

The morbidity and mortality associated with infection with HIV-1 together with the uncertainty of vaccine efficacy generally precludes the use of conventional vaccine strategies employing live attenuated or even killed preparations which have been successful with other viruses (except perhaps in individuals already infected, as has been practiced with rabies virus). A safer approach is to construct immunogens containing the different categories of epitopes that induce protective immunity against the virus and its infected target cells. This work must be based on continuous refinement of the definition of such epitopes.

With such information in hand, the sequences responsible for these epitopes can be assembled into a single gene, excluding unwanted sequences. The protein produced by such a gene might then generate a safe and effective immunogen.

To be successful, the candidate vaccine must be presented to the immune system so as to evoke the most favorable response. The requirements for antibody production are somewhat distinct from those needed for generation of cytotoxic T cells. Antibodies originate from B cells but the process requires T cell help. Helper T cells ($CD4^+$ T cells) recognize antigens in association with class II MHC. In most cases this occurs only after the immunogen is taken up by antigen-presenting cells and processed (proteolyticaly digested) within endocytic vesicles to a form which can associate with class II MHC (Figure 3, left side). Consequently immunogens designed to produce antibody responses should contain epitopes that are recognized by the B cell compartment (B cell epitopes) as well as those needed to activate helper T cells. These can either be contained with the immunogen or provided as carrier molecules.

A different principle is used to generate cytotoxic T cells ($CD8^+$ CTL). In this instance, the antigen must associate with class I MHC in order to be recognized (Figure 3, right side). Current knowledge indicates that a different processing pathway is involved and that the most efficient (but not the only) way to achieve this may be when the antigen is actually produced within the presenting cells. Approaches to induce cytotoxic T cells therefore, include innocuous virus vectors or bacterial vaccine strains which can be engineered to carry the genes which encode the desired antigen (Figure 3, right side) within the cell. Under ideal circumstances, both processed (class I MHC) and unprocessed antigen (for uptake and presentation with class II MHC) could be achieved through use of such vectors to yield a full range of humoral and cellular responses.

4. Remaining Challenges in HIV Vaccine Development

The encouraging developments reviewed above notwithstanding, challenges standing before the development of an HIV vaccine are prodigious. Ways to present HIV antigens to the human immune system so as to evoke protective humoral and cellular immunity are still being perfected. Much more must be done along the lines of development of safe and effective adjuvants or constructions of innocuous replicating vectors into which critical HIV target epitopes can be incorporated. Also, more must be known about the role of secretory immunity in a protective response against transmission of HIV across mucosal surfaces.

New and better models for testing of vaccine efficacy must also become available. The difficulty with current animal models focuses on the fact that the chimpanzee is the only species which can be routinely infected with HIV-1. However, HIV-1 has so far not induced disease in this species. Thus, it is impossible to know whether a vaccine that would reduce but not eliminate infection in the chimpanzee is effective against disease. The reverse side of this question is whether a successful vaccine for HIV requires complete protection against infection. That being the case the chimp might be a suitable model but one would still be hampered by the vastly insufficient numbers of experimental animals.

To counter this, other animal models are being sought. Successful transplantation of the human immune system in immunodeficient mice holds promise and such animals have already proved to be infectable with HIV-1. Whether the full spectrum of immune response necessary to resist HIV-1 infection can be induced in them remains to be determined. Until such models are perfected, one must rely on the cognate simian immunodeficiency viruses; however, one must keep in mind that SIV is not HIV and subhuman primates are not the equal of man.

The ultimate test of a vaccine is through clinical trials in man. Some trials under way have already raised a number of questions that require resolution. One is the distinction between the immune response to the vaccine and a response to HIV infection. If the vaccine is represented by a limited number of virus components, the distinction is easy, but if it is more complex, the distinction will be much less clear. Use of attenuated or killed virus as an immunogen would make the distinction very difficult. This question also relates to a number of sociopolitical problems deriving from the stigma of HIV-1 seropositivity and how it affects employment, travel, procurement of life and health insurance, etc. In addition to discriminatory tests, clear documentation of participation in a vaccine trial is obligatory. It follows for all the reasons mentioned above that the degree and specificity of immunological monitoring necessary for HIV vaccine testing will have no precedent.

V. Summary

A number of issues distinguish the problems associated with retrovirus vaccine development from those against other pathogens. This is particularly true for HIV. Major considerations are safety issues in that infection with HIV probably results in a uniformly fatal outcome. Vaccines that have any possibility of being "infectious," even in a small number of cases, should be used only as a last resort. It is also not known at this time to what extent immune responses to HIV contribute to the disease process. Special problems with HIV vaccines also relate to complexities of how to measure the consequences of immunization, due both to the long and variable eclipse period between infection and disease and the low attack rate of virus infection in known risk groups. Safety and efficacy testing will thus be very difficult to assess critically. An important point is whether the primary effect of the vaccine should be to delay or reduce disease symptoms. With many viral infections, this would be an eminently acceptable goal but with HIV one must consider the dangers associated with large numbers of disease-free carriers, capable of transmitting infection. Finally, one must be concerned about the availability of a suitable number of volunteers for entry into vaccine trials and this is complicated by sociopolitical factors as well as ethical and liability issues. A vaccine candidate suitable for clinical trials in man should be (1) a potent immunogen without toxic side effects both alone and subsequent to virus infection in a suitable animal model and (2) successful in preventing infection and disease from occurring upon challenge with diverse isolates of both free and cell-associated HIV.

There will be a number of by-products emanating from research on HIV vaccines. One will certainly learn which immune responses are most effective against the virus. Passive transfer of immunity can be brought to bear on individuals already infected, particularly in cases of congenital transmission of HIV infection (e.g., from mother to child) or in acute cases of HIV exposure (e.g., needlesticks). Such approaches can be applied if the protective immune elements can be identified and if safety can be guaranteed. In this sense the goals of prevention and immune intervention are congruent. Finally, HIV vaccine research has and will continue to stimulate development of vaccines to other retroviruses, particularly human retroviruses associated with various forms of adult T cell leukemia, neurological disorders, and yet undiscovered disease forms.

Acknowledgments

For their contributions of unpublished data the following individuals are gratefully acknowledged: Ruth Ruprecht (Dana Farber Cancer Center), Edward Hoover (Colorado State University), Dante Marciani (Cambridge Bioscience), Oswald and William Jarrett (University of Glasgow), Arsene Burny (University of Brussels), Ronald Montelaro (Louisiana State University), Preston Marks and Murray Gardner (University of California at Davis), and Thomas Palker and Barton F. Haynes (Duke University).

Bibliography

Bolognesi, D. P. (1990). Immunobiology of the HIV envelope and its relationship to vaccine strategies. *In* "Molecular Biology and Medicine." **7**, 1–15. Academic Press, New York.

Desrosiers, R. C., Wyand, M. S., Kodama, T., Ringler, D. J., Arthur, L. O., Sehgal, P. K., Letvin, N. L., King, N. W., and Daniel, M. D. (1989). Vaccine protection against simian immunodeficiency virus infection. *Proc. Natl. Acad. Sci. U.S.A.* **86,** 6353–6357.

Haase, A. T. (1986). Pathogenesis of lentivirus infections. *Nature* (*London*) **322,** 130–136.

Koff, W. C., and Hoth, D. F. (1988). Development and testing of AIDS vaccines. *Science* **241,** 426–432.

Matthews, T. J., and Bolognesi, D. P. (1988). AIDS vaccines. *Sci. Amer.* **259,** 120–127.

Murphey-Corb, M., Martin, L. N., Davison-Fairburn, B., Montelaro, R. C., Miller, M., West, M., Ohkawa, S., Baskin, G. B., Zhang, J., Putney, S. D., Allison, A. C., and Eppstein, D. A. (1989). A formalin-inactivated whole SIV vaccine confers protection in macaques. *Science* **246,** 1293–1297.

Nakamura, H., Hayami, M., Ohta, Y., Ishikawa, K., Tsujimoto, H., Kiyokawa, T., Yoshida, M., Sasagawa, A., and Honjo, S. (1987). Protection of cynomolgus monkeys against infection by human T-cell leukemia virus type-1 by immunization with viral env gene products produced in *Escherichia coli*. *Int. J. Cancer* **40,** 403–407.

Osterhaus, A., Weijer, K., UytdeHaag, F., Knell, P., Jarrett, O., Akerblom, L., and Morein, B. (1989). Serological responses in cats vaccinated with FeLV ISCOM and an inactivated FeLV vaccine. *Vaccine* **7,** 137–141.

Portetelle, D., Burny, A., Desmettre, P., Mammerickx, M., and Paoletti, E. (1990). "Development of a Specific Serological Test and an Efficient Subunit Vaccine to Control Bovine Leukemia Virus Infection." The International Association of Biological Standardization, Vaccine, (in press).

Schafer, W., and Bolognesi, D. P. (1977). Mammalian C-type oncornaviruses: Relationships between viral structural and cell-surface antigens and their possible significance in immunological defense mechanisms. *Contemp. Top. Immunobiol.* **6,** 127–167.

Varmus, H. (1989). Retroviruses. *Science* **240,** 1427–1435.

Wong-Staal, F., and Gallo, R. C. (1985). Human T-lymphotropic retroviruses. *Nature* (*London*) **317,** 395–403.

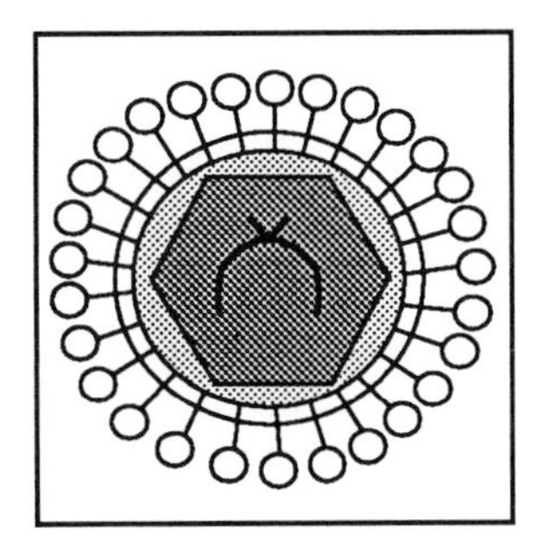

Retroviruses as Vectors for Gene Transfer

RALPH DORNBURG, *University of Medicine and Dentistry, New Jersey–Robert Wood Johnson Medical School*

HOWARD M. TEMIN, *McArdle Laboratory for Cancer Research, University of Wisconsin*

Glossary

Antigen Structure (e.g., a protein or polysaccharide) that can be recognized by an antibody
Antisense RNA RNA transcript of a gene in the opposite orientation to the gene (complementary to the normal mRNA)
cDNA Complementary DNA copy transcribed from an mRNA template rather than a DNA template
Enhancer DNA sequence that increases the efficiency of transcription from a promoter
Long terminal repeats Sequences located at the 5′ and 3′ ends of retroviral DNA
Promoter DNA sequence that is recognized by RNA polymerases and drives gene expression
Oncogene Gene whose expression results in the malignant transformation of a cell
Protooncogene Cellular gene (most probably involved in differentiation and/or cell division), from which an oncogene evolved
Transfection Introduction of genetic material into a cell by experimental procedures
Vector Genetically engineered DNA construct to introduce and express a gene in a target cell

RETROVIRUSES ARE RNA viruses that replicate through a DNA intermediate which is integrated into the host cell genome. Retroviruses are widespread in nature and can be associated with different forms of malignant tumors. Thus, they are also termed RNA tumor viruses. Many tumorigenic retroviruses contain a gene (i.e., an oncogene) in their genome, in addition to or substituting for their replication genes. The expression of the oncogene results in the malignant transformation of the infected cells. Oncogenes are of cellular origin and were picked up by retroviruses in the course of earlier infections. Thus, retroviruses act as vehicles to transfer modified cellular genes from cell to cell and from organism to organism. This ability, as well as the high efficiency of retrovirus replication, makes these viruses useful for the genetic engineering of transfer vectors to study a large variety of biological processes. Moreover, because of their properties, retroviral vectors are also being used in human gene therapy and for marking human tumor cells.

I. Retroviral Life Cycle

In general the retroviral life cycle resembles that of many other animal viruses, in that it can be divided into virus attachment and entrance into cells, a synthesis period, and a period of virus assembly and release. However, the molecular mechanisms of replication are unique for this family of viruses (and a few related viruses) and follows a rather complicated, but highly efficient, pathway.

A. Virions, Genome, and Taxonomy

Retrovirus virions are medium-sized particles with a diameter of about 100 nm, consisting of a core structure surrounded by a lipid bilayer membrane (Fig. 1). In electron micrographs the shape of the core varies among different retroviruses, ranging from spherical to rodlike forms. The core consists of viral proteins called gag (*g*roup-specific *a*nti*g*en), the viral DNA polymerase and integration protein, two identical copies of the viral genomic RNA, and associated tRNA. The lipid bilayer of the virion is derived from the host cell membrane and contains

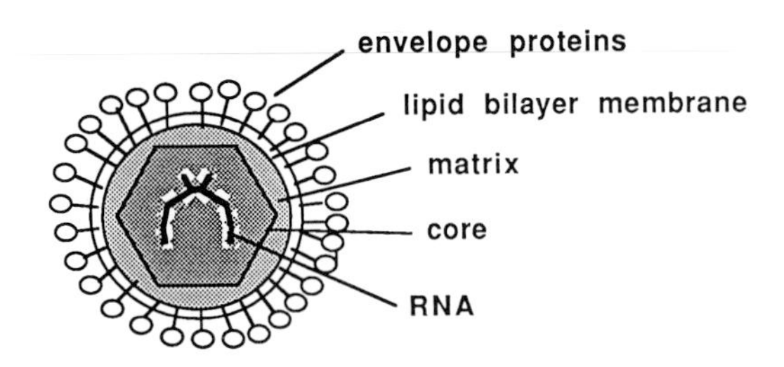

FIGURE 1 A retroviral particle. Two genomic RNA molecules are encapsidated in a core structure. The core is surrounded by a lipid bilayer membrane. Envelope proteins are embedded in the bilayer membrane. For a more detailed explanation see the text.

viral envelope proteins which look like spikes in electron micrographs. These proteins recognize and interact with host cell surface receptors and enable the virus to penetrate the cell.

The RNA genome of retroviruses resembles a eukaryotic mRNA, has a 5′ cap, and contains a poly(A) tail. It has the same polarity as the viral mRNAs, and therefore it is called plus-stranded RNA. In simple retroviruses it contains three genes coding for proteins (Fig. 2). The *gag* gene usually codes for four core proteins: the matrix protein, which is mainly found in the matrix and probably forms the junction between the core and the lipid bilayer; the capsid protein, which is the major component of the core; the nucleocapsid protein, which is tightly bound to the RNA; and the protease protein. They are also designated with the letter "p" and a number that reflects the molecular weight in thousands (e.g., p 20). In all retroviruses the core proteins are derived from a single polypeptide precursor by proteolytic cleavage by the viral protease PR. The *pol* gene codes for the viral polymerase—an RNA-dependent DNA polymerase (i.e., reverse transcriptase)—and the viral integration protein.

DNA polymerase and integration protein are first translated with the core proteins into a single polypeptide and then are cleaved into the two active enzymes by the viral protease. The products of the *env* gene are two envelope proteins, surface protein and transmembrane protein, which are found on and in the lipid bilayer membrane, respectively. They are also derived from a single precursor protein, cleaved by a cellular enzyme, and glycosylated, as are cellular membrane proteins. Hence, they are also termed glycoproteins (e.g., gp70); the number reflects the molecular weight in thousands. Recently, it has been reported that some retroviruses, such as the human T cell leukemia viruses and the human immunodeficiency virus, carry additional genes other than *gag, pol,* and *env*. Some of these genes have essential functions in regulatory viral gene expression.

In addition to the protein coding genes, the retroviral RNA genome contains several regulatory sequences (also called *cis*-acting sequences) required for efficient viral replication. These sequences are located primarily at the ends of the genome.

In modern molecular biology the nucleotide sequence of the retroviral genome serves as the yardstick for classification. Therefore, the most commonly studied retroviruses are organized into six groups: the avian leukosis–sarcoma viruses, the avian reticuloendotheliosis viruses and mammalian leukemia–viruses, the mouse mammary tumor viruses, the primate Type D viruses, the human T cell leukemia-related viruses, and the lentiviruses. Sequence comparisons on the nucleotide and amino acid levels have led to the establishment of evolutionary trees.

B. Life Cycle

To enter a cell, the virus first has to attach to the cell membrane of the target cell (Fig. 3). This attachment is mediated by the surface protein, which recognizes a specific receptor of the target cell. After attachment the virus is absorbed by endocytosis (human immunodeficiency virus can enter directly). Fusion of the membranes leads to the release of the core structure into the cytoplasm of the infected cell. Now the virus has access to the cellular nucleoside triphosphates and starts to copy its RNA genome into a double-stranded DNA. In this complicated process all enzymatic reactions are carried out by the viral reverse transcriptase (Fig. 4).

As a result of this replication process, the protein coding regions of the retroviral genome are flanked by long terminal repeats (LTRs). The LTRs carry specific sequences (i.e., short inverted repeats) at their ends which are essential for the efficient integration of the provirus into the host genome: Sequences at the 5′ end of the left LTR and sequences at the 3′ end of the right LTR form the attachment site recognized by the viral integration protein which carries out this process. Thus, integration is sequence specific with regard to the viral DNA. However, no sequence specificity is known regarding the nucleotide sequence of the integration site in the host genome, although "hot spots" of integration have been reported. As a result of the integra-

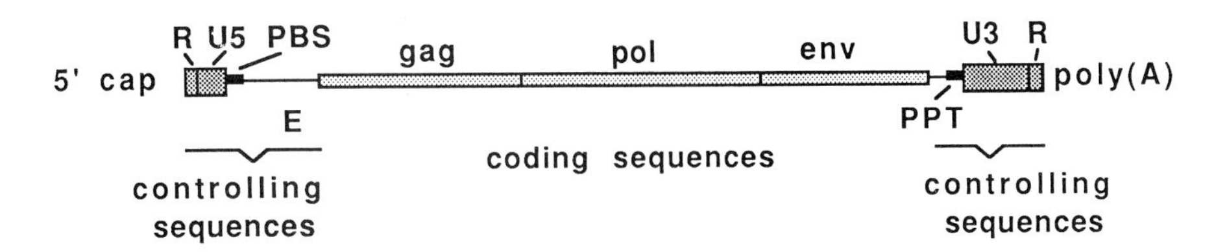

FIGURE 2 Organization of a C-type retrovirus RNA genome. The genome contains three protein coding genes. Regulatory sequences are located mainly at the ends. For a more detailed explanation see the text. U3, U5, Unique in all retroviral RNAs at the 3′ and 5′ ends, respectively; R, repeated region; PPT, purine-rich region; E, encapsidation sequence.

tion, four to six nucleotides of the chromosomal DNA are duplicated, depending on the viral integration protein. The integrated DNA is called the provirus.

After integration of the provirus, RNA transcription produces genomic retroviral RNAs. RNA transcription is performed by cellular RNA polymerase II and is driven by a promoter and enhancers present in the U3 region of the LTRs. The 3′ end RNA processing is regulated by viral sequences present

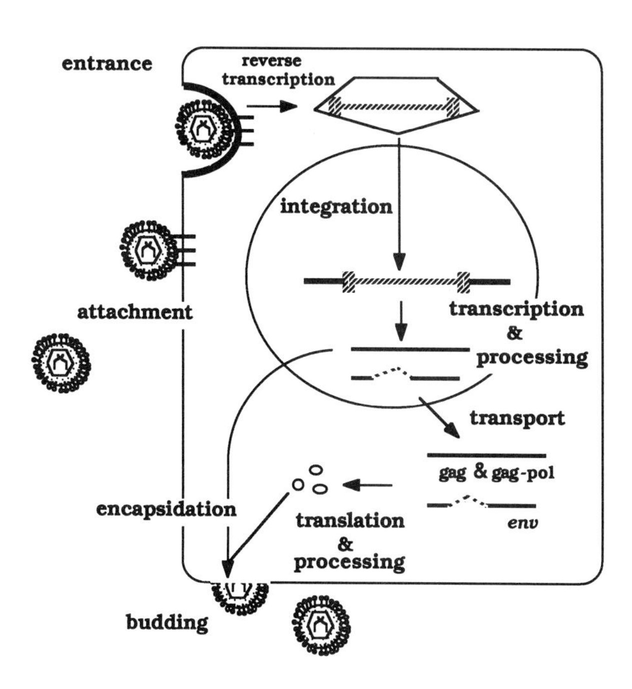

FIGURE 3 Retrovirus life cycle. After entrance of the virion into the cell, the retroviral RNA is copied into a double-stranded DNA, which is integrated into the host genome. DNA transcription and RNA processing results in genomic RNA, as well as retroviral mRNAs. Virus assembly takes place at the cell membrane and results in the budding of retroviral virions. For a more detailed explanation see the text.

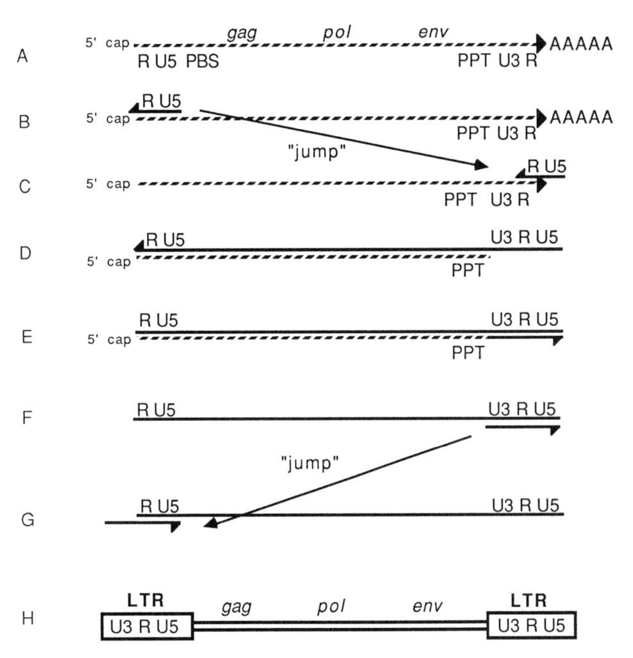

FIGURE 4 Reverse transcription of the retroviral RNA. (A) Retroviral genomic RNA. (B) A short cDNA copy (solid line) of the viral RNA is synthesized using as a primer a cellular tRNA which hybridizes to the primer binding site (PBS) located near the 5′ end of the viral RNA. (C) Next the RNA moiety of the resulting double-stranded DNA–RNA hybrid is removed, allowing the short single-stranded cDNA copy to "jump" to a repeated region (R) at the 3′ end of the second virion RNA molecule and to hybridize to it. (D) This cDNA serves as a primer for the cDNA synthesis of the complete viral genome. As a result of this jump, a region designated U5 (because it is unique in all retroviral RNAs as the 5′ end) is now attached to the 3′ end of the single-stranded cDNA copy. Next the RNA moiety of the DNA–RNA hybrid is cut next to a purine-rich region (PPT). (E) This polypurine tract serves as a primer for the second-strand DNA synthesis. (F) Then all of the remaining RNA is degraded. (G) The double strand is melted, enabling the short second strand to jump to the other end of the first cDNA strand. (H) The single-stranded regions are filled in to complete the synthesis of a double-stranded DNA molecule. In the course of this DNA synthesis, a region called U3 (unique at the 3′ end of retroviruses) is duplicated. As a result of this replication mechanism, the protein coding regions are flanked by long terminal repeats (LTRs). All enzymatic reactions are carried out by the viral reverse transcriptase.

in the (repeat) R or U3 (depending on the particular virus) and U5 regions. In any case the RNA transcript starts at the first nucleotide of R in the left LTR and ends with the last nucleotide of R in the right LTR. As a result RNA transcripts are identical to the original genomic viral RNA. gag proteins and reverse transcriptase are translated from genomic RNA. Expression of the envelope proteins is from spliced genomic RNAs and follows the pathway of cellular membrane proteins. [*See* DNA AND GENE TRANSCRIPTION.]

To complete the viral life cycle, genomic viral RNAs are encapsidated by gag proteins to form core structures. The selective encapsidation of genomic viral RNAs is mediated by specific encapsidation sequences (called E in avian retroviruses or ψ in murine retroviruses) located 3′ from the primer binding site on the viral RNA. Core structures interact with viral envelope proteins, which are embedded in the cell membrane. As a result of this interaction, virus particles bud from infected cells to give progeny virus. Finally, the proteolytic cleavage of precursor proteins takes place immediately after the budding process, resulting in virions ready to infect fresh target cells.

As a result of this replication process, retroviruses often do not kill or lyse the infected cells. Instead, the cell machinery is continuously used for virus production. Moreover, after the provirus becomes a part of the cell genome, both daughter cells carry the provirus and produce retrovirus particles. Retroviral replication can be efficient: Many thousands of virus particles can be produced per day from a single cell infected with one provirus.

C. Endogenous Retroviruses

Retroviruses are also able to infect germ-line cells or their precursors in preimplantation embryos. As a result the progeny carry retroviral proviruses in all body cells. Such retroviruses are called endogenous, to distinguish them from those resulting from exogenous infections. Endogenous retroviruses are found in all vertebrates (including humans) investigated thoroughly.

There are also other cellular DNA sequences with partial nucleotide sequence homology to retroviruses. These and other repeated cell sequences probably arose by the reverse transcription of RNAs. They are estimated to make up about 10% of the mammalian genome.

II. Natural Retroviral Vectors

Many retroviruses contain another gene (i.e., an oncogene) in their genomes. Expression of the oncogene results in the malignant transformation of the infected cells. Such retroviruses are called highly oncogenic retroviruses. They arose by recombination of the viral genome with cellular protooncogenes. As a result of this process, in most oncogenic retroviruses the viral oncogene substitutes for parts of the protein coding regions of the viral genome (Fig. 5). Thus, the viral oncogene product is often a fusion protein of virus and cell sequences. Oncogenes are always expressed from transcripts originating from and terminating in the viral LTRs. Most highly oncogenic retroviruses cannot synthesize all of the proteins necessary for retroviral replication. Thus, they are replication–defective. [*See* ONCOGENE AMPLIFICATION IN HUMAN CANCER.]

Highly oncogenic retroviruses are natural gene transfer vehicles (i.e., vectors) which carry a nonviral gene. Defective highly oncogenic viruses can only infect fresh target cells if their host cell is also infected with a wild-type virus, called a "helper" virus. The helper virus provides those viral proteins the defective retrovirus cannot synthesize as a result of the deletion and substitution in its genome. However, highly oncogenic retroviruses still contain all of the *cis*-acting sequences necessary for retroviral replication. Thus, their genomic retroviral RNA is encapsidated into virions (provided by the helper virus) to form infectious retroviral particles.

Retroviruses without an oncogene can also induce tumors. It has been documented in several cases that insertion of a provirus near a protooncogene can result in the uncontrolled and/or increased

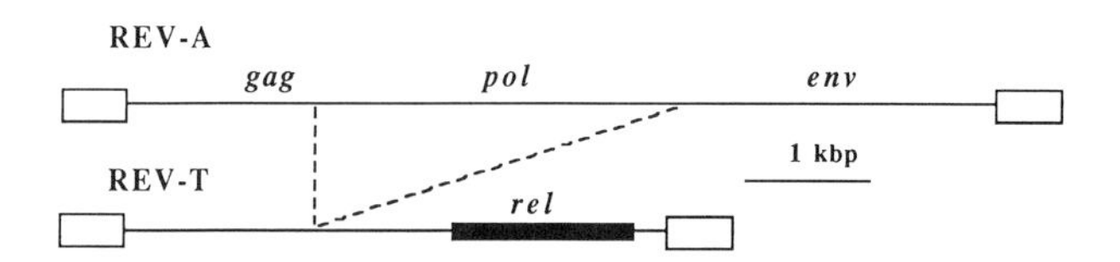

FIGURE 5 A highly oncogenic retrovirus. A provirus of a replication-competent avian retrovirus, reticuloendotheliosis virus A (REV-A), is shown at the top. Recombination with a cellular protooncogene (called c-*rel*) resulted in the highly oncogenic retrovirus REV-T. In most highly oncogenic retroviruses the oncogene substitutes for viral protein coding sequences. Retroviruses that carry an oncogene are naturally occurring gene transfer vectors.

gene expression of that oncogene, leading ultimately to transformation of the infected cell.

III. Retroviral Vectors

A. Vectors, Helper Cell, and Experimental Design

Oncogenic retroviruses are used as a model in the construction of retroviral gene transfer systems, which usually consist of two components: the retroviral vector which contains the gene of interest replacing retroviral protein coding sequences and a helper cell which supplies the retroviral proteins for the encapsidation of the vector genome.

Retroviral vectors have been mainly derived from chicken and murine retroviruses and are constructed in several molecular cloning steps. First, a retrovirus provirus is cloned in a bacterial plasmid, in which it can be amplified to obtain large quantities of DNA for genetic engineering purposes. Next the viral protein coding genes are removed and replaced by the gene(s) of interest. Figure 6 shows some examples of retroviral vectors. Genes can be inserted in the same or reverse orientation to the vector. They can be expressed by the LTR promoter, additional internal (inducible) promoters,

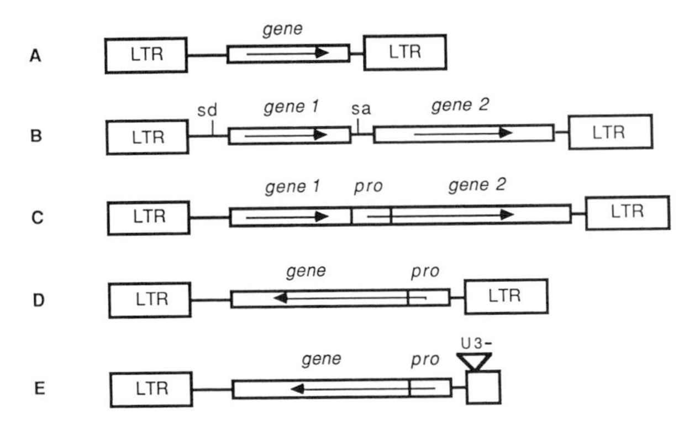

FIGURE 6 Retroviral vectors. All kinds of genes can be inserted into a retroviral vector in the same orientation (A–C) or in the reverse orientation to the vector (D and E). Genes can be expressed from the long terminal repeat (LTR) promoter (A–C), spliced mRNAs (B), or internal promoters (C–E). Retroviral vector constructs with a deleted U3 region (U3−) in the right LTR result in a provirus without an LTR promoter after one round of replication. Gene expression is only performed from the internal promoter (E). sd, Splice donor site; sa, splice acceptor site; pro, internal promoter. The vector constructs are in bacterial plasmids. The plasmid sequences which abut the LTRs are not shown.

and/or spliced RNAs. Genes inserted into the reverse orientation must be expressed from an internal promoter. The inserted genes can contain introns. Introns of genes in the same orientation as the vector are lost as a consequence of retroviral replication. In many vectors selectable marker genes (usually genes for resistance to antibiotics) have been inserted to obtain high virus titers (see Section III).

Helper cells carry a defective retrovirus provirus which is constitutively expressing retroviral proteins, but cannot encapsidate its own genomic RNAs, as a result of deletion of the encapsidation sequence in the viral nucleic acid. Helper cells have also been made by the transfection of plasmid constructs into established cell lines to express retroviral proteins from nonretroviral promoters. Such helper cells avoid the risk of recombination between the retroviral vector and the helper sequences.

Formation of a helper-free retrovirus vector for gene transfer is outlined in Fig. 7. Helper cells are transfected with retroviral vector DNA constructs which carry appropriate encapsidation sequences. Thus, RNA transcripts of such constructs are packaged into virions provided by the helper cell. The virus produced from the helper cell is called "helper free," since it contains no replication-competent helper virus. If the vector has a selectable gene, helper cells can be selected for the expression of the marker gene. As a result each selected helper cell produces virions containing the RNA transcript of the transfected vector. Supernatant tissue culture medium is used to infect fresh target cells. Titers of up to 10^7 infectious virus particles per milliliter of tissue culture medium have been obtained. The efficiency of introducing a gene by other methods is several orders of magnitude lower than that by retroviral infection.

B. Experiments with Retroviral Vectors

A large variety of eukaryotic, bacterial, and viral genes have been inserted and expressed in retroviral vectors. So far, there is no report that any gene could not be expressed.

Retroviral vectors have been used mainly in tissue culture experiments to investigate gene expression and the effects of gene products (e.g., oncogenes or protooncogenes) in particular cells. They were also used to introduce genes for the production of antisense RNAs to investigate the effect of

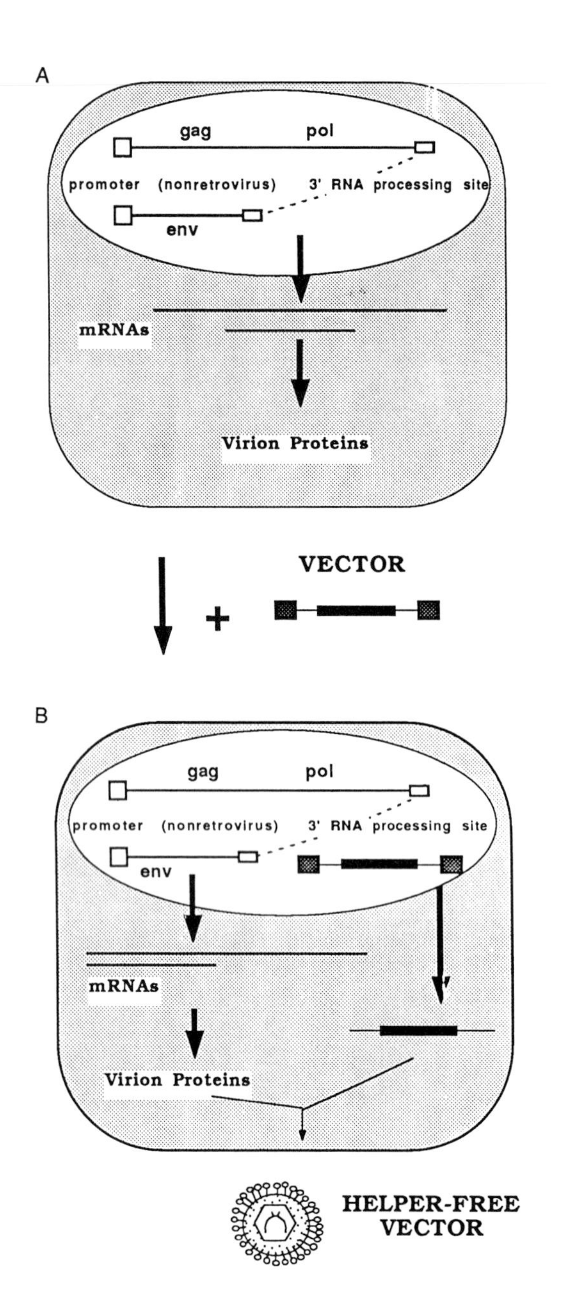

FIGURE 7 Formation of a helper-free retroviral vector for gene transfer. (A) In helper cell retroviral proteins are expressed from different plasmid DNAs. These RNA transcripts do not contain encapsidation sequences. Thus, they are not encapsidated into retroviral particles. (B) Such helper cells are transfected with a retroviral vector plasmid construct. The RNA transcript of the retroviral vector contains a encapsidation sequence and, therefore, is encapsidated into virus proteins supplied by the helper cell. Supernatant tissue culture medium is used to infect fresh target cells.

antisense RNAs on gene expression, particularly in cells infected with other viruses.

Several experiments with retroviral vectors have been performed in early embryos (mainly in mice) to tag chromosomal locations of developmental genes. The idea of these experiments is to destroy the genes involved in cellular differentiation by the insertion of a vector provirus. Animals in which the insertion has taken place show a characteristic developmental defect. Molecular cloning of the chromosomal sequences surrounding the integrated provirus has led to the discovery of developmental genes. Infection of murine embryos with retroviral vectors was also used for cell lineage studies, because all cells derived from a single infected progenitor cell carry the vector provirus at the same location.

In other experiments bone marrow cells (i.e., stem cells) infected or marked, with retroviral vectors have been injected into lethally irradiated mice to study hematopoiesis and the development of the immune system. The marked stem cells undergo differentiation in the injected animal. Comparison of the location of integrated vector proviruses in different fully differentiated cells from such animals led to insights into how cells of the immune system develop. Cell transplantation experiments were also performed to study possible application of retroviral vectors in gene therapy. For example, the gene for human blood clotting Factor VIII was introduced by a retroviral into rat fibroblasts in tissue culture. Cells expressing Factor VIII were transplanted in rats. Human Factor VIII was detected in the blood of rats. In addition, the infection of preimplantation embryos with retroviral vectors is considered for use in farm animals as a substitute for gene transfer by microinjections.

C. Problems and Limitations

As a consequence of the properties of retroviruses, there are several limitations in the use of retroviral vectors. There appears to be a maximum limit on the size of the RNA that can be encapsidated into a retrovirus virion. Many eukaryotic genes contain large introns and/or are regulated by sequences far upstream or downstream from the coding sequences. Thus, several genes can be too large to be inserted in a retroviral vector. Moreover, gene regulation from internal promoters can interfere with virus production, or the LTR promoter can suppress or stimulate the internal promoter. Attempts have been made to overcome this shortcoming with the construction of "suicide vectors" (see Fig. 6E). In addition, retroviral vectors often behave in unpredictable ways. Many sequences inserted into a retroviral vector appear to be unstable, and there is sometimes a strong selection of spontaneous mutations against particular constructs.

D. Retroviral Vectors in Human Gene Therapy

Somatic gene therapy could be valuable in treating human single-gene diseases. Today, approximately 3000 human diseases are known which result from a single-defective gene. Some diseases are rare; others occur with relatively high frequencies (e.g., blood clotting Factor VIII deficiency or sickle cell anemia). It is estimated that 1–2% of newborns are affected with single-gene disorders. The present therapies are mainly based on special diets, the injection of proteins, and/or blood transfusions (e.g., for the supplement of blood clotting factors). However, in many cases such therapies are ineffective and unsatisfactory. Thus, there is a desire to cure such genetic disorders by introducing a functional gene into the body cells of such patients. The functional gene could be expressed even while the inherited defective gene persists.

Gene therapy experiments with retroviral vectors have been performed in animals (mostly mice and rats) with genes for globin and blood clotting factors. Although the results of these experiments have been promising, many questions are being raised concerning gene therapy in humans. In addition to the technical problems of expression with retroviral vectors discussed above, the questions concern safety, efficiency, and social effects. For example, can the insertion of a retroviral vector into the genome lead, in some cells, to protooncogene gene expression and to the development of tumors (see Section II)? Is there a chance that infectious viruses will be produced by recombination with retroviral sequences in the helper cell which can infect other people and/or germ-line cells? What are the costs in relation to the efficiency of the treatment? A committee of the National Institutes of Health (Human Gene Therapy Subcommittee of NIH Recombinant DNA Advisory Committee) has been discussing these and related questions since 1984. However, now these questions have been resolved satisfactorily so that retrovirus-mediated gene therapy of somatic cells and retrovirus-mediated cancer therapy are to begin.

Bibliography

Dornburg, R., and Temin, H. M. (1988). Retroviral vector system for the study of cDNA gene formation. *Mol. Cell. Biol.* **8,** 2328.

Dougherty, J. P., and Temin, H. M. (1986). High mutation rate of a spleen necrosis virus-based retrovirus vector. *Mol. Cell. Biol.* **6,** 4387.

Eglitis, M. A., and Anderson, W. F. (1988). Retroviral vectors for introduction of genes into mammalian cells. *BioTechniques* **6,** 608.

Miller, A. D. (1990). Retrovirus packaging cells. *Hu. Gene Ther.* **1,** 2. NIH. (1990). The revised "Points to Consider" document. *Hu. Gen. Ther.* **1,** 93.

Temin, H. M. (1986). Retrovirus vectors for gene transfer: Efficient integration into and expression of exogenous DNA in vertebrate cell genomes. *In* "Gene Transfer" (R. Kucherlapati, ed.). Plenum, New York.

Temin, H. M. (1989). Retrovirus vectors: Promise and reality. *Science* **246,** 983.

Temin, H. M. (1990). Safety and societal issues of using retroviral vectors for human gene therapy. *Hu. Gene. Ther.* **1,** in press.

Varmus, H. (1988). Retroviruses. *Science* **240,** 1427.

Varmus, H., and Brown, P. (1988). Retroviruses. *In* "Mobile DNA" (M. Howe and D. Berg, eds.). Am. Soc. Microbiol., Washington, D. C.

Weiss, R., Teich, N., Varmus, H., and Coffin, J. (eds.) (1985). "RNA Tumor Viruses: Molecular Biology of Tumor Viruses," 2nd ed. Cold Spring Harbor Lab., Cold Spring Harbor, New York.

RNases

Ribonucleases, Metabolic Involvement

DUANE C. EICHLER, *University of South Florida*

Glossary

Ataxia Loss of muscular coordination
Eosinophil White blood cell, a type of leukocyte
Interferon Substance capable of inducing a state of resistance in response to infection by a virus
Intrathecal Within the spinal cord
Retrovirus Virus which contains genomic RNA and whose life cycle has a stage in which its RNA is copied to DNA
Ribonucleoprotein particle RNA with bound protein

RIBONUCLEASES (RNases) are enzymes which hydrolyze phosphodiester linkages in RNA. Because RNA can be perceived as the more transient of the nucleic acid molecules when compared to DNA, enzymes which degrade or affect RNA structure can play a key role in the metabolism of RNA. However, far more knowledge and understanding has been gained about the enzymatic capacity of cellular RNases than is known about their actual *in vivo* function. In general, ribonucleases can be considered to participate in one of two roles in RNA metabolism: either in the removal of an RNA species by degradation, or in the maturation of an RNA species by affecting the formation of a functionally active RNA from an inactive precursor.

I. Background

Like other polymer-degrading enzymes, ribonucleases may be distinguished and therefore classified by their mode of attack, mechanism of cleavage, or by their substrate specificity. For example, the term exoribonuclease defines an enzyme which requires recognition of an RNA terminus in order to initiate attack, and includes not only enzymes which release mononucleotides as degradation proceeds from a terminus but also enzymes which release oligonucleotides as well. Because RNA has direction, known as polarity, the direction of attack of an exonuclease may be used to further distinguish ribonucleases as either a $5' \rightarrow 3'$ or $3' \rightarrow 5'$ exonuclease. The corollary, endoribonuclease, refers to a nuclease which does not require a free terminus to initiate attack. Ribonucleases also show specificity for RNA structure and therefore can be distinguished as single or double strand-specific ribonucleases. Finally, ribonucleases can be distinguished by their mode of attack at the phosphodiester linkage. Ribonucleases which require the 2′-OH group of the ribose moiety to initiate attack at the adjacent phosphodiester bond are termed phosphotransferase-type RNases (or cyclizing RNases). By necessity, this group of RNases shows absolute specificity for RNA and the products of hydrolysis contain either a 3′ phosphate group or 2′,3′-cyclic phosphate termini. Also, this group often includes RNases which demonstrate base specificity as well. For example, pancreatic RNase is a phosphotransferase-type RNase specific for pyrimidine bases and is one of the most extensively studied of all known ribonucleases. Ribonucleases which utilize water as the nucleophilic agent to attack the phosphodiester linkage are termed phosphodiesterase-type RNases. Cleavage of the phosphodiester bond by a phosphodiesterase-type RNase results in products with either a 3′- or 5′-terminal phosphate depending on which side of the phosphodiester bond cleavage was initiated. Phosphodiesterase-type RNases also include enzymes which can cleave both RNA and DNA, for example, spleen phosphodiesterase. A third mode of attack distinguishes the phosphorylase-type RNases that utilize inorganic phosphate as the nucleophilic agent to attack the phosphodiester linkage. Although this third group may be quite interesting, ribonucleases of this type have not been very well characterized in mammalian tissue.

Categorization of ribonucleases by their specificity and/or mechanism of action may tell us something about the enzymatic properties of the ribonuclease, but it does not address the metabolic role of the ribonuclease. Rather than listing groupings of RNases based on their catalytic properties, this article will focus on metabolic roles of ribonucleases. In this regard, cleavage of a phosphodiester bond by a ribonuclease can be considered as playing one of two possible roles metabolically. Ribonuclease cleavage may result simply in the degradation of an RNA species where the resulting products of degradation are used to replenish ribonucleotide pools, or cleavage may, instead, affect the maturation of an RNA species resulting in formation of a biologically functional RNA molecule. In addition, ribonucleases may have metabolic roles that do not seem to directly relate to their capacity to hydrolyze RNA. Examples of some of the metabolic roles of ribonucleases will be discussed in the body of this text. The examples chosen are not so much meant to be all inclusive but rather to provide a more general understanding of the metabolic role of ribonucleases in human tissue and physiology.

The purpose of this article is to bring the nonspecialist an overall view on the status of information relating to ribonuclease function in human tissue. Descriptions of specific ribonucleases will be limited; therefore references at the end of this article are, in part, intended to provide a resource for more detailed information relating to known RNases. For convenience, specific enzymes which are discussed in this article will be referred to by their most common names.

II. Pyrimidine-Specific Human Ribonucleases

A. Pancreatic-Like Ribonucleases

In mammalian tissue, two major groups of pyrimidine-specific ribonucleases have been distinguished as secretory and nonsecretory RNases. Secretory RNases, exemplified by pancreatic ribonuclease, are mainly found in secretory organs such as pancreas and submaxillary glands. The secretory RNases have been extensively studied and a considerable volume of literature exists concerned with their characterization. This group of pyrimidine-specific RNases is distinguished by their preferential hydrolysis of the homoribopolymers poly(C) and poly(U), and a pH optimum for the hydrolysis of RNA at pH 7.5. In contrast, nonsecretory RNases are generally less well characterized and exist in considerable amounts in tissue such as liver, kidney, and spleen. Nonsecretory RNases hydrolyze RNA most effectively at pH 6.5, and are less active toward poly(C) than secretory RNases. Both groups, however, still share extensive amino acid sequence homology and such common properties as the endonucleolytic cleavage of RNA and 2′,3′-cyclic pyrimidine products.

Although considerable work has been carried out with pyrimidine-specific RNases, their specific metabolic role in human tissue has not been well established. In most cases, our understanding suggests that the pyrimidine-specific RNases are predominantly involved in RNA degradative processes rather than RNA maturation, but the types of RNA and the actual involvement are still unclear. The regulation of their activity, particularly in tissue, seems often to be correlated with the presence of specific protein inhibitors which bind to the RNase, rendering it inactive. The best studied of these inhibitors is human placental RNase inhibitor. These protein inhibitors form a very tight complex with pancreatic-like RNases but can be rapidly inactivated by sulfhydryl modification reagents such as *N*-ethylmaleimide, restoring RNase activity. *In vivo*, inactivation of the specific RNase inhibitor has been suggested as a way of controlling the latent form (active form) of nonsecretory RNase activity. Additional signals which may further affect substrate selection *in vivo* have not been elucidated.

In addition to providing information on the structure and function of pyrimidine-specific ribonucleases, their study has stimulated attempts to correlate levels of ribonuclease released into the circulation or excreted in the urine that might serve as diagnostic markers for a variety of diseases. However, the multiplicity of these enzymes and the variety of possible origins from which they may be derived have made it difficult to perform qualitative and/or quantitative analysis to establish strict tissue correlations required for diagnostic purposes.

B. Angiogenin

The primary structure of tumor-derived angiogenin, a protein that stimulates formation of blood vessels, was shown to be very similar to that of the pancreatic ribonucleases. In keeping with this similarity, the angiogenic protein was also shown to possess ribonuclease activity. The ribonuclease activity of angiogenin, however, could be distinguished from that of human pancreatic RNase, particularly in its limited ability to cleave either 18S or 28S ribosomal RNA and the lack of activity under conditions of a standard pancreatic RNase assay. Nevertheless, both the RNase activity and angiogenin activity of this protein are inhibited by human placental RNase inhibitor, suggesting that similar tissue-specific inhibitors may regulate the activity of angiogenin *in vivo*.

C. Eosinophil-Derived Neurotoxin and Eosinophil Cationic Protein

When injected intrathecally into rabbits, both the eosinophil-derived neurotoxin and the eosinophil cationic protein will produce the Gordon phenomenon, a neurologic syndrome characterized in the rabbit by stiffness, ataxia, muscle weakness, and muscle wasting. Interestingly, both the eosinophil-derived neurotoxin and eosinophil cationic protein were first linked by their structural similarity with human pancreatic RNases; however, further analysis suggests that they are better related to the nonsecretory ribonucleases. In addition, the observed amino acid similarities also suggest that eosinophil-derived neurotoxin, eosinophil cationic protein, and ribonucleases may be part of a multigene family that has evolved by the process of gene duplication, giving rise to the variety of proteins in the pyrimidine-specific ribonuclease family. The findings of homology between tumor-derived angiogenin and pancreatic-like RNases also indicate that the angiogenin protein belongs to this family.

III. Ribonuclease Involvement in RNA Stability

Although transcriptional regulation may often be the major determinant of gene expression, the stability of an RNA molecule will ultimately determine how long it can function. For this reason the rate at which an RNA is degraded can serve as a major control point in the regulation of gene expression. The wide diversity in decay rates of RNA seen in both procaryotes and eukaryotes appears to be due to recognition of some structural features in the individual RNAs. For example, special structures at the 3′-termini of many messenger RNAs (mRNAs) appear to provide protection against rapid exonucleolytic digestion, but the selectivity of the decay process appears to be determined to a large extent by interactions between endonucleases or other factors and internal mRNA structures.

The existence of RNA stability factors supports the idea that RNA decay rates may be determined by multiple interactions, especially of two kinds. First, sequence within the RNA would determine the "intrinsic" or "unregulated half-life," independent of regulatory factors. These sequences might influence the susceptibility of the RNA to ribonucleases. In contrast, some sequences within the RNA might interact with regulatory factors that modify the intrinsic half-life. *In vivo*, it may well be a combination of both intrinsic and regulated half-life that ultimately work together to determine the overall decay rate of an RNA molecule.

A. mRNA Decay

The half-lives of eukaryotic mRNAs vary greatly. While globin mRNA has a half-life in excess of 50 hr, the half-life of the c-*myc* oncogene mRNA is less than 20 min. Covalent modifications of mRNAs [5′ caps and poly(A)] appear to be necessary for the stability of most mRNAs. In particular, 5′ caps appear to protect mRNAs from degradation by a 5′ → 3′ exoribonuclease. Although controversial, most published evidence supports the idea that poly(A) effects mRNA stability by affecting the rate of degradation in mammalian cells. For example, some mRNAs lose their poly(A) before they are degraded and the addition of poly(A) to some deadenylated mRNAs stabilizes them.

Recent studies indicate that the presence of certain 5′ or 3′ untranslated sequences, specific mRNA degradative activities, and cellular ribonuclease inhibitors may be involved in regulating mRNA turnover. Specific sequences within the 5′ untranslated regions of certain mRNAs appear to protect these mRNAs from exoribonuclease attack or to target them for rapid degradation. The specific RNases responsible for mRNA turnover have not been identified, although there is some recent evidence to suggest that eukaryotic mRNAs may be degraded in a 3′ to 5′ manner by exonucleases analogous to the prokaryotic enzymes RNase II and polynucleotide phosphorylase. Several studies suggest that the rate-limiting step in mRNA degradation is endonucleolytic cleavage near the 3′ end that leaves the mRNA susceptible to 3′ exonucleolytic attack. Some features of mRNA decay in mammalian cells suggest the involvement of a ribosome-bound ribonuclease that responds to very specific signals; however, further progress will require the identification of this ribonuclease whose function is controlled by translational processes. In addition, there are cases in which rates of specific mRNA turnover are modulated by hormones and other physiological effectors such as estrogen, growth hormone, cellular differentiation, regulation of DNA synthesis, interferon, cyclic AMP, and factors associated with viral infection.

B. RNase Inhibitor

Many mammalian tissues have been found to contain an inhibitor of neutral RNase activity (see Section II,A). Thus, equilibrium of the association and dissociation between RNases and their inhibitors may play a role in regulating the rate of decay of mRNAs. HeLa cells possess a factor which resembles human placental RNase inhibitor in that the mRNA protecting activity is effective against pancreatic-like RNase activities and that treatment of the extract with *N*-ethylmaleimide completely destroys the protective activity. Interestingly, though, purified human placental RNase inhibitor was unable to inhibit the ribonuclease activity responsible for mRNA decay in the cytoplasmic extract of HeLa cells. [*See* HELA CELLS.]

C. The (2′,5′)-Oligoadenylate Synthetase-Nuclease System

When exposed to viruses or certain chemical inducers, animal cells produce interferons which strongly inhibit translation. Interferons exhibit a wide range of biological and biochemical effects. Their interaction with specific receptors determines the biochemical events and their modification of cellular functions. This is a complex process that is just beginning to be dissected. Two interferon-induced, double-stranded RNA-dependent enzymes have been identified that may play important roles in the regulation of viral and cellular macromolecular synthesis and degradation. They are the (2′,5′)-oligoadenylate synthetase (2′,5′-A_n synthetase) and the p1/eIF-2α protein kinase. Both the interferon-induced enzymes must be activated following induction, and both synthetic and natural double-stranded RNAs (dsRNA) can fulfill the activation requirement *in vitro*. Three enzymes have been shown to play important roles in the 2′,5′-A_n system; a synthetase that catalyzes the formation of the novel oligonucleotides possessing 2′,5′-phosphodiester bonds, and endoribonuclease (RNase L) that is activated by certain 2′,5′-A_n structures, and a phosphodiesterase that catalyzes the hydrolysis of oligonucleotides possessing 2′,5′-phosphodiester bonds. So far, the only biochemical function of 2′,5′-A_n, the unusual product of the interferon-induced synthetase, is the activation of an endoribonuclease that is present as a latent protein in both untreated and interferon-treated cells. This latent RNase L catalyzes the cleavage of both viral and cellular RNA on the 3′ side of -UpXp- sequences (predominantly UA, UG, and UU) to yield products with -UpXp at the 3′ termini. Many different types of single-stranded RNAs are cleaved, including synthetic polyuridylic acid and various natural RNAs; however, ribosomal RNA is, perhaps, the best characterized type of RNA observed to be cleaved both *in vitro* and *in vivo*. [*See* INTERFERONS.]

RNase L has an estimated molecular weight of 75 to 85 kDa and binds 2′,5′-A_n. This enzyme has been identified in cytoplasmic extracts prepared from a variety of mammalian sources. The level of latent 2′,5′-A_n-dependent RNase L present in most types of cells in culture differs less than twofold between untreated and interferon-treated cells. However, culture conditions have been described in which the functional activity of the latent RNase L is regulated by parameters other than the availability of 2′,5′-A_n. For example, in NIH 3T3 cells, the 2′,5′-A_n-dependent RNase L can be independently induced by interferon or by growth arrest. This activity has also been observed to be regulated during cell differentiation.

IV. RNA-Processing Ribonucleases

In eukaryotic cells, most nuclear transcripts are synthesized as larger precursors that must undergo a series of cleavage and/or modification reactions in order to form a smaller functional RNA species. In some cases, the shortening of the primary transcripts results from direct cleavage of phosphodiester bonds, in other cases the shortening results from a series of intramolecular rearrangements that collectively result in excision of the nonexpressed sequences (introns) and ligation of the flanking expressed regions (exons). Some processing steps appear to be completely dependent on the occurrence of earlier steps and may be governed by the order in which the substrate recognition signals become available. Recognition sites may also be composed of unique combinations of secondary and/or tertiary structure, and may result from packaging in the form of a ribonucleoprotein particle (RNP). It is also important to consider that in many instances a large proportion of the primary transcript is discarded during posttranscriptional processing steps. A ribonuclease, therefore, may play either a degradative or processing role, and perhaps in some instances both. The involvement of ribonucleases in the maturation of eukaryotic transcripts is of primary importance, particularly since these cleavage events may serve as critical sites for the overall maintenance of RNA expression in mammalian tissue.

Difficulties in specifically defining processing ribonucleases has been overcome in some cases by our ability to develop adequate *in vitro* systems which mimic the *in vivo* situation. Additionally, recent developments in molecular biology have provided efficient cloning and *in vitro* transcriptional systems to provide defined precursor transcripts, and preparations of cellular extracts which will process these *in vitro*-synthesized transcripts. Antibodies which specifically cross-react with components of nuclear processing systems obtained from patients suffering from the autoimmune disease systemic lupus erythematosus (SLE), have provided extremely powerful tools to elucidate the structure of processing complexes. Our knowledge of RNA processing has therefore expanded rapidly in the last few years, and a step-by-step analysis of the components and mechanisms involved in RNA processing has progressed rapidly.

A. Ribosomal RNA Processing

Synthesis, processing, and assembly of ribosomal RNA (rRNA) occurs in nucleoli. Transcription of mammalian ribosomal DNA by RNA polymerase I yields a large primary transcript of 13.9 kb, which is subsequently cleaved to yield the mature 18S, 5.8S, and 28S rRNA (ribosomal RNA) sequences. The temporal sequence of processing events *en route* to the formation of functional rRNA species has been extensively studied, and the processing events which release the 5′ and 3′ termini of the mature 18S, 5.8S, and 28S rRNA species often occur in a polar fashion, from the 5′ to the 3′ end of the transcript. The precise order and intermediates generated may vary with growth and between species. However, at least six sites are cleaved for the sequential production of the lower molecular weight mature ribosomal RNA species. Several of the cleavage sites have been identified and mapped *in vivo*. In addition to the major cleavages, trimming processes may also be involved in the formation of mature ribosomal RNA products. Precursor rRNA is, however, also cleaved rapidly at positions other than those which release mature ends of rRNA species. One very early cleavage takes place in a region approximately 400 nucleotides downstream of the transcription initiation site in humans. This cleavage was suggested as the primary processing event in mammalian rRNA because its cleavage, followed by a cleavage approximately 600 residues upstream from the 3′ terminus of the ribosomal RNA precursor, yields a 45S rRNA precursor detected *in vivo*.

Processing of the ribosomal RNA precursor is also a nonconservative process since nonutilizable products of rRNA maturation are degraded. For example, if protein synthesis is inhibited (e.g., by puromycin), synthesis and partial processing of precursor RNA still continues, but little or no mature rRNA emerges from the nucleolus. In resting lymphocytes the synthesis of ribosomal RNA can appear to exceed the capacity to be utilized in ribosome assembly. Thus, in both cases, substantial turnover or "wastage" of precursor rRNA occurs in the nucleolus, suggesting that processing of ribosomal RNA not only involves the specific cleavage of larger precursor RNA molecules for production of smaller mature species, but also the degradation of cleavage products during biogenesis as well as that of nonutilizable transcripts.

Although the sites of rRNA processing are reasonably well defined, the mechanism and proteins involved have proved elusive. Development of an *in vitro* system for characterizing these ribosomal RNA maturation processes, similar to those used to analyze the processing of mRNA precursors, has lagged behind. There are only a few reports where limited processing of precursor rRNA has been achieved *in vitro*; first, with processing coupled to transcription systems, and later with transcripts made *in vitro* and then incubated in cellular extracts. A single-stranded endonuclease which specifically recognizes and cleaves at the +650 site in mouse ribosomal RNA (comparable to the +414 site in human ribosomal RNA) represents the only defined enzyme species directly implicated in ribosomal RNA processing.

B. Maturation of mRNA

Because one of the first characteristics used to distinguish them from other RNAs in the nucleus was the heterogeneity of their sizes, RNA polymerase II transcripts in the nucleus were known as heterogeneous nuclear RNA (hnRNA) molecules. Many of these transcripts leave the nucleus as messenger RNA (mRNA) molecules and, as they are being synthesized, they are covalently modified at both their 5′ end and their 3′ end in ways that clearly distinguish them from transcripts made by other RNA polymerases. In addition, many of the transcripts contain noncoding regions, called introns, which are not found in the mature functional mRNAs, and which are removed by a process called splicing during the biogenesis of the mRNA. Development of *in vitro* systems which mimic the *in vivo* ability to remove the intron sequence while splicing together the coding sequences facilitated the rapid elucidation of the components involved. Unlike normal cleavage reactions, splicing reactions involve the intramolecular rearrangement of phosphodiester bonds. These reactions are catalyzed by a structure called a splicesome which assembles the boundaries of the intron sequence and coding sequences (exons) into a complex which permits the specific rearrangements to occur. The splicesome is composed of small ribonucleoprotein particles called snRNPs. The order and assembly of the snRNPs into the splicesome and their participation in splicing has, in part, been worked out; however, the actual catalytic participants have not. Noncoding sequences which are removed during splicing are very rapidly degraded, presumably due to the uncapped nature of their 5′ ends. These results suggest the active participation of a ribonuclease which recognizes free, uncapped 5′ ends to initiate degradation.

The 5′ end of the newly synthesized RNA molecule is first "capped" by addition of a methylated guanosine nucleotide. Capping occurs almost immediately, and it involves the condensation of the triphosphate group of a molecule of GTP with a diphosphate left at the 5′ end of the initial transcript. The 3′ ends of most RNA polymerase II transcripts are defined not by termination of transcription, but by a second modification in which the growing transcript is cleaved at a specific site and a poly(A) tail is added by a separate polymerase to the cut 3′ end. In the case of most eukaryotic mRNAs, this occurs posttranscriptionally in a two-step reaction involving cleavage at the poly(A) site and subsequent addition of approximately 200 adenosine residues to the newly made 3′ end. A hexamer sequence (AAUAAA) 18 to 30 nucleotides upstream of the poly(A) addition site is indispensable *in vivo* and *in vitro*. A less well defined GU-rich and U-rich sequence downstream of the poly(A) site has also been shown to affect 3′ end formation. However, it has not been possible to demonstrate that the downstream sequences play a role specific for each pre-mRNA species. Both the poly(A) polymerase activity and the ribonuclease activity responsible for providing the available 3′ end do not separate on extensive purification, suggesting that the component which contains the polymerase activity is also required for the cleavage reaction. Similar to the splicing reactions, a variety of experiments have produced evidence indicating that a snRNP is involved in the cleavage–polyadenylation reaction. RNA fragments containing the AAUAAA sequence have been immunoprecipitated by anti-snRNP antibodies and by antibodies directed against the trimethyl cap structure of U-type small nuclear RNAs. The function of the poly(A) tail is not known for certain, but it may play a role in the export of mature mRNA from the nucleus and, as mentioned in Section III,A, there is also evidence that polyadenylation of mRNA helps to stabilize at least some mRNA molecules.

C. RNA Polymerase III Products

Like RNA polymerase II, RNA polymerase III acts on a number of different cellular genes. It synthesizes ribosomal 5S RNA, all transfer RNAs (tRNAs), and certain other small RNAs. Similar to other nuclear RNAs, the tRNAs are synthesized in precursor forms that are processed posttranscriptionally to yield the mature tRNA molecules. The precursor tRNAs are longer than the final products and the extra polyribonucleotides are removed from both the 3′ and 5′ ends. The maturation of tRNA also entails changes in the purine and pyrimidine bases. Similarly, the 5S rRNA is synthesized as a precursor and cleaved to yield an approximate 150-nucleotide product.

In eukaryotes, most tRNA genes are transcribed as monomers and only the few extra nucleotides on either side of the mature sequence are removed in processing. RNase P removes, with great accuracy, the extra nucleotides from the transcripts of tRNA genes to yield the correct 5′ terminus of the mature tRNAs. Although this activity was first defined in bacterial cells, recent evidence has supported the presence of this activity in human cells as well. It is a particularly interesting enzyme because the catalytic element of the enzyme is RNA, not protein. The reaction catalyzed by RNase P is not a transesterification (rearrangement reactions) like the splicing reactions involved in mRNA processing. No covalent linkage is formed between the enzyme and the substrate, nor are any intramolecular bonds formed transiently in the substrate during the reaction. Studies of RNase P from eukaryotic cells have proceeded much more slowly than similar studies carried out in bacteria because the enzymatic activity is more labile and less abundant. Recently, anti-RNase P antibodies, derived from patients with SLE, were used to identify the RNA species which copurifies with RNase P derived from human cells.

V. Ribonucleases Specific for RNA/DNA Hybrid Structures

One of the unique specificities observed for some ribonucleases is the recognition of the RNA strand of an RNA/DNA hybrid structure. Enzymatic activities capable of specifically degrading the RNA complement of the hybrid RNA/DNA duplex have been designated as RNase H type activities, where the letter "H" refers to hybrid. RNA/DNA hybrid structures are typically transient structures and such hybrid structures are observed during transcription and during DNA replication at both the origin of replication and at the replication fork. Also, the unique pathway of replication of retroviruses necessitates the formation of an RNA/DNA hybrid structure in order for the virus to convert its genetic information from RNA to DNA.

A. Cellular DNA Replication

During replication, DNA polymerase synthesizes continuously along the parental template strand in the 5′ → 3′ direction; however, on the lagging strand DNA polymerase requires a primer to synthesize each short, newly replicated DNA fragment. A process is therefore required not only to produce the primer but also to remove the base-paired primer to permit repair and ligation of the newly synthesized fragments. Synthesis of this primer involves an enzyme called DNA primase, which uses ribonucleoside triphosphates to synthesize short RNA primers. These primers are about 10 nucleotides long in eukaryotes, and they are made at intervals on the lagging strand, where they are elongated by the DNA polymerase to begin each newly replicated fragment (Okazaki fragment). The synthesis of each Okazaki fragment ends when the DNA polymerase runs into the RNA primer attached to the 5′ end of the previous adjacent fragment. A possible role for RNase H, then, is in the removal of this primer hydrolyzing the RNA component of RNA/DNA hybrids. When the RNase H activity removes the RNA primer, it is replaced with DNA. DNA ligase then joins the 3′ ends of the adjacent DNA fragments to produce a continuous DNA chain from the many newly synthesized DNA fragments made on the lagging strand. [*See* NUCLEOTIDE AND NUCLEIC ACID SYNTHESIS, CELLULAR ORGANIZATION.]

Despite the extensive investigations, the cellular functions of the various RNase H activities have not been clearly established. However, RNase H activity has been reported to rise and fall in concert with replication in whole cells consistent with its role in DNA replication. Cellular RNase H activity has an apparent molecular weight of 70 to 90 kDa and has been observed to associate with larger protein complexes which are involved in replication. In contrast to the viral RNase H activity, the mammalian RNase H activity is an endonuclease releasing di- to oligonucleotides containing 5′-terminal phosphate groups.

B. Retrovirus Replication

Ribonuclease H also turns out to be a ubiquitous activity of retrovirus and is an integral component of the virion RNA-dependent DNA polymerase known as reverse transcriptase. The infecting RNA of retroviruses serves as a template for the synthesis of a single-stranded DNA complement by the viral reverse transcriptase. Reverse transcriptase has three associated enzymatic activities. It copies an RNA molecule to yield double-stranded DNA/RNA. It copies a single strand of DNA to form double-stranded DNA, and it degrades RNA in the DNA/RNA hybrid. The latter activity is the associated RNase H activity. Recent evidence suggests that the DNA polymerase and RNase H activities reside within separate domains of a single polypeptide chain. Viral RNase H is an exonuclease, which can apparently carry out the processive attack in either $5' \rightarrow 3'$ or $3' \rightarrow 5'$ directions, liberating oligonucleotides two to eight residues in length with 5′-terminal phosphate groups. RNase H is also associated with other classes of tumor viruses but in some instances its viral origin has not been firmly established.

C. Transcription

Despite extensive investigation, the cellular functions of the various RNase H activities have not been clearly established. One particular role for RNase H activity, not initially anticipated, is its possible involvement in transcription. Recent studies have shown that there are a large number of genetic elements and protein factors needed in conjunction with purified RNA polymerase II for accurate and efficient transcription of different genes. Although the majority of known factors are implicated in correct initiation of transcription, additional factors appear to be important in the elongation and termination phases of transcription as well. Previous studies have shown that transcriptional elongation carried out by highly purified mammalian RNA polymerase II is often defective, in that nascent transcripts remain associated with the DNA template strand, with continual displacement of the nontranscribed DNA strand. In contrast, bacterial RNA polymerases invariably displace their transcripts under similar conditions and with similar templates. These results suggested that the "renaturase" function of the bacterial RNA polymerase was missing from the mammalian enzyme. Just recently, such an activity was detected and partially purified from HeLa cells. In the presence of the partially purified renaturase, RNA polymerase II synthesizes undiminished amounts of RNA transcripts, all of which are efficiently displaced from the DNA template strand. When this activity was characterized, it was found to be inseparable from an RNase H activity. The mechanism of RNase H, or renaturase, suggests that this reaction may involve displacement of the RNA transcript after formation of a short single-stranded region of DNA on the template strand. *In vitro*, this displacement was followed by hydrolysis of the hybrid transcript oligonucleotide by the RNase H activity. Since RNA polymerase II transcripts *in vivo* require the presence of the 5′-triphosphate terminus for capping (the addition of 7-methylguanosine to the 5′ end of the transcript), the *in vivo* role of this RNase H activity is, therefore, still in question. [*See* DNA AND GENE TRANSCRIPTION.]

Bibliography

Apirion, D. (ed.) (1984). "Processing of RNA." CRC Press, Boca Raton, Florida.

Barnard, E. A. (1969). Ribonucleases. *Annu. Rev. Biochem.* **38,** 677–732.

Brawerman, G. (1989). mRNA Decay: Finding the right targets. *Cell* **57,** 9–10.

Deutscher, M. P. (1988). The metabolic role of RNases. *Trends Biochem. Sci.* **13,** 136–139.

Gerhard, C. and Keller, W. (1989). Poly(A) polymerase purified from Hela cell nuclear extract is required for both cleavage and polyadenylation of pre-mRNA *in vitro*. *Mol. Cell. Biol.* **9,** 193–203.

Gleich, G. J., Loegering, D. A., Bell, M. P., Checkel, J. L., Ackerman, S. J., and McKean, D. J. (1986). Biochemical and functional similarities between human eosinophil-derived neurotoxin and eosinophil cationic protein: Homology with ribonuclease. *Proc. Natl. Acad. Sci. U.S.A.* **83,** 3146–3150.

Gold, H. A., Craft, J., Hardin, J. A., Bartkiewicz, M., and Altman, S. (1988). Antibodies in human serum that precipitate ribonuclease P. *Proc. Natl. Acad. Sci. U.S.A.* **85,** 5483–5487.

Kane, C. M. (1988). Renaturase and ribonuclease H: A novel mechanism that influences transcript displacement by RNA polymerase II *in vitro*. *Biochemistry* **27,** 3187–3196.

Pestka, S., Langer, J. A., Zoon, K. C., Samuel, C. E. (1987). Interferons and their actions. *Annu. Rev. Biochem.* **56,** 727–777.

Shapiro, R., Fett, J. W., Strydom, D. J., and Vallee, B. L. (1986). Isolation and characterization of a human colon carcinoma-secreted enzyme with pancreatic ribonuclease-like activity. *Biochemistry* **25,** 7255–7264.
Sierakowska, H., and Shugar, D. (1977). Mammalian nucleolytic enzymes. *Prog. Nucl. Acid Res. Mol. Biol.* **20,** 59–130.
Stolle, C. A., and Benz, E. J., Jr. (1988). Cellular factor affecting the stability of β-globin mRNA. *Gene* **62,** 65–74.
Weickmann, J. L., and Glitz, D. G. (1982). Human ribonucleases: Quantitation of pancreatic-like enzymes in serum, urine, and organ preparations. *J. Biol. Chem.* **257,** 8705–8710.

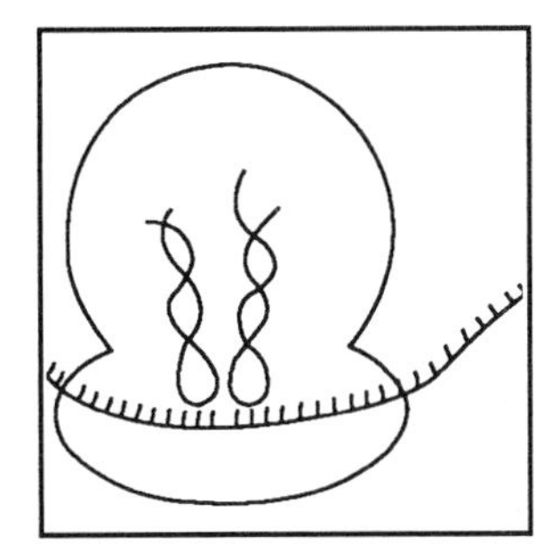

Ribosomes

IRA G. WOOL, *The University of Chicago*

Glossary

Genome Complete set of genes in the organism
Homology Implies that two proteins are derived from a common ancestral gene. Related proteins have a certain percentage of similar or identical amino acid residues at the same positions but are not necessarily homologous
Ribosomes Ribonucleoprotein particles that bind mRNA, aminoacyl-tRNA, and initiation, elongation, and termination factors, and catalyze peptide bond formation; they are the cellular machines for the synthesis of the proteins encoded in mRNAs, which latter are transcripts of genes
Transcription Process in which the enzyme RNA polymerase copies the DNA of a gene into the complementary RNA that then serves as either stable transcripts (transfer and ribosomal RNAs) or as messenger RNAs which encode proteins
Translation Conversion of the amino acid sequence encoded in mRNA into a protein; the process occurs on ribosomes

RIBOSOMES ARE ribonucleoprotein cellular organelles that link the genotype to the phenotype by translating a sequence of nucleotides in a transcript of a gene into a sequence of amino acids in a protein. They catalyze protein synthesis in all of the organisms in our biosphere; indeed, they are at one and the same time, universal, essential, and complicated. The ribosomal proteins and the ribosomal nucleic acids provide the specific binding sites (for messenger RNA, transfer RNA, and initiation, elongation, and termination factors) and the catalytic activities required for the formation of a peptide bond and the synthesis of a protein. The grand imperative is to know the structure of ribosomes so as to be able to account for these functions. This has as a prerequisite knowledge of the chemistry of the constituents. Human ribosomes are composed of two subunits which are designated by their sedimentation coefficients: the smaller is 40S and the larger 60S. The subunits associate—they are held together by noncovalent bonds, perhaps by magnesium salt bridges—and form the functional 80S ribosome. The 40S subunit has a single molecule of RNA, designated 18S rRNA, and 30 to 35 proteins; the 60S subunit has 3 molecules of RNA, 5S, 5.8S, and 28S, and 40 to 45 proteins. A great deal is known about prokaryotic (principally *Escherichia coli*) ribosomes: the primary structures of the nucleic acids and of the proteins; the secondary structures, from comparative sequence analysis, and preliminary proposals for the tertiary folding of the rRNAs; the binding sites for the ribosomal proteins on the rRNAs; the topography of the proteins from neutron scattering and from immune electron microscopy; and there are even preliminary, low-resolution X-ray crystallographic data for the structure of the ribosomal subunits. Far less is known of eukaryotic ribosomes because their structure is more complicated, because of the difficulty of applying genetic analysis, and because of the lack of a means for reconstituting ribosomal subunits. Nonetheless, progress is being made, especially with ribosomes from yeast and from rat. Comparatively little has been done directly with ribosomes from humans but what we know supports the contention

Encyclopedia of Human Biology, Volume 6.

that they closely resemble in both structure and function the particles from other mammals. For this reason this article draws heavily on studies of rat ribosomes.

I. Structure of Ribosomes

It has become very important to obtain a solution to the structure of ribosomes because knowledge of the structure is believed, with cause, essential for a rational, molecular account of the function of the organelle in protein synthesis. For a solution of the structure, a requisite is the sequences of nucleotides and of amino acids in the constituent nucleic acids and proteins.

A. The Sequences of Nucleotides and the Secondary Structure of Ribosomal RNAs

The sequences of nucleotides in the four species of RNA—5S, 5.8S, 18S, and 28S—have been determined for human ribosomes and for a large number of other eukaryotic species. Indeed, the sequences of nucleotides for some 400 18S-like and about 75 28S-like rRNAs have now been determined. This library of sequences provides the data for the determination of the secondary structure using comparative sequence analysis. In this procedure, one begins with the assumption that two rRNAs, say *E. coli* 16S and human 18S, have the same secondary structure. The primary structures are aligned using regions of sequence identity as a guide and putative helices are constructed. Compensated base changes in the nucleotide sequences in the helices is taken as evidence for the structure, whereas uncompensated changes leading to mismatches are evidence against it. Generally, at least two independent examples of compensated base changes are required to establish the existence of a helix. In principle, the phylogenetic method amounts to having the data from a preexisting genetic experiment in which all the organisms in the biosphere are considered pseudorevertants of the various mutations in rRNA genes that have occurred during evolution. Most of the structure derived by phylogenetic comparison has now been confirmed by direct experimental test, principally by modification of the rRNA in intact ribosomes with chemical reagents followed by identification of the altered nucleotides by the extension of primers complementary to nucleotides in specified regions of the RNA with reverse transcriptase. Modified bases, generally in single-stranded sequences, interrupt transcription by the enzyme. The conclusion is that the rRNAs from all species, including humans, can be folded into the same secondary structure.

B. The Sequences of Amino Acids in Ribosomal Proteins

An effort is being made to determine the sequences of amino acids in all of the proteins in a single mammalian species, the rat. Eighty-four proteins have been isolated from rat ribosomes, but it is not certain that all of them are constituents of the particle. The actual number of proteins in mammalian ribosomes, including those from humans, is thought to be between 70 and 80. Human ribosomal proteins are relatively small—they range in mass from about 7000 to 37,000 Da; they are nearly all basic, most having p*I* values in the range of 10–12.

To date the sequences of amino acids in 35 rat and 11 human ribosomal proteins have been determined; related rat and human ribosomal proteins diverge at only a few residues if at all.

There are features of the ribosomal proteins and of their cDNAs that are noteworthy:

1. At the start of the 5′ noncoding region of eukaryotic ribosomal protein mRNAs there is a pyrimidine sequence that may play a role in the regulation of their translation, i.e., the pyrimidine sequences may be cis-acting translation promoters.
2. The hybridization of the cDNAs to restriction enzyme digests of genomic DNA indicates that mammalian ribosomal protein genes are present in multiple copies, generally 7–20; moreover, they are dispersed in the genome. However, in no instance has it been shown that more than one of the genes is functional. The presumption is that for each ribosomal protein the genome contains only one gene that is expressed, that the other copies are nonfunctional pseudogenes. However, this presumption derives from a detailed analysis of only a limited number of families. [*See* GENES; GENOME, HUMAN.]

II. Function of Ribosomes

The synthesis of proteins on ribosomes involves a series of reiterative cycles—the overall process is divided for convenience of description into three parts, initiation, elongation, and termination. It is easiest to begin with elongation since it is prototypical for the others. [*See* DNA SYNTHESIS.]

The growing nascent polypeptide is attached by an ester linkage to a transfer RNA (tRNA) just as amino acids are attached to tRNA in aminoacyl-tRNA (aa-tRNA). The latter is the substrate for protein synthesis. The peptidyl-tRNA (pep-tRNA) alternately occupies one of two adjacent sites on the ribosome termed P for peptidyl and A for aminoacyl. The A and P sites are the crucial postulates of a paradigm proposed by Watson in 1964 to explain the biochemistry of protein synthesis; a coherent exposition would be difficult without invocation of the A and P sites, however, the molecules that comprise their structure remains uncertain.

The initial step in elongation is recognition: It is the process wherein a trinucleotide codon in the mRNA residing in the recognition region of the A site, i.e., the decoding domain, specifies the binding of a cognate aa-tRNA by base pairing with the complementary anticodon. The aa-tRNAs are synthesized in a coupled reaction: first, the amino acids are activated by formation of an enzyme-bound aminoacyladenylate with the liberation of inorganic pyrophosphate (PP_i); second, the aminoacyl group is attached to a cognate transfer RNA by a high-energy ester linkage and enzyme and AMP are released. These reactions are catalyzed by a set of 20 enzymes (aminoacyl-tRNA synthetases), one for each amino acid, and require ATP. This is the reaction most critical for the fidelity of protein synthesis—it is essential that the correct amino acid is attached to a cognate tRNA.

The recognition step requires an accessory nonribosomal protein termed elongation factor 1 (EF-1). EF-1 forms a ternary complex with aa-tRNA and GTP and this complex binds noncovalently to the ribosome in a codon-specific reaction. The aa-tRNA is positioned in the A site, GTP is hydrolzyed, and the EF-1·GDP complex and P_i are released.

During peptidyl transfer, which is the second step in elongation, the nascent polypeptide in the P site replaces the ester bond to tRNA by a peptide bond with the α-amino group of the aa-tRNA in the A site; the peptide is increased in length by one amino acid and is transferred from the P to the A site. This reaction is catalyzed by an activity termed peptidyl transferase; the activity is located in the 60S ribosomal subunits but its molecular identity is not known. The energy for the synthesis of this peptide bond is derived from the high group-transfer potential of the ester linkage in the aa-tRNA. Peptidyl transfer follows immediately on the binding of aa-tRNA to the A site.

The final step in elongation is translocation. The pep-tRNA is transferred from the A to the P site and the mRNA is dragged along with it. The deacylated tRNA left by the peptidyl transfer reaction is displaced from the ribosome and the next codon is brought into apposition with the A site. The ribosome is now ready to repeat the process in a recursive manner until the synthesis of the protein is completed. Translocation is catalyzed by a binary complex of GTP with a second elongation factor, EF-2. A possible mechanism for translocation is considered in some detail later.

The initiation of protein synthesis has the same biochemical scenario as elongation. The prime purpose served by the initiation reactions is the proper framing of the mRNA on the ribosome so that the initial AUG codon is translated in phase. The initiation codon specifies a methionyl residue and that amino acid is esterified to a tRNA that is used only to begin the synthesis of a protein. The initiator methionyl-tRNA forms a ternary complex with GTP and a particular initiation factor, termed eIF-2, rather than with EF-1 as occurs during elongation; the ternary complex binds to the 40S ribosomal subunit and then mRNA and 60S subunits join the complex. The initiation met-tRNA is base paired to the AUG codon nearest the 5′ end of the mRNA and positioned in the P site. These latter reactions are promoted by a number of other initiation factors. The 80S ribosomal initiation complex is now competent to begin the elongation cycle.

Termination is the process by which the completed protein is released from the tRNA that bore the carboxyl-terminal amino aci. Chain termination requires a specific codon (UAA, UAG, or UGA) at the 3′ end of the reading frame in the mRNA and a specific factor termed release factor (RF). The hydrolysis of the ester linkage between the tRNA and the completed polypeptide is catalyzed by peptidyl transferase, the same enzyme that is responsible for peptide bond formation. The enzyme can form a peptide bond with an amino group as the nucleophilic agent attacking the ester bond of aa-tRNA or the enzyme can catalyze hydrolysis with water as the nucleophilic agent. It appears that the interaction of RF with the termination codon in the A site not only activates peptidyl transferase but favors water as the nucleophilic agent; the result is the release of the nascent peptide.

A. Internal Duplications in Ribosomal Proteins

A number of rat ribosomal proteins have duplications of amino acid sequences. For example, protein L7 has 5 repeats of a segment of 12 amino acids arranged in tandem near the NH_2 terminus. The repeats are very basic; four to six of the residues are lysyl or arginyl and there are no acidic amino acids.

The occurrence of multiple, related, generally basic repeats in ribosomal proteins insinuates that they have functional significance, but there is no indication as yet what this might be. Possibilities that suggest themselves are that they play a role in the interaction with RNA (ribosomal, transfer, or messenger) or that they are involved in directing the proteins to the nucleolus for assembly of ribosomes. There is evidence that information specifying the localization of proteins in the nucleus is encoded in short sequences of amino acids, although it is not yet possible to derive a consensus sequence nor to formulate general rules for the structure of the peptide. In the best characterized examples, entry into the nucleus is contingent on a consecutive sequence of several basic amino acids preceded by a prolyl residue. For example, the first 21 amino acids of yeast ribosomal protein L3, which are needed for entry into the nucleus, contains a sequence of this type. Several of the repeats in rat ribosomal proteins also have these characteristics. Although there is no definitive experimental evidence available to evaluate the proposal the repeats have sufficient similarity to known nuclear localization sequences as to require consideration of the possibility that they serve the same function.

III. Evolution of Ribosomes

It has been apparent for some time that individual ribosomal proteins from different eukaryotic species are derived from common ancestral genes. The sequences of amino acids are closest for related mammalian ribosomal proteins; for example, rat, mouse, and human proteins L32 have exactly the same sequence. The homologies are also obvious when the sequences of amino acids in rat and yeast ribosomal proteins, the two eukaryotic species for which the largest sets of data have been collected (35 for rat and 25 for yeast), are compared. Despite the two species being evolutionarily distant eukaryotes, 17 can be correlated. The perceptage of identities in the alignments range from 40 to 80. The data are sufficient to provide confidence that most if not all of the eukaryotic ribosomal proteins are homologous and that it will be possible to establish a protein-to-protein correlation for all species.

Until recently common wisdom has had it that with only a few exceptions there are no close sequence similarities between eukaryotic and prokaryotic ribosomal proteins and it was suggested that a relationship might be found only in their three-dimensional structure. The thinking with regard to this issue was strongly influenced by the results of the comparison of the structure of ribosomal RNAs. The motif for these molecules is conservation of secondary structure rather than primary sequence. There are conserved nucleotide sequences, but these tend to be (although not exclusively) in non-helical regions, and there is evidence that they are important for function. It is therefore possible that only the functionally important amino acids in ribosomal proteins, which need not be contiguous in the sequence, will be conserved. This may account for the difficulty in establishing relationships between eukaryotic and prokaryotic ribosomal proteins.

Intuition leads one to surmise that ribosomes arose on a single occasion and, hence, that the proteins as well as the rRNAs of eukaryotic ribosomes are likely to be related to their prokaryotic progenitors. The problem is how to trace the chemical spoor and unravel the mechanism by which the proteins evolved. A strong argument can be made for homology of eukaryotic and prokaryotic ribosomal acidic proteins. The ribosomes of all the species that have been analyzed have a protein or proteins homologous to *E. coli* L7/L12; the human equivalents are P0, P1, and P2. There is convincing evidence that a few additional eukaryotic ribosomal proteins are related to *E. coli* ribosomal proteins.

There is another aspect to the problem of the evolution of ribosomes. Eukaryotic ribosomes are larger; they contain a greater number of proteins and an additional molecule of RNA. The latter is, however, a difference more apparent than real; prokaryotes lack 5.8S rRNA but have a related sequence at the 5′ end of 23S rRNA—prokaryotes have not gone to the trouble of processing the 5.8S-like RNA out of the primary rRNA transcript, whereas eukaryotes have. The individual protein and RNA molecules in eukaryotic ribosomes are, on the average, larger than those in prokaryotes. There is no explanation for these differences. Moreover, it is a paradox since eukaryotic and prokaryotic ribosomes perform precisely the same function (the catalysis of protein synthesis) and, more importantly, they do so by the same biochemical means (albeit the initiation of protein synthesis is more complicated in animal cells). The accretion of extra proteins and the increase in size of the proteins and of the RNAs in eukaryotic ribosomes may imply that eukaryotic ribosomes have functions lacking in prokaryotic particles related, perhaps, to the regulation of the translation of mRNA.

One can, of course, look at the problem the other way around and ask why prokaryotic ribosomes are smaller. One answer is that they had to be streamlined. During log-phase growth as much as 35% of the total protein in a bacterium is ribosomal; the bacterium might have difficulty supporting a particle with as much protein as eukaryotic ribosomes have. Human ribosomes are not only larger than those of bacteria but 50% rather than 35% of the mass is protein. The assumption is that ribosomes evolved once, and that prokaryotic ribosomes at one time had a greater number of proteins, a number comparable to that in eukaryotes. Prokaryotic ribosomes then responded to selective pressure by discarding proteins and reducing the size of those that were retained—the advantage being a greater number of smaller ribosomes capable of performing the same function as their larger progenitors. This would have happened after the divergence of primitive eukaryotes.

One possibility is that the earliest cells, progenitors of prokaryotes and eukaryotes, had nuclei and eukaryotic-like ribosomes and that a number of the proteins were needed, not for protein synthesis, but to manage the complicated traffic between nucleus and cytoplasm. Ribosomal proteins are synthesized in the cytoplasm and then transported to the nucleus where they are assembled on nascent rRNA transcripts; after the processing of the rRNA, which most likely occurs during assembly, the ribosomal subunits must pass through nuclear pores to get to the cytoplasm. Thus, eukaryotic ribosomal proteins are likely to contain amino acid sequences that serve as "zip codes" for nuclear localization; ribosomes may also have proteins whose sole function is to facilitate transnuclear membrane traffic. Obviously, bacteria no longer require either the nuclear localization amino acid sequences or the transport proteins.

Another rationalization derives from the recent discoveries concerning RNA by Thomas Cech. The initial finding was that ribosomal RNA has the capacity to mediate self-splicing and self-ligation, that is to say, the removal of a sequence of nucleotides (an intron) from an rRNA, and the religation of the ends so as to reestablish the integrity of the molecule. Before this observation nucleic acids were considered to be relatively inert chemically—capable of serving in information transfer by providing a template for polymerization reactions (as with messenger and transfer RNAs) and capable of providing the scaffolding for proteins in RNP particles and organelles (as in ribosomes) but not of catalysis. The latter was assumed to be the sole province of proteins. Cech's experiments have shown that RNA can be both the repository of information to be transmitted, and that it can have enzymatic activity. This makes it possible, even likely, that the precellular biological world was dominated by RNA; that RNA preceded DNA and protein. It is of relevance that Cech's work lends support to the idea that primoridal ur-ribosomes had only RNA, as had been proposed early on by Francis Crick and Leslie Orgel. The concept is that the basic biochemistry of protein synthesis (the binding of aminoacyl-tRNA, peptide bond formation, and translocation) is an intrinsic property of ribosomal RNA; that the ribosomal proteins, a later evolutionary embellishment, facilitate the folding and the maintenance of an optimal configuration of the ribosomal RNA and in this way confer on protein synthesis speed and accuracy. Perhaps then eukaryotic ribosomes, which have far more RNA than prokaryotes (approximately 1300 more nucleotides), need more proteins to tune their nucleic acids. There is still another scenario: If the ur-ribosome had only RNA and divergence of eukaryotes and prokaryotes occurred before the emergence of a complete set of ribosomal proteins then one would not expect in the amino acid sequences the close resemblance that is apparent in the structure of the rRNAs; rather one would expect homology only of those ribosomal proteins (like the acidic proteins and a few others) that had evolved before divergence.

IV. The Biogenesis of Ribosomes: The Regulation of the Synthesis of the Molecular Components and the Assembly of the Particles

The assembly of ribosomes is an extraordinarily complex process. Biogenesis of human ribosomes requires the synthesis of equimolar amounts of the four rRNAs and of 70 to 80 proteins. (Human ribosomes are presumed, in the absence of definitive evidence, to have molar amounts of most, if not all, proteins). In addition, the synthesis of the rRNAs and of the ribosomal proteins must be precisely balanced. What is more, human cells have as many as a million ribosomes. The coordinate synthesis in the nucleolus of 5.8S, 18S, and 28S rRNAs is easily accomplished since their genes are in a single large transcription unit. However, the synthesis of 5S rRNA occurs outside of the nucleolus so its transcription must be coordinated with that of the other rRNAs and, in addition, 5S rRNA must be delivered to the nucleolus. The coordination of the synthesis of the ribosomal proteins is more complicated than that of the rRNAs: the transcription of 70 to 80 unlinked genes must be balanced. In exponentially growing cells the balanced synthesis of the various ribosomal proteins is due primarily to their mRNAs being present in similar amounts and to their being translated with similar efficiencies. Thus, coordination of synthesis would seem to be determined in the first instance, and most importantly, by regulation of transcription of the individual ribosomal protein mRNAs, although there is evidence also for regulation of pre-mRNA processing and of translation. This implies that the source of the coordination resides in cis-acting elements of the promoters of ribosomal protein genes and in trans-acting factors.

In *E. coli*, the 53 ribosomal protein genes are organized into 20 operons (transcription units) the largest has 11 genes, the smallest a single one. Almost half of these genes, all of which are present in single copies, are at one locus (*Str*) on the chromosome; the remainder are scattered over the genome. The ribosomal protein operons are complex transcriptional units; some also contain nonribosomal protein genes and some have two promoters. In bacteria the main control of ribosomal protein formation is not transcription of mRNAs but is mediated by autogenous regulation of their translation. This regulation exploits the arrangement of the genes in operons and specific interactions of the ribosomal proteins with RNAs. Certain ribosomal proteins, when synthesized in excess, function as repressors of the expression of their own operons by binding to a control region on the polycistronic mRNA rather than to rRNA. This follows from similarities in the structure of the control region of the polycistronic mRNA and the repressor ribosomal protein-binding site on rRNA; of course the affinity for the latter must be greater than for the former. As to the synthesis of rRNA, there probably is regulation (repression) of transcription by nontranslating ribosomes and perhaps by ppGpp.

Information on the organization of ribosomal protein genes in eukaryotes is just beginning to accumulate. What is certain is that both their structure and the regulation of their expression is different than in prokaryotes. The number of copies of the genes for individual ribosomal proteins varies among eukaryotes. In yeast there are either 1 or 2 copies, most often 2, of the genes that have been analyzed (more than 20 of them); where there are 2 they are both functional. In *Xenopus* there are two to six genes; no nontranscribed ribosomal protein genes have been identified. In mammals there are 7 to 20 copies of each but only 1 is transcribed.

In eukaryotes the genes are not clustered; not even multiple copies of the same gene are clustered, far less genes for different ribosomal proteins. Indeed, different genes for the same ribosomal protein may be on different chromosomes.

A distinctive feature of the structure of ribosomal protein genes in mammals (most of the analyses have been of mouse genes but the few human genes studied are very similar) is the lack of a canonical TATA box, which normally determines where transcription begins. In the region where one would be expected there is a 6- to 7-base pair element that contains five or six AT pairs. Presumably, some aspect of this sequence pattern, or the novel organization of the cap region, or both, assumes the function usually served by the TATA box, i.e., to position RNA polymerase II for the initiation of transcription. Although precise initiation of transcription in the absence of a canonical TATA box is rare it does occur; it has been described for a few viral and cellular genes. The latter are, like the ribosomal protein genes, members of the "housekeeping" class. Conceivably, novel promoter structure may confer special regulatory properties on this class of genes.

Another striking characteristic of mammalian ribosomal protein genes is the structure of the cap sites, i.e., the region where transcription is initiated. It is embedded in a ≥12-nucleotide stretch of pyrimidines flanked by blocks of greater than 80% GC content. The cap site pyrimidine tract has the motif 5′-CTTCCYTYYTC-3′; initiation of transcription is at the C at position 4 or 5. The promoter region of mammalian ribosomal protein genes contains, in addition, a number of cis-acting control elements, indeed, maximal expression requires 200 bp in the 5′ flanking region. Within this region there are at least five discrete elements that affect expression and which bind nuclear factors. Although the general patterns of the structure of the promoters in individual mammalian ribosomal protein genes are similar, they do have distinct characteristics.

As has already been indicated, the regulation of transcription is likely to be embedded in the structure of the promoters of mammalian ribosomal protein genes; more importantly this is also likely to be the site of coordination of the transcription of the 70 to 80 ribosomal protein genes. The assumption, until proved otherwise, must be that coordination is mediated by a set of common or overlapping trans-acting factors affecting in some complex way a pattern of cis-acting sequences in the promoters. The analysis to date has provided information on the architecture of these promoters but we still lack knowledge of the exact pattern of the control elements, of whether there is a common architecture for all mammalian ribosomal protein gene promoters, and of the identity of the factors that bind to cis-acting sequences.

V. The Function of Ribosome Domains

Until recently little was known of the function of individual ribosomal components or even of ribosomal domains. None of the ribosomal proteins or nucleic acids has activity when separated from the particle. To circumvent this impediment advantage has been taken of the activity of toxins and antibiotics. The value that derives from an analysis of their mechanism of action is in concentrating our attention on regions of the ribosome where our efforts to comprehend functional correlates of structure are likely to be rewarded.

α-Sarcin is a small, basic, cytotoxic protein produced by the mold *Aspergillus giganteus* that inhibits protein synthesis by inactivating ribosomes. The inhibition is the result of the hydrolysis of a phosphodiester bond on the 3′ side of residue G-4325 which is in a single-stranded loop 459 residues from the 3′ end of 28S rRNA. The cleavage site is embedded in a purine-rich, single-stranded segment of 14 nucleotides that is nearly universal. This is one of the most strongly conserved regions of rRNA and, indeed, the ribosomes of all the organisms that have been tested, including the producing fungus, are sensitive to the toxin. α-Sarcin catalyzes the hydrolysis of only the one phosphodiester bond and this single break accounts entirely for its cytotoxicity. This remarkable specificity is peculiar to α-sarcin; treatment of ribosomes with other ribonucleases causes extensive digestion of rRNA.

The finding that cleavage of a single phosphodiester bond inactivates the ribosome implies that this sequence in the α-sarcin domain is crucial for function since ribosomes ordinarily survive mild treatment with nucleases despite many nicks in their RNA; indeed, some organisms physiologically divide their 28S rRNA into domains. Thus, intact rRNA per se is not essential for protein synthesis. The presumption that the α-sarcin region of 28S rRNA is critical for ribosome function has gained considerable reinforcement from the elucidation of the mechanism of action of ricin. Ricin, which is among the most toxic substances known (a single molecule is sufficient to kill a cell), is an RNA *N*-glycosidase and the single base in 28S rRNA that is depurinated is A-4324, i.e., the nucleotide adjacent to the α-sarcin cut site.

Although the molecular details of the function of the α-sarcin/ricin domain are not known there are good reasons to suspect that it is involved in EF-1-dependent binding of aminoacyl-tRNA to ribosomes and EF-2-catalyzed GTP hydrolysis and translocation. This supposition follows from the findings that these are the partial reactions most adversely affected by α-sarcin and by ricin, respectively, and that cleavage at the α-sarcin site interferes only with the binding of elongation factors. The most convincing evidence for this interpretation comes from the demonstration that elongation factors footprint in the α-sarcin/ricin domain. Elongation factor Tu (EF-Tu) protects only four of the nucleotides in prokaryotic 23S rRNA against chemical modification and these correspond in eukaryotic 28S rRNA to A-4324 (ricin), G-4325 (α-sarcin), and G-4319 and A-4329, which latter are also in the universal sequence. EF-G also protects only four nucleotides and three are the same as the ones protected by EF-Tu: the bases that correspond to G-4319, A-4324, and G-4325.

An attempt is being made to determine how α-sarcin recognizes a single phosphodiester bond in a particular domain in rRNA; this is part of an effort, unfortunately but necessarily, oblique, to understand how ribosomal proteins, which like α-sarcin are small and basic, recognize specific sites in rRNA. At the same time information is being sought on the contribution that the α-sarcin site RNA makes to ribosome function.

The structure of the α-sarcin domain RNA encompasses a helical stem of 7 base pairs with a single bulged nucleotide and a single-stranded loop of

17 nucleotides. Recognition by α-sarcin of the toxin domain RNA requires the stem (but not the bulged nucleotide) and the single-stranded loop in which the sequence of at least 14 nucleotides (the universal sequence) affect binding and enzymatic activity; the stem needs only 3 base pairs and the identity of the Watson–Crick interactions does not have a large influence. Perhaps most important is the observation that the tetranucleotide GAG(sarcin)A in the loop must have the correct geometry with respect to the stem.

It is assumed that the structure of the α-sarcin domain in 28S rRNA is more complex than is depicted in the usual two-dimensional cartoons and, furthermore, that it is capable of undergoing reversible alterations. What suggests that the structure of this domain is complex is the enormous amount of ricin required to cleave the glycosidic bond in naked 28S rRNA or in the synthetic RNA. Depurination of A-4324 in 28S rRNA resident in the ribosome occurs at a ricin : substrate ratio of 0.001, whereas with naked 28S rRNA the ratio is 10, i.e., 10,000 times greater. The reason for this incredible discrepancy is not known but the most likely explanation is that the ordered structure of the domain is different in ribosomes than it is in naked RNA. We take note that the single-stranded loop is large (17 nucleotides) and, hence, unlikely to exist as such; it might well participate in a tertiary interaction with other regions of 28S rRNA.

What suggests that the domain might undergo transitions in its structure is an observation of the effect of α-sarcin on the association of 5.8S rRNA with 28S rRNA. In the large subunit of eukaryotic ribosomes these two nucleic acids are noncovalently but stably associated. This interaction, which involves approximately 40 hydrogen bonds in 2 separate contact regions, is destabilized by the α-sarcin-catalyzed cleavage which is at a site more than 4000 nucleotides away. Dissociation of 5.8S rRNA from 28S rRNA, which ordinarily requires treatment with 4–6 *M* urea, occurs spontaneously after α-sarcin action on ribosomes.

It is difficult to provide a coherent physical chemical explanation for this phenomenon. The cleavage at G-4325 appears to initiate a propagated change in the secondary or tertiary structure, or both, that in some way is transmitted through the molecule and leads to the collapse of 28S rRNA. This collapse could account for the destabilization of the 5.8S·28S rRNA complex and to the loss of function of the ribosomes. It might interfere also with the operation of an rRNA switch, i.e., with the ability to disrupt and reestablish a crucial tertiary interaction that involves the single-stranded region of the α-sarcin domain.

There are other observations that support this conjecture. Oligodeoxynucleotides complementary to the universal sequence in the loop of the α-sarcin domain will not bind to ribosomes suspended in buffer, suggesting, but by no means proving, that the structure is not simply single stranded. It is most important that if ribosomes are catalyzing protein synthesis they will bind the complementary oligodeoxynucleotide (manifest as sensitivity to ribonuclease H), suggesting a reversible change in the structure of the domain. Further support for the proposal comes from experiments with inhibitors of protein synthesis (cycloheximide, guanylyl-imidodiphosphate [GMPPNP], and sparsomycin) in which α-sarcin is used as a probe of structure. The results of these experiments also indicate that there is a conformational change in the RNA during translocation since ribosomes are sensitive to α-sarcin only when peptidyl-tRNA is in the A site prior to translocation. The conformational transition, perhaps the making and breaking of a tertiary interaction, could provide the motive force for changes in the structure of ribosomal 60S subunits and might underlie the movement required for the translocation of peptidyl-tRNA from the A to the P site and for the movement of mRNA one codon after each of the reiterative rounds in translation. In this paradigm it is the elongation factors that initiate the reversible transition or switch in rRNA structure that propels translocation either directly or indirectly through the binding or the hydrolysis of GTP. Cleavage at G-4325 in 28S rRNA by α-sarcin or depurination at A-4324 by ricin might abolish the capacity to reversibly switch structures and in this way account for the catastrophic effect of the toxins on ribosome function.

VI. Coda

With respect to the structure and function of human ribosomes all that remains are the difficult tasks: The tertiary folding patterns of the rRNAs, the three-dimensional structure of the proteins, and ultimately the structure of the subunits at atomic resolution. These are formidable problems but progress in studies of macromolecular structure has been so astonishing in the last several years as to encourage optimism. As to the function of the organelle we *only* lack information on how a peptide bond is made and how movement of mRNA and of peptidyl-tRNA is catalyzed. It is difficult to be optimistic that solutions to these problems will soon be had. But we live in hope.

Bibliography

Perez-Bercoff, R. (ed.). (1982). "Protein Biosynthesis in Eukaryotes," Nato Advance Study Institutes Series. Plenum, New York.

Hardesty, B., and Kramer, G. (eds.) (1985). "Structure, Function, and Genetics of Ribosomes." Springer Verlag, New York.

Hill, W. E. (ed.) (1990). "Structure, Function and Evolution of Ribosomes." American Society of Microbiology, Washington, D.C.

Wool, I. G. (1979). The structure and function of eukaryotic ribosomes. *Annu. Rev. Biochem.* **48,** 719–754.

Noller, H. F. (1984). Structure of ribosomal RNA. *Annu. Rev. Biochem.* **53,** 119–162.

Wittmann, H. F. (1982). Components of bacterial ribosomes. *Annu. Rev. Biochem.* **51,** 155–183.

Wittmann, H. F. (1983). Architecture of prokaryotic ribosomes. *Annu. Rev. Biochem.* **52,** 35–65.

Raué, H. A., Klootwijk, J., and Musters, W. (1988). Evolutionary conservation of structure and function of ribosomal RNA. *Prog. Biophys. Molec. Biol.* **51,** 77-129.

Mager, W. H. (1988). Control of ribosomal protein gene expression. *Biochim. Biophys. Acta* **949,** 1–15.

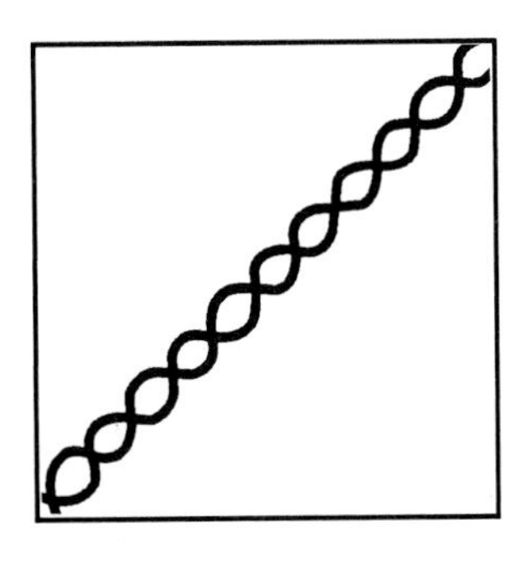

RNA Replication

CLAUDE A. VILLEE, *Harvard University*

Glossary

Codons and anticodons The specific sequence of three nucleotides in messenger RNA (codon) and the complementary sequence of three nucleotides in transfer RNA (anticodon) in which genetic information is stored and transferred, determines the base-pairing between specific codons and specific anticodons and this in turn determines the order in which amino acids are synthesized into peptide chains

Exons and introns The initial RNA product of the transcription of DNA is a large molecule termed heterogeneous nuclear RNA. This is cut and spliced by specific enzymes to yield a much shorter messenger RNA composed of only a portion of the original heterogeneous nuclear RNA. The pieces of RNA that are included in the final messenger RNA are termed exons. The other, unused, portions are termed introns, or intervening sequences. It is not at all clear what function, if any, these intervening sequences may have. For the heterogeneous nuclear RNA to be converted into a functional messenger RNA, the introns must be deleted from the molecule and the exons must be precisely spliced together to form a continuous message that codes for specific protein

Helix-destabilizing proteins Two strands that constitute DNA must be physically separated in order for replication to proceed; the unwinding of the double strand is catalyzed by DNA helicases, and the separated strands are then bound by helix-destabilizing proteins, which bind to single-strand DNA and prevent the reestablishment of the double helix until each strand has been copied

Okazaki fragments Relatively short DNA chains formed on the lagging strand; each is initiated by a separate primer and then is extended toward the 5′ end of the previously synthesized fragment by DNA polymerase

Retroviruses Have a genomic RNA that is a single plus strand, which may serve as a template to direct the formation of a DNA molecule, which, in turn, may act as a template for making messenger RNA; first the virion RNA is copied into a single strand of DNA, which then forms a complementary second DNA strand; this double-strand DNA is integrated into the chromosomal DNA of the infected cell and the integrated DNA is then transcribed by the cell's own machinery into RNA that either acts as a viral messenger RNA or becomes enclosed in a virus; the so-called central dogma of biology that information is transferred only in the direction of DNA → RNA → protein had to be amended after the discovery of retroviruses in which information is transferred from RNA to DNA

Ribozymes New class of biological catalysts that, as nucleic acids, can act as enzymes; the RNA that composes the introns may have the ability to splice itself without the assistance of protein catalysts

Sense strands and antisense strands Of the two strands that comprise the double helix of a DNA molecule, only sense strand contains a sequence of nucleotides that can be read out to form a protein. The complementary strand, termed the antisense strand, has a sequence of nucleotides that, if read out, would give either a garbled or a totally lacking messenger RNA

THE REPLICATION OF RNA is a process basically similar to that of the replication of DNA and the transcription of DNA to form RNA. In all of these processes, one polynucleotide strand serves as a template, and specific polymerases use specific nucleotide triphosphates to add nucleotides to the strand in a specific order. The specific base added in sequence is determined by the complementary base-pairing of the bases in the initial template strand and the newly forming strand. The mechanism by which double-strand DNA undergoes replication to form two double helices has been understood for a long time. That RNA can serve as a template to form complementary DNA strands was recognized somewhat later, and, more recently, it was realized that RNA itself can undergo replication. It can serve as a template for the formation of a complementary strand of RNA. The complementary pairing of purine and pyrimidine bases is basic to the transfer of genetic information and the synthesis of specific proteins composed of sequences of amino acids. Specific base-pairing is determined by the nature of the hydrogen bonds joining the bases. Two bonds join adenine and thymine, and three bonds join guanine and cytosine. The pairs are always A-T and G-C.

Our understanding of the process of RNA replication has developed relatively recently whereas our understanding of DNA replication developed much earlier. In fact, the knowledge of DNA replication made possible our appreciation of the mechanisms involved in RNA replication. Thus it is appropriate to precede the discussion of RNA replication with a brief presentation of our current knowledge of DNA replication.

I. Nucleic Acid Structure

The nucleic acids are of two types: deoxyribonucleic acid (DNA) and ribonucleic acid (RNA). Each consists of large molecules composed of four kinds of nucleotides: adenylate (A), thymidylate (T), guanylate (G), or cytidylate (C). In RNA, the T is replaced by U (uridylate). Each nucleotide consists of a nitrogenous base (adenine, thymine, guanine, and cytosine) plus a sugar (deoxyribose in DNA and ribose in RNA) plus orthophosphate. The nucleic acids are of fundamental importance in biology because of their role in transmitting biological information from one generation to the next and in transcribing that genetic information in the synthesis of proteins. RNA serves as the genetic material by which information is transferred from one generation to the next in certain small viruses. All other viruses and all bacteria, plants, and animals use DNA as the genetic material.

A. DNA Structure

The DNA molecule consists of a very long double helix. The two chains that make up the helix extend in opposite directions and are paired by the so-called Chargaff rules—i.e., a nitrogenous base in one chain is paired to a base in the other chain such that an A (adenine) in one chain always pairs to a thymine (T) in the other chain and a G (guanine) pairs with C (cytidine). The specific base-pairing is determined by the nature of the hydrogen bonds joining the bases: two bonds join A and T and three bonds join G and C. To fit into the double helix, one large base, A or G, must pair with a smaller base, T or C. An A-G pair would be too large and a C-T pair would be too small to fit in the available space. In DNA replication (as shown subsequently), the two chains separate momentarily and, with the aid of a number of proteins (some of which are enzymes), a new chain is formed by base-pairing to each of the original chains. The product is two DNA molecules, each composed of a double helix. [*See* DNA AND GENE TRANSCRIPTION.]

B. RNA Structure

RNA functions primarily in translating the genetic information in DNA into the sequence of amino acids that will comprise the protein coded by that sequence of nucleotides in the gene. There are three distinct types of RNA, each with a particular function in bringing about the synthesis of specific proteins. Ribosomal RNA (rRNA) serves as an integral part of the ribosomes (structures inside the cell on which protein synthesis occurs). Transfer RNA (tRNA) consists of relatively small nucleic acid molecules, each with a specific binding area of three nucleotides (the anticodon) and with a specific amino acid attached to the opposite end of the tRNA molecule. The third type of RNA, messenger RNA (mRNA), is produced by the transcription of the sequence of nucleotides in DNA to form a comparable sequence of ribonucleotides in RNA. Each set of three nucleotides in a molecule of mRNA constitutes a codon, which undergoes specific base-pairing with the three nucleotides in the anticodon of tRNA. This ensures that the amino acids will be lined up in the proper sequence, yielding the correct protein molecule.

Because similar codes are used in both DNA and RNA, in certain retroviruses, such as the acquired immunodeficiency disease virus, information is transferred from RNA, which is the genetic material of the retroviruses, to DNA, which becomes an intermediate in the synthesis of the next generation of RNA. Information is transferred directly from RNA to RNA in certain RNA viruses such as the influenza virus or the poliomyelitis virus.

II. DNA is the Genetic Material

Experiments have shown that the transforming principle of pneumococci is DNA and not protein. Analyses of a variety of cells then showed that eggs and sperm, which are haploid (containing one set of genes), also contain only half as much DNA as somatic diploid cells (two sets of genes). In 1952, radiolabeled precursors demonstrated that viruses multiplying within bacteria (bacteriophages) inject their DNA, but not their protein, into the bacterial cell. The injected DNA undergoes replication, resulting in the presence of many DNA molecules within the bacterial cell. Despite these findings, most biologists did not accept the idea that the genetic material was DNA. However, in 1953, Watson and Crick published their famous paper in *Nature* describing their model of DNA as a double helix with the nitrogenous bases on the inside of the helix. This immediately suggested a mechanism by which DNA could undergo replication, and very quickly biologists accepted the theory that DNA was the genetic material and that it was a double helix of two nucleotide chains.

This model explained many previous findings such as Chargaff's analyses demonstrating that in a great variety of cells the amounts of A and T were equal, and also the amount of G and the amount of C were equal. It explained the X-ray crystallographic pictures of DNA molecules taken by Wilkins and Franklin.

A. Information Transfer: DNA to DNA

The above model also suggested a mechanism by which the information in DNA could be copied precisely. Because nucleotides pair with each other in a complementary fashion, i.e., A to T and G to C, each of the nucleotide strands in the DNA molecule could serve as a template for the synthesis of the opposite strand. When the hydrogen bonds joining the two strands are broken, the two chains can separate. Each chain can then pair with complementary nucleotides to form the corresponding strand. This results in two DNA double helices, both of which are identical to the original one. Each consists of one original strand from the parent molecule and one newly synthesized complementary strand.

The Watson–Crick model also suggested a mechanism by which DNA could undergo a mutation. For a long time it had been known that genes can undergo mutations (sudden, heritable changes) that are inherited in subsequent generations. The Watson–Crick double-helix model suggested that a mutation could simply involve a change in the sequence of bases in the DNA. If the DNA is copied by a mechanism using complementary base-pairing, then any change in the sequence in the bases on one strand would result in a new sequence of complementary bases being paired during the next cycle of replication. The new sequence of bases would then be transmitted to the daughter molecules by the copying mechanism that had been used to copy the original genetic material.

1. DNA Replication is Semiconservative

The replication of DNA is termed semiconservative because each of the original strands is conserved in one of the daughter strands and constitutes one half of the daughter helix. Direct evidence that the replication process is semiconservative was provided by experiments using the bacterium *Escherichia coli.* For several generations, bacteria were grown in a medium containing heavy nitrogen (^{15}N), labeling the parent strands of DNA in the bacterium. The presence of the ^{15}N atoms increased the density of the DNA molecules so that they can be distinguished by appropriate techniques from strands containing regular nitrogen (^{14}N). Some of the bacteria containing the ^{15}N-labeled DNA were transferred to a medium containing the usual nitrogen isotope, ^{14}N, and permitted to undergo additional cell divisions. The newly synthesized DNA strands were less dense because they incorporated ^{14}N bases. After one generation, the DNA molecules of the cells had a density intermediate between bacteria containing the ^{15}N bases and normal bacteria containing only ^{14}N bases. After a further cycle of cell division, the DNA in the density gradient sedimented at levels indicating that about half consisted of hybrid DNA helices containing equal amounts of ^{15}N and ^{14}N; the remaining half contained only ^{14}N DNA. From these results, each strand of the parental double helix was shown to be conserved in a different daughter molecule just as predicted by the semiconservative replication model.

Because the two strands of DNA in the double helix are physically intertwined, they must be separated for replication to proceed. The separation of the strands has proved to be a complex process. Replication can occur in these very long DNA molecules only if the strain of the unwinding strands is relieved. The unwinding is catalyzed by DNA helicases, enzymes that move along the helix, unwinding the strands as they move. The separated strands are bound by helix-destabilizing proteins. These bind to single-strand DNA, preventing the reestablishment of the double helix until each strand has been copied.

The DNA polymerases that link together the nucleotide subunits in the replication of DNA add nucleotides to the 3′ end of a polynucleotide strand that is paired to the strand being copied. The substrates for the DNA polymerases are deoxyribonucleoside triphosphates. As the nucleotides are joined, two of the phosphates are removed, and this provides the energy to drive the synthetic reaction. A new polynucleotide chain is elongated by the addition of the 5′ phosphate group of the next nucleotide subunit to the 3′ hydroxyl sugar at the end of the growing strand; thus, DNA synthesis proceeds in a 5′ → 3′ direction. DNA polymerases can catalyze the addition of nucleotides only at the 3′ end of an existing DNA strand. This leads to the question of how the synthesis of DNA can be initiated when the two strands are separated. This is accomplished by utilizing a short piece of an RNA primer that is synthesized by an aggregate of proteins called a primosome. The RNA primer pairs with a single-strand DNA template at the point of initiation of replication. DNA polymerase can then synthesize the new chain by adding nucleotides to the 3′ end of the RNA primer. When DNA synthesis has proceeded to an appropriate extent, the RNA primer is degraded by specific enzymes and that space is filled in by DNA polymerase. [*See* DNA SYNTHESIS.]

2. DNA Replication: Leading and Lagging Strands

One of the initial puzzles regarding the replication mechanism stemmed from the complementary DNA strands extending in opposite directions, but DNA synthesis can proceed only in the 5′ → 3′ direction. The strand being copied is being read in a 3′ → 5′ direction; hence, it would follow that only one strand can be copied at a time. Experiments demonstrate clearly that DNA replication begins at specific sites on the DNA molecule, called origins of replication, and that both strands are replicated at the same time. One, termed the leading strand, is formed continuously after the process has been initiated; its complementary lagging strand is synthesized in short pieces, which are subsequently joined to make a complete DNA chain. DNA is synthesized beginning at a Y-shaped structure called a replication fork. The lagging strand is synthesized in relatively short fragments (Okazaki fragments), each of which is initiated by a separate primer and is then extended toward the 5′ end of the previously synthesized fragment by a DNA polymerase. DNA polymerases are complex enzymes that serve several functions. As the growing fragment approaches the fragment synthesized previously, one part of DNA polymerase degrades the previous RNA primer, allowing other polymerases to fill in the gap between the two fragments. These are then linked by DNA ligase, which joins the 3′ end of one end of the fragment to the 5′ end of another by a phosphodiester bond.

When double-strand DNA is separated, two fork-like structures are created. The molecule is replicated in both directions from the origin of replication. In eukaryotic chromosomes, each of which is composed of a linear DNA molecule, there usually are multiple origins of replication. Each replication fork proceeds until it meets one coming from the opposite direction. This results in the formation of a chromosome containing two DNA double helices. [*See* CHROMOSOMES.]

B. Information Transfer: DNA to RNA

The transcription of DNA to form RNA and the replication of RNA are processes that are basically similar to that of DNA replication. In all of these, one polynucleotide strand serves as a template and specific polymerases use specific nucleoside triphosphates to add specific nucleotides. The specific base added in sequence is determined by the complementary base-pairing of the bases in the initial strand and the forming strand.

The synthesis of proteins using the genetic information present in DNA involves two stages. In the first stage, the information present in the specific sequence of nucleotides in one of the DNA strands (called the minus strand) is copied and a complementary plus strand of mRNA is produced. This transcription process is very similar to the process by which DNA is replicated: the plus mRNA strand is formed by complementary base-pairing with the minus strand of DNA. The genetic information in DNA is transcribed to yield mRNA.

C. Information Transfer: RNA to Protein

In the second stage of protein synthesis, information that has been transcribed into mRNA is used to determine the amino acid sequence in the protein. Clearly, this process involves the conversion of the nucleic acid code of the mRNA into an amino acid code of the protein; hence, it is termed a process of translation. The information present in the mRNA is in the form of a genetic code composed of three bases forming a codon. Each group of three bases in mRNA determines the presence of a specific amino acid. Thus, each codon in the mRNA specifies one of the amino acids in the protein. The codon that specifies the amino acid tryptophan is UGG. Any of six different codons (CGU, CGC, CGA, CGG, AGA, and AGG) may specify the amino acid arginine.

III. Protein Synthesis: mRNA and tRNA

The information contained in the codons of mRNA is translated by a very complex process that takes place in eukaryotic cells within the ribosomes. The recognition and decoding of the codons in the mRNA is accomplished by tRNAs. The anticodon of the tRNA molecules recognizes a codon in mRNA by complementary base-pairing. The amino acid specified by the anticodon is attached to the other end of the tRNA molecule.

The synthesis of a specific protein guided by the genetic information located in the DNA requires the joining of amino acids in the correct order by chemical bonds. This occurs on the ribosomes. The complex process of protein synthesis involves first the attaching of the ribosomes to the 5′ end of the mRNA. The ribosome then travels along the mRNA, joining the codon of the mRNA with the anticodon of the tRNA, so that the amino acids are lined up in the proper sequence.

A. Sense and Antisense Strands

DNA is a double helix, composed of two complementary anti-parallel chains and only one of the two chains, termed the minus or sense strand, normally undergoes transcription to form mRNA. The other DNA strand, the antisense strand, is usually not transcribed but may undergo transcription by DNA dependent RNA polymerases in certain cells to yield a plus or antisense RNA. The RNA polymerase that catalyzes the transcription process recognizes a specific promotor sequence of bases at the 5′ end. The mRNA is synthesized by the addition of nucleotides one at a time, to the 3′ end of the growing molecule.

The antisense RNA has been shown to regulate the expression of certain genes in both prokaryotes and eukaryotes. The antisense RNA can pair with mRNA, thus inhibiting the translation of the mRNA. It seems possible that antisense RNAs could be prepared and introduced into human cells where they could inactivate specific genes. Thus, antisense RNAs could be used in the treatment of cancer and viral diseases such as AIDS. The antisense strand of DNA, the one which is normally not transcribed, is complementary in sequence to the sense or template strand. Thus it is identical in base sequence to the product, RNA, but it is composed of T's instead of U's.

The promotor sequences of different genes may be different, and this can determine which genes will be transcribed at any given time. A promotor sequence of bacteria is typically about 40 bases in length and located about 8 bases upstream (i.e., toward the 5′ end), at the point that RNA transcription begins. Three of the specific codons in mRNA serve as stop signals for the RNA polymerase and bring about the termination of transcription.

B. Modification and Processing of mRNA: Introns and Exons

The mRNAs of bacteria can be used directly after they have been transcribed without any further processing. In contrast, the mRNA molecules of eukaryotic organisms undergo posttranscriptional modifications and processing. A molecule of 7-methyl guanylate, an unusual nucleotide, is added as a cap at the 5′ end of the mRNA chain. This may be the basis of the greater stability of eukaryotic mRNAs, which have half-lives ranging as long as 24 hr, whereas the half-life of prokaryotic mRNAs is about 15 min. A long tail of polyadenylic acid, composed of 100–200 adenine nucleotides, is joined to the 3′ end of the mRNA.

A third step in the modification of eukaryotic mRNA molecules involves cutting and splicing the mRNA at specific sites. Recent evidence has revealed the presence of interrupted coding sequences in eukaryotic DNA. The interrupted sequences may be quite long; they do not code for the amino acids present in the final protein product. These noncoding regions within the gene are called intervening sequences, or introns, as opposed to exons (which are expressed sequences and parts of the protein-coding sequence). The number of introns present in a gene can be quite variable. The Beta-globin gene, which produces one of the components of hemoglobin, contains two introns. The ovalbumin gene, which determines one of the proteins in egg white, contains 7 introns and the gene for another egg white protein, conalbumin, contains 16. The combined lengths of the introns may be considerably longer than the combined protein-coding exon sequences. The ovalbumin gene contains about 7,700 base-pairs, whereas the sum of its coding sequences is only 1,859 base-pairs.

The transcription of a gene containing both introns and exons yields a large RNA transcript termed heterogeneous nuclear RNA (hnRNA). For the hnRNA to be converted into a functional mRNA, the introns must be deleted from the molecule and the exons must be spliced together to form a continuous message that codes for a specific protein. The splicing reactions are mediated by special base sequences within and to either side of the introns. The splicing may involve the association of small nuclear ribonucleoprotein complexes, which bind to the introns and catalyze the cleavage and splicing reactions. In some species, the RNA within the intron may have the ability to splice itself without the assistance of protein catalysts. These nucleic acids can act as enzymes. This new class of biological catalysts has been termed ribozymes. The final product, the functional mRNA, has had its introns removed and the exons spliced together; it has a 7-methyl guanylic cap at the 5′ end and a poly-A tail at the 3′ end. It is then ready to pass out of the nucleus into the cytoplasm and, when attached to a ribosome, serve as the blueprint for the synthesis of a protein with its specific sequence of amino acids.

IV. Viral DNA and RNA

Much of what we know about genetic machinery in general and the replication and transcription processes in particular has been derived from studies of viruses. These small (most of them are too small to be seen in a light microscope) and relatively simple organisms have lent themselves to a great variety of studies of replication. The viruses that infect animal cells exhibit a tremendous variety of shapes, sizes, and genetic strategies.

The generalization gradually emerged that all viruses contain nucleic acids. This further suggested that viruses and genetic material had similar functions. These speculations received confirmation by studies of bacterial viruses or bacteriophages. As shown by Hershey and Chase, only the bacteriophage DNA and not the bacteriophage protein enters the bacterial host cell and initiates replication in the host cell. This leads to the production of several hundred progeny viruses. Viruses thus are genetic elements enclosed by a protective coat and are able to move from one cell to another. They have been termed mobile genes by some investigators. As studies proceeded, it became clear that some viruses have RNA, but not DNA. Thus, in these organisms it was concluded that RNA must be the viral genetic component. This theory was confirmed by the finding that purified RNA preparations from tobacco mosaic virus were infectious even in the absence of any tmv protein. Subsequently, many other viruses, such as polio, influenza, and measles, were found to contain RNA, but not DNA. Thus, the potential of RNA for carrying genetic information and undergoing replication cannot be denied. Viral DNA or RNA codes not only for the coding proteins of the virus, but also for the enzymes that are required to replicate the viral nucleic acid. The smaller DNA viruses, such as the monkey SV-40 virus, and the very small bacteriophage ϕX174 contain much less genetic information and must rely to a much greater extent on enzymes from the host cells to carry out the synthesis of proteins and DNA. These viruses do contain coding for enzymes that initiate their own DNA synthesis selectively. For a virus to be successful when it invades the cell, it must override the cellular control signals that would otherwise prevent the viral DNA from doubling more than once in each cell cycle.

Viruses originally were classified by the names of the diseases they caused or the animals or plants that they infected; however, many different kinds of viruses can produce the same symptoms and appear to be the same disease states. A dozen or more different viruses can produce red eyes, runny noses, and sneezing. Viruses are now classified according to the sequence of reactions by which the mRNA is produced. In this classification system, a viral mRNA is termed a plus strand, and its complementary sequence, which cannot function as mRNA, is termed a minus strand. Furthermore, a strand of DNA complementary to a viral mRNA is termed a minus strand. Production of a plus strand of mRNA requires that a minus strand of RNA or of

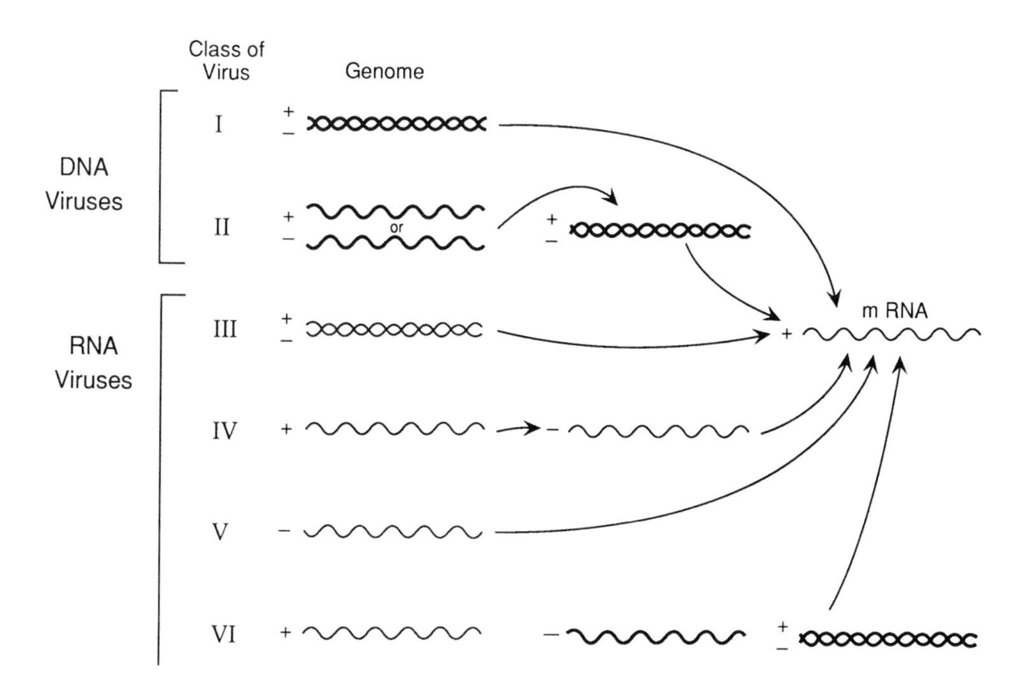

FIGURE 1 Classification of animal viruses. Heavy lines indicate DNA; light lines, RNA. Class I: The minus strand of DNA is transcribed to yield mRNA. Class II: The single strand of DNA is replicated to yield double-strand DNA, and its minus strand is translated to yield mRNA. Class III: The minus strand of the genome is copied to yield a minus strand of RNA, which is then copied to form the plus strand of RNA. Class V: The minus strand of the single-strand RNA in the genome is copied directly to yield a plus strand of mRNA. Class VI: The single-strand RNA genome is copied to form a minus strand of DNA. This is copied to yield double-strand DNA. The minus strand of this double-strand DNA is copied to form a plus strand of mRNA.

DNA be used as the template. This permits the identification of six classes of animal viruses. In each of these classes, the nucleic acid of the virion (the infective particle) ultimately becomes the mRNA of the virus (Fig. 1).

A. Classes of Animal Viruses

Class I viruses have double-stranded DNA. The adenoviruses and the SV-40 virus are Class I viruses. The DNA of these viruses typically enters the nucleus of the host cell where the enzymes that normally are responsible for producing cellular mRNA are diverted to producing viral mRNA. The pox viruses, another group of Class I viruses, are large viruses that have their own enzymes for making mRNA; they undergo replication in the cell cytoplasm.

The parvoviruses, members of Class II, are simple viruses that contain a single strand of DNA. Some parvoviruses enclose within their capsules both plus and minus strands of DNA, but they are present in separate virions. Others enclose only a minus strand within the capsid, and this is copied within the cell into double-strand DNA, which is then copied to yield mRNA.

B. Information Transfer: RNA to RNA, RNA Replication

The other four classes of animal viruses contain RNA genomes. A wide range of animals, including human beings, are infected by viruses in each of these four classes. Viruses of Class III contain a double strand of RNA. The minus RNA strand acts as a template for the synthesis of a plus strand of mRNA. The virions of Class III viruses have segmented genomes containing 8–12 double-strand RNA segments, each of which codes for a specific polypeptide. These viruses contain a complete set of enzymes that can produce mRNA. The viruses of Class IV contain a single plus strand of RNA. Because the viral genomic RNA is identical to the mRNA, the virion (genomic) RNA by itself can initiate the process of infection when introduced into a cell. The mRNA is copied into a minus strand, which then produces more plus strands. We can recognize two types of Class IV viruses. The RNA molecule in the virion of poliomyelitis virus serves as the mRNA to encode all of the viral proteins. The individual proteins are first synthesized as a single, very long polypeptide strand, which is subsequently cleaved to yield the various functional proteins. The mRNA of these viruses is the same length as the genome RNA. Viruses of Class IVb, called togaviruses because the virions are surrounded by a lipid envelope, synthesize at least two forms of mRNA in the host cell. One of these mRNAs is the same length as the virion RNA, whereas the other corresponds to the third of the virion RNA at the 3′ end. Class IVb includes many rare insect-borne viruses that cause encephalitis in human beings.

Class V viruses contain single minus strands of RNA. The RNA in the virion has a base sequence complementary to that of the mRNA. Thus, the virion contains a template for making mRNA but that template does not itself encode proteins.

We can distinguish two subdivisions of Class V. Class Va viruses have a genome that is a single molecule of RNA. A virus-specific polymerase contained in the virion synthesizes several different mRNAs from different parts of this single RNA template strand. Each of the Class Va viral mRNAs encodes one protein. Class Vb viruses, exemplified by influenza virus, have segmented genomes. Each segment is a template for the synthesis of a single mRNA. As with Class Va viruses, the virion contains a virus-specific RNA polymerase required to produce the mRNA. The minus strands of Class V nucleic acids alone, i.e., in the absence of the virus-specific polymerase, are not infectious. The influenza virus RNA polymerase initiates the transcription of each mRNA by a unique mechanism. The polymerase begins mRNA synthesis by borrowing 12–15 nucleotides from the 5′ end of the cellular mRNA or mRNA precursor in the nucleus. This oligonucleotide serves as a primer for the replication of RNA catalyzed by the viral RNA polymerase. The individual mRNAs made by Class Vb viruses generally encode single proteins, but some of the mRNAs can produce two distinct proteins (by reading different sequences of triplets within the same mRNA).

The multiplication of RNA viruses involves the formation of complementary strands. Many RNA viruses studied, such as polio virus, have a single-strand RNA polynucleotide chain. Other viruses have double helical RNA viral chromosomes; this is seen in the reoviruses that infect a great variety of organisms. RNA replication is mediated by specific RNA-dependent RNA polymerases (called replicases), which are coded for in the viral RNA chromosome.

C. Retroviruses: Information Transfer from RNA to DNA

The viruses of Class VI, called retroviruses, have a genomic RNA that is a single plus strand. This serves as a template that directs the formation of a DNA molecule. This, in turn, acts as the template for making the mRNA. First, an enzyme present in the virion, reverse transcriptase, transcribes the virion RNA into a single strand of DNA, which serves as a template for the synthesis of a complementary second strand. The double-strand DNA is integrated into the chromosomal DNA of the infected cell as a provirus. Finally, the provirus is transcribed by the cell's own machinery into RNA that either acts as a viral mRNA or becomes enclosed in a virus. This completes the retrovirus cycle. If the retrovirus contains cancer genes, the cell that it infects may be transformed into a tumor cell. [*See* RETROVIRUSES AS VECTORS FOR GENE TRANSFER.]

Bibliography

Alberts, B., Bray, D., Lewis, J., Raff, M., Roberts, K., and Watson, J. D. (1983). "Molecular Biology of the Cell," Garland Publishing, Inc., New York.

Darnell, J. E., Jr. (1985). RNA. *Sci. Am.* October. **253,** 75.

Darnell, J., Lodish, H., and Baltimore, D. (1986). "Molecular Cell Biology." Scientific American Books, New York.

Felsenfeld, G. (1985). DNA, *Sci. Am.* October. **253,** 90.

Gallo, R. C. (1986). The first human retrovirus. *Sci. Am.* December. **255,** 88.

Hogle, J. M., Chow, M., and Filman, D. J. (1987). The structure of polio virus. *Sci. Am.* March. **256,** 101.

Lewin, B. (1987). "Genes," 3rd ed. John Wiley, New York.

Watson, J. D., Hopkins, N. H., Roberts, J. W., Steitz, J. A., and Weiner, V. M. (1987). "Molecular Biology of the Gene," 4th ed. Benjamin Cummings, Menlo Park, California.

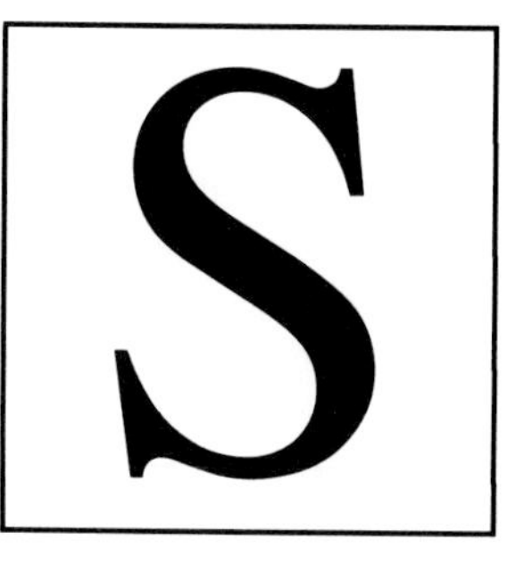

Salivary Glands and Saliva

STEVEN D. BRADWAY, MICHAEL J. LEVINE, *State University of New York at Buffalo*

Glossary

Acylation In the context of this chapter, refers to the covalent incorporation of fatty acids into protein

Cariogenic Substance that produces carious lesions in teeth

Hypotonic Solution that has a lower osmotic pressure (contains less solutes) than serum

Disulfide cross-linked proteins Proteins which contain either intermolecular or intramolecular covalent bonds between the sulfur groups of their cysteine amino acids

Electrolyte In the context of this chapter, refers to the inorganic ions found in serum and saliva: sodium, chloride, potassium, calcium, bicarbonate, phosphate, etc.

Genetic polymorphism Differential expression of the same gene or gene family in different individuals

Glycosylation Covalent coupling of carbohydrate to protein

Parenchyma Functional cells of an exocrine gland; in the context of this chapter, refers to the acinar and ductal cells of salivary glands

Phosphorylation Addition of phosphate groups onto proteins

Structural modulation Alteration of a salivary component by chemicals or enzymes after it has been secreted into the oral cavity

SALIVA IS A COMPLEX EXOCRINE secretion, which coats all oral surfaces and plays a major role in maintaining the homeostasis of the oral cavity. A constant supply of saliva (1–2 liters/day at a flow rate of 1–3 ml/min) is produced by four major groups of glands, which generate characteristic secretions in response to a diverse group of stimuli from both the oral and extraoral environment. The proteins and glycoprotein components of saliva are synthesized by conventional biosynthetic mechanisms in acinar cells to form primary saliva. As the primary saliva passes through the secretory ducts, some electrolytes are readsorbed by cells lining the ducts, and the secretion is discharged into the oral cavity as a hypotonic solution. Many salivary components, by virtue of their physical characteristics, are well adapted to form protective films called pellicles on oral surfaces. These pellicles function to lubricate and moisten oral tissues, mediate selective microbial attachment, and act as barriers against noxious substances. Other salivary components and electrolytes maintain oral pH, aid in digestion, exert antimicrobial activity, and mediate mineralization processes. The significance of these functions can be seen in salivary dysfunction, where decreased or absent salivary flow is associated with rampant tooth decay, yeast infections, and inflammation of the oral mucous membranes. Over 300 therapeutic pharmacologic agents as well as local and systemic diseases are known to adversely affect salivary function. Indeed, salivary dysfunction is increasingly recognized as a significant clinical problem in the therapy of many patient groups. Extensive research to characterize the functional characteristics of saliva has been performed over the last 20 years, and the information obtained is now being applied to create functional salivary replacements for individuals with salivary dysfunction.

I. Salivary Gland Anatomy

A. Gross Anatomy

There are three sets of anatomically distinct major salivary glands and a number of minor salivary glands (Fig. 1). The largest of these glands, the parotid gland, has a superficial and deep lobe, which grossly forms an inverted three-dimensional triangle positioned at the posterior border of the mandible. Secretions of the parotid gland are carried to the mouth through Stensen's duct, which has an orifice in the buccal mucosa adjacent to the maxillary second molar. The submandibular gland is round in shape (≈2–3 cm in diameter) and is found just below and inside the lower angle of the mandible. The submandibular duct (Wharton's duct) empties through orifices under the most anterior frenum of the tongue. The sublingual gland is composed of a major and multiple minor lobes, which lie just under the mucosal lining of the floor of the mouth. The major lobe drains through a common duct (Bartholin's duct) below the tongue, and the minor lobes drain through individual ducts in the lingual fold close to the base of the tongue. All of the major glands are innervated by parasympathetic and sympathetic fibers of the autonomic nerve system.

The fourth group of salivary organs are collectively referred to as the minor salivary glands. These reside in the submucosa of the oral mucous membranes and are given regional names such as the labial, buccal, and palatine glands to designate their location in the lips, cheek, and palate, respectively. The glandular parenchyma of the minor salivary glands is small, unilobular, and generally empties into the oral cavity through a single common duct. The ducts of minor salivary glands are often associated with cells of the immune system called the gland-associated lymphoid tissue.

B. Microanatomy

Salivary glands, like other exocrine organs, have a glandular parenchyma encapsulated by fibrous connective tissue, which partially divides the gland into lobes. The functional structure of the parenchyma is composed of a highly branched system of progressively smaller epithelial ducts that end in a semicircular cluster of acinar cells. The organelles of the acinar cells reflect their synthetic function and consist of abundant rough endoplasmic reticulum, Golgi apparatus, and secretory granules. The acinar cells can be functionally and histologically segregated into serous and mucous types. Serous cells have small, dense secretory granules confined to the apical portion of the cell, whereas the secretory granules of mucous cells are large, lucent structures that occupy most of the cellular cytoplasm. The parotid gland is composed primarily of serous cells, whereas the submandibular and sublingual glands are composed of mixed populations of serous and mucous cells. In contrast, the minor salivary glands contain purely mucous acinar cells. In the mixed acinar populations, the mucous cells form a slightly elongated acinus, which ends in a semicircular cap of serous cells (serous demilune). Adjacent acinar cells are connected at discrete areas by an intercellular seal called the junctional complex, which is comprised of tight junctions, adhering junctions, and desmosomes. An additional cellular connection, the gap junction, provides a porous junction, which allows communication between the acinar cells. The intercellular space not occupied by the junctional complex is termed the intercellular canaliculi. The basal membranes of the acinar cells are highly folded as an adaptation for ion and fluid transport. The acinar cell complex is capped with contractile cells called myoepithelial cells, and the entire complex of cells is surrounded by capillaries.

The salivary gland ducts are classified in order of ascending size as intercalated, striated, and main excretory ducts. The ductal cells possess highly-folded basal membranes, which increase the basal surface of these cells. This is thought to be a functional adaptation for the transport of electrolytes from primary saliva. These cells also perform a minor secretory function and contain small numbers of secretory organelles. The ductal cells are connected by junctional complexes and gap junctions but are also separated by intercellular canaliculi. In combination, the acinar and ductal cells form a semipermeable membrane, termed the saliva–blood barrier, that separates acinar and ductal lumen from the glandular stroma.

II. Salivary Gland Physiology

Saliva like other excretory secretions is derived from two separate functional events: (1) the synthesis, storage, and secretion of salivary macromolecules and (2) the fluid and electrolyte transport from the salivary capillary beds. Both synthesis and fluid

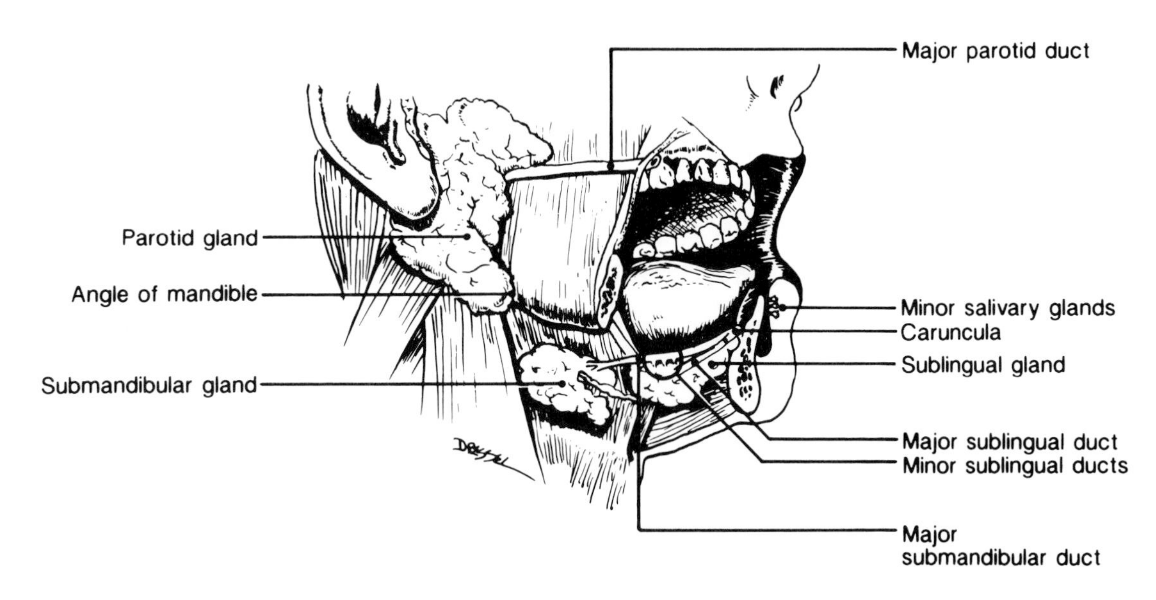

FIGURE 1 Gross anatomy of the salivary glands.

transport are mediated by acinar cell mechanisms, and the products of these events are combined in the acinar lumen to produce an isotonic secretion called primary saliva. Electrolytes are then readsorbed and/or secreted into the primary saliva in the salivary duct system to finally produce a hypotonic secretion. However, the concentrations of proteins, water, and electrolytes are also dependent on the time of day as well as the duration and type of the secretory stimulus. Integration of this system is primarily controlled by the autonomic nerve system, which is stimulated by taste and mechanical receptors in the mouth. This control can be modulated by neural influences (anxiety, fear, etc.) from the central nerve system as well as by hormones and drugs, which interact with receptors on the parenchymal cells of the salivary gland.

A. Protein Synthesis and Secretion

The synthesis, packaging, and secretion of proteins by salivary parenchymal cells proceeds by the same mechanisms of transcription and translation found in other exocrine cell types. In this process, a "message" in the form of mRNA is transcribed from genes in the acinar cell nucleus and exported to the cell cytoplasm. There, the message directs the synthesis of a protein (i.e., is "translated") into the lumen of the rough endoplasmic reticulum, a hollow cytoplasmic organelle. During and after synthesis, the protein can be further modified in a process termed posttranslational modification, by the covalent addition of carbohydrate (glycosylation), sulfate (sulfation), phosphate (phosphorylation), and/or lipid (acylation). Concomitant with protein processing, the rough endoplasmic reticulum undergoes a continuous transformation, first into the smooth endoplasmic reticulum and then into an organelle termed the Golgi apparatus. The Golgi apparatus finally gives rise to spherical structures, called secretory granules, that contain a mixture of the processed proteins, which have been highly concentrated in preparation for secretion as salivary products. Following stimulation of the salivary glands, the secretory granules migrate to the luminal surface of the acinar cell, fuse with the plasma membrane, and empty their contents into the acinar lumen. [*See* DNA SYNTHESIS.]

B. Fluid and Electrolyte Transport

The fluid phase of saliva is derived from a two-step process. Initially, water and electrolytes are transported from extraglandular interstitial fluid into the acinar lumen to produce primary saliva. The electrolyte concentration of the primary saliva is then modified in the ductal system to produce a hypotonic secretion. Salivary fluid and electrolytes are derived from capillary-derived interstitial fluid. Fluid transport into the salivary gland is thought to function in agreement with the solute-solvent coupling hypothesis, which can be conceptualized as three chambers defined by two permeable barriers (Fig. 2). The interstitial tissue surrounding the acini represents the first chamber, which contains serum transudates, whereas the acinar lumen and salivary

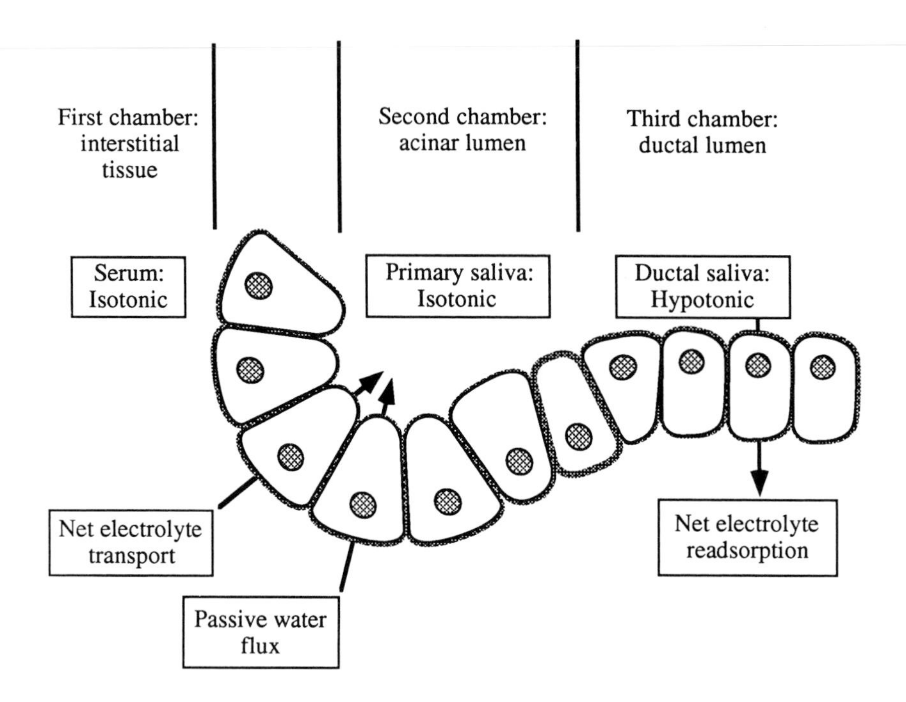

FIGURE 2 Fluid and electrolyte transport.

duct system comprise the second and third chambers, respectively. The acinar cells act as a complex barrier that serves two functions: (1) a semipermeable membrane allowing free passage of water but not salt as well as (2) a pump for the active transport of electrolytes into the acinar lumen. During secretion, electrolytes, pumped from the interstitial fluid, raise the osmotic pressure in the acinar lumen and attract a passive influx of water. This combined action creates a force that drives the primary saliva against the hydrostatic pressure (the second "barrier") created by saliva already in the ductal system.

As unstimulated or resting saliva passes through the salivary duct system, ion concentrations relative to serum are modified such that sodium, chloride, and bicarbonate are decreased and potassium is increased. Sodium is readsorbed across the luminal membrane of the duct cells by an energy-requiring mechanism in a direct or indirect exchange for potassium. As the sodium is pumped into the extraductal stroma, it attracts chloride across the ductal barrier by a transcellular route through the junctional complex. On the other hand, with glandular stimulation, sodium and chloride concentrations increase to near serum levels, bicarbonate ion concentration increases to levels far above those of serum, and potassium ion concentration decreases slightly. As salivary flow increases, the ductal transport mechanisms become saturated and allow sodium and chloride ions to be discharged with the salivary secretion. Collectively, these processes result in a hypotonic secretion (Table I).

C. Neural Control of Salivary Secretion

Protein secretion and fluid transport are triggered by the synergistic influence of both sympathetic and parasympathetic neurons of the autonomic nerve system. Neural impulses, however, cannot cross the plasma membrane and must be converted to intercellular chemical messages by the process of signal transduction. Salivary glands possess two functionally distinct signal transduction mechanisms, the adenylate cyclase and the phosphatidylinositol systems. These can be segregated into three components: (1) an extracellular receptor for neural transmitters, (2) an enzyme system, which synthesizes intracellular chemical messengers, and (3) coupling proteins, which regulate the receptor–enzyme interaction.

Separate functions of acinar and ductal cells appear to be activated by different types of autonomic neurons, which activate the synthesis of specific intracellular messengers. Stimulation of β-adrenergic (sympathetic) neurons, for instance, initiates protein synthesis by activating the adenylate cyclase system to produce the messenger cyclic adenosine

TABLE I Composition of Major Human Salivary Secretions[a]

Electrolytes (meq/liter)	Constituent	Parotid[b]	Submandibular[b]	Sublingual[c]	Plasma
	Potassium	21.0/24.0	17.0/14.4	13.2	4.0
	Sodium	36.0/1.3	45.0/3.3	32.7	140.0
	Chloride	28.0/22.0	25.0/12.0	26.2	105.0
	Bicarbonate	30.0/1.1	18.0/4.0	10.9	27.0
	Calcium	1.6/1.1	2.4/1.56	2.1	5.0
	Magnesium	0.12/0.16	0.04/0.07	?	2.0
	Phosphate	3.7/9.0	5.5/5.6	4.1	2.0
Organics[c] (mg/100 ml)	**Constituent**	**Parotid**	**Submandibular/sublingual**		**Plasma**
	Protein	221.0	132.0		7,000
	Lipids	8.0	8.0		600
	Carbohydrate[d]	31.0	15.0		100–140

[a] Ferguson (1989), Levine *et al.* (1978), and Slomiany *et al.* (1982).
[b] Mean values for stimulated/unstimulated saliva.
[c] Mean values for stimulated salivas.
[d] Carbohydrate content of glycoproteins.

monophosphate (cAMP). In contrast, stimulation of α_1-adrenergic (sympathetic) and parasympathetic neurons results in fluid and electrolyte transport following the synthesis of the intercellular messenger inositol triphosphate (IP_3) by the phosphatidylinositol system. Similarly, ductal readsorption and secretion of salivary electrolytes are controlled by the effect of second messengers on the ductal epithelium following the selective stimulation of sympathetic as well as parasympathetic neurons. [*See* AUTONOMIC NERVOUS SYSTEM.]

III. Composition of Saliva

The total protein, carbohydrate, and lipid concentrations of the major human salivary secretions is considerably less than that in serum (Table I). Salivary gland secretions contain at least 40 proteins and glycoproteins, many of which can be grouped into at least seven families whose members are genetically and structurally related (Table II). In addition, there are singular components that may have relatives in other exocrine secretions. The secretions of all salivary gland groups, combined with oral microflora, gingival secretions, and sloughed or desquamated oral epithelium, result in a mixture termed whole saliva. Only the glandular products will be discussed in the following sections.

A. Families of Salivary Proteins

Multiple members of families of salivary proteins within the same individual have been shown to result from a small group of gene transcripts by posttranscriptional and/or posttranslational modifications. The family of proline-rich proteins (PRPs) have been well studied and serve to illustrate this phenomenon. To date more than 20 PRPs have been grouped into three major categories: acidic, basic, and glycosylated species. All are related by their high content of proline, glycine, and glutamine, but each group is distinguished by repeated amino acid sequences or posttranslational modifications that endow them with a distinct isoelectric point (acidic or basic) and/or carbohydrate content. The PRPs are now thought to be derived from the transcripts of six gene loci through differential gene splicing. These proteins may then undergo additional posttranslational modifications in which carbohydrate as well as phosphate groups are added. In addition, some of the acidic PRPs are thought to undergo proteolytic processing in the salivary duct by the enzyme kallikrein.

Other salivary families are composed of groups of proteins that share similar structural and functional characteristics. For instance, human salivary mucins exist as two distinct, highly-glycosylated molecules designated MG1 and MG2. MG1 is composed of multiple, disulfide-linked subunits comprising a molecule of >1,000 kDa. The MG1 oligosaccharides may be sialylated (i.e., terminate with a negatively charged sialic acid residue) or sulfated and are located in densely glycosylated regions, which are separated from nonglycosylated or "naked" peptide regions. MG1 also contains a small amount of covalently linked fatty acids (i.e., acyl-

TABLE II Salivary Families

Family name	Functions	Biochemical composition
Mucins	1. Selective clearance and adherence of microflora 2. Tissue-coating and formation of intraoral pellicles 3. Lubrication at hard and soft tissue interfaces 4. Microbial nutrient source 5. Digestion and taste 6. Complexing with lysozyme, cystatin, and sIgA	Glycoproteins
Acidic and basic proline-rich proteins	1. Selective clearance and adherence of microflora 2. Tissue-coating and formation of intraoral pellicles 3. Lubrication at hard and soft tissue inferfaces 4. Microbial nutrient source 5. Modulation of mineralization processes on tooth surfaces 6. Complexing with albumin	Phosphoproteins and glycoproteins
Cystatins	1. Antimicrobial activity 2. Complexing with other salivary molecules to coat oral surfaces 3. Modulation of mineralization processes on tooth surfaces 4. Thiol protease inhibition	Proteins and phosphoproteins
Histatins and statherin	1. Antimicrobial activity 2. Tissue-coating and formation of intraoral pellicles 3. Modulation of mineralization processes 4. Buffering of salivary pH	Proteins and phosphoproteins
Amylase	1. Digestion of complex carbohydrates 2. Tissue-coating and formation of intraoral pellicles 3. Selective clearance and adherence of microflora 4. Antimicrobial activity 5. Digestion and taste 6. Complexing with other salivary molecules to coat oral surfaces	Proteins and glycoproteins
Carbonic anhydrases	1. Reduce salivary pH by catalyzing the formation of bicarbonate from carbon dioxide	Glycoproteins
Salivary peroxidases	1. Catalysis of the formation of products that are toxic to some oral bacteria	Glycoproteins

ated) and possesses hydrophobic domains in the naked core regions. In contrast, MG2 is a smaller molecule of 120–130 kDa, which contains a single polypeptide chain uniformly glycosylated with small oligosaccharide chains to give this molecule a "bottle brush" configuration. Another family of glycoproteins, the amylases, are synthesized as two subfamilies composed of glycosylated and nonglycosylated isoenzymes. Each of these subfamilies contains isoenzymes, which may have only slight charge variations produced by posttranslational deamidation of glutamine residues in the parent molecule. Two additional families of enzymes, the salivary peroxidases and the carbonic anhydrases, each contain members that differ slightly in physicochemical characteristics but perform the same catalytic functions. Other families of salivary proteins such as the cystatins and histatins are categorized by common compositional characteristics. Cystatins, for instance, contain a characteristic content of cysteine amino acids, whereas histatins are rich in histidine residues.

B. Singular Salivary Components

The singular salivary components (Table III) include kallikrein, lactoferrin, lysozyme, secretory IgA (sIgA), and fibronectin. With the exception of sIgA, these components are synthesized primarily by ductal cells; however, some are synthesized in lesser quantities by acinar cells. Secretory IgA is synthesized as monomeric IgA by plasma cells surrounding the acinar cells. The salivary acinar cells then import monomeric IgA, process it, and secrete the modified immunoglobulin, secretory sIgA, into saliva. Kallikrein is a serine protease that is se-

TABLE III Individual Salivary Molecules

Molecule	Functions	Biochemical composition
Fibronectin	1. Tissue-coating and formation of intraoral pellicles 2. Mediates microbial adherence	Glycoprotein
Kallikrein	1. Posttranslational processing of proline-rich proteins and cystatins	Glycoprotein
Lactoferrin	1. Antimicrobial activity 2. Complexing with other salivary molecules to coat oral surfaces	Glycoprotein
Lysozyme	1. Antimicrobial activity 2. Complexing with other salivary molecules to coat oral surfaces	Protein
Secretory IgA	1. Complexing with other salivary molecules to coat oral surfaces 2. Mediates clearance and adherence of microflora 3. Antimicrobial activity	Glycoprotein

creted by the striated ducts of the major salivary glands. Both lactoferrin and lysozyme are antimicrobial agents that share identity with similar agents in tears, gastric mucosal secretions, and specific granules of neutrophils. Fibronectin, a substance that mediates intercellular adherence interactions in most human tissues, also appears to be present in saliva.

IV. Function of Saliva

In general, salivary constituents play a protective role either as individuals or in molecular complexes. Each of these constituents also appears to possess unique physicochemical characteristics that allow them to express their biologic function either free in solution or when adsorbed to oral surfaces. The functional characteristic(s) of individual salivary components are dependant on their structural characteristics; thus, alteration of their structure by host and/or microbial enzymes alters their functional characteristics. The collective result of all of these functions serves to (1) maintain the microbial ecology of the oral cavity, (2) prepare food for swallowing and digestion, and/or (3) preserve the integrity of oral tissues.

A. Tissue-Coating

Salivary components adsorb to tooth enamel, dental materials, microbial surfaces, and epithelial cells with a selectivity that depends on the physicochemical characteristics of the individual adsorbent surfaces as well as that of the salivary component. Adsorbed components, in turn, can complex with other salivary constituents to form protective films called pellicles, which function as lubricants, permeability barriers against acids, moisture retainers, and modulators of microbial adherence. Complexing among salivary components may act to concentrate and possibly enhance the functional characteristics of an individual salivary component. For instance, salivary mucins coat surfaces and also form complexes with antimicrobial factors such as sIgA, lysozyme, and cystatins. This may serve to localize and concentrate these substances on oral surfaces and increase antimicrobial activity. Additionally, proline-rich glycoprotein, a salivary lubricant, provides enhanced lubrication when complexed with human serum albumin. Other functional characteristics of salivary pellicles depend on the carbohydrate moieties of their salivary constituents. Moisture retention, for instance, is primarily mediated by the carbohydrate moieties. Highly glycosylated components such as mucin and proline-rich glycoprotein also provide better lubrication than that of less glycosylated salivary components such as lactoferrin, amylase, and secretory IgA. Enzymatic alteration of salivary proteins prior to or after they are adsorbed to oral surfaces (i.e., structural modulation) may also affect their functional characteristics. For instance, a relatively nonpathogenic bacteria, *Streptococcus sanguis*, utilizes a sialic acid-binding adhesion on its surface to bind to oral surfaces coated with salivary glycoproteins containing terminal sialic acid residues. Other oral bacteria produce an enzyme, neuraminidase, which may cleave these sialic acid residues to expose underlying galactose residues. Subsequently, putative oral pathogens such as *Actinomyces viscosis* or *Streptococcus mutans*, which possess galactose-binding proteins are then able to adhere to these surfaces and initiate a disease process.

B. Antimicrobial Activity

Salivary peroxidase, lysozyme, lactoferrin, histatins, and sIgA all exert antimicrobial activity. The salivary peroxidase system utilizes dietary thiocyanate ions (SCN^-) and bacterial hydrogen peroxide to synthesize hypothiocyanate, a substance that reversibly inhibits bacterial growth and metabolism. Lysozyme cleaves carbohydrate linkages in the cell wall of gram-positive bacteria, making them susceptible to changes in osmotic pressure and causing them to burst in the hypotonic environment of the oral cavity. Lactoferrin is a noncatalytic iron-binding molecule that exerts bacteriostatic effects by binding iron, an essential bacterial nutrient. Additional studies suggest that lactoferrin may possess a direct, iron-independent, bacterocidal effect on various strains of streptococci. The histatins can inhibit the viability of the oral pathogen *Candida albicans* and can also inhibit the growth of *S. mutans*. Secretory IgA (sIgA) is the primary component of the oral mucosal immune system. sIgA is thought to produce antimicrobial action by specifically binding oral microbes to prevent them from adhering to and colonizing oral surfaces.

C. Posttranslational Processing

In vitro studies have shown that salivary kallikrein cleaves the long forms of the acidic proline-rich proteins to produce a shorter form of these proteins and a small C-terminal peptide. Similarly, the N-terminal portion of salivary cystatins may also be processed. The immunohistochemical localization of kallikrein suggests that this cleavage takes place in the ductal system. The biological significance of these events remains to be determined.

D. Digestion

Digestion is primarily an enzymatically mediated function of the lower alimentary canal. Saliva contains two enzymes that contribute to this process. The first is α-amylase, one of the most abundant salivary components, which hydrolyzes dietary starch into smaller fragments. Lipase, the second enzyme, is secreted by the minor salivary glands of Von Ebner at the base of the tongue and is thought to play a role in the initial digestion of lipids. In addition, saliva assists in the hydration and dispersion of the food particles during the mastication or chewing process. This aids in the formation and lubrication of the food bolus in preparation for passage through the esophagus and provides a fluid medium that is beneficial to the processes involved in taste.

E. Buffering Capacity

Food residue left on teeth after meals is rapidly converted by oral bacteria to organic acids which can cause decalcification of tooth structure and subsequent tooth decay (dental caries). Saliva provides several buffering mechanisms to counteract this process. Salivary bicarbonate, the primary buffering agent in saliva, is produced in the salivary ductal cells. Bicarbonate may also be formed directly in the oral cavity by the action of carbonic anhydrases. Once in the oral cavity, bicarbonate ions may be complexed to salivary mucins, which are adsorbed to oral surfaces. This may enhance the protective nature of mucins by producing a buffered barrier against acid penetration to oral mucosa and tooth enamel. Other salivary components may act as buffers by virtue of their amino acid content. Histidine-rich peptides (i.e., histatins) contain a high content of the basic amino acid histidine, which may act to neutralize acidic by-products of bacterial metabolism.

F. Mineralization Processes

Tooth enamel is composed of a relatively insoluble calcium-phosphate mineral termed hydroxyapatite. Under normal conditions of ionic strength and pH, this mineral will slowly dissolve in a saliva that is devoid of its protein constituents. However, with normal salivary flow and composition and minimal exposure to bacterial acids, individuals do not lose their teeth to dissolution over long periods of time. In fact, decalcified enamel associated with an early carious lesion will remineralize if the tooth surface is regularly cleaned and allowed to come in contact with saliva. Recent evidence suggests that the acidic residues, including phosphate, of salivary phosphoproteins may bind calcium in saliva to produce a much higher concentration of calcium than would otherwise be possible. This supersaturation constitutes a thermodynamic driving force that fa-

TABLE IV Factors That Affect Salivary Gland Function[a]

Category	Condition	Effect on salivary gland
Factors that directly affect the gland	Ductal obstruction	Pressure atrophy of the acinar cells
	Acute and chronic inflammation	Immune-mediated destruction of glandular parenchyma
	Head and neck radiation therapy	Destruction of the glandular parenchyma by radiation
	Trauma	Neural or parenchymal destruction
	Benign and malignant neoplasm	Neoplastic infiltration of the glandular parenchyma
Systemic diseases	Sjogrens's syndrome	Autoimmune destruction of the glandular parenchyma
	Graft versus host disease	Inflammatory destruction of the glandular parenchyma
	Viral infection (such as mumps)	Inflammatory destruction of the glandular parenchyma
	Cystic fibrosis	Increase salivary viscosity, altered electrolyte readsorption
	Dehydration in diabetic acidosis, uremia, etc.	Decreased salivary flow

[a] Adopted from Mandel (1980) and Baum *et al.* (1985).

vors the formation of calcium phosphate salts (i.e., remineralization) on the surface of the teeth. [*See* DENTAL AND ORAL BIOLOGY, ANATOMY.]

V. Salivary Dysfunction and Therapy

A. Etiologies of Salivary Dysfunction

Some variations in salivary flow rates and composition are part of normal salivary physiology. In contrast, sustained or permanent alterations in these parameters are considered pathologic and are referred to as salivary dysfunction. Salivary glands may be affected by local or systemic factors, which result in a transient or permanent glandular dysfunction, possibly glandular destruction. Factors that cause salivary gland dysfunction include (1) physical obstruction of salivary duct, (2) destruction of glandular parenchyma, and/or (3) pharmacologic alteration of secretory mechanisms (Tables IV and V).

Ductal obstruction often completely blocks the salivary duct but does not stop acinar secretion. This results in an increased intraductal pressure, which can lead to atrophy of the acinar cells and loss of glandular function. Primary factors that lead to obstruction of the salivary duct are the formation of mineralized stones and/or trauma. These conditions are often treated by removal of the obstruction or repair of the salivary duct. Obstructive dysfunction may also be a manifestation of systemic disease such as cystic fibrosis or tumor involvement of the salivary duct. Cystic fibrosis is a genetically inherited disease that is associated, in part, with a highly viscous secretion in mucin-producing glands, which results in obstruction of the salivary duct by "mucin plugs." Salivary gland or nonsalivary head and neck tumors can also compress or involve the salivary gland ducts. In both of these cases, treatment of the primary disease usually dictates the resolution of the salivary dysfunction. [*See* CYSTIC FIBROSIS, MOLECULAR GENETICS.]

Destruction of the salivary gland parenchyma may take place as the result of autoimmune disease or radiation therapy, or as a consequence of infection. Autoimmune involvement of salivary glands is most commonly associated with the primary and secondary forms of sicca syndrome. The primary form of sicca syndrome is characterized by involvement of the lacrimal and salivary glands, which is accompanied by symptoms of both dry eyes and dry mouth. The secondary form (Sjogrens's syndrome) involves at least one additional connective tissue disease, which may include systemic lupus erythematosus, scleroderma, chronic hepatobiliary disease, and Raynaud's phenomenon. Both forms of sicca syndrome are characterized by a localized inflammatory infiltrate that results in swelling of the salivary gland and progressive destruction of the

TABLE V Effects of Pharmacologic Agents on Salivary Glands[a]

Effect on salivary gland	Drug category	Representative drug
Increased salivary flow	Centrally acting drugs	Strychnine
		Reserpine
	Parasympathomimetic drugs	Muscarine
		Neostigmine
		Physostigmine
	Sympathomimetic drugs	Norepinepherine
		Epinepherine
		Ephederine
Decreased salivary flow	Centrally acting drugs	General anesthetics
		Barbituates
	Parasympatholytic	Atropine
		Scopolamine
		Antihistamines
	Sympatholytic	Phentolamine
		Ergotamine
		Clorporomazine
	Ganglionic blockers	Psychoactive drugs
		Hexamethonium
Altered salivary composition	Cardiac glycosides	Digitalis
	Cancer chemotherapy drugs	Methotrexate
		Vincristine
		Cytoxan

[a] Adopted from Mandel, 1980; Baum *et al.* 1985; Sreebny & Broich, 1987.

glandular parenchyma. While sicca syndrome patients may retain partial salivary gland function, this disease often progresses to glandular destruction and almost complete loss of function. An immune-related glandular dysfunction similar in histological appearance to sicca syndrome has also been reported in patients receiving bone marrow transplants for diseases such as leukemia. In these patients, all immune component cells are killed by whole-body radiation after which a patient receives new immune cells in the form of a normal marrow transplant. In some patients, the transplanted immune cells recognize the new host as foreign and mount an immune response against the salivary glands among other tissues. This response has been associated with symptoms of dry mouth and decreased secretion of sIgA in minor salivary glands.

All of the salivary glands can also be damaged by the doses of radiation used in the treatment of head and neck tumors with each gland varying in its sensitivity to radiation. Immediately following radiation therapy, <5% of parotid gland function remains. Due to a higher resistance to radiation, the submandibular-sublingual glands initially retain approximately 25% of their function. However, their residual function decreases to <5% after 3 years.

Infection of salivary glands may be the result of a primary systemic infection (e.g., mumps) or localized retrograde infections, which are secondary to other glandular dysfunction. Mumps is a viral infection transmitted through direct or indirect contact with saliva containing the mumps virus. The primary target of this virus is the parotid glands, but it may also infect the gonads, pancreas, central nervous system, and heart. Mumps is characterized by parotid gland swelling and focal necrosis of the glandular parenchyma. This is a transient self-limiting disease that runs its clinical course in 3–7 days and does not result in permanent glandular dysfunction. Localized retrograde infections are generally secondary to reduced salivary flow. The flow of saliva through the salivary duct provides a natural countercurrent barrier against the migration of oral bacteria into the salivary gland. During decreased salivary flow, bacteria can migrate up the salivary ducts and produce acute or chronic suppurative infections of the glandular parenchyma. This type of infection can occur with any condition that produces reduced salivary flow and commonly occurs during reduced salivary flow associated with dehydration in infant, elderly, or debilitated patients.

Many pharmacologic agents are known to affect the autonomic nerve system by altering central nervous function by affecting either nervous impulses at autonomic ganglia and/or postganglionic synapses. At the postganglionic level, sympathomimetic and parasympathomimetic drugs generally increase salivary flow by stimulating fluid transport and acinar secretion, whereas sympatholytic and parasympatholytic drugs decrease salivary flow (Table V). Many drugs used to treat serious conditions such as Parkinson's disease, various psychoses, and hypertension produce decreased salivary flow or possibly altered salivary composition. However, in a therapeutic context, stimulatory drugs such as pilocarpine have recently been tested in clinical trials to treat patients with decreased salivary flow.

B. Therapy of Salivary Dysfunction

In many cases, patients with significant salivary flow report a perception of dry mouth, whereas others with little salivary flow have no sensation of dryness. Thus, the treatment for salivary dysfunction should only be initiated after determining the extent of salivary flow. In many instances, a differential diagnosis should also include radiographic and biopsy procedures as well as salivary flow measurements. The information obtained will enable the clinician to determine if patients with deficient salivary function can be stimulated to produce additional saliva (i.e., intrinsic therapy) or if the patient will require topical replacement of lost salivary flow (i.e., extrinsic therapy). Intrinsic therapy is based on a pharmacologic approach in which agents such as dilute citric acid, pilocarpine, and bromhexine are used to stimulate any residual function in the salivary glands. Therapy with drugs such as pilocarpine produces inherent side effects that have prevented this method of therapy from progressing beyond the experimental stage. The extrinsic approach is the most widely practiced method of therapy and is predicated upon periodic oral rinsing with salivary replacements. The most commonly used preparations are formulated from carboxymethylcellulose and the sugar alcohols, xylitol and sorbitol. However, these preparations exhibit low oral retention as well as poor taste and mucosa-irritating characteristics. A more therapeutic preparation should have sustained effects that would include lubrication, tissue-coating and moistening properties, and selected antimicrobial function. Research directed at the chemical and physical characterization of salivary proteins and glycoproteins now suggests that discrete portions of these molecules endow them with their functional characteristics. This knowledge is now being used to design and synthesize composite molecules that will contain multiple protective functions. This approach should result in saliva replacements that have enhanced functional characteristics compared to authentic saliva.

Bibliography

Abdel-Latif, A. A. (1986). Calcium-mobilizing receptors, polyphosphoinositides, and the generation of second messengers. *Pharm. Rev.* **38,** 227–272.

Arnold, R. R., Brewer, M., and Gauthier, J. J. (1980). Bacterial activity of human lactoferrin: Sensitivity of a variety of microorganisms. *Infect. Immun.* **28,** 893–898.

Baum, B. J. (1987). "Regulation of salivary secretion." *In* "The Salivary System," Chapter 6 (L. M. Sreebny, ed.). CRC Press, Boca Raton, Florida.

Baum, B. J., Bodner, L., Fox, P. C., Izutsu, K. T., Pizzo, P. A., and Wright, W. E. (1985). Therapy-induced dysfunction of salivary glands: Implications for oral health. *Spec. Care Dent.* **5,** 274–277.

Cohen, R. E., and Levine, M. J. (1989). Salivary Glycoproteins. *In* "Human Saliva: Clinical Chemistry and Microbiology," Vol. 1, Chapter 4 (J. O. Tenovuo, ed.). CRC Press, Boca Raton, Florida.

Ferguson, D. B. (1989). Salivary electrolytes. *In* "Human Saliva: Clinical Chemistry and Microbiology," Vol. 1, Chapter 3 (J. O. Tenovuo, ed.). CRC Press, Boca Raton, Florida.

Gardner, E. A., Gray, D. J., and O'Rahilly, R. (1986). "Anatomy: A Regional Study of Human Structure," 5th ed., Chapters 58–59. W. B. Saunders, Philadelphia.

Goedert, M., Naby, J. I., and Emson, P. C. (1982). The origin of substance-P in the rat submandibular gland and its major duct. *Brain Res.* **252,** 327–333.

Iacono, V. J., MacKay, B. J., Direnzo, S., and Pollock, J. J. (1980). Selective antibacterial properties of lysozyme for oral microorganisms. *Infect. Immun.* **29,** 623–632.

Izutsu, K. T. (1987). Salivary fluid production in health and disease. *In* "The Salivary System," Chapter 5 (L. M. Sreebny, ed.). CRC Press, Boca Raton, Florida.

Knauf, H., Lubcke, R., Kreutz, W., and Sachs, G. (1982). Interrelationship of ion transport in rat submaxillary duct epithelium. *Am. J. Physiol.* **242,** F132–F139.

Levine, M. J., Aguirre, A., Hatton, M. N., and Tabak, L. A. (1987). Artificial salivas: Present and future. *J. Dent. Res.* (Special Issue) **66,** 693–698.

Levine, M. J., Herzberg, M. C., Ellison, S. A., Shomers, J. P., and Sadowski, G. A. (1978). Biochemical and immunological comparison of monkey (*Macaca arc-*

toides) and human salivary secretions. *Comp. Biochem. Physiol.* **60B,** 423–431.
Mandel, I. D. (1980). Sialochemistry in disease and clinical situations affecting salivary glands. *CRC Crit. Rev. Clin. Lab. Sci.* **12,** 321–366.
Minaguchi, K., and Bennick, A. (1989). Genetics of human salivary proteins. *J. Dent. Res.* **68,** 2–15.
Oppenheim, F. G., Xu, T., McMillian, F. M., Levitz, S. M., Diamond, R. D., Offner, G. D., and Troxler, R. F. (1988). Histatins, a novel family of histidine-rich proteins in human parotid secretion: Isolation, characterization, primary structure and fungistatic effects on *Candida albicans*. *J. Biol. Chem.* **16,** 472–477.
Sicher, H., and DuBrul, L. E. (1970). "Oral Anatomy," 5th ed., pp. 191–196. C. V. Mosby, St. Louis.
Slomiany, B. L., Murty, V. L. N., and Aono, M., Slomiany, A., and Mandel, I. D. (1982). Lipid composition of human parotid and submandibular saliva from caries-resistant and caries-susceptible adults. *Arch. Oral Biol.* **27,** 803–808.
Sreebny, L. M., and Broich, G. (1987). "Xerostomia (Dry Mouth)." *In* "The Salivary System, Chapter 9 (L. M. Sreebny, ed.). CRC Press, Boca Raton, Florida.
Sreebny, L. M., and Valdini, A. (1987). Xerostomia: A neglected symptom. *Arch. Intern. Med.* **147,** 1333–1337.
Tenovuo, J., and Pruitt, K. M. (1984). Relationship of the human salivary peroxidase system to oral health. *J. Oral Path.* **13,** 573–584.

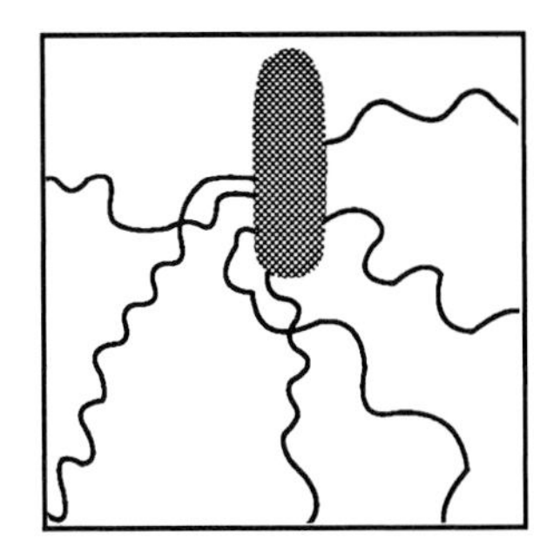

Salmonella

M. A. ROUF, *University of Wisconsin*

Glossary

Carrier Apparently healthy person harboring a pathogen

Facultative anaerobe Organism that is able to grow in the presence or absence of oxygen

Phagovar Subdivision of a serovar based on the sensitivity to a series of bacteriophages at appropriate dilutions

Plasmid Small piece of circular DNA besides bacterial chromosomal DNA

Salmonellosis Infection of the gastrointestinal tract by *Salmonella*

Serovar (serotype) Subdivision of a species based on its antigenic analysis

THE SALMONELLAE ARE pathogenic for humans, causing enteric fever, gastroenteritis, and septicemia. They also infect many animal species besides humans. Animals are the main reservoirs of *Salmonella* species. Over 2,100 serovars of *Salmonella* have been identified based on "O" and "H" and Vi antigens. The serovars and species, although equated, are not the same. The O antigen, also known as somatic antigen, is the outer portion of the endotoxin complex. The H antigen (flagellar antigen) is used in conjunction with O antigen to identify serovars. The Vi antigens primarily occur in *Salmonella typhi*. Each serovar of *Salmonella* is characterized by its unique structure and arrangement of simple sugars in the O-specific polysaccharide chain.

Salmonella cells enter the animal body by ingestion of contaminated food or drink or via person-to-person contact. The infection begins in the intestinal tract. Disease manifestations by *Salmonella* depend on the virulence and the type of the serovars, general health of the animal, and the number of viable cells ingested. Typhoid fever is caused by *S. typhi*. It is a systemic infectious disease resulting in prolonged fever. Paratyphoid fever is relatively mild and of shorter duration compared with typhoid fever. Although paratyphoid fever may be caused by any serovars of *Salmonella,* the serovars most often responsible are *S. paratyphi* A, *S. schottmulleri, S. hirschfeldii,* and *S. senadi*. The most effective treatment for enteric fevers is administration of antibiotics. Typhoid vaccines are available and offer important protection. Any of the over 2,000 serovars that are ubiquitous in nature can cause gastroenteritis. Most commonly *S. typhimurium* and *S. enteritidis* are responsible for food-borne infection. Poultry, meat, and eggs and recently milk and cheese are the main sources of transmission of nontyphoidal salmonellosis. Septicemia is most often caused by *S. choleraesuis,* followed by *S. typhimurium*.

I. Characteristics and Classification

Salmonella is a genus of bacterium that belongs to the family Enterobacteriaceae. The organisms are facultatively anaerobic, Gram-negative, nonsporforming, straight rods, and 0.7–1.5 in width and 2–5 millimicron in length. Most are motile by means of peritrichous flagella (flagella arising out of entire cell surface). Colonies in typical culture media are 2–4 mm in diameter.

The typhoid bacillus (*S. typhi*) was the first organism in this genus to be observed in tissue of dead patients by Eberth in 1880. The organism at that time was known as *Eberthella*. The genus is now named after an American bacteriologist, D. E. Salmon.

In 1884, Gaffky was able to culture typhoid bacteria from mesenteric lymph nodes, and Pfeiffer cultured them from a fecal specimen. In 1885, Salmon and Smith isolated *Salmonella* from cases of swine fever and named the organism *Bacillus cholera suis*. Swine fever, however, is caused by a virus; the authors mistakenly thought it was a bacterial disease. *S. enteritidis* was isolated by Gaertner in 1888 from patients who ate contaminated meat and developed food poisoning.

As most etiological agents of various infectious diseases were isolated and identified during the period leading up to the 1930s, the number of organisms in this genus increased rapidly in number, creating a confusion in naming of the species, which even exists today. Presently, the bacteria in the genus *Salmonella* are classified according to antigenic relationship. Over 2,100 serovars have been described, and new ones are being added from time to time. Although antigenic relationship is useful in classifying *Salmonella,* it is taxonomically unsound.

The salmonellae are pathogenic for humans, causing enteric fever (typhoid and paratyphoid fever), gastroenteritis, and septicemia. They also infect many animal species besides humans. Animals are the main reservoirs of *Salmonella* species.

A. Taxonomy

Based on biochemical characteristics, the genus *Salmonella* is presently subdivided into five subgenera, namely subgenera I, II, III, IV, and V (Table I). The subgenus V includes organisms that grow in the presence of potassium cyanide (KCN) (as those of subgenus IV) but have other biochemical differences from all other subgenera. They are further divided into over 2,100 serovars (serotypes) based on the Kauffman-White schema. Each is given a specific epithet. For example, *S. typhi* is the name of a serovar, not the name of a species, although it has a specific epithet. The use of a species name to describe a serovar (serotype) has lead to much confusion and misconception of equating serovars to species, a definite taxonomic term.

The naming of *Salmonella* is controlled by an in-

TABLE I Differential Characteristics of the "Subgenera" of the Genus *Salmonella*[a]

	Subgenus				
	I	II	III	IV	V[b]
β-galactosidase (ONPG test)	−	− or x	+	−	+
Acid production from:					
Lactose	−	−	+ or x	−	−
Dulcitol	+	+	−	−	+
Mucate	+	+	d	−	+
Galacturonate[c]	−	+	d	+	+
Utilization of:					
Malonate	−	+	+	−	−
d-tartrate	+	− or x	− or x	− or x	−
Gelatin hydrolysis (film method)	−	+	+	+	−
Growth in presence of KCN	−	−	−	+	+
Habitat of the majority of strains:					
Warm-blooded animals	+	−	−	−	−
Cold-blooded animals and environment	−	+	+	+	+

[a] Symbols: +, positive for 90% or more of strains in 1–2 days; d, positive for 11–89% of strains in 1–2 days; −, positive for 0–10% of strains in 1–2 days; x, late and irregularly positive (3–7 days). The temperature for all reactions is 37°C.

[b] L. Le Minor, M. Veron, and M. Popoff, 1982, *Ann. Microbiol.* (*Inst. Pasteur*) **133B,** 223–243.

[c] From Le Minor *et al.* (1979). Monophasic serovars of "Subgenus" III are galacturonate-negative; diphasic serovars are positive.

[Reprinted, with permission, from L. Le Minor, 1984, Genus III *Salmonella, in* "Bergey's Manual of Systematic Bacteriology," Vol. 1 (N. R. Krieg and J. G. Holt, eds.), p. 427.

ternational agreement. According to this system, a serovar is named after the place where it was first isolated (e.g., *S. bangkok, S. dakar, S. karachi, S. london, S. miami*).

The classification of *Salmonella* remains unsatisfactory. In Bergey's Manual of Systematic Bacteriology (1984), it is stated that employing specific names for *Salmonella* serovars is extremely useful, although serovars and species should not be regarded as equivalent. It is further stated that DNA relatedness data of organisms distinctly belonging to the five subgroups (subgenera) show that they belong to a single genetic *Salmonella* species. It is, therefore, proposed as "a single diverse species with five subspecies." The suggestion is made to designate "*S. enterica*" as the only species. It was also suggested that in medical bacteriology Latin binomials (generic and specific names) could be used for serovars in subgenus I and for the named serovars of the subgenera II and IV. Serovars in subgenera II, III, and IV, which have not been named, could be listed as subspecies followed by antigenic formulae. Based on DNA relatedness, the editorial board of Bergey's Manual of Systematic Bacteriology, without any specific recommendation, stated that the genus *Salmonella* should consist of a single species, *S. choleraesuis,* with six subspecies as suggested by LeMinor, Veron, and Popoff. The following organisms, respectively, were proposed as the type species for the six subspecies: *S. choleraesuis, S. salamae, S. arizonae, S. diarizonae, S. houtenae,* and *S. bongori*.

The sensitivity of *Salmonella* isolates to various bacteriophages is also used to classify them into phagovars. A serovar can be divided into biovars based on different sugar fermentation patterns. Biovars may serve as markers to identify pathogens in epidemic outbreaks.

B. Physiological and Biochemical Characteristics

The salmonellae have simple nutritional requirements. They grow on most routine bacteriological growth media without enrichment or special supplement. They are facultatively anaerobic (i.e., they do not require oxygen for growth but grow well both in presence or in absence of oxygen). Their optimum growth temperature is 37°C, but they grow reasonably well at room temperature.

Salmonella does not ferment lactose except for some organisms in the subgenus *Salmonella* III (*arizonae*). Also, they usually do not ferment sucrose, salicin, inositol, and amygdalin. Glucose is fermented usually with the production of gas, except in *S. typhi,* which does not produce gas either from glucose or other carbohydrates. Typically, H_2S is produced and citrate utilized as the sole source of carbon. They are indole and urease-negative and unable to liquefy gelatin. Lysine and ornithine decarboxylase reactions are usually positive. The differential characteristics of the various subgenera are given in Table I.

C. Antigenic Differentiation

All *Salmonella* are identified antigenically by a simple agglutination test using intact cells and monospecific antisera. The salmonellae possess three major antigens called the "O", "H", and "Vi" antigens. Over 2,100 serovars of *Salmonella* have been identified based on different combinations of these antigens. The detection of Vi antigen, which occurs primarily in *S. typhi* and a few other serovars, can be used as a screening test for typhoid carriers.

The O antigen, also known as somatic antigen, is the outer portion of the endotoxin complex. The endotoxin complex of *Salmonella* and other Gram-negative organisms is a lipopolysaccharide (Fig. 1) that joins the outer membrane structure to the peptidoglycan cell wall. Although the endotoxin component of the cell wall is of low toxicity, it may play an important role in the pathogenesis of Gram-negative bacteria. Endotoxins can evoke fever (pyrogenic), cause disturbance in capillary permeability and associated changes affecting circulation and blood pressure, and activate serum complement, kinin, and the clotting systems, among a multitude of effects. This endotoxin is made up of three covalently linked parts: (1) the outer O-specific chain, (2) the middle oligosaccharide R-core that links to the O-specific chain, and (3) the inner lipid A layer bound to the cell wall.

The O-specific chain (O antigen) is heat-stable and composed of oligosaccharides (groups of sugars), which extend like whiskers from the membrane surface to the outer environment. The O chains are made up of repeating units of identical oligosaccharides. Each serovar of *Salmonella* is characterized by its unique structure and arrangement of simple sugars in the O-specific chain. Often, but not always, the terminal sugar residue of the O-specific chain is immunodominant. The smooth virulent "strains" of *Salmonella* possess a full complement of O sugar repeating units, while

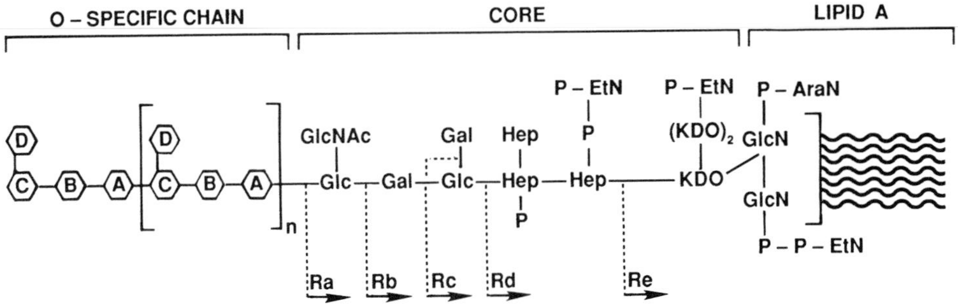

FIGURE 1 General structure of a *Salmonella lipopolysaccharide* (LPS). Key: A–D, sugar residues; Gal, D-glucose; GlcN, D-glucosamine; GlcNAC, *N*-acetyl-D-glucosamine; Hep, L-glycero-D-*manno*-heptose; KDO, 2-keto-3-deoxy-D-*manno*-octonate; AraN, 4-amino-L-arabinose; P, phosphate; EtN, ethanolamine; ~, hydroxy and nonhydroxy fatty acids; Ra–Re, incomplete R-form LPS. [Reprinted, with permission, from O. Luderitz, C. Galanos, H. Mayer, and E. Th. Rietschel, 1986, *in* Proceedings of the 10th International Convocation on Immunology (H. Kohler and P. T. Lo Verde, eds.), p. 78, John Wiley & Sons, Inc., New York.]

the "rough" avirulent or less virulent *Salmonella* lack a complete sequence of O sugar repeating units. The O antigens are designated by numbers 1–67. "Smooth" and "rough" derive from the appearance of colonies on agar.

Besides the importance of the O-specific chain (O antigen) of the endotoxin in determining the specificity of serovars, R-core is also an important entity in that antibodies directed against R-core may protect against infection by a wide variety of Gram-negative bacteria. This is probably due to the fact that the R-core structure has less diversity than O-chains and is common to other Enterobacteriaceae. In addition to the *Salmonella* R-core, other core types have also been identified in Enterobacteriaceae.

Because *Salmonella* are motile by peritrichous flagella (Fig. 2), flagellar antigen, also known as H antigen, has been used in conjunction with O antigen to identify serovars. There are two classes of H antigen designated as phase 1 and phase 2. The phase 1 antigen is specific, occurs in only a few serovars, and is designated by a small letter, a–z. Recently identified phase 1 antigens are designated as Z_1, Z_2, Z_3, . . . , Z_{60}, for lack of additional letters available for designations. There are fewer kinds of phase 2 antigens but they are much more widely distributed than phase 1 antigens. Many *Salmonella* have common phase 2 antigen; therefore, they exhibit antigenic relationships. The phase 2 antigens are designated by arabic numerals.

The Vi (capsular) antigen is a surface antigen most commonly found in *S. typhi* and *S. hirschfeldii* (*S. paratyphi* C). The Vi antigen can be removed from cells by hot saline or trichloracetic acid treatment. Although several types of Vi antigens are recognized, they are immunologically related. Presently, the specific name given to a *Salmonella* is in reality a designation for a particular serovar. While most serovars are named binomially, some are designated only by their antigenic formulas. A serovar with antigenic formulae 6,7 : r : 1,7 represents O antigen 6 and 7 : phase 1 H antigen r : phase 2 H antigens 1 and 7 and is named *S. colindale*. In the Kauffman-White schema, serovars with common O antigens are arranged into groups. Table II lists the names, O-antigen groups, and antigenic formulas of some common salmonellae infecting humans. For a detailed listing of all serovars, see Bergey's Manual of Systematic Bacteriology (1984).

D. Bacteriophage Typing

A number of *Salmonella* serovars, which are indistinguishable by serological or biochemical test, can be divided into phage type based on the sensitivity of cultures to a series of bacteriophages. Phages are bacterial viruses that infect bacteria and eventually cause lysis of the bacteria. *S. typhi, S. hirschfeldii, S. paratyphi* A, *S. schottmulleri, S. typhimurium* and, to a much lesser extent, other serovars have been typed using various bacteriophages.

The preponderance of one or another of the phage types of *S. typhi* may characterize a geographical origin and could help determine the origin of an epidemic.

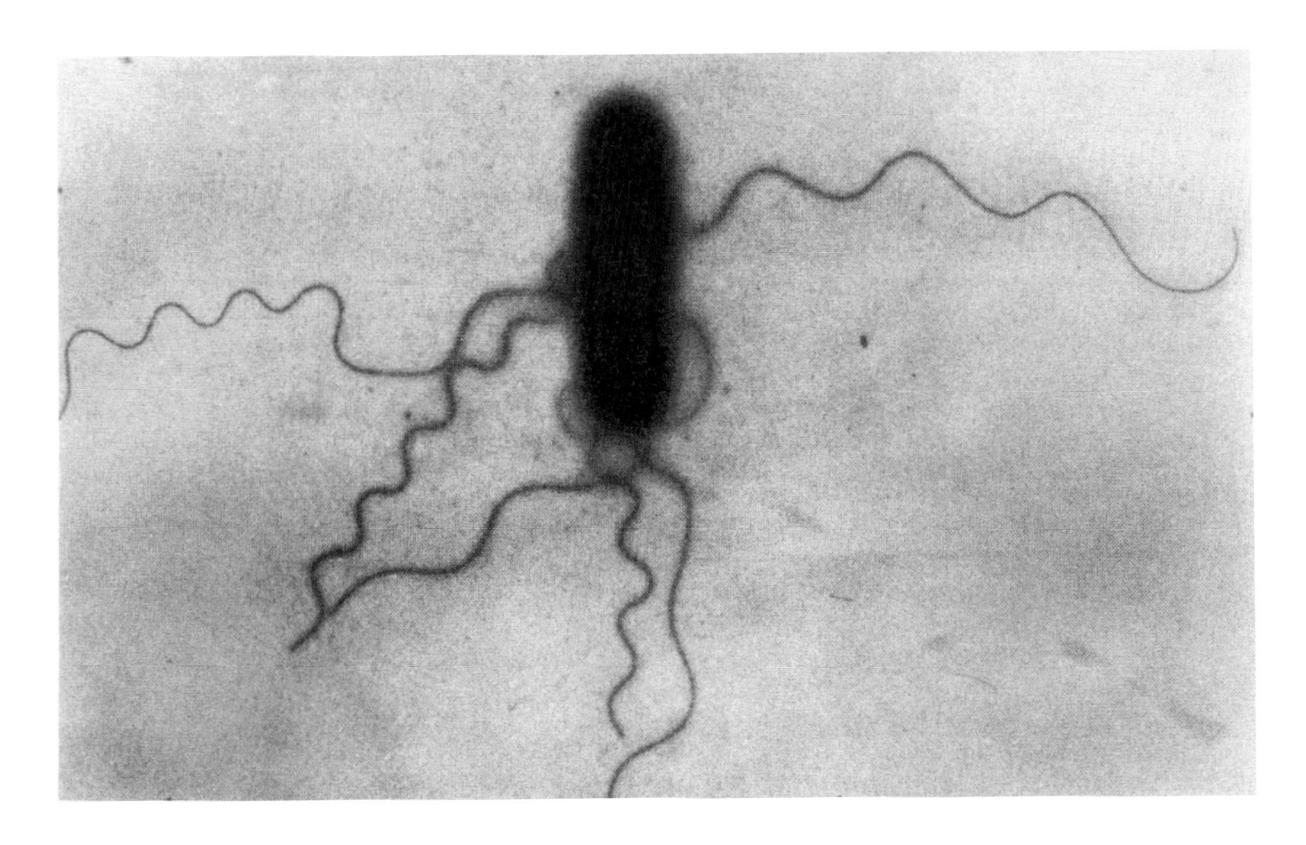

FIGURE 2 A negatively stained cell of *Salmonella* sp. showing peritrichous flagellation. [Courtesy of W. L. Dentler; Reprinted, with permission, from V. T. Schuhardt and T. W. Huber, 1978, *in* "Pathogenic Microbiology," p. 262, J. B. Lippincott Company, Philadelphia.]

E. Habitat

Although *Salmonella* serovars are widely distributed in nature, they are obligate parasites (i.e., live inside animal hosts) and are not found elsewhere. They may be adapted to single host or ubiquitous human and animal pathogens, and some may be of still unknown pathogenicity. They may be primary

TABLE II Antigenic Grouping and Formulas of Some Common *Salmonella*

Serovar	O group	Somatic antigen	Flagellar (H) antigen Phase 1	Flagellar (H) antigen Phase 2	Disease
S. paratyphi	A	1, 2, 12	a	1, 5	Enteric fever (paratyphoid), gastroenteritis
S. schottmuelleri	B	1, 4, 5, 12	b	1, 2	Enter fever (paratyphoid, gastroenteritis
S. typhimurium	B	1, 4, 5, 12	i	1, 2	Most frequent agent of gastroenteritis, septicemia
S. hirschfeldi	C	6, 7 (Vi)	c	1, 5	Enteric fever (paratyphoid, endocarditis)
S. choleraesuis	C	6, 7	c	1, 5	Enteric fever (paratyphoid), gastroenteritis, septicemia
S. typhi	D	9, 12 (Vi)	d	—	Enteric fever (typhoid)
S. enteritidis	D	1, 9, 12	g m	1, 7	Gastroenteritis

etiological agents or secondary invaders. They are found in many ecosystems that have been polluted by humans and animals. A large number of *Salmonella* are parasites of lower animals such as rodents, birds, reptiles, and even insects.

Some serovars are strictly adapted to one particular host. For example, *S. typhi, S. sendai,* and *S. paratyphi* A are adapted to humans, *S. abortusovis* to sheep, and *S. gallinarium* (*S. pullorum*) to poultry. Many *Salmonella* are also ubiquitous and can be found in a variety of animals at any given time. A majority of the serovars show no host specificity, and most common in this group is *S. typhimurium.*

II. Pathogenicity for Humans

Salmonellosis is an important infectious disease of humans. Countries with poor sanitation and hygiene have the largest incidence of salmonellosis. Salmonellosis can be caused by any of the 2,100 serovars; however, 10 serovars account for about 73% of the infections. *Salmonella* infection may range from mild gastroenteritis to enteric fever to severe fatal septicemia. Millions of people worldwide are affected annually by *Salmonella* infection. Worldwide financial loss from salmonellosis is in the billions of dollars.

A. Mechanism of Pathogenicity

Various events leading to infection are summarized in Figure 3. *Salmonella* cells enter the animal body by ingestion of contaminated food or drink or via person-to-person contact. The infection begins in the intestinal tract. Disease manifestations by *Salmonella* depend on the virulence and the type of the serovars, general health of the animal, and the number of viable cells ingested. Approximately 10^5–10^9 organisms are needed to produce disease in >50% of individuals. However, some serovars under certain circumstances may produce disease with a much lower number of organisms.

Following ingestion of *Salmonella,* organisms pass through the stomach, where they are subjected to gastric acidity. Conditions within the stomach regulate the number of bacteria that enter the small intestine. In the acid-deficient stomach, a larger number of viable bacteria are discharged into the small intestine, thereby increasing the possibility of disease production. From the stomach, bacterial cells are transported to the ileum and colon, where the cells adhere to the target cells by various mechanisms that are not completely understood. The adhesion process is influenced by bacterial motility and chemoreactants produced by intestinal cells (which bring *Salmonella* cells into closer contact with mucosal receptors). Some salmonellae such as *S. typhi* produce adhesion, free of O, H, and Vi antigen, that is made of 85% cell-surface protein and Rd_1P^+ lipopolysaccharide, which specifically binds to mannoselike receptors on epithelial cells. Several outer membrane proteins are also known to be virulence factors in mice infected with *S. typhimurium.* Production of adherence factors may also be enhanced by specific plasmids. For example, in *S. typhimurium,* a 60-megadalton plasmid, has been found to be responsible for adhesive and invasive properties.

Studies of invasion of guinea pig ileum by *S. typhimurium* indicate that bacterial adherence to the brush border of the cells leads to the damage or degeneration of microvilli at the site of attachment (Fig. 4). This degeneration enables bacteria to be engulfed into the cells by a process termed receptor-mediated endocytosis (RME). The epithelial cells that are involved in this process are known as M cells, and they overlay Peyer's patches (lymphoid organs) and other mucosal lymphoid follicles. Only viable bacteria are transported by this process, which is known as translocation. Translocation is an important initial step in *Salmonella* pathogenicity. It is interesting to note that other intestinal bacteria are not taken up along with *Salmonella.* After colonization is established, the organisms invade the intestinal epithelium and proliferate within the epithelial cells and lymphoid tissue.

After RME, the organisms are found in vacuoles within the epithelial cells. Those salmonellae causing typhoidlike disease seem to be transported through the cells in the vacuoles and then enter the lamina propria, which is under the epithelium, via the basal cell membrane. Once in that area, the inflammatory response consists primarily of infiltration by monocytic cells. Finally, organisms colonize the cells of the reticuloendothelial system, and presumably, when a certain population is reached, organisms break out into the bloodstream (secondary bacteremia), producing enteric fever symptoms. *Salmonella* serovars that cause gastroenteritis remain localized in the mucosa. Further spread of nontyphoidal salmonellosis is prevented by polymorphonuclear cell (PMN) response.

Most *Salmonella* elicit an inflammatory response

FIGURE 3 Mechanisms of pathogenicity of *Salmonella*.

and ulceration in the *lamina propria,* with fluid secretion and diarrhea. Generally, there are two types of *Salmonella* diarrhea: (1) toxigenic diarrhea caused by enterotoxin (chlorealike toxin), and (2) acute *Salmonella* diarrhea, which occurs during the invasion process. Additionally, some salmonellae produce Shiga-like cytotoxin (yet to be isolated), which leads to the ulceration of intestinal epithelial tissues. *Salmonella* enterotoxin activates adenylate cyclase (AC) causing elevation of cyclic adenosine monophosphate (cAMP). That in turn leads to ion flux disruption of the intestinal epithelial cells and cascade events leading to secretion of chloride and bicarbonate ions and inhibition of sodium absorption, resulting in diarrhea. Salmonellae that do not produce enterotoxin cause diarrhea by invading intestinal epithelia and stimulating an inflammatory reaction, which evokes fluid secretion and diarrhea. However, some invasive strains do not evoke fluid secretion and others secrete fluid before inflammation. Thus, it seems that more than one mechanism is involved in this type of diarrhea.

Endotoxin is a potent inflammatory inducing factor and a strong chemotactic factor. The attracted PMNs phagocytize bacterial cell, and endotoxin is released from the dead bacteria. The released endotoxin causes damage to surrounding cells and in-

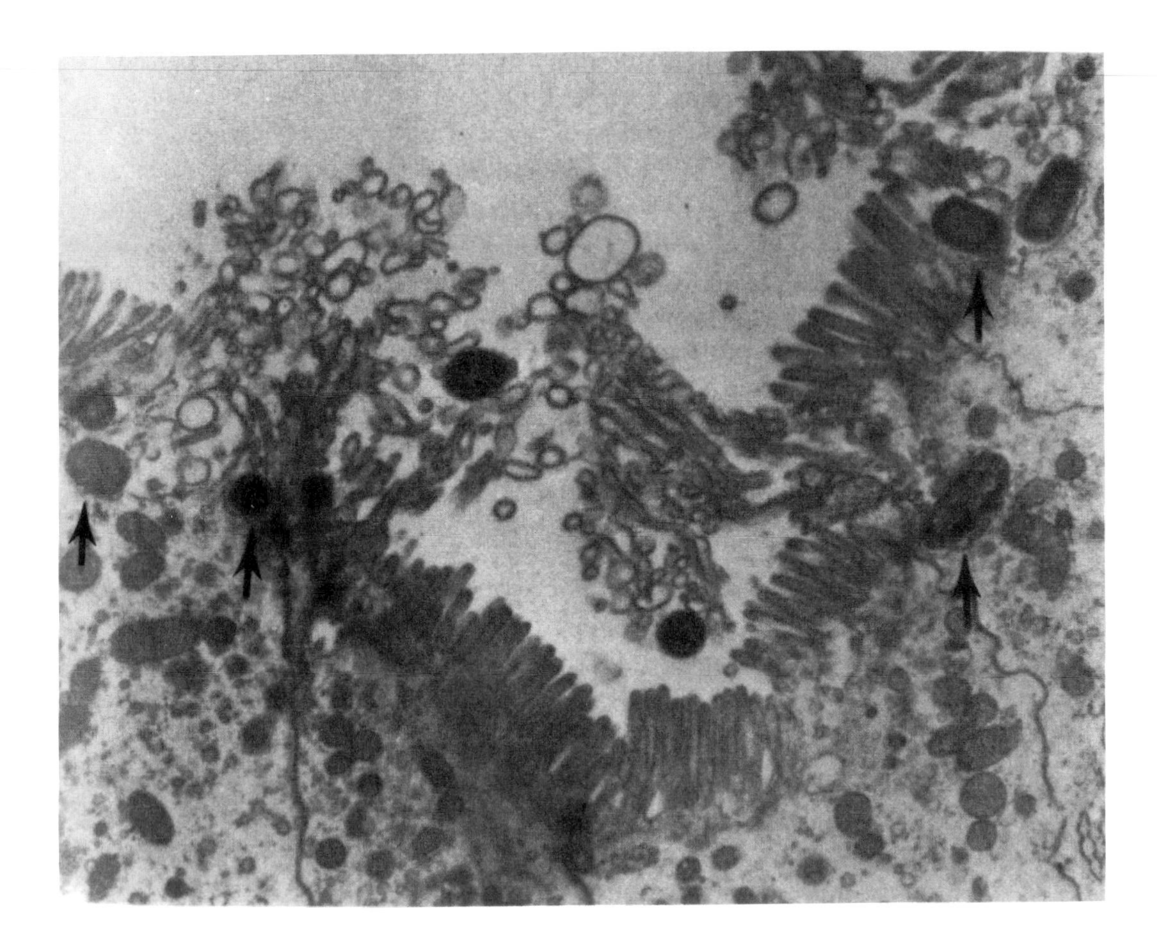

duces release of prostaglandins, mediators of inflammation, and platelet-activating factors (PAFs). The prostaglandins from inflamed tissue also activate adenylate cyclase, leading to elevation of cAMP, which leads to fluid secretion, as explained above.

PAF is a mediator of allergic and inflammatory reactions produced by various blood cells. It acts as a vasodilator, elevates vascular permeability, and stimulates the release of lysosomal enzymes. It is also a potent ulcerogenic necrotizing enterocolitis factor. Following neutrophil aggregation, other vasoconstrictors such as thromboxane A_2 and noradrenaline, free radicals, and lysosomal enzymes are released in association with PAF, causing local tissue ulceration. The result of these actions leads to severe colitis with diarrhea that may be associated with bloody mucous stool. Blood leaks out of capillaries as a result of alteration of intestinal permeability, which causes marked inhibition of fluid absorption and active colonic secretion. [*See* PLATELET-ACTIVATING FACTOR PAF-ACETHER.]

FIGURE 4 Electron photomicrograph showing the invasion by *S. typhimurium* of the epithelial lining of small intestine in guinea pig. Arrows point to invading *Salmonella* organisms. [Reprinted, with permission, from A. Takeuchi, 1975, *in* "Microbiology," p. 176, American Society for Microbiology, Washington, D.C.]

B. Dominant Serovars in Various Diseases

Salmonella serovars may be adapted to a particular host or can be ubiquitous. Often, a particular serovar is involved in producing the specific symptoms of a disease. However, a serovar can occasionally produce symptoms of any of the various diseases caused by *Salmonella*. For example, *S. typhi* causes typhoid fever and *S. enteritidis* gastroenteritis; *S. paratyphi, S. schottmuller, S. hirschfeldii,* and *S. choleraesuis* all can cause paratyphoid fever and gastroenteritis; *S. typhimurium* causes gastroenteritis and septicemia; *S. choleraesuis* can also cause septicemia or focal infections.

The salmonellae can be divided into three groups on the basis of their host preference. The first group includes those highly adapted to humans (*S. typhi, S. paratyphi,* and *S. senadi*) and commonly causing enteric fever. The second group includes those primarily adapted to a particular animal host. Of the organisms in this group, *S. dublin* and *S. choleraesuis* are also pathogenic to humans. Infection by the latter can be quite severe in children. The third group includes unadapted serovars. This group includes over 2,000 serovars that are ubiquitous in nature and which seemingly attack humans and other animals with equal facility. Many serovars in this group cause gastroenteritis and account for 85% of all *Salmonella* infections in the United States. In developing countries, *Salmonella* is the major cause of bacterial diarrhea.

C. Enteric Fever—Typhoid

Typhoid fever is caused by *S. typhi.* It is a systemic infectious disease resulting in a prolonged fever. Humans are the only reservoir for *S. typhi,* which is pathogenic only for humans. The organism is transmitted by food or drink contaminated with human excreta. It is prevalent in developing countries with improper sewage disposal, contaminated water, and poor hygienic habits. In developed countries, foods contaminated by carriers are primarily responsible for transmissions of the organisms. Flies and other insects also spread the organism from feces to food. Occasional transmission results from homosexual activity (anal–oral route), by direct contact in children, and by improper use of toilet paper and contamination of hands with feces. The disease is most severe in children and older adults. In the United States, over 400 cases of typhoid per year have been reported to the Center for Disease Control (CDC), Atlanta, under the surveillance program of the last 10 years. Throughout much of the world, typhoid is endemic, and at times it becomes epidemic, causing suffering to millions and death to many.

1. Symptoms and Diagnosis

The incubation period for typhoid fever depends on virulence and the number of organisms ingested. It is usually 7–14 days but may be as short as 3 or as long as 60 days. The clinical symptoms are gradual onset of continuous fever, headache, aches and pains, enlargement of the spleen, inflammation of the intestine, tender abdomen, rose-spot rash, leucopenia, arthralgias, diarrhea, pharyngitis, constipation, anoxemia, abdominal pain, and tenderness. Normally, diarrhea is not a typical symptom of typhoid fever. Recent findings in Asian countries indicate diarrhea in patients with positive blood culture for *S. typhi* or *S. paratyphi* A, but without any other pathogen capable of producing diarrhea. This diarrhea is watery and contains large quantities of leukocytes and protein. Although not a typical symptom, diarrhea in typhoid may also occur as a result of double infection from a diarrheogenic pathogen. Untreated, the fever rises, reaching a plateau in 2–3 days and remains high (39.4–40°C) for another 1–2 wk, falling to normal in 4–5 wk. The bacteria pass from the intestine to the mesenteric lymph nodes, where they multiply and eventually reach the bloodstream. They may be found in urine, the gallbladder, bile ducts, and bone marrow. Complications in untreated patients may result in intestinal perforation and acute cholecystis, hepatitis, pneumonia, or abscess formations and neuropsychiatric disturbances. Bacterial localization in different organs (metastatic infection) can appear during acute or convalescent periods of infection, or months to years after the infection. The most common sites of infections are the bones, bone marrow, and joints, but any part of the body may be involved. Osteomyelitis (bone infection) may develop as long as 6–7 yr after typhoid fever. Relapses occur in about 10–20% of patients. The disease symptoms and the treatment are the same as the first infection. Occasionally, a second relapse occurs.

Initial diagnosis may be based on continuous fever and other symptoms, but ultimate laboratory diagnosis is based on the isolation of the *Salmonella* serovar from the blood. Within the first 10 days, the majority of the cases show positive blood culture, which drops off during the later stages of infection. Stool cultures are positive during the third to fifth week. Urine cultures are also positive at the same time as stool cultures, however at a much lower percentage. The organism can also be isolated from bone marrow and rose spots. A persistent positive stool culture in a patient beyond 3–6 mo indicates a 90% possibility of the patient becoming a carrier.

2. Treatment and Control

The most effective treatment for typhoid fever is administration of antibiotics. Ampicillin, chloramphenicol, and trimethoprim-sulfamethoxazole (TMP-SMZ) are primarily used. Chloramphenicol is the drug of choice for treating typhoid fever. In recent times, isolates of *Salmonella* have increasingly shown drug resistance to antibiotics due to the R plasmids, which carry genes for resistance, and can transfer their resistance to sensitive strains. They are responsible for the rapid rise of multiple-drug resistance. *S. typhi* resistant to ampicillin, chloramphenicol, and TMP-SMZ have been observed in many parts of the world. Cefotaxime, a third-generation cephalosporin, has recently been used with success to treat typhoid and nontyphoid salmonelloses and may be a prime candidate for the treatment of multiresistant systemic salmonelloses. Proper treatment and disposal of sewage and appropriate purification of water are the two major means of controlling typhoid and *Salmonella* outbreaks in general. Also, pasteurization of milk and other drinks, proper hygienic habits, and control of carriers from working in the food industry where they come in contact with food would help reduce the transmission of *Salmonella*. *Salmonella* surveillance programs would also be helpful in controlling epidemics. In appropriately treated situations, the mortality rate is negligible. [*See* ANTIBIOTICS.]

3. Typhoid Vaccine

Presently, typhoid vaccines are made from *S. typhi* whole cells killed with acetone to preserve the Vi antigen; it has been more effective than the formalin-fixed vaccine. Typhoid vaccines have been used for immunization since the turn of the century. Vaccination alters the character of the epidemics and lessens the severity of the illness. The present whole-cell vaccine has adverse side effects such as fever and aches and does not provide absolute protection. It also needs to be administered periodically. Even with these drawbacks, the vaccine offers the best available protection. Two different vaccines are presently in field trials. One administered orally consists of live *S. typhi,* made avirulent by chemical mutagens. The other is made from purified capsular material of bacterial cells and is administered by injection. Both of these vaccines offer better protection than the present commercially available vaccine. A recent field trial of oral vaccine in Egypt showed 96% protection against typhoid; however, this high level of protection has not been duplicated in Chile under World Health Organization trials. The vaccine given in Chile was "enteric-coated" rather than the lyophilized material given with bicarbonate in Egypt. Injectable typhoid vaccine made from capsular polysaccharide of *S. typhi* has undergone tests for several years in Nepal. It seems to provide 70–78% protection for 17 mo or more. Presently, no vaccine is available for nontyphoidal salmonellosis.

D. Enteric Fever—Paratyphoid

Paratyphoid fever is relatively mild and of shorter duration compared with typhoid fever. It has the same symptoms as typhoid fever, and both are known as enteric fever. Although paratyphoidal enteric fever may be caused by any serovars of *Salmonella,* the serovars most often responsible are *S. paratyphi* A, *S. schottmulleri* (*S. paratyphi* B), *S. hirschfeldii* (*S. paratyphi* C), and *S. senadi*. Paratyphoid is often marked by the sudden onset of chills, but otherwise the pathogenesis and the manifestations are very similar to typhoid. The incidence of paratyphoid, compared with typhoid based on hospital records, is about 1 : 10. Proper clinical diagnosis involves isolation, identification, and serological typing of the organisms. Treatment is similar to that for typhoid fever.

E. Gastroenteritis

Any of the over 2,000 serovars that are ubiquitous in nature can cause gastroenteritis. Most commonly, *S. typhimurium* and *S. enteritidis* are responsible for gastroenteritis. The disease symptoms appear within 12–48 hr and often overnight after the ingestion of food contaminated with *Salmonella.* A relatively large inoculum is usually required (10^8–10^9 cells) to produce clinical symptoms, but infection may occur at smaller inoculum. The disease symptoms are sudden onset of nausea, acute vomiting, abdominal cramp and pain, watery diarrhea occasionally containing mucous, blood, or both. A slight fever of 38.3–38.9°C is seen in over half the cases. The peak incidence occurs in children under the age of 6 yr and in persons over 50 yr of age. In newborn and young children, symptoms are diverse and range from grave typhoidlike illness and septicemia to asymptomatic infection. The mortality rate in very young children may be as high as 20% from *Salmonella* sepsis. The disease is self-limiting, usually mild, lasting 1–4 days, and in a majority of cases treatment is not required. However, at times it can be debilitating and protracted. Fatalities are rare except in nursing homes (7–8%) and among infants (6–7%). In a pediatric ward, the infection is transmitted by the hands of personnel.

Salmonellosis may also occur in patients with an

established diagnosis of acquired immunodeficiency syndrome, or it may occur during the first manifestation of this disorder. While the incidence of enteric fever has gone down dramatically, the salmonellosis of the gastroenteritis type has gone up astronomically. The Committee on *Salmonella* of the National Research Council in 1969 conservatively estimated that 2 million cases of salmonellosis occur in the United States per year. The present estimate is 4–5 million cases per year. The true magnitude of the problem is impossible to determine. The rapid rise of food-borne gastroenteritis is blamed, among other things, on the use of antibiotics in farm animals, which has given rise to antibiotic-resistant *Salmonella* serovars.

1. Various Food and Feed Sources Implicated

The major source of the *Salmonella* problem is food of animal origin. Poultry, meat, and eggs are the main sources of nontyphoidal salmonellosis, accounting for about 36% of the outbreaks. A miscellaneous variety of processed foods are also known to contain *Salmonella*. Many cases of salmonellosis in which the source of infection was not established may have been food-borne. [*See* Food Microbiology and Hygiene.]

The main source of *Salmonella* problems is poultry, which accounts for 17% of salmonellosis. The U.S. Department of Agriculture reports that 37% of chicken carcasses carry *Salmonella*. This figure has remained constant for more than a decade. Chickens eat each other's droppings, and in this way a large percentage of animals become inoculated. Four percent of beef and 12% of pork is also contaminated with *Salmonella*.

Salmonellosis associated with egg and egg products led to the passage of the Egg Products Inspection Act in 1970 by the United States. Since this legislation, the CDC has not received any report of major outbreaks associated with bulk egg products. However, recent findings indicate the possibility of contamination of Grade A eggs with *S. enteritidis* through the hen's ovary as a major source of infections. Presently, eggs and egg products still constitute a significant source of salmonellosis.

Raw or improperly pasteurized milk, milk products, and powdered milk are other sources of *Salmonella* infections. In 1985, a major outbreak of *Salmonella,* the biggest in U.S. history, which caused food poisoning to about 200,000 people in the United States, occurred from improperly pasteurized milk. Various cheeses made from unpasteurized or improperly pasteurized milk have also been involved in *Salmonella* infections. Miscellaneous products such as dried yeast, smoked fish, bakery products, cream-filled desserts, ice cream, sauces and salad dressings, bread mix, coconut, frog legs, spices, gelatin, sandwiches in vending machines, vegetables, etc. have been incriminated at times as the vehicle of *Salmonella* transmission and gastroenteritis.

Another major source of salmonellosis is improper kitchen hygiene, such as improperly cleaned cutting boards, and the contamination of fruits, salads, vegetables, beef, ham, etc. from chicken drippings happening either at the grocery store or at home.

By-products of animal origin, milk and milk products, egg products, and protein concentrates of vegetable origin are some of the numerous sources of *Salmonella* in animal feed. Food-borne salmonellosis can be controlled only by eradication of *Salmonella* from feed and farm environments, pasteurization of raw foods of animal origin, and consumer education to the danger of salmonellosis. The U.S. Department of Agriculture and the FDA believe that eradication of *Salmonella* from feed and feed ingredients is too costly for the gained benefit. Therefore, feed will continue to be a source of *Salmonella* infection to animals.

Most gastroenteritis symptoms subside promptly in healthy individuals and require no treatment. If needed, the illness is treated symptomatically except in high risk patients who are treated with antibiotics.

2. Detection of *Salmonella* in Food

Detection of *Salmonella* in food is complicated by a lower population of *Salmonella* and other bacteria in food, injury of surviving cells from processing, and the inherent variability encountered in the analysis of various types of food. Presently, detection of *Salmonella* in food involves pre-enrichment in nutritive nonselective media, selective enrichment into a special broth, and finally selective plating onto suitable solid media. Suspicious colonies are cultivated further and examined for typical reactions.

F. Septicemia

Although any *Salmonella* serovar can cause septicemia or localized disease or focal infections (liver, gallbladder, etc.), *S. choleraesuis* infection is most frequent, followed by *S. typhimurium*. The clinical symptoms vary with the serovar involved. In general, from the intestinal tract, the organisms reach the bloodstream in a manner similar to that in *S. typhi* and multiply in the bloodstream producing recurrent high fever, chill, loss of appetite, etc. From the bloodstream, organisms may reach various parts of the body and produce pneumonia, myocardial abscesses, osteomyelitis, meningitis, arthritis, etc. They may also invade the endocardium and the endothelium of large arteries, thereby causing intravascular lesions. Septicemia by *Salmonella* can occur with or without bacteremia. In bacteremia, the organism can be isolated from blood samples but not from stools.

S. choleraesuis is most prone to invade the bloodstream, even when it causes other clinical forms of salmonellosis. Isolation of this organism from the blood usually indicates the presence of an abscess. The fatality rate in nontyphoidal bacteremia is high, from 15 to 20%. The illness is particularly severe in children. Ampicillin or chloramphenicol are effective in treating bacteremia.

III. *Salmonella* Infection of Animals (Other Than Humans)

Salmonella infections of cattle and poultry are of common occurrence. About 10% of about 2,100 serovars are routinely isolated from animals. Although some serovars show specificity to certain animals producing characteristic clinical symptoms, the majority do not. Infection symptoms may vary from acute septicemia to diarrhea to abortion. Many infections are asymptomatic.

The most common cattle isolates are *S. typhimurium, S. dublin,* and *S. newport*. In calves, *S. typhimurium* is associated with serious clinical symptoms of acute salmonellosis. *Salmonella* infection in cows may also cause abortion either during or after the acute phase of illness, or even without frank clinical symptoms of illness. *S. abortusequi, S. abortusovis,* and *S. choleraesuis* may cause severe illness and abortion in horses, sheep, and pigs, respectively. Fur-bearing animals such as dogs, cats, rodents, ferrets, bats, and marsupials are also infected with *Salmonella,* producing a variety of salmonellosis symptoms. Poultry constitutes the greatest host reservoir for *Salmonella* and an important indirect source of salmonellosis in humans. *S. pullorum* once caused pullorum disease in chicks 1–14 days old, resulting in death to almost 100% of the hatch. *S. pullorum,* which can be transported by eggs, was eradicated by slaughter of infected adult birds, whose infection had been detected by blood tests. Fowl typhoid caused by *S. gallinarium* also used to cause severe financial loss in poultry operation, and it has also been eliminated by blood-testing programs. The occurrence of salmonellae among cold-blooded vertebrates such as snakes, turtles, and lizards has been known for many years. Pet turtles have been the cause of concern in the last decade. In 1975, the FDA prohibited the distribution of red-eared turtles in the United States due to documentation of thousands of cases of salmonellosis per year from this source.

IV. Detection of *Salmonella* in Clinical Specimen

Diagnosis of salmonellosis ultimately depends on the isolation and identification of *Salmonella* from clinical specimens. The organisms may be isolated from circulating blood in earlier phases of enteric fever and septicemia. In later stages organisms are found in stool, urine, etc.

Stool specimens are normally inoculated into an enrichment medium such as gram negative broth or selenite F broth and incubated for 12–18 hr at 35°C. Several drops of this enrichment broth is then used to inoculate moderately selective agars such as Hektoen enteric or XLD agar, or highly selective media such as bismuth sulfite and brilliant green agar.

Most other clinical specimens are inoculated into a supportive medium such as blood, chocolate or MacConkey agar. After incubation, the media should be checked for growth by Gram stain and streaked on suitable media as described above for isolation of *Salmonella*. Suspected colonies from various media are then identified using routine biochemical tests for *Enterobacteriaceae* and *Salmonella*. It is important to do differential tests for the identification of related organisms. Next, tentatively identified organisms are further identified by serological typing using O antisera for serogroups A to E, H antisera for flagellar antigens, and Vi antiserum as warranted. Further identification of *Salmonella,* including the complete serological identification, is normally carried out by a specialized laboratory.

IV. Carrier Status

Pathogenic microbes can be transmitted to healthy people by a person who has the disease, one who has recovered and carries the organism, or one who has no clinical symptoms and was asymptomatically infected. Such a person is called a carrier. In case of enteric fever, *Salmonella* is present in the patient's stool in the later stages, and a small percentage of these patients can discharge the organisms for a period of 3–10 wk after the onset of the illness. A certain small portion of these people in turn continue to excrete *Salmonella* for from several years to throughout their entire life. About 3% of untreated or recovered typhoid patients are believed to be chronic carriers. Carriers harbor a high population of *S. typhi,* up to 10^{10} organisms per gram of feces. A persistent fecal carrier usually has infection in the gallbladder. Urinary carriers are also found in certain parts of the world. In many of the asymptomatic carriers, bacteria persist for a relatively short period and the excretion of *Salmonella* is intermittent. Food handlers constitute a high percentage of carriers. Infants frequently become long-term carriers and excrete *Salmonella* longer than adults. It is estimated that there are about 2,000 typhoid carriers in the United States, most of them elderly females who have chronic biliary disease. The proportion of typhoid carriers must be high in developing countries, but it varies from different geographic locations and for different time periods.

Duration of carrier state is variable in different age groups and in both humans and animals. Most serovars causing gastroenteritis do not usually establish a chronic carrier state. There are no general estimates of the number of carriers of unadapted serovars; it is estimated to be very high.

All types of carriers can be treated with antibiotics to remove the foci of infection; however, until the underlying anatomical problems are corrected by surgical intervention, antibiotic treatment is of little value.

V. Epidemiology of *Salmonella*

Salmonellosis is one of the most important epidemiological problems in both developed and in developing countries, the former from nontyphoidal salmonellosis and the latter from both typhoidal and nontyphoidal salmonellosis. Although all the serovars of *Salmonella* are potential pathogens for humans, only a handful of serovars are most frequently involved in salmonellosis. The 10 most frequently reported serovars from human and nonhuman sources reported to the CDC in 1987 represent 73.1 and 57.5% of the total isolates (Table III) from human and nonhuman sources, respectively.

S. typhimurium is the most frequently isolated serovar from both human and nonhuman sources. Salmonellosis, excluding typhoid, is most prevalent

TABLE III Ten Most Frequently Reported *Salmonella* Serotypes from Human Sources Reported to the CDC in 1987 and from Nonhuman Sources Reported to the CDC and USDA in 1987

	Human 1987			Non-human 1987		
Rank	Serotype	Number	Percent	Serotype	Number	Percent
1	typhimurium[a]	10,462	23.5	typhimurium[a]	1,246	13.5
2	enteritidis	6,950	15.6	heidelberg	1,124	12.2
3	heidelberg	5,714	12.8	choleraesuis[b]	667	7.2
4	newport	2,858	6.4	reading	438	4.8
5	hadar	2,170	4.9	hadar	352	3.8
6	infantis	1,136	2.5	senftenberg	331	3.6
7	agona	1,080	2.4	newport	302	3.3
8	montevideo	1,037	2.3	montevideo	301	3.3
9	thompson	635	1.4	enteritidis	270	2.9
10	braenderup	548	1.2	anatum	266	2.9
	Subtotal	32,590	73.1		5,297	57.5
	Total	44,609			9,208	

[a] Typhimurium includes var. copenhagen.
[b] Choleraesuis includes var. kunzendorf.
[Courtesy of Center for Disease Control, Atlanta.]

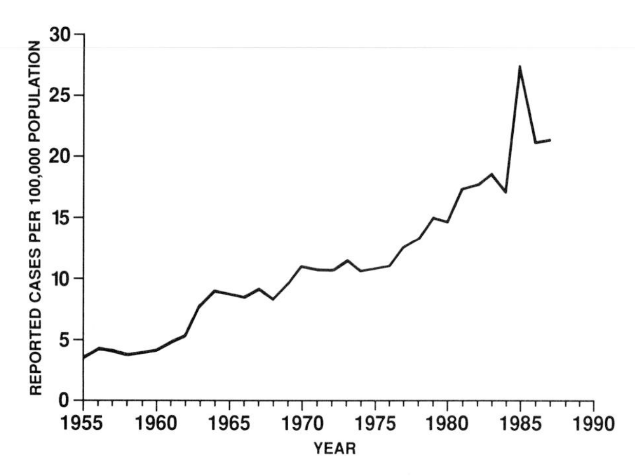

FIGURE 5 Salmonellosis, by year, United States, 1955–1987. [Courtesy of Center for Disease Control, Atlanta.]

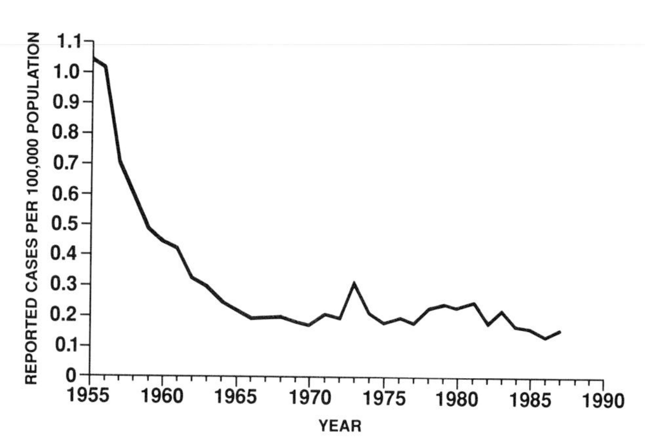

FIGURE 6 Typhoid fever, by year, United States, 1955–1987. [Courtesy of Center for Disease Control, Atlanta.]

in infants <1 yr old and children from 1 to 4 yr of age. This group constitutes about 32% of the reported cases in the United States. Older adults, ≥60 yr old, represent about 11% of the cases. The incidence of nontyphoidal salmonellosis as reported to the CDC has about doubled in the last 10 years from 13 to 21 per 100,000 population (Fig. 5). However, the real estimate is impossible to make and there are conservatively estimated to be 4–5 million cases per year in the United States alone.

The major source of this problem in humans is food of animal origin; predominantly poultry, beef, and pork. Inadequate cooking or processing of contaminated products, cross-contamination of working surfaces in the kitchen environment, and a lack of human hygiene (and, more specifically, lack of consumer education) are most contributory to the salmonellosis problem.

S. typhi, the causative organism for typhoid, is highly adapted to humans, and humans are the only source of this organism. Human carriers excrete the typhoid bacillus in stool and occasionally in urine. The bacillus can then be transmitted to healthy individuals via contamination of food and drink. In the rural areas of developing countries, typhoid epidemics usually result from fecal contamination of drinking water. Typhoid fever in general is on the decline throughout the world due to better drinking water, sanitary disposal of wastes, and the processing of milk. In the United States, the incidence of typhoid fever has gone down from 1/100,000 in 1955 to 0.2/100,000 population in 1987 (Fig. 6). The sources of infection in the United States are already infected people, carriers, or contraction of infection while overseas. Healthy carriers in food-handling situations are the most important link in the chain of transmission of this infection. Milk can also be a very important source of *Salmonella* infection. For example, a recent massive outbreak of salmonellosis by antimicrobial-resistant *Salmonella* was traced to pasteurized milk that affected about 200,000 people in Illinois. Flies and pests—rats, mice, cockroaches, ants, etc.—can also spread the organism from feces to food.

Bibliography

Committee on Salmonella (1969). "An Evaluation of the *Salmonella* Problem," publication No. 1683. National Academy of Sciences, Washington, D.C.

LeMinor, L. (1984). Genus III *Salmonella. In* "Bergey's Manual of Systematic Bacteriology," Vol. I (N. R. Krieg and J. G. Holt, eds.), pp. 427–458. Williams and Wilkins, Baltimore, Maryland.

Lund, B. M., Sussman, M., Jones, D., and Stringer, M. F. (eds.) (1988). "Enterobacteriaceae in the Environment and as Pathogens." The Society for Applied Bacteriology Symposium Series No. 17. Nottingham, England, U.K. July 7–9, 1987. Blackwell Scientific Publications, Oxford, England.

Rubin, R. H., and Weinstein, L. (1977). "Salmonellosis: Microbiologic, Pathologic and Clinical Features." Stratton International Medical Book Corporation, New York.

Silliker, J. H. (1980). Status of *Salmonella*—Ten years later. *J. Food Prot.* **43,** 307–313.

Soe, G. B., and Overturf, G. D. (1987). Treatment of typhoid fever and other systemic salmonelloses with cefotaxime, ceftriaxone, cefoperazone, and other newer cephalosporins. *Rev. Infect. Dis.* **9,** 719–736.

Salt Preference in Humans

GARY K. BEAUCHAMP, *Monell Chemical Senses Center*

Glossary

Addison's disease Primary failure or insufficiency of the adrenal cortex to secrete hormones

Bartters syndrome Rare condition characterized by renal potassium wasting and elevated renin and aldosterone levels, with normal or low blood pressure

Diuretic Agent that increases the volume and flow of urine, often with associated electrolytes

SALT (NaCl) is ubiquitous in diets in developed countries. A major reason for this is that people prefer salted foods to the same foods without added salt. Although little evidence exists for genetic determination of individual differences in consumption and preferred level of salt, more research in this area is necessary. Considerable data support the view that the optimal level of salt in the diet is determined in part by the level an individual is currently consuming; increasing or decreasing customary salt intake, as long as the salt is tasted, increases or decreases, respectively, the preferred level of salt in food. While these data are consistent with a hypothesis that optimal salt preferences are learned, other data, from both animal models and human developmental studies, suggest that salt preference has an innate component. Furthermore, early experience with low or high salt diets may have a long-term impact on preferred salt levels. A preference for salt, like a preference for sweets, has an innate basis that can be modified by individual experience.

I. Introduction

Because sodium is an essential nutrient, it is not surprising that regulatory systems have evolved to ensure discovery, recognition, and consumption of sufficient sodium to meet that requirement. A nexus in these systems is the sense of taste. Salty (NaCl) taste sensation is thought to be one of a small number of primary taste qualities, others being sweet, bitter, sour, and perhaps a few others including those involved in detecting amino acids. Most humans express preferences for familiar salty foods over those same foods without salt. Consequently, many of the foods available for purchase contain relatively high levels of salt added by the manufacturers to increase acceptability. For many foods, added salt appears not only to make the food taste more salty but to modify the perception of the food in other, less well-understood ways. Thus, bread contains a relatively high level of salt; unsalted bread is quite unpalatable. Yet, few perceive bread as salty or even notice the salt when their attention is called to it. Preferences for salty foods may not only involve liking of salty tastes but a liking for the tastes and flavors as modified by added salt. [*See* TONGUE AND TASTE.]

The major medical issue concerning salt intake is the apparent relationship between high salt consumption and hypertension. While some animal model studies have linked salt preference with the genetic tendency to develop elevated blood pressure or hypertension, few reliable data indicate a positive relationship between individual differences in human salt sensitivity or preference and the probability of developing hypertension. Although a few studies are positive, the bulk of the data indicates that hypertensive individuals are neither more nor less sensitive to the taste of salt and their level of preference for salt is not different from those individuals with normal blood pressure. However, because many individuals consume salt at levels in the range thought by many investigators studying hypertension to be excessive, high salt preference may be medically counterproductive for those susceptible to hypertension. For this reason alone, it is important to understand the factors controlling and modifying salt preferences and salt intake. [*See* HYPERTENSION.]

II. Determinants of Human Salt Preference

That humans have a preference or liking for salt, particularly in food, is obvious. The factors that influence and control this preference and the ways it can be modified are, however, complex and not fully understood. The determinants of salt preference are divided here into physiological, genetic, psychological, and developmental factors or influences. In reality, these categories of explanation interact and are not mutually exclusive.

A. Physiological Factors

Most adults in the United States consume 6–12 g NaCl/day (100–200 meq), an order of magnitude or more above physiologically determined requirements. Although it seems unlikely that this level of consumption would reflect some sort of physiological need, that has been suggested.

If people on very low-sodium diets were actually in a chronically depleted state, they might be expected to crave salt. The history of salt traders, the levying of taxes on salt, and the frequency with which wars have been fought over salt all attest to the intense desire for this substance. Events in India represent only one example of this history. During the British occupation, salt was so heavily taxed that the poor in some states had to part with a substantial portion of their income just to purchase what they felt to be sufficient salt. Huge fortifications were built to stifle salt smuggling and Gandhi used the issue of salt monopolies to strike at the heart of the British Empire. All of these historical observations, not only from India but from Africa, South America, and elsewhere, indicate that when access to salt it restricted, people will go to great lengths and pay large sums to get it. This suggests that the desire for salt is a powerful motivating force and that a craving for salt ought to exist when low-sodium diets are consumed, either chronically or acutely.

Arguing against this suggestion is the observation that some people live on as little as 1 g/day of NaCl or less without apparent harm and perhaps without craving. At the extreme, the Yanomamo Indians are reported to consume approximately 0.06 g Na/day, as determined by urinary excretion (extra loss in sweat is possible). Other unacculturated peoples also have been reported to chronically consume very low-sodium diets. In most cases, when these peoples are faced with the opportunity to consume diets high in salt, their salt consumption increases, but evidence that this is due to a particular craving for salt is equivocal.

Indeed, exceedingly little direct human evidence indicates that even an extremely low-sodium diet and sodium depletion are followed by a greatly increased desire for salt. Evidence from clinical studies (Table I) indicates that excessive salt preference, even under conditions of presumably extreme sodium loss (e.g., Addison's disease, Bartters syndrome), is rare in humans and, when reported, is almost always of childhood onset (see below).

Acute experimental studies of salt depletion have also found little evidence for elevated salt preference or salt appetite. When human volunteers have been placed on very low-sodium diets and depleted of sodium in other ways (e.g., heavy sweating, treatment with diuretics), they have not expressed a strong hunger for salt. However, when asked in one study to rate how much they would like to eat foods varying in salt content, sodium-depleted subjects did express a greater desire for salty foods, and in taste tests they tended to prefer higher levels of salt in food when depleted. These data do not conclusively demonstrate a physiologically determined heightened salt appetite following sodium depletion in adult humans because a psychological explanation is also possible: extremely low-sodium diets are bland and unpalatable, when contrasted with the previously experienced food, and it may not be the salt taste these subjects desire so much as the more flavorful food associated with saltiness. However, because preference for sweet foods and sucrose in food declined during depletion, a simple desire for more orosensory stimulation cannot explain these results. This issue needs further investigation.

Related work has shown that normal subjects administered a hydrochlorothiozide diuretic that causes excretion of sodium exhibited a compensatory increase in sodium intake. Apparently, sub-

TABLE I Clinical Reports of Salt Appetite

Subject(s), description	Age at onset of salt appetite	Amount of NaCl consumed	Medical diagnosis	Comments
A. 15-yr-old diabetic boy (three other diabetic children referred to—one seemed also to have a salt appetite)	Unknown, <15 yr	60–90 g/day	Diabetes	Consumed salt to satisfy "an abnormal craving for salt." Excess salt consumption resulted in elevated blood pressure.
B. 20-yr-old woman	"Present since early childhood"	130–195 g/day (self-report), 138 g/day (clinically determined)	Primary pulmonary arteriosclerosis	First noticed appetite for salt when large enough to climb into chair to reach it. Carried rock salt with her at all times. Refused to go on low-salt diet for >4 days. Threshold for salt taste low relative to control values.
C. 3.5-yr-old boy	First noticed at 1 yr old	Unknown, probably substantially >20 g/day	Corticoadrenal insufficiency	First began eating pure salt at about 18 mo of age. "Salt" among first words learned. Did not like sweets.
D. 10 of 64 patients with Addison's disease	Unknown	Unknown	Addison's disease	"Increased desire for salt and salty foods." According to Richter, one patient covered food with salt and eats very salty foods, salting such times as oranges and lemons; he disliked sweets.
E. 19% of patients with Addison's disease	Unknown	Unknown	Addison's disease	Salt appetite *not* listed among nine diagnostic signs and symptoms of disease. Not one of the 12 adult cases described had elevated salt appetite listed as symptom.
F. 36-yr-old male	Apparently recent	0.25 kg salty olives each day	Addison's disease	Also exhibited an appetite for licorice sweets, known to contain a palliative for Addison's disease. In another case, an adult craved licorice but denied craving salt.
G. The Southwood-Gannon family ($n = 20$ total)	Unknown, disease onset in childhood	Unknown	Genetically determined periodic paralysis with normalkalemia	No direct evidence of salt appetite. Many family members consumed a large quantity of salt daily, perhaps as a therapeutic agent as attacks are moderated.
H. 5-yr-old boy	"The patient had always been a poor eater and craved salt"	Unknown	Bartters syndrome	Another child was not noted to have salt appetite. Most often clinical descriptions of Bartters syndrome do not mention salt appetite; most involve observations on adults.
I. 8-yr-old boy	Unknown, <8 yr	Unknown	Pleoconial myopathy	Desired salt; covered food with it and "frequently will eat a teaspoon of salt directly." A 12-yr-old sibling also exhibited salt hunger. No relationship among sodium loading, sodium deprivation, aldosterone antagonism, and attacks.

TABLE I *(continues)*

TABLE I *(continued)*

Subject(s), description	Age at onset of salt appetite	Amount of NaCl consumed	Medical diagnosis	Comments
J. 13-yr-old male	"Present all his life"	Unknown	Mitochondrial myopathy	Ate sandwiches of bread and salt and salted bananas. Mother had to forbid him from pouring salt from store containers directly onto food.
K. 16 of 43 children with sickle-cell disease	Unknown, "as early as 2.5 yr"	Unknown	Sickle-cell hemoglobinopathies	Six had "abnormal" and 10 had "exceedingly increased salt appetite" relative to rest of family. Latter category included children who salted apples, oranges, and peaches, as well as one 2.5-yr-old who salted rock candy, licked off salt, and discarded candy.
L. 33-yr-old female	Adult, recent	"By the shakerful"	Iron deficiency	Iron replacement therapy resulted in a cessation of salt eating.

[Adapted from Beauchamp *et al.*, in press.]

jects expressed no obvious desire for increased salt and, in fact, it was not possible to determine the source of the increased intake. Interestingly, this effect was not observed with the diuretic amiloride, so it would seem that simple sodium loss could not account for the observed results.

It would appear that sodium depletion should be the adequate and necessary condition stimulating a heightened salt preference. However, depletion of other nutrients may also stimulate avid salt consumption. Some research suggests that calcium and protein depletion stimulates salt appetite in experimental animals. It is generally believed that individuals in cultures in which a primarily vegetarian diet is consumed need and crave more salt than do primarily meat-eating peoples. The traditional explanation for this has been that meat-eaters naturally obtain substantial amounts of sodium in their diet. Although it is certainly true that meat diets generally contain more sodium than most vegetable diets, little direct evidence (as reviewed above) indicates that the low-sodium diets that might exist in vegetarian cultures lead to an elevated desire for salt. Vegetarian diets in some populations are not only low in salt but may also be low in protein; perhaps that, as well as or instead of sodium content of the diet, underlies an increased desire for salt.

B. Genetic Factors

Genetic factors could account for differences between groups and between individuals in their salt requirements, intakes, and preferences. A number of animal model studies implicate genetic control over salt intake and preference. For example, both inbred strains of rats and inbred strains of mice differ in their salt preference. However, for humans, the evidence favoring a genetic influence is not yet strong.

It has been speculated that Blacks may have a different mechanism than Whites for handling sodium. In particular, they may be less efficient in dealing with excess salt. This could contribute to the high prevalence of hypertension among Blacks when salt is readily available. Although one report indicates that black adolescents and adults prefer higher levels of salt than do whites, little evidence indicates a race difference in salt intake or avidity. Additionally, it is problematic whether or not genetic differences would necessarily underlie a race different in preference.

More direct evidence for a role for genes could come from studies comparing identical and fraternal twins. Twin studies so far have failed to find a heritable component to salt taste preference. The studies conducted to date cannot be considered definitive, however, due to methodological deficiencies. Consequently, a genetic explanation for individual differences in salt taste preference remains a possibility. More work is needed both in human populations and with animal models.

C. Psychological Factors

Experimental studies have examined the effects of alterations in consumption of salt on salt taste preference. In general, the amount of salt in food required to make the food taste best (optimal level) has been measured. A series of studies has demonstrated that if salt intake is reduced by about 50% over a period of time, the optimal level of salt in foods declines. The amount of salt required to optimize food flavor is decreased to about 65–75% of the prediet amount. Thus, if the optimal level of salt in soup for an individual was 1% while he or she was on a diet containing 150 meq Na/day, decreasing sodium consumption to about 75 meq/day would result, on average, in a reduction in optimal salt levels in soup to 0.75%. Importantly, this change is gradual, taking probably 1–2 mo to reach asymptote.

For several reasons, this change in preference probably is due to the altered (decreased) experience of tasting salty foods rather than to a physiological response to the change in amount of sodium the body must handle. First, the gradual nature of the effect, as noted above, would suggest experience plays the major role, because physiological responses to changes in sodium consumption (e.g., changes in the hormones of sodium balance such as renin and aldosterone) are quite rapid, occurring within hours or days. Second, one study reported that if salt consumption was increased by requiring subjects to add approximately 10 g of salt to their food, preferred levels of salt in food increased. However, if the same amount of additional salt was consumed as tablets and, thus, not tasted, there were no changes in taste preferences. Apparently, a sensory adaptation to different levels of salt accompanies a change in salt intake. To some extent, people like the level of salt they taste rather than, or in addition to, choosing the intensity of the taste of salt they like.

The third reason for believing that this change in optimal salt level following dietary change is a psychological phenomenon comes from another study in which dietary salt was reduced by 50%, but continued *ad libitum* use of table salt was permitted. Under these conditions, individuals use of added salt increased almost fourfold but did not approach compensation for the amount removed. In fact, they only replaced 20% of the decrement so that there was an overall decrease of approximately 40% in sodium intake. In this instance, salt was presumably added "to taste," implying that although the salt content of their regular food was in a range that the subjects found palatable, it was unnecessarily high, probably because it was dispersed throughout the food, rather than all being on its surface and easily accessible to the taste receptors. This study suggested a novel technique to reduce salt intake in that individuals were able to use as much salt as they wished while still reducing total salt intake. For this to be practical, however, there should be wide availability of all categories of lowered sodium foods. In terms of control of salt preference, the importance of this research is that there were no changes in taste preference, even though salt intake was decreased 40%. By adding salt to the outside of their food, individuals apparently obtained a sufficient salty taste experience to prevent any preference changes.

In sum, a substantial number of studies support the conclusion that changes in salt taste preferences following changing of salt intake are mediated by hedonic–cognitive expectations based on current dietary experience. Physiological changes associated with alterations in the amount of NaCl available to the body do not appear to play a role.

D. Developmental Factors

It is in infancy and childhood that the strongest claims for an experiential influence on human salt taste preference have been made. Interestingly, both high-salt intake and extremely low-salt intake in infancy have been postulated to increase later salt preference. However, actual data concerning the early development of salt preference and how experiences may serve to modify or perhaps permanently establish heightened preferences are conflicting and confusing.

Newborn infants are either indifferent to or avoid moderate to high concentrations of saline solution relative to water. By the time children are 2–3 or more yr of age, preferences for salty foods over those same foods without salt are common. These two observations have led some to conclude that salt preference is learned, although that conclusion does not necessarily follow. In fact, several studies indicate that a preference for salt solutions over plain water, although not evident in infants <4 mo of age, is evident in infants 4–23 mo of age. A developmental change in response to salt in human infants may represent, in part at least, postnatal maturation of the ability to taste salt. This interpretation is consistent with a substantial body of evidence from animal models that demonstrates that the neurophysiological response to salt exhibits a postnatal developmental change. In both rats and sheep, responses to salt in newborn animals are less robust than responses in adults.

Even if there is a postnatal maturation of salt taste perception and preference, experience with salt could modulate this preference in infants just as it does with adults. There are reports of a correlation between exposure to salted food and relative preference for saltiness in that food; however, these data are not conclusive as they are based on data from very few infants. Further longitudinal studies on the development and early modification of salt taste preference are needed. In particular, the issue of the relationship between level of exposure to salt and salt preference should be explored at several ages to understand the origin of the very high salt preferences. In this regard, a number of studies now suggest that young children and adolescents actually prefer higher levels of salt in food than do adults. It is unclear why this is the case, although it is possibly due to altered nutritional requirements during growth.

Contrasting with the hypothesis that high-salt intake in infancy and childhood may elevate later preferences, some animal experimental data and human clinical reports suggest that sodium depletion early in life may have profound effects on later salt preference. A body of evidence, primarily derived from studies with rats, now strongly indicates that the salt taste system is plastic during its development. Salt taste perception, as measured electrophysiologically, pharmacologically, and behaviorally, matures postnatally. Also, early restriction of dietary sodium alters subsequent response to NaCl. Finally, an episode of sodium depletion induced over a period of 1 or several days by a very high dose of diuretic combined with salt-free diets produces permanent elevation in salt intake weeks or months after recovery; this effect appears especially robust if the deletion occurs during early development.

The mechanisms responsible for these effects likely have both peripheral and central components. Changes in peripheral activity of salt-sensitive neurons may be involved. The persistent elevation of salt appetite following extreme sodium depletion, however, seems more likely to be of central origin, possibly a consequence of the elevation of renin and angiotensin that follows sodium depletion.

The analysis of available human clinical studies suggests that early sodium depletion may have profound effects on subsequent salt appetite in humans as well (see Table I). Evidence that sodium depletion (occurring in clinical entities such as Bartters syndrome and Addison's disease) produces a craving for salt that might be homologous with salt appetite induced in some experimental animals following severe depletion is not extensive. However, when human salt appetite had been reported, the onset was almost invariably in childhood. The most dramatic case was of a 3.5-yr-old boy with adrenocortical insufficiency causing an obsessive salt appetite. This child died due to sodium depletion following his placement on a controlled hospital diet. Prior to this he had consumed salt in very large amounts sufficient to replace the massive sodium loss that occurred in the absence of adrenal hormones that ensure sodium conservation. Several other similar cases of early childhood onset of salt appetite have also been reported (Table I). The absence of persuasive evidence for an adult onset of salt appetite suggests that in humans too, early depletion may have especially potent effects on later salt taste perception and preference. This is an area ripe for further research.

III. Conclusions

Humans prefer to consume substantially more salt than is necessary for the maintenance of normal physiology. In salt-sensitive hypertensives, this may lead to illness, and it also may contribute to the worsening of other illness (e.g., congestive heart failure, hepatic cirrhosis with fluid retention, hypertensives on antihypertensive medication). Some sensory and behavioral factors responsible for this high consumption have been identified, but a full explanation of the mechanisms is still lacking. To the extent that one understands the determinants of salt intake, one may be better able to provide guidance for successful reduction of dietary salt intake if this is medically indicted.

Acknowledgment

Preparation of this essay was supported by the National Institutes of Health, Grant Number DC 00882, and the Clinical Research Center, Grant Number RR 00040.

Bibliography

Beauchamp, G. K. (1987). The human preference for excess salt. *Am. Sci.* **75,** 27–34.

Beauchamp, G. K., Bertino, M., and Engelman, K. (1990). Human salt preference. *In* "Chemical Senses: Appetite and Nutrition" (M. I. Friedman, M. Tordoff, and M. R. Kare, eds.). Marcel Dekker, New York (*in press*).

Denton, D. (1982). "The Hunger for Salt." Springer-Verlag, Berlin.

Mattes, R. D. (1984). Salt taste and hypertension: A critical review of the literature. *J. Chronic Dis.* **37,** 195–208.

Shephard, R. (1988). Sensory influence on salt, sugar and fat intake. *In* "Nutrition Research Reviews," vol. 1. (J. W. T. Dickerson, A. G. Low, D. J. Milward, and R. H. Smith, eds.). Cambridge University Press, Cambridge.

Satiating Power of Food

JOHN E. BLUNDELL, *BioPsychology Group, University of Leeds*

PETER J. ROGERS, *Agriculture and Food Research Council, Institute of Food Research*

I. Food Habits, Eating Patterns, and Satiating Power
II. Satiation and Satiety
III. Hunger
IV. The Satiety Cascade
V. How to Measure Satiating Power
VI. Effect of Food Characteristics on Satiating Power
VII. Physiological Mechanisms Mediating Satiating Power
VIII. Implications

Glossary

Caloric compensation Capacity of the system to adjust food intake (by increasing or decreasing ingestion) in response to prior over- or underfeeding

Eating profile Pattern of eating episodes (meals, snacks, etc.) reflecting food ingested over a day, week, or longer

Hunger (1) Abstract term referring to an underlying drive to eat; (2) Subjective sensation associated with a desire to obtain and eat food

Palatability Degree of perceived pleasantness of a food. May vary from time to time according to physiological state or social circumstances

Prandial Associated with a meal; e.g., preprandial (before a meal), postprandial (after a meal)

Satiation Process of bringing an episode of eating to an end

Satiety State of inhibition over eating brought about by the consequences of food ingestion

THE SATIATING POWER OF FOOD refers to the capacity of any food material to suppress the experience of hunger and to take away the desire for further eating. The satiating power of food is therefore one important factor which limits the total amount of food consumed. Satiating power is sometimes referred to as satiating efficiency. This power is achieved by certain properties of the food itself engaging with various physiological and biochemical mechanisms within the body which are concerned with the processing of food once it has been ingested. Satiating power is therefore an effect resulting from a variety of biological processes and is an important component of the healthy regulation of appetite.

Information concerning the satiating power of food items is important for theoretical and practical reasons. First, knowledge of how the composition of food alters energy intake and food selection throws light upon the mechanisms of appetite control and helps us to better understand the complex processes involved in the regulation of human appetite. Second, this knowledge can be employed to develop a coherent strategy of nutritional intake for everyday use in the home, at work, and in the clinic. Knowledge about the effects on appetite control exerted by particular components of food can help industry to provide appropriate foods for specific requirements and can allow the consumer to rationally select a suitable diet. [*See* APPETITE.]

I. Food Habits, Eating Patterns, and Satiating Power

One major objective in appetite and nutrition research is to examine ways in which the properties of foods influence their long-term pattern of consumption and therefore affect the nutritional status (physical well being) of individuals. This enterprise is vital since it is possible that strongly held positive beliefs about a food, or very attractive taste qualities, may stimulate and maintain consumption but, if the commodity is not nutritionally valuable, then long-term effects may not be beneficial. These long-term patterns of eating are maintained by two sets of factors. The first involves the enduring habits, attitudes, and opinions about the value and suitability of foods and a general liking for them (factors concerning economics and availability obviously also determine eating patterns). A second set of factors arises from the events immediately surrounding the consumption of foods—the so called periprandial circumstances. These involve the impact that foods have upon the sense organs (particularly receptors in the nose and mouth) and upon other biological processes involving digestion and absorption.

The satiating power of foods is generated mainly through the second set of factors but with some influence from the first.

II. Satiation and Satiety

For technical precision and conceptual clarity it is useful to describe the distinction between satiation and satiety. Both terms may be assigned workable operational definitions (i.e., definitions which depend upon measurable events). Satiation can be regarded as the process which develops during the course of eating and which eventually brings a period of eating to an end. Thus it involves an increase in the tendency to inhibit eating behavior and ultimately ends in the temporary termination of eating. Accordingly satiation can be defined according to the measured size of the eating episode (volume or weight of food, or caloric value of the energy content). Satiety is defined as the state of inhibition over further eating which follows the end of an eating episode and which arises from the consequences of food ingestion. The intensity of satiety can be measured most clearly by duration of time lapsing until eating is recommenced, or by the amount consumed at the next eating episode. Other measures include the degree of suppression of hunger induced by the food eaten or the degree of aversion (or decline in preference) for specific types of foods. In the view of some researchers satiation and satiety can be referred to as intrameal satiety and intermeal satiety.

Clearly the processes of satiation and satiety operate conjointly to determine the pattern of eating episodes. Accordingly, reference to satiation and satiety must pay some attention to the measurement of meals, snacks, and other units of ingestion which make up a person's repertoire of eating activities. The satiating power of food will be one important component influencing the eating profile. At the same time it is clear that the satiating power of food can be expressed through the process of satiation or the state of satiety.

III. Hunger

What is hunger and how is it implicated in the satiating power of foods? Hunger is used in two ways in research associated with the control of appetite. On the one hand it is used as a term for a drive or state of motivation; in this sense it stands for an impulsion to action. Naturally the primary target of this urge is food. However, the notion of a drive has a rather unusual status since it is a force which is not measured directly but which is inferred from other events (physiological or behavioral) which can be monitored. Used in this way the hunger drive occupies an important place in the motivational theory of appetite. However, hunger also represents a subjective experience or feeling which is associated with the desire to obtain and eat food. This is how the layman understands the term hunger and it is how hunger is defined by the social rules of language. This is the way that hunger is used in everyday social discourse. The term is given understanding through a social consensus; that is, use of the term is generally understood as implying a willingness or desire to eat.

A number of sources of information make up the subjective feeling of hunger. They include physical sensations in the body and head such as sensations of stomach emptiness, light headedness, stomach cramps, mild nausea, tightening of the throat, and so on. Hunger feeling may also be influenced by certain cues such as the presence of food, the arrival of meal times, and the stimulation of thoughts about food or eating. The particular patterns of physical sensations, environmental stimuli (events), and mental thoughts which contribute to the overall feeling of hunger may vary markedly from person to person. However, if someone is asked to indicate the strength of the feeling of hunger (or desire to eat) they can do it reliably by using a rating scale. These scales are simple instruments which allow people to transform the intensity of a subjective feeling into an objective point or category on a scale.

From a functional point of view hunger achieves

a purpose as that nagging irritating feeling whose presence constantly serves to stimulate thoughts about food and eating. Hunger is useful and reminds us that the body needs food. In this way hunger can be seen to possess a clear biological function. Some theorists would argue that hunger *causes* eating but this is a very strong view that makes an uncompromising statement about causality. It is safer to argue that hunger is one important component which precedes the onset of eating and is strongly associated with consumption.

How is hunger related to the satiating power of food? Since the subjective feeling of hunger is a factor in motivating people to eat, hunger can be used as one measure of satiating power. When someone has gone without food for several hours the feeling of hunger will be strong and will promote thoughts of food and help to initiate eating. As eating proceeds the subjective intensity of hunger will decline and fall to a low level during the development of satiation. Hunger remains low during the course of satiety but begins to grow again as satiety weakens. Therefore the time course of hunger during satiety can be used to evaluate the satiating power of food.

IV. The Satiety Cascade

Eating food has the capacity to take away hunger and, after satiation has occurred, further eating is inhibited for a period. What mechanisms are responsible for these processes? It is clear that the mechanisms involved in terminating eating and in maintaining inhibition range from those which occur when food is initially sensed, to the effects of metabolites on bodily tissues following digestion and absorption (across the wall of the intestine and into the blood stream). By definition, satiety is not an instantaneous event but occurs over a considerable time period; it is therefore useful to distinguish different phases of satiety which can be associated with different mechanisms. This concept is illustrated in Fig. 1. Four mediating processes are identified: sensory, cognitive, postingestive, and postabsorptive. These maintain inhibition over eating (and hunger) during the early and late phases of satiety. Sensory effects are generated through the smell, taste, temperature, and texture of food and it is likely that these factors help to bring eating to a halt and inhibit eating (of foods with similar sensory characteristics) in the short term. Such a mecha-

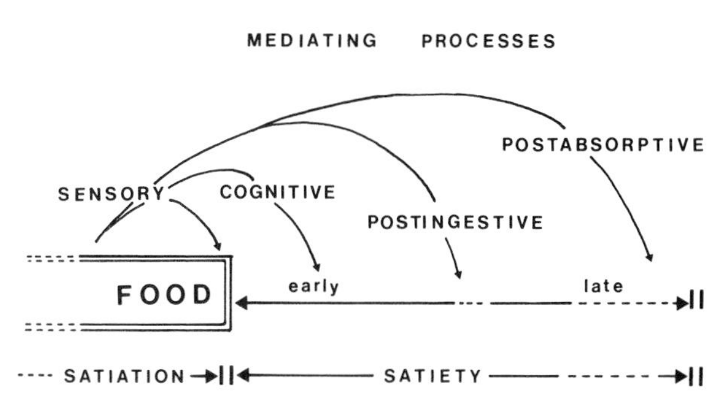

FIGURE 1 Conceptualization of the satiety cascade illustrating how different features of ingested food contribute to the time course of satiety.

nism is embodied in the idea of sensory-specific satiety. Cognitive effects represent the beliefs held about the properties of foods and their presumed effect upon the eater. The category identified here as postingestive processes includes a number of possible actions, including gastric distension and rate of gastric emptying, the release of hormones such as cholecystokinin from the duodenum, and the stimulation of physicochemically specific receptors along the gastrointestinal tract. The postabsorptive phase of satiety includes those mechanisms arising from the action of metabolites after absorption across the intestine and into the blood system. This category embraces the actions of chemicals such as glucose and the amino acids, which may act directly upon the brain after crossing the blood–brain barrier or which may influence the brain indirectly via neural inputs following stimulation of peripheral chemoreceptors.

The approximate anticipated moment of action of these mediating processes is shown in Fig. 1 but of course the mechanisms will overlap and their effects will be integrated to produce a combined effect. It should also be kept in mind that the psychobiological system for appetite control has the capacity to learn, that is, to form associations between the sensory and postabsorptive characteristics of foods. This means that it will be useful to distinguish between the unconditioned effects of food, i.e., those in which the natural biological consequences of food processing in the gut are reflected upon satiety, and the conditioned effects which come into play due to the links developed between sensory aspects of food—particularly those which are tasted—and the later metabolic effects generated by the same food. The sensory characteristics (or cues) therefore come to predict the impact which the food will later exert. This means that the sensory factors themselves (conditioned stimuli) can be used to mediate the satiating efficiency or power of food. However, the potency of this mechanism depends upon the stability and reliability of the relationship between tastes (sensory cues) and physiological effects (metabolic consequences) of food. When there is distortion or random variation between sensory characteristics and nutritional properties then the conditioned mechanisms of the satiety cascade are greatly weakened.

All of these factors add to the importance of measuring the strength of satiety at various times after the end of ingestion, in order to throw light upon the effect of individual mediating processes. Good experimental designs will be required to analyze the operations of the satiety cascade. However, it is clear that the effects of foods upon the satiety cascade have important consequences for appetite control and for the subjective feelings associated with eating. It is therefore likely that the expression of the satiety cascade will have implications for food acceptability.

V. How to Measure Satiating Power

The ability to assess the satiating effectiveness of foods depends upon having valid monitoring procedures and appropriate experimental designs. It will also be useful to monitor the motivation to eat along with actual caloric intake, the distribution of eating episodes, and the selection of particular nutrients. In addition, since the overall satiating capacity of food will depend upon facilitatory and inhibitory effects, it will be helpful to measure the various parameters before, during, and after eating. The temporal profiles of motivation (e.g., hunger) and behavior (e.g., eating) should provide the most complete description of satiation and satiety.

As an episode of eating comes to an end it would be widely agreed that motivational measures have declined to low values. Motivation here means those subjective feelings of willingness to eat or to continue eating and the most obvious of these is the sensation of hunger. Accordingly by measuring various motivational parameters before and after eating it is possible to establish the profile of changes in behavior and sensations which are correlated with the initiation and termination of eating. The assessment of these is described more comprehensively elsewhere but some of the details most relevant to the satiating power of foods are set out below.

1. Visual analog ratings made on 100-mm scales (word anchored at either end) in the following order: First, "How strong is your desire to eat?" (Very strong–Very weak); second, "How hungry do you feel?" (As hungry as I have ever felt–Not at all hungry); third, "How full do you feel?" (Very full–Not at all full); and fourth, "How much food do you think you could eat?" (A large amount–Nothing at all). This last rating is called prospective consumption.
2. Rating of the pleasantness of a meal can be checked by a further scale asking "How pleasant have you found the food?" (Very pleasant–Not at all pleasant). Similarly, individual food items can be assessed for their pleasantness to indicate how the appreciation of the taste of particular types of foods alters during the course of consumption, and afterward. This measure throws light upon the so called sensory-specific satiety of foods in which the change in satiating value of a specific food is linked to its sensory qualities.
3. A food preference checklist in which subjects tick off the foods they would like to eat from a list of 32 common food items: The items represent eight high-protein, eight high-carbohydrate, eight high-fat foods, all of which are portioned so that the energy content is between 180–220 kcal, and eight low-calorie foods such as fruit or vegetables. The number of items checked correlates well (correlation coefficients of 0.7–0.8) with total amount of food eaten and the separate macronutrient scores correlate to approximately the same degree with nutrient selection.
4. Forced choice food preference test in which subjects are forced to choose between a high-carbohydrate and high-protein item: This procedure is designed to reveal a specific high preference or craving for carbohydrate or protein.
5. Food intake in a test meal: Clearly one of the most important measures of the satiating power of a food is the impact of that food upon subsequent consumption. Consequently it is necessary to measure accurately and objectively the amount of food eaten. This is done most conveniently by administering a "test meal" at a specified interval after the consumption of the food being evaluated. This test meal usually contains a variety of different foods varying in nutritional content whose weight and composition are known exactly. The foods are presented in abundance and from the amount of foods remaining after consumption the exact energy value and nutrient content of the eaten food can be derived.
6. Diary records: In addition to the direct measurement of intake in a test meal, longer term assessment can be computed from the records of food consumed kept by subjects in a food diary. These

estimates will not be as accurate as those measures taken directly from test meals but they provide information about the pattern of energy and nutrient intake when subjects are free to eat when and where they wish.

Taken together these procedures are designed to measure strength of motivation to eat, the direction of food preferences, the prominence of particular physical sensations, and the frequency and sizes of eating episodes. When used in conjunction with the technique of temporal tracking, these techniques provide a body of information concerning the psychological state which precedes a meal, which accompanies eating and which follows consumption.

This procedure for the analysis of motivational variables accompanying eating provides a profile of characteristics illustrated in Fig. 2. This diagram also includes some information derived from certain related procedures. This information illustrates the way in which certain measurable parameters of motivation have been demonstrated to change as food is eaten. Since food is the natural agent which induces satiation, the constellation of features surrounding the end of a meal can be used to characterize the existence of satiation. Similarly, the continued tracking of these parameters following the termination of eating would disclose the features of the state of satiety and of recovery from satiation. At the start of a meal the motivational process of hunger is characterized by strong hunger sensations and a low value for fullness, a high hedonic rating for food, a large score on the food preference checklist, and a relative preference for protein over carbohydrate. At the end of a meal, satiation is characterized by low reported sensations of hunger, high level of fullness, low checklist score, and a relative preference for carbohydrate over protein.

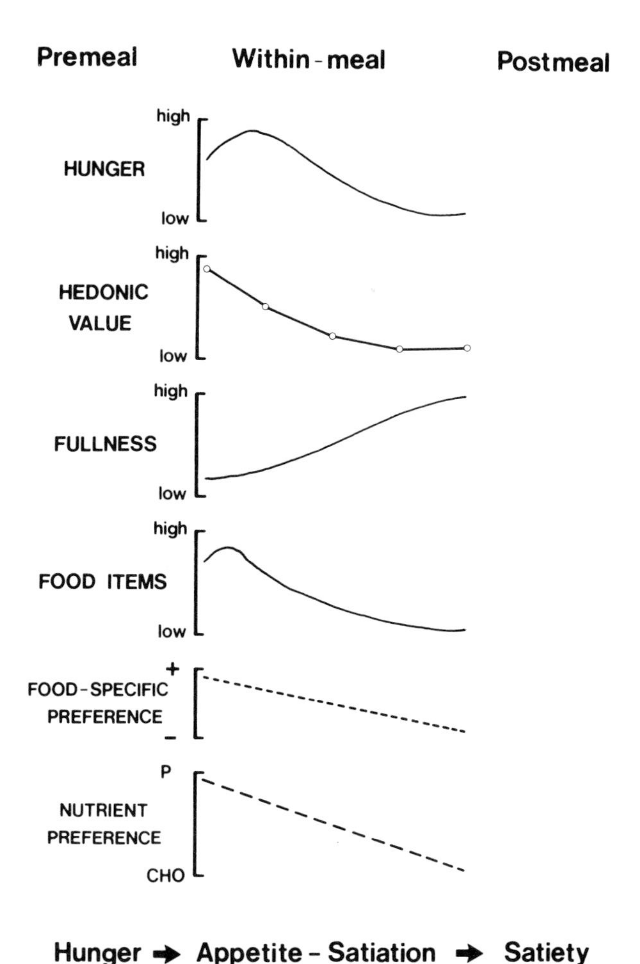

FIGURE 2 Adjustments in motivational parameters which accompany eating during the course of a meal. P, Protein; CHO, carbohydrate.

The profile of variables revealed by this battery of procedures provides a baseline for evaluating the consequences of various types of experimental manipulations of eating. Agents believed to intensify satiation should clearly influence the constellation of changes associated with the end of a meal. Changing the properties of particular foods would be expected to produce shifts in the profile of variables and to change immediate or later intake.

One other aspect of measuring satiating power involves the development and use of appropriate experimental designs. Different types of experiments may be required to demonstrate the pre- or postabsorptive effects of foods or to distinguish between sensory and postingestive actions (see Fig. 1). A general plan of experiments to examine satiating power is set out in Fig. 3. This diagram illustrates how nutritional manipulations of one meal (labeled MEAL) can be tracked during that meal and the consequences followed in subsequent test meals or by means of diary records of eating patterns and meal profiles. These designs incorporate the necessity of validating the changes in ratings and checklist response by measuring actual intake of energy and the selection of particular nutrients. For example, if the total number of calories consumed at a meal is experimentally manipulated while the appearance and pleasantness of the food remains unchanged, then the cognitive effects on motivational parameters may prevent any instantaneous change in these variables. However, an effect should become apparent as the postingestive effects of the food take effect.

One further important feature of designs to mea-

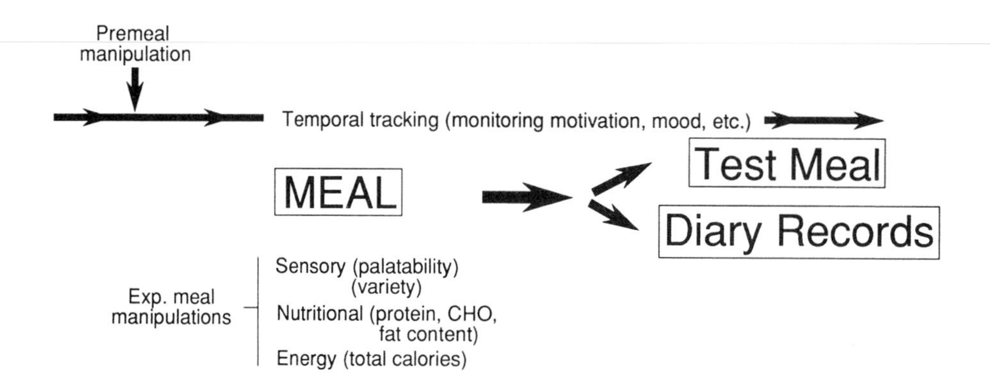

FIGURE 3 Experimental strategies for investigating the satiating power of foods.

sure satiating power is the use of a manipulation termed a preload. This term refers to an accurately prepared sample of a food or drink which is presented to the subject at a specified time, often 1 hr before a test of consumption (test meal). The preload can vary in volume, number of calories, and all the other parameters which may influence satiating power. It is often a convenient way of isolating a potentially interesting variable by preparing a sample in which that variable is covertly manipulated with other variables kept constant. In this way the effect of this prominent variable (for example, the amount of protein or dietary fiber or sweetness) in the food can be more readily evaluated.

VI. Effect of Food Characteristics on Satiating Power

What are the features of food which give it the capacity to suppress hunger and to inhibit further intake of food? Some properties of food which are of considerable significance are the level of perceived pleasantness (or palatability), the energy value (calories), the proportion of the macronutrients protein, carbohydrate, and fat, and the quantity of indigestible material present (or dietary fiber). In order to describe the consequences of manipulating any one of these features it is of course necessary to hold the others constant. For example, if palatability is experimentally manipulated then total calories and the proportion of macronutrients should be held at constant values. When working with real foods this methodological requirement can present the experimenter with some difficult nutritional problems. Together with these components there must also be considered those aspects of the satiety cascade—sensory factors and cognitions—already referred to.

A. Caloric Value of Food

There are good logical and biological reasons for considering that the total energy content (caloric value) of a consumed food will play a major role in satiating power. Many experiments have demonstrated that this is the case and this principle is embodied in the so-called energostatic theory of appetite control. However, generally speaking the caloric value of a food becomes available only following the processes of digestion and absorption (or during the processes themselves) when the energy-yielding substances become available to receptive tissues and the chemosensitive processes in the cells of those tissues. Therefore in order to assess the full extent to which the total caloric value of a food contributes to satiating power sufficient time must elapse for the calories to be detected and registered by the body. It requires carefully controlled laboratory experiments to determine the satiating power of calories themselves. This is because it is probably the case that the appetite control system has evolved so that the taste of a food can be a predictor of caloric value (see Section IV). There are prominent neural pathways which facilitate the development of this association. Therefore under normal circumstances the body can predict the satiating power of a food (and therefore display accurate and appropriate control of appetite) from the taste itself. Of course in scientific investigations of caloric value it is therefore necessary to eliminate taste as a signal of satiating power by presenting to the subject (normally as a preload) two or more food samples identical in taste, texture, and physical properties but differing in caloric value. When these studies are properly carried out caloric content of food is shown to be the most important contributor to satiating power.

B. Macronutrient Content: Protein, Fat, and Carbohydrate

Do the macronutrients protein, carbohydrate, and fat contribute equally to satiating power? The answer is not obvious. One reason is that these macronutrients are not of equal caloric density, therefore comparisons among them should be made on the basis of their individual caloric contributions rather than upon weight. Protein and carbohydrate are worth approximately 4 kcal/g while fat is 9 kcal/g. Investigations of the action of macronutrients can be made by presenting preloads in which one macronutrient is held constant and the other two are systematically varied. For example, when fat is held constant and protein and carbohydrate adjusted it is generally found that protein provides greater satiating power than does carbohydrate. Therefore high-protein meals would be expected to give rise to intense and prolonged satiety. Comments to this effect can be found in long-established textbooks but good experiments confirming the issue have been carried out only recently. [*See* PROTEINS (NUTRITION).]

There is currently a great deal of interest in the contribution of dietary fat to satiating power. This is because the current dietary intake of fat in technologically highly developed societies is rather high (often more than 45% of total calories) and it is widely believed that this excessive intake is an important factor in the development and maintenance of obesity. It is certainly true that, owing to the high caloric density of fat, it is possible to consume a huge number of calories for a relatively small amount of high-fat food products. It is partly for this reason that the nutrition industry is currently actively engaged in the development of fat replacers or fat substitutes. These substances, some of which have now almost reached the market place, provide most of the textural and sensory qualities of true fats but contain far fewer, if any, calories per gram. Accordingly it is a matter of some importance to establish the satiating power of the calories represented as fat. At the present time insufficient well-controlled and well-designed studies have been done to provide an unequivocal answer. However, it is beginning to look as if fat may be less satiating calorie for calorie than either carbohydrate or protein. However, this effect may depend upon the time of day that the fat is consumed. If this is confirmed by subsequent experiments then it would imply that a contribution to the alleviation of obesity could be made by replacing at least some of the fat in foods by a substitute or "fake fat." [*See* FATS AND OILS (NUTRITION); OBESITY.]

C. Dietary Fiber

The contribution of dietary fiber to satiating power is another currently disputed issue. Do high-fiber foods possess high satiating power? Once again the answer to this question can only be realized through good experimentation. Initially, it is worth noting that dietary fiber refers to those components of food which resist digestion and therefore do not contribute to the energy value of food. This usually means the cell walls of plants and the outer coats or husks of seeds and fruits. (The energy value of fiber itself is presently under considerable scrutiny.) Assuming, for the moment, that fiber contributes few, if any, calories, experiments to assess its impact on satiating power can use three major manipulations. First, preloads or meals can be composed of those foods which are naturally high in fiber. Second, foods can be produced in which the level of fiber has been deliberately raised (e.g., bread containing guar gum, an insoluble type of fiber). Third, the fiber can be isolated and delivered separately as tablets or capsules. Considering the results from all three types of studies there is evidence that fiber augments the satiating power of food and under certain circumstances it has effects on both satiation and satiety. It should be added that the mechanisms underlying these effects are not known but fiber could exert an effect on satiating power through action in the mouth, stomach, or intestine as well as by altering postabsorptive plasma levels of nutrients. In addition the action of fiber is heavily dependent on the amount delivered; it is very unlikely that any effect on satiating power is achieved by small amounts of fiber but significant effects are observed with 30 or more grams per day. Finally, rather than considering that fiber itself provides satiating power, it is probably more accurate to argue that fiber intensifies the effect on satiating power conferred by the calories contained in the food. [*See* DIETARY FIBER, CHEMISTRY AND PROPERTIES.]

D. Sensory Factors

The satiety cascade (Fig. 1) described earlier indicates that the early detection of the physical properties of foods by sensory receptors makes some contribution to satiating power. Certainly the taste and texture of foods can have a rapid effect on our willingness to continue eating, and if the taste happens to be aversive or unpleasant then eating will probably cease. Consequently, sensory factors allow us to determine if food is acceptable for consumption; if acceptability reaches a particular threshold then the food will be eaten and its satiating power could then be determined by the factors already mentioned. In addition sensory factors provide signals of information about the nutritional content of food and in this way they mediate satiating power by representing the genuine agents of satiating power such as calorie value or protein content. However, a more singular role for sensory factors has been proposed through an agency termed "sensory-specific satiety." This notion suggests that the specific taste (or smell, texture, etc.) of a food item reduces the willingness to consume more of that item (relative to other food items of similar pleasantness but not yet eaten) by reducing its pleasantness in relation to other foods. It appears to be widely accepted that this phenomenon can account for overconsumption when a variety of foods are on offer and indeed it has been experimentally demonstrated that subjects in laboratory conditions do eat more when presented with a variety of different food items rather than many items of the same type. However, some experts have remarked that "eating a food in a state of satiety induced by the immediately preceding intake of a sensorily different food is a trivial experience within human meals." Sensory-specific satiety does appear to describe a certain feature of human eating but it should not be confused with the phenomenon of satiating power. Manipulating the range of tastes available can influence the amount consumed at any sitting and can therefore delay the onset of the process of satiation, but once the food has been consumed then satiating power is engendered by the major components of caloric value and macronutrient content.

E. Palatability

This is a term which connotes a preference or a liking for a food and this idea suggests that this factor should exert some effect on satiating power. Indeed it is known that food confers both facilitatory and inhibitory effects upon appetite. The facilitatory effects are those which transiently augment the desire for food and which can stimulate consumption. The palatability of food is one such factor and it has been demonstrated that food which is perceived as being very pleasant can enhance the sensation of hunger during eating and hasten the return of hunger after a meal. High palatability also generates effects on physiology (plasma levels of glucose) which are related to the sensations of hunger.

F. Sweetness

Sweetness is a potent dimension of food and a powerful psychobiological phenomenon. There is an innate preference for the sweet taste which is present in newborn infants and sweetness is a prized characteristic of foods in many human societies. The sweet taste per se appears to have a stimulating effect on appetite since sweetening power (in the form of high intensity or artificial sweeteners) added to a bland-tasting food or beverage generally tends to augment the feeling of hunger. This manipulation is called the addition procedure and would be expected to work against the satiating power of other properties of food. Many sweet-tasting foods are energy rich (often sweet carbohydrates). Some energy-containing material may be removed and replaced by a high-intensity sweetening agent which has the effect of maintaining sweetening power but reducing the caloric or energy value. This manipulation, which is called the substitution procedure, means that the satiating power of food will almost certainly be reduced. However, because of the powerful effects of sweetness upon physiology, behavior, and subjective states other effects may be possible. The overall biological potency of sweetness suggests that it should be manipulated with caution. [*See* SWEETNESS, CHEMISTRY.]

G. Cognitions

This is the technical name given to beliefs and attitudes and in this context it indicates the feelings a person has concerning the expected impact of the eaten food. For example, someone eating a food which they believe is very high in energy value may display a strengthening of the satiating power of the food conferred by the actual nutritional content. Some cleverly devised laboratory studies have indicated that cognitions can be equal in power to the effects of strictly biological properties. Therefore under certain conditions it is possible that cognitions could exert a significant influence upon food consumption; the outcome would of course depend upon whether the cognitions complemented or opposed the satiating properties of the nutritional components. However, because of the capacity of physiological mechanisms to detect the "true" satiating power of food (conferred by the nutritional properties) it is likely that the contribution of cognitions would not be long lasting. (This action of cognitions upon satiating power is of course separate from the exertion of conscious control over food intake when under extreme circumstances, e.g., to achieve political goals, individuals may cease eating and starve themselves to death. [*See* ATTITUDES AS DETERMINANTS OF FOOD CONSUMPTION.]

H. Summary

It can be seen that a number of characteristics of foods contribute to satiating power. These characteristics include both sensory and nutritional components of the food material. Some factors strengthen satiating power while others act in opposition. Accordingly in assessing the potential satiating power of a new food it would be important to know, for example, the total caloric content, the amount of protein, the presence of dietary fiber, and the overall sweetness and palatability. On the basis of this knowledge some estimate could be made of the likely satiating power. However, scientific methods and experimental designs should be used wherever possible to confirm the validity of estimates.

VII. Physiological Mechanisms Mediating Satiating Power

It is clear that the components of food which give rise to the satiating power achieve this via an interaction with physiological mechanisms as the food is sensed, consumed, digested, and the products absorbed. The complexity of the physiological processes involved in satiating power is formidable and these same processes form the physiological basis for appetite control. This is not the place for a detailed examination of these processes but the diagram in Fig. 4 illustrates some of the most important components. The lower part of Fig. 4 sets out the satiety cascade already described (see Fig. 1), the middle part summarizes some peripheral physiological events, and the upper part shows in simple form how the peripheral information engages with brain neurotransmitter systems. Current understanding of the physiology of satiety takes account of peripheral information reaching the brain via neural routes (particularly via the various branches of the vagus nerve) and via humoral routes (chemicals and metabolites reaching significant brain receptor sites by being carried in the blood). Figure 4 illustrates how processing of consumed food is mediated by physical changes (gastric factors) or the release of hormones such as cholecystokinin (CCK) or insulin. Much of this preabsorptive information reaches the nucleus of the solitary tract (NST) of the brain by nerve pathways. From here it is currently believed that the information is relayed forward to the hypothalamus. One key zone appears to be the paraventricular nucleus (PVN) where a number of neurotransmitters (amines and peptides) have potent actions. Products of digestion such as glucose, free fatty acids (FFA), or amino acids (AA), may influence brain mechanisms directly (by being carried to the brain in the blood supply) or indirectly (via nerve pathways responding to oxidative metabolism in the liver). Certain other agents such as the corticosteroids or hypothesized appetite agents such as "satietin" may affect the appetite control systems in the brain and could therefore modulate the satiety response to foods. The so-called cephalic phase mechanisms are activated by sensory properties of foods such as taste and smell and this information is detected by the brain and then fed downward to release hormones in the gastrointestinal (GI) tract in anticipation of the ingestion of food. This simple diagram provides a conceptual picture to indicate the physiological basis for the satiating power of food. The changes in hunger and fullness which food engenders are mediated by peripheral physiological events and by brain processes.

VIII. Implications

At the present time there is a great deal of interest in the effect of food on behavior and psychological processes. Eating patterns and feelings of hunger are of special interest. In addition strong cultural pressure encourages people (particularly women) to lose weight and to aspire to socially defined ideals of slimness. It is often argued that one way to

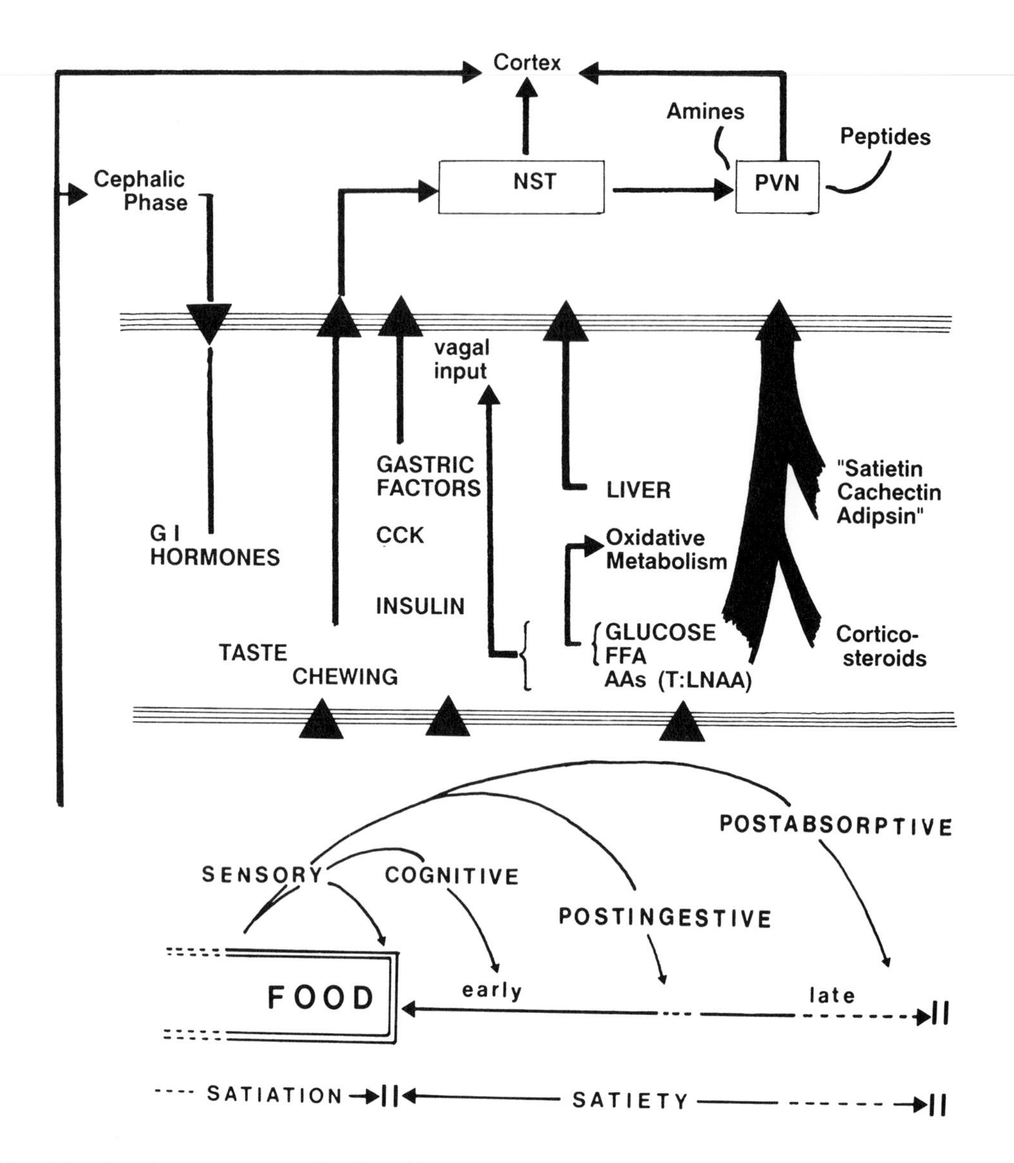

FIGURE 4 Diagram showing how some physiological events in the peripheral system and in the brain could be associated with the operations of the satiety cascade. NST, Nucleus of the solitary tract; PVN, paraventricular nucleus; GI, gastrointestinal; CCK, cholecystokinin; FFA, free fatty acid; AA, amino acid; T : LNAA, ratio of T (tryptophan) to LNAA (large neutral amino acids).

achieve this objective is to consume foods with reduced energy value. Advances in food technology have placed in the market place a wide variety of low-calorie or "lite" food products. These are widely believed to be a useful aid to weight control. It is therefore of interest to inquire about the satiating power of low-calorie foods. Is it possible to decrease the caloric value of a food product yet leave satiating power undiminished? Research to date suggests that this is not generally the case. When calories are withdrawn from a food (to decrease the energy value) then a weakening of satiating power would be expected. When more information is available about the details of the physiological processes underlying satiety then certain nutritional "tricks" could be incorporated into foods in order to amplify satiating power or to maintain satiating power when calories are withdrawn. Certainly during this present period in the late part of the twentieth century the introduction of new nutritional products into the food supply may have considerable implications for the satiating power of the foods we consume. In turn this may influence the extent to which good appetite control can be achieved. The effect of nutritional changes in our diet must of course be evaluated alongside various cultural changes in attitudes to eating, health, and body shape. It can be concluded that eating food with adequate satiating power, which therefore has the capacity to control the irritation of hunger without producing weight gain, is a significant component of maintaining good appetite control. In turn this contributes to good psychological and physical health.

Bibliography

Blundell, J. E., and Burley, V. J. (1987). Satiation, satiety and the action of fibre on food intake. *Int. J. Obes.* **11,** Suppl. 1, 9–25.

Blundell, J. E., Hill, A. J., and Rogers, P. J. (1988). Hunger and the satiety cascade—their importance for food acceptance in the late 20th century. *In* "Food Acceptability" (D. M. H. Thompson, ed.), pp. 232–250. Elsevier, Amsterdam and New York.

Kissileff, H. R., Gruss, L. P., Thornton, J., and Jordan, H. A. (1984). The satiating efficiency of foods. *Physiol. Behav.* **32,** 319–332.

Le Magnen, J. (1985). "Hunger." Cambridge University Press, Cambridge.

Rogers, P. J., and Blundell, J. E. (1989). Separating the actions of sweetness and calories: Effects of saccharin and carbohydrates on hunger and food intake in human subjects. *Physiol. Behav.* **45,** 1093–1099.

Shepherd, R. (ed.) (1989). "Handbook of the Psychophysiology of Human Eating," p. 383. Wiley, New York.

Van Itallie, T. B., and Vanderweele, D. A. (1981). The phenomenon of satiety. *In* "Recent Advances in Obesity Research III" (P. Bjorntorp, M. Cairella, and A. N. Howard, eds.), pp. 278–289. Libbey, London.

SEM

Scanning Electron Microscopy

DAVID C JOY, *University of Tennessee*

Glossary

Convergence angle Cone angle formed by electron beam as focused onto the specimen
Electron gun Device to produce high-energy electrons for imaging
Pixel Smallest element of image on the display or record screen
Raster Rectangular pattern of scan lines generated during imaging
Scintillator Material that emits light when struck by electrons
Working distance Free space between the sample and the microscope lens

THE SCANNING ELECTRON MICROSCOPE (SEM) is the most widely used form of electron microscope in the health and biological sciences. The SEM is popular because it combines some of the simplicity of the optical microscope with much of the performance of the more expensive and complex transmission electron microscope. Most important of all, the SEM can look at real, solid specimens such as whole cells, pieces of tissue, and bones and even complete organisms such as viruses and bacteria. Its ability to produce high-resolution, high depth-of-field images of three-dimensional surfaces makes it an instrument of unique value.

I. Introduction

The SEM was originally developed in Germany in the 1930s by Knoll and Von Ardenne. Later, important improvements were made by Zworykin, Hillier, and Snyder at the RCA Research Laboratories in the United States in the 1940s. The design and performance of their instrument anticipated much that was found in later microscopes, but their success was ultimately limited by the poor vacuum conditions under which they had to work. The current form of the instrument is the result of the work of Oatley and his students at Cambridge University between 1948 and 1965. The first commercial SEM, the Cambridge "Stereoscan," was produced by Cambridge Instruments in the United Kingdom in 1965. Today nearly a dozen companies manufacture scanning microscopes for the international market with prices varying from US$40,000 to in excess of US$500,000.

II. Principles of the SEM

Figure 1 shows schematically the basic principle of the SEM. Two electron beams are used simultaneously: The incident beam strikes the specimen to be examined; the second electron beam strikes a cathode ray tube (CRT) viewed by the operator. As a result of the impact of the incident beam on the specimen, a variety of electron and photon emissions are produced. The chosen signal is collected,

Encyclopedia of Human Biology, Volume 6.

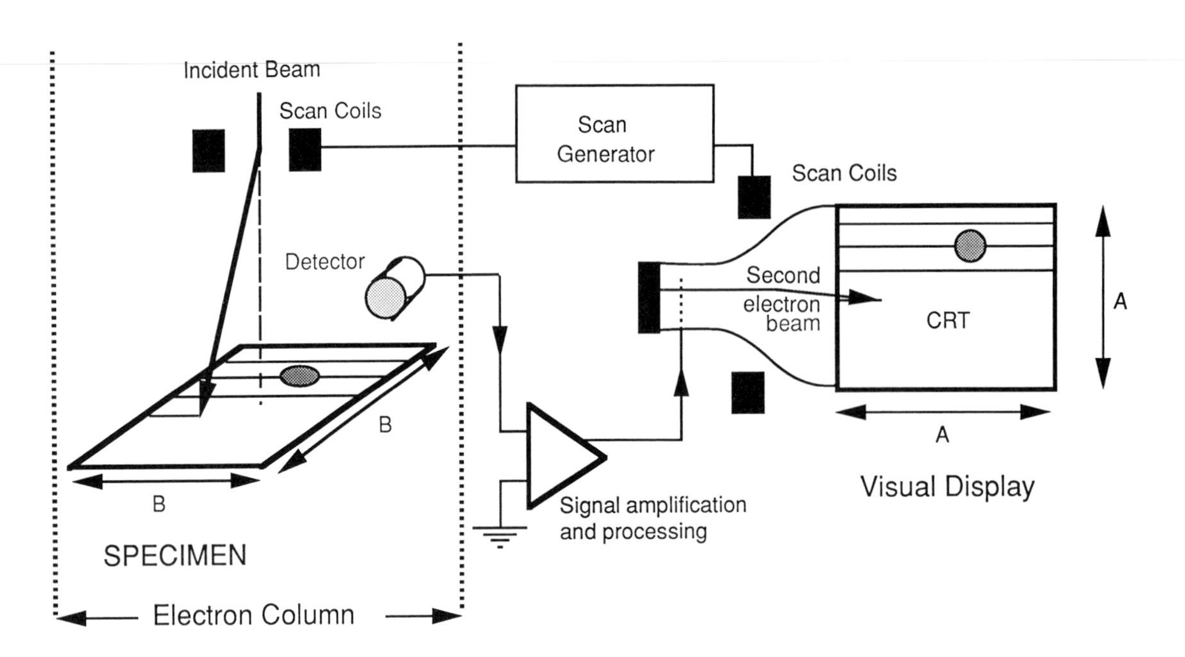

FIGURE 1 Schematic illustration of the principle of operation of the SEM.

detected, amplified, and used to modulate the brightness of the second electron beam, so that a big collected signal produces a bright spot on the CRT while a small signal produces a dimmer spot. The two beams are scanned synchronously so that for every point scanned on the specimen there is a corresponding point on the CRT. Typically, the beams scan square patterns on both the specimen and the CRT. They start at the top left-hand corner of the area, scan a "line" of points parallel to the top edge, and then, when they reach the end of the line, they fly back to the starting edge and scan a second line, and so on until the whole square area has been "rastered." Each complete image is conventionally called a frame. If the display area of the CRT tube is $A \times A$ in size and the area scanned on the specimen is $B \times B$ in size, then variations in the signal from the specimen will be mapped onto the CRT as variations in brightness with a linear magnification of A/B. Thus, a magnified map or image of the specimen is produced without the need for any imaging lenses.

This method of imaging offers several important advantages:

(1) Magnification is achieved in a purely geometric manner and can be varied by simply changing the dimensions of the area scanned on the specimen.

(2) Any emission that can be stimulated from the specimen under the impact of the incident electron beam—e.g., electrons, X rays, visible photons, heat, sound—can be collected, detected, and used to form an image. The SEM is therefore not restricted to imaging with radiations that can be focused by lenses.

(3) Several different types of image can be produced and displayed simultaneously from the same area of the sample, enabling different types of information to be correlated; only a suitable detector, amplifier, and display screen for each signal of interest are necessary. Furthermore, these signals can be mixed with each other to generate new types of imaging information.

(4) Because the picture on the screen is formed from an electrical signal, which varies with the position of the beam and hence with time, the image can be electronically processed to control or enhance contrast.

A consequence of this arrangement is that a fundamental limit to the imaging performance is set by the display CRT screen. The smallest feature that can be discerned on the CRT is equal to the size of the electron spot on the display screen. Conventionally it is assumed that 1,000 scan lines, each containing 1,000 picture elements, or pixels, make up each image frame scanned. Each picture is, therefore, formed from a pattern of 1,000 $\times$ 1,000 (i.e., 1 million) pixels. When the SEM is operating at a magnification of M, then the resolution in the image (i.e., the smallest detail on the specimen that can be observed) is equal to the pixel size divided by the magnification. Because the size of the spot on the CRT is typically 100–200 μm (0.2 mm), magnifications of a few hundred times the resolution are limited to 1 μm or so. Only at high magnifications is the resolution limited by more fundamental electron-optical considerations.

III. Components of the SEM

The main components of an SEM are contained in two units: the electron column, which contains the electron beam scanning the specimen, and the display console, which contains the second electron beam, which impinges on the CRT. The high-energy electron beam incident on the specimen is generated by an electron gun, two basic types of which are in current use. The first (Fig. 2A) is the thermionic gun in which electrons are obtained by heating a tungsten or lanthanum hexaboride cathode or filament to between 1,500 and 3,000 K. The cathode is held negative at the required accelerating voltage E_0 with respect to the grounded anode of the gun so that the negatively charged electrons are accelerated from the cathode and leave the anode with an energy E_0 kiloelectron volts (keV). Thermionic guns are in wide use because they may safely be run in vacuums of 10^{-5} Pa (i.e., 10^{-7} Torr) or worse. The alternative source (Fig. 2B) is the field emission gun in which a sharply pointed wire of tungsten is held close to an extraction anode to which is applied a potential of several thousand volts. Electrons tunnel out of the tungsten wire, which can be at room temperature, into the vacuum and are then accelerated as in the thermionic gun toward the anode. Field emission guns require an atomically clean emitter surface, thus they must be operated under ultrahigh vacuum conditions, typically in a vacuum of 10^{-7} Pa (i.e., 10^{-9} Torr) or better. For either emitter the entire length of the electron column traveled by the electron beam from the gun to the specimen chamber must also be pumped to an adequate vacuum using oil-diffusion, turbo-molecular or ion pumps individually or in combination.

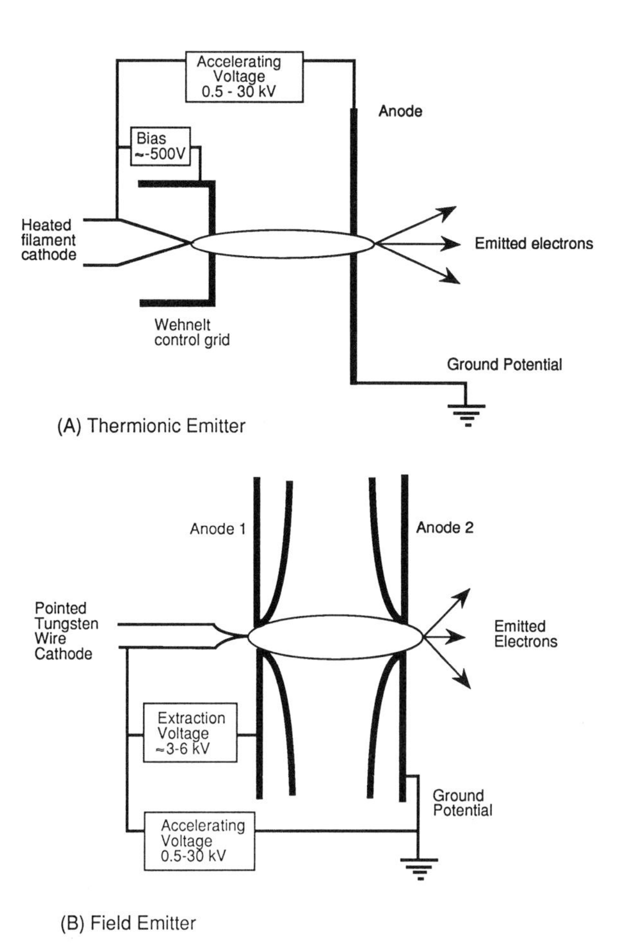

FIGURE 2 Electron sources used in the SEM. A. Thermionic emitter using a tungsten or LaB_6 cathode. B. Field emitter.

IV. Performance Limits of the SEM

The performance of the SEM depends on a number of related factors, perhaps the most important of which is the output of the electron source. The source is quantified by its brightness, β, which is the current density (amps/m^2) it delivers into unit solid angle (steradian). The brightness increases linearly with the accelerating voltage of the microscope but also varies greatly from one type of source to another. At a given energy, a field emission gun is between 10 and 100 times as bright as an LaB_6 thermionic emitter, which is, in turn, between 3 and 10 times brighter than a tungsten thermionic emitter. Typically at 20 keV, a field emission gun has a brightness in excess of 10^{12} amps/m^2/sterad.

The diameter of the electron beam from the gun is reduced or demagnified by passing it through two or more lenses before it reaches the sample surface (Fig. 3). An electron lens, which consists of a coil of wire carrying a current, focuses the electron beam in exactly the same way as a glass lens focuses light, but it has the convenient property that the focal length can be varied by changing the magnitude of the current flowing through the solenoid. By varying the excitations of the lenses, the beam diameter

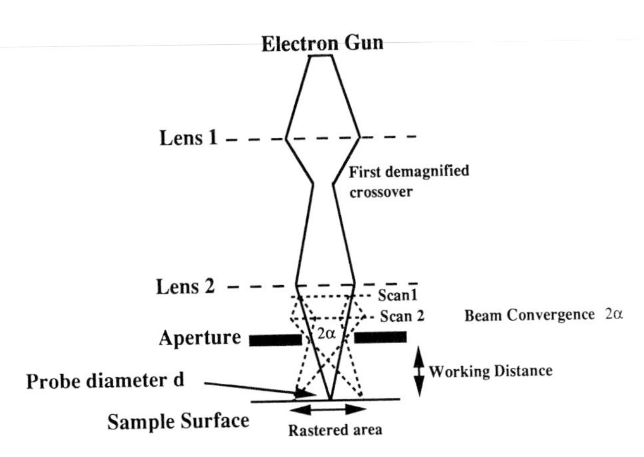

FIGURE 3 Diagram illustrating the ray paths taken by electron traveling in the column of the SEM.

at the specimen can be set to any desired value from the source size downward.

The electron beam is usually scanned by first deflecting it across the optical axis in one direction, and then immediately deflecting it in the opposite sense through twice the angle. This arrangement ensures that all of the scanned rays pass through a single point. An aperture can then be placed at this position to define the beam convergence angle α, where α will be equal to the diameter of aperture divided by twice the working distance (see Fig. 3).

The gun brightness is constant throughout the electron optical system, and hence the value β measured for the focused probe of electrons impinging on the specimen is the same as the value that would be measured at the source. If the probe diameter is d, if the incident beam current is I_B, and if the convergence angle of the beam is α, then by definition the brightness β at the sample, and hence at the gun, is:

$$\beta = \frac{4\,I_B}{\pi^2 d^2 \alpha^2}\,. \tag{1}$$

For typical operation α is fixed by the choice of aperture size, and β is constant, thus:

$$I_B = \left(\frac{\pi^2}{4}\right)(\beta \cdot \alpha^2) d^2, \tag{2}$$

which shows that as the diameter (d) of the probe is made smaller, the current contained in the probe falls as d^2. The spatial resolution of the SEM (i.e., the smallest detail on the sample that can be observed) cannot be significantly less than the probe diameter (d). The so-called brightness equation, equation (2), therefore indicates that the available source brightness will set a limit to the resolution of the microscope because, as discussed below, a certain minimum incident current is required to form an image.

The depth of field, D_f, of the image, defined as the vertical focusing range outside of which the image resolution is visibly degraded, is given as:

$$D_f \approx \frac{\text{pixel size}}{\alpha}\,. \tag{3}$$

Because α is between 10^{-3} and 10^{-2} radians, the depth of field is typically several hundred times the pixel size. At low magnifications, therefore, the depth of field can be of the order of several millimeters, giving the SEM an unrivaled ability to image complex surface topography and to produce images with a pronounced three-dimensional quality to them. Notably, however, this fortunate effect results from the fact that electron-optical lenses must be "stopped down" to very small apertures (i.e., small values of α) in order to work, and thus very few of the electrons leaving the source actually reach the specimen. An exact optical analogy would be the "pinhole" camera, which also has a large depth of field but requires long exposure times.

The image in the SEM is built up from an electrical signal, which varies with time. If the average signal level is S, and if as the beam scans across some feature the signal changes by some amount, ∂S, then the feature is said to have a "contrast" level C given by:

$$C = \frac{\delta S}{S}\,. \tag{4}$$

Changes in the signal can also occur because of statistical fluctuations in the incident beam current and in the efficiency with which the various emission processes take place within the specimen. Thus, repeated measurements across the same feature on a specimen will give signal intensities that will vary randomly around some mean value. These inherent statistical variations constitute a "noise" contribution to the image, which therefore has a finite "signal-to-noise" ratio. For image information to be visible, the magnitude of the signal change ∂S occurring at the specimen must exceed the magnitude of the random fluctuations by a factor of five times or so. This leads to the concept of threshold current I_{TH}, which is the minimum incident beam current required to observe a feature of contrast level, C. I_{TH} is given by the relationship:

$$I_{TH} = \frac{4.10^{-12}}{C^2\tau} \text{ amps}, \quad (5)$$

where τ is the time in seconds required to record one frame of the image (assumed to contain 10^6 pixels). The observation of low-contrast features therefore needs high-beam currents or long exposure times. Typically, images are produced in 1 sec or less, but images to be photographed are recorded for 30–100 sec to improve the signal-to-noise ratio. The threshold current requirement sets a fundamental limit to the performance of the SEM in all modes of operation. A comparison of equation (5) with equation (2) shows that because the beam current varies rapidly with the beam diameter the resolution limit of the SEM will depend on the contrast of the feature being observed.

V. Electron-Solid Interactions

The interaction of an electron beam with a solid specimen produces a wide variety of emissions, all of which are potentially useful for imaging. Each of these signals is produced with a different efficiency, comes from a different volume of the sample, and carries different information about the specimen. Figure 4 shows schematically the energy distribution of the electrons produced by an incident beam of energy E_0. The distribution displays two peaks: one peak at an energy close to that of the incident beam and a second peak at a much lower energy. The high-energy peak is made up of electrons that have been "backscattered," or reflected, by the sample. The fractional yield of backscattered electrons, i.e., the number of backscattered electrons per incident electrons, is called the backscatter yield and usually is identified as η. The average energy of these electrons is about 0.5–0.6 of the incident energy E_0, and the backscatter yield η is typically 0.2–0.4.

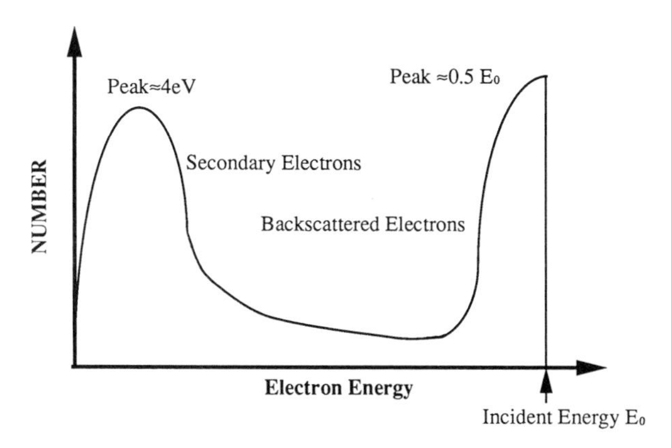

FIGURE 4 The energy distribution of electrons emitted from a solid specimen irradiated by an electron beam of energy E_0.

The lower-energy peak, which lies within the energy range 0–50 eV, is made up of what is usually called the secondary electron signal. As before, we can define a relative secondary yield, i.e., number of secondary electrons per number of incident electrons, identified as δ. The average energy of the secondary electrons is about 4 eV, independent of the incident beam energy, but the secondary yield δ varies rapidly with accelerating energy being of the order of 0.1 or less for most materials at 30 keV, but of the order of unity for energies of 1 or 2 keV. Because the secondaries are low in energy, they cannot travel more than a few nanometers through the sample to reach the surface and escape, so the secondary signal images the surface region of the specimen.

The total current flowing into and out of the specimen must balance to zero, thus:

$$I_B = \eta\, I_B + \delta I_B + I_{SE}, \quad (6)$$

where I_B is the incident beam current and I_{SE} is the current flowing in the ground connection to the sample. This "specimen current" contains information about both the secondary and backscatter signals and can form the basis of an important imaging mode. Equation (6) shows that if $\delta + \eta$ is unity, then no current flows to earth. At this condition, the incident electron beam is neither injecting charge into the specimen nor extracting charge from it. When the sample being examined is an electrical conductor, this situation is not of much significance, but when the specimen is not a good electrical conductor then no current can flow to earth, so any excess (or deficit) charge is retained in the specimen. At high-incident electron energies (greater than a few kiloelectron volts), the total yield $\delta + \eta$ is less than unity, so charge is injected into the specimen, which therefore charges negatively. This reduces the effective incident energy of the beam so $(\delta + \eta)$ increases, but charging will continue until the effective incident energy reaches a value E2 at which $\delta + \eta$ becomes unity. At this energy, each electron in produces, on average, one electron out, so no further charge is deposited and the specimen potential stabilizes. Therefore, at this E2 energy it is possible to form an image from even an insulating material without the need to make the surface an electrical conductor by coating it with metal. Typically, E2 is of the order of 1–3 keV for materials such as polymers or ceramics, but low-density materials such as dried, unstained, biological tissue can have E2 values as low as a few hundred electron volts. Because operation at the E2 energy avoids the necessity of coating the sample, there is increasing interest in low-voltage scanning microscopy in the biological field.

VI. Secondary Electron Imaging

Secondary electrons are the most popular choice of interaction with which to form an image. There are two main reasons for this:

(1) Because they are low in energy, it is possible to collect most of the secondaries produced by the sample by biasing the detector to a modest positive potential so that it attracts electrons to itself. Efficient collection is possible even when the detector is not in the line of sight of the sample.

(2) The main contrast mechanism associated with secondary electrons produces images that are readily interpretable by analogy with reflected light images in the macroscopic world.

Figure 5 shows a schematic drawing of the detector that is now standard for secondary electron detection. The detector is based on a disc of scintillator, which emits light under the impact of electrons. The light travels along a light pipe, through a vacuum window, and into a photomultiplier where it is converted back into an electrical current. Because the amount of light produced by the scintillator directly depends on the energy of the electron that strikes it, secondary electrons, which have only a few electron volts of energy, would produce only a very little signal. To increase the efficiency therefore a bias of +10 kV is applied to the front face of the scintillator to accelerate all electrons to at least

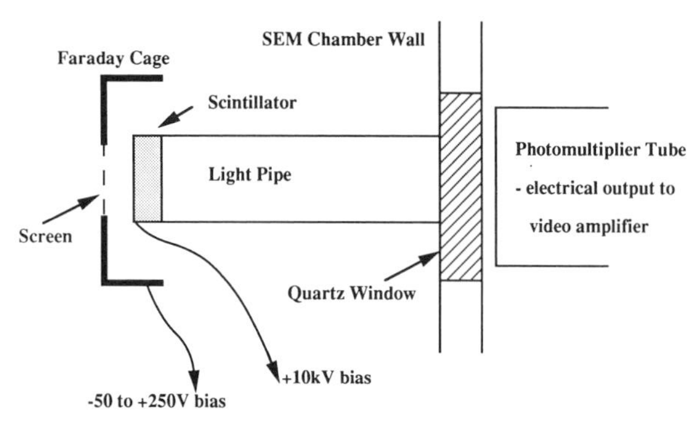

FIGURE 5 The Everhart-Thornley secondary electron detector.

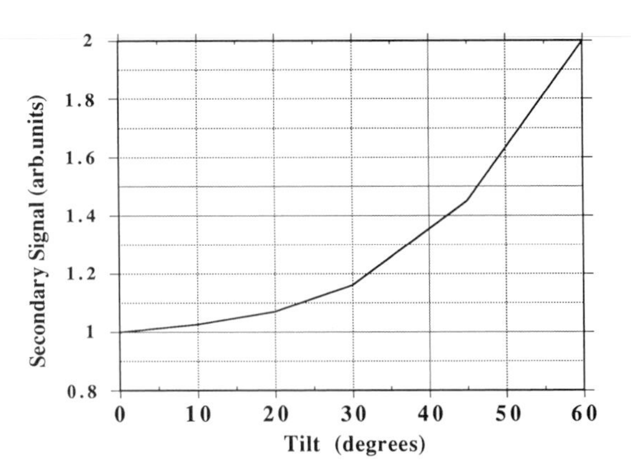

FIGURE 6 Variation in secondary electron yield with θ, the angle of incidence of the beam.

this energy. This high potential may, however, distort or deflect a low-energy incident beam so the scintillator is shielded by a Faraday cage, made of open mesh wire and carrying a potential of only +200 V or so. This is still sufficient to attract and collect 50% or more of the available secondary electrons.

The dominant imaging mechanism for secondary electrons is topographic contrast. It has been estimated that >90% of all scanning micrographs rely on this mode. The effect arises because an increase in θ, the angle of incidence between the beam and the surface normal, will lead to an increase in the yield of secondary electrons as shown in Figure 6. This can be understood by noting that as θ is increased, the fraction of secondary electrons produced within the escape region of the surface also increases. If an electron beam moves over a rough surface, then the local angle of incidence between the beam and the surface normal will change and produce a corresponding change in the secondary signal.

Figure 7 shows examples of this type of imaging, from cultured rat basophilic leukemia cells that have been fixed but not metal-coated, at magnifications ranging from a value (600×) that is comparable with that usually associated with an optical microscope to an upper value (100,000×) that is equivalent to that obtained on a transmission electron microscope. Figure 8 shows another pair of examples of this type of image, this time from human prostate cancer cells that have been coated with a 3-nm film of gold-palladium. Again the benefits of a wide magnification range, three-dimensional view, and high depth of field are evident. Perhaps surprising is the ease with which this type

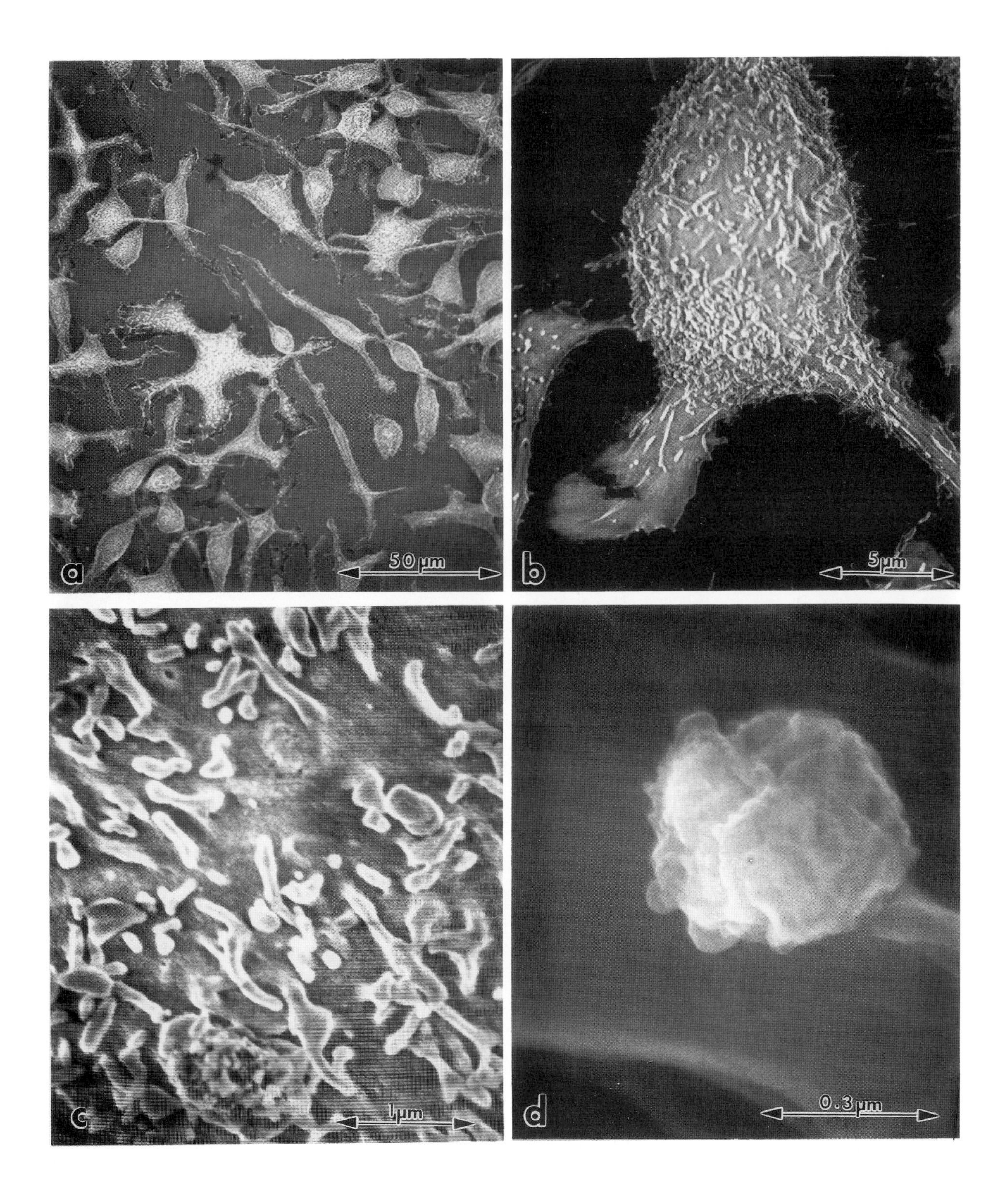

FIGURE 7 Images of rat basophilic leukemia cells at original magnifications of (a) 600×, (b) 6,000×, (c) 25,000×, and (d) 100,000×. Micrographs recorded at a beam energy of 5 keV on a Hitachi S-900 field emission SEM.

of image can be understood and interpreted. In fact, the view of a surface obtained by the secondary electron image in the SEM is analogous to that which would be obtained if the observer were to

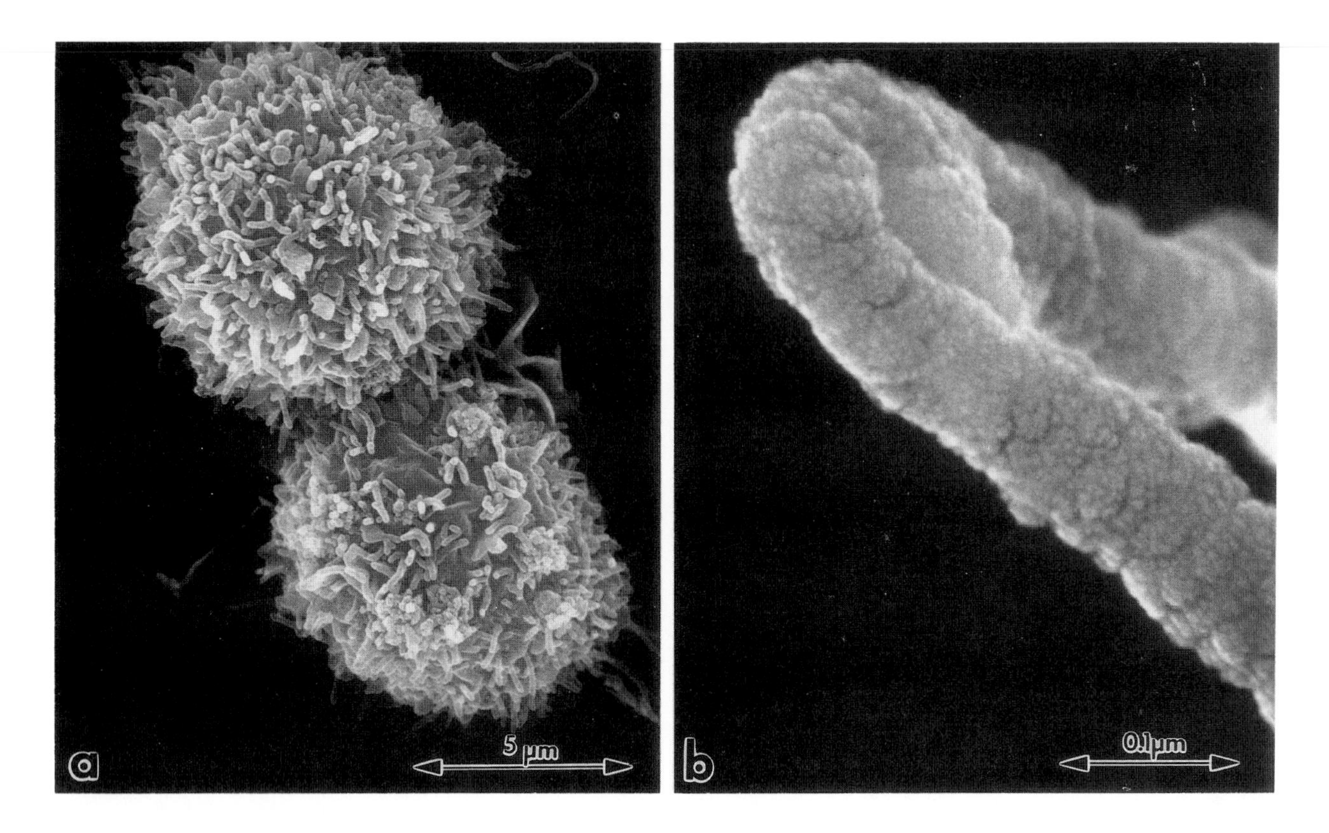

FIGURE 8 Secondary electron images of human prostate cancer cells at original magnifications of (a) 6,000× and (b) 180,000×. The specimen was coated with a 3-nm-thick film of gold-palladium. The fine structure visible at image (b) is the grain of the coating. Image recorded at 20 keV on a Hitachi S-900 field emission SEM. [Micrographs courtesy of Dr. Carolyn Joy, University of Tennessee.]

look down the column at the specimen illuminated by a light source placed at the detector. Faces at a high angle of inclination to the beam and facing the detector will be bright, while surfaces normal to the beam will be darker. Most microscopists also rotate the micrograph so as to place the illumination at the top of the picture; then all bright surfaces are tilted up and facing toward the detector, while darker surfaces are horizontal or facing away from the detector. Our brain, based on a lifetime of experience in using such clues, can than reconstruct the surface topography with a high degree of confidence. This, coupled with the three-dimensional quality that comes from the high depth of field, explains much of the popularity of the SEM and secondary electron imaging.

The spatial resolution of the secondary electron image under optimum conditions can approach 1 nm (10 Å) with modern, high-performance field emission SEMs. In general, however, the spatial resolution of biological images is limited by the preservation of detail in the tissue after it has been stabilized, dried, and fixed. The typical biological material is mostly water, so it cannot simply be placed in the SEM and examined because the rapid drying that would occur in the vacuum would lead to gross shrinkage and distortion. Instead, the water must first be exchanged carefully with alcohol, and then the alcohol removed by freeze-drying, critical-point drying, or vacuum-aided chemical drying. Because dry tissue is very fragile, it is also usually necessary to fix the sample—either chemically or by freezing—before it is observed. Each of these essential steps can, however, produce artifacts in the specimen and obscure or destroy fine detail. Thus, high-resolution secondary imaging involves as a prerequisite painstaking care with the protocols of specimen preparation.

VII. Backscattered Imaging Modes

Backscattered electrons are those incident electrons that have been scattered through an angle

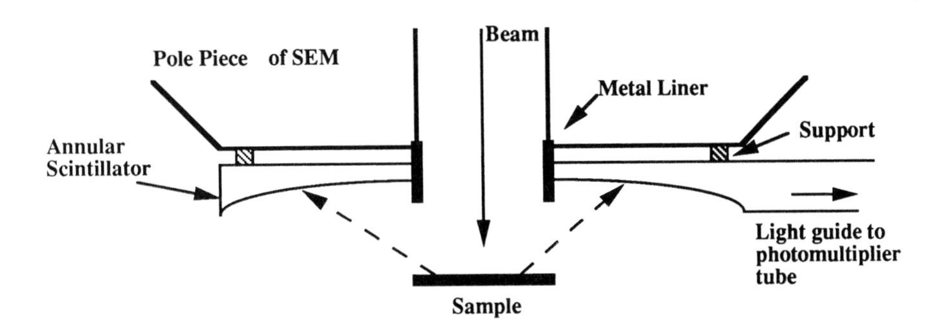

FIGURE 9 Typical detector for backscattered electrons using a large annular scintillator placed above the specimen and concentric with the beam axis.

>90° within the sample and, thus, can leave it again. They typically have an energy that is of the order of 0.5 of their original incident energy. Consequently, the backscattered electrons can emerge from considerable depths within the specimen, an estimate for this escape depth, R_{BS}, being

$$R_{BS} = \frac{70\ E^{1.67}}{\rho}\ \text{nm}, \quad (7)$$

where ρ is the density and E is the beam energy in kiloelectron volts. For a beam energy of 20 keV, backscattered electrons can carry information about regions 5–10 μm below the surface of dry tissue compared with just a few nanometers for the secondary electrons. However, because the diameter over which the backscattered electrons emerge from the surface is of the same order as R_{BS}, the spatial resolution of the image will be worse than that of the secondary signal.

Backscattered electron imaging modes are complementary to secondary modes and produce unique information of their own; however, they have so far attracted somewhat less attention because of the problem of signal collection. While the low-energy secondary electrons are readily collected by the application of a small bias field, the high-energy backscattered electrons travel in straight lines from the specimen and must therefore be collected by placing a suitable detector in the path of the electrons. A typical arrangement (Fig. 9) places a ring of scintillator concentrically around the beam and directly above the specimen. No bias is applied to the scintillator so only high-energy backscattered electrons produce an output. For beam energies over a few kiloelectron volts, this type of arrangement is highly efficient (>50%).

The yield η of backscattered electrons varies with both the atomic number of the target and the energy of the incident electrons. For energies over about 5 keV, the backscattering yield is almost independent of the accelerating voltage, and η can be approximated by the function:

$$\eta = -0.0254 + 0.016Z - 0.000186Z^2, \quad (8)$$

where Z is the atomic number of the target. η varies between about 0.05 for carbon (Z = 6) and about 0.5 for gold (Z = 79). The yield is also somewhat dependent on the topography of the surface, but for a large backscattered detector placed symmetrically above the specimen surface this effect is small enough that such backscattered images contain little topographic information.

Because the yield is simply related to the atomic number Z, the backscattered image displays contrast that is directly related to the atomic number of the area sampled by the electron probe. This technique has found several important applications in biology. For example, regions of tissue or cells can be identified by heavy metal (e.g., osmium, silver, tungsten, gold) stains, which bind preferentially to specific features of interest. While the metal might be invisible in the secondary image because of the strength of the topographic contrast, the large difference in effective atomic number between the stain and the predominantly carbon matrix provides a high-contrast signal in the backscattered image mode. A comparison of the secondary and backscattered images then permits the stain to be localized. An extension of this technique—cell-surface labeling—uses small (2–20 nm diameter) particles of colloidal gold either bound to a chemical entity, which has a specific affinity for some site or sites on a cell surface, or to an antibody. Even though the gold particles are small, they are readily visible in the backscattered image and, hence, by again comparing secondary and backscattered images, the binding sites for the labeled compound can be identified. Figure 10 shows an example of this type of

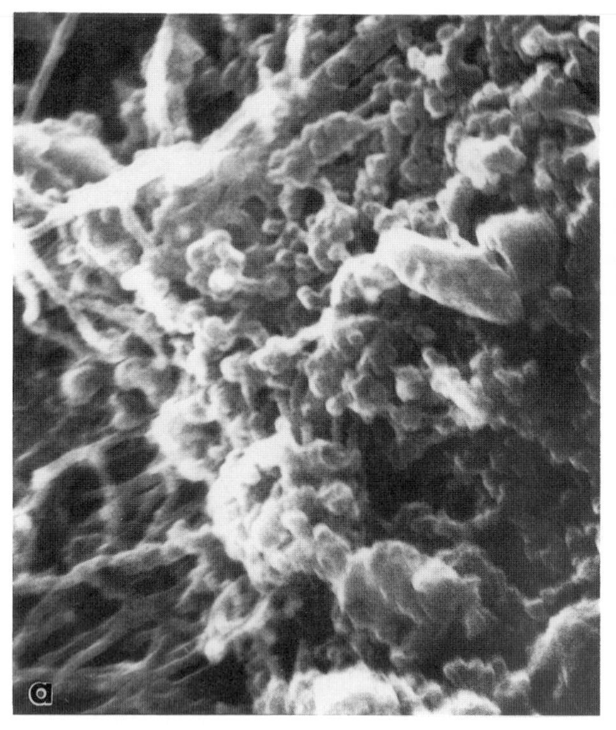

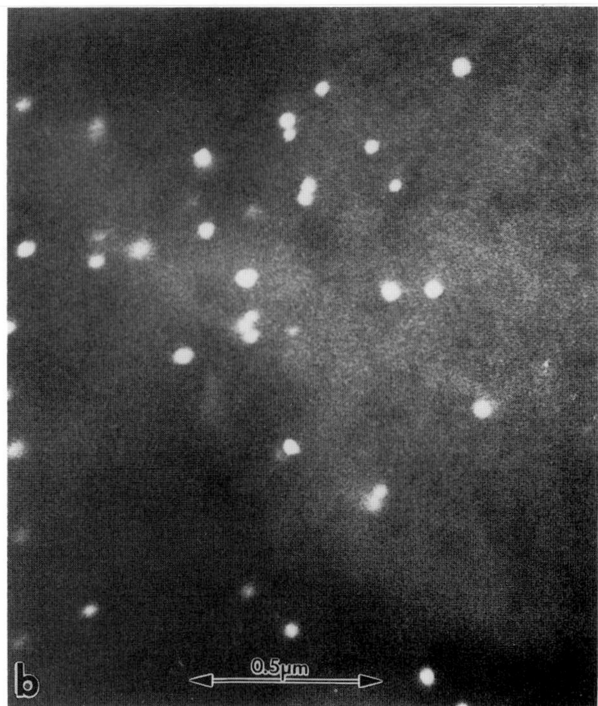

FIGURE 10 (a) Secondary and (b) corresponding backscattered electron images of 12 nm colloidal gold-labeled prolactin on the zona pellucida of a porcine oocyte. Images recorded at 20 keV energy and a magnification of 60,000× using a Hitachi S-900 field emission SEM. [Micrographs courtesy of Dr. D. Wininger, University of Tennessee.]

imaging in which the receptor sites for prolactin on porcine oocytes are identified by the 12-nm gold labels visible as bright dots against the dark carbon background.

VIII. Unwanted Beam Interactions

Although interactions between the sample and the beam of incident electrons provides the information carrying contrast of the image, other interactions are equally potent but less beneficial. In particular, the electron beam is a powerful source for ionizing radiation. To give an idea of how significant this effect is, the specimen in the SEM is subjected to as much radiation as an individual standing 30 m from a 10 megaton H-bomb. This is very harmful to biological materials because ionization can lead to the breaking and cross-linkage of bonds. This radiation damage is a severe limitation to the ability of the SEM to efficiently image such materials. A deposited dose of only 1,000 e^-/nm^2 (160 coulombs/m^2) is sufficient to destroy the crystal structure of a soft protein, yet this is less than one-tenth of the dose normally required to take a single high-resolution SEM image. An X-ray spectrum requires a greater dose by a further factor of 100 times. At the moment, there is no way of protecting against such damage although, for some samples, the rate of damage is known to be reduced by a factor of two to five times if the specimen can be held at very low (below 10 K) temperatures. Therefore, it is necessary to plan microscopy so as to minimize the electron dose to which any part of the sample is exposed—e.g., by working at lower magnifications, shooting blind on previously unexposed areas, and accepting noisier images and spectra—and to be aware of the kinds of artifacts that radiation damage can cause.

IX. Conclusion

Although the SEM is a simple instrument, its mode of operation gives it some unique advantages. It combines the low-magnification capability and simplicity of the optical microscope with the high resolution of the transmission electron microscope and produces images of solid specimens, which are both readily interpretable and aesthetically pleasing. Thus, the SEM is a tool of growing importance in the biomedical sciences.

Bibliography

Glauert, A. (ed.) (1974). "Practical Methods in Electron Microscopy." North Holland Press, Amsterdam.

Goldstein, J. I., Newbury, D. E., Echlin, P. E., Joy, D. C., and Fiori, C. E. (1981). "Scanning Electron Microscopy and X-ray Microanalysis." Plenum Press, New York.

Joy, D. C., Romig, A. D., and Goldstein, J. I. (1986). "Principles of Analytical Electron Microscopy." Plenum Press, New York.

Newbury, D. E., Joy, D. C., Echlin, P., Fiori, C. E., and Goldstein, J. I. (1986). "Advanced Scanning Electron Microscopy and X-ray Microanalysis." Plenum Press, New York.

Oatley, C. (1982). The early development of the SEM. *J. Appl. Phys.* **53,** R1–13.

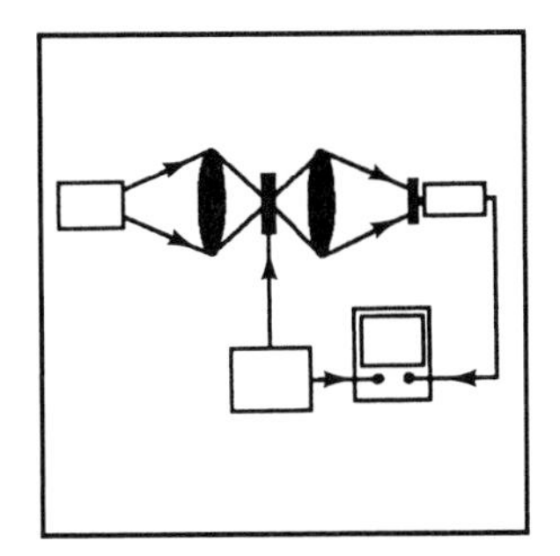

Scanning Optical Microscopy

C. J. R. SHEPPARD, *University of Sydney*

Glossary

Autofocus technique Method of producing projections in confocal microscopy by selecting the peak signal in an appropriate direction

Beam scanning Type of scanning optical microscope in which the beam is scanned by, for example, galvomirrors. The beam thus transverses the optical system off-axis during scanning

Confocal microscopy Technique of scanning optical microscopy, often performed in a fluorescence mode, but also in brightfield reflection or transmission. Confocal microscopy results in optical sectioning of thick objects, and improved resolution and contrast

Differential phase contrast Technique of scanning optical microscopy, achieved by using a detector split into two halves, giving an image of phase gradients in the object. It can be performed in either reflection or transmission

Extended focus technique Method of producing projections in confocal microscopy by averaging the signal in an appropriate direction

On-axis scanning Type of scanning optical microscope in which the specimen (or objective lens) is scanned mechanically so that the beam always travels along the axis of the instrument. This avoids off-axis aberrations and shading.

SCANNING OPTICAL MICROSCOPY is a recently introduced method which exhibits a range of advantages compared with conventional optical microscopy, while retaining the noninvasive nature of optical microscopy. Thus specimens can be observed with the minimum of preparation, and need not be exposed to a vacuum environment as in electron microscopy. Living tissue can be observed in its natural watery condition. Specimens can be observed without staining, taking advantage of electronic contrast enhancement or methods of phase imaging. Alternatively, specific components can be stained using immunogold or immunofluorescence techniques. Perhaps the most well known at present of the various scanning methods available is confocal fluorescence microscopy, which is considered in detail in Section III,C. [*See* OPTICAL MICROSCOPIC METHODS IN CELL BIOLOGY.]

I. Introduction

A. Advantages of Scanning

Broadly, the advantages of scanning optical microscopy stem from two main properties. First is the fact that the image is measured in the form of an electronic signal, which allows a whole range of electronic image processing techniques, both analog and digital, to be employed. These include image enhancement techniques such as frame averaging, contrast enhancement, edge enhancement, and image subtraction to show changes or movement, image restoration techniques for resolution enhancement, and noise reduction and image analysis techniques such as feature recognition and cell sizing and counting.

Second is the property that imaging in a scanning microscope is achieved by illuminating the object with a finely focused light spot. This allows a number of novel optical imaging modes to be employed such as confocal imaging or differential phase contrast, but also introduces the possibility of imaging modes in which the incident light spot produces some related effect in the specimen which can be monitored to produce an image. In addition, by restricting the size of the photosensitive detector the noise level is reduced and the measurement accuracy can be greatly improved.

B. Methods of Scanning

In a scanning microscope the object is illuminated with a focused light spot which is scanned relative to the object (Fig. 1). This can be achieved either by scanning the light spot, or by scanning the object itself. Most commercial instruments at present scan the light spot. The two most widely used methods employ galvomirror scanners for both the x and y scan, or a Nipkow disk as is used in the tandem scanning microscope. Galvomirrors are either of the resonant variety which can oscillate at high speed but are fixed frequency devices, or more usually of the feedback-stabilized type, which can scan at a line frequency of about 1 kHz. Nipkow disk scanners, in which a disk with an array of holes is rotated, allow the use of a white light source because they have multiple apertures. Their main disadvantages are that signal level can be low, and that a television camera must be used in order to produce an electronic image signal. Alternative beam-scanning systems include polygon mirror scanners, which achieve high scanning speeds but are difficult to synchronize, and acoustooptic scanners, which allow TV scanning rates but suffer from chromatic variations which rule them out for fluorescence applications without use of special imaging geometries.

The main advantage of object scanning is that the optical system is then completely unchanging during scanning so that the imaging properties are unvarying across the field. Beam-scanning systems, on the other hand, can experience brightness variations across the field, a fall-off in resolution at the edges of the field, and also noticeable curvature of field. Object scanning is thus preferable for quantitative work or in image reconstruction. The disadvantages of object scanning are that imaging can be rather slow, taking perhaps a few seconds to record a single frame, and that use of electrical probes is more difficult. The speed consideration is perhaps not so important when it is realized that often long scan times are necessary in order to collect sufficient light to produce a low-noise image. We can perhaps anticipate a revival of object-scanning methods in the future.

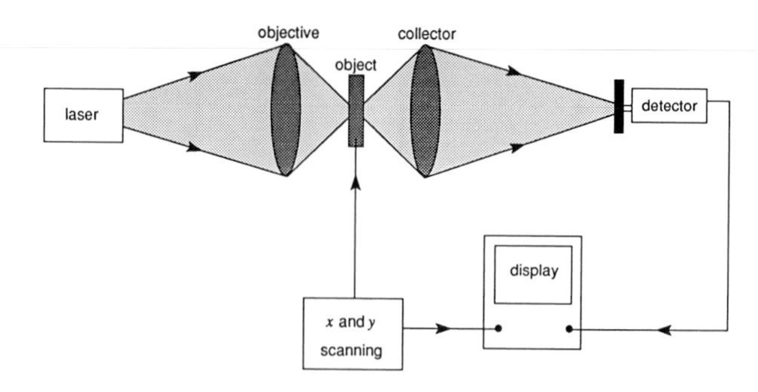

FIGURE 1 A schematic diagram of a scanning optical microscope.

The final method of scanning to be considered is mechanical scanning of the objective lens. This again has the advantage that the light travels on-axis through the optical system, thereby giving good quantitative imaging. On the other hand, speed is again rather low.

C. Design of Scanning Systems

A number of alternative arrangements are used for beam-scanning microscopes. Usually a microscope eyepiece, or lens of similar design, is used to produce a focus which subsequently illuminates the objective lens. In order that the beam fills the objective lens aperture, the axis of rotation of the scanning mirrors must be situated close to the plane of the entrance pupil of the eyepiece. A single mirror placed here and scanned in both the x and y directions is the simplest design. Alternatively, two separated galvomirrors can be used, coupled by a telecentric system consisting of two lenses or mirrors. Finally, two close-coupled galvomirrors can be used if the axis of rotation of one is offset so that it both translates and rotates.

In an object-scanning system, the optical system is simplified as the system is operated in an on-axis condition. It is found that it is possible to optimize performance by small corrections to the effective tube length of the objective. Mechanical object scanning is usually achieved using electromechanical devices. One feature which must receive some attention is the necessity of ensuring that the plane of scanning is accurately located. This can be achieved by mounting the specimen stage on leaf springs or stretched wires.

D. Alignment

An optical system gives its best performance only if it is accurately aligned. In particular a confocal microscope must be aligned correctly or it will exhibit artifacts, especially with a small pinhole size. First of all the beam-expanding system must be adjusted and the beam collimated so that the objective is used at its correct tube length. The best way of aligning the system is to examine a planar object with a pinhole of large diameter. Without refocusing the object the pinhole size is reduced and its axial and transverse position adjusted for maximum signal.

E. Photodetectors

Many types of detectors may be used for scanning optical microscopy. If signal level is high, as for example in transmission or in reflection from a surface such as bone or tooth, a simple photodiode may be employed. For confocal fluorescence applications signal level is low, and is reduced further as the thickness of the optical section is reduced. Then highly sensitive detectors such as photomultiplier tubes, perhaps cooled, and with photon counting are advantageous.

The relevant properties of a detector are quantum efficiency, sensitivity, dynamic range, and linearity. Conventional photocathodes have a quantum efficiency of only about 20%, but this figure can be raised to above 30% by using negative electron affinity photocathodes. Silicon detectors of either the photodiode, avalanche photodiode, or charge coupled device (CCD) variety can have quantum efficiencies greater than 80%. Avalanche photodiodes in principle combine high quantum efficiency with the ability to photon count with pulse height analysis for rejection of dark current. However, at present there are difficult practical problems in their use. Charge-coupled devices combine high quantum efficiency with the facility for integration within the device and the possibility of cooling to improve noise performance. CCDs are usually of the area or linear array variety, and in some cases additional information can be extracted from this spatial information.

F. Choice of Objectives

At present there are no objective lenses specially designed for scanning microscopy, and so objectives must be chosen from commercially available types. For laser microscopy the objective must be corrected for only one wavelength (or two, including a fluorescent wavelength), which should allow lenses of increased numerical aperture or working distance to be developed. Similarly in microscopes with on-axis scanning, off-axis aberrations are unimportant, giving further flexibility in the design. For example, an immersion lens, designed for one wavelength and with a limited field of view for on-axis scanning, with numerical aperture of 1.4 and working distance of 1 mm, seems feasible.

Achromat lenses have some advantages over apochromat lenses for some applications as they have lower loss, as well as being considerably cheaper. We have found fluorite lenses to be a good compromise for on-axis scanning. However, for beam scanning the off-axis aberrations and field curvature are too large. Again with beam-scanning confocal microscopes, there is a need for high-aperture lenses (to give good collection efficiency) of low magnification. These are not necessary with on-axis microscopes as then one objective can be used to cover the whole range of magnifications, simply by altering the amplitude of scan. Confocal microscopes are more sensitive to aberrations than conventional instruments, particularly for imaging in the depth direction. So incorporation of a correction collar is useful. Finally, for many biological studies water immersion objectives are preferable, because with uncovered specimens this removes both the reflection and aberrations produced by the cover glass–specimen interface.

G. Image Processing for Scanning Microscopy

First of all it is worth pointing out that many of the early scanning laser microscopes did not use a computer for image manipulation, but rather used analog electronics together with a long-persistence display. The main impetus for employing digital techniques was to provide image storage to improve the real-time observation of images, but in fact very high-quality images with a resolution of several thousands of lines can be produced by photographing directly the cathode ray tube display. Similarly many forms of image processing, such as contrast enhancement, filtering, image addition, and subtraction, can be achieved using analog methods. The use of digital methods of course greatly extends the flexibility of the system, but nevertheless analog-processing facilities are worth retaining in order to improve the dynamic range of the recorded data.

A single 512 × 512 image contains a quarter of a megabyte of information. This means that a three-dimensional (3D) image consisting of many sections requires large memory and also processing time for 3D manipulation. For this reason images which consist mainly of "empty space" can be stored and processed alternatively using a vector scan method in which only the nonzero contributions to the image are stored with their coordinates. Actually, good stereoscopic effects can be obtained from few sections, and it is also possible to store projections directly, rather than the sections themselves, which can also greatly reduce the quantity of data.

Most scanning microscopes generate a single image in a few seconds, which means that with a commercial imaging system a slow-scan input is necessary. Alternatively, personal computers nowadays have large enough memories to permit direct storage of the images in their random access memory (RAM). Personal computers can provide most of the functions necessary for scanning microscopes, but are not fast enough for real-time manipulation of 3D images.

A range of standard image enhancement methods can be used with advantage in scanning optical microscopy. These include contrast enhancement by linear stretching or histogram equalization, edge enhancement by filtering in either the spatial or Fourier domain, low-pass or median filtering to reduce noise in images, image averaging, and so on. Most of these methods can be employed with 3D as well as 2D data sets.

Once a 3D image has been stored projections in arbitrary directions can be produced by image rotation. This is a computationally intensive process, and an alternative is to stack sections with an appropriate pixel offset between adjacent sections. The sections may be stacked by summation, corresponding to the extended focus method described in Section III,B, or by selecting the peak signal, corresponding to the autofocus method.

II. Imaging Modes of Scanning Microscopy

A. Introduction

In a scanning microscope the object is illuminated with a focused spot of light. Such an arrangement is extremely versatile as an image can be generated from a wide range of different effects of this illuminating spot. Examples include photoelectron imaging, photoacoustic imaging, photothermal imaging, and photodesorption studies. A further method of great application in the semiconductor industry involves detecting the current (or voltage) generated by the incident radiation. This technique could also have applications in the biological area. Here we discuss in more detail two particular imaging modes. The first involves the detection of light at a different wavelength from that incident. The second uses detector arrays to give differential phase contrast.

B. Spectroscopic Methods

An advantage of using scanning methods for spectroscopic imaging is that imaging, performed with the incident radiation, is separated from wavelength selection and analysis of the emitted radiation, thus simplifying system design and resulting in superior performance. The detection system, because it does not have to image, may also have greater sensitivity. This is of great advantage in fluorescence microscopy, which also results in the further advantage that the resolution is determined by the shorter incident wavelength, rather than the longer fluorescence wavelength.

Fluorescence, or luminescence, microscopy can give information concerning spatial variations in excitation states, binding energies, band structure, molecular configuration, structural defects, and the concentration of different atomic and molecular species.

Use of a pulsed laser allows investigation of transient effects such as the lifetime of excited states, and capture and emission cross-sections. Other examples of spectroscopy which may be performed using scanning techniques include absorption spectroscopy, Raman spectroscopy, resonance Raman spectroscopy, coherent anti-Stokes Raman spectroscopy (CARS), two-photon fluorescence, photoelectron spectroscopy, and photoacoustic spectroscopy.

C. Differential Phase Contrast

Image formation in a scanning microscope (nonconfocal) is in principle identical to that in a conventional microscope. However, one difference is that the detector sensitivity distribution of a scanning microscope can be made of negative strength, which is not true for the source intensity distribution of a conventional microscope. An example is the arrangement of Fig. 2, in which the detector is split into two halves. In the absence of a specimen each half receives an equal signal, so that if they are subtracted one from the other there is no net signal.

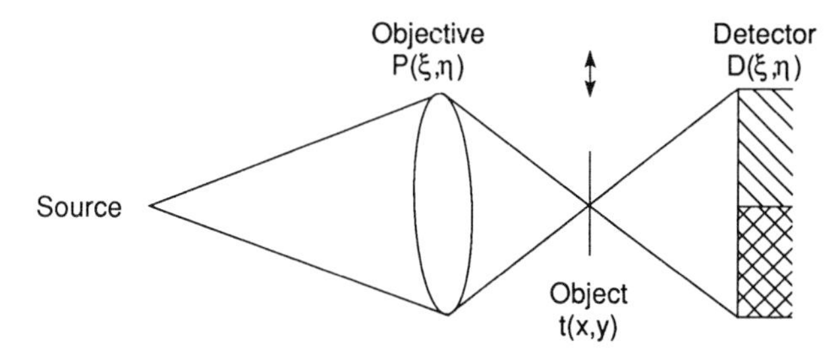

FIGURE 2 The split detector method for achieving differential phase contrast.

If an object consisting of a phase wedge (i.e., a prism) is inserted, the beam on the detector is deflected, and a net signal results which is proportional to the phase gradient. Adding the two signals gives a conventional brightfield image. The split detector technique allows the observation of weak phase structure. It is extremely sensitive: for example, it can show up the edge of a monomolecular film. It has various advantages over the alternative Nomarski differential interference contrast (DIC) method, most notably that the phase information can be more easily extracted. By electronic integration of the phase gradient information, an absolute phase image can also be produced. The method can also be used in reflection, to show up surface topography. By altering the configuration of the detector the system can be optimized for different properties, for example weak phase gradients or fine detail.

III. Confocal Microscopy

A. Confocal Imaging

In a conventional microscope the object is illuminated using a large-area incoherent source via the condenser, and each point of the object imaged by the objective lens. It is the objective which is responsible for determining the resolution of the system. If the image is now measured point by point by a detector of small aperture the image is unchanged, providing the detector is small enough. In a scanning microscope, however, we have a point source and a large-area detector, rather than a large-area source and a point detector. Reciprocity argues that image formation is identical in a scanning microscope and a conventional microscope. However, now it is the first lens (also termed the objective, but sometimes the projector) which determines the resolution.

In the confocal microscope we use both a point source and a point detector, achieved by placing a pinhole in front of the detector, so that in this case both lenses take part equally in the image formation process and the resolution is improved by a factor of about 1.4. Other advantages of confocal microscopy are, first, that out-of-focus information is rejected by the pinhole so that an optical section is imaged and, second, that unwanted scattered light is also rejected by the pinhole. Although the system shown in Fig. 1 is a confocal transmission system commercial systems usually operate in the reflection mode. This makes the operation of the system much easier as the point detector can be arranged to coincide with the image of the point source in the beam splitter, resulting in coincidence of the illuminating and detection spots.

B. Optical Sectioning and Three-Dimensional Microscopy

Light emanating from regions of the specimen separated from the focal plane are defocused at the pinhole plane and hence rejected, thus resulting in an optical sectioning effect. By scanning in the depth direction sequential sections can be studied and a complete three-dimensional image built up. The 3D data can, for example, be stored in a computer for subsequent processing and display.

Of course, 3D images are difficult to display directly so instead we can extract sections oriented in some arbitrary direction. For example *xz* images are sections parallel to the system axis. We can also produce projections in an arbitrary direction, in which the depth information is suppressed. Projections can be produced either by summation of sections, resulting in an extended focus image, or alternatively by detecting the peak signal in depth, giving an autofocus image. Either approach can be achieved using digital or analog methods, and an extended focus image can also be generated directly by photographic integration on a cathode ray tube. Two projections produced at slightly differing angles either by the extended focus or autofocus method give a stereoscopic image pair.

As well as recording the peak signal in depth, as in the autofocus method, we can record the depth position in order to locate its position in depth. This depth information can be displayed as a surface profile image in gray levels or color, or can be combined with the peak information to produce color-coded images or reconstructed views. The sensitivity of the depth measurement is about 50 nm, or can be better than 0.1 nm using confocal interference techniques.

Projections in different directions can be computed and stored for animated display, but at present information content of a full 3D image is too great to permit real-time rotation.

By using a flat reflecting surface such as a mirror as specimen the imaging performance in the depth direction can be investigated by axial scanning and observation of the resulting defocus signal. This can alternatively be achieved by tilting the mirror slightly, scanning in a transverse plane, and examining the line-scan signal, or by producing an *xz* image and again extracting a line scan. The defocus signal gives much information about the performance of the optical system. Ideally it should be a smooth, narrow response with weak sidelobes. In practice it is made broader as a result of the finite size of the pinhole and often degraded by lack of alignment or the presence of aberrations. In particular the presence of strong sidelobes or auxiliary peaks is very detrimental to three-dimensional imaging performance. Confocal microscopy is very sensitive to the presence of small amounts of aberration such as spherical aberration or astigmatism, which can be introduced by the use of incorrect cover glass thickness, or the effects of the mounting medium or focusing deep into the specimen. In these cases observation of the defocus signal allows optimization of the imaging performance by altering the tube length at which the objective is used or by insertion of correction lenses.

C. Confocal Fluorescence Microscopy

At present most commercial confocal microscopes are designed for fluorescence operation, so that to many people confocal and fluorescence are almost synonymous. However, it should be stressed that there are many useful nonfluorescent applications using confocal brightfield techniques. The main advantages of the fluorescence mode of confocal microscopy is that specific stains can be introduced, but also high-quality 3D images can be produced without problems arising from coherent optical noise (speckle).

As described earlier, resolution of a scanning fluorescence microscope is superior to that in a conventional fluorescence microscope. There is a further improvement for confocal fluorescence microscopy, so that for example if the fluorescence wavelength is 1.5 times the primary wavelength the cut-off in the spatial frequency response is 1.5 times as great in a scanning compared with a conventional fluorescence microscope, but 2.5 times as great in a confocal fluorescence system. The price one pays for this dramatic improvement in resolution is the decreased signal strength of the confocal arrangement.

Most commercial confocal fluorescence systems use an argon ion laser to provide a main line at 488 nm which is close to the absorption peak of fluorescein isothiocyanate. The other strong line of the argon ion laser at 514 nm can be used to excite rhodamine or Texas Red. An alternative laser for fluorescence work is the helium cadmium laser, which gives lines at 442 and 325 nm. The former can be used with acridine orange, fluorescein, and Feulgen-Schiff staining, the latter with Hoechst or DAPI. Either helium-neon (633 nm) or krypton ion (647 nm) lasers can be used to excite chlorophyll *b*. The krypton ion laser can give a large number of lines throughout the visible and ultraviolet. A further laser of interest is the frequency-doubled neodymium YAG laser at 532 nm.

D. Effects of Pinhole or Slit Size

As the pinhole of a scanning microscope is reduced in size the imaging performance more nearly approximates true confocal imaging, but the signal strength also decreases so that in practice some compromise setting is necessary. The important parameter is the size of the pinhole compared with the diameter of the Airy disk in the pinhole plane, so that the absolute size of the pinhole needed to obtain a specific effect depends on the geometry of the system. We introduce a normalized pinhole radius v_d given by

$$v_d = \frac{2\pi r_d \,\mathrm{NA}}{\lambda\, M}$$

where r_d is the true radius of the pinhole, NA is the numerical aperture of the objective, λ is the wavelength, and M is the magnification between the pinhole and object planes. Use of a pinhole with v_d equal to about 3 gives a good compromise in performance.

As the pinhole size is increased the resolution in both transverse and axial directions is degraded. For the axial direction resolution for a planar object decreases monotonically with increasing pinhole size, and for large pinholes the width of the response increases linearly. On the other hand, for a pointlike object axial resolution does not become significantly worse for v_d greater than about 10. This explains how 3D restoration techniques can be used for some forms of object with conventional fluorescence microscopes.

Some designs of microscope use a slit aperture rather than a circular pinhole. While this can give an increased signal relative to a pinhole of diameter equal to the slit width, the resolution is also degraded and overall no benefit is gained in this respect. Nevertheless slit apertures can be used to advantage in various ways to obtain real-time image formation.

E. Confocal Brightfield Reflection Microscopy

Confocal microscopy in the brightfield (nonfluorescence) reflection mode has many applications in the materials science and industrial areas, and can also be exploited in biological studies. The tandem scanning microscope is often used in a brightfield mode as signal levels are frequently low in the fluorescence mode. Reflected brightfield has the advantage of giving extremely sharp depth imaging, sharper than in confocal fluorescence, but in some cases there can be difficulties in interpretation as a result of coherent noise (speckle). This is not usually a problem when the specimen consists of reflecting surfaces, but can be when it is necessary to investigate refractive index variations within a semitransparent object. The coherent noise is reduced by using the autofocus, or more effectively by the extended focus, technique and in some cases the visibility of the image features can be improved by filtering.

A further important technique of confocal reflection microscopy is the use of immunogold probes. In this case the gold particles are spatially well separated so that coherent interactions are not important, and the gold particles can be located in three-dimensional space. Scanning microscopes can be very sensitive for detecting scattering from these gold particles, and preparations with particles as small as 5 nm in diameter without silver enhancement have been imaged. In many cases, however, the lower limit to the size of the particles is set by the requirement that the scattering, which varies with the fourth power of the diameter, is strong compared with scattering by the refractive index variations in the specimen itself.

F. Confocal Transmission Microscopy

Although much of the early work on confocal microscopy was performed in the transmission mode, this is now not nearly as frequently used as the confocal fluorescence technique. The reason for this is that there is extreme difficulty in aligning, and maintaining alignment during scanning, caused by refractive effects in the specimen. This can be to some degree alleviated by using a double-pass method, where the transmitted light is reflected back through the specimen and detected using the usual reflected light detector. This method can also be used to increase the signal in confocal fluorescence microscopy.

Confocal transmission retains the resolution improvement of confocal imaging, and also results in an improvement in depth imaging, but not to such a marked degree as in confocal reflection. Confocal transmission microscopy can also be performed in a differential phase-contrast mode, and by mixing with a coherent reflection signal can result in further improvement in depth imaging.

G. Confocal Interference Methods

Scanning techniques are well suited to interference methods, and in particular confocal interference microscopy is a powerful technique. This is possible because confocal imaging is a coherent process, and is further simplified experimentally by the long coherence length of lasers. Confocal methods have the major advantage for interference microscopy that the shape of the reference beam wavefront is immaterial, only its phase and amplitude at the detector pinhole being of importance, removing the requirement for matched optics and making alignment much less critical. Furthermore, it is possible to combine the system with multiple detectors and real-time processing to extract image information. Confocal interference microscopy can be used for investigation of refractive index or surface height variations, or for obtaining the signal phase for restoration of images or measurement of system aberrations.

Confocal interference microscopy can be performed in transmission, using a Mach-Zehnder arrangement, or in reflection using a Michelson geometry. High sensitivity can be achieved using either phase-shifting or heterodyne technique. The heterodyne technique also exhibits an imaging property, which allows nonconfocal images to be formed without an objective lens being necessary. This may prove useful in focusing deep into a specimen which would be impossible with a real lens. Alternatively combining illumination of the object by a focused spot and heterodyne detection can result in confocal imaging without the use of a physical pinhole.

As described earlier the axial imaging performance of a confocal microscope can be investigated by observation of the variation with defocus in signal using a reflecting plane as specimen (the defocus signal). Confocal imaging is a coherent technique, so that if only the intensity of the defocus signal is measured the phase information is lost. However, if interference methods are used to extract the phase and amplitude, the aberrations of the imaging system can be determined by a simple Fourier transformation of the defocus signal. This is useful for optimization of the imaging system, or to provide information for image reconstruction.

Bibliography

Inoué, S. (1986). "Video Microscopy." Plenum, New York.

Pluta, M. (1988). "Advanced Light Microscopy. Specialized Methods," Vol. 2. Elsevier, Amsterdam and New York.

Sheppard, C. J. R. (1987). Scanning optical microscopy. *Adv. Opt. Electr. Microsc.* **10,** 1–98.

Wilson, T. (ed.) (1990). "Confocal Microscopy." Academic Press, New York.

Wilson, T., and Sheppard, C. J. R. (1984). "Theory and Practice of Scanning Optical Microscopy." Academic Press, New York.

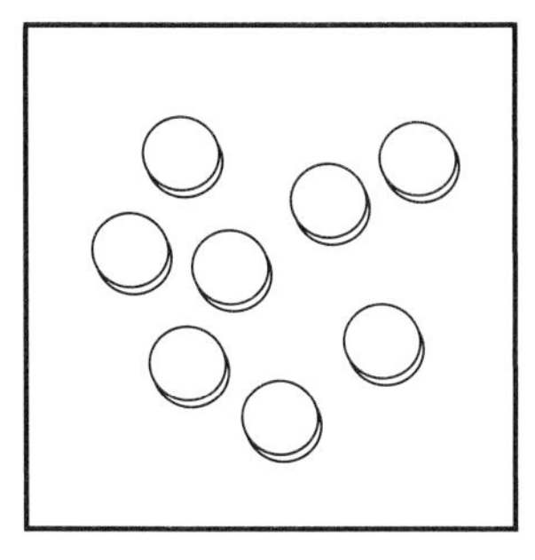

Schizophrenia, Psychosocial Treatment

KIM T. MUESER, *Medical College of Pennsylvania at Eastern Pennsylvania Psychiatric Institute*

ROBERT P. LIBERMAN, *University of California at Los Angeles; Brentwood VA Medical Center*

Glossary

Behavioral family therapy Intervention provided to families to improve their ability to manage the illness effectively and minimize the negative impact of the illness on the family unit

Expressed emotion Stressful communication of negative affect from relatives to the family member with schizophrenia, which is hypothesized to increase the risk of those patients to symptomatic relapses

Milieu therapy Remediation of social deficits in psychiatric patients through modification of the immediate environment to create a "therapeutic community" conducive to patients' acceptance of social responsibility

Psychodynamic treatment Individual treatment approach oriented to improving social functioning by fostering insights into unconscious processes underlying past and current interpersonal relationships

Relapse Return of schizophrenic symptoms to a patient whose symptoms were previously in remission, or the exacerbation of symptoms in a chronically symptomatic patient

Social learning theory Employed in behavioral approaches to psychosocial interventions, a set of principles assumed to govern the acquisition of socially appropriate behavior in the natural environment; behaviors are learned through a combination of observation of others (modeling), positive social reinforcement for certain behaviors (reward), and negative reinforcement for certain behaviors

Social skills training Intervention developed to improve the social competence of persons through behavioral rehearsal (practice) of the skills

Stress–vulnerability–coping skills model Postulates that the outcome of schizophrenia is determined by a dynamic balance among biological vulnerability, environmental stress, and the ability of patients to cope effectively with the effect of stress

Token economy Environmentally based treatment approach whereby desirable social behaviors are rewarded by the provision of "tokens," which are exchangeable for material goods and privileges, and undesirable behaviors are suppressed by "fining" patients through tokens

THE PSYCHOSOCIAL TREATMENT of schizophrenia is coupled with judicious types and doses of antipsychotic medications to improve the course and outcome of this chronic illness. The utility of an array of psychosocial interventions is based on evidence that factors in the environment, such as interpersonal stressors, influence the course and outcome of schizophrenia. Psychosocial interventions are aimed at remediating social impairments resulting from the illness and decreasing socioenvironmental stressors impinging on the patient. To achieve this, a diversity of techniques have been developed that are applied to either the patient, the family, or the environment in which the patient resides. Relevant to the central goal of improving the outcome of schizophrenia, psychosocial treatments also are aimed at reducing the stress and burden on relatives who are directly responsible for the patient's care and whose behavior may have favorable or adverse impacts on the course of the illness itself.

I. Introduction

A. The Course of the Illness

Schizophrenia is a severe psychotic illness, most often with an onset in late adolescence or early adulthood and a chronic course throughout adult life. The social functioning of afflicted patients is always impaired, but characteristic symptoms of the illness (e.g., hallucinations, delusions, incoherence) fluctuate over time, requiring periodic hospitalizations for most patients. While the pathogenesis of schizophrenic symptoms is assumed to be biological in nature, psychosocial treatments are based on evidence that socioenvironmental factors influence the onset and course of illness. [*See* SCHIZOPHRENIC DISORDERS.]

Psychosocial interventions for schizophrenia can be guided by a multidimensional, interactive model of the disorder: the stress–vulnerability–coping skills model. According to this model, symptoms and their associated social impairments are the result of stressors impinging on a person's enduring biological vulnerability. The noxious effects of these stressors are modulated by a person's social competence and the amount of social support available to him or her. The appearance or worsening of schizophrenic symptoms and disabilities may be caused by changes in the environment, behavior, and biology of an individual, such as the following: (1) stressful life events or daily levels of tension intervene and overwhelm the individual's ability to cope in social and instrumental roles (e.g., critical or overinvolved family relationships); (2) the individual's social support network weakens or diminishes (e.g., family member dies, therapist terminates, patient leaves home); (3) social problem-solving skills that were previously in the patient's repertoire are lost as a result of disuse, poor motivation, or reinforcement of the sick role; (4) the underlying biological vulnerability increases or is physiologically stressed (e.g., patient abuses alcohol or psychotomimetic drugs).

Thus, the symptomatology and social functioning of persons with a biological vulnerability for schizophrenia, at any point in time, are determined by the amount and type of life stressors, on the one hand, and the social problem-solving capacities of these persons and the availability of social support, on the other hand. Either too much environmental change, stress, or ambient tension or not enough coping skills and social support can lead to a symptom exacerbation and loss of social and occupational functioning.

B. Role of Psychosocial Treatment

The significance of the stress–vulnerability–coping skills model of symptom formation lies in the emphasis given to the active role of the patient's coping skills and support system, both of which suggest targeted objectives and modalities for therapeutic intervention. The clinician can prescribe neuroleptic drugs to buffer the underlying biological vulnerability. When necessary, environmental modification can be employed to ameliorate the negative effects of stressors on a vulnerable person. In such instances, hospitalization can be beneficial because the patient is removed temporarily from stressors in family and community settings. Alternatively, treatment can emphasize strengthening the patient's social support network, as is the case in family and group therapies and self-help clubs. Finally, increasing the patient's resilience through training in social and problem-solving skills can also reduce the likelihood of symptom relapse.

Psychosocial treatments are an essential component of rehabilitation for schizophrenics. Although treatment with neuroleptic medications reduces the risk of relapse, approximately 35–40% of schizophrenic patients who are compliant with neuroleptic medications still relapse within a year. Furthermore, the side effects of neuroleptics reduce adherence, even in patients responsive to the beneficial effects of the drugs. Most importantly, drugs cannot teach life and coping skills, nor can they improve the quality of a person's life, except indirectly through suppression of symptoms. Most schizophrenic patients need to learn or relearn social and personal skills for surviving in the community and reducing the risk of symptom relapses. Many patients also require psychosocial interventions to reduce their abuse of alcohol or drugs, which may worsen the illness.

A final role for psychosocial intervention is to remediate basic social impairments due to the illness. Social inadequacy correlates with poor symptomatic and behavioral outcomes, as well as rehospitalization. Interpersonal problem-solving appears to be markedly deficient in schizophrenic persons, especially in limitations in generating alternative ways of responding to situational challenges. Deficits in social skills may include misperception of relevant social cues and poor cognitive processing of these cues, leading to inadequate generation of response alternatives and inappropriate behavioral responses to others in the situation. Poor social problem-solving may partly reflect core attentional and psychophysiological impairments in schizophrenia and are important targets for modification in psychosocial treatment programs.

II. Taxonomy of Psychosocial Interventions

Psychosocial interventions can be organized according to their focus, locus, and modus, as well as to their goals and objectives. The focus can be on the individual as in one-to-one therapy, group therapy, family therapy, or a total milieu. The locus of therapeutic and rehabilitative efforts can be the hospital, private office, clinic, mental health center, natural home, board-and-care home, or social club. The modus, or method, of intervention can derive from one or more explicit or implicit orientation, such as behavioral, psychodynamic, family system, client-centered, or supportive therapy. It is recognized that modalities of treatment may overlap considerably, and that much of the therapeutic impact of any psychiatric or medical treatment derives from nonspecific effects that are inherent in therapy that is offered in a credible, hopeful, and positive manner. Focus, locus, and modus of treatment may change over time as the specific problems and needs of the patient and available resources change. Also, at any point in time, multiple foci, loci, and modalities may be harnessed to implement a comprehensive rehabilitation plan.

A wide spectrum of goals and objectives are reflected in the variety of psychosocial interventions provided to persons with schizophrenia. Some treatments aim primarily to maintain a person at a marginal level of functioning by minimizing stress and risk of relapse. Other treatments provide crisis counseling and assertive outreach to help patients meet immediate survival needs in the community. Still others attempt to build social and independent living skills. Thus, the process of designing, employing, and evaluating psychosocial interventions requires reference to at least four mutually exclusive domains of attributes; for example, family therapy (focus) may be provided in the home (locus), with a behavioral orientation (modus) that aims to improve the problem-solving and communication skills (goals) of the patient and relatives. Psychosocial treatments of schizophrenia will be reviewed according to the focus of each intervention: individual treatment, milieu therapy, group therapy, and family therapy.

III. Individual Treatment

A. Traditional Psychotherapy

A range of different psychotherapies have been tried for schizophrenic patients, but most traditional approaches have met with limited success in improving the outcome of the illness. The clinical efficacy of psychodynamic treatment for schizophrenia has been debated since soon after the development of psychoanalysis in the early twentieth century. Recent controlled research on exploratory psychodynamic treatment for schizophrenic outpatients has overcome past methodological weaknesses by providing treatment by experienced clinicians over extended periods of time, and by utilizing objective diagnostic criteria and standard outcome measures of symptomatology, occupational and social functioning, and recidivism. The results of these studies uniformly demonstrate that psychodynamic therapies, at best, confer little benefit to schizophrenic patients and may actually worsen the course of the illness and adaptive functioning of some patients. The results of long-term psychodynamic interventions conducted in carefully crafted residential milieus for schizophrenia have not been more encouraging than in outpatient treatment studies. Thus, exploratory, insight-oriented psychodynamic therapies do not currently have an empirically supported role in the residential, inpatient, or outpatient treatment of schizophrenia. Early applications of cognitive psychotherapy for schizophrenia have produced promising results, but controlled outcome studies are lacking. Social skills training, which can be conducted either individually or in groups of patients, will be described under group therapy.

B. Case Management

The comprehensive treatment of schizophrenia must include a range of medical and psychiatric services to remediate primary symptoms, to rehabilitate social and self-care skills, to provide basic medical, housing, and social services, and to maintain continuity of care through various stages of the illness. Case management is the process of linking these needs for treatment and community support with resources available in the hospital or community. Assisting schizophrenic individuals to maximize their use of existing resources can enable them to increase their independence and the quality of their lives. Case management also helps to ensure accountability, accessibility to resources, efficiency, and continuity of care. The requirements of effective case management are summarized in Table I.

An important facet of case management for schizophrenia is the development of social supports that can act as a buffer against the aversive effects of environmental stress. Schizophrenics are particularly vulnerable to disruptions in their social networks. Patients recovering from an acute phase of the illness often fail to re-establish their former network because of the social stigma of the illness and deficits in social skills. Because a fundamental aspect of schizophrenia is that patients have difficulty or are unable to identify and articulate many of their most basic needs, the case manager assumes this vital role as coordinator of psychosocial and somatic treatments. In some treatment programs, the process of case management is extended beyond the traditional inpatient or outpatient treatment setting to locales in the community where patients reside. Such assertive outreach treatments are reviewed next.

C. Assertive Outreach Programs

To address the problem of schizophrenic patients' inability to advocate for themselves, assertive outreach programs have been developed to aid patients' functioning in the community. The hallmark of these approaches is the locus of treatment, which is almost always in the community, at locales such as homes, streets, day programs, and community rehabilitation facilities and residences where the basic focus is on helping patients meet their basic living needs. In addition, intervention is provided on an as-needed basis, by an interdisciplinary team, 7 days a week, with intensive case management, crisis intervention, and patient advocacy as the main modalities of treatment. The most widely disseminated and emulated program has been the Training in Community Living (TCL) program in Madison, Wisconsin. In controlled studies, the TCL program has been found to reduce symptoms and rates of rehospitalization and to modestly increase the capacity to work.

The dramatic reduction in rehospitalization rates for schizophrenic patients treated in an Australian replication of the TCL program is depicted in Figure 1. In addition, cost–benefit analyses have shown that assertive outreach programs are more economical than traditional case management approaches for recidivistic patients. However, when

TABLE I Requirements of Effective Case Management

- Client identification and outreach
- Individual assessment
- Service planning
- Linkage with required services
- Monitoring of service delivery
- Patient advocacy

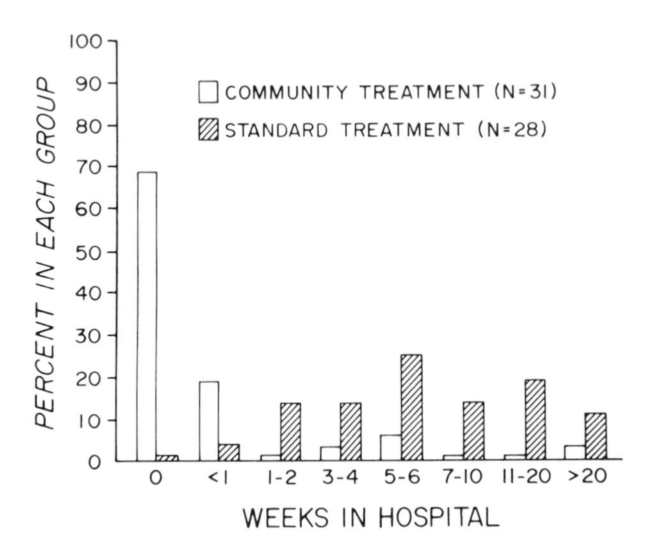

FIGURE 1 Fifty-nine schizophrenic patients presented for admission to a psychiatric hospital were randomly assigned to two groups: community treatment, and standard hospital plus aftercare. Community treatment patients were not admitted to the hospital, if possible, and were provided comprehensive treatment and 24-hour crisis services. These patients achieved superior clinical outcomes, and along with their relatives were more satisfied with their treatment than standard-care patients. Patients in community treatment spent substantially less time in the hospital, a difference that was highly statistically significant ($P < 0.001$). [Adapted from Hoult J. Reynolds I, 1984, Schizophrenia: A comparative trial of community-oriented and hospital-oriented psychiatric care, *Acta Psychiatr. Scand.* **69,** 359.]

the demonstration program was terminated, the differences favoring the experimental patients evaporated rapidly. This finding underscores the need to sustain psychosocial support indefinitely, just as neuroleptic drugs often must be maintained on a long-term basis. It remains to be seen whether or not the use of social learning principles in such community support programs can improve acquisition, durability, and generalization of living skills.

The TCL model has been successfully adapted for even more chronic patients who have histories of multiple hospitalizations and difficulty sustaining community tenure. For example, the Bridge Program was developed as a community outreach treatment based in the urban environment of Chicago for the most highly recidivistic psychiatric patients. Only patients with at least three admissions in the past year were admitted to the program. Results from the program suggested that it effectively lowered rehospitalization rates. Seventy-one members treated at the Bridge Program for 1 year had their hospitalizations reduced from an average of 3.3 in the year preceding the program to 1.9 the following year, with a corresponding drop in the average time spent in the hospital from 106.9 days to 43.1 days. The economic savings by avoiding costly inpatient treatment more than paid for the program, with savings of almost $6,000 per patient per year.

IV. Milieu Therapy

A. Elements of Effective Milieus

The locus for milieu therapy is a living, learning, or working environment. Examples include the inpatient psychiatric unit, day hospitals, psychosocial rehabilitation clubs, board-and-care homes, community-based residential alternatives to hospitals, and sheltered workshops. The defining characteristics of treatment and rehabilitation milieus are the use of a team to provide treatment and the large amount of time spent by the patient in the environment. Recent adaptations of milieu therapy have included 24-hour-per-day programs that are situated in the community locales frequented by patients and that provide support, case management, and training in living skills.

Milieu therapy, or the therapeutic community, may be based on any one of a number of modalities ranging from structured behavior therapy to spontaneous, humanistically oriented approaches. Most programs encompass the following attributes: (1) emphasis on group and social interaction, (2) rules and expectations that are mediated by peer pressure for normalization of adaptation, (3) blurring of the patient role by viewing patients as responsible human beings, (4) emphasis on patients' rights for involvement in setting goals, for freedom of movement, and for informality of relationships with staff, and (5) emphasis on interdisciplinary participation and goal-oriented, clear communications.

Data has emerged that describes the elements of effective therapeutic milieus with schizophrenics. These elements can be divided into those related to milieu structure and those related to treatment procedures. Structural elements associated with favorable outcomes include (1) small size of milieu and patient census, (2) high staff–patient ratios and staff stability, (3) heterogeneity of patient population, with an optimal mix of two-thirds higher functioning of acutely ill patients and one-third lower functioning of chronic patients, and (4) clarity and consistency in status and roles among staff. Other programmatic elements that appear to be related to successful treatment are (1) active participation by patients and nursing staff, (2) administrative commitment to short stays (i.e., 3 months or less for the average hospital episode, and (3) outplacement of patients requiring long-term, custodial care.

Treatment process variables correlated with good outcome include high levels of staff–patient interaction, with the focus of interactions on adaptive, practical aspects of everyday behavior, rather than on symptomatic and psychodynamic issues. Treatment units that set clearly defined and time-limited goals with patients are more effective, as are units that organize and schedule prosocial activities for most waking hours. A variety of psychosocial treatment models have been developed and tested and are considered more effective than standard or comparison approaches. These models include the social learning–token economy program and therapeutic communities such as the Fountain House social club (described later).

B. Social Learning–Token Economy

For chronic, treatment-refractory, long-stay schizophrenics who have resisted all efforts at deinstitutionalization, the psychosocial strategy of choice is social learning–token economy. Utilizing behavioral assessment and therapy, highly trained paraprofessionals and nursing staff have been successful in remediating the bizarre symptoms and social and self-care deficits of most chronic schizophrenic patients. In a rigorous study conducted on the token economy, a social learning–token economy program is compared with an equally intensive therapeutic milieu, and a customary care group.

Over 100 severely debilitated and chronically institutionalized patients were treated by a paraprofessional staff who rotated between the units to control for nonspecific personalities of the therapists. Staff–patient ratios were similar to those used in custodial institutions. The social learning program employed a highly specific token economy with many hours of structured educational activities throughout the day and evening. Patients were rewarded with tokens for engaging in socially adaptive behaviors, which could then be exchanged for material goods or privileges. The therapeutic milieu followed principles of peer pressure and democratic decision-making. Both programs were offered in 28-bed units in a regional psychiatric hospital. Most of the patients received inpatient care for at least 1 year and then obtained aftercare for 6 months after their discharge into the community.

A multimodal assessment battery revealed impressive and clear-cut results favoring the social learning–token economy approach. Improved functioning, enabling long-term community tenure, occurred in 97% of the social learning patients. The therapeutic milieu program was less effective, but its 71% release and community maintenance rate was still a favorable outcome when compared with the 45% rate of patients released from custodial care and living in the community for 18 months or longer. The success in sustaining patients in the community was mirrored by the significant clinical and behavioral improvements, which even produced a minority of patients who, by direct behavioral observational ratings, could not be distinguished from a normal population. After only 14 weeks of treatment, every resident in the social learning program showed dramatic improvements in overall functioning, regardless of usual prognostic indicators, such as duration of hospitalization.

By the end of the second year of programming, fewer than 25% of residents in either experimental program were on maintenance psychotropic drugs. Two clinically significant conclusions reached by the study were that (1) most older chronic mental patients, when provided active and structured psychosocial therapies, did not require maintenance neuroleptic drugs, and (2) clinical improvements resulted, not from *how much* attention and enthusiasm was offered to patients by staff, but *how* that attention was given. Patients in the therapeutic milieu program received more overall attention but improved less than their counterparts in the social learning program.

Social learning principles have been extended into the community where loci of treatment include day hospitals and community support programs. One such program, associated with a veterans administration hospital, offered structured and scheduled classes in which social, vocational, and survival skills were taught. Rehospitalization rates were only 10% for the patients participating in the behavior-oriented educational program, whereas the rates were 53% in a comparison group receiving traditional aftercare services. In a similar program located in a community mental health center, specific goal-setting and active behavioral training of social and community living skills led to increasing numbers of clinical goals attained during a 24-month follow-up period, whereas matched patients from a more traditional day hospital showed decrements in goal attainment over the 24-month period. Treatment procedures based on this program have been widely adopted by community mental health centers throughout the United States.

C. Therapeutic Communities

Some efforts to develop a milieu conducive to improving schizophrenics' social and independent living skills have resulted in therapeutic communities in which patients are involved in the administration and ongoing operation of the program. An early effort to demonstrate that a comprehensive learning-based therapeutic community could be effective with chronic mental patients used an approach in which the hospital was a base for initial training of patients in interpersonal skills, effective decision-making, and group governance and cohesion. Following this training, the patients were transferred as a group to living quarters in the community. These quarters resembled a lodge, and the patients were assisted in gradually assuming full responsibility for its maintenance and operation. They bought and prepared food, kept financial records of income and expenditures, and consulted community-based physicians and agencies for their medical, psychiatric, and support needs. Live-in staff faded themselves out of the picture until the lodge residents were functioning autonomously. In addition, the patients earned money by starting a local business that provided janitorial services, yard work, general hauling, and painting. A 40-months follow-up indicated that the patients trained to live and work in a community lodge sustained significantly greater time outside the hospital and in gainful employment.

Another approach to the therapeutic community, termed psychosocial rehabilitation, emerged during the late 1940s when ex-patients began to meet together in a social club in New York City to satisfy their needs for acceptance and emotional support. Emphasizing self-help, mutual interdependence, and reliance on assets, the movement led to the establishment of Fountain House, which has spawned hundreds of similar programs. The main assumption of this approach is that patients have a fundamental right to work and that employment facilitates community adjustment and reduces symptoms. Employment opportunities are provided both in the clubhouse (e.g., food preparation, switchboard) and by transitional jobs available in the community, with no limitation on the length of participation in the program. These transitional jobs are opportunities for club members to work temporarily en route to full-time employment elsewhere or to work on a longer-term basis in the entry-level position.

An 18-month follow-up evaluation of club members working in transitional jobs revealed that 16% were employed independently on a full-time basis, and an additional 45% continued part-time work in the transitional program or were attending school or other training programs. Only 2% were in a psychiatric hospital at the time of the 18-month follow-up. Another evaluation of the psychosocial club model found that 38% of members were rehospitalized during a 2-year follow-up period, in comparison to a 60% rehospitalization rate for a contrast group. Members of the club also had significantly lower rehospitalization rates 5 years later; those who were hospitalized from the club spent 40% fewer days in the hospital than did the rehospitalized control subjects. Despite these positive results, methodological problems with self-selection, lack of diagnostic clarity, and nonrandomly assigned control groups limit the conclusions that can be drawn from these studies.

V. Group Therapy

A. Modes of Therapy

There are as many schools of group psychotherapy as there are of individual therapy. Group therapy can be characterized by its theoretical and operational qualities and procedures. Some group therapies are highly structured and employ a behavioral orientation, other groups are unstructured with a psychodynamic and insight-oriented focus, and still others are primarily supportive in nature. The locus for most therapy groups is the hospital, clinic, or private office, although multiple family groups have met in homes and storefronts, and groups emphasizing peer support and normalization often meet in community centers, schools, and churches.

The goals of therapy groups overlap considerably, with varying degrees of emphasis placed on insight, behavior change, and skill development; social support and maintenance; and participation in recreational activities. What cuts across most modalities is the group leader's use of the naturally developing interactional dynamics in groups, such as cohesion, to strengthen the group process and to improve the outcomes. A large body of evidence collected from groups serving a spectrum of patient populations suggests that cohesion has a generically favorable impact on group therapy, an impact that is similar to the therapeutic alliance between patient and therapist in individual therapy.

With the trend toward brief inpatient hospitalizations for schizophrenic patients, followed by continuing care in the community, almost all group psychotherapy takes place in the aftercare period. Exploratory and psychodynamic group therapy during the inpatient period may worsen the clinical state of patients who are still floridly psychotic and, thus, vulnerable to overstimulation and hyperarousal from their treatment environment. Group therapy is likely to be more beneficial when offered after symptoms, such as delusions and hallucinations, have been controlled, and if it focuses on practical, everyday problems of living experienced by patients trying to adjust to the community. Most outpatient groups aim at supporting a patient's stabilization and community tenure, assisting the patient in coping with stressful life events, and facilitating efforts at longer-term rehabilitation.

While controlled research studies have shown that insight-oriented group therapy is not efficacious for schizophrenics, the beneficial effects of more socially interactive groups on outcome criteria such as symptomatology, rehospitalization, or vocational or social adjustment remains to be established. The consensus of clinicians, buttressed by controlled studies, is that employing group therapy during the aftercare, outpatient phase of treatment is more effective than during inpatient treatment. Group therapy formats have been adapted broadly for use in providing such services as social skills training, medication evaluations, occupational and recreational therapy, patient and family education, patient government, self-help, and mutual support.

B. Social Skills Training

1. Training Methods

The most highly structured form of group therapy for schizophrenic patients is social skills training. The goals are explicit, the session agendas are usually planned in advance, the procedures follow written guidelines often derived from a manual, and *in vivo*[1] practice and homework assignments are emphasized. Social skills can be defined as those interpersonal behaviors required (1) to attain instrumental goals necessary for community survival and independence and (2) to establish, maintain, and deepen supportive and socially rewarding relationships. Schizophrenia disrupts one or more of the affective, cognitive, verbal, and behavioral domains of functioning and thereby impairs a person's potential for enjoying and sustaining interpersonal relationships, which are the essence of the social quality of life. Recurring schizophrenic disorders pose enduring social disruptions for affected persons. These disorders involve symptoms that adversely affect the schizophrenic patients' social quality of life and also evoke impairments that hamper learning or relearning adaptive social behaviors. Applying behavior analysis principles to identify and remediate deficits in social behaviors, clinicians have developed treatment packages, termed social skills training, that have proved effective with schizophrenic patients.

In virtually all published reports of social skills training, role playing (a simulated social encounter) is the vehicle used both to assess patients' pretreatment social competence and to train targeted behavioral excesses or deficits during treatment. Training scenes are selected either on the basis of the individual's past difficulties or from problem situations that have been found to apply to most patients. Training sessions vary in length from 15 to 120 minutes, depending on the number of patients participating and on their level of functioning. Although the group format provides vicarious learning opportunities through observation of other patients' behavior, as well as from amplified reinforcement from peers, the group experience is sometimes supplemented by individual training; such training allows more intensive focus on a single patient's behavior and provides an opportunity for more practice within sessions.

Participants in the role playing include the target patient(s), a respondent, and the therapist. A combination of focused instructions, modeling (demonstration), feedback, and social reinforcement are applied as a "package" to remediate deficits in social behavior. Modeling and feedback are provided by group participants or through videotape playback. Target behaviors selected for change usually include both nonverbal and paralinguistic behaviors (e.g., eye contact, voice loudness or intonation, response latency, smiles) and content behaviors (e.g., requests for change, highlighting the importance of a need, empathic responses, compliance, hostile comments, irrelevant remarks). The efficacy of social skills training lies in the specific, functional, and goal-oriented nature of the behaviors that are targeted for change. The sequence of steps in a social skills training session is outlined in Table II. Social skills training techniques have been packaged in modules—comprising a trainer's manual, patient's workbook, and demonstration videocassette—to increase their dissemination and utilization by mental health and rehabilitation practitioners.

2. Research Findings

Extensive research has examined the effects of social skills training with schizophrenic patients. Many of the studies used single-case experimental designs, which inherently limit the generality of the findings. Some studies are also limited in external validity because they failed to adequately define the patient population and control for medication effects; however, there is convergence across studies with respect to several findings. First, schizophrenic patients in a treatment setting can be trained to improve social skills in specific situations. Second, moderate generalization of acquired skills to similar situations can be expected from the training. When patients are encouraged to use the skills they have learned in their natural living, learning, and working environments and when they are reinforced by their peers, relatives, and caregivers for employing their skills, generalization is promoted.

Clearly, evidence indicates that, given the reinforcing learning environment structure by behav-

1. Practice in actual social situations, such as in the community or conversing directly with a treatment provider (e.g., physician).

TABLE II Procedures for Structured Social Skills Training in Groups

1. Specify the interpersonal problem in each group member in turn by asking the following:
   a. What emotion, need, or communication is lacking or not being appropriately expressed?
   b. With whom does the patient want and need to improve social contact?
   c. What are the patient's short- and long-term goals?
   d. What are the patient's rights and responsibilities?
   e. Where and when does the problem occur?
2. For each patient in the group formulate a scene that simulates or recapitulates the features of the problem situation. The scene should include the following characteristics:
   a. Constructed as a positive goal
   b. Functional for the patient
   c. Frequently occurring in the patient's life
   d. Specific
   e. Consistent with the patient's rights and responsibilities
   f. Attainable
3. Observe while the patient and surrogate role players (other group members) rehearse the scene. During this "dry run," therapists should position themselves close to the action so that they can make an assessment.
4. Identify the asserts, deficits, and excesses in the patient's performance during the dry run. Praise assets and efforts and solicit positive feedback from other group members.
5. Assess and train "receiving" and "processing" skills by asking the patient:
   a. What did the other person say?
   b. What was the other person feeling?
   c. What were the patient's short-term goals?
   d. What were the patient's long-term goals?
   e. Did the patient obtain these goals?
   f. What other alternatives could the patient use in this situation?
   g. Would one of these alternatives help the patient reach their goals?
6. Employ modeling, using therapist or other group members to demonstrate potentially effective alternatives using expressive and adaptive "sending" skills.
7. Highlight the desired behaviors being modeled and review "receiving," "processing," and "sending" skills, and ask other group members and targeted patients to repeat this procedure.
8. Rerun the scene with the patient, giving positive feedback to reinforce progress and effort and soliciting further positive feedback from other group members.
9. Use coaching and nonverbal prompts to shape behavioral changes and improvements in small increments, starting at the patient's present level.
10. Focus on all dimensions of social competence in training "sending" skills:
    a. Topical content and choice of words and phrases
    b. Nonverbal behaviors
    c. Timing, reciprocity, and listening skills
    d. Effective alternatives
11. Generalize the improvements in competence by the following procedures:
    a. Repeating practice and overlearning
    b. Selecting specific, attainable, and functional goals and scenes
    c. Providing positive feedback for successful transfer of skills to real-life situations
    d. Prompting the patient to use self-evaluation and self-reinforcement
    e. Fading the structure and frequency of the training
    f. "Programming" for generalization in the natural environment

ioral principles, most schizophrenics can acquire or relearn social and conversational skills. Learning, however, occurs tediously or little when patients are still floridly ill and are highly distractable. Third, participants in the training consistently report decreases in social anxiety after training. Fourth, follow-up evaluations indicate that durability of acquired social skills depends on the duration of training; thus, overlearning that stems from repeated practice promotes retention, but retention is unlikely to occur for brief training of less than 2–3 months of twice weekly sessions. Last, social skills training has been found to significantly lower the rate of relapse in controlled studies and to improve a variety of indices of social functioning.

There are also findings that limit the applicability

of social skills training. Generalization of complex conversational skills is less likely than that of briefer, more discrete verbal and nonverbal responses. Because complex behaviors are more critical for generating social support in the community, methods have been developed to improve the learning and durability of conversational skills; these training methods focus on problem-solving and perceptual and information-processing skills. Another challenge to the broad use of social skills training is the limited evidence that generalization of skills occurs in the natural environment of persistently psychotic patients in the absence of prompting and planned supportive reinforcement.

VI. Family Therapy

Early theories regarding the etiology of schizophrenia postulated that the family played an important role in the development of the illness. As evidence mounted that the illness was biological in nature and transmitted genetically rather than interpersonally, the focus of research and treatment shifted to examining family factors that influence the course of the illness. In a series of carefully designed, cross-culturally replicated studies, critical, intrusive, or emotionally overinvolved attitudes and feelings of relatives—termed high expressed emotion—have consistently been found to be a powerful predictor of relapse in schizophrenia. This communication of negative affect from family members to the patient, which increases patients' risk of relapse, is a consequence, in part, of the burden relatives experience while caring for a chronically ill patient. This burden has increased dramatically since the discovery of neuroleptic medications in the 1950s, which allowed the majority of schizophrenics to be treated in the community.

Several modes of family therapy have been designed and empirically tested for their ability to change the emotional climate of the family, reduce the burden of the illness on relatives, and improve the outcome of the illness. These new, behavior-oriented, psychoeducational approaches to family therapy have a distinctively different rationale than the earlier clinical studies that implicated family structure, communication, and relationships in the etiology of schizophrenia. In fact, as part of the initial contacts with the family group, the newer family therapies emphasize that no etiological link exists between family relations and the development of schizophrenia. Instead, the stress–vulnerability–coping skills model of schizophrenia is described to explain how the stress of an already established major mental illness can place burdens on patient and relatives alike, thereby raising tension levels in the family. Given the fragile coping capacity of the index patient, the increased stress reverberating throughout the ambient family emotional climate can lead to a relapse of symptoms. The dynamic interplay among family stress, patient vulnerability and symptomatology, and burden of the illness on the family is illustrated in Figure 2.

A common feature of effective approaches to family therapy is an emphasis on educating the patient and family members about the nature of schizophrenia and its available treatment. Time is spent on demystifying the varied symptoms, signs, and prognoses associated with the disorder and on translating the neurobiological underpinnings into

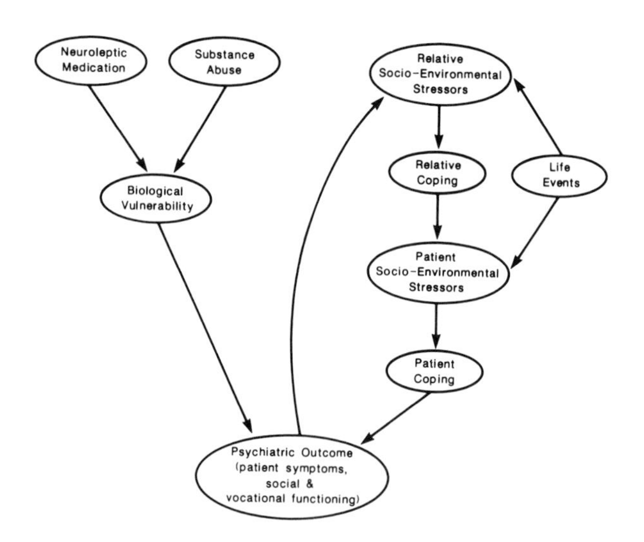

FIGURE 2 The stress–vulnerability–coping skills model of schizophrenia is illustrated. The outcome of the illness (symptoms, social and vocational functioning) is determined by patients' biological vulnerability and their ability to cope with environmental stressors impinging on them. Biological vulnerability may be increased by the abuse of illicit substances (e.g., amphetamines, cocaine) and decreased by neuroleptic medications. The ability of patients to cope with socioenvironmental stressors, such as life events or a hostile emotional climate in the family, impacts on the illness. Just as stress has a negative impact on the patient, severe stress on the family, such as that caused by a floridly ill patient at home, can overwhelm the coping efforts of the family, leading to greater stress on the patient and an increased risk of symptom relapse. Social skills training focuses on improving patients' ability to cope effectively with stress. Behavioral family therapy focuses on decreasing tension in the home by improving family members' ability to cope with stressors and to manage the illness of schizophrenia.

TABLE III Component Interventions Comprising Behavioral Family Management

Behavioral analysis of all members of the family
Defining individual and family-wide problems
Setting goals for each individual and family system
Identifying reinforcers
Pinpointing assets and resources within individuals, family, and community
Education of all family members about schizophrenia and currently available treatment and rehabilitation modalities
Training in communication skills
Expressing positive feelings to others and acknowledging when others do or say something positive toward you
Active, reflective listening
Making positive requests and asking for what you want
Expressing negative feelings in constructive ways
Training in problem-solving skills
Be specific and objective in describing the problem
Express how you feel directly and subjectively about the problem
Listen to each other actively and reflectively as the problem is described and feelings are expressed
Help each other generate alternatives and options in dealing with the problem
Weigh the potential consequences or outcomes (risks and benefits, pros and cons) of each alternative
Choose a reasonable alternative
Decide how to implement the alternative
Behavioral interventions for specific problems
Contingency management for negative symptoms
Job-finding skills training
Friendship skills training
Independent living skills training

lay terminology. The role of neuroleptic drugs in the treatment and prophylaxis of schizophrenia is highlighted, and an effort is made to improve adherence to the pharmacotherapeutic regimen. Some therapists prefer to meet with the relatives alone for initial sessions and to invite the patient to join in later when acute symptomatology has been controlled and the span of attention increased. Therapists provide educational and other interventions with individual families in multiple family groups, and during day-long or evening "survival skills workshops."

The methods of family therapy run along a continuum in terms of the systematic use of behavioral learning principles and the use of family systems theory; however, they have in common educating relatives and patients about the illness, teaching family members skills for communicating effectively and solving problems together, and using the principles of social skills training (Table III). In contrast to purely educational family treatments, the behavioral method does not assume that stressful negative affective communication by relatives can be lowered by the provision of information alone and aims instead to improve the quality of family interactions. As in social skills training with patient groups, behavioral family therapy sessions are highly structured with preplanned agendas and homework assignments to program generalization of learned skills. Assessments are routinely conducted on all family members and the family as a unit, with the principal goal of improving each member's ability to achieve their own personal goals.

Most recently developed family therapies for schizophrenia have been shown to produce highly significant reductions in relapse rates. In families where relatives are high in expressed emotion and patients are at increased risk for relapse, the base rate of relapse can be expected to fall in the 50–60% range. Educational and behavioral therapies that employ skills-training procedures have reduced relapse rates substantially, even while using lower doses of neuroleptic drugs. Recently, four controlled clinical studies have documented the efficacy of family therapy for schizophrenia over traditional treatment of patients living with high expressed emotion relatives. The relapse rates of these studies are summarized in Table IV. In each study, schizophrenic patients who received family therapy had significantly fewer relapses over 2 years following a hospitalization than patients who received routine treatment, including neuroleptic medications. Research has not yet compared two

TABLE IV Two-Year Cumulative Relapse Rates (in percentages) of Schizophrenics Receiving Family Therapy versus Routine Treatment

Theoretical orientation	Family therapy	Routine treatment	Principal investigator
Educational–behavioral	17	83	Falloon, 1985
Educational–supportive	14	78	Leff, 1985
Educational–family systems	32	67	Hogarty, 1987
Educational–behavioral	33	59	Tarrier, 1989

different types of family therapy, nor is it known whether or not the benefits of therapy persist after it has been discontinued. However, the strong effects of family therapy on relapse rates, replicated across different modes of therapy, indicate that family therapy may be one of the most potent psychosocial treatments for schizophrenia.

VII. Integrating Psychosocial and Pharmacological Treatments

Evidence from many studies supports the conclusion that, when combined with rationally prescribed neuroleptic drugs, properly designed psychosocial treatment offers greater protection against relapse and higher levels of social adjustment than drugs or psychosocial treatment alone. Neuroleptic drugs have a primary effect on cognitive disorganization, hallucinations, and delusions but have less impact on impairments in psychosocial functioning. The opposite seems to be the case with social and psychosocial therapies. In combination, their beneficial impact on the comprehensive needs of the schizophrenic patient is additive.

The critical time to offer psychosocial treatment is during the aftercare period, when the patient needs assistance in surmounting the problems and stressors of readjusting to family and community. Psychosocial treatment is most helpful for patients who are in reasonably good states of partial or full remission from florid symptoms and who have reached stable levels of maintenance medication. During acute flare-ups of symptomatology, treatment should be aimed at reducing ambient stress on the patient, reducing levels of social and physical stimulation, and assisting the patient to integrate and understand the symptoms as part of an illness process.

Psychosocial treatment should be long-term. Its benefits do not become apparent before 12 months and are even greater after 2 years. It is likely that indefinite psychosocial support, guidance, and training are optimal for most chronic schizophrenic patients. As neuroleptic drugs are most effective in maintaining symptomatic improvement when continued indefinitely, it is not surprising that psychosocial rehabilitation efforts are similarly optimized by continuity.

The most effective psychosocial treatment—whether provided by individual therapy, group or family therapy, day hospital, or inpatient milieu therapy—contains elements of practicality, concrete problem-solving of everyday challenges, low-key socialization and recreation, engagement of attainable tasks, and specific goal orientation. A continuing positive relationship is central in the overall strategy for treating the schizophrenic patient, no matter how much drug or psychosocial treatment contributes to the overall plan. This relationship may be with the prescribing psychiatrist or with a paraprofessional case manager.

Psychosocial treatment providers must routinely assess alcohol and drug abuse and, when possible, intervene to reduce its negative influence on the course of illness. Schizophrenic patients are prone to "self-medicating" their symptoms and anhedonia (incapacity for experiencing happiness) by abusing psychoactive substances, particularly stimulants. The result of substance abuse is often the precipitation of relapses requiring rehospitalization. Psychosocial treatment should aim to decrease patient's vulnerability to substance abuse by methods including: education regarding the effects of substance abuse on the symptoms of schizophrenia and likelihood of relapse, social skills training to improve assertive behavior in refusing peer pressure to abuse drugs or alcohol, helping patients develop alternative leisure activities, expanding patients' social support networks, and linking patients and family members with self-help organizations such as Alcoholics Anonymous.

Finally, psychosocial treatment should focus on stressors in the environment and deficits in personal characteristics that seem to play specific roles in relapse and community maladjustment. Schizophrenic relapse is common even when drug compliance is firmly established. Nor is there any evidence that a patient's level of manifest psychopathology at hospitalization or discharge predicts subsequent relapse. The best explanation, based on converging lines of evidence from empirical studies, is that the patient's personal assets and deficits, the social environment, and the type of psychosocial therapy are the most powerful influences on relapse, even in the face of reliably administered maintenance medication.

Bibliography

Anderson, C. M., Reiss, D. J., and Hogarty, G. E. (1986). "Schizophrenia and the Family." Guilford Press, New York.

Bellack, A. S. (ed.) (1989). "A Guide to the Treatment of Schizophrenia." Plenum Press, New York.
Falloon, I. R. H., Boyd, J. L., and McGill, D. W. (1984). "Family Care of Schizophrenia." Guilford Press, New York.
Liberman, R. P. (ed.) (1988). "Psychiatric Rehabilitation of Chronic Mental Patients." American Psychiatric Press, Washington, D.C.
Liberman, R. P., DeRisi, W. J., and Mueser, K. T. (1989). "Social Skills Training for Psychiatric Patients." Pergamon Press, New York.
McFarlane, W. R. (ed.) (1983). "Family Therapy in Schizophrenia." Guilford Press, New York.
Mueser, K. T., and Berenbaum, H. (1990). Psychodynamic treatment of schizophrenia: Is there a future? *Psychol. Med.* **20,** 253–262.
Mueser, K. T., Yarnold, P. R., Levinson, D. F., Singh, H., Bellack, A. S., Kee, K., Morrison, R. L., and Yadalam, K. G. (1990). Prevalence of substance abuse in schizophrenia: Demographic and clinical correlates. *Schizophr. Bull.* **16,** 31–56.
Paul, G. L., and Lentz, R. J. (1977). "The Psychosocial Treatment of the Chronic Mental Patient." Harvard University Press, Cambridge, Massachusetts.
Stein, L. I., Test, M. A. (eds.) (1985). "The Training in Community Living Model: A Decade of Experience." Jossey-Bass Inc., San Francisco.

Schizophrenic Disorders

ERMIAS SELESHI AND JAMES W. MAAS, *University of Texas Health Science Center at San Antonio*

Glossary

Anticholinergic Blocking effect of nervous system function mediated by the neurotransmitter acetylcholine

Extrapyramidal system Part of the central nervous system controlling certain aspects of voluntary motor behavior

Neuroleptic Drug possessing antipsychotic effect. The term means to "grasp the neuron" from its action on the nervous system

Paranoia Specific fear or suspicion not founded on reality, often associated with other disturbances of thought

Pathognmonic Characteristic that is a unique or specific identifying feature of a disease

Psychosis Disturbances of thought characterized by false beliefs (delusions), perceptions (hallucinations), and loss of order in thought processes

Psychotomimetic Capacity or potential of inducing psychotic symptoms

Tardive dyskinesia (TD) Abnormal involuntary movements of voluntary muscles such as the tongue, face, neck, diaphragm, and limbs, most commonly associated with prolonged use of neuroleptic drugs

SCHIZOPHRENIA is a disorder of behavior characterized by disturbance of perception, thought processes, and reality testing. The term, as originally coined, meant split mindedness, to describe the dissociation between thought, emotion, and behavior. This led to the popular misconception of split personality, equating schizophrenia to the condition currently known as multiple personality disorder. An estimated 1% of the population of the United States are victims of this disorder, which costs approximately 2% of the Gross National Product, mostly because of recurrent hospitalizations and secondary loss of the productive capacity of patients. Only a tiny fraction of research funds are allocated to the study of schizophrenia when compared with other less common and less costly diseases. Social stigma and gradual socioeconomic decline of affected individuals and their families give the distorted impression that the disease is more prevalent among the poor and minority ethnic groups. Available data, however, indicate that schizophrenia has no socioeconomic or ethno-cultural boundaries. Aggregate data obtained from empirical observations, biological studies, and the wide variability in the severity, course, and treatment response indicate that schizophrenia is the manifestation of a group of heterogeneous disorders. Genetic and other environmental factors may predispose an individual to develop the disorder or modify its clinical manifestation.

I. Clinical Features

A. Early Description and Terminology

The term *schizophrenia* was introduced in 1911 by Eugene Bleuler, one of the two European psychiatrists who are credited for the early description of the clinical picture and phenomenology of the disorder. However, many of the other frequently used terminologies associated with schizophrenia (e.g., *paranoia, hebephrenia,* and *dementia precox*) did appear in the latter half of the 19th century. The Swiss psychiatrist, Emil Kraepelin, applied the term *dementia precox* to describe various psychotic states with an early age of onset in the teens or twenties and a progressive decline of mental function, resulting in a clinical state reminiscent of dementia. He distinguished schizophrenia from manic depressive illness, which is characterized by episodic psychotic disturbance with symptom-free intervals and a lack of progressive mental deterioration. In addition to providing a detailed description of the symptoms of schizophrenia, Kraepelin wrote of a small percentage of patients who may experience complete recovery or run a benign favorable course.

Bleuler's notions differed from those of Kraepelin in several ways. He applied psychoanalytic interpretations to the behaviors manifested by schizophrenic patients and focused on the severity and duration of the illness rather than its clinical course. He proposed four cardinal diagnostic features of psychotic disturbance (i.e., looseness of associations of thought content, autism, disturbance of affect, and ambivalence) often referred to as the "four A's" of Bleuler. He categorized hallucinations, delusions, and catatonia as nonspecific accessory symptoms. Thus Bleuler's criteria were broader and more inclusive. Both pioneering workers believed in an underlying organic cause of schizophrenia.

B. Clinical Diagnosis

None of the symptoms of schizophrenia are specific or pathognomonic for the illness. Thus diagnosis is based on the presence and duration of a cluster of symptoms, on the clinical course over time, and when possible, by documentation of certain personality traits or prodromal changes in behavior antedating the onset of symptoms. The disorder usually begins in late teens or early twenties. Unlike in its predecessor, in the current Diagnostic and Statistical Manual Revised Version (DSM-III-R), the uncommon onset after the fifth decade of life is accepted. The age of onset and clinical course have both therapeutic and prognostic implications.

The cardinal features of psychotic symptoms in schizophrenic disorders include disturbances of thought, perception, language, psychomotor function, mood, and affect. Additionally, eating disorders and so-called soft neurologic signs could be present. Soft neurologic signs cannot be attributed to a specific area of brain abnormality and are therefore referred to as nonlocalizing. They include asymmetric coordination difficulties with fine motor tasks (e.g., clumsiness in counting the fingers rapidly on one side of the body). The core category of symptoms is referred to as negative symptoms and will be discussed later. These contrast with the positive symptoms of thought and perceptual disturbances. Impairment of judgment and insight is often associated both with acute psychotic symptoms and the residual phase of the illness.

The disturbance of thought can be broken down into form and content. Form refers to the overall structure of the thought process. The schizophrenic patient may manifest looseness of associations, whereby the verbalization of the patient's thoughts lack logical order of cause and effect. Topics are abruptly changed and bear no clear association to the issue being discussed. The patient's trends of thought may abruptly be interrupted for a few seconds, and the patient may be unable to continue from the point of interruption. Language function can also be affected. The patient may use complex terms inappropriately (stilted speech), create new words (neologism), repeat words or statements several times (echolalia), or be verbose without conveying much information (poverty of content). The ability to understand and interpret abstract concepts becomes impaired, and ideas are oversimplified (concrete thinking). Although memory and orientation is preserved, the acutely psychotic patient may manifest impairment of attention.

The disturbance in the content of thought is broadly referred to as delusion, in which the patient displays firm beliefs that have no basis in reality. These may take the form of paranoia, whereby the patient believes he or she is being persecuted by familiar persons, institutions, government agencies (especially law enforcement), or by alien beings, such as extraterrestrial creatures, or the devil. The delusion can be extremely elaborate and often bizarre, taking on a science fiction quality in which patients may describe how they are being spied on through various equipment (i.e., TV, radio, ovens, etc.) in their surroundings. Not uncommonly, real life events, such as being arrested or being fired from a job because of abnormal behavior, get incorporated into the delusional system and reinforce the belief. Another form of delusion is ideas of reference, in which patients infer a special significance about events around them. These may include the ways others communicate with them, the arrangement of furniture in a room, contents of news broadcasts, or prevailing weather conditions. Patients may believe that their ideas and thoughts are removed from their minds (thought withdrawal) or broadcast on news media (thought broadcasting). In addition, they may believe that alien thoughts or feelings are being forced on them (thought insertion) or that they possess telepathic powers to understand and influence the behavior of others. At times the content of delusional material takes on a grandiose quality. In such instances, patients believe they possess tremendous wealth and physical or political power, have a special religious or political mission, or have the ability to influence events around them (magical thinking). Somatic delusions refer to the patient's belief in suffering from terminal illness, deformity of the body, or destruction of internal organs. In summary, delusional thought contents can take on many forms, from a well-organized and elaborated system with some internal logic to bizarre and fragmented beliefs much more diverse than what has been highlighted here.

Perceptual disturbances consist of perversion of sensory modalities in the absence of actual environmental stimuli. These symptoms are referred to as hallucinations and may involve any of the five senses (i.e., hearing, touch, vision, smell, or taste). Auditory hallucinations are the most common form, and the last three are more frequently associated with an organic brain disorder. A single patient may have more than one form of hallucination simultaneously or at different points in time. The hallucinations may be elaborated and be part of the patient's delusional system. As an example, a patient may believe that he or she receives verbal instructions from controlling forces, such as the President or God. Frequently patients report one or more continuous, critical, praising, or threatening voices in reference to the patient (i.e., a running commentary). Voices instructing patients to carry out certain tasks are called *command hallucinations*. Not uncommonly, patients may only hear single words, such as their names, muffled voices, foreign languages, and animal or machinery noises. Visual hallucinations consist of either fully formed animated appearances of humans, animals, or alien beings or fragments of such (i.e., faces, skeletons, shadows, flying objects, etc.). Patients do not usually readily admit the presence of delusions or hallucinations, thus inferences are made based on their outward behavior. They may carry on a conversation or appear to be responding to an unidentifiable stimulus.

Various types of deviant psychomotor behavior are encountered in schizophrenia. Catatonia is a type of motor behavior in which patients assume a still position for long periods of time (catatonic stupor) or are extremely agitated and in constant vigorous movement (catatonic excitement). Catatonic patients may resist attempts to move their limbs (negativism) or maintain any position to which a limb is moved (waxy flexibility). Catatonic features are less frequent now than they used to be, probably because of the advent of antipsychotic drugs and early treatment of patients. Other abnormal motor behaviors that are encountered include stereotyped, repetitive gesticulations and echopraxia, in which the patient will imitate movements and positions assumed by the examiner. Although signs and symptoms of depression or anxiety may be associated with schizophrenia, in general the mood and affect displayed by the patients may not be reflective of the thought content, and this is referred to as inappropriate affect. Patients may show little or no emotion while talking about issues that would be expected to generate strong feelings (e.g., fear, sadness, happiness) in the average person. This is described as blunting or flattening of affect.

Nonspecific behavioral changes that are frequently encountered include a decline in the patient's personal appearance, hygiene, and social grace. These features are often associated with the so-called negative symptoms of schizophrenia, encompassing social withdrawal, lack of motivation, poverty of speech, and an emotional distancing from others. These groups of symptoms are responsible for a major part of the disabling effect of the illness, as they impair patients in their educational, occupational, and social functioning. More significantly, unlike the positive symptoms, the negative symptoms respond poorly to treatment. Impairment in executing sound judgment, as well as insight about the illness, are a common feature in both the acutely psychotic and chronic schizophrenic patient.

C. Course and Long-Term Outcome

Schizophrenia runs a chronic relapsing and remitting course. In general, unlike mood disorders, there is no return to premorbid level of function during periods of remission. It is not uncommon to observe symptoms of depression when the patient is recovering from an acute psychotic episode. Ten percent of schizophrenics die from suicide, and four times as many attempt it during their lifetime. Unlike the popular misconception, homicide is not more frequent among schizophrenics than in the general population. Although treatment noncompliance is probably one of the most common causes of relapse, other factors may be contributory. Environmental stress and concurrent use of alcohol or other illicit drugs play some role. [*See* DEPRESSION; SUICIDE.]

There appears to be a cumulative effect of deterioration during the first few years of recurrent relapse, which stabilizes as time goes on. Positive symptoms become less marked although up to one-third of patients will continue to experience them in a substantial way. The progressive deterioration, associated with emergence of the functionally and socially disabling negative symptoms, is sometimes referred to as deficit state. Socioeconomic decline of the schizophrenic patient has given rise to the downward drift theory. This phenomenon may be explained by the affliction of young patients, before they achieve their educational and vocational potentials, and subsequent failure of the majority to become self-sufficient because of recurrent psychotic relapses. Many gravitate downward into poverty, and a sizable portion of the homeless in urban centers in the United States are schizophrenics.

The disorder, however, is not invariably bleak in outcome. Various studies have indicated that up to a quarter of patients achieve significant recovery and lead relatively normal lives, and an equal number function with mild to moderate symptoms. The remaining 50%, nonetheless, are significantly impaired for life. Recovery and overall improvement have been reported among chronically ill, deinstitutionalized patients over a long period of prospective follow-up. In general, factors such as older age and acute manner of onset associated with apparent precipitant factors when symptoms initially appear, favorable premorbid function, significant positive, depressive and paranoid symptoms, as well as being married with a good social support system, confer favorable outcome. Family history of schizophrenia, history of birth injury, and younger age of onset with a chronic, relatively unremitting course are associated with poor prognosis.

II. Classification

Over the years many terms have been applied to describe clinical subtypes of schizophrenia primarily based on clinical presentation. Some of these have prognostic significance in their treatment response or long-term course. Paranoid schizophrenia is characterized by prominent positive symptoms of paranoid delusions. Catatonic schizophrenia or catatonia is the subtype with the peculiar psychomotor symptoms of stupor or excitement. Both paranoid and catatonic categories have a favorable outcome in their response to treatment. The patient with hebephrenic schizophrenia manifests gross disorganization of behavior and thought processes with a disinhibited, inappropriate, and silly affect. Simple schizophrenia, however, has prominent features of negative symptoms with minimal and occasional, if any, hallucinations and delusions. The hebephrenic and simple subtypes carry poor prognosis. The term *latent or borderline schizophrenia* has been applied to patients with enduring odd personality traits and occasional thought disorders. Pseudoneurotic schizophrenics have prominent anxiety symptoms with some underlying thought disorders, phobias, and sexual identity conflicts.

The DSM-III-R introduced in 1987 is the current guideline for the diagnosis of psychiatric disorders in the United States. However, there are other closely related guidelines, such as the Research Diagnostic Criteria and the Schedule for Affective Disorders and Schizophrenia, primarily used in clinical research settings. In addition to the presence of a constellation of symptoms, the DSM-III-R requires temporal criteria to be met to subtype diagnostic groups, as well as to classify the course of illness. Prodromal and residual symptoms of change in behavior, as well as impairment of function, are also included in the scheme. To diagnose schizophrenia, a combination of positive symptoms and prodromal or residual features should be documented for at least 6 months. The course of the illness is classified as subchronic, if less than 2 years, or chronic, if longer. The clinical picture over the long course is further subclassified depending on periods of improvement or worsening of symptoms. Five subtypes of the disorder are noted. The paranoid and catatonic types closely resemble the subtypes in older criteria. The disorganized type is equivalent to hebephrenia. Patients with the undifferentiated type manifest prominent positive symptoms or grossly disorganized behavior and do not fulfill criteria for other subtypes. Unlike the old subtype of simple schizophrenia, positive symptoms are prominent in the undifferentiated group. The fifth subtype, residual type, is used for patients who show little or no positive symptoms but display an apparent impairment from negative symptoms. The criteria emphasize that organic causes, including drug abuse and other psychiatric diagnoses such as manic-depressive illness, be excluded before the diagnosis of schizophrenia is made.

III. Genetics

There is compelling evidence that genetic predisposition confers vulnerability to the development of schizophrenia. The pattern of inheritance is not clearly understood but is believed to be a Mendelian pattern with incomplete penetrance. The search for a chromosomal marker has been elusive to date. Recent restriction fragment length polymorphism studies in families with multigeneration expression of the disease suggested a locus on chromosome 5. This finding, however, has not been replicated by other workers. Many family and twin studies have been carried out both in the United States and Europe, supporting the contribution of genetic factors. The findings indicate that even in adopted apart siblings and twins raised in separate environments, the risk of developing the disease remains the same as those raised by their biological parents. Increasing genetic load confers a higher likelihood of becoming schizophrenic. The risk of a nontwin sibling of a schizophrenic patient is 8%, compared with the 1% prevalence rate in the general population. This rate rises to 12% if a parent or a dizygotic twin is affected. The child of two schizophrenic parents has a 40% chance of becoming schizophrenic, while the concordance rate in monozygotic twins is 50%.

IV. Epidemiology

The wide diversity of diagnostic criteria applied has resulted in a wide range of incidence and prevalence rates of schizophrenia worldwide. The estimated incidence of schizophrenia in the United States is one per 1,000 population, and the lifetime prevalence is approximately 1%. However, only a small fraction of the schizophrenic patient population receives any form of treatment, and the majority of the patients receiving treatment require hospitalized care. Although there is no difference in prevalence between the sexes, the peak age of onset in men, 15–25 years old, is 10 years younger than in women. The disease rarely manifests itself during the first or after the fifth decade of life. Some reported differences in incidence rates among races and socioeconomic classes are attributed to probable biased application of diagnostic criteria and caused by the downward drift concept. Significant correlation with population density in large but not in medium or small urban centers has been applied to support the role of environmental stress as a cause for schizophrenia. Yet demographic factors that may promote the aggregation of patients in large cities have not been adequately addressed. Higher incidence of schizophrenia among new immigrants in urban centers has raised a question about the possible contribution of cultural factors in the etiology of the disease. Similarly, the higher incidence of the disorder in industrialized nations and the growing incidence in developing nations undergoing industrial changes have been explored.

The birth of a disproportionately high percentage of schizophrenic patients during the winter months has led to the development of environmental etiologic factor theories such as viral infections and nutritional deficiencies. There is to date no substantial evidence to support these theories. Schizophrenics as a group have a higher incidence of birth complications than the general population and have an as yet unexplained higher rate of mortality from various medical disorders.

V. Metapsychological Theories

The psychoanalytic school pioneered by Sigmund Freud has its central focus on the disturbance of ego function. According to Freud, ego function disintegrates and returns to a more primitive stage of development. The term *primitive* refers to earlier stages of life of the individual. The structural theory, formulated by Freud, divides psychic apparatus organization into the id, ego, and superego. The ego, driven by narcissistic needs, develops in response to demands placed on it by the environment, represented in part by the superego. In return, it has the executive function of monitoring the id, which is governed by innate drives of sex and aggression. Sustained, overwhelming conflict between id impulses and superego values are believed to be the source of psychological symptoms. The ego's failure to master and resolve these conflicts in an appropriate way results in its disintegration and return to an infantile level in which reality testing is impaired. Freud also proposed that specific symptoms were reflective of underlying conflicts (e.g., paranoia representing ambiguity of sexual identity). More recent theories emanating from this school suggest that various symptoms have symbolic significance, representing the fears and wishes of the individual, and the regressive phenomenon may be an adaptive event for reorganization of the psyche in creating new reality. [*See* PSYCHOANALYTIC THEORY.]

The interpersonal theorists focus on the influence of parental figures. The schizophrenic patient fails to master the separation-individuation phase of development when the infant starts identifying itself as a separate entity from the mother and incorporates object constancy. This development enables the infant to feel secure in the mother's availability whenever she is not within visual range. The failure of development in this area is believed to be generated by ambiguous and conflicting messages the mother may convey to her child. Harry Sullivan postulated that the behavior of overanxious mothers may generate the same anxious feelings in their babies who later develop schizophrenia.

Various family theories have been advanced as a cause of emotional disturbance in the growing child that later becomes schizophrenic. Learning theorists believe that schizophrenic patients develop irrational perceptions and reactions by imitating emotionally disturbed parents. The double-bind concept refers to a setting in which emotionally disturbed families place the child in a constant situation of making decisions between ambiguous choices, producing confusion. Another theory about the balance of relationships among members of a family relates to the abnormal over-closeness of the child with the parent of the opposite sex or the parent playing the dominant role in the family. The pattern and style of expression of various emotions among members of a family has also been postulated to influence the expression of schizophrenic symptoms.

Despite the compelling evidence supporting an underlying, genetically determined organic cause of schizophrenia, many psychodynamic and psychosocial models have been proposed to explain disease etiology as well as means of therapeutic intervention. Although few traditional centers focus on a psychotherapeutic approach as a primary modality of treatment, most currently use a combined approach that includes the use of antipsychotic drugs. Successful comprehensive treatment programs incorporate some aspects of the preceding theoretical models in their attempt to rehabilitate patients to various levels of independent functioning within the community. The effort centers on modifying maladaptive interpersonal relationships by fostering and rewarding appropriate behaviors, while discouraging dysfunctional ones. [*See* SCHIZOPHRENIA, PSYCHOSOCIAL TREATMENT.]

VI. Current Biological Treatment

First of all, it must be appreciated that there is no definite cure for schizophrenia, and the syndrome can only be controlled, not cured. It should be also recognized that in our present state of knowledge we are not aware of the etiology of schizophrenia or psychosis. In fact, there is some debate as to what the relation is between schizophrenia and psychosis. Some psychiatrists, for example, feel that the neuroleptic drugs are useful in the treatment of both schizophrenia and psychosis, whereas others are of the opinion that they are helpful only in managing psychosis. The latter point out that the neuroleptic drugs will control positive symptoms and not the negative symptoms. Thus, if the drugs selectively treat the positive symptoms, leaving the core negative symptoms untouched, any attempt to construct an etiologic theory of schizophrenia based on response to drugs rests on shaky ground. This is because the dopamine (DA) blocking effect of neuroleptic drugs is central to the so-called dopamine hypothesis of schizophrenia.

The care of the psychotic patient, nonetheless, was revolutionized as of 35 years ago when the neuroleptics were first introduced to the North American continent. These drugs adequately control agitation, delusions, hallucinations, and other miscellaneous thought disorders of patients, whether they are in a schizophrenic, manic, organic, or drug-induced clinical setting. Recent work with the pharmacology of these agents indicates that they bind to the so-called D2 receptor in the brain, which is believed to be responsible for the manifestation of psychosis, and not the D1 receptor. Both are subtypes of DA receptors. More recent studies, however, indicate that there may be interactions between the D1 and D2 receptors, and thus the drugs may indirectly affect the D1 receptor.

The drugs are now most often referred to as neuroleptics or antipsychotics, but in the early history of their use they were also called major tranquilizers. They are divided into several chemical classes. The phenothiazine class of drugs is divided into three subgroups (i.e., the aliphatic, piperidine and piperazine derivatives). Chlorpromazine

(Thorazine) is the most prominent member of the aliphatic group and one of the first drugs used. Thioridazine (Mellaril) is a commonly used piperidine derivative. Examples of piperazine phenothiazines include trifluoperazine (Stelazine) and fluphenazine (Prolixin). Haloperidol (Haldol), a butyrophenone class of neuroleptic, is probably the most commonly prescribed neuroleptic today. Thiothixene (Navane) belongs to the thioxanthene class. Molindone (Moban) is a dihydroindolone and loxapine (Loxitane) is a dibenzoxazepine. In addition to the above conventional classes, other so-called atypical neuroleptics, because of their unconventional biochemical and behavioral effects, are available. Clozapine (Clozaril) is an example of these atypical neuroleptics with promising clinical applications. This drug has just been approved in the United States for use in schizophrenics resistant or unresponsive to conventional neuroleptics. Because of a 2–4% incidence of agranulocytosis (i.e., a marked reduction in the number of white blood cells), strict guidelines for frequent blood testing and patient monitoring must be followed. This will make the use of clozapine quite costly. However, as well as being effective in a substantial proportion of patients who have failed to respond to standard neuroleptics, clozapine may improve some negative symptoms. It also appears to be free of both the acute and long-term neurologic side effects associated with the other agents.

The antipsychotic drugs are also classified according to their clinical potency. This is based on comparison of potency with 100 mg of chlorpromazine equivalents and appears to parallel the degree of DA receptor blocking activity of the drugs. The drugs are usually given orally in pill or liquid form and take about 100 minutes to achieve peak plasma level. Parenteral preparations for intramuscular (IM) injection are available for many of the drugs. Administration by injection results in peak plasma levels within 30 minutes and is used in emergent situations when rapid tranquilization of a severely agitated or violent patient is desired. Haloperidol and fluphenazine are available in oil-based depot injection forms. These preparations are slowly absorbed from the tissue and can be given at 2–4-week intervals. They are useful in ensuring compliance, instead of oral preparations that have to be taken on a daily basis. There is no convincing evidence at the present time that any one neuroleptic is better than another in equivalent doses. The choice of drug is usually based on side-effect profile and desired clinical response. The high-potency drugs such as haloperidol or fluphenazine are less sedating but are associated with a high incidence of extrapyramidal side effects (EPS). However, the low potency drugs such as chlorpromazine or thioridazine produce significant sedation and postural hypotension (i.e., a significant drop of blood pressure in the upright position) and possess potent anticholinergic action. This property may actually have a protective effect from EPS, as anticholinergic drugs are frequently prescribed with the high-potency drugs to treat EPS.

Treatment of the schizophrenic patient can be divided into the acute and chronic phases. Objectives of treatment during the acute phase include controlling agitation and acute psychotic symptoms. The type of drug, dose, and route of administration depends on the prevailing clinical state. If marked agitation or violent behavior is encountered, a more sedating agent is given. The first few doses may be given by IM injection if needed. The autonomic side effects, especially postural hypotension, may limit the use of these drugs at adequate doses in the elderly or in those with underlying cardiovascular problems. Although the high-potency drugs are less sedating, the EPS can be severe and intolerable especially at higher doses. It can, however, be effectively controlled with the use of anticholinergic drugs such as benztropine (Cogentin), trihexyphenidyl (Artane), or diphenhydramine (Benadryl). These agents are usually given by mouth but can be administered by injection for rapid effect. As emergence of side effects may interfere in the patient's acceptance of the treatment, they should be promptly and effectively treated to ensure long-term compliance. In general, most patients develop tolerance to these side effects within 8–12 weeks of initiating neuroleptic treatment and the anticholinergic agents can be gradually stopped. A few patients require long-term use of the drugs, and an occasional patient may begin to manifest the side effects after months of neuroleptic use. The antipsychotic effects of the neuroleptics may take 6 weeks or longer to be maximally effective. Therefore, an aggressive increase of neuroleptic dose, more than is recommended, is probably not appropriate during the first few weeks of treatment. Such dose escalation is associated with an increasing incidence of the unpleasant and troublesome side effects without significantly benefitting the target symptoms. Not surprisingly, rapid neuroleptization (i.e., a treatment strategy of giving high doses of haloperidol or other high potency agents by IM injection at one to two hourly intervals to achieve rapid tranquilization of the acutely psychotic patient) has lost favor in recent years.

The objectives of the long-term or maintenance phase of treatment are control of psychotic symptoms with lower maintenance dose of neuroleptics. In addition to minimizing the acute dose–related EPS, the cumulative exposure to neuroleptics, presumed to be responsible for tardive dyskinesia (TD), will be controlled. If psychotic symptoms have been completely absent for at least 6 months on maintenance dose after the treatment of an acute episode, attempts are made to gradually decrease and discontinue the neuroleptic, as long as the patient remains symptom-free. Although a higher incidence of relapse is likely with low-dose therapy, recurrences of psychotic symptoms are usually rapidly controlled with dose adjustments. Studies of blood neuroleptic level measurement for dose determination and clinical response monitoring have produced conflicting results and have generally been disappointing. This is partly due to the multitude of metabolites of undetermined significance these drugs have, as well as problems with the various laboratory techniques used to assay drug concentrations. In general, the findings of the various studies support the presence of a minimum threshold blood drug concentration that should be present to achieve clinical response. Higher drug concentrations are associated with increasing incidence of side effects.

Several neuroleptic treatment–related side effects should be noted. The acute ones take several forms. The EPS involves the extrapyramidal part of the central nervous system, where a high concentration of DA neurons are localized. It usually has features of Parkinsonism including muscle rigidity, bradykinesia (i.e., slowness of movements), and gait disturbances. Akithsia is a peculiar sense of motor restlessness. The patient is unable to remain still or relax for any length of time. It is often confused with agitation and mismanaged by giving more of the offending agent. Beta blocker drugs, such as propranolol, are useful in controlling this symptom. A unique form of EPS is acute dystonia, which is a painful sustained spasm of neck, limb, extraocular, and sometimes oropharyngeal and laryngeal muscles. It is an emergent condition that responds to parenteral anticholinergic drug administration. Approximately 15–20% of long-term neuroleptic-treated schizophrenic patients develop TD. This is usually a cosmetically undesirable, but sometimes a physically disabling, movement disorder. It manifests as an uncontrollable stereotyped movement of the tongue and mouth (chewing, lip pursing), face (grimacing), neck and spine (contorting movements), or the hands and feet (writhing). The condition affects women more frequently than men. Although most patients with TD have an average of 3–5 years of chronic neuroleptic exposure, TD has been reported in some cases after only a few months of neuroleptic use. It is believed to be related to increased sensitivity of DA receptors deprived of normal DA stimulation. Its onset is usually insidious and may first be observed during neuroleptic dose reduction, oftentimes prompting dose escalation, which temporarily masks the manifestation. There is no effective treatment for TD, but improvement occurs over time after discontinuation of neuroleptic use. Although there is potential for full recovery, the condition is irreversible in many patients. The neuroleptic malignant syndrome is a life-threatening medical emergency of undetermined cause, associated with both acute and maintenance treatment of high-potency neuroleptics. It is characterized by marked elevation of body temperature, severe muscle rigidity, profuse sweating, elevated muscle enzymes and white blood cell counts, and alteration of consciousness. Elevated blood myoglobin level may cause renal failure. The treatment is symptomatic, including lowering body temperature, use of the muscle relaxant agent dantrolene, and administration of the DA agonist drug bromocriptine. The symptoms can persist from 10 days to several weeks, depending on the duration of action of the offending neuroleptic.

Adjunct treatments to neuroleptics in schizophrenics include lithium carbonate, antidepressants, benzodiazepines, and the anticonvulsant drug carbamazepine. Although these agents can be of some benefit in selected patients with concurrent specific symptoms, such as mood disturbance or anxiety, they are of equivocal value on schizophrenic symptoms. Other forms of therapy for schizophrenia include electroconvulsive therapy (ECT), psychosurgery, and insulin coma. The last one is obsolete and is only of historical interest. Psychosurgery is used rarely to control violent or compulsive behavior, when all other means fail. The surgery involves destruction of the frontal lobe or its connections. ECT was once widely used in catatonic patients, and although it appears to control psychotic symptoms, the benefits are short-lived. There is some recent resurgence of interest in ECT use to treat schizophrenia.

VII. Current Biological Theories

The etiology of schizophrenia basically remains unknown. In a situation in which a chronic and destructive disease such as schizophrenia exists, it is only natural that there would be a great number of theories or hypotheses about its origin. Some of these are interesting but have no support whatsoever. Others fall into the "where there's smoke, there's fire" idea that there is supportive evidence of the possibility of what may be wrong but no firm evidence. Unfortunately, there is no hypothesis for which there is absolutely firm and substantial data, both from clinical and basic science observations, to indicate the type of underlying abnormality in schizophrenia. An example of the first of these types of theories is the vitamin C deficiency theory, which was proposed by Dr. Linus Pauling. However, tests of this hypothesis have only yielded equivocal results.

In the second category, there is suggestive evidence, but no firm data, to clearly support theories that deal with neurotransmitters, with constitutional factors, and with structural problems. One of the neurotransmitter theories involves serotonin. Serotonin is a molecule that is a natural transmitter in the brain and has many structural similarities to the psychotomimetic drug lysergic acid diethylamide (LSD). For this reason, attention has long been focused on the possibility that serotonin might be handled abnormally in some way in the central nervous system of schizophrenics, although there is no agreement or firm evidence to support this. Furthermore, the clinical picture that one obtains with LSD is somewhat different than with schizophrenia. Visual hallucinations are common with LSD, whereas they are uncommon with schizophrenic illness.

The neurotransmitter for which there is the most convincing evidence at the basic science level is DA. This is a neurotransmitter that is a precursor of norepinephrine (NE). In certain regions of the brain there is no enzyme necessary to convert DA to NE, and DA itself becomes the primary neurotransmitter. Probably the best known example of the effects of a deficiency of DA would be Parkinson's disease. Work with schizophrenia was given a particular impetus by the fact that amphetamine can produce a paranoid-like psychosis. It causes the release of DA and NE. The balance tipped in favor of DA, rather than NE, when it was found that almost all the classical neuroleptics used to treat schizophrenia bind to the D2 receptor with a potency that has a relation to the doses used clinically. In contrast, the potency with which the drugs bind to NE or serotonin receptors is different from their clinical potency. Furthermore, certain experiments with animals can produce behaviors that are analogous to stereotypy in the schizophrenic when the DA neuronal systems of these animals are manipulated. Interest in the DA system was occasioned by a report some 20 years ago from a group from Sweden led by Carlsson. He and his co-workers gave neuroleptics to mice and then measured the metabolic products of DA in the brain. They found that when neuroleptics were administered to the mice, there was an increase in the metabolites of DA, and they postulated that this was because of a blockade of DA receptors in the brain, with concurrent feedback that caused more DA to be released. This basic *in vivo* study has been replicated now many times, and there is little question to its validity, in that the administration of neuroleptics is associated with increased DA metabolites. Whether this effect is mediated by receptors on the cell body or by feedback loops is perhaps more controversial and is not really necessary to the theory's credibility. This observation evolved into the so-called DA hypothesis of schizophrenia. Unfortunately, although the theory is attractive and has much support from animal research, it has little evidence from actual clinical studies. Most clinical studies have not revealed any difference from the norm in cerebrospinal fluid (CSF) concentrations of homovanillic acid (HVA), the major metabolite of DA in the brain.

There are distinct neuronal DA systems within the brain (i.e., tubero-infundibular, mesocortical, mesolimbic, and nigrostriatal). The binding of neuroleptics to receptors in the brain can be determined in the laboratory or can be determined *in vivo* by positron emission tomography (PET), a novel neuro-imaging technique. Because a substantial portion of the DA-like ligands or their antagonists need to bind to receptors to be visualized by PET, the question has been left open as to whether there may be deficit in the mesolimbic and mesocortical systems, which cannot be visualized because of the low density of DA receptors in these areas. Alternatively, perhaps there may not be any DA receptors in the cortex and the limbic system. There are two opposing results from PET studies, one from Johns Hopkins University and the other from the Karolinska Institute in Stockholm. The group at Johns Hopkins has shown an increased density of D2 receptors in brains of schizophrenics, even in those who were not on, or have never been on, medication. Whereas at the Karolinska Institute, it was demonstrated that the number of receptors is the same in both healthy controls and patients. Although both groups used PET, different mathematical models were used to arrive at what constitutes D2 receptor density. Hence, the controversy is not resolved, and we will have to wait for newer techniques or different models of measuring binding to answer this important question. Neuroanatomically, it has been postulated that the deficit in the schizophrenic patient may be in the mesocortical or mesolimbic system. This was originally thought to be the case because the neostriatum, which is visualized by PET, seems to be concerned mostly with motor functions or the control of movement. However, more recent work has suggested that the caudate, a part of the neostriatum, may be implicated in cortical functioning; hence, it may be that lesions in the schizophrenic brain will be found in the striatum.

There is stronger clinical evidence supporting a role for NE in psychotic states than for DA. This neurotransmitter is absent in the striatum but is found widely distributed throughout the entire cortex. However, the results from animal studies are less convincing for NE than for DA. For example, the potency with which receptors in the central nervous system bind NE-like ligands or neuroleptics has no relation to their clinical potency as antipsychotic drugs. Three separate studies have reported elevated CSF NE levels in psychotic patients. A study of levels of NE in neuroanatomic structures indicates that there are elevated NE levels in limbic structures of paranoid patients. Lastly, there is evidence that 3-methoxy-4-hydroxyphenylglycol (MHPG) levels in urine and plasma are elevated in schizophrenic patients and the level of MHPG correlates directly with the severity of psychotic symptoms. MHPG is a major metabolite of NE in brain. It is possible that both DA and NE brain systems are involved in the genesis of psychosis and that NE and DA are functionally related. However, the linkage of the two systems is at present poorly understood.

Various structural abnormalities have been reported in postmortem studies of the brain of schizophrenic patients, as well as by the modern imaging techniques of computerized axial tomography (CAT) and magnetic resonance imaging (MRI). The postmortem findings are varied and inconsistent. They include atrophy of various parts of the brain including the neostriatum, the cerebellum, and temporal lobes, as well as enlargement of the cerebral ventricles. It is unclear if any of the changes in brain tissue are a reflection of the disorder, changes associated with neuroleptic exposure, or postmortem artifacts. Similarly, CAT and MRI findings are nonspecific and not universally found. Abnormal findings are more frequently associated with the so-called type II schizophrenia, or deficit state, than with type I, which is characterized by positive symptoms. In conclusion, despite the availability of these powerful tools, some theoretical framework is needed to search systematically and identify specific findings. [*See* MAGNETIC RESONANCE IMAGING.]

Bibliography

Buchsbaum, M. S., and Haier, R. J. (1987). Functional and anatomical brain imaging: Impact on schizophrenia research. *Schizophr Bull* **13,** 115–132.

Kane, J. M. (1987). Treatment of schizophrenia. *Schizophr Bull* **13,** 133–156.

Kaplan, H. I., and Sadock, B. J., eds. (1989). "Comprehensive Textbook of Psychiatry," 5th ed. Williams and Wilkins, Baltimore, Maryland.

Kleinman, E. J., Casanova, M. F., and Jaskiw, G. E. (1988). The neuropathology of schizophrenia. *Schizophr Bull* **14,** 209–222.

McGlashan, T. H., and Carpenter, W. T., Jr., eds. (1988). Long-term followup studies of schizophrenia. *Schizophr Bull* **14,** 497–673.

Meltzer, H. Y. (1987). Biological studies in schizophrenia. *Schizophr Bull* **13,** 77–111.

Sherrington, R., Brynjjolfsson, J., Petursson, H., et al. (1988). Localization of a susceptibility locus for schizophrenia on chromosome 5. *Nature* **336,** 164–167.

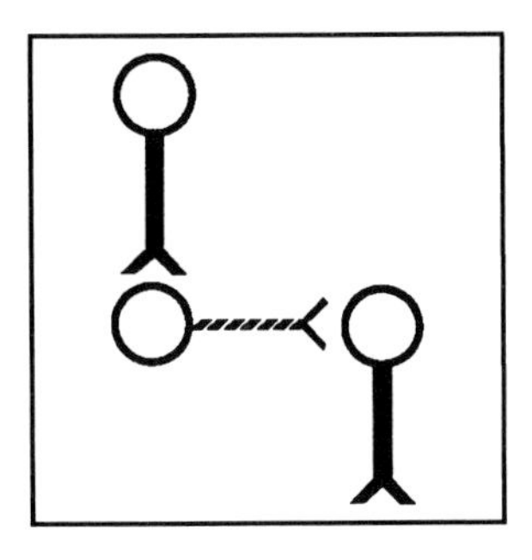

Seizure Generation, Subcortical Mechanisms

KAREN GALE, *Georgetown University Medical Center*

Glossary

Amino acid neurotransmitters Amino acids released by specific neurons in order to transmit information across synaptic junctions to the postsynaptic target neuron. Some are excitatory (causing depolarization of the postsynaptic target, thereby increasing the chance of its firing), such as glutamate and aspartate. Others are inhibitory (causing hyperpolarization of the postsynaptic target, thereby reducing the chance of its firing) including γ-aminobutyric acid (GABA) and glycine. Each amino acid neurotransmitter exerts its action via highly specific and selective receptors located on the postsynaptic target neuron. The "NMDA receptor" refers to a subtype of excitatory amino acid receptor which is selectively activated by the drug *N*-methyl-D-aspartate

Basal ganglia Includes corpus striatum (caudate nucleus, putamen, and globus pallidus or entopeduncular nucleus) and substantia nigra. The caudate and putamen (neostriatum) receive input from widespread regions of the cortex and, after integrating this input, send projections out via relays in entopeduncular nucleus and substantia nigra. The basal ganglia are important for the initiation of movements, regulation of posture, and sensory–motor integration

Brainstem (also referred to as hindbrain) Part of the brain containing the midbrain, pons, and medulla. This part of the brain is especially important for controlling general arousal and regulating states of consciousness, waking and sleeping, breathing, and heart rate

Clonic convulsion Seizure manifestation consisting of repetitive, rhythmic jerking of the limbs, trunk, head, and/or facial muscles

Electroencephalographic (EEG) recording Recording of variations in electrical potentials of the brain by means of electrodes placed on the surface of the scalp, skull, dura, or within deep structures of the brain. This activity is amplified and then recorded by the movement of a pen on a paper chart passing beneath it. The resulting record represents the summation of a variety of complex electrical potentials in the cortical or subcortical areas being monitored

Forebrain Part of the brain containing the telencephalon (cerebral cortex, corpus striatum, limbic system) and the diencephalon (thalamus and hypothalamus). In general, the most recently evolved areas of the brain are located in the forebrain

Limbic system Group of anatomically interconnected brain regions that regulates emotional experience and expression. Major forebrain components include hippocampal formation, septum, amygdala, olfactory bulbs, hypothalamus, piriform and entorhinal cortex, and nucleus accumbens. Also known as the "visceral brain," this system subserves functions essential for individual and species survival such as feeding, fight and flight, mating, and care of offspring. This system also plays a crucial role in learning and memory and goal-directed behavior

Maximal electroshock Electroconvulsive shock, when applied at a current strength above that sufficient to cause tonic hindlimb extension in all subjects (usually rats or mice), is referred to as *maximal electroshock*

Subcortical All brain structures other than the cerebral cortex (i.e., neocortex). While the cerebellar cortex is not properly "subcortical," it will be included in the present discussion

Tonic convulsion One type of motor manifestation of a generalized seizure, it consists of coordinated contractions of the musculature, especially evident in the limbs. Often it involves the contraction of extensor muscles leading to *tonic extension* of the forelimbs and/or hindlimbs

I. Introduction

The abnormal paroxysmal discharge of neurons in the central nervous system is referred to as a *seizure* and is usually detected by electroencephalographic (EEG) recording. If a seizure is confined to a group of neurons within a circumscribed brain region, it may not necessarily produce any behavioral symptoms. When seizure activity is conducted to distant brain regions it is said to be propagated and the manifestations of the seizure depend upon the particular neural circuits involved. Most forms of seizures involve activation of motor systems and are referred to as convulsive seizures, as distinct from seizures that cause impairment of consciousness without motor activation. Motor manifestations of convulsive seizures range from automatisms and spasms to tonic or clonic jerking, often accompanied by impaired consciousness. The precise motor components and their pattern of expression is closely related to the brain areas engaged in the seizure. Seizures can be categorized by their behavioral symptoms and by the nature and extent of the brain areas involved in their generation.

When seizures occur recurrently and spontaneously, the term *epilepsy* is used to describe the disorder. Epilepsy develops as a secondary consequence of various chronic neurological conditions, especially when damage and scarring of brain tissue is present as in the case of cerebral palsy, brain tumors, head injury, or cerebrovascular insults. The anatomical site of origin of the seizure is referred to as a focus and, when a defined focus exists, the pathways for seizure conduction are determined by the anatomic connections of that focus. [*See* EPILEPSY.]

In the normal brain, acute convulsive seizures can be provoked by any of a number of conditions. Specific drugs are capable of inducing seizures, and such drugs are generally referred to as chemoconvulsants. Severe hypoglycemia (as can be produced by excess insulin), hypoxia, sleep deprivation, nutritional deficiencies (especially in vitamin B_6 or magnesium), fever, poisoning, and electroconvulsive shock all can precipitate seizure activity. Convulsive seizures are also one component of a withdrawal syndrome that occurs when continuous long-term use of alcohol, barbiturates, benzodiazepines, and certain other sedative–hypnotic drugs is abruptly discontinued.

This article will focus on convulsive seizures as they occur in animal models, because it is only in experimental animals that we have had the ability to probe and analyze subcortical brain regions that influence seizure initiation and/or propagation. While it is not known to what degree seizure mechanisms in animals are equatable with those in humans, it is likely that significant similarities exist. This assertion is supported by the fact that many of the same acute conditions that provoke convulsive seizures in humans do the same in experimental animals, and the drugs that are anticonvulsant in humans are also effective in many of the experimental animal seizure models. Moreover, prolonged seizures have been documented to cause damage to brain tissue in both humans and experimental animals and the anatomic pattern of this damage exhibits some cross-species similarities.

Based on studies in experimental animal models, a number of different brain regions have been implicated in the initiation and modulation of convulsive seizure activity. The particular brain regions and pathways that are engaged by a seizure vary according to how the seizure is elicited. For any seizure there are brain regions responsible for (1) initiating the seizure, (2) propagating the seizure activity, and (3) controlling or suppressing seizure activity once it has been initiated. At present, we do not have a complete neuroanatomical map of all these components for any particular seizure model. At best, we have a partial list of subcortical brain regions that have been experimentally identified as contributing to seizure initiation, propagation, and/or control in one or more experimental seizure models in animals.

II. Forebrain Regions Involved in Seizure Initiation: The Limbic System

The subcortical brain regions that have received the greatest attention in connection with convulsive seizures are those associated with the limbic system. In particular, the amygdala and hippocampus are part of circuits involved with the initiation of certain types of convulsive seizures. These seizures, characterized by automatisms and/or clonic movements of the mouth, face, and upper extremities have been referred to as "limbic motor seizures" since they are accompanied by afterdischarges and electrographic seizure activity throughout the limbic system. Electrical stimulation of either amygdala or hippocampus (among other forebrain regions) when applied repeatedly over a period of several days can lead to *kindling*, the process by which a subconvulsant stimulus, when applied repeatedly, can come to trigger a convulsion. Once an animal has been kindled (for example by electrical stimulation in the amygdala), the animal will reliably exhibit convulsive seizures when stimulated (in the amygdala) and this response will be maintained indefinitely. Moreover, an animal kindled from one limbic region (e.g., amygdala), will require very few stimulations in another limbic region in order for the kindling to be transferred to that second region (e.g., hippocampus). The behavioral characteristics of these seizures are relatively constant, regardless of the limbic structure (septum, olfactory bulb, entorhinal cortex, amygdala, or hippocampus) used for the stimulation. This suggests that a common seizure-generating pathway or circuit can be engaged by stimulation of any of a number of sites in the limbic system. No single limbic region appears necessary for producing limbic motor seizures, because attempts to interfere with these seizures by destroying various individual structures have been generally unsuccessful. Lesion studies must be viewed with extreme caution, however, because over hours or days, intact brain tissue adjacent to or connected with the damaged region often can change its activity to compensate for the damage. In some instances, this occurs because inhibitory neurons are damaged, thereby disinhibiting other groups of neurons outside the lesioned area. Therefore, a particular brain region may in fact play a crucial role in seizure initiation and/or propagation even though seizure activity continues to occur following its removal. One way to minimize this problem is to use focal drug injections into the brain area of interest in order to rapidly and reversibly suppress the activity of the area. Anticonvulsant effects can then be evaluated before many of the adaptive and compensatory changes set in. [*See* HIPPOCAMPAL FORMATION; LIMBIC MOTOR SYSTEM.]

Recently, a very small region within the deep prepiriform cortex has been identified that triggers limbic motor seizures when stimulated by the application of minute quantities of certain drugs. This functionally defined area, the area tempestas, is regulated by inhibitory and excitatory amino acid neurotransmitters. Blockade of inhibitory neurotransmission [mediated by γ-aminobutyric acid (GABA)] or augmentation of excitatory neurotransmission (mediated by glutamate or aspartate) within this site in one hemisphere initiates bilaterally synchronous motor and electrographic seizures that resemble those evoked by kindling of limbic structures. The area tempestas is most likely part of a seizure-generating circuit that connects the piriform cortex, entorhinal cortex, hippocampus, and amygdala. This area can also be kindled rapidly, requiring relatively few electrical stimulations (as is true for amygdala) and exhibiting an especially rapid rate of kindling transfer to/from amygdala. Moreover, there is evidence suggesting that drug-induced inhibition of transmission within the area tempestas can significantly elevate the threshold for seizures triggered from amygdala in kindled rats. It is suspected that the area tempestas influences limbic system excitability via projections to the piriform cortex and entorhinal cortex, regions known to be particularly prone to the generation of epileptic seizure activity.

Whereas the area tempestas may be concerned with the *initiation* of limbic seizure activity, the substantia innominata is a region concerned with the *expression* of the motor components of limbic seizures. Inhibition of activity within this brain region prevents the convulsions evoked by amygdala stimulation in kindled animals but does not prevent the afterdischarge recorded electrically from the amygdala. It has been suggested that the substantia innominata serves as a relay between the amygdala and cortical areas participating in the organization of motor components of the seizure response. Additionally, the substantia innominata sends projections to midbrain regions that can influence motor systems via descending pathways.

III. Forebrain and Midbrain Regions Involved in Seizure Propagation and Control: The Basal Ganglia and Related Nuclei

The major components of the basal ganglia, i.e., the caudate-putamen (striatum), globus pallidus (entopeduncular nucleus in particular), and substantia nigra, have all been demonstrated to influence seizure susceptibility in numerous experimental seizure models. The best studied of these structures is the substantia nigra, the only region that has been examined in connection with as many as 10 different experimental seizure models. Treatments that either increase inhibitory transmission (mediated by the neurotransmitter GABA) in substantia nigra or block excitatory transmission (mediated by glutamate or certain neuropeptides such as substance P) in this nucleus in both hemispheres, prevent or attenuate convulsive seizures induced by several chemoconvulsants (pilocarpine, kainic acid, bicuculline, and flurothyl, among others), by maximal electroshock, by kindling of amygdala, by drug application into the area tempestas, and by acoustic stimulation in rats susceptible to audiogenic seizures due to either ethanol withdrawal or genetic determinants. In addition, inhibition within the substantia nigra decreases susceptibility to nonconvulsive electrographic seizures that occur either spontaneously (in genetically predisposed strains of animals) or as a consequence of treatment with drugs such as pentylenetetrazol in low doses.

The integrity of the substantia nigra is not required for seizure induction, indicating that this structure is not part of a crucial seizure-conducting pathway. Instead, the substantia nigra is probably part of a seizure-suppressing circuit that becomes engaged by seizure discharge. Viewed in this way, the substantia nigra and associated nuclei of the basal ganglia act to maintain a homeostatic balance of brain excitability by creating a resistance to seizure spread and generalization.

Certain key structures of the forebrain and midbrain may work in concert with the substantia nigra in this seizure-resisting capacity. The striatum is one major source of neural input to the substantia nigra, whereas the superior colliculus (deep layers) is one important target of neural projections coming from the nigra. One of the most prominent pathways connecting the striatum and substantia nigra is inhibitory and utilizes GABA as its transmitter. Likewise, the substantia nigra sends an inhibitory GABA-containing projection to the superior colliculus. Consequently, it might be expected that the stimulation of the neurons in the striatum that give rise to the GABA inputs to the substantia nigra, would be anticonvulsant (see Fig. 1) because this would enhance GABA transmission in the substantia nigra. The experimental evidence supports this proposal; electrical or drug-induced excitatory stimulation in striatum tends to exert an anticonvulsant action in the few experimental seizure models that have been examined so far.

Within substantia nigra, inhibitory GABAergic transmission acts to suppress the activity of output projections to superior colliculus. This relationship predicts that blockade of GABA transmission in the nigral projection target area of superior colliculus should be anticonvulsant. Again, the experimental evidence is consistent with expectations. Blockade of GABA receptor-mediated transmission in the deep layers of superior colliculus is anticonvulsant against both clonic and tonic forms of convulsive seizure activity, as well as against nonconvulsive spike-and-wave electrographic seizure discharge. This illustrates a fundamental principle of central nervous system organization: disinhibition. Inhibitory transmission within the substantia nigra acts to reduce the activity of the nigral outputs to the colliculus (which are themselves inhibitory), resulting in the withdrawal of inhibition, or *disinhibition*, of neuronal targets in the superior colliculus (see Fig. 1). Presently, it is not understood how neuronal activity within the superior colliculus acts to impede seizure progression. Some of the colliculus neurons that are activated by nigral disinhibition may be those that project to the brainstem reticular formation where they can relay with both ascending and descending neural pathways. Stimulation of certain regions of the reticular formation is capable of interfering with the development of a synchronized pattern of neuronal discharge in cortex. By engaging these desynchronizing influences of the reticular formation, the superior colliculus (Fig. 1) could disrupt the generation of seizures, which require synchronized neuronal bursting.

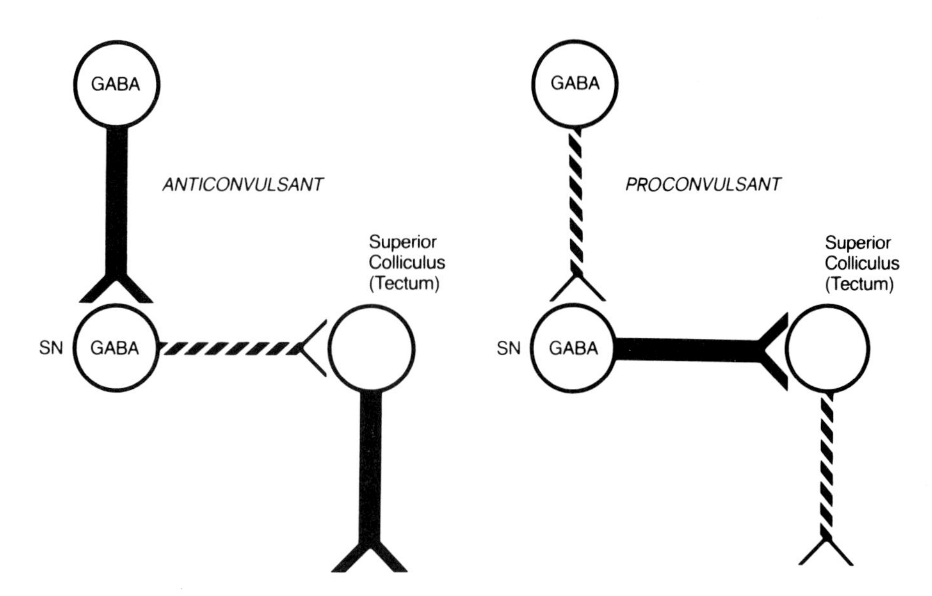

FIGURE 1 Serial inhibitory neuronal links in the basal ganglia. From the striatum to the substantia nigra (SN), a GABA-containing neural pathway projects. From substantia nigra to superior colliculus (tectum), there is also a GABA-containing neural pathway. Within the substantia nigra, the first of these pathways (striato-nigral) inhibits the second (nigro-tectal). When there is more GABA released by the first pathway, there is more inhibition in substantia nigra and, consequently, the activity of the nigro-tectal pathway is suppressed (broken lines). The suppression of this second pathway results in the removal of inhibition (i.e., disinhibition) in the superior colliculus. This is associated with anticonvulsant effects (see text). The opposite effect (i.e., the effect of decreasing the activity of the striatonigral GABA pathway, is portrayed on the right side of the figure). Heavy solid lines indicate increased neural firing. Broken lines indicate decreased firing.

The entopeduncular nucleus, also known as the medial segment of the globus pallidus, is similar to the substantia nigra in terms of function, morphology, and neuroanatomic connections. Together with the substantia nigra, this nucleus relays neuronal outputs from the striatum. It is therefore not surprising that the same treatments that suppress seizure propagation when placed in the substantia nigra, exert a similar action in the entopeduncular nucleus. This brain region has not yet been studied in the multiplicity of seizure models that have demonstrated a regulatory role for substantia nigra. Consequently, it is not known whether this brain region can influence seizure propagation regardless of the site of origin or type of seizure, or whether it is involved with a specific subset of convulsive seizures.

The influence of the basal ganglia on seizure generalization is not confined to the motor manifestations of the seizures. In instances in which electrographic signs of seizures have been monitored, anticonvulsant manipulations of basal ganglia structures induce a corresponding suppression of electrographic seizure discharge in cortical and subcortical structures. This is consistent with a role of basal ganglia circuits in the control of cortical excitability and the regulation of synchronization of neuronal discharge in widespread regions of the forebrain.

IV. Diencephalic Regions Involved in the Propagation of Pentylenetetrazol-Induced Seizures

Seizures induced in animals by the chemoconvulsant agent, pentylenetetrazol (PTZ), have been used extensively as one model for screening potential antiepileptic drugs. Metabolic mapping studies have shown that the neural projections from the mammillary bodies of the hypothalamus to the anterior thalamus exhibit markedly increased activity during PTZ-induced seizures. Lesions of this mammilothalamic pathway can abolish both the behavioral and electrographic expression of PTZ seizures, as can focal microinjections of drugs that enhance GABA transmission in the anterior thalamus. The mammilothalamic tract is believed to be part of a subcortical pathway that interconnects the thalamus with the midbrain tegmentum, a route that may be significant for the behavioral expression of the seizures. The cortical projections to and from the thalamus are most likely responsible for the cortical electrographic manifestations of the seizures. It is noteworthy that lesions of the thalamus have been found to suppress cortical seizure discharge in other seizure models in which seizure activity is elicited by focal drug or electrical stimulation of cortex itself.

V. Hindbrain Regions Implicated in Seizure Initiation

Convulsive seizures can be elicited by appropriate stimulation of brainstem regions, but these seizures are readily distinguished from those elicited by stimulation of forebrain structures. In the rodent, forebrain-evoked seizures usually involve clonic movements of the face, neck and forelimbs, whereas hindbrain-evoked seizures lack this feature and instead consist of explosive running and bouncing clonus and/or tonic extension of forelimbs and hindlimbs. Electrical stimulation of the reticular formation of the midbrain, pons, or medulla can elicit self-sustained tonic convulsive seizures. These seizures do not necessarily engage forebrain circuitry and, moreover, their expression does not require intact connections with the forebrain. Animals with complete transection of the brain (precollicular transections) disconnecting forebrain from hindbrain circuits, are fully capable of exhibiting tonic seizures as well as explosive running and bouncing clonus in response to brainstem electrical stimulation, systemic injection of chemoconvulsants such as PTZ, or electroconvulsive shock applied via corneal or ear-clip electrodes. On the other hand, the seizure responses associated with forebrain limbic circuitry (see above) cannot be elicited in animals with precollicular transections that disconnect the forebrain from the hindbrain.

A brainstem site from which running/bouncing clonic seizures can be triggered is the inferior colliculus. Electrical stimulation of the inferior colliculus in the rat as well as focal application of drugs that stimulate excitatory amino acid transmission in this site, elicit sudden, explosive bouts of leaping, running, and bouncing on all four limbs. Blockade of GABA transmission in the inferior colliculus will have the same convulsive effect. This behavior resembles that associated with sound-induced (audiogenic) seizures in susceptible rodent strains.

The inferior colliculus, as part of the auditory sensory pathways, is probably a crucial afferent relay station for converting acoustic stimulation into convulsive discharge and subconvulsant stimulation of this region can render otherwise normal animals susceptible to audiogenic seizures. Direct chemical or electrical stimulation of this brain region does not typically evoke *tonic* convulsions, unless ascending noradrenergic projections have been compromised by lesions or depletion of norepinephrine.

VI. Hindbrain Regions Required for Seizure Expression

Lesions that produce bilateral damage in the pontine reticular formation block tonic convulsions induced by electroshock and chemoconvulsants. At the same time, these lesions do not interfere with the facial and forelimb clonus associated with limbic seizures. A key structure damaged by the pontine lesions is the nucleus reticularis pontis oralis (RPO). Selective lesions of the RPO will also prevent the tonic convulsive components of chemoconvulsant and electroshock seizures. Together with the stimulation effects described above, these observations demonstrate that the brainstem is both necessary and sufficient for the initiation and development of tonic convulsions.

VII. Hindbrain Projections Associated with Autonomic Activity: Influence on Seizure Susceptibility

Afferent vagal activation can influence cortical excitability, producing electroencephalographic synchronization or desynchronization depending upon the stimulation parameters. Recently, it has been demonstrated in both animals and humans that electrical stimulation of the afferent vagus can attenuate a variety of seizures, ranging from limbic motor seizures to tonic convulsions. The nucleus tractus solitarius (NTS) is the major relay for vagal afferents, and this nucleus, located in the brainstem, probably mediates the anticonvulsant action of vagal stimulation. The NTS has widespread influence on subcortical and cortical circuits, most likely exerted via projections to the brainstem reticular formation (especially the parabrachial area), hypothalamus and amygdala. It remains to be determined which of the projections from NTS are crucial for influencing seizure susceptibility and seizure progression.

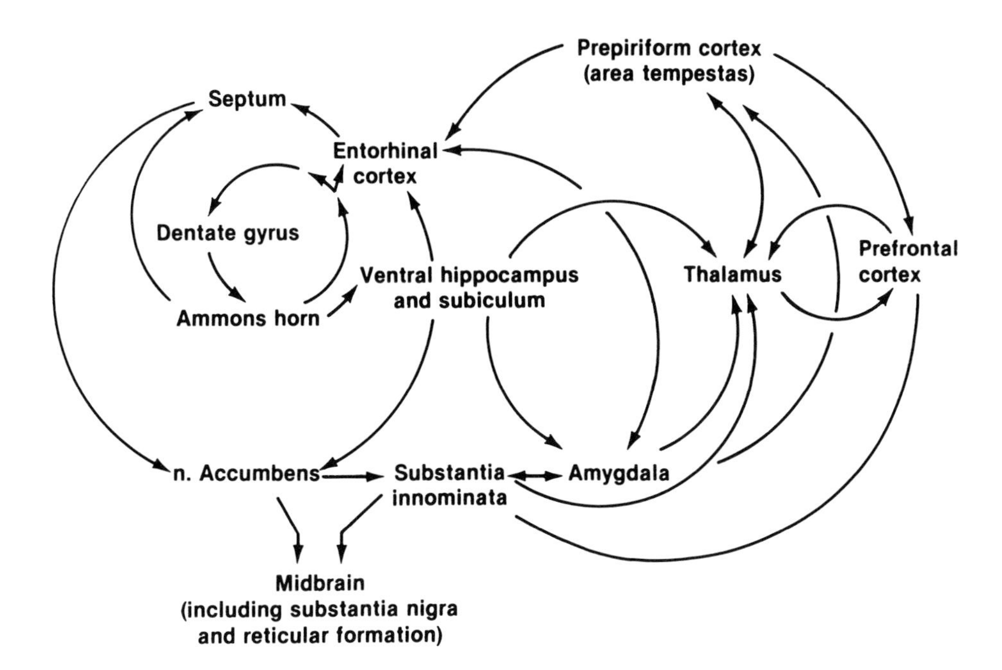

FIGURE 2 Schematic diagram of some limbic system regions and their connections that are involved in seizure generation and propagation.

VIII. Cerebellum

Removal of the cerebellum selectively inhibits tonic components of convulsive seizures, such as tonic hindlimb extension produced by maximal electroshock. Cerebellectomy does not interfere with the expression of clonic seizure components, and may actually facilitate focal seizure discharge in cortex and certain types of clonic seizure activity.

Consistent with the results of cerebellar removal, damage to the cerebellar peduncles of both hemispheres also blocks tonic convulsions. The cerebellar peduncles contain the major efferent projections from the cerebellum to the brainstem reticular formation and other motor nuclei.

IX. Insights from Metabolic and Electrographic Mapping Studies

The information discussed above derives largely from studies utilizing selective stimulation or lesions of specific brain regions or pathways. Another approach to elucidating brain regions involved in seizure generation involves the use of radiolabeled 2-deoxyglucose (2-DG). This method for estimating brain glucose utilization by contact autoradiographic analysis of brain sections, allows for a survey of all brain regions in any given subject. In addition to its use in animal studies, this metabolic mapping procedure has been applied to human seizures by using positron emission tomography (PET) scanning to detect the distribution of radiolabeled 2-DG. In the human studies, analyses concentrate on patterns of metabolic alterations in cortical structures, because the resolution of most PET scans is not sufficient to allow a detailed evaluation of subcortical regions. The contact autoradiographic analyses used for the animal studies, however, is sufficient to resolve even some of the smallest subcortical nuclei.

During seizures evoked by stimulation of structures within the limbic system, the pattern of 2-DG uptake depends on the severity of the seizure. Partial seizures are associated with increased 2-DG uptake in the closest target regions to which the stimulated brain region directly projects. During full seizures, in which a complete pattern of convulsive seizure activity is manifest bilaterally, 2-DG uptake is increased in hippocampus, amygdala, nucleus accumbens, substantia nigra, entorhinal cortex, substantia innominata, and the anterior and periventricular nuclei of the thalamus (see Fig. 2 for some of the interconnections between these regions).

Seizure-evoked activation of 2-DG uptake limited to the hippocampus is associated with staring and behavioral arrest. When bilateral activation of amygdala, substantia nigra, and thalamic nuclei occurs, strong facial and forelimb clonus with rearing is usually evident. A consistent finding from all such studies is that seizures propagate along known neuroanatomical pathways. Moreover, the neuroanatomical pathways that are metabolically activated during a seizure correspond to the circuitry connected with the stimulated region, the type of seizure evoked (e.g., limbic), and the type of convulsive behavior produced.

When seizures are elicited by stimulation of discrete regions of neocortex, the resulting focal motor seizures are associated with a pattern of 2-DG uptake and behavior that is distinct from the limbic seizures described above. In this case, the basal ganglia (caudate-putamen, globus pallidus, substantia nigra) and paraventricular and ventral thalamic nuclei are involved, even with mild and unilateral seizure activity. Larger areas of thalamus and contralateral frontal cortex become involved when seizures are more severe, and activation of medial thalamic nuclei are associated with bilateral seizure spread. It has been suggested that when the seizure discharge activates a sufficient neuronal population in medial thalamus, transsynaptic target areas in medial-frontal or orbito-frontal cortex can become bilaterally recruited.

The results obtained using metabolic mapping are generally in agreement with electrographic recording from subcortical structures. Although electrographic recording can only sample activity from a limited number of brain regions in any one subject, it has the advantage of providing information concerning the temporal sequence of seizure spread which is not provided by metabolic mapping. Typically, the first limbic system region to exhibit electrographic signs of seizure discharge is the hippocampus, perhaps because this region is especially susceptible to developing synchronous bursting patterns of neuronal discharge. Although the hippocampus can readily generate electrographic seizure activity, seizures evoked within this brain region do not easily spread to engage other brain regions. As long as the seizure discharge is confined to hippocampal circuits, there may be little or no signs of motor seizure activity. Thus, the fact that in various experimental models, as well as in human seizure disorders, the earliest signs of seizure activity are often detected in hippocampus does not necessarily mean that this is the site of seizure initiation. In fact, experimental evidence indicates that the hippocampus usually manifests the first signs of electrographic seizure discharge even when seizures are evoked from elsewhere in the brain (e.g., the area tempestas in the deep prepiriform cortex). Thus, the hippocampus appears to be an easily recruited target of epileptogenic stimuli, and hippocampal circuits may participate in the amplification of the propagated seizure discharge.

X. Effect of Seizures on Immediate Early Gene (IEG) Expression in Brain Regions

Activation of neural cells can lead to the induction of new programs of gene expression as part of the adaptive and trophic processes that continually occur in the brain. For one class of genes, increased transcription occurs within minutes of stimulation; this class is referred to as immediate early genes (IEGs). These genes, which include c-fos, c-jun, jun-B, and zif/268 among others, are thought to control the expression of late response genes whose products influence specific structural and functional aspects of the neurons in which they are expressed. Some of the IEGs encode transcription factors which bind to DNA and can facilitate (or suppress) expression of nearby genes. [*See* DNA AND GENE TRANSCRIPTION.]

Convulsive seizures provoke an increase in expression of the c-fos gene and other IEGs in neural cells in the brain. It appears that, at least for certain neuronal populations, induction of IEG expression can be used as a marker for increased excitatory activation of neuronal cells during convulsive seizures.

With seizures evoked from area tempestas or by systemic treatment with chemoconvulsants such as kainic acid, pentylenetetrazol and picrotoxin, the most pronounced increases in mRNA for c-fos are seen in the piriform cortex, amygdala, entorhinal cortex, olfactory bulbs, and throughout the hippocampus (Color Plate 24). A similar pattern of increase occurs with mRNAs for zif/268, c-jun and jun-B. There is a consistent lack of detectable increase in IEG expression in superior and inferior colliculus, geniculate bodies, substantia nigra, and cerebellar cortex in these circumstances. The increases in IEG expression reach a peak at 30 minutes after seizure initiation; by four hours, mRNA levels for the IEGs return to baseline.

Based on the electrographic and behavioral characterization of convulsive seizures evoked from the inferior colliculus, we would not expect to see marked activation of IEG expression in forebrain areas in association with these seizures. Stimulation of inferior colliculus by focal application of bicuculline evokes explosive running–bouncing convulsions in the absence of forebrain electrographic seizure activity and this type of convulsion, as discussion earlier, is dependent upon hindbrain, but not forebrain, circuitry for its expression. In fact, little or no elevation of IEG expression is detected in hippocampus, entorhinal cortex, or piriform cortex following convulsions evoked from inferior colliculus. In general, no marked increases in IEG expression are evident in any forebrain area thus far surveyed. Similarly, there are no apparent increases in c-fos expression in forebrain regions in response to audiogenic seizures; this is another seizure model predominantly associated with hindbrain mechanisms. Interestingly, in the audiogenic seizure model, several subcortical auditory nuclei, including inferior colliculus, exhibited marked increases in c-fos expression. As these nuclei do not exhibit increased c-fos expression in association with seizures evoked by chemical stimulation of inferior colliculus, it is likely that they are part of circuits involved in processing of the seizure-eliciting sensory (i.e., auditory) input and are not substrates involved in seizure propagation or expression.

It is interesting to compare the results obtained with 2DG accumulation and IEG expression. With seizures associated with the forebrain, both techniques reveal marked activation of the olfactory bulb, piriform and entorhinal cortex, and hippocampus. These areas are interconnected via excitatory synaptic links, which, based on these mapping studies, appear to be highly activated by the seizures. There are also several regions in which marked seizure-evoked increases in 2DG accumulation occur without obvious changes in IEG expression. For example, the substantia nigra and selected thalamic nuclei have dramatic increases in the 2DG signal while appearing unchanged when examined for expression of c-fos or other IEGs. In these instances, increased 2DG accumulation is probably associated with the activation of inhibitory inputs into these structures, such as the inhibitory GABA-containing terminals innervating the substantia nigra. Increased inhibitory input to a region would not be expected to evoke an increase in IEG expression in the post-synaptic cells, thus accounting for the discrepant patterns with the two techniques.

XI. Neurotransmitter Mechanisms

It has become clear that no single neurotransmitter can be related to the genesis of seizures. Rather, neurotransmitters and neuromodulators interact in a complex pattern to determine seizure susceptibility and the characteristics of seizure spread. This interaction is anatomically specific and depends on the configuration and identity of the synapses in a given brain locus. γ-Amino-butyric acid (GABA), the most prevalent inhibitory transmitter in brain, exerts an anticonvulsant influence in certain brain regions whereas in other regions this same neurotransmitter acts in a proconvulsant fashion. Likewise, glutamate, the most prevalent excitatory transmitter in brain, has both seizure-inducing and seizure-suppressing actions in different brain areas.

At the same time, pharmacological studies have demonstrated that the overall net effect of enhancing GABA transmission in the central nervous system (CNS) is to reduce convulsive seizure susceptibility and, conversely, a general reduction in GABA transmission can evoke convulsive seizures. In humans, the drug γ-vinyl-GABA, or vigabatrin, an agent that interferes with the metabolic breakdown of GABA, and thereby elevates the GABA content in brain, has clinically effective anticonvulsant activity. Conversely, isoniazid (a drug used in the treatment of tuberculosis) interferes with GABA synthesis and, consequently, can cause convulsive seizures. It is believed that the anticonvulsant action of the benzodiazepines (e.g., valium) in both experimental animals and humans is due to the ability of these drugs to enhance GABA-mediated neurotransmission. On the other hand, general (systemic) treatment with drugs that block glutamate transmission (especially those drugs that block *N*-methyl-D-aspartate (NMDA)-sensitive glutamate receptors) is anticonvulsant, whereas the dominant action of drugs that stimulate glutamate receptors is to induce seizure activity.

Another neurotransmitter that has been found to have a general influence on seizure susceptibility is norepinephrine. Depletion of this transmitter, or lesions of a major norepinephrine-containing cell group, the locus coeruleus, usually cause a decrease in seizure threshold. Likewise, drugs which block CNS receptors for norepinephrine (α receptors in particular) can lower seizure threshold in experimental animals as well as in humans. This may account for the increased seizure susceptibility occasionally observed in patients treated chronically with certain antipsychotic drugs such as phenothiazines.

Opioid drugs have complex influences on seizure susceptibility. Depending upon the mechanism of seizure induction, morphine and other compounds that stimulate the μ opiate receptors exert either anticonvulsant or proconvulsant actions: anticonvulsant effects are obtained against electroshock seizures, whereas proconvulsant effects can be seen with most chemoconvulsant seizures. Stimulation of μ opiate receptors can induce electrographic seizure discharge in hippocampus, and when placed directly into the area tempestas or into the ventral hippocampus, convulsive seizures are evoked. When placed into the substantia nigra, morphine and related opioid peptides produce anticonvulsant actions against electroshock seizures.

In experimental animals, high doses of drugs that stimulate muscarinic cholinergic transmission tend to induce convulsive seizures. One such agent, pilocarpine, has been used to create an animal model of limbic motor seizures. These seizures are especially sensitive to blockade by inhibition of activity in the area tempestas (see earlier discussion). Since muscarinic stimulation in the area tempestas can evoke convulsive seizures, it is possible that this region is an important site of origin of seizures evoked by pilocarpine or similar agents. In addition, stimulation of muscarinic receptors by drug treatment directly in amygdala can be used instead of electrical stimulation to produce kindling of this region.

Finally, it should be recognized that convulsive seizures do not necessarily require activation of brain circuitry for their occurrence. It is possible to induce a type of tonic convulsion by a direct action on spinal cord neurotransmission. This is best exemplified by the convulsant drug, strychnine, which acts by blocking receptors for glycine, an inhibitory neurotransmitter in the spinal cord. Strychnine induces tonic convulsions without the appearance of cerebral electrographic seizure discharge and without any other convulsive components.

Bibliography

Avoli, M., Gloor, P., Kostopoulos, G., and Naquet, R., eds., (1990). "Generalized Epilepsy, Neurobiological Approaches." Birkhauser, Boston.

Dichter, M., ed., (1988). "Mechanisms of Epileptogenesis: From Membranes to Man." Plenum, New York.

Gale, K. (1989). "GABA in epilepsy: The pharmacologic basis." *Epilepsia* **30**(*Suppl* 3), S1–S11.

Gale, K. (1985). Mechanisms of seizure control mediated by gamma-aminobutyric acid: role of the substantia nigra. *Fed. Proc.* **44,** 2414–2424.

Wada, J. A. (ed.) (1981). "Kindling," Vol. II, Raven Press, New York.

Wada, J. A. (ed.) (1986). "Kindling," Vol. III, Raven Press, New York.

Wada, J. A. (ed.) (1990). "Kindling," Vol. IV, Plenum Press, New York.

Se

Selenium in Nutrition

GERALD F. COMBS, JR., *Cornell University*

Glossary

Keshan disease Juvenile cardiomyopathy associated with severe selenium deficiency and prevalent in areas of endemic selenium deficiency in China

Lipid peroxidation Process of oxidative degradation of polyunsaturated lipids, involving the generation in the vicinity of their 1,4-pentadiene structures of carbon-centered free radicals followed by the attack on those centers by molecular oxygen and the subsequent formation of chain cleavage products

Selenium Group VIA element that is an essential nutrient for animals and humans in which it functions as the active center of the antioxidant enzyme glutathione peroxidase

Selenium-dependent glutathione peroxidase Enzyme that catalyzes the reduction of peroxides (e.g., H_2O_2, lipid hydroperoxides) using reducing equivalents from the reduced form of glutathione; this enzyme contains selenium in a selenocysteinyl residue at its active center and is the only known metabolically active form of that element

Vitamin E Membrane-resident factor with antioxidant function due to its ability to reduce free radicals species without propagating lipid peroxidation

NOT UNTIL THE late 1950s was the element selenium (Se) thought to play a role in normal metabolism. Until that time, the biomedical significance of this metalloid had been recognized only for its toxic properties. However, in 1957 it was discovered that trace amounts of Se could alleviate necrotic liver disease and capillary leakage in vitamin E-deficient animals, suggesting that Se spared the need for that fat-soluble vitamin. Research in the 1960s culminated in the recognition that Se is more than just a vitamin E–sparing factor, that it is an essential nutrient required for the antioxidant enzyme glutathione peroxidase. Since that time, an increasing understanding has emerged of the metabolic functions and health implications of this trace element. At present, Se is generally regarded as having prime importance in the metabolic protection from oxidative stress with special relevance to disease of the heart muscle and drug metabolism. In addition, the results of animal tumor model studies and several human epidemiological investigations support the hypothesis that Se may have a role in the protection against carcinogenesis. It is the purpose of this review to summarize present understanding of the role of Se in human nutrition and health.

I. Selenium in Foods and Human Diets

A. Selenium in Foods

The Se contents of foods widely vary due mainly to the amount of biologically available Se in the environment (e.g., the soluble Se content of the soil for plant species; the biologically available Se content of the diet for animal species). Because of the intimate relation between plants and animals in food chains, the Se contents of foods from both plant and animal origins tend to be greatly influenced by the local soil Se environment. Thus, foods of all types tend to show geographic patterns of variation in Se content reflecting, in general, local soil Se conditions. Because almost all Se in plant and animal tissues is bound to proteins, the Se contents of foods also tend to be correlated with its protein content.

Variation in the Se content of foods because of geochemical differences is readily seen by comparing the Se contents of like foods from different countries. For example, whole wheat grain may contain more than 2 ppm Se (air-dry basis) if produced in the Dakotas, but as little as 0.11 ppm Se if produced in New Zealand, and only 0.005 ppm Se if produced in Shaanxi Province, China. On a global basis, foods with the lowest Se contents are found in the low-Se regions of China, in particular, the provinces of Heilongjiang, northern Shaanxi and Sichuan. Ironically, foods containing the greatest concentrations of Se have also been found in the same country, although in different locales.[1]

The Se contents of foods of animal origin largely depend on the Se intakes of livestock. Food animals raised in regions with feeds of low-Se content deposit relatively low concentrations of the mineral in their edible tissues and products (e.g., milk, eggs), whereas animals raised with relatively high-Se nutriture yield food products with much greater Se concentrations. Because of the needs of livestock for Se to prevent debilitating deficiency syndromes, Se (usually in the form of Na_2SeO_3) is used as a feed supplement in animal agriculture in many parts of the world. This practice became widespread in North America and Europe only within the past 10–15 years and has reduced what would otherwise be a stronger geographic variation in the Se contents of animal food products.

Within the normal ranges of Se supplementation of livestock diets, muscle meats from most species tend to contain 0.3–0.4 ppm Se (fresh weight basis). Organ meats usually accumulate greater concentrations of Se; the livers of most species generally contain about four times as much Se as skeletal muscle, and the kidneys of steers, lambs, and swine have been found to accumulate 10–16 times the amounts in muscle.

B. Selenium in Human Diets

Because of differences in geography, agronomic practices, food availability, and preferences, most of which are difficult to quantify, evaluations of Se intakes of specific human population groups are often not precise. General comparisons can be made, however, of the Se contents of different food supplies by using the average Se concentrations determined within specific major classes of foods in different locales. Table I presents the typical Se contents of the major classes of foods from several countries. These values are, for the most part, based on actual analyses of foods from each country.

The average per capita daily Se intakes of adults in various countries of the world have been estimated (Table II). They widely vary among different regions, the lowest being only 7–11 μg Se/person/day in the areas of Se-responsive human diseases in China. In contrast, the Se intakes of residents of countries with well-recognized endemic Se deficiency disorders in livestock (i.e., Finland and New Zealand) are estimated to be at least threefold those in the Se-deficient regions of China. Residents of the so-called low-Se parts of the United States (e.g., Ohio, southeastern seaboard) have estimated Se intakes approximately two- to fivefold those of Finns or New Zealanders.

The most important sources of Se in the diets for most people are cereals, meats, and fish. Dairy products and eggs contribute small amounts of Se to the total intakes in most countries, although these can represent large percentages of the total Se intakes in countries where the rest of the diet provides little Se (e.g., Finland, New Zealand). Vegetables and fruits are uniformly low in Se (when expressed on a fresh weight basis) and provide only small amounts (less than 8% of the total intake) of the mineral in most human diets.

Core foods for Se in American diets have been identified in a USDA study that compiled data from the USDA 1977–78 Nationwide Food Consumption Survey with published values for the Se contents of American foods. That analysis revealed that five foods (beef, white bread, pork, chicken, and eggs) contributed ca. 50% of the total Se in the "typical" American diet and that 80% of the total dietary Se was provided by a core of only 22 foods (Table III).

Differences in patterns of food consumption, whether general ones caused by cultural influences or specific ones caused by personal preferences and food availability, can significantly affect Se intake. In view of this potential variation in Se intake with different dietary habits, it is not surprising to find this actually to be the case for individuals able to select their own patterns of food consumption from a variety of foods. In a study of the Se intakes of

1. Endemic selenosis of animals and humans has been identified in a few mountainous communities in Hubei and southern Shaanxi provinces.

TABLE I Typical Se Contents of Major Classes of Foods from Several Countries

Food	USA	England	W. Germany	Finland	N. Zealand	China			Venezuela
						Se-deficient	Moderate	High Se[b]	
Cereal products	0.03–0.66	0.02–0.53	0.03–0.88	0.005–0.12	0.004–0.09	0.005–0.02	0.017–0.11	1.06–6.9	0.132–0.51
Vegetables	0.001–0.10	0.01–0.09	0.04–0.10	0.001–0.02	0.001–0.02	0.002–0.02	0.002–0.09	0.34–45.7	0.002–2.98
Fruits	0.002–0.01	0.005–0.01	0.002–0.04	0.002–0.03	0.001–0.004	0.001–0.003	0.005–0.04	—	0.005–0.06
Red meats	0.05–0.27	0.05–0.14	0.13–0.28	0.01–0.07	0.01–0.04	0.01–0.03	0.05–0.25	—	0.17–0.83
Organ meats	0.43–1.90	0.20–2.46	0.09–0.95	0.06–1.71	0.05–2.03	0.05–0.10	0.05–1.00	—	0.36–0.83
Poultry	0.04–0.15	0.05–0.15	0.05–0.15	0.05–0.10	0.05–0.10	0.02–0.06	0.05–0.10	—	0.10–0.70
Fish	0.19–1.9	0.10–0.61	0.24–0.53	0.18–0.98	0.03–0.31	0.03–0.20	0.10–0.60	—	0.32–0.93
Milk products	0.01–0.24	0.01–0.08	0.01–0.10	0.01–0.09	0.003–0.025	0.002–0.01	0.01–0.03	—	0.11–0.43
Eggs	0.06–0.20	0.05–0.20	0.05–0.20	0.10–0.20	0.24–0.98	0.02–0.06	0.05–0.15	—	0.50–1.5

[a] I.e., Keshan disease–endemic areas.
[b] I.e., areas of endemic selenosis.

TABLE II Estimated Per Capita Daily Intakes of Se (μg) of Adults and Infants in Several Countries

Age group	USA	England	W. Germany	Finland	New Zealand	China			Venezuela
						Se-deficient[a]	Moderate	High Se[b]	
Adults	60–220	50–120	60–150	30–100	30–80	7–11	60–120	750–4,990	200–350
Infants (3 mos.)	12–18	10–15	18–24	5–10	7–9	2–3	12–18	180–250	—

[a] I.e., Keshan disease–endemic areas.
[b] I.e., areas of endemic selenosis.

TABLE III Core Foods for Se in American Diets

Rank	Food	Contribution to total Se intake	
		Individual %	Cumulative %
1	Beef	17.2	17.2
2	White bread	14.2	31.4
3	Pork/ham	8.2	39.6
4	Chicken	6.5	46.2
5	Eggs	4.8	51.0
6	White rolls	4.0	55.0
7	Whole wheat bread	3.3	58.3
8	Noodles, pasta, etc.	3.0	61.3
9	Whole milk	2.8	64.1
10	Canned tuna	2.1	66.2
11	Milk, 2% butterfat	1.8	68.0
12	White rice	1.7	69.7
13	Macaroni and cheese	1.3	71.0
14	Luncheon meat	1.2	72.2
15	Spaghetti w/meat sauce	1.1	73.3
16	Mayonnaise	1.1	74.4
17	Meat loaf (beef)	1.0	75.4
18	Hamburger on buns	1.0	76.4
19	Oatmeal	1.0	77.4
20	Cracked wheat bread	1.0	78.4
21	Rye bread	1.0	79.4
22	Turkey	0.9	80.3

free-living individuals in Maryland, the mean daily intake of Se was found to be 81 μg per person; however, 17% of diets provided less than 50 μg Se/person/day, while 5% provided more than 150 μg Se/person/day.

Differences in Se consumption by women in different parts of the world are reflected as differences in the Se contents of human milk and, therefore, in the Se intakes of breast-fed infants (Table II). Large differences in the Se nutriture (i.e., estimated intakes from 7.5 to 212 μg/infant/day) are seen. However, it is apparent that the differences in Se intakes are generally much less than those of adults in the same countries. Limited analyses of the Se contents of human milk in the low-Se areas of China indicate that the Se intakes of nursing infants, particularly in areas of endemic Keshan disease, may be as little as 33–40% of those of breast-fed infants in Finland and New Zealand.

C. Bioavailability of Dietary Selenium

The use of dietary Se involves several physiological and metabolic processes that convert a portion of ingested Se to certain metabolically critical forms that are necessary for normal physiological function. Ingested Se (normally in the form of Se-containing proteins and inorganic compounds) is subject to several potential losses en route to the metabolic production of its metabolically active form(s). These losses include those associated with the digestion and enteric absorption of ingested Se; thus, Se compounds that are insoluble under conditions of the lumenal environment of the small intestine, as well as Se-containing proteins of low digestibility, will pass through the animal to be eliminated in the feces. It is probable that under normal circumstances, there is only a small enterohepatic circulation of absorbed Se and, therefore, that fecal Se represents only that amount of ingested Se that was not absorbed. In general, the apparent absorption of Se in foods, inorganic compounds, and Se amino acids is good (ca. 70%); however, it is highly variable both between and within single sources.

Not all absorbed Se is physiologically important. Some is metabolized to methylated forms that are readily excreted. For example, trimethyl selenonium cation comprises 10–20% of urinary Se, and the volatile dimethyl selenide is readily excreted across the lung in expired air, although the production of the latter species is probably only important under conditions of high intakes of Se. In addition to these forms, Se is normally metabolized to several species, most notably the Se analogues of the S-containing amino acids and, hence, to Se-containing polypeptides and proteins. Of these, the Se-dependent glutathione peroxidase (SeGSHpx) is the only physiologically critical species known. However, several other Se proteins with undefined function have been identified. One or more of these may prove to be physiologically important; however, because Se is metabolized in many cases indiscriminately as an analogue of S, it is likely that at least some of these Se proteins are physiologically inert, serving only as reserves of Se to be made available for metabolism to critical forms as they turn over.

In general, the bioavailability of Se compounds can be described as follows: (1) The more reduced (and insoluble) inorganic forms of Se have low bioavailabilities; (2) the common selenoamino acids (i.e., selenomethionine, selenocysteine) and Se in most plant materials have reasonably good bioavailabilities (i.e., approaching that of sodium selenite); and (3) Se in many animal products has low to moderate bioavailability. In addition, other factors can significantly influence the bioavailability of Se, by affecting the use of ingested Se in either the digestion/absorption or the metabolism/excretion phases (Table IV). Factors (e.g., vitamin E and other antioxidants) that increase the enteric absorption of Se and/or increase the metabolism of absorbed Se to the physiologically critical forms will positively affect Se bioavailability. Alternatively, factors (e.g., heavy metals) that decrease the enteric absorption of Se and/or increase the metabolism of Se to more readily excreted forms (e.g., the methylated forms) negatively affect Se bioavailability.

II. Biochemical Functions of Selenium

Although the nutritional role of Se has always been associated with that of vitamin E and, hence, with antioxidant activity, the metabolic basis of this nu-

TABLE IV Factors Affecting the Bioavailability of Dietary Selenium

Type of factor	Good Se bioavailability	Poor Se bioavailability
Chemical form of Se	Oxidation to selenate or selenite Presence as Se amino acids (Se-cysteine, Se-methionine)	Reduction to insoluble selenide or elemental Se
Food sources of Se	Se-enriched yeasts Wheat Most plant products	Meat and fish products Soy protein products
Dietary factors	Restricted food intake Vitamins E, C, and A (at high levels) Supplemental methionine Synthetic antioxidants (e.g., ethoxyquin)	Heavy metals (e.g., Cd, Hg) Deficiencies of pyridoxine, vitamin E, riboflavin, methionine Excess sulfur, arsenic

tritional role remained unknown until the early 1970s when Se was discovered to be an essential part of the antioxidant enzyme SeGSHpx. That enzyme catalyzes the reduction of hydrogen peroxide and fatty acid hydroperoxides using reducing equivalents of reduced GSH. Human SeGSHpx has been purified and has been found to be an homologous tetramer, each subunit of which contains one Se atom in the form of selenocysteine located at the active site.

The metabolic function of SeGSHpx is as part of the cellular defense system against damage induced by free radicals, particularly those of oxygen. The addition of electrons to molecular oxygen can yield a number of radical intermediates, some of which can interact with membrane lipids to alter membrane structure and function. Two such radicals, superoxide anion ($O_2^{\cdot-}$) and peroxide ($HOO^{\cdot}$), can be formed by endogenous enzyme systems and by such exogenous agents as drugs and pesticides. Although superoxide itself is relatively unreactive with the polyunsaturated fatty acids (PUFAs) of membrane phospholipids, in the presence of ferric iron (Fe^{+3}), it can interact with peroxide to form highly reactive species such as hydroxyl radical (HO^-) and singlet oxygen ($^1O_2$). The latter can initiate lipid peroxidation by abstracting hydrogen atoms from PUFAs, causing reactions that branch and become autocatalytic. Whereas the formation of peroxides and free radicals is essential to certain physiological functions (e.g., the bactericidal activities of polymorphonuclear leukocytes and macrophages; intermediary steps in prostaglandin synthesis), uncontrolled lipid peroxidation is detrimental to the cell.

Cellular defense against damage caused by free radical chain reactions involves attacks on these processes at several different stages. The proximal reactive oxygen species, superoxide, is converted by a family of enzymes called *superoxide dismutases* to hydrogen peroxide. That product is then converted either to water by SeGSHpx, or to water and O_2 by the heme-enzyme catalase. Because SeGSHpx is the more widely distributed of these enzymes within the cell (i.e., it is found in the cytosol and mitochondrial matrix space, whereas catalase is limited to the peroxisomes), it is most important in removing hydrogen peroxide and, thus, diminishing the amount of that species available to react with superoxide to initiate free radical reactions. Free radicals that are formed can be scavenged by vitamin E, which can donate an hydrogen atom to quench the radical. This explains the long-observed relation of vitamin E and Se, and why either one is sufficient to prevent a variety of deficiency diseases of animals. [*See* VITAMIN E.]

Substrates of SeGSHpx include hydroperoxides of free fatty acids, as well as hydrogen peroxide, which suggests that SeGSHpx may also serve in the antioxidant defense system by interrupting the autocatalytic phase of lipid peroxidation. However, SeGSHpx cannot reduce lipid peroxides when they are acylated (i.e., in phospholipids of biological membranes); therefore, peroxidized fatty acids must be rendered free of their glyceryl components by a membrane-associated lipase for SeGSHpx to function in this manner. Otherwise, it acts only to reduce hydrogen peroxide in preventing the initiation of lipid peroxidation.

Nutritional Se deficiency results in decreased SeGSHpx activities in all tissues. These changes are accompanied by changes in other hepatic enzymes and substrates involved in the metabolism and detoxification of drugs and other foreign compounds. The latter changes are not necessarily of the same magnitude or direction for all enzymes. For example, in vitamin E–adequate mice the activities of certain cytosolic enzymes (e.g., some GSH-S-transferases, GSSG reductase) were elevated by long-term Se deprivation, while others (e.g., GSH-thioltransferase and GSH-sulfotransferase) were simultaneously decreased. Similarly, some microsomal enzyme activities were increased (e.g., cytochrome P_{450}-dependent hydroperoxidase, heme oxygenase, UDP-glucuronyl transferase), some were reduced (e.g., NADPH-cytochrome P_{450}-reductase, flavin-monooxygenase), and others were unchanged (e.g., NADH-cytochrome-b_5-reductase, anilinehydroxylase, aminopyrine-N-demethylase). Glutathione-S-transferase activities have been shown in several species to be elevated in Se deficiency.

These changes in drug metabolism can increase the toxicities of some compounds but decrease the toxicities of others. In the first class are such compounds as nitrofurantoin (an antibacterial agent used for treatment of urinary tract infections) and paraquat (a dipyridinium herbicide), both of which are metabolized to free radicals that generate reactive oxygen species. Dietary Se protects against the acute toxicities of these types of compounds through its function as SeGSHpx in removing H_2O_2 and, thus, preventing lipid peroxidation that would otherwise be stimulated by these pro-oxidants.

Other compounds (e.g., iodipamide, acetaminophen, aflatoxin B_1) are rendered less toxic by nutritional Se deficiency. The protection is thought to be due to the increases in GSH-S-transferase activities and GSH concentrations that occur in Se deficiency. At least one of the GSH-S-transferases also has GSHpx activity, and it has been suggested that the increases in this enzyme in Se deficiency may compensate for the losses in SeGSHpx. In addition, GSH-S-transferases serve important detoxification functions by conjugating foreign compounds with GSH, thereby promoting their excretion.

Several Se-containing proteins in addition to SeGSHpx have been identified in animal tissues. Whether any has enzymic activity or any other physiologic function is not known. The levels of one selenoprotein found in rat plasma and liver (i.e., "selenoprotein-P") reflect changes in Se nutriture and have been proposed as a possible transporter of Se; it may prove useful as an index of nutritional Se status.

III. Evaluating Selenium Status of Humans

Analyses of the Se concentrations of human tissues demonstrate a strong geographic variation in Se status correlating with the geographic variation in the Se contents of food supplies (Table V). Populations with the lowest apparent intakes of Se (i.e., the Se-deficient regions of China, New Zealand, Finland) also have the lowest concentrations of Se in their blood. Blood Se concentrations have revealed substantial geographic variation within the United States: In the 1960s the mean Se concentrations of whole blood from blood banks in 19 cities was found to vary from 157 to 256 ppb. That variation appeared to relate to differences in local intakes of Se, as the cities ranking at the extremes of that range were located in areas known to be either high (e.g., northern Great Plains states) or low (northeastern states) with respect to Se. The levels of Se and SeGSHpx in whole blood show a good correlation at blood Se concentrations less than 100 ppb; however, at higher levels (such as are generally found in Americans), the correlation is relatively weak, suggesting that Se resides in forms other than active SeGSHpx in increasing amounts in individuals with adequate or better Se intakes.

Nutritional Se status can be assessed on the basis of the Se contents of hair or nails, which Chinese studies have shown to be highly correlated with the level of Se intake. This approach has the obvious advantage of ease of sample availability, storage, and preparation; however, in the case of hair sampling, standardization of the scalp location and amount of hair is important to ensure reproducibility. It is also important to screen subjects to ascertain who may have recently used Se-containing antidandruff shampoos, as their hair will bear Se residues that do not reflect nutritional Se status. The renal clearance of Se is important in the homeostasis of the element; therefore, measurement of urinary Se excretion can provide useful information for the assessment of nutritional Se status in humans. The urinary excretion of Se is generally greater than the fecal excretion of the element; urine Se content is a function of the level and form of Se intake, the nature of the diet, as well as the Se status of the individual. Urinary Se excretion is closely correlated with the Se concentrations of blood or plasma but can, therefore, be influenced by the Se consumed with a recent meal.

Plasma Se concentrations are 10–45% lower in pregnant women than in nonpregnant women. It has been suggested that this difference may be a factor in the etiology of pre-eclampsia; however, blood Se levels in normal and pre-eclamptic pregnancies have been found to be similar. It is likely that at least a portion of the decrease in plasma and whole blood Se are due to the hemodilution of pregnancy

TABLE V Typical Blood Se Concentrations (ng/ml) of Residents of Several Countries

						China			
Tissue	USA	England	W. Germany	Finland	N. Zealand	Se-deficient[a]	Moderate	High Se[b]	Venezuela
Plasma	65–180	80–110	32–88	46–109	43–54	—	—	—	—
Whole blood	75–265	—	—	56–130	—	10–21	33–136	3,200–3,480	355–813

[a] I.e., Keshan disease-endemic areas.
[b] I.e., areas of endemic selenosis.

and the demands of the growing fetus for Se. Blood Se concentration rapidly returns to nonpregnancy levels after delivery. Changes in plasma Se concentration during pregnancy have been associated with changes of similar magnitude in the plasma activity of SeGSHpx.

Premature infants are generally born with Se status comparable with those of full-term infants at the time of delivery. However, because hospitalized premature infants are frequently maintained with low-Se parenteral nutrient solutions or are fed low-Se formula diets for as long as several weeks, they frequently show progressive declines in blood Se concentrations and SeGSHpx activities. Because they also have low plasma levels of vitamin E, their low Se status with its attendant reduction in antioxidant protection may have significant physiological consequences (e.g., increased adverse drug reactions).

IV. Consequences of Low Selenium Status

Two diseases of children are recognized as being associated with severe nutritional Se deficiency: a cardiomyopathy named *Keshan disease*, and a chondrodystrophy named *Kaschin-Beck disease*. Both diseases occur in rural areas in the belt of endemic Se deficiency of China. The endemic distributions of both diseases are similar but not always identical; both correspond to the distribution of endemic Se deficiency in foods.

Keshan disease has been diagnosed in more than a dozen Chinese provinces, almost exclusively among people living in the mountainous areas where the soil is low in Se (i.e., <125 ppb Se, of which less than 2.5% is water-soluble). Accordingly, locally produced foods are exceedingly low in Se content (e.g., grains generally contain <40 ppb Se). Se/vitamin E–deficiency diseases of livestock (e.g., "white muscle disease"[2] in lambs, "mulberry heart disease"[3] in pigs) are also endemic in these areas. Humans living in those areas typically show the lowest tissue Se levels of any free-living populations anywhere on earth (e.g., blood Se <25 ppb; hair Se <100 ppb).[4] Keshan disease has been more prevalent among farming families in rural districts than among residents of urban centers; this difference probably relates to the more monotonous dietary habits of rural residents and to their stronger dependency on food produced in their immediate locales.

Keshan disease is a multifocal myocarditis occurring primarily in children between the ages of 2 and 10 years and, to a lesser extent, among women of child-bearing age. Infants are rarely affected; their intakes of Se from breast milk, even in these deficient areas, appears to be ca. 3 μg/day, which is thought to be sufficient to protect them. The first signs of the disease are normally not observed until after children are weaned to solid foods, at which time their intakes drop to ca. 1.5 μg Se/child/day. Keshan disease shows marked seasonal and annual variations in incidence, suggesting that other factors must be involved in its etiology.

Keshan disease is manifest as acute or chronic insufficiency of cardiac function, cardiac enlargement, gallop rhythm, cardiac arrhythmias, and electrocardiographic (EKG) and radiographic abnormalities. Subjects may show cardiogenic shock or congestive heart failure; embolic episodes from cardiac thromboses have been reported. Four clinical subtypes of Keshan disease have been identified: an acute type with sudden onset of dizziness, malaise, substernal discomfort, and dyspnea in otherwise apparently healthy children with no history of cardiac disorders; a chronic type, characterized by chronic congestive heart failure with varying degrees of cardiac insufficiency; a subacute type, with signs and symptoms varying with the degree of cardiac insufficiency but with an accelerated course; and a latent type characterized by normal heart function with mild cardiac dilation and associated EKG changes. The case fatality of Keshan disease in China was greater than 80% in the 1940s but has been reduced in recent years to ca. 30%, primarily as the result of better medical care.

Hypotheses for the etiology of Keshan disease have proposed that it may be caused by food- and/or water-borne toxins (e.g., mycotoxins, heavy metals, nitrite), certain infectious agents (e.g., cardiophilic viruses), or specific nutrient deficiencies (e.g., Se, Mo, Mg, riboflavin). Available experimental evidence provides support for only the last two of these hypotheses. Intervention studies have demonstrated that Se can be effective in the prevention of Keshan disease. Improvements in the Se status of children have been achieved by either the

2. I.e., a skeletal myopathy.
3. I.e., a cardiomyopathy.
4. I.e., less than half of the corresponding values for New Zealanders or Finns.

use of table salt fortified with Se (e.g., 10–15 ppm as Na_2SeO_3) or oral tablets containing Na_2SeO_3; both vehicles have been used to produce dramatic reductions in the incidence of Keshan disease. Although the etiology of Keshan disease is still not fully elucidated, it is clear that Se deficiency is at least a contributing factor and is responsible for defining the endemic distribution of the disease.

Cardiomyopathies associated with low Se status have been reported in a few cases outside of China. However, it cannot be determined at this time the extent to which Se deficiency may have been involved in the pathogenesis of those cases. Certainly, low Se status is not a general feature of cardiomyopathy patients in the United States.

Kaschin-Beck disease is an osteoarthropathy with an endemic distribution of corresponding to that of Keshan disease. The disease primarily affects the epiphyseal cartilage, articular cartilage, and epiphyseal growth plates of growing bones. The long bones are most frequently affected; however, the cartilage tissue associated with any bone in the skeleton may be involved. Affected cartilage shows atrophy and necrosis with repair and disturbance in endochondral ossification. The most striking histological feature is chondronecrosis with proliferation of surviving chondrocytes in clusters, categorized as a coagulation necrosis. The condition results in enlarged joints (especially of the fingers, toes, and knees), shortened fingers, toes, and extremities, and, in severe cases, dwarfism.

Studies of the effects of Se supplementation in the prevention and therapy of Kaschin-Beck disease are few but have yielded some encouraging results. Two controlled Se intervention studies showed that the Se tablets reduced the severity of the disease and facilitated improvement of prevalent cases. The etiology of Kaschin-Beck disease is presently unclear. Its coincident distribution with that of severe Se deficiency and the reported effectiveness of supplemental Se in reducing its incidence and/or severity suggest that Se may play some role in its etiology.

Altered Se status has been found consistently in association with relatively few human diseases. Reduced blood Se levels (by 21–60%) have been measured in patients with alcoholic liver disease and cirrhosis; however, these effects may be without significant biological consequences inasmuch as they have not been found to be associated with reductions in blood SeGSHpx. Patients with neuronal ceroid lipofucsinosis, acrodermatitis enteropathica, Kwashiorkor, and chronic renal failure have been found to show reductions in blood Se by 17–45%, whereas patients with muscular dystrophy have been found with increased (by ca. 40%) plasma Se levels. Multiple sclerosis patients have had reductions of blood SeGSHpx by as much as 23%; yet their plasma Se concentrations appear not to be affected. The plasma Se concentrations of patients with Down's syndrome are consistently reported to be abnormally low (by ca. 27%), although these individuals also show increased (by ca. 55%) erythrocyte SeGSHpx activities. Many of these apparent differences in Se status between diseased and healthy persons may relate to differences in diet (i.e., Se intake). Such an effect is to be expected in such cases as Kwashiorkor, in which inadequate intake of protein, the vehicle for most dietary Se, results in the disease. No evidence indicates abnormalities in Se metabolism in any of these diseases.

Finnish studies have shown low serum Se levels to be associated with increased risk to cardiovascular disease. The apparent protective effect was associated with increased concentrations of high-density lipoprotein-bound cholesterol (by 43%) in serum.

Low Se status, as indicated by low concentrations of the element in blood cells and/or plasma, has been identified in infants with the inborn errors of amino acid metabolism, maple syrup urine disease, or phenylketonuria. However, low Se status is not related directly to any of these metabolic diseases. Because these children are fed purified diets low in protein and, hence, low in Se, they show serum Se levels (e.g., as low as 5 ppb) with erythrocyte SeGSHpx activities of only 10–20% of those of healthy children.

Parenteral nutrition fluids based on amino acid mixtures contain negligible amounts of Se (e.g., <5 μg Se/1,000 kcal) if they are not supplemented with the element. Therefore, it is not surprising that low Se status has been observed in patients maintained with total parenteral nutrition (TPN) for extended periods of time. To prevent this from happening, it is recommended that TPN solutions be supplemented to provide 25–30 μg Se/day.

The quantitative dietary requirements of humans for Se are subject to some debate for the reason that unlike most other essential nutrients, no clear-cut deficiency syndrome has been identified in deficient humans. Nevertheless, there is general agreement that most humans require ca. 0.85 μg Se/kg body weight/day to support maximal tissue activities of

TABLE VI 1989 National Research Council Recommended Dietary Allowances for Se

Group	Age (yrs.)	μg Se/day
Infants	0–0.5	10
	0.5–1	15
Children	1–6	20
	7–10	30
Males	11–14	40
	15–18	50
	19+	70
Females	11–14	45
	15–18	50
	19+	55
Pregnant		65
Lactating		75

SeGSHpx. For most adults, this corresponds to intakes of 55–70 μg Se/person/day (see Table VI).

V. Selenium and Cancer

Several Se compounds have been demonstrated to inhibit or retard carcinogenesis in a variety of experimental animal models. The available information concerning experimental carcinogenesis as affected by either nutritional or pharmacological levels of Se reveals different general effects that may relate to dose-related differences in Se metabolism. Numerous studies have evaluated the effects of pharmacological levels of Se on experimental carcinogenesis using several systems, including chemical, viral, and transplantable tumor models. Of those, two-thirds have found that high-level Se treatment reduced the development of tumors at least moderately (i.e., by 15–35% from control levels) and, in most cases, substantially (i.e., by more than 35%). Only a few studies have found Se treatments of animals to be without effect on tumor outcome. Taken collectively, these results indicate that Se may have less antitumor efficacy when used at levels less than the equivalent of 1 ppm in the diet. Although relatively few studies have evaluated the effect of nutritionally deficient levels of Se on carcinogenesis in experimental animal models, the available results indicate that dietary deficiencies of Se generally do not affect carcinogenesis of the colon or liver in the rat or of the mammary gland in the mouse but may actually enhance carcinogenesis of the mammary gland in the rat and of the skin in the mouse.

Selenium status has been evaluated as a putative factor in the etiologies of several human cancers. Ecological studies have tended to show that Se status and cancer mortality were inversely correlated. These have concerned cancers at several sites: lymphomas, peritoneum, lung, breast, colon, rectum, and liver. A recent ecological study of Se status and cancer mortality in the United States indicated that intermediate- and high-Se counties had lower rates than low-Se counties of cancers of all sites combined and, specifically, in both sexes, of the lung, colon, rectum, bladder, esophagus, and pancreas. Inverse associations were also found for cancers of the breast, ovary, and cervix. Cancers of the liver and stomach, Hodgkin's disease, and leukemia showed positive associations with forage Se level in either sex.

A number of case-control studies have been conducted to test the hypothesis that Se status may be related to cancer risk. In these studies the Se status of cancer patients is compared with that of healthy persons with the same general characteristics. One study conducted in eastern North Carolina found that subjects ranking in the lowest decile of plasma Se concentration (with a mean of 84 ppb) had a relative risk of nonmelanoma skin cancer[5] greater than 3 in comparison to subjects ranked in the highest decile (mean plasma Se of 221 ppb). Further, when cases and controls were stratified on high versus low plasma retinol and total carotenoids, the results indicated that low plasma Se was associated with elevated risk of skin cancer in patients low with respect to either of those variables.

A prospective case-control study was conducted to evaluate the relation of serum Se level and total cancer mortality in a random population sample of more than 8,100 persons in eastern Finland. During a 6-year period of observation, the subjects that developed cancers were found to have had significantly lower serum Se concentrations at the beginning of the study than those of matched controls that did not develop cancer (50.5 ± 1.1 ppb versus 54.3 ± 1.0 ppb). Another prospective case-control study found that the relative risk of subjects in the lowest quintile of serum Se (i.e., <115 ppb) was twice that of subjects in the highest quartile (i.e., >154 ppb). In addition, the relative risk in low-Se subjects appeared to be greatest if they were also relatively low in vitamins A and E (as indicated by low plasma concentrations of retinol and α-to-

5. E.g., basal cell or squamous cell carcinomas.

copherol). A third prospective case-control study showed that the mean serum Se concentrations were lower in terminal cancer patients than in matched controls (i.e., 53.7 ppb versus 60.9 ppb). The relative risk of death to cancer was 5.8 among subjects in the lowest tertile of serum Se concentration when compared with subjects in the highest tertile. Again, an interaction of Se and vitamin E was detected; subjects in the lowest tertiles with respect to serum concentrations of both nutrients had a relative risk of fatal cancer of 11.4 compared with subjects in the highest tertiles.

Some epidemiological investigations of Se and cancer have failed to detect such relations. Nevertheless, the *plausibility* of the hypothesis that low Se status may increase human cancer risk is supported by the results of the ecological correlational studies and cross-sectional case-control studies discussed above. It should be noted that this hypothesis is different from that derived from studies with animal tumor models, which would hold that pharmacologic levels of Se may be anticarcinogenic. In fact, experimental animal studies have indicated that there are likely to be multiple mechanisms by which Se can affect carcinogenesis, although these, too, are yet to be elucidated. Therefore, it is likely that the dose-response relation between Se and tumor development is not linear (i.e., that the nature of the low-Se status effect that is suggested by human epidemiological studies may be quite different from the high-dose Se effect indicated by the results of animal studies).

VI. Selenosis

The potential for Se toxicity was first recognized well before the nutritional role of the element was discovered. In the 1930s Se was found to be a cause of a dermatologic disorder ("alkali disease") that occurred in cattle and horses in northern Nebraska and South Dakota. Investigations revealed that those animals grazed on pasture plants that accumulated high levels of Se from those seleniferous soils. Although Se toxicity has subsequently been studied in several animal species, only a few cases of human exposure to hazardous levels of Se have been reported. Most of these cases have involved occupational exposures (e.g., of workers in copper smelters or Se rectifier plants) caused by the inhalation of Se aerosols. Some cases, however, have involved the oral consumption of high levels of Se in various forms: selenous acid (H_2SeO_3) in gun bluing preparations,[6] sodium selenate (Na_2SeO_4), selenium dioxide (SeO_2), extremely high-Se foods in a particular locale of endemic selenosis in China, a particular nut (coco de mono) that appears to accumulate Se, and an over-the-counter supplement that was erroneously formulated with excessive Se. These cases have demonstrated that acute exposure to high levels of Se can produce hypotension (resulting from vasodilation), respiratory distress, and a garlic-like odor of the breath (caused by the production of the respiratory metabolite dimethylselenide). They also show that chronic exposures can produce gastrointestinal disturbances (e.g., dyspepsia, diarrhea, anorexia) and garlic breath as the major signs; more severe cases show skin eruptions, pathological nails, and hair loss.

The lethality of acute Se intoxication has been established in animals and has been documented in a few human cases. In two cases (a 3-year-old boy who drank gun bluing, and a 17-year-old man who consumed an unknown amount of SeO_2), the subjects died within only a few hours of exposure. In one case (a 52-year-old woman who drank gun bluing), the subject was revived only to die 8 days later of respiratory failure. Reports of nonfatal acute selenosis (including one of a 2-year-old girl who drank gun bluing) indicate that the signs and symptoms of the toxicity can be reversible on cessation of Se exposure and that complete recovery without sequelae can be expected.

Naturally occurring chronic selenosis was identified in the 1960s among residents of Enshi County, Hubei Province, China. It appears to have resulted from exceedingly high concentrations of Se in the local food supplies and, in fact, throughout that environment. In that particular area, the local soils were found to contain nearly 8 ppm Se, and coal (the ash of which was used to amend the soil) was found to contain as much as 84,000 ppm Se. Consequently, locally produced foods contained the highest concentrations of Se ever reported: corn, 6.33 ppm Se; rice, 1.48 ppm Se. Even the water, which leached through seleniferous coal seams, contained unusually high concentrations of Se (e.g., 54 ppb Se). In the five most heavily affected villages, morbidity was ca. 50%. Almost all residents showed signs, the most common of which were losses of

6. These can contain as much as 2% H_2SeO_3, i.e., >12,000 ppm Se.

hair and nails. Some also showed skin lesions (e.g., erythema, edema, eruptions, intense itching), hepatomegaly, polyneuritis (e.g., peripheral anesthesia, acroparesthesia, pain in the extremities, convulsions, partial paralysis, motor impairment, hemiplegia), and gastrointestinal disturbances. One death was attributed to selenosis (a postmortem evaluation of that case was not made). In a village that had a history of high prevalence of these signs and symptoms, it was estimated that local residents consumed 3,200–6,690 μg Se/person/day. It should be noted that this level of intake is approximately 100 times the nutritionally significant level for Se. Recent studies in these regions have indicated that risk of selenosis may be significant among individuals with chronic intakes greater than 750–850 μg Se from that seleniferous food supply.

Signs of chronic selenosis have also been reported among users of certain oral Se supplements. Although the consumption of Na_2SeO_3 at rates as great as 1 mg Se/person/day for at least short periods of time appears to produce no toxic signs or symptoms, clear signs of selenosis have been reported in a man who took an oral supplement that provided that level of Se daily for 2 years. That subject had garlic breath and thickened and fragile nails, but both symptoms subsided when he stopped taking the Se supplement.

Thirteen cases of selenosis were identified among consumers of a commercial Se supplement that through a manufacturing error, contained an average of 27 mg Se/tablet. The cases showed variable signs according to the number to tablets of the supplement actually consumed. The individual that consumed the greatest amount of total Se in this group developed total alopecia, severe nausea and vomiting, severe diarrhea, and garlic breath. Some cases showed irritability, fatigue, and paresthesia; peripheral neuropathy was documented in two cases.

Selenosis was reported in a woman who had used a SeS_2-containing antidandruff shampoo two to three times weekly for 8 months. Her symptoms appeared one day within an hour after using the shampoo: eruptive scalp and mild nonrhythmical tremors of the arms and hands progressing to generalized tremors of increasing severity. Within 2 hours she had a metallic taste of the mouth and garlic breath. Two days later she was weak and anorectic; the lethargy persisted for a couple of days during which time she was nauseous and had porphyrinuria. She recovered gradually thereafter, having ceased using the shampoo; her recovery was complete in 2 weeks.

Present knowledge of the toxicology of Se in humans is incomplete. Although the proximal biochemical lesion(s) concerned in Se toxicity are not clear, it is thought that these involve the oxidation and/or binding of critical sulfhydryl groups in proteins and/or nonprotein thiols by Se species present in excessive concentrations. It has been proposed that the signs of Se toxicity are related to changes in intracellular concentrations of reduced GSH and/or other nonprotein sulfhydryls.

In humans, garlic breath and dermatologic and gastrointestinal signs are the earliest indicators of Se intoxication. The lowest Se dosage found to be associated with the manifestation of these signs is ca. 1,000 μg Se/day, which was achieved by the daily use of an oral supplement of 900 μg Se (as 2 mg Na_2SeO_3) and a daily intake of ca. 100 μg Se from other dietary sources. It is likely that level may be near the threshold dose for toxic responses, as the signs associated with it were mild and were not manifest for nearly 2 years of exposure. Therefore, it has been suggested that for adults, levels of 550 and 750 μg Se/person/day be used as upper limits for safe exposure to inorganic (e.g., Na_2SeO_3) and organic (e.g., Se amino acids) Se compounds, respectively. Because foods do not contain such high concentrations of Se except in unusual circumstances, the risks of exposure to potentially intoxicating amounts of the element are limited to its improper use as a food supplement (i.e., formulation errors) and the improper consumption of seleniferous products not intended for human consumption (e.g., accidental consumption of Se reagents).

Bibliography

Combs, G. F., Jr. (1988). Selenium in foods. *In* "Advances in Food Research," vol. 32. (C. O. Chichester and B. S. Schweigert, eds.), pp. 85–113. Academic Press, New York.

Combs, G. F., Jr. (1989). Selenium. *In* "Nutrition and Cancer Prevention: Investigating the Role of Micronutrients" (T. E. Moon and M. S. Micozzi, eds.), pp. 389–420. Marcel Dekker, New York.

Combs, G. F., Jr., and Combs, S. B. (1986). "The Role of Selenium in Nutrition." Academic Press, New York.

Combs, G. F., Jr., Spallholz, J. E., Levander, O. A., and Oldfield, J. E., eds. (1987). "Selenium in Biology and Medicine," vols. A and B. AVI Publishing Co., Westport, Connecticut.

Levander, O. A. (1986). Selenium. *In* "Trace Elements in Human and Animal Nutrition," 5th ed., vol. 2. (W. Mertz, ed.), pp. 209–279. Academic Press, New York.

Levander, O. A. (1987). A global view of human selenium nutrition. *Annu. Rev. Nutr.* **7,** 227.

Schubert, A., Holden, J. M., and Wolf, W. R. (1987). Selenium content of a core group of foods based on a critical evaluation of published analytical data. *J. Am. Diet. Assoc.* **87,** 285–299.

Wendel, A., ed. (1989). "Selenium in Biology and Medicine." Springer-Verlag, New York.

Yang, G. Q., Wang, S., Zhou, R., and Sun, R. (1983). Endemic selenium intoxication of humans in China. *Am. J. Clin. Nutr.* **37,** 872–881.

Yang, G. Q., Yin, S., Zhou, R., Gu, L., Yan, B., Liu, Y., and Liu, Y. (1989). Studies of safe maximal daily dietary selenium intake in a seleniferous area in China. II. Relations between Se-intake and the manifestations of clinical signs and certain biochemical alterations in blood and urine. *J. Trace Elem. Electrolytes Health Dis.* **3,** 123–130.

Yang, G. Q., Zhou, R., Yin, S., Gu, L., Yan, B., Liu, Y., and Li, X. (1989). Studies of safe maximal daily dietary selenium intake in a seleniferous area in China. I. Selenium intake and tissue selenium levels of the inhabitant. *J. Trace Elem. Electrolytes Health Dis.* **3,** 77–87.

Self-Monitoring

MARK SNYDER, SARAH A. MEYERS, *University of Minnesota*

Glossary

Attitude–action relationship Amount of similarity between someone's stated attitudes or beliefs and the way he or she behaves

Self-concept Feelings, beliefs, and attitudes a person holds about himself or herself

Self-monitoring Degree to which people control the images of self they present to others during social interactions, either on the basis of cues from social situations or information from personal characteristics

Self-monitoring inventory Self-report measure designed to assess the degree to which a person engages in self-monitoring

Situational specificity Tendency to be different or consistent as a function of how different or similar the situations are in which the person finds himself or herself

SELF-MONITORING CONCERNS THE DEGREE to which people monitor or control the images of self they present in social interactions. High self-monitors are especially concerned with the images they project and, thus, tend to tailor their behavior to the demands of the situation. Low self-monitors are more concerned with expressing their personal beliefs and opinions; therefore, their behavior is more consistent among different situations because it reflects the same set of attitudes. High and low self-monitors differ with regard to many domains of social life, including their behavior in social situations, the kinds of friendships they form, the ways they develop romantic attachments, their job or career preferences, and their reactions to advertising. Although these differences are shaped by environmental events, evidence also suggests a genetic component to the construct of self-monitoring.

I. Introduction

Everyday, people ask themselves "Who am I really?" in hope of discovering the one true self cloaked by the many roles they play. Some people have no trouble answering this question. They seem to be able to look inward and "know" themselves. Others spend much of their lives searching for the self. Self-help books and encounter weekends have become popular tools used in the pursuit of self-identity, self-esteem, and self-respect. Yet, no matter how difficult the search may be, very few people ever even question the assumption that each person has a true self—one that uniquely distinguishes him or her from all others, that gives meaning to experience, that provides continuity to life. Nevertheless, researchers who study the self are directly challenging the assumption that each person has one unique, special, true self. These researchers argue that rather than one self, some people may have many selves. Moreover, for some people, apparently the self is the product of their relationships with other people rather than an integral part of their own personalities. Despite conventional wisdom, there may be striking gaps and contradictions between the public appearances and the private realities of the self.

These gaps and contradictions between the selves we display to others and the self only we can know have served as the foundation for a new way of examining the self: the self as actor. Poets and psychologists alike have used the metaphor that life is theater and each person is really only an actor playing a part. People assume many different parts over the course of their lifetimes. Each new situation is really a new backdrop on the stage and, as such, requires the actor to ascertain which role is most appropriate. This metaphor holds true, at least in part, for most people.

However, for some people, "acting" is almost a way of life. Some people are particularly skilled in the ways they express and present themselves in social situations—at parties, at professional meetings, in any circumstance in which they might choose to create a specific appearance. These people are capable of observing their own behavior and modifying it when indications from others suggest they are not having the desired effect. These persons are called high self-monitors because of the great extent to which they monitor or control the selves they project to others in social interaction. In contrast, low self-monitors tend to express what they think and feel, rather than mold and fashion their behavior to fit the situation.

High self-monitors, then, are people who are very concerned with acting in ways appropriate to each situation in which they find themselves. They look to the situation and to others in the situation for cues. These cues help them monitor (i.e., regulate and control) their self-presentation. Low self-monitors do not look outward for cues, they look inward. They look to their opinions, beliefs, attitudes, and mood to determine the way they will present themselves.

One frequently asked question is how to tell if people are high or low self-monitors. To determine any specific person as a high or low self-monitor, a Self-Monitoring Inventory is used. This questionnaire consists of a set of true–false statements such as "At parties and social gatherings, I do not attempt to do or say things that others will like," or "When I am uncertain how to act in social situations, I look to the behavior of others for cues." High self-monitors tend to answer true to questions such as the following:

I would probably make a good actor.
In different situations and with different people, I often act like very different persons.
I'm not always the person I appear to be.

Low self-monitors tend to agree with these kinds of questions:

I have trouble changing my behavior to suit different people and different situations.
I can only argue for ideas that I already believe.
I would not change my opinions (or the way I do things) in order to please someone else or win their favor.

Two versions of the Self-Monitoring Inventory are available: an 18-item and a 25-item scale. In addition, the Self-Monitoring Inventory has been translated into a number of different languages including Japanese, Arabic, German, Spanish, and Polish. Information on how to administer and score the inventory can be found in Snyder's 1987 book "Public Appearances/Private Realities: The Psychology of Self-Monitoring."

How do the differences measured by the Self-Monitoring Inventory express themselves in daily life? The following sections attempt to answer this question by examining the development of self-monitoring, friendship patterns, romantic relationships, job and career choices, and responses to advertising.

II. Situational Specificity and Conformity

If high self-monitors select a way to present themselves based on the situation, then their behavior should differ only insofar as the circumstances in which they find themselves change. The behavior of high self-monitors should be consistent to the extent that these situations are consistent. To some, this seems to imply that high self-monitors are conformists. This is not necessarily the case. The norms of behavior in some situations do require conformity. However, the norms of other situations require autonomy, as shown, for example, by a comparison of people's behavior in two situations. The first was a public situation in which group discussions were videotaped for use in psychology classes. In such public situations, the norms for behavior were for autonomy, because presumably psychology classes viewing this videotape would think relatively poorly of someone who simply conformed to the rest of the group. The second situation was a more private one. The only people the participants were concerned about were the other discussion group members; thus, the norms for behavior favored group consensus or conformity, possibly as a strategy to develop bonds of cohesion and solidarity within the group.

High self-monitors were very aware of the differences between these situations. They conformed in the private discussions where conformity was the preferred norm, and they acted autonomously in the public discussions where autonomy was the norm. Low self-monitors did not differ between the groups. They were autonomous or conforming depending on whether they considered themselves to be autonomous or conformists, not depending on which situation they were in.

This ability to conform or act independently based on the demands of the situation indicates that the high self-monitor has considerable flexibility in choosing his or her behavior. Being flexible in this way allows the high self-monitor to select the appropriate presentation of self from a wider repertoire of alternatives. Rather than having a limited number of responses, the high self-monitor has the ability to present himself or herself in a wide variety of ways as the situation requires.

III. Attitude–Action Relationship

Although high self-monitors have the advantage of being more flexible in their self-presentation, this advantage does have associated costs. Because they frequently bend to the will of the situation, high self-monitors' behavior may not always reveal their private feelings, beliefs, or intentions. Low self-monitors hold the edge in this area: they tend to act consistently with their inner beliefs and attitudes.

Personality psychologists are particularly interested in the relationship between people's dispositions (their personalities, their attitudes, etc.) and actions. The reason for trying to identify someone's personality traits is to be able to predict and understand that person's behavior. If the person is disposed to be patient, then he or she ought to be patient in almost all situations. However, this may not be the case for high self-monitors. Because high self-monitors act according to the demands of situations, they may be doing what the situation calls for rather than acting on their own personalities and attitudes. Thus, the link between their attitudes and their actions or behaviors ought to be minimal. To predict the behavior of high self-monitors, information about the situation should be more helpful than information about their personality.

One test, for example, is to examine the consistency between people's attitudes toward affirmative action and their decisions as jurors. A mock trial was held involving allegations of sex discrimination in a hiring decision. Jurors were presented with information concerning the qualifications of two biologists, Ms. Harrison and Mr. Sullivan, who applied for an appointment as an assistant professor of biology. The university offered the position to Mr. Sullivan, and Ms. Harrison sued the university for sex discrimination. After considering the arguments on both sides, the jurors reached verdicts and wrote essays explaining their decisions.

Low self-monitors reached verdicts consistent with their attitudes: Those who supported affirmative action rendered verdicts favorable to Ms. Harrison; those who opposed affirmative action rendered verdicts unfavorable to Ms. Harrison. High self-monitors' verdicts were not related to their attitudes toward affirmative action.

This consistency between the dispositions and behaviors of low self-monitors illustrates one important way in which their behavior can and does communicate their feelings, beliefs, and other personal qualities. If a low self-monitor has a strong opinion about some issue, others are destined to know about it. Furthermore, low self-monitors are more consistent among different situations and over time; thus, a low self-monitor who believes in civil rights can be expected to express that opinion in a number of different situations, not just to a specific group of friends. The habit of accurately expressing and communicating one's attitudes and feelings may serve low self-monitors well in many areas of life, especially those calling for self-disclosure and, potentially, psychological intimacy and closeness.

IV. Sense of Self

As you have seen, high and low self-monitors have two different and distinct behavioral styles. They act in different ways, highs according to situations and lows according to dispositions. These contrasting behavioral styles are accompanied by different underlying notions that high and low self-monitors have about the self.

One's sense of self is his or her identity. How does a person construe his or her self? High self-monitors are thought to have a pragmatic sense of self, in which their identity is based on the situation and is very flexible and changes as the situation shifts. In this sense, high self-monitors can be said to have a self for each situation. Their self is whoever they are at the moment. Low self-monitors are thought to have a principled self, in which their identity is construed in terms of inner characteristics and attributes. Rather than having a different self for different situations, they can be said to have one self for all situations, i.e., a sense of self as coherent, consistent, and stable.

How are these senses of self linked to people's behavior? Before engaging in any activity or behavior, people ask themselves how they want to be perceived—what kind of image they want to give to others. High self-monitors answer this question using their pragmatic sense of self. This pragmatic self answers the question by saying behavior should be based on the cues inherent in the situation. Low self-monitors answer this question using their principled sense of self. The principled self answers the question by saying behavior should be a reflection of who one really is. [*See* DEVELOPMENT OF THE SELF.]

V. Development of Self-Monitoring

How does someone get to be a high or a low self-monitor? What roles do the biological–genetic factors and environmental–socialization influences play in the development of self-monitoring? Surprisingly, self-monitoring does not vary meaningfully with many of the indicators thought to be important in child-rearing patterns and practices: social class, economic status, regional origins, geographical movement, or religious affiliation. Perhaps even more surprising is the fact that self-monitoring seems to have genetic roots.

To assess the genetic component of personality, psychologists often compare identical twins, fraternal twins, and randomly selected pairs of individuals. Identical twins share all of their genetic material, thus, if self-monitoring has a genetic basis, they should belong to the same self-monitoring category. Randomly selected pairs of individuals share no genetic material; therefore, any two individuals should have a fifty–fifty chance of having the same self-monitoring propensity. Fraternal twins share half of their genetic material and thus should be the same in self-monitoring approximately 75% of the time. It was, in fact, found that identical twins (as determined by blood tests) belonged to the same category of self-monitoring, either both high or both low, 99% of the time. Fraternal twins were found to be the same in self-monitoring 74% of the time. And, randomly selected pairs were the same only 55% of the time.

This evidence supports the hypothesis that self-monitoring has a biological–genetic etiology. But what does that mean? First, it does not mean that people are born high or low self-monitors only to rigidly remain that way for the rest of their lives. Rather than being born a high or a low self-monitor, it appears that some people are born with a predisposition to become a high or a low self-monitor. Personal experiences and socialization are what translates that potential into a reality. Having a predisposition to become a high self-monitor may increase the likelihood that the person opts to be in situations in which the talents and qualities of a high self-monitor are specifically encouraged and in which they may be preferable. Similarly, someone who is predisposed to become a low self-monitor may select situations in which his or her talents are useful.

The basic idea here is that minor differences present during infancy grow to be major differences in self-monitoring orientations as the child matures. For example, one of the few notable ways in which infants differ by temperament is in how adaptable they are to changes in their environment. Some children are relatively rigid to change, whereas others show greater adaptability and greater social responsiveness. Children who are especially responsive socially, and have the greater versatility this may bring, also may be more likely to search out new situations and to develop different responses to the numerous situations they face. These people may be in the process of becoming high self-monitors. Children who are relatively rigid may be more likely to remain in situations corresponding to their particular interests and preferences. They may be on the road to developing into low self-monitors.

To examine the self-monitoring propensities of children of differing ages, two additional versions of the self-monitoring inventory have been created.

One is a measure suitable for children between the ages of $1\frac{1}{2}$ and 4 years, to be completed by an adult, such as a parent or a teacher, who knows the child well. Another is a measure suitable for children between the ages of 6 and 13 years. This "junior" measure is very similar to the adult version. The items simply were altered to be more applicable to a child. For example, one item on the scale reads "I sometimes wear some kinds of clothes just because my friends are wearing that kind."

One example of the use of the Self-Monitoring Inventory created for very young children had children between 3 and 5 years old label the emotions in a set of photographs. They found that even children as young as 3 years were sensitive to this kind of social information. Furthermore, differences in this sensitivity were related to self-monitoring scores: young high self-monitors were more sensitive than young low self-monitors.

Self-monitoring scores on the Junior Self-Monitoring Inventory have been related to the amount of social comparison these children engage in. To assess social comparison, children were asked to answer a number of questions about their opinions on several topics. The children were given information about how other students had answered these questions, and they could consult these answers at any time while they themselves completed the opinion survey. The researchers found that the high self-monitors spent more time consulting this information than did the low self-monitors.

VI. Friendships

One particularly striking difference in the ways high and low self-monitors live their lives is in how they select their friends. Low self-monitors tend to spend their time with people they like; high self-monitors, on the other hand, tend to spend time with people who are skilled at specific activities. For example, a high self-monitor may have one friend with whom he or she plays tennis and another with whom he or she plays chess. These two games require very different skills, so the high self-monitor probably would not opt to play chess with someone who is good at tennis, but not good at chess. A low self-monitor, however, would play both games with the same person, not because the person was necessarily good at both games, but simply because the person was a friend.

A study was conducted in which college students were asked to describe all of their friends and the specific activities they engaged in with those friends. Each friend was then paired with each activity, and the students were asked to rate the likelihood that they would participate in that activity with that friend. Low self-monitors usually preferred the same friends for most or even all of the activities, whereas high self-monitors typically paired their friends with activities, such that specific people were selected for specific activities.

This experiment brings up the question of what people do when faced with the choice between a proficient partner or a likable partner. For instance, if you have the opportunity to play tennis with someone who is very likable but who is only a mediocre tennis player, or to go sailing with an excellent sailor whom you do not like a lot, which would you choose? People presented with this dilemma decided according to their self-monitoring tendencies. High self-monitors preferred to engage in an activity with an expert, whereas low self-monitors preferred to spend the time with someone they liked. In fact, 80% of the high self-monitors selected activity partners on the basis of skill level rather than likability. Of the low self-monitors, 67% made their choices on the basis of likability rather than the person's skill.

Because low self-monitors elect to spend time with people on the basis of how much they like the other person, low self-monitors should have a high probability of forming very close friendships. High self-monitors are less likely to form close bonds of this kind because they select their friends on the basis of skills in specific activities. These differences have implications for the way high and low self-monitors form romantic attachments.

VII. Romantic Relationships

Similar differences in the selection of partners is found in the realm of romantic relationships. Here the differences focus more on physical appearances for high self-monitors and personality traits for low self-monitors. College-aged men were presented with information about a number of prospective dating partners. The information contained a photograph and a sketch of the woman's personality. The researchers measured the amount of time each man spent looking at the two types of information (the photograph and the personality information) and found that high self-monitors spent proportionally more time looking at the picture than low self-monitors, and that low self-monitors spent proportionally more time examining the personality sketches than did high self-monitors.

To examine the choices high and low self-monitors made when they had to select between an attractive physical appearance and a pleasant personality, college-aged men were presented with information on two prospective dates. The first woman was physically attractive but did not have a very pleasant personality: she was moody, withdrawn, and self-centered. The second woman had a pleasant personality, but was relatively unattractive. Sixty-nine percent of the high self-monitors selected the physically attractive woman, whereas 81% of the low self-monitors selected the woman with the pleasant personality.

Given these differences in the ways high and low self-monitors choose dating partners, what kinds of differences exist in the relationships they actually form? High self-monitors tend to hold relatively uncommitted orientations toward their relationships. Among college students not in long-term relationships, high self-monitors express a greater willingness to participate in social activities with people other than their current dating partners. They also report a willingness to terminate a relationship in favor of a new one. This is reflected in their tendency to date more partners over the course of a year than do low self-monitors. Low self-monitors report relatively less willingness to spend time with people other than their current dating partners. They usually reject the opportunity to end a current relationship in favor of a new dating partner. And, among those in steady relationships, the relationships of low self-monitors are much more long-lasting than those of high self-monitors.

A similar difference between high and low self-monitors extends to their sexual relations with their dating partners. High self-monitors tend to have rather unrestricted views about sexuality in comparison to low self-monitors; they report having sexual relations with more different partners than do low self-monitors, and they report a willingness to engage in sexual relations on a rather casual basis. Low self-monitors, on the other hand, believe that sexual activity should be confined to relationships in which bonds of closeness and intimacy have already developed.

Differences in the dating relationships of high and low self-monitors may indicate potential differences in the marriages of high and low self-monitors. Because high self-monitors tend to prefer dating physically attractive partners or skilled activity partners, they may also prefer to marry people who fit these descriptions. Low self-monitors may opt to marry people who have similar attitudes and beliefs as themselves, as they select these types of people as dating partners. The difference in high and low self-monitors' willingness to spend time with people outside their dating relationships may also carry over into their married life. Low self-monitors may abandon many of their old activities and friends in favor of their new responsibilities as a spouse. In contrast, high self-monitors may retain relatively more of their outside commitments.

Marital complaints may reflect the different conceptions high and low self-monitors have of marriage. For example, one commonly heard complaint in a marriage counselor's office is "My spouse just is not the same person I married." This simple statement may mean very different things to high and low self-monitors. If high self-monitors have selected a spouse on the same basis as research suggests they choose a dating partner (namely, someone who is very attractive or who is good at particular activities), this statement may mean the spouse is no longer as attractive as when the wedding took place or the spouse no longer prefers the same kinds of activities as before. If a low self-monitor has selected a spouse on the same basis as they do in research on choices of dating partners (namely, similar feelings and beliefs), the claim of change on the part of the spouse may mean the spouse no longer holds those same beliefs. Knowing which meaning is intended may be important to marriage counselors if they are to help the couple resolve their differences.

VIII. Jobs and Careers

One of the biggest choices people make in their lives is what job they want or what career they will pursue. Because we have already seen how high and low self-monitors select different situations, different friends, and different romantic relationship styles, it seems logical to expect they may also choose different professions. Low self-monitors may select jobs or careers that reflect their personal attitudes and talents. High self-monitors may select jobs or careers that allow them to make use of their ability to be different people in different situations.

One such type of job is known as "boundary-spanning" positions.

Boundary-spanning positions are jobs in which the employee must act as a go-between for groups that may not be able to communicate with each other effectively face to face. Researchers have found that high self-monitors excel in boundary-spanning positions such as field representative at a large franchise organization. In their study, high self-monitors tended to be more successful than low self-monitors in boundary-spanning jobs. Although they did not examine issues of satisfaction or job promotion, one would expect to find high self-monitors more satisfied with boundary-spanning jobs than low self-monitors. In addition, one would expect high self-monitors to receive more promotions than low self-monitors in this area. In fact, researchers examining employees of an insurance company found high self-monitors received more promotions.

All of this research would seem to indicate that high self-monitors are more likely than low self-monitors to hold jobs with which they are satisfied and to be promoted. This conclusion may not be warranted. The jobs and careers that have been studied to date tend to be those that high self-monitors excel at. This does not mean there are not jobs and careers better suited to low self-monitors. Low self-monitors in positions suited to them should perform better at those jobs and be more satisfied with them than high self-monitors.

Research has examined the performance of both high and low self-monitors in leadership positions. High self-monitors emerge as leaders in circumstances involving high levels of verbal interaction. Low self-monitors express their leadership potential when they work in relatively homogeneous groups. Furthermore, there are two styles of management: task-oriented and relationship-oriented. The task-oriented leader is one who is decisive and direct. The relationship-oriented leader is tolerant and understanding. Low self-monitors work best when their personal inclinations toward task- or relationship-oriented leadership match the needs of the leadership position.

IX. Advertising and Consumer Behavior

Issues concerning the world of paid work naturally lead to questions about what people do with the fruits of their labors. How do people decide how to spend the money they earn? Questions such as this are examined by the field of consumer psychology. One of the major focal points in this area concerns the effectiveness of advertising.

Two different schools of thought exist within the advertising profession: the soft-sell and the hard-sell. Advertisers who believe the soft-sell approach is most effective create ads that appeal to images they want the consumer to associate with the products. Those who agree with the hard-sell approach create ads that make claims about the quality of the product. Researchers have found that these two advertising approaches appeal differently to high and low self-monitors.

The responses of high and low self-monitors to these two approaches was compared by creating ads with the same picture but different written copy. For example, a picture of a bottle of Canadian Club whisky sitting on blueprints for a house was paired with the text "You're not just moving in, you're moving up" for the image-oriented ad and with the text "When it comes to great taste, everyone draws the same conclusion" for the quality-oriented ad. High self-monitors preferred the image-oriented ad, whereas low self-monitors preferred the quality-oriented ad.

How do these ad preferences translate into actual purchase selections at the store? Although high self-monitors prefer the image-oriented ads, they will only buy products associated with images they want to project at particular times in their lives. Similarly, although low self-monitors prefer quality-oriented ads, they will purchase the highest quality product only if they like the item. Thus, high self-monitors may purchase the car that looks flashy and sporty rather than the better handling sports car, and they may drink the imported premium beer that gives a special status to its drinker rather than a cheaper domestic beer that tastes the same. Likewise, low self-monitors may eat the most nutritious breakfast cereal even though there isn't a star athlete on the box, and they may purchase the most energy-efficient refrigerator even though it doesn't have a designer-styled finish.

X. Summary and Conclusions

High and low self-monitors differ in a number of important ways. High self-monitors are particularly concerned with behaving appropriately from situation to situation and from role to role. They determine which behaviors are appropriate by looking at the cues inherent in the situation. Low self-monitors are interested in behaving according to their dispositional characteristics and personal opinions and attitudes. As a result, they are consistent among different situations and over time.

These self-monitoring styles have important implications for various aspects of people's lives, including friendships and romantic relationships. High self-monitors tend to spend time with people who are skilled at certain activities, whereas low self-monitors prefer to spend time with people they consider likable. Similarly, high self-monitors prefer to date people who are physically attractive, whereas low self-monitors prefer people who have pleasant personalities.

Another difference in the lives of high and low self-monitors pertains to the jobs or careers they have. High self-monitors excel at positions in which their flexibility can be exploited. One such job is a boundary-spanning position in which the person serves as a go-between for two groups that otherwise cannot communicate effectively. Low self-monitors excel at jobs in which their dispositions and opinions can be utilized.

Given all of these differences between high and low self-monitors, which is better? At first glance, most people think the high self-monitor is preferable to the low, that somehow the low self-monitor is lacking something the high self-monitor has. After further consideration, some people change their mind and believe the low self-monitoring style is preferable to that of the high. In fact, both styles have advantages and disadvantages. High self-monitors have a special flexibility that enables them to modify their behavior in very different situations. They tend to excel in boundary-spanning positions and receive more promotions in such roles. And they prefer to spend their free time with friends who are skilled at particular activities. Low self-monitors display consistency between their beliefs and their actions. They prefer to spend time with people they really like, and they have longer-lasting and more intimate dating relationships.

Because each self-monitoring style has its own pluses and minuses, there is no evidence that, on balance and overall, life's pleasures and pains occur disproportionately in the lives of high or low self-monitors. Thus, there is no reason to grant more favored status to one or the other lifestyle. Moreover, it takes but a moment's reflection to realize that the answer to the question "Which is better—to be high or low in self-monitoring" is largely a matter of semantics. For example, the high self-monitor could be favorably described as "flexible and adaptable" or less favorably described as "unreliable and deceptive." Similarly, the low self-monitor could be favorably described as "consistent and reliable" or less favorably described as "rigid and stubborn." Thus, value judgments should be avoided. Instead, until there is convincing reason to believe otherwise, we should assume that neither the high nor the low self-monitor is better than the other; rather, both interactional styles should be seen as having their own advantages and disadvantages, and as representing alternative lifestyles, each with its own potential assets and liabilities.

Bibliography

Snyder, M. (1979). Self-monitoring processes. *In* "Advances in Experimental Social Psychology," Vol. 12 (L. Berkowitz, ed.). Academic Press, New York.

Snyder, M. (1987). "Public Appearances/Private Realities: The Psychology of Self-Monitoring." Freeman, New York.

Snyder, M. (1980). "The many me's of the self-monitor." *Psychol. Today* **March,** 32–34, 36, 39–40, 92.

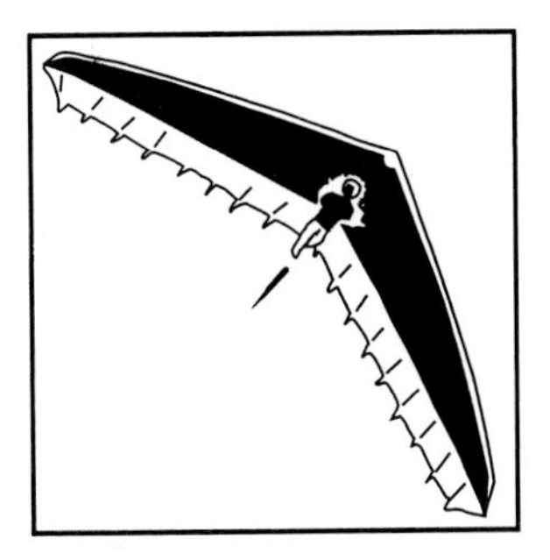

Sensation-Seeking Trait

MARVIN ZUCKERMAN, *University of Delaware*

Glossary

Augmenting–reducing (of the cortical evoked potential) Tendency of the amplitude of the cortical evoked potential to increase (augment) or decrease (reduce) as a function of the intensity of a stimulus; individuals differ reliably in this function and may be characterized as augmenters or reducers

Cortisol One of the corticosteroids from the adrenal cortex that regulates the release of glycogen from the liver into the bloodstream, fat metabolism, striated muscle strength, blood pressure, and lymphoid tissue (antiinflammatory effect); it is released by stress acting through the adrenocorticotropic hormone from the pituitary gland

Estrogens (estradiol, estrone) Hormones produced by the ovaries in the female, controlling maturation of the female reproductive system and development of secondary sex characteristics; they function in vascularization of the vaginal tissue and the lubrication of the vagina during stimulation but are not an essential factor in sexual arousability; some estrogen is produced in males, largely through conversion from androgens

Monoamine oxidase Enzyme involved in the catabolic deamination of the monoamines: norepinephrine, dopamine, and serotonin

Orienting reflex Involuntary physiological response to the first presentation of a stimulus that tends to habituate or diminish in strength with subsequent presentations of the same stimulus

Testosterone Hormone produced by the testes in the male, controlling maturation of the male reproductive system, sperm production, and secondary sex characteristics; also found in females where the major source is the adrenal cortex; affects sex drive and sexual arousability in both sexes

SENSATION SEEKING IS A PERSONALITY trait defined by the disposition to seek varied, novel, complex, and intense sensations and experiences and to take physical and social risks for the sake of such experiences. Operationally, the sensation seeking scale (SSS) or other test varients related to the construct (e.g., Monotony Avoidance, Venturesomeness and Arousal Seeking scales) are used to define the personality dimension or groups of subjects characterized as "high" or "low" sensation seekers.

I. Theory and Trait Description

The theory and initial trait assessment of sensation seeking emerged in the context of experimental research on sensory deprivation in humans. The wide variation in responses to periods of sensory deprivation or invariant sensory environments suggested the possibility that people have optimal levels of stimulation and arousal that influence their behavior and preferences in many areas of life activities. A trait of sensation seeking was defined (see glossary) and a questionnaire, the SSS, was designed to assess the general trait.

Further development of the SSS revealed four stable factors based on the item content of the scale. These factors have been replicated in many different populations around the world. Three of the factors describe different modes or styles of sensation seeking while the fourth factor, Boredom Suscepti-

Encyclopedia of Human Biology, Volume 6.

bility, represents the reaction to a lack of varied stimulation. The factors may be defined as follows:

1. *Thrill and Adventure Seeking* (TAS) consists of items expressing a desire to engage in sports or activities involving an above-average level of physical danger or risk such as mountain climbing, parachute jumping, etc.
2. *Experience Seeking* (ES) contains items describing the seeking of new experiences by living in a nonconforming life-style with unconventional friends and/or an involvement with travel, art, music, and drugs.
3. *Disinhibition* (Dis) was named for the items describing the disinhibiting of behavior in the social sphere by drinking, partying, and seeking variety in sexual partners.
4. *Boredom Susceptibility* (BS) items describe an aversion for repetitive experience of any kind, routine work, or even dull or predictable people. Other items indicate a restless reaction when things are unchanging.

Form V of the SSS has a total score based on the sum of the four factor-derived scales in place of a general scale used in previous versions.

II. Phenomenal Correlates

The validity of a personality scale is usually assessed in terms of its predictions or correlations with phenomena outside of itself or other self-report tests. The SSS was first applied to the prediction of responses to experimental sensory deprivation. Male college student subjects were exposed to two 8-hour sessions of confinement in a sound-proof room on two different occasions. One condition involved sensory deprivation (darkness and quiet), whereas the other involved social isolation and confinement but the room was lighted and minimal visual and auditory stimuli were provided. The order of conditions was counterbalanced. Nonpurposive movements were measured using a pressure transducer connected to the air mattress on which they were reclining. Sensation seekers (highs on SSS) were found to show more restlessness than low-sensation seekers in both kinds of confinement conditions as a function of time in the monotonous environment. In another experiment, sensation seekers tended to respond more for the reward of visual stimuli, again as a function of time in the invariant environment.

Highs in a college population reported having had more varied sexual activities with more partners than lows. They also reported using alcohol more heavily and admitted more use of illegal drugs than lows. Studies of drug use in normal and drug-abusing populations show that sensation seeking correlates not only with drug use but also with the variety of drugs used. Some evidence indicates that sensation seekers are somewhat more attracted to amphetamine and hallucinogenic drugs than to suppressant drugs, but users of a wide range of illegal drugs, including opiates, tend to be higher sensation seekers than nonusers. The relationships with alcohol and drug use are obtained even when items referring to a desire to use, or actual use of, these substances are removed from the scales. High-sensation seekers are also more likely than lows to smoke tobacco. Food preferences are related to sensation seeking: sensation seekers like foreign, crunchy, and spicy foods, whereas lows tend to prefer bland, soft, and sweet foods. Vegetarians are lower sensation seekers than gourmets.

High sensation seekers tend to volunteer for unusual types of experiments or activities such as hypnosis, sensory deprivation, or meditation on the expectation that they will have interesting experiences. However, when these activities turn out to be boring (completely predictable, or nonarousing types of experiences) they persist less than lows. Highs like designs that are novel, complex, and symmetrical, whereas lows show a relatively greater preference for familiar, simple, and symmetrical designs. Highs like paintings that are impressionistic, ambiguous, or surrealistic, whereas lows prefer quiet pastoral scenes. Highs tend to watch less television than lows. When offered a choice of channels to watch, highs tend to switch from channel to channel, whereas lows watch one program. Highs like films with explicit sexual and violent themes, whereas lows tend to avoid sexual and morbid themes in films and reading. Highs like rock or jazz; lows like more bland popular or film sound track types of music.

Highs are attracted to occupations involving interactions with people, whereas lows tend to be attracted to solitary business or clerical occupations. Female highs are attracted to nontraditional occupations, whereas female lows prefer the more traditionally feminine ones. Highs more than lows are dissatisfied in nonstimulating occupations, such as assembly-line jobs. Highs given boring tasks in the laboratory report more dissatisfaction than lows.

Highs tend to engage in risky sports involving

speed (auto racing), unusual sensations (parachuting, scuba diving), or skiing. Among novice skiers, those higher on sensation seeking tend to have more accidents, indicating their propensity for risk-taking before they have developed the requisite skills. Highs also tend to engage in risky body contact sports more than lows. Other sports such as bowling, tennis, or gymnastics are equally likely to be practiced by highs and lows.

III. Psychopathology

Sensation seeking is a normal dimension of personality unrelated to traits of anxiety or neuroticism but positively related to traits such as impulsivity, social dominance, surgency, sociability, autonomy, and exhibitionism and negatively related to traits such as deference, nuturance, and orderliness. Most highs or lows on sensation seeking do not have psychiatric disorders.

A problem with using test results from subjects who are currently imprisoned or hospitalized is that their current scores on the tests may reflect the current state of the disorders or the influence of incarceration rather than their personalities in normal conditions and states. Data were obtained from a large scale survey of 2,115 persons in the general community. As part of a demographic survey, they were asked about past treatments and hospitalizations for psychological problems and their diagnosis. The drugs reportedly prescribed were used as an additional check on their self-reported diagnoses. While 17% of the sample reported having received some diagnoses, 11% of the total sample were reliably classifiable within the categories listed in Table I. Each subject in each diagnostic group was matched for sex, age, and education (as closely as possible) with a control subject from the same pool of survey respondents who reported no history of psychological problems. Those who matched the controls were blind as to the diagnoses and SSS scores of the subjects with positive histories and the SSS scores of the controls.

A subgroup of bipolar disorders (manic-depressive) scored significantly higher than matched controls on the total SSS, ES, and BS, but unipolar major depressive disorders did not differ from their matched controls. The results are congruent with psychometric results showing consistent correla-

TABLE I Comparisons of Diagnostic Groups and Matched Control Groups

Groups	Mean sensation-seeking scores								
	Ns[a]	TAS	ES	Dis	BS	Total	Age	Ed[b]	% Male
Bipolar (manic-depressive)	19	8.2	7.6	6.7	5.4	27.8	30.3	5.0	37
Controls	19	5.6	6.1	5.2	2.0	21.0	30.5	4.7	37
t test		3.53**	2.28*	1.54	2.01	4.79***			
Major depressive	39	6.5	7.1	5.9	4.5	24.0	31.0	5.9	28
Controls	39	7.2	7.1	6.1	4.2	24.6	31.5	5.9	28
t test		1.25	0.00	0.26	0.36	0.64			
Antisocial substance abuse	17	9.0	8.4	7.2	4.7	29.2	30.8	5.0	65
Controls	17	7.7	6.2	5.4	3.7	22.7	29.7	5.1	65
t test		2.44*	3.10**	2.52*	0.91	2.75*			
Neurotic	100	6.2	7.2	6.1	5.0	25.1	35.8	6.3	34
Controls	100	6.8	7.1	5.6	4.5	24.1	35.7	6.2	34
t test		1.58	0.24	1.23	1.42	0.37			
Personality trait disorders	15	7.4	7.7	5.9	5.3	26.3	33.8	5.5	27
Controls	15	7.1	6.6	5.8	4.0	23.5	34.0	5.5	27
t test		0.26	0.99	0.12	1.62	0.95			
Schizophrenic	20	6.7	7.0	5.0	4.8	23.4	28.5	5.2	45
Controls	20	6.6	6.9	5.9	4.0	23.3	27.8	5.2	45

[a] Ns, Numbers of subjects.
[b] Education (Ed.) is on an 8-point scale: 1 = grade school, 2 = 1–3 years of high school, 3 = high school graduate, 4 = 1–3 years of college, 5 = college graduate, 6 = some graduate school, 7 = master's degree, 8 = doctorate degree.
* $P < 0.05$, ** $P < 0.01$, *** $P < 0.001$.
From M. Zuckerman, and M. Neeb, 1979, Sensation seeking and psychopathology, *Psychiatry Res.* **1**, 255. Copyright Elsevier/North Holland Biomedical Press. With permission.

tions between the hypomania score on the Minnesota Multiphasic Personality Inventory in normal and patient populations, as well as a number of common biological correlates of sensation seeking and bipolar disorders. A group loosely labelled antisocial personality, including sociopathic and substance abusers, also scored higher than their controls on the total SSS and TAS, ES, and Dis subscales. These results are consistent with previous findings showing that prisoners diagnosed as psychopathic scored higher than nonpsychopathic prisoners on all of the SSS scales. They are also consistent with many studies showing a relationship between sensation seeking and drug and alcohol abuse. No differences were found among neurotics, personality trait disorders (other than antisocial types), or schizophrenics and their respective control groups. The lack of difference between neurotics and controls is consistent with the absence of correlation between scales of anxiety or neuroticism and sensation seeking in normal and abnormal populations. One type of schizophrenic, chronic behaviorally retarded types, have been reported to have low SSS scores, but these subjects were hospitalized at the time of testing.

IV. Biological Correlates

A large-scale study on twins shows a relatively high degree of heritability for sensation seeking compared with other personality traits. Because we do not inherit traits as such, but only their biological bases, the data suggest that we might expect to find some biological bases for the sensation-seeking trait. Such correlates have been found and many have proven to be replicable.

A. Orienting Reflex

The orienting reflex (OR) in humans is measured as a physiological response to a novel stimulus (see glossary). The amplitude of electodermal response (EDR), measured as skin resistance to a current applied across electrodes on the palmar surface of the hand or fingers, provides a sensitive index of changes in interest or emotions elicited by stimuli. High-sensation seekers showed a stronger EDR than lows to the first presentation of a simple but novel stimulus but did not differ from lows on subsequent presentations of the stimulus as response habituated in both groups. Although the phenomena proved difficult to replicate with simple, meaningless stimuli, auditory or visual stimuli with strong content appealing to sensation seekers yielded consistent electrodermal OR differences between highs and lows. Using heart rate (HR), more consistent differences have been obtained between high and low scorers on the Disinhibition (Dis) subscale. When heart rate is measured beat-by-beat following a stimulus, the extent of the initial deceleration provides a measure of OR. With an auditory stimulus of moderate intensity, high-Dis subjects tend to show stronger ORs. Lows showed either weaker ORs or a defensive or startle response, as indicated by an acceleration of HR to the same stimulus. Stimulus novelty and intensity interact in producing the reaction related to Dis. Figure 1 shows the contrasting HR reactions of high and low disinhibitors to a 70 dB tone.

B. Augmenting–Reducing of the Cortical Evoked Potential

To obtain a measure of the cortical response to a stimulus, a brief stimulus, such as a light flash or a tone, is presented to the subject a number of times

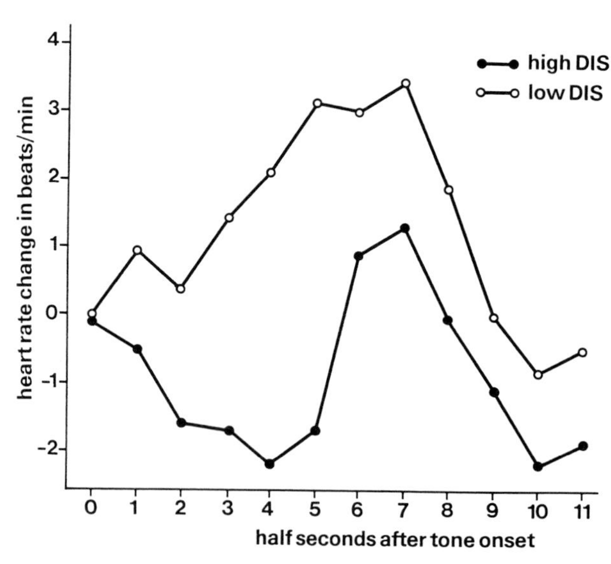

FIGURE 1 Mean heart rate changes during 5.5 sec following stimulus offset for subjects with high and low disinhibition (Dis) scores averaged over the first three trials. Stimulus was a 1,000-Hz 80-dB tone. High versus low Dis × trials × time interaction significant, $P < 0.05$. [Reprinted, with permission, from J. F. Orlebeke and J. A. Feji, 1979, The orienting reflex as a personality correlate, *in* "The Orienting Reflex in Humans," H. D. Kimmel, E. H. van Olst, and J. F. Orlebeke (eds.), Copyright 1979 by Lawrence Erlbaum Associates.]

while the electroencephalogram (EEG) is recorded. The half second periods of the EEG subsequent to the stimulus are averaged by a computer at selective points in time. The resulting wave form represents the evoked potential (EP) for that individual in response to the specific stimulus. Studies of identical twins have shown a very high similarity of the complex wave forms indicating a high heritability for the EP. The augmenting–reducing paradigm compares the relationships between the amplitude of an early EP component, at approximately 100–140 msec after stimulus presentation, reflecting the initial impact of the stimulus on the cortex, and the intensities of stimuli (see Glossary).

A significant relationship has been found in many studies between the Disinhibition subscale of the SSS and the augmenting–reducing continuum of visual and auditory EPs. Figure 2 shows the results of a study using the visual EP, and Figure 3 shows the results of a study using the auditory EP. In both studies, the high disinhibitors tend to show an increasing cortical reaction proportional to increasing intensities of the stimuli. In contrast, the low disinhibitors show little increase in EP amplitude with increases in stimulus intensity, and, in the case of the visual stimuli, they show a marked reduction in amplitude at the brightest intensity of stimulation. Studies in humans and cats (who show a similar individual variability in the phenomena) show that augmenting–reducing reflects brain and not peripheral receptor differences.

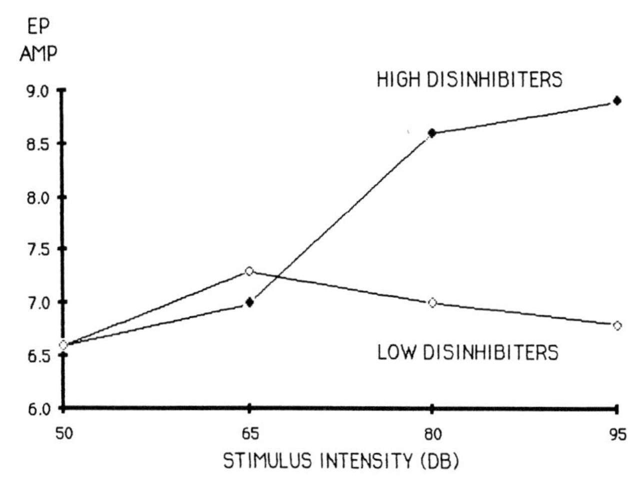

FIGURE 3 Mean auditory EP amplitudes (μVs) for groups scoring high and low on the Dis subscale of the SSS, form V, at each of four stimulus intensities (50, 65, 80, 95 dB). High versus low Dis × stimulus intensity interaction significant, F (df 3, 111) = 3.25, $P < 0.05$. [Reprinted, with permission, from M. Zuckerman, R. F. Simons, and P. G. Como, 1988, Sensation seeking and stimulus intensity as modulators of cortical, cardiovascular, and electrodermal response: A cross-modality study, *Personal. Individ. Diff.* **9**, 368.]

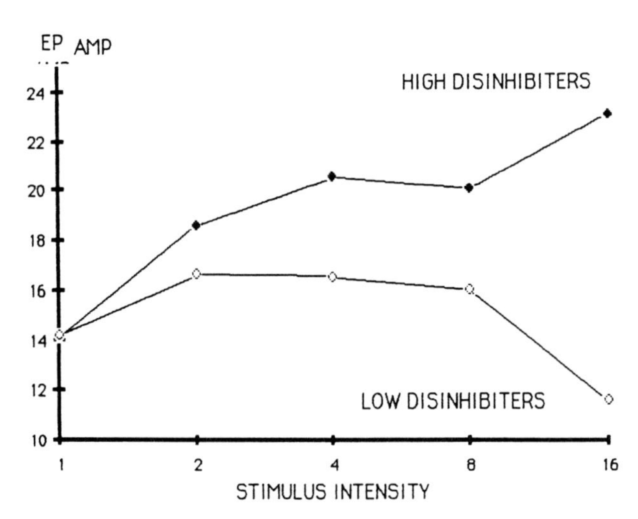

FIGURE 2 Mean visual EP amplitudes (in arbitrary millimeter deflection units; each unit = 0.42 μV) for groups scoring high and low on the Dis subscale of the SSS, form IV, at each of five stimulus intensity settings for the Grass photostimulator. High versus low Dis × stimulus intensity interaction significant, F (df 4/104) = 2.83, $P < 0.05$. [Reprinted, with permission, from M. Zuckerman, T. Murtaugh, and J. Siegel, 1974, Sensation seeking and cortical augmenting–reducing, *Psychophysiology* **11**, 539. Copyright 1974 by The Society for Psychophysiological Research.]

Cats who are classified as augmenters by the EP method show analogous types of behavior to human sensation seekers, tending to be active and exploratory, to approach novel stimuli, and to show strong response for reward. Reducer cats show slow adaptation to a novel environment and fear and avoidance of novel stimuli, but they learn better than augmenter cats on a task requiring inhibition of response in order to obtain reward. The correlation of the slope measure of EP augumenting and days to learn a task requiring the animal to maintain a low rate of response for reward was +0.73 ($P < 0.05$); the higher the slope (augmenting) of a cat, the longer it took to learn to delay its responses. The impulsivity shown by augmenter cats and the constraint shown by reducers is also consistent with the finding that augmenting–reducing of the EP is related to the trait of impulsivity as well as sensation seeking in humans.

TABLE II Gonadal Hormone[a] Levels of Male Subjects[b] Scoring in High and Low Ranges of the SSS Disinhibition (Dis) Subscale

	High Dis		Low Dis		
Hormone	Mean	SD	Mean	SD	*t*
Testosterone[c]	943.6	210.3	711.1	278.3	2.98*
17-estradiol[d]	38.0	11.1	27.2	6.4	3.78**
Estrone[d]	40.1	18.1	24.6	11.4	3.23*
Progesterone[d]	56.4	25.3	52.5	16.7	0.57

[a] Obtained from plasma, venous puncture.
[b] ns = 40 high Dis and 40 low Dis subjects.
[c] ng/100 ml.
[d] pg/ml.
* $P < 0.01$, ** $P < 0.001$.
From R. Daitzman, and M. Zuckerman, 1979, Disinhibitory sensation seeking, personality and gonadal hormones, *Personal. Individ. Diff.* **1**, 103. With permission.

C. Gonadal Hormones

The relationships between gonadal (sex) hormones and sensation seeking have been studied in males. Females have not been extensively studied because of the problems of variation due to the menstrual cycle. Both testosterone and estrogens in males correlated with sensation seeking, particularly with the Dis subscale. Table II shows the mean values for plasma testosterone, estradiol, estrone, and progesterone in males scoring high or low on the Dis subscale of the SSS. High-Dis males had significantly higher levels of testosterone, estradiol, and estrone than low-Dis males. In addition to sensation seeking, testosterone tended to correlate with sociability, impulsivity, and extent of heterosexual experience in males.

D. Other Hormones

Cortisol in cerebrospinal fluid (CSF) has been found to be negatively correlated with sensation seeking. Cortisol tends to be elevated in severe depressions, and low levels of cortisol in CSF may be related to the lack of inhibition in the disinhibitory type of sensation seeking. Similarly, thyroid function is lower in high- than in low-sensation seekers. Thyroid is related to tension and anxiety and thus may inhibit behavior that raises arousal levels.

E. Monoamine Oxidase

Monoamine oxidase (MAO) is an enzyme found in neurons of the monoamine systems in brain. It regulates the levels of neurotransmitters in these systems by breaking down the neurotransmitter after it is taken up into the cell from the synaptic cleft. It regulates the levels of neurotransmitters and, therefore, the degree of neural transmission by keeping a balance between production and amount stored in the neuron. In living humans, type B MAO is measured from blood platelets. Levels of platelet MAO show high heritability and are related to individual differences in activity in newborn infants. MAO has been found to be negatively correlated with general sensation seeking in a number of studies. Although the typical correlations are not high ($r = -0.25$), they are usually significant and always in a negative direction. High-sensation seekers tend to have low levels of MAO. This is consistent with higher levels of MAO in women than in men and the tendency of both brain and platelet MAO to rise with age. Sensation seeking is higher in men than in women and scores on the SSS fall with age. Studies of high and low MAO individuals in the general population show that low MAO types tend to be more sociable, less law-abiding, and more prone to use drugs (Table III). Drug users, alcoholics, and bipolar disorders have low levels of MAO, consistent with their high levels of sensation seeking. MAO is related to EP augmenting–reducing: augmenters have low levels of MAO relative to reducers. Studies of high and low MAO monkeys in natural colonies show analogous relationships to social behavior. Low

TABLE III Significant Differences Between Low and High (Upper and Lower 10%) Platelet MAO Subjects[a]

	Males		Females	
MAO	Low	High	Low	High
No. hrs. socializing on average weekday and weekend	13.2	8.5**	15.2	11.4*
Seen psychologist/psychiatrist for personal problems	37%	12%*	44%	29%*
Convictions for offenses other than traffic violations	37%	6%*	0%	0%

[a] The investigators mention other differences for which significance levels but no incidence figures are given: More low MAO males than high MAO males had used illegal drugs ($P = 0.008$), including depressants ($P < 0.05$) and stimulants ($P < 0.05$), and were currently using more stimulants ($P < 0.05$) and hallucinogens ($P < 0.05$). Low MAO males smoked more cigarettes than highs ($P < 0.01$).
* ($P < 0.05$).
** ($P < 0.01$).
From R. D. Coursey, M. S. Buchsbaum, and D. L. Murphy, 1979, Platelet MAO activity and evoked potentials in the identification of subjects biologically at risk for psychiatric disorders. *British Journal of Psychiatry* **134**, 372.

TABLE IV Partial Correlations, Removing Age Effects, Between Mean Behavior and Platelet MAO Activity in Rhesus Monkeys Living in Natural Conditions

Activities	Males (n = 17)	Females (n = 26)	All subjects (n = 43)
Forage	−0.14	−0.04	−0.08
Move	−0.22	−0.18	−0.20
Lookout	0.13	0.03	0.07
Play	−0.71*	0.26	−0.17
Give grooming	−0.28	0.16	−0.01
Receive grooming	−0.21	−0.41*	−0.34*
Self-grooming	0.69**	0.33	0.49***
Inactive	0.15	0.01	0.06
Rest–sleep	0.32	0.40	0.37*
Dominant–agonistic	−0.55*	−0.14	−0.31*
Submissive	0.38	0.09	0.20
Social contact	−0.54*	−0.18	−0.33*
Alone	0.59*	0.45*	0.51***

*$P < 0.05$, **$P < 0.01$, ***$P < 0.001$.
From D. E. Redmond, Jr., D. L. Murphy, and J. Baulu, 1979, Platelet monoamine oxidase activity correlates with social affiliative and agonistic behaviors in normal rhesus monkeys. Psychosom. Med. **41**, 87. Copyright by American Psychosomatic Society, with permission

MAO monkeys are more playful, aggressive, and sexually active; high MAO monkeys are passive and relatively unsociable (Table IV).

F. Monoamines: Their Metabolites and Enzymes

The findings relating MAO to sensation seeking suggested that there might be a relationship between at least one of the monoamine systems and the trait. The primary monoamine systems in brain are identified with one of the neurotransmitters norepinephrine (NE), dopamine, or serotonin. Although it is impossible to assess brain levels of the neurotransmitters in humans, their activity may be gauged by levels in CSF, plasma, and urine. Ordinarily, the metabolites of the monoamines are assessed, although CSF levels of NE may furnish a direct index of activity in one NE brain system that sends efferents down the spinal cord. In one study, CSF NE was negatively correlated with general sensation seeking ($r = -0.51$; partial r controlled for age, height, and weight = -0.49; both r's significant; $P < 0.01$). Consistent with this finding were negative correlations between plasma levels of the enzyme dopamine-beta-hydroxylase (DBH) and sensation seeking in this study ($r = -0.44$; partial $r = -0.60$; $P < 0.05$) and others. DBH is the enzyme converting dopamine to NE in the NE neuron. A low level of the enzyme would inhibit the production of NE. As yet, no correlations have been found between the metabolites of serotonin and dopamine in CSF and sensation seeking; however, the serotonin metabolite is low in impulsive persons.

V. A Psychobiological Model

The approach to the psychobiology of sensation seeking has thus far been largely correlational. Experimental work on the role of biological factors in determining behavior in humans is difficult. It is unethical to produce irreversible brain alterations in humans in order to study the effects on behavior and personality; however, using data from biological psychiatry, where potent drugs are used to ameliorate disturbed behavior, and from comparative experimental work with animals, where the brain may be selectively lesioned or stimulated, it may be possible to fit the correlational data from normal humans into a biological model.

The strong OR found in high-sensation seekers is a product of novelty because differences between highs and lows disappear when stimuli are repeated. The strong OR trait may reflect the disposition in a nervous system adapted to process novelty and approach novel stimuli. This is a characteristic of sensation seekers at the behavioral level. They constantly seek novel information and tend to avoid repetitious or redundant information. In low-sensation seekers, novelty tends to be threatening, and in the face of novel stimuli they tend to manifest defensive physiological responses as if in preparation for flight or planning escape.

In contrast to the OR that is a response to novelty at low to moderate levels of intensity, the augmenting or reducing of the EP is a reaction to the intensity of stimulation, because it is defined by the relationship between EP amplitude and intensity of stimulation. Augmenters seem to have a "strong nervous system," to use a Pavlovian theoretical term. The task performance of augmenters is less affected than reducers by stressful distractions. Reducing is essentially a protective reaction, but it may result in cortical inefficiency as a consequence of cortical inhibition in response to a high intensity of stimulation. Augmenting could also represent a vulnerability in many situations because the reflexive protection against high levels of stimulation is lacking. In behavioral terms, the link between augmenting and disinhibition can be seen in the tolerance of high-sensation seekers for loud music and highly arousing stimuli of all kinds. The vulnerability of augmenters can be seen in their disposition to affective disorders of the bipolar type. In manic states, bipolars seek intense stimulation without relent, and the stimulation serves to further excite them. The combination of low MAO and augmenting in persons at risk for this kind of affective disorder suggests that the lack of regulation of one or more of the monoamine systems predisposes them to the failure of cortical inhibition mechanisms and the loss of behavioral control seen in mania.

Gonadal hormones seem to play an important role in activating social, sexual, and sensation-seeking behavior in men. One of the causal paths for this effect may be their lowering effect on MAO. Although MAO is consistently linked to a variety of sensation seeking-related behaviors, from activity in newborn infants to interest and participation in mountain climbing, the role of this enzyme is not clear. Although it regulates three monoamine systems, these systems have different roles in behavior. Serotonin tends to be related to behavioral control in animals, and low levels in humans are associated with impulsive, aggressive, and suicidal behavior. The Norepinephrine (NE) system seems to serve to arouse the cortex in response to signals of biological significance. Dopamine in the nucleus accumbens (a region of the basal forebrain involved in the initiation of locomotor activity) is associated with intrinsic reward; whether released by direct stimulation or stimulant drugs such as amphetamine or cocaine, it produces a state of energy and euphoria in humans and maintains self-stimulation through electrodes placed in that area of the brain in rats. The direction of the MAO relationship with sensation seeking only makes sense if, as many assume, it is a better marker for activity in the serotonin-using neurons than the other two systems. However, if sensation seekers suffered from a low base level of activity in the NE and dopamine systems, but were highly responsive to novel stimuli (which activate the NE arousal system and the OR), they might engage in exciting, risky activities, seek novel and intense sensations, and take stimulant drugs to activate these systems to some optimal level of arousal of these neurotransmitters. If low-sensation seekers were near their optimal level, further stimulation might push them into the high range of activity where the same neurotransmitters may produce anxiety and even panic (as sometimes happens to drug abusers who have developed a tolerance and are taking high doses of stimulants). Both sensation seeking and drug use may be attempts to find an optimal level of neurotransmitter activity somewhere between states of boredom and anxiety.

Environmental factors obviously must play some role in producing high or low levels of the trait, but studies on twins suggest that it is not the shared family environment but the environment that is specific for each member of a family that influences sensation seeking. The peer environment, for instance, may be more influential than the level of stimulation in the home shared by all family members. This is not to say that *prolonged* periods of low or high stimulation may not readjust the inherited component of sensation seeking. Subjects coming out of sensory deprivation experiments of several days or more show reduced cortical activity and a lack of motivation to seek stimulation. Paradoxically, most subjects simply want to go home and be alone. However, such extremes of stimulation are not that common during development. Peer or parental modeling or reinforcement may influence the phenotypic forms that sensation seeking will take, more than the basic tendency. A youth growing up in the inner city may find that drugs and crime are the only kinds of sensation seeking available, whereas one raised in more affluent surroundings can also seek sensation in exciting sports, fast cars, rock concerts, travel, and other such diversions. While socioeconomic class and education show some influence on subscales like TAS and ES, particularly in women, the Dis scale seems to be relatively free of such influence. Partying, sex, and alcohol are the age-old forms of social sensation seeking. Such needs may be modified by attitudinal and religious training, but it is just as likely that sensation seeking will modify learned attitudes if they are not in accord with the basic disposition. There is much to learn about the interaction of genotype and environment in producing the consistent differences in behavior that we call personality. This is clearly a two-way interaction. We select what we need from the range of environments and are in turn influenced by what we select as well as that which we cannot control.

Bibliography

Zuckerman, M. (1979). "Sensation Seeking: Beyond the Optimal Level of Arousal." Lawrence Erlbaum Associates, Hillsdale, New Jersey.

Zuckerman, M. (ed.) (1983). "Biological Bases of Sensation Seeking, Impulsivity and Anxiety." Lawrence Erlbaum Associates, Hilldale, New Jersey.

Zuckerman, M. (1983). Sensation seeking: A biosocial dimension of personality. *In* "Physiological Correlates of Human Behavior: Vol. 3 Individual Differences" (A. Gale and J. Edwards, eds.). Academic Press, New York.

Zuckerman, M. (1984). Sensation seeking: A comparative approach to a human trait. *Behav. Brain Sci.* **7,** 175.

Zuckerman, M. (1985). Biological foundations of the sensation seeking temperament. *In* "The Biological Bases of Personality and Behavior," Vol. I (J. Strelau, F. H. Farley, and A. Gale, eds.). Hemisphere, Washington, D.C.

Zuckerman, M. (1988). Brain monoamine systems and personality. *In* "Neurobiological Approaches to Human Diseases" (D. Hellhammer, I. Florin, and H. Weiner, eds.). Hans Huber Publishers, Toronto.

Zuckerman, M. (1990). The psychophysiology of sensation seeking. *J. Pers.* **58,** 313–345.

Zuckerman, M., Ballenger, J. C., and Post, R. M. (1984). The neurobiology of some dimensions of personality. *In* "International Review of Neurobiology," Vol. 25 (J. R. Smythies and R. J. Bradley, eds.). Academic Press, New York.

Zuckerman, M., Buchsbaum, M. S., and Murphy, D. L. (1980). Sensation seeking and its biological correlates. *Psychol. Bull.* **88,** 198.

Sensory-Motor Behavioral Organization and Changes in Infancy

HENRIETTE BLOCH, *Laboratoire de Psycho-Biologie de l'Enfant*

Glossary

Allometry Different parts of the body do not grow up and increase their weight at the same speed, so their relation is called *allometric*

Altricial This word characterizes the species born immature

Automatic walking By handling a newborn under its armpits, upright, with the soles of its feet in contact with a plane surface, a forward stepping movement is triggered. This stepping reflex, called *automatic walking,* disappears spontaneously around the third month

Behavioral state "Temporary stable conditions of neural and autonomic functions, known as sleep and wakefulness" (Prechtl). With behavioral criteria such as open/closed eyes, breath regular/irregular rhythm, gross movements/no movement, vocalization/no vocalization, and concomittant cues of polygraphic recording, five behavioral states are distinguished: (1) quiet sleep; (2) agitated sleep; (3) quiet alertness; (4) agitated alertness; (5) cries

Biomechanical constraint Any limitation to moving due to mechanical arrangement among body components, such as muscles, joints, skeleton elements, fate deposition

Cephalocaudal law Universal principle of development in organized species. Development begins from the cephalic segment and then, step by step, reaches the extremities

Circular reaction Organized act by which any positive effect leads the subsequent re-elicitation

Gestational age Fetus or infant age computed from the last menses of the mother

Habituation When a stimulation perdures or is repeated, the initial response it provokes decreases with time. Such a decline differs from saturation or forgetting because the initial rate of response can be recovered with partial changes in the stimulation or a delay

Moro reflex Reflexive extension of the arms, followed by an abduction on the chest. This reaction responds to a sudden lost of support. It seems to have a vestibular origin

Myelination Discontinuous sheath of myeline encircles progressively many nerve fibers in the vertebrate central nervous system and trains a faster conduction of nervous impulse

Posture Antigravity positioning of the whole body and positioning of its mobile parts with respect to each other

Reversibility Main property of a logical structure in which invariants are formed by reciprocal ties between two actions or intellectual operations oriented in opposite directions

Saccadic eye movement Jump of the gaze. To fixate a target, an adult directs his/her gaze by an initial jump of the eyes, matched to the target distance. The newborn's jump is of standard amplitude. So, a series of jumps is necessary to point a fixed target or to track and suit a mobile

LIKE OTHER mammals, the human infant arrives at birth still immature. Nevertheless, it is equipped with instruments that can ensure relations with the external world. Its sense organs are morphologically developed, and its sensory systems, although they are at different levels of maturation, are all functional. Their functioning implies a motor participation: Every sense organ can change its orientation either by a self, local movement or by a movement of the head, which contains all the sense organs except the tactile ones. If anything prevents a sense organ from moving, perception will be altered. It has been shown in kittens that by cutting off the ophthalmic branch of the trigeminal nerve, one eye can be impeded to move in the orbit, and that trains alterations in the functional array of cells in the striate cortex and causes defects in the binocular visual field.

Moreover, the neonate's motility includes more than rapid stereotyped reflex reactions and mass agitation. Slow, aimed movements and general, isolated movements are performed; some of them appear gracious and delicate, as manipulation of fingers, which can be observed in wakefulness and in the so-called agitated sleep. Touching a part of the body and reaching to some object in the immediate environment may carry out a gaze shift toward the touched or reached surface. Therefore, we can assume that sensory and motor activity are linked in several ways from the first days of life. Observations on preterm-born infants and on fetuses show that such linkages are prewired. Neurophysiological studies conducted with animals invite us to think that early sensory-motor behavior is controlled by a common structure in the midbrain brainstem, the colliculus. It may be that some other structures in receptive and motor cortices and in the cerebellum take part in the command process. The image of the impulsive newborn living in chaotic confusion has been shattered; such a view was only due to our ignorance. However, the human neonate is altricial. Many limitations and constraints bear heavily on its sensory and motor behaviors (i.e., fragile behavioral states, biomechanical constraints due to allometry, and coaction effects caused by a possible unclear differentiation between systems). The organization of early behavior appears different in the newborn and the young infant and from what it will be later in childhood and adulthood.

To understand the initial features and the mechanisms of transition, we first must examine separately the panel of sensory and motor activity and then their possible coordinations.

I. The Neonate Repertoire

A. Newborn Sensory Activity

Sensory activity is devoted to gathering and processing information from the external environment. A primary condition to do that is to detect separate stimulations, which requires us to specify where the stimulus is and what it is. The orientation response can be obtained from the first hours after delivery, and it is observed in healthy prematures as early as 28 weeks gestational age. In visual perception, which is the most often studied modality, peripheral receptors of the retina are activated and trigger a head-eye turning toward the source of stimulation. The gaze can stare at the visual target, but holding the fixation seems to be difficult if not impossible, because of the foveal immaturity and a lack of control of the rapid, little eye movements that accompany ocular fixation. These are considered strong indications that human newborn perception could be very poor. However, the neonate can distinguish a shape on the ground, when the shape is a closed, structured, and complex one and when the contrast between the ground and the shape is high enough. Perceptual habituation is also possible for auditory, olfactory, gustative, and visual stimulus. This allows us to assume that external information is brought and processed. However, many parameters of perceptual behavior are yet unknown.

Perceptual discriminations are now studied systematically. Some of them appear better in neonates than later in infancy, such as speech sounds and odors, and some of them seem rather less sharp. Perceptual exploration and the organization of activity seem also greatly different than at later ages. For instance, visual tracking is jumpy and is called saccadic; and visual pursuit of any mobile, discontinuous. The visual field is less extended in the neonate than later ages. So, we can summarize these evidences in concluding that the perceptual world in the beginning of life is very different from the adult's. However, it does exist. [*See* PERCEPTION.]

B. Motor Activity

Motor activity of the newborn involves several categories of movements and positionings. Any sort of movement refers to the postural framework. Posture is defined as the spatial orientation of the whole body or the limbs for holding an attitude or executing a movement. In vertebrates, posture is performed by the action of muscular systems on the skeleton. Every positioning and movement can be described as a part of any postural organization and in relation to it. Postural maturation has far to be achieved at birth. Nevertheless, human fetuses and neonates can adopt preferential postures that might

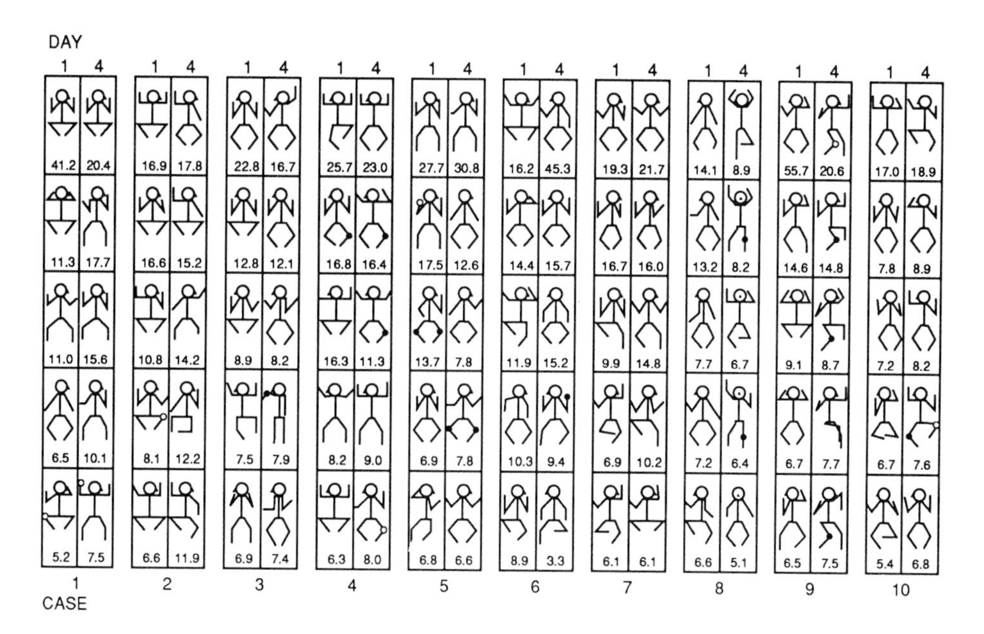

FIGURE 1 The five most frequently occurring postures per baby on day 1 and day 4 after birth. (From Cioni, Ferrari, and Prechtl, (1989). Posture and spontaneous motility in fullterm infants. *Early Hum. Dev.* **18,** 247–262, with permission.)

have an influence on some aspects of later motor development.

Spontaneous motility can be observed with different frequencies in every behavioral state as shown in Fig. 1. A perceptual stimulus can induce more than one movement, depending on its signification and value. Rhythmical reactions often are given as immediate responses. Adapted responses are slower than later in the life.

Part of the motor repertoire present at birth is going to disappear in the first postnatal months: Many reflexes (e.g., the Moro reflex, the grasping, the automatic walking) are called *archaic reflexes*. Some of them can be considered to have influence on later, more integrated responses, while some of them will not. Grasping, for example, was first considered as an archaic, rudimentary form of prehension; but recent studies about early hand-reaching have convinced us that grasping has to be viewed as acting against reaching and smooth prehension. However, automatic walking seems to exert influence on later autonomous locomotion. Training in automatic walking accelerates the accession to autonomous walk. So the importance of early reflexes for motor development is controversial.

Biomechanical constraints are heavy at the beginning of extrauterine life. The weight of the head, which is comparatively greater than the trunk and the limbs weight, pushes the body to flexion and provokes head inclinations, even falls. Antigravitational reactions are not formed. So the newborn has only a weak control of its own posture. Its postures are arranged by the adult caregiver. Despite this obvious immature aspect, preferential positions adopted by neonates are in continuity with fetal positions. The preferential positions are asymetric and seem to be triggered by the asymetric tonic neck reflex. Therefore, the spontaneous motility, which has to be referred to posture, is different in quantity and in quality among the newborns. Nevertheless it reveals some lateral differences and invites us to think that the two hemispheres are not involved in the same way in any motor act.

The motor repertoire of the newborn is, however, considerable, but some movements that are spontaneous can be used as aimed movements in a coordinated act. This is the case for head movement: In supine position, a baby can turn its head in response to a perceptual stimulation, and head movement is a part of the orienting reaction. However, it is not performed simultaneously with eye movement, and it can be impeded by an eccentric eye displacement, as if it is concerned by a musculo-external muscular locking.

C. Newborn Sensory-Motor Coordinated Behaviors

Because of many evidences of prewired sensory and motor abilities, newborn capacity to couple sensory-motor acts is currently explored in some

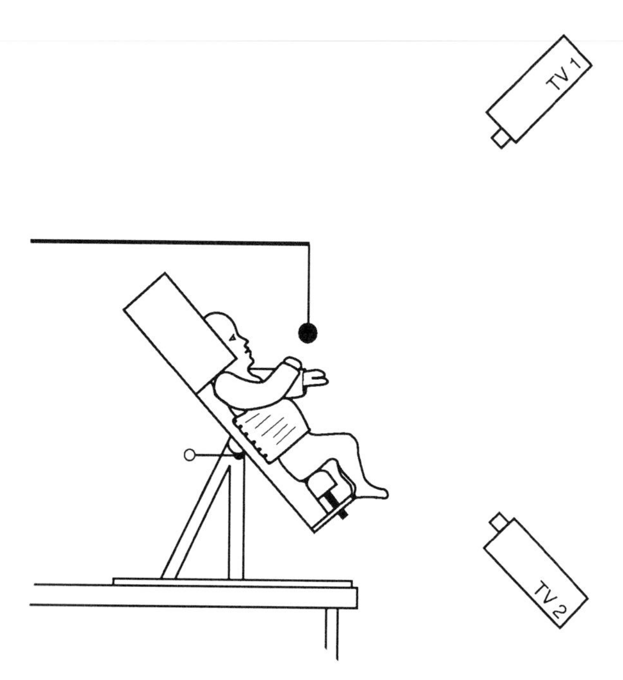

FIGURE 2 Experimental situation for studying newborn eye-hand coordination. (From Hofsten, (1982). Eye-hand coordination in the newborn. *Dev. Psychol.* **18,** 450–461, with permission.)

observational and experimental settings. Such studies, although, unfortunately, still not numerous enough, provide useful information about available conditions and suitable criteria of early couplings. Attention is presently paid to eye-head, eye-hand, and hand-mouth coordinations. Observations in every-day-life environments show that such coordinations are neither obvious nor frequent when the infant is lying in its cradle. However, they can be observed when the newborn is held in a sitting position, with the head upright. So, posture appears as a determinant condition for coordinated behavior.

As studied in experimental designs as in Fig. 2, with respect to good postural condition, neonate coordinated behavior is attested by a global synchrony between perception and movement toward any object in the proximal environment. However, it has specific features that raise questions and controversies:

1. Movement is not always triggered by the perception of the target. In eye-head coordination, head movement can precede a gaze shift; hand movement toward an object can be performed with the eyes closed.
2. Attainment of the target, considered as the goal, is neither frequent nor accurate. The reaching movement of the head or of the hand stops before encountering the target. In hand-mouth coordination, it has been shown that the hand can touch some parts of the face before or without going to the mouth, which stays open.
3. Perception and movement seem not to be closely and continuously coupled. For instance, velocities and amplitude of head movement are not in harmony with eye movement. Some accelerations and decelerations can occur in head-turning. In hand-reaching, the baby does not visually control its hand, except when one hand is in the vicinity of the target.
4. Perceptual pointing seems to be better than the associated movement. The newborn is able to point accurately on a visible target by moving its eyes, despite saccadic and rigid eye movement. There is only one pattern of eye movement; on the other hand, there are several patterns of head and hand movement.

II. Bases and Determinants of Sensory-Motor Development

The 1-year-old infant's behavior appears both very different from the newborn's and more similar to child and adult behavior: The infant can stay standing up, it begins to walk, it recognize persons, it remembers places of objects, it searches for and hides objects, it can compare and associate several objects and build some new configurations with them, etc. Such large differences have led us and still lead us to consider development in the first year as made through successive qualitative changes rather than a linear, quantitative progression. At the beginning of the 20th century, the very commonplace opinion was that this development is mainly sensory-motor and implies some changes in nervous functioning and in the postural framework. The latter were easier to study than the internal changes.

A. Postural Maturation and Its Consequences

Posture develops in humans within the first 2 years after birth, until body positions are generally mastered, following a cephalocaudal law and a proximo-distal principle. Four stages have been distinguished from the neonate prone position:

1. Around the third month, the infant becomes able to hold its head upright. It is able to prevent forward and lateral inclinations or "falls" of the head and can turn the head into a horizontal plane without participation of the torso and the shoulders. The main conse-

quence of this change is that the line of the gaze can remain constant.

2. The sitting position is attained at 4–5 months. When the infant can stay seated without any back support, that "liberates" its arms. So, the space of manual prehension extends, and may be structured with a stable reference to the body median vertical axis.

3. The infant begins to stand up at about the ninth month. First, it stands up holding a solid stationary object such as a wall or a chair; then it can stay up without any material support. This stage appears as a transition to walking.

4. At last, the infant begins to walk by itself at about 13–15 months, first leaning on a support and then without any support. The autonomous locomotion represents achievement of a general motor coordination and implies a stable organization of movements. It greatly enlarges the space of action.

This postural development includes some other local transitions and steps, such as creeping and crawling, which show a continuity. Cultural influences can shorten a stage, but they cannot modify the successive order of the main stages. However, sensory or motor handicaps can be responsible for severe retardations. For instance, blind-born infants attempt the sitting and standing position and walking later than normal infants when they do not receive any compensatory help, such as an ultrasonic guide, but they attempt these stages in the same order as the normal infants. Therefore, the postural changes in human infancy appear similar to the postural changes in all the superior vertebrates and are considered maturational changes. They can be considered organizers of the environmental space: Every sense organ has a specific receiving area, which catches the stimulations from a spatial limited part of the surroundings. Movement makes this part mobile, and a whole coordinated system of movement makes the successive parts homogeneous. [*See* MOTOR SKILLS, ACQUISITION.]

B. Changes in Nervous Functioning

The infant central nervous system differs from that of the child and the adult in its morphology and its functioning. It is the seat of large, rapid, and antagonist transformations, which are not yet totally clear, during the first years of life. All the structures of the brain are formed at birth, but they include populations of cells and synapses that can be more or less numerous than they will be later; their spatial array is also different, and some of them seem not to be implicated in early behaviors (e.g., the associative, prefrontal, and frontal cortices). Two major developmental nervous processes have been predicted to cause changes in sensory-motor organizations and behaviors: myelination, which was viewed during a long time as a single, general criterion of nervous maturation, and then, corticalization.

1. Myelination

Myelination of several motor and sensory pathways begins before birth in humans but continues after birth. The pyramidal pathways, which play the main role in motor-coordinated acts and in the spatial orientation of the head and the body, are not myelinated at birth. Their myelination begins at about the normal term of birth and accelerates at about the third month. It is not achieved before the end of the second postnatal year and occurs earlier in the corticospinal paths than in the cortico-motoneural ways that command voluntary movements. Among the sensory systems, the visual system has been the most studied. Although it develops after the others, its rhythm of maturation is rapid after birth. Nevertheless, it is far from a full myelination during the first months postnatally. The subcortical pathway, which goes from peripheral retina to the superior colliculus, is fully myelinated at 3 months, having started at the sixth month of gestation. It contrast, the central retina-to-cortex tract begins to myelinate only after birth; the myelination increases rapidly up to 4 months and then decelerates, but is not achieved before the end of the second year. Other input connections to the primary visual cortex remain poorly myelinated up to 3–4 months. The myelination of extrastriate areas and intracortical neurons takes a long course, as long as 7 years.

According to the myelination process, voluntary movements, such as involved in prehension, could not be performed before the fourth postnatal month. Such requirements of prehensive behavior, such as the continuous control of movement from its initiation and the anticipatory shaping of the hand, in relation to the perceived size and shape of the object to be reached, would not be possible before this time.

2. Corticalization

Corticalization is an evolutionary process that characterizes development in primates and is the most important in humans. Regarding sensory-motor capacities, it presents the following features:

- An enlargement of the lateral parts of the cerebellum, which are involved in the motor control of limbs;
- An increasing of the precentral cortex, which is involved in the motor control of manuality;
- A functional specification of the parietal associative cortex, which commands the accurate visual guidance of hand movements; and
- A large development of the associative areas of the frontal cortex, which are concerned with the processing of external information and self-movement-produced information [*See* CORTEX.]

The corticalization process is commonly supposed to be important in the first months after birth, with a shift from subcortical to cortical control of perceptual responses to external stimulations and any influence on some sensory-motor coordinations. According to this view, perceptual analysis of form, which is performed by the foveal vision, would truly begin at 2 months; the eye-head coordination would also appear and would be controlled by the corticofugal pathways and not by the colliculus. Changes in the oculomotor organization of eye movements, namely, in the saccadic components, have been observed at this time.

The corticalization seems to be obviously the main factor of the onset of autonomous walking, which forms a sort of synthesis of voluntary movement, with a control of alternative leg movement, parallelism of the lower limbs, vertical equilibrium of the body, and a control of direction and distance. So the autonomous locomotion appears as a goal-directed behavior compared with the newborn automatic walking.

C. Open Questions

Questions arising from general bases and determinants of sensory-motor development in human infancy are currently discussed. First, these bases and determinants do not allow us to understand all the changes observed in behavior. Second, the sensory-motor repertoire in the first few months appears more rich and organized than should be presumed from the myelination and the corticalization processes. That does not mean that these processes have no influence on behavior, but that they are not its only determinants and that the relations are not simple and direct. The temporal coincidences between neural and behavioral changes provide some opposite evidences: For instance, if we accept that eye movements are first controlled by the colliculus and cannot include a foveal fixation, we cannot explain the visual habituation phenomenon in newborns, which requires a fixation holding a long time, often more than 3–5 seconds; we cannot explain some fine discriminations of form features performed by newborns as distinction between curvilinear and linear contours of visual forms. If we accept that any aimed arm-hand movement can be generated before the pyramidal pathways are myelinated, we cannot understand the neonate reaching behavior.

III. Course and Processes of Sensory-Motor Development

Sensory-motor development has been studied, with different approaches, since almost the first century. However, its course and its processes remain still partly unclear. Causes of obscurities and discrepancies seem to be due to three causes: First, hypotheses were different for motor and for perceptual development and induced separate research. Motor changes in infancy and childhood were approached with maturationist hypotheses, and perceptual changes, with cognitive and learning ones.

Second, it is difficult to detect and describe successive changes without any model or theory about their directionality or without a reference to a stable state. The reference taken in motor studies was more often biological and phylogenetical. The aim was to explain the changes into a hierarchical continuity from embryology. The studies of perception referred to adult perceptual behavior and were used to understand how a stable world is built.

Third, for understanding sensory-motor coupled development, the research of a single principle predominated for a long time. Three general theories have been purposed.

Because some questions remained in suspense within such general, theoretical frames, current research has moved away from them and tried to accumulate more empirical data, without any explicit *a priori* model. During the past 20 years, research was focused on collecting data and inventing efficient methods of investigation. Sometimes some theoretical points were discussed or revisited. Presently, it seems that psychologists again feel a need for theoretical overviews. Such a pendulum of historical movement imposes to review general models, to summarize the data, and at last, to consider current interpretations.

A. Review of General Theories of Sensory-Motor Development

All the general theories that have been proposed consider the sensory-motor development as a part of ontogenesis and as the first level in a constructive process.

1. Baldwin's Model of Accommodation

Baldwin was the first psychologist who paid attention to the relation between perception and action from birth. He described the course of sensory-motor development as a series of changes into "circular reactions." What he called circular reactions are sensory-motor coupled responses to external stimulations, which appear and during the first year of life. Movement would be first, such as reflexive in the newborn. Any spontaneous movement can have a positive or a negative effect and can be felt by the infant as pleasant or annoying. According to the Thorndyke's law of effect, any act that has a positive or pleasant effect would be repeated. For instance, sucking is a reflexive reaction for ingesting milk. Then sucking movement would be repeated in a vacuum and would constitute a "primary circular reaction."

Later, a "secondary circular reaction" would appear when two parts of the body can be associated in a single act: When the hand of the baby encounters a visible bauble by accident, the sound its movement produces would train the act again. The infant learns that its own movement has a consequence on an external object, so it can learn to adjust its movement to replicate the sound. Such an adjustment reflects an accommodation. The secondary circular reaction appears around the fourth month postnatally and implies a coordination between vision and hand-reaching.

A third step would be when the infant becomes able to associate more than two systems. A "tertiary circular reaction" would appear at about 8–10 months, with complex or sequential actions and several objects.

Such a model allies sensory-motor development to a learning process. It supposes that the human infant has a "hedonic"-oriented capacity as soon as birth. That has been confirmed by Lipsitt in his research on neonate sucking behavior.

2. Piagetian Theory of Sensory-Motor Development

Piaget started from Baldwin's description of circular reactions. However, he did not accept that they give evidence of accommodation. He considered that sensory-motor coordinations allow the infant to create means–goal relationships, which develop through active assimilation: The infant would make a mental work for linking its own actions and the consequences they have. When it is placed in a new situation, it can distinguish known and unknown components and apply a strategy in acting. The successive behaviors toward a hidden object provide evidences about an assimilation, for instance, when a 9-month-old infant searches for a hidden object at the place where it was found previously and not at the place where the object disappeared.

Therefore, sensory-motor development represents the first stage of cognitive development. This stage would be achieved when the fundamental organization that governs both action and cognition is built. Piaget demonstrated that such a structure is formed when the infant is able to walk, because autonomous walking implies what he called "reversibility." According to this theory, logic would be first in action, then it would pass into reasoning on concrete situations. Active assimilation could be only inferred from changes into a repeated act. Sensory-motor learning would not only be a gain in performance but would involve a transformation of a mental structure that controls the behavioral response.

3. Gibsons' Theory

Gibson and Gibson assumed that sensory-motor development is not a process of enrichment but rather depends on perceptual differentiation. First, they considered that the perceptual systems can take and treat simultaneously a lot of information specified in any sensorial array, which contains object and space features and has an ecologic meaning. This information is not referred to the modality of input, but to the physical environment. Second, they claimed that there are functional equivalences between exteroceptive and proprioceptive information. So, intermodal relation can exist and perception is intrinsically meaningful. Perception-action couplings are determined by what the Gibsons called "affordances." Affordances are properties of objects or events in the surroundings that respond to the needs of the perceiver. They are together physical and psychological and they are, as told by the Gibsons, "ecologic." These properties are immediately translated in action terms. The perception of a nipple implies the act of sucking, and by sucking a nipple, the infant learns its elasticity, its texture, its form, and its size. Objects can be distinguished from the possibility of action they offer. A young infant does not know that a projected slide cannot be taken in hand, so it reaches for the picture as for a "real" object. By doing that it learns to discriminate pictures and solid objects; so it learns to reach only for solids.

B. Data on the Course of Sensory-Motor Development

The course of sensory-motor development does not appear as a linear progression. In many cases, it includes stops and apparent regressions.

The example of hand-reaching is illustrative: (1) The neonate reacts to contact of any object placed in the palm of its hand by a grasping response. Its fingers close up the object by an immediate, sudden, and sharp reflexive movement. This reaction is described as stereotyped and serves as an item in the postnatal first neurological examination.

In the first days after birth, reaching movement has been shown toward distal visible objects. Movement is an arm-hand slow projection, without any grasping. The hand remains open during the movement. More often, it stops before encountering the object, although the infant is looking at it. Such a movement happens only when the infant is in a quiet alert state and is promoted when infant is seated, with the head upright. Timing of movement is slow (about 1 second for reaching the vicinity of an object put at 20 cm from the body); velocity is not regular but shows accelerations and decelerations. At about the end of the first month, the frequency of such reaching hand movements decreases. The movement is made with the hand closed; when the hand contacts the object, there is no manual exploration. In this period, movement can be described as ballistic. At about 13 months, another change occurs: The arm-hand movement becomes again slower, visual control of it can be observed, the hand opens before the encounter of the object, and there is an anticipatory shaping according to the size of the object. However, the speed of reaching is often too rapid in the final phase, and the hand pushes the object and cannot easily ensure a prehension. Contact is performed by trials and errors. Constant directionality and distance of the hand-aimed movement are observed at about the sixth month.

This development has a U shape and that leads us to consider that the successive stages are attained through reorganizations. The decline of the reaching behavior between the second and the fourth months may be due to an increasing importance of visual search and attention, which would inhibit the movement. Because of different rhythms of maturation in sensory and motor systems, it can be supported that earliest sensory-motor behaviors cannot be harmoniously coupled. A good synchrony would suppose reciprocal control and take time to be built.

Continuity from grasping to prehension is not obvious. On the contrary, grasping appears as a transitory constraint, which is not a component of the early reaching movement. That can be interpreted as a change in the composition of elements, with a change in the function of behavior, or as an effect of inhibition by higher nervous centers.

(2) Development of prehension is not achieved at 6 months postnatally. Several later changes will occur: The reaching movement becomes less rigid, with smooth final bracking, the manual gesture to take an object becomes more and more accurate, and more direct than tentative. The thumb–index finger grip, which appears at about 9–10 months, allows the child to take very small objects. In the same time, visual guidance is alleged, and visual pointing very precise. However, the little child remains unhandy for several years. Child performance in sensory-motor ordinary tasks, such as catching a drop, lacing shoes, buttoning up a coat, and in sensory-motor skills, such as sewing, drawing, and writing, increases up to 7 or 9 years with active practice and may also improve by modeling. However, performance is still inferior to the adult performance. Qualitative differences can be shown in the organization of such behaviors, between 7-year-old children and adults, namely, in their spatio-temporal unfolding. Nevertheless, the commonplace opinion is that the course of sensory-motor development ends at about 2 years, because basic local and general sensory and motor structures are stabilized and are homogeneously connected. So, the changes that happen later are considered as gains provided by some kinds of learning. Discussions are generally centered on acquisition of motor skills and focus on learning processes in this field.

However, some currently conducted research induces some doubt about that. In studying running and jumping child abilities, it has been shown that some postural and dynamical constraints are responsible for specific features and determine transient compositions of movements in relation to perceptual control, which are not overtopped with only practice. We need more information about them. However, they lead us to consider that the course of sensory-motor development is longer than we thought and could mean that sensory-motor development does not entirely depend on cognitive structures and functioning that emerges around the second year but would continue in a parallel direction to cognitive development.

C. Concluding Remarks

For summarizing what we know about sensory-motor development, the following statements can be made:

1. Despite sensory and motor systems that do not mature at the same speed and rhythm, perception-action couplings are performed very early in the neonatal period. They appear asymetric and transitional and are followed by several deep changes.
2. Several kinds of transition are observed in infancy and are presumed in childhood. Some of them involve a change of function: The development from early reaching to prehensive behavior examplifies such a change. Early reaching seems not to be devoted to manual prehension or to tactile exploration, but rather to help stabilize vision. Some changes appear as structural ones, when connection between perceptual and motor components become closer and when new nervous networks can be inferred from behavioral analysis. Some changes can be both functional and structural.
3. The general theories assigned a cognitive aim to sensory-motor development and suggested that sensory-motor organizations and behaviors are more and more controlled by cognitive goals. However, some changes appear more adaptive than truly cognitive, and such a single principle of development cannot be currently supported. Even if it is obvious that the well-organized perception-action coupled responses serve to master the external world in relation to body stabilization, many kinds and models of mastering have to be considered.
4. New perspectives are now open in psychobiological studies of development, with the dynamic systems approach and with a renewal of functionalist approaches that use precise control procedures for detecting the aim of behavior and for relating subjective goal to proceedings of attainment.

Bibliography

Blass, E. M., ed. (1986). "Handbook of Behavioural Neurobiology: Developmental Processes in Psychobiology and Neurobiology." Plenum, New York.

Bloch, H. (1989). On early coordinations and their future. *In* "Transition Mechanisms in Child Development" (A. de Ribeaupierre, ed.), pp. 259–282. Cambridge University Press, Cambridge, New York.

Lipsitt, L. P., ed. (1976). "Developmental Psychobiology." L. Erlbaum, Hillsdale, New Jersey.

Pick, H. L. (1984). Cognition and action in development: A tutorial discussio. *In* "Cognition and Motor Processes" (W. Prinz and A. F. Sanders, ed.). Springer Verlag, Berlin.

von Hofsten, C. (1989). Transition mechanisms in sensory-motor development. *In* "Transition Mechanisms in Child Development" (A. de Ribeaupierre, ed.), pp. 233–258. Cambridge University Press, Cambridge, New York.

Serotonin in the Nervous System

JOAN M. LAKOSKI AND BERNARD HABER, *The University of Texas Medical Branch at Galveston*

Glossary

5-Hydroxyindole acetic acid (5-HIAA) Principle metabolite formed by deamination via the enzyme monoamine oxidase

Receptor subtypes Multiple membrane-associated recognition sites for a neurotransmitter

Serotonin 5-Hydroxytryptamine (5-HT) is an indolealkyl amine that is localized in neurons where it functions as a neurotransmitter

Serotonin agonist Compound recognized as serotonin-like by the receptor

Serotonin antagonist Compounds that block the interaction with the receptor

Tryptophan Essential amino acid transported into brain tissues, where it is the primary substrate for the synthesis of serotonin

Tryptophan hydroxylase Rate-limiting enzyme in the synthesis of serotonin that hydroxylates the precursor substrate tryptophan

SEROTONIN (5-HT; 5-hydroxytryptamine) is an endogenously produced indolealkyl amine located in platelets, mast cells, enterochromaffin cells, and in specific neurons of the central nervous system. Within the nervous system, serotonin is located in discrete neuronal cell groups located in the midbrain and pons. These parts of the brain provide both ascending and descending neuronal projections, which can be either excitatory or inhibitory, to innervate most all nervous system structures. The neurochemistry, pharmacology, and physiology of serotonin support its role as a major neurotransmitter. Recent data obtained with a variety of radioligands have identified multiple selective receptor subtypes (5-HT_{1A}, 5-HT_{1B}, 5-HT_{1C}, 5-HT_2, 5-HT_3). A critical role for serotonin in the nervous system is intimately linked to the regulation of processes including sleep, pain, anxiety, depression, migraine, as well as abused substances such as hallucinogens and stimulants. Serotonin continues to emerge as a key regulator in central processes regulating cardiovascular function, neuroendocrine regulation of hormone secretion, and the development and aging of the nervous system. The future understanding of the physiological specificity of serotonergic effects will be provided by detailed knowledge of the receptor subtype structures within the cell membrane and their linkage to signal transduction mechanisms.

I. Distribution of Serotonin

Serotonin was first isolated in the blood in 1948. Subsequently, in the mid-1960s, serotonin-containing neurons were first discovered and mapped in the central nervous system with the application of the Falck–Hillarp fluorescence immunohistochemical technique. Characteristic of the serotonergic system, these neurons are grouped in clusters of cells lying in or near the midline region of the midbrain and pons (upper brain stem). Nine serotonin-containing nuclei (termed B_1–B_9; collectively called "raphe nuclei") have been characterized as the principal groups of 5-HT neurons. These neurons, in turn, send axons to most all regions of the nervous system; the more caudal nuclei (B_1–B_3) send theirs to the spinal cord, whereas the more rostral

Encyclopedia of Human Biology, Volume 6.

nuclei (B_4–B_9) provide ascending innervation to the pituitary, hypothalamus, thalamus, amygdala, hippocampus, and cortical brain regions.

II. Serotonin Receptor Subtypes

A. Neurochemical Profile

Serotonin is synthesized in the nervous system from the amino acid tryptophan via the enzyme tryptophan hydroxylase to form the intermediate compound 5-hydroxytryptophan; this product is then decarboxylated to form serotonin. This neurotransmitter is metabolized primarily by monoamine oxidase (type A) to form 5-hydroxyindole acetic acid (5-HIAA); levels of this metabolite are often assayed in the cerebrospinal fluid or circulating blood in order to assess the clinical status of serotonergic function in the brain. It is important to note that serotonin cannot enter the brain due to its inability to cross the blood–brain barrier. Rather, the precursor tryptophan can be actively taken up into nervous system tissues to then serve as the substrate for the formation of this neurotransmitter.

In its function as a chemical signaling substance or neurotransmitter in the nervous system, serotonin is synthesized in the nerve terminal, stored in vesicles located in the cytoplasm, and is released upon stimulation provided by a nerve impulse into the synaptic cleft. In this extracellular domain located between two adjacent neurons, serotonin then binds to a selective recognition site, or receptor, in the plasma membrane of the adjacent neuron to complete the process of chemical neurotransmission across the synapse. Importantly, the action of serotonin released into the synaptic cleft is terminated either by being taken up again into the nerve terminal from which it was originally released, or into adjacent glia. The transport of serotonin back into the nerve terminal from which it was released is mediated through the action of specific membrane carrier or uptake pumps located on the presynaptic nerve membrane. Once inside the neuron, serotonin may be stored in vesicles for reuse or metabolically degraded to an inactive form.

B. Pharmacological Profile

Since the early 1950s it has been recognized that 5-HT receptors may exist in multiple forms. With the use of radioligand binding techniques in the 1970s, the use of ligands with high affinity for 5-HT receptors began to reveal a marked pattern of heterogeneous distribution of binding sites for serotonin in the central nervous system. Initially, two central serotonin binding sites were proposed: a 5-HT_1 site labeled by [^{3}H]5-HT and a 5-HT_2 site labeled by [^{3}H]spiperone, with the hallucinogen LSD possessing similar affinity for both ligands. Since these observations, evidence has emerged for 5-HT_1, 5-HT_2, and 5-HT_3 receptor subtypes.

The 5-HT_1 family of receptors are characterized by a nanomolar affinity for 5-HT and 5-carboxyamidotryptamine and only micromolar affinity for 5-HT_2 and 5-HT_3 antagonists. This receptor class can be further subdivided into receptor subtypes, including 5-HT_{1A}, 5-HT_{1B}, 5-HT_{1C}, and 5-HT_{1D}. Each of these receptor subtypes is characterized by a distinct pharmacologic profile and pattern of distribution in the brain. The aminotetralin 8-hydroxy-2-(di-*n*-propylamino)tetralin (8-OH-DPAT) has a 1000-fold higher affinity for the 5-HT_{1A} receptor than serotonin. Thus, [^{3}H]8-OH-DPAT is used as a radioligand for 5-HT_{1A} receptors and has revealed the highest density of binding to 5-HT_{1A} is in the hippocampus and dorsal raphe nucleus, while that of 5-HT_{1B} is in the substantia nigra and globus pallidus. The 5-HT_{1C} receptor is structurally similar to a 5-HT_2 receptor and has the highest density of binding sites in the choroid plexus of the cerebral ventricles.

With the introduction of the quinazolinediane derivative ketanserin, a selective 5-HT_2 antagonist, characterization of the 5-HT_2 binding site was carried out. The highest density of 5-HT_2 sites is found in the medial prefrontal region of the cortex as identified in a variety of mammalian brain tissues, including humans. In general, this binding site is characterized by a relatively low affinity for serotonin agonists but a high affinity for 5-HT_2 antagonists. Recently several phenyalkyamines, such as DOM, have been identified as potent 5-HT_2 agonists.

The recent introduction of selective 5-HT_3 antagonists, including MDL 72222, ICS 205-903 and GR 38032F, and the selective 5-HT_3 agonist 2-methyl serotonin, has prompted extensive studies of 5-HT_3 binding sites in the central nervous system. Central 5-HT_3 binding sites have been identified in the brain with high concentrations of sites in cortical regions and in the area postrema of the brainstem. Such distribution of binding sites may prove to underlie an apparent therapeutic role of antagonists specific for this receptor as anti-emetic drugs.

C. Signal Transduction Mechanisms

Functional correlates of the activation of the multiple 5-HT recognition sites defined via receptor binding techniques also require consideration. The identification of the biochemical transducing systems which are associated with each site has lent further support to the concept of multiple 5-HT receptor subtypes.

One major signaling pathway involving adenylate cyclase activation can be stimulated by 5-HT to induce an increase in the formation of cyclic AMP, a second messenger which in turn activates specific protein kinases. Serotonin-induced stimulation of adenylate cyclase activity is the signaling mechanism that is coupled to 5-HT_{1A}-mediated responses. In contrast, actions of serotonin linked to 5-HT_{1B} and 5-HT_{1D} receptors are negatively coupled to adenylate cyclase, such that activation of these receptors will result in a decrease in the formation of cyclic AMP.

A second major signaling pathway which mediates the actions of serotonin on nervous system function is the phospholipase C/phosphoinositide hydrolysis mechanism which utilizes inositol lipids located in the plasma membrane for signal transduction. Ultimately, transduction of this receptor-mediated signal will include effects on calcium ions and formation of diacylglycerol which stimulates protein kinase C. This phosphoinositide signaling pathway is also involved in 5-HT-mediated responses with selectivity for 5-HT_{1C} and 5-HT_2 receptor subtypes. A number of 5-HT_2 antagonists block serotonin-induced activation of the phosphoinositide hydrolysis system. Interestingly, as the molecular structure of the 5-HT_{1C} and 5-HT_2 receptors have become available through cloning techniques, the structural similarity of these receptors are striking.

The 5-HT_1 and the 5-HT_2 receptors are members of the G protein superfamily. Molecular cloning techniques show that 5-HT receptor genes encode single-subunit proteins whose structure is characterized by seven transmembrane domains and the ability to activate G protein-dependent processes, such as adenylate cyclase or phosphoinositide signaling mechanisms.

The signaling mechanism of the 5-HT_3-mediated response is distinctly different from the above. This receptor is a member of the ligand-gated ion channel superfamily, such that 5-HT_3 receptor activation is directly coupled to ion flow through a Na^+ channel. [*See* CELLULAR SIGNALING.]

D. Physiological Functions

Multiple physiological effects of 5-HT were first identified in studies of smooth muscle contraction in the gut. In the central nervous system, the technique of drug microiontophoresis as applied to different brain regions established the presence of both inhibitory and excitatory responses to the application of serotonin. Autoreceptors, which are inhibited by serotonin, have been identified in midbrain raphe cell groups, with dorsal raphe responses most sensitive to 5-HT_{1A} agonists and median raphe responses most sensitive to 5-HT_{1B} agonists. *In vitro* brain slice preparations coupled with intracellular recording techniques have identified hyperpolarizing effects, depolarizing effects, or direct ion-gated depolarizing, rapidly desensitizing effects on resting membrane potential in association with 5-HT_{1A}-, 5-HT_2-, and 5-HT_3-mediated responses, respectively.

III. Role of Serotonin in Human Disease

A. Pain and Nociception

Central serotonin neurons have been implicated in pain and the actions of analgesic drugs. It has been recognized for sometime that serotonergic neurons project to and comprise components of ascending and descending neuronal pathways known to mediate nociceptive responsiveness. In addition, reciprocal projections between 5-HT neurons and neurons containing the peptide enkephalin, as well as colocalization of these neurotransmitters, may underlie nociception. In general, direct or indirect enhancement of serotonergic function has been demonstrated to have antinociceptive activity in animal models. For example, 5-HT reuptake inhibitors have been demonstrated to enhance morphine analgesia. While clinical usefulness of serotonergic related compounds is at present limited with respect to the therapeutic treatment of pain, this neuronal system has a critical role in the central regulation of pain pathways. [*See* PAIN.]

B. Sleep

Serotonin has long been implicated in the complex behavior of sleep. In general, enhancement of 5-HT activity has been reported to increase total sleep time while the inhibition of 5-HT synthesis with *p*-chloroamphetamine results in insomnia. Dose-dependent effects of the 5-HT reuptake inhibitor fluoxetine have also been reported to lengthen the latency to the onset of the rapid eye movement (REM) stage of sleep in animals. [*See* SLEEP.]

C. Depressive Disorders

A role for serotonin in depressive disorders has emerged from numerous clinical studies of decreases in 5-HT and its metabolite, 5-HIAA, seen in brain tissue from depressed patients who have committed suicide. Similarly, monoamine oxidase inhibitors and 5-HT reuptake inhibitors which effectively increase central levels of 5-HT are therapeutically useful antidepressants. Several more selective 5-HT reuptake inhibitors have been recently developed, including fluoxetine, sertraline, citalopram, and paroxetine, which block neuronal reuptake of 5-HT without effects on dopamine or noradrenergic reuptake processes. These compounds have been found to be effective antidepressant drugs with the added advantage of lacking several side effects often associated with use of tricyclic antidepressant drugs. Thus, the effectiveness of selective 5-HT reuptake inhibitors as antidepressant drugs support the hypothesis that enhancement of synaptically available 5-HT will ameliorate symptoms of depression. [*See* DEPRESSION: NEUROTRANSMITTERS AND RECEPTORS.]

D. Anxiety

Serotonin has long been implicated for a role in mediating anxiety. While early studies hypothesized a direct interaction of central 5-HT with the anxiolytic actions of benzodiazepines (such as valium), a role for two 5-HT receptor subtypes has emerged. Both behavioral and clinical studies have identified the selective 5-HT_2 antagonist ritanserin to have efficacy as an anxiolytic agent. However, most attention has focused on buspirone, a nonbenzodiazepine anxiolytic drug with high affinity for the 5-HT_{1A} receptor. While several 5-HT_{1A} agonists, including 8-OH-DPAT, gepirone, and ipsapirone, have all been demonstrated to have similar anxiolytic properties, the actual neuronal mechanisms and pathways mediating their effects remain to be established.

E. Obesity and Eating Disorders

Brain serotonergic neurons have been demonstrated to mediate, in part, the hypothalamic regulation of food intake. Drugs which act either directly or indirectly to increase 5-HT stimulation decrease food intake in rats. Such drugs include serotonin-releasing drugs such as *p*-chloroamphetamine and (±)fenfluramine, which have been marketed as anti-obesity drugs in the clinical setting. In addition, the selective 5-HT uptake inhibitors fluoxetine and zimelidine have been demonstrated to decrease body weight in nondepressed obese patients. Conversely, several direct acting 5-HT_{1A} agonists, including 8-OH-DPAT, buspirone, and ipsapirone, have recently been reported to increase food intake in rats. Clearly, serotonin has a role in mediating the responses associated with food intake; however, the neuronal circuitry and receptors mediating the regulation of food intake remain unestablished. [*See* EATING DISORDERS; OBESITY.]

F. Psychopharmacology of Hallucinogens

Historically, the original classification of multiple 5-HT responses by Gaddum included both "D" and "M" receptors, with the former being antagonized by LSD (lysergic acid diethylamide). This observation, coupled with the known structural similarity of 5-HT to the psychedelic drug LSD, has prompted extensive investigation into the role of serotonergic neurons in mediating hallucinations. As identified in numerous biochemical and physiologic studies, the central effects of LSD include direct effects on 5-HT_{1A} autoreceptor function and on 5-HT_2 receptors. Most recently, the selective 5-HT_2 antagonists, such as ritanserin, have proved successful in blocking the biochemical, physiological, and behaviorial effects of LSD in animal studies. Such interactions contribute to our understanding of the complex and poorly understood phenomenon of hallucinations and may prove to be a neuronal basis which is common to the actions of all hallucinogens.

G. Psychopharmacology of Stimulants

Central nervous system stimulants, including amphetamine and cocaine, produce numerous effects which range from euphoria to convulsions. Clearly established in mediating the central effects of stimulants are effects on dopamine, noradrenergic, and serotonergic neuronal systems. Specifically, amphetamine, which acts by releasing stores of a given neurotransmitter, will enhance the availability of 5-HT in the synapse and ultimately prolong the agonist effect. While acting by a somewhat difference mechanism, cocaine also enhances the action of available 5-HT, thereby acting as an indirect serotonin agonist. Specifically, in addition to known actions at blocking Na^+ channels, cocaine blocks the reuptake of 5-HT into the presynaptic terminal. Cocaine has also been recently demonstrated to alter the cell firing of spontaneously active 5-HT neurons recorded in the dorsal raphe nucleus as well as modulate the neuroendocrine regulation of several hormones under tonic serotonergic regulation. While the receptor subtypes which may mediate the actions of cocaine in the central nervous system are as yet unresolved, vasoconstrictive actions of serotonin in peripheral tissues, which are mediated via a 5-HT_3 receptor, are attenuated by cocaine. Indeed, the structure of cocaine has been successfully modified to yield new potent and selective 5-HT_3 antagonists, such as MDL 72222 and ICS 205-930.

H. Central Cardiovascular Function

Serotonin has the ability to constrict blood vessels, amplify the vasoconstrictor responses to other compounds, as well as enhance blood platelet aggregation. These actions, coupled with the known serotonergic innervation of cerebral vasculature, have helped to identify a role for serotonin in normal and pathogenic cardiovascular function.

Several centrally acting 5-HT agonists and peripheral 5-HT antagonists have been suggested to be effective hypertensive agents. Several 5-HT_1 agonists have been reported to decrease blood pressure. However, the antihypertensive properties of ketanserin, a 5-HT_2 receptor antagonist, are due not to actions on a 5-HT receptor but rather an α-adrenergic receptor. Another potential therapeutic use of 5-HT_2 antagonists with respect to cardiovascular dysfunction is emerging in the treatment of thrombotic obstructions of the aorta. Likewise, 5-HT_3 receptor density is most strikingly high in brainstem regions, e.g., the area postrema, with critical roles in the central pathways regulating cardiovascular function.

I. Etiology of Migraine

A role for 5-HT in the pathogenesis of migraines has been implicated by the effectiveness of 5-HT antagonists in the treatment of migraine. Serotonin receptors are located on sensory nerves, in central pain pathways, and in cerebral blood vessels which may be the site of their antimigraine action via either 5-HT_{1A}, 5-HT_2, or 5-HT_3 receptors. Effective antimigraine drugs, such as methysergide and ergotamine, have high affinities for a 5-HT_{1A} receptor. However, these compounds may also block vascular 5-HT_2 receptors and/or act as partial agonists which potentiate serotonin-induced contraction of intracranial blood vessels. However, recent reports have identified 5-HT_3 antagonists to be effective in the treatment of migraine, thereby implicating a role for this selective receptor subtype in the etiology of migraine. [*See* HEADACHE.]

IV. Neuroendocrine Regulation

The regulation of the secretion of hormones from the pituitary is a complex process that involves the interaction of the hypothalamic–pituitary–adrenal and/or gonadal axis. In its role as a neurotransmitter in the nervous system, serotonin-containing neuronal systems are one of several neurotransmitter systems identified to have specific modulatory roles in the regulation of a variety of neuroendocrine functions. Not only has a role for serotonin been identified in the regulation of neuroendocrine function in depressive disorders, this neurotransmitter is emerging as a key factor in the normal processes that regulate reproductive processes such as ovulation, menstral cyclicity, sexual behavior, and fertility. [*See* NEUROENDOCRINOLOGY.]

A. Anterior Pituitary Hormones

The secretion of hormones from the anterior pituitary, including luteinizing hormone (LH), prolactin, growth hormone (GH), thyroid stimulating hormone (TSH), and adrenocorticotropin hormone (ACTH), is closely regulated by specific hypophyseal hormones. These enter into the portal circulation at the median eminence of the hypothalamus to regulate the production of a given hormone by the pituitary. For the above listed hormones, these hypophyseal hormones include luteinizing-releasing hormone (Gn-RH), "prolactin releasing factor" (PRF), sommatostatin, thyroid releasing hormone (TRH), and corticotropin releasing factor (CRF), respectively. These are, in turn, closely regulated by a variety of neurotransmitters located in the hypothalamus and in extrahypothalamic brain regions. The specific roles of serotonin in regulating neuroendocrine function in the nervous system are mediated via the modulation of these hypophyseal hormones. [*See* HYPOTHALAMUS.]

Serotonin has been demonstrated to have an inhibitory role on the secretion of LH. Electrical stimulation of serotonergic neurons in the dorsal raphe nucleus as well as intraventricular administration of 5-HT suppresses LH in both intact male and ovariectomized female rats. Administration of a 5-HT synthesis inhibitor or a selective 5-HT neurotoxin also attenuates proestrus increases in LH as well as ovulation in the rat. While the neuronal site of action of 5-HT in mediating these responses is not yet clear, the presence of dense serotonergic innervation in the median eminence and in the anterior pituitary has been identified. With respect to identification of a role for specific receptor subtypes, evidence for a multiplicity of serotonergic influences on the release of LH in the rat supports both stimulatory and inhibitory roles of both 5-HT_1 and 5-HT_2 receptors in a steroid hormone-dependent manner. Indeed, estrogen has been demonstrated to alter 5-HT receptor binding characteristics with specific decreases in 5-HT_1 and 5-HT_2 binding sites reported in various brain regions following hormone pretreatment.

The regulation of prolactin secretion, while under inhibitory regulation from dopamine neurons located in the arcuate nucleus of the hypothalamus, is under a stimulatory influence by serotonergic mechanisms. The stimulatory effects of 5-HT on prolactin release occur at the level of the hypothalamus. Lesion of the paraventricular nucleus will block the serotonergic stimulation of this hormone. The specific neuronal pathways mediating this response include the midbrain raphe nuclei, as destruction of 5-HT cell bodies in the dorsal raphe nucleus decreases circulating levels of prolactin. With respect to the role of receptor subtypes which may mediate these effects, 5-HT_{1A} agonists (8-OH-DPAT, gepirone, ipsapirone) and 5-HT_2 agonists {TFMPP [1-(3-trifluormethylphenyl)-perazine], MK-212} have been reported to stimulate prolactin release. These serotonergic agonists may ultimately prove useful as probes in clinical neuroendocrine challenge studies.

Serotonergic neurons outside of the dorsal raphe nucleus have been demonstrated to stimulate the secretion of ACTH and β-endorphin in the rat. The 5-HT_{1A} agonist, 8-OH-DPAT, the 5-HT_{1C} agonist *m*-chlorophenylpiperazine (m-CPP), and the 5-HT_2 agonist 1-(2,5-dimethoxy-4-iodophenyl)-2-aminopropane (DOI) all produce marked, dose-dependent increases in ACTH levels. In addition, recent evidence suggests that 5-HT_2 receptors mediate CRF secretion from the hypothalamus. Thus, stimulation of the hypothalamic-pituitary-adrenal axis occurs by multiple 5-HT receptor subtypes.

B. Posterior Pituitary Hormones

The hormones released from the posterior pituitary, which include vasopressin and oxytocin, are synthesized in the supraoptic and paraventricular hypothalamic nuclei. Serotonergic innervation to the paraventricular nucleus has been demonstrated to provide a facilitatory effect on the secretion of vasopressin. The stimulatory input on vasopressin secretion is provided by the 5-HT neurons which originate in the dorsal raphe nucleus. To date, the relative contributions of various 5-HT receptor subtypes to vasopression secretion are unclear.

C. Other Endocrine Glands

Serotonin has also been implicated in the regulation of endocrine glands other than the pituitary. Such endocrine glands are either controlled via circulation of blood-borne substances or regulated via direct neuronal innervation. The juxtaglomerulosa cells of the kidney, which secrete renin in response to changes in sympathetic output, are one such endocrine gland. The secretion of renin is centrally regulated by serotonin, such that a stimulatory effect on renin secretion is mediated via the 5-HT neurons of the dorsal raphe nucleus. Stimulation of 5-HT_2 receptors is consistent with a stimulatory effect on renin secretion, while 5-HT_{1A} agonists have little apparent effect in regulating this neuroendocrine response.

V. Development and Aging

Serotonin is one of the earliest developing neurotransmitter systems (circuits) in the brain. Serotonergic neurons influence the growth of other neurons with which they come into contact, either speeding up, slowing down, or having no effect, depending on the type of cell they interact with during development. Thus, this neuroactive substance plays an important role in the development of neural circuits where serotonin will be later used as a chemical signal important for brain function.

Serotonin has also been found in early embryos where it is present outside the nervous system in regions undergoing active morphogenesis (formation of tissues and body structures), such as the heart, intestine, and craniofacial region (nose, face, ears), where congenital malformations are most common. Animal studies suggest that certain psychoactive drugs, including some antidepressants, tranquilizers, and antipsychotics, which act by altering the functions of serotonergic neurons, can cause structural and/or functional abnormalities in the brain when given to pregnant mothers during critical periods of gestation. Animal studies suggest that manipulations of serotonergic systems may have permanent effects on the developing nervous system.

The role of serotonin in aging continues to emerge as a strikingly important one with respect to age-related changes in central nervous system function. Receptor binding studies have revealed significant decreases in the number of 5-HT binding sites with age in the brain regions involved with cognitive processes such as the cortex and hippocampus. Furthermore, significant age-related alterations in the synthesis and turnover of 5-HT have been identified in the brain as well as changes in the cellular physiologic activity of serotonin neurons. These observations confirm that a key role of serotonin is in mediating aged-related changes in Alzheimer's disease or other cognitive decline.

VI. Summary and Future Directions

The role of serotonin as a major neurotransmitter substance in the central nervous system is clearly established. Serotonin contributes to the regulation of numerous biological states such as sleep, eating, perception of pain and affective disorders such as anxiety, depression, and the mood-altering properties of several widely abused drugs and migraine. Recent strides have been taken in identifying multiple recognition sites for 5-HT in the brain as well as understanding their functional role in the nervous system. The future understanding of the physiological specificity of serotonergic effects will continue to expand as our knowledge of the molecular structure of these multiple receptor subtypes emerges. Ultimately, this knowledge of the role of serotonin in the nervous system will provide new therapeutic approaches toward the treatment of human disease.

Acknowledgments

This work was supported by NIA Grant AG-06017 and NIDA Grant DA-04296 (J.M.L.) and a grant from Sigma Tau S.p.A. (B.H.). J.M.L. is a recipient of a Research Career Development Award from the National Institute on Aging.

Bibliography

Aghajanian, G. K., Sprouse, J. S., and Rasmussen, K. (1987). Physiology of the midbrain serotonin system. *In* "Psychopharmacology: The Third Generation of Progress" (H. Y. Meltzer, ed.), p. 141. Raven, New York.

Bradley, P. B., Engel, G., Feniuk, W., Fozard, J. R., Humphrey, P. P. A., Middlemiss, D. N., Mylecharane, E. J., Richardson, B. P., and Saxena, P. R. (1986). Proposals for the classification and nomenclature of functional receptors for 5-hydroxytryptamine. *Neuropharmacology* **25,** 563.

Fozard, J. R. (1987). 5-HT: The enigma variations. *Trends Pharmacol.* **8,** 501.

Fuller, R. W. (1988). The pharmacology and therapeutic potential of serotonin receptor agonists and antagonists. *Adv. Drug Res.* **17,** 350.

Glennon, R. A. (1987). Central serotonin receptors as targets for drug research. *J. Med. Chem.* **30,** 1.

Hartig, P. R. (1989). Molecular biology of 5-HT receptors. *Trends Pharmacol.* **10,** 64.

Sanders-Bush, E., and Conn, P. J. (1987). Neurochemistry of serotonin neuronal systems: Consequences of serotonin receptor activation. *In* "Psychopharmacology: The Third Generation of Progress" (H. Y. Meltzer, ed.), p. 95. Raven, New York.

Schmidt, A. K., and Peroutka, S. J. (1989). 5-Hydroxytryptamine receptor "families." *FASEB J.* **3,** 2242.

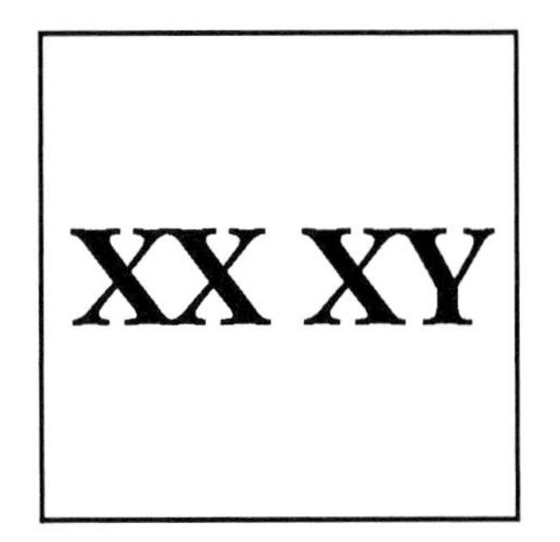

Sex Differences, Biological

SUSUMU OHNO, *Beckman Research Institute of the City of Hope*

Glossary

Inhibin Gonadal peptite hormone that acts on the hypothalamus-pituitary axis
Estradiol Female steroid hormone
H-Y Antigen gene Minor male-specific histocompatibility antigen gene
MHC Major histocompatibility gene
Müllerian (Paranephric duct) Embryonic duct system that develops into female internal reproductive organs
Tdf Testis determining gene
Testosterone Male steroid hormone

A CLEAR GENETIC DIFFERENCE between man and woman responsible for sexual dimorphism resides solely in the presence or absence of the male-specific Y chromosome. However, most of the small number of genes present on the Y chromosome have their counterparts on the X.

A great majority (but not all) of mammalian species are sexually dimorphic, adult males being noticeably taller and heavier than adult females. In these species, adult males are more often than not endowed with variously conspicuous signs of masculinity; e.g., antlers of the stag and mane of the lion. Our own species is no exception. According to the *Statistical Abstract of the United States, 1986,* the average height and weight of men between 25 and 34 years of age were 176.7 cm and 78.5 kg during the period 1976–80, whereas those of women were 163.0 cm and 64.4 kg. These differentials are about the same as one sees between stallions and mares. In view of the fact that in certain seal species, adult males weigh nearly 10 times more than adult females, sexual dimorphism of our own species is not outrageously conspicuous as far as height and weight differentials are concerned.

I. Is There a Male Specific Gene on the Y?

If one is to seek genes that are present exclusively in one sex, they can only be sought on the Y chromosome. The X and the Y in the common ancestor of all mammals were homologous, with the same genetic content. The X chromosome has since been conserved so that whatever gene resides on the human X chromosome automatically resides on the *X chromosome* of all other mammalian species. In contrast, the *Y chromosome* underwent extensive genetic degeneration, accumulating nongenic DNA base sequences and becoming reduced in size. It follows then that the number of functioning genes still residing on the Y must necessarily be small, and that the majority of them must still find their counterparts (alleles) on the X.

The X and the Y still pair with each other and exchange genetic materials during meiosis of spermatocytes in the testis. This very short, homologous segment resides at the tip of the short arm of the X as well as the Y. While the short arm of the X is quite substantial, that of the Y is barely visible under the light microscope. The still functioning Y-linked genes appear to be concentrated in the vicinity of this tiny homologous segment, while their counterparts on the X may or may not be in the vicinity. Those in the vicinity of the pairing segment escape the X-inactivation mechanism that inactivates one of the two X's in female somatic cells. Such Y-homologous genes on the X include the one for Xg blood group and the other for steroid sulphatase. In the case of the steroid sulphatase gene, its counterpart on the Y has apparently become a functionless pseudogene in humans, but not in a number of strains of the laboratory mouse. In the case of the Xg blood group locus, its counterpart on the Y may correspond to the 12E7 antigen locus, although it is not definitely established.

While these two genes have nothing to do with the mechanism of sex determination in our own species or in other mammals, steadily accumulating evidence suggests that the all important Y-linked *testis determining gene* (*Tdf*), which decides the fate of embryonic indifferent gonads to be a pair of testes instead of ovaries, also resides in the vicinity of the pairing segment on the Y. This gene was discovered by studying the inheritance of the Xg gene, which is polymorphic; individuals typing either as Xga + Xga −. Normal XX daughters of the mating between an Xga + father and an Xga − mother should uniformly type as Xga + for they inherit one Xga + X from the father and one Xga − X from the mother. It was noted, however, that the above expectation was frequently violated by abnormal XX male progeny who managed to organize testes in an apparent absence of the Y. They more often than not typed as Xga −, like the mother. It was thus reasoned that the cause of such an *XX male condition* is found in an abnormal genetic exchange between the X and the Y (Fig. 1) that occurred in the testes of the father. An exchange occurred at a point just beyond the small homologous segment. As a result, the X lost Xga + allele by donating it to the Y, in exchange gaining Tdf from the Y. Such XX males should always carry Y-derived fragments containing Tdf, but of variable lengths at the short arm tip of their paternally derived X chromosome. Even the shortest Y fragment should still contain Tdf. Following this rationale, a gene thought to be the Y-linked Tdf was identified in XX men. The protein encoded by a gene in this region contains repeated zinc-binding finger peptide domains, as is characteristic of a group of DNA-binding proteins. There is, however, no conclusive evidence that this protein corresponds to the male-determining gene. The Tdf gene has a counterpart on the X, not in the vicinity of its pairing segment, but deeper in its short arm. This X-linked Tdf is present in women as well as in men, but it is not known whether it is still equally active in the two sexes. During the course of mammalian evolution, if this Y-linked Tdf has diverged sufficiently from its counterpart on the X to acquire the unique function of organizing the tes-

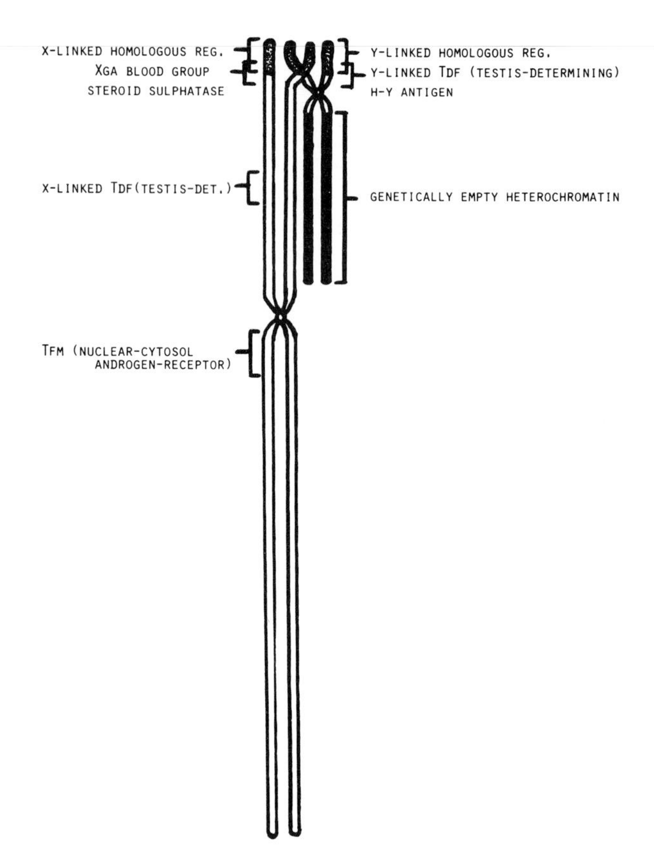

FIGURE 1 The human X (left) and Y (right) chromosomes are schematically illustrated. Each chromosome is divided into its short and long arms by a small circle that denotes a centromere, the attachment site to spindle fibres during cell division. The X chromosome is of a substantial size containing roughly 1.8×10^8 base pairs of DNA. It carries about as many genes as other chromosomes (autosomes) of comparable size such as chromosomes 11 and 12. The number of genes carried by the X should, therefore, approach 6,000. Most of these genes, however, have nothing whatsoever to do with sex determination and sex differentiation.

The Y chromosome, in contrast, is a minute element of variable sizes, usually one-third of the X. A bulk of it is made of heterochromatin, shown as solid black, which contains no gene. The variable amounts of heterochromatin carried by individual Y chromosomes explain their size differences.

The shaded area at the tip of the short arm of the X and Y is the homologous segment with which the X and Y pair with each other and regularly exchange genetic materials between the two during meiosis (reduction division) of spermatocytes in the testis of adult males. In the vicinity of the homologous segment of the X are two genes; one for Xga blood group antigen, and the other for steroid sulfatase. In somatic cells of females, one of the two X chromosomes is inactivated, but the above two genes are not. In the vicinity of the homologous segment of the Y are two genes, the encoding Tdf (testis determining factor) and the other the H-Y minor histocompatibility antigen. Tdf is also located on the X somewhere near the middle of its short arm. The long arm of the X also carries a gene important for sexual development; it encodes the nuclear-cytosol androgen-receptor protein, which mediates all the divergent responses of various target cells to testosterone and 5α-dihydrotestosterone.

tis, the human Y-linked Tdf gene should show greater homology to the Y-linked Tdf gene of all other mammalian species than to its own X-linked Tdf gene. Apparently, this is not the case. It seems, therefore, that humans have both an X-linked and a Y-linked Tdf gene. If so, sex is determined by a dose of Tdf. The X-linked Tdf gene, for its location toward the middle of the X short arm, would be subjected to X-inactivation, endowing normal XX females with only one active dose of the Tdf gene, which is not sufficient for testicular development. Males, on the other hand are endowed with two doses of the Tdf gene, one Y-linked and the other X-linked.

In a more recent report, it has been claimed that X-linked Tdf is not subjected to the X-inactivation mechanism in females. This is rather strange from its position on the X. At any rate, if so, there is essentially no difference between males with one dose each of X-linked and Y-linked Tdf and females with two doses of X-linked Tdf. Indeed, this zinc-binding finger DNA protein may not be Tdf after all. In Australian marsupials (kangaroos and their allies), Y-linked Tdf has been translocated from the Y to an autosome (ordinary chromosome); so has X-linked Tdf been translocated to another autosome. Yet, the marsupial Y without this gene is as strong a male determiner as the Y of true (placental) mammals; XXY developing as males and XO developing as females. Thus, we are still in the dark as to the male-determining capacity of the Y.

Another possible male-specific gene is H-Y, which encodes a minor histocompatibility antigen. By immunological tests, it was shown that H-Y antigen is present in some, but not all, of the aforementioned XX men, suggesting that it is not the sex-determining gene. Findings in mice indicate that the H-Y antigen is involved in germ cell maturation in the testis.

In humans, as with other mammals, the Y chromosome is an important and autonomous determiner of maleness, irrespective of the X's, in fact, XXY and even XXXXY are males. There are XY women, but they are usually devoid of gonads (*gonadal dysgenesis*). In contrast, fertile XY females exist in some rodents.

II. The Testis and the Ovary as Homologous Organs

The chromosomal sex (normally male XY and normally female XX) determines the fate of embryonic indifferent gonads, but subsequent manifestations of maleness or femaleness are the hormonal consequences of the above decision. At first glance, the sperm-producing *testis* and the egg-producing *ovary* appear as two very different organs. Nevertheless, both are derived from the same elements, thus, they are homologous organs in the manner of a small intestine versus a large intestine, rather than, say, liver versus lung.

The development of the main reproductive organs is intimately associated with a stage in the development of the kidneys. Mammalian embryos develop three kinds of kidneys in succession, the last one persisting as the functional kidney. Development of the gonads is associated with the *mesonephros,* and to a certain extent with the pronephros. A pair of gonadal ridges, from which future testes or ovaries will evolve, develop at the inner sides of the mesonephros, very close to the base of the mesentery.

The all-important *primordial germ cells* arise outside the embryo in its yolk sac and migrate toward the newly formed gonadal ridges mainly through the mesentery. As soon as the migration of primordial germ cells to the gonadal ridges is completed, the fate of gonads is determined, for normally male XY embryos of our own species organize the testis between the 43d and 49th days of gestation. *Primordial germ cells* become encased within tubular structures (future seminiferous tubules) together with one somatic element: *Sertoli cells,* whereas the other somatic element left outside the tubules become *Leydig's cells* (Fig. 2). Soon, these two somatic elements begin hormonal activities, thus determining the fates of two pairs of duct systems related to the kidney: The wolffian and müllerian. *Wolffian ducts* would regress without a trace unless exposed to testosterone synthesized by the Leydig's cells of newly organized testes. In the presence of testosterone, this duct system would differentiate into epidydimis, ductus deferens, and seminal vesicles. Although the concentration of circulating testosterone is very low until the 112th day of gestation, masculinization of wolffian ducts can proceed because they apparently receive testosterone directly from the testis. In fact, rare individuals manage to develop a testis only on one side, generally the right. In such an individual, wolffian duct derivatives persist only on the testicular side. Leydig's cells of newly organized testis have yet one more assigned function; to induce the *urogenital sinus* into a penis and scrotum, instead of a vagina. The circulating testosterone level of male em-

FIGURE 2 Schematic illustration of the human embryonic testis (left) and fetal ovary (right) as homologous organs. Large cells with clear cytoplasm are germ cells that have migrated from the yolk sac; primordial germ cells of the male (left) and oogonia and oocytes of the female (right). Cells with black cytoplasm are Sertoli cells of the testis (left) and future granulosa cells of the ovary (right). Cells with shaded cytoplasm are Leydig's cells of the testis (left) and future theca cells of the ovary (right).

Other somatic elements are shown without nuclei. Outlined ducts shown at the base of both the testis and the ovary are rete tubules.

After puberty, rete tubules of the testis perform the important function of collecting and transporting spermatozoa made in seminiferous tubules toward the epidydimis. Rete tubules of the ovary, on the other hand, have no known function.

In the male gonad, testicular cords begin to take shape as early as the 49th day of gestation and establish themselves as definitive seminiferous tubules by the 60th day.

Their ovarian counterparts, ovigerous cords, on the other hand, are not formed until the 100th day of gestation; they are not as convoluted as testicular cords and are open-ended at the top. Nevertheless, ovigerous cords contain germ cells and future granulosa cells in the same manner as testicular cords encase germ cells and Sertoli cells.

In the testis, the germ cells remain primordial until the neonatal stage where they differentiate into definitive spermatogonia, which remain quiescent until puberty (about the 12th year). In the ovary, in contrast, germ cells begin differentiation as soon as ovigerous cords are formed. After a brief period of multiplication as oogonia, they differentiate into oocytes and enter the first meiotic prophase. As they do so, they move more deeply into the ovigerous cords. Meiosis is interrupted at the diplotene stage and each oocyte then surrounds itself with a single layer of follicular (granulosa) cells. This formation of primordial follicles breaks up the ovigerous cords structure from the deepest part.

bryos of that stage is not high enough to saturate the androgen receptors that mediate its effect in the urogenital sinus. For this reason, cells of the urogenital sinus are endowed with the enzyme *5α-reductase,* which converts testosterone to *5α-dihydrotestosterone*. The aforementioned androgen receptor has a considerably higher binding affinity for the latter compound. If there is a genetic defect of this enzyme in XY persons, their external genitalia look more like a vagina, although they have testes with completely masculinized wolffian duct derivatives. These individuals are initially raised as girls. With the approach of puberty, their circulating testosterone levels become high enough to masculinize the urogenital sinus without the help of 5α-reductase. Because the "girl" becomes a man at age 12, this trait is known as "penis at 12 syndrome." Interestingly, the presence of this enzyme in the urogenital sinus derivatives appears to render the vaginal area of women susceptible to testosterone as well. The circulating testosterone level of adult women is roughly 1/20th of that of adult men. Yet, the growth of pubic hair in women is apparently dependent upon this miniscule amount of testosterone, which becomes significant when converted *in situ* to 5α-dihydrotestosterone.

While the wolffian (mesonephric) duct system perishes without testosterone, differentiation of the *mullerian duct* system (paranephric of probably pronephros origin) to female reproductive tracts, fallopian tubules, uterus, and upper part of the vagina, is an automatic process that requires no hormonal induction. There are no mullerian duct derivatives in normal men because the mullerian duct system is destroyed during embryonic development. This destruction is carried out by the Sertoli cells of the newly organized testes. The glycoprotein excreted by Sertoli cells for this purpose is MIS (*mullerian inhibiting substance*), which is about 550 amino acid residues long. This protein is related to transforming growth factor "β" as well as to β-chain of inhibin.

In view of the above-noted specific function assigned to Leydig and Sertoli cells of newly formed testes, it may appear that these two cell types are truly male or testis specific, having no counterparts in the female reproductive organ, the ovary. However, this is not the case. When a pair of gonadal ridges begins to differentiate toward the ovary instead of toward the testis, they too form tubular structures. These ovigerous cords are not as conspicuous nor as convoluted as testicular cords, which give rise to siminiferous tubules. Yet, the ovigerous cords, which open toward the surface, contain primordial germ cells and one somatic cell type: the *follicular cells,* which after puberty are called granulosa cells. The other somatic cell type left outside the ovigerous cord becomes *theca cells*.

While male germ cells in the testis remain inactive until the onset of puberty, primordial germ cells in the ovary soon cease their multiplication and enter into meiosis. Each oocyte, upon completing the diplotene stage of the first meiotic division, becomes encased by a single layer of follicular cells, thus becoming the *primordial ovarian follicle*. It is this formation of follicles that breaks up the structure of the ovigerous cords, as shown in Fig. 2. Thus, testicular Sertoli cells have their counterparts in ovarian follicular (or granulosa) cells, and testicular Leydig cells in ovarian theca cells. Indeed, ovarian *granulosa cells* also produce MIS, although at a later period, after the mullerian duct system became indestructible by MIS. As already noted, inhibin is related to MIS and the function of this hormone is to prevent, by negative feedback, the pituitary secretion of FSH (follicle stimulating hormone). This too is produced by testicular Sertoli cells and ovarian granulosa cells alike. Ovarian follicular cells are rich in one microsomal enzyme, *aromatase,* which converts male hormone testosterone to the female hormone *estradiol*. In women of reproductive age, the aromatase function is essential because each menstrual cycle is initiated by a rise in estradiol concentration. However, testicular Sertoli cells are also rich in aromatase. Nevertheless, the testis does not become an estradiol-producing organ because it is sequestered inside seminiferous tubules. Thus, testosterone secreted by Leydig cells is not accessible to them. This separation between Sertoli cells and Leydig cells breaks down when the former become malignant and begin to propagate profusely; testicular Sertoli cell tumors are rather common in dogs. When they occur, the testis becomes the feminizing estradiol-producing organ.

If ovarian theca cells are the equivalent of testicular Leydig cells, can the ovary become the testosterone-producing organ? They probably do, as seen in female rabbits immunized with testosterone to obtain the antibodies that are used in clinical radioimmunoassay of testosterone. The ovaries of these rabbits begin to produce as much testosterone as male rabbits, probably because all the testosterone they produce is inactivated by antibodies. Indeed, testicular Leydig's cells and ovarian theca cells are homologous to each other, as are testicular Sertoli cells and ovarian granulosa cells.

Thus, the testis and the ovary should be viewed as homologous organs representing two sides of the same coin. In fact, many fish species readily exchange the ovary with the testis or vice versa in their lifetime. The most notable is the perch-like tropical coral fish *Anthias squaminpinnis*. This is a conspicuously, sexually dimorphic species (a rather unusual event in fish). The female is bright yellow, while the mature male is brightly reddish wearing a blue sash across his chest. Each group consists of only one male and a number of females. When this male is killed or removed, the most dominant of the females becomes a male, changing the ovary to the testis and the yellowish coloration to red with a blue sash within a matter of a few weeks.

III. The Female as the Primordial Sex and the Male as a Modified Female

As far as mammals are concerned, the basic plan of development is to be feminine. It will be recalled that differentiation of müllerian (paranephric) ducts to fallopian tubules and uterus of the female reproductive tract is an automatic process that requires no intervention, and so is the urogenital sinus's differentiation toward the vagina. In sharp contrast, wolffian (mesonephric) ducts would perish if left alone, their differentiation to male reproductive tracts requiring testosterone, while differentiation of the urogenital sinus to penis and scrotum requires not only testosterone but also its *"in situ"* conversion to 5α-dihydrotestosterone.

Even primordial germ cells are inherently planned to be eggs rather than sperm. It will be recalled that primordial germ cells originate outside the embryo in the yolk sac. During their migration toward the gonadal ridges, which emerge on the surface of the mesonephros, a number of them mismigrate and home into the nearby adrenals. In certain inbred strains of laboratory mice, these mismigrated primordian germ cells are numerous and readily identifiable. In the adrenals, even XY germ cells of male embryos behave as eggs and undergo up to the diplotene stage of the first meiotic prophase; then each surrounds itself with a single layer of adrenal cortical cells, forming a misguided primordial follicle.

For professional biologists, the sex starts with differentiation of gonads, but for others, gonads are of little interest. What matters are external appearances and behaviors. From animal experiments it seems likely that if given ample amounts of testosterone at three critical stages of development, XX individuals with the ovary would develop as men both in appearance and behavior. The critical stages are: (1) 50th to 80th day of gestation to develop wolffian ducts into male reproductive tracts and the urogenital sinus to penis and scrotum; (2) days around the parturition to *imprint* the central nervous system, to insure the later manifestation of masculine behavior; and (3) 12th year onward to gain the height and weight of a man and to maintain masculine symbols (e.g., beards) and behavior. A number of reports have described sex differences in spatial perception and certain mathematical abilities. If they exist, they are not likely to be the direct consequences of a genetic difference but of the presence or absence of paranatal imprinting on the central nervous system. [*See* SEX DIFFERENCES, PSYCHOLOGICAL.]

Conversely, if protected from the effect of one's own testosterone, XY individuals with testes should develop as females. The effect of testosterone as well as 5α-dihydrotestosterone on target cells is mediated by a *nuclear-cytosol androgen-receptor protein,* which is encoded by a gene on the X chromosome (Fig. 1). When this locus becomes dysfunctional by mutation, the entire body of XY individuals becomes totally nonresponsive to the normal amount of testosterone produced by their testes. This is known as the complete form of *testicular feminization* syndrome. An affected individual has neither male nor female reproductive tracts because the wolffian ducts, unable to respond to testosterone, perish, while mullerian ducts are actively destroyed by the hormone MIS, secreted by testicular Sertoli cells. Not being able to respond to 5α-dihydrotestosterone converted from testosterone *in situ,* the urogenital sinus develops as the vagina. Thus, the external appearances of such XY individuals are female. After puberty, testosterone increases in amount but it is not utilized by the target cells; it is then converted to the female hormone estradiol by the already mentioned aromatase, which is present in extragonadal tissue, noticeably in fat tissue. As these individuals are normally responsive to estradiol they develop well-shaped breasts as well as hips. In fact, these XY individuals with the complete form of testicular feminization syndrome are invariably more beautifully feminine than their normal XX sisters. Such is the consequence of the female being the primordial sex.

One tell-tale sign of the complete form of testicular feminization is the total absence of pubic and axillary hair. This suggests that even in sexually mature normal women, the hair development is maintained by testosterone. Although the circulating level of testosterone in adult women is roughly 1/20th of that of adult men, the skin of the above-noted area contains 5α-reductase, which converts testostene to 5α-dihydrotestosterone. The already-mentioned androgen receptor demonstrates a considerably higher binding affinity for 5α-dihydrotestosterone than for testosterone itself, as also noted. Testosterone can, therefore, induce hair formation at these sites. The fact that a portion of the signs of sexual maturity by normal women is dependent upon a male hormone is another reflection of the little differences separating man from woman. In this connection, it is possible that the paranatal masculine imprinting of the central nervous system is carried out by the female hormone, estradiol, generated from testosterone by a local aromatase. Evidence obtained on experimental animals suggest this possibility.

Bibliography

Cate, R. L., Mattaliano, R. J., Hession, C., Tizard, R., Farber, N. M., Cheung, A., Ninfa, E. G., Frey, A. Z., Gash, D. J., and Chow, E. W. (1986). Isolation of the bovine and human genes for müllerian inhibiting substance and expression of human gene in animal cells, *Cell* **45,** 685–89.

Ohno, S. (1987). Conservation *in toto* of the mammalian X-linkage group as a frozen accident, *Chromosomes Today* **9,** 147–53.

Page, D. C., Mosher, R., Simpson, E. M., Fisher, E.M.C., Mardon, G., Pollack, J., McGillvray, B., de la Chapelle, A., and Brown, L. G. (1987). The sex-determining region of the human Y chromosome encodes a finger protein, *Cell* **51,** 1091–1104.

Schneider-Gädlicke, A., Beer-Romero, P., Brown, L. G., Nussbaum, R., and Page, D. C. (1989). ZFX has a gene structure similar to ZFY, the putative human sex determinant, and escapes X-inactivation, *Cell* **57,** 1247–58.

Simpson, E., Chandler, P., Goulmy, E., Page, D. C., Disteche, C., and Ferguson-Smith, M. A. (1987). Separation of the genetic loci for the H-Y antigen and testis determination on human Y chromosome, *Nature* **326,** 876–78.

Sinclair, A. H., Foster, J. W., Spencer, J. A., Page, D. C., Palmer, M., Goodfellow, P. N., and Marshall Graves, J. A. (1988). Sequences homologous to ZFY, a candidate human sex-determining gene, are autosomal in marsupials, *Nature* **336,** 780–83.

Takahashi, M., Hayashi, M., Manganaro, T. F., and Donahoe, P. K. (1986). The ontogeny of müllerian inhibiting substance in granulosa cells of the bovine ovarian follicle, *Biological Reproduction* **35,** 447–53.
Wieacker, P., Griffin, J. E., Wienker, T., Lopez, J. M., Wilson, J. D., and Breckwoldt, M. (1987). Linkage analysis with RFLPs in families with androgen resistance syndromes: evidence for close linkage between the androgen receptor locus and the DXS1 segment, *Human Genetics* **76,** 248–52.

Sex Differences, Biocultural

ROBERTA L. HALL, *Oregon State University*

Glossary

Gathering and hunting economy Economic organization that dominated the evolutionary history of the hominids from 4 million to 12,000 years ago; plant and animal resources were obtained through gathering, scavenging of dead animals, and hunting; patterned tools are associated with cultures of the last 2 million years

Genotype Total complement of an individual's genetic material

Hominid All species, living and extinct, of the genera *Australopithecus* or *Homo*; they share a common ancestor with the chimpanzee and gorilla but are not ancestral to either

Prehistory The period, prior to 5,000 years ago, for which no written records exist; it is studied by archaeology, which analyzes material remains of human activity

Sex ratio Number of males to 100 females at particular points in the life cycle; primary refers to the ratio at conception, secondary to the ratio at birth, and tertiary to the ratio at some other specified time

Sexual dimorphism Patterned differences between males and females of one species; may include color, size, shape, body composition, anatomical features, etc.

ANTHROPOLOGISTS HAVE NOTED that some cultures emphasize distinctions between the sexes, while others mute them. At various times in history, the biological differences between males and females have been considered either as major determinants of individual and social patterns or as minor phenomena. A biocultural perspective considers the evolution of male and female characteristics within a social context. The values of a society affect behaviors such as nutrition, education, and levels of physical activity; these in turn may have effects on the development of sex-specific characteristics. In addition, a society's values influence the cultural interpretation of differences between the sexes.

I. Introduction

A biocultural perspective cannot define for all time the core differences between men and women. Instead, this perspective suggests that absolute definition is not an appropriate goal. Certainly, a biocultural perspective will not provide a blueprint of proper behavioral patterns for the two sexes, nor will it provide answers to ethical questions. However, a biocultural perspective examines diverse solutions to the problem of male–female definition that have been developed by various species and societies. By going beyond the boundaries of one culture, one time period, or species, it puts the problem of sex definition within the domain of research so that the subject can be approached by the intellect rather than by the emotions.

II. Animal Insights

At its most inclusive level, sexual reproduction encompasses all types of reproduction in which an individual offspring receives genetic contributions from more than one parent. Whereas microbial species that meet this definition minimally do not have defined male or female roles, organs, or individuals, more complex sexual species can be categorized by whether the reproductive cell provides only genetic material (male) or provides both nourishment and genetic material (female). Some organisms, including many common plants, such as corn, contain both male and female organs and functions; some groups, for example aphids (insects that prey on plants), have the capacity to reproduce either sexually or asexually; and other species, including all primates, develop individuals as either male or female by structuring their reproductive capacity to one or the other style. Except among mammals, who bear their young alive and provide care for their young during infancy, the female parent is not always a care-giver. Both parents in many species of birds devote much of their energy to providing for their young; among other vertebrates, care-giving by either sex is rare. Examples of paternal care are uncommon and tend to be repeated in the literature (e.g., the male sea horse tends its young). Parental care is simply one among many possible behavioral adaptations that may be advantageous to a species and hence become part of its innate behavioral repertoire. Lacking a complementary anatomical program such as exists in mammals, parental care is neither obligatory nor inherently superior.

Among mammals, females not only house and nourish the developing embryo but feed the offspring after birth. While involvement of the mother is obligatory, the involvement of the male is optional. However, from an evolutionary standpoint, a male that participates in reproduction will have to follow the behavioral pattern of its species. Most bears, for example, do not have a paternal role. Male and female bears have a courting period in which individuals permit another adult of their species to become intimate for several days, but their normal relationship with other adults is avoidance or conflict. Indeed, adult male bears threaten the security of cubs. Wolves, who are members of the same taxonomic order as bears, have developed the opposite strategy. Male adults, the father and other related males of the same pack, provide food for the pack's mother and for wolf pups. In addition, they act as baby-sitters and teach skills such as hunting. These examples illustrate the wide range of options open to mammalian species, but this does not mean that the same range is open to individual bears or wolves. Females and males who participate in the evolutionary process (i.e., who produce offspring for the next generation) must perform their respective roles, which appear to be genetically programmed to a large extent but, to some extent, vary with individual social experience. Problems that captive animals have with reproduction illustrate the dependence of sexual behavior on maintenance of a species' socioecological pattern.

We do not know specifically how a male is programmed to act like a father. Male wolves and coyotes are good fathers, but male dogs characteristically reserve their loyalty for their human masters. Female coyotes and wolves who are sexually stimulated to court with dogs pay a stiff price when their suitor abandons them and they must try to raise puppies alone. How much of the paternal behavior of the canids is controlled by its genes and how much by its own life experience? Some flexibilities must exist in canids, because coyote family behavior differs greatly according to whether or not they are protected. Where hunted, pups are reared for a short time but family groups are neither large nor permanent. Where coyotes are protected, family ties persist. Both males and females adapt their behavior to necessity; while pack life may be the preferred option within the coyote repertoire, it is not obligatory where the environment demands a solitary existence.

III. Primate Patterns

Social carnivores such as the canids provide the best nonprimate analogue for human social patterns, but biologists and anthropologists seeking models for the evolution of human sex differences have usually turned to the order Primates, in which *Homo sapiens* belongs. However, great diversity in reproduction, sex roles, and sexual dimorphism is found among the primates. Scholars have long debated which groups make the most appropriate model. Some have focused on chimpanzees and gorillas on the grounds that the groups that share the most recent common ancestor with humans are most appropriate. Interestingly, these two groups differ greatly between themselves in social organization, including sex roles, and in male–female size differences. Chimpanzee males and females show no greater overall size differences than human males and females. The two species of chimps (*Pan troglodytes*, the common chimp, and *Pan paniscus*, known as the pygmy chimp but actually not much smaller than the other species) differ in sex roles. Common chimp troops live in territories within which females mate with various males who are organized into a shifting hierarchy. Females are more likely than males to move from troop to troop, and they tend to take their young with them; females and their grown and adolescent offspring form one part of the social structure, while liaisons among males form another. Pygmy chimps are reported to have a different and more humanlike structure with more female bonding; more sexual and affiliative behavior, even between members of the same sex; and occasional use of the face to face copulatory position. It is unfortunate for many reasons that so few chimpanzees—particularly *P. paniscus*—exist in the wild. If more did, one benefit would be further studies elucidating the range of social behaviors in free-ranging chimps. Of all primates, these two species might be expected to show the greatest number of sex role options. Biochemical studies suggest that the common and pygmy chimps have been separate species for 1.5 million years, and it would be valuable to know the degree to which their sex differences are encoded. Such an understanding might provide models for the role of sex differences in the evolution of prehistoric hominids (see next section).

Three subspecies of gorilla are recognized. Although they differ in size, all show a large degree of sex differences. Gorillas live in troops consisting of one elder male, a band of females, and their young. Their social pattern is markedly different from that of both chimpanzee species.

A third great ape species, the orangutan of Southeast Asia, has yet a different form of social structure in the wild. Although apparently quite socially responsive when reared in captivity, orangs in the wild tend to be solitary as adults, with the exception of the courtship period. The basic social group is the adult female and young. Biochemical studies show orangs are less closely related to chimps, gorillas, and humans than these three are to each other. Orangs, gorillas, and chimpanzees are alike in placing almost all of the responsibility for rearing infants on the mother. While male chimps and gorillas may provide defense against attack by outsiders, the burdens of feeding and providing emotional security rest with the female. In chimps and gorillas, older siblings may be of some help to infants, but not predictably or consistently. One theory of human origins holds that the human innovation of the father, responsible for providing food and protection, permitted human ancestors to reproduce faster and more effectively and gave human ancestors the edge over other apes. While this theory is too simplistic, it does call attention to problems that the great apes have in maintaining their populations when habitat is lost and adult females are killed by humans.

Some anthropologists have argued that the great apes are inappropriate social models for humans. This argument is based on the idea that, in splitting from the apes, human ancestors left the forests and chose semi-open or savanna habitat, and, accordingly, adjusted their behavior. By this reasoning, primates of the savanna offer more useful models of early human sex differences. Baboons in Africa and macaque species in Asia, both highly social, intelligent, and successful monkeys, have thereby received considerable attention. They and other monkeys of Africa and Asia have diversified greatly in the last several millions of years and now occupy a wide range of habitats from the cold and snowy mountains of northern Japan to the lush tropics and arid rocky deserts of Africa. Many monkeys live in multimale groups in which adult males play a role in rearing young. Differences exist between closely related species, but in general females and their young form the core of society while males tend to move, usually every several years, from troop to troop. In some arid regions, the baboon foraging unit consists of a single adult male and a small group of females and their young. The adult male guides their foraging and defends the females against predators and against being taken over by other males. This organization has been quite inaccurately called a harem—a luxury for the well-to-do male it definitely is not! Rather, it is an adaptation to a harsh environment.

In richer savanna areas, baboons live in large troops that consist of many males and females of all ages. In these troops, males and females show consistent differences in growth patterns, reproductive styles, and social behavior. Females are smaller than males and mature at a younger age. The longer growth period of males means that they grow larger and can thus defend the periphery of the troop; if they survive to adulthood they start their reproductive careers later. Meanwhile, females start to produce offspring at a young age. Reductions in the nutritional needs of females due to small size are balanced by demands due to reproduction (every second year) and lactation. Several decades ago studies of troops of savanna baboons portrayed their societies as male-dominated and rigidly hierarchical; however, long-term studies have established that females are the core of baboon life. These studies show intertroop mobility, mainly of adolescent and adult males, and have identified social mechanisms for changes in troop hierarchies. In addition, long-term studies of monkey troops, like those of great apes, have identified considerable personality differences among members of the same troop.

While each troop's environmental and historical situation may require its males and females to have certain behavioral traits, not all individuals comply. The existence of variation permits a population to shift its social norms to meet new environmental or historical constraints. Studies of our primate relatives have suggested the roots of human diversity but have not provided a blueprint of male and female behavior. [*See* PRIMATES.]

IV. Prehistoric Hominids

Identification of an individual's sex from its preserved skeleton, which often is quite fragmentary, is never certain. This lack of certainty is the primary impediment to developing models of the history of hominid sexual dimorphism. Teeth are the most frequently preserved fossil remains of hominid skeletons. Fortunately, teeth, particularly the canines, show consistent male–female size differences in nonhuman primates and, to a lesser but statistically significant extent, in modern people as well. Still, identification of fossil specimens by sex (or by species) is seldom agreed upon by all experts, so disputes about primary data categories persist.

One of the most important early hominid groups, identified at two sites in Eastern Africa, is *Australopithecus afarensis*, dating from 3 to 4 million years ago. Donald Johanson has identified two sexes, the male considerably larger than the female, at these sites. A contrary hypothesis holds that the finds represent two species of different size. The Johanson model suggests the possibility of divergent sex roles, and perhaps different growth and maturity patterns for males and females, such as were noted in some contemporary monkeys. Questions suggested by the analysis include whether this implies a female-centered social structure and intertroop promiscuity, like that of the monkeys, or a society like that of the gorilla, dominated by an elder male. Or, some other alternative may have been the social structure: perhaps closer to some modern cultures that live in small bands in which one family may include one male, several wives, and all their children. What role, if any, males played in child-rearing is a frequently asked, but unanswerable, question.

The australopithecine period of 1.5–5 million years ago ends several hundred thousand years after the first fossils of our own genus, *Homo*, are known. *Homo erectus* specimens are found in the period from 1.8 to 0.3 million years ago, but their spare skeletal remains are insufficient to establish sex differences in size. Behavioral models of the species in East Asia, Africa, and Europe are based on the archaeological record, primarily on stone tools and evidence of hunting. Behaviors of males and females in the *H. erectus* period have been hypothesized primarily from ethnographic patterns described over the past 100 years in the technologically simplest gathering and hunting cultures of *H. sapiens*. Such speculation (or model-building) is acceptable if used as a hypothesis, but there is a danger that such descriptive pictures will take on a false reality and subsequently will be used to make unsubstantiated inferences or to "explain" the modern data from which they have been derived.

The Neanderthal population that occupied Western Europe from about 125,000–40,000 years ago is one of the best known and most argued about skeletal populations in prehistory. In contrast to *H. erectus* samples, enough data exist to at least hold conversations about Neanderthal body size and shape and about sex differences. Compared with modern people, Neanderthals were short and very muscular. Both males and females appear to have had broad pelvises, but explanations of this finding differ. One hypothesis is that their locomotory adaptations differed from ours; i.e., that Neanderthals were less efficient in their stride. A second idea is that the broad pelvis is simply an aspect of their robusticity, which evolved as an adaptation both to extreme cold and to facilitate combat at close range with large mammals such as the cave bear and mammoth. A third idea is that females were selected for bearing robust young with large heads; a variant on this hypothesis proposed a gestation period of perhaps 11 months with precocious young, more physiologically mature than modern infants, presumably as an adaptation to the extreme climatic conditions of Europe during the Wurm glaciation. A nineteenth century hypothesis that argued Neanderthals were not upright was shelved decades ago. Notably absent from Neanderthal theories is the petite female, protected by one or more robust males, for the skeletal remains indicate that both male and female Neanderthals were robust and muscular. From the Neanderthal period on, size sexual dimorphism is considered to vary in the degree to which it varies in modern people. As the next section illustrates, this variation offers some insights into general principles of human growth and development.

V. Gathering and Hunting Humans Compared with Early Farmers

Study of the sexual dimorphism in prehistoric populations of *H. sapiens* is possible to a degree far exceeding that in premodern hominids, due in large part to confidence in the primary data. Within *H. sapiens*, our own species, it is possible to use recognized criteria to determine the sex of a skeletal individual with >90% confidence of accuracy. Although modern *H. sapiens* appeared in Asia and in Europe at least 40,000 years ago, and in Africa at least 100,000 years ago, there are no large samples of >15,000 years antiquity. Large samples are required in the study of sex differences to be sure the data are not skewed by individual anomalies.

Thus, when we discuss the degree of sexual dimorphism in modern *H. sapiens*, we really are referring only to the most recent populations of *Homo sapiens sapiens*, and only to some selected samples, luckily preserved and studied. With this *caveat*, based on studies of North American populations of the last several thousand years and of Eurasian populations of <10,000 years antiquity, a few generalizations can be made.

Sexual dimorphism in skeletal populations of modern people appears to depend on a number of factors. Important factors include the type of economy (gathering and hunting, village farming, preindustrial states in which farming was the main occupation, etc.); the degree of class structure, or lack of it, and the social class of the sample; and the ecological and nutritional status of the skeletal population. A hunting life-style appears to select for larger, more robust males if all other things, such as the ecological richness of the habitat, are equal. A farming life-style tends to select less for robusticity than for physical perseverance. Farming permits, and in fact encourages, increased population density, but when drought or other climatic changes occur it selects for small body size, because this reduces caloric needs and allows survival when food is scarce. Research indicates that females tend to be more programmed in their growth patterns and that males tend to be more susceptible to increase in size when nutrition is plentiful, and to reduction in growth when it is not. Because adult females in nonindustrialized cultures, and certainly females in prehistoric societies, could expect to spend most of their reproductive years either pregnant or lactating, selection has tended to buttress the female against the nutritional stress of childbearing and nursing. One way to do this is to restrain total growth; a second way concerns body composition.

VI. Body Composition

Body composition of contemporary male and female adults varies most notably in the percentages of muscle and fat. Whereas young adult males average 15% of body weight in fat, females average about 27%; in muscle, the comparison is between 52 (males) and 40% (females). Hormonal control of fat patterns in females has been shown to represent nutrient reserves required in successful completion of pregnancy and lactation.

Although useful in comparing some male–female traits, these blanket average statements fail to take behavior and biological variability into account. Recent research has focused on the trained female athlete who reduces her fat composition and increases relative muscle composition. The institution of a training regimen during adolescence can postpone menarche, whereas institution of an extensive exercise program after menarche can produce periods of amenorrhea. These observations have contributed to understanding the role of relative fat in triggering the onset of menarche and of understanding how the trend toward taller adults and a younger age at the adolescent growth spurt have been produced by overnutrition.

While sex differences in body composition become acute during the adolescent growth spurt, they originate long before adolescence. Variation occurs among individuals as well as between different body sites where fat and muscle tissue are deposited. Reduction of fat tissue in females in industrialized societies, such as during periods of famine in war-torn countries, may lead to a reduction in the birth rate. This process is due to prolonged periods of amenorrhea in a substantial fraction of the female population. Such a condition may also be self-induced, as in the case of anorexic women. In populations of hunter–gatherers, of which few remained in the late twentieth century, such conditions may be related to prolonged lactation or to undernutrition. Restoration of a critical percentage of fat appears to be required before menstrual cycling resumes.

Young boys, by contrast, may lose muscle mass under conditions of malnutrition. In modern society, girls are not alone in experiencing eating disorders, such as anorexia and bulimia, in which the affected persons have complicity in their bodies' inability to take in and digest food; boys involved in wrestling have been known to use artificial means to reduce their weight in order to qualify in a lower weight class. Like bulimics, they willingly vomit food and lose nourishment. Additional research is needed to determine the long-term consequences of these behaviors in adolescence. [*See* EATING DISORDERS.]

The consequences of taking steroids to achieve the opposite effect, i.e., to put on muscle and weight, have been documented, but so long as dire consequences have only a probabilistic basis they fail to discourage all young men from steroid intake.

In cosmopolitan industrial societies, enhancement of muscle tissues is considered fashionable by both men and women. This slight exaggeration of a male pattern of body composition is favored over enhancement of the female pattern of fat deposition. A biocultural perspective relates this fashion to the abundance of food and food storage technology; it is no longer necessary or even healthful to store calories as fat. Success, symbolized in the appearance of strength and apparent youthfulness, counts more than stored food, perhaps in part because the symbol is the scarcer commodity.

VII. Environmental and Cultural Variation

Over the past century many unplanned experiments that involve rapid changes in the environment and culture of populations have occurred. Franz Boas was one of the first anthropologists to seize the opportunity presented by migration to North America of peoples from all over the world to compare the physical traits of migrants with those of people who remained in their original territory. He found that migrants tended to have proportionately broader heads; later studies have found an increase in body size, often occurring to a greater extent in males than in females. Many sophisticated studies in recent decades have compared migrants with two groups of stay-at-homes: those who continue to practice traditional culture and those who adopt a Western life-style. Predictably, the change in life-style is more important than migration as such.

As discussed in the preceding section, attitudes toward body size and shape undergo change as access to nutrition changes. Peasant cultures tend to value fat deposition as an indication of stored wealth and as a hedge against hard times, but in a cosmopolitan industrialized culture fat is superfluous. Activity patterns differ; in most nonindustrial cultures, women perform hard physical labor and develop muscles, whereas women in industrialized societies have to exercise during leisure time to achieve muscularity. In these societies, personal choice concerning daily activities, as well as in selection of food, assumes a major role, along with the genotype, in determining the body's shape.

Many scholars of women studies have pointed out that just as women's average levels of physical activity and body shapes have changed with the advent of industrial society, so have attitudes toward women's roles. The "helpless female" concept did not arise in traditional culture, where women's work was essential to family survival. This syndrome appears to be a short-lived cultural phenomenon, an aristocratic status symbol that the middle class could afford for only a few decades. Environmental and cultural changes will continue to test the degree of flexibility of the human frame.

VIII. Cognitive and Behavioral Differences

All studies of cognitive differences between the sexes show that great overlap exists between males and females; group averages for either sex offer little predictive power regarding the performance of any one individual. Still, some patterned behavior differences have been identified and deserve mention. Following behavioral patterns, a second type of cognitive study focuses on sex differences in the anatomy of the brain; a third type looks at the influence of hormones, which may have an effect on brain development or act directly on behavior.

Sensory and motor abilities of boys and girls differ in that girls tend to have greater auditory acuity, whereas boys excel in visual keenness, particularly in tracking moving objects. These traits were studied in the development of the verbal skills of speaking and reading, in which girls tend to outperform boys, and of spatial and mathematical skills, in which boys tend to outperform girls. Because reading is considered an essential skill for all citizens, most boys are strongly encouraged to succeed as readers. The same pressure is rarely applied to girls to assure they achieve mathematical competence. The role of socialization in developing or maintaining these divergent patterns was hotly debated in the mid-twentieth century with no final resolution. While unlikely to be a total cause of observed sex differences, socialization probably serves a contributory role. If, as suggested above, boys are forced to devote extra effort to gain essential skills in reading, but girls are not forced to gain essential skills in mathematics, socialization pressures contribute to the disadvantaging of females in certain professions. [*See* SEX DIFFERENCES, PSYCHOLOGICAL.]

Anatomical comparisons of male and female brain structure are in their infancy. Researchers have focused on differences between the brain's two hemispheres and on links between them. Language processing tends to be localized in the left hemisphere and rapid pattern analysis in the right. However, research has not shown that one sex shows more development in one hemisphere while the other favors the opposite. Some studies have indicated greater laterality in males—i.e., that a greater proportion of males than females have brains in which one hemisphere dominates the other—but this finding is not universal. An innovative research project undertaken in the late 1980s found the area of fibers that connect the two halves of the brain to be larger in women than in men. Researchers related this finding to the greater verbal fluency of women, but until further studies are done these results are tentative and any inferences are speculative.

Hormonal effects on sex development have been studied experimentally in animal models and in humans with chromosomal abnormalities. Though these studies demonstrate that hormones produce significant effects, they do not provide ready models for subtle behavioral variation in genetically normal individuals. Studies conducted in 1988 showed that variation of hormone levels in healthy women is correlated with performance of tasks involving motor skills and spatial reasoning. During times when the female sex hormones estrogen and progesterone were highest, subjects tended to perform better on fine motor tasks, but their performance of spatial skills improved when these hormones were low. In addition to confirmatory tests, studies of the performance on similar tasks by men in relation to circulating levels of various hormones are needed before these preliminary findings can serve as a basis for interpreting male–female variation in motor and spatial skills.

IX. Sex Ratio and Mortality

In most populations, the number of newborn male children exceeds that of female children by 5–8%. The sex ratio at birth is termed the secondary sex ratio, while the proportion at conception, the primary sex ratio of males to females, is believed to be considerably higher. R. A. Fisher's theory of the sex ratio proposes that evolution works to produce an equal sex ratio at the age of procreation; thus, given the higher mortality of male embryos, infants, and children, there must be an excess of them at conception. The degree of vulnerability of males in comparison to females varies among societies. It may either be enhanced or compensated for by cultural practices such as male participation in warfare or other occupational sex differences with attendant levels of risk, marital patterns such as polygyny, differential age at marriage for males and females, and the existence of social institutions such as monasteries and nunneries. In recent decades, much attention has been focused on the greater life expectancy of females and its relationship to their lower level or later onset of chronic diseases of old age. Hormonal and genetic factors appear to account for many of these sex differences but behavioral factors also contribute. Significantly, outside of societies in which many young men perish in warfare, the greater longevity of females appears to be due to their less than maximal fertility, a relatively recent development. By contrast, the very high fertility of some village farming communities puts women at greater health risk than men, not only by childbirth but by producing stresses that predispose to other disorders.

In some traditional farming communities, both males and females married at a young age and immediately started a family in which children made an economic contribution. Newly formed families often moved to the frontier, and by this means vast areas of Eurasian forest and steppe became farmland several thousand years ago. In the Americas, immigrants produced a similar ecological change within 100 years. In these agrarian societies, the two sexes usually had different roles but both were constrained and their life cycles were relatively symmetric. In societies in which new land was not available other patterns held. For example, many traditional African societies practiced polygyny, but a male had to prove himself by accumulating wealth and prestige before marriage, whereas females married young. The different age rules for marriage produced asymmetry due to attrition of the older males, but polygyny supplied a balance.

In contrast to traditional societies, industrialized society supplies no social customs to maintain a balanced adult sex ratio. Age at marriage varies widely from decade to decade and apparently is affected by political conditions such as war, by erratic economic conditions, and by wide pendulum swings in attitudes toward women. Whereas in traditional societies members of neither sex are permitted to set their life course, males in modern society tend to be granted this right more consistently than women. Although the trend in the twentieth century has been toward increasing the choices offered women, particularly in the workplace, the pattern is an erratic one involving both advances and set backs. Modern society is capable of rapid swings in public policy, which can be brought about by well-organized political forces. An example of shifting attitudes toward women and the choices they are allowed to make in their own lives can be found in changes in the legal status of abortions in the United States. Not illegal until the late nineteenth century, abortions were legalized by a Supreme Court decision in 1973, but court actions, following political pressure during the 1980s, produced restrictions on this female province.

X. Individual Variability

The process of sexual reproduction ensures that each individual (excluding monozygotic twins) is genetically unique. Individual variability originates in the genotype but can be fostered, or muted, socially. Traditional societies are known for channeling the abilities of their people; as anthropologist Ruth Benedict termed it, they are "personalities writ large." Benedict studied the processes whereby traditional cultures shape the perspectives and behavior of their members at a time when her own culture (U.S. society between the two world wars) was experiencing uncertainties about its own attitude toward individuality and the changing roles of males and females. In her work she became concerned about the misfit, the aberrant individual whose personality or talents simply do not incline toward the culture into which he/she is born. Benedict's hope was that modern society could accommodate diverse personalities more than traditional societies had done.

Is Benedict's hope feasible? A biocultural perspective on differences between males and females shows that biological differences between the two groups are real, but that the extent of the differences depends on both environmental factors (such as nutrition) and behavioral factors (such as activity level). A historical biocultural review shows that interpretations of these patterns do change within a culture over time. Historically, many different perspectives have waxed and waned, but few cultures at any time have been truly tolerant of aberrant personalities.

So far in human history it appears that the dominant perspective toward differences between males and females at any one time rests on the economic, political and historical traditions of the culture more than on scientific understanding. Scientific data are used by many politicians in attempts to prove points, rather than in a dispassionate attempt to seek the truth. Indeed, most ethicists would argue that scientific understanding is not an appropriate base on which to rest values or legislation. Though science can help us to perceive the natural phenomena—in this case the biology and behavior of males and females and their interactions—it is people who must strive to treat individuals ethically. If a society does evolve that treats males and females fairly, and allows both sexes the opportunity for innovative personal growth, it will be because its members have chosen to value individual productivity more than they value uniformity.

Acknowledgment

I am grateful to Lizbeth Gray, Dee Baer, and Don Hall for their helpful comments on a draft of this manuscript, and to Claire Younger for preparing it.

Bibliography

Frisch, R. E. (ed.) (1990). "Adipose Tissue and Reproduction." S. Karger, Basel.

Hall, R. L. (ed.) (1982). "Sexual Dimorphism in Homo Sapiens: A Question of Size." Praeger Scientific, New York.

Hall, R. L., Draper, P., Hamilton, M. E., McGuinness, D., Otten, C. M., and Roth, E. A. (1985). "Male–Female Differences: A Bio-Cultural Perspective." Praeger Scientific, New York.

Halvorson, H. O., and Monroy, A. (eds.) (1985). "The Origin and Evolution of Sex." Alan R. Liss, New York.

Hrdy, S. B. (1981). "The Woman that Never Evolved." Harvard University Press, Cambridge.

Kimura, D. (1987). Are men's and women's brains really different? *Can. Psychol.* **28(2),** 133–147.

Pickford, M., and Chiarelli, B. (eds.) (1986). "Sexual Dimorphism in Living and Fossil Primates." Il Sedicesimo, Firenze, Italy.

Sex Differences, Psychological

LEE WILLERMAN, *University of Texas at Austin*

Glossary

Dimorphism Differences in form or structure between members of the same species

PSYCHOLOGICAL SEX DIFFERENCES are broadly construed to refer to behavioral differences between the sexes. While the unmistakable accent is on difference, considerable overlap exists for most of the behaviors to be described, and personal characteristics other than sex usually provide a firmer foundation for making socially important decisions that affect individuals.

Sexual dimorphism is generated by a testis-determining factor located on the short arm of the Y chromosome producing maleness. The absence of this factor results in a female. Differences in the brain accompany the more conspicuous sex differences in body structure and size. Women appear less prone to aphasic disturbances following damage to their left hemisphere, except for a small region in the left anterior frontal lobe where damage is more likely to produce aphasia in women. The sexes also differ in dendritic arborization patterns in the visual cortex and hippocampus, areas not previously thought to be sexually dimorphic. There is sex dimorphism in the size of certain nuclei in the anterior hypothalamus, and only the hypothalamus of females responds to rising plasma titers of estradiol with the cyclic release of luteinizing hormone–releasing hormone. Anterior hypothalamic nuclei are targets of the organizing influence of prenatal testosterone and are involved in regulating sexual behavior in lower animals.

Observations such as this have led to theories that the brain of the two sexes is somewhat differently organized. Experiments in some strains of rats reveal that male fetuses whose mothers were stressed during pregnancy often had a smaller amount of tissue in the preoptic area of the hypothalamus; as adults they were less effective copulators and had lower testosterone levels than nonstressed control animals. The relatively smaller preoptic area in stressed males is believed to arise from a 1-day delay in the onset of a normal testosterone surge during the last trimester of pregnancy. These findings confirm the centrality of the hypothalamus and the importance of testosterone in regulating sexual behavior, although it is unwarranted to generalize from results such as these to all aspects of sexual behavior. For example, these stressed male fetuses often display more feminine behavior than control males, but they do not show a sexual *preference* for other male rats.

If psychological sex differences were a direct result of biological sex differences, controversy surrounding the following findings would be considerably lessened. The shifting magnitude and, occasionally, even the direction of sex differences on many psychological measures over time make it difficult to draw hard and fast conclusions, as does the joint effect of biology and culture. Some experts argue that even posing questions about "built-in versus acquired" sex differences misses the mark by ignoring their interaction. Their analogy is to first language learning, the capacity for which is innate, but the specific language learned depending entirely on the language spoken in the home. The more traditional approach argues that variations in constitutional factors can be pitted against variations in environmental factors to estimate the relative importance of biology versus social environment in a given population. This relative importance approach is more dominant, although the joint effects of constitution and environment are drawing increasing attention.

Initial biological differences often are correlated with socialization practices that prepare children for their adult sex roles. These treatment differences make it difficult to determine the degree to which constitution alone produces predictable variations in later behavior. With only two sexes, explanatory power is greatly constrained, because any sex-related biological or rearing difference becomes a potential candidate to explain any psychological sex difference. So great is the sheer number of these potential candidates that hypotheses can easily proliferate. For example, toddlers' predilection for rough-and-tumble play could be a rehearsal strategy for adult intermale competition (analogous to the playful pouncing of kittens in preparation for later predation) and/or the result of the higher rates of rough-and-tumble father–child interactions observed with male, as opposed to female, infants.

I. Childhood Differences

Gender identity, the ineffable sense of one's own femaleness or maleness, evolves early in life and, in most normal children, seems to be established by 4 yr of age. Exceptions occur when children with gender identity disorders are brought to professional attention having repudiated the manifestations of their sex—for example, asserting that genitalia will appear or disappear, or reporting feelings of being trapped in the body of the wrong sex. Most of these children eventually develop a gender identity congruent with their genetic sex, but perhaps two-thirds of such boys become homosexually oriented despite their development of a masculine gender identity (the rates for extremely masculine girls are less certain). Gender identity, however, should not be thought of as dichotomously male or female. A significant fraction of people, not all of whom are homosexuals or transsexuals, have dreams or fantasies of being the other sex or wonder about having a sex-change operation, suggesting the existence of intermediate gender identities.

Gender role refers to the outward manifestation of gender identity (i.e., the behaviors conventionally associated with being male or female). Distress about gender role adequacy is much more common than gender identity disturbance, predictably so in teenagers if puberty or dating skills are delayed, and in adults if the inability to fulfill stereotypic gender roles is accentuated (e.g., when previously employed men lose their jobs, or when women want to but cannot conceive).

Some sex dimorphisms are inevitable (e.g., men inseminating, women bearing children), others are completely arbitrary (e.g., dresses versus trousers), and still others are intermediate dimorphisms usually more common in one sex or the other but expressible by either sex under appropriate circumstances (e.g., sleeping mothers being more sensitive than fathers to an infant's crying). Among the most psychologically interesting of these latter differences are energy expenditure, physical competitiveness, fighting off predators in defense of territory, and erotic arousal by visual imagery (more commonly observed in males) and retrieving, cuddling, and rehearsal play with dolls or playmates (more frequently observed in females).

By 3 yr of age, most toddlers can distinguish the two sexes in pictures, although some may be uncertain about their own sex remaining forever constant or the hegemony of anatomy in sex determination. They also show preferences for interacting with same-sex playmates, the patterns of interaction being quite different within the sexes. In comparison to girls, boys more often prefer rough-and-tumble activities and mock fighting. Boys also accede less often to protests from girls not to take their toys, even at 3 yr of age.

Sex-segregated subcultures of the preschool years persist into the early school years, but the source of that persistence is unknown. It is possible to identify masculine and feminine preschoolers of both sexes, i.e., children whose behaviors tend to resemble the stereotype of one sex or the other. Nevertheless, most children, regardless of masculinity or femininity, play mainly with same-sex peers, despite occasional efforts by adults to encourage mixed-sex interactions. Sex segregation in the early school years is sustained in part by peers who tease about being in love if another child shows an interest in someone of the opposite sex.

It has been difficult to demonstrate compellingly that differential parental socialization of the sexes or even extrafamilial social influences such as the media cause, rather than correlate with, the gender segregation and play style differences. The absence of decisive evidence on this matter has complicated environmental theories about the origins of psychological sex differences. Biological theories about the play style differences probably are more established because of behavioral homologies in lower animals.

Whatever the causes, sex-related preferences in early childhood appear to affect the matrix of experiences encountered and influence later mate choice. For example, virtually no marriages exist between classmates in Israeli kibbutzim, where like-aged children live in mixed-sex group homes from infancy on. Classmates often marry members of the same kibbutz, but they nearly always come from younger or older classes, the early familiarity apparently discouraging romantic interests within their age cohort. Exceptions occur if one of the children enters the cohort after 6 yr of age, suggesting that very early playmate experiences have important functions in later mate selection. If early familiarity discourages later sexual interests, then these results may also help to explain the absence of widespread sibling incest in ordinary families.

II. Interest and Personality Differences

Sex differences in interest patterns are among the largest and most stable psychological sex differences known. These differences are telling because they reflect preferences rather than obligations and, therefore, reveal more about what people really like to do rather than what they must do. Studies of reading preferences in elementary and high school students indicate that boys are 3 times more likely than girls to prefer adventure and mystery stories, whereas girls are 10 times more likely to prefer stories about home and school life. These differences persist into adulthood, when women are much more likely than men to read romance novels. However, sex differences in interests tend to wane slightly in middle-age with males moving in the feminine direction.

Despite decades of increasing educational and occupational opportunities for women in industrialized societies, interest differences do not seem to have diminished appreciably. For example, the Strong-Campbell Interest Inventory was constructed by asking adults in many occupations about their likes and dislikes (e.g., decorating a room with flowers, cabinetmaking), to see whether or not interest pattern differences relate to the likelihood of success in various occupations (they do). The most recent restandardization in the 1970s revealed sex differences on 149 of the 325 items, and the differences are still similar to those indicated in the original 1930s standardization. In one item, "liking to decorate a room with flowers," 79% of women lawyers in the original standardization indicated a liking as opposed to 21% of the men lawyers; in the most recent survey, 62% of the women lawyers and 15% of the men lawyers indicated a similar liking, a modest decrease in overall liking, but nevertheless a similar large difference between the sexes.

Most readers could have guessed the direction of the sex difference in liking to decorate with flowers. Early masculinity–femininity scales of the 1930s were empirically constructed from such items to identify psychological differences between the sexes, the assumption being that masculinity–femininity was a single bipolar dimension; high femininity perforce meant low masculinity, and vice versa.

More recently, perhaps motivated by increasing career flexibility for women, experts devised independent scales of masculinity and femininity from the perspective of two separate personality dimensions, thus allowing people to be high or low on both, rather than masculine *or* feminine. These new masculinity–femininity scales focus on socially desirable personality traits such as emotional expressiveness, nurturance, and interpersonal orientation in women, and instrumentality (task as opposed to person orientation) and competitiveness in men. Despite meaningful average differences between the sexes on these personality scales, a tight linkage among gender-congruent items does not exist—masculine men often express tender feelings, and feminine women often enjoy the competition in careers formerly occupied by men. The merit of these scales, however, was to demonstrate the wide variety of possible psychological adaptations to one's gender.

Perhaps the absence of a tight linkage among personality items is not so surprising if one assumes that sexual selection affected physiology or behaviors more directly related to reproductive success—physical attractiveness and nurturance in women, size and aggression in men. Other sex-related adaptations probably arose because of their correlations with the primary targets of sexual selection (e.g., a greater preference for decorating a room with flowers because "home" has more significance for women, or men more often finding the idea of sex without love erotic because of their relatively low necessary investment in the costs of gestation and postnatal nurturance).

Recent cross-cultural research has shown fairly consistent patterns of sex differences in mate preference that seem related to signs of reproductive potential in women and the potential of men to provide resources for mother and child. Questionnaires completed by respondents of both sexes in 37 industrialized and preindustrialized societies revealed, in nearly every case, that women more than men preferred ambitious and financially secure potential spouses, whereas men valued more highly than women a good-looking spouse. In every society, both sexes independently agreed about the desirability of men marrying younger women, confirming the social worth of youth in female reproductive potential and male resource capacity in mate choice.

Differences between the sexes on most self-report personality inventories are quite modest if scales of masculinity–femininity are excluded. The recent massive restandardization of the 462-item California Personality Inventory (CPI), one of the most popular instruments of this genre for the normal adolescent and adult population, found no substantial sex differences for 19 of 20 personality traits. Among the personality traits measured were dominance, empathy, and sociability, and none revealed mean differences between the sexes of more than one-fifth of a standard deviation, suggesting nearly complete overlap of the distributions of scores for the two sexes. The CPI feminity scale, however, revealed that females scored nearly two standard deviations higher than males, a difference of this magnitude indicating that the average female scores at the 95th percentile of the male distribution. High scorers of either sex on this scale are rated as sympathetic, sensitive to criticism, interpreting events from a personal point of view, and reporting feelings of vulnerability; low scorers of either sex are viewed as decisive, action-oriented, taking the initiative, and unsentimental. Descriptions such as these seem to capture stereotypes about the major personality differences between the sexes.

III. Cognitive Differences

A review of cognitive differences between the sexes must acknowledge that tests are designed by people who first must decide on the types of items to include in the domain of interest. Most tests do not ask about tracking game, surviving in the desert, or making dyes from plants, perfectly legitimate items in preindustrialized societies. Conventional intelligence tests attempt to exclude items that rely on material explicitly taught in school. The tests generally include items that call for problem-solving, reasoning, and abstraction. Verbal intelligence test items might ask a child about the erroneous thinking in "Mary's jeans were so tight that she had to put her pants on over her head" or explain how sewing machines and automobiles are alike. By contrast, pure achievement test items ask about material explicitly taught in school, although some achievement and intelligence test items do resemble each other when they require reasoning based on school-related subject matter. Nonverbal intelligence tests require no specific knowledge of language; completing jigsaw puzzles and constructing block designs after model pictures are exemplars and can even be pantomimed for people who are congenitally deaf or come from preliterate societies.

A. General Intelligence

Evaluating sex differences in general intelligence using standardized tests is usually not meaningful, because most tests are designed to be equally applicable to both sexes, and items initially showing sex differences in percent passing were either excluded from final versions or counterbalanced by items with the opposite pattern of sex difference. Men do perform relatively better than women on items that require elementary geographic or scientific knowledge (e.g., Why does water boil in the mountains at a lower temperature than in the valleys?), but this is most likely to be a result of discrepant interest patterns. Perhaps one tends to remember the gas laws only if they jibe in some way with interests or are reinforced in specific courses. A widely used nonverbal intelligence test (Raven Progressive Matrices) ignored sex differences in its construction and found none afterward, suggesting that sex differences in general intelligence are either negligible or nonexistent.

B. Specific Abilities and Scholastic Achievement

Tests designed to tap specific abilities reveal moderately interpretable differences between the sexes. Females outdo males on tests of verbal fluency (e.g., in one minute say as many words as possible that begin with the letter "R") and perceptual speed (e.g., quickly cross out all the words beginning with the letter "e" in a long string of words). Achievement tests tapping written language, grammar, and *arithmetic computation* also show a female disadvantage.

None of these differences, however, has attracted quite so much attention as the conspicuous male advantage on *mathematical reasoning* tests, such as the Scholastic Aptitude Test (SAT) mathematics section. The sex difference is most apparent at the highest levels of achievement with males comprising 96% of those with the maximum possible mathematics score of 800. The cause of the *mean* advantage is not clear, but sex differentials in high school dropout rates and low ability males being less likely to take the test are partial explanations; however, no selection differential can account for the remarkable overrepresentation of males at the very highest level, and many hypotheses have been proposed to explain the result. One hypothesis has focused on the greater nonconformity of boys in school (i.e., they are less well behaved), which paradoxically gives them an advantage in solving math problems that require nonconventional reasoning instead of direct application of explicitly taught and well-practiced strategies. Because males now show a slight advantage on the SAT verbal test as well, where nonconformity would not seem to have been advantageous, that hypothesis seems doubtful.

The male advantage on the SAT mathematics test is accentuated at the top of the score distribution because males are also more variable. Greater male variability needs to be explained, and the current biological focus is on genetic and hormonal differences between the sexes. Nonbiological approaches focus on differential expectancies about the role of mathematics in one's future (e.g., attitudes toward mathematics or taking different courses), but they have not gotten very far in accounting for the difference. The same could be said for sex differences in self-perceptions (e.g., females tending to attribute success to luck and failure to personal inadequacy), and thus a tendency exists, *faute de mieux,* to find biological approaches more appealing.

Males perform better than females on tests of spatial visualization, especially when three-dimensional mental rotation is required. The reason for this advantage also is not clear, but an analogous sex difference in learning mazes that require some form of spatial problem-solving is found in several rodent species. Human females with congenital adrenal hyperplasia (CAH), involving overproduction of androgen from the adrenals prior to birth, appear to be better than normal females at three-dimensional rotation. Because the overproduction of androgen is curtailed by cortisone treatment beginning soon after birth, partial prenatal masculinization of the CAH female brain is a viable hypothesis for this advantage. A subsidiary hypothesis is that interest and activity patterns of females with high spatial visualization scores (or males with low spatial visualization scores) may be more like the other sex. Thus, it might be the interest–activity patterns that provide test-relevant experiences that influence spatial visualization, rather than something in the hormonal environment directly affecting spatial visualization.

Studies offer modest support for this hypothesis. Some species of voles show differences in ranging behavior between the sexes, although all voles are monomorphic. Male voles of high-ranging species show better spatial maze-learning ability than males in low-ranging species. Young human males are also more likely to range farther from their caretakers, suggesting some experiential basis to the sex difference in spatial ability. Both normal and CAH females with masculine activity patterns tend to do better on spatial visualization tasks, but the effect is too small to eradicate the overall mean sex difference in spatial visualization scores, and the direction of effects has not been determined; does higher spatial visualizing ability lead to more masculine interest–ranging patterns, or vice versa?

Lower spatial visualization scores in men with idiopathic hypogonadotropic hypogonadism (IHH) would seem to reinforce the role of pubertal testosterone in the male spatial visualization advantage, especially because these men show no deficits in verbal ability. IHH is characterized by a failure to enter puberty because of a hypothalamic deficiency in gonadotropin-releasing hormone. Although IHH males appear to have a near-normal prenatal hormonal environment, this is not certain, and none of these males was tested prior to his teenage years. Moreover, studies of mathematically precocious preadolescents reveal a disproportionate number of males scoring at the very highest levels, suggesting hormonal or experiential influences prior to the pubertal testosterone surge. In short, evidence exists for prenatal and pubertal testosterone influences on spatial ability; their relative influence, however, remains uncertain.

Other evidence for an early biological influence on spatial and mathematical ability comes from the study of females with Turner syndrome, a disorder caused by the complete or partial absence of one X chromosome. Even as children, many show a significant deficit in spatial and mathematical ability despite normal verbal intelligence. These females have low levels of estrogen because of defective or absent ovaries and do not enter puberty unless estrogen supplements are provided. Although several somatic problems are associated with this condition, including very short stature, affected females are usually within the normal range of mental health and are heterosexually oriented. No brain defects have yet been identified, and a prenatal hormonal deficiency effect on spatial and numerical ability is quite plausible, although the specific hormonal mechanism has not been clarified.

IV. Sexual Orientation

Aside from gender identity, the largest psychological gender difference known is in the sex of the preferred mate. All mammalian species are heterosexual, and there is even controversy about whether homosexual orientation exists in lower mammals. The debate is not about whether or not homosexual behavior naturally occurs—it does, but mostly when males are blocked from access to females. Studies of mate choice in some lower mammals reveal that early castration of males or testosterone administration to females can partially or completely reverse sexual orientation, so that hormonal factors must figure prominently.

Heterosexual and homosexual orientation refer, respectively, to the tendency to become erotically aroused by the opposite or same sex. Sexual orientation is not solely determined by sex acts with opposite or same-sexed partners; heterosexual teenage boys often engage in mutual masturbation, for example, but it is friction rather than homosexual fantasy that sparks their sexual arousal.

Current research suggests that the matrix out of which homosexuality emerges is evident early in life. Gender-incongruent behavior even in preschool children, especially extreme femininity in boys, appears to augur later homosexuality; tomboyism in girls is not nearly as good a predictor of later lesbianism because it occurs with a high base rate even among future heterosexuals. Although male homosexuality runs in families (little evidence indicates that female homosexuality is familial), a genetic explanation seems doubtful because homosexuality is associated with reduced reproductive fitness, and putative genes for this orientation should have been eliminated from the population. Various, and as yet unsupported, alternatives to salvage a genetic hypothesis include a gene that increases the fitness of relatives by "helper at the nest" behaviors or a gene that reduces fetal wastage in heterosexual female carriers. Social environmental explanations have been unsatisfactory thus far because distinctive peculiarities in the early lives of homosexuals have not been discovered. Recollections of childhood can be faulty, however, and large, prospective longitudinal studies that begin with monitoring events during pregnancy and observing the development of the children hold more promise. Current hypotheses emphasize prenatal endocrine factors, perhaps arising from stresses to the mother or fetus reducing androgen levels in males, and excess prenatal androgen increasing the risk for eventual lesbianism in females. Unfortunately, trustworthy evidence for these hypotheses is lacking in humans. Whereas some studies have reported that heterosexual and homosexual males differed in their luteinizing-hormone levels following an estrogen challenge, with homosexual males tending to respond somewhat like females, other studies have failed replicate and no consensus has been achieved about the reliability or meaning of these results.

Other evidence that neuroendocrine factors contribute to sexual orientation comes from people whose prenatal hormonal milieu has been altered by accidents of nature. But describing the psychological characteristics of people exposed to unusual prenatal hormonal environments can be complicated, because phenotypic, and not genetic, sex determines whether children are reared as boys or girls. Testicular feminization syndrome provides an example of appearance as being decisive in the sex of rearing, while not shedding much light on the role of prenatal hormonal factors in shaping sexual orientation. This syndrome arises because an autosomal recessive gene leads to a reduced or complete inability of receptors in the brain and elsewhere to respond to circulating testosterone. In genetic males, the outcome is a feminine phenotype, feminine gender identity, and sexual attraction to males. Because they are reared as girls—subjectively and phenotypically female except for the absence of ovaries and uterus and the presence of undescended testes—it is misleading to regard their sexual attraction to males as homosexual.

A more informative condition is CAH, which produces nearly comparable levels of prenatal androgen in affected female and in normal male fetuses. The external consequence in the affected female is an enlarged clitoris, correctable by surgery; excess secretion of androgen from the adrenals after birth is controlled by cortisone administration. Though reared as females, they are often tomboys, and some become lesbians. This outcome points to a significant role for the prenatal hormonal milieu in influencing sexual orientation. Because only a minority become lesbians, however, uncertainty remains about the possible influence of other factors such as the timing or degree of prenatal androgen excess or postnatal experiences on sexual orientation.

Another interesting syndrome in this context involves a genetic inability to produce 5-alpha-reductase, an enzyme that converts testosterone to dihydrotestosterone (DHT). DHT is necessary for the early masculinization of the external genitalia but has no known function in the brain. Genetic males with 5-alpha-reductase deficiency have ambiguous external genitalia, and more than half are reared unquestioningly as girls. However, a gonadal testosterone surge at puberty causes a deepening of the voice, muscularity, the descent of undescended testes, and phallic enlargement so that erection and intromission can occur. Virtually all of these girls develop erotic interests in females, eventually cohabiting and foresaking their earlier identity for a masculine gender identity and gender role. Because this disorder runs in families, parental knowledge about the risk for the disorder could have biased the girls' rearing, but parents typically report shock and amazement when their child exhibits sexual interests in females. Because nearly all such females developed erotically in ways consistent with their prenatal testosterone levels despite unambiguous rearing as girls, hormonal influences are implied to play a powerful role in sexual orientation. It is premature, however, to argue strongly on behalf of any one hypothesis to explain the origins of homosexuality, and multifactorial causation is a distinct possibility.

V. Sex Differences in Psychopathology

There are many sex differences in the rates of psychopathology. In childhood, the rates of attention-deficit hyperactivity disorder, conduct disorder, dyslexia, stuttering, some forms of mental retardation, and early infantile autism are much more prevalent in boys. These disorders run in families, although most cases occur sporadically. Explanatory hypotheses have focused on X-linked genes or sex chromosome anomalies, sex differences in prenatal vulnerability to anoxia or infection, mother–male fetus antibody reactions, and sex differences in child-rearing practices.

The fragile X syndrome, occurring in 1/2,000 male births, provides an example of a disorder that produces an excess of mental retardation in males and may also account for some of the male excess among those with early infantile autism. Identified by a constriction on the X chromosome in lymphocytes cultured in a medium deficient in folic acid and thymidine, this syndrome produces mental retardation in affected males (a few have normal IQs) and a much more variable outcome ranging from mild retardation to normal intelligence in females. In both sexes, an elongated face and large ears are associated with the fragile X syndrome, and enlarged testicles are common in men. Although most people with fragile X syndrome lack remarkable behavioral features, with the exceptions of mental retardation and perhaps hyperactivity, 12 to 25% of males with early infantile autism have the fragile X anomaly, suggesting behavioral pleiotropy of the fragile site. [*See* FRAGILE X SYNDROME.]

Among adults, men are much more likely than women to have a history of antisocial personality disorder and alcohol abuse or dependence, with a sex ratio ranging from 4 : 1 to 6 : 1 in epidemiologic surveys. The male excess for hyperactivity, alcoholism, and antisocial personality disorder has been attributed to interactions of genetic factors with sex differences in prenatal vulnerability and to social factors. Disorders with by far the greatest male preponderance are paraphilias such as fetishism. Paraphilias are characterized by recurrent intense sexual urges and fantasies that require nonhuman objects, humiliation to self or others, or nonconsenting adults or children for sexual satisfaction. Fetishism specifically involves intense preoccupations with nonliving objects (e.g., gloves, panties) for sexual satisfaction. In comparison to thousands of male fetishists, only three female fetishists have ever been reported, and all three were homosexually or bisexually oriented, implying the possibility of early hormonal factors figuring in the etiology. A sex difference of such magnitude is truly remarkable and an explanation could bear on sex differences in other realms, including the preponderance of males (whether homosexual or heterosexual) as purchasers of pornography.

Women are more likely than men to have a history of major depressive episode (2 : 1 or 3 : 1), agoraphobia (3 : 1), and simple phobia (2 : 1). Other disorders show smaller, less reliable sex differences in prevalence. Included among these are nonalcoholic drug abuse and dependence, with men predominating, and dysthymia (chronic depression), somatization (an unusually wide variety of inexplicable bodily complaints), panic disorder, and obsessive–compulsive disorder, with women predominating. Many hypotheses have been proposed for the female excess, including feelings of powerlessness, less reluctance to admit vulnerability and fear, a greater tendency to internalize stress, and a sex bias in diagnostic criteria. All of these hypotheses have found support to various degrees, but more refined hypotheses are required to handle the sex ratio heterogeneity from disorder to disorder. [*See* DEPRESSION; MENTAL DISORDERS; NONNARCOTIC DRUG USE AND ABUSE.]

Many of the disorders with sex differences are heritable; for example, adoption and/or twin studies indicate that hyperactivity, antisocial personality disorder, major depression, alcoholism, and somatization have heritable components. Except for a putative X-linked dominant form of manic-depressive disorder and X-linked mental retardation, no psychopathological studies indicate X-linkage, suggesting that sex differences in rates for specific disorders could arise from sex differences in vulnerability thresholds (e.g., female stutterers are rare, but, if affected, must have required a greater dose of the relevant genes because they are more likely than affected men to produce similarly affected offspring). Another possibility is sex-specific symptom expression of the same diathesis (e.g., antisocial males often have sisters with somatization disorder).

VI. The Future of Sex Differences

Modern society no longer requires that physical strength be the final arbiter in the division of labor. Very few occupations are necessarily limited to one sex, and as external barriers to equal opportunity diminish, occupational choices will be made more often on the basis of personal interests and talent. Sex differences in occupational interests have remained fairly stable for several decades and, if the past is a guide, we should not expect much change in preferences for various occupations.

Because the onus of reproduction falls largely on women, any correlates of differential parental investment such as different attitudes about sexual behavior might be difficult to eliminate. Perhaps emblematic of these differences are findings in a large British survey that while three-quarters of unmarried college women think that the idea of a sex orgy is disgusting, only 18% of a corresponding group of men do and that 73% of the men versus 35% of the women believe it is all right to seduce people who are old enough to know what they are doing. Another study reported that a third of men said they might rape if there were no chance of getting caught. Differences such as these can be perennial sources of intersexual conflict, at least for a large proportion of people.

It seems likely that sex differences in preference and style during sexual intercourse also will not change much, with conflicts arising that often affect other aspects of the relationship. Women frequently complain that their partners have a narrow concept of affection, moving too speedily toward intromission while short-circuiting whole-body caressing. On the other hand, men often complain that their mates are sexually witholding. Observational studies of sexual behavior in homosexual couples provide insights into how the sexes differ in preferred styles of sexual intercourse because they need not make as many compromises as heterosexual couples. Results from these studies indicate that homosexual males tend to spend relatively less time in foreplay in comparison to lesbians and tend to leave the bed more quickly after orgasm, in contrast to lesbians, who often fall asleep holding one another. Moreover, cohabiting male homosexual couples are also more likely than lesbian couples to be involved in sexual infidelities. Some of these sex differences could also be a source of tension within heterosexual couples. [*See* SEXUAL BEHAVIOR, HUMAN.]

Whatever their origins, differences in sexual behavior appear to be fairly sturdy and not necessarily rational from the perspective of an intelligent automaton. For example, the mammalian sex difference in longevity is tied to testosterone since male neonatal castration lengthens life span and testosterone administration to females shortens life span. This difference in longevity, about 7 yr in humans, combined with an average difference of about 3 yr in age of spouses, means that the average woman will be widowed for 10 yr. If other things were equal, a demographic reversal in the average age at marriage would increase the duration of couple companionship.

Sex dimorphism with respect to body size predicts overall patterns of mating across many species: monomorphic species monogamous, dimorphic species polygynous or promiscuous. In the monomorphic gibbon, pairs mate for life; in the extremely dimorphic gorilla, the dominant male has a harem, and most males are perforce heterosexually celibate. In a moderately dimorphic species such as humans, the relevant evolution is unmistakably closer to a heritage of intermale competition, but our smaller degree of dimorphism suggests some divergence, as does the humanly distinctive pubertal growth of breasts in the absence of pregnancy.

Human males also have moderately sized testes relative to other primates, which means a limited reservoir of sperm. Consequently, human males have not been equipped by evolution for closely spaced bouts of sperm-producing intercourse and the sex drive is probably somewhat less predictably imperious. Interpersonal ambiguities arising from being intermediate on these evolutionary factors probably figure importantly in some tensions between the sexes. These considerations may lead one to ask how relations between the sexes and psychological sex differences might alter, absent of any changes in brain biology, if the sex difference in strength were reversed or if men bore children.

Acknowledgment

The author thanks Professors David B. Cohen, Stephen Finn, Judith Langlois, and John C. Loehlin for reviewing earlier versions of this article.

Bibliography

Benbow, C. (1988). Sex differences in mathematical reasoning ability in intellectually talented preadolescents: Their nature, effects, and possible causes. *Behav. Brain Sci.* **11,** 160.

Buss, D. (1989). Sex differences in human mate preferences: Evolutionary hypotheses tested in 37 cultures. *Behav. Brain Sci.* **12,** 1.

Campbell, D. P., and Hanson, J.-I. C. (1981). "Manual for the Strong-Campbell Interest Inventory," 3rd ed. Stanford University Press, Stanford California.

Gough, H. G. (1987). "California Psychological Inventory: Administrator's Guide." Consulting Psychologists Press, Palo Alto, California.

Ilai, D., and Willerman, L. (1989). Sex differences in WAIS-R item performance. *Intelligence* **13,** 225.

Kimura, D. (1987). Are men's and women's brains really different? *Can. Psychol.* **28,** 133.

Maccoby, E. E. (1988). Gender as a social category. *Dev. Psychol.* **24,** 755.

Reinisch, J. M., Rosenblum, L. A., and Sanders, S. A. (eds.) (1987). "Masculinity/Femininity: Basic Perspectives." Oxford University Press, New York.

Stewart, J. (ed.) (1988). Sexual differentiation and gender-related behaviors. *Psychobiology* **16,** whole issue.

Wilder, G. Z., and Powell, K. (1989). Sex differences in test performance: A survey of the literature. *College Board Report* No. 89-3.

Willerman, L., and Cohen, D. B. (1990). "Psychopathology." McGraw-Hill, New York.

Sexual Behavior, Human

JOHN D. BALDWIN AND JANICE I. BALDWIN, *University of California at Santa Barbara*

Glossary

Erotic stimulus Any stimulus that can elicit the sexual reflex, be it a conditioned stimulus or the unconditioned stimulus that is biologically associated with the sexual reflex

Myotonia Heightened muscle tension, adequate to cause various subcomponents of the sexual response

Pavlovian conditioning Process by which stimuli that are paired with and predictive of a reflexive response become conditioned stimuli that can elicit certain aspects of that reflexive response

Sexual dysfunction Deviation from the normal functioning of the sexual reflex that impairs a person's ability to have gratifying sexual responses

Sexual orientation Degree to which a person is sexually attracted to the other sex, the same sex, both sexes, or neither sex

Vasocongestion Increased blood flow into and congestion of the blood vessels that activate parts of the sexual response

HUMAN SEXUAL BEHAVIOR includes all the voluntary and reflexive activities—and their underlying physiological mechanisms—that are involved in sexual reproduction, though these activities are not always used in ways that lead to pregnancy and childbirth. The reflexive and inborn facets of sexual behavior are not adequate for assuring successful reproduction; and considerable learning is needed for an individual to become reproductively capable. That learning can be influenced significantly by social, cultural, and individual variables, leading to considerable variability in human sexual activities.

I. The Sexual Response of Healthy Adult Males and Females

The human sexual response has been described in several ways. One of the more widely accepted descriptions divides the continuous series of changes of the sexual response into four general phases: excitement, plateau, orgasm, and resolution. Another categorization is desire, excitement, and orgasm, with desire occurring before biological arousal. Since many people have problems of sexual desire (i.e., a lack of interest in sexual fantasies and activity), sexual desire is an important topic; and it will be considered before the four phases described above.

A. Desire

There is considerable variation among people in the amount of sexual activity that they desire. People with no sexual desire are sometimes described as "asexual" or "nonsexual." People with low sexual desire may think about sex from time to time, but they have little interest in sexual fantasies or activity and may not become easily excited in sexual situations. People with high sexual desire think about sex often and may even become sexually aroused in nonsexual situations. Desire for sexual activity is not an essential prerequisite for the remaining phases of the sexual response: People can—and some do—progress through the biological phases of the sexual response without having sexual desire.

There is no single "normal" level of sexual desire. Two people with high sexual desire may be very happy with their sexual life having sexual interactions every day; and two people with low sexual desire may have an equally satisfactory sex life, engaging in intercourse only once a month. Levels of sexual desire are only considered to be problematic when they lead to personal or interpersonal problems, for example, when two sexual partners have different levels of sexual desire, i.e., "discrepant sexual desire." The greater the discrepancy between two people's level of sexual desire, the greater their sexual problems can be. The partner who wants to have frequent sexual interactions may feel very frustrated by the lack of interest that his or her partner shows; and the person with low sexual desire may feel pressured to engage in sex more than he or she desires. Such problems can create considerable tensions in a relationship.

B. Excitement

Any of a variety of stimuli can lead to sexual excitement. Sometimes people become excited without planning it or desiring it. Other times, a clear sexual desire leads them to seek out the stimuli that elicit sexual excitement. The central nervous system is structured from birth such that touch to the genitals leads to sexual excitement, if it is of the appropriate pressure and movement pattern. Visual, auditory, olfactory, and fantasy stimuli can also take on the ability to trigger sexual excitement.

In the female, the first sign of sexual excitement is vaginal lubrication; and in the male it is penile erection. These early signs of excitement are easily reversible if there is no further sexual stimulation. However, with continued effective stimulation, there can be further developments of sexual excitement: In the female, the inner two-thirds of the vagina expands, the uterus lifts up, the major lips (or labia majora) pull back and the minor lips (or labia minora) and clitoris increase in size. In the male, the testes enlarge and are elevated somewhat closer to the body.

C. Plateau

With continued effective sexual stimulation, people reach a higher level of sexual arousal, the plateau phase. In the female, the outer third of the vagina swells, narrowing this portion of the vaginal barrel, while the inner two-thirds expands more, and complete elevation of the uterus occurs. The clitoris retracts under the clitoral hood (but is still sensitive to stimulation around it). In the male, the testes continue to swell, becoming 50 to 100% larger than normal. They are pulled closer to the body. Usually at this time, the Cowper's glands secrete a fluid that neutralizes the acidity of the urethra. The fluid may also contain sperm left from the last ejaculation or nocturnal emission.

D. Orgasm

After sufficient effective sexual stimulation, orgasm occurs. In both the female and male, the length and intensity of orgasm can vary, depending on the level of sexual arousal, the type of stimulation obtained during orgasm, health, age, and other variables. Typically, orgasm lasts 3 to 15 sec.

In the female, the primary orgasmic response consists of rhythmical contractions of the muscles surrounding the outer third of the vagina. The first several contractions occur at approximately every 0.8 sec; but subsequent contractions are spaced further apart and are less intense. These contractions are experienced as pleasurable, often being described as waves, surges, or pulses of pleasure. Females have the potential for multiple orgasms: if they continue to desire and receive effective stimulation, they may alternate between plateau and orgasm several times in close succession.

In the male, orgasm is divided into two clearly recognizable stages: ejaculatory inevitability and ejaculation. Ejaculatory inevitability occurs a few seconds prior to ejaculation. At this time, contractions of the prostate, vas deferens, and seminal vesicles cause semen to be released into the urethra. After this several-second-long process, the semen is expelled from the male's body by rhythmical contractions of the perineal muscles, which contract at intervals of 0.8 sec for the first several contractions, though later contractions are spaced further apart and gradually become less intense.

E. Resolution

After orgasm is over, the various physical changes that develop during excitement, plateau, and orgasm reverse, and the body gradually returns to the unaroused condition. Resolution from the plateau phase takes longer if there is no orgasm: The occurrence of orgasm hastens the changes during resolution.

After ejaculation, males enter a period called the refractory period, in which further stimulation does not produce a sexual response. The refractory period can last from a few minutes to many hours, depending largely on a person's age—being longer in older males. After the refractory period is over, sexual stimulation can once again lead to erection, rising sexual arousal, and orgasm.

II. The Mechanisms of the Sexual Response

A. Reflexes

The sexual response is based on complex reflex mechanisms that are mediated by the sacral, lumbar, and parts of the thoracic vertebrae in the lower spinal cord. The primary neural receptors that trigger the reflex come from the genitals (or nearby erogenous areas). The unconditioned stimuli that activate the reflex are touch to the genitals. The reflex mechanisms also send signals to the cerebral cortex and are responsive to signals from the cerebral cortex. Through cerebral mediation and Pavlovian conditioning, additional stimuli, such as fantasies, can come to elicit the sexual response. Finally, the sexual reflex can be suppressed by any of a variety of aversive stimuli, such as worry, guilt, anxiety; and much of sexual therapy focuses on helping people reduce the aversive components of their sexual lives, while increasing the aspects that are sexually arousing and exciting.

B. Vasocongestion and Myotonia

During the early phases of sexual stimulation, sacral reflexes are activated that cause increased blood flow to the genitals. The resulting increase in blood in the genital area is called vasocongestion; and it causes an increase in the size of various parts of the genitals. Upon early stimulation, the penis becomes larger, firmer, and more erect. In the female, vasocongestion around the vaginal barrel causes transudation—or lubrication—on the inside walls of the vagina. In addition, vasocongestion causes the labia minora to increase in size.

Among the changes mediated by myotonia (muscle tension) are heightened muscle tension in various parts of the body, the elevation of the testicles close to the body, and, with further stimulation, there are muscular contractions of the auxiliary organs that operate during ejaculatory inevitability to prepare the ejaculate for expulsion from the body. Myotonia is also responsible for the muscle contractions at orgasm that are experienced as waves of pleasure and cause the ejaculation of the semen in the male.

C. Cerebral Cortex

The neural mechanisms that connect the reflex centers to the cerebral cortex connect with the sensory centers, allowing people to feel touch and pressure in the genitals. In addition, some of these neural systems innervate the pleasure and pain (i.e., reinforcement and punishment) centers in the brain. This assures that certain forms of sexual stimulation are experienced as pleasurable; and people usually learn to repeat these activities, due to positive reinforcement. Other forms of sexual stimulation—for example, strong pressure to the gonads—are painful and people learn to avoid these forms of stimulation. These reinforcement and punishment mechanisms help assure that most people, if they have appropriate life experiences, learn how to obtain effective stimulation and avoid aversive stimulation. The muscle contractions at the time of orgasm are usually experienced as especially pleasurable.

Nerves that connect the cortex to the sexual reflex mechanisms in the lower spinal cord allow for the "psychogenic" stimulation of the sexual reflex. Namely, when people see or fantasize about erotic stimuli, the cortical activity can activate the sexual reflexes. For some people, psychogenic stimulation produces only a mild sexual response, such as partial penile erection or slight lubrication of the vagina; but in other people, the response can be quite strong, sometimes leading to orgasm. There is a great deal of variation among people in the types of fantasy and erotic stimuli that produce such psychogenic responses, and these differences are due to differences in prior sexual learning experiences, especially those involving Pavlovian conditioning. [*See* CORTEX.]

D. Pavlovian Conditioning

The sexual reflex is among the reflexes capable of being conditioned via Pavlovian conditioning. Namely, when a neutral stimulus is paired repeatedly with the unconditioned stimulus of appropriate touch to the genitals, that neutral stimulus gradually becomes a conditioned stimulus capable of eliciting the sexual reflex. In common parlance, these conditioned stimuli are called erotic stimuli or sexual "turn-ons." Usually, the more frequently that an erotic stimulus or turn-on has been paired with the unconditioned stimulus of touch to the genitals, the more capable the conditioned stimulus is of eliciting sexual responses.

Almost any stimulus that is frequently paired with and is predictive of the sexual reflex can become an erotic stimulus or sexual turn-on through Pavlovian conditioning. The specific stimuli that people respond to as sexually exciting differ from person to person, depending on each individual's prior sexual experience. Some people learn to respond to fantasies or pictures of nude bodies as erotic stimuli, if these have been paired with effective sexual stimulation in the past. Others come to find "talking dirty" to be sexually arousing, if such talk has been a common part of their exciting sexual experiences. Although many people would be offended by such language and do not understand how anyone could find it sexually exciting, the mechanisms of Pavlovian conditioning allow us to understand that almost *any* stimulus can become an erotic stimulus or sexual turn-on, if it is frequently paired with and is predictive of the sexual response.

When people learn to imagine their erotic stimuli (even in the absence of any external stimuli), they gain fantasy access to these stimuli. Some people cultivate sexual fantasies because of the pleasure and sexual arousal that they bring. During sexual intercourse, fantasies can add to the total excitatory sexual stimulation and enhance the sexual response. Fantasies also serve to distract people from some of the stimuli that can inhibit the sexual response (e.g., guilt, fears about the adequacy of one's sexual performance, or fears of not being attractive enough.)

Any of a variety of stimuli can become associated with sexual anxiety, depending on each individual's prior history of Pavlovian conditioning. For one person, fears of becoming pregnant can interfere with the sexual response; whereas another person has no fears of pregnancy. Given the pervasive social attention to physical attractiveness—especially for females—some people may fear removing their clothes to make love because they fear that their partner will not think that their body is attractive enough. A given individual's sexual fears can be understood only in terms of that individual's prior learning experiences. [*See* CONDITIONING.]

E. Social and Cultural Influences

Each individual is strongly influenced by his or her location in the social matrix. People who grow up in homes that accept sexuality as a natural part of life are likely to receive little or no socialization that induces guilt or fear about sexuality; and they may learn many positive thoughts and activities that help them have a gratifying sexual life (with the minimum of problems). On the other hand, people who grow up in homes where sexuality is never discussed other than in ways that clearly associate it with sin, shame, or guilt may learn many negative associations with sexuality, creating a condition called erotophobia. Such people often feel sufficiently guilty about sex that they avoid sex education material and do not learn effective means of birth control. This can leave them very vulnerable to sexual accidents if their phobia about sex is overcome by curiosity about sex or pressure from another person to experiment with sex. While erotophobic people can become pregnant and contract sexually transmitted diseases as easily as other people, their lack of knowledge about sex leaves them more vulnerable to such mishaps since they have not planned ahead for sex and have less knowledge about ways to avoid the problems.

III. Sexual Orientation

Although most people are attracted to and sexually aroused by people of the other sex, heterosexuality is not universal in humans or in other species. In almost all societies that have been studied, homosexuality has been found; and there is also evidence of bisexual and nonsexual people in many studies.

The diversity of sexual orientation has been viewed as reflecting the type of erotic stimuli that excite a person's sexual thoughts, feelings, and desires. Figure 1 shows that a person can have varying degrees of sexual responsiveness to other-sex and same-sex stimuli. People who have little or no sexual response to either same- or other-sex stimuli are labeled nonsexual. In a society that places as much emphasis on sexuality as ours does, there can be negative associations with this word; and people sometimes feel uncomfortable with their lack of sexual interest or their choosing to be celibate at

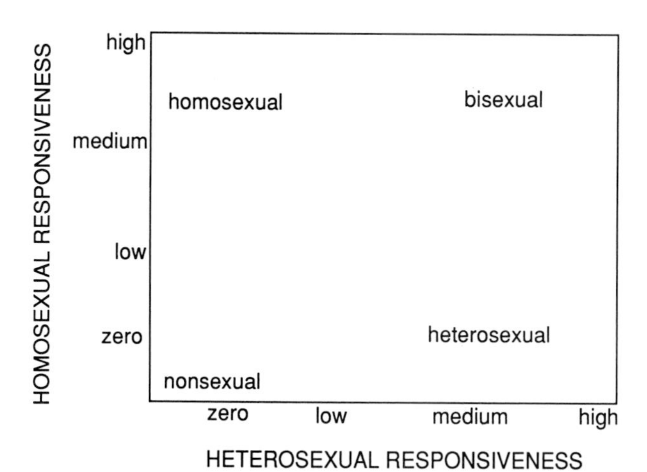

FIGURE 1 Four variations on sexual orientation (heterosexual, homosexual, bisexual, and nonsexual) are related to an individual's levels of hetero- and homosexual responsiveness—i.e., desire and preference for hetero- and homosexual partners.

certain parts of life. However, some people never become interested in sexual activity; and others may pass through periods of their lives, such as after a divorce or the death of a spouse, in which they have no sexual interest or desire and want to lead celibate lives.

Individuals whose primary erotic and amorous emotional interest is in members of the other sex, even though that interest may not be overtly expressed, are called heterosexuals. However, it should be clear from the figure that there is actually a range of variation within the heterosexual population: Some people respond to a few heterosexual stimuli as sexually interesting and arousing, whereas other people are attracted to and sexually excited by a broad range of heterosexual stimuli. Thus heterosexual sexual desire can range from low to medium to high, sometimes changing across an individual's lifetime, depending on many variables.

Individuals whose primary erotic and amorous emotional interest is in members of the same sex, even though that interest may not be overtly expressed, are called homosexuals. Again, it should be clear from the figure that there is actually a range of variation within the homosexual population: Some people respond to a few homosexual stimuli as sexually interesting and arousing, whereas other people are attracted to and sexually excited by a broad range of same-sex stimuli. Thus homosexual sexual desire can range from low to medium to high, sometimes changing across an individual's lifetime, in response to many variables. A person can be a homosexual and never engage in sexual relations with a member of the same sex. However, many homosexuals can respond sexually to the other sex though they find that their emotional and psychological attachment is to the same sex.

People who respond erotically to both male and female stimuli and enjoy engaging in sex with both sexes are labeled bisexuals. Their level of sexual interest can range from low to medium to high—changing across their lifetimes. Not surprisingly, there is a great deal of variation within this category; and too few studies have been done to date to provide an accurate overview of the range of bisexual lifestyles and activities.

Although most people believe that individuals are *either* heterosexual or homosexual, Figure 1 reveals a much greater range of patterns of sexual orientation. In spite of the fact that there has been a great deal of research on the causes of various sexual orientations, there are at present no conclusive answers. It is possible that there are biological factors that predispose individuals to be heterosexual, homosexual, bisexual, or nonsexual. But environmental factors, personal life experiences, and various forms of learning, including Pavlovian conditioning, could also be part of the explanation. More research is needed on this topic.

IV. Patterns of Sexual Behavior

The sexual reflex functions from the first days of life: Soon after birth, many boys have penile erections; and some girls experience vaginal lubrication and clitoral tumescence within the first day of birth. Even before birth, ultrasound studies suggest that erections may occur in baby boys. Sexual behavior, in the form of masturbation, often emerges in the first year. There is a considerable range of variation among children in early sexual behavior, depending in part on their parents' response to the behavior. Some parents punish and suppress early sexual behavior; whereas others allow considerable sexual exploration. Due to the hormonal influences of puberty, sexual stimulation becomes more pleasurable, and teenagers typically become increasingly interested in sexuality. During the teen and early adult years, the diversity of people's sexual activities tends to increase as they explore sexuality; but by mid or late adulthood, many people show a declining interest in and diversity of sexual activities.

A. Masturbation

During early infancy, babies explore their environments and their own bodies, touching and looking at many parts of their anatomy with as much interest as they have in the world around them. When they touch the genitals, the sexual response may be elicited, along with the pleasurable feelings associated with it. The pleasurable sensations reinforce touching the genitals, and young children may learn habits of engaging in genital stimulation. Of course, children who are swaddled, kept completely dressed most of the time, or punished for genital exploration may not learn to touch their genitals. Punishment may in fact condition inhibitions and fears about touching themselves. However, there are documented cases of babies learning to masturbate to orgasm before 1 year of age (though boys do not ejaculate any fluid before puberty).

Sometimes children, inquisitive to learn what other bodies look like, engage in mutual sexual exploration; and this can also lead to touching that elicits sexual responses. Such sexual exploration can lead to sexual interactions, including same-sex and/or other-sex exploration. While parents may fear that same-sex play may lead to homosexuality, this is not usually the case.

In cultures where people sleep together in a common room or hut, children may see their parents making love and later imitate coital activities during play with other children. There is a great deal of variability among cultures in childhood sexuality, but usually less variation within a cultural group. In most Western industrial societies, childhood sexuality is usually suppressed, since punishment of sexual exploration is commonplace. In societies where parents do not inhibit their children, sexual exploration and play can continue throughout childhood: There is no biologically determined "latency period" for sexuality.

As genital stimulation and sexual arousal become more pleasurable during puberty, adolescents find it more rewarding to think about and/or to do. Boys are much more likely than girls to learn to masturbate to orgasm during the adolescent years, reflecting, in part, the stronger punishment and social inhibitions that are placed on girls.

B. Fantasy

Thoughts about sex are most likely to become erotic fantasies when they are paired with the sexual arousal induced by masturbation or coitus; but sexual thoughts can also become very exciting by being treated as a "forbidden fruit"—something that is exciting, mysterious, but not permitted before marriage. Fantasies of the forbidden fruit are tempting because of their prohibited status. Fantasies that are paired with sexual stimulation—masturbation or coitus—are exciting because they become conditioned erotic stimuli (through Pavlovian conditioning). Teens and adults who have masturbated to images of the other sex become sexually aroused by seeing people who somewhat resemble those images. The more that people masturbate to different images, the more erotic stimuli they learn to respond to; thus there are numerous stimuli that can elicit their sexual arousal. People who develop an extensive and exciting fantasy life respond to many stimuli as sexually interesting, which increases their desire for further sexual experience. In contrast, people with little masturbatory or coital experience often have few sexually exciting thoughts and low sexual desire, since little they see or hear leads to sexual arousal.

C. Petting and Foreplay

Most modern Western societies allow teenagers to explore intimate contact before marriage, though it has not always been this way. Because physical touch to the skin, lips, hair, and other body parts are all pleasurable, it is not surprising that couples enjoy discovering the sensations of touching, fondling, kissing, and hugging. In such circumstances, pressure and touch to the genital areas are quite likely; and the sexual pleasures of such contact reinforce the learning of activities that may be labeled as petting or foreplay, depending on the context and other variables. The learning of these erotic forms of stimulation may be inhibited in societies that make young people feel guilty about doing such things; or they can be facilitated in societies that inundate their youth with pictures, movies, and videos of people doing sexual activities. Societal messages also influence the next step of sexual exploration: Societies with clear messages that sex must be reserved until marriage may suppress further sexual exploration until marriage (or at least engagement); but such prohibitions are not very strong in most Western, industrial societies.

D. Sexual Positions

As petting becomes more intimate and clothing is shed, it is easy for most people to discover coitus.

For young people who have been told (or shown in the media) that there are "certain best sexual positions," they would be very likely to explore those positions. Without cultural guidelines, patterns of sexual exploration are more random and unpredictable.

There are numerous sexual positions; and people may discover them through sexual exploration or by learning about them from books, friends, movies, therapists, or other sources. Erotophobes and people who have been made to feel guilty about sex may feel inhibited about exploring various sexual positions. However, less inhibited people often explore a variety of sexual positions and learn which ones feel good, which are unexciting, and which are painful. Because people commonly cease exploring and settle down into a few favorite positions after they have explored a range of possibilities, older people usually engage in less sexual exploration than do younger people. If married couples report that they are losing interest in sex because it has become too routine, sexual therapists and marriage counselors sometimes give couples books that show multiple sexual positions and encourage them to try novel ones. There are several good books, such as "Sexual Awareness" by McCarthy and McCarthy, "How to Make Love to the Same Person for the Rest of Your Life, and Still Love It" by O'Connor, and "Making Love" by Raley, that contain pictures and information on how to improve sexual enjoyment and pleasure.

V. Sex Differences in Motivation for Sexual Interaction

In most cultures, males tend to have a greater sexual interest and motivation then females. From the perspective of sociobiology, it is possible to argue that these sex differences are innate, since males with high sexual interest and motivation can reproduce dozens if not hundreds of their own kind but females cannot. However, the difference could also arise from social–psychological factors. For example, women learn that careless sexual exploration leads to the loss of virginity and perhaps premarital pregnancy (which have almost always hurt a woman's chances of marriage) and even sterility (induced by sexually transmitted diseases). These undesirable side effects of sexual behavior often condition fears and anxiety about sex that can inhibit female masturbation, fantasy, sexual desire, and sexual experimentation.

VI. Sexuality and the Life Cycle

A. Premarital Sex

Many cultures have strict rules against premarital sex. Since most people throughout history have not had effective means of birth control, premarital sexual activity often led to pregnancy; and young unmarried mothers usually could not find marriage partners. Males often did not want to marry a pregnant woman or a woman with a child because the males wanted to devote their childrearing efforts to raising their own children, not someone else's. In such cultures, virginity can be very important; and a man may give a woman back to her family if she is found (or suspected) not to be a virgin. Naturally, there is much variation among cultures, and some societies allow a certain amount of premarital sex, perhaps with a fiancé, just before marriage.

With the development of increasingly effective forms of birth control, the decreasing influence of strict religious codes, and other recent changes, many modern industrialized nations have experienced significant increases in premarital sexuality. In the United States, this trend has lead to an increasing amount of teenage pregnancy and teenage abortion, in part because many parents and teachers have been reluctant to provide good birth control information to teenagers. In Western Europe and Canada, where rates of teenage premarital sexual activity are similar to those in the United States, the teen pregnancy and abortion rates are considerably lower. For example, in the United States, the pregnancy rate is 96 per 1000 15- to 19-year-olds, compared to 35 in Sweden and 14 in the Netherlands. In modern Western European countries, teenagers are given better sexual education and better access to birth control services than in the United States. There is also less of a traditional religious influence.

While the prevalence of premarital sexual intercourse has gradually increased, the age of onset has gradually declined. For example, Alfred Kinsey, a pioneer sex researcher, found that half of the women he interviewed some 40 years ago had intercourse before marriage; but Morton Hunt found that 81% of 18- to 24-year-old married women had done so in the early 1970s. In 1971, only 46% of unmarried 19-year-old women had engaged in intercourse; but this percentage increased to 55% in 1976 and 69% in 1979. In the 1980s, this trend for increasingly early sexuality is believed to have leveled off. Between 1970 and 1986, the typical age of marriage has been postponed by over 2 years, shifting from

20.6 years to 23.3 years in females and 22.5 to 25.1 years in males. Thus, many young people are spending a longer portion of their lives as sexually active but unmarried. Some individuals spend this time with a small number of partners, whereas others have numerous partners.

B. Cohabitation

Although cohabitation is less common in the United States than in France or Sweden, increasing numbers of people are living with a partner without being married. In the United States, approximately 4% of unmarried adults are cohabiting, in contrast to 1% in 1960. Those who cohabit tend to be more liberal, more likely to use drugs, and less religious than people who do not cohabit. However, cohabitation is becoming increasingly routine before marriage, though there is no evidence that it helps people make better choices of marital partners. Most cohabiting unions do not lead to marriage, dissolving relatively quickly. This is preliminary evidence that this living arrangement may be one factor reducing the number of people marrying and increasing the risk of divorce among those who cohabited before marrying.

C. Marriage

Most married people report that sexual relations are most fulfilling in the first year after marriage and that the quality of their sexual experience declines thereafter. In their 20s and 30s, married couples typically engage in coitus two or three times a week. The frequency of intercourse usually declines gradually as people age. People also report a decline in communication, listening, respect, and romantic love after marriage. These changes are not inevitable, but they do indicate that many people do not devote much attention to making their marriages work. They also suggest that many people do not have the time, effort, or skill to solve their marital and sexual problems and/or that the multiple other demands of adult life—jobs, careers, children, socializing—are so demanding or rewarding that the marriage becomes neglected.

Couples commonly engage in noncoital sex play before penile–vaginal intercourse. Foreplay typically lasts some 5 to 15 min, though the time varies from couple to couple and from time to time for any given couple. Various stimulative techniques are used in foreplay, including touching, caressing, and kissing of the genitals, breasts, lips, and other body parts. Couples use a variety of positions for engaging in intercourse. While the male superior, face-to-face position is very common, couples are increasingly using the female superior position. Side-by-side positions and combinations of female superior and side-by-side positions are also used. Rear-entry positions are less commonly explored. Coitus usually lasts approximately 10 min.

Many couples engage in oral–genital stimulation. Ninety percent of people married less than 25 years were found to have engaged in oral sex during the preceding year. Anal intercourse is not a common practice for most couples: for married people between 18 and 35, approximately 50% of men and 25% of women had engaged in anal intercourse on at least one occasion.

D. Extramarital Sex

Although marriage vows require faithfulness, studies reveal that a considerable number of people violate the traditional marital contract. A survey in the 1970s found that 50% of married men and 20% of married women had at least one extramarital relationship by the age of 45. However, the rates for younger women were similar to those of men. Recent magazine surveys of their readers indicate higher rates of extramarital sex, in one 60% of the women and 75% of men had extramarital sexual experience. There is historical and cross-cultural evidence that extramarital sex has been common at other times and in other societies. Although most individuals attempt to keep their extramarital affairs secret, they do not always succeed; and the discovery that there has been extramarital sex can be very damaging to the quality of a marriage. Such a major breach of trust may make it difficult for the cheated partner to completely trust the other again.

Some people recognize that secret affairs are common yet potentially deleterious to marriage and advocate "open marriage" as an alternative. Built on the assumption that most people want both a committed relationship and freedom, open marriage is designed to allow a couple to see other people and decide if this should include sexual relations or not. The incidence of open marriage is unknown. One study found that in 15% of marriages, both partners agreed that extramarital sex was acceptable under certain circumstances.

VII. Sexual Dysfunctions

It has been estimated that approximately 50% of couples will have a major sexual dysfunction, such as erectile or orgasmic dysfunction, at some time during their lives. Many people do not seek professional help or even read self-help books; and a substantial number of these people do not overcome their sexual problems. This is unfortunate, since modern sexual therapies and medical interventions can help people even after they have had a sexual problem for a decade or more. While perhaps 25 to 35% of problems have an organic basis, most problems result from having too little effective stimulation to elicit the sexual reflex and/or experiencing guilt, anxiety, feelings of inadequacy, or other negative emotions that suppress the sexual reflex. Therapy consists of diagnosis and treatment of any physical condition and/or providing correct sexual information, teaching effective sexual techniques, and resolving the negative emotions that suppress the sexual reflex.

A. Female Problems and Therapies

The most common female sexual dysfunction is orgasmic dysfunction, or the inability to have orgasms. This can be primary (if a woman has never had an orgasm) or secondary (if a woman once had orgasms but no longer does). Situational orgasmic dysfunction indicates that a woman can have orgasms in certain situations but not others (e.g., during masturbation but not during coitus). Other common and closely related problems are low sexual desire, difficulty become excited, difficulty staying excited, and not having orgasms as often or as easily as desired. Although orgasmic dysfunction is usually not caused by organic factors, a variety of physical problems, such as neurological or pelvic disorders, can result in anorgasmia. The therapies for psychologically induced orgasmic problems focus on increasing communication, giving women and men useful knowledge about female sexuality, and teaching the sexual skills needed to assure that the female receives effective sexual stimulation. Therapies typically proceed in gradual steps from exercises in nonsexual body touching (called sensate focus) to gradually adding more sexual stimulation. Counseling also focuses on the specific types of myths, fears, anxieties, or guilts that a woman might have that are suppressing her sexual responsiveness and enjoyment.

Many women experience dyspareunia—or painful intercourse—from time to time; and some, often. Dyspareunia can result from any of dozens of causes, including inadequate lubrication, overly vigorous and overly lengthy penile stimulation, vaginal infections (such as monilia, gardnerella, or trichomoniasis), pelvic infections, irritations or infections of the vulva, and thinning of the vaginal walls. Psychological factors can also result in painful intercourse. Only after a careful analysis of the specific causes that are relevant to a given individual's problem is a therapy arranged to solve that problem.

A less common female problem is vaginismus, in which the muscles surrounding the outer third of the vagina contract involuntarily and either completely prevent penile intromission or make it painful. This strong, involuntary muscular response can be due to any of a variety of aversive experiences related to sex, including a history of painful intercourse, past physical or sexual assault, a religious socialization stressing sexual guilt, and fear or dislike of her partner. The woman receives counseling to help her cope with the emotional causes of the problem, such as dealing with memories of her childhood sexual assault. To help her overcome the physical response, therapists give the woman a small dilator and have her insert it in her vagina—in privacy at home—after she does a series of muscle relaxation exercises. Once she can successfully insert a small dilator, she is given a slightly larger dilator with which to practice. Over multiple therapy sessions, as she becomes used to the exercises, she is given larger dilators until she can accept one the size of a penis. Once she has learned to keep her vaginal muscles relaxed with the large dilator, she is gradually reintroduced to sexual intercourse in ways that help her remain relaxed during early intromission.

B. Male Problems and Therapies

When a man cannot attain an erection firm enough for intromission, he is said to have erectile dysfunction. This problem is sometimes called impotence, but therapists prefer the less threatening and more accurate label of erectile dysfunction. The most common organic causes for erectile dysfunction are diabetes and alcoholism. Other causes include spinal cord injuries, multiple sclerosis, and hormone deficiencies. Psychological causes for this weak sexual response can usually be traced to the presence of anxiety, guilt, or other negative emotions. The therapy for psychological causes parallels the therapy for female orgasmic dysfunction: Therapists give the man and his partner information about male sexuality in order to overcome myths and anxieties people have about the male sexual response. The male and his partner are taught more effective means of providing sexual stimulation for him, and they receive counseling that can reduce whatever negative emotions may be suppressing the sexual reflex. For example, it is common for males to have "performance anxiety" during intercourse, since they know that the penis must be firm to attain intromission; therefore, therapists have developed various exercises to reduce performance anxiety. The sensate focus exercises assigned as homework during the beginning of therapy take the pressure off the male, since there is no need to perform sexually; and if he does have an erection, he is not allowed to use it. As therapy progresses, the clients are introduced to "teasing," in which the man's partner alternates between stimulating the penis to erection, then allowing the erection to disappear by ceasing stimulation. After having his erection appear then disappear on numerous occasions, the man begins to lose his performance anxiety, since he learns that the sexual reflex will occur quite automatically with adequate stimulation. As his performance anxiety declines, he has fewer problems with attaining erection.

Men who can have intercourse but cannot ejaculate inside a woman's body are described as having ejaculatory incompetence. A variation of this problem is retarded ejaculation, in which a man takes a great deal of stimulation and time before he can ejaculate. These problems can result from organic causes, with drugs and neurological problems being the most common. However, most cases result from psychological factors, such as a very negative sexual socialization or traumatic sexual experiences. In addition to sensate focus exercises, the therapy for psychological problems uses stages of successive approximation that start with the man masturbating by himself to orgasm. Next, his partner joins him to watch and learn from him. After the man can successfully ejaculate in front of his partner and is comfortable with this, his partner learns to masturbate the man to ejaculation. Finally, the partner stimulates the man to the point of ejaculation, then inserts the penis into her vagina before ejaculation. After a few repetitions of this, most men adjust to normal intercourse. Counseling can help reduce any fears, guilt, or anxiety a man might have about ejaculating inside the partner's vagina.

The most common sexual problem for young men is premature ejaculation, or reaching orgasm so quickly that he and/or his partner are deprived of important sexual gratification and find this problematic. Premature ejaculation can be characterized as a lack of voluntary control over the timing of ejaculation. This dysfunction is rarely caused by organic problems. It often emerges when a male learns sexual habits of attaining sexual stimulation very rapidly, causing the sexual reflex to progress to orgasm in a very brief period of time. There are several forms of therapy, which focus on slowing down the sexual reflex. The "stop–start" method teaches the man to alternate between starting sexual stimulation and stopping. During the stop phases, his sexual arousal level will decline a little; and by inserting enough stop periods, he will be able to extend his sexual response to a much longer duration. The "tease–squeeze" method consists of slowing a man's sexual response by applying a rather firm squeeze to the base of the glans or the base of the shaft of the penis. The squeeze reduces a man's sexual arousal and erection temporarily, then he can resume stimulation. These therapies are usually performed with a cooperative partner.

Dyspareunia—or painful intercourse—in men is not common; but infections or inflammations of the penis, urethra, prostate, or testes can be among the causes leading to painful intercourse. Psychological influences may also be involved. After careful diagnosis to determine the cause, specific therapies are designed.

Bibliography

Baldwin, John D., and Baldwin, Janice I. (1988). The socialization of homosexuality and heterosexuality in a non-Western society. *Arch. Sex. Behav.* **18,** 13–29.

Bell, Alan P., Weinberg, Martin S., Hammersmith, Sue Kiefer (1981). "Sexual Preference: Its Development in Men and Women." Indiana University Press, Bloomington, Indiana.

Heiman, Julia, and LoPiccolo, Joseph (1988). "Becoming Orgasmic: A Sexual Growth Program for Women." Prentice Hall, New York.

Kaplan, Helen S. (1987). "The Illustrated Manual of Sex Therapy," 2nd Ed. Brunner-Mazel, New York.

Katchadourian, Herant A. (1989). "Fundamentals of Human Sexuality" 5th Ed. Holt, Rinehart and Winston, Chicago.

Leiblum, Sandra R., and Rosen, Raymond C. (eds.) (1988). "Sexual Desire Disorders." Guilford Press, New York.

Masters, William H., Johnson, Virginia E., and Kolodny, Robert C. (1988). "Human Sexuality," 3rd Ed. Scott, Foresman, and Company, Glenville, Illinois.

Silverstein, Judith (1986). "Sexual Enhancement for Men." Vantage Press, New York.

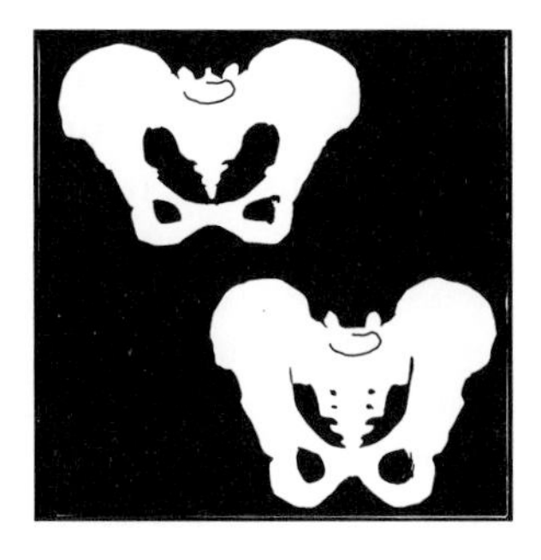

Sexuality, Anthropological and Evolutionary Perspectives

BECKY A. SIGMON, *University of Toronto*

Glossary

Biological sex Phenomenon of being male or female based on biological factors such as sex chromosomes—XX for female and XY for male

Gender Each of us has a gender assignment that is given at birth, based on the appearance of the genitalia; our gender identity is our personal assessment of being male or female; and our gender role is a set of socially defined expectations about behavior appropriate for people of a particular gender

Polygamy Marriage structure based on having more than one mate. It is referred to as polyandry if a wife has more than one husband, or polygny if a husband has more than one wife

Sexual dimorphism Physical differences in the anatomy and skeletal structure of males and females, usually based on quantitative measures such as size, amount, and robusticity of a trait

Taboos Rules in a society that state what is forbidden to those who live within the society

ANTHROPOLOGISTS study variation in people's physical and behavioral adaptations. They seek to describe and explain the nature of that variation and to look for patterns and reasons for similarities or differences in cultures of the world. Human sexuality is one aspect of the physical and behavioral nature of people. Anthropologists study the evolutionary aspects of human sexuality such as the adaptations that evolved to ensure the perpetuation of the species. They use research from nonhuman primate societies as comparative models in attempting to understand the origins of sexual asymmetry. Biological studies of human sexuality provide information that can be insightful into understanding the multiple behavioral dimensions of sexual practices. Finally, how a human society has organized aspects of human sexuality into the social structure of the culture is a major concern; cross-cultural comparisons are extremely enlightening in this regard.

I. Evolutionary Perspectives on the Origin and Uniqueness of Sexuality in Humans

A. Palaeoanthropology

One might think it farfetched to consider questions about human sexuality in relation to ancient fossilized bones, but this is indeed what occurred when the oldest woman of our lineage was discovered in 1937 at a limestone quarry called Sterkfontein, in the Republic of South Africa. Our earliest identifiable female ancestor was nicknamed "Mrs. Ples," a shorter version of her scientific name *Plesianthropus transvaalensis*, and she lived about 2 million years ago. How did the scientist, Dr. Robert Broom, know the sex of the fossil bones that he discovered? The question could be answered by any university student studying anthropology. The skeleton of males and females looks different, a concept that is known as sexual dimorphism. The pelvis is particularly different in females because it is adapted for giving birth. And it was Mrs. Ples' pelvis that was found to identify her as female. [*See* BONY PELVIS OF ARCHAIC HOMO SAPIENS.]

Another famous female ancestor named "Lucy" was discovered in 1974 at Hadar, Ethiopia. Her sexual identification was also made on the basis of her pelvis, and she may have been an ancestor of "Mrs. Ples" because she lived nearly a million years earlier in time. Sexual dimorphism has been in the hominid lineage at least as long as our ancestors walked upright on their two lower limbs.

There are also some famous fossilized male skeletal remains. An adolescent boy was uncovered in Kenya in 1984 who is known to have lived about 1.5 million years ago. He was only a teenager of about 12 years of age at the time of his death. Had he lived to become an adult man, it is estimated that he would have been nearly 6 feet in stature. Most of his race at that time were presumed to be shorter by 4–6 inches. The discovery of a nearly complete skeleton of this adolescent boy (a *Homo erectus* type) provided scientists with information on male growth patterns.

Size differences between males and females is also considered to be a sexually dimorphic trait. In general, in modern human populations, within a given cultural group, males are larger than females by 5–12%. The data from studies of the fossilized skeletal remains of early hominids who lived from 1 to 3 million years ago suggest that the amount of sexual dimorphism was equivalent to or greater than that which occurs in modern human populations. This information is extremely insightful when compared with data that have been gathered from studies of nonhuman primate species.

In nonhuman primates, the presence of sexual dimorphism in size can be used as an indicator of the type of mating strategy that is practiced in a primate species. The absence of this dimorphism is found in species with monogamous mating. The presence of sexual dimorphism is an indication of polygamous mating strategies. What this suggests for early hominid groups that show sexual dimorphism is the tendency toward polygamy (i.e., for males and females to mate with several or more of the opposite sex). An alternative point of view, however, emphasizes that pair bonding and a monogamous system of mating was critical to the emergence of early human societies.

In comparing early hominid societies with those of nonhuman primates, two major features make humans unique. The first is that females have the capability of always being sexually receptive, and the second is that ovulation and thus estrus becomes more "concealed." There are two opinions on the extent of this latter evolutionary change. Some authors believe that estrus—the period just before and during the time of ovulation when mammalian females express a desire to mate—has been lost in human females, while other authors maintain that the tendency is still present. Let us look further at nonhuman primate sexual behavior.

Some of the following data have been recorded by primatologists. (1) In monogamous primates, both the male and female of the pair are about the same size. Both defend their territory against intruders. The breeding female in monogamous matings is intolerant of the presence of other breeding females. (2) In polygynous primates, there is sexual dimorphism in size and other features of body shape. It is the larger of the two, the male, that tends to act as a "defense" for the troop. (3) In general, primate males offer some form of care and protection to "known" mothers and their infants. "Known" in this respect refers to those females with whom the male has copulated. In many other cases, males practice infanticide on offspring of females who are "unknown." (4) The presence of a primate male may affect the hormonal balance of the female and may cause her to ovulate prematurely. (5) Rape is unknown in all primates except for orangutans and humans. (6) In primate female–female interactions, the effect of the dominant female on another may result in reproductive inhibition of the latter. This can occur through delay of maturation, inhibition of ovulation and menstruation, or spontaneous abortion. The causes are not clear, although stress is considered to be one factor. (7) Males compete for access to females in estrus. Female primates in general exhibit competition with other females. (8) Differences in female fertility seems related to differential use of resources and differential fertility according to rank, for example, the dominant female, or the better-fed female, or the "first wife" tends to have more offspring. (9) Females are sexually aggressive, actively seek out multiple males to copulate with, and to some extent are selective in matings. [*See* PRIMATES.]

B. The Meaning for Human Evolution

How do all these data on nonhuman primate research relate to the evolution and adaptation of human sexuality? First, *natural selection is operating on males and on females*. Both are competing for natural resources and ultimately for survival and successful reproduction. Those who have adaptations that enhance these features will be those whose genes are maintained in the population.

Turning now to those features that males and females possess, we will look at them in terms of their adaptability. First, why are there size differences between human males and females? This seems puzzling in view of the fact that there seems to be an optimum body size for a given environment and for the life style of the animal in making use of that environment. Studies on sexual dimorphism in mammals have suggested that differences in body size are probably related to the reproductive functions in pregnancy and lactation. It is estimated that up to 80% of the adult female life may be spent being pregnant or feeding an infant. She may spend an estimated 25% more energy during pregnancy and 50% more during lactation above the requirements for maintenance of her own body. In effect, she is then using more from the environment that what is her actual body weight. Parallels from comparative mammalian and particularly primate studies probably apply to some extent to human groups. Perhaps in addition there have been behavioral and more subtle biological responses that have evolved in both human males and females that serve to balance out these morphological differences.

Second, because she has the capacity for "situation-dependent" sexual receptivity, females can mate with more males in a given period of time than males can mate with females. Once a male has copulated and ejaculated, he is usually temporarily impotent, although there are cases of multi-orgasmic men who can ejaculate several or more times within a given intercourse situation. However, the female can continue further copulations with additional males. It would seem that the advantage that this gives the female is to bond more males to her, and that because a male has "known" her through copulating with her, he will tend to be supportive of her and, by association, also to her offspring. Because nonhuman primate males are known to kill the offspring of unknown females, this is clearly an evolutionary device to perpetuate her offspring.

Other evolutionary tendencies in primate sexuality would be selection for (1) sexually aggressive females and for sexually aggressive males, for (2) males who are bigger and stronger yet supportive and nurturing within limits described above, for (3) males that do practice infanticide of the young of unknown females, and for (4) competition among females (i.e., competition based on the premise of getting more males to bond to her than to other females).

Selection appears to have affected the male also in terms of his own investment in passing on his genes to the succeeding generations. The males most likely to father offspring would be those who can identify and thereby protect those infants who carry their genes. In nonhuman primates, this may be based more on biological features whereas in humans it may additionally be based on cultural features discussed in Section III,E. In nonhuman primates, there is a tendency for males to kill the offspring of unknown females, which gets rid of the genes of competitive fathers. This gives the male a better chance of inseminating a new female with his own seed and passing on his genes. The tendency to protect these females is selective because it is those offspring that will have a better chance to survive. Following through with this argument, those males who are larger and stronger, as well as more biologically fit, will have the greater chance of continuing their genetic lineage. Obviously, this would also result in intrasex (male–male) competition of some sort. The dominant male tends to be the main consort during the peak of estrus in the female (i.e., he has the greatest access to her during the time that she is ovulating), thus providing a greater probability that he inseminates her and that the resulting offspring will carry his genes.

Because there is a tendency toward concealed ovulation in human females, it may occur that neither the female nor the male are aware of the optimal time for fertilization. How, then, can the male maximize the chance that it will be his genes that fertilize the ovum? Because it is highly probable that males and females are copulating with multiple partners, the best response for the male to make in this regard is to be protective of all the females with whom he has copulated and possibly impregnated. Perhaps this is the manner in which the earliest hominids behaved. *Or* perhaps the earliest hominids did not have concealed ovulation, but rather continued to have breeding seasons, with detectable ovulation that occurs during the peak of estrus. *And*, following this reasoning, the evolution of concealed ovulation was a much later development in human evolution. This would also suggest that it corresponded with a development in the human male to have some means of awareness of this prime fertile period in females, thus ensuring the perpetuation of his genes.

Unfortunately, in human evolution, this is the kind of information that we cannot know directly. Primatologists do think that there is an association between concealed ovulation in human females and continual female sexual receptivity, which has resulted in a cultural adaptation consisting of the invention and implementation of female sexual sanctions. Since fatherhood is uncertain in connection with polygamous females, if one removes the possibility that the female can mate with other males, it increases the possibility that the sole consort is the father of the offspring.

Female sexual sanctions come in many varieties. The earliest evidence that we have for female sexual sanctions can be traced to the early Egyptians. The practice of female circumcision including removal of the clitoris and labial fusions, which surgically close the vaginal opening, have been found in Egyptian female mummies. How much earlier in time sexual sanctions occurred is not known.

It has also been suggested that concealed ovulation could be an evolutionary adaptation that prevents the human female from knowing when she might conceive, thus preventing her from choosing not to conceive. This possibility is also suggestive of a late evolutionary development for concealed ovulation—in that as intelligence, rational thought, and symbolic behavior developed, there also evolved biological checks for the perpetuation of the species, and this may have been one.

II. Biological Dimensions of Human Sexuality

A. Making a Baby

Let us begin with sexual intercourse. The evolutionary gift to humans to encourage reproduction is sexual pleasure and the energy associated with sexual drive. Humans might be more reluctant to invest the enormous amounts of time and energy required in the making of and caring for offspring if sexual pleasure did not exist. When human females are in a position of choosing whether to make baby after baby, they have stopped short of the maximum number of which they are capable. When they can, women tend to select to control their reproductive capacities. An example is the popularity of the contraceptive pill since its introduction in the 1960s.

Returning to sexual union between male and female, we know that one possible consequence of this is the making of a baby. The intensity of coupling is frequently increased especially notable in our own society by the mutually shared phrase "Let's make a baby." Evolution meant it to be so. The phenomenon of "falling in love," referred to by Tennov as *limerence* in her book *Love and Limerence*, provides a cohesive bond, which she suggests serves the evolutionary function of keeping the couple bonded together on an average of 2 years. This period of time is sufficient for the couple to make a baby and give it dual parental nurturance during its first year, thus increasing its chances of survival.

As intercourse continues, the sexual union continues to give sensory pleasure to both partners. At any point either or both male and female may experience an orgasm, which is an intensely delightful feeling at the culmination or climax of this pleasure, preceded by a series of muscular contractions. In the female, these contractions occur in the clitoris, vagina, and uterus and may or may not result in an ejaculation. The location of orgasm in females has recently been a topic of interest to sexologists. Looking at data from women's own testimonials, a woman's orgasm includes erotic and pleasurable sensations that can be felt in the clitoris, the vagina, especially in the area of the G-spot, and the back of the vagina at the cervix; sexual sensations from these areas can also result in pleasurable uterine contractions. It would seem that women have a unified sex organ in which all parts can react during sexual arousal, no matter which part is being stimulated. [*See* SEXUAL BEHAVIOR, HUMAN.]

The point that the female orgasm may or may not result in an ejaculation is an interesting example of the occurrence of different cultural patterns and differences in cultural knowledge. The recent acknowledgment (within the past decade or so) of the female ejaculate in Western cultures was preceded by the assumption by sexologists that females do not experience an ejaculation during orgasm. However, the knowledge that there is a female ejaculation has existed for millennia in ancient China. It was described by Taoists as the third of three "waters" emitted by a female from an area that is now referred to as the G-spot (after Grafenberg), an area in the anterior wall of the vagina that is an erotic zone with erectile tissue of the same embryological origin as the male prostate gland, which is an equivalent erotic zone.

In the male, orgasm is typically associated with ejaculation, although the two phenomena may also be independent events. It begins with rhythmic penile contractions, resulting in the first phase of ejaculation called *emission*, in which the seminal fluid is forced into the base of the urethra through the contractions of the vas deferens, seminal vesicles, and prostate gland. This produces the sensation of "coming" or "ejaculatory inevitability." The second phase has two variants. One type results in the contraction of the urethral bulb and the penis, resulting in expulsion, during which the seminal fluid is expelled with great force. The second type occurs with the alternative or additional stimulation of the prostate gland, resulting in ejaculation in which the seminal fluid flows out rather than spurts; it is like the "pushing down" feeling described by women with stimulation of their G-spot, which seems reasonable considering the tissues are of the same embryological origin.

Again, different cultural patterns prevail, even in the phenomenon of ejaculation. The Taoists of ancient China and the Hindus of India advocate orgasm without the second phase of ejaculation, the expulsion of the seminal fluid. Western physiologists state that what this does is to store the seminal fluid in the urethra until the next voiding of the urine, when it is voided with the urine. The Western attitude reflects another cultural pattern in thinking, which tends to be more "scientific." Representing a different cultural perspective, they seem to be missing a major point in Taoist philosophy, which advocates conserving the semen and energy associated with ejaculating it. Taoists and the Hindus of India refer to the inner orgasm of the organs, as compared with the outer orgasm described above when the ejaculate is released through the urethra. The pressure that is built up through vasocongestion and muscular contractions is not released outward or out of the body by those who practice the "inner orgasm"; it is directed both symbolically and to some extent dissipated biologically inside the body, causing effects that the Taoists refer to as the "organ orgasm."

This inner orgasm is also advocated for females (emphasizing sexual symmetry) and is carried out in a similar manner. At the near point when the female thinks she is about to experience orgasm, she concentrates on controlling this outward release of tension and directs it inwardly, symbolically, as well as practically, because the typical definition of the orgasm is not a state that she will have allowed herself to reach. The contrasting Western view of orgasm is the release of muscular tension built up during sexual intercourse.

The following is a summary of the biologically adaptive aspects of sexual intercourse that include orgasm. (1) The sensory aspect in sexual intercourse including the climax or orgasm makes intercourse pleasurable and thus functions in the making of further offspring and perpetuation of the species. (2) It is a strong bonding agent between couples because of the mutual sharing of this pleasure and the presumed accompanying caring that both feel, which is usually a part of it. (3) It may be different in males and females: males usually require a resting (refractory) period after ejaculation (from 20 minutes to 24 hours) before they can experience another orgasm/ejaculation, although this is not always so. Females can have intercourse and (possibly) orgasm multiple times in a short period of time, but when females ejaculate, they too usually require a resting period afterward. (4) It has health benefits. It is thought to serve in keeping the reproductive organs of females and of males healthy; one way is through "use" from the exercise of the body parts themselves, which improves their functioning, and another (based on Western beliefs) is through relieving the vasocongestion that can become built up. (5) The muscular contractions that occur both before and more intensely during orgasm act in males to force the sperm into the female reproductive tract, and in the female these contractile waves of the uterus and vaginal muscles aid in carrying the sperm into the fallopian tubes. For fertilization to occur, the sperm must have arrived in the fallopian tubes at the same time as the egg, because it is here that fertilization must occur.

B. Will the Baby Be a Boy Or Girl, Or Other?

The first 8 weeks of embryonic life is critical in the making of the normal male and female. A brief review of events that can go wrong at this time include (1) sex chromosomal aneuploidies, which result in abnormal sex chromosomal types ranging from a sterile female with only one X chromosome to XXX females who are usually normal in appearance and fertility to XXY and XYY males who vary in morphology and fertility from normal to abnormal. Although these are the most common, other variations are known: (2) improper fetal development from week 5–8, resulting from a failure of hormonal production at the appropriate time, which may cause a genetic XY fetus to develop female genitalia and some internal reproductive organs; (3) abnormal exposure of a genetic XX female to male hormones, causing the fetus to develop male external genitalia; (4) inherited disorders in which the timing or amount of fetal sex hormones is abnormal; and (5) a variety of other genetic mistakes.

What this adds up to is that there is no guarantee that those who are XY necessarily become male and those who are XX become female. It is interesting to consider these sex chromosomal variants in anthropological perspective. There is an expected frequency of occurrence of these chromosomal variants, and no population is immune in this regard. Presumably such variants have arisen in every human population at some time. Some human populations have a third or a fourth sex category for people who look different or express behavioral preferences other than standard male or female into which they were classified at birth based on which type of genitalia they had or seemed to have had. Perhaps this group of sex chromosomal variants might have occasionally fit into this "other" sex classification for a third gender type. The important point to note is that there is a biological basis for having more than a dual classification of sex. Some cultures have acknowledged this, whereas others have not. This point will be taken up in the cross-cultural comparison.

C. Different Patterns of Growth and Development for Boys and Girls

Patterns of growth are inherited, they vary from individual to individual, they are predictable, and they differ in males and females. The actual chronological age of a person is different from the developmental or physiological age, which is not determined by age in years. The latter is determined by the maturity of the skeletal and dental growth and is commonly referred to as "skeletal age."

Until adulthood, boys are behind girls in skeletal growth. This pattern is evident even at birth when baby boys are about 4 weeks behind girls in this measure. From birth to adolescence they remain about 20% behind girls in skeletal age.

At adolescence, which begins in girls at the skeletal ages of about 12–13 and in boys at the skeletal ages of about 14–15, differences in a number of body features begin to appear. These differences can be summarized as follows. After the adolescent growth spurt (i.e., a period of intensive growth in males and females), males in comparison to females have (1) larger hearts and lungs relative to their size, (2) greater ability to neutralize the chemical products of muscular exercise including an increase in the alkali reserve, which enables the blood to absorb more lactic acid and other substances produced by muscles during exercise without a change in the pH, (3) an increase in red blood cells and thus (4) greater oxygen-carrying capacity in the blood because of the increased amount of hemoglobin, (5) larger size, and (6) greater strength potential. Subcutaneous adipose or fatty tissue distribution and amount differs also, in that limb fat in males—but not in females—tends to decrease, and trunk fat in females—but not in males—tends to increase. Finally, the growth spurt itself at puberty is different in that it occurs later in males, allowing for additional or extra growth time, and its intensity is greater in males, resulting in more exaggerated effects of the changes that are occurring in both sexes. Facial changes are occurring also, but with greater intensity in males (e.g., the development of the "Adam's apple," which deepens the voice, and the generally more robust development in the face of heavier brows, jaws, and more projecting noses). The result of these changes at puberty produces the phenomenon we have referred to earlier as "sexual dimorphism." As a result of this phenomenon in the human species, a variety of cultural adaptations have arisen. Here I will turn to a few examples of cultural patterns that have emerged with respect to dealing with human sexuality. [*See* PUBERTY.]

III. Human Sexuality Within a Social Context

A. The Beginning of Cultural Rules

Animals behave according to certain biological dictates that are, to some extent, species-specific. Those relating to reproductive behavior are quite species-specific. These "biological dictates" are like cues that an animal follows when in a situation involving other members of its species, or even members of other species. This phenomenon allows for a somewhat predictable, orderly progression of events in the life of the animal.

When the earliest humans began to develop culture, they also began to "override" the innate biological dictates. Cultural means of adaptation began to be used as a substitute for biological means. From this point forward, human sexual behavior is definable primarily from the social context within which an individual is born and grows up and from which he or she identifies and models his or her behavior. The society of which he or she is a part affects every aspect of his or her sexual behavior, indirectly or directly.

The earliest examples of sexual behavior in humans come from prehistoric times. Cave art and figurines are found throughout Europe and other parts of the Old World from as early as 20,000 or more years ago and portray exaggerated and often enlarged forms of the human reproductive organs. Reproduction in animals that they used for food and clothing, as well as fertility in their own groups, was apparently a major interest of these early human societies. However, it is not until historic times with the development of written language that one can begin to study, directly, the cultural practices relating to sexual behavior.

We will now consider how societies classify sex in people.

B. Sex and Gender Classification

At birth, all children are assigned a gender of male or female, based on the appearance of their genitalia. But the appearance of the genitalia may not always reflect the biological sex, and it may also not reflect the gender orientation of the individual. Some societies make allowances for individuals to change their gender role to the opposite from that of their physical sex; this change can occur also because parents wish to change the gender role of their child.

This phenomenon of a third gender is generally referred to as "berdache," although it was originally a term applied to certain aboriginal North American males. The derivation of the term "berdache" is from a French term "bardash" meaning male prostitute—which is obviously a culturally limited term because the early explorers had no equivalent name for a third gender.

Berdache-like people have been found to exist in a number of societies throughout the world. Basically, the berdache are persons who have been allowed to assume the gender role opposite to that assigned them at birth. They can be physical males or physical females, although different names may be applied to females who change their gender role. When this change is initiated by the individual at an adult stage of life, it may be recognized by the society through a ceremony at which the person receives social sanction for the new gender classification. The person may receive a new name and can at that time assume the new role.

There is a great deal of variability in cultural and individual interpretations of this third gender role. A society like the Bellacoola of the Northwest Coast of North America who believed that a berdache was selected by a hermaphrodite god, regarded this role with awe and respect. Among the Siberian Chuckchee, the role was institutionalized; it was a ceremonial role with privileges and responsibilities and was highly regarded. However, among other cultures such as the Zuni of the American Southwest it was not institutionalized and was merely tolerated as a variant of human behavior.

Individual variation in assuming the third gender ranged from partial changes to a total switch-over to the opposite gender role in which, for example, a physical male assumed the female gender and the feminine role including marrying a person of the same physical sex but of opposite gender orientation. Partial changes to the new gender role might include only changing a name and changing clothing to that of the assumed gender role.

Eunuchs add another special gender category. Historically they have occupied a unique and special position in certain cultures including China, the Byzantine Empire, the Islam state, and others. A eunuch is a male who has had his testes removed, and in many cases also his penis has been amputated. Eunuchs have occupied unique positions in society because they were considered "incomplete men" who could neither reproduce, carry on sexual relations with women, nor take the highest positions of leadership in countries. Because they were not a threat to other males or to females with power, they were used as guards in harems, informers to rulers, and trusted and loyal servants who often held positions of power and prestige. Their importance in certain societies is evident by the data that estimates that the number of eunuchs in China even in 1912 was 100,000.

In summary, human biological sexuality may be divided into more than a dual classification of male and female. There are two genders, male and female. There are two gender roles, masculine and feminine. And in the latter case, there are some degrees of overlap in gender roles, which some societies have allowed for by the creation of a third gender role that individuals may adopt if they or their parents feel the desire to do so. The eunuch, although physically induced and representing an asexual category, seems to fill a related role in some societies.

C. Socialization and Sanctions over the Sexuality of the Young

Most males and females who have been so classified from birth onward will follow the customs that have been laid out in their society. Children do not question what they suppose is normal in their society.

They receive their values and ideas of proper behavior from what the adult members of the society are doing, and they assume that this is what is normal. The child's views of what is proper or improper, what is erotic or not erotic, and what is regarded as "good" or "bad" are already formed by the time he or she has reached adolescence. What children in one society learn may be very different from what those in another society learn.

Societies are always interested in the sexual behavior of their younger members. The form of permissiveness, control, and restrictiveness of the sexuality of their children is tied in with the basic running of a culture and is influenced by its economy, technology, wealth, and social structure. There is an extremely wide range of variation among cultures in what is allowed and what is restricted among the young.

1. Initiation Rituals at Puberty

Puberty is the period of transition of the child into an individual with reproductive potential. This time may be marked in different cultures by a variety of cultural practice including public or private ceremony, body mutilation, and the application of new rules of behavior. Adolescent puberty rites serve to inform the society that one of its younger members is ready to assume the new role that involves participation in the reproductive aspects of adulthood.

Initiation rituals at adolescence occur in females with the onset of their first menstruation and in males at a certain predetermined age. Because there is less of a clear-cut physical marker in a boy's "coming of age," the ceremony is usually a public one involving circumcision of young boys. One study of 182 societies summarized the following on the occurrence of adolescent initiation ceremonies: 80 societies had no adolescent ceremony, 17 had it for males only, 39 had it for girls only, and 46 had ceremonies for both.

What kind of explanation can one use to interpret this data? Some anthropologists think there is an association between the presence of puberty rituals and the economic/political structure of the society. Those societies without adolescent rituals tend to be the larger ones with intensive agriculture and complex forms of social organization. Cultures practicing puberty rites use them as ritualized communication about gender status, suggesting that gender plays an important role in the organization of the society. These societies would include hunger/gatherer bands and horticultural groups without plows or irrigation. However, when means of identification of self comes from sources other than gender status (e.g., occupation, class, or voluntary associations), then identification by gender lessens—and ritualization of its significance to the society decreases or disappears altogether. Ritualization then becomes incorporated into the other means by which the person identifies—e.g. ceremonies at school, membership in the church or in clubs, etc.

Body mutilation, as mentioned, often takes the form, in males, of penile circumcision. Both or either sex may undergo tatooing and scarification. Where genital mutilation is practiced for females, it may take three different forms. Sunna circumcision involves the cutting away of the hood and glans of the clitoris. Excision–clitoridectomy is the removal of the clitoris in its entirety as well as much of the labia minora. Excision–infibulation is the entire removal of the clitoris and labia minora, after which the entire area is scraped so that scar tissue will form; then the two sides of the vulva are sewn together so that they will fuse and block the vaginal opening. In certain African societies, a woman is not considered marriageable unless this operation is performed. After marriage, this sealed opening is reopened to permit intercourse and child delivery. Afterward, the woman may or may not be reinfibulated. An estimate by Hosken in 1976 in *Women and Health* suggests a figure of about 20 m females in the world today who have undergone genital mutilation of one form or another. This operation is usually performed by women on the adolescent female in a private ceremony at home. Numerous medical complications, as well as death, may occur particularly from the excision–infibulation type of operation.

2. New Rules of Behavior at Adolescence

When children are near the age of puberty, many societies alter their sleeping arrangements. Some allow their young men and women to live together in a communal hut or dwelling. During this period, before marriage, males and females may sleep together, engage in intercourse with any or all the members of the opposite sex, and generally experience a care-free existence for a period of time. One of these "permissive" societies include the Muria Gond of India.

Most cases of adolescent segregation involve young males moving away from the parental home, and young females remaining there but possibly being relocated to an area farther away from her father or brothers and nearer the mother. The Nyakyusa in Tanzania segregate their adolescent boys into age villages just on the periphery of the main village.

D. Marriage

For the purpose of cross-cultural research, it is difficult to define marriage because although something like marriage is found in every culture, the rules and practices related to it differ considerably. Marriages may be arranged by parents or relatives, or they may be undertaken with the initiative of the two partners involved. Usually it is assumed that marriage gives a man and woman cultural approval for living together, having intercourse, and producing children. Marriage definitions include not only members of the opposite sex, but members of the same sex as well. This phenomenon was observed in the case of the berdache, or the third gender category, and it may be observed in modern North American society between two biological males referred to as homosexual or gay males and between two biological females referred to as lesbians.

Although monogamous marriages (one husband and one wife) are the most common, other arrangements are also known. Polygamy is the state of plural marriage. *Polyandry* is the more precise term referring to one female with more than one husband; *polygyny* refers to the state of one husband being married to more than one wife.

A review of 862 societies showed that 16% were monogamous. Of these, most allowed serial monogamy, in which a spouse can remarry after divorce or death of the other spouse; only a few forbid remarriage. Of the polygamous societies, 83% were polygynous and only four of these 862 societies had polyandry. Even in those cultures that allow polygyny, the more common pattern is monogamy because only a minority of men in a society have sufficient economic resources to provide for more than one family at a time.

Other variations in marriage rules concern who can marry whom. Cultures that practice endogamy require that persons find a marriage partner within their own village or band. Groups that practice exogamy require that individuals marry partners outside of their own immediate group. Marriage within kinship categories is a further variation. Many cultures in Southeast Asia or the Pacific Ocean encourage men to marry women in the category of mother's brother's daughters. Contrariwise, many North African cultures have a preference for men marrying father's brother's daughters. Other kinship restrictions may be found as well, such as a variety of cousin-marriage rules. The origins of, and the reasons for, these practices and rules must be studied in the context of each culture. Often they may be related to the political, economic, or social structure.

Finally, as has been shown already, there is variation among cultures regarding sexual intercourse before marriage. Some groups allow sexual play and experimentation before marriage, whereas others try to control the sexuality, especially of the young females. In the latter case, female virginity may be prized or even required before marriage. Male virginity is usually less important to most cultures. However, among the Hindus of India, female virginity was not esteemed, and often mothers would break a daughter's hymen and enlarge the vulva, which made her more marriageable.

E. After Marriage; Extramarital Sex Norms in Cross-Cultural Perspective

Study of this topic shows that cultural patterns serve a purpose in the perpetuation of the reproductive success of the society. In a study of 116 cultures, it was found that 63 approved of extramarital sex for husbands, 13 cultures permitted it for wives, and the remainder condemned it for both. In the latter group, the punishment differed for men and women engaging in extramarital sex. In general, wives are penalized more, and the punishment might be extreme such as beating, mutilating, or even death. Husbands tend to be punished only mildly or not at all. One explanation for this double standard is that control over female sexuality is a way for males to ensure that they are the father of the offspring of their wives. There is much consistency, cross-culturally, in management of sexuality for married women, but there is a certain amount of inconsistency in relative constraints on the two married partners.

Despite the potential punishment, women in many cultures carry on extramarital affairs, even at the threat of mutilation or death. From 56 societies that were studied, women in 41 of them acknowledged having extramarital affairs. This phenomenon has been explained as increasing the reproductive success of the woman. Like her nonhuman primate relatives, the human female through extramarital affairs could bond more males to her for added security and protection, and there would be a greater guarantee of her getting the superior genotype for the father of her offspring.

A similar explanation can be given for men in those societies that permit extramarital sex for females. Among the Chukchee, the Nama Hottentots, Masai, and Toda, there exists the custom of wife exchange. With the Toda this takes the form of kin-based extramarital sex exchange. Both of these forms of permissiveness can further the bonds of mutual support among the members of the society. In the kin-based type it can ensure that genes of a relative, if not the husband himself, are passed on in the lineage. The last type of female extramarital sex relationship that occurs, for example, in the Huron, Kazak, and Lesu, is that involving a form of "wife-bartering" in which the husband receives compensation from his wife's lover in the form of money, labor, or property. Such a cultural pattern could provide concrete benefits to the family unit, thus providing a greater chance of offspring success.

Evolutionary explanations like this provide us with insight as to the origins and perpetuation of the variety of cultural practices regarding sexual behavior that can be found in human societies.

F. Menstrual Ritual: Blood, Sex, Power, and Fear

A young Ashanti girl of West Africa is being honored at a public ceremony because she has reached the "hunter" state of "having killed an elephant." Gifts are exchanged; it is a time of rejoicing because the girl is ready to take on the role of contributing to the society through childbearing; yet it is a time of anxiety because her arrival at this state means the death of a child of a departed ancestor. It is a time of continuing with the natural rhythm of life giving and life taking. The Ashanti compare the female reproductive function to that of a hunter and warrior. They see a symmetry in the nature and the functions of males and females.

In New Guinea, the Arapesh seek sexual balance through incision of the boy's penis at puberty. The drawing of the boy's blood symbolizes the feeding of the fetus, which is done by the "good blood," which is what the mother provides for the fetus. The female's menstrual blood, however, is regarded as dangerous, and menstruation is healthy for the female because it discharges these dangerous fluids from the female's body, including those fluids received from men in intercourse.

The Bellacolla of the Northwest Coast of North America symbolized menstrual blood with the powers of Nature—both could destroy as well as protect. Because menstruation was involuntary, it was thought to be necessary to control this "power" through taboos. Bellacoola subsistence depended primarily on fishing. Menstruating women were not allowed to bathe in the river nor be in the vicinity of the fishing area nor repair nets. Other taboos included avoidance of menstruating women by the man who was going hunting or going to war—he believed that this would endanger his life and sap his energy.

Food taboos for the menstruating female can be found in a number of other cultures in which food supplies are inconstant (e.g., Innuit, Habbe of West Sudan, Bukka, and Melanesian Ifaluk Islanders). Isolation of the female after her first menstruation is another custom practiced by cultures; its duration varies in length from a few days to a few years. For example, the Kolosh Indians of Alaska confined the pubescent girl for 1 year in a hut, darkened except for one small opening. The most extreme case of isolation is recorded for the Carrier Indians of British Columbia where the newly menstruating female was sent away from the village to live in the wilderness for 3–4 years.

A summary of menstrual taboos in 110 societies showed that eight had no menstrual restrictions, 51 had one or two restrictions, and 51 had three to five restrictions. When the environment provides ample or plentiful resources for the people, as in the Mbuti, the Semang Negritos, or the Andamese, a female's first menstruation is a period of rejoicing and celebrating the harmony with nature. This blood is seen as a fertility symbol; it is benign and represents a balancing between the people and their environment. However, when nature's provisions are not constant nor always plentiful, menstrual blood symbolizes power and requires controlling or it will unwittingly destroy. Females are separated from food sources, and this serves to relieve the people's fear of their vulnerability to nature.

As the society's balance with nature is further disrupted (e.g., with the use of technology and with global expansions), this principle changes and control over the "inward female world," as symbolizing nature, becomes even more prevalent.

G. Incest

Rules against incest-sexual activity between close relatives or between people that the culture defines as having too close a relationship can be found in every human society. The exact definition of which relatives are too close for sexual relations varies from group to group. It may include parents and children, brothers and sisters, or two people who are unrelated but have been breast-fed by the same wet nurse. Ritual mating between brothers and sisters may be allowable in certain instances (e.g., a man of the Lele culture of Africa is allowed to have sexual relations with his clan sister before he becomes chief, thus placing him above the moral code for normal members of the society). When sexual relations occur between brother and sister, it is usually within a special class in the society who are considered to be "immune" from the usual incest rules.

Incest rules against first-cousin matings vary, depending on the kinship structure of the society. In some societies a man who marries his father's brother's daughter or his mother's sister's daughter would be committing incest; but if he married his mother's brother's daughter or his father's sister's daughter, it is regarded as normal. Both have the same biological relationship—children of siblings.

Each culture has its own particular kinship structure, and it is within this structure that incest regulations are set up. Cultures seem to derive two benefits from incest rules. The biological advantage is the prevention of inbreeding, thus reducing the chance of the recombining of deleterious genetic traits. A second benefit is that it provides a more stabilized social structure. Incestuous relationships between members of a family or a kinship group could confuse the roles of the individuals and disrupt the authority structure of the family and society.

H. Reproduction

The primary purpose of sexual behavior and intercourse is for the perpetuation of members of a species. Sexual behavior has its origins in mating behavior. This is as true for humans as it is for the rest of the animal kingdom. Before mating, animals must somehow inform possible partners of their availability and their intentions. In other animals, this involves courtship display and a variety of other species-specific behavior. In humans, this information comes about through adolescent initiation ceremonies, "coming of age" parties, newspaper or other forms of advertisements, and a number of diverse "mating" rules, which may differ considerably from society to society. These rules substitute, to some extent, for the more innate species-specific mating behavior of other animals. The culture defines who can have sexual access to whom.

When reproduction becomes less of an issue in sexual relations, the culture may define and accept or tolerate same-sex relationships. Homosexual behavior occurs in a number of societies. It is formalized into tradition in cultures like the Azande of Africa, a polygynous society in which the number of available women is reduced because of their being married to one man. Azande boys aged 12–20 may be taken as temporary "wives" of older men. This is primarily a device for providing sexual outlet plus economic help as the "boy-wives" perform a variety of domestic functions such as washing and cooking. Another example comes from what one author (Bullough 1976, in *Sexual Variance in Society and History*) describes as a "sex-positive" religion which attaches positive attributes to sex. Islams, he claims, are relatively tolerant of homosexuality both because of this sex-positive attitude and because Islam is a sexually segregating religion where males and females have reduced access to each other, resulting in greater reliance on same-sex relationships. In summary, there is a wide range of variation for homosexual behavior in cultures from simply tolerance to its being culturally defined as a normal pattern in the social structure.

Producing children is the means by which a human society perpetuates itself. This requires a female and male, the biological mother and father of a child. The culture defines the roles of mother and father, and these do not necessarily always correspond to the actual biological parents, the genitors. Surrogate parents may play these roles, but the culture defines the behavior associated with mothering and fathering.

It will always be clear who is the biological mother, but the identity of the biological father is not always certain. The biological mother is associated with the child because it develops within her and she gives birth to it; in addition, she is the only parent who can nourish it with milk from her breasts. The mother–child relationship is a basic unit of the social system of all societies. The newborn usually remains with the biological mother who nurses it and provides much of the early caretaking. In some societies like the Bushmen (San) of the Kalahari Desert of Africa, the mother may nurse the child up to 4 years. In many societies, when the biological mother is not capable of caring for her newborn, it may be adopted by another individual of the society, and another female may ''wet-nurse'' it. Shared breast-feeding of children occurs in many human societies.

The culture also defines the male role of ''father'' and appropriate fathering behavior. Margaret Mead in her book *Male and Female* discussed human fatherhood as a social invention. She says, ''Somewhere at the dawn of human history, some social invention was made under which males starting nurturing females and their young.'' Stressing that humans are the only primate group in which the male contributes to feeding the female (although other primate males may protect females and their young, they do not provision her with food), she states that the human society teaches its young males that this is their job and duty. The fact that there are so many variations in the fathering pattern adds to the proof that this behavior is not deeply biological but is learned, she suggests.

Mead states further that ''when we survey all known human societies, we find everywhere some form of the family, some set of permanent arrangements by which males assist females in caring for children while they are young.'' The usual pattern is for the male to perform this role with his sexual partner and her children. Studies of human societies show that it may be irrelevant whose children they are, because the mother's sexual partner can and does act as father irrespective of his being the biological parent.

The role of father and fathering behavior are important cultural concerns. Comparisons of cultures show that their definitions of ''father'' may mean either genitor or ''social father.'' For example, even when the genitor or biological father is known, it is the mother's brother in the Nayar of Malabar who plays the role of father to the child, whereas the biological father plays a more passive but supportive role like a benign uncle. This is because the kinship system is matrilineal and inheritance is through the mother's lineage, thereby making mother's brother the ''social father.'' The real genitor passes his inheritance to the children of his sister.

Another custom concerning fatherhood is the phenomenon of ''couvade,'' in which a man mimics the pregnant female's behavior. The practices in couvade vary among cultures and may include just food taboos, or they may involve a man's mimicking the woman undergoing childbirth. It is an empathetic symbolic behavior that acts to legitimize a man's claim of fatherhood and sets up a bond between father and unborn child.

Variation in kinship structures of societies results in variable definitions of the role of father and his expected fathering behavior, although the latter usually involves some kind of caretaking function of children who may or may not be his own biological offspring. Because the mother–child association is biologically clear, it becomes a primary stabilizing unit in the social structure of a culture. There are cases of adoptive mothers or shared breast-feeding of children. However, the adoption of children by an adult of either sex represents a different situation. Human societies characteristically take care of their young, if they are economically able to do so, and this then becomes a trait of the society itself, not just the role of one or another of the adult sexes.

I. Menopause, Aging, and Sexual Response

Reproduction in women is biologically possible between puberty, or shortly thereafter, and menopause. The period of menopause in females, which is a gradual diminishing of the menstrual cycle until it ceases and no more ova are produced, is not noted ceremonially by the culture. This release from reproductive capabilities may become a time in which the female has great influence in the culture, or it may be regarded as a period in which she experiences decreased respect because of her loss in reproductive capability. Cultures vary in their attitude to the postmenopausal female.

A decrease in sperm fertilizing capacity occurs later in men than the cessation of ova production occurs in women. There may be a reduction in the sperm count of men, therefore also of their fertilizing capacity, by their mid-sixties onward, compared with the earlier loss of female fertility in her early fifties.

However, the interest in sexual relationships at this time, past 50, does not necessarily diminish. Most of the sexual reaction of people at this time is related to behavior that they have learned within their cultural context. Biologically, the major problem that arises is the reduction by about one-sixth of the female hormone estrogen; this may cause a thinning of the vaginal walls and a decrease in genital lubrication, both important for comfortable intercourse. The interest in sexual relationships continues, and there is no biological reason that would terminate this interest. Evidence from people in these age categories indicates that the sexual desires continue to exist, as do satisfying sexual relationships, and where there is a will, there is a way.

IV. Summary

Humans evolved the capacity for cultural behavior in which they could devise their own rules of behavior rather than yielding just to biological dictates. This is particularly evident in the area of sexuality. This article is an attempt to tie together all those areas that have influenced human sexuality including the role that evolution and biology have played and continue to play in human sexuality. Anthropologists seek an understanding of how cultures have taken biological information and shaped it into certain patterns of behavior that define how people are allowed and expected to express their sexuality.

Bibliography

Bullough, V. L. (1976). "Sexual Variance in Society and History." John Wiley & Sons, New York.

Frayser, S. G. (1985). "Varieties of Sexual Experience: An Anthropological Perspective on Human Sexuality." Human Area Relations File, New Haven.

Harmatz, M. G., and Novak, M. A. (1983). "Human Sexuality." Harper & Row, New York.

Hrdy, S. B. (1981). "The Woman That Never Evolved." Harvard University Press, Cambridge, Massachusetts.

Katchadourian, H. A. (1989). "Fundamentals of Human Sexuality." Holt, Rinehart & Winston, New York.

Masters, W. H., Johnson, V. E., and Kolodny, R. C. (1988). "Human Sexuality." Scott, Foresman, Glenview, Illinois.

Mead, M. (1949). "Male and Female." W. Morrow, New York.

Tanner, J. M. (1963). "Growth At Adolescence." Blackwell, Oxford.

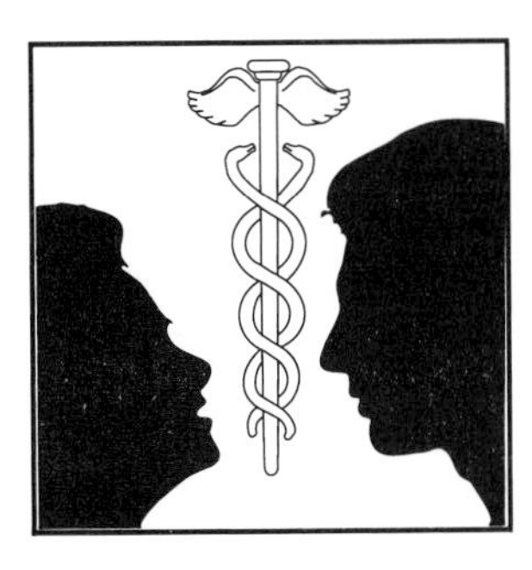

Sexually Transmitted Diseases (Public Health)

WILLARD CATES, JR., *Centers for Disease Control*

Glossary

AIDS Acquired immunodeficiency syndrome
CDC Centers for Disease Control
HIV Human immunodeficiency virus
HPV Human papillomavirus
HSV Herpes simplex virus
NGU Non-gonococcal urethritis
PPNG β-lactamase-producing *Neisseria gonorrhoeae*
STD Sexually transmitted diseases

SEXUALLY TRANSMITTED DISEASES (STD) are infections spread from person to person through intimate human sexual contact. Few fields of human biology and public health are more dynamic than that of STD control. During the past decade, this discipline has evolved from one emphasizing the traditional venereal diseases of gonorrhea and syphilis, to one concerned with the bacterial and viral syndromes associated with *Chlamydia trachomatis*, herpes simplex virus (HSV), and human papillomavirus (HPV), to one preoccupied with the fatal systemic infections caused by human immunodeficiency virus (HIV).

This expanded spectrum of infections and new responsibilities has been a two-edged sword. On one hand, it stretched limited resources allocated to prevention/control of STD; however, it also accelerated the acceptance of constructive new approaches to reducing STD transmission among high-risk groups. This article discusses (1) the reasons why we are so concerned with STD; (2) the general factors affecting STD control; (3) the trends in sexually transmitted infections; (4) prevention and control strategies for STD; and (5) future directions in STD.

I. Why Are STD So Important Now?

Except for HIV, many of these infections have been around for ages, but they have achieved recent prominence as STD For several reasons.

1. Laboratory diagnostic techniques have improved. New diagnostic approaches facilitated epidemic investigations, elucidating the extent, method of transmission, and clinical consequences of STD.
2. The population-at-risk for STD rose. The number of young adults in Western countries increased faster than the total population during the 1960s and 1970s because of the aging of the baby-boom generation. Moreover, this cohort was more sexually active than its predecessors.
3. The composition of the STD "core" population changed. During the 1970s, many homosexual men exercised greater sexual liberties than in previous eras. In the 1980s, the influence of illicit drugs, especially crack cocaine, has expanded the number of persons exchanging sex for drugs.
4. The incidence of the newer STD increased. For example, in the United States, sexually transmitted viral infections are increasing; estimates of symptomatic genital herpes and genital HPV range up to 1 million new cases annually.
5. A higher proportion of infections with multiple modes of transmission are being transmitted sexually. As sexual behavior has changed, hepatitis A and B viruses, cytomegalovirus, and some enteric pathogens have become frequent sexually transmitted agents, especially among those engaging in anal sex.

6. The STD have been associated with incurable and fatal conditions. The acquired immunodeficiency syndrome (AIDS), HPV-associated genital cancers, and chronic recurrent genital herpes have captured public attention.
7. The key impact of STD on maternal and child health is now apparent. Estimates of the cost of pelvic infection and its sequelae (tubal infertility and ectopic pregnancy) account for over $2 billion annually in the United States. Those interested in improving reproductive health realize their programs must include activities to prevent and control STD.
8. International travel disseminates STD into a global problem. The rapid spread of HIV and the progressive growth of plasmid-mediated gonococcal resistance are tragic examples of modern STD that have been widely transmitted by world travelers.

II. Key Factors Affecting Public Health

A. Population at Risk

The size of the sexually active population at risk for STD peaked in 1985 and will decrease by 1990. Thus, times are opportune to have an impact on STD trends. During the past three decades, various sociosexual changes in the United States dramatically influenced those in danger of transmitting or acquiring STD. In 1970, an estimated 50 million sexually experienced persons were between the ages of 15 and 34 years; in 1985, that number was 69 million; by 1990, it will decline to 67 million.

Several factors have influenced this population at risk: (1) the baby boomers' coming to, and passing through, the most active sexual years; (2) the percentage of young persons who were sexually active increased during the 1970s and 1980s; and (3) sexual behaviors of specific high-risk STD "core" populations—primarily homosexual men and those using illicit drugs—have varied. In the 1980s, public perceptions of the AIDS and herpes risks have apparently led many Americans to change their sexual practices to reduce exposure to STD. These behavioral changes have measurably influenced STD trends among homosexual males, but less impact has been observed in teenage and low-income, inner city, minority heterosexual populations.

B. Governmental Responsibilities for STD Control

National, regional, and local tiers of government have different responsibilities for STD control. At the Federal level, in the United States, the Centers for Disease Control (CDC) coordinate developing and implementing STD control strategies and undertake epidemiologic research. The National Institutes of Health (NIH) support both basic science and applied clinical investigations.

State health departments have the statutory authority for control of communicable diseases, including STD. States (and the largest metropolitan areas) receive Federal project grants (numbering 63 in 1988) from CDC for STD control activities, including disease reporting and program evaluation. Local health departments are charged with providing direct clinical services that include diagnosis, treatment, patient counseling, and sex partner notification activities.

These differences require Federal, state, and local health officials to cooperate closely to ensure an integrated STD program. Crucial factors include identification of local priorities based on well-defined epidemiologic indicators, and application of control strategies with the greatest potential for preventing disease. Moreover, each of the three health tiers has had to increasingly create a matrix of activities that bear on preventing STD—school health programs have required collaboration with education officials, drug-abuse programs with law enforcement officials, and so on.

C. Resources for STD Control

Paradoxically, even with recent funding increases, the Federal budget allocated to preventing STD has not kept pace with the worsening problem. After adjusting for effects of inflation, the peak year for funding STD control programs was 1947. In that year, over $130 million (in 1988 dollars) was focused on the control of a single STD, syphilis. By 1973, even with the boost of a national gonorrhea control program, *total* Federal grant resources for these two diseases was $64 million. By 1988, this aggregate amount had further declined to $54 million, and was spread across a full spectrum of STD.

Although Federal resources directed against AIDS have increased markedly, major supplementation of categorical funds for the other STD is unlikely. Thus, to gain overall leverage on the range of sexually transmitted infections, several simultaneous approaches are necessary: (1) lessons learned in primary prevention of HIV transmission must be rapidly applied to other STD; (2) resources from other governmental programs, namely, education, mental health, family planning, and maternal and child health, must be marshaled to complement STD categorical funds; and (3) the private medical community must see prevention/control of STD as a basic part of primary care.

D. Facilities Providing STD Services

Many perceive STD control efforts as concentrated in the network of the approximately 3500 public health clinics throughout the country; however, an equal number of patients seeking treatment for STD apparently visit the private sector as the public health facilities. Approximately 5 million persons are seen each year at public STD clinics for such conditions as nongonococcal urethritis, gonorrhea, genital herpes, trichomoniasis, and genital warts. A similar number of persons visit private physicians for these same complaints. This latter estimate may be low since it does not include visits to hospital-based outpatient facilities.

E. Public Interest in STD

Because the public's perception of the STD problem influences both policymakers and program planners, it has a strong impact on the resources available for STD control. However, the public's interest in STD has varied. It has ignored (or denied) the importance of these infections in some years, and then has overreacted in others. For example, the news media first focused on the problem of genital herpes in 1982, nearly 10 years after the initial rise in number of cases. A media lag of about 1 year occurred with AIDS. Currently, STD, and especially AIDS, make headlines nearly every day.

Different interest groups also affect policy. One private voluntary organization, the American Social Health Association, has as its mission to increase attention on preventing STD. Other organizations are now recognizing the necessity of controlling STD at both the national level and in their own communities. Recent networking among private health-interest organizations, AIDS-service groups, community-based foundations serving minorities, and the Federal/state/local governments will help improve more efficient use of this increasing public interest in STD.

III. Trends in Sexually Transmitted Infections

A. Persistent Viral Infections

1. Human Immunodeficiency Virus

The epidemiology of HIV—and its fatal sequelae AIDS—is well known. In fact, even before the virus was discovered, epidemiologic analysis of those with AIDS allowed development of landmark AIDS prevention guidelines in March 1983 that are still relevant today. Risky behaviors had been identified, a virus was felt to be the causative agent, routes of transmission were understood, and the "core" population of asymptomatic infected persons capable of transmitting the agent was assumed. [*See* AIDS EPIDEMIC (PUBLIC HEALTH).]

2. Genital Herpes

Genital herpes, though now less publicized than AIDS, still accounts for sizeable morbidity. It is the main cause of genital ulcers in the United States, accounting for at least 10 times more cases than syphilis. The total number of physician-patient consultations for genital herpes increased 15-fold between 1966 and 1987, from 30,000 to almost 450,000 (Fig. 1). First office visits—a more likely indicator of first genital infections—also increased sixfold over this same period, from 18,000 in 1966 to 112,000 in 1987. Adults aged 20–29 continued to account for most consultations; women visited physicians' offices more frequently than men for genital herpes.

These data must be interpreted cautiously for several reasons: (1) recent media attention—especially since 1982—may have increased both physicians' and patients' awareness of the signs and symptoms of genital herpes, thus inflating the numbers of patients seen in recent years; (2) a patient treated for genital herpes by a physician for the first time may not actually represent a newly diagnosed case; (3) asymptomatic infections are not reported; and (4) many of those with symptomatic genital herpes probably did not seek health care at all or did so through public clinics.

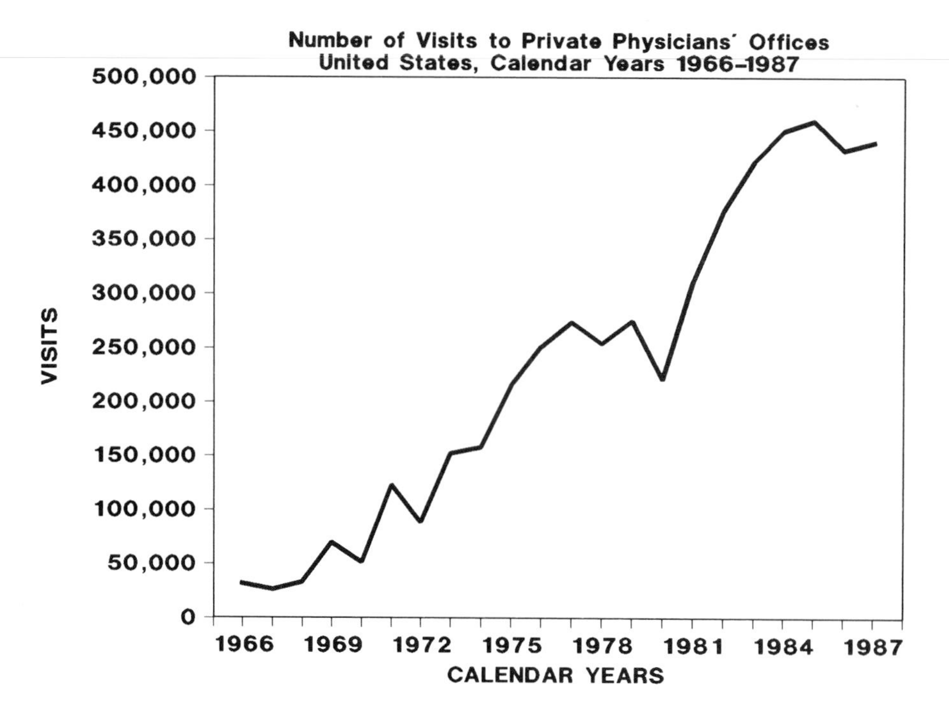

FIGURE 1 Genital herpes simplex virus infections.

The most recent finding about the epidemiology of genital herpes is that symptomatic infections are merely a tip of the iceberg—both for disease magnitude and viral transmission. Only one-fourth of those with antibodies to HSV-2 give histories compatible with genital herpes infection. Blacks are more likely to have HSV-2 antibodies than whites. In both races, HSV-2 antibody prevalence was slightly higher in women than in men.

Asymptomatic transmission of primary genital herpes infections is a striking feature. Three-fourths of those who had been the sources of infection for patients with documented primary HSV-2 infections gave no histories of genital lesions at the time of contact. Although all sources had HSV antibodies (indicative of prior infection), only one-third were aware that they had had any disease compatible with genital herpes.

Infection with HSV has been linked to higher risks of HIV transmission in homosexual men. Presumably this is due to the virus's causing genital ulcers (both recognized and unrecognized), which act as the portal of entry (or egress) for HIV. Because of the magnitude of symptomatic HSV infection in the United States, public health officials are turning attention back to potential approaches for HSV control, including HSV vaccine and prophylactic acyclovir (a drug effective against HSV symptoms) in high-risk symptomatic persons.

A serious consequence of genital herpes infection is neonatal herpes. Because it is not a reportable disease, we have no national data with which to calculate incidence. In several regions, the incidence of neonatal herpes reported between 1966 and 1988 has risen more than fourfold. This increase can be attributed to the rise in both symptomatic and asymptomatic genital herpes. Whether these reports represent a true increase in incidence of neonatal herpes or else reflect an improved ability to diagnose the disease is unknown. [*See* HERPESVIRUSES.]

3. Genital Human Papillomavirus (HPV) Infections

The epidemiology of genital HPV infections is similar to that of genital herpes—except the number of affected people is nearly threefold higher and the key consequence associated with this infection, cervical neoplasia, more severe. Using external genital warts as the index for HPV infections, the number of physician-patient consultations for this condition increased over ninefold between 1966 and 1987, from 179,000 to 1,860,000 (Fig. 2). First visits also increased nearly tenfold over the same period from 54,000 in 1966 to 517,000 in 1987. Persons aged 20–24 had more frequent genital wart consultations than did patients in other age groups; visits for women outnumbered those for men. As with genital

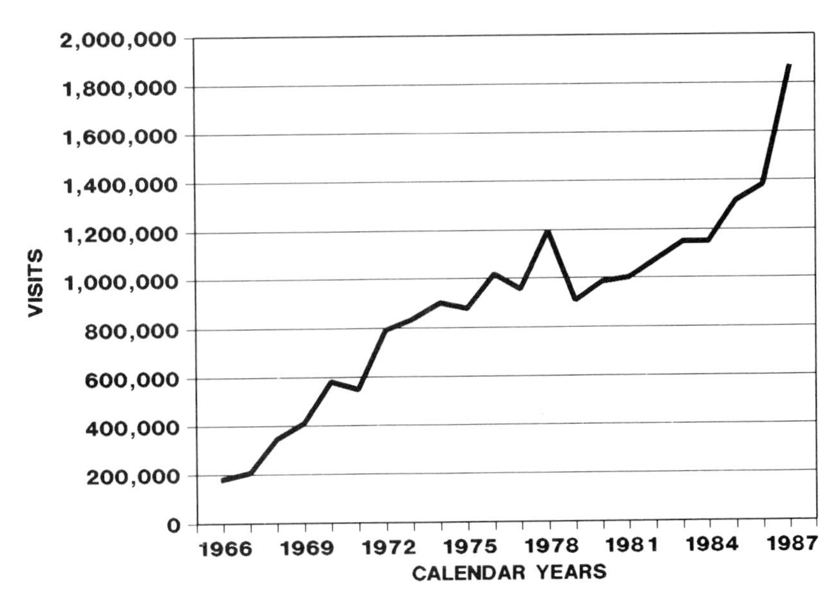

FIGURE 2 Genital warts: number of visits to private physicians' offices.

herpes, these data from physician practices have limitations in their interpretation.

Clinically apparent genital wart infections have also increased at STD clinics. From 1980 to 1985, in Columbus, Ohio, the frequency of patients diagnosed with genital warts rose by 138% in white women, 225% in black women, 116% in heterosexual men, and 27% in homosexual men. The disease was twice as frequent among white women compared to black women. The highest infection rates in women occurred among those 19 years old and younger; among men, infection rates increased until age 29 years, after which disease rates declined.

Like herpes, genital warts represent only the symptomatic tip of the iceberg of HPV infections. As physician awareness and availability of diagnostic methods increase, subclinical papillomavirus infections of the male and female genital tract are becoming commonly recognized. At present, no serologic test is available, and the virus cannot be recovered through tissue culture. Subclinical infection may be diagnosed by the presence of specific cells on a cytologic smear or tissue biopsy, HPV DNA sequences detected by sophisticated studies, HPV antigen detected by immunoperoxidase stains, or certain morphologic features on magnified examination of the cervix.

During 1983–88, evidence of cervical HPV lesions was found in 1–3% of all Papanicolaou smears in U.S. Planned Parenthood clinics, especially in women under 20 years of age. Moreover, routine cervical cytologic examination alone is inadequate for diagnosing cervical HPV infection. Human papillomavirus DNA sequences have been detected in 5–10% of women attending STD clinics who had normal Pap smears and normal colposcopic examinations. Colposcopic evidence of cervical HPV infection has been observed in 25–30% of STD clinic attendees.

Men also have sizable levels of asymptomatic HPV infection. A large majority of male partners of women with histologic evidence of condyloma acuminata have histologic evidence of penile condyloma. Most have asymptomatic infection that can be detected colposcopically after application of 5% acetic acid to normal penile epithelium. Thus, HPV infections of the genital tract are probably the most common STD. [*See* PAPILLOMAVIRUSES AND NEOPLASTIC TRANSFORMATION.]

4. Hepatitis B Virus (HBV) Infections

Nationwide, the incidence of hepatitis B has increased steadily over the last decade in spite of both

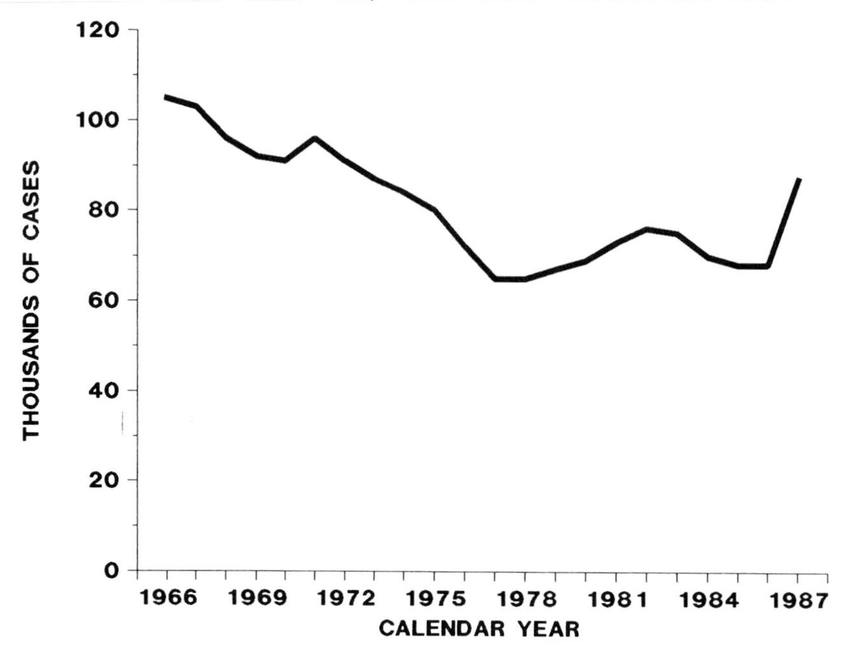

FIGURE 3 Syphilis, United States, calendar years 1966–87.

effective blood screening programs and the availability of a vaccine. New infections have risen from about 200,000 in 1978 to 300,000 in 1987. Approximately half of those infected suffer symptomatic acute hepatitis. However, the more serious concerns involve the effects of chronic HBV infection. Between 6 and 10% of those infected with HBV become chronic carriers. Chronic active hepatitis develops in more than 25% of carriers and often progresses to cirrhosis and hepatocellular carcinoma.

Most HBV infections in the United States with known routes of transmission result from sexual exposure. Unfortunately, our vaccination programs have focused primarily on three risk groups—health-care workers who are exposed to blood; staff and residents of institutions for the developmentally disabled; and staff and patients in hemodialysis units. These groups, however, account for less than 10% of acute hepatitis B cases. The risk groups that account for most cases—IV drug users, persons acquiring the disease through heterosexual exposure, and homosexual men—are not being reached effectively by current hepatitis B vaccine programs.

The behavior changes of homosexual men to reduce their risk of HIV have already led to striking decreases in the number of hepatitis B cases (along with the other STD) among this group. However, the number of cases of hepatitis B caused by heterosexual exposure has increased over the past several years, not surprisingly, primarily among inner city minority heterosexuals. Of similar concern is the large rise in the proportion of hepatitis B patients with a history of intravenous drug use. Because of the implications of HBV for both chronic hepatitis and hepatocellular carcinoma, the recent trends of sexual transmission carry long-term risks to STD core populations.

B. Bacterial Infections

1. Syphilis

Syphilis remains an important sexually transmitted organism because of its public health heritage, its association with HIV transmission, its escalating rate among inner city, minority heterosexuals, and its capacity for prevention. Since the introduction of penicillin in the late 1940s, the number of primary/secondary syphilis cases in the United States has declined by 99%. However, in recent years, infectious syphilis trends have followed a roller-coaster course (Fig. 3). In males, the number of reported cases has been affected by homosexual behavior: steadily increasing during the 1970s, but decreasing in the first half of the 1980s; presumably, this decline reflects behavior changes to reduce the risk of transmitting HIV, which in turn affects the other STD.

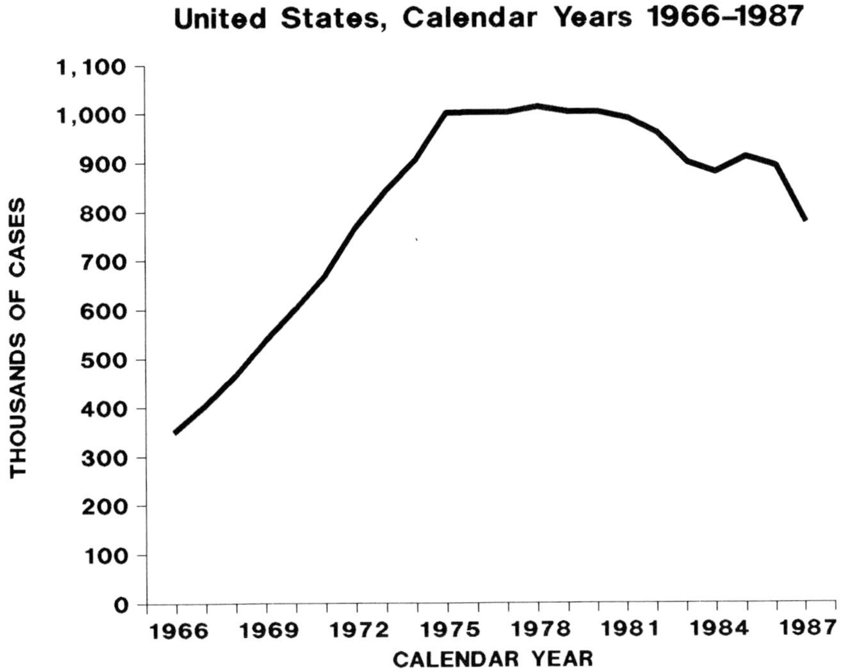

FIGURE 4 Gonorrhea, United States, calendar years 1966–87.

During the second half of the 1980s, infectious syphilis increased dramatically, to its highest level in 30 years. All of the recent increase in primary/secondary syphilis has occurred in low-income, inner city minority heterosexual populations. An important contributor to this rise has been the exchange of sexual services for drugs, especially crack cocaine. The increasing syphilis rate in this traditional STD "core" population has important implications: rises in heterosexual adult syphilis predict similar trends in congenital syphilis; community health education messages—generated by concerns about HIV—to reduce risky sexual behavior have not yet permeated the minority, heterosexual community; and because of the association of genital ulcers with HIV transmission, controlling syphilis infections provides an additional opportunity to reduce HIV spread in these communities.

Trends in congenital syphilis (CS) reflect recent heterosexual rates. While steady drops in incidence of CS occurred in the 1950s and 1960s, substantial increases have been reported in recent years. Part of the rise observed in 1984 may be attributed to broader surveillance definitions—particularly by expanding the syndrome to include stillbirths. However, the increase observed from 1985 on is less attributable to changes in reporting activity. The recent rise in CS incidence suggests increased vertical transmission may be related to underutilization and inadequacy of prenatal care. With escalating rates of female syphilis occurring in many areas of the United States, it is particularly important to provide early, high-quality prenatal care to the high-risk populations, and to encourage serologic testing in both the first and third trimester for those women who may have been exposed to syphilis during their pregnancy.

2. Gonorrhea

An examination of recent gonorrhea trends reveals two major themes: a sustained decrease of penicillin-sensitive organisms; and the continued increase in the number and variety of antibiotic resistant strains. From 1975 to 1988, reported gonorrhea in the United States has declined 30%, with nearly all the decrease occurring since 1981 (Fig. 4). Both males and females have shared in the decline, with the rate of decrease occurring slightly faster in males. Although reported cases account for only an estimated 50% of actual disease incidence, these data have been invaluable in monitoring trends over time. Moreover, trends in gonorrhea visits to private physicians show declines similar to the reported data, beginning even earlier in the 1970s.

The behavioral responses of homosexual men to HIV prevention recommendations have apparently affected gonorrhea trends. During the 1980s, in many areas of the country that measure STD by

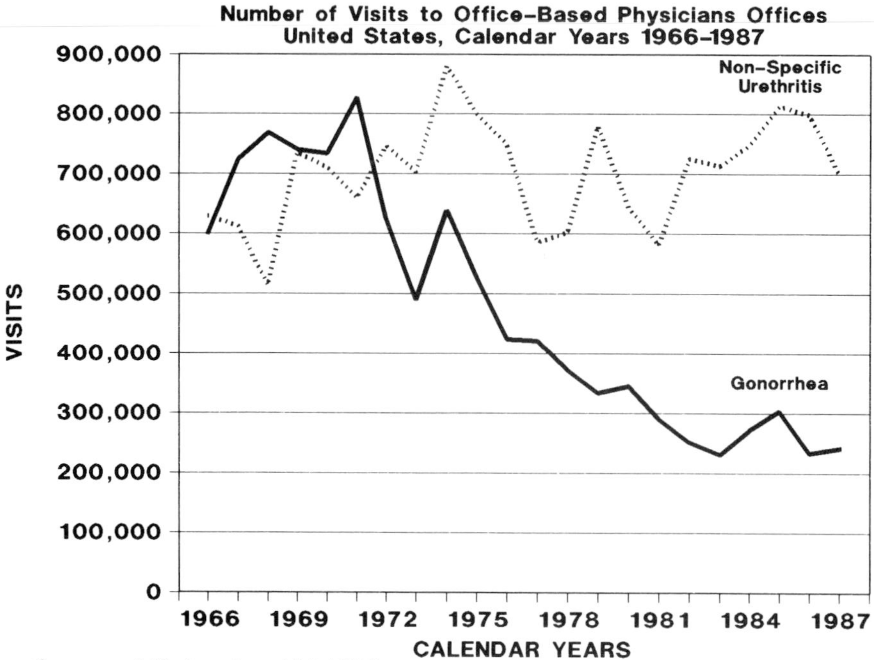

FIGURE 5 Nonspecific urethritis and gonorrhea for males. Number of visits to office-based physicians offices. United States Calendar, Years 1966–1987.

sexual preference, gonorrhea has declined faster in gay males than in other groups. The effect of a decreasing trend in gonococcal infections of homosexual men influences interpretation of the recent heterosexual morbidity trends. Because national gonorrhea data cannot be separated by sexual preference, the sharper decline among the gay population has accelerated the decrease in overall gonorrhea during the 1980s.

Unfortunately, gonorrhea trends in teenagers are disturbing. Teenagers, and especially white teenage females, have not shared in the recent gonorrhea decrease. Because the proportion of sexually active teenagers has apparently not risen in the 1980s, the stable disease rates among the teenage female population mean that gonorrhea control programs designed to lower disease incidence are apparently not yet reaching this key risk group.

Gonococcal antibiotic resistance clouds the generally encouraging trends in reported gonorrhea morbidity. Resistance to antimicrobial agents by the gonococcus has been evolving since the availability of sulfonamides in the 1930s and penicillin in the 1940s. However, the discovery of β-lactamase-producing *N. gonorrhoeae* (PPNG) in 1976 marked the beginning of an accelerated trend toward greater antibiotic resistance. Since the emergence of PPNG, clinically significant resistance has been described for the three most widely used classes of drugs—the penicillins, tetracyclines, and aminoglycosides (spectinomycin). Furthermore, the variety of mechanisms involved is cause for increasing concern, both plasmid-mediated and chromosomally mediated antibiotic resistance have expanded.

Since 1980, and especially in the last four years, the incidence of PPNG has increased dramatically. In 1988, PPNG strains accounted for 4.4% of national gonococcal morbidity, and current data indicate that this proportion will increase. In earlier epidemics, the majority of PPNG cases were linked to disease importation; however, by 1985, endemic foci of disease took hold. Areas initially with severe endemic PPNG problems included Florida, New York City, Los Angeles, and Houston. By 1988, 14 states had levels of 3% or greater PPNG, the threshold CDC has characterized as hyperendemic.

3. Chlamydia

Genital infections caused by *C. trachomatis* are the most common bacterial sexually transmitted syndromes in the United States today. Nongonococcal urethritis (NGU) in men is caused by *C. trachomatis* about 40% of the time. In 1972, NGU surpassed gonorrhea as the most common diagnosis for patient visits to private physicians offices; the gap has widened in recent years, with NGU now being twice as common as gonococcal urethritis. This pattern of increasing NGU (contrasted with the decreasing gonococcal urethritis) is consistent with trends in other developed countries.

Besides its role in male infections, *C. trachomatis* also plays an important role in causing mucopurulent cervicitis—the female equivalent of NGU. This condition predisposes either to acute pelvic inflammatory disease in nonpregnant women, or to infant and puerperal infections in pregnant women. The usual prevalence has ranged from 7 to 12%; those at highest risk are unwed teenagers living in urban areas, precisely the group at highest risk both for other STD and for adverse pregnancy outcomes.

In 1986, chlamydia was estimated to cause over 4 million infections. Approximately 2.6 million infections occurred in women, 1.8 million in men, and one-quarter million in infants. The ratio of *C. trachomatis* to *N. gonorrhoeae* infection was found to be influenced by at least five variables besides gender and age. These included race, pregnancy status, choice of contraception, the proportion of infections without symptoms, and sexual preference. Considerably higher chlamydia to gonorrhea ratios were found among whites, pregnant women, oral contraceptive users, and asymptomatic individuals; lower ratios were found among homosexual men.

Efforts to control chlamydia have been hampered by the *relative* difficulties, compared to gonorrhea, of diagnosis and treatment. National chlamydia control guidelines were developed by CDC in 1985. To make maximum use of limited STD funds, recommendations for chlamydia control were primarily based on treating syndromes rather than specific infections. However, broader strategies, where feasible, have proven successful. In areas with the capacity to provide such traditional STD strategies as diagnosis, screening, and partner notification services for chlamydia, apparent declines in sexually transmitted chlamydial infections have occurred.

IV. Prevention and Control Strategies for STD

These recent trends in STD have stimulated new directions in prevention and control programs. Previous successful approaches to controlling syphilis and gonorrhea have relied on diagnosis, therapy, and partner notification. However, in the future more emphasis will be needed on primary prevention of STD, through behavioral messages and vaccine development. Many simultaneous activities are necessary to reduce STD. For purposes of simplification, I will group the strategies into six somewhat overlapping categories: health promotion, clinical care, disease detection, disease treatment, patient counseling, and partner notification.

A. Health Promotion

Health education messages are, in general, an integral part of STD intervention activities. Specific community health education efforts, if presented properly, can supplement other strategies by encouraging primary preventive behaviors in healthy persons at risk. Traditionally, judgmental messages have not been a popular or effective method of changing behavior. In fact, stigmatizing infected individuals by widespread social disapproval may even hinder disease control through delaying care. In today's world of AIDS concerns, public health officials are continually trying to balance their messages between creating excessive fear or excessive reassurance.

To have any chance of success, community health education must stress the benefits of preventive action and promote individual decision making. For example, in our modern society, messages that have emphasized safer sexual practices (uninfected partners, no receptive anal intercourse, use of condoms) rather than proscribing coital activity altogether may have already had an impact. In the United States, because of increasing concern with HIV, over half of unmarried persons who believed themselves at risk reported changing their sexual behavior to avoid this disease. As discussed earlier, changing sexual behavior among homosexual men in response to concern about AIDS has also affected other STD.

Health education interventions also affect secondary prevention, by reducing risks of complications in those already infected and by limiting the transmission of infection after symptoms appear. Recent experience with use of videotapes to promote such actions as follow-up cultures, treatment compliance, and condom use have been encouraging. These messages are generally given to STD patients as they are being counseled about the appropriate actions they should take both to ensure a cure and also to prevent a reinfection.

School education strategies to increase the students' knowledge of the full spectrum of STD (including AIDS) have recently been encouraged by the Surgeon General. Unfortunately, our past heritage of teaching about "VD" in high schools has generally consisted of didactic biomedical lectures concentrating only on syphilis and gonorrhea. Moreover, when delivered in health science rather than sex education texts, the information failed to stress appropriate preventive skills and behaviors. To correct this situation, prototype school AIDS and STD curriculum materials for both teachers and students in grades 6 through 12 have been developed and field tested. These curricula use a self-instructional format and emphasize behavioral skill building messages. It will facilitate systematic STD/AIDS education in family life/sex education courses throughout the country.

National hotlines for both STD and AIDS provide both general information and specific answers to questions of all callers. In 1988, almost 80,000 calls were answered on the STD hotline. Over 30,000 persons requested referral to confidential medical services in either the public or private sector. Knowing where to turn for STD diagnosis and treatment plays a crucial role in preventing both transmission and complications, and is a major component of community education efforts.

In the future, those concerned with preventing STD need to make better use of the mass media to convey effective health promotion messages. Awareness of herpes and AIDS is widespread largely because of the attention given these conditions by the media. A national mail campaign to improve the public's knowledge of AIDS was conducted in 1988. Several large newspapers even employ AIDS reporters. Teenagers are a special audience for the broadcast media; they spend an average of 23 hours a week listening to radio or watching television. Messages are increasingly being aired to encourage condom use in high-risk settings and to publicize hotline numbers for further questions.

B. STD Clinical Care

Medical schools in the United States have been slow to respond to the increasing magnitude of the STD problem. In 1980, less than one in six U.S. medical schools had a specific STD clinic available for STD training. A survey conducted in 1985 showed some improvement, but still only one in five medical schools provided even half its students with STD clinical training. Paradoxically, this occurred just when U.S. physicians need STD training the most; the majority enter specialties requiring knowledge of STD diagnosis and treatment. In 1988, nearly two-thirds of first-year residents chose internal medicine, pediatrics, obstetrics/gynecology, family practice, emergency medicine, and dermatology. Without training in STD problems, the medical community cannot be effectively mobilized to support control programs.

Thus, an integral part of recent STD control strategies in the United States has been to train health-care providers in the rapidly changing field of STD. This training must involve all specialties affecting STD control—clinicians, laboratory workers, managers, field investigators, and health educators. For each group, both formal education systems and ongoing inservice training opportunities are essential.

The training initiatives have taken several approaches. To respond to immediate needs, clinical guidelines were developed by CDC to assist STD facilities to improve patient management. National STD treatment recommendations are regularly updated to reflect changing diagnostic capabilities, antibiotic susceptibilities, and pharmacologic innovations.

Another training initiative in the United States has involved creating regional STD prevention/training centers. Each center is based on integration of a university medical school with a model public STD clinic for purposes of providing training for mid-career clinicians, as well as for medical, nursing, and paramedical students. By 1988, 10 such multidisciplinary centers were in operation, with an eleventh opening in 1989; over 13,000 students have been trained in these facilities since 1979.

Finally, training support has been provided directly to medical schools. To create a cadre of medically qualified clinicians who have a career commitment to the STD discipline, interest in this field must be developed early during clinical training. These individuals could subsequently establish their own STD academic programs, thus multiplying the effect of the training efforts. As of 1988, STD research and/or STD clinical training centers exist in 18 medical schools in the United States.

Physicians interested in STD will need to complement their traditional diagnostic and therapeutic skills with training in additional biosocial sciences—psychology, sociology, and epidemiology. Most clinicians are incapable of taking an adequate sexual history; furthermore, inquiring about the intimate details of their patient's sexual partners does not come naturally. These skills must be taught. Regarding psychosocial training, the increasing emphasis on behavioral factors in STD control is part of the growing awareness in medicine generally of the importance of behavior in both producing and preventing disease. The role of lifestyle in STD is obvious; the transition from curable bacterial SDS to incurable viral STD makes it particularly important that physicians shift their emphasis toward education of individuals to voluntarily modify their behavior.

C. Disease Detection

Early disease detection is crucial to STD intervention strategies. Case finding methods include clinical diagnosis based on symptoms and signs, confirmatory laboratory diagnostic testing in patients with symptoms or signs suggestive of an STD, targeted laboratory diagnostic testing in individuals at high risk for having STD, broad application of diagnostic screening without regard to likelihood of STD, and examination of sex partners of individuals with STD.

Accurate diagnosis is the intervention cornerstone for the early detection strategies of STD control. Whether for making specific diagnoses in those with symptoms, or for screening of persons without symptoms, diagnostic tests for STD should ideally be rapid, inexpensive, simple, and accurate. The usual considerations for assessing diagnostic techniques—sensitivity, specificity, and predictive value—have a slightly different interpretation for STD control than for screening of chronic conditions, largely because STD treatment is generally shorter and safer than therapy for other conditions. Moreover, as discussed earlier, curing STD in one individual frequently prevents disease in others. Consequently, achieving high sensitivity by reducing false negatives takes on an increased importance because missed cases place the infected person at continued risk of more serious complications, and result in further disease dissemination.

Achieving high specificity by reducing false positives is less important from the public health perspective when treatment is associated with minimal morbidity, cost, and inconvenience. From the perspective of the individual with an STD, especially HIV infection, however, the human and emotional costs of erroneously stigmatizing anyone means specificity cannot be neglected. In addition, the public health costs of interviewing and tracing large numbers of partners of patients with false-positive tests would drain limited STD resources.

Advances in laboratory techniques have led to many major initiatives in STD control. Serologic capabilities improved syphilis case finding, and selective culture media allowed gonorrhea screening and diagnostic testing. Development of more rapid and less expensive immunodiagnostic tests for chlamydia and HIV have stimulated strategies for controlling these infections. The decision to initiate or abandon a new diagnostic test should be dictated both by the prevalence of the disease in the population being tested and by the sensitivity and specificity of the test being considered. The predictive values of test results are crucial to clinical and public health decisions.

D. Disease Treatment

Once the diagnosis is suspected, treatment should be inexpensive, simple, safe, and effective. Early and adequate treatment of patients and their sexual partners is an effective means of preventing the community spread of STD. To promulgate successful treatment regimens for specific diagnoses, the U.S. Public Health Service has established treatment recommendations as a standard part of its control strategy. Initially, these recommendations covered syphilis and gonorrhea, but have been expanded in the 1980s to include 18 other sexually transmitted organisms and syndromes.

Selective prophylactic (preventive or epidemiological) treatment also has a major role in STD control strategies. In certain instances, waiting to confirm the specific diagnosis prior to initiating therapy is inappropriate. Rather, based on epidemiological indications, antibiotics should be administered to high-risk individuals when the diagnosis is considered likely, even without clinical signs of infection and before proof of infection by laboratory methods. Thus, selected groups of patients with a high likelihood of infection are identified by epidemiological analyses—usually because of history of exposure—and treated before confirmation of their infection status. This interrupts the chain of transmission and prevents complications that might occur between the time of testing and treatment, insures treatment for infected individuals with false-negative laboratory tests, and guarantees treatment for those who might not return when notified of positive tests.

The benefits of this approach are generally thought to outweigh the costs of treating a percentage of uninfected persons, especially of those exposed to gonorrhea, syphilis, or chlamydial infection. Most recently, this approach of selective preventive treatment has effectively limited outbreaks of syphilis, PPNG, and chancroid in metropolitan areas. The same philosophy underlies the recommendation for giving tetracycline concurrently with penicillin to patients with confirmed gonococcal infection, since a relatively high proportion are liable to be harboring coexistent *C. trachomatis*.

E. Patient Counseling

Because of HIV and other incurable STD, counseling (i.e., educating) patients to facilitate changes in their behavior has taken a new importance. STD counselors actively encourage changing patient behaviors to both reduce risks of reinfection or complications and to limit spread of STD in the community. The behaviors sought include (1) responding to disease suspicion by promptly seeking appropriate medical evaluation; (2) taking oral medications as directed; (3) returning for follow-up tests when applicable; (4) assuring examination of sexual partners; (5) avoiding future sexual exposure while infectious; and (6) preventing exposure by using barrier protection in high-risk settings.

Risk-reduction counseling to prevent acquisition of STD is increasingly becoming ingrained as a standard part of STD clinical care, whether provided in public STD clinics, other public health facilities, or private physician offices. The expansion of the STD field to include persistent viral infections has lessened the role for prompt treatment and simultaneously raised the need for primary prevention. The concept of "safer sex" has captured worldwide attention because of HIV infection.

Risk-reduction counseling is much more than emphasizing condom use. Patients need to understand the importance of knowing the risk behaviors of their partners, as well as which sexual practices reduce the potential risk of infection. They should be counseled about social skills essential to negotiating safer behaviors with future partners. Counselors need to adopt nonjudgmental attitudes in discussing potential lifestyle changes. Unrealistic recommendations will either be ignored or will lead to only short-term changes. Patients must find the counseling messages comprehensible, acceptable, and attainable.

Counseling has additional meaning when involved with HIV testing programs. Persons wishing an HIV antibody test should have pretest counseling to establish any risk behaviors for HIV, understand the meaning and implications of test results, allow proper informed consent, and prepare for dealing with the results of the test. Posttest counseling should be delivered by well-trained professionals, one-on-one, face-to-face, not by letter or telephone. Information imparted should build altruistic motivations in seropositive persons and self-protective motivations in seronegative. Moreover, HIV infected individuals should be referred for appropriate medical and psychosocial follow up.

F. Partner Notification

Traditionally, STD control programs in the United States have emphasized active intervention by the health providers to interview the patient, to locate the named sexual partners, and to assure that these individuals are evaluated and treated. The privacy of original patients and contacts is rigorously protected. During the 1970s, in large part due to the expanded spectrum of infections, this process of active intervention was modified in many settings to include a more simplified approach. Instead of relying solely on the health worker, the patient is often encouraged to assume responsibility for locating and referring all of his or her sexual partners.

This self-referral method actively involves patients in the disease control effort, is inexpensive, is normally acceptable to patients, and reserves scarce staff time for other activities. Potential shortcomings of patient referral methods, however, include limited effectiveness, difficulty in evaluating its outcomes, and nonproductivity with noncompliant patients. Under most circumstances, patient referral will be the most easily implemented approach to partner notification. Active referral by the health provider is more labor-intensive, time consuming, and expensive; therefore, active provider referral is restricted to high yield cases or to high risk "core" environments. Special situations where this more costly strategy has been useful include: (1) introduction of a serious disease (e.g., syphilis or PPNG infection) into a community previously unaffected; (2) men and women with repeated STD infections; (3) female consorts of infectious syphilis cases; (4) STD infections in children; and (5) partners of persons infected with HIV who might not otherwise realize they have been exposed to infection.

V. Future Directions

First, reducing the transmission of HIV and controlling its progression to AIDS will become even more integral to the STD field. Activities will evolve over time from emphasis on patient and professional education to actual disease intervention through testing and confidential partner notification. Skills will increasingly involve supportive counseling to assist those dealing with consequences of persistent/fatal infections and the need for maintaining safer sexual behaviors.

Second, other facilities beyond STD clinics will increasingly provide STD clinical care to the high-risk populations, e.g., drug treatment centers, adolescent health, maternal and child health, and family planning clinics. Diagnosis and treatment of STD in these settings will be funded by non-STD public and private resources and will be justified as part of essential medical services provided to these population groups.

Third, health professionals and patients alike will increasingly learn about the medical problems of STD, the need for compliance with therapeutic recommendations, and the responsibility toward sexual partners. The ability to take and provide an accurate sexual history will become part of routine medical examinations.

Fourth, the private medical sphere will increasingly assure that modern diagnostic and therapeutic methods are applied to the patient population seen by this group. STD care will be paid for by private insurance, and understood by patients and providers alike to be cost-effective. With STD, curative medicine equals preventive medicine.

Fifth, acquisition of new computer-based skills will be essential, particularly for modern data collection and their systematic analysis. STD clinical and managerial decisions will be based on these data to obtain the most cost-beneficial results.

Bibliography

Barnes, R. C., and Holmes, K. K. (1984). Epidemiology of gonorrhea: current perspectives, *Epidemiol. Rev.* **6,** 1.

Brandt, A. M. (1985). "No Magic Bullet: A Social History of Venereal Disease in the United States since 1880." Oxford University Press, New York.

Cates, W., Jr. (1986). Priorities for sexually transmitted diseases in the late 1980s and beyond, *Sex. Transm. Dis.* **13,** 114.

Cates, W., Jr., and Holmes, K. K. (1986). Sexually transmitted diseases, *in* "Maxcy-Rosenau Public Health and Preventive Medicine," 12th ed. (J. M. Last, J. Fielding, and J. Chin, eds.). Appleton-Century-Crofts, East Norwalk, Conn.

Guinan, M. E., Wolinsky, S. M., and Reichman, R. C. (1985). Epidemiology of genital herpes simplex virus infection, *Epidemiol. Rev.* **7,** 127.

Holmes, K. K., Mardh, P.-A., Sparling, P. F., Wiesner, P. J., Cates, W. Jr., Lemon, S. M., and Stamm, W. E., eds. (1990). "Sexually Transmitted Diseases," 2d ed. McGraw-Hill, New York.

Koop, C. E. (1986). "The Surgeon General's Report on AIDS." U.S. Public Health Service, Washington, D.C.

Koutsky, L. A., Galloway, D. A., and Holmes, K. K. (1988). Epidemiology of genital human papillomavirus infection, *Epidemiol. Rev.* **10,** 122.

Thompson, S. E., and Washington, A. E. (1983). Epidemiology of sexually transmitted *Chlamydia trachomatis* infections, *Epidemiol. Rev.* **5,** 96.

Wentworth, B. B., and Judson, F. N., eds. (1986). "Laboratory Methods for the Diagnosis of Sexually Transmitted Diseases." American Public Health Association, Washington, D.C.

World Health Organization (1986). "WHO Expert Committee on Venereal Diseases: Sixth Report." World Health Organization, Geneva.